THE NU AFRICAN CALENDAR

VOLUME THREE: CALENDAR FACSIMILES

From MP 449 B.S.E. to MP 588 B.S.E.
(circa A1965 C.E. to A2104 C.E.

Forget about the Mayan Calendar and the end of the world that was suppose to happen in 2012 C.E. In this project I have introduced a exciting and original work that may prove to be the most astonishing, far-reaching and creative work of calendrical science with its philosophic, epicological, mathematical and technical foundations grounded in the experience of the Nu African People within the crucible of American grand-domestication (i.e. slavery, Jim Crow/Zip Coon, indirect rule, racialism, discrimination, and murder under the color of law). Despite the hype about the supposed end of the world in 2012 (based on the end-date of the Mayan Calendar) and auto-apocalypse (still to come!); the Nu African Cal- endar extends that time into the 22nd Century, that is, to 2104 C.E., in a note of optimism, based on the fact that Jesus declared: "But of that day or hour no one knows, not even the angels in heaven, nor the Son, but the Father alone" (Mark 13:32, NAS). To that end I have created the first modern black civil numeric perpetual calendar since the invention of the ancient Ta-Merry [Egyptian] Calendar whose base year is 4236 B.C.E. The ancient Ta-Merry Calendar was the first every civil-numeric calendar used by man. The ancient Ta-Merry Calendar had its development rooted in the history, epicology, and experien- tial needs of the black people living in the Nile Valley. Likewise, the Nu African Calendar is rooted in the epicology and experiential phenomena, grounded in the history, experience, and needs of the Nu Af- rican (i.e. African-American) People in the American Hemisphere. Below the reader is provided with a comprehensive series of interlinear Nu African Calendars (from 1965 to 2104 C.E.) that have the Grego- rian Calendar, Ta-Merry Calendar, and Ethiopian Calendar acronymically keyed to it. Further the reader will see that the base year of the Nu African Calendar is February 21, 1517 C.E. (reflecting the date- point of Shahidi Malcolm X's martyrdom on Februrary 21, 1965). Although we wanted to begin these calendars at year A1954 (the end of legal segregation in the South) logically since this Nu African Calen- dar is a memorial to Shahidi Malcolm X and the base year is based around his assassination February 21, 1965 these calendars will begin around about c. 1964 leap year sequence inclusive of events in 1965 and the brief Saasa Period and beyond ending with A2104 C.E. For Nu African Calendars from 1860 to the leap year sequence inclusive of the year 1965, the year of Malcolm's assassination the reader is invited to read my project work Our Ancestors Have Names. Finally the reader will see the unique way the Nu Af- rican Calendar has incorporated the black national holiday of Kwanzaa into its calendrical apparatus fall- ing precisely on the date points this Nu African holiday is generally celebrated. It is hoped the reader will enjoy the unique experience that this project and work offers to those interested in the epicological and experiential phenomena of the Nu African People as given unique expression in this modern black calen- darical system.

ACRONYMIC KEY TO THE TA-MERRY CALENDAR MONTHS

KEY	*TA-MERRY*	*COPTIC*
THT	TEHUTI	THOTH
PAP	PA-IPT	PAOPI
HTH	HET-HERU	HATHOR
KHK	KA-HR-RA	KHOIAK
TBT	TA-^{C}BET	TOBI
MKH	M^{C}KHIR	MEKHIR
AMN	AMEN HOTEP	PHAMENOTH
RNT	RENUTET	PHARMUTHI
KHN	KHENSU	PAKHUN
PIN	PA-INT	PAONI
APP	APAP	EPEP
MSR	MESUT-RA	MESORE

ACRONYMIC KEY THE TA-MERRY *H'ORISHA*

ASR	AUSAR	OSIRIS
HRU	HERU	HORUS
STH	SETH	SET
AST	AUSET	ISIS
NBH	NEBET-HET	NEPTHYS
ANU	ANUBIS	ANPU

ACRONYMIC KEY TO THE ETHIOPIAN CALENDAR MONTHS

KEY	*LATIN*	*AMHARIC*
MAS	MASKARAM	መስከረም
THi	THIKIMT	ጥቅምት
HID	HIDAR	ሕዳር
TAH	TAHSAS	ታሕሳስ
TIR	TIR	ጥር
YEK	YEKATIT	የካቲት
MAG	MAGABIT	መጋቢት
MIY	MIYAZIA	ሚያዝያ
GIN	GINBOT	ግንቦት
SAN	SANE	ሰኔ
HAM	HAMLE	ሐምሌ
NAH	NAHASE	ነሐሴ
PAG	PAGUMEN	ጳጉሜን
QUA	QUAGGIMi[1026]	ቋጊሚ

1026 There are only 13 months in the *Ethiopian Calendar* (EC) and just like the *Egyptian Calendar* it has a short 13th month of 5 or 6 days. *Pagumen and Quaggimi represent the same short month in the Ethiopian Calendar.* You will find that some modern Ethiopian Calendars use either *Pagumen* or *Quaggimi* as the short month days in the 13th month. This can be confusing. In the calendars below *Pagumen* will be used for the *5 day short month*; and *Quaggimi* will be used for the *6 day short month* in each EC calendar common year that comes *just prior* to a Gregorian Leap Year. In the Nu African Calendar this will occur in the *Month of Uwezo*. We have marked *Uwezo* with a different month of color to represent this calendrical nuance. See *Excursus 6* for a fuller explanation of the *Ethiopian Calendar*.

ANUP-THE DOG OF TIME AND ETERNITY

By the maya calendar the conclusion of the "fifth sun" was 2012, i.e. the end of the world! Doom's day notions all in a swirl, some even make your hair curl! The nu African calendar say that time goes past to 2100 + 4 and more, by that time will it matter for us who keeps score? Gate-keepers for the grim-ripper! Some writers say the maya calendar right; we just lost sight; turning their day into night; but by what we say, we mean no slight; massey and diop no dought right, giving us the true epicological insight. The black holy fathers said long ago; the end of years no man can know, nevertheless they did show; time is the brother of eternity, granting him to man with pain and tears; they even spoke of millions of years; their great year was twenty thousand + six, even that the great maya could not put a fix; they used someone else's science, we bring all first claims into compliance; we say the black holy fathers taught even the maya what they knew, we won't even get into how they even flew! The black holy fathers even knew about the ***"speed of light",***  = ***Khakhet er shuit = "swith more than light", that's right!*** *The aztex called the maya "tonalamatl" 260 day sacred calendar the "book of day", they like the black holy fathers had a "haab year" of 360 rays + 5 [xma kaba kin] and that's no jive; u yail haab = without name = sorrowful days of labor, our end to this we can savor, they even had a 18,000 + 980 day calendar round, but we know the black holy father's calendar was yet more sound and now we have the Nu African Calenar unbound! D.S.M.*

End of the *Mayan Calendar* Round

Beginning of the *Nu African Calendar* Unbound

MAAFA 448 B.S.E./ASIENTO 1964-65 C.E./
KAM 6200-01 S.M./EC 1956-57 A.M. (ዓ.ም.)

AFU 448/1964-65/6200-01/1956-57						
Ashe	Ojobo	Iwa	Araorun	Ile	Itutu	Erinle
1	2	3	4	5	6	7
F21	F22	F23	F24	F25	F26	F27
RNT7	RNT8	RNT9	RNT10	RNT11	RNT12	RNT13
YEK13	YEK14	YEK15	YEK16	YEK17	YEK18	YEK19
8	9	10	11	12	13	14
F28	F29	M1	M2	M3	M4	M5
RNT14	RNT15	RNT16	RNT17	RNT18	RNT19	RNT20
YEK 20	YEK21	YEK22	YEK23	YEK24	YEK25	YEK26
15	16	17	18	19	20	21
M6	M7	M8	M9	M10	M11	M12
RNT21	RNT22	RNT23	RNT24	RNT25	RNT26	RNT27
YEK 27	YEK28	YEK29	YEK30	MAG1	MAG2	MAG3
22	23	24	25	26	27	28
M13	M14	M15	M16	M17	M18	M19
RNT28	RNT29	RNT30	KHN1	KHN2	KHN3	KHN4
MAG4	MAG5	MAG6	MAG7	MAG8	MAG9	MAG10
SHARIFU 109						
M20						
KHN5						
MAG11						

KAWAIDA 448/1964-65/6200-01/1956-57						
Ashe	Ojobo	Iwa	Araorun	Ile	Itutu	Erinl
1	2	3	4	5	6	7
A18	A19	A20	A21	A22	A23	A24
PIN4	PIN5	PIN6	PIN7	PIN8	PIN9	PIN10
MIY10	MIY11	MIY12	MIY13	MIY14	MIY15	MIY16
8	9	10	11	12	13	14
A25	A26	A27	A28	A29	A30	MA1
PIN11	PIN12	PIN13	PIN14	PIN15	PIN16	PIN17
MIY17	MIY18	MIY19	MIY20	MIY21	MIY22	MIY2
15	16	17	18	19	20	21
MA2	MA3	MA4	MA5	MA6	MA7	MA
PIN18	PIN19	PIN20	PIN21	PIN22	PIN23	PIN2
MIY24	MIY25	MIY26	MIY27	MIY28	MIY29	MIY
22	23	24	25	26	27	28
MA9	MA10	MA11	MA12	MA13	MA14	MA
PIN25	PIN26	PIN27	PIN28	PIN29	PIN30	AP
GIN1	GIN2	GIN3	GIN4	GIN5	GIN6	GI

UHURU 448/1964-65/6200-01/1956-57						
Ashe	Ojobo	Iwa	Araorun	Ile	Itutu	Erinle
1	2	3	4	5	6	7
M21	M22	M23	M24	M25	M26	M27
KHN6	KHN7	KHN8	KHN9	KHN10	KHN11	KHN12
MAG12	MAG13	MAG14	MAG15	MAG16	MAG17	MAG18
8	9	10	11	12	13	14
M28	M29	M30	M31	A1	A2	A3
KHN13	KHN14	KHN15	KHN16	KHN17	KHN18	KHN19
MAG19	MAG20	MAG21	MAG22	MAG23	MAG24	MAG25
15	16	17	18	19	20	21
A4	A5	A6	A7	A8	A9	A10
KHN20	KHN21	KHN222	KHN23	KHN24	KHN25	KHN26
MAG26	MAG27	MAG28	MAG29	MAG30	MIY1	MIY2
22	23	24	25	26	27	28
A11	A12	A13	A14	A15	A16	A17
KHN27	KHN28	KHN29	KHN30	PIN1	PIN2	PIN3
MIY3	MIY4	MIY5	MIY6	MIY7	MIY8	MIY9

MABABA WEUSI 448/1964-65/6200-01/1956-57						
Ashe	Ojobo	Iwa	Araorun	Ile	Itutu	Erinl
1	2	3	4	5	6	7
MA16	MA17	MA18	MA19	MA20	MA21	MA2
APP2	APP3	APP4	APP5	APP6	APP7	APP
GIN8	GIN9	GIN10	GIN11	GIN12	GIN13	GIN
8	9	10	11	12	13	14
MA23	MA24	MA25	MA26	MA27	MA28	MA2
APP9	APP10	APP11	APP12	APP13	APP14	APP
GIN15	GIN16	GIN17	GIN18	GIN19	GIN20	GIN
15	16	17	18	19	20	21
MA30	MA31	J1	J2	J3	J4	J5
APP16	APP17	APP18	APP19	APP20	APP21	APP
GIN22	GIN23	GIN24	GIN25	GIN26	GIN27	GIN
22	23	24	25	26	27	28
J6	J7	J8	J9	J10	J11	J1
APP23	APP24	APP25	APP26	APP27	APP28	APP
GIN29	GIN30	SAN1	SAN2	SAN3	SAN4	SA

MWANAMKE WEUSI 448/1964-65/6200-01/1956-57

Ashe	Ojobo	Iwa	Araorun	Ile	Itutu	Erinle
1	2	3	4	5	6	7
J13	J14	J15	J16	J17	J18	J19
APP30	MSR1	MSR2	MSR3	MSR4	MSR5	MSR6
SAN6	SAN7	SAN8	SAN9	SAN10	SAN11	SAN12
8	9	10	11	12	13	14
J20	J21	J22	J23	J24	J25	J26
MSR7	MSR8	MSR9	MSR10	MSR11	MSR12	MSR13
SAN13	SAN14	SAN15	SAN16	SAN17	SAN18	SAN19
15	16	17	18	19	20	21
J27	J28	J29	J30	JU1	JU2	JU3
MSR14	MSR15	MSR16	MSR17	MSR18	MSR19	MSR20
SAN20	SAN21	SAN22	SAN23	SAN24	SAN25	SAN26
22	23	24	25	26	27	28
JU4	JU5	JU6	JU7	JU8	JU9	JU10
MSR21	MRS22	MSR23	MSR24	MSR25	MSR26	MSR27
SAN27	SAN28	SAN29	SAN30	HAM1	HAM2	HAM3

TAIFA WEUSI 448/1964-65/6200-01/1956-57

Ashe	Ojobo	Iwa	Araorun	Ile	Itutu	Erinle
1	2	3	4	5	6	7
AU8	AU9	AU10	AU11	AU12	AU13	AU14
THT20	THT21	THT22	THT23	THT24	THT25	THT26
NAH2	NAH3	NAH4	NAH5	NAH6	NAH7	NAH8
8	9	10	11	12	13	14
AU15	AU16	AU17	AU18	AU19	AU20	AU21
THT27	THT28	THT29	THT30	PAP1	PAP1	PAP2
NAH9	NAH10	NAH11	NAH12	NAH13	NAH14	NAH15
15	16	17	18	19	20	21
AU22	AU23	AU24	AU25	AU26	AU27	AU28
PAP3	PAP4	PAP5	PAP6	PAP7	PAP8	PAP9
NAH16	NAH17	NAH18	NAH19	NAH20	NAH21	NAH22
22	23	24	25	26	27	28
AU29	AU30	AU31	S1	S2	S3	S4
PAP10	PAP11	PAP12	PAP13	PAP14	PAP15	PAP16
NAH23	NAH24	NAH25	NAH26	NAH27	NAH28	NAH29

MTU WEUSI 448/1964-65/6200-01/1956-57

Ashe	Ojobo	Iwa	Araorun	Ile	Itutu	Erinle
1	2	3	4	5	6	7
JU11	JU12	JU13	JU14	JU15	JU16	JU17
MSR28	MSR29	MSR30	*ASR1*	*HRU2*	*STH3*	*AST4*
HAM4	HAM5	HAM6	HAM7	HAM8	HAM9	HAM10
8	9	10	11	12	13	14
JU18	JU19	JU20	JU21	JU22	JU23	JU24
NBH5	*ANU6*	THT1	THT2	THT3	THT4	THT5
HAM11	HAM12	HAM13	HAM14	HAM15	HAM16	HAM17
15	16	17	18	19	20	21
JU25	JU26	JU27	JU28	JU29	JU30	JU31
THT6	THT7	THT8	THT9	THT10	THT11	THT12
HAM18	HAM19	HAM20	HAM21	HAM22	HAM23	HAM24
22	23	24	25	26	27	28
AU1	AU2	AU3	AU4	AU5	AU6	AU7
THT13	THT14	THT15	THT16	THT17	THT18	THT19
HAM25	HAM26	HAM27	HAM28	HAM29	HAM30	NAH1

UWEZO 448/1964-65/6200-01/1956-57

Ashe	Ojobo	Iwa	Araorun	Ile	Itutu	Erinle
1	2	3	4	5	6	7
S5	S6	S7	S8	S9	S10	S11
PAP17	PAP18	PAP19	PAP20	PAP21	PAP22	PAP23
NAH30	*PAG1*	*PAG2*	*PAG3*	*PAG4*	*PAG5*	MAS1
8	9	10	11	12	13	14
S12	S13	S14	S15	S16	S17	S18
PAP24	PAP25	PAP26	PAP27	PAP28	PAP29	PAP30
MAS2	MAS3	MAS4	MAS5	MAS6	MAS7	MAS8
15	16	17	18	19	20	21
S19	S20	S21	S22	S23	S24	S25
HTH1	HTH2	HTH3	HTH4	HTH5	HTH6	HTH7
MAS9	MAS10	MAS11	MAS12	MAS13	MAS14	MAS15
22	23	24	25	26	27	28
S26	S27	S28	S29	S30	O1	O2
HTH8	HTH9	HTH10	HTH11	HTH12	HTH13	HTH14
MAS16	MAS17	MAS18	MAS19	MAS20	MAS21	MAS22

EKUNDU 448/1964-65/6200-01/1956-57						
Ashe	Ojobo	Iwa	Araorun	Ile	Itutu	Erinle
1	2	3	4	5	6	7
O3	O4	O5	O6	O7	O8	O9
HTH15	HTH16	HTH17	HTH18	HTH19	HTH20	HTH21
MAS23	MAS24	MAS25	MAS26	MAS27	MAS28	MAS29
8	9	10	11	12	13	14
O10	O11	O12	O13	O14	O15	O16
HTH22	HTH23	HTH24	HTH25	HTH26	HTH27	HTH28
MAS30	THi1	THi2	THi3	THi4	THi5	THi6
15	16	17	18	19	20	21
O17	O18	O19	O20	O21	O22	O23
HTH29	HTH30	KHK1	KHK2	KHK3	KHK4	KHK5
THi7	THi8	This9	THi10	THi11	THi12	THi13
22	23	24	25	26	27	28
O24	O25	O26	O27	O28	O29	O30
KHK6	KHK7	KHK8	KHK9	KHK10	KHK11	KHK12
Thi14	THi15	THi16	THi17	THi18	THi19	THi20

KIJANI 448/1964-65/6200-01/1956-57						
Ashe	Ojobo	Iwa	Araorun	Ile	Itutu	Erinle
1	2	3	4	5	6	7
N28	N29	N30	D1	D2	D3	D4
TBT11	TBT12	TBT13	TBT14	TBT15	TBT16	TBT17
HID19	HID20	HID21	HID22	HID23	HID24	HID25
8	9	10	11	12	13	14
D5	D6	D7	D8	D9	D10	D11
TBT18	TBT19	TBT20	TBT21	TBT22	TBT23	TBT24
HID26	HID27	HID28	HID29	HID30	TAH1	TAH2
15	16	17	18	19	20	21
D12	D13	D14	D15	D16	D17	D18
TBT25	TBT26	TBT27	TBT28	TBT29	TBT30	MKH1
TAH3	TAH4	TAH5	TAH6	TAH7	TAH8	TAH9
22	23	24	25	26	27	28
D19	D20	D21	D22	D23	D24	D25
MKH2	MKH3	MKH4	MKH5	MKH6	MKH7	MKH8
TAH10	TAH11	TAH12	TAH13	TAH14	TAH15	TAH16

EUSI 448/1964-65/6200-01/1956-57						
Ashe	Ojobo	Iwa	Araorun	Ile	Itutu	Erinle
1	2	3	4	5	6	7
O31	N1	N2	N3	N4	N5	N6
KHK13	KHK14	KHK15	KHK16	KHK17	KHK18	KHK19
THi21	THi22	THi23	THi24	THi25	THi26	THi27
8	9	10	11	12	13	14
N7	N8	N9	N10	N11	N12	N13
KHK20	KHK21	KHK22	KHK23	KHK24	KHK25	KHK26
THi28	THi29	THi30	HID1	HID2	HID3	HID4
15	16	17	18	19	20	21
N14	N15	N16	N17	N18	N19	N20
KHK27	KHK28	KHK29	KHK30	TBT1	TBT2	TBT3
HID5	HID6	HID7	HID8	HID9	HID10	HID11
22	23	24	25	26	27	28
N21	N22	N23	N24	N25	N26	N27
TBT4	TBT5	TBT6	TBT7	TBT8	TBT9	TBT10
HID12	HID13	HID14	HID15	HID16	HID17	HID18

KWANZA 448/1964-65/6200-01/1956-57						
Umoja	Kujichagulia	Ujima	Ujamaa	Nia	Kuumba	Imani
1	2	3	4	5	6	7
D26	D27	D28	D29	D30	D31	JA1
MKH9	MKH10	MKH11	MKH12	MKH13	MKH14	MKH15
TAH17	TAH18	TAH19	TAH20	TAH21	TAH22	TAH23
8	9	10	11	12	13	14
JA2	JA3	JA4	JA5	JA6	JA7	JA8
MKH16	MKH17	MKH18	MKH19	MKH20	MKH21	MKH22
TAH24	TAH25	TAH26	TAH27	TAH28	TAH29	TAH30
15	16	17	18	19	20	21
JA9	JA10	JA11	JA12	JA13	JA14	JA15
MKH23	MKH24	MKH25	KMH26	MKH27	MKH28	MKH29
TIR1	TIR2	TIR3	TIR4	TIR5	TIR6	TIR7
22	23	24	25	26	27	28
JA16	JA17	JA18	JA19	JA20	JA21	JA22
MKH30	AMN1	AMN2	AMN3	AMN4	AMN5	AMN6
TIR8	TIR9	TIR10	TIR11	TIR12	TIR13	TIR14

SHAHIDI 448/1964-65/6200-01/1956-57						
Ashe	Ojobo	Iwa	Araorun	Ile	Itutu	Erinle
1	2	3	4	5	6	7
JA23	JA24	JA25	JA26	JA27	JA28	JA29
AMN7	AMN8	AMN9	AMN10	AMN11	AMN12	AMN13
TIR15	TIR16	TIR17	TIR18	TIR19	TIR20	TIR21
8	9	10	11	12	13	14
JA30	JA31	F1	F2	F3	F4	F5
MN14	AMN15	AMN16	AMN17	AMN18	AMN19	AMN20
IR22	TIR23	TIR24	TIR25	TIR26	TIR27	TIR28
5	16	17	18	19	20	21
6	F7	F8	F9	F10	F 11	F12
MN21	AMN22	AMN23	AMN24	AMN25	AMN26	AMN28
IR29	TIR30	YEK1	YEK2	YEK3	YEK4	YEK5
22	23	24	25	26	27	28
F13	F14	F15	F16	F17	F18	F19
MN29	AMN30	RNT1	RNT2	RNT3	RNT4	RNT5
YEK6	YEK7	YEK8	YEK9	YEK10	YEK11	YEK12
HAHIDI 1						
20						
RNT6						
YEK13						

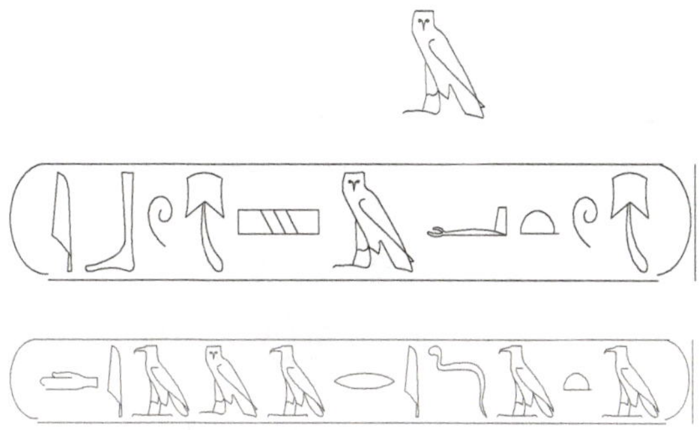

MAAFA 449 B.S.E./ASIENTO 1965-66 C.E./
KAM 6201-02 S.M./EC 1957-58 A.M. (ዓ.ም.)

AFU 449/1965-66/6201-02/1957-58						
Ashe	Ojobo	Iwa	Araorun	Ile	Itutu	Erinle
1	2	3	4	5	6	7
F21	F22	F23	F24	F25	F26	F27
RNT7	RNT8	RNT9	RNT10	RNT11	RNT12	RNT13
YEK14	YEK15	YEK16	YEK17	YEK18	YEK19	YEK20
8	9	10	11	12	13	14
F28	M1	M2	M3	M4	M5	M6
RNT14	RNT15	RNT16	RNT17	RNT18	RNT19	RNT20
YEK21	YEK22	YEK23	YEK24	YEK25	YEK26	YEK27
15	16	17	18	19	20	21
M7	M8	M9	M10	M11	M12	M13
RNT21	RNT22	RNT23	RNT24	RNT25	RNT26	RNT27
YEK28	YEK29	YEK30	MAG1	MAG2	MAG3	MAG4
22	23	24	25	26	27	28
M14	M15	M16	M17	M18	M19	M20
RNT28	RNT29	RNT30	KHN1	KHN2	KHN3	KHN4
MAG5	MAG6	MAG7	MAG8	MAG9	MAG10	MAG11

KAWAIDA 449/1965-66/6201-02/1957-58						
Ashe	Ojobo	Iwa	Araorun	Ile	Itutu	Erinle
1	2	3	4	5	6	7
A18	A19	A20	A21	A22	A23	A24
PIN3	PIN4	PIN5	PIN6	PIN7	PIN8	PIN9
MIY10	MIY11	MIY12	MIY13	MIY14	MIY15	MIY16
8	9	10	11	12	13	14
A25	A26	A27	A28	A29	A30	MA1
PIN10	PIN11	PIN12	PIN13	PIN14	PIN15	PIN16
MIY17	MIY18	MIY19	MIY20	MIY21	MIY22	MIY23
15	16	17	18	19	20	21
MA2	MA3	MA4	MA5	MA6	MA7	MA8
PIN17	PIN18	PIN19	PIN20	PIN21	PIN22	PIN23
MIY24	MIY25	MIY26	MIY27	MIY28	MIY29	MIY30
22	23	24	25	26	27	28
MA9	MA10	MA11	MA12	MA13	MA14	MA15
PIN24	PIN25	PIN26	PIN27	PIN28	PIN29	PIN30
GIN1	GIN2	GIN3	GIN4	GIN5	GIN6	GIN7

UHURU 449/1965-66/6201-02/1957-58						
Ashe	Ojobo	Iwa	Araorun	Ile	Itutu	Erinle
1	2	3	4	5	6	7
M21	M22	M23	M24	M25	M26	M27
KHN5	KHN6	KHN7	KHN8	KHN9	KHN10	KHN11
MAG12	MAG13	MAG14	MAG15	MAG16	MAG17	MAG18
8	9	10	11	12	13	14
M28	M29	M30	M31	A1	A2	A3
KHN12	KHN13	KHN14	KHN15	KHN16	KHN17	KHN18
MAG19	MAG20	MAG21	MAG22	MAG23	MAG24	MAG25
15	16	17	18	19	20	21
A4	A5	A6	A7	A8	A9	A10
KHN19	KHN20	KHN21	KHN22	KHN23	KHN24	KHN25
MAG26	MAG27	MAG28	MAG29	MAG30	MIY1	MIY2
22	23	24	25	26	27	28
A11	A12	A13	A14	A15	A16	A17
KHN26	KHN27	KHN28	KHN29	KHN30	PIN1	PIN2
MIY3	MIY4	MIY5	MIY6	MIY7	MIY8	MIY9

MABABA WEUSI 449/1965-66/6201-02/1957-58						
Ashe	Ojobo	Iwa	Araorun	Ile	Itutu	Erinle
1	2	3	4	5	6	7
MA16	MA17	MA18	MA19	MA20	MA21	MA22
APP1	APP2	APP3	APP4	APP5	APP6	APP7
GIN8	GIN9	GIN10	GIN11	GIN12	GIN13	GIN14
8	9	10	11	12	13	14
MA23	MA24	MA25	MA26	MA27	MA28	MA29
APP8	APP9	APP10	APP11	APP12	APP13	APP14
GIN15	GIN16	GIN17	GIN18	GIN19	GIN20	GIN21
15	16	17	18	19	20	21
MA30	MA31	J1	J2	J3	J4	J5
APP15	APP16	APP17	APP18	APP19	APP20	APP21
GIN22	GIN23	GIN24	GIN25	GIN26	GIN27	GIN28
22	23	24	25	26	27	28
J6	J7	J8	J9	J10	J11	J12
APP22	APP23	APP24	APP25	APP26	APP27	APP28
GIN29	GIN30	SAN1	SAN2	SAN3	SAN4	SAN5

MWANAMKE WEUSI 449/1965-66/6201-02/1957-58

Ashe	Ojobo	Iwa	Araorun	Ile	Itutu	Erinle
1	2	3	4	5	6	7
J13	J14	J15	J16	J17	J18	J19
APP29	APP30	MSR1	MSR2	MSR3	MSR4	MSR5
SAN6	SAN7	SAN8	SAN9	SAN10	SAN11	SAN12
8	9	10	11	12	13	14
J20	J21	J22	J23	J24	J25	J26
MSR6	MSR7	MSR8	MSR9	MSR10	MSR11	MSR12
SAN13	SAN14	SAN15	SAN16	SAN17	SAN18	SAN19
15	16	17	18	19	20	21
J27	J28	J29	J30	JU1	JU2	JU3
MSR13	MSR14	MSR15	MSR16	MSR17	MSR18	MSR19
SAN20	SAN21	SAN22	SAN23	SAN24	SAN25	SAN26
22	23	24	25	26	27	28
JU4	JU5	JU6	JU7	JU8	JU9	JU10
MSR20	MRS21	MSR22	MSR23	MSR24	MSR25	MSR26
SAN27	SAN28	SAN29	SAN30	HAM1	HAM2	HAM3

MTU WEUSI 449/1965-66/6201-02/1957-58

Ashe	Ojobo	Iwa	Araorun	Ile	Itutu	Erinle
1	2	3	4	5	6	7
JU11	JU12	JU13	JU14	JU15	JU16	JU17
MSR27	MSR28	MSR29	MSR30	*ASR1*	*HRU2*	*STH3*
HAM4	HAM5	HAM6	HAM7	HAM8	HAM9	HAM10
8	9	10	11	12	13	14
JU18	JU19	JU20	JU21	JU22	JU23	JU24
AST4	*NBH5*	THT1	THT2	THT3	THT4	THT5
HAM11	HAM12	HAM13	AM14	HAM15	HAM16	HAM17
15	16	17	18	19	20	21
JU25	JU26	JU27	JU28	JU29	JU30	JU31
THT6	THT7	THT8	THT9	THT10	THT11	THT12
HAM18	HAM19	HAM20	HAM21	HAM22	HAM23	HAM24
22	23	24	25	26	27	28
AU1	AU2	AU3	AU4	AU5	AU6	AU7
THT13	THT14	THT15	THT16	THT17	THT18	THT19
HAM25	HAM26	HAM27	HAM28	HAM29	HAM30	NAH1

TAIFA WEUSI 449/1965-66/6201-02/1957-58

Ashe	Ojobo	Iwa	Araorun	Ile	Itutu	Erinle
1	2	3	4	5	6	7
AU8	AU9	AU10	AU11	AU12	AU13	AU14
THT20	THT21	THT22	THT23	THT24	THT25	THT26
NAH2	NAH3	NAH4	NAH5	NAH6	NAH7	NAH8
8	9	10	11	12	13	14
AU15	AU16	AU17	AU18	AU19	AU20	AU21
THT27	THT28	THT29	THT30	PAP1	PAP1	PAP2
NAH9	NAH10	NAH11	NAH12	NAH13	NAH14	NAH15
15	16	17	18	19	20	21
AU22	AU23	AU24	AU25	AU26	AU27	AU28
PAP3	PAP4	PAP5	PAP6	PAP7	PAP8	PAP9
NAH16	NAH17	NAH18	NAH19	NAH20	NAH21	NAH22
22	23	24	25	26	27	28
AU29	AU30	AU31	S1	S2	S3	S4
PAP10	PAP11	PAP12	PAP13	PAP14	PAP15	PAP16
NAH23	NAH24	NAH25	NAH26	NAH27	NAH28	NAH29

UWEZO 449/1965-66/6201-02/1957-58

Ashe	Ojobo	Iwa	Araorun	Ile	Itutu	Erinle
1	2	3	4	5	6	7
S5	S6	S7	S8	S9	S10	S11
PAP17	PAP18	PAP19	PAP20	PAP21	PAP22	PAP23
NAH30	*PAG1*	*PAG2*	*PAG3*	*PAG4*	*PAG5*	MAS1
8	9	10	11	12	13	14
S12	S13	S14	S15	S16	S17	S18
PAP24	PAP25	PAP26	PAP27	PAP28	PAP29	PAP30
MAS2	MAS3	MAS4	MAS5	MAS6	MAS7	MAS8
15	16	17	18	19	20	21
S19	S20	S21	S22	S23	S24	S25
HTH1	HTH2	HTH3	HTH4	HTH5	HTH6	HTH7
MAS9	MAS10	MAS11	MAS12	MAS13	MAS14	MAS15
22	23	24	25	26	27	28
S26	S27	S28	S29	S30	O1	O2
HTH8	HTH9	HTH10	HTH11	HTH12	HTH13	HTH14
MAS16	MAS17	MAS18	MAS19	MAS20	MAS21	MAS22

EKUNDU 449/1965-66/6201-02/1957-58						
Ashe	Ojobo	Iwa	Araorun	Ile	Itutu	Erinle
1	2	3	4	5	6	7
O3	O4	O5	O6	O7	O8	O9
HTH15	HTH16	HTH17	HTH18	HTH19	HTH20	HTH21
MAS23	MAS24	MAS25	MAS26	MAS27	MAS28	MAS29
8	9	10	11	12	13	14
O10	O11	O12	O13	O14	O15	O16
HTH22	HTH23	HTH24	HTH25	HTH26	HTH27	HTH28
MAS30	THi1	THi2	THi3	THi4	THi5	THi6
15	16	17	18	19	20	21
O17	O18	O19	O20	O21	O22	O23
HTH29	HTH30	KHK1	KHK2	KHK3	KHK4	KHK5
THi7	THi8	THi9	THi10	THi11	THi12	THi13
22	23	24	25	26	27	28
O24	O25	O26	O27	O28	O29	O30
KHK6	KHK7	KHK8	KHK9	KHK10	KHK11	KHK12
THi14	THi15	THi16	THi17	THi18	THi19	THi20

KIJANI 449/1965-66/6201-02/1957-58						
Ashe	Ojobo	Iwa	Araorun	Ile	Itutu	Erinle
1	2	3	4	5	6	7
N28	N29	N30	D1	D2	D3	D4
TBT11	TBT12	TBT13	TBT14	TBT15	TBT16	TBT17
HID19	HID20	HID21	HID22	HID23	HID24	HID25
8	9	10	11	12	13	14
D5	D6	D7	D8	D9	D10	D11
TBT18	TBT19	TBT20	TBT21	TBT22	TBT23	TBT24
HID26	HID27	HID28	HID29	HID30	TAH1	TAH2
15	16	17	18	19	20	21
D12	D13	D14	D15	D16	D17	D18
TBT25	TBT26	TBT27	TBT28	TBT29	TBT30	MKH1
TAH3	TAH4	TAH5	TAH6	TAH7	TAH8	TAH9
22	23	24	25	26	27	28
D19	D20	D21	D22	D23	D24	D25
MKH2	MKH3	MKH4	MKH5	MKH6	MKH7	MKH8
TAH10	TAH11	TAH12	TAH13	TAH14	TAH15	TAH16

EUSI 449/1965-66/6201-02/1957-58						
Ashe	Ojobo	Iwa	Araorun	Ile	Itutu	Erinle
1	2	3	4	5	6	7
O31	N1	N2	N3	N4	N5	N6
KHK13	KHK14	KHK15	KHK16	KHK17	KHK18	KHK19
THi21	THi22	THi23	THi24	THi25	THi26	THi27
8	9	10	11	12	13	14
N7	N8	N9	N10	N11	N12	N13
KHK20	KHK21	KHK22	KHK23	KHK24	KHK25	KHK26
THi28	THi29	THi30	HID1	HID2	HID3	HID4
15	16	17	18	19	20	21
N14	N15	N16	N17	N18	N19	N20
KHK27	KHK28	KHK29	KHK30	TBT1	TBT2	TBT3
HID5	HID6	HID7	HID8	HID9	HID10	HID11
22	23	24	25	26	27	28
N21	N22	N23	N24	N25	N26	N27
TBT4	TBT5	TBT6	TBT7	TBT8	TBT9	TBT10
HID12	HID13	HID14	HID15	HID16	HID17	HID18

KWANZA 449/1965-66/6201-02/1957-58						
Umoja	Kujichagulia	Ujima	Ujamaa	Nia	Kuumba	Imani
1	2	3	4	5	6	7
D26	D27	D28	D29	D30	D31	JA1
MKH9	MKH10	MKH11	MKH12	MKH13	MKH14	MKH15
TAH17	TAH18	TAH19	TAH20	TAH21	TAH22	TAH23
8	9	10	11	12	13	14
JA2	JA3	JA4	JA5	JA6	JA7	JA8
MKH16	MKH17	MKH18	MKH19	MKH20	MKH21	MKH22
TAH24	TAH25	TAH26	TAH27	TAH28	TAH29	TAH30
15	16	17	18	19	20	21
JA9	JA10	JA11	JA12	JA13	JA14	JA15
MKH23	MKH24	MKH25	KMH26	MKH27	MKH28	MKH29
TIR1	TIR2	TIR3	TIR4	TIR5	TIR6	TIR7
22	23	24	25	26	27	28
JA16	JA17	JA18	JA19	JA20	JA21	JA22
MKH30	AMN1	AMN2	AMN3	AMN4	AMN5	AMN6
TIR8	TIR9	TIR10	TIR11	TIR12	TIR13	TIR14

SHAHIDI 448/1964-65/6200-01/1956-57						
Ashe	Ojobo	Iwa	Araorun	Ile	Itutu	Erinle
1	2	3	4	5	6	7
JA23	JA24	JA25	JA26	JA27	JA28	JA29
AMN7	AMN8	AMN9	AMN10	AMN11	AMN12	AMN13
TIR15	TIR16	TIR17	TIR18	TIR19	TIR20	TIR21
8	9	10	11	12	13	14
JA30	JA31	F1	F2	F3	F4	F5
AMN14	AMN15	AMN16	AMN17	AMN18	AMN19	AMN20
TIR22	TIR23	TIR24	TIR25	TIR26	TIR27	TIR28
15	16	17	18	19	20	21
F6	F7	F8	F9	F10	F 11	F12
AMN21	AMN22	AMN23	AMN24	AMN25	AMN26	AMN28
TIR29	TIR30	YEK1	YEK2	YEK3	YEK4	YEK5
22	23	24	25	26	27	28
F13	F14	F15	F16	F17	F18	F19
AMN29	AMN30	RNT1	RNT2	RNT3	RNT4	RNT5
YEK6	YEK7	YEK8	YEK9	YEK10	YEK11	YEK12
SHAHIDI 1						
F20						
RNT6						
YEK13						

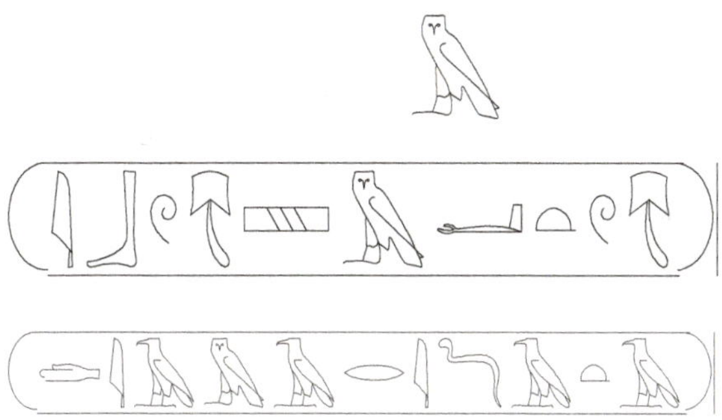

MAAFA 450 B.S.E./ASIENTO 1966-67 C.E./
KAM 6202-03 S.M./EC 1958-59 A.M. (ዓ.ም.)

AFU 450/1966-67/6202-03/1958-59						
Ashe	Ojobo	Iwa	Araorun	Ile	Itutu	Erinle
1	2	3	4	5	6	7
F21	F22	F23	F24	F25	F26	F27
RNT7	RNT8	RNT9	RNT10	RNT11	RNT12	RNT13
YEK14	YEK15	YEK16	YEK17	YEK18	YEK19	YEK20
8	9	10	11	12	13	14
F28	M1	M2	M3	M4	M5	M6
RNT14	RNT15	RNT16	RNT17	RNT18	RNT19	RNT20
YEK21	YEK22	YEK23	YEK24	YEK25	YEK26	YEK27
15	16	17	18	19	20	21
M7	M8	M9	M10	M11	M12	M13
RNT21	RNT22	RNT23	RNT24	RNT25	RNT26	RNT27
YEK28	YEK29	YEK30	MAG1	MAG2	MAG3	MAG4
22	23	24	25	26	27	28
M14	M15	M16	M17	M18	M19	M20
RNT28	RNT29	RNT30	KHN1	KHN2	KHN3	KHN4
MAG5	MAG6	MAG7	MAG8	MAG9	MAG10	MAG11

KAWAIDA 450/1966-67/6202-03/1958-59						
Ashe	Ojobo	Iwa	Araorun	Ile	Itutu	Erinle
1	2	3	4	5	6	7
A18	A19	A20	A21	A22	A23	A24
PIN3	PIN4	PIN5	PIN6	PIN7	PIN8	PIN9
MIY10	MIY11	MIY12	MIY13	MIY14	MIY15	MIY16
8	9	10	11	12	13	14
A25	A26	A27	A28	A29	A30	MA1
PIN10	PIN11	PIN12	PIN13	PIN14	PIN15	PIN16
MIY17	MIY18	MIY19	MIY20	MIY21	MIY22	MIY23
15	16	17	18	19	20	21
MA2	MA3	MA4	MA5	MA6	MA7	MA8
PIN17	PIN18	PIN19	PIN20	PIN21	PIN22	PIN23
MIY24	MIY25	MIY26	MIY27	MIY28	MIY29	MIY30
22	23	24	25	26	27	28
MA9	MA10	MA11	MA12	MA13	MA14	MA15
PIN24	PIN25	PIN26	PIN27	PIN28	PIN29	PIN30
GIN1	GIN2	GIN3	GIN4	GIN5	GIN6	GIN7

UHURU 450/1966-67/6202-03/1958-59						
Ashe	Ojobo	Iwa	Araorun	Ile	Itutu	Erinle
1	2	3	4	5	6	7
M21	M22	M23	M24	M25	M26	M27
KHN5	KHN6	KHN7	KHN8	KHN9	KHN10	KHN11
MAG12	MAG13	MAG14	MAG15	MAG16	MAG17	MAG18
8	9	10	11	12	13	14
M28	M29	M30	M31	A1	A2	A3
KHN12	KHN13	KHN14	KHN15	KHN16	KHN17	KHN18
MAG19	MAG20	MAG21	MAG22	MAG23	MAG24	MAG25
15	16	17	18	19	20	21
A4	A5	A6	A7	A8	A9	A10
KHN19	KHN20	KHN21	KHN22	KHN23	KHN24	KHN25
MAG26	MAG27	MAG28	MAG29	MAG30	MIY1	MIY2
22	23	24	25	26	27	28
A11	A12	A13	A14	A15	A16	A17
KHN26	KHN27	KHN28	KHN29	KHN30	PIN1	PIN2
MIY3	MIY4	MIY5	MIY6	MIY7	MIY8	MIY9

MABABA WEUSI 450/1966-67/6202-03/1958-59						
Ashe	Ojobo	Iwa	Araorun	Ile	Itutu	Erinle
1	2	3	4	5	6	7
MA16	MA17	MA18	MA19	MA20	MA21	MA22
APP1	APP2	APP3	APP4	APP5	APP6	APP7
GIN8	GIN9	GIN10	GIN11	GIN12	GIN13	GIN14
8	9	10	11	12	13	14
MA23	MA24	MA25	MA26	MA27	MA28	MA29
APP8	APP9	APP10	APP11	APP12	APP13	APP14
GIN15	GIN16	GIN17	GIN18	GIN19	GIN20	GIN21
15	16	17	18	19	20	21
MA30	MA31	J1	J2	J3	J4	J5
APP15	APP16	APP17	APP18	APP19	APP20	APP21
GIN22	GIN23	GIN24	GIN25	GIN26	GIN27	GIN28
22	23	24	25	26	27	28
J6	J7	J8	J9	J10	J11	J12
APP22	APP23	APP24	APP25	APP26	APP27	APP28
GIN29	GIN30	SAN1	SAN2	SAN3	SAN4	SAN5

MWANAMKE WEUSI 450/1966-67/6202-03/1958-59

Ashe	Ojobo	Iwa	Araorun	Ile	Itutu	Erinle
1	2	3	4	5	6	7
J13	J14	J15	J16	J17	J18	J19
APP29	APP30	MSR1	MSR2	MSR3	MSR4	MSR5
SAN6	SAN7	SAN8	SAN9	SAN10	SAN11	SAN12
8	9	10	11	12	13	14
J20	J21	J22	J23	J24	J25	J26
MSR6	MSR7	MSR8	MSR9	MSR10	MSR11	MSR12
SAN13	SAN14	SAN15	SAN16	SAN17	SAN18	SAN19
15	16	17	18	19	20	21
J27	J28	J29	J30	JU1	JU2	JU3
MSR13	MSR14	MSR15	MSR16	MSR17	MSR18	MSR19
SAN20	SAN21	SAN22	SAN23	SAN24	SAN25	SAN26
22	23	24	25	26	27	28
JU4	JU5	JU6	JU7	JU8	JU9	JU10
MSR20	MRS21	MSR22	MSR23	MSR24	MSR25	MSR26
SAN27	SAN28	SAN29	SAN30	HAM1	HAM2	HAM3

MTU WEUSI 450/1966-67/6202-03/1958-59

Ashe	Ojobo	Iwa	Araorun	Ile	Itutu	Erinle
1	2	3	4	5	6	7
JU11	JU12	JU13	JU14	JU15	JU16	JU17
MSR27	MSR28	MSR29	MSR30	*ASR1*	*HRU2*	*STH3*
HAM4	HAM5	HAM6	HAM7	HAM8	HAM9	HAM10
8	9	10	11	12	13	14
JU18	JU19	JU20	JU21	JU22	JU23	JU24
AST4	*NBH5*	THT1	THT2	THT3	THT4	THT5
HAM11	HAM12	HAM13	AM14	HAM15	HAM16	HAM17
15	16	17	18	19	20	21
JU25	JU26	JU27	JU28	JU29	JU30	JU31
THT6	THT7	THT8	THT9	THT10	THT11	THT12
HAM18	HAM19	HAM20	HAM21	HAM22	HAM23	HAM24
22	23	24	25	26	27	28
AU1	AU2	AU3	AU4	AU5	AU6	AU7
THT13	THT14	THT15	THT16	THT17	THT18	THT19
HAM25	HAM26	HAM27	HAM28	HAM29	HAM30	NAH1

TAIFA WEUSI 450/1966-67/6202-03/1958-59

Ashe	Ojobo	Iwa	Araorun	Ile	Itutu	Erinle
1	2	3	4	5	6	7
AU8	AU9	AU10	AU11	AU12	AU13	AU14
THT20	THT21	THT22	THT23	THT24	THT25	THT26
NAH2	NAH3	NAH4	NAH5	NAH6	NAH7	NAH8
8	9	10	11	12	13	14
AU15	AU16	AU17	AU18	AU19	AU20	AU21
THT27	THT28	THT29	THT30	PAP1	PAP1	PAP2
NAH9	NAH10	NAH11	NAH12	NAH13	NAH14	NAH15
15	16	17	18	19	20	21
AU22	AU23	AU24	AU25	AU26	AU27	AU28
PAP3	PAP4	PAP5	PAP6	PAP7	PAP8	PAP9
NAH16	NAH17	NAH18	NAH19	NAH20	NAH21	NAH22
22	23	24	25	26	27	28
AU29	AU30	AU31	S1	S2	S3	S4
PAP10	PAP11	PAP12	PAP13	PAP14	PAP15	PAP16
NAH23	NAH24	NAH25	NAH26	NAH27	NAH28	NAH29

UWEZO 450/1966-67/6202-03/1958-59

Ashe	Ojobo	Iwa	Araorun	Ile	Itutu	Erinle
1	2	3	4	5	6	7
S5	S6	S7	S8	S9	S10	S11
PAP17	PAP18	PAP19	PAP20	PAP21	PAP22	PAP23
NAH30	*PAG1*	*PAG2*	*PAG3*	*PAG4*	*PAG5*	MAS1
8	9	10	11	12	13	14
S12	S13	S14	S15	S16	S17	S18
PAP24	PAP25	PAP26	PAP27	PAP28	PAP29	PAP30
MAS2	MAS3	MAS4	MAS5	MAS6	MAS7	MAS8
15	16	17	18	19	20	21
S19	S20	S21	S22	S23	S24	S25
HTH1	HTH2	HTH3	HTH4	HTH5	HTH6	HTH7
MAS9	MAS10	MAS11	MAS12	MAS13	MAS14	MAS15
22	23	24	25	26	27	28
S26	S27	S28	S29	S30	O1	O2
HTH8	HTH9	HTH10	HTH11	HTH12	HTH13	HTH14
MAS16	MAS17	MAS18	MAS19	MAS20	MAS21	MAS22

EKUNDU 450/1966-67/6202-03/1958-59						
he	Ojobo	Iwa	Araorun	Ile	Itutu	Erinle
	2	3	4	5	6	7
	O4	O5	O6	O7	O8	O9
H15	HTH16	HTH17	HTH18	HTH19	HTH20	HTH21
S23	MAS24	MAS25	MAS26	MAS27	MAS28	MAS29
	9	10	11	12	13	14
.0	O11	O12	O13	O14	O15	O16
H22	HTH23	HTH24	HTH25	HTH26	HTH27	HTH28
S30	THi1	THi2	THi3	THi4	THi5	THi6
	16	17	18	19	20	21
.7	O18	O19	O20	O21	O22	O23
H29	HTH30	KHK1	KHK2	KHK3	KHK4	KHK5
i7	THi8	THi9	THi10	THi11	THi12	THi13
	23	24	25	26	27	28
4	O25	O26	O27	O28	O29	O30
K6	KHK7	KHK8	KHK9	KHK10	KHK11	KHK12
i14	THi15	THi16	THi17	THi18	THi19	THi20

KIJANI 450/1966-67/6202-03/1958-59						
Ashe	Ojobo	Iwa	Araorun	Ile	Itutu	Erinle
1	2	3	4	5	6	7
N28	N29	N30	D1	D2	D3	D4
TBT11	TBT12	TBT13	TBT14	TBT15	TBT16	TBT17
HID19	HID20	HID21	HID22	HID23	HID24	HID25
8	9	10	11	12	13	14
D5	D6	D7	D8	D9	D10	D11
TBT18	TBT19	TBT20	TBT21	TBT22	TBT23	TBT24
HID26	HID27	HID28	HID29	HID30	TAH1	TAH2
15	16	17	18	19	20	21
D12	D13	D14	D15	D16	D17	D18
TBT25	TBT26	TBT27	TBT28	TBT29	TBT30	MKH1
TAH3	TAH4	TAH5	TAH6	TAH7	TAH8	TAH9
22	23	24	25	26	27	28
D19	D20	D21	D22	D23	D24	D25
MKH2	MKH3	MKH4	MKH5	MKH6	MKH7	MKH
TAH10	TAH11	TAH12	TAH13	TAH14	TAH15	TAH16

EUSI 450/1966-67/6202-03/1958-59						
he	Ojobo	Iwa	Araorun	Ile	Itutu	Erinle
	2	3	4	5	6	7
1	N1	N2	N3	N4	N5	N6
K13	KHK14	KHK15	KHK16	KHK17	KHK18	KHK19
i21	THi22	THi23	THi24	THi25	THi26	THi27
	9	10	11	12	13	14
'	N8	N9	N10	N11	N12	N13
K20	KHK21	KHK22	KHK23	KHK24	KHK25	KHK26
i28	THi29	THi30	HID1	HID2	HID3	HID4
	16	17	18	19	20	21
4	N15	N16	N17	N18	N19	N20
K27	KHK28	KHK29	KHK30	TBT1	TBT2	TBT3
D5	HID6	HID7	HID8	HID9	HID10	HID11
	23	24	25	26	27	28
1	N22	N23	N24	N25	N26	N27
T4	TBT5	TBT6	TBT7	TBT8	TBT9	TBT10
D12	HID13	HID14	HID15	HID16	HID17	HID18

KWANZA 450/1966-67/6202-03/1958-59						
Umoja	Kujichagulia	Ujima	Ujamaa	Nia	Kuumba	Imani
1	2	3	4	5	6	7
D26	D27	D28	D29	D30	D31	JA1
MKH9	MKH10	MKH11	MKH12	MKH13	MKH14	MKH1
TAH17	TAH18	TAH19	TAH20	TAH21	TAH22	TAH23
8	9	10	11	12	13	14
JA2	JA3	JA4	JA5	JA6	JA7	JA8
MKH16	MKH17	MKH18	MKH19	MKH20	MKH21	MKH2
TAH24	TAH25	TAH26	TAH27	TAH28	TAH29	TAH30
15	16	17	18	19	20	21
JA9	JA10	JA11	JA12	JA13	JA14	JA15
MKH23	MKH24	MKH25	KMH26	MKH27	MKH28	MKH2
TIR1	TIR2	TIR3	TIR4	TIR5	TIR6	TIR7
22	23	24	25	26	27	28
JA16	JA17	JA18	JA19	JA20	JA21	JA22
MKH30	AMN1	AMN2	AMN3	AMN4	AMN5	AMN6
TIR8	TIR9	TIR10	TIR11	TIR12	TIR13	TIR14

SHAHIDI 450/1966-67/6202-03/1958-59						
Ashe	Ojobo	Iwa	Araorun	Ile	Itutu	Erinle
1	2	3	4	5	6	7
JA23	JA24	JA25	JA26	JA27	JA28	JA29
AMN7	AMN8	AMN9	AMN10	AMN11	AMN12	AMN13
TIR15	TIR16	TIR17	TIR18	TIR19	TIR20	TIR21
8	9	10	11	12	13	14
JA30	JA31	F1	F2	F3	F4	F5
AMN14	AMN15	AMN16	AMN17	AMN18	AMN19	AMN20
TIR22	TIR23	TIR24	TIR25	TIR26	TIR27	TIR28
15	16	17	18	19	20	21
F6	F7	F8	F9	F10	F 11	F12
AMN21	AMN22	AMN23	AMN24	AMN25	AMN26	AMN28
TIR29	TIR30	YEK1	YEK2	YEK3	YEK4	YEK5
22	23	24	25	26	27	28
F13	F14	F15	F16	F17	F18	F19
AMN29	AMN30	RNT1	RNT2	RNT3	RNT4	RNT5
YEK6	YEK7	YEK8	YEK9	YEK10	YEK11	YEK12
SHAHIDI 1						
F20						
RNT6						
YEK13						

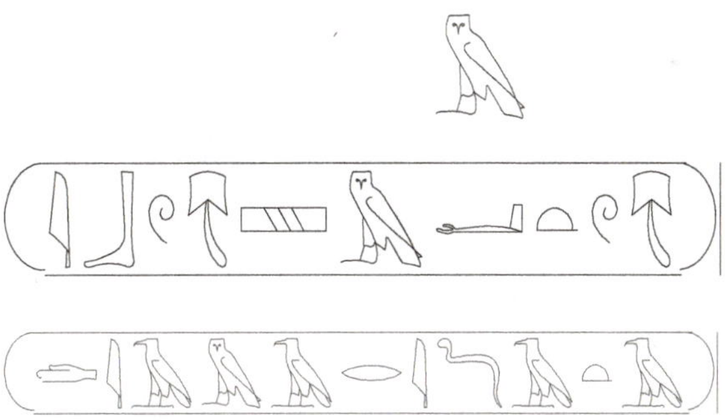

MAAFA 451 B.S.E./ASIENTO 1967-68 C.E./
KAM 6203-04 S.M./EC 1959-60 A.M. (ዓ.ም.)

AFU 451/1967-68/6203-04/1959-60						
Ashe	Ojobo	Iwa	Araorun	Ile	Itutu	Erinle
1	2	3	4	5	6	7
F21	F22	F23	F24	F25	F26	F27
RNT7	RNT8	RNT9	RNT10	RNT11	RNT12	RNT13
YEK14	YEK15	YEK16	YEK17	YEK18	YEK19	YEK20
8	9	10	11	12	13	14
F28	M1	M2	M3	M4	M5	M6
RNT14	RNT15	RNT16	RNT17	RNT18	RNT19	RNT20
YEK21	YEK22	YEK23	YEK24	YEK25	YEK26	YEK27
15	16	17	18	19	20	21
M7	M8	M9	M10	M11	M12	M13
RNT21	RNT22	RNT23	RNT24	RNT25	RNT26	RNT27
YEK28	YEK 29	YEK30	MAG1	MAG2	MAG3	MAG4
22	23	24	25	26	27	28
M14	M15	M16	M17	M18	M19	M20
RNT28	RNT29	RNT30	KHN1	KHN2	KHN3	KHN4
MAG5	MAG6	MAG7	MAG8	MAG9	MAG10	MAG11

KAWAIDA 451/1967-68/6203-04/1959-60						
Ashe	Ojobo	Iwa	Araorun	Ile	Itutu	Erinle
1	2	3	4	5	6	7
A18	A19	A20	A21	A22	A23	A24
PIN3	PIN4	PIN5	PIN6	PIN7	PIN8	PIN9
MIY10	MIY11	MIY12	MIY13	MIY14	MIY15	MIY16
8	9	10	11	12	13	14
A25	A26	A27	A28	A29	A30	MA1
PIN10	PIN11	PIN12	PIN13	PIN14	PIN15	PIN16
MIY17	MIY18	MIY19	MIY20	MIY21	MIY22	MIY23
15	16	17	18	19	20	21
MA2	MA3	MA4	MA5	MA6	MA7	MA8
PIN17	PIN18	PIN19	PIN20	PIN21	PIN22	PIN23
MIY24	MIY25	MIY26	MIY27	MIY28	MIY29	MIY30
22	23	24	25	26	27	28
MA9	MA10	MA11	MA12	MA13	MA14	MA15
PIN24	PIN25	PIN26	PIN27	PIN28	PIN29	PIN30
GIN1	GIN2	GIN3	GIN4	GIN5	GIN6	GIN7

UHURU 451/1967-68/6203-04/1959-60						
Ashe	Ojobo	Iwa	Araorun	Ile	Itutu	Erinle
1	2	3	4	5	6	7
M21	M22	M23	M24	M25	M26	M27
KHN5	KHN6	KHN7	KHN8	KHN9	KHN10	KHN11
MAG12	MAG13	MAG14	MAG15	MAG16	MAG17	MAG18
8	9	10	11	12	13	14
M28	M29	M30	M31	A1	A2	A3
KHN12	KHN13	KHN14	KHN15	KHN16	KHN17	KHN18
MAG19	MAG20	MAG21	MAG22	MAG23	MAG24	MAG25
15	16	17	18	19	20	21
A4	A5	A6	A7	A8	A9	A10
KHN19	KHN20	KHN21	KHN22	KHN23	KHN24	KHN25
MAG26	MAG27	MAG28	MAG29	MAG30	MIY1	MIY2
22	23	24	25	26	27	28
A11	A12	A13	A14	A15	A16	A17
KHN26	KHN27	KHN28	KHN29	KHN30	PIN1	PIN2
MIY3	MIY4	MIY5	MIY6	MIY7	MIY8	MIY9

MABABA WEUSI 451/1967-68/6203-04/1959-60						
Ashe	Ojobo	Iwa	Araorun	Ile	Itutu	Erinle
1	2	3	4	5	6	7
MA16	MA17	MA18	MA19	MA20	MA21	MA22
APP1	APP2	APP3	APP4	APP5	APP6	APP7
GIN8	GIN9	GIN10	GIN11	GIN12	GIN13	GIN14
8	9	10	11	12	13	14
MA23	MA24	MA25	MA26	MA27	MA28	MA29
APP8	APP9	APP10	APP11	APP12	APP13	APP14
GIN15	GIN16	GIN17	GIN18	GIN19	GIN20	GIN21
15	16	17	18	19	20	21
MA30	MA31	J1	J2	J3	J4	J5
APP15	APP16	APP17	APP18	APP19	APP20	APP21
GIN22	GIN23	GIN24	GIN25	GIN26	GIN27	GIN28
22	23	24	25	26	27	28
J6	J7	J8	J9	J10	J11	J12
APP22	APP23	APP24	APP25	APP26	APP27	APP28
GIN29	GIN30	SAN1	SAN2	SAN3	SAN4	SAN5

MWANAMKE WEUSI 451/1967-68/6203-04/1959-60

Ashe	Ojobo	Iwa	Araorun	Ile	Itutu	Erinle
1	2	3	4	5	6	7
J13	J14	J15	J16	J17	J18	J19
APP29	APP30	MSR1	MSR2	MSR3	MSR4	MSR5
SAN6	SAN7	SAN8	SAN9	SAN10	SAN11	SAN12
8	9	10	11	12	13	14
J20	J21	J22	J23	J24	J25	J26
MSR6	MSR7	MSR8	MSR9	MSR10	MSR11	MSR12
SAN13	SAN14	SAN15	SAN16	SAN17	SAN18	SAN19
15	16	17	18	19	20	21
J27	J28	J29	J30	JU1	JU2	JU3
MSR13	MSR14	MSR15	MSR16	MSR17	MSR18	MSR19
SAN20	SAN21	SAN22	SAN23	SAN24	SAN25	SAN26
22	23	24	25	26	27	28
JU4	JU5	JU6	JU7	JU8	JU9	JU10
MSR20	MRS21	MSR22	MSR23	MSR24	MSR25	MSR26
SAN27	SAN28	SAN29	SAN30	HAM1	HAM2	HAM3

TAIFA WEUSI 451/1967-68/6203-04/1959-60

Ashe	Ojobo	Iwa	Araorun	Ile	Itutu	Erinle
1	2	3	4	5	6	7
AU8	AU9	AU10	AU11	AU12	AU13	AU14
THT20	THT21	THT22	THT23	THT24	THT25	THT26
NAH2	NAH3	NAH4	NAH5	NAH6	NAH7	NAH8
8	9	10	11	12	13	14
AU15	AU16	AU17	AU18	AU19	AU20	AU21
THT27	THT28	THT29	THT30	PAP1	PAP1	PAP2
NAH9	NAH10	NAH11	NAH12	NAH13	NAH14	NAH15
15	16	17	18	19	20	21
AU22	AU23	AU24	AU25	AU26	AU27	AU28
PAP3	PAP4	PAP5	PAP6	PAP7	PAP8	PAP9
NAH16	NAH17	NAH18	NAH19	NAH20	NAH21	NAH22
22	23	24	25	26	27	28
AU29	AU30	AU31	S1	S2	S3	S4
PAP10	PAP11	PAP12	PAP13	PAP14	PAP15	PAP16
NAH23	NAH24	NAH25	NAH26	NAH27	NAH28	NAH29

MTU WEUSI 451/1967-68/6203-04/1959-60

Ashe	Ojobo	Iwa	Araorun	Ile	Itutu	Erinle
1	2	3	4	5	6	7
JU11	JU12	JU13	JU14	JU15	JU16	JU17
MSR27	MSR28	MSR29	MSR30	*ASR1*	*HRU2*	*STH3*
HAM4	HAM5	HAM6	HAM7	HAM8	HAM9	HAM10
8	9	10	11	12	13	14
JU18	JU19	JU20	JU21	JU22	JU23	JU24
AST4	*NBH5*	THT1	THT2	THT3	THT4	THT5
HAM11	HAM12	HAM13	HAM14	HAM15	HAM16	HAM17
15	16	17	18	19	20	21
JU25	JU26	JU27	JU28	JU29	JU30	JU31
THT6	THT7	THT8	THT9	THT10	THT11	THT12
HAM18	HAM19	HAM20	HAM21	HAM22	HAM23	HAM24
22	23	24	25	26	27	28
AU1	AU2	AU3	AU4	AU5	AU6	AU7
THT13	THT14	THT15	THT16	THT17	THT18	THT19
HAM25	HAM26	HAM27	HAM28	HAM29	HAM30	NAH1

UWEZO 451/1967-68/6203-04/1959-60

Ashe	Ojobo	Iwa	Araorun	Ile	Itutu	Erinle
1	2	3	4	5	6	7
S5	S6	S7	S8	S9	S10	S11
PAP17	PAP18	PAP19	PAP20	PAP21	PAP22	PAP23
NAH30	*QUA1*	*QUA2*	*QUA3*	*QUA4*	*QUA5*	*QUA6*
8	9	10	11	12	13	14
S12	S13	S14	S15	S16	S17	S18
PAP24	PAP25	PAP26	PAP27	PAP28	PAP29	PAP30
MAS1	MAS2	MAS3	MAS4	MAS5	MAS6	MAS7
15	16	17	18	19	20	21
S19	S20	S21	S22	S23	S24	S25
HTH1	HTH2	HTH3	HTH4	HTH5	HTH6	HTH7
MAS8	MAS9	MAS10	MAS11	MAS12	MAS13	MAS14
22	23	24	25	26	27	28
S26	S27	S28	S29	S30	O1	O2
HTH8	HTH9	HTH10	HTH11	HTH12	HTH13	HTH14
MAS15	MAS16	MAS17	MAS18	MAS19	MAS20	MAS21

EKUNDU 451/1967-68/6203-04/1959-60						
Ashe	Ojobo	Iwa	Araorun	Ile	Itutu	Erinle
1	2	3	4	5	6	7
O3	O4	O5	O6	O7	O8	O9
HTH15	HTH16	HTH17	HTH18	HTH19	HTH20	HTH21
MAS22	MAS23	MAS 24	MAS25	MAS26	MAS27	MAS28
8	9	10	11	12	13	14
O10	O11	O12	O13	O14	O15	O16
HTH22	HTH23	HTH24	HTH25	HTH26	HTH27	HTH28
MAS29	MAS30	THi1	THi2	THi3	THi4	THi5
15	16	17	18	19	20	21
O17	O18	O19	O20	O21	O22	O23
HTH29	HTH30	KHK1	KHK2	KHK3	KHK4	KHK5
THi6	THi7	THi8	THi9	THi10	THi11	THi12
22	23	24	25	26	27	28
O24	O25	O26	O27	O28	O29	O30
KHK6	KHK7	KHK8	KHK9	KHK10	KHK11	KHK12
THi13	THi14	THi15	THi16	THi17	THi18	THi19

KIJANI 451/1967-68/6203-04/1959-60						
Ashe	Ojobo	Iwa	Araorun	Ile	Itutu	Erinle
1	2	3	4	5	6	7
N28	N29	N30	D1	D2	D3	D4
TBT11	TBT12	TBT13	TBT14	TBT15	TBT16	TBT17
HID18	HID19	HID20	HID21	HID22	HID23	HID24
8	9	10	11	12	13	14
D5	D6	D7	D8	D9	D10	D11
TBT18	TBT19	TBT20	TBT21	TBT22	TBT23	TBT24
HID25	HID26	HID27	HID28	HID29	HID30	TAH1
15	16	17	18	19	20	21
D12	D13	D14	D15	D16	D17	D18
TBT25	TBT26	TBT27	TBT28	TBT29	TBT30	MKH1
TAH2	TAH3	TAH4	TAH5	TAH6	TAH7	TAH8
22	23	24	25	26	27	28
D19	D20	D21	D22	D23	D24	D25
MKH2	MKH3	MKH4	MKH5	MKH6	MKH7	MKH8
TAH9	TAH10	TAH11	TAH12	TAH13	TAH14	TAH15

EUSI 451/1967-68/6203-04/1959-60						
Ashe	Ojobo	Iwa	Araorun	Ile	Itutu	Erinle
1	2	3	4	5	6	7
O31	N1	N2	N3	N4	N5	N6
KHK13	KHK14	KHK15	KHK16	KHK17	KHK18	KHK19
THi20	THi21	THi22	THi23	THi24	THi25	THi26
8	9	10	11	12	13	14
N7	N8	N9	N10	N11	N12	N13
KHK20	KHK21	KHK22	KHK23	KHK24	KHK25	KHK26
THi27	THi28	THi29	THi30	HID1	HID2	HID3
15	16	17	18	19	20	21
N14	N15	N16	N17	N18	N19	N20
KHK27	KHK28	KHK29	KHK30	TBT1	TBT2	TBT3
HID4	HID5	HID6	HID7	HID8	HID9	HID10
22	23	24	25	26	27	28
N21	N22	N23	N24	N25	N26	N27
TBT4	TBT5	TBT6	TBT7	TBT8	TBT9	TBT10
HID11	HID12	HID13	HID14	HID15	HID16	HID17

KWANZA 451/1967-68/6203-04/1959-60						
Umoja	Kujichagulia	Ujima	Ujamaa	Nia	Kuumba	Imani
1	2	3	4	5	6	7
D26	D27	D28	D29	D30	D31	JA1
MKH9	MKH10	MKH11	MKH12	MKH13	MKH14	MKH15
TAH16	TAH17	TAH18	TAH19	TAH20	TAH21	TAH22
8	9	10	11	12	13	14
JA2	JA3	JA4	JA5	JA6	JA7	JA8
MKH16	MKH17	MKH18	MKH19	MKH20	MKH21	MKH22
TAH23	TAH24	TAH25	TAH26	TAH27	TAH28	TAH29
15	16	17	18	19	20	21
JA9	JA10	JA11	JA12	JA13	JA14	JA15
MKH23	MKH24	MKH25	KMH26	MKH27	MKH28	MKH29
TAH30	TIR1	TIR2	TIR3	TIR4	TIR5	TIR6
22	23	24	25	26	27	28
JA16	JA17	JA18	JA19	JA20	JA21	JA22
MKH30	AMN1	AMN2	AMN3	AMN4	AMN5	AMN6
TIR7	TIR8	TIR9	TIR10	TIR11	TIR12	TIR13

SHAHIDI 451/1967-68/6203-04/1959-60						
Ashe	Ojobo	Iwa	Araorun	Ile	Itutu	Erinle
1	2	3	4	5	6	7
JA23	JA24	JA25	JA26	JA27	JA28	JA29
AMN7	AMN8	AMN9	AMN10	AMN11	AMN12	AMN13
TIR14	TIR15	TIR16	TIR17	TIR18	TIR19	TIR20
8	9	10	11	12	13	14
JA30	JA31	F1	F2	F3	F4	F5
AMN14	AMN15	AMN16	AMN17	AMN18	AMN19	AMN20
TIR21	TIR22	TIR23	TIR24	TIR25	TIR26	TIR27
15	16	17	18	19	20	21
F6	F7	F8	F9	F10	F 11	F12
AMN21	AMN22	AMN23	AMN24	AMN25	AMN26	AMN28
TIR28	TIR29	TIR30	YEK1	YEK2	YEK3	YEK4
22	23	24	25	26	27	28
F13	F14	F15	F16	F17	F18	F19
AMN29	AMN30	RNT1	RNT2	RNT3	RNT4	RNT5
YEK5	YEK6	YEK7	YEK8	YEK9	YEK10	YEK11
SHAHIDI 2						
F20						
RNT6						
YEK12						

MAAFA 452 B.S.E./ASIENTO 1968-69 C.E./
KAM 6204-05 S.M./EC 1960-61 A.M. (ዓ.ም.)

AFU 452/1968-69/6204-05/1960-61						
Ashe	Ojobo	Iwa	Araorun	Ile	Itutu	Erinle
1	2	3	4	5	6	7
21	F22	F23	F24	F25	F26	F27
NT7	RNT8	RNT9	RNT10	RNT11	RNT12	RNT13
EK13	YEK14	YEK15	YEK16	YEK17	YEK18	YEK19
	9	10	11	12	13	14
28	F29	M1	M2	M3	M4	M5
NT14	RNT15	RNT16	RNT17	RNT18	RNT19	RNT20
EK 20	YEK21	YEK22	YEK23	YEK24	YEK25	YEK26
5	16	17	18	19	20	21
16	M7	M8	M9	M10	M11	M12
NT21	RNT22	RNT23	RNT24	RNT25	RNT26	RNT27
EK 27	YEK28	YEK29	YEK30	MAG1	MAG2	MAG3
22	23	24	25	26	27	28
13	M14	M15	M16	M17	M18	M19
NT28	RNT29	RNT30	KHN1	KHN2	KHN3	KHN4
AG4	MAG5	MAG6	MAG7	MAG8	MAG9	MAG10
HARIFU 110						
20						
HN5						
AG11						

KAWAIDA 452/1968-69/6204-05/1960-61						
Ashe	Ojobo	Iwa	Araorun	Ile	Itutu	Erinle
1	2	3	4	5	6	7
A18	A19	A20	A21	A22	A23	A24
PIN4	PIN5	PIN6	PIN7	PIN8	PIN9	PIN10
MIY10	MIY11	MIY12	MIY13	MIY14	MIY15	MIY16
8	9	10	11	12	13	14
A25	A26	A27	A28	A29	A30	MA1
PIN11	PIN12	PIN13	PIN14	PIN15	PIN16	PIN17
MIY17	MIY18	MIY19	MIY20	MIY21	MIY22	MIY23
15	16	17	18	19	20	21
MA2	MA3	MA4	MA5	MA6	MA7	MA8
PIN18	PIN19	PIN20	PIN21	PIN22	PIN23	PIN24
MIY24	MIY25	MIY26	MIY27	MIY28	MIY29	MIY30
22	23	24	25	26	27	28
MA9	MA10	MA11	MA12	MA13	MA14	MA15
PIN25	PIN26	PIN27	PIN28	PIN29	PIN30	APP1
GIN1	GIN2	GIN3	GIN4	GIN5	GIN6	GIN7

UHURU 452/1968-69/6204-05/1960-61						
Ashe	Ojobo	Iwa	Araorun	Ile	Itutu	Erinle
1	2	3	4	5	6	7
M21	M22	M23	M24	M25	M26	M27
HN6	KHN7	KHN8	KHN9	KHN10	KHN11	KHN12
AG12	MAG13	MAG14	MAG15	MAG16	MAG17	MAG18
8	9	10	11	12	13	14
M28	M29	M30	M31	A1	A2	A3
HN13	KHN14	KHN15	KHN16	KHN17	KHN18	KHN19
AG19	MAG20	MAG21	MAG22	MAG23	MAG24	MAG25
15	16	17	18	19	20	21
A4	A5	A6	A7	A8	A9	A10
HN20	KHN21	KHN222	KHN23	KHN24	KHN25	KHN26
MAG26	MAG27	MAG28	MAG29	MAG30	MIY1	MIY2
22	23	24	25	26	27	28
A11	A12	A13	A14	A15	A16	A17
HN27	KHN28	KHN29	KHN30	PIN1	PIN2	PIN3
IY3	MIY4	MIY5	MIY6	MIY7	MIY8	MIY9

MABABA WEUSI 452/1968-69/6204-05/1960-61						
Ashe	Ojobo	Iwa	Araorun	Ile	Itutu	Erinle
1	2	3	4	5	6	7
MA16	MA17	MA18	MA19	MA20	MA21	MA22
APP2	APP3	APP4	APP5	APP6	APP7	APP8
GIN8	GIN9	GIN10	GIN11	GIN12	GIN13	GIN14
8	9	10	11	12	13	14
MA23	MA24	MA25	MA26	MA27	MA28	MA29
APP9	APP10	APP11	APP12	APP13	APP14	APP15
GIN15	GIN16	GIN17	GIN18	GIN19	GIN20	GIN21
15	16	17	18	19	20	21
MA30	MA31	J1	J2	J3	J4	J5
APP16	APP17	APP18	APP19	APP20	APP21	APP22
GIN22	GIN23	GIN24	GIN25	GIN26	GIN27	GIN28
22	23	24	25	26	27	28
J6	J7	J8	J9	J10	J11	J12
APP23	APP24	APP25	APP26	APP27	APP28	APP29
GIN29	GIN30	SAN1	SAN2	SAN3	SAN4	SAN5

MWANAMKE WEUSI 452/1968-69/6204-05/1960-61

Ashe	Ojobo	Iwa	Araorun	Ile	Itutu	Erinle
1	2	3	4	5	6	7
J13	J14	J15	J16	J17	J18	J19
APP30	MSR1	MSR2	MSR3	MSR4	MSR5	MSR6
SAN6	SAN7	SAN8	SAN9	SAN10	SAN11	SAN12
8	9	10	11	12	13	14
J20	J21	J22	J23	J24	J25	J26
MSR7	MSR8	MSR9	MSR10	MSR11	MSR12	MSR13
SAN13	SAN14	SAN15	SAN16	SAN17	SAN18	SAN19
15	16	17	18	19	20	21
J27	J28	J29	J30	JU1	JU2	JU3
MSR14	MSR15	MSR16	MSR17	MSR18	MSR19	MSR20
SAN20	SAN21	SAN22	SAN23	SAN24	SAN25	SAN26
22	23	24	25	26	27	28
JU4	JU5	JU6	JU7	JU8	JU9	JU10
MSR21	MRS22	MSR23	MSR24	MSR25	MSR26	MSR27
SAN27	SAN28	SAN29	SAN30	HAM1	HAM2	HAM3

TAIFA WEUSI 452/1968-69/6204-05/1960-61

Ashe	Ojobo	Iwa	Araorun	Ile	Itutu	Erinle
1	2	3	4	5	6	7
AU8	AU9	AU10	AU11	AU12	AU13	AU14
THT20	THT21	THT22	THT23	THT24	THT25	THT26
NAH2	NAH3	NAH4	NAH5	NAH6	NAH7	NAH8
8	9	10	11	12	13	14
AU15	AU16	AU17	AU18	AU19	AU20	AU21
THT27	THT28	THT29	THT30	PAP1	PAP1	PAP2
NAH9	NAH10	NAH11	NAH12	NAH13	NAH14	NAH15
15	16	17	18	19	20	21
AU22	AU23	AU24	AU25	AU26	AU27	AU28
PAP3	PAP4	PAP5	PAP6	PAP7	PAP8	PAP9
NAH16	NAH17	NAH18	NAH19	NAH20	NAH21	NAH22
22	23	24	25	26	27	28
AU29	AU30	AU31	S1	S2	S3	S4
PAP10	PAP11	PAP12	PAP13	PAP14	PAP15	PAP16
NAH23	NAH24	NAH25	NAH26	NAH27	NAH28	NAH29

MTU WEUSI 452/1968-69/6204-05/1960-61

Ashe	Ojobo	Iwa	Araorun	Ile	Itutu	Erinle
1	2	3	4	5	6	7
JU11	JU12	JU13	JU14	JU15	JU16	JU17
MSR28	MSR29	MSR30	*ASR1*	*HRU2*	*STH3*	*AST4*
HAM4	HAM5	HAM6	HAM7	HAM8	HAM9	HAM10
8	9	10	11	12	13	14
JU18	JU19	JU20	JU21	JU22	JU23	JU24
NBH5	*ANU6*	THT1	THT2	THT3	THT4	THT5
HAM11	HAM12	HAM13	HAM14	HAM15	HAM16	HAM17
15	16	17	18	19	20	21
JU25	JU26	JU27	JU28	JU29	JU30	JU31
THT6	THT7	THT8	THT9	THT10	THT11	THT12
HAM18	HAM19	HAM20	HAM21	HAM22	HAM23	HAM24
22	23	24	25	26	27	28
AU1	AU2	AU3	AU4	AU5	AU6	AU7
THT13	THT14	THT15	THT16	THT17	THT18	THT19
HAM25	HAM26	HAM27	HAM28	HAM29	HAM30	NAH1

UWEZO 452/1968-69/6204-05/1960-61

Ashe	Ojobo	Iwa	Araorun	Ile	Itutu	Erinle
1	2	3	4	5	6	7
S5	S6	S7	S8	S9	S10	S11
PAP17	PAP18	PAP19	PAP20	PAP21	PAP22	PAP23
NAH30	*PAG1*	*PAG2*	*PAG3*	*PAG4*	*PAG5*	MAS1
8	9	10	11	12	13	14
S12	S13	S14	S15	S16	S17	S18
PAP24	PAP25	PAP26	PAP27	PAP28	PAP29	PAP30
MAS2	MAS3	MAS4	MAS5	MAS6	MAS7	MAS8
15	16	17	18	19	20	21
S19	S20	S21	S22	S23	S24	S25
HTH1	HTH2	HTH3	HTH4	HTH5	HTH6	HTH7
MAS9	MAS10	MAS11	MAS12	MAS13	MAS14	MAS15
22	23	24	25	26	27	28
S26	S27	S28	S29	S30	O1	O2
HTH8	HTH9	HTH10	HTH11	HTH12	HTH13	HTH14
MAS16	MAS17	MAS18	MAS19	MAS20	MAS21	MAS22

EKUNDU 452/1968-69/6204-05/1960-61

he	Ojobo	Iwa	Araorun	Ile	Itutu	Erinle
	2	3	4	5	6	7
	O4	O5	O6	O7	O8	O9
TH15	HTH16	HTH17	HTH18	HTH19	HTH20	HTH21
AS23	MAS24	MAS25	MAS26	MAS27	MAS28	MAS29
	9	10	11	12	13	14
10	O11	O12	O13	O14	O15	O16
H22	HTH23	HTH24	HTH25	HTH26	HTH27	HTH28
AS30	THi1	THi2	THi3	THi4	THi5	THi6
	16	17	18	19	20	21
17	O18	O19	O20	O21	O22	O23
H29	HTH30	KHK1	KHK2	KHK3	KHK4	KHK5
i7	THi8	This9	THi10	THi11	THi12	THi13
	23	24	25	26	27	28
4	O25	O26	O27	O28	O29	O30
K6	KHK7	KHK8	KHK9	KHK10	KHK11	KHK12
i14	THi15	THi16	THi17	THi18	THi19	THi20

KIJANI 452/1968-69/6204-05/1960-61

Ashe	Ojobo	Iwa	Araorun	Ile	Itutu	Erinle
1	2	3	4	5	6	7
N28	N29	N30	D1	D2	D3	D4
TBT11	TBT12	TBT13	TBT14	TBT15	TBT16	TBT17
HID19	HID20	HID21	HID22	HID23	HID24	HID25
8	9	10	11	12	13	14
D5	D6	D7	D8	D9	D10	D11
TBT18	TBT19	TBT20	TBT21	TBT22	TBT23	TBT24
HID26	HID27	HID28	HID29	HID30	TAH1	TAH2
15	16	17	18	19	20	21
D12	D13	D14	D15	D16	D17	D18
TBT25	TBT26	TBT27	TBT28	TBT29	TBT30	MKH1
TAH3	TAH4	TAH5	TAH6	TAH7	TAH8	TAH9
22	23	24	25	26	27	28
D19	D20	D21	D22	D23	D24	D25
MKH2	MKH3	MKH4	MKH5	MKH6	MKH7	MKH8
TAH10	TAH11	TAH12	TAH13	TAH14	TAH15	TAH16

EUSI 452/1968-69/6204-05/1960-61

he	Ojobo	Iwa	Araorun	Ile	Itutu	Erinle
	2	3	4	5	6	7
1	N1	N2	N3	N4	N5	N6
K13	KHK14	KHK15	KHK16	KHK17	KHK18	KHK19
i21	THi22	THi23	THi24	THi25	THi26	THi27
	9	10	11	12	13	14
7	N8	N9	N10	N11	N12	N13
K20	KHK21	KHK22	KHK23	KHK24	KHK25	KHK26
i28	THi29	THi30	HID1	HID2	HID3	HID4
	16	17	18	19	20	21
4	N15	N16	N17	N18	N19	N20
K27	KHK28	KHK29	KHK30	TBT1	TBT2	TBT3
D5	HID6	HID7	HID8	HID9	HID10	HID11
2	23	24	25	26	27	28
1	N22	N23	N24	N25	N26	N27
T4	TBT5	TBT6	TBT7	TBT8	TBT9	TBT10
D12	HID13	HID14	HID15	HID16	HID17	HID18

KWANZA 452/1968-69/6204-05/1960-61

Umoja	Kujichagulia	Ujima	Ujamaa	Nia	Kuumba	Imani
1	2	3	4	5	6	7
D26	D27	D28	D29	D30	D31	JA1
MKH9	MKH10	MKH11	MKH12	MKH13	MKH14	MKH15
TAH17	TAH18	TAH19	TAH20	TAH21	TAH22	TAH23
8	9	10	11	12	13	14
JA2	JA3	JA4	JA5	JA6	JA7	JA8
MKH16	MKH17	MKH18	MKH19	MKH20	MKH21	MKH22
TAH24	TAH25	TAH26	TAH27	TAH28	TAH29	TAH30
15	16	17	18	19	20	21
JA9	JA10	JA11	JA12	JA13	JA14	JA15
MKH23	MKH24	MKH25	KMH26	MKH27	MKH28	MKH29
TIR1	TIR2	TIR3	TIR4	TIR5	TIR6	TIR7
22	23	24	25	26	27	28
JA16	JA17	JA18	JA19	JA20	JA21	JA22
MKH30	AMN1	AMN2	AMN3	AMN4	AMN5	AMN6
TIR8	TIR9	TIR10	TIR11	TIR12	TIR13	TIR14

SHAHIDI 452/1968-69/6204-05/1960-61						
Ashe	Ojobo	Iwa	Araorun	Ile	Itutu	Erinle
1	2	3	4	5	6	7
JA23	JA24	JA25	JA26	JA27	JA28	JA29
AMN7	AMN8	AMN9	AMN10	AMN11	AMN12	AMN13
TIR15	TIR16	TIR17	TIR18	TIR19	TIR20	TIR21
8	9	10	11	12	13	14
JA30	JA31	F1	F2	F3	F4	F5
AMN14	AMN15	AMN16	AMN17	AMN18	AMN19	AMN20
TIR22	TIR23	TIR24	TIR25	TIR26	TIR27	TIR28
15	16	17	18	19	20	21
F6	F7	F8	F9	F10	F 11	F12
AMN21	AMN22	AMN23	AMN24	AMN25	AMN26	AMN28
TIR29	TIR30	YEK1	YEK2	YEK3	YEK4	YEK5
22	23	24	25	26	27	28
F13	F14	F15	F16	F17	F18	F19
AMN29	AMN30	RNT1	RNT2	RNT3	RNT4	RNT5
YEK6	YEK7	YEK8	YEK9	YEK10	YEK11	YEK12
SHAHIDI 3						
F20						
RNT6						
YEK13						

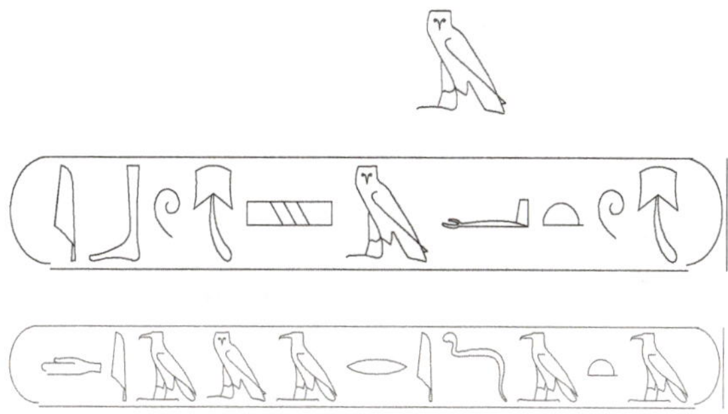

AFU 453/1969-70/6205-06/1961-62						
shhe	Ojobo	Iwa	Araorun	Ile	Itutu	Erinle
1	2	3	4	5	6	7
21	F22	F23	F24	F25	F26	F27
NT7	RNT8	RNT9	RNT10	RNT11	RNT12	RNT13
EK14	YEK15	YEK16	YEK17	YEK18	YEK19	YEK20
	9	10	11	12	13	14
28	M1	M2	M3	M4	M5	M6
NT14	RNT15	RNT16	RNT17	RNT18	RNT19	RNT20
EK21	YEK22	YEK23	YEK24	YEK25	YEK26	YEK27
5	16	17	18	19	20	21
17	M8	M9	M10	M11	M12	M13
NT21	RNT22	RNT23	RNT24	RNT25	RNT26	RNT27
EK28	YEK29	YEK30	MAG1	MAG2	MAG3	MAG4
2	23	24	25	26	27	28
14	M15	M16	M17	M18	M19	M20
NT28	RNT29	RNT30	KHN1	KHN2	KHN3	KHN4
AG5	MAG6	MAG7	MAG8	MAG9	MAG10	MAG11

KAWAIDA 453/1969-70/6205-06/1961-62						
Ashe	Ojobo	Iwa	Araorun	Ile	Itutu	Erinle
1	2	3	4	5	6	7
A18	A19	A20	A21	A22	A23	A24
PIN3	PIN4	PIN5	PIN6	PIN7	PIN8	PIN9
MIY10	MIY11	MIY12	MIY13	MIY14	MIY15	MIY16
8	9	10	11	12	13	14
A25	A26	A27	A28	A29	A30	MA1
PIN10	PIN11	PIN12	PIN13	PIN14	PIN15	PIN16
MIY17	MIY18	MIY19	MIY20	MIY21	MIY22	MIY23
15	16	17	18	19	20	21
MA2	MA3	MA4	MA5	MA6	MA7	MA8
PIN17	PIN18	PIN19	PIN20	PIN21	PIN22	PIN23
MIY24	MIY25	MIY26	MIY27	MIY28	MIY29	MIY30
22	23	24	25	26	27	28
MA9	MA10	MA11	MA12	MA13	MA14	MA15
PIN24	PIN25	PIN26	PIN27	PIN28	PIN29	PIN30
GIN1	GIN2	GIN3	GIN4	GIN5	GIN6	GIN7

UHURU 453/1969-70/6205-06/1961-62						
shhe	Ojobo	Iwa	Araorun	Ile	Itutu	Erinle
	2	3	4	5	6	7
21	M22	M23	M24	M25	M26	M27
HN5	KHN6	KHN7	KHN8	KHN9	KHN10	KHN11
AG12	MAG13	MAG14	MAG15	MAG16	MAG17	MAG18
8	9	10	11	12	13	14
28	M29	M30	M31	A1	A2	A3
HN12	KHN13	KHN14	KHN15	KHN16	KHN17	KHN18
AG19	MAG20	MAG21	MAG22	MAG23	MAG24	MAG25
15	16	17	18	19	20	21
4	A5	A6	A7	A8	A9	A10
HN19	KHN20	KHN21	KHN22	KHN23	KHN24	KHN25
AG26	MAG27	MAG28	MAG29	MAG30	MIY1	MIY2
22	23	24	25	26	27	28
11	A12	A13	A14	A15	A16	A17
HN26	KHN27	KHN28	KHN29	KHN30	PIN1	PIN2
IY3	MIY4	MIY5	MIY6	MIY7	MIY8	MIY9

MABABA WEUSI 453/1969-70/6205-06/1961-62						
Ashe	Ojobo	Iwa	Araorun	Ile	Itutu	Erinle
1	2	3	4	5	6	7
MA16	MA17	MA18	MA19	MA20	MA21	MA22
APP1	APP2	APP3	APP4	APP5	APP6	APP7
GIN8	GIN9	GIN10	GIN11	GIN12	GIN13	GIN14
8	9	10	11	12	13	14
MA23	MA24	MA25	MA26	MA27	MA28	MA29
APP8	APP9	APP10	APP11	APP12	APP13	APP14
GIN15	GIN16	GIN17	GIN18	GIN19	GIN20	GIN21
15	16	17	18	19	20	21
MA30	MA31	J1	J2	J3	J4	J5
APP15	APP16	APP17	APP18	APP19	APP20	APP21
GIN22	GIN23	GIN24	GIN25	GIN26	GIN27	GIN28
22	23	24	25	26	27	28
J6	J7	J8	J9	J10	J11	J12
APP22	APP23	APP24	APP25	APP26	APP27	APP28
GIN29	GIN30	SAN1	SAN2	SAN3	SAN4	SAN5

MWANAMKE WEUSI 453/1969-70/6205-06/1961-62						
Ashe	Ojobo	Iwa	Araorun	Ile	Itutu	Erinle
1	2	3	4	5	6	7
J13	J14	J15	J16	J17	J18	J19
APP29	APP30	MSR1	MSR2	MSR3	MSR4	MSR5
SAN6	SAN7	SAN8	SAN9	SAN10	SAN11	SAN12
8	9	10	11	12	13	14
J20	J21	J22	J23	J24	J25	J26
MSR6	MSR7	MSR8	MSR9	MSR10	MSR11	MSR12
SAN13	SAN14	SAN15	SAN16	SAN17	SAN18	SAN19
15	16	17	18	19	20	21
J27	J28	J29	J30	JU1	JU2	JU3
MSR13	MSR14	MSR15	MSR16	MSR17	MSR18	MSR19
SAN20	SAN21	SAN22	SAN23	SAN24	SAN25	SAN26
22	23	24	25	26	27	28
JU4	JU5	JU6	JU7	JU8	JU9	JU10
MSR20	MRS21	MSR22	MSR23	MSR24	MSR25	MSR26
SAN27	SAN28	SAN29	SAN30	HAM1	HAM2	HAM3

MTU WEUSI 453/1969-70/6205-06/1961-62						
Ashe	Ojobo	Iwa	Araorun	Ile	Itutu	Erinle
1	2	3	4	5	6	7
JU11	JU12	JU13	JU14	JU15	JU16	JU17
MSR27	MSR28	MSR29	MSR30	*ASR1*	*HRU2*	*STH3*
HAM4	HAM5	HAM6	HAM7	HAM8	HAM9	HAM10
8	9	10	11	12	13	14
JU18	JU19	JU20	JU21	JU22	JU23	JU24
AST4	*NBH5*	THT1	THT2	THT3	THT4	THT5
HAM11	HAM12	HAM13	AM14	HAM15	HAM16	HAM17
15	16	17	18	19	20	21
JU25	JU26	JU27	JU28	JU29	JU30	JU31
THT6	THT7	THT8	THT9	THT10	THT11	THT12
HAM18	HAM19	HAM20	HAM21	HAM22	HAM23	HAM24
22	23	24	25	26	27	28
AU1	AU2	AU3	AU4	AU5	AU6	AU7
THT13	THT14	THT15	THT16	THT17	THT18	THT19
HAM25	HAM26	HAM27	HAM28	HAM29	HAM30	NAH1

TAIFA WEUSI 453/1969-70/6205-06/1961-62						
Ashe	Ojobo	Iwa	Araorun	Ile	Itutu	Erinle
1	2	3	4	5	6	7
AU8	AU9	AU10	AU11	AU12	AU13	AU14
THT20	THT21	THT22	THT23	THT24	THT25	THT26
NAH2	NAH3	NAH4	NAH5	NAH6	NAH7	NAH8
8	9	10	11	12	13	14
AU15	AU16	AU17	AU18	AU19	AU20	AU21
THT27	THT28	THT29	THT30	PAP1	PAP1	PAP2
NAH9	NAH10	NAH11	NAH12	NAH13	NAH14	NAH15
15	16	17	18	19	20	21
AU22	AU23	AU24	AU25	AU26	AU27	AU28
PAP3	PAP4	PAP5	PAP6	PAP7	PAP8	PAP9
NAH16	NAH17	NAH18	NAH19	NAH20	NAH21	NAH22
22	23	24	25	26	27	28
AU29	AU30	AU31	S1	S2	S3	S4
PAP10	PAP11	PAP12	PAP13	PAP14	PAP15	PAP16
NAH23	NAH24	NAH25	NAH26	NAH27	NAH28	NAH29

UWEZO 453/1969-70/6205-06/1961-62						
Ashe	Ojobo	Iwa	Araorun	Ile	Itutu	Erinle
1	2	3	4	5	6	7
S5	S6	S7	S8	S9	S10	S11
PAP17	PAP18	PAP19	PAP20	PAP21	PAP22	PAP23
NAH30	*PAG1*	*PAG2*	*PAG3*	*PAG4*	*PAG5*	MAS1
8	9	10	11	12	13	14
S12	S13	S14	S15	S16	S17	S18
PAP24	PAP25	PAP26	PAP27	PAP28	PAP29	PAP30
MAS2	MAS3	MAS4	MAS5	MAS6	MAS7	MAS8
15	16	17	18	19	20	21
S19	S20	S21	S22	S23	S24	S25
HTH1	HTH2	HTH3	HTH4	HTH5	HTH6	HTH7
MAS9	MAS10	MAS11	MAS12	MAS13	MAS14	MAS15
22	23	24	25	26	27	28
S26	S27	S28	S29	S30	O1	O2
HTH8	HTH9	HTH10	HTH11	HTH12	HTH13	HTH14
MAS16	MAS17	MAS18	MAS19	MAS20	MAS21	MAS22

EKUNDU 453/1969-70/6205-06/1961-62						
Ashe	Ojobo	Iwa	Araorun	Ile	Itutu	Erinle
1	2	3	4	5	6	7
O3	O4	O5	O6	O7	O8	O9
HTH15	HTH16	HTH17	HTH18	HTH19	HTH20	HTH21
MAS23	MAS24	MAS25	MAS26	MAS27	MAS28	MAS29
8	9	10	11	12	13	14
O10	O11	O12	O13	O14	O15	O16
HTH22	HTH23	HTH24	HTH25	HTH26	HTH27	HTH28
MAS30	THi1	THi2	THi3	THi4	THi5	THi6
15	16	17	18	19	20	21
O17	O18	O19	O20	O21	O22	O23
HTH29	HTH30	KHK1	KHK2	KHK3	KHK4	KHK5
THi7	THi8	THi9	THi10	THi11	THi12	THi13
22	23	24	25	26	27	28
O24	O25	O26	O27	O28	O29	O30
KHK6	KHK7	KHK8	KHK9	KHK10	KHK11	KHK12
THi14	THi15	THi16	THi17	THi18	THi19	THi20

KIJANI 453/1969-70/6205-06/1961-62						
Ashe	Ojobo	Iwa	Araorun	Ile	Itutu	Erinle
1	2	3	4	5	6	7
N28	N29	N30	D1	D2	D3	D4
TBT11	TBT12	TBT13	TBT14	TBT15	TBT16	TBT17
HID19	HID20	HID21	HID22	HID23	HID24	HID25
8	9	10	11	12	13	14
D5	D6	D7	D8	D9	D10	D11
TBT18	TBT19	TBT20	TBT21	TBT22	TBT23	TBT24
HID26	HID27	HID28	HID29	HID30	TAH1	TAH2
15	16	17	18	19	20	21
D12	D13	D14	D15	D16	D17	D18
TBT25	TBT26	TBT27	TBT28	TBT29	TBT30	MKH1
TAH3	TAH4	TAH5	TAH6	TAH7	TAH8	TAH9
22	23	24	25	26	27	28
D19	D20	D21	D22	D23	D24	D25
MKH2	MKH3	MKH4	MKH5	MKH6	MKH7	MKH8
TAH10	TAH11	TAH12	TAH13	TAH14	TAH15	TAH16

EUSI 453/1969-70/6205-06/1961-62						
Ashe	Ojobo	Iwa	Araorun	Ile	Itutu	Erinle
1	2	3	4	5	6	7
O31	N1	N2	N3	N4	N5	N6
KHK13	KHK14	KHK15	KHK16	KHK17	KHK18	KHK19
THi21	THi22	THi23	THi24	THi25	THi26	THi27
8	9	10	11	12	13	14
N7	N8	N9	N10	N11	N12	N13
KHK20	KHK21	KHK22	KHK23	KHK24	KHK25	KHK26
THi28	THi29	THi30	HID1	HID2	HID3	HID4
15	16	17	18	19	20	21
N14	N15	N16	N17	N18	N19	N20
KHK27	KHK28	KHK29	KHK30	TBT1	TBT2	TBT3
HID5	HID6	HID7	HID8	HID9	HID10	HID11
22	23	24	25	26	27	28
N21	N22	N23	N24	N25	N26	N27
TBT4	TBT5	TBT6	TBT7	TBT8	TBT9	TBT10
HID12	HID13	HID14	HID15	HID16	HID17	HID18

KWANZA 453/1969-70/6205-06/1961-62						
Umoja	Kujichagulia	Ujima	Ujamaa	Nia	Kuumba	Imani
1	2	3	4	5	6	7
D26	D27	D28	D29	D30	D31	JA1
MKH9	MKH10	MKH11	MKH12	MKH13	MKH14	MKH15
TAH17	TAH18	TAH19	TAH20	TAH21	TAH22	TAH23
8	9	10	11	12	13	14
JA2	JA3	JA4	JA5	JA6	JA7	JA8
MKH16	MKH17	MKH18	MKH19	MKH20	MKH21	MKH22
TAH24	TAH25	TAH26	TAH27	TAH28	TAH29	TAH30
15	16	17	18	19	20	21
JA9	JA10	JA11	JA12	JA13	JA14	JA15
MKH23	MKH24	MKH25	KMH26	MKH27	MKH28	MKH29
TIR1	TIR2	TIR3	TIR4	TIR5	TIR6	TIR7
22	23	24	25	26	27	28
JA16	JA17	JA18	JA19	JA20	JA21	JA22
MKH30	AMN1	AMN2	AMN3	AMN4	AMN5	AMN6
TIR8	TIR9	TIR10	TIR11	TIR12	TIR13	TIR14

SHAHIDI 453/1969-70/6205-06/1961-62						
Ashe	Ojobo	Iwa	Araorun	Ile	Itutu	Erinle
1	2	3	4	5	6	7
JA23	JA24	JA25	JA26	JA27	JA28	JA29
AMN7	AMN8	AMN9	AMN10	AMN11	AMN12	AMN13
TIR15	TIR16	TIR17	TIR18	TIR19	TIR20	TIR21
8	9	10	11	12	13	14
JA30	JA31	F1	F2	F3	F4	F5
MN14	AMN15	AMN16	AMN17	AMN18	AMN19	AMN20
TIR22	TIR23	TIR24	TIR25	TIR26	TIR27	TIR28
5	16	17	18	19	20	21
6	F7	F8	F9	F10	F 11	F12
MN21	AMN22	AMN23	AMN24	AMN25	AMN26	AMN28
IR29	TIR30	YEK1	YEK2	YEK3	YEK4	YEK5
22	23	24	25	26	27	28
F13	F14	F15	F16	F17	F18	F19
MN29	AMN30	RNT1	RNT2	RNT3	RNT4	RNT5
EK6	YEK7	YEK8	YEK9	YEK10	YEK11	YEK12
HAHIDI 4						
20						
NT6						
EK13						

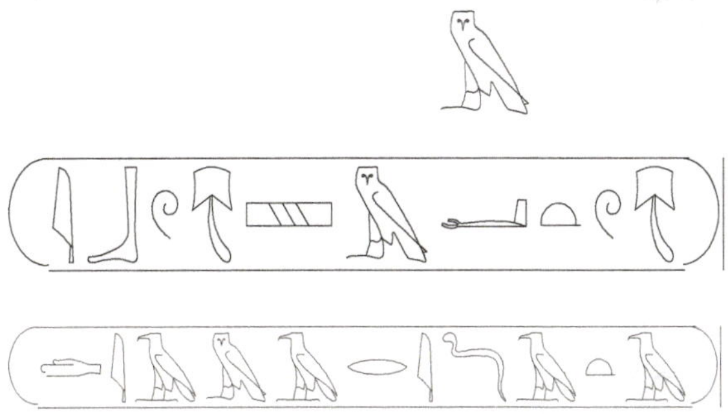

MAAFA 454 B.S.E./ASIENTO 1970-71 C.E./ KAM 6206-07 S.M./EC 1962-63 A.M. (ዓ.ም.)

AFU 454/1970-71/6206-07/1962-63

she	Ojobo	Iwa	Araorun	Ile	Itutu	Erinle
	2	3	4	5	6	7
1	F22	F23	F24	F25	F26	F27
NT7	RNT8	RNT9	RNT10	RNT11	RNT12	RNT13
EK14	YEK15	YEK16	YEK17	YEK18	YEK19	YEK20
	9	10	11	12	13	14
28	M1	M2	M3	M4	M5	M6
NT14	RNT15	RNT16	RNT17	RNT18	RNT19	RNT20
EK21	YEK22	YEK23	YEK24	YEK25	YEK26	YEK27
	16	17	18	19	20	21
I7	M8	M9	M10	M11	M12	M13
NT21	RNT22	RNT23	RNT24	RNT25	RNT26	RNT27
EK28	YEK29	YEK30	MAG1	MAG2	MAG3	MAG4
2	23	24	25	26	27	28
I14	M15	M16	M17	M18	M19	M20
NT28	RNT29	RNT30	KHN1	KHN2	KHN3	KHN4
AG5	MAG6	MAG7	MAG8	MAG9	MAG10	MAG11

KAWAIDA 454/1970-71/6206-07/1962-63

Ashe	Ojobo	Iwa	Araorun	Ile	Itutu	Eri
1	2	3	4	5	6	7
A18	A19	A20	A21	A22	A23	A2
PIN3	PIN4	PIN5	PIN6	PIN7	PIN8	PIN
MIY10	MIY11	MIY12	MIY13	MIY14	MIY15	MI
8	9	10	11	12	13	14
A25	A26	A27	A28	A29	A30	MA
PIN10	PIN11	PIN12	PIN13	PIN14	PIN15	PIN
MIY17	MIY18	MIY19	MIY20	MIY21	MIY22	MIY
15	16	17	18	19	20	21
MA2	MA3	MA4	MA5	MA6	MA7	M
PIN17	PIN18	PIN19	PIN20	PIN21	PIN22	PI
MIY24	MIY25	MIY26	MIY27	MIY28	MIY29	M
22	23	24	25	26	27	2
MA9	MA10	MA11	MA12	MA13	MA14	M
PIN24	PIN25	PIN26	PIN27	PIN28	PIN29	P
GIN1	GIN2	GIN3	GIN4	GIN5	GIN6	G

UHURU 454/1970-71/6206-07/1962-63

she	Ojobo	Iwa	Araorun	Ile	Itutu	Erinle
	2	3	4	5	6	7
I21	M22	M23	M24	M25	M26	M27
HN5	KHN6	KHN7	KHN8	KHN9	KHN10	KHN11
AG12	MAG13	MAG14	MAG15	MAG16	MAG17	MAG18
	9	10	11	12	13	14
I28	M29	M30	M31	A1	A2	A3
HN12	KHN13	KHN14	KHN15	KHN16	KHN17	KHN18
AG19	MAG20	MAG21	MAG22	MAG23	MAG24	MAG25
5	16	17	18	19	20	21
4	A5	A6	A7	A8	A9	A10
HN19	KHN20	KHN21	KHN22	KHN23	KHN24	KHN25
AG26	MAG27	MAG28	MAG29	MAG30	MIY1	MIY2
2	23	24	25	26	27	28
11	A12	A13	A14	A15	A16	A17
HN26	KHN27	KHN28	KHN29	KHN30	PIN1	PIN2
IY3	MIY4	MIY5	MIY6	MIY7	MIY8	MIY9

MABABA WEUSI 454/1970-71/6206-07/1962-63

Ashe	Ojobo	Iwa	Araorun	Ile	Itutu	Eri
1	2	3	4	5	6	7
MA16	MA17	MA18	MA19	MA20	MA21	M.
APP1	APP2	APP3	APP4	APP5	APP6	AI
GIN8	GIN9	GIN10	GIN11	GIN12	GIN13	GI
8	9	10	11	12	13	14
MA23	MA24	MA25	MA26	MA27	MA28	M.
APP8	APP9	APP10	APP11	APP12	APP13	AI
GIN15	GIN16	GIN17	GIN18	GIN19	GIN20	GI
15	16	17	18	19	20	21
MA30	MA31	J1	J2	J3	J4	J
APP15	APP16	APP17	APP18	APP19	APP20	
GIN22	GIN23	GIN24	GIN25	GIN26	GIN27	(
22	23	24	25	26	27	28
J6	J7	J8	J9	J10	J11	J1
APP22	APP23	APP24	APP25	APP26	APP27	A
GIN29	GIN30	SAN1	SAN2	SAN3	SAN4	SA

MWANAMKE WEUSI 454/1970-71/6206-07/1962-63

Ashe	Ojobo	Iwa	Araorun	Ile	Itutu	Erinle
1	2	3	4	5	6	7
J13	J14	J15	J16	J17	J18	J19
APP29	APP30	MSR1	MSR2	MSR3	MSR4	MSR5
SAN6	SAN7	SAN8	SAN9	SAN10	SAN11	SAN12
8	9	10	11	12	13	14
J20	J21	J22	J23	J24	J25	J26
MSR6	MSR7	MSR8	MSR9	MSR10	MSR11	MSR12
SAN13	SAN14	SAN15	SAN16	SAN17	SAN18	SAN19
15	16	17	18	19	20	21
J27	J28	J29	J30	JU1	JU2	JU3
MSR13	MSR14	MSR15	MSR16	MSR17	MSR18	MSR19
SAN20	SAN21	SAN22	SAN23	SAN24	SAN25	SAN26
22	23	24	25	26	27	28
JU4	JU5	JU6	JU7	JU8	JU9	JU10
MSR20	MRS21	MSR22	MSR23	MSR24	MSR25	MSR26
SAN27	SAN28	SAN29	SAN30	HAM1	HAM2	HAM3

TAIFA WEUSI 454/1970-71/6206-07/1962-63

Ashe	Ojobo	Iwa	Araorun	Ile	Itutu	Erinle
1	2	3	4	5	6	7
AU8	AU9	AU10	AU11	AU12	AU13	AU14
THT20	THT21	THT22	THT23	THT24	THT25	THT26
NAH2	NAH3	NAH4	NAH5	NAH6	NAH7	NAH8
8	9	10	11	12	13	14
AU15	AU16	AU17	AU18	AU19	AU20	AU21
THT27	THT28	THT29	THT30	PAP1	PAP1	PAP2
NAH9	NAH10	NAH11	NAH12	NAH13	NAH14	NAH15
15	16	17	18	19	20	21
AU22	AU23	AU24	AU25	AU26	AU27	AU28
PAP3	PAP4	PAP5	PAP6	PAP7	PAP8	PAP9
NAH16	NAH17	NAH18	NAH19	NAH20	NAH21	NAH22
22	23	24	25	26	27	28
AU29	AU30	AU31	S1	S2	S3	S4
PAP10	PAP11	PAP12	PAP13	PAP14	PAP15	PAP16
NAH23	NAH24	NAH25	NAH26	NAH27	NAH28	NAH29

MTU WEUSI 454/1970-71/6206-07/1962-63

Ashe	Ojobo	Iwa	Araorun	Ile	Itutu	Erinle
1	2	3	4	5	6	7
JU11	JU12	JU13	JU14	JU15	JU16	JU17
MSR27	MSR28	MSR29	MSR30	*ASR1*	*HRU2*	*STH3*
HAM4	HAM5	HAM6	HAM7	HAM8	HAM9	HAM10
8	9	10	11	12	13	14
JU18	JU19	JU20	JU21	JU22	JU23	JU24
AST4	*NBH5*	THT1	THT2	THT3	THT4	THT5
HAM11	HAM12	HAM13	AM14	HAM15	HAM16	HAM17
15	16	17	18	19	20	21
JU25	JU26	JU27	JU28	JU29	JU30	JU31
THT6	THT7	THT8	THT9	THT10	THT11	THT12
HAM18	HAM19	HAM20	HAM21	HAM22	HAM23	HAM24
22	23	24	25	26	27	28
AU1	AU2	AU3	AU4	AU5	AU6	AU7
THT13	THT14	THT15	THT16	THT17	THT18	THT19
HAM25	HAM26	HAM27	HAM28	HAM29	HAM30	NAH1

UWEZO 454/1970-71/6206-07/1962-63

Ashe	Ojobo	Iwa	Araorun	Ile	Itutu	Erinle
1	2	3	4	5	6	7
S5	S6	S7	S8	S9	S10	S11
PAP17	PAP18	PAP19	PAP20	PAP21	PAP22	PAP23
NAH30	*PAG1*	*PAG2*	*PAG3*	*PAG4*	*PAG5*	MAS1
8	9	10	11	12	13	14
S12	S13	S14	S15	S16	S17	S18
PAP24	PAP25	PAP26	PAP27	PAP28	PAP29	PAP30
MAS2	MAS3	MAS4	MAS5	MAS6	MAS7	MAS8
15	16	17	18	19	20	21
S19	S20	S21	S22	S23	S24	S25
HTH1	HTH2	HTH3	HTH4	HTH5	HTH6	HTH7
MAS9	MAS10	MAS11	MAS12	MAS13	MAS14	MAS15
22	23	24	25	26	27	28
S26	S27	S28	S29	S30	O1	O2
HTH8	HTH9	HTH10	HTH11	HTH12	HTH13	HTH14
MAS16	MAS17	MAS18	MAS19	MAS20	MAS21	MAS22

EKUNDU 454/1970-71/6206-07/1962-63						
Ashe	Ojobo	Iwa	Araorun	Ile	Itutu	Erinle
1	2	3	4	5	6	7
O3	O4	O5	O6	O7	O8	O9
HTH15	HTH16	HTH17	HTH18	HTH19	HTH20	HTH21
MAS23	MAS24	MAS25	MAS26	MAS27	MAS28	MAS29
8	9	10	11	12	13	14
O10	O11	O12	O13	O14	O15	O16
HTH22	HTH23	HTH24	HTH25	HTH26	HTH27	HTH28
MAS30	THi1	THi2	THi3	THi4	THi5	THi6
15	16	17	18	19	20	21
O17	O18	O19	O20	O21	O22	O23
HTH29	HTH30	KHK1	KHK2	KHK3	KHK4	KHK5
THi7	THi8	THi9	THi10	THi11	THi12	THi13
22	23	24	25	26	27	28
O24	O25	O26	O27	O28	O29	O30
KHK6	KHK7	KHK8	KHK9	KHK10	KHK11	KHK12
THi14	THi15	THi16	THi17	THi18	THi19	THi20

KIJANI 454/1970-71/6206-07/1962-63						
Ashe	Ojobo	Iwa	Araorun	Ile	Itutu	Erinle
1	2	3	4	5	6	7
N28	N29	N30	D1	D2	D3	D4
TBT11	TBT12	TBT13	TBT14	TBT15	TBT16	TBT17
HID19	HID20	HID21	HID22	HID23	HID24	HID25
8	9	10	11	12	13	14
D5	D6	D7	D8	D9	D10	D11
TBT18	TBT19	TBT20	TBT21	TBT22	TBT23	TBT24
HID26	HID27	HID28	HID29	HID30	TAH1	TAH2
15	16	17	18	19	20	21
D12	D13	D14	D15	D16	D17	D18
TBT25	TBT26	TBT27	TBT28	TBT29	TBT30	MKH1
TAH3	TAH4	TAH5	TAH6	TAH7	TAH8	TAH9
22	23	24	25	26	27	28
D19	D20	D21	D22	D23	D24	D25
MKH2	MKH3	MKH4	MKH5	MKH6	MKH7	MKH8
TAH10	TAH11	TAH12	TAH13	TAH14	TAH15	TAH16

EUSI 454/1970-71/6206-07/1962-63						
Ashe	Ojobo	Iwa	Araorun	Ile	Itutu	Erinle
1	2	3	4	5	6	7
O31	N1	N2	N3	N4	N5	N6
KHK13	KHK14	KHK15	KHK16	KHK17	KHK18	KHK19
THi21	THi22	THi23	THi24	THi25	THi26	THi27
8	9	10	11	12	13	14
N7	N8	N9	N10	N11	N12	N13
KHK20	KHK21	KHK22	KHK23	KHK24	KHK25	KHK26
THi28	THi29	THi30	HID1	HID2	HID3	HID4
15	16	17	18	19	20	21
N14	N15	N16	N17	N18	N19	N20
KHK27	KHK28	KHK29	KHK30	TBT1	TBT2	TBT3
HID5	HID6	HID7	HID8	HID9	HID10	HID11
22	23	24	25	26	27	28
N21	N22	N23	N24	N25	N26	N27
TBT4	TBT5	TBT6	TBT7	TBT8	TBT9	TBT10
HID12	HID13	HID14	HID15	HID16	HID17	HID18

KWANZA 454/1970-71/6206-07/1962-63						
Umoja	Kujichagulia	Ujima	Ujamaa	Nia	Kuumba	Imani
1	2	3	4	5	6	7
D26	D27	D28	D29	D30	D31	JA1
MKH9	MKH10	MKH11	MKH12	MKH13	MKH14	MKH15
TAH17	TAH18	TAH19	TAH20	TAH21	TAH22	TAH23
8	9	10	11	12	13	14
JA2	JA3	JA4	JA5	JA6	JA7	JA8
MKH16	MKH17	MKH18	MKH19	MKH20	MKH21	MKH22
TAH24	TAH25	TAH26	TAH27	TAH28	TAH29	TAH30
15	16	17	18	19	20	21
JA9	JA10	JA11	JA12	JA13	JA14	JA15
MKH23	MKH24	MKH25	KMH26	MKH27	MKH28	MKH29
TIR1	TIR2	TIR3	TIR4	TIR5	TIR6	TIR7
22	23	24	25	26	27	28
JA16	JA17	JA18	JA19	JA20	JA21	JA22
MKH30	AMN1	AMN2	AMN3	AMN4	AMN5	AMN6
TIR8	TIR9	TIR10	TIR11	TIR12	TIR13	TIR14

SHAHIDI 454/1970-71/6206-07/1962-63						
Ashe	Ojobo	Iwa	Araorun	Ile	Itutu	Erinle
1	2	3	4	5	6	7
JA23	JA24	JA25	JA26	JA27	JA28	JA29
AMN7	AMN8	AMN9	AMN10	AMN11	AMN12	AMN13
TIR15	TIR16	TIR17	TIR18	TIR19	TIR20	TIR21
8	9	10	11	12	13	14
JA30	JA31	F1	F2	F3	F4	F5
AMN14	AMN15	AMN16	AMN17	AMN18	AMN19	AMN20
TIR22	TIR23	TIR24	TIR25	TIR26	TIR27	TIR28
15	16	17	18	19	20	21
F6	F7	F8	F9	F10	F 11	F12
AMN21	AMN22	AMN23	AMN24	AMN25	AMN26	AMN28
TIR29	TIR30	YEK1	YEK2	YEK3	YEK4	YEK5
22	23	24	25	26	27	28
F13	F14	F15	F16	F17	F18	F19
AMN29	AMN30	RNT1	RNT2	RNT3	RNT4	RNT5
YEK6	YEK7	YEK8	YEK9	YEK10	YEK11	YEK12
SHAHIDI 5						
F20						
RNT6						
YEK13						

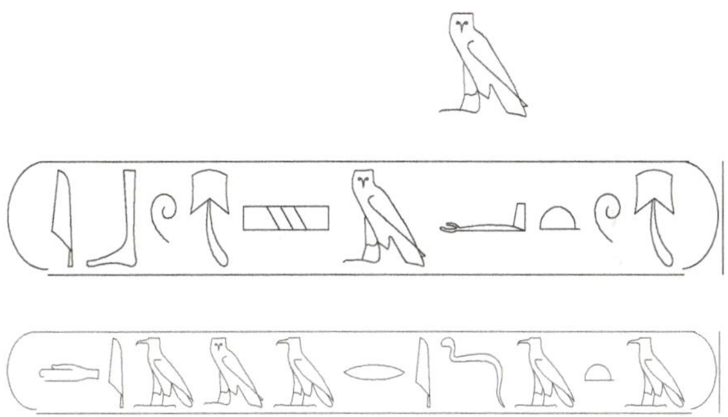

MAAFA 455 B.S.E./ASIENTO 1971-72 C.E./
KAM 6207-08 S.M./EC 1963-64 A.M. (ዓ.ም.)

AFU 455/1971-72/6207-08/1963-64

Ashe	Ojobo	Iwa	Araorun	Ile	Itutu	Erinle
1	2	3	4	5	6	7
F21	F22	F23	F24	F25	F26	F27
RNT7	RNT8	RNT9	RNT10	RNT11	RNT12	RNT13
YEK14	YEK15	YEK16	YEK17	YEK18	YEK19	YEK20
8	9	10	11	12	13	14
F28	M1	M2	M3	M4	M5	M6
RNT14	RNT15	RNT16	RNT17	RNT18	RNT19	RNT20
YEK21	YEK22	YEK23	YEK24	YEK25	YEK26	YEK27
15	16	17	18	19	20	21
M7	M8	M9	M10	M11	M12	M13
RNT21	RNT22	RNT23	RNT24	RNT25	RNT26	RNT27
YEK28	YEK 29	YEK30	MAG1	MAG2	MAG3	MAG4
22	23	24	25	26	27	28
M14	M15	M16	M17	M18	M19	M20
RNT28	RNT29	RNT30	KHN1	KHN2	KHN3	KHN4
MAG5	MAG6	MAG7	MAG8	MAG9	MAG10	MAG11

KAWAIDA 455/1971-72/6207-08/1963-64

Ashe	Ojobo	Iwa	Araorun	Ile	Itutu	Erinle
1	2	3	4	5	6	7
A18	A19	A20	A21	A22	A23	A24
PIN3	PIN4	PIN5	PIN6	PIN7	PIN8	PIN9
MIY10	MIY11	MIY12	MIY13	MIY14	MIY15	MIY16
8	9	10	11	12	13	14
A25	A26	A27	A28	A29	A30	MA1
PIN10	PIN11	PIN12	PIN13	PIN14	PIN15	PIN16
MIY17	MIY18	MIY19	MIY20	MIY21	MIY22	MIY23
15	16	17	18	19	20	21
MA2	MA3	MA4	MA5	MA6	MA7	MA8
PIN17	PIN18	PIN19	PIN20	PIN21	PIN22	PIN23
MIY24	MIY25	MIY26	MIY27	MIY28	MIY29	MIY30
22	23	24	25	26	27	28
MA9	MA10	MA11	MA12	MA13	MA14	MA15
PIN24	PIN25	PIN26	PIN27	PIN28	PIN29	PIN30
GIN1	GIN2	GIN3	GIN4	GIN5	GIN6	GIN7

UHURU 455/1971-72/6207-08/1963-64

Ashe	Ojobo	Iwa	Araorun	Ile	Itutu	Erinle
1	2	3	4	5	6	7
M21	M22	M23	M24	M25	M26	M27
KHN5	KHN6	KHN7	KHN8	KHN9	KHN10	KHN11
MAG12	MAG13	MAG14	MAG15	MAG16	MAG17	MAG18
8	9	10	11	12	13	14
M28	M29	M30	M31	A1	A2	A3
KHN12	KHN13	KHN14	KHN15	KHN16	KHN17	KHN18
MAG19	MAG20	MAG21	MAG22	MAG23	MAG24	MAG25
15	16	17	18	19	20	21
A4	A5	A6	A7	A8	A9	A10
KHN19	KHN20	KHN21	KHN22	KHN23	KHN24	KHN25
MAG26	MAG27	MAG28	MAG29	MAG30	MIY1	MIY2
22	23	24	25	26	27	28
A11	A12	A13	A14	A15	A16	A17
KHN26	KHN27	KHN28	KHN29	KHN30	PIN1	PIN2
MIY3	MIY4	MIY5	MIY6	MIY7	MIY8	MIY9

MABABA WEUSI 455/1971-72/6207-08/1963-64

Ashe	Ojobo	Iwa	Araorun	Ile	Itutu	Erinle
1	2	3	4	5	6	7
MA16	MA17	MA18	MA19	MA20	MA21	MA22
APP1	APP2	APP3	APP4	APP5	APP6	APP7
GIN8	GIN9	GIN10	GIN11	GIN12	GIN13	GIN14
8	9	10	11	12	13	14
MA23	MA24	MA25	MA26	MA27	MA28	MA29
APP8	APP9	APP10	APP11	APP12	APP13	APP14
GIN15	GIN16	GIN17	GIN18	GIN19	GIN20	GIN21
15	16	17	18	19	20	21
MA30	MA31	J1	J2	J3	J4	J5
APP15	APP16	APP17	APP18	APP19	APP20	APP21
GIN22	GIN23	GIN24	GIN25	GIN26	GIN27	GIN28
22	23	24	25	26	27	28
J6	J7	J8	J9	J10	J11	J12
APP22	APP23	APP24	APP25	APP26	APP27	APP28
GIN29	GIN30	SAN1	SAN2	SAN3	SAN4	SAN5

MWANAMKE WEUSI 455/1971-72/6207-08/1963-64						
Ashe	Ojobo	Iwa	Araorun	Ile	Itutu	Erinle
1	2	3	4	5	6	7
J13	J14	J15	J16	J17	J18	J19
APP29	APP30	MSR1	MSR2	MSR3	MSR4	MSR5
SAN6	SAN7	SAN8	SAN9	SAN10	SAN11	SAN12
8	9	10	11	12	13	14
J20	J21	J22	J23	J24	J25	J26
MSR6	MSR7	MSR8	MSR9	MSR10	MSR11	MSR12
SAN13	SAN14	SAN15	SAN16	SAN17	SAN18	SAN19
15	16	17	18	19	20	21
J27	J28	J29	J30	JU1	JU2	JU3
MSR13	MSR14	MSR15	MSR16	MSR17	MSR18	MSR19
SAN20	SAN21	SAN22	SAN23	SAN24	SAN25	SAN26
22	23	24	25	26	27	28
JU4	JU5	JU6	JU7	JU8	JU9	JU10
MSR20	MRS21	MSR22	MSR23	MSR24	MSR25	MSR26
SAN27	SAN28	SAN29	SAN30	HAM1	HAM2	HAM3

TAIFA WEUSI 455/1971-72/6207-08/1963-64						
Ashe	Ojobo	Iwa	Araorun	Ile	Itutu	Erinle
1	2	3	4	5	6	7
AU8	AU9	AU10	AU11	AU12	AU13	AU14
THT20	THT21	THT22	THT23	THT24	THT25	THT26
NAH2	NAH3	NAH4	NAH5	NAH6	NAH7	NAH8
8	9	10	11	12	13	14
AU15	AU16	AU17	AU18	AU19	AU20	AU21
THT27	THT28	THT29	THT30	PAP1	PAP1	PAP2
NAH9	NAH10	NAH11	NAH12	NAH13	NAH14	NAH15
15	16	17	18	19	20	21
AU22	AU23	AU24	AU25	AU26	AU27	AU28
PAP3	PAP4	PAP5	PAP6	PAP7	PAP8	PAP9
NAH16	NAH17	NAH18	NAH19	NAH20	NAH21	NAH22
22	23	24	25	26	27	28
AU29	AU30	AU31	S1	S2	S3	S4
PAP10	PAP11	PAP12	PAP13	PAP14	PAP15	PAP16
NAH23	NAH24	NAH25	NAH26	NAH27	NAH28	NAH29

MTU WEUSI 455/1971-72/6207-08/1963-64						
Ashe	Ojobo	Iwa	Araorun	Ile	Itutu	Erinle
1	2	3	4	5	6	7
JU11	JU12	JU13	JU14	JU15	JU16	JU17
MSR27	MSR28	MSR29	MSR30	*ASR1*	*HRU2*	*STH3*
HAM4	HAM5	HAM6	HAM7	HAM8	HAM9	HAM10
8	9	10	11	12	13	14
JU18	JU19	JU20	JU21	JU22	JU23	JU24
AST4	*NBH5*	THT1	THT2	THT3	THT4	THT5
HAM11	HAM12	HAM13	HAM14	HAM15	HAM16	HAM17
15	16	17	18	19	20	21
JU25	JU26	JU27	JU28	JU29	JU30	JU31
THT6	THT7	THT8	THT9	THT10	THT11	THT12
HAM18	HAM19	HAM20	HAM21	HAM22	HAM23	HAM24
22	23	24	25	26	27	28
AU1	AU2	AU3	AU4	AU5	AU6	AU7
THT13	THT14	THT15	THT16	THT17	THT18	THT19
HAM25	HAM26	HAM27	HAM28	HAM29	HAM30	NAH1

UWEZO 455/1971-72/6207-08/1963-64						
Ashe	Ojobo	Iwa	Araorun	Ile	Itutu	Erinle
1	2	3	4	5	6	7
S5	S6	S7	S8	S9	S10	S11
PAP17	PAP18	PAP19	PAP20	PAP21	PAP22	PAP23
NAH30	*QUA1*	*QUA2*	*QUA3*	*QUA4*	*QUA5*	*QUA6*
8	9	10	11	12	13	14
S12	S13	S14	S15	S16	S17	S18
PAP24	PAP25	PAP26	PAP27	PAP28	PAP29	PAP30
MAS1	MAS2	MAS3	MAS4	MAS5	MAS6	MAS7
15	16	17	18	19	20	21
S19	S20	S21	S22	S23	S24	S25
HTH1	HTH2	HTH3	HTH4	HTH5	HTH6	HTH7
MAS8	MAS9	MAS10	MAS11	MAS12	MAS13	MAS14
22	23	24	25	26	27	28
S26	S27	S28	S29	S30	O1	O2
HTH8	HTH9	HTH10	HTH11	HTH12	HTH13	HTH14
MAS15	MAS16	MAS17	MAS18	MAS19	MAS20	MAS21

EKUNDU 455/1971-72/6207-08/1963-64						
Ashe	Ojobo	Iwa	Araorun	Ile	Itutu	Erinle
1	2	3	4	5	6	7
O3	O4	O5	O6	O7	O8	O9
HTH15	HTH16	HTH17	HTH18	HTH19	HTH20	HTH21
MAS22	MAS23	MAS 24	MAS25	MAS26	MAS27	MAS28
8	9	10	11	12	13	14
O10	O11	O12	O13	O14	O15	O16
HTH22	HTH23	HTH24	HTH25	HTH26	HTH27	HTH28
MAS29	MAS30	THi1	THi2	THi3	THi4	THi5
15	16	17	18	19	20	21
O17	O18	O19	O20	O21	O22	O23
HTH29	HTH30	KHK1	KHK2	KHK3	KHK4	KHK5
THi6	THi7	THi8	THi9	THi10	THi11	THi12
22	23	24	25	26	27	28
O24	O25	O26	O27	O28	O29	O30
KHK6	KHK7	KHK8	KHK9	KHK10	KHK11	KHK12
THi13	THi14	THi15	THi16	THi17	THi18	THi19

KIJANI 455/1971-72/6207-08/1963-64						
Ashe	Ojobo	Iwa	Araorun	Ile	Itutu	Erinle
1	2	3	4	5	6	7
N28	N29	N30	D1	D2	D3	D4
TBT11	TBT12	TBT13	TBT14	TBT15	TBT16	TBT17
HID18	HID19	HID20	HID21	HID22	HID23	HID24
8	9	10	11	12	13	14
D5	D6	D7	D8	D9	D10	D11
TBT18	TBT19	TBT20	TBT21	TBT22	TBT23	TBT24
HID25	HID26	HID27	HID28	HID29	HID30	TAH1
15	16	17	18	19	20	21
D12	D13	D14	D15	D16	D17	D18
TBT25	TBT26	TBT27	TBT28	TBT29	TBT30	MKH1
TAH2	TAH3	TAH4	TAH5	TAH6	TAH7	TAH8
22	23	24	25	26	27	28
D19	D20	D21	D22	D23	D24	D25
MKH2	MKH3	MKH4	MKH5	MKH6	MKH7	MKH8
TAH9	TAH10	TAH11	TAH12	TAH13	TAH14	TAH15

EUSI 455/1971-72/6207-08/1963-64						
Ashe	Ojobo	Iwa	Araorun	Ile	Itutu	Erinle
1	2	3	4	5	6	7
O31	N1	N2	N3	N4	N5	N6
KHK13	KHK14	KHK15	KHK16	KHK17	KHK18	KHK19
THi20	THi21	THi22	THi23	THi24	THi25	THi26
8	9	10	11	12	13	14
N7	N8	N9	N10	N11	N12	N13
KHK20	KHK21	KHK22	KHK23	KHK24	KHK25	KHK26
THi27	THi28	THi29	THi30	HID1	HID2	HID3
15	16	17	18	19	20	21
N14	N15	N16	N17	N18	N19	N20
KHK27	KHK28	KHK29	KHK30	TBT1	TBT2	TBT3
HID4	HID5	HID6	HID7	HID8	HID9	HID10
22	23	24	25	26	27	28
N21	N22	N23	N24	N25	N26	N27
TBT4	TBT5	TBT6	TBT7	TBT8	TBT9	TBT10
HID11	HID12	HID13	HID14	HID15	HID16	HID17

KWANZA 455/1971-72/6207-08/1963-64						
Umoja	Kujichagulia	Ujima	Ujamaa	Nia	Kuumba	Imani
1	2	3	4	5	6	7
D26	D27	D28	D29	D30	D31	JA1
MKH9	MKH10	MKH11	MKH12	MKH13	MKH14	MKH15
TAH16	TAH17	TAH18	TAH19	TAH20	TAH21	TAH22
8	9	10	11	12	13	14
JA2	JA3	JA4	JA5	JA6	JA7	JA8
MKH16	MKH17	MKH18	MKH19	MKH20	MKH21	MKH22
TAH23	TAH24	TAH25	TAH26	TAH27	TAH28	TAH29
15	16	17	18	19	20	21
JA9	JA10	JA11	JA12	JA13	JA14	JA15
MKH23	MKH24	MKH25	KMH26	MKH27	MKH28	MKH29
TAH30	TIR1	TIR2	TIR3	TIR4	TIR5	TIR6
22	23	24	25	26	27	28
JA16	JA17	JA18	JA19	JA20	JA21	JA22
MKH30	AMN1	AMN2	AMN3	AMN4	AMN5	AMN6
TIR7	TIR8	TIR9	TIR10	TIR11	TIR12	TIR13

SHAHIDI 455/1971-72/6207-08/1963-64						
Ashe	Ojobo	Iwa	Araorun	Ile	Itutu	Erinle
1	2	3	4	5	6	7
JA23	JA24	JA25	JA26	JA27	JA28	JA29
AMN7	AMN8	AMN9	AMN10	AMN11	AMN12	AMN13
TIR14	TIR15	TIR16	TIR17	TIR18	TIR19	TIR20
8	9	10	11	12	13	14
JA30	JA31	F1	F2	F3	F4	F5
MN14	AMN15	AMN16	AMN17	AMN18	AMN19	AMN20
IR21	TIR22	TIR23	TIR24	TIR25	TIR26	TIR27
5	16	17	18	19	20	21
6	F7	F8	F9	F10	F 11	F12
MN21	AMN22	AMN23	AMN24	AMN25	AMN26	AMN28
IR28	TIR29	TIR30	YEK1	YEK2	YEK3	YEK4
22	23	24	25	26	27	28
F13	F14	F15	F16	F17	F18	F19
MN29	AMN30	RNT1	RNT2	RNT3	RNT4	RNT5
EK5	YEK6	YEK7	YEK8	YEK9	YEK10	YEK11
HAHIDI 6						
20						
NT6						
EK12						

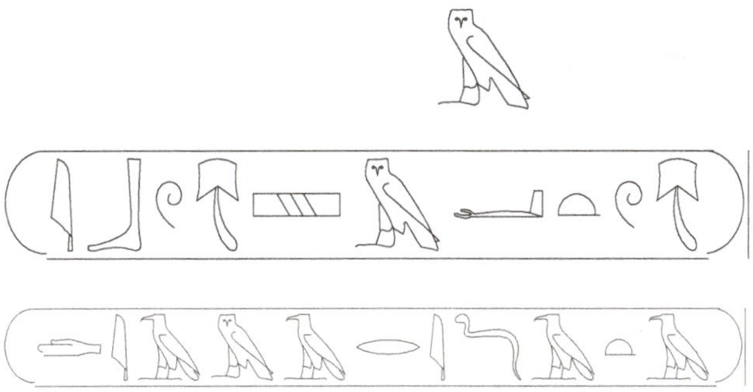

MAAFA 456 B.S.E./ASIENTO 1972-73 C.E./
KAM 6208-09 S.M./EC 1964-65 A.M. (ዓ.ም.)

AFU 456/1972-73/6208-09/1964-65

Ashe	Ojobo	Iwa	Araorun	Ile	Itutu	Erinle
1	2	3	4	5	6	7
F21	F22	F23	F24	F25	F26	F27
RNT7	RNT8	RNT9	RNT10	RNT11	RNT12	RNT13
YEK13	YEK14	YEK15	YEK16	YEK17	YEK18	YEK19
8	9	10	11	12	13	14
F28	F29	M1	M2	M3	M4	M5
RNT14	RNT15	RNT16	RNT17	RNT18	RNT19	RNT20
YEK 20	YEK21	YEK22	YEK23	YEK24	YEK25	YEK26
15	16	17	18	19	20	21
M6	M7	M8	M9	M10	M11	M12
RNT21	RNT22	RNT23	RNT24	RNT25	RNT26	RNT27
YEK 27	YEK28	YEK29	YEK30	MAG1	MAG2	MAG3
22	23	24	25	26	27	28
M13	M14	M15	M16	M17	M18	M19
RNT28	RNT29	RNT30	KHN1	KHN2	KHN3	KHN4
MAG4	MAG5	MAG6	MAG7	MAG8	MAG9	MAG10
SHARIFU 111						
M20						
KHN5						
MAG11						

KAWAIDA 456/1972-73/6208-09/1964-65

Ashe	Ojobo	Iwa	Araorun	Ile	Itutu	Erinle
1	2	3	4	5	6	7
A18	A19	A20	A21	A22	A23	A24
PIN4	PIN5	PIN6	PIN7	PIN8	PIN9	PIN10
MIY10	MIY11	MIY12	MIY13	MIY14	MIY15	MIY16
8	9	10	11	12	13	14
A25	A26	A27	A28	A29	A30	MA1
PIN11	PIN12	PIN13	PIN14	PIN15	PIN16	PIN17
MIY17	MIY18	MIY19	MIY20	MIY21	MIY22	MIY23
15	16	17	18	19	20	21
MA2	MA3	MA4	MA5	MA6	MA7	MA8
PIN18	PIN19	PIN20	PIN21	PIN22	PIN23	PIN24
MIY24	MIY25	MIY26	MIY27	MIY28	MIY29	MIY30
22	23	24	25	26	27	28
MA9	MA10	MA11	MA12	MA13	MA14	MA15
PIN25	PIN26	PIN27	PIN28	PIN29	PIN30	APP1
GIN1	GIN2	GIN3	GIN4	GIN5	GIN6	GIN7

UHURU 456/1972-73/6208-09/1964-65

Ashe	Ojobo	Iwa	Araorun	Ile	Itutu	Erinle
1	2	3	4	5	6	7
M21	M22	M23	M24	M25	M26	M27
KHN6	KHN7	KHN8	KHN9	KHN10	KHN11	KHN12
MAG12	MAG13	MAG14	MAG15	MAG16	MAG17	MAG18
8	9	10	11	12	13	14
M28	M29	M30	M31	A1	A2	A3
KHN13	KHN14	KHN15	KHN16	KHN17	KHN18	KHN19
MAG19	MAG20	MAG21	MAG22	MAG23	MAG24	MAG25
15	16	17	18	19	20	21
A4	A5	A6	A7	A8	A9	A10
KHN20	KHN21	KHN222	KHN23	KHN24	KHN25	KHN26
MAG26	MAG27	MAG28	MAG29	MAG30	MIY1	MIY2
22	23	24	25	26	27	28
A11	A12	A13	A14	A15	A16	A17
KHN27	KHN28	KHN29	KHN30	PIN1	PIN2	PIN3
MIY3	MIY4	MIY5	MIY6	MIY7	MIY8	MIY9

MABABA WEUSI 456/1972-73/6208-09/1964-65

Ashe	Ojobo	Iwa	Araorun	Ile	Itutu	Erinle
1	2	3	4	5	6	7
MA16	MA17	MA18	MA19	MA20	MA21	MA22
APP2	APP3	APP4	APP5	APP6	APP7	APP8
GIN8	GIN9	GIN10	GIN11	GIN12	GIN13	GIN14
8	9	10	11	12	13	14
MA23	MA24	MA25	MA26	MA27	MA28	MA29
APP9	APP10	APP11	APP12	APP13	APP14	APP15
GIN15	GIN16	GIN17	GIN18	GIN19	GIN20	GIN21
15	16	17	18	19	20	21
MA30	MA31	J1	J2	J3	J4	J5
APP16	APP17	APP18	APP19	APP20	APP21	APP22
GIN22	GIN23	GIN24	GIN25	GIN26	GIN27	GIN28
22	23	24	25	26	27	28
J6	J7	J8	J9	J10	J11	J12
APP23	APP24	APP25	APP26	APP27	APP28	APP29
GIN29	GIN30	SAN1	SAN2	SAN3	SAN4	SAN5

MWANAMKE WEUSI 456/1972-73/6208-09/1964-65						
Ashe	**Ojobo**	**Iwa**	**Araorun**	**Ile**	**Itutu**	**Erinle**
1	2	3	4	5	6	7
J13	J14	J15	J16	J17	J18	J19
APP30	MSR1	MSR2	MSR3	MSR4	MSR5	MSR6
SAN6	SAN7	SAN8	SAN9	SAN10	SAN11	SAN12
8	9	10	11	12	13	14
J20	J21	J22	J23	J24	J25	J26
MSR7	MSR8	MSR9	MSR10	MSR11	MSR12	MSR13
SAN13	SAN14	SAN15	SAN16	SAN17	SAN18	SAN19
15	16	17	18	19	20	21
J27	J28	J29	J30	JU1	JU2	JU3
MSR14	MSR15	MSR16	MSR17	MSR18	MSR19	MSR20
SAN20	SAN21	SAN22	SAN23	SAN24	SAN25	SAN26
22	23	24	25	26	27	28
JU4	JU5	JU6	JU7	JU8	JU9	JU10
MSR21	MRS22	MSR23	MSR24	MSR25	MSR26	MSR27
SAN27	SAN28	SAN29	SAN30	HAM1	HAM2	HAM3

TAIFA WEUSI 456/1972-73/6208-09/1964-65						
Ashe	**Ojobo**	**Iwa**	**Araorun**	**Ile**	**Itutu**	**Erinle**
1	2	3	4	5	6	7
AU8	AU9	AU10	AU11	AU12	AU13	AU14
THT20	THT21	THT22	THT23	THT24	THT25	THT26
NAH2	NAH3	NAH4	NAH5	NAH6	NAH7	NAH8
8	9	10	11	12	13	14
AU15	AU16	AU17	AU18	AU19	AU20	AU21
THT27	THT28	THT29	THT30	PAP1	PAP1	PAP2
NAH9	NAH10	NAH11	NAH12	NAH13	NAH14	NAH15
15	16	17	18	19	20	21
AU22	AU23	AU24	AU25	AU26	AU27	AU28
PAP3	PAP4	PAP5	PAP6	PAP7	PAP8	PAP9
NAH16	NAH17	NAH18	NAH19	NAH20	NAH21	NAH22
22	23	24	25	26	27	28
AU29	AU30	AU31	S1	S2	S3	S4
PAP10	PAP11	PAP12	PAP13	PAP14	PAP15	PAP16
NAH23	NAH24	NAH25	NAH26	NAH27	NAH28	NAH29

MTU WEUSI 456/1972-73/6208-09/1964-65						
Ashe	**Ojobo**	**Iwa**	**Araorun**	**Ile**	**Itutu**	**Erinle**
1	2	3	4	5	6	7
JU11	JU12	JU13	JU14	JU15	JU16	JU17
MSR28	MSR29	MSR30	*ASR1*	*HRU2*	*STH3*	*AST4*
HAM4	HAM5	HAM6	HAM7	HAM8	HAM9	HAM10
8	9	10	11	12	13	14
JU18	JU19	JU20	JU21	JU22	JU23	JU24
NBH5	*ANU6*	THT1	THT2	THT3	THT4	THT5
HAM11	HAM12	HAM13	HAM14	HAM15	HAM16	HAM17
15	16	17	18	19	20	21
JU25	JU26	JU27	JU28	JU29	JU30	JU31
THT6	THT7	THT8	THT9	THT10	THT11	THT12
HAM18	HAM19	HAM20	HAM21	HAM22	HAM23	HAM24
22	23	24	25	26	27	28
AU1	AU2	AU3	AU4	AU5	AU6	AU7
THT13	THT14	THT15	THT16	THT17	THT18	THT19
HAM25	HAM26	HAM27	HAM28	HAM29	HAM30	NAH1

UWEZO 456/1972-73/6208-09/1964-65						
Ashe	**Ojobo**	**Iwa**	**Araorun**	**Ile**	**Itutu**	**Erinle**
1	2	3	4	5	6	7
S5	S6	S7	S8	S9	S10	S11
PAP17	PAP18	PAP19	PAP20	PAP21	PAP22	PAP23
NAH30	*PAG1*	*PAG2*	*PAG3*	*PAG4*	*PAG5*	MAS1
8	9	10	11	12	13	14
S12	S13	S14	S15	S16	S17	S18
PAP24	PAP25	PAP26	PAP27	PAP28	PAP29	PAP30
MAS2	MAS3	MAS4	MAS5	MAS6	MAS7	MAS8
15	16	17	18	19	20	21
S19	S20	S21	S22	S23	S24	S25
HTH1	HTH2	HTH3	HTH4	HTH5	HTH6	HTH7
MAS9	MAS10	MAS11	MAS12	MAS13	MAS14	MAS15
22	23	24	25	26	27	28
S26	S27	S28	S29	S30	O1	O2
HTH8	HTH9	HTH10	HTH11	HTH12	HTH13	HTH14
MAS16	MAS17	MAS18	MAS19	MAS20	MAS21	MAS22

EKUNDU 456/1972-73/6208-09/1964-65						
Ashe	Ojobo	Iwa	Araorun	Ile	Itutu	Erinle
1	2	3	4	5	6	7
O3	O4	O5	O6	O7	O8	O9
HTH15	HTH16	HTH17	HTH18	HTH19	HTH20	HTH21
MAS23	MAS24	MAS25	MAS26	MAS27	MAS28	MAS29
8	9	10	11	12	13	14
O10	O11	O12	O13	O14	O15	O16
HTH22	HTH23	HTH24	HTH25	HTH26	HTH27	HTH28
MAS30	THi1	THi2	THi3	THi4	THi5	THi6
15	16	17	18	19	20	21
O17	O18	O19	O20	O21	O22	O23
HTH29	HTH30	KHK1	KHK2	KHK3	KHK4	KHK5
THi7	THi8	This9	THi10	THi11	THi12	THi13
22	23	24	25	26	27	28
O24	O25	O26	O27	O28	O29	O30
KHK6	KHK7	KHK8	KHK9	KHK10	KHK11	KHK12
Thi14	THi15	THi16	THi17	THi18	THi19	THi20

KIJANI 456/1972-73/6208-09/1964-65						
Ashe	Ojobo	Iwa	Araorun	Ile	Itutu	Erinle
1	2	3	4	5	6	7
N28	N29	N30	D1	D2	D3	D4
TBT11	TBT12	TBT13	TBT14	TBT15	TBT16	TBT17
HID19	HID20	HID21	HID22	HID23	HID24	HID25
8	9	10	11	12	13	14
D5	D6	D7	D8	D9	D10	D11
TBT18	TBT19	TBT20	TBT21	TBT22	TBT23	TBT24
HID26	HID27	HID28	HID29	HID30	TAH1	TAH2
15	16	17	18	19	20	21
D12	D13	D14	D15	D16	D17	D18
TBT25	TBT26	TBT27	TBT28	TBT29	TBT30	MKH1
TAH3	TAH4	TAH5	TAH6	TAH7	TAH8	TAH9
22	23	24	25	26	27	28
D19	D20	D21	D22	D23	D24	D25
MKH2	MKH3	MKH4	MKH5	MKH6	MKH7	MKH8
TAH10	TAH11	TAH12	TAH13	TAH14	TAH15	TAH16

EUSI 456/1972-73/6208-09/1964-65						
Ashe	Ojobo	Iwa	Araorun	Ile	Itutu	Erinle
1	2	3	4	5	6	7
O31	N1	N2	N3	N4	N5	N6
KHK13	KHK14	KHK15	KHK16	KHK17	KHK18	KHK19
THi21	THi22	THi23	THi24	THi25	THi26	THi27
8	9	10	11	12	13	14
N7	N8	N9	N10	N11	N12	N13
KHK20	KHK21	KHK22	KHK23	KHK24	KHK25	KHK26
THi28	THi29	THi30	HID1	HID2	HID3	HID4
15	16	17	18	19	20	21
N14	N15	N16	N17	N18	N19	N20
KHK27	KHK28	KHK29	KHK30	TBT1	TBT2	TBT3
HID5	HID6	HID7	HID8	HID9	HID10	HID11
22	23	24	25	26	27	28
N21	N22	N23	N24	N25	N26	N27
TBT4	TBT5	TBT6	TBT7	TBT8	TBT9	TBT10
HID12	HID13	HID14	HID15	HID16	HID17	HID18

KWANZA 456/1972-73/6208-09/1964-65						
Umoja	Kujichagulia	Ujima	Ujamaa	Nia	Kuumba	Imani
1	2	3	4	5	6	7
D26	D27	D28	D29	D30	D31	JA1
MKH9	MKH10	MKH11	MKH12	MKH13	MKH14	MKH1
TAH17	TAH18	TAH19	TAH20	TAH21	TAH22	TAH2
8	9	10	11	12	13	14
JA2	JA3	JA4	JA5	JA6	JA7	JA
MKH16	MKH17	MKH18	MKH19	MKH20	MKH21	MKH2
TAH24	TAH25	TAH26	TAH27	TAH28	TAH29	TAH3
15	16	17	18	19	20	21
JA9	JA10	JA11	JA12	JA13	JA14	JA1
MKH23	MKH24	MKH25	KMH26	MKH27	MKH28	MKH2
TIR1	TIR2	TIR3	TIR4	TIR5	TIR6	TIR7
22	23	24	25	26	27	28
JA16	JA17	JA18	JA19	JA20	JA21	JA2
MKH30	AMN1	AMN2	AMN3	AMN4	AMN5	AMN
TIR8	TIR9	TIR10	TIR11	TIR12	TIR13	TIR1

SHAHIDI 456/1972-73/6208-09/1964-65						
Ashe	Ojobo	Iwa	Araorun	Ile	Itutu	Erinle
1	2	3	4	5	6	7
JA23	JA24	JA25	JA26	JA27	JA28	JA29
AMN7	AMN8	AMN9	AMN10	AMN11	AMN12	AMN13
TIR15	TIR16	TIR17	TIR18	TIR19	TIR20	TIR21
8	9	10	11	12	13	14
JA30	JA31	F1	F2	F3	F4	F5
AMN14	AMN15	AMN16	AMN17	AMN18	AMN19	AMN20
TIR22	TIR23	TIR24	TIR25	TIR26	TIR27	TIR28
15	16	17	18	19	20	21
F6	F7	F8	F9	F10	F 11	F12
AMN21	AMN22	AMN23	AMN24	AMN25	AMN26	AMN28
TIR29	TIR30	YEK1	YEK2	YEK3	YEK4	YEK5
22	23	24	25	26	27	28
F13	F14	F15	F16	F17	F18	F19
AMN29	AMN30	RNT1	RNT2	RNT3	RNT4	RNT5
YEK6	YEK7	YEK8	YEK9	YEK10	YEK11	YEK12
SHAHIDI 7						
F20						
RNT6						
YEK13						

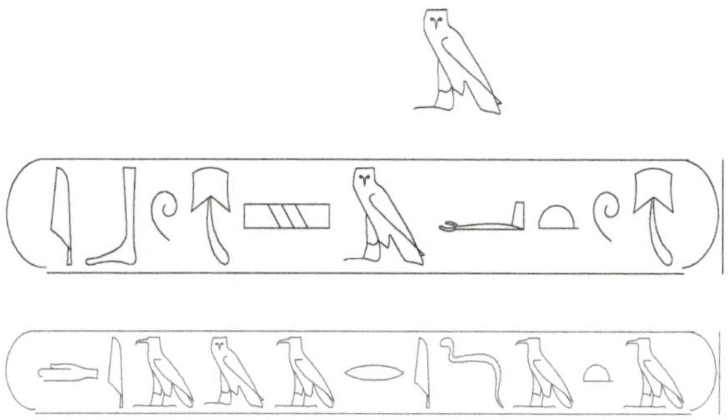

MAAFA 457 B.S.E./ASIENTO 1973-74 C.E./
KAM 6209-10 S.M./EC 1965-66 A.M. (ዓ.ም.)

AFU 457/1973-74/6209-10/1965-66						
Ashe	Ojobo	Iwa	Araorun	Ile	Itutu	Erinle
1	2	3	4	5	6	7
F21	F22	F23	F24	F25	F26	F27
RNT7	RNT8	RNT9	RNT10	RNT11	RNT12	RNT13
YEK14	YEK15	YEK16	YEK17	YEK18	YEK19	YEK20
	9	10	11	12	13	14
F28	M1	M2	M3	M4	M5	M6
RNT14	RNT15	RNT16	RNT17	RNT18	RNT19	RNT20
YEK21	YEK22	YEK23	YEK24	YEK25	YEK26	YEK27
15	16	17	18	19	20	21
M7	M8	M9	M10	M11	M12	M13
RNT21	RNT22	RNT23	RNT24	RNT25	RNT26	RNT27
YEK28	YEK29	YEK30	MAG1	MAG2	MAG3	MAG4
22	23	24	25	26	27	28
M14	M15	M16	M17	M18	M19	M20
RNT28	RNT29	RNT30	KHN1	KHN2	KHN3	KHN4
MAG5	MAG6	MAG7	MAG8	MAG9	MAG10	MAG11

KAWAIDA 457/1973-74/6209-10/1965-66						
Ashe	Ojobo	Iwa	Araorun	Ile	Itutu	Erinle
1	2	3	4	5	6	7
A18	A19	A20	A21	A22	A23	A24
PIN3	PIN4	PIN5	PIN6	PIN7	PIN8	PIN9
MIY10	MIY11	MIY12	MIY13	MIY14	MIY15	MIY16
8	9	10	11	12	13	14
A25	A26	A27	A28	A29	A30	MA1
PIN10	PIN11	PIN12	PIN13	PIN14	PIN15	PIN16
MIY17	MIY18	MIY19	MIY20	MIY21	MIY22	MIY23
15	16	17	18	19	20	21
MA2	MA3	MA4	MA5	MA6	MA7	MA8
PIN17	PIN18	PIN19	PIN20	PIN21	PIN22	PIN23
MIY24	MIY25	MIY26	MIY27	MIY28	MIY29	MIY30
22	23	24	25	26	27	28
MA9	MA10	MA11	MA12	MA13	MA14	MA15
PIN24	PIN25	PIN26	PIN27	PIN28	PIN29	PIN30
GIN1	GIN2	GIN3	GIN4	GIN5	GIN6	GIN7

UHURU 457/1973-74/6209-10/1965-66						
Ashe	Ojobo	Iwa	Araorun	Ile	Itutu	Erinle
1	2	3	4	5	6	7
M21	M22	M23	M24	M25	M26	M27
KHN5	KHN6	KHN7	KHN8	KHN9	KHN10	KHN11
MAG12	MAG13	MAG14	MAG15	MAG16	MAG17	MAG18
8	9	10	11	12	13	14
M28	M29	M30	M31	A1	A2	A3
KHN12	KHN13	KHN14	KHN15	KHN16	KHN17	KHN18
MAG19	MAG20	MAG21	MAG22	MAG23	MAG24	MAG25
15	16	17	18	19	20	21
A4	A5	A6	A7	A8	A9	A10
KHN19	KHN20	KHN21	KHN22	KHN23	KHN24	KHN25
MAG26	MAG27	MAG28	MAG29	MAG30	MIY1	MIY2
22	23	24	25	26	27	28
A11	A12	A13	A14	A15	A16	A17
KHN26	KHN27	KHN28	KHN29	KHN30	PIN1	PIN2
MIY3	MIY4	MIY5	MIY6	MIY7	MIY8	MIY9

MABABA WEUSI 457/1973-74/6209-10/1965-66						
Ashe	Ojobo	Iwa	Araorun	Ile	Itutu	Erinle
1	2	3	4	5	6	7
MA16	MA17	MA18	MA19	MA20	MA21	MA22
APP1	APP2	APP3	APP4	APP5	APP6	APP7
GIN8	GIN9	GIN10	GIN11	GIN12	GIN13	GIN14
8	9	10	11	12	13	14
MA23	MA24	MA25	MA26	MA27	MA28	MA29
APP8	APP9	APP10	APP11	APP12	APP13	APP14
GIN15	GIN16	GIN17	GIN18	GIN19	GIN20	GIN21
15	16	17	18	19	20	21
MA30	MA31	J1	J2	J3	J4	J5
APP15	APP16	APP17	APP18	APP19	APP20	APP21
GIN22	GIN23	GIN24	GIN25	GIN26	GIN27	GIN28
22	23	24	25	26	27	28
J6	J7	J8	J9	J10	J11	J12
APP22	APP23	APP24	APP25	APP26	APP27	APP28
GIN29	GIN30	SAN1	SAN2	SAN3	SAN4	SAN5

MWANAMKE WEUSI 457/1973-74/6209-10/1965-66						
Ashe	Ojobo	Iwa	Araorun	Ile	Itutu	Erinle
1	2	3	4	5	6	7
J13	J14	J15	J16	J17	J18	J19
APP29	APP30	MSR1	MSR2	MSR3	MSR4	MSR5
SAN6	SAN7	SAN8	SAN9	SAN10	SAN11	SAN12
8	9	10	11	12	13	14
J20	J21	J22	J23	J24	J25	J26
MSR6	MSR7	MSR8	MSR9	MSR10	MSR11	MSR12
SAN13	SAN14	SAN15	SAN16	SAN17	SAN18	SAN19
15	16	17	18	19	20	21
J27	J28	J29	J30	JU1	JU2	JU3
MSR13	MSR14	MSR15	MSR16	MSR17	MSR18	MSR19
SAN20	SAN21	SAN22	SAN23	SAN24	SAN25	SAN26
22	23	24	25	26	27	28
JU4	JU5	JU6	JU7	JU8	JU9	JU10
MSR20	MRS21	MSR22	MSR23	MSR24	MSR25	MSR26
SAN27	SAN28	SAN29	SAN30	HAM1	HAM2	HAM3

MTU WEUSI 457/1973-74/6209-10/1965-66						
Ashe	Ojobo	Iwa	Araorun	Ile	Itutu	Erinle
1	2	3	4	5	6	7
JU11	JU12	JU13	JU14	JU15	JU16	JU17
MSR27	MSR28	MSR29	MSR30	*ASR1*	*HRU2*	*STH3*
HAM4	HAM5	HAM6	HAM7	HAM8	HAM9	HAM10
8	9	10	11	12	13	14
JU18	JU19	JU20	JU21	JU22	JU23	JU24
AST4	*NBH5*	THT1	THT2	THT3	THT4	THT5
HAM11	HAM12	HAM13	AM14	HAM15	HAM16	HAM17
15	16	17	18	19	20	21
JU25	JU26	JU27	JU28	JU29	JU30	JU31
THT6	THT7	THT8	THT9	THT10	THT11	THT12
HAM18	HAM19	HAM20	HAM21	HAM22	HAM23	HAM24
22	23	24	25	26	27	28
AU1	AU2	AU3	AU4	AU5	AU6	AU7
THT13	THT14	THT15	THT16	THT17	THT18	THT19
HAM25	HAM26	HAM27	HAM28	HAM29	HAM30	NAH1

TAIFA WEUSI 457/1973-74/6209-10/1965-66						
Ashe	Ojobo	Iwa	Araorun	Ile	Itutu	Erinle
1	2	3	4	5	6	7
AU8	AU9	AU10	AU11	AU12	AU13	AU14
THT20	THT21	THT22	THT23	THT24	THT25	THT26
NAH2	NAH3	NAH4	NAH5	NAH6	NAH7	NAH8
8	9	10	11	12	13	14
AU15	AU16	AU17	AU18	AU19	AU20	AU21
THT27	THT28	THT29	THT30	PAP1	PAP1	PAP2
NAH9	NAH10	NAH11	NAH12	NAH13	NAH14	NAH15
15	16	17	18	19	20	21
AU22	AU23	AU24	AU25	AU26	AU27	AU28
PAP3	PAP4	PAP5	PAP6	PAP7	PAP8	PAP9
NAH16	NAH17	NAH18	NAH19	NAH20	NAH21	NAH22
22	23	24	25	26	27	28
AU29	AU30	AU31	S1	S2	S3	S4
PAP10	PAP11	PAP12	PAP13	PAP14	PAP15	PAP16
NAH23	NAH24	NAH25	NAH26	NAH27	NAH28	NAH29

UWEZO 457/1973-74/6209-10/1965-66						
Ashe	Ojobo	Iwa	Araorun	Ile	Itutu	Erinle
1	2	3	4	5	6	7
S5	S6	S7	S8	S9	S10	S11
PAP17	PAP18	PAP19	PAP20	PAP21	PAP22	PAP23
NAH30	*PAG1*	*PAG2*	*PAG3*	*PAG4*	*PAG5*	MAS1
8	9	10	11	12	13	14
S12	S13	S14	S15	S16	S17	S18
PAP24	PAP25	PAP26	PAP27	PAP28	PAP29	PAP30
MAS2	MAS3	MAS4	MAS5	MAS6	MAS7	MAS8
15	16	17	18	19	20	21
S19	S20	S21	S22	S23	S24	S25
HTH1	HTH2	HTH3	HTH4	HTH5	HTH6	HTH7
MAS9	MAS10	MAS11	MAS12	MAS13	MAS14	MAS15
22	23	24	25	26	27	28
S26	S27	S28	S29	S30	O1	O2
HTH8	HTH9	HTH10	HTH11	HTH12	HTH13	HTH14
MAS16	MAS17	MAS18	MAS19	MAS20	MAS21	MAS22

EKUNDU 457/1973-74/6209-10/1965-66						
Ashe	Ojobo	Iwa	Araorun	Ile	Itutu	Erinle
1	2	3	4	5	6	7
O3	O4	O5	O6	O7	O8	O9
HTH15	HTH16	HTH17	HTH18	HTH19	HTH20	HTH21
MAS23	MAS24	MAS25	MAS26	MAS27	MAS28	MAS29
8	9	10	11	12	13	14
O10	O11	O12	O13	O14	O15	O16
HTH22	HTH23	HTH24	HTH25	HTH26	HTH27	HTH28
MAS30	THi1	THi2	THi3	THi4	THi5	THi6
15	16	17	18	19	20	21
O17	O18	O19	O20	O21	O22	O23
HTH29	HTH30	KHK1	KHK2	KHK3	KHK4	KHK5
THi7	THi8	THi9	THi10	THi11	THi12	THi13
22	23	24	25	26	27	28
O24	O25	O26	O27	O28	O29	O30
KHK6	KHK7	KHK8	KHK9	KHK10	KHK11	KHK12
THi14	THi15	THi16	THi17	THi18	THi19	THi20

KIJANI 457/1973-74/6209-10/1965-66						
Ashe	Ojobo	Iwa	Araorun	Ile	Itutu	Erinle
1	2	3	4	5	6	7
N28	N29	N30	D1	D2	D3	D4
TBT11	TBT12	TBT13	TBT14	TBT15	TBT16	TBT17
HID19	HID20	HID21	HID22	HID23	HID24	HID25
8	9	10	11	12	13	14
D5	D6	D7	D8	D9	D10	D11
TBT18	TBT19	TBT20	TBT21	TBT22	TBT23	TBT24
HID26	HID27	HID28	HID29	HID30	TAH1	TAH2
15	16	17	18	19	20	21
D12	D13	D14	D15	D16	D17	D18
TBT25	TBT26	TBT27	TBT28	TBT29	TBT30	MKH1
TAH3	TAH4	TAH5	TAH6	TAH7	TAH8	TAH9
22	23	24	25	26	27	28
D19	D20	D21	D22	D23	D24	D25
MKH2	MKH3	MKH4	MKH5	MKH6	MKH7	MKH8
TAH10	TAH11	TAH12	TAH13	TAH14	TAH15	TAH16

EUSI 457/1973-74/6209-10/1965-66						
Ashe	Ojobo	Iwa	Araorun	Ile	Itutu	Erinle
1	2	3	4	5	6	7
O31	N1	N2	N3	N4	N5	N6
KHK13	KHK14	KHK15	KHK16	KHK17	KHK18	KHK19
THi21	THi22	THi23	THi24	THi25	THi26	THi27
8	9	10	11	12	13	14
N7	N8	N9	N10	N11	N12	N13
KHK20	KHK21	KHK22	KHK23	KHK24	KHK25	KHK26
THi28	THi29	THi30	HID1	HID2	HID3	HID4
15	16	17	18	19	20	21
N14	N15	N16	N17	N18	N19	N20
KHK27	KHK28	KHK29	KHK30	TBT1	TBT2	TBT3
HID5	HID6	HID7	HID8	HID9	HID10	HID11
22	23	24	25	26	27	28
N21	N22	N23	N24	N25	N26	N27
TBT4	TBT5	TBT6	TBT7	TBT8	TBT9	TBT10
HID12	HID13	HID14	HID15	HID16	HID17	HID18

KWANZA 457/1973-74/6209-10/1965-66						
Umoja	Kujichagulia	Ujima	Ujamaa	Nia	Kuumba	Imani
1	2	3	4	5	6	7
D26	D27	D28	D29	D30	D31	JA1
MKH9	MKH10	MKH11	MKH12	MKH13	MKH14	MKH15
TAH17	TAH18	TAH19	TAH20	TAH21	TAH22	TAH23
8	9	10	11	12	13	14
JA2	JA3	JA4	JA5	JA6	JA7	JA8
MKH16	MKH17	MKH18	MKH19	MKH20	MKH21	MKH22
TAH24	TAH25	TAH26	TAH27	TAH28	TAH29	TAH30
15	16	17	18	19	20	21
JA9	JA10	JA11	JA12	JA13	JA14	JA15
MKH23	MKH24	MKH25	KMH26	MKH27	MKH28	MKH29
TIR1	TIR2	TIR3	TIR4	TIR5	TIR6	TIR7
22	23	24	25	26	27	28
JA16	JA17	JA18	JA19	JA20	JA21	JA22
MKH30	AMN1	AMN2	AMN3	AMN4	AMN5	AMN6
TIR8	TIR9	TIR10	TIR11	TIR12	TIR13	TIR14

SHAHIDI 457/1973-74/6209-10/1965-66						
Ashe	Ojobo	Iwa	Araorun	Ile	Itutu	Erinle
1	2	3	4	5	6	7
JA23	JA24	JA25	JA26	JA27	JA28	JA29
AMN7	AMN8	AMN9	AMN10	AMN11	AMN12	AMN13
TIR15	TIR16	TIR17	TIR18	TIR19	TIR20	TIR21
8	9	10	11	12	13	14
JA30	JA31	F1	F2	F3	F4	F5
AMN14	AMN15	AMN16	AMN17	AMN18	AMN19	AMN20
TIR22	TIR23	TIR24	TIR25	TIR26	TIR27	TIR28
15	16	17	18	19	20	21
F6	F7	F8	F9	F10	F 11	F12
AMN21	AMN22	AMN23	AMN24	AMN25	AMN26	AMN28
TIR29	TIR30	YEK1	YEK2	YEK3	YEK4	YEK5
22	23	24	25	26	27	28
F13	F14	F15	F16	F17	F18	F19
AMN29	AMN30	RNT1	RNT2	RNT3	RNT4	RNT5
YEK6	YEK7	YEK8	YEK9	YEK10	YEK11	YEK12
SHAHIDI 8						
F20						
RNT6						
YEK13						

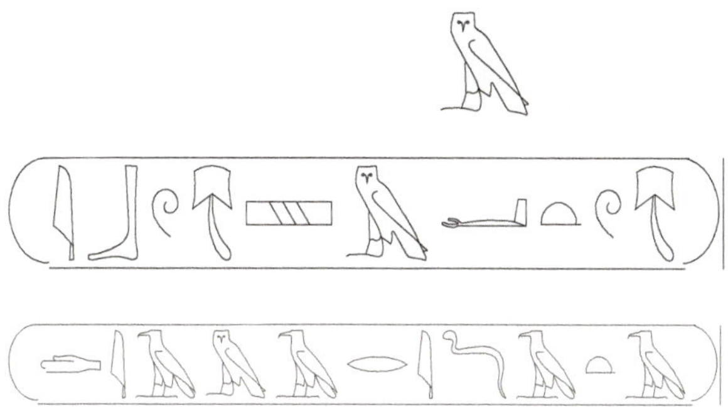

MAAFA 458 B.S.E./ASIENTO 1974-75 C.E./
KAM 6210-11 S.M./EC 1966-67 A.M. (ዓ.ም.)

AFU 458/1974-75/6210-11/1966-67

he	Ojobo	Iwa	Araorun	Ile	Itutu	Erinle
	2	3	4	5	6	7
	F22	F23	F24	F25	F26	F27
T7	RNT8	RNT9	RNT10	RNT11	RNT12	RNT13
K14	YEK15	YEK16	YEK17	YEK18	YEK19	YEK20
	9	10	11	12	13	14
8	M1	M2	M3	M4	M5	M6
T14	RNT15	RNT16	RNT17	RNT18	RNT19	RNT20
K21	YEK22	YEK23	YEK24	YEK25	YEK26	YEK27
	16	17	18	19	20	21
7	M8	M9	M10	M11	M12	M13
T21	RNT22	RNT23	RNT24	RNT25	RNT26	RNT27
K28	YEK29	YEK30	MAG1	MAG2	MAG3	MAG4
	23	24	25	26	27	28
14	M15	M16	M17	M18	M19	M20
T28	RNT29	RNT30	KHN1	KHN2	KHN3	KHN4
G5	MAG6	MAG7	MAG8	MAG9	MAG10	MAG11

KAWAIDA 458/1974-75/6210-11/1966-67

Ashe	Ojobo	Iwa	Araorun	Ile	Itutu	Erinle
1	2	3	4	5	6	7
A18	A19	A20	A21	A22	A23	A24
PIN3	PIN4	PIN5	PIN6	PIN7	PIN8	PIN9
MIY10	MIY11	MIY12	MIY13	MIY14	MIY15	MIY16
8	9	10	11	12	13	14
A25	A26	A27	A28	A29	A30	MA1
PIN10	PIN11	PIN12	PIN13	PIN14	PIN15	PIN16
MIY17	MIY18	MIY19	MIY20	MIY21	MIY22	MIY23
15	16	17	18	19	20	21
MA2	MA3	MA4	MA5	MA6	MA7	MA8
PIN17	PIN18	PIN19	PIN20	PIN21	PIN22	PIN23
MIY24	MIY25	MIY26	MIY27	MIY28	MIY29	MIY30
22	23	24	25	26	27	28
MA9	MA10	MA11	MA12	MA13	MA14	MA15
PIN24	PIN25	PIN26	PIN27	PIN28	PIN29	PIN30
GIN1	GIN2	GIN3	GIN4	GIN5	GIN6	GIN7

UHURU 458/1974-75/6210-11/1966-67

he	Ojobo	Iwa	Araorun	Ile	Itutu	Erinle
	2	3	4	5	6	7
21	M22	M23	M24	M25	M26	M27
N5	KHN6	KHN7	KHN8	KHN9	KHN10	KHN11
G12	MAG13	MAG14	MAG15	MAG16	MAG17	MAG18
	9	10	11	12	13	14
28	M29	M30	M31	A1	A2	A3
N12	KHN13	KHN14	KHN15	KHN16	KHN17	KHN18
G19	MAG20	MAG21	MAG22	MAG23	MAG24	MAG25
	16	17	18	19	20	21
4	A5	A6	A7	A8	A9	A10
N19	KHN20	KHN21	KHN22	KHN23	KHN24	KHN25
G26	MAG27	MAG28	MAG29	MAG30	MIY1	MIY2
	23	24	25	26	27	28
1	A12	A13	A14	A15	A16	A17
N26	KHN27	KHN28	KHN29	KHN30	PIN1	PIN2
Y3	MIY4	MIY5	MIY6	MIY7	MIY8	MIY9

MABABA WEUSI 458/1974-75/6210-11/1966-67

Ashe	Ojobo	Iwa	Araorun	Ile	Itutu	Erinle
1	2	3	4	5	6	7
MA16	MA17	MA18	MA19	MA20	MA21	MA22
APP1	APP2	APP3	APP4	APP5	APP6	APP7
GIN8	GIN9	GIN10	GIN11	GIN12	GIN13	GIN14
8	9	10	11	12	13	14
MA23	MA24	MA25	MA26	MA27	MA28	MA29
APP8	APP9	APP10	APP11	APP12	APP13	APP14
GIN15	GIN16	GIN17	GIN18	GIN19	GIN20	GIN21
15	16	17	18	19	20	21
MA30	MA31	J1	J2	J3	J4	J5
APP15	APP16	APP17	APP18	APP19	APP20	APP21
GIN22	GIN23	GIN24	GIN25	GIN26	GIN27	GIN28
22	23	24	25	26	27	28
J6	J7	J8	J9	J10	J11	J12
APP22	APP23	APP24	APP25	APP26	APP27	APP28
GIN29	GIN30	SAN1	SAN2	SAN3	SAN4	SAN5

MWANAMKE WEUSI	458/1974-75/6210-11/1966-67					
Ashe	Ojobo	Iwa	Araorun	Ile	Itutu	Erinle
1	2	3	4	5	6	7
J13	J14	J15	J16	J17	J18	J19
APP29	APP30	MSR1	MSR2	MSR3	MSR4	MSR5
SAN6	SAN7	SAN8	SAN9	SAN10	SAN11	SAN12
8	9	10	11	12	13	14
J20	J21	J22	J23	J24	J25	J26
MSR6	MSR7	MSR8	MSR9	MSR10	MSR11	MSR12
SAN13	SAN14	SAN15	SAN16	SAN17	SAN18	SAN19
15	16	17	18	19	20	21
J27	J28	J29	J30	JU1	JU2	JU3
MSR13	MSR14	MSR15	MSR16	MSR17	MSR18	MSR19
SAN20	SAN21	SAN22	SAN23	SAN24	SAN25	SAN26
22	23	24	25	26	27	28
JU4	JU5	JU6	JU7	JU8	JU9	JU10
MSR20	MRS21	MSR22	MSR23	MSR24	MSR25	MSR26
SAN27	SAN28	SAN29	SAN30	HAM1	HAM2	HAM3

TAIFA WEUSI	458/1974-75/6210-11/1966-67					
Ashe	Ojobo	Iwa	Araorun	Ile	Itutu	Erinl
1	2	3	4	5	6	7
AU8	AU9	AU10	AU11	AU12	AU13	AU14
THT20	THT21	THT22	THT23	THT24	THT25	THT
NAH2	NAH3	NAH4	NAH5	NAH6	NAH7	NAH
8	9	10	11	12	13	14
AU15	AU16	AU17	AU18	AU19	AU20	AU21
THT27	THT28	THT29	THT30	PAP1	PAP1	PAP
NAH9	NAH10	NAH11	NAH12	NAH13	NAH14	NAH15
15	16	17	18	19	20	21
AU22	AU23	AU24	AU25	AU26	AU27	AU28
PAP3	PAP4	PAP5	PAP6	PAP7	PAP8	PA
NAH16	NAH17	NAH18	NAH19	NAH20	NAH21	NAH
22	23	24	25	26	27	28
AU29	AU30	AU31	S1	S2	S3	S4
PAP10	PAP11	PAP12	PAP13	PAP14	PAP15	PAP
NAH23	NAH24	NAH25	NAH26	NAH27	NAH28	NAH2

MTU WEUSI	458/1974-75/6210-11/1966-67					
Ashe	Ojobo	Iwa	Araorun	Ile	Itutu	Erinle
1	2	3	4	5	6	7
JU11	JU12	JU13	JU14	JU15	JU16	JU17
MSR27	MSR28	MSR29	MSR30	*ASR1*	*HRU2*	*STH3*
HAM4	HAM5	HAM6	HAM7	HAM8	HAM9	HAM10
8	9	10	11	12	13	14
JU18	JU19	JU20	JU21	JU22	JU23	JU24
AST4	*NBH5*	THT1	THT2	THT3	THT4	THT5
HAM11	HAM12	HAM13	AM14	HAM15	HAM16	HAM17
15	16	17	18	19	20	21
JU25	JU26	JU27	JU28	JU29	JU30	JU31
THT6	THT7	THT8	THT9	THT10	THT11	THT12
HAM18	HAM19	HAM20	HAM21	HAM22	HAM23	HAM24
22	23	24	25	26	27	28
AU1	AU2	AU3	AU4	AU5	AU6	AU7
THT13	THT14	THT15	THT16	THT17	THT18	THT19
HAM25	HAM26	HAM27	HAM28	HAM29	HAM30	NAH1

UWEZO	458/1974-75/6210-11/1966-67					
Ashe	Ojobo	Iwa	Araorun	Ile	Itutu	Erinl
1	2	3	4	5	6	7
S5	S6	S7	S8	S9	S10	S11
PAP17	PAP18	PAP19	PAP20	PAP21	PAP22	PAP2
NAH30	*PAG1*	*PAG2*	*PAG3*	*PAG4*	*PAG5*	MAS1
8	9	10	11	12	13	14
S12	S13	S14	S15	S16	S17	S18
PAP24	PAP25	PAP26	PAP27	PAP28	PAP29	PAP3
MAS2	MAS3	MAS4	MAS5	MAS6	MAS7	MAS8
15	16	17	18	19	20	21
S19	S20	S21	S22	S23	S24	S25
HTH1	HTH2	HTH3	HTH4	HTH5	HTH6	HTH
MAS9	MAS10	MAS11	MAS12	MAS13	MAS14	MAS15
22	23	24	25	26	27	28
S26	S27	S28	S29	S30	O1	O2
HTH8	HTH9	HTH10	HTH11	HTH12	HTH13	HTH1
MAS16	MAS17	MAS18	MAS19	MAS20	MAS21	MAS2

EKUNDU 458/1974-75/6210-11/1966-67						
Ashe	Ojobo	Iwa	Araorun	Ile	Itutu	Erinle
1	2	3	4	5	6	7
O3	O4	O5	O6	O7	O8	O9
HTH15	HTH16	HTH17	HTH18	HTH19	HTH20	HTH21
MAS23	MAS24	MAS25	MAS26	MAS27	MAS28	MAS29
8	9	10	11	12	13	14
O10	O11	O12	O13	O14	O15	O16
HTH22	HTH23	HTH24	HTH25	HTH26	HTH27	HTH28
MAS30	THi1	THi2	THi3	THi4	THi5	THi6
15	16	17	18	19	20	21
O17	O18	O19	O20	O21	O22	O23
HTH29	HTH30	KHK1	KHK2	KHK3	KHK4	KHK5
THi7	THi8	THi9	THi10	THi11	THi12	THi13
22	23	24	25	26	27	28
O24	O25	O26	O27	O28	O29	O30
KHK6	KHK7	KHK8	KHK9	KHK10	KHK11	KHK12
THi14	THi15	THi16	THi17	THi18	THi19	THi20

KIJANI 458/1974-75/6210-11/1966-67						
Ashe	Ojobo	Iwa	Araorun	Ile	Itutu	Erinle
1	2	3	4	5	6	7
N28	N29	N30	D1	D2	D3	D4
TBT11	TBT12	TBT13	TBT14	TBT15	TBT16	TBT17
HID19	HID20	HID21	HID22	HID23	HID24	HID25
8	9	10	11	12	13	14
D5	D6	D7	D8	D9	D10	D11
TBT18	TBT19	TBT20	TBT21	TBT22	TBT23	TBT24
HID26	HID27	HID28	HID29	HID30	TAH1	TAH2
15	16	17	18	19	20	21
D12	D13	D14	D15	D16	D17	D18
TBT25	TBT26	TBT27	TBT28	TBT29	TBT30	MKH1
TAH3	TAH4	TAH5	TAH6	TAH7	TAH8	TAH9
22	23	24	25	26	27	28
D19	D20	D21	D22	D23	D24	D25
MKH2	MKH3	MKH4	MKH5	MKH6	MKH7	MKH8
TAH10	TAH11	TAH12	TAH13	TAH14	TAH15	TAH16

EUSI 458/1974-75/6210-11/1966-67						
Ashe	Ojobo	Iwa	Araorun	Ile	Itutu	Erinle
1	2	3	4	5	6	7
O31	N1	N2	N3	N4	N5	N6
KHK13	KHK14	KHK15	KHK16	KHK17	KHK18	KHK19
THi21	THi22	THi23	THi24	THi25	THi26	THi27
8	9	10	11	12	13	14
N7	N8	N9	N10	N11	N12	N13
KHK20	KHK21	KHK22	KHK23	KHK24	KHK25	KHK26
THi28	THi29	THi30	HID1	HID2	HID3	HID4
15	16	17	18	19	20	21
N14	N15	N16	N17	N18	N19	N20
KHK27	KHK28	KHK29	KHK30	TBT1	TBT2	TBT3
HID5	HID6	HID7	HID8	HID9	HID10	HID11
22	23	24	25	26	27	28
N21	N22	N23	N24	N25	N26	N27

KWANZA 458/1974-75/6210-11/1966-67						
Umoja	Kujichagulia	Ujima	Ujamaa	Nia	Kuumba	Imani
1	2	3	4	5	6	7
D26	D27	D28	D29	D30	D31	JA1
MKH9	MKH10	MKH11	MKH12	MKH13	MKH14	MKH15
TAH17	TAH18	TAH19	TAH20	TAH21	TAH22	TAH23
8	9	10	11	12	13	14
JA2	JA3	JA4	JA5	JA6	JA7	JA8
MKH16	MKH17	MKH18	MKH19	MKH20	MKH21	MKH22
TAH24	TAH25	TAH26	TAH27	TAH28	TAH29	TAH30
15	16	17	18	19	20	21
JA9	JA10	JA11	JA12	JA13	JA14	JA15
MKH23	MKH24	MKH25	KMH26	MKH27	MKH28	MKH29
TIR1	TIR2	TIR3	TIR4	TIR5	TIR6	TIR7
22	23	24	25	26	27	28
JA16	JA17	JA18	JA19	JA20	JA21	JA22

SHAHIDI 458/1974-75/6210-11/1966-67						
Ashe	Ojobo	Iwa	Araorun	Ile	Itutu	Erinle
1	2	3	4	5	6	7
JA23	JA24	JA25	JA26	JA27	JA28	JA29
AMN7	AMN8	AMN9	AMN10	AMN11	AMN12	AMN13
TIR15	TIR16	TIR17	TIR18	TIR19	TIR20	TIR21
8	9	10	11	12	13	14
JA30	JA31	F1	F2	F3	F4	F5
AMN14	AMN15	AMN16	AMN17	AMN18	AMN19	AMN20
TIR22	TIR23	TIR24	TIR25	TIR26	TIR27	TIR28
15	16	17	18	19	20	21
F6	F7	F8	F9	F10	F 11	F12
AMN21	AMN22	AMN23	AMN24	AMN25	AMN26	AMN28
TIR29	TIR30	YEK1	YEK2	YEK3	YEK4	YEK5
22	23	24	25	26	27	28
F13	F14	F15	F16	F17	F18	F19
AMN29	AMN30	RNT1	RNT2	RNT3	RNT4	RNT5
YEK6	YEK7	YEK8	YEK9	YEK10	YEK11	YEK12
SHAHIDI 9						
F20						
RNT6						
YEK13						

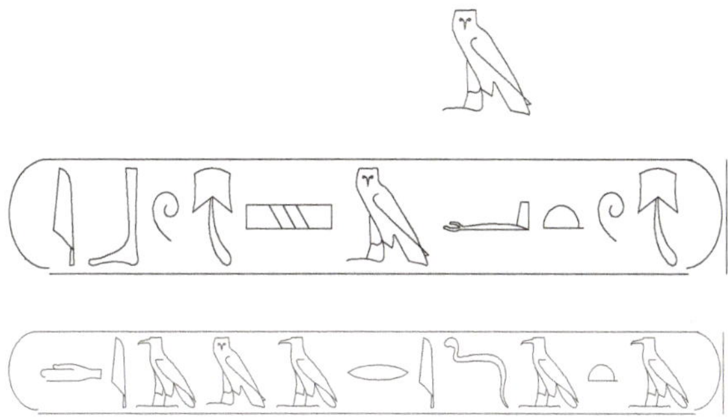

MAAFA 459 B.S.E./ASIENTO 1975-76 C.E./ KAM 6211-12 S.M./EC 1967-68 A.M. (ዓ.ም.)

AFU 459/1975-76/6211-12/1967-68

Ashe	Ojobo	Iwa	Araorun	Ile	Itutu	Erinle
1	2	3	4	5	6	7
F21	F22	F23	F24	F25	F26	F27
RNT7	RNT8	RNT9	RNT10	RNT11	RNT12	RNT13
YEK14	YEK15	YEK16	YEK17	YEK18	YEK19	YEK20
8	9	10	11	12	13	14
F28	M1	M2	M3	M4	M5	M6
RNT14	RNT15	RNT16	RNT17	RNT18	RNT19	RNT20
YEK21	YEK22	YEK23	YEK24	YEK25	YEK26	YEK27
15	16	17	18	19	20	21
M7	M8	M9	M10	M11	M12	M13
RNT21	RNT22	RNT23	RNT24	RNT25	RNT26	RNT27
YEK28	YEK 29	YEK30	MAG1	MAG2	MAG3	MAG4
22	23	24	25	26	27	28
M14	M15	M16	M17	M18	M19	M20
RNT28	RNT29	RNT30	KHN1	KHN2	KHN3	KHN4
MAG5	MAG6	MAG7	MAG8	MAG9	MAG10	MAG11

KAWAIDA 459/1975-76/6211-12/1967-68

Ashe	Ojobo	Iwa	Araorun	Ile	Itutu	Erinle
1	2	3	4	5	6	7
A18	A19	A20	A21	A22	A23	A24
PIN3	PIN4	PIN5	PIN6	PIN7	PIN8	PIN9
MIY10	MIY11	MIY12	MIY13	MIY14	MIY15	MIY16
8	9	10	11	12	13	14
A25	A26	A27	A28	A29	A30	MA1
PIN10	PIN11	PIN12	PIN13	PIN14	PIN15	PIN16
MIY17	MIY18	MIY19	MIY20	MIY21	MIY22	MIY23
15	16	17	18	19	20	21
MA2	MA3	MA4	MA5	MA6	MA7	MA8
PIN17	PIN18	PIN19	PIN20	PIN21	PIN22	PIN23
MIY24	MIY25	MIY26	MIY27	MIY28	MIY29	MIY30
22	23	24	25	26	27	28
MA9	MA10	MA11	MA12	MA13	MA14	MA15
PIN24	PIN25	PIN26	PIN27	PIN28	PIN29	PIN30
GIN1	GIN2	GIN3	GIN4	GIN5	GIN6	GIN7

UHURU 459/1975-76/6211-12/1967-68

Ashe	Ojobo	Iwa	Araorun	Ile	Itutu	Erinle
1	2	3	4	5	6	7
M21	M22	M23	M24	M25	M26	M27
KHN5	KHN6	KHN7	KHN8	KHN9	KHN10	KHN11
MAG12	MAG13	MAG14	MAG15	MAG16	MAG17	MAG18
8	9	10	11	12	13	14
M28	M29	M30	M31	A1	A2	A3
KHN12	KHN13	KHN14	KHN15	KHN16	KHN17	KHN18
MAG19	MAG20	MAG21	MAG22	MAG23	MAG24	MAG25
15	16	17	18	19	20	21
A4	A5	A6	A7	A8	A9	A10
KHN19	KHN20	KHN21	KHN22	KHN23	KHN24	KHN25
MAG26	MAG27	MAG28	MAG29	MAG30	MIY1	MIY2
22	23	24	25	26	27	28
A11	A12	A13	A14	A15	A16	A17
KHN26	KHN27	KHN28	KHN29	KHN30	PIN1	PIN2
MIY3	MIY4	MIY5	MIY6	MIY7	MIY8	MIY9

MABABA WEUSI 459/1975-76/6211-12/1967-68

Ashe	Ojobo	Iwa	Araorun	Ile	Itutu	Erinle
1	2	3	4	5	6	7
MA16	MA17	MA18	MA19	MA20	MA21	MA22
APP1	APP2	APP3	APP4	APP5	APP6	APP7
GIN8	GIN9	GIN10	GIN11	GIN12	GIN13	GIN14
8	9	10	11	12	13	14
MA23	MA24	MA25	MA26	MA27	MA28	MA29
APP8	APP9	APP10	APP11	APP12	APP13	APP14
GIN15	GIN16	GIN17	GIN18	GIN19	GIN20	GIN21
15	16	17	18	19	20	21
MA30	MA31	J1	J2	J3	J4	J5
APP15	APP16	APP17	APP18	APP19	APP20	APP21
GIN22	GIN23	GIN24	GIN25	GIN26	GIN27	GIN28
22	23	24	25	26	27	28
J6	J7	J8	J9	J10	J11	J12
APP22	APP23	APP24	APP25	APP26	APP27	APP28
GIN29	GIN30	SAN1	SAN2	SAN3	SAN4	SAN5

MWANAMKE WEUSI 459/1975-76/6211-12/1967-68						
Ashe	Ojobo	Iwa	Araorun	Ile	Itutu	Erinle
1	2	3	4	5	6	7
J13	J14	J15	J16	J17	J18	J19
APP29	APP30	MSR1	MSR2	MSR3	MSR4	MSR5
SAN6	SAN7	SAN8	SAN9	SAN10	SAN11	SAN12
8	9	10	11	12	13	14
J20	J21	J22	J23	J24	J25	J26
MSR6	MSR7	MSR8	MSR9	MSR10	MSR11	MSR12
SAN13	SAN14	SAN15	SAN16	SAN17	SAN18	SAN19
15	16	17	18	19	20	21
J27	J28	J29	J30	JU1	JU2	JU3
MSR13	MSR14	MSR15	MSR16	MSR17	MSR18	MSR19
SAN20	SAN21	SAN22	SAN23	SAN24	SAN25	SAN26
22	23	24	25	26	27	28
JU4	JU5	JU6	JU7	JU8	JU9	JU10
MSR20	MRS21	MSR22	MSR23	MSR24	MSR25	MSR26
SAN27	SAN28	SAN29	SAN30	HAM1	HAM2	HAM3

TAIFA WEUSI 459/1975-76/6211-12/1967-68						
Ashe	Ojobo	Iwa	Araorun	Ile	Itutu	Erinle
1	2	3	4	5	6	7
AU8	AU9	AU10	AU11	AU12	AU13	AU14
THT20	THT21	THT22	THT23	THT24	THT25	THT26
NAH2	NAH3	NAH4	NAH5	NAH6	NAH7	NAH8
8	9	10	11	12	13	14
AU15	AU16	AU17	AU18	AU19	AU20	AU21
THT27	THT28	THT29	THT30	PAP1	PAP1	PAP2
NAH9	NAH10	NAH11	NAH12	NAH13	NAH14	NAH15
15	16	17	18	19	20	21
AU22	AU23	AU24	AU25	AU26	AU27	AU28
PAP3	PAP4	PAP5	PAP6	PAP7	PAP8	PAP9
NAH16	NAH17	NAH18	NAH19	NAH20	NAH21	NAH22
22	23	24	25	26	27	28
AU29	AU30	AU31	S1	S2	S3	S4
PAP10	PAP11	PAP12	PAP13	PAP14	PAP15	PAP16
NAH23	NAH24	NAH25	NAH26	NAH27	NAH28	NAH29

MTU WEUSI 459/1975-76/6211-12/1967-68						
Ashe	Ojobo	Iwa	Araorun	Ile	Itutu	Erinle
1	2	3	4	5	6	7
JU11	JU12	JU13	JU14	JU15	JU16	JU17
MSR27	MSR28	MSR29	MSR30	*ASR1*	*HRU2*	*STH3*
HAM4	HAM5	HAM6	HAM7	HAM8	HAM9	HAM10
8	9	10	11	12	13	14
JU18	JU19	JU20	JU21	JU22	JU23	JU24
AST4	*NBH5*	THT1	THT2	THT3	THT4	THT5
HAM11	HAM12	HAM13	HAM14	HAM15	HAM16	HAM17
15	16	17	18	19	20	21
JU25	JU26	JU27	JU28	JU29	JU30	JU31
THT6	THT7	THT8	THT9	THT10	THT11	THT12
HAM18	HAM19	HAM20	HAM21	HAM22	HAM23	HAM24
22	23	24	25	26	27	28
AU1	AU2	AU3	AU4	AU5	AU6	AU7
THT13	THT14	THT15	THT16	THT17	THT18	THT19
HAM25	HAM26	HAM27	HAM28	HAM29	HAM30	NAH1

UWEZO 459/1975-76/6211-12/1967-68						
Ashe	Ojobo	Iwa	Araorun	Ile	Itutu	Erinle
1	2	3	4	5	6	7
S5	S6	S7	S8	S9	S10	S11
PAP17	PAP18	PAP19	PAP20	PAP21	PAP22	PAP23
NAH30	*QUA1*	*QUA2*	*QUA3*	*QUA4*	*QUA5*	*QUA6*
8	9	10	11	12	13	14
S12	S13	S14	S15	S16	S17	S18
PAP24	PAP25	PAP26	PAP27	PAP28	PAP29	PAP30
MAS1	MAS2	MAS3	MAS4	MAS5	MAS6	MAS7
15	16	17	18	19	20	21
S19	S20	S21	S22	S23	S24	S25
HTH1	HTH2	HTH3	HTH4	HTH5	HTH6	HTH7
MAS8	MAS9	MAS10	MAS11	MAS12	MAS13	MAS14
22	23	24	25	26	27	28
S26	S27	S28	S29	S30	O1	O2
HTH8	HTH9	HTH10	HTH11	HTH12	HTH13	HTH14
MAS15	MAS16	MAS17	MAS18	MAS19	MAS20	MAS21

EKUNDU 459/1975 -76/6211-12/1967-68

Ashe	Ojobo	Iwa	Araorun	Ile	Itutu	Erinle
1	2	3	4	5	6	7
O3	O4	O5	O6	O7	O8	O9
HTH15	HTH16	HTH17	HTH18	HTH19	HTH20	HTH21
MAS23	MAS24	MAS25	MAS26	MAS27	MAS28	MAS29
8	9	10	11	12	13	14
O10	O11	O12	O13	O14	O15	O16
HTH22	HTH23	HTH24	HTH25	HTH26	HTH27	HTH28
MAS30	THi1	THi2	THi3	THi4	THi5	THi6
15	16	17	18	19	20	21
O17	O18	O19	O20	O21	O22	O23
HTH29	HTH30	KHK1	KHK2	KHK3	KHK4	KHK5
THi7	THi8	THi9	THi10	THi11	THi12	THi13
22	23	24	25	26	27	28
O24	O25	O26	O27	O28	O29	O30
KHK6	KHK7	KHK8	KHK9	KHK10	KHK11	KHK12
THi14	THi15	THi16	THi17	THi18	THi19	THi20

KIJANI 459/1975 -76/6211-12/1967-68

Ashe	Ojobo	Iwa	Araorun	Ile	Itutu	Erinle
1	2	3	4	5	6	7
N28	N29	N30	D1	D2	D3	D4
TBT11	TBT12	TBT13	TBT14	TBT15	TBT16	TBT17
HID19	HID20	HID21	HID22	HID23	HID24	HID25
8	9	10	11	12	13	14
D5	D6	D7	D8	D9	D10	D11
TBT18	TBT19	TBT20	TBT21	TBT22	TBT23	TBT24
HID26	HID27	HID28	HID29	HID30	TAH1	TAH2
15	16	17	18	19	20	21
D12	D13	D14	D15	D16	D17	D18
TBT25	TBT26	TBT27	TBT28	TBT29	TBT30	MKH1
TAH3	TAH4	TAH5	TAH6	TAH7	TAH8	TAH9
22	23	24	25	26	27	28
D19	D20	D21	D22	D23	D24	D25
MKH2	MKH3	MKH4	MKH5	MKH6	MKH7	MKH8
TAH10	TAH11	TAH12	TAH13	TAH14	TAH15	TAH16

EUSI 459/1975 -76/6211-12/1967-68

Ashe	Ojobo	Iwa	Araorun	Ile	Itutu	Erinle
1	2	3	4	5	6	7
O31	N1	N2	N3	N4	N5	N6
KHK13	KHK14	KHK15	KHK16	KHK17	KHK18	KHK19
THi21	THi22	THi23	THi24	THi25	THi26	THi27
8	9	10	11	12	13	14
N7	N8	N9	N10	N11	N12	N13
KHK20	KHK21	KHK22	KHK23	KHK24	KHK25	KHK26
THi28	THi29	THi30	HID1	HID2	HID3	HID4
15	16	17	18	19	20	21
N14	N15	N16	N17	N18	N19	N20
KHK27	KHK28	KHK29	KHK30	TBT1	TBT2	TBT3
HID5	HID6	HID7	HID8	HID9	HID10	HID11
22	23	24	25	26	27	28
N21	N22	N23	N24	N25	N26	N27

KWANZA 459/1975 -76/6211-12/1967-68

Umoja	Kujichagulia	Ujima	Ujamaa	Nia	Kuumba	Imani
1	2	3	4	5	6	7
D26	D27	D28	D29	D30	D31	JA1
MKH9	MKH10	MKH11	MKH12	MKH13	MKH14	MKH15
TAH17	TAH18	TAH19	TAH20	TAH21	TAH22	TAH23
8	9	10	11	12	13	14
JA2	JA3	JA4	JA5	JA6	JA7	JA8
MKH16	MKH17	MKH18	MKH19	MKH20	MKH21	MKH22
TAH24	TAH25	TAH26	TAH27	TAH28	TAH29	TAH30
15	16	17	18	19	20	21
JA9	JA10	JA11	JA12	JA13	JA14	JA15
MKH23	MKH24	MKH25	KMH26	MKH27	MKH28	MKH29
TIR1	TIR2	TIR3	TIR4	TIR5	TIR6	TIR7
22	23	24	25	26	27	28
JA16	JA17	JA18	JA19	JA20	JA21	JA22

SHAHIDI 459/1975-76/6211-12/1967-68						
Ashe	Ojobo	Iwa	Araorun	Ile	Itutu	Erinle
1	2	3	4	5	6	7
JA23	JA24	JA25	JA26	JA27	JA28	JA29
AMN7	AMN8	AMN9	AMN10	AMN11	AMN12	AMN13
TIR14	TIR15	TIR16	TIR17	TIR18	TIR19	TIR20
8	9	10	11	12	13	14
JA30	JA31	F1	F2	F3	F4	F5
AMN14	AMN15	AMN16	AMN17	AMN18	AMN19	AMN20
TIR21	TIR22	TIR23	TIR24	TIR25	TIR26	TIR27
15	16	17	18	19	20	21
F6	F7	F8	F9	F10	F 11	F12
AMN21	AMN22	AMN23	AMN24	AMN25	AMN26	AMN28
TIR28	TIR29	TIR30	YEK1	YEK2	YEK3	YEK4
22	23	24	25	26	27	28
F13	F14	F15	F16	F17	F18	F19
AMN29	AMN30	RNT1	RNT2	RNT3	RNT4	RNT5
YEK5	YEK6	YEK7	YEK8	YEK9	YEK10	YEK11
SHAHIDI 10						
F20						
RNT6						
YEK12						

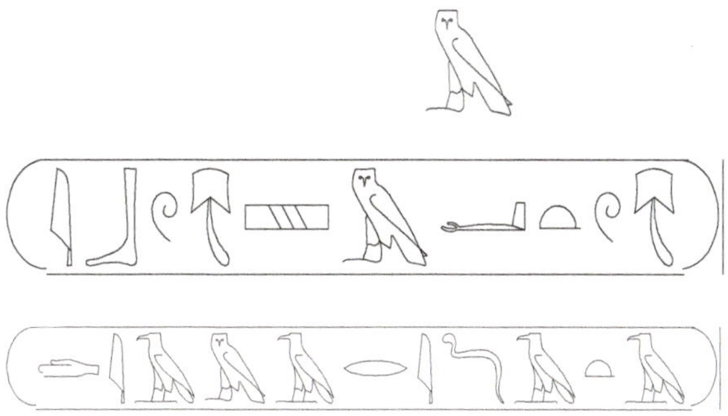

MAAFA 460 B.S.E./ASIENTO 1976-77 C.E./
KAM 6212-13 S.M./EC 1968-69 A.M. (ዓ.ም.)

AFU 460/1976-77/6212-13/1968-69						
Ashe	Ojobo	Iwa	Araorun	Ile	Itutu	Erinle
1	2	3	4	5	6	7
F21	F22	F23	F24	F25	F26	F27
RNT7	RNT8	RNT9	RNT10	RNT11	RNT12	RNT13
YEK13	YEK14	YEK15	YEK16	YEK17	YEK18	YEK19
8	9	10	11	12	13	14
F28	F29	M1	M2	M3	M4	M5
RNT14	RNT15	RNT16	RNT17	RNT18	RNT19	RNT20
YEK 20	YEK21	YEK22	YEK23	YEK24	YEK25	YEK26
15	16	17	18	19	20	21
M6	M7	M8	M9	M10	M11	M12
RNT21	RNT22	RNT23	RNT24	RNT25	RNT26	RNT27
YEK 27	YEK28	YEK29	YEK30	MAG1	MAG2	MAG3
22	23	24	25	26	27	28
M13	M14	M15	M16	M17	M18	M19
RNT28	RNT29	RNT30	KHN1	KHN2	KHN3	KHN4
MAG4	MAG5	MAG6	MAG7	MAG8	MAG9	MAG10
SHARIFU 112						
M20						
KHN5						
MAG11						

KAWAIDA 460/1976-77/6212-13/1968-69						
Ashe	Ojobo	Iwa	Araorun	Ile	Itutu	Erinle
1	2	3	4	5	6	7
A18	A19	A20	A21	A22	A23	A24
PIN4	PIN5	PIN6	PIN7	PIN8	PIN9	PIN10
MIY10	MIY11	MIY12	MIY13	MIY14	MIY15	MIY16
8	9	10	11	12	13	14
A25	A26	A27	A28	A29	A30	MA1
PIN11	PIN12	PIN13	PIN14	PIN15	PIN16	PIN17
MIY17	MIY18	MIY19	MIY20	MIY21	MIY22	MIY23
15	16	17	18	19	20	21
MA2	MA3	MA4	MA5	MA6	MA7	MA8
PIN18	PIN19	PIN20	PIN21	PIN22	PIN23	PIN24
MIY24	MIY25	MIY26	MIY27	MIY28	MIY29	MIY30
22	23	24	25	26	27	28
MA9	MA10	MA11	MA12	MA13	MA14	MA15
PIN25	PIN26	PIN27	PIN28	PIN29	PIN30	APP1
GIN1	GIN2	GIN3	GIN4	GIN5	GIN6	GIN7

UHURU 460/1976-77/6212-13/1968-69						
Ashe	Ojobo	Iwa	Araorun	Ile	Itutu	Erinle
1	2	3	4	5	6	7
M21	M22	M23	M24	M25	M26	M27
KHN6	KHN7	KHN8	KHN9	KHN10	KHN11	KHN12
MAG12	MAG13	MAG14	MAG15	MAG16	MAG17	MAG18
8	9	10	11	12	13	14
M28	M29	M30	M31	A1	A2	A3
KHN13	KHN14	KHN15	KHN16	KHN17	KHN18	KHN19
MAG19	MAG20	MAG21	MAG22	MAG23	MAG24	MAG25
15	16	17	18	19	20	21
A4	A5	A6	A7	A8	A9	A10
KHN20	KHN21	KHN222	KHN23	KHN24	KHN25	KHN26
MAG26	MAG27	MAG28	MAG29	MAG30	MIY1	MIY2
22	23	24	25	26	27	28
A11	A12	A13	A14	A15	A16	A17
KHN27	KHN28	KHN29	KHN30	PIN1	PIN2	PIN3
MIY3	MIY4	MIY5	MIY6	MIY7	MIY8	MIY9

MABABA WEUSI 460/1976-77/6212-13/1968-69						
Ashe	Ojobo	Iwa	Araorun	Ile	Itutu	Erinle
1	2	3	4	5	6	7
MA16	MA17	MA18	MA19	MA20	MA21	MA22
APP2	APP3	APP4	APP5	APP6	APP7	APP8
GIN8	GIN9	GIN10	GIN11	GIN12	GIN13	GIN14
8	9	10	11	12	13	14
MA23	MA24	MA25	MA26	MA27	MA28	MA29
APP9	APP10	APP11	APP12	APP13	APP14	APP15
GIN15	GIN16	GIN17	GIN18	GIN19	GIN20	GIN21
15	16	17	18	19	20	21
MA30	MA31	J1	J2	J3	J4	J5
APP16	APP17	APP18	APP19	APP20	APP21	APP22
GIN22	GIN23	GIN24	GIN25	GIN26	GIN27	GIN28
22	23	24	25	26	27	28
J6	J7	J8	J9	J10	J11	J12
APP23	APP24	APP25	APP26	APP27	APP28	APP29
GIN29	GIN30	SAN1	SAN2	SAN3	SAN4	SAN5

MWANAMKE WEUSI 460/1976-77/6212-13/1968-69						
Ashe	Ojobo	Iwa	Araorun	Ile	Itutu	Erinle
1	2	3	4	5	6	7
J13	J14	J15	J16	J17	J18	J19
APP30	MSR1	MSR2	MSR3	MSR4	MSR5	MSR6
SAN6	SAN7	SAN8	SAN9	SAN10	SAN11	SAN12
8	9	10	11	12	13	14
J20	J21	J22	J23	J24	J25	J26
MSR7	MSR8	MSR9	MSR10	MSR11	MSR12	MSR13
SAN13	SAN14	SAN15	SAN16	SAN17	SAN18	SAN19
15	16	17	18	19	20	21
J27	J28	J29	J30	JU1	JU2	JU3
MSR14	MSR15	MSR16	MSR17	MSR18	MSR19	MSR20
SAN20	SAN21	SAN22	SAN23	SAN24	SAN25	SAN26
22	23	24	25	26	27	28
JU4	JU5	JU6	JU7	JU8	JU9	JU10
MSR21	MRS22	MSR23	MSR24	MSR25	MSR26	MSR27
SAN27	SAN28	SAN29	SAN30	HAM1	HAM2	HAM3

TAIFA WEUSI 460/1976-77/6212-13/1968-69						
Ashe	Ojobo	Iwa	Araorun	Ile	Itutu	Erinle
1	2	3	4	5	6	7
AU8	AU9	AU10	AU11	AU12	AU13	AU14
THT20	THT21	THT22	THT23	THT24	THT25	THT26
NAH2	NAH3	NAH4	NAH5	NAH6	NAH7	NAH8
8	9	10	11	12	13	14
AU15	AU16	AU17	AU18	AU19	AU20	AU21
THT27	THT28	THT29	THT30	PAP1	PAP1	PAP2
NAH9	NAH10	NAH11	NAH12	NAH13	NAH14	NAH15
15	16	17	18	19	20	21
AU22	AU23	AU24	AU25	AU26	AU27	AU28
PAP3	PAP4	PAP5	PAP6	PAP7	PAP8	PAP9
NAH16	NAH17	NAH18	NAH19	NAH20	NAH21	NAH22
22	23	24	25	26	27	28
AU29	AU30	AU31	S1	S2	S3	S4
PAP10	PAP11	PAP12	PAP13	PAP14	PAP15	PAP16
NAH23	NAH24	NAH25	NAH26	NAH27	NAH28	NAH29

MTU WEUSI 460/1976-77/6212-13/1968-69						
Ashe	Ojobo	Iwa	Araorun	Ile	Itutu	Erinle
1	2	3	4	5	6	7
JU11	JU12	JU13	JU14	JU15	JU16	JU17
MSR28	MSR29	MSR30	*ASR1*	*HRU2*	*STH3*	*AST4*
HAM4	HAM5	HAM6	HAM7	HAM8	HAM9	HAM10
8	9	10	11	12	13	14
JU18	JU19	JU20	JU21	JU22	JU23	JU24
NBH5	*ANU6*	THT1	THT2	THT3	THT4	THT5
HAM11	HAM12	HAM13	HAM14	HAM15	HAM16	HAM17
15	16	17	18	19	20	21
JU25	JU26	JU27	JU28	JU29	JU30	JU31
THT6	THT7	THT8	THT9	THT10	THT11	THT12
HAM18	HAM19	HAM20	HAM21	HAM22	HAM23	HAM24
22	23	24	25	26	27	28
AU1	AU2	AU3	AU4	AU5	AU6	AU7
THT13	THT14	THT15	THT16	THT17	THT18	THT19
HAM25	HAM26	HAM27	HAM28	HAM29	HAM30	NAH1

UWEZO 460/1976-77/6212-13/1968-69						
Ashe	Ojobo	Iwa	Araorun	Ile	Itutu	Erinle
1	2	3	4	5	6	7
S5	S6	S7	S8	S9	S10	S11
PAP17	PAP18	PAP19	PAP20	PAP21	PAP22	PAP23
NAH30	*PAG1*	*PAG2*	*PAG3*	*PAG4*	*PAG5*	MAS1
8	9	10	11	12	13	14
S12	S13	S14	S15	S16	S17	S18
PAP24	PAP25	PAP26	PAP27	PAP28	PAP29	PAP30
MAS2	MAS3	MAS4	MAS5	MAS6	MAS7	MAS8
15	16	17	18	19	20	21
S19	S20	S21	S22	S23	S24	S25
HTH1	HTH2	HTH3	HTH4	HTH5	HTH6	HTH7
MAS9	MAS10	MAS11	MAS12	MAS13	MAS14	MAS15
22	23	24	25	26	27	28
S26	S27	S28	S29	S30	O1	O2
HTH8	HTH9	HTH10	HTH11	HTH12	HTH13	HTH14
MAS16	MAS17	MAS18	MAS19	MAS20	MAS21	MAS22

EKUNDU 460/1976-77/6212-13/1968-69						
e	Ojobo	Iwa	Araorun	Ile	Itutu	Erinle
	2	3	4	5	6	7
	O4	O5	O6	O7	O8	O9
15	HTH16	HTH17	HTH18	HTH19	HTH20	HTH21
23	MAS24	MAS25	MAS26	MAS27	MAS28	MAS29
	9	10	11	12	13	14
	O11	O12	O13	O14	O15	O16
22	HTH23	HTH24	HTH25	HTH26	HTH27	HTH28
30	THi1	THi2	THi3	THi4	THi5	THi6
	16	17	18	19	20	21
	O18	O19	O20	O21	O22	O23
29	HTH30	KHK1	KHK2	KHK3	KHK4	KHK5
	THi8	This9	THi10	THi11	THi12	THi13
	23	24	25	26	27	28
	O25	O26	O27	O28	O29	O30
6	KHK7	KHK8	KHK9	KHK10	KHK11	KHK12
4	THi15	THi16	THi17	THi18	THi19	THi20

KIJANI 460/1976-77/6212-13/1968-69						
Ashe	Ojobo	Iwa	Araorun	Ile	Itutu	Erinle
1	2	3	4	5	6	7
N28	N29	N30	D1	D2	D3	D4
TBT11	TBT12	TBT13	TBT14	TBT15	TBT16	TBT17
HID19	HID20	HID21	HID22	HID23	HID24	HID25
8	9	10	11	12	13	14
D5	D6	D7	D8	D9	D10	D11
TBT18	TBT19	TBT20	TBT21	TBT22	TBT23	TBT24
HID26	HID27	HID28	HID29	HID30	TAH1	TAH2
15	16	17	18	19	20	21
D12	D13	D14	D15	D16	D17	D18
TBT25	TBT26	TBT27	TBT28	TBT29	TBT30	MKH1
TAH3	TAH4	TAH5	TAH6	TAH7	TAH8	TAH9
22	23	24	25	26	27	28
D19	D20	D21	D22	D23	D24	D25
MKH2	MKH3	MKH4	MKH5	MKH6	MKH7	MKH8
TAH10	TAH11	TAH12	TAH13	TAH14	TAH15	TAH16

EUSI 460/1976-77/6212-13/1968-69						
e	Ojobo	Iwa	Araorun	Ile	Itutu	Erinle
	2	3	4	5	6	7
	N1	N2	N3	N4	N5	N6
13	KHK14	KHK15	KHK16	KHK17	KHK18	KHK19
1	THi22	THi23	THi24	THi25	THi26	THi27
	9	10	11	12	13	14
	N8	N9	N10	N11	N12	N13
20	KHK21	KHK22	KHK23	KHK24	KHK25	KHK26
8	THi29	THi30	HID1	HID2	HID3	HID4
	16	17	18	19	20	21
	N15	N16	N17	N18	N19	N20
27	KHK28	KHK29	KHK30	TBT1	TBT2	TBT3
5	HID6	HID7	HID8	HID9	HID10	HID11
	23	24	25	26	27	28
	N22	N23	N24	N25	N26	N27
4	TBT5	TBT6	TBT7	TBT8	TBT9	TBT10
12	HID13	HID14	HID15	HID16	HID17	HID18

KWANZA 460/1976-77/6212-13/1968-69						
Umoja	Kujichagulia	Ujima	Ujamaa	Nia	Kuumba	Imani
1	2	3	4	5	6	7
D26	D27	D28	D29	D30	D31	JA1
MKH9	MKH10	MKH11	MKH12	MKH13	MKH14	MKH15
TAH17	TAH18	TAH19	TAH20	TAH21	TAH22	TAH23
8	9	10	11	12	13	14
JA2	JA3	JA4	JA5	JA6	JA7	JA8
MKH16	MKH17	MKH18	MKH19	MKH20	MKH21	MKH22
TAH24	TAH25	TAH26	TAH27	TAH28	TAH29	TAH30
15	16	17	18	19	20	21
JA9	JA10	JA11	JA12	JA13	JA14	JA15
MKH23	MKH24	MKH25	KMH26	MKH27	MKH28	MKH29
TIR1	TIR2	TIR3	TIR4	TIR5	TIR6	TIR7
22	23	24	25	26	27	28
JA16	JA17	JA18	JA19	JA20	JA21	JA22
MKH30	AMN1	AMN2	AMN3	AMN4	AMN5	AMN6
TIR8	TIR9	TIR10	TIR11	TIR12	TIR13	TIR14

SHAHIDI 460/1976-77/6212-13/1968-69						
Ashe	Ojobo	Iwa	Araorun	Ile	Itutu	Erinle
1	2	3	4	5	6	7
JA23	JA24	JA25	JA26	JA27	JA28	JA29
AMN7	AMN8	AMN9	AMN10	AMN11	AMN12	AMN13
TIR15	TIR16	TIR17	TIR18	TIR19	TIR20	TIR21
8	9	10	11	12	13	14
JA30	JA31	F1	F2	F3	F4	F5
AMN14	AMN15	AMN16	AMN17	AMN18	AMN19	AMN20
TIR22	TIR23	TIR24	TIR25	TIR26	TIR27	TIR28
15	16	17	18	19	20	21
F6	F7	F8	F9	F10	F 11	F12
AMN21	AMN22	AMN23	AMN24	AMN25	AMN26	AMN28
TIR29	TIR30	YEK1	YEK2	YEK3	YEK4	YEK5
22	23	24	25	26	27	28
F13	F14	F15	F16	F17	F18	F19
AMN29	AMN30	RNT1	RNT2	RNT3	RNT4	RNT5
YEK6	YEK7	YEK8	YEK9	YEK10	YEK11	YEK12
SHAHIDI 11						
F20						
RNT6						
YEK13						

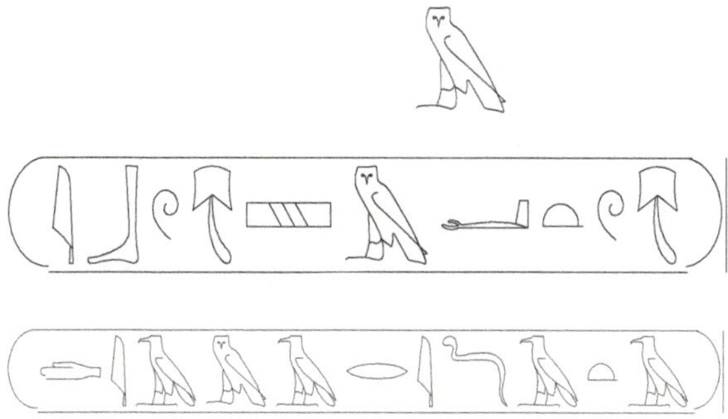

MAAFA 461 B.S.E./ASIENTO 1977-78 C.E./
KAM 6213-14 S.M./EC 1969-70 A.M. (ዓ.ም.)

AFU 461/1977-78/6213-14/1969-70						
she	Ojobo	Iwa	Araorun	Ile	Itutu	Erinle
1	2	3	4	5	6	7
21	F22	F23	F24	F25	F26	F27
NT7	RNT8	RNT9	RNT10	RNT11	RNT12	RNT13
EK14	YEK15	YEK16	YEK17	YEK18	YEK19	YEK20
	9	10	11	12	13	14
28	M1	M2	M3	M4	M5	M6
NT14	RNT15	RNT16	RNT17	RNT18	RNT19	RNT20
EK21	YEK22	YEK23	YEK24	YEK25	YEK26	YEK27
5	16	17	18	19	20	21
17	M8	M9	M10	M11	M12	M13
NT21	RNT22	RNT23	RNT24	RNT25	RNT26	RNT27
EK28	YEK29	YEK30	MAG1	MAG2	MAG3	MAG4
2	23	24	25	26	27	28
114	M15	M16	M17	M18	M19	M20
NT28	RNT29	RNT30	KHN1	KHN2	KHN3	KHN4
AG5	MAG6	MAG7	MAG8	MAG9	MAG10	MAG11

KAWAIDA 461/1977-78/6213-14/1969-70						
Ashe	Ojobo	Iwa	Araorun	Ile	Itutu	Erinle
1	2	3	4	5	6	7
A18	A19	A20	A21	A22	A23	A24
PIN3	PIN4	PIN5	PIN6	PIN7	PIN8	PIN9
MIY10	MIY11	MIY12	MIY13	MIY14	MIY15	MIY16
8	9	10	11	12	13	14
A25	A26	A27	A28	A29	A30	MA1
PIN10	PIN11	PIN12	PIN13	PIN14	PIN15	PIN16
MIY17	MIY18	MIY19	MIY20	MIY21	MIY22	MIY23
15	16	17	18	19	20	21
MA2	MA3	MA4	MA5	MA6	MA7	MA8
PIN17	PIN18	PIN19	PIN20	PIN21	PIN22	PIN23
MIY24	MIY25	MIY26	MIY27	MIY28	MIY29	MIY30
22	23	24	25	26	27	28
MA9	MA10	MA11	MA12	MA13	MA14	MA15
PIN24	PIN25	PIN26	PIN27	PIN28	PIN29	PIN30
GIN1	GIN2	GIN3	GIN4	GIN5	GIN6	GIN7

UHURU 461/1977-78/6213-14/1969-70						
she	Ojobo	Iwa	Araorun	Ile	Itutu	Erinle
1	2	3	4	5	6	7
121	M22	M23	M24	M25	M26	M27
HN5	KHN6	KHN7	KHN8	KHN9	KHN10	KHN11
AG12	MAG13	MAG14	MAG15	MAG16	MAG17	MAG18
8	9	10	11	12	13	14
128	M29	M30	M31	A1	A2	A3
HN12	KHN13	KHN14	KHN15	KHN16	KHN17	KHN18
AG19	MAG20	MAG21	MAG22	MAG23	MAG24	MAG25
15	16	17	18	19	20	21
A4	A5	A6	A7	A8	A9	A10
HN19	KHN20	KHN21	KHN22	KHN23	KHN24	KHN25
AG26	MAG27	MAG28	MAG29	MAG30	MIY1	MIY2
22	23	24	25	26	27	28
11	A12	A13	A14	A15	A16	A17
HN26	KHN27	KHN28	KHN29	KHN30	PIN1	PIN2
IY3	MIY4	MIY5	MIY6	MIY7	MIY8	MIY9

MABABA WEUSI 461/1977-78/6213-14/1969-70						
Ashe	Ojobo	Iwa	Araorun	Ile	Itutu	Erinle
1	2	3	4	5	6	7
MA16	MA17	MA18	MA19	MA20	MA21	MA22
APP1	APP2	APP3	APP4	APP5	APP6	APP7
GIN8	GIN9	GIN10	GIN11	GIN12	GIN13	GIN14
8	9	10	11	12	13	14
MA23	MA24	MA25	MA26	MA27	MA28	MA29
APP8	APP9	APP10	APP11	APP12	APP13	APP14
GIN15	GIN16	GIN17	GIN18	GIN19	GIN20	GIN21
15	16	17	18	19	20	21
MA30	MA31	J1	J2	J3	J4	J5
APP15	APP16	APP17	APP18	APP19	APP20	APP21
GIN22	GIN23	GIN24	GIN25	GIN26	GIN27	GIN28
22	23	24	25	26	27	28
J6	J7	J8	J9	J10	J11	J12
APP22	APP23	APP24	APP25	APP26	APP27	APP28
GIN29	GIN30	SAN1	SAN2	SAN3	SAN4	SAN5

MWANAMKE WEUSI 461/1977-78/6213-14/1969-70

Ashe	Ojobo	Iwa	Araorun	Ile	Itutu	Erinle
1	2	3	4	5	6	7
J13	J14	J15	J16	J17	J18	J19
APP29	APP30	MSR1	MSR2	MSR3	MSR4	MSR5
SAN6	SAN7	SAN8	SAN9	SAN10	SAN11	SAN12
8	9	10	11	12	13	14
J20	J21	J22	J23	J24	J25	J26
MSR6	MSR7	MSR8	MSR9	MSR10	MSR11	MSR12
SAN13	SAN14	SAN15	SAN16	SAN17	SAN18	SAN19
15	16	17	18	19	20	21
J27	J28	J29	J30	JU1	JU2	JU3
MSR13	MSR14	MSR15	MSR16	MSR17	MSR18	MSR19
SAN20	SAN21	SAN22	SAN23	SAN24	SAN25	SAN26
22	23	24	25	26	27	28
JU4	JU5	JU6	JU7	JU8	JU9	JU10
MSR20	MRS21	MSR22	MSR23	MSR24	MSR25	MSR26
SAN27	SAN28	SAN29	SAN30	HAM1	HAM2	HAM3

TAIFA WEUSI 461/1977-78/6213-14/1969-70

Ashe	Ojobo	Iwa	Araorun	Ile	Itutu	Erinle
1	2	3	4	5	6	7
AU8	AU9	AU10	AU11	AU12	AU13	AU14
THT20	THT21	THT22	THT23	THT24	THT25	THT26
NAH2	NAH3	NAH4	NAH5	NAH6	NAH7	NAH8
8	9	10	11	12	13	14
AU15	AU16	AU17	AU18	AU19	AU20	AU21
THT27	THT28	THT29	THT30	PAP1	PAP1	PAP2
NAH9	NAH10	NAH11	NAH12	NAH13	NAH14	NAH15
15	16	17	18	19	20	21
AU22	AU23	AU24	AU25	AU26	AU27	AU28
PAP3	PAP4	PAP5	PAP6	PAP7	PAP8	PAP9
NAH16	NAH17	NAH18	NAH19	NAH20	NAH21	NAH22
22	23	24	25	26	27	28
AU29	AU30	AU31	S1	S2	S3	S4
PAP10	PAP11	PAP12	PAP13	PAP14	PAP15	PAP16
NAH23	NAH24	NAH25	NAH26	NAH27	NAH28	NAH29

MTU WEUSI 461/1977-78/6213-14/1969-70

Ashe	Ojobo	Iwa	Araorun	Ile	Itutu	Erinle
1	2	3	4	5	6	7
JU11	JU12	JU13	JU14	JU15	JU16	JU17
MSR27	MSR28	MSR29	MSR30	*ASR1*	*HRU2*	*STH3*
HAM4	HAM5	HAM6	HAM7	HAM8	HAM9	HAM10
8	9	10	11	12	13	14
JU18	JU19	JU20	JU21	JU22	JU23	JU24
AST4	*NBH5*	THT1	THT2	THT3	THT4	THT5
HAM11	HAM12	HAM13	AM14	HAM15	HAM16	HAM17
15	16	17	18	19	20	21
JU25	JU26	JU27	JU28	JU29	JU30	JU31
THT6	THT7	THT8	THT9	THT10	THT11	THT12
HAM18	HAM19	HAM20	HAM21	HAM22	HAM23	HAM24
22	23	24	25	26	27	28
AU1	AU2	AU3	AU4	AU5	AU6	AU7
THT13	THT14	THT15	THT16	THT17	THT18	THT19
HAM25	HAM26	HAM27	HAM28	HAM29	HAM30	NAH1

UWEZO 461/1977-78/6213-14/1969-70

Ashe	Ojobo	Iwa	Araorun	Ile	Itutu	Erinle
1	2	3	4	5	6	7
S5	S6	S7	S8	S9	S10	S11
PAP17	PAP18	PAP19	PAP20	PAP21	PAP22	PAP23
NAH30	*PAG1*	*PAG2*	*PAG3*	*PAG4*	*PAG5*	MAS1
8	9	10	11	12	13	14
S12	S13	S14	S15	S16	S17	S18
PAP24	PAP25	PAP26	PAP27	PAP28	PAP29	PAP30
MAS2	MAS3	MAS4	MAS5	MAS6	MAS7	MAS8
15	16	17	18	19	20	21
S19	S20	S21	S22	S23	S24	S25
HTH1	HTH2	HTH3	HTH4	HTH5	HTH6	HTH7
MAS9	MAS10	MAS11	MAS12	MAS13	MAS14	MAS15
22	23	24	25	26	27	28
S26	S27	S28	S29	S30	O1	O2
HTH8	HTH9	HTH10	HTH11	HTH12	HTH13	HTH14
MAS16	MAS17	MAS18	MAS19	MAS20	MAS21	MAS22

EKUNDU 461/1977-78/6213-14/1969-70

Ashe	Ojobo	Iwa	Araorun	Ile	Itutu	Erinle
1	2	3	4	5	6	7
O3	O4	O5	O6	O7	O8	O9
HTH15	HTH16	HTH17	HTH18	HTH19	HTH20	HTH21
MAS23	MAS24	MAS25	MAS26	MAS27	MAS28	MAS29
8	9	10	11	12	13	14
O10	O11	O12	O13	O14	O15	O16
HTH22	HTH23	HTH24	HTH25	HTH26	HTH27	HTH28
MAS30	THi1	THi2	THi3	THi4	THi5	THi6
15	16	17	18	19	20	21
O17	O18	O19	O20	O21	O22	O23
HTH29	HTH30	KHK1	KHK2	KHK3	KHK4	KHK5
THi7	THi8	This9	THi10	THi11	THi12	THi13
22	23	24	25	26	27	28
O24	O25	O26	O27	O28	O29	O30
KHK6	KHK7	KHK8	KHK9	KHK10	KHK11	KHK12
Thi14	THi15	THi16	THi17	THi18	THi19	THi20

KIJANI 461/1977-78/6213-14/1969-70

Ashe	Ojobo	Iwa	Araorun	Ile	Itutu	Erinle
1	2	3	4	5	6	7
N28	N29	N30	D1	D2	D3	D4
TBT11	TBT12	TBT13	TBT14	TBT15	TBT16	TBT17
HID19	HID20	HID21	HID22	HID23	HID24	HID25
8	9	10	11	12	13	14
D5	D6	D7	D8	D9	D10	D11
TBT18	TBT19	TBT20	TBT21	TBT22	TBT23	TBT24
HID26	HID27	HID28	HID29	HID30	TAH1	TAH2
15	16	17	18	19	20	21
D12	D13	D14	D15	D16	D17	D18
TBT25	TBT26	TBT27	TBT28	TBT29	TBT30	MKH1
TAH3	TAH4	TAH5	TAH6	TAH7	TAH8	TAH9
22	23	24	25	26	27	28
D19	D20	D21	D22	D23	D24	D25
MKH2	MKH3	MKH4	MKH5	MKH6	MKH7	MKH8
TAH10	TAH11	TAH12	TAH13	TAH14	TAH15	TAH16

EUSI 461/1977-78/6213-14/1969-70

Ashe	Ojobo	Iwa	Araorun	Ile	Itutu	Erinle
1	2	3	4	5	6	7
O31	N1	N2	N3	N4	N5	N6
KHK13	KHK14	KHK15	KHK16	KHK17	KHK18	KHK19
THi21	THi22	THi23	THi24	THi25	THi26	THi27
8	9	10	11	12	13	14
N7	N8	N9	N10	N11	N12	N13
KHK20	KHK21	KHK22	KHK23	KHK24	KHK25	KHK26
THi28	THi29	THi30	HID1	HID2	HID3	HID4
15	16	17	18	19	20	21
N14	N15	N16	N17	N18	N19	N20
KHK27	KHK28	KHK29	KHK30	TBT1	TBT2	TBT3
HID5	HID6	HID7	HID8	HID9	HID10	HID11
22	23	24	25	26	27	28
N21	N22	N23	N24	N25	N26	N27
TBT4	TBT5	TBT6	TBT7	TBT8	TBT9	TBT10
HID12	HID13	HID14	HID15	HID16	HID17	HID18

KWANZA 461/1977-78/6213-14/1969-70

Umoja	Kujichagulia	Ujima	Ujamaa	Nia	Kuumba	Imani
1	2	3	4	5	6	7
D26	D27	D28	D29	D30	D31	JA1
MKH9	MKH10	MKH11	MKH12	MKH13	MKH14	MKH15
TAH17	TAH18	TAH19	TAH20	TAH21	TAH22	TAH23
8	9	10	11	12	13	14
JA2	JA3	JA4	JA5	JA6	JA7	JA8
MKH16	MKH17	MKH18	MKH19	MKH20	MKH21	MKH22
TAH24	TAH25	TAH26	TAH27	TAH28	TAH29	TAH30
15	16	17	18	19	20	21
JA9	JA10	JA11	JA12	JA13	JA14	JA15
MKH23	MKH24	MKH25	KMH26	MKH27	MKH28	MKH29
TIR1	TIR2	TIR3	TIR4	TIR5	TIR6	TIR7
22	23	24	25	26	27	28
JA16	JA17	JA18	JA19	JA20	JA21	JA22
MKH30	AMN1	AMN2	AMN3	AMN4	AMN5	AMN6
TIR8	TIR9	TIR10	TIR11	TIR12	TIR13	TIR14

SHAHIDI 461/1977-78/6213-14/1969-70						
Ashe	Ojobo	Iwa	Araorun	Ile	Itutu	Erinle
1	2	3	4	5	6	7
JA23	JA24	JA25	JA26	JA27	JA28	JA29
AMN7	AMN8	AMN9	AMN10	AMN11	AMN12	AMN13
TIR15	TIR16	TIR17	TIR18	TIR19	TIR20	TIR21
8	9	10	11	12	13	14
JA30	JA31	F1	F2	F3	F4	F5
AMN14	AMN15	AMN16	AMN17	AMN18	AMN19	AMN20
TIR22	TIR23	TIR24	TIR25	TIR26	TIR27	TIR28
15	16	17	18	19	20	21
F6	F7	F8	F9	F10	F 11	F12
AMN21	AMN22	AMN23	AMN24	AMN25	AMN26	AMN28
TIR29	TIR30	YEK1	YEK2	YEK3	YEK4	YEK5
22	23	24	25	26	27	28
F13	F14	F15	F16	F17	F18	F19
AMN29	AMN30	RNT1	RNT2	RNT3	RNT4	RNT5
YEK6	YEK7	YEK8	YEK9	YEK10	YEK11	YEK12
SHAHIDI 12						
F20						
RNT6						
YEK13						

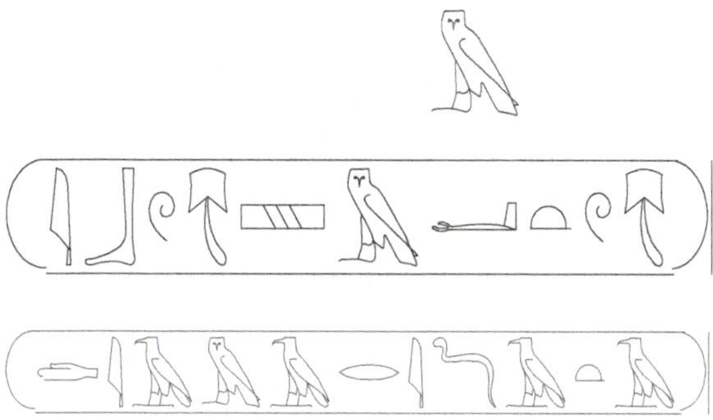

MAAFA 462 B.S.E./ASIENTO 1978-79 C.E./ KAM 6214-15 S.M./EC 1970-71 A.M. (ዓ.ም.)

AFU 462/1978-79/6214-15/1970-71						
Ashe	Ojobo	Iwa	Araorun	Ile	Itutu	Erinle
1	2	3	4	5	6	7
F21	F22	F23	F24	F25	F26	F27
RNT7	RNT8	RNT9	RNT10	RNT11	RNT12	RNT13
YEK14	YEK15	YEK16	YEK17	YEK18	YEK19	YEK20
8	9	10	11	12	13	14
F28	M1	M2	M3	M4	M5	M6
RNT14	RNT15	RNT16	RNT17	RNT18	RNT19	RNT20
YEK21	YEK22	YEK23	YEK24	YEK25	YEK26	YEK27
15	16	17	18	19	20	21
M7	M8	M9	M10	M11	M12	M13
RNT21	RNT22	RNT23	RNT24	RNT25	RNT26	RNT27
YEK28	YEK29	YEK30	MAG1	MAG2	MAG3	MAG4
22	23	24	25	26	27	28
M14	M15	M16	M17	M18	M19	M20
RNT28	RNT29	RNT30	KHN1	KHN2	KHN3	KHN4
MAG5	MAG6	MAG7	MAG8	MAG9	MAG10	MAG11

KAWAIDA 462/1978-79/6214-15/1970-71						
Ashe	Ojobo	Iwa	Araorun	Ile	Itutu	Erinle
1	2	3	4	5	6	7
A18	A19	A20	A21	A22	A23	A24
PIN3	PIN4	PIN5	PIN6	PIN7	PIN8	PIN9
MIY10	MIY11	MIY12	MIY13	MIY14	MIY15	MIY16
8	9	10	11	12	13	14
A25	A26	A27	A28	A29	A30	MA1
PIN10	PIN11	PIN12	PIN13	PIN14	PIN15	PIN16
MIY17	MIY18	MIY19	MIY20	MIY21	MIY22	MIY23
15	16	17	18	19	20	21
MA2	MA3	MA4	MA5	MA6	MA7	MA8
PIN17	PIN18	PIN19	PIN20	PIN21	PIN22	PIN2
MIY24	MIY25	MIY26	MIY27	MIY28	MIY29	MIY3
22	23	24	25	26	27	28
MA9	MA10	MA11	MA12	MA13	MA14	MA1
PIN24	PIN25	PIN26	PIN27	PIN28	PIN29	PIN3
GIN1	GIN2	GIN3	GIN4	GIN5	GIN6	GIN

UHURU 462/1978-79/6214-15/1970-71						
Ashe	Ojobo	Iwa	Araorun	Ile	Itutu	Erinle
1	2	3	4	5	6	7
M21	M22	M23	M24	M25	M26	M27
KHN5	KHN6	KHN7	KHN8	KHN9	KHN10	KHN11
MAG12	MAG13	MAG14	MAG15	MAG16	MAG17	MAG18
8	9	10	11	12	13	14
M28	M29	M30	M31	A1	A2	A3
KHN12	KHN13	KHN14	KHN15	KHN16	KHN17	KHN18
MAG19	MAG20	MAG21	MAG22	MAG23	MAG24	MAG25
15	16	17	18	19	20	21
A4	A5	A6	A7	A8	A9	A10
KHN19	KHN20	KHN21	KHN22	KHN23	KHN24	KHN25
MAG26	MAG27	MAG28	MAG29	MAG30	MIY1	MIY2
22	23	24	25	26	27	28
A11	A12	A13	A14	A15	A16	A17
KHN26	KHN27	KHN28	KHN29	KHN30	PIN1	PIN2
MIY3	MIY4	MIY5	MIY6	MIY7	MIY8	MIY9

MABABA WEUSI 462/1978-79/6214-15/1970-71						
Ashe	Ojobo	Iwa	Araorun	Ile	Itutu	Erinle
1	2	3	4	5	6	7
MA16	MA17	MA18	MA19	MA20	MA21	MA22
APP1	APP2	APP3	APP4	APP5	APP6	APP7
GIN8	GIN9	GIN10	GIN11	GIN12	GIN13	GIN14
8	9	10	11	12	13	14
MA23	MA24	MA25	MA26	MA27	MA28	MA29
APP8	APP9	APP10	APP11	APP12	APP13	APP1
GIN15	GIN16	GIN17	GIN18	GIN19	GIN20	GIN21
15	16	17	18	19	20	21
MA30	MA31	J1	J2	J3	J4	J5
APP15	APP16	APP17	APP18	APP19	APP20	API
GIN22	GIN23	GIN24	GIN25	GIN26	GIN27	GIN
22	23	24	25	26	27	28
J6	J7	J8	J9	J10	J11	J12
APP22	APP23	APP24	APP25	APP26	APP27	APP
GIN29	GIN30	SAN1	SAN2	SAN3	SAN4	SAN5

MWANAMKE WEUSI 462/1978-79/6214-15/1970-71						
Ashe	Ojobo	Iwa	Araorun	Ile	Itutu	Erinle
1	2	3	4	5	6	7
J13	J14	J15	J16	J17	J18	J19
APP29	APP30	MSR1	MSR2	MSR3	MSR4	MSR5
SAN6	SAN7	SAN8	SAN9	SAN10	SAN11	SAN12
8	9	10	11	12	13	14
J20	J21	J22	J23	J24	J25	J26
MSR6	MSR7	MSR8	MSR9	MSR10	MSR11	MSR12
SAN13	SAN14	SAN15	SAN16	SAN17	SAN18	SAN19
15	16	17	18	19	20	21
J27	J28	J29	J30	JU1	JU2	JU3
MSR13	MSR14	MSR15	MSR16	MSR17	MSR18	MSR19
SAN20	SAN21	SAN22	SAN23	SAN24	SAN25	SAN26
22	23	24	25	26	27	28
JU4	JU5	JU6	JU7	JU8	JU9	JU10
MSR20	MRS21	MSR22	MSR23	MSR24	MSR25	MSR26
SAN27	SAN28	SAN29	SAN30	HAM1	HAM2	HAM3

TAIFA WEUSI 462/1978-79/6214-15/1970-71						
Ashe	Ojobo	Iwa	Araorun	Ile	Itutu	Erinle
1	2	3	4	5	6	7
AU8	AU9	AU10	AU11	AU12	AU13	AU14
THT20	THT21	THT22	THT23	THT24	THT25	THT26
NAH2	NAH3	NAH4	NAH5	NAH6	NAH7	NAH8
8	9	10	11	12	13	14
AU15	AU16	AU17	AU18	AU19	AU20	AU21
THT27	THT28	THT29	THT30	PAP1	PAP1	PAP2
NAH9	NAH10	NAH11	NAH12	NAH13	NAH14	NAH15
15	16	17	18	19	20	21
AU22	AU23	AU24	AU25	AU26	AU27	AU28
PAP3	PAP4	PAP5	PAP6	PAP7	PAP8	PAP9
NAH16	NAH17	NAH18	NAH19	NAH20	NAH21	NAH22
22	23	24	25	26	27	28
AU29	AU30	AU31	S1	S2	S3	S4
PAP10	PAP11	PAP12	PAP13	PAP14	PAP15	PAP16
NAH23	NAH24	NAH25	NAH26	NAH27	NAH28	NAH29

MTU WEUSI 462/1978-79/6214-15/1970-71						
Ashe	Ojobo	Iwa	Araorun	Ile	Itutu	Erinle
1	2	3	4	5	6	7
JU11	JU12	JU13	JU14	JU15	JU16	JU17
MSR27	MSR28	MSR29	MSR30	*ASR1*	*HRU2*	*STH3*
HAM4	HAM5	HAM6	HAM7	HAM8	HAM9	HAM10
8	9	10	11	12	13	14
JU18	JU19	JU20	JU21	JU22	JU23	JU24
AST4	*NBH5*	THT1	THT2	THT3	THT4	THT5
HAM11	HAM12	HAM13	AM14	HAM15	HAM16	HAM17
15	16	17	18	19	20	21
JU25	JU26	JU27	JU28	JU29	JU30	JU31
THT6	THT7	THT8	THT9	THT10	THT11	THT12
HAM18	HAM19	HAM20	HAM21	HAM22	HAM23	HAM24
22	23	24	25	26	27	28
AU1	AU2	AU3	AU4	AU5	AU6	AU7
THT13	THT14	THT15	THT16	THT17	THT18	THT19
HAM25	HAM26	HAM27	HAM28	HAM29	HAM30	NAH1

UWEZO 462/1978-79/6214-15/1970-71						
Ashe	Ojobo	Iwa	Araorun	Ile	Itutu	Erinle
1	2	3	4	5	6	7
S5	S6	S7	S8	S9	S10	S11
PAP17	PAP18	PAP19	PAP20	PAP21	PAP22	PAP23
NAH30	*PAG1*	*PAG2*	*PAG3*	*PAG4*	*PAG5*	MAS1
8	9	10	11	12	13	14
S12	S13	S14	S15	S16	S17	S18
PAP24	PAP25	PAP26	PAP27	PAP28	PAP29	PAP30
MAS2	MAS3	MAS4	MAS5	MAS6	MAS7	MAS8
15	16	17	18	19	20	21
S19	S20	S21	S22	S23	S24	S25
HTH1	HTH2	HTH3	HTH4	HTH5	HTH6	HTH7
MAS9	MAS10	MAS11	MAS12	MAS13	MAS14	MAS15
22	23	24	25	26	27	28
S26	S27	S28	S29	S30	O1	O2
HTH8	HTH9	HTH10	HTH11	HTH12	HTH13	HTH14
MAS16	MAS17	MAS18	MAS19	MAS20	MAS21	MAS22

EKUNDU 462/1978-79/6214-15/1970-71						
Ashe	Ojobo	Iwa	Araorun	Ile	Itutu	Erinle
1	2	3	4	5	6	7
O3	O4	O5	O6	O7	O8	O9
HTH15	HTH16	HTH17	HTH18	HTH19	HTH20	HTH21
MAS23	MAS24	MAS25	MAS26	MAS27	MAS28	MAS29
8	9	10	11	12	13	14
O10	O11	O12	O13	O14	O15	O16
HTH22	HTH23	HTH24	HTH25	HTH26	HTH27	HTH28
MAS30	THi1	THi2	THi3	THi4	THi5	THi6
15	16	17	18	19	20	21
O17	O18	O19	O20	O21	O22	O23
HTH29	HTH30	KHK1	KHK2	KHK3	KHK4	KHK5
THi7	THi8	THi9	THi10	THi11	THi12	THi13
22	23	24	25	26	27	28
O24	O25	O26	O27	O28	O29	O30
KHK6	KHK7	KHK8	KHK9	KHK10	KHK11	KHK12
THi14	THi15	THi16	THi17	THi18	THi19	THi20

KIJANI 462/1978-79/6214-15/1970-71						
Ashe	Ojobo	Iwa	Araorun	Ile	Itutu	Erinle
1	2	3	4	5	6	7
N28	N29	N30	D1	D2	D3	D4
TBT11	TBT12	TBT13	TBT14	TBT15	TBT16	TBT17
HID19	HID20	HID21	HID22	HID23	HID24	HID25
8	9	10	11	12	13	14
D5	D6	D7	D8	D9	D10	D11
TBT18	TBT19	TBT20	TBT21	TBT22	TBT23	TBT24
HID26	HID27	HID28	HID29	HID30	TAH1	TAH2
15	16	17	18	19	20	21
D12	D13	D14	D15	D16	D17	D18
TBT25	TBT26	TBT27	TBT28	TBT29	TBT30	MKH1
TAH3	TAH4	TAH5	TAH6	TAH7	TAH8	TAH9
22	23	24	25	26	27	28
D19	D20	D21	D22	D23	D24	D25
MKH2	MKH3	MKH4	MKH5	MKH6	MKH7	MKH8
TAH10	TAH11	TAH12	TAH13	TAH14	TAH15	TAH16

EUSI 462/1978-79/6214-15/1970-71						
Ashe	Ojobo	Iwa	Araorun	Ile	Itutu	Erinle
1	2	3	4	5	6	7
O31	N1	N2	N3	N4	N5	N6
KHK13	KHK14	KHK15	KHK16	KHK17	KHK18	KHK19
THi21	THi22	THi23	THi24	THi25	THi26	THi27
8	9	10	11	12	13	14
N7	N8	N9	N10	N11	N12	N13
KHK20	KHK21	KHK22	KHK23	KHK24	KHK25	KHK26
THi28	THi29	THi30	HID1	HID2	HID3	HID4
15	16	17	18	19	20	21
N14	N15	N16	N17	N18	N19	N20
KHK27	KHK28	KHK29	KHK30	TBT1	TBT2	TBT3
HID5	HID6	HID7	HID8	HID9	HID10	HID11
22	23	24	25	26	27	28
N21	N22	N23	N24	N25	N26	N27
TBT4	TBT5	TBT6	TBT7	TBT8	TBT9	TBT10
HID12	HID13	HID14	HID15	HID16	HID17	HID18

KWANZA 462/1978-79/6214-15/1970-71						
Umoja	Kujichagulia	Ujima	Ujamaa	Nia	Kuumba	Imani
1	2	3	4	5	6	7
D26	D27	D28	D29	D30	D31	JA1
MKH9	MKH10	MKH11	MKH12	MKH13	MKH14	MKH15
TAH17	TAH18	TAH19	TAH20	TAH21	TAH22	TAH23
8	9	10	11	12	13	14
JA2	JA3	JA4	JA5	JA6	JA7	JA8
MKH16	MKH17	MKH18	MKH19	MKH20	MKH21	MKH22
TAH24	TAH25	TAH26	TAH27	TAH28	TAH29	TAH30
15	16	17	18	19	20	21
JA9	JA10	JA11	JA12	JA13	JA14	JA15
MKH23	MKH24	MKH25	KMH26	MKH27	MKH28	MKH29
TIR1	TIR2	TIR3	TIR4	TIR5	TIR6	TIR7
22	23	24	25	26	27	28
JA16	JA17	JA18	JA19	JA20	JA21	JA22
MKH30	AMN1	AMN2	AMN3	AMN4	AMN5	AMN6
TIR8	TIR9	TIR10	TIR11	TIR12	TIR13	TIR14

SHAHIDI 462/1978-79/6214-15/1970-71						
Ashe	Ojobo	Iwa	Araorun	Ile	Itutu	Erinle
1	2	3	4	5	6	7
JA23	JA24	JA25	JA26	JA27	JA28	JA29
AMN7	AMN8	AMN9	AMN10	AMN11	AMN12	AMN13
TIR15	TIR16	TIR17	TIR18	TIR19	TIR20	TIR21
8	9	10	11	12	13	14
JA30	JA31	F1	F2	F3	F4	F5
AMN14	AMN15	AMN16	AMN17	AMN18	AMN19	AMN20
TIR22	TIR23	TIR24	TIR25	TIR26	TIR27	TIR28
15	16	17	18	19	20	21
F6	F7	F8	F9	F10	F 11	F12
AMN21	AMN22	AMN23	AMN24	AMN25	AMN26	AMN28
TIR29	TIR30	YEK1	YEK2	YEK3	YEK4	YEK5
22	23	24	25	26	27	28
F13	F14	F15	F16	F17	F18	F19
AMN29	AMN30	RNT1	RNT2	RNT3	RNT4	RNT5
YEK6	YEK7	YEK8	YEK9	YEK10	YEK11	YEK12
SHAHIDI 13						
F20						
RNT6						
YEK13						

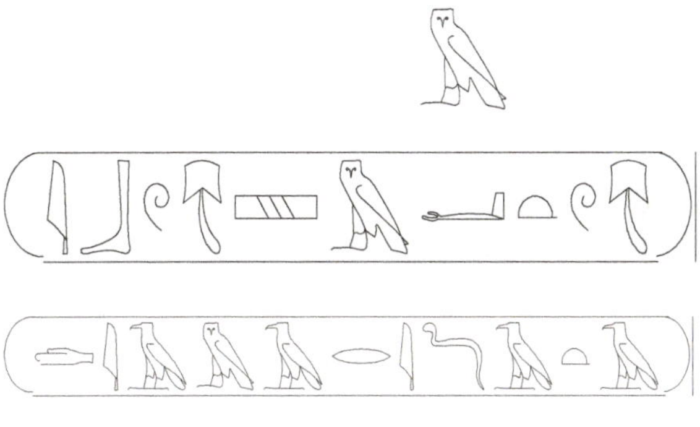

MAAFA 463 B.S.E./ASIENTO 1979-80 C.E./
KAM 6215-16 S.M./EC 1971-72 A.M. (ዓ.ም.)

AFU 463/1979-80/6215-16/1971-72						
Ashe	Ojobo	Iwa	Araorun	Ile	Itutu	Erinle
1	2	3	4	5	6	7
F21	F22	F23	F24	F25	F26	F27
RNT7	RNT8	RNT9	RNT10	RNT11	RNT12	RNT13
YEK14	YEK15	YEK16	YEK17	YEK18	YEK19	YEK20
8	9	10	11	12	13	14
F28	M1	M2	M3	M4	M5	M6
RNT14	RNT15	RNT16	RNT17	RNT18	RNT19	RNT20
YEK21	YEK22	YEK23	YEK24	YEK25	YEK26	YEK27
15	16	17	18	19	20	21
M7	M8	M9	M10	M11	M12	M13
RNT21	RNT22	RNT23	RNT24	RNT25	RNT26	RNT27
YEK28	YEK 29	YEK30	MAG1	MAG2	MAG3	MAG4
22	23	24	25	26	27	28
M14	M15	M16	M17	M18	M19	M20
RNT28	RNT29	RNT30	KHN1	KHN2	KHN3	KHN4
MAG5	MAG6	MAG7	MAG8	MAG9	MAG10	MAG11

KAWAIDA 463/1979-80/6215-16/1971-72						
Ashe	Ojobo	Iwa	Araorun	Ile	Itutu	Erinle
1	2	3	4	5	6	7
A18	A19	A20	A21	A22	A23	A24
PIN3	PIN4	PIN5	PIN6	PIN7	PIN8	PIN9
MIY10	MIY11	MIY12	MIY13	MIY14	MIY15	MIY16
8	9	10	11	12	13	14
A25	A26	A27	A28	A29	A30	MA1
PIN10	PIN11	PIN12	PIN13	PIN14	PIN15	PIN16
MIY17	MIY18	MIY19	MIY20	MIY21	MIY22	MIY23
15	16	17	18	19	20	21
MA2	MA3	MA4	MA5	MA6	MA7	MA8
PIN17	PIN18	PIN19	PIN20	PIN21	PIN22	PIN23
MIY24	MIY25	MIY26	MIY27	MIY28	MIY29	MIY30
22	23	24	25	26	27	28
MA9	MA10	MA11	MA12	MA13	MA14	MA15
PIN24	PIN25	PIN26	PIN27	PIN28	PIN29	PIN30
GIN1	GIN2	GIN3	GIN4	GIN5	GIN6	GIN7

UHURU 463/1979-80/6215-16/1971-72						
Ashe	Ojobo	Iwa	Araorun	Ile	Itutu	Erinle
1	2	3	4	5	6	7
M21	M22	M23	M24	M25	M26	M27
KHN5	KHN6	KHN7	KHN8	KHN9	KHN10	KHN11
MAG12	MAG13	MAG14	MAG15	MAG16	MAG17	MAG18
8	9	10	11	12	13	14
M28	M29	M30	M31	A1	A2	A3
KHN12	KHN13	KHN14	KHN15	KHN16	KHN17	KHN18
MAG19	MAG20	MAG21	MAG22	MAG23	MAG24	MAG25
15	16	17	18	19	20	21
A4	A5	A6	A7	A8	A9	A10
KHN19	KHN20	KHN21	KHN22	KHN23	KHN24	KHN25
MAG26	MAG27	MAG28	MAG29	MAG30	MIY1	MIY2
22	23	24	25	26	27	28
A11	A12	A13	A14	A15	A16	A17
KHN26	KHN27	KHN28	KHN29	KHN30	PIN1	PIN2
MIY3	MIY4	MIY5	MIY6	MIY7	MIY8	MIY9

MABABA WEUSI 463/1979-80/6215-16/1971-72						
Ashe	Ojobo	Iwa	Araorun	Ile	Itutu	Erinle
1	2	3	4	5	6	7
MA16	MA17	MA18	MA19	MA20	MA21	MA22
APP1	APP2	APP3	APP4	APP5	APP6	APP7
GIN8	GIN9	GIN10	GIN11	GIN12	GIN13	GIN14
8	9	10	11	12	13	14
MA23	MA24	MA25	MA26	MA27	MA28	MA29
APP8	APP9	APP10	APP11	APP12	APP13	APP14
GIN15	GIN16	GIN17	GIN18	GIN19	GIN20	GIN21
15	16	17	18	19	20	21
MA30	MA31	J1	J2	J3	J4	J5
APP15	APP16	APP17	APP18	APP19	APP20	APP21
GIN22	GIN23	GIN24	GIN25	GIN26	GIN27	GIN28
22	23	24	25	26	27	28
J6	J7	J8	J9	J10	J11	J12
APP22	APP23	APP24	APP25	APP26	APP27	APP28
GIN29	GIN30	SAN1	SAN2	SAN3	SAN4	SAN5

MWANAMKE WEUSI 463/1979-80/6215-16/1971-72						
Ashe	Ojobo	Iwa	Araorun	Ile	Itutu	Erinle
1	2	3	4	5	6	7
J13	J14	J15	J16	J17	J18	J19
APP29	APP30	MSR1	MSR2	MSR3	MSR4	MSR5
SAN6	SAN7	SAN8	SAN9	SAN10	SAN11	SAN12
8	9	10	11	12	13	14
J20	J21	J22	J23	J24	J25	J26
MSR6	MSR7	MSR8	MSR9	MSR10	MSR11	MSR12
SAN13	SAN14	SAN15	SAN16	SAN17	SAN18	SAN19
15	16	17	18	19	20	21
J27	J28	J29	J30	JU1	JU2	JU3
MSR13	MSR14	MSR15	MSR16	MSR17	MSR18	MSR19
SAN20	SAN21	SAN22	SAN23	SAN24	SAN25	SAN26
22	23	24	25	26	27	28
JU4	JU5	JU6	JU7	JU8	JU9	JU10
MSR20	MRS21	MSR22	MSR23	MSR24	MSR25	MSR26
SAN27	SAN28	SAN29	SAN30	HAM1	HAM2	HAM3

TAIFA WEUSI 463/1979-80/6215-16/1971-72						
Ashe	Ojobo	Iwa	Araorun	Ile	Itutu	Erinle
1	2	3	4	5	6	7
AU8	AU9	AU10	AU11	AU12	AU13	AU14
THT20	THT21	THT22	THT23	THT24	THT25	THT26
NAH2	NAH3	NAH4	NAH5	NAH6	NAH7	NAH8
8	9	10	11	12	13	14
AU15	AU16	AU17	AU18	AU19	AU20	AU21
THT27	THT28	THT29	THT30	PAP1	PAP1	PAP2
NAH9	NAH10	NAH11	NAH12	NAH13	NAH14	NAH15
15	16	17	18	19	20	21
AU22	AU23	AU24	AU25	AU26	AU27	AU28
PAP3	PAP4	PAP5	PAP6	PAP7	PAP8	PAP9
NAH16	NAH17	NAH18	NAH19	NAH20	NAH21	NAH22
22	23	24	25	26	27	28
AU29	AU30	AU31	S1	S2	S3	S4
PAP10	PAP11	PAP12	PAP13	PAP14	PAP15	PAP16
NAH23	NAH24	NAH25	NAH26	NAH27	NAH28	NAH29

MTU WEUSI 463/1979-80/6215-16/1971-72						
Ashe	Ojobo	Iwa	Araorun	Ile	Itutu	Erinle
1	2	3	4	5	6	7
JU11	JU12	JU13	JU14	JU15	JU16	JU17
MSR27	MSR28	MSR29	MSR30	*ASR1*	*HRU2*	*STH3*
HAM4	HAM5	HAM6	HAM7	HAM8	HAM9	HAM10
8	9	10	11	12	13	14
JU18	JU19	JU20	JU21	JU22	JU23	JU24
AST4	*NBH5*	THT1	THT2	THT3	THT4	THT5
HAM11	HAM12	HAM13	HAM14	HAM15	HAM16	HAM17
15	16	17	18	19	20	21
JU25	JU26	JU27	JU28	JU29	JU30	JU31
THT6	THT7	THT8	THT9	THT10	THT11	THT12
HAM18	HAM19	HAM20	HAM21	HAM22	HAM23	HAM24
22	23	24	25	26	27	28
AU1	AU2	AU3	AU4	AU5	AU6	AU7
THT13	THT14	THT15	THT16	THT17	THT18	THT19
HAM25	HAM26	HAM27	HAM28	HAM29	HAM30	NAH1

UWEZO 463/1979-80/6215-16/1971-72						
Ashe	Ojobo	Iwa	Araorun	Ile	Itutu	Erinle
1	2	3	4	5	6	7
S5	S6	S7	S8	S9	S10	S11
PAP17	PAP18	PAP19	PAP20	PAP21	PAP22	PAP23
NAH30	*QUA1*	*QUA2*	*QUA3*	*QUA4*	*QUA5*	*QUA6*
8	9	10	11	12	13	14
S12	S13	S14	S15	S16	S17	S18
PAP24	PAP25	PAP26	PAP27	PAP28	PAP29	PAP30
MAS1	MAS2	MAS3	MAS4	MAS5	MAS6	MAS7
15	16	17	18	19	20	21
S19	S20	S21	S22	S23	S24	S25
HTH1	HTH2	HTH3	HTH4	HTH5	HTH6	HTH7
MAS8	MAS9	MAS10	MAS11	MAS12	MAS13	MAS14
22	23	24	25	26	27	28
S26	S27	S28	S29	S30	O1	O2
HTH8	HTH9	HTH10	HTH11	HTH12	HTH13	HTH14
MAS15	MAS16	MAS17	MAS18	MAS19	MAS20	MAS21

EKUNDU 463/1979-80/6215-16/1971-72						
Ashe	Ojobo	Iwa	Araorun	Ile	Itutu	Erinle
1	2	3	4	5	6	7
O3	O4	O5	O6	O7	O8	O9
HTH15	HTH16	HTH17	HTH18	HTH19	HTH20	HTH21
MAS22	MAS23	MAS 24	MAS25	MAS26	MAS27	MAS28
8	9	10	11	12	13	14
O10	O11	O12	O13	O14	O15	O16
HTH22	HTH23	HTH24	HTH25	HTH26	HTH27	HTH28
MAS29	MAS30	THi1	THi2	THi3	THi4	THi5
15	16	17	18	19	20	21
O17	O18	O19	O20	O21	O22	O23
HTH29	HTH30	KHK1	KHK2	KHK3	KHK4	KHK5
THi6	THi7	THi8	THi9	THi10	THi11	THi12
22	23	24	25	26	27	28
O24	O25	O26	O27	O28	O29	O30
KHK6	KHK7	KHK8	KHK9	KHK10	KHK11	KHK12
THi13	THi14	THi15	THi16	THi17	THi18	THi19

KIJANI 463/1979-80/6215-16/1971-72						
Ashe	Ojobo	Iwa	Araorun	Ile	Itutu	Erinle
1	2	3	4	5	6	7
N28	N29	N30	D1	D2	D3	D4
TBT11	TBT12	TBT13	TBT14	TBT15	TBT16	TBT17
HID18	HID19	HID20	HID21	HID22	HID23	HID24
8	9	10	11	12	13	14
D5	D6	D7	D8	D9	D10	D11
TBT18	TBT19	TBT20	TBT21	TBT22	TBT23	TBT24
HID25	HID26	HID27	HID28	HID29	HID30	TAH1
15	16	17	18	19	20	21
D12	D13	D14	D15	D16	D17	D18
TBT25	TBT26	TBT27	TBT28	TBT29	TBT30	MKH1
TAH2	TAH3	TAH4	TAH5	TAH6	TAH7	TAH8
22	23	24	25	26	27	28
D19	D20	D21	D22	D23	D24	D25
MKH2	MKH3	MKH4	MKH5	MKH6	MKH7	MKH8
TAH9	TAH10	TAH11	TAH12	TAH13	TAH14	TAH15

EUSI 463/1979-80/6215-16/1971-72						
Ashe	Ojobo	Iwa	Araorun	Ile	Itutu	Erinle
1	2	3	4	5	6	7
O31	N1	N2	N3	N4	N5	N6
KHK13	KHK14	KHK15	KHK16	KHK17	KHK18	KHK19
THi20	THi21	THi22	THi23	THi24	THi25	THi26
8	9	10	11	12	13	14
N7	N8	N9	N10	N11	N12	N13
KHK20	KHK21	KHK22	KHK23	KHK24	KHK25	KHK26
THi27	THi28	THi29	THi30	HID1	HID2	HID3
15	16	17	18	19	20	21
N14	N15	N16	N17	N18	N19	N20
KHK27	KHK28	KHK29	KHK30	TBT1	TBT2	TBT3
HID4	HID5	HID6	HID7	HID8	HID9	HID10
22	23	24	25	26	27	28
N21	N22	N23	N24	N25	N26	N27
TBT4	TBT5	TBT6	TBT7	TBT8	TBT9	TBT10
HID11	HID12	HID13	HID14	HID15	HID16	HID17

KWANZA 463/1979-80/6215-16/1971-72						
Umoja	Kujichagulia	Ujima	Ujamaa	Nia	Kuumba	Imani
1	2	3	4	5	6	7
D26	D27	D28	D29	D30	D31	JA1
MKH9	MKH10	MKH11	MKH12	MKH13	MKH14	MKH15
TAH16	TAH17	TAH18	TAH19	TAH20	TAH21	TAH22
8	9	10	11	12	13	14
JA2	JA3	JA4	JA5	JA6	JA7	JA8
MKH16	MKH17	MKH18	MKH19	MKH20	MKH21	MKH22
TAH23	TAH24	TAH25	TAH26	TAH27	TAH28	TAH29
15	16	17	18	19	20	21
JA9	JA10	JA11	JA12	JA13	JA14	JA15
MKH23	MKH24	MKH25	KMH26	MKH27	MKH28	MKH29
TAH30	TIR1	TIR2	TIR3	TIR4	TIR5	TIR6
22	23	24	25	26	27	28
JA16	JA17	JA18	JA19	JA20	JA21	JA22
MKH30	AMN1	AMN2	AMN3	AMN4	AMN5	AMN6
TIR7	TIR8	TIR9	TIR10	TIR11	TIR12	TIR13

SHAHIDI 463/1979-80/6215-16/1971-72						
Ashe	Ojobo	Iwa	Araorun	Ile	Itutu	Erinle
1	2	3	4	5	6	7
JA23	JA24	JA25	JA26	JA27	JA28	JA29
AMN7	AMN8	AMN9	AMN10	AMN11	AMN12	AMN13
TIR14	TIR15	TIR16	TIR17	TIR18	TIR19	TIR20
8	9	10	11	12	13	14
JA30	JA31	F1	F2	F3	F4	F5
AMN14	AMN15	AMN16	AMN17	AMN18	AMN19	AMN20
TIR21	TIR22	TIR23	TIR24	TIR25	TIR26	TIR27
15	16	17	18	19	20	21
F6	F7	F8	F9	F10	F 11	F12
AMN21	AMN22	AMN23	AMN24	AMN25	AMN26	AMN28
TIR28	TIR29	TIR30	YEK1	YEK2	YEK3	YEK4
22	23	24	25	26	27	28
F13	F14	F15	F16	F17	F18	F19
AMN29	AMN30	RNT1	RNT2	RNT3	RNT4	RNT5
YEK5	YEK6	YEK7	YEK8	YEK9	YEK10	YEK11
SHAHIDI 14						
F20						
RNT6						
YEK12						

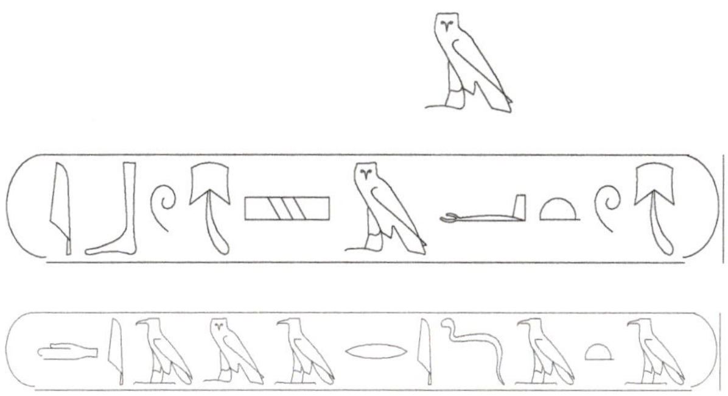

MAAFA 464 B.S.E./ASIENTO 1980-81 C.E./
KAM 6216-17 S.M./EC 1972-73 A.M. (ዓ.ም.)

AFU 464/1980-81/6216-17/1972-73

Ashe	Ojobo	Iwa	Araorun	Ile	Itutu	Erinle
1	2	3	4	5	6	7
F21	F22	F23	F24	F25	F26	F27
RNT7	RNT8	RNT9	RNT10	RNT11	RNT12	RNT13
YEK13	YEK14	YEK15	YEK16	YEK17	YEK18	YEK19
8	9	10	11	12	13	14
F28	F29	M1	M2	M3	M4	M5
RNT14	RNT15	RNT16	RNT17	RNT18	RNT19	RNT20
YEK 20	YEK21	YEK22	YEK23	YEK24	YEK25	YEK26
15	16	17	18	19	20	21
M6	M7	M8	M9	M10	M11	M12
RNT21	RNT22	RNT23	RNT24	RNT25	RNT26	RNT27
YEK 27	YEK28	YEK29	YEK30	MAG1	MAG2	MAG3
22	23	24	25	26	27	28
M13	M14	M15	M16	M17	M18	M19
RNT28	RNT29	RNT30	KHN1	KHN2	KHN3	KHN4
MAG4	MAG5	MAG6	MAG7	MAG8	MAG9	MAG10
SHARIFU 113						
M20						
KHN5						
MAG11						

KAWAIDA 464/1980-81/6216-17/1972-73

Ashe	Ojobo	Iwa	Araorun	Ile	Itutu	Erinle
1	2	3	4	5	6	7
A18	A19	A20	A21	A22	A23	A24
PIN4	PIN5	PIN6	PIN7	PIN8	PIN9	PIN10
MIY10	MIY11	MIY12	MIY13	MIY14	MIY15	MIY16
8	9	10	11	12	13	14
A25	A26	A27	A28	A29	A30	MA1
PIN11	PIN12	PIN13	PIN14	PIN15	PIN16	PIN17
MIY17	MIY18	MIY19	MIY20	MIY21	MIY22	MIY23
15	16	17	18	19	20	21
MA2	MA3	MA4	MA5	MA6	MA7	MA8
PIN18	PIN19	PIN20	PIN21	PIN22	PIN23	PIN24
MIY24	MIY25	MIY26	MIY27	MIY28	MIY29	MIY30
22	23	24	25	26	27	28
MA9	MA10	MA11	MA12	MA13	MA14	MA15
PIN25	PIN26	PIN27	PIN28	PIN29	PIN30	APP1
GIN1	GIN2	GIN3	GIN4	GIN5	GIN6	GIN7

UHURU 464/1980-81/6216-17/1972-73

Ashe	Ojobo	Iwa	Araorun	Ile	Itutu	Erinle
1	2	3	4	5	6	7
M21	M22	M23	M24	M25	M26	M27
KHN6	KHN7	KHN8	KHN9	KHN10	KHN11	KHN12
MAG12	MAG13	MAG14	MAG15	MAG16	MAG17	MAG18
8	9	10	11	12	13	14
M28	M29	M30	M31	A1	A2	A3
KHN13	KHN14	KHN15	KHN16	KHN17	KHN18	KHN19
MAG19	MAG20	MAG21	MAG22	MAG23	MAG24	MAG25
15	16	17	18	19	20	21
A4	A5	A6	A7	A8	A9	A10
KHN20	KHN21	KHN222	KHN23	KHN24	KHN25	KHN26
MAG26	MAG27	MAG28	MAG29	MAG30	MIY1	MIY2
22	23	24	25	26	27	28
A11	A12	A13	A14	A15	A16	A17
KHN27	KHN28	KHN29	KHN30	PIN1	PIN2	PIN3
MIY3	MIY4	MIY5	MIY6	MIY7	MIY8	MIY9

MABABA WEUSI 464/1980-81/6216-17/1972-73

Ashe	Ojobo	Iwa	Araorun	Ile	Itutu	Erinle
1	2	3	4	5	6	7
MA16	MA17	MA18	MA19	MA20	MA21	MA22
APP2	APP3	APP4	APP5	APP6	APP7	APP8
GIN8	GIN9	GIN10	GIN11	GIN12	GIN13	GIN14
8	9	10	11	12	13	14
MA23	MA24	MA25	MA26	MA27	MA28	MA29
APP9	APP10	APP11	APP12	APP13	APP14	APP15
GIN15	GIN16	GIN17	GIN18	GIN19	GIN20	GIN21
15	16	17	18	19	20	21
MA30	MA31	J1	J2	J3	J4	J5
APP16	APP17	APP18	APP19	APP20	APP21	APP22
GIN22	GIN23	GIN24	GIN25	GIN26	GIN27	GIN28
22	23	24	25	26	27	28
J6	J7	J8	J9	J10	J11	J12
APP23	APP24	APP25	APP26	APP27	APP28	APP29
GIN29	GIN30	SAN1	SAN2	SAN3	SAN4	SAN5

MWANAMKE WEUSI 464/1980-81/6216-17/1972-73

Ashe	Ojobo	Iwa	Araorun	Ile	Itutu	Erinle
1	2	3	4	5	6	7
J13	J14	J15	J16	J17	J18	J19
APP30	MSR1	MSR2	MSR3	MSR4	MSR5	MSR6
SAN6	SAN7	SAN8	SAN9	SAN10	SAN11	SAN12
8	9	10	11	12	13	14
J20	J21	J22	J23	J24	J25	J26
MSR7	MSR8	MSR9	MSR10	MSR11	MSR12	MSR13
SAN13	SAN14	SAN15	SAN16	SAN17	SAN18	SAN19
15	16	17	18	19	20	21
J27	J28	J29	J30	JU1	JU2	JU3
MSR14	MSR15	MSR16	MSR17	MSR18	MSR19	MSR20
SAN20	SAN21	SAN22	SAN23	SAN24	SAN25	SAN26
22	23	24	25	26	27	28
JU4	JU5	JU6	JU7	JU8	JU9	JU10
MSR21	MRS22	MSR23	MSR24	MSR25	MSR26	MSR27
SAN27	SAN28	SAN29	SAN30	HAM1	HAM2	HAM3

TAIFA WEUSI 464/1980-81/6216-17/1972-73

Ashe	Ojobo	Iwa	Araorun	Ile	Itutu	Erinle
1	2	3	4	5	6	7
AU8	AU9	AU10	AU11	AU12	AU13	AU14
THT20	THT21	THT22	THT23	THT24	THT25	THT26
NAH2	NAH3	NAH4	NAH5	NAH6	NAH7	NAH8
8	9	10	11	12	13	14
AU15	AU16	AU17	AU18	AU19	AU20	AU21
THT27	THT28	THT29	THT30	PAP1	PAP1	PAP2
NAH9	NAH10	NAH11	NAH12	NAH13	NAH14	NAH15
15	16	17	18	19	20	21
AU22	AU23	AU24	AU25	AU26	AU27	AU28
PAP3	PAP4	PAP5	PAP6	PAP7	PAP8	PAP9
NAH16	NAH17	NAH18	NAH19	NAH20	NAH21	NAH22
22	23	24	25	26	27	28
AU29	AU30	AU31	S1	S2	S3	S4
PAP10	PAP11	PAP12	PAP13	PAP14	PAP15	PAP16
NAH23	NAH24	NAH25	NAH26	NAH27	NAH28	NAH29

MTU WEUSI 464/1980-81/6216-17/1972-73

Ashe	Ojobo	Iwa	Araorun	Ile	Itutu	Erinle
1	2	3	4	5	6	7
JU11	JU12	JU13	JU14	JU15	JU16	JU17
MSR28	MSR29	MSR30	*ASR1*	*HRU2*	*STH3*	*AST4*
HAM4	HAM5	HAM6	HAM7	HAM8	HAM9	HAM10
8	9	10	11	12	13	14
JU18	JU19	JU20	JU21	JU22	JU23	JU24
NBH5	*ANU6*	THT1	THT2	THT3	THT4	THT5
HAM11	HAM12	HAM13	HAM14	HAM15	HAM16	HAM17
15	16	17	18	19	20	21
JU25	JU26	JU27	JU28	JU29	JU30	JU31
THT6	THT7	THT8	THT9	THT10	THT11	THT12
HAM18	HAM19	HAM20	HAM21	HAM22	HAM23	HAM24
22	23	24	25	26	27	28
AU1	AU2	AU3	AU4	AU5	AU6	AU7
THT13	THT14	THT15	THT16	THT17	THT18	THT19
HAM25	HAM26	HAM27	HAM28	HAM29	HAM30	NAH1

UWEZO 464/1980-81/6216-17/1972-73

Ashe	Ojobo	Iwa	Araorun	Ile	Itutu	Erinle
1	2	3	4	5	6	7
S5	S6	S7	S8	S9	S10	S11
PAP17	PAP18	PAP19	PAP20	PAP21	PAP22	PAP23
NAH30	*PAG1*	*PAG2*	*PAG3*	*PAG4*	*PAG5*	MAS1
8	9	10	11	12	13	14
S12	S13	S14	S15	S16	S17	S18
PAP24	PAP25	PAP26	PAP27	PAP28	PAP29	PAP30
MAS2	MAS3	MAS4	MAS5	MAS6	MAS7	MAS8
15	16	17	18	19	20	21
S19	S20	S21	S22	S23	S24	S25
HTH1	HTH2	HTH3	HTH4	HTH5	HTH6	HTH7
MAS9	MAS10	MAS11	MAS12	MAS13	MAS14	MAS15
22	23	24	25	26	27	28
S26	S27	S28	S29	S30	O1	O2
HTH8	HTH9	HTH10	HTH11	HTH12	HTH13	HTH14
MAS16	MAS17	MAS18	MAS19	MAS20	MAS21	MAS22

EKUNDU 464/1980-81/6216-17/1972-73						
Ashe	Ojobo	Iwa	Araorun	Ile	Itutu	Erinle
1	2	3	4	5	6	7
O3	O4	O5	O6	O7	O8	O9
HTH15	HTH16	HTH17	HTH18	HTH19	HTH20	HTH21
MAS23	MAS24	MAS25	MAS26	MAS27	MAS28	MAS29
8	9	10	11	12	13	14
O10	O11	O12	O13	O14	O15	O16
HTH22	HTH23	HTH24	HTH25	HTH26	HTH27	HTH28
MAS30	THi1	THi2	THi3	THi4	THi5	THi6
15	16	17	18	19	20	21
O17	O18	O19	O20	O21	O22	O23
HTH29	HTH30	KHK1	KHK2	KHK3	KHK4	KHK5
THi7	THi8	This9	THi10	THi11	THi12	THi13
22	23	24	25	26	27	28
O24	O25	O26	O27	O28	O29	O30
KHK6	KHK7	KHK8	KHK9	KHK10	KHK11	KHK12
Thi14	THi15	THi16	THi17	THi18	THi19	THi20

KIJANI 464/1980-81/6216-17/1972-73						
Ashe	Ojobo	Iwa	Araorun	Ile	Itutu	Erinle
1	2	3	4	5	6	7
N28	N29	N30	D1	D2	D3	D4
TBT11	TBT12	TBT13	TBT14	TBT15	TBT16	TBT17
HID19	HID20	HID21	HID22	HID23	HID24	HID25
8	9	10	11	12	13	14
D5	D6	D7	D8	D9	D10	D11
TBT18	TBT19	TBT20	TBT21	TBT22	TBT23	TBT24
HID26	HID27	HID28	HID29	HID30	TAH1	TAH2
15	16	17	18	19	20	21
D12	D13	D14	D15	D16	D17	D18
TBT25	TBT26	TBT27	TBT28	TBT29	TBT30	MKH1
TAH3	TAH4	TAH5	TAH6	TAH7	TAH8	TAH9
22	23	24	25	26	27	28
D19	D20	D21	D22	D23	D24	D25
MKH2	MKH3	MKH4	MKH5	MKH6	MKH7	MKH8
TAH10	TAH11	TAH12	TAH13	TAH14	TAH15	TAH16

EUSI 464/1980-81/6216-17/1972-73						
Ashe	Ojobo	Iwa	Araorun	Ile	Itutu	Erinle
1	2	3	4	5	6	7
O31	N1	N2	N3	N4	N5	N6
KHK13	KHK14	KHK15	KHK16	KHK17	KHK18	KHK19
THi21	THi22	THi23	THi24	THi25	THi26	THi27
8	9	10	11	12	13	14
N7	N8	N9	N10	N11	N12	N13
KHK20	KHK21	KHK22	KHK23	KHK24	KHK25	KHK26
THi28	THi29	THi30	HID1	HID2	HID3	HID4
15	16	17	18	19	20	21
N14	N15	N16	N17	N18	N19	N20
KHK27	KHK28	KHK29	KHK30	TBT1	TBT2	TBT3
HID5	HID6	HID7	HID8	HID9	HID10	HID11
22	23	24	25	26	27	28
N21	N22	N23	N24	N25	N26	N27
TBT4	TBT5	TBT6	TBT7	TBT8	TBT9	TBT10
HID12	HID13	HID14	HID15	HID16	HID17	HID18

KWANZA 464/1980-81/6216-17/1972-73						
Umoja	Kujichagulia	Ujima	Ujamaa	Nia	Kuumba	Imani
1	2	3	4	5	6	7
D26	D27	D28	D29	D30	D31	JA1
MKH9	MKH10	MKH11	MKH12	MKH13	MKH14	MKH15
TAH17	TAH18	TAH19	TAH20	TAH21	TAH22	TAH23
8	9	10	11	12	13	14
JA2	JA3	JA4	JA5	JA6	JA7	JA8
MKH16	MKH17	MKH18	MKH19	MKH20	MKH21	MKH22
TAH24	TAH25	TAH26	TAH27	TAH28	TAH29	TAH30
15	16	17	18	19	20	21
JA9	JA10	JA11	JA12	JA13	JA14	JA15
MKH23	MKH24	MKH25	KMH26	MKH27	MKH28	MKH29
TIR1	TIR2	TIR3	TIR4	TIR5	TIR6	TIR7
22	23	24	25	26	27	28
JA16	JA17	JA18	JA19	JA20	JA21	JA22
MKH30	AMN1	AMN2	AMN3	AMN4	AMN5	AMN6
TIR8	TIR9	TIR10	TIR11	TIR12	TIR13	TIR14

SHAHIDI 464/1980-81/6216-17/1972-73						
Ashe	Ojobo	Iwa	Araorun	Ile	Itutu	Erinle
1	2	3	4	5	6	7
JA23	JA24	JA25	JA26	JA27	JA28	JA29
AMN7	AMN8	AMN9	AMN10	AMN11	AMN12	AMN13
TIR15	TIR16	TIR17	TIR18	TIR19	TIR20	TIR21
8	9	10	11	12	13	14
JA30	JA31	F1	F2	F3	F4	F5
AMN14	AMN15	AMN16	AMN17	AMN18	AMN19	AMN20
TIR22	TIR23	TIR24	TIR25	TIR26	TIR27	TIR28
15	16	17	18	19	20	21
F6	F7	F8	F9	F10	F 11	F12
AMN21	AMN22	AMN23	AMN24	AMN25	AMN26	AMN28
TIR29	TIR30	YEK1	YEK2	YEK3	YEK4	YEK5
22	23	24	25	26	27	28
F13	F14	F15	F16	F17	F18	F19
AMN29	AMN30	RNT1	RNT2	RNT3	RNT4	RNT5
YEK6	YEK7	YEK8	YEK9	YEK10	YEK11	YEK12
SHAHIDI 15						
F20						
RNT6						
YEK13						

**MAAFA 465 B.S.E./ASIENTO 1981-82 C.E./
KAM 6217-18 S.M./EC 1973-74 A.M. (ዓ.ም.)**

AFU 465/1981-82/6217-18/1973-74						
Ashe	Ojobo	Iwa	Araorun	Ile	Itutu	Erinle
1	2	3	4	5	6	7
F21	F22	F23	F24	F25	F26	F27
RNT7	RNT8	RNT9	RNT10	RNT11	RNT12	RNT13
YEK14	YEK15	YEK16	YEK17	YEK18	YEK19	YEK20
	9	10	11	12	13	14
F28	M1	M2	M3	M4	M5	M6
RNT14	RNT15	RNT16	RNT17	RNT18	RNT19	RNT20
YEK21	YEK22	YEK23	YEK24	YEK25	YEK26	YEK27
15	16	17	18	19	20	21
M7	M8	M9	M10	M11	M12	M13
RNT21	RNT22	RNT23	RNT24	RNT25	RNT26	RNT27
YEK28	YEK29	YEK30	MAG1	MAG2	MAG3	MAG4
22	23	24	25	26	27	28
M14	M15	M16	M17	M18	M19	M20
RNT28	RNT29	RNT30	KHN1	KHN2	KHN3	KHN4
MAG5	MAG6	MAG7	MAG8	MAG9	MAG10	MAG11

KAWAIDA 465/1981-82/6217-18/1973-74						
Ashe	Ojobo	Iwa	Araorun	Ile	Itutu	Erinle
1	2	3	4	5	6	7
A18	A19	A20	A21	A22	A23	A24
PIN3	PIN4	PIN5	PIN6	PIN7	PIN8	PIN9
MIY10	MIY11	MIY12	MIY13	MIY14	MIY15	MIY16
8	9	10	11	12	13	14
A25	A26	A27	A28	A29	A30	MA1
PIN10	PIN11	PIN12	PIN13	PIN14	PIN15	PIN16
MIY17	MIY18	MIY19	MIY20	MIY21	MIY22	MIY23
15	16	17	18	19	20	21
MA2	MA3	MA4	MA5	MA6	MA7	MA8
PIN17	PIN18	PIN19	PIN20	PIN21	PIN22	PIN23
MIY24	MIY25	MIY26	MIY27	MIY28	MIY29	MIY30
22	23	24	25	26	27	28
MA9	MA10	MA11	MA12	MA13	MA14	MA15
PIN24	PIN25	PIN26	PIN27	PIN28	PIN29	PIN30
GIN1	GIN2	GIN3	GIN4	GIN5	GIN6	GIN7

UHURU 465/1981-82/6217-18/1973-74						
Ashe	Ojobo	Iwa	Araorun	Ile	Itutu	Erinle
1	2	3	4	5	6	7
M21	M22	M23	M24	M25	M26	M27
KHN5	KHN6	KHN7	KHN8	KHN9	KHN10	KHN11
MAG12	MAG13	MAG14	MAG15	MAG16	MAG17	MAG18
8	9	10	11	12	13	14
M28	M29	M30	M31	A1	A2	A3
KHN12	KHN13	KHN14	KHN15	KHN16	KHN17	KHN18
MAG19	MAG20	MAG21	MAG22	MAG23	MAG24	MAG25
15	16	17	18	19	20	21
A4	A5	A6	A7	A8	A9	A10
KHN19	KHN20	KHN21	KHN22	KHN23	KHN24	KHN25
MAG26	MAG27	MAG28	MAG29	MAG30	MIY1	MIY2
22	23	24	25	26	27	28
A11	A12	A13	A14	A15	A16	A17
KHN26	KHN27	KHN28	KHN29	KHN30	PIN1	PIN2
MIY3	MIY4	MIY5	MIY6	MIY7	MIY8	MIY9

MABABA WEUSI 465/1981-82/6217-18/1973-74						
Ashe	Ojobo	Iwa	Araorun	Ile	Itutu	Erinle
1	2	3	4	5	6	7
MA16	MA17	MA18	MA19	MA20	MA21	MA22
APP1	APP2	APP3	APP4	APP5	APP6	APP7
GIN8	GIN9	GIN10	GIN11	GIN12	GIN13	GIN14
8	9	10	11	12	13	14
MA23	MA24	MA25	MA26	MA27	MA28	MA29
APP8	APP9	APP10	APP11	APP12	APP13	APP14
GIN15	GIN16	GIN17	GIN18	GIN19	GIN20	GIN21
15	16	17	18	19	20	21
MA30	MA31	J1	J2	J3	J4	J5
APP15	APP16	APP17	APP18	APP19	APP20	APP21
GIN22	GIN23	GIN24	GIN25	GIN26	GIN27	GIN28
22	23	24	25	26	27	28
J6	J7	J8	J9	J10	J11	J12
APP22	APP23	APP24	APP25	APP26	APP27	APP28
GIN29	GIN30	SAN1	SAN2	SAN3	SAN4	SAN5

MWANAMKE WEUSI 465/1981-82/6217-18/1973-74

Ashe	Ojobo	Iwa	Araorun	Ile	Itutu	Erinle
1	2	3	4	5	6	7
J13	J14	J15	J16	J17	J18	J19
APP29	APP30	MSR1	MSR2	MSR3	MSR4	MSR5
SAN6	SAN7	SAN8	SAN9	SAN10	SAN11	SAN12
8	9	10	11	12	13	14
J20	J21	J22	J23	J24	J25	J26
MSR6	MSR7	MSR8	MSR9	MSR10	MSR11	MSR12
SAN13	SAN14	SAN15	SAN16	SAN17	SAN18	SAN19
15	16	17	18	19	20	21
J27	J28	J29	J30	JU1	JU2	JU3
MSR13	MSR14	MSR15	MSR16	MSR17	MSR18	MSR19
SAN20	SAN21	SAN22	SAN23	SAN24	SAN25	SAN26
22	23	24	25	26	27	28
JU4	JU5	JU6	JU7	JU8	JU9	JU10
MSR20	MRS21	MSR22	MSR23	MSR24	MSR25	MSR26
SAN27	SAN28	SAN29	SAN30	HAM1	HAM2	HAM3

MTU WEUSI 465/1981-82/6217-18/1973-74

Ashe	Ojobo	Iwa	Araorun	Ile	Itutu	Erinle
1	2	3	4	5	6	7
JU11	JU12	JU13	JU14	JU15	JU16	JU17
MSR27	MSR28	MSR29	MSR30	*ASR1*	*HRU2*	*STH3*
HAM4	HAM5	HAM6	HAM7	HAM8	HAM9	HAM10
8	9	10	11	12	13	14
JU18	JU19	JU20	JU21	JU22	JU23	JU24
AST4	*NBH5*	THT1	THT2	THT3	THT4	THT5
HAM11	HAM12	HAM13	AM14	HAM15	HAM16	HAM17
15	16	17	18	19	20	21
JU25	JU26	JU27	JU28	JU29	JU30	JU31
THT6	THT7	THT8	THT9	THT10	THT11	THT12
HAM18	HAM19	HAM20	HAM21	HAM22	HAM23	HAM24
22	23	24	25	26	27	28
AU1	AU2	AU3	AU4	AU5	AU6	AU7
THT13	THT14	THT15	THT16	THT17	THT18	THT19
HAM25	HAM26	HAM27	HAM28	HAM29	HAM30	NAH1

TAIFA WEUSI 465/1981-82/6217-18/1973-74

Ashe	Ojobo	Iwa	Araorun	Ile	Itutu	Erinle
1	2	3	4	5	6	7
AU8	AU9	AU10	AU11	AU12	AU13	AU14
THT20	THT21	THT22	THT23	THT24	THT25	THT26
NAH2	NAH3	NAH4	NAH5	NAH6	NAH7	NAH8
8	9	10	11	12	13	14
AU15	AU16	AU17	AU18	AU19	AU20	AU21
THT27	THT28	THT29	THT30	PAP1	PAP1	PAP2
NAH9	NAH10	NAH11	NAH12	NAH13	NAH14	NAH15
15	16	17	18	19	20	21
AU22	AU23	AU24	AU25	AU26	AU27	AU28
PAP3	PAP4	PAP5	PAP6	PAP7	PAP8	PAP9
NAH16	NAH17	NAH18	NAH19	NAH20	NAH21	NAH22
22	23	24	25	26	27	28
AU29	AU30	AU31	S1	S2	S3	S4
PAP10	PAP11	PAP12	PAP13	PAP14	PAP15	PAP16
NAH23	NAH24	NAH25	NAH26	NAH27	NAH28	NAH29

UWEZO 465/1981-82/6217-18/1973-74

Ashe	Ojobo	Iwa	Araorun	Ile	Itutu	Erinle
1	2	3	4	5	6	7
S5	S6	S7	S8	S9	S10	S11
PAP17	PAP18	PAP19	PAP20	PAP21	PAP22	PAP23
NAH30	*PAG1*	*PAG2*	*PAG3*	*PAG4*	*PAG5*	MAS1
8	9	10	11	12	13	14
S12	S13	S14	S15	S16	S17	S18
PAP24	PAP25	PAP26	PAP27	PAP28	PAP29	PAP30
MAS2	MAS3	MAS4	MAS5	MAS6	MAS7	MAS8
15	16	17	18	19	20	21
S19	S20	S21	S22	S23	S24	S25
HTH1	HTH2	HTH3	HTH4	HTH5	HTH6	HTH7
MAS9	MAS10	MAS11	MAS12	MAS13	MAS14	MAS15
22	23	24	25	26	27	28
S26	S27	S28	S29	S30	O1	O2
HTH8	HTH9	HTH10	HTH11	HTH12	HTH13	HTH14
MAS16	MAS17	MAS18	MAS19	MAS20	MAS21	MAS22

EKUNDU 465/1981-82/6217-18/1973-74					
Ojobo	Iwa	Araorun	Ile	Itutu	Erinle
2	3	4	5	6	7
O4	O5	O6	O7	O8	O9
HTH16	HTH17	HTH18	HTH19	HTH20	HTH21
MAS24	MAS25	MAS26	MAS27	MAS28	MAS29
9	10	11	12	13	14
O11	O12	O13	O14	O15	O16
HTH23	HTH24	HTH25	HTH26	HTH27	HTH28
THi1	THi2	THi3	THi4	THi5	THi6
16	17	18	19	20	21
O18	O19	O20	O21	O22	O23
HTH30	KHK1	KHK2	KHK3	KHK4	KHK5
THi8	THi9	THi10	THi11	THi12	THi13
23	24	25	26	27	28
O25	O26	O27	O28	O29	O30
KHK7	KHK8	KHK9	KHK10	KHK11	KHK12
THi15	THi16	THi17	THi18	THi19	THi20

KIJANI 465/1981-82/6217-18/1973-74						
Ashe	Ojobo	Iwa	Araorun	Ile	Itutu	Erinle
1	2	3	4	5	6	7
N28	N29	N30	D1	D2	D3	D4
TBT11	TBT12	TBT13	TBT14	TBT15	TBT16	TBT17
HID19	HID20	HID21	HID22	HID23	HID24	HID25
8	9	10	11	12	13	14
D5	D6	D7	D8	D9	D10	D11
TBT18	TBT19	TBT20	TBT21	TBT22	TBT23	TBT24
HID26	HID27	HID28	HID29	HID30	TAH1	TAH2
15	16	17	18	19	20	21
D12	D13	D14	D15	D16	D17	D18
TBT25	TBT26	TBT27	TBT28	TBT29	TBT30	MKH1
TAH3	TAH4	TAH5	TAH6	TAH7	TAH8	TAH9
22	23	24	25	26	27	28
D19	D20	D21	D22	D23	D24	D25
MKH2	MKH3	MKH4	MKH5	MKH6	MKH7	MKH8
TAH10	TAH11	TAH12	TAH13	TAH14	TAH15	TAH16

EUSI 465/1981-82/6217-18/1973-74					
Ojobo	Iwa	Araorun	Ile	Itutu	Erinle
2	3	4	5	6	7
N1	N2	N3	N4	N5	N6
KHK14	KHK15	KHK16	KHK17	KHK18	KHK19
THi22	THi23	THi24	THi25	THi26	THi27
9	10	11	12	13	14
N8	N9	N10	N11	N12	N13
KHK21	KHK22	KHK23	KHK24	KHK25	KHK26
THi29	THi30	HID1	HID2	HID3	HID4
16	17	18	19	20	21
N15	N16	N17	N18	N19	N20
KHK28	KHK29	KHK30	TBT1	TBT2	TBT3
HID6	HID7	HID8	HID9	HID10	HID11
23	24	25	26	27	28
N22	N23	N24	N25	N26	N27
TBT5	TBT6	TBT7	TBT8	TBT9	TBT10
HID13	HID14	HID15	HID16	HID17	HID18

KWANZA 465/1981-82/6217-18/1973-74						
Umoja	Kujichagulia	Ujima	Ujamaa	Nia	Kuumba	Imani
1	2	3	4	5	6	7
D26	D27	D28	D29	D30	D31	JA1
MKH9	MKH10	MKH11	MKH12	MKH13	MKH14	MKH15
TAH17	TAH18	TAH19	TAH20	TAH21	TAH22	TAH23
8	9	10	11	12	13	14
JA2	JA3	JA4	JA5	JA6	JA7	JA8
MKH16	MKH17	MKH18	MKH19	MKH20	MKH21	MKH22
TAH24	TAH25	TAH26	TAH27	TAH28	TAH29	TAH30
15	16	17	18	19	20	21
JA9	JA10	JA11	JA12	JA13	JA14	JA15
MKH23	MKH24	MKH25	KMH26	MKH27	MKH28	MKH29
TIR1	TIR2	TIR3	TIR4	TIR5	TIR6	TIR7
22	23	24	25	26	27	28
JA16	JA17	JA18	JA19	JA20	JA21	JA22
MKH30	AMN1	AMN2	AMN3	AMN4	AMN5	AMN6
TIR8	TIR9	TIR10	TIR11	TIR12	TIR13	TIR14

SHAHIDI 465/1981-82/6217-18/1973-74						
Ashe	Ojobo	Iwa	Araorun	Ile	Itutu	Erinle
1	2	3	4	5	6	7
JA23	JA24	JA25	JA26	JA27	JA28	JA29
AMN7	AMN8	AMN9	AMN10	AMN11	AMN12	AMN13
TIR15	TIR16	TIR17	TIR18	TIR19	TIR20	TIR21
8	9	10	11	12	13	14
JA30	JA31	F1	F2	F3	F4	F5
AMN14	AMN15	AMN16	AMN17	AMN18	AMN19	AMN20
TIR22	TIR23	TIR24	TIR25	TIR26	TIR27	TIR28
15	16	17	18	19	20	21
F6	F7	F8	F9	F10	F 11	F12
AMN21	AMN22	AMN23	AMN24	AMN25	AMN26	AMN28
TIR29	TIR30	YEK1	YEK2	YEK3	YEK4	YEK5
22	23	24	25	26	27	28
F13	F14	F15	F16	F17	F18	F19
AMN29	AMN30	RNT1	RNT2	RNT3	RNT4	RNT5
YEK6	YEK7	YEK8	YEK9	YEK10	YEK11	YEK12
SHAHIDI 16						
F20						
RNT6						
YEK13						

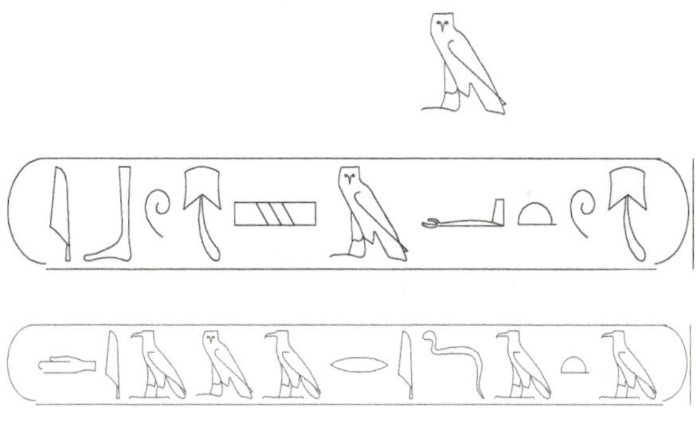

MAAFA 466 B.S.E./ASIENTO 1982-83 C.E./
KAM 6218-19 S.M./EC 1974-75 A.M. (ዓ.ም.)

AFU 466/1982-83/6218-19/1974-75						
Ashe	Ojobo	Iwa	Araorun	Ile	Itutu	Erinle
1	2	3	4	5	6	7
F21	F22	F23	F24	F25	F26	F27
RNT7	RNT8	RNT9	RNT10	RNT11	RNT12	RNT13
YEK14	YEK15	YEK16	YEK17	YEK18	YEK19	YEK20
8	9	10	11	12	13	14
F28	M1	M2	M3	M4	M5	M6
RNT14	RNT15	RNT16	RNT17	RNT18	RNT19	RNT20
YEK21	YEK22	YEK23	YEK24	YEK25	YEK26	YEK27
15	16	17	18	19	20	21
M7	M8	M9	M10	M11	M12	M13
RNT21	RNT22	RNT23	RNT24	RNT25	RNT26	RNT27
YEK28	YEK29	YEK30	MAG1	MAG2	MAG3	MAG4
22	23	24	25	26	27	28
M14	M15	M16	M17	M18	M19	M20
RNT28	RNT29	RNT30	KHN1	KHN2	KHN3	KHN4
MAG5	MAG6	MAG7	MAG8	MAG9	MAG10	MAG11

KAWAIDA 466/1982-83/6218-19/1974-75						
Ashe	Ojobo	Iwa	Araorun	Ile	Itutu	Erinle
1	2	3	4	5	6	7
A18	A19	A20	A21	A22	A23	A24
PIN3	PIN4	PIN5	PIN6	PIN7	PIN8	PIN9
MIY10	MIY11	MIY12	MIY13	MIY14	MIY15	MIY16
8	9	10	11	12	13	14
A25	A26	A27	A28	A29	A30	MA1
PIN10	PIN11	PIN12	PIN13	PIN14	PIN15	PIN16
MIY17	MIY18	MIY19	MIY20	MIY21	MIY22	MIY23
15	16	17	18	19	20	21
MA2	MA3	MA4	MA5	MA6	MA7	MA8
PIN17	PIN18	PIN19	PIN20	PIN21	PIN22	PIN23
MIY24	MIY25	MIY26	MIY27	MIY28	MIY29	MIY30
22	23	24	25	26	27	28
MA9	MA10	MA11	MA12	MA13	MA14	MA15
PIN24	PIN25	PIN26	PIN27	PIN28	PIN29	PIN30
GIN1	GIN2	GIN3	GIN4	GIN5	GIN6	GIN7

UHURU 466/1982-83/6218-19/1974-75						
Ashe	Ojobo	Iwa	Araorun	Ile	Itutu	Erinle
1	2	3	4	5	6	7
M21	M22	M23	M24	M25	M26	M27
KHN5	KHN6	KHN7	KHN8	KHN9	KHN10	KHN11
MAG12	MAG13	MAG14	MAG15	MAG16	MAG17	MAG18
8	9	10	11	12	13	14
M28	M29	M30	M31	A1	A2	A3
KHN12	KHN13	KHN14	KHN15	KHN16	KHN17	KHN18
MAG19	MAG20	MAG21	MAG22	MAG23	MAG24	MAG25
15	16	17	18	19	20	21
A4	A5	A6	A7	A8	A9	A10
KHN19	KHN20	KHN21	KHN22	KHN23	KHN24	KHN25
MAG26	MAG27	MAG28	MAG29	MAG30	MIY1	MIY2
22	23	24	25	26	27	28
A11	A12	A13	A14	A15	A16	A17
KHN26	KHN27	KHN28	KHN29	KHN30	PIN1	PIN2
MIY3	MIY4	MIY5	MIY6	MIY7	MIY8	MIY9

MABABA WEUSI 466/1982-83/6218-19/1974-75						
Ashe	Ojobo	Iwa	Araorun	Ile	Itutu	Erinle
1	2	3	4	5	6	7
MA16	MA17	MA18	MA19	MA20	MA21	MA22
APP1	APP2	APP3	APP4	APP5	APP6	APP7
GIN8	GIN9	GIN10	GIN11	GIN12	GIN13	GIN14
8	9	10	11	12	13	14
MA23	MA24	MA25	MA26	MA27	MA28	MA29
APP8	APP9	APP10	APP11	APP12	APP13	APP14
GIN15	GIN16	GIN17	GIN18	GIN19	GIN20	GIN21
15	16	17	18	19	20	21
MA30	MA31	J1	J2	J3	J4	J5
APP15	APP16	APP17	APP18	APP19	APP20	APP21
GIN22	GIN23	GIN24	GIN25	GIN26	GIN27	GIN28
22	23	24	25	26	27	28
J6	J7	J8	J9	J10	J11	J12
APP22	APP23	APP24	APP25	APP26	APP27	APP28
GIN29	GIN30	SAN1	SAN2	SAN3	SAN4	SAN5

MWANAMKE WEUSI 466/1982-83/6218-19/1974-75

Ashe	Ojobo	Iwa	Araorun	Ile	Itutu	Erinle
1	2	3	4	5	6	7
J13	J14	J15	J16	J17	J18	J19
APP29	APP30	MSR1	MSR2	MSR3	MSR4	MSR5
SAN6	SAN7	SAN8	SAN9	SAN10	SAN11	SAN12
8	9	10	11	12	13	14
J20	J21	J22	J23	J24	J25	J26
MSR6	MSR7	MSR8	MSR9	MSR10	MSR11	MSR12
SAN13	SAN14	SAN15	SAN16	SAN17	SAN18	SAN19
15	16	17	18	19	20	21
J27	J28	J29	J30	JU1	JU2	JU3
MSR13	MSR14	MSR15	MSR16	MSR17	MSR18	MSR19
SAN20	SAN21	SAN22	SAN23	SAN24	SAN25	SAN26
22	23	24	25	26	27	28
JU4	JU5	JU6	JU7	JU8	JU9	JU10
MSR20	MRS21	MSR22	MSR23	MSR24	MSR25	MSR26
SAN27	SAN28	SAN29	SAN30	HAM1	HAM2	HAM3

TAIFA WEUSI 466/1982-83/6218-19/1974-75

Ashe	Ojobo	Iwa	Araorun	Ile	Itutu	Erinle
1	2	3	4	5	6	7
AU8	AU9	AU10	AU11	AU12	AU13	AU14
THT20	THT21	THT22	THT23	THT24	THT25	THT26
NAH2	NAH3	NAH4	NAH5	NAH6	NAH7	NAH8
8	9	10	11	12	13	14
AU15	AU16	AU17	AU18	AU19	AU20	AU21
THT27	THT28	THT29	THT30	PAP1	PAP1	PAP2
NAH9	NAH10	NAH11	NAH12	NAH13	NAH14	NAH15
15	16	17	18	19	20	21
AU22	AU23	AU24	AU25	AU26	AU27	AU28
PAP3	PAP4	PAP5	PAP6	PAP7	PAP8	PAP9
NAH16	NAH17	NAH18	NAH19	NAH20	NAH21	NAH22
22	23	24	25	26	27	28
AU29	AU30	AU31	S1	S2	S3	S4
PAP10	PAP11	PAP12	PAP13	PAP14	PAP15	PAP16
NAH23	NAH24	NAH25	NAH26	NAH27	NAH28	NAH29

MTU WEUSI 466/1982-83/6218-19/1974-75

Ashe	Ojobo	Iwa	Araorun	Ile	Itutu	Erinle
1	2	3	4	5	6	7
JU11	JU12	JU13	JU14	JU15	JU16	JU17
MSR27	MSR28	MSR29	MSR30	*ASR1*	*HRU2*	*STH3*
HAM4	HAM5	HAM6	HAM7	HAM8	HAM9	HAM10
8	9	10	11	12	13	14
JU18	JU19	JU20	JU21	JU22	JU23	JU24
AST4	*NBH5*	THT1	THT2	THT3	THT4	THT5
HAM11	HAM12	HAM13	AM14	HAM15	HAM16	HAM17
15	16	17	18	19	20	21
JU25	JU26	JU27	JU28	JU29	JU30	JU31
THT6	THT7	THT8	THT9	THT10	THT11	THT12
HAM18	HAM19	HAM20	HAM21	HAM22	HAM23	HAM24
22	23	24	25	26	27	28
AU1	AU2	AU3	AU4	AU5	AU6	AU7
THT13	THT14	THT15	THT16	THT17	THT18	THT19
HAM25	HAM26	HAM27	HAM28	HAM29	HAM30	NAH1

UWEZO 466/1982-83/6218-19/1974-75

Ashe	Ojobo	Iwa	Araorun	Ile	Itutu	Erinle
1	2	3	4	5	6	7
S5	S6	S7	S8	S9	S10	S11
PAP17	PAP18	PAP19	PAP20	PAP21	PAP22	PAP23
NAH30	*PAG1*	*PAG2*	*PAG3*	*PAG4*	*PAG5*	MAS1
8	9	10	11	12	13	14
S12	S13	S14	S15	S16	S17	S18
PAP24	PAP25	PAP26	PAP27	PAP28	PAP29	PAP30
MAS2	MAS3	MAS4	MAS5	MAS6	MAS7	MAS8
15	16	17	18	19	20	21
S19	S20	S21	S22	S23	S24	S25
HTH1	HTH2	HTH3	HTH4	HTH5	HTH6	HTH7
MAS9	MAS10	MAS11	MAS12	MAS13	MAS14	MAS15
22	23	24	25	26	27	28
S26	S27	S28	S29	S30	O1	O2
HTH8	HTH9	HTH10	HTH11	HTH12	HTH13	HTH14
MAS16	MAS17	MAS18	MAS19	MAS20	MAS21	MAS22

EKUNDU 466/1982-83/6218=19/1974-75

Ashe	Ojobo	Iwa	Araorun	Ile	Itutu	Erinle
1	2	3	4	5	6	7
O3	O4	O5	O6	O7	O8	O9
HTH15	HTH16	HTH17	HTH18	HTH19	HTH20	HTH21
MAS23	MAS24	MAS25	MAS26	MAS27	MAS28	MAS29
8	9	10	11	12	13	14
O10	O11	O12	O13	O14	O15	O16
HTH22	HTH23	HTH24	HTH25	HTH26	HTH27	HTH28
MAS30	THi1	THi2	THi3	THi4	THi5	THi6
15	16	17	18	19	20	21
O17	O18	O19	O20	O21	O22	O23
HTH29	HTH30	KHK1	KHK2	KHK3	KHK4	KHK5
THi7	THi8	This9	THi10	THi11	THi12	THi13
22	23	24	25	26	27	28
O24	O25	O26	O27	O28	O29	O30
KHK6	KHK7	KHK8	KHK9	KHK10	KHK11	KHK12
Thi14	THi15	THi16	THi17	THi18	THi19	THi20

KIJANI 466/1982-83/6218=19/1974-75

Ashe	Ojobo	Iwa	Araorun	Ile	Itutu	Erinle
1	2	3	4	5	6	7
N28	N29	N30	D1	D2	D3	D4
TBT11	TBT12	TBT13	TBT14	TBT15	TBT16	TBT17
HID19	HID20	HID21	HID22	HID23	HID24	HID25
8	9	10	11	12	13	14
D5	D6	D7	D8	D9	D10	D11
TBT18	TBT19	TBT20	TBT21	TBT22	TBT23	TBT24
HID26	HID27	HID28	HID29	HID30	TAH1	TAH2
15	16	17	18	19	20	21
D12	D13	D14	D15	D16	D17	D18
TBT25	TBT26	TBT27	TBT28	TBT29	TBT30	MKH1
TAH3	TAH4	TAH5	TAH6	TAH7	TAH8	TAH9
22	23	24	25	26	27	28
D19	D20	D21	D22	D23	D24	D25
MKH2	MKH3	MKH4	MKH5	MKH6	MKH7	MKH8
TAH10	TAH11	TAH12	TAH13	TAH14	TAH15	TAH16

EUSI 466/1982-83/6218=19/1974-75

Ashe	Ojobo	Iwa	Araorun	Ile	Itutu	Erinle
1	2	3	4	5	6	7
O31	N1	N2	N3	N4	N5	N6
KHK13	KHK14	KHK15	KHK16	KHK17	KHK18	KHK19
THi21	THi22	THi23	THi24	THi25	THi26	THi27
8	9	10	11	12	13	14
N7	N8	N9	N10	N11	N12	N13
KHK20	KHK21	KHK22	KHK23	KHK24	KHK25	KHK26
THi28	THi29	THi30	HID1	HID2	HID3	HID4
15	16	17	18	19	20	21
N14	N15	N16	N17	N18	N19	N20
KHK27	KHK28	KHK29	KHK30	TBT1	TBT2	TBT3
HID5	HID6	HID7	HID8	HID9	HID10	HID11
22	23	24	25	26	27	28
N21	N22	N23	N24	N25	N26	N27
TBT4	TBT5	TBT6	TBT7	TBT8	TBT9	TBT10
HID12	HID13	HID14	HID15	HID16	HID17	HID18

KWANZA 466/1982-83/6218=19/1974-75

Umoja	Kujichagulia	Ujima	Ujamaa	Nia	Kuumba	Imani
1	2	3	4	5	6	7
D26	D27	D28	D29	D30	D31	JA1
MKH9	MKH10	MKH11	MKH12	MKH13	MKH14	MKH15
TAH17	TAH18	TAH19	TAH20	TAH21	TAH22	TAH23
8	9	10	11	12	13	14
JA2	JA3	JA4	JA5	JA6	JA7	JA8
MKH16	MKH17	MKH18	MKH19	MKH20	MKH21	MKH22
TAH24	TAH25	TAH26	TAH27	TAH28	TAH29	TAH30
15	16	17	18	19	20	21
JA9	JA10	JA11	JA12	JA13	JA14	JA15
MKH23	MKH24	MKH25	KMH26	MKH27	MKH28	MKH29
TIR1	TIR2	TIR3	TIR4	TIR5	TIR6	TIR7
22	23	24	25	26	27	28
JA16	JA17	JA18	JA19	JA20	JA21	JA22
MKH30	AMN1	AMN2	AMN3	AMN4	AMN5	AMN6
TIR8	TIR9	TIR10	TIR11	TIR12	TIR13	TIR14

SHAHIDI 466/1982-83/6218-19/1974-75						
Ashe	Ojobo	Iwa	Araorun	Ile	Itutu	Erinle
1	2	3	4	5	6	7
JA23	JA24	JA25	JA26	JA27	JA28	JA29
AMN7	AMN8	AMN9	AMN10	AMN11	AMN12	AMN13
TIR15	TIR16	TIR17	TIR18	TIR19	TIR20	TIR21
8	9	10	11	12	13	14
JA30	JA31	F1	F2	F3	F4	F5
AMN14	AMN15	AMN16	AMN17	AMN18	AMN19	AMN20
TIR22	TIR23	TIR24	TIR25	TIR26	TIR27	TIR28
15	16	17	18	19	20	21
F6	F7	F8	F9	F10	F 11	F12
AMN21	AMN22	AMN23	AMN24	AMN25	AMN26	AMN28
TIR29	TIR30	YEK1	YEK2	YEK3	YEK4	YEK5
22	23	24	25	26	27	28
F13	F14	F15	F16	F17	F18	F19
AMN29	AMN30	RNT1	RNT2	RNT3	RNT4	RNT5
YEK6	YEK7	YEK8	YEK9	YEK10	YEK11	YEK12
SHAHIDI 17						
F20						
RNT6						
YEK13						

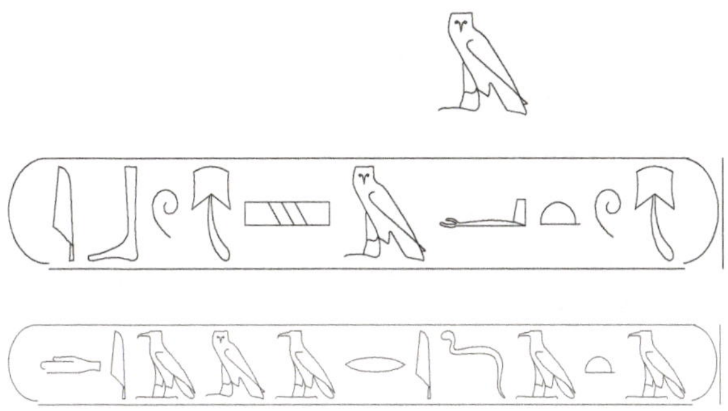

MAAFA 467 B.S.E./ASIENTO 1983-84 C.E./
KAM 6219-20 S.M./EC 1975-76 A.M. (ዓ.ም.)

AFU 467/1983-84/6219-20/1975-76

Ashe	Ojobo	Iwa	Araorun	Ile	Itutu	Erinle
1	2	3	4	5	6	7
F21	F22	F23	F24	F25	F26	F27
RNT7	RNT8	RNT9	RNT10	RNT11	RNT12	RNT13
YEK14	YEK15	YEK16	YEK17	YEK18	YEK19	YEK20
8	9	10	11	12	13	14
F28	M1	M2	M3	M4	M5	M6
RNT14	RNT15	RNT16	RNT17	RNT18	RNT19	RNT20
YEK21	YEK22	YEK23	YEK24	YEK25	YEK26	YEK27
15	16	17	18	19	20	21
M7	M8	M9	M10	M11	M12	M13
RNT21	RNT22	RNT23	RNT24	RNT25	RNT26	RNT27
YEK28	YEK 29	YEK30	MAG1	MAG2	MAG3	MAG4
22	23	24	25	26	27	28
M14	M15	M16	M17	M18	M19	M20
RNT28	RNT29	RNT30	KHN1	KHN2	KHN3	KHN4
MAG5	MAG6	MAG7	MAG8	MAG9	MAG10	MAG11

UHURU 467/1983-84/6219-20/1975-76

Ashe	Ojobo	Iwa	Araorun	Ile	Itutu	Erinle
1	2	3	4	5	6	7
M21	M22	M23	M24	M25	M26	M27
KHN5	KHN6	KHN7	KHN8	KHN9	KHN10	KHN11
MAG12	MAG13	MAG14	MAG15	MAG16	MAG17	MAG18
8	9	10	11	12	13	14
M28	M29	M30	M31	A1	A2	A3
KHN12	KHN13	KHN14	KHN15	KHN16	KHN17	KHN18
MAG19	MAG20	MAG21	MAG22	MAG23	MAG24	MAG25
15	16	17	18	19	20	21
A4	A5	A6	A7	A8	A9	A10
KHN19	KHN20	KHN21	KHN22	KHN23	KHN24	KHN25
MAG26	MAG27	MAG28	MAG29	MAG30	MIY1	MIY2
22	23	24	25	26	27	28
A11	A12	A13	A14	A15	A16	A17
KHN26	KHN27	KHN28	KHN29	KHN30	PIN1	PIN2
MIY3	MIY4	MIY5	MIY6	MIY7	MIY8	MIY9

KAWAIDA 467/1983-84/6219-20/1975-76

Ashe	Ojobo	Iwa	Araorun	Ile	Itutu	Erinle
1	2	3	4	5	6	7
A18	A19	A20	A21	A22	A23	A24
PIN3	PIN4	PIN5	PIN6	PIN7	PIN8	PIN9
MIY10	MIY11	MIY12	MIY13	MIY14	MIY15	MIY16
8	9	10	11	12	13	14
A25	A26	A27	A28	A29	A30	MA1
PIN10	PIN11	PIN12	PIN13	PIN14	PIN15	PIN16
MIY17	MIY18	MIY19	MIY20	MIY21	MIY22	MIY23
15	16	17	18	19	20	21
MA2	MA3	MA4	MA5	MA6	MA7	MA8
PIN17	PIN18	PIN19	PIN20	PIN21	PIN22	PIN23
MIY24	MIY25	MIY26	MIY27	MIY28	MIY29	MIY30
22	23	24	25	26	27	28
MA9	MA10	MA11	MA12	MA13	MA14	MA15
PIN24	PIN25	PIN26	PIN27	PIN28	PIN29	PIN30
GIN1	GIN2	GIN3	GIN4	GIN5	GIN6	GIN7

MABABA WEUSI 467/1983-84/6219-20/1975-76

Ashe	Ojobo	Iwa	Araorun	Ile	Itutu	Erinle
1	2	3	4	5	6	7
MA16	MA17	MA18	MA19	MA20	MA21	MA22
APP1	APP2	APP3	APP4	APP5	APP6	APP7
GIN8	GIN9	GIN10	GIN11	GIN12	GIN13	GIN14
8	9	10	11	12	13	14
MA23	MA24	MA25	MA26	MA27	MA28	MA29
APP8	APP9	APP10	APP11	APP12	APP13	APP14
GIN15	GIN16	GIN17	GIN18	GIN19	GIN20	GIN21
15	16	17	18	19	20	21
MA30	MA31	J1	J2	J3	J4	J5
APP15	APP16	APP17	APP18	APP19	APP20	APP21
GIN22	GIN23	GIN24	GIN25	GIN26	GIN27	GIN28
22	23	24	25	26	27	28
J6	J7	J8	J9	J10	J11	J12
APP22	APP23	APP24	APP25	APP26	APP27	APP28
GIN29	GIN30	SAN1	SAN2	SAN3	SAN4	SAN5

MWANAMKE WEUSI 467/1983-84/6219-20/1975-76

Ashe	Ojobo	Iwa	Araorun	Ile	Itutu	Erinle
1	2	3	4	5	6	7
J13	J14	J15	J16	J17	J18	J19
APP29	APP30	MSR1	MSR2	MSR3	MSR4	MSR5
SAN6	SAN7	SAN8	SAN9	SAN10	SAN11	SAN12
8	9	10	11	12	13	14
J20	J21	J22	J23	J24	J25	J26
MSR6	MSR7	MSR8	MSR9	MSR10	MSR11	MSR12
SAN13	SAN14	SAN15	SAN16	SAN17	SAN18	SAN19
15	16	17	18	19	20	21
J27	J28	J29	J30	JU1	JU2	JU3
MSR13	MSR14	MSR15	MSR16	MSR17	MSR18	MSR19
SAN20	SAN21	SAN22	SAN23	SAN24	SAN25	SAN26
22	23	24	25	26	27	28
JU4	JU5	JU6	JU7	JU8	JU9	JU10
MSR20	MRS21	MSR22	MSR23	MSR24	MSR25	MSR26
SAN27	SAN28	SAN29	SAN30	HAM1	HAM2	HAM3

TAIFA WEUSI 467/1983-84/6219-20/1975-76

Ashe	Ojobo	Iwa	Araorun	Ile	Itutu	Erinle
1	2	3	4	5	6	7
AU8	AU9	AU10	AU11	AU12	AU13	AU14
THT20	THT21	THT22	THT23	THT24	THT25	THT26
NAH2	NAH3	NAH4	NAH5	NAH6	NAH7	NAH8
8	9	10	11	12	13	14
AU15	AU16	AU17	AU18	AU19	AU20	AU21
THT27	THT28	THT29	THT30	PAP1	PAP1	PAP2
NAH9	NAH10	NAH11	NAH12	NAH13	NAH14	NAH15
15	16	17	18	19	20	21
AU22	AU23	AU24	AU25	AU26	AU27	AU28
PAP3	PAP4	PAP5	PAP6	PAP7	PAP8	PAP9
NAH16	NAH17	NAH18	NAH19	NAH20	NAH21	NAH22
22	23	24	25	26	27	28
AU29	AU30	AU31	S1	S2	S3	S4
PAP10	PAP11	PAP12	PAP13	PAP14	PAP15	PAP16
NAH23	NAH24	NAH25	NAH26	NAH27	NAH28	NAH29

MTU WEUSI 467/1983-84/6219-20/1975-76

Ashe	Ojobo	Iwa	Araorun	Ile	Itutu	Erinle
1	2	3	4	5	6	7
JU11	JU12	JU13	JU14	JU15	JU16	JU17
MSR27	MSR28	MSR29	MSR30	*ASR1*	*HRU2*	*STH3*
HAM4	HAM5	HAM6	HAM7	HAM8	HAM9	HAM10
8	9	10	11	12	13	14
JU18	JU19	JU20	JU21	JU22	JU23	JU24
AST4	*NBH5*	THT1	THT2	THT3	THT4	THT5
HAM11	HAM12	HAM13	HAM14	HAM15	HAM16	HAM17
15	16	17	18	19	20	21
JU25	JU26	JU27	JU28	JU29	JU30	JU31
THT6	THT7	THT8	THT9	THT10	THT11	THT12
HAM18	HAM19	HAM20	HAM21	HAM22	HAM23	HAM24
22	23	24	25	26	27	28
AU1	AU2	AU3	AU4	AU5	AU6	AU7
THT13	THT14	THT15	THT16	THT17	THT18	THT19
HAM25	HAM26	HAM27	HAM28	HAM29	HAM30	NAH1

UWEZO 467/1983-84/6219-20/1975-76

Ashe	Ojobo	Iwa	Araorun	Ile	Itutu	Erinle
1	2	3	4	5	6	7
S5	S6	S7	S8	S9	S10	S11
PAP17	PAP18	PAP19	PAP20	PAP21	PAP22	PAP23
NAH30	*QUA1*	*QUA2*	*QUA3*	*QUA4*	*QUA5*	*QUA6*
8	9	10	11	12	13	14
S12	S13	S14	S15	S16	S17	S18
PAP24	PAP25	PAP26	PAP27	PAP28	PAP29	PAP30
MAS1	MAS2	MAS3	MAS4	MAS5	MAS6	MAS7
15	16	17	18	19	20	21
S19	S20	S21	S22	S23	S24	S25
HTH1	HTH2	HTH3	HTH4	HTH5	HTH6	HTH7
MAS8	MAS9	MAS10	MAS11	MAS12	MAS13	MAS14
22	23	24	25	26	27	28
S26	S27	S28	S29	S30	O1	O2
HTH8	HTH9	HTH10	HTH11	HTH12	HTH13	HTH14
MAS15	MAS16	MAS17	MAS18	MAS19	MAS20	MAS21

EKUNDU 467/1983-84/6219-20/1975-76						
Ashe	Ojobo	Iwa	Araorun	Ile	Itutu	Erinle
1	2	3	4	5	6	7
O3	O4	O5	O6	O7	O8	O9
HTH15	HTH16	HTH17	HTH18	HTH19	HTH20	HTH21
MAS22	MAS23	MAS 24	MAS25	MAS26	MAS27	MAS28
8	9	10	11	12	13	14
O10	O11	O12	O13	O14	O15	O16
HTH22	HTH23	HTH24	HTH25	HTH26	HTH27	HTH28
MAS29	MAS30	THi1	THi2	THi3	THi4	THi5
15	16	17	18	19	20	21
O17	O18	O19	O20	O21	O22	O23
HTH29	HTH30	KHK1	KHK2	KHK3	KHK4	KHK5
THi6	THi7	THi8	THi9	THi10	THi11	THi12
22	23	24	25	26	27	28
O24	O25	O26	O27	O28	O29	O30
KHK6	KHK7	KHK8	KHK9	KHK10	KHK11	KHK12
THi13	THi14	THi15	THi16	THi17	THi18	THi19

KIJANI 467/1983-84/6219-20/1975-76						
Ashe	Ojobo	Iwa	Araorun	Ile	Itutu	Erinle
1	2	3	4	5	6	7
N28	N29	N30	D1	D2	D3	D4
TBT11	TBT12	TBT13	TBT14	TBT15	TBT16	TBT17
HID18	HID19	HID20	HID21	HID22	HID23	HID24
8	9	10	11	12	13	14
D5	D6	D7	D8	D9	D10	D11
TBT18	TBT19	TBT20	TBT21	TBT22	TBT23	TBT24
HID25	HID26	HID27	HID28	HID29	HID30	TAH1
15	16	17	18	19	20	21
D12	D13	D14	D15	D16	D17	D18
TBT25	TBT26	TBT27	TBT28	TBT29	TBT30	MKH1
TAH2	TAH3	TAH4	TAH5	TAH6	TAH7	TAH8
22	23	24	25	26	27	28
D19	D20	D21	D22	D23	D24	D25
MKH2	MKH3	MKH4	MKH5	MKH6	MKH7	MKH8
TAH9	TAH10	TAH11	TAH12	TAH13	TAH14	TAH15

EUSI 467/1983-84/6219-20/1975-76						
Ashe	Ojobo	Iwa	Araorun	Ile	Itutu	Erinle
1	2	3	4	5	6	7
O31	N1	N2	N3	N4	N5	N6
KHK13	KHK14	KHK15	KHK16	KHK17	KHK18	KHK19
THi20	THi21	THi22	THi23	THi24	THi25	THi26
8	9	10	11	12	13	14
N7	N8	N9	N10	N11	N12	N13
KHK20	KHK21	KHK22	KHK23	KHK24	KHK25	KHK26
THi27	THi28	THi29	THi30	HID1	HID2	HID3
15	16	17	18	19	20	21
N14	N15	N16	N17	N18	N19	N20
KHK27	KHK28	KHK29	KHK30	TBT1	TBT2	TBT3
HID4	HID5	HID6	HID7	HID8	HID9	HID10
22	23	24	25	26	27	28
N21	N22	N23	N24	N25	N26	N27
TBT4	TBT5	TBT6	TBT7	TBT8	TBT9	TBT10
HID11	HID12	HID13	HID14	HID15	HID16	HID17

KWANZA 467/1983-84/6219-20/1975-76						
Umoja	Kujichagulia	Ujima	Ujamaa	Nia	Kuumba	Imani
1	2	3	4	5	6	7
D26	D27	D28	D29	D30	D31	JA1
MKH9	MKH10	MKH11	MKH12	MKH13	MKH14	MKH15
TAH16	TAH17	TAH18	TAH19	TAH20	TAH21	TAH22
8	9	10	11	12	13	14
JA2	JA3	JA4	JA5	JA6	JA7	JA8
MKH16	MKH17	MKH18	MKH19	MKH20	MKH21	MKH22
TAH23	TAH24	TAH25	TAH26	TAH27	TAH28	TAH29
15	16	17	18	19	20	21
JA9	JA10	JA11	JA12	JA13	JA14	JA15
MKH23	MKH24	MKH25	KMH26	MKH27	MKH28	MKH29
TAH30	TIR1	TIR2	TIR3	TIR4	TIR5	TIR6
22	23	24	25	26	27	28
JA16	JA17	JA18	JA19	JA20	JA21	JA22
MKH30	AMN1	AMN2	AMN3	AMN4	AMN5	AMN6
TIR7	TIR8	TIR9	TIR10	TIR11	TIR12	TIR13

SHAHIDI 467/1983-84/6219-20/1975-76						
Ashe	Ojobo	Iwa	Araorun	Ile	Itutu	Erinle
1	2	3	4	5	6	7
JA23	JA24	JA25	JA26	JA27	JA28	JA29
AMN7	AMN8	AMN9	AMN10	AMN11	AMN12	AMN13
TIR14	TIR15	TIR16	TIR17	TIR18	TIR19	TIR20
8	9	10	11	12	13	14
JA30	JA31	F1	F2	F3	F4	F5
AMN14	AMN15	AMN16	AMN17	AMN18	AMN19	AMN20
TIR21	TIR22	TIR23	TIR24	TIR25	TIR26	TIR27
15	16	17	18	19	20	21
F6	F7	F8	F9	F10	F 11	F12
AMN21	AMN22	AMN23	AMN24	AMN25	AMN26	AMN28
TIR28	TIR29	TIR30	YEK1	YEK2	YEK3	YEK4
22	23	24	25	26	27	28
F13	F14	F15	F16	F17	F18	F19
AMN29	AMN30	RNT1	RNT2	RNT3	RNT4	RNT5
YEK5	YEK6	YEK7	YEK8	YEK9	YEK10	YEK11
SHAHIDI 18						
F20						
RNT6						
YEK12						

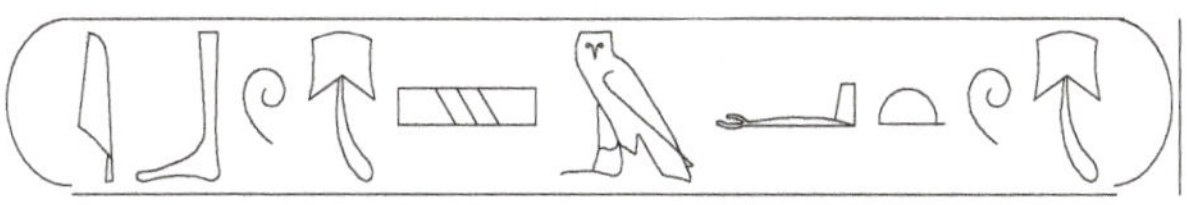

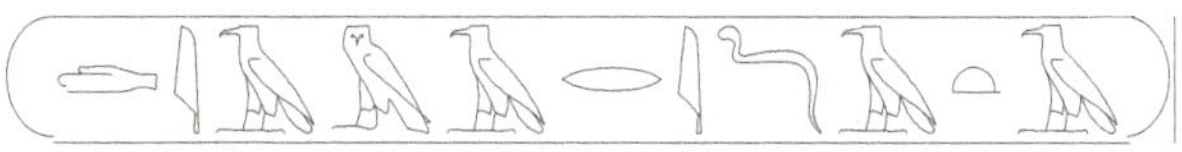

MAAFA 468 B.S.E./ASIENTO 1984-85 C.E./ KAM 6220-21 S.M./EC 1976-77 A.M. (ዓ.ም.)

AFU 468/1984-85/6220-21/1976-77

Ashe	Ojobo	Iwa	Araorun	Ile	Itutu	Erinle
1	2	3	4	5	6	7
F21	F22	F23	F24	F25	F26	F27
RNT7	RNT8	RNT9	RNT10	RNT11	RNT12	RNT13
YEK13	YEK14	YEK15	YEK16	YEK17	YEK18	YEK19
8	9	10	11	12	13	14
F28	F29	M1	M2	M3	M4	M5
RNT14	RNT15	RNT16	RNT17	RNT18	RNT19	RNT20
YEK 20	YEK21	YEK22	YEK23	YEK24	YEK25	YEK26
15	16	17	18	19	20	21
M6	M7	M8	M9	M10	M11	M12
RNT21	RNT22	RNT23	RNT24	RNT25	RNT26	RNT27
YEK 27	YEK28	YEK29	YEK30	MAG1	MAG2	MAG3
22	23	24	25	26	27	28
M13	M14	M15	M16	M17	M18	M19
RNT28	RNT29	RNT30	KHN1	KHN2	KHN3	KHN4
MAG4	MAG5	MAG6	MAG7	MAG8	MAG9	MAG10
SHARIFU 114						
M20						
KHN5						
MAG11						

KAWAIDA 468/1984-85/6220-21/1976-77

Ashe	Ojobo	Iwa	Araorun	Ile	Itutu	Erinle
1	2	3	4	5	6	7
A18	A19	A20	A21	A22	A23	A24
PIN4	PIN5	PIN6	PIN7	PIN8	PIN9	PIN10
MIY10	MIY11	MIY12	MIY13	MIY14	MIY15	MIY16
8	9	10	11	12	13	14
A25	A26	A27	A28	A29	A30	MA1
PIN11	PIN12	PIN13	PIN14	PIN15	PIN16	PIN17
MIY17	MIY18	MIY19	MIY20	MIY21	MIY22	MIY23
15	16	17	18	19	20	21
MA2	MA3	MA4	MA5	MA6	MA7	MA8
PIN18	PIN19	PIN20	PIN21	PIN22	PIN23	PIN24
MIY24	MIY25	MIY26	MIY27	MIY28	MIY29	MIY30
22	23	24	25	26	27	28
MA9	MA10	MA11	MA12	MA13	MA14	MA15
PIN25	PIN26	PIN27	PIN28	PIN29	PIN30	APP1
GIN1	GIN2	GIN3	GIN4	GIN5	GIN6	GIN7

UHURU 468/1984-85/6220-21/1976-77

Ashe	Ojobo	Iwa	Araorun	Ile	Itutu	Erinle
1	2	3	4	5	6	7
M21	M22	M23	M24	M25	M26	M27
KHN6	KHN7	KHN8	KHN9	KHN10	KHN11	KHN12
MAG12	MAG13	MAG14	MAG15	MAG16	MAG17	MAG18
8	9	10	11	12	13	14
M28	M29	M30	M31	A1	A2	A3
KHN13	KHN14	KHN15	KHN16	KHN17	KHN18	KHN19
MAG19	MAG20	MAG21	MAG22	MAG23	MAG24	MAG25
15	16	17	18	19	20	21
A4	A5	A6	A7	A8	A9	A10
KHN20	KHN21	KHN222	KHN23	KHN24	KHN25	KHN26
MAG26	MAG27	MAG28	MAG29	MAG30	MIY1	MIY2
22	23	24	25	26	27	28
A11	A12	A13	A14	A15	A16	A17
KHN27	KHN28	KHN29	KHN30	PIN1	PIN2	PIN3
MIY3	MIY4	MIY5	MIY6	MIY7	MIY8	MIY9

MABABA WEUSI 468/1984-85/6220-21/1976-77

Ashe	Ojobo	Iwa	Araorun	Ile	Itutu	Erinle
1	2	3	4	5	6	7
MA16	MA17	MA18	MA19	MA20	MA21	MA22
APP2	APP3	APP4	APP5	APP6	APP7	APP8
GIN8	GIN9	GIN10	GIN11	GIN12	GIN13	GIN14
8	9	10	11	12	13	14
MA23	MA24	MA25	MA26	MA27	MA28	MA29
APP9	APP10	APP11	APP12	APP13	APP14	APP15
GIN15	GIN16	GIN17	GIN18	GIN19	GIN20	GIN21
15	16	17	18	19	20	21
MA30	MA31	J1	J2	J3	J4	J5
APP16	APP17	APP18	APP19	APP20	APP21	APP22
GIN22	GIN23	GIN24	GIN25	GIN26	GIN27	GIN28
22	23	24	25	26	27	28
J6	J7	J8	J9	J10	J11	J12
APP23	APP24	APP25	APP26	APP27	APP28	APP29
GIN29	GIN30	SAN1	SAN2	SAN3	SAN4	SAN5

MWANAMKE WEUSI 468/1984-85/6220-21/1976-77

Ashe	Ojobo	Iwa	Araorun	Ile	Itutu	Erinle
1	2	3	4	5	6	7
J13	J14	J15	J16	J17	J18	J19
APP30	MSR1	MSR2	MSR3	MSR4	MSR5	MSR6
SAN6	SAN7	SAN8	SAN9	SAN10	SAN11	SAN12
8	9	10	11	12	13	14
J20	J21	J22	J23	J24	J25	J26
MSR7	MSR8	MSR9	MSR10	MSR11	MSR12	MSR13
SAN13	SAN14	SAN15	SAN16	SAN17	SAN18	SAN19
15	16	17	18	19	20	21
J27	J28	J29	J30	JU1	JU2	JU3
MSR14	MSR15	MSR16	MSR17	MSR18	MSR19	MSR20
SAN20	SAN21	SAN22	SAN23	SAN24	SAN25	SAN26
22	23	24	25	26	27	28
JU4	JU5	JU6	JU7	JU8	JU9	JU10
MSR21	MRS22	MSR23	MSR24	MSR25	MSR26	MSR27
SAN27	SAN28	SAN29	SAN30	HAM1	HAM2	HAM3

TAIFA WEUSI 468/1984-85/6220-21/1976-77

Ashe	Ojobo	Iwa	Araorun	Ile	Itutu	Erinle
1	2	3	4	5	6	7
AU8	AU9	AU10	AU11	AU12	AU13	AU14
THT20	THT21	THT22	THT23	THT24	THT25	THT26
NAH2	NAH3	NAH4	NAH5	NAH6	NAH7	NAH8
8	9	10	11	12	13	14
AU15	AU16	AU17	AU18	AU19	AU20	AU21
THT27	THT28	THT29	THT30	PAP1	PAP1	PAP2
NAH9	NAH10	NAH11	NAH12	NAH13	NAH14	NAH15
15	16	17	18	19	20	21
AU22	AU23	AU24	AU25	AU26	AU27	AU28
PAP3	PAP4	PAP5	PAP6	PAP7	PAP8	PAP9
NAH16	NAH17	NAH18	NAH19	NAH20	NAH21	NAH22
22	23	24	25	26	27	28
AU29	AU30	AU31	S1	S2	S3	S4
PAP10	PAP11	PAP12	PAP13	PAP14	PAP15	PAP16
NAH23	NAH24	NAH25	NAH26	NAH27	NAH28	NAH29

MTU WEUSI 468/1984-85/6220-21/1976-77

Ashe	Ojobo	Iwa	Araorun	Ile	Itutu	Erinle
1	2	3	4	5	6	7
JU11	JU12	JU13	JU14	JU15	JU16	JU17
MSR28	MSR29	MSR30	*ASR1*	*HRU2*	*STH3*	*AST4*
HAM4	HAM5	HAM6	HAM7	HAM8	HAM9	HAM10
8	9	10	11	12	13	14
JU18	JU19	JU20	JU21	JU22	JU23	JU24
NBH5	*ANU6*	THT1	THT2	THT3	THT4	THT5
HAM11	HAM12	HAM13	HAM14	HAM15	HAM16	HAM17
15	16	17	18	19	20	21
JU25	JU26	JU27	JU28	JU29	JU30	JU31
THT6	THT7	THT8	THT9	THT10	THT11	THT12
HAM18	HAM19	HAM20	HAM21	HAM22	HAM23	HAM24
22	23	24	25	26	27	28
AU1	AU2	AU3	AU4	AU5	AU6	AU7
THT13	THT14	THT15	THT16	THT17	THT18	THT19
HAM25	HAM26	HAM27	HAM28	HAM29	HAM30	NAH1

UWEZO 468/1984-85/6220-21/1976-77

Ashe	Ojobo	Iwa	Araorun	Ile	Itutu	Erinle
1	2	3	4	5	6	7
S5	S6	S7	S8	S9	S10	S11
PAP17	PAP18	PAP19	PAP20	PAP21	PAP22	PAP23
NAH30	*PAG1*	*PAG2*	*PAG3*	*PAG4*	*PAG5*	MAS1
8	9	10	11	12	13	14
S12	S13	S14	S15	S16	S17	S18
PAP24	PAP25	PAP26	PAP27	PAP28	PAP29	PAP30
MAS2	MAS3	MAS4	MAS5	MAS6	MAS7	MAS8
15	16	17	18	19	20	21
S19	S20	S21	S22	S23	S24	S25
HTH1	HTH2	HTH3	HTH4	HTH5	HTH6	HTH7
MAS9	MAS10	MAS11	MAS12	MAS13	MAS14	MAS15
22	23	24	25	26	27	28
S26	S27	S28	S29	S30	O1	O2
HTH8	HTH9	HTH10	HTH11	HTH12	HTH13	HTH14
MAS16	MAS17	MAS18	MAS19	MAS20	MAS21	MAS22

EKUNDU 468/1984-85/6220-21/1976-77						
Ashe	Ojobo	Iwa	Araorun	Ile	Itutu	Erinle
1	2	3	4	5	6	7
O3	O4	O5	O6	O7	O8	O9
HTH15	HTH16	HTH17	HTH18	HTH19	HTH20	HTH21
MAS23	MAS24	MAS25	MAS26	MAS27	MAS28	MAS29
8	9	10	11	12	13	14
O10	O11	O12	O13	O14	O15	O16
HTH22	HTH23	HTH24	HTH25	HTH26	HTH27	HTH28
MAS30	THi1	THi2	THi3	THi4	THi5	THi6
15	16	17	18	19	20	21
O17	O18	O19	O20	O21	O22	O23
HTH29	HTH30	KHK1	KHK2	KHK3	KHK4	KHK5
THi7	THi8	This9	THi10	THi11	THi12	THi13
22	23	24	25	26	27	28
O24	O25	O26	O27	O28	O29	O30
KHK6	KHK7	KHK8	KHK9	KHK10	KHK11	KHK12
Thi14	THi15	THi16	THi17	THi18	THi19	THi20

KIJANI 468/1984-85/6220-21/1976-77						
Ashe	Ojobo	Iwa	Araorun	Ile	Itutu	Erinle
1	2	3	4	5	6	7
N28	N29	N30	D1	D2	D3	D4
TBT11	TBT12	TBT13	TBT14	TBT15	TBT16	TBT17
HID19	HID20	HID21	HID22	HID23	HID24	HID25
8	9	10	11	12	13	14
D5	D6	D7	D8	D9	D10	D11
TBT18	TBT19	TBT20	TBT21	TBT22	TBT23	TBT24
HID26	HID27	HID28	HID29	HID30	TAH1	TAH2
15	16	17	18	19	20	21
D12	D13	D14	D15	D16	D17	D18
TBT25	TBT26	TBT27	TBT28	TBT29	TBT30	MKH1
TAH3	TAH4	TAH5	TAH6	TAH7	TAH8	TAH9
22	23	24	25	26	27	28
D19	D20	D21	D22	D23	D24	D25
MKH2	MKH3	MKH4	MKH5	MKH6	MKH7	MKH8
TAH10	TAH11	TAH12	TAH13	TAH14	TAH15	TAH16

EUSI 468/1984-85/6220-21/1976-77						
Ashe	Ojobo	Iwa	Araorun	Ile	Itutu	Erinle
1	2	3	4	5	6	7
O31	N1	N2	N3	N4	N5	N6
KHK13	KHK14	KHK15	KHK16	KHK17	KHK18	KHK19
THi21	THi22	THi23	THi24	THi25	THi26	THi27
8	9	10	11	12	13	14
N7	N8	N9	N10	N11	N12	N13
KHK20	KHK21	KHK22	KHK23	KHK24	KHK25	KHK26
THi28	THi29	THi30	HID1	HID2	HID3	HID4
15	16	17	18	19	20	21
N14	N15	N16	N17	N18	N19	N20
KHK27	KHK28	KHK29	KHK30	TBT1	TBT2	TBT3
HID5	HID6	HID7	HID8	HID9	HID10	HID11
22	23	24	25	26	27	28
N21	N22	N23	N24	N25	N26	N27
TBT4	TBT5	TBT6	TBT7	TBT8	TBT9	TBT10
HID12	HID13	HID14	HID15	HID16	HID17	HID18

KWANZA 468/1984-85/6220-21/1976-77						
Umoja	Kujichagulia	Ujima	Ujamaa	Nia	Kuumba	Imani
1	2	3	4	5	6	7
D26	D27	D28	D29	D30	D31	JA1
MKH9	MKH10	MKH11	MKH12	MKH13	MKH14	MKH15
TAH17	TAH18	TAH19	TAH20	TAH21	TAH22	TAH23
8	9	10	11	12	13	14
JA2	JA3	JA4	JA5	JA6	JA7	JA8
MKH16	MKH17	MKH18	MKH19	MKH20	MKH21	MKH22
TAH24	TAH25	TAH26	TAH27	TAH28	TAH29	TAH30
15	16	17	18	19	20	21
JA9	JA10	JA11	JA12	JA13	JA14	JA15
MKH23	MKH24	MKH25	KMH26	MKH27	MKH28	MKH29
TIR1	TIR2	TIR3	TIR4	TIR5	TIR6	TIR7
22	23	24	25	26	27	28
JA16	JA17	JA18	JA19	JA20	JA21	JA22
MKH30	AMN1	AMN2	AMN3	AMN4	AMN5	AMN6
TIR8	TIR9	TIR10	TIR11	TIR12	TIR13	TIR14

SHAHIDI 468/1984-85/6220-21/1976-77						
Ashe	Ojobo	Iwa	Araorun	Ile	Itutu	Erinle
1	2	3	4	5	6	7
JA23	JA24	JA25	JA26	JA27	JA28	JA29
AMN7	AMN8	AMN9	AMN10	AMN11	AMN12	AMN13
TIR15	TIR16	TIR17	TIR18	TIR19	TIR20	TIR21
8	9	10	11	12	13	14
JA30	JA31	F1	F2	F3	F4	F5
AMN14	AMN15	AMN16	AMN17	AMN18	AMN19	AMN20
TIR22	TIR23	TIR24	TIR25	TIR26	TIR27	TIR28
15	16	17	18	19	20	21
F6	F7	F8	F9	F10	F 11	F12
AMN21	AMN22	AMN23	AMN24	AMN25	AMN26	AMN28
TIR29	TIR30	YEK1	YEK2	YEK3	YEK4	YEK5
22	23	24	25	26	27	28
F13	F14	F15	F16	F17	F18	F19
AMN29	AMN30	RNT1	RNT2	RNT3	RNT4	RNT5
YEK6	YEK7	YEK8	YEK9	YEK10	YEK11	YEK12
SHAHIDI 19						
F20						
RNT6						
YEK13						

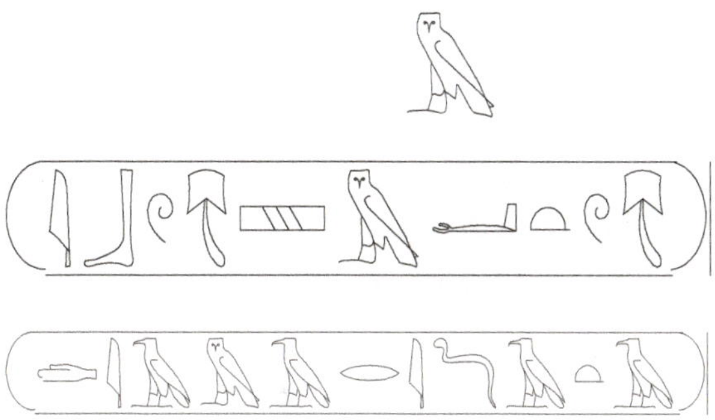

MAAFA 469 B.S.E./ASIENTO 1985-86 C.E./
KAM 6221-22 S.M./EC 1977-78 A.M. (ዓ.ም.)

AFU 469/1985-86/6221-22/1977-78						
Ashe	Ojobo	Iwa	Araorun	Ile	Itutu	Erinle
1	2	3	4	5	6	7
F21	F22	F23	F24	F25	F26	F27
RNT7	RNT8	RNT9	RNT10	RNT11	RNT12	RNT13
YEK14	YEK15	YEK16	YEK17	YEK18	YEK19	YEK20
8	9	10	11	12	13	14
F28	M1	M2	M3	M4	M5	M6
RNT14	RNT15	RNT16	RNT17	RNT18	RNT19	RNT20
YEK21	YEK22	YEK23	YEK24	YEK25	YEK26	YEK27
15	16	17	18	19	20	21
M7	M8	M9	M10	M11	M12	M13
RNT21	RNT22	RNT23	RNT24	RNT25	RNT26	RNT27
YEK28	YEK29	YEK30	MAG1	MAG2	MAG3	MAG4
22	23	24	25	26	27	28
M14	M15	M16	M17	M18	M19	M20
RNT28	RNT29	RNT30	KHN1	KHN2	KHN3	KHN4
MAG5	MAG6	MAG7	MAG8	MAG9	MAG10	MAG11

KAWAIDA 469/1985-86/6221-22/1977-78						
Ashe	Ojobo	Iwa	Araorun	Ile	Itutu	Erinle
1	2	3	4	5	6	7
A18	A19	A20	A21	A22	A23	A24
PIN3	PIN4	PIN5	PIN6	PIN7	PIN8	PIN9
MIY10	MIY11	MIY12	MIY13	MIY14	MIY15	MIY16
8	9	10	11	12	13	14
A25	A26	A27	A28	A29	A30	MA1
PIN10	PIN11	PIN12	PIN13	PIN14	PIN15	PIN16
MIY17	MIY18	MIY19	MIY20	MIY21	MIY22	MIY23
15	16	17	18	19	20	21
MA2	MA3	MA4	MA5	MA6	MA7	MA8
PIN17	PIN18	PIN19	PIN20	PIN21	PIN22	PIN23
MIY24	MIY25	MIY26	MIY27	MIY28	MIY29	MIY30
22	23	24	25	26	27	28
MA9	MA10	MA11	MA12	MA13	MA14	MA15
PIN24	PIN25	PIN26	PIN27	PIN28	PIN29	PIN30
GIN1	GIN2	GIN3	GIN4	GIN5	GIN6	GIN7

UHURU 469/1985-86/6221-22/1977-78						
Ashe	Ojobo	Iwa	Araorun	Ile	Itutu	Erinle
1	2	3	4	5	6	7
M21	M22	M23	M24	M25	M26	M27
KHN5	KHN6	KHN7	KHN8	KHN9	KHN10	KHN11
MAG12	MAG13	MAG14	MAG15	MAG16	MAG17	MAG18
8	9	10	11	12	13	14
M28	M29	M30	M31	A1	A2	A3
KHN12	KHN13	KHN14	KHN15	KHN16	KHN17	KHN18
MAG19	MAG20	MAG21	MAG22	MAG23	MAG24	MAG25
15	16	17	18	19	20	21
A4	A5	A6	A7	A8	A9	A10
KHN19	KHN20	KHN21	KHN22	KHN23	KHN24	KHN25
MAG26	MAG27	MAG28	MAG29	MAG30	MIY1	MIY2
22	23	24	25	26	27	28
A11	A12	A13	A14	A15	A16	A17
KHN26	KHN27	KHN28	KHN29	KHN30	PIN1	PIN2
MIY3	MIY4	MIY5	MIY6	MIY7	MIY8	MIY9

MABABA WEUSI 469/1985-86/6221-22/1977-78						
Ashe	Ojobo	Iwa	Araorun	Ile	Itutu	Erinle
1	2	3	4	5	6	7
MA16	MA17	MA18	MA19	MA20	MA21	MA22
APP1	APP2	APP3	APP4	APP5	APP6	APP7
GIN8	GIN9	GIN10	GIN11	GIN12	GIN13	GIN14
8	9	10	11	12	13	14
MA23	MA24	MA25	MA26	MA27	MA28	MA29
APP8	APP9	APP10	APP11	APP12	APP13	APP14
GIN15	GIN16	GIN17	GIN18	GIN19	GIN20	GIN21
15	16	17	18	19	20	21
MA30	MA31	J1	J2	J3	J4	J5
APP15	APP16	APP17	APP18	APP19	APP20	APP21
GIN22	GIN23	GIN24	GIN25	GIN26	GIN27	GIN28
22	23	24	25	26	27	28
J6	J7	J8	J9	J10	J11	J12
APP22	APP23	APP24	APP25	APP26	APP27	APP28
GIN29	GIN30	SAN1	SAN2	SAN3	SAN4	SAN5

MWANAMKE WEUSI 469/1985-86/6221-22/1977-78						
Ashe	Ojobo	Iwa	Araorun	Ile	Itutu	Erinle
1	2	3	4	5	6	7
J13	J14	J15	J16	J17	J18	J19
APP29	APP30	MSR1	MSR2	MSR3	MSR4	MSR5
SAN6	SAN7	SAN8	SAN9	SAN10	SAN11	SAN12
8	9	10	11	12	13	14
J20	J21	J22	J23	J24	J25	J26
MSR6	MSR7	MSR8	MSR9	MSR10	MSR11	MSR12
SAN13	SAN14	SAN15	SAN16	SAN17	SAN18	SAN19
15	16	17	18	19	20	21
J27	J28	J29	J30	JU1	JU2	JU3
MSR13	MSR14	MSR15	MSR16	MSR17	MSR18	MSR19
SAN20	SAN21	SAN22	SAN23	SAN24	SAN25	SAN26
22	23	24	25	26	27	28
JU4	JU5	JU6	JU7	JU8	JU9	JU10
MSR20	MRS21	MSR22	MSR23	MSR24	MSR25	MSR26
SAN27	SAN28	SAN29	SAN30	HAM1	HAM2	HAM3

TAIFA WEUSI 469/1985-86/6221-22/1977-78						
Ashe	Ojobo	Iwa	Araorun	Ile	Itutu	Erinle
1	2	3	4	5	6	7
AU8	AU9	AU10	AU11	AU12	AU13	AU14
THT20	THT21	THT22	THT23	THT24	THT25	THT26
NAH2	NAH3	NAH4	NAH5	NAH6	NAH7	NAH8
8	9	10	11	12	13	14
AU15	AU16	AU17	AU18	AU19	AU20	AU21
THT27	THT28	THT29	THT30	PAP1	PAP1	PAP2
NAH9	NAH10	NAH11	NAH12	NAH13	NAH14	NAH15
15	16	17	18	19	20	21
AU22	AU23	AU24	AU25	AU26	AU27	AU28
PAP3	PAP4	PAP5	PAP6	PAP7	PAP8	PAP9
NAH16	NAH17	NAH18	NAH19	NAH20	NAH21	NAH22
22	23	24	25	26	27	28
AU29	AU30	AU31	S1	S2	S3	S4
PAP10	PAP11	PAP12	PAP13	PAP14	PAP15	PAP16
NAH23	NAH24	NAH25	NAH26	NAH27	NAH28	NAH29

MTU WEUSI 469/1985-86/6221-22/1977-78						
Ashe	Ojobo	Iwa	Araorun	Ile	Itutu	Erinle
1	2	3	4	5	6	7
JU11	JU12	JU13	JU14	JU15	JU16	JU17
MSR27	MSR28	MSR29	MSR30	*ASR1*	*HRU2*	*STH3*
HAM4	HAM5	HAM6	HAM7	HAM8	HAM9	HAM10
8	9	10	11	12	13	14
JU18	JU19	JU20	JU21	JU22	JU23	JU24
AST4	*NBH5*	THT1	THT2	THT3	THT4	THT5
HAM11	HAM12	HAM13	AM14	HAM15	HAM16	HAM17
15	16	17	18	19	20	21
JU25	JU26	JU27	JU28	JU29	JU30	JU31
THT6	THT7	THT8	THT9	THT10	THT11	THT12
HAM18	HAM19	HAM20	HAM21	HAM22	HAM23	HAM24
22	23	24	25	26	27	28
AU1	AU2	AU3	AU4	AU5	AU6	AU7
THT13	THT14	THT15	THT16	THT17	THT18	THT19
HAM25	HAM26	HAM27	HAM28	HAM29	HAM30	NAH1

UWEZO 469/1985-86/6221-22/1977-78						
Ashe	Ojobo	Iwa	Araorun	Ile	Itutu	Erinle
1	2	3	4	5	6	7
S5	S6	S7	S8	S9	S10	S11
PAP17	PAP18	PAP19	PAP20	PAP21	PAP22	PAP23
NAH30	*PAG1*	*PAG2*	*PAG3*	*PAG4*	*PAG5*	MAS1
8	9	10	11	12	13	14
S12	S13	S14	S15	S16	S17	S18
PAP24	PAP25	PAP26	PAP27	PAP28	PAP29	PAP30
MAS2	MAS3	MAS4	MAS5	MAS6	MAS7	MAS8
15	16	17	18	19	20	21
S19	S20	S21	S22	S23	S24	S25
HTH1	HTH2	HTH3	HTH4	HTH5	HTH6	HTH7
MAS9	MAS10	MAS11	MAS12	MAS13	MAS14	MAS15
22	23	24	25	26	27	28
S26	S27	S28	S29	S30	O1	O2
HTH8	HTH9	HTH10	HTH11	HTH12	HTH13	HTH14
MAS16	MAS17	MAS18	MAS19	MAS20	MAS21	MAS22

EKUNDU 469/1985-86/6221-22/1977-78						
Ashe	Ojobo	Iwa	Araorun	Ile	Itutu	Erinle
1	2	3	4	5	6	7
O3	O4	O5	O6	O7	O8	O9
HTH15	HTH16	HTH17	HTH18	HTH19	HTH20	HTH21
MAS23	MAS24	MAS25	MAS26	MAS27	MAS28	MAS29
8	9	10	11	12	13	14
O10	O11	O12	O13	O14	O15	O16
HTH22	HTH23	HTH24	HTH25	HTH26	HTH27	HTH28
MAS30	THi1	THi2	THi3	THi4	THi5	THi6
15	16	17	18	19	20	21
O17	O18	O19	O20	O21	O22	O23
HTH29	HTH30	KHK1	KHK2	KHK3	KHK4	KHK5
THi7	THi8	THi9	THi10	THi11	THi12	THi13
22	23	24	25	26	27	28
O24	O25	O26	O27	O28	O29	O30
KHK6	KHK7	KHK8	KHK9	KHK10	KHK11	KHK12
THi14	THi15	THi16	THi17	THi18	THi19	THi20

KIJANI 469/1985-86/6221-22/1977-78						
Ashe	Ojobo	Iwa	Araorun	Ile	Itutu	Erinle
1	2	3	4	5	6	7
N28	N29	N30	D1	D2	D3	D4
TBT11	TBT12	TBT13	TBT14	TBT15	TBT16	TBT17
HID19	HID20	HID21	HID22	HID23	HID24	HID25
8	9	10	11	12	13	14
D5	D6	D7	D8	D9	D10	D11
TBT18	TBT19	TBT20	TBT21	TBT22	TBT23	TBT24
HID26	HID27	HID28	HID29	HID30	TAH1	TAH2
15	16	17	18	19	20	21
D12	D13	D14	D15	D16	D17	D18
TBT25	TBT26	TBT27	TBT28	TBT29	TBT30	MKH1
TAH3	TAH4	TAH5	TAH6	TAH7	TAH8	TAH9
22	23	24	25	26	27	28
D19	D20	D21	D22	D23	D24	D25
MKH2	MKH3	MKH4	MKH5	MKH6	MKH7	MKH8
TAH10	TAH11	TAH12	TAH13	TAH14	TAH15	TAH16

EUSI 469/1985-86/6221-22/1977-78						
Ashe	Ojobo	Iwa	Araorun	Ile	Itutu	Erinle
1	2	3	4	5	6	7
O31	N1	N2	N3	N4	N5	N6
KHK13	KHK14	KHK15	KHK16	KHK17	KHK18	KHK19
THi21	THi22	THi23	THi24	THi25	THi26	THi27
8	9	10	11	12	13	14
N7	N8	N9	N10	N11	N12	N13
KHK20	KHK21	KHK22	KHK23	KHK24	KHK25	KHK26
THi28	THi29	THi30	HID1	HID2	HID3	HID4
15	16	17	18	19	20	21
N14	N15	N16	N17	N18	N19	N20
KHK27	KHK28	KHK29	KHK30	TBT1	TBT2	TBT3
HID5	HID6	HID7	HID8	HID9	HID10	HID11
22	23	24	25	26	27	28
N21	N22	N23	N24	N25	N26	N27
TBT4	TBT5	TBT6	TBT7	TBT8	TBT9	TBT10
HID12	HID13	HID14	HID15	HID16	HID17	HID18

KWANZA 469/1985-86/6221-22/1977-78						
Umoja	Kujichagulia	Ujima	Ujamaa	Nia	Kuumba	Imani
1	2	3	4	5	6	7
D26	D27	D28	D29	D30	D31	JA1
MKH9	MKH10	MKH11	MKH12	MKH13	MKH14	MKH15
TAH17	TAH18	TAH19	TAH20	TAH21	TAH22	TAH23
8	9	10	11	12	13	14
JA2	JA3	JA4	JA5	JA6	JA7	JA8
MKH16	MKH17	MKH18	MKH19	MKH20	MKH21	MKH22
TAH24	TAH25	TAH26	TAH27	TAH28	TAH29	TAH30
15	16	17	18	19	20	21
JA9	JA10	JA11	JA12	JA13	JA14	JA15
MKH23	MKH24	MKH25	KMH26	MKH27	MKH28	MKH29
TIR1	TIR2	TIR3	TIR4	TIR5	TIR6	TIR7
22	23	24	25	26	27	28
JA16	JA17	JA18	JA19	JA20	JA21	JA22
MKH30	AMN1	AMN2	AMN3	AMN4	AMN5	AMN6
TIR8	TIR9	TIR10	TIR11	TIR12	TIR13	TIR14

SHAHIDI 469/1985-86/6221-22/1977-78						
Ashe	Ojobo	Iwa	Araorun	Ile	Itutu	Erinle
1	2	3	4	5	6	7
JA23	JA24	JA25	JA26	JA27	JA28	JA29
AMN7	AMN8	AMN9	AMN10	AMN11	AMN12	AMN13
TIR15	TIR16	TIR17	TIR18	TIR19	TIR20	TIR21
8	9	10	11	12	13	14
JA30	JA31	F1	F2	F3	F4	F5
AMN14	AMN15	AMN16	AMN17	AMN18	AMN19	AMN20
TIR22	TIR23	TIR24	TIR25	TIR26	TIR27	TIR28
15	16	17	18	19	20	21
F6	F7	F8	F9	F10	F 11	F12
AMN21	AMN22	AMN23	AMN24	AMN25	AMN26	AMN28
TIR29	TIR30	YEK1	YEK2	YEK3	YEK4	YEK5
22	23	24	25	26	27	28
F13	F14	F15	F16	F17	F18	F19
AMN29	AMN30	RNT1	RNT2	RNT3	RNT4	RNT5
YEK6	YEK7	YEK8	YEK9	YEK10	YEK11	YEK12
SHAHIDI 20						
F20						
RNT6						
YEK13						

MAAFA 470 B.S.E./ASIENTO 1986-87 C.E./
KAM 6222-23 S.M./EC 1978-79 A.M. (ዓ.ም.)

AFU 470/1986-87/6222-23/1978-79						
Ashe	Ojobo	Iwa	Araorun	Ile	Itutu	Erinle
1	2	3	4	5	6	7
F21	F22	F23	F24	F25	F26	F27
RNT7	RNT8	RNT9	RNT10	RNT11	RNT12	RNT13
YEK14	YEK15	YEK16	YEK17	YEK18	YEK19	YEK20
8	9	10	11	12	13	14
F28	M1	M2	M3	M4	M5	M6
RNT14	RNT15	RNT16	RNT17	RNT18	RNT19	RNT20
YEK21	YEK22	YEK23	YEK24	YEK25	YEK26	YEK27
15	16	17	18	19	20	21
M7	M8	M9	M10	M11	M12	M13
RNT21	RNT22	RNT23	RNT24	RNT25	RNT26	RNT27
YEK28	YEK29	YEK30	MAG1	MAG2	MAG3	MAG4
22	23	24	25	26	27	28
M14	M15	M16	M17	M18	M19	M20
RNT28	RNT29	RNT30	KHN1	KHN2	KHN3	KHN4
MAG5	MAG6	MAG7	MAG8	MAG9	MAG10	MAG11

KAWAIDA 470/1986-87/6222-23/1978-79						
Ashe	Ojobo	Iwa	Araorun	Ile	Itutu	Erinle
1	2	3	4	5	6	7
A18	A19	A20	A21	A22	A23	A24
PIN3	PIN4	PIN5	PIN6	PIN7	PIN8	PIN9
MIY10	MIY11	MIY12	MIY13	MIY14	MIY15	MIY16
8	9	10	11	12	13	14
A25	A26	A27	A28	A29	A30	MA1
PIN10	PIN11	PIN12	PIN13	PIN14	PIN15	PIN16
MIY17	MIY18	MIY19	MIY20	MIY21	MIY22	MIY23
15	16	17	18	19	20	21
MA2	MA3	MA4	MA5	MA6	MA7	MA8
PIN17	PIN18	PIN19	PIN20	PIN21	PIN22	PIN23
MIY24	MIY25	MIY26	MIY27	MIY28	MIY29	MIY30
22	23	24	25	26	27	28
MA9	MA10	MA11	MA12	MA13	MA14	MA15
PIN24	PIN25	PIN26	PIN27	PIN28	PIN29	PIN30
GIN1	GIN2	GIN3	GIN4	GIN5	GIN6	GIN7

UHURU 470/1986-87/6222-23/1978-79						
Ashe	Ojobo	Iwa	Araorun	Ile	Itutu	Erinle
1	2	3	4	5	6	7
M21	M22	M23	M24	M25	M26	M27
KHN5	KHN6	KHN7	KHN8	KHN9	KHN10	KHN11
MAG12	MAG13	MAG14	MAG15	MAG16	MAG17	MAG18
8	9	10	11	12	13	14
M28	M29	M30	M31	A1	A2	A3
KHN12	KHN13	KHN14	KHN15	KHN16	KHN17	KHN18
MAG19	MAG20	MAG21	MAG22	MAG23	MAG24	MAG25
15	16	17	18	19	20	21
A4	A5	A6	A7	A8	A9	A10
KHN19	KHN20	KHN21	KHN22	KHN23	KHN24	KHN25
MAG26	MAG27	MAG28	MAG29	MAG30	MIY1	MIY2
22	23	24	25	26	27	28
A11	A12	A13	A14	A15	A16	A17
KHN26	KHN27	KHN28	KHN29	KHN30	PIN1	PIN2
MIY3	MIY4	MIY5	MIY6	MIY7	MIY8	MIY9

MABABA WEUSI 470/1986-87/6222-23/1978-79						
Ashe	Ojobo	Iwa	Araorun	Ile	Itutu	Erinle
1	2	3	4	5	6	7
MA16	MA17	MA18	MA19	MA20	MA21	MA22
APP1	APP2	APP3	APP4	APP5	APP6	APP7
GIN8	GIN9	GIN10	GIN11	GIN12	GIN13	GIN14
8	9	10	11	12	13	14
MA23	MA24	MA25	MA26	MA27	MA28	MA29
APP8	APP9	APP10	APP11	APP12	APP13	APP14
GIN15	GIN16	GIN17	GIN18	GIN19	GIN20	GIN21
15	16	17	18	19	20	21
MA30	MA31	J1	J2	J3	J4	J5
APP15	APP16	APP17	APP18	APP19	APP20	APP21
GIN22	GIN23	GIN24	GIN25	GIN26	GIN27	GIN28
22	23	24	25	26	27	28
J6	J7	J8	J9	J10	J11	J12
APP22	APP23	APP24	APP25	APP26	APP27	APP28
GIN29	GIN30	SAN1	SAN2	SAN3	SAN4	SAN5

MWANAMKE WEUSI 470/1986-87/6222-23/1978-79						
Ashe	Ojobo	Iwa	Araorun	Ile	Itutu	Erinle
1	2	3	4	5	6	7
J13	J14	J15	J16	J17	J18	J19
APP29	APP30	MSR1	MSR2	MSR3	MSR4	MSR5
SAN6	SAN7	SAN8	SAN9	SAN10	SAN11	SAN12
8	9	10	11	12	13	14
J20	J21	J22	J23	J24	J25	J26
MSR6	MSR7	MSR8	MSR9	MSR10	MSR11	MSR12
SAN13	SAN14	SAN15	SAN16	SAN17	SAN18	SAN19
15	16	17	18	19	20	21
J27	J28	J29	J30	JU1	JU2	JU3
MSR13	MSR14	MSR15	MSR16	MSR17	MSR18	MSR19
SAN20	SAN21	SAN22	SAN23	SAN24	SAN25	SAN26
22	23	24	25	26	27	28
JU4	JU5	JU6	JU7	JU8	JU9	JU10
MSR20	MRS21	MSR22	MSR23	MSR24	MSR25	MSR26
SAN27	SAN28	SAN29	SAN30	HAM1	HAM2	HAM3

TAIFA WEUSI 470/1986-87/6222-23/1978-79						
Ashe	Ojobo	Iwa	Araorun	Ile	Itutu	Erinle
1	2	3	4	5	6	7
AU8	AU9	AU10	AU11	AU12	AU13	AU14
THT20	THT21	THT22	THT23	THT24	THT25	THT26
NAH2	NAH3	NAH4	NAH5	NAH6	NAH7	NAH8
8	9	10	11	12	13	14
AU15	AU16	AU17	AU18	AU19	AU20	AU21
THT27	THT28	THT29	THT30	PAP1	PAP1	PAP2
NAH9	NAH10	NAH11	NAH12	NAH13	NAH14	NAH15
15	16	17	18	19	20	21
AU22	AU23	AU24	AU25	AU26	AU27	AU28
PAP3	PAP4	PAP5	PAP6	PAP7	PAP8	PAP9
NAH16	NAH17	NAH18	NAH19	NAH20	NAH21	NAH22
22	23	24	25	26	27	28
AU29	AU30	AU31	S1	S2	S3	S4
PAP10	PAP11	PAP12	PAP13	PAP14	PAP15	PAP16
NAH23	NAH24	NAH25	NAH26	NAH27	NAH28	NAH29

MTU WEUSI 470/1986-87/6222-23/1978-79						
Ashe	Ojobo	Iwa	Araorun	Ile	Itutu	Erinle
1	2	3	4	5	6	7
JU11	JU12	JU13	JU14	JU15	JU16	JU17
MSR27	MSR28	MSR29	MSR30	*ASR1*	*HRU2*	*STH3*
HAM4	HAM5	HAM6	HAM7	HAM8	HAM9	HAM10
8	9	10	11	12	13	14
JU18	JU19	JU20	JU21	JU22	JU23	JU24
AST4	*NBH5*	THT1	THT2	THT3	THT4	THT5
HAM11	HAM12	HAM13	AM14	HAM15	HAM16	HAM17
15	16	17	18	19	20	21
JU25	JU26	JU27	JU28	JU29	JU30	JU31
THT6	THT7	THT8	THT9	THT10	THT11	THT12
HAM18	HAM19	HAM20	HAM21	HAM22	HAM23	HAM24
22	23	24	25	26	27	28
AU1	AU2	AU3	AU4	AU5	AU6	AU7
THT13	THT14	THT15	THT16	THT17	THT18	THT19
HAM25	HAM26	HAM27	HAM28	HAM29	HAM30	NAH1

UWEZO 470/1986-87/6222-23/1978-79						
Ashe	Ojobo	Iwa	Araorun	Ile	Itutu	Erinle
1	2	3	4	5	6	7
S5	S6	S7	S8	S9	S10	S11
PAP17	PAP18	PAP19	PAP20	PAP21	PAP22	PAP23
NAH30	*PAG1*	*PAG2*	*PAG3*	*PAG4*	*PAG5*	MAS1
8	9	10	11	12	13	14
S12	S13	S14	S15	S16	S17	S18
PAP24	PAP25	PAP26	PAP27	PAP28	PAP29	PAP30
MAS2	MAS3	MAS4	MAS5	MAS6	MAS7	MAS8
15	16	17	18	19	20	21
S19	S20	S21	S22	S23	S24	S25
HTH1	HTH2	HTH3	HTH4	HTH5	HTH6	HTH7
MAS9	MAS10	MAS11	MAS12	MAS13	MAS14	MAS15
22	23	24	25	26	27	28
S26	S27	S28	S29	S30	O1	O2
HTH8	HTH9	HTH10	HTH11	HTH12	HTH13	HTH14
MAS16	MAS17	MAS18	MAS19	MAS20	MAS21	MAS22

EKUNDU 470/1986-87/6222-23/1978-79

Ashe	Ojobo	Iwa	Araorun	Ile	Itutu	Erinle
1	2	3	4	5	6	7
O3	O4	O5	O6	O7	O8	O9
HTH15	HTH16	HTH17	HTH18	HTH19	HTH20	HTH21
MAS23	MAS24	MAS25	MAS26	MAS27	MAS28	MAS29
8	9	10	11	12	13	14
O10	O11	O12	O13	O14	O15	O16
HTH22	HTH23	HTH24	HTH25	HTH26	HTH27	HTH28
MAS30	THi1	THi2	THi3	THi4	THi5	THi6
15	16	17	18	19	20	21
O17	O18	O19	O20	O21	O22	O23
HTH29	HTH30	KHK1	KHK2	KHK3	KHK4	KHK5
THi7	THi8	THi9	THi10	THi11	THi12	THi13
22	23	24	25	26	27	28
O24	O25	O26	O27	O28	O29	O30
KHK6	KHK7	KHK8	KHK9	KHK10	KHK11	KHK12
THi14	THi15	THi16	THi17	THi18	THi19	THi20

KIJANI 470/1986-87/6222-23/1978-79

Ashe	Ojobo	Iwa	Araorun	Ile	Itutu	Erinle
1	2	3	4	5	6	7
N28	N29	N30	D1	D2	D3	D4
TBT11	TBT12	TBT13	TBT14	TBT15	TBT16	TBT17
HID19	HID20	HID21	HID22	HID23	HID24	HID25
8	9	10	11	12	13	14
D5	D6	D7	D8	D9	D10	D11
TBT18	TBT19	TBT20	TBT21	TBT22	TBT23	TBT24
HID26	HID27	HID28	HID29	HID30	TAH1	TAH2
15	16	17	18	19	20	21
D12	D13	D14	D15	D16	D17	D18
TBT25	TBT26	TBT27	TBT28	TBT29	TBT30	MKH1
TAH3	TAH4	TAH5	TAH6	TAH7	TAH8	TAH9
22	23	24	25	26	27	28
D19	D20	D21	D22	D23	D24	D25
MKH2	MKH3	MKH4	MKH5	MKH6	MKH7	MKH8
TAH10	TAH11	TAH12	TAH13	TAH14	TAH15	TAH16

EUSI 470/1986-87/6222-23/1978-79

Ashe	Ojobo	Iwa	Araorun	Ile	Itutu	Erinle
1	2	3	4	5	6	7
O31	N1	N2	N3	N4	N5	N6
KHK13	KHK14	KHK15	KHK16	KHK17	KHK18	KHK19
THi21	THi22	THi23	THi24	THi25	THi26	THi27
8	9	10	11	12	13	14
N7	N8	N9	N10	N11	N12	N13
KHK20	KHK21	KHK22	KHK23	KHK24	KHK25	KHK26
THi28	THi29	THi30	HID1	HID2	HID3	HID4
15	16	17	18	19	20	21
N14	N15	N16	N17	N18	N19	N20
KHK27	KHK28	KHK29	KHK30	TBT1	TBT2	TBT3
HID5	HID6	HID7	HID8	HID9	HID10	HID11
22	23	24	25	26	27	28
N21	N22	N23	N24	N25	N26	N27
TBT4	TBT5	TBT6	TBT7	TBT8	TBT9	TBT10
HID12	HID13	HID14	HID15	HID16	HID17	HID18

KWANZA 470/1986-87/6222-23/1978-79

Umoja	Kujichagulia	Ujima	Ujamaa	Nia	Kuumba	Imani
1	2	3	4	5	6	7
D26	D27	D28	D29	D30	D31	JA1
MKH9	MKH10	MKH11	MKH12	MKH13	MKH14	MKH15
TAH17	TAH18	TAH19	TAH20	TAH21	TAH22	TAH23
8	9	10	11	12	13	14
JA2	JA3	JA4	JA5	JA6	JA7	JA8
MKH16	MKH17	MKH18	MKH19	MKH20	MKH21	MKH22
TAH24	TAH25	TAH26	TAH27	TAH28	TAH29	TAH30
15	16	17	18	19	20	21
JA9	JA10	JA11	JA12	JA13	JA14	JA15
MKH23	MKH24	MKH25	KMH26	MKH27	MKH28	MKH29
TIR1	TIR2	TIR3	TIR4	TIR5	TIR6	TIR7
22	23	24	25	26	27	28
JA16	JA17	JA18	JA19	JA20	JA21	JA22
MKH30	AMN1	AMN2	AMN3	AMN4	AMN5	AMN6
TIR8	TIR9	TIR10	TIR11	TIR12	TIR13	TIR14

SHAHIDI 470/1986-87/6222-23/1978-79						
Ashe	Ojobo	Iwa	Araorun	Ile	Itutu	Erinle
1	2	3	4	5	6	7
JA23	JA24	JA25	JA26	JA27	JA28	JA29
AMN7	AMN8	AMN9	AMN10	AMN11	AMN12	AMN13
TIR15	TIR16	TIR17	TIR18	TIR19	TIR20	TIR21
8	9	10	11	12	13	14
JA30	JA31	F1	F2	F3	F4	F5
AMN14	AMN15	AMN16	AMN17	AMN18	AMN19	AMN20
TIR22	TIR23	TIR24	TIR25	TIR26	TIR27	TIR28
15	16	17	18	19	20	21
F6	F7	F8	F9	F10	F 11	F12
AMN21	AMN22	AMN23	AMN24	AMN25	AMN26	AMN28
TIR29	TIR30	YEK1	YEK2	YEK3	YEK4	YEK5
22	23	24	25	26	27	28
F13	F14	F15	F16	F17	F18	F19
AMN29	AMN30	RNT1	RNT2	RNT3	RNT4	RNT5
YEK6	YEK7	YEK8	YEK9	YEK10	YEK11	YEK12
SHAHIDI 21						
F20						
RNT6						
YEK13						

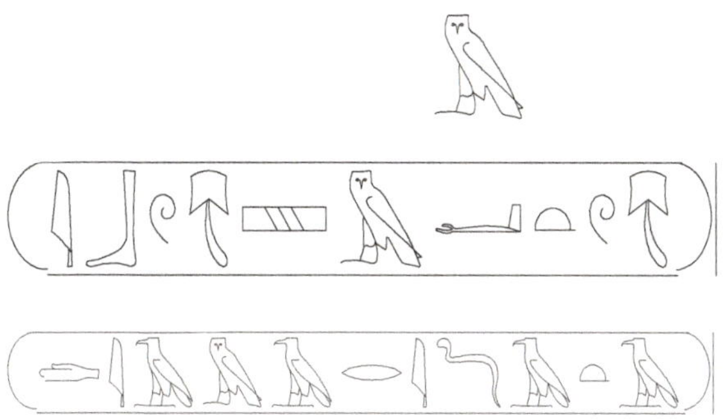

**MAAFA 471 B.S.E./ASIENTO 1987-88 C.E./
KAM 6223-24 S.M./EC 1979-80 A.M. (ዓ.ም.)**

AFU 471/1987-88/6223-24/1979-80						
Ashe	Ojobo	Iwa	Araorun	Ile	Itutu	Erinle
1	2	3	4	5	6	7
F21	F22	F23	F24	F25	F26	F27
RNT7	RNT8	RNT9	RNT10	RNT11	RNT12	RNT13
YEK14	YEK15	YEK16	YEK17	YEK18	YEK19	YEK20
8	9	10	11	12	13	14
F28	M1	M2	M3	M4	M5	M6
RNT14	RNT15	RNT16	RNT17	RNT18	RNT19	RNT20
YEK21	YEK22	YEK23	YEK24	YEK25	YEK26	YEK27
15	16	17	18	19	20	21
M7	M8	M9	M10	M11	M12	M13
RNT21	RNT22	RNT23	RNT24	RNT25	RNT26	RNT27
YEK28	YEK 29	YEK30	MAG1	MAG2	MAG3	MAG4
22	23	24	25	26	27	28
M14	M15	M16	M17	M18	M19	M20
RNT28	RNT29	RNT30	KHN1	KHN2	KHN3	KHN4
MAG5	MAG6	MAG7	MAG8	MAG9	MAG10	MAG11

KAWAIDA 471/1987-88/6223-24/1979-80						
Ashe	Ojobo	Iwa	Araorun	Ile	Itutu	Erinle
1	2	3	4	5	6	7
A18	A19	A20	A21	A22	A23	A24
PIN3	PIN4	PIN5	PIN6	PIN7	PIN8	PIN9
MIY10	MIY11	MIY12	MIY13	MIY14	MIY15	MIY16
8	9	10	11	12	13	14
A25	A26	A27	A28	A29	A30	MA1
PIN10	PIN11	PIN12	PIN13	PIN14	PIN15	PIN16
MIY17	MIY18	MIY19	MIY20	MIY21	MIY22	MIY23
15	16	17	18	19	20	21
MA2	MA3	MA4	MA5	MA6	MA7	MA8
PIN17	PIN18	PIN19	PIN20	PIN21	PIN22	PIN23
MIY24	MIY25	MIY26	MIY27	MIY28	MIY29	MIY30
22	23	24	25	26	27	28
MA9	MA10	MA11	MA12	MA13	MA14	MA15
PIN24	PIN25	PIN26	PIN27	PIN28	PIN29	PIN30
GIN1	GIN2	GIN3	GIN4	GIN5	GIN6	GIN7

UHURU 471/1987-88/6223-24/1979-80						
Ashe	Ojobo	Iwa	Araorun	Ile	Itutu	Erinle
1	2	3	4	5	6	7
M21	M22	M23	M24	M25	M26	M27
KHN5	KHN6	KHN7	KHN8	KHN9	KHN10	KHN11
MAG12	MAG13	MAG14	MAG15	MAG16	MAG17	MAG18
8	9	10	11	12	13	14
M28	M29	M30	M31	A1	A2	A3
KHN12	KHN13	KHN14	KHN15	KHN16	KHN17	KHN18
MAG19	MAG20	MAG21	MAG22	MAG23	MAG24	MAG25
15	16	17	18	19	20	21
A4	A5	A6	A7	A8	A9	A10
KHN19	KHN20	KHN21	KHN22	KHN23	KHN24	KHN25
MAG26	MAG27	MAG28	MAG29	MAG30	MIY1	MIY2
22	23	24	25	26	27	28
A11	A12	A13	A14	A15	A16	A17
KHN26	KHN27	KHN28	KHN29	KHN30	PIN1	PIN2
MIY3	MIY4	MIY5	MIY6	MIY7	MIY8	MIY9

MABABA WEUSI 471/1987-88/6223-24/1979-80						
Ashe	Ojobo	Iwa	Araorun	Ile	Itutu	Erinle
1	2	3	4	5	6	7
MA16	MA17	MA18	MA19	MA20	MA21	MA22
APP1	APP2	APP3	APP4	APP5	APP6	APP7
GIN8	GIN9	GIN10	GIN11	GIN12	GIN13	GIN14
8	9	10	11	12	13	14
MA23	MA24	MA25	MA26	MA27	MA28	MA29
APP8	APP9	APP10	APP11	APP12	APP13	APP14
GIN15	GIN16	GIN17	GIN18	GIN19	GIN20	GIN21
15	16	17	18	19	20	21
MA30	MA31	J1	J2	J3	J4	J5
APP15	APP16	APP17	APP18	APP19	APP20	APP21
GIN22	GIN23	GIN24	GIN25	GIN26	GIN27	GIN28
22	23	24	25	26	27	28
J6	J7	J8	J9	J10	J11	J12
APP22	APP23	APP24	APP25	APP26	APP27	APP28
GIN29	GIN30	SAN1	SAN2	SAN3	SAN4	SAN5

MWANAMKE WEUSI 471/1987-88/6223-24/1979-80						
Ashe	Ojobo	Iwa	Araorun	Ile	Itutu	Erinle
1	2	3	4	5	6	7
J13	J14	J15	J16	J17	J18	J19
APP29	APP30	MSR1	MSR2	MSR3	MSR4	MSR5
SAN6	SAN7	SAN8	SAN9	SAN10	SAN11	SAN12
8	9	10	11	12	13	14
J20	J21	J22	J23	J24	J25	J26
MSR6	MSR7	MSR8	MSR9	MSR10	MSR11	MSR12
SAN13	SAN14	SAN15	SAN16	SAN17	SAN18	SAN19
15	16	17	18	19	20	21
J27	J28	J29	J30	JU1	JU2	JU3
MSR13	MSR14	MSR15	MSR16	MSR17	MSR18	MSR19
SAN20	SAN21	SAN22	SAN23	SAN24	SAN25	SAN26
22	23	24	25	26	27	28
JU4	JU5	JU6	JU7	JU8	JU9	JU10
MSR20	MRS21	MSR22	MSR23	MSR24	MSR25	MSR26
SAN27	SAN28	SAN29	SAN30	HAM1	HAM2	HAM3

TAIFA WEUSI 471/1987-88/6223-24/1979-80						
Ashe	Ojobo	Iwa	Araorun	Ile	Itutu	Erinle
1	2	3	4	5	6	7
AU8	AU9	AU10	AU11	AU12	AU13	AU14
THT20	THT21	THT22	THT23	THT24	THT25	THT26
NAH2	NAH3	NAH4	NAH5	NAH6	NAH7	NAH8
8	9	10	11	12	13	14
AU15	AU16	AU17	AU18	AU19	AU20	AU21
THT27	THT28	THT29	THT30	PAP1	PAP1	PAP2
NAH9	NAH10	NAH11	NAH12	NAH13	NAH14	NAH15
15	16	17	18	19	20	21
AU22	AU23	AU24	AU25	AU26	AU27	AU28
PAP3	PAP4	PAP5	PAP6	PAP7	PAP8	PAP9
NAH16	NAH17	NAH18	NAH19	NAH20	NAH21	NAH22
22	23	24	25	26	27	28
AU29	AU30	AU31	S1	S2	S3	S4
PAP10	PAP11	PAP12	PAP13	PAP14	PAP15	PAP16
NAH23	NAH24	NAH25	NAH26	NAH27	NAH28	NAH29

MTU WEUSI 471/1987-88/6223-24/1979-80						
Ashe	Ojobo	Iwa	Araorun	Ile	Itutu	Erinle
1	2	3	4	5	6	7
JU11	JU12	JU13	JU14	JU15	JU16	JU17
MSR27	MSR28	MSR29	MSR30	*ASR1*	*HRU2*	*STH3*
HAM4	HAM5	HAM6	HAM7	HAM8	HAM9	HAM10
8	9	10	11	12	13	14
JU18	JU19	JU20	JU21	JU22	JU23	JU24
AST4	*NBH5*	THT1	THT2	THT3	THT4	THT5
HAM11	HAM12	HAM13	HAM14	HAM15	HAM16	HAM17
15	16	17	18	19	20	21
JU25	JU26	JU27	JU28	JU29	JU30	JU31
THT6	THT7	THT8	THT9	THT10	THT11	THT12
HAM18	HAM19	HAM20	HAM21	HAM22	HAM23	HAM24
22	23	24	25	26	27	28
AU1	AU2	AU3	AU4	AU5	AU6	AU7
THT13	THT14	THT15	THT16	THT17	THT18	THT19
HAM25	HAM26	HAM27	HAM28	HAM29	HAM30	NAH1

UWEZO 471/1987-88/6223-24/1979-80						
Ashe	Ojobo	Iwa	Araorun	Ile	Itutu	Erinle
1	2	3	4	5	6	7
S5	S6	S7	S8	S9	S10	S11
PAP17	PAP18	PAP19	PAP20	PAP21	PAP22	PAP23
NAH30	*QUA1*	*QUA2*	*QUA3*	*QUA4*	*QUA5*	*QUA6*
8	9	10	11	12	13	14
S12	S13	S14	S15	S16	S17	S18
PAP24	PAP25	PAP26	PAP27	PAP28	PAP29	PAP30
MAS1	MAS2	MAS3	MAS4	MAS5	MAS6	MAS7
15	16	17	18	19	20	21
S19	S20	S21	S22	S23	S24	S25
HTH1	HTH2	HTH3	HTH4	HTH5	HTH6	HTH7
MAS8	MAS9	MAS10	MAS11	MAS12	MAS13	MAS14
22	23	24	25	26	27	28
S26	S27	S28	S29	S30	O1	O2
HTH8	HTH9	HTH10	HTH11	HTH12	HTH13	HTH14
MAS15	MAS16	MAS17	MAS18	MAS19	MAS20	MAS21

EKUNDU 471/1987-88/6223-24/1979-80						
Ashe	Ojobo	Iwa	Araorun	Ile	Itutu	Erinle
1	2	3	4	5	6	7
O3	O4	O5	O6	O7	O8	O9
HTH15	HTH16	HTH17	HTH18	HTH19	HTH20	HTH21
MAS22	MAS23	MAS 24	MAS25	MAS26	MAS27	MAS28
8	9	10	11	12	13	14
O10	O11	O12	O13	O14	O15	O16
HTH22	HTH23	HTH24	HTH25	HTH26	HTH27	HTH28
MAS29	MAS30	THi1	THi2	THi3	THi4	THi5
15	16	17	18	19	20	21
O17	O18	O19	O20	O21	O22	O23
HTH29	HTH30	KHK1	KHK2	KHK3	KHK4	KHK5
THi6	THi7	THi8	THi9	THi10	THi11	THi12
22	23	24	25	26	27	28
O24	O25	O26	O27	O28	O29	O30
KHK6	KHK7	KHK8	KHK9	KHK10	KHK11	KHK12
THi13	THi14	THi15	THi16	THi17	THi18	THi19

KIJANI 471/1987-88/6223-24/1979-80						
Ashe	Ojobo	Iwa	Araorun	Ile	Itutu	Erinle
1	2	3	4	5	6	7
N28	N29	N30	D1	D2	D3	D4
TBT11	TBT12	TBT13	TBT14	TBT15	TBT16	TBT17
HID18	HID19	HID20	HID21	HID22	HID23	HID24
8	9	10	11	12	13	14
D5	D6	D7	D8	D9	D10	D11
TBT18	TBT19	TBT20	TBT21	TBT22	TBT23	TBT24
HID25	HID26	HID27	HID28	HID29	HID30	TAH1
15	16	17	18	19	20	21
D12	D13	D14	D15	D16	D17	D18
TBT25	TBT26	TBT27	TBT28	TBT29	TBT30	MKH1
TAH2	TAH3	TAH4	TAH5	TAH6	TAH7	TAH8
22	23	24	25	26	27	28
D19	D20	D21	D22	D23	D24	D25
MKH2	MKH3	MKH4	MKH5	MKH6	MKH7	MKH8
TAH9	TAH10	TAH11	TAH12	TAH13	TAH14	TAH15

EUSI 471/1987-88/6223-24/1979-80						
Ashe	Ojobo	Iwa	Araorun	Ile	Itutu	Erinle
1	2	3	4	5	6	7
O31	N1	N2	N3	N4	N5	N6
KHK13	KHK14	KHK15	KHK16	KHK17	KHK18	KHK19
THi20	THi21	THi22	THi23	THi24	THi25	THi26
8	9	10	11	12	13	14
N7	N8	N9	N10	N11	N12	N13
KHK20	KHK21	KHK22	KHK23	KHK24	KHK25	KHK26
THi27	THi28	THi29	THi30	HID1	HID2	HID3
15	16	17	18	19	20	21
N14	N15	N16	N17	N18	N19	N20
KHK27	KHK28	KHK29	KHK30	TBT1	TBT2	TBT3
HID4	HID5	HID6	HID7	HID8	HID9	HID10
22	23	24	25	26	27	28
N21	N22	N23	N24	N25	N26	N27
TBT4	TBT5	TBT6	TBT7	TBT8	TBT9	TBT10
HID11	HID12	HID13	HID14	HID15	HID16	HID17

KWANZA 471/1987-88/6223-24/1979-80						
Umoja	Kujichagulia	Ujima	Ujamaa	Nia	Kuumba	Imani
1	2	3	4	5	6	7
D26	D27	D28	D29	D30	D31	JA1
MKH9	MKH10	MKH11	MKH12	MKH13	MKH14	MKH15
TAH16	TAH17	TAH18	TAH19	TAH20	TAH21	TAH22
8	9	10	11	12	13	14
JA2	JA3	JA4	JA5	JA6	JA7	JA8
MKH16	MKH17	MKH18	MKH19	MKH20	MKH21	MKH22
TAH23	TAH24	TAH25	TAH26	TAH27	TAH28	TAH29
15	16	17	18	19	20	21
JA9	JA10	JA11	JA12	JA13	JA14	JA15
MKH23	MKH24	MKH25	KMH26	MKH27	MKH28	MKH29
TAH30	TIR1	TIR2	TIR3	TIR4	TIR5	TIR6
22	23	24	25	26	27	28
JA16	JA17	JA18	JA19	JA20	JA21	JA22
MKH30	AMN1	AMN2	AMN3	AMN4	AMN5	AMN6
TIR7	TIR8	TIR9	TIR10	TIR11	TIR12	TIR13

SHAHIDI 471/1987-88/6223-24/1979-80						
Ashe	**Ojobo**	**Iwa**	**Araorun**	**Ile**	**Itutu**	**Erinle**
1	2	3	4	5	6	7
JA23	JA24	JA25	JA26	JA27	JA28	JA29
AMN7	AMN8	AMN9	AMN10	AMN11	AMN12	AMN13
TIR14	TIR15	TIR16	TIR17	TIR18	TIR19	TIR20
8	9	10	11	12	13	14
JA30	JA31	F1	F2	F3	F4	F5
AMN14	AMN15	AMN16	AMN17	AMN18	AMN19	AMN20
TIR21	TIR22	TIR23	TIR24	TIR25	TIR26	TIR27
15	16	17	18	19	20	21
F6	F7	F8	F9	F10	F 11	F12
AMN21	AMN22	AMN23	AMN24	AMN25	AMN26	AMN28
TIR28	TIR29	TIR30	YEK1	YEK2	YEK3	YEK4
22	23	24	25	26	27	28
F13	F14	F15	F16	F17	F18	F19
AMN29	AMN30	RNT1	RNT2	RNT3	RNT4	RNT5
YEK5	YEK6	YEK7	YEK8	YEK9	YEK10	YEK11
SHAHIDI 22						
F20						
RNT6						
YEK12						

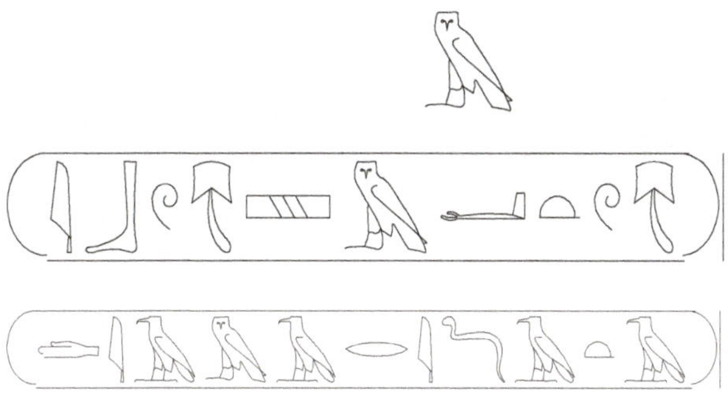

MAAFA 472 B.S.E./ASIENTO 1988-89 C.E./
KAM 6224-25 S.M./EC 1980-81 A.M. (ዓ.ም.)

AFU 472/1988-89/6224-25/1980-81						
Ashe	Ojobo	Iwa	Araorun	Ile	Itutu	Erinle
1	2	3	4	5	6	7
F21	F22	F23	F24	F25	F26	F27
RNT7	RNT8	RNT9	RNT10	RNT11	RNT12	RNT13
YEK13	YEK14	YEK15	YEK16	YEK17	YEK18	YEK19
8	9	10	11	12	13	14
F28	F29	M1	M2	M3	M4	M5
RNT14	RNT15	RNT16	RNT17	RNT18	RNT19	RNT20
YEK 20	YEK21	YEK22	YEK23	YEK24	YEK25	YEK26
15	16	17	18	19	20	21
M6	M7	M8	M9	M10	M11	M12
RNT21	RNT22	RNT23	RNT24	RNT25	RNT26	RNT27
YEK 27	YEK28	YEK29	YEK30	MAG1	MAG2	MAG3
22	23	24	25	26	27	28
M13	M14	M15	M16	M17	M18	M19
RNT28	RNT29	RNT30	KHN1	KHN2	KHN3	KHN4
MAG4	MAG5	MAG6	MAG7	MAG8	MAG9	MAG10
SHARIFU 115						
M20						
KHN5						
MAG11						

KAWAIDA 472/1988-89/6224-25/1980-81						
Ashe	Ojobo	Iwa	Araorun	Ile	Itutu	Erinle
1	2	3	4	5	6	7
A18	A19	A20	A21	A22	A23	A24
PIN4	PIN5	PIN6	PIN7	PIN8	PIN9	PIN10
MIY10	MIY11	MIY12	MIY13	MIY14	MIY15	MIY16
8	9	10	11	12	13	14
A25	A26	A27	A28	A29	A30	MA1
PIN11	PIN12	PIN13	PIN14	PIN15	PIN16	PIN17
MIY17	MIY18	MIY19	MIY20	MIY21	MIY22	MIY23
15	16	17	18	19	20	21
MA2	MA3	MA4	MA5	MA6	MA7	MA8
PIN18	PIN19	PIN20	PIN21	PIN22	PIN23	PIN24
MIY24	MIY25	MIY26	MIY27	MIY28	MIY29	MIY30
22	23	24	25	26	27	28
MA9	MA10	MA11	MA12	MA13	MA14	MA15
PIN25	PIN26	PIN27	PIN28	PIN29	PIN30	APP1
GIN1	GIN2	GIN3	GIN4	GIN5	GIN6	GIN7

UHURU 472/1988-89/6224-25/1980-81						
Ashe	Ojobo	Iwa	Araorun	Ile	Itutu	Erinle
1	2	3	4	5	6	7
M21	M22	M23	M24	M25	M26	M27
KHN6	KHN7	KHN8	KHN9	KHN10	KHN11	KHN12
MAG12	MAG13	MAG14	MAG15	MAG16	MAG17	MAG18
8	9	10	11	12	13	14
M28	M29	M30	M31	A1	A2	A3
KHN13	KHN14	KHN15	KHN16	KHN17	KHN18	KHN19
MAG19	MAG20	MAG21	MAG22	MAG23	MAG24	MAG25
15	16	17	18	19	20	21
A4	A5	A6	A7	A8	A9	A10
KHN20	KHN21	KHN222	KHN23	KHN24	KHN25	KHN26
MAG26	MAG27	MAG28	MAG29	MAG30	MIY1	MIY2
22	23	24	25	26	27	28
A11	A12	A13	A14	A15	A16	A17
KHN27	KHN28	KHN29	KHN30	PIN1	PIN2	PIN3
MIY3	MIY4	MIY5	MIY6	MIY7	MIY8	MIY9

MABABA WEUSI 472/1988-89/6224-25/1980-81						
Ashe	Ojobo	Iwa	Araorun	Ile	Itutu	Erinle
1	2	3	4	5	6	7
MA16	MA17	MA18	MA19	MA20	MA21	MA22
APP2	APP3	APP4	APP5	APP6	APP7	APP8
GIN8	GIN9	GIN10	GIN11	GIN12	GIN13	GIN14
8	9	10	11	12	13	14
MA23	MA24	MA25	MA26	MA27	MA28	MA29
APP9	APP10	APP11	APP12	APP13	APP14	APP15
GIN15	GIN16	GIN17	GIN18	GIN19	GIN20	GIN21
15	16	17	18	19	20	21
MA30	MA31	J1	J2	J3	J4	J5
APP16	APP17	APP18	APP19	APP20	APP21	APP22
GIN22	GIN23	GIN24	GIN25	GIN26	GIN27	GIN28
22	23	24	25	26	27	28
J6	J7	J8	J9	J10	J11	J12
APP23	APP24	APP25	APP26	APP27	APP28	APP29
GIN29	GIN30	SAN1	SAN2	SAN3	SAN4	SAN5

MWANAMKE WEUSI 472/1988-89/6224-25/1980-81						
Ashe	Ojobo	Iwa	Araorun	Ile	Itutu	Erinle
1	2	3	4	5	6	7
J13	J14	J15	J16	J17	J18	J19
APP30	MSR1	MSR2	MSR3	MSR4	MSR5	MSR6
SAN6	SAN7	SAN8	SAN9	SAN10	SAN11	SAN12
8	9	10	11	12	13	14
J20	J21	J22	J23	J24	J25	J26
MSR7	MSR8	MSR9	MSR10	MSR11	MSR12	MSR13
SAN13	SAN14	SAN15	SAN16	SAN17	SAN18	SAN19
15	16	17	18	19	20	21
J27	J28	J29	J30	JU1	JU2	JU3
MSR14	MSR15	MSR16	MSR17	MSR18	MSR19	MSR20
SAN20	SAN21	SAN22	SAN23	SAN24	SAN25	SAN26
22	23	24	25	26	27	28
JU4	JU5	JU6	JU7	JU8	JU9	JU10
MSR21	MRS22	MSR23	MSR24	MSR25	MSR26	MSR27
SAN27	SAN28	SAN29	SAN30	HAM1	HAM2	HAM3

TAIFA WEUSI 472/1988-89/6224-25/1980-81						
Ashe	Ojobo	Iwa	Araorun	Ile	Itutu	Erinle
1	2	3	4	5	6	7
AU8	AU9	AU10	AU11	AU12	AU13	AU14
THT20	THT21	THT22	THT23	THT24	THT25	THT26
NAH2	NAH3	NAH4	NAH5	NAH6	NAH7	NAH8
8	9	10	11	12	13	14
AU15	AU16	AU17	AU18	AU19	AU20	AU21
THT27	THT28	THT29	THT30	PAP1	PAP1	PAP2
NAH9	NAH10	NAH11	NAH12	NAH13	NAH14	NAH15
15	16	17	18	19	20	21
AU22	AU23	AU24	AU25	AU26	AU27	AU28
PAP3	PAP4	PAP5	PAP6	PAP7	PAP8	PAP9
NAH16	NAH17	NAH18	NAH19	NAH20	NAH21	NAH22
22	23	24	25	26	27	28
AU29	AU30	AU31	S1	S2	S3	S4
PAP10	PAP11	PAP12	PAP13	PAP14	PAP15	PAP16
NAH23	NAH24	NAH25	NAH26	NAH27	NAH28	NAH29

MTU WEUSI 472/1988-89/6224-25/1980-81						
Ashe	Ojobo	Iwa	Araorun	Ile	Itutu	Erinle
1	2	3	4	5	6	7
JU11	JU12	JU13	JU14	JU15	JU16	JU17
MSR28	MSR29	MSR30	*ASR1*	*HRU2*	*STH3*	*AST4*
HAM4	HAM5	HAM6	HAM7	HAM8	HAM9	HAM10
8	9	10	11	12	13	14
JU18	JU19	JU20	JU21	JU22	JU23	JU24
NBH5	*ANU6*	THT1	THT2	THT3	THT4	THT5
HAM11	HAM12	HAM13	HAM14	HAM15	HAM16	HAM17
15	16	17	18	19	20	21
JU25	JU26	JU27	JU28	JU29	JU30	JU31
THT6	THT7	THT8	THT9	THT10	THT11	THT12
HAM18	HAM19	HAM20	HAM21	HAM22	HAM23	HAM24
22	23	24	25	26	27	28
AU1	AU2	AU3	AU4	AU5	AU6	AU7
THT13	THT14	THT15	THT16	THT17	THT18	THT19
HAM25	HAM26	HAM27	HAM28	HAM29	HAM30	NAH1

UWEZO 472/1988-89/6224-25/1980-81						
Ashe	Ojobo	Iwa	Araorun	Ile	Itutu	Erinle
1	2	3	4	5	6	7
S5	S6	S7	S8	S9	S10	S11
PAP17	PAP18	PAP19	PAP20	PAP21	PAP22	PAP23
NAH30	*PAG1*	*PAG2*	*PAG3*	*PAG4*	*PAG5*	MAS1
8	9	10	11	12	13	14
S12	S13	S14	S15	S16	S17	S18
PAP24	PAP25	PAP26	PAP27	PAP28	PAP29	PAP30
MAS2	MAS3	MAS4	MAS5	MAS6	MAS7	MAS8
15	16	17	18	19	20	21
S19	S20	S21	S22	S23	S24	S25
HTH1	HTH2	HTH3	HTH4	HTH5	HTH6	HTH7
MAS9	MAS10	MAS11	MAS12	MAS13	MAS14	MAS15
22	23	24	25	26	27	28
S26	S27	S28	S29	S30	O1	O2
HTH8	HTH9	HTH10	HTH11	HTH12	HTH13	HTH14
MAS16	MAS17	MAS18	MAS19	MAS20	MAS21	MAS22

EKUNDU 472/1988-89/6224-25/1980-81						
Ashe	Ojobo	Iwa	Araorun	Ile	Itutu	Erinle
1	2	3	4	5	6	7
O3	O4	O5	O6	O7	O8	O9
HTH15	HTH16	HTH17	HTH18	HTH19	HTH20	HTH21
MAS23	MAS24	MAS25	MAS26	MAS27	MAS28	MAS29
8	9	10	11	12	13	14
O10	O11	O12	O13	O14	O15	O16
HTH22	HTH23	HTH24	HTH25	HTH26	HTH27	HTH28
MAS30	THi1	THi2	THi3	THi4	THi5	THi6
15	16	17	18	19	20	21
O17	O18	O19	O20	O21	O22	O23
HTH29	HTH30	KHK1	KHK2	KHK3	KHK4	KHK5
THi7	THi8	This9	THi10	THi11	THi12	THi13
22	23	24	25	26	27	28
O24	O25	O26	O27	O28	O29	O30
KHK6	KHK7	KHK8	KHK9	KHK10	KHK11	KHK12
Thi14	THi15	THi16	THi17	THi18	THi19	THi20

KIJANI 472/1988-89/6224-25/1980-81						
Ashe	Ojobo	Iwa	Araorun	Ile	Itutu	Erinle
1	2	3	4	5	6	7
N28	N29	N30	D1	D2	D3	D4
TBT11	TBT12	TBT13	TBT14	TBT15	TBT16	TBT17
HID19	HID20	HID21	HID22	HID23	HID24	HID25
8	9	10	11	12	13	14
D5	D6	D7	D8	D9	D10	D11
TBT18	TBT19	TBT20	TBT21	TBT22	TBT23	TBT24
HID26	HID27	HID28	HID29	HID30	TAH1	TAH2
15	16	17	18	19	20	21
D12	D13	D14	D15	D16	D17	D18
TBT25	TBT26	TBT27	TBT28	TBT29	TBT30	MKH1
TAH3	TAH4	TAH5	TAH6	TAH7	TAH8	TAH9
22	23	24	25	26	27	28
D19	D20	D21	D22	D23	D24	D25
MKH2	MKH3	MKH4	MKH5	MKH6	MKH7	MKH8
TAH10	TAH11	TAH12	TAH13	TAH14	TAH15	TAH16

EUSI 472/1988-89/6224-25/1980-81						
Ashe	Ojobo	Iwa	Araorun	Ile	Itutu	Erinle
1	2	3	4	5	6	7
O31	N1	N2	N3	N4	N5	N6
KHK13	KHK14	KHK15	KHK16	KHK17	KHK18	KHK19
THi21	THi22	THi23	THi24	THi25	THi26	THi27
8	9	10	11	12	13	14
N7	N8	N9	N10	N11	N12	N13
KHK20	KHK21	KHK22	KHK23	KHK24	KHK25	KHK26
THi28	THi29	THi30	HID1	HID2	HID3	HID4
15	16	17	18	19	20	21
N14	N15	N16	N17	N18	N19	N20
KHK27	KHK28	KHK29	KHK30	TBT1	TBT2	TBT3
HID5	HID6	HID7	HID8	HID9	HID10	HID11
22	23	24	25	26	27	28
N21	N22	N23	N24	N25	N26	N27
TBT4	TBT5	TBT6	TBT7	TBT8	TBT9	TBT10
HID12	HID13	HID14	HID15	HID16	HID17	HID18

KWANZA 472/1988-89/6224-25/1980-81						
Umoja	Kujichagulia	Ujima	Ujamaa	Nia	Kuumba	Imani
1	2	3	4	5	6	7
D26	D27	D28	D29	D30	D31	JA1
MKH9	MKH10	MKH11	MKH12	MKH13	MKH14	MKH15
TAH17	TAH18	TAH19	TAH20	TAH21	TAH22	TAH23
8	9	10	11	12	13	14
JA2	JA3	JA4	JA5	JA6	JA7	JA8
MKH16	MKH17	MKH18	MKH19	MKH20	MKH21	MKH22
TAH24	TAH25	TAH26	TAH27	TAH28	TAH29	TAH30
15	16	17	18	19	20	21
JA9	JA10	JA11	JA12	JA13	JA14	JA15
MKH23	MKH24	MKH25	KMH26	MKH27	MKH28	MKH29
TIR1	TIR2	TIR3	TIR4	TIR5	TIR6	TIR7
22	23	24	25	26	27	28
JA16	JA17	JA18	JA19	JA20	JA21	JA22
MKH30	AMN1	AMN2	AMN3	AMN4	AMN5	AMN6
TIR8	TIR9	TIR10	TIR11	TIR12	TIR13	TIR14

SHAHIDI 472/1988-89/6224-25/1980-81						
Ashe	Ojobo	Iwa	Araorun	Ile	Itutu	Erinle
1	2	3	4	5	6	7
JA23	JA24	JA25	JA26	JA27	JA28	JA29
AMN7	AMN8	AMN9	AMN10	AMN11	AMN12	AMN13
TIR15	TIR16	TIR17	TIR18	TIR19	TIR20	TIR21
8	9	10	11	12	13	14
JA30	JA31	F1	F2	F3	F4	F5
AMN14	AMN15	AMN16	AMN17	AMN18	AMN19	AMN20
TIR22	TIR23	TIR24	TIR25	TIR26	TIR27	TIR28
15	16	17	18	19	20	21
F6	F7	F8	F9	F10	F 11	F12
AMN21	AMN22	AMN23	AMN24	AMN25	AMN26	AMN28
TIR29	TIR30	YEK1	YEK2	YEK3	YEK4	YEK5
22	23	24	25	26	27	28
F13	F14	F15	F16	F17	F18	F19
AMN29	AMN30	RNT1	RNT2	RNT3	RNT4	RNT5
YEK6	YEK7	YEK8	YEK9	YEK10	YEK11	YEK12
SHAHIDI 23						
F20						
RNT6						
YEK13						

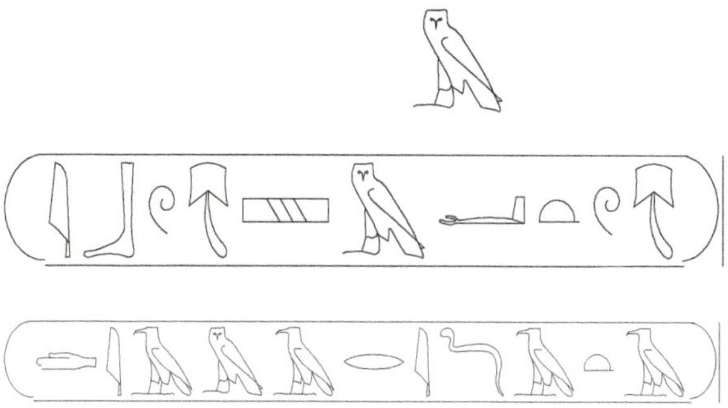

MAAFA 473 B.S.E./ASIENTO 1989-90 C.E./
KAM 6225-26 S.M./EC 1981-82 A.M. (ዓ.ም.)

AFU 473/1989-90/6225-26/1981-82						
Ashe	Ojobo	Iwa	Araorun	Ile	Itutu	Erinle
1	2	3	4	5	6	7
F21	F22	F23	F24	F25	F26	F27
RNT7	RNT8	RNT9	RNT10	RNT11	RNT12	RNT13
YEK14	YEK15	YEK16	YEK17	YEK18	YEK19	YEK20
8	9	10	11	12	13	14
F28	M1	M2	M3	M4	M5	M6
RNT14	RNT15	RNT16	RNT17	RNT18	RNT19	RNT20
YEK21	YEK22	YEK23	YEK24	YEK25	YEK26	YEK27
15	16	17	18	19	20	21
M7	M8	M9	M10	M11	M12	M13
RNT21	RNT22	RNT23	RNT24	RNT25	RNT26	RNT27
YEK28	YEK29	YEK30	MAG1	MAG2	MAG3	MAG4
22	23	24	25	26	27	28
M14	M15	M16	M17	M18	M19	M20
RNT28	RNT29	RNT30	KHN1	KHN2	KHN3	KHN4
MAG5	MAG6	MAG7	MAG8	MAG9	MAG10	MAG11

KAWAIDA 473/1989-90/6225-26/1981-82						
Ashe	Ojobo	Iwa	Araorun	Ile	Itutu	Erinle
1	2	3	4	5	6	7
A18	A19	A20	A21	A22	A23	A24
PIN3	PIN4	PIN5	PIN6	PIN7	PIN8	PIN9
MIY10	MIY11	MIY12	MIY13	MIY14	MIY15	MIY16
8	9	10	11	12	13	14
A25	A26	A27	A28	A29	A30	MA1
PIN10	PIN11	PIN12	PIN13	PIN14	PIN15	PIN16
MIY17	MIY18	MIY19	MIY20	MIY21	MIY22	MIY23
15	16	17	18	19	20	21
MA2	MA3	MA4	MA5	MA6	MA7	MA8
PIN17	PIN18	PIN19	PIN20	PIN21	PIN22	PIN23
MIY24	MIY25	MIY26	MIY27	MIY28	MIY29	MIY30
22	23	24	25	26	27	28
MA9	MA10	MA11	MA12	MA13	MA14	MA15
PIN24	PIN25	PIN26	PIN27	PIN28	PIN29	PIN30
GIN1	GIN2	GIN3	GIN4	GIN5	GIN6	GIN7

UHURU 473/1989-90/6225-26/1981-82						
Ashe	Ojobo	Iwa	Araorun	Ile	Itutu	Erinle
1	2	3	4	5	6	7
M21	M22	M23	M24	M25	M26	M27
KHN5	KHN6	KHN7	KHN8	KHN9	KHN10	KHN11
MAG12	MAG13	MAG14	MAG15	MAG16	MAG17	MAG18
8	9	10	11	12	13	14
M28	M29	M30	M31	A1	A2	A3
KHN12	KHN13	KHN14	KHN15	KHN16	KHN17	KHN18
MAG19	MAG20	MAG21	MAG22	MAG23	MAG24	MAG25
15	16	17	18	19	20	21
A4	A5	A6	A7	A8	A9	A10
KHN19	KHN20	KHN21	KHN22	KHN23	KHN24	KHN25
MAG26	MAG27	MAG28	MAG29	MAG30	MIY1	MIY2
22	23	24	25	26	27	28
A11	A12	A13	A14	A15	A16	A17
KHN26	KHN27	KHN28	KHN29	KHN30	PIN1	PIN2
MIY3	MIY4	MIY5	MIY6	MIY7	MIY8	MIY9

MABABA WEUSI 473/1989-90/6225-26/1981-82						
Ashe	Ojobo	Iwa	Araorun	Ile	Itutu	Erinle
1	2	3	4	5	6	7
MA16	MA17	MA18	MA19	MA20	MA21	MA22
APP1	APP2	APP3	APP4	APP5	APP6	APP7
GIN8	GIN9	GIN10	GIN11	GIN12	GIN13	GIN14
8	9	10	11	12	13	14
MA23	MA24	MA25	MA26	MA27	MA28	MA29
APP8	APP9	APP10	APP11	APP12	APP13	APP14
GIN15	GIN16	GIN17	GIN18	GIN19	GIN20	GIN21
15	16	17	18	19	20	21
MA30	MA31	J1	J2	J3	J4	J5
APP15	APP16	APP17	APP18	APP19	APP20	APP21
GIN22	GIN23	GIN24	GIN25	GIN26	GIN27	GIN28
22	23	24	25	26	27	28
J6	J7	J8	J9	J10	J11	J12
APP22	APP23	APP24	APP25	APP26	APP27	APP28
GIN29	GIN30	SAN1	SAN2	SAN3	SAN4	SAN5

MWANAMKE 473/1989-90/6225-26/1981-82

Ashe	Ojobo	Iwa	Araorun	Ile	Itutu	Erinle
1	2	3	4	5	6	7
J13	J14	J15	J16	J17	J18	J19
APP30	MSR1	MSR2	MSR3	MSR4	MSR5	MSR6
SAN6	SAN7	SAN8	SAN9	SAN10	SAN11	SAN12
8	9	10	11	12	13	14
J20	J21	J22	J23	J24	J25	J26
MSR7	MSR8	MSR9	MSR10	MSR11	MSR12	MSR13
SAN13	SAN14	SAN15	SAN16	SAN17	SAN18	SAN19
15	16	17	18	19	20	21
J27	J28	J29	J30	JU1	JU2	JU3
MSR14	MSR15	MSR16	MSR17	MSR18	MSR19	MSR20
SAN20	SAN21	SAN22	SAN23	SAN24	SAN25	SAN26
22	23	24	25	26	27	28
JU4	JU5	JU6	JU7	JU8	JU9	JU10
MSR21	MRS22	MSR23	MSR24	MSR25	MSR26	MSR27
SAN27	SAN28	SAN29	SAN30	HAM1	HAM2	HAM3

TAIFA WEUSI 473/1989-90/6225-26/1981-82

Ashe	Ojobo	Iwa	Araorun	Ile	Itutu	Erinle
1	2	3	4	5	6	7
AU8	AU9	AU10	AU11	AU12	AU13	AU14
THT20	THT21	THT22	THT23	THT24	THT25	THT26
NAH2	NAH3	NAH4	NAH5	NAH6	NAH7	NAH8
8	9	10	11	12	13	14
AU15	AU16	AU17	AU18	AU19	AU20	AU21
THT27	THT28	THT29	THT30	PAP1	PAP1	PAP2
NAH9	NAH10	NAH11	NAH12	NAH13	NAH14	NAH15
15	16	17	18	19	20	21
AU22	AU23	AU24	AU25	AU26	AU27	AU28
PAP3	PAP4	PAP5	PAP6	PAP7	PAP8	PAP9
NAH16	NAH17	NAH18	NAH19	NAH20	NAH21	NAH22
22	23	24	25	26	27	28
AU29	AU30	AU31	S1	S2	S3	S4
PAP10	PAP11	PAP12	PAP13	PAP14	PAP15	PAP16
NAH23	NAH24	NAH25	NAH26	NAH27	NAH28	NAH29

MTU WEUSI 473/1989-90/6225-26/1981-82

Ashe	Ojobo	Iwa	Araorun	Ile	Itutu	Erinle
1	2	3	4	5	6	7
JU11	JU12	JU13	JU14	JU15	JU16	JU17
MSR28	MSR29	MSR30	*ASR1*	*HRU2*	*STH3*	*AST4*
HAM4	HAM5	HAM6	HAM7	HAM8	HAM9	HAM10
8	9	10	11	12	13	14
JU18	JU19	JU20	JU21	JU22	JU23	JU24
NBH5	*ANU6*	THT1	THT2	THT3	THT4	THT5
HAM11	HAM12	HAM13	HAM14	HAM15	HAM16	HAM17
15	16	17	18	19	20	21
JU25	JU26	JU27	JU28	JU29	JU30	JU31
THT6	THT7	THT8	THT9	THT10	THT11	THT12
HAM18	HAM19	HAM20	HAM21	HAM22	HAM23	HAM24
22	23	24	25	26	27	28
AU1	AU2	AU3	AU4	AU5	AU6	AU7
THT13	THT14	THT15	THT16	THT17	THT18	THT19
HAM25	HAM26	HAM27	HAM28	HAM29	HAM30	NAH1

UWEZO 473/1989-90/6225-26/1981-82

Ashe	Ojobo	Iwa	Araorun	Ile	Itutu	Erinle
1	2	3	4	5	6	7
S5	S6	S7	S8	S9	S10	S11
PAP17	PAP18	PAP19	PAP20	PAP21	PAP22	PAP23
NAH30	*PAG1*	*PAG2*	*PAG3*	*PAG4*	*PAG5*	MAS1
8	9	10	11	12	13	14
S12	S13	S14	S15	S16	S17	S18
PAP24	PAP25	PAP26	PAP27	PAP28	PAP29	PAP30
MAS2	MAS3	MAS4	MAS5	MAS6	MAS7	MAS8
15	16	17	18	19	20	21
S19	S20	S21	S22	S23	S24	S25
HTH1	HTH2	HTH3	HTH4	HTH5	HTH6	HTH7
MAS9	MAS10	MAS11	MAS12	MAS13	MAS14	MAS15
22	23	24	25	26	27	28
S26	S27	S28	S29	S30	O1	O2
HTH8	HTH9	HTH10	HTH11	HTH12	HTH13	HTH14
MAS16	MAS17	MAS18	MAS19	MAS20	MAS21	MAS22

EKUNDU 473/1989-90/6225-26/1981-82						
Ashe	Ojobo	Iwa	Araorun	Ile	Itutu	Erinle
1	2	3	4	5	6	7
)3	O4	O5	O6	O7	O8	O9
TH15	HTH16	HTH17	HTH18	HTH19	HTH20	HTH21
IAS23	MAS24	MAS25	MAS26	MAS27	MAS28	MAS29
8	9	10	11	12	13	14
O10	O11	O12	O13	O14	O15	O16
TH22	HTH23	HTH24	HTH25	HTH26	HTH27	HTH28
IAS30	THi1	THi2	THi3	THi4	THi5	THi6
15	16	17	18	19	20	21
O17	O18	O19	O20	O21	O22	O23
TH29	HTH30	KHK1	KHK2	KHK3	KHK4	KHK5
THi7	THi8	THi9	THi10	THi11	THi12	THi13
2	23	24	25	26	27	28
)24	O25	O26	O27	O28	O29	O30
HK6	KHK7	KHK8	KHK9	KHK10	KHK11	KHK12
THi14	THi15	THi16	THi17	THi18	THi19	THi20

KIJANI 473/1989-90/6225-26/1981-82						
Ashe	Ojobo	Iwa	Araorun	Ile	Itutu	Erinle
1	2	3	4	5	6	7
N28	N29	N30	D1	D2	D3	D4
TBT11	TBT12	TBT13	TBT14	TBT15	TBT16	TBT17
HID19	HID20	HID21	HID22	HID23	HID24	HID25
8	9	10	11	12	13	14
D5	D6	D7	D8	D9	D10	D11
TBT18	TBT19	TBT20	TBT21	TBT22	TBT23	TBT24
HID26	HID27	HID28	HID29	HID30	TAH1	TAH2
15	16	17	18	19	20	21
D12	D13	D14	D15	D16	D17	D18
TBT25	TBT26	TBT27	TBT28	TBT29	TBT30	MKH1
TAH3	TAH4	TAH5	TAH6	TAH7	TAH8	TAH9
22	23	24	25	26	27	28
D19	D20	D21	D22	D23	D24	D25
MKH2	MKH3	MKH4	MKH5	MKH6	MKH7	MKH8
TAH10	TAH11	TAH12	TAH13	TAH14	TAH15	TAH16

EUSI 473/1989-90/6225-26/1981-82						
Ashe	Ojobo	Iwa	Araorun	Ile	Itutu	Erinle
1	2	3	4	5	6	7
)31	N1	N2	N3	N4	N5	N6
HK13	KHK14	KHK15	KHK16	KHK17	KHK18	KHK19
THi21	THi22	THi23	THi24	THi25	THi26	THi27
8	9	10	11	12	13	14
N7	N8	N9	N10	N11	N12	N13
HK20	KHK21	KHK22	KHK23	KHK24	KHK25	KHK26
THi28	THi29	THi30	HID1	HID2	HID3	HID4
15	16	17	18	19	20	21
I14	N15	N16	N17	N18	N19	N20
HK27	KHK28	KHK29	KHK30	TBT1	TBT2	TBT3
IID5	HID6	HID7	HID8	HID9	HID10	HID11
22	23	24	25	26	27	28
I21	N22	N23	N24	N25	N26	N27
TBT4	TBT5	TBT6	TBT7	TBT8	TBT9	TBT10
IID12	HID13	HID14	HID15	HID16	HID17	HID18

KWANZA 473/1989-90/6225-26/1981-82						
Umoja	Kujichagulia	Ujima	Ujamaa	Nia	Kuumba	Imani
1	2	3	4	5	6	7
D26	D27	D28	D29	D30	D31	JA1
MKH9	MKH10	MKH11	MKH12	MKH13	MKH14	MKH15
TAH17	TAH18	TAH19	TAH20	TAH21	TAH22	TAH23
8	9	10	11	12	13	14
JA2	JA3	JA4	JA5	JA6	JA7	JA8
MKH16	MKH17	MKH18	MKH19	MKH20	MKH21	MKH22
TAH24	TAH25	TAH26	TAH27	TAH28	TAH29	TAH30
15	16	17	18	19	20	21
JA9	JA10	JA11	JA12	JA13	JA14	JA15
MKH23	MKH24	MKH25	KMH26	MKH27	MKH28	MKH29
TIR1	TIR2	TIR3	TIR4	TIR5	TIR6	TIR7
22	23	24	25	26	27	28
JA16	JA17	JA18	JA19	JA20	JA21	JA22
MKH30	AMN1	AMN2	AMN3	AMN4	AMN5	AMN6
TIR8	TIR9	TIR10	TIR11	TIR12	TIR13	TIR14

SHAHIDI 473/1989-90/6225-26/1981-82						
Ashe	Ojobo	Iwa	Araorun	Ile	Itutu	Erinle
1	2	3	4	5	6	7
JA23	JA24	JA25	JA26	JA27	JA28	JA29
AMN7	AMN8	AMN9	AMN10	AMN11	AMN12	AMN13
TIR15	TIR16	TIR17	TIR18	TIR19	TIR20	TIR21
8	9	10	11	12	13	14
JA30	JA31	F1	F2	F3	F4	F5
MN14	AMN15	AMN16	AMN17	AMN18	AMN19	AMN20
TIR22	TIR23	TIR24	TIR25	TIR26	TIR27	TIR28
5	16	17	18	19	20	21
6	F7	F8	F9	F10	F 11	F12
MN21	AMN22	AMN23	AMN24	AMN25	AMN26	AMN28
IR29	TIR30	YEK1	YEK2	YEK3	YEK4	YEK5
22	23	24	25	26	27	28
F13	F14	F15	F16	F17	F18	F19
MN29	AMN30	RNT1	RNT2	RNT3	RNT4	RNT5
EK6	YEK7	YEK8	YEK9	YEK10	YEK11	YEK12
HAHIDI 24						
20						
NT6						
EK13						

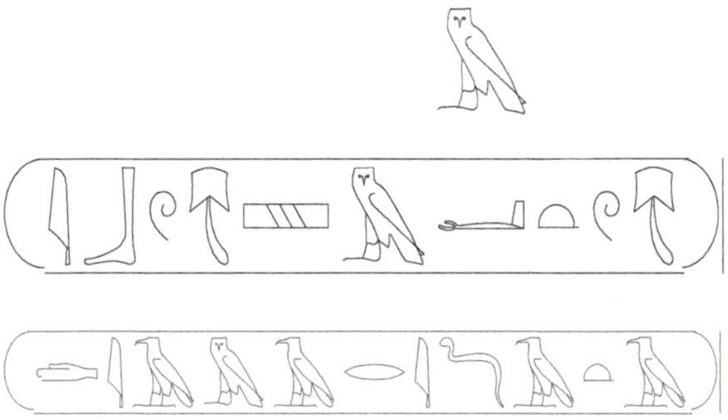

MAAFA 474 B.S.E./ASIENTO 1990-91 C.E./
KAM 6226-27 S.M./EC 1982-83 A.M. (ዓ.ም.)

AFU 474/1990-91/6226-27/1982-83						
Ashe	Ojobo	Iwa	Araorun	Ile	Itutu	Erinle
1	2	3	4	5	6	7
F21	F22	F23	F24	F25	F26	F27
RNT7	RNT8	RNT9	RNT10	RNT11	RNT12	RNT13
YEK14	YEK15	YEK16	YEK17	YEK18	YEK19	YEK20
8	9	10	11	12	13	14
F28	M1	M2	M3	M4	M5	M6
RNT14	RNT15	RNT16	RNT17	RNT18	RNT19	RNT20
YEK21	YEK22	YEK23	YEK24	YEK25	YEK26	YEK27
15	16	17	18	19	20	21
M7	M8	M9	M10	M11	M12	M13
RNT21	RNT22	RNT23	RNT24	RNT25	RNT26	RNT27
YEK28	YEK29	YEK30	MAG1	MAG2	MAG3	MAG4
22	23	24	25	26	27	28
M14	M15	M16	M17	M18	M19	M20
RNT28	RNT29	RNT30	KHN1	KHN2	KHN3	KHN4
MAG5	MAG6	MAG7	MAG8	MAG9	MAG10	MAG11

KAWAIDA 474/1990-91/6226-27/1982-83						
Ashe	Ojobo	Iwa	Araorun	Ile	Itutu	Erinle
1	2	3	4	5	6	7
A18	A19	A20	A21	A22	A23	A24
PIN3	PIN4	PIN5	PIN6	PIN7	PIN8	PIN9
MIY10	MIY11	MIY12	MIY13	MIY14	MIY15	MIY16
8	9	10	11	12	13	14
A25	A26	A27	A28	A29	A30	MA1
PIN10	PIN11	PIN12	PIN13	PIN14	PIN15	PIN16
MIY17	MIY18	MIY19	MIY20	MIY21	MIY22	MIY23
15	16	17	18	19	20	21
MA2	MA3	MA4	MA5	MA6	MA7	MA8
PIN17	PIN18	PIN19	PIN20	PIN21	PIN22	PIN23
MIY24	MIY25	MIY26	MIY27	MIY28	MIY29	MIY30
22	23	24	25	26	27	28
MA9	MA10	MA11	MA12	MA13	MA14	MA15
PIN24	PIN25	PIN26	PIN27	PIN28	PIN29	PIN30
GIN1	GIN2	GIN3	GIN4	GIN5	GIN6	GIN7

UHURU 474/1990-91/6226-27/1982-83						
Ashe	Ojobo	Iwa	Araorun	Ile	Itutu	Erinle
1	2	3	4	5	6	7
M21	M22	M23	M24	M25	M26	M27
KHN5	KHN6	KHN7	KHN8	KHN9	KHN10	KHN11
MAG12	MAG13	MAG14	MAG15	MAG16	MAG17	MAG18
8	9	10	11	12	13	14
M28	M29	M30	M31	A1	A2	A3
KHN12	KHN13	KHN14	KHN15	KHN16	KHN17	KHN18
MAG19	MAG20	MAG21	MAG22	MAG23	MAG24	MAG25
15	16	17	18	19	20	21
A4	A5	A6	A7	A8	A9	A10
KHN19	KHN20	KHN21	KHN22	KHN23	KHN24	KHN25
MAG26	MAG27	MAG28	MAG29	MAG30	MIY1	MIY2
22	23	24	25	26	27	28
A11	A12	A13	A14	A15	A16	A17
KHN26	KHN27	KHN28	KHN29	KHN30	PIN1	PIN2
MIY3	MIY4	MIY5	MIY6	MIY7	MIY8	MIY9

MABABA WEUSI 474/1990-91/6226-27/1982-83						
Ashe	Ojobo	Iwa	Araorun	Ile	Itutu	Erinle
1	2	3	4	5	6	7
MA16	MA17	MA18	MA19	MA20	MA21	MA22
APP1	APP2	APP3	APP4	APP5	APP6	APP7
GIN8	GIN9	GIN10	GIN11	GIN12	GIN13	GIN14
8	9	10	11	12	13	14
MA23	MA24	MA25	MA26	MA27	MA28	MA29
APP8	APP9	APP10	APP11	APP12	APP13	APP14
GIN15	GIN16	GIN17	GIN18	GIN19	GIN20	GIN21
15	16	17	18	19	20	21
MA30	MA31	J1	J2	J3	J4	J5
APP15	APP16	APP17	APP18	APP19	APP20	APP21
GIN22	GIN23	GIN24	GIN25	GIN26	GIN27	GIN28
22	23	24	25	26	27	28
J6	J7	J8	J9	J10	J11	J12
APP22	APP23	APP24	APP25	APP26	APP27	APP28
GIN29	GIN30	SAN1	SAN2	SAN3	SAN4	SAN5

MWANAMKE WEUSI 474/1990-91/6226-27/1982-83						
Ashe	Ojobo	Iwa	Araorun	Ile	Itutu	Erinle
1	2	3	4	5	6	7
J13	J14	J15	J16	J17	J18	J19
APP29	APP30	MSR1	MSR2	MSR3	MSR4	MSR5
SAN6	SAN7	SAN8	SAN9	SAN10	SAN11	SAN12
8	9	10	11	12	13	14
J20	J21	J22	J23	J24	J25	J26
MSR6	MSR7	MSR8	MSR9	MSR10	MSR11	MSR12
SAN13	SAN14	SAN15	SAN16	SAN17	SAN18	SAN19
15	16	17	18	19	20	21
J27	J28	J29	J30	JU1	JU2	JU3
MSR13	MSR14	MSR15	MSR16	MSR17	MSR18	MSR19
SAN20	SAN21	SAN22	SAN23	SAN24	SAN25	SAN26
22	23	24	25	26	27	28
JU4	JU5	JU6	JU7	JU8	JU9	JU10
MSR20	MRS21	MSR22	MSR23	MSR24	MSR25	MSR26
SAN27	SAN28	SAN29	SAN30	HAM1	HAM2	HAM3

TAIFA WEUSI 474/1990-91/6226-27/1982-83						
Ashe	Ojobo	Iwa	Araorun	Ile	Itutu	Erinle
1	2	3	4	5	6	7
AU8	AU9	AU10	AU11	AU12	AU13	AU14
THT20	THT21	THT22	THT23	THT24	THT25	THT26
NAH2	NAH3	NAH4	NAH5	NAH6	NAH7	NAH8
8	9	10	11	12	13	14
AU15	AU16	AU17	AU18	AU19	AU20	AU21
THT27	THT28	THT29	THT30	PAP1	PAP1	PAP2
NAH9	NAH10	NAH11	NAH12	NAH13	NAH14	NAH15
15	16	17	18	19	20	21
AU22	AU23	AU24	AU25	AU26	AU27	AU28
PAP3	PAP4	PAP5	PAP6	PAP7	PAP8	PAP9
NAH16	NAH17	NAH18	NAH19	NAH20	NAH21	NAH22
22	23	24	25	26	27	28
AU29	AU30	AU31	S1	S2	S3	S4
PAP10	PAP11	PAP12	PAP13	PAP14	PAP15	PAP16
NAH23	NAH24	NAH25	NAH26	NAH27	NAH28	NAH29

MTU WEUSI 474/1990-91/6226-27/1982-83						
Ashe	Ojobo	Iwa	Araorun	Ile	Itutu	Erinle
1	2	3	4	5	6	7
JU11	JU12	JU13	JU14	JU15	JU16	JU17
MSR27	MSR28	MSR29	MSR30	*ASR1*	*HRU2*	*STH3*
HAM4	HAM5	HAM6	HAM7	HAM8	HAM9	HAM10
8	9	10	11	12	13	14
JU18	JU19	JU20	JU21	JU22	JU23	JU24
AST4	*NBH5*	THT1	THT2	THT3	THT4	THT5
HAM11	HAM12	HAM13	AM14	HAM15	HAM16	HAM17
15	16	17	18	19	20	21
JU25	JU26	JU27	JU28	JU29	JU30	JU31
THT6	THT7	THT8	THT9	THT10	THT11	THT12
HAM18	HAM19	HAM20	HAM21	HAM22	HAM23	HAM24
22	23	24	25	26	27	28
AU1	AU2	AU3	AU4	AU5	AU6	AU7
THT13	THT14	THT15	THT16	THT17	THT18	THT19
HAM25	HAM26	HAM27	HAM28	HAM29	HAM30	NAH1

UWEZO 474/1990-91/6226-27/1982-83						
Ashe	Ojobo	Iwa	Araorun	Ile	Itutu	Erinle
1	2	3	4	5	6	7
S5	S6	S7	S8	S9	S10	S11
PAP17	PAP18	PAP19	PAP20	PAP21	PAP22	PAP23
NAH30	*PAG1*	*PAG2*	*PAG3*	*PAG4*	*PAG5*	MAS1
8	9	10	11	12	13	14
S12	S13	S14	S15	S16	S17	S18
PAP24	PAP25	PAP26	PAP27	PAP28	PAP29	PAP30
MAS2	MAS3	MAS4	MAS5	MAS6	MAS7	MAS8
15	16	17	18	19	20	21
S19	S20	S21	S22	S23	S24	S25
HTH1	HTH2	HTH3	HTH4	HTH5	HTH6	HTH7
MAS9	MAS10	MAS11	MAS12	MAS13	MAS14	MAS15
22	23	24	25	26	27	28
S26	S27	S28	S29	S30	O1	O2
HTH8	HTH9	HTH10	HTH11	HTH12	HTH13	HTH14
MAS16	MAS17	MAS18	MAS19	MAS20	MAS21	MAS22

EKUNDU 474/1990-91/6226-27/1982-83						
Ashe	Ojobo	Iwa	Araorun	Ile	Itutu	Erinle
1	2	3	4	5	6	7
O3	O4	O5	O6	O7	O8	O9
HTH15	HTH16	HTH17	HTH18	HTH19	HTH20	HTH21
MAS23	MAS24	MAS25	MAS26	MAS27	MAS28	MAS29
8	9	10	11	12	13	14
O10	O11	O12	O13	O14	O15	O16
HTH22	HTH23	HTH24	HTH25	HTH26	HTH27	HTH28
MAS30	THi1	THi2	THi3	THi4	THi5	THi6
15	16	17	18	19	20	21
O17	O18	O19	O20	O21	O22	O23
HTH29	HTH30	KHK1	KHK2	KHK3	KHK4	KHK5
THi7	THi8	THi9	THi10	THi11	THi12	THi13
22	23	24	25	26	27	28
O24	O25	O26	O27	O28	O29	O30
KHK6	KHK7	KHK8	KHK9	KHK10	KHK11	KHK12
THi14	THi15	THi16	THi17	THi18	THi19	THi20

KIJANI 474/1990-91/6226-27/1982-83						
Ashe	Ojobo	Iwa	Araorun	Ile	Itutu	Erinle
1	2	3	4	5	6	7
N28	N29	N30	D1	D2	D3	D4
TBT11	TBT12	TBT13	TBT14	TBT15	TBT16	TBT17
HID19	HID20	HID21	HID22	HID23	HID24	HID25
8	9	10	11	12	13	14
D5	D6	D7	D8	D9	D10	D11
TBT18	TBT19	TBT20	TBT21	TBT22	TBT23	TBT24
HID26	HID27	HID28	HID29	HID30	TAH1	TAH2
15	16	17	18	19	20	21
D12	D13	D14	D15	D16	D17	D18
TBT25	TBT26	TBT27	TBT28	TBT29	TBT30	MKH1
TAH3	TAH4	TAH5	TAH6	TAH7	TAH8	TAH9
22	23	24	25	26	27	28
D19	D20	D21	D22	D23	D24	D25
MKH2	MKH3	MKH4	MKH5	MKH6	MKH7	MKH8
TAH10	TAH11	TAH12	TAH13	TAH14	TAH15	TAH16

EUSI 474/1990-91/6226-27/1982-83						
Ashe	Ojobo	Iwa	Araorun	Ile	Itutu	Erinle
1	2	3	4	5	6	7
O31	N1	N2	N3	N4	N5	N6
KHK13	KHK14	KHK15	KHK16	KHK17	KHK18	KHK19
THi21	THi22	THi23	THi24	THi25	THi26	THi27
8	9	10	11	12	13	14
N7	N8	N9	N10	N11	N12	N13
KHK20	KHK21	KHK22	KHK23	KHK24	KHK25	KHK26
THi28	THi29	THi30	HID1	HID2	HID3	HID4
15	16	17	18	19	20	21
N14	N15	N16	N17	N18	N19	N20
KHK27	KHK28	KHK29	KHK30	TBT1	TBT2	TBT3
HID5	HID6	HID7	HID8	HID9	HID10	HID11
22	23	24	25	26	27	28
N21	N22	N23	N24	N25	N26	N27
TBT4	TBT5	TBT6	TBT7	TBT8	TBT9	TBT10
HID12	HID13	HID14	HID15	HID16	HID17	HID18

KWANZA 474/1990-91/6226-27/1982-83						
Umoja	Kujichagulia	Ujima	Ujamaa	Nia	Kuumba	Imani
1	2	3	4	5	6	7
D26	D27	D28	D29	D30	D31	JA1
MKH9	MKH10	MKH11	MKH12	MKH13	MKH14	MKH15
TAH17	TAH18	TAH19	TAH20	TAH21	TAH22	TAH23
8	9	10	11	12	13	14
JA2	JA3	JA4	JA5	JA6	JA7	JA8
MKH16	MKH17	MKH18	MKH19	MKH20	MKH21	MKH22
TAH24	TAH25	TAH26	TAH27	TAH28	TAH29	TAH30
15	16	17	18	19	20	21
JA9	JA10	JA11	JA12	JA13	JA14	JA15
MKH23	MKH24	MKH25	KMH26	MKH27	MKH28	MKH29
TIR1	TIR2	TIR3	TIR4	TIR5	TIR6	TIR7
22	23	24	25	26	27	28
JA16	JA17	JA18	JA19	JA20	JA21	JA22
MKH30	AMN1	AMN2	AMN3	AMN4	AMN5	AMN6
TIR8	TIR9	TIR10	TIR11	TIR12	TIR13	TIR14

SHAHIDI 474/1990-91/6226-27/1982-83						
Ashe	Ojobo	Iwa	Araorun	Ile	Itutu	Erinle
1	2	3	4	5	6	7
JA23	JA24	JA25	JA26	JA27	JA28	JA29
AMN7	AMN8	AMN9	AMN10	AMN11	AMN12	AMN13
TIR15	TIR16	TIR17	TIR18	TIR19	TIR20	TIR21
8	9	10	11	12	13	14
JA30	JA31	F1	F2	F3	F4	F5
AMN14	AMN15	AMN16	AMN17	AMN18	AMN19	AMN20
TIR22	TIR23	TIR24	TIR25	TIR26	TIR27	TIR28
15	16	17	18	19	20	21
F6	F7	F8	F9	F10	F 11	F12
AMN21	AMN22	AMN23	AMN24	AMN25	AMN26	AMN28
TIR29	TIR30	YEK1	YEK2	YEK3	YEK4	YEK5
22	23	24	25	26	27	28
F13	F14	F15	F16	F17	F18	F19
AMN29	AMN30	RNT1	RNT2	RNT3	RNT4	RNT5
YEK6	YEK7	YEK8	YEK9	YEK10	YEK11	YEK12
SHAHIDI 25						
F20						
RNT6						
YEK13						

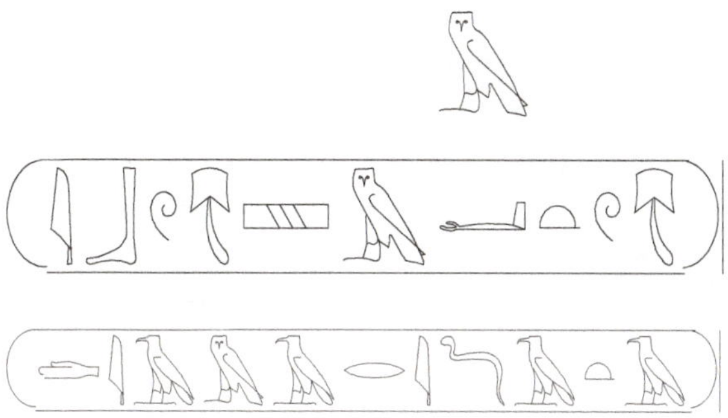

MAAFA 475 B.S.E./ASIENTO 1991-92 C.E./
KAM 6227-28 S.M./EC 1983-84 A.M. (ዓ.ም.)

AFU 475/1991-92/6227-28/1983-84

Ashe	Ojobo	Iwa	Araorun	Ile	Itutu	Erinle
1	2	3	4	5	6	7
F21	F22	F23	F24	F25	F26	F27
RNT7	RNT8	RNT9	RNT10	RNT11	RNT12	RNT13
YEK14	YEK15	YEK16	YEK17	YEK18	YEK19	YEK20
8	9	10	11	12	13	14
F28	M1	M2	M3	M4	M5	M6
RNT14	RNT15	RNT16	RNT17	RNT18	RNT19	RNT20
YEK21	YEK22	YEK23	YEK24	YEK25	YEK26	YEK27
15	16	17	18	19	20	21
M7	M8	M9	M10	M11	M12	M13
RNT21	RNT22	RNT23	RNT24	RNT25	RNT26	RNT27
YEK28	YEK 29	YEK30	MAG1	MAG2	MAG3	MAG4
22	23	24	25	26	27	28
M14	M15	M16	M17	M18	M19	M20
RNT28	RNT29	RNT30	KHN1	KHN2	KHN3	KHN4
MAG5	MAG6	MAG7	MAG8	MAG9	MAG10	MAG11

KAWAIDA 475/1991-92/6227-28/1983-84

Ashe	Ojobo	Iwa	Araorun	Ile	Itutu	Erinle
1	2	3	4	5	6	7
A18	A19	A20	A21	A22	A23	A24
PIN3	PIN4	PIN5	PIN6	PIN7	PIN8	PIN9
MIY10	MIY11	MIY12	MIY13	MIY14	MIY15	MIY16
8	9	10	11	12	13	14
A25	A26	A27	A28	A29	A30	MA1
PIN10	PIN11	PIN12	PIN13	PIN14	PIN15	PIN16
MIY17	MIY18	MIY19	MIY20	MIY21	MIY22	MIY23
15	16	17	18	19	20	21
MA2	MA3	MA4	MA5	MA6	MA7	MA8
PIN17	PIN18	PIN19	PIN20	PIN21	PIN22	PIN23
MIY24	MIY25	MIY26	MIY27	MIY28	MIY29	MIY30
22	23	24	25	26	27	28
MA9	MA10	MA11	MA12	MA13	MA14	MA15
PIN24	PIN25	PIN26	PIN27	PIN28	PIN29	PIN30
GIN1	GIN2	GIN3	GIN4	GIN5	GIN6	GIN7

UHURU 475/1991-92/6227-28/1983-84

Ashe	Ojobo	Iwa	Araorun	Ile	Itutu	Erinle
1	2	3	4	5	6	7
M21	M22	M23	M24	M25	M26	M27
KHN5	KHN6	KHN7	KHN8	KHN9	KHN10	KHN11
MAG12	MAG13	MAG14	MAG15	MAG16	MAG17	MAG18
8	9	10	11	12	13	14
M28	M29	M30	M31	A1	A2	A3
KHN12	KHN13	KHN14	KHN15	KHN16	KHN17	KHN18
MAG19	MAG20	MAG21	MAG22	MAG23	MAG24	MAG25
15	16	17	18	19	20	21
A4	A5	A6	A7	A8	A9	A10
KHN19	KHN20	KHN21	KHN22	KHN23	KHN24	KHN25
MAG26	MAG27	MAG28	MAG29	MAG30	MIY1	MIY2
22	23	24	25	26	27	28
A11	A12	A13	A14	A15	A16	A17
KHN26	KHN27	KHN28	KHN29	KHN30	PIN1	PIN2
MIY3	MIY4	MIY5	MIY6	MIY7	MIY8	MIY9

MABABA WEUSI 475/1991-92/6227-28/1983-84

Ashe	Ojobo	Iwa	Araorun	Ile	Itutu	Erinle
1	2	3	4	5	6	7
MA16	MA17	MA18	MA19	MA20	MA21	MA22
APP1	APP2	APP3	APP4	APP5	APP6	APP7
GIN8	GIN9	GIN10	GIN11	GIN12	GIN13	GIN14
8	9	10	11	12	13	14
MA23	MA24	MA25	MA26	MA27	MA28	MA29
APP8	APP9	APP10	APP11	APP12	APP13	APP14
GIN15	GIN16	GIN17	GIN18	GIN19	GIN20	GIN21
15	16	17	18	19	20	21
MA30	MA31	J1	J2	J3	J4	J5
APP15	APP16	APP17	APP18	APP19	APP20	APP21
GIN22	GIN23	GIN24	GIN25	GIN26	GIN27	GIN28
22	23	24	25	26	27	28
J6	J7	J8	J9	J10	J11	J12
APP22	APP23	APP24	APP25	APP26	APP27	APP28
GIN29	GIN30	SAN1	SAN2	SAN3	SAN4	SAN5

MWANAMKE WEUSI 475/1991-92/6227-28/1983-84

Ashe	Ojobo	Iwa	Araorun	Ile	Itutu	Erinle
1	2	3	4	5	6	7
J13	J14	J15	J16	J17	J18	J19
APP29	APP30	MSR1	MSR2	MSR3	MSR4	MSR5
SAN6	SAN7	SAN8	SAN9	SAN10	SAN11	SAN12
8	9	10	11	12	13	14
J20	J21	J22	J23	J24	J25	J26
MSR6	MSR7	MSR8	MSR9	MSR10	MSR11	MSR12
SAN13	SAN14	SAN15	SAN16	SAN17	SAN18	SAN19
15	16	17	18	19	20	21
J27	J28	J29	J30	JU1	JU2	JU3
MSR13	MSR14	MSR15	MSR16	MSR17	MSR18	MSR19
SAN20	SAN21	SAN22	SAN23	SAN24	SAN25	SAN26
22	23	24	25	26	27	28
JU4	JU5	JU6	JU7	JU8	JU9	JU10
MSR20	MRS21	MSR22	MSR23	MSR24	MSR25	MSR26
SAN27	SAN28	SAN29	SAN30	HAM1	HAM2	HAM3

TAIFA WEUSI 475/1991-92/6227-28/1983-84

Ashe	Ojobo	Iwa	Araorun	Ile	Itutu	Erinle
1	2	3	4	5	6	7
AU8	AU9	AU10	AU11	AU12	AU13	AU14
THT20	THT21	THT22	THT23	THT24	THT25	THT26
NAH2	NAH3	NAH4	NAH5	NAH6	NAH7	NAH8
8	9	10	11	12	13	14
AU15	AU16	AU17	AU18	AU19	AU20	AU21
THT27	THT28	THT29	THT30	PAP1	PAP1	PAP2
NAH9	NAH10	NAH11	NAH12	NAH13	NAH14	NAH15
15	16	17	18	19	20	21
AU22	AU23	AU24	AU25	AU26	AU27	AU28
PAP3	PAP4	PAP5	PAP6	PAP7	PAP8	PAP9
NAH16	NAH17	NAH18	NAH19	NAH20	NAH21	NAH22
22	23	24	25	26	27	28
AU29	AU30	AU31	S1	S2	S3	S4
PAP10	PAP11	PAP12	PAP13	PAP14	PAP15	PAP16
NAH23	NAH24	NAH25	NAH26	NAH27	NAH28	NAH29

MTU WEUSI 475/1991-92/6227-28/1983-84

Ashe	Ojobo	Iwa	Araorun	Ile	Itutu	Erinle
1	2	3	4	5	6	7
JU11	JU12	JU13	JU14	JU15	JU16	JU17
MSR27	MSR28	MSR29	MSR30	*ASR1*	*HRU2*	*STH3*
HAM4	HAM5	HAM6	HAM7	HAM8	HAM9	HAM10
8	9	10	11	12	13	14
JU18	JU19	JU20	JU21	JU22	JU23	JU24
AST4	*NBH5*	THT1	THT2	THT3	THT4	THT5
HAM11	HAM12	HAM13	HAM14	HAM15	HAM16	HAM17
15	16	17	18	19	20	21
JU25	JU26	JU27	JU28	JU29	JU30	JU31
THT6	THT7	THT8	THT9	THT10	THT11	THT12
HAM18	HAM19	HAM20	HAM21	HAM22	HAM23	HAM24
22	23	24	25	26	27	28
AU1	AU2	AU3	AU4	AU5	AU6	AU7
THT13	THT14	THT15	THT16	THT17	THT18	THT19
HAM25	HAM26	HAM27	HAM28	HAM29	HAM30	NAH1

UWEZO 475/1991-92/6227-28/1983-84

Ashe	Ojobo	Iwa	Araorun	Ile	Itutu	Erinle
1	2	3	4	5	6	7
S5	S6	S7	S8	S9	S10	S11
PAP17	PAP18	PAP19	PAP20	PAP21	PAP22	PAP23
NAH30	*QUA1*	*QUA2*	*QUA3*	*QUA4*	*QUA5*	*QUA6*
8	9	10	11	12	13	14
S12	S13	S14	S15	S16	S17	S18
PAP24	PAP25	PAP26	PAP27	PAP28	PAP29	PAP30
MAS1	MAS2	MAS3	MAS4	MAS5	MAS6	MAS7
15	16	17	18	19	20	21
S19	S20	S21	S22	S23	S24	S25
HTH1	HTH2	HTH3	HTH4	HTH5	HTH6	HTH7
MAS8	MAS9	MAS10	MAS11	MAS12	MAS13	MAS14
22	23	24	25	26	27	28
S26	S27	S28	S29	S30	O1	O2
HTH8	HTH9	HTH10	HTH11	HTH12	HTH13	HTH14
MAS15	MAS16	MAS17	MAS18	MAS19	MAS20	MAS21

EKUNDU 475/1991-92/6227-28/1983-84

Ashe	Ojobo	Iwa	Araorun	Ile	Itutu	Erinle
1	2	3	4	5	6	7
O3	O4	O5	O6	O7	O8	O9
HTH15	HTH16	HTH17	HTH18	HTH19	HTH20	HTH21
MAS22	MAS23	MAS 24	MAS25	MAS26	MAS27	MAS28
8	9	10	11	12	13	14
O10	O11	O12	O13	O14	O15	O16
HTH22	HTH23	HTH24	HTH25	HTH26	HTH27	HTH28
MAS29	MAS30	THi1	THi2	THi3	THi4	THi5
15	16	17	18	19	20	21
O17	O18	O19	O20	O21	O22	O23
HTH29	HTH30	KHK1	KHK2	KHK3	KHK4	KHK5
THi6	THi7	THi8	THi9	THi10	THi11	THi12
22	23	24	25	26	27	28
O24	O25	O26	O27	O28	O29	O30
KHK6	KHK7	KHK8	KHK9	KHK10	KHK11	KHK12
THi13	THi14	THi15	THi16	THi17	THi18	THi19

KIJANI 475/1991-92/6227-28/1983-84

Ashe	Ojobo	Iwa	Araorun	Ile	Itutu	Erinle
1	2	3	4	5	6	7
N28	N29	N30	D1	D2	D3	D4
TBT11	TBT12	TBT13	TBT14	TBT15	TBT16	TBT17
HID18	HID19	HID20	HID21	HID22	HID23	HID24
8	9	10	11	12	13	14
D5	D6	D7	D8	D9	D10	D11
TBT18	TBT19	TBT20	TBT21	TBT22	TBT23	TBT24
HID25	HID26	HID27	HID28	HID29	HID30	TAH1
15	16	17	18	19	20	21
D12	D13	D14	D15	D16	D17	D18
TBT25	TBT26	TBT27	TBT28	TBT29	TBT30	MKH1
TAH2	TAH3	TAH4	TAH5	TAH6	TAH7	TAH8
22	23	24	25	26	27	28
D19	D20	D21	D22	D23	D24	D25
MKH2	MKH3	MKH4	MKH5	MKH6	MKH7	MKH8
TAH9	TAH10	TAH11	TAH12	TAH13	TAH14	TAH15

EUSI 475/1991-92/6227-28/1983-84

Ashe	Ojobo	Iwa	Araorun	Ile	Itutu	Erinle
1	2	3	4	5	6	7
O31	N1	N2	N3	N4	N5	N6
KHK13	KHK14	KHK15	KHK16	KHK17	KHK18	KHK19
THi20	THi21	THi22	THi23	THi24	THi25	THi26
8	9	10	11	12	13	14
N7	N8	N9	N10	N11	N12	N13
KHK20	KHK21	KHK22	KHK23	KHK24	KHK25	KHK26
THi27	THi28	THi29	THi30	HID1	HID2	HID3
15	16	17	18	19	20	21
N14	N15	N16	N17	N18	N19	N20
KHK27	KHK28	KHK29	KHK30	TBT1	TBT2	TBT3
HID4	HID5	HID6	HID7	HID8	HID9	HID10
22	23	24	25	26	27	28
N21	N22	N23	N24	N25	N26	N27
TBT4	TBT5	TBT6	TBT7	TBT8	TBT9	TBT10
HID11	HID12	HID13	HID14	HID15	HID16	HID17

KWANZA 475/1991-92/6227-28/1983-84

Umoja	Kujichagulia	Ujima	Ujamaa	Nia	Kuumba	Imani
1	2	3	4	5	6	7
D26	D27	D28	D29	D30	D31	JA1
MKH9	MKH10	MKH11	MKH12	MKH13	MKH14	MKH15
TAH16	TAH17	TAH18	TAH19	TAH20	TAH21	TAH22
8	9	10	11	12	13	14
JA2	JA3	JA4	JA5	JA6	JA7	JA8
MKH16	MKH17	MKH18	MKH19	MKH20	MKH21	MKH22
TAH23	TAH24	TAH25	TAH26	TAH27	TAH28	TAH29
15	16	17	18	19	20	21
JA9	JA10	JA11	JA12	JA13	JA14	JA15
MKH23	MKH24	MKH25	KMH26	MKH27	MKH28	MKH29
TAH30	TIR1	TIR2	TIR3	TIR4	TIR5	TIR6
22	23	24	25	26	27	28
JA16	JA17	JA18	JA19	JA20	JA21	JA22
MKH30	AMN1	AMN2	AMN3	AMN4	AMN5	AMN6
TIR7	TIR8	TIR9	TIR10	TIR11	TIR12	TIR13

SHAHIDI 475/1991-92/6227-28/1983-84						
Ashe	Ojobo	Iwa	Araorun	Ile	Itutu	Erinle
1	2	3	4	5	6	7
JA23	JA24	JA25	JA26	JA27	JA28	JA29
AMN7	AMN8	AMN9	AMN10	AMN11	AMN12	AMN13
TIR14	TIR15	TIR16	TIR17	TIR18	TIR19	TIR20
8	9	10	11	12	13	14
JA30	JA31	F1	F2	F3	F4	F5
AMN14	AMN15	AMN16	AMN17	AMN18	AMN19	AMN20
TIR21	TIR22	TIR23	TIR24	TIR25	TIR26	TIR27
15	16	17	18	19	20	21
F6	F7	F8	F9	F10	F 11	F12
AMN21	AMN22	AMN23	AMN24	AMN25	AMN26	AMN28
TIR28	TIR29	TIR30	YEK1	YEK2	YEK3	YEK4
22	23	24	25	26	27	28
F13	F14	F15	F16	F17	F18	F19
AMN29	AMN30	RNT1	RNT2	RNT3	RNT4	RNT5
YEK5	YEK6	YEK7	YEK8	YEK9	YEK10	YEK11
SHAHIDI 26						
F20						
RNT6						
YEK12						

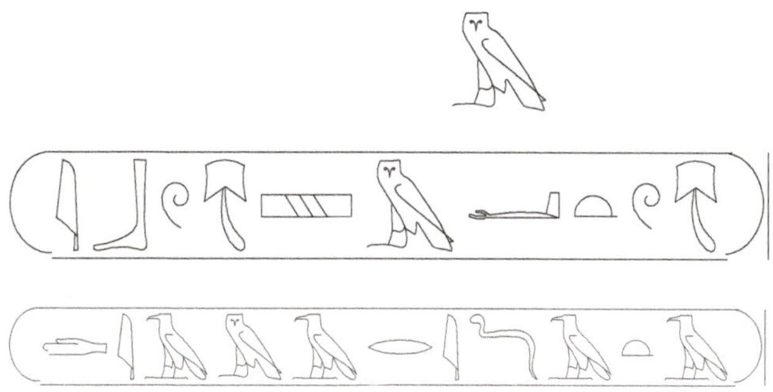

MAAFA 476 B.S.E./ASIENTO 1992-93 C.E./ KAM 6228-29 S.M./EC 1984-85 A.M. (ዓ.ም.)

AFU 476/1992-93/6228-29/1984-85

Ashe	Ojobo	Iwa	Araorun	Ile	Itutu	Erinle
1	2	3	4	5	6	7
F21	F22	F23	F24	F25	F26	F27
RNT7	RNT8	RNT9	RNT10	RNT11	RNT12	RNT13
YEK13	YEK14	YEK15	YEK16	YEK17	YEK18	YEK19
8	9	10	11	12	13	14
F28	F29	M1	M2	M3	M4	M5
RNT14	RNT15	RNT16	RNT17	RNT18	RNT19	RNT20
YEK 20	YEK21	YEK22	YEK23	YEK24	YEK25	YEK26
15	16	17	18	19	20	21
M6	M7	M8	M9	M10	M11	M12
RNT21	RNT22	RNT23	RNT24	RNT25	RNT26	RNT27
YEK 27	YEK28	YEK29	YEK30	MAG1	MAG2	MAG3
22	23	24	25	26	27	28
M13	M14	M15	M16	M17	M18	M19
RNT28	RNT29	RNT30	KHN1	KHN2	KHN3	KHN4
MAG4	MAG5	MAG6	MAG7	MAG8	MAG9	MAG10
SHARIFU 116						
M20						
KHN5						
MAG11						

KAWAIDA 476/1992-93/6228-29/1984-85

Ashe	Ojobo	Iwa	Araorun	Ile	Itutu	Erinle
1	2	3	4	5	6	7
A18	A19	A20	A21	A22	A23	A24
PIN4	PIN5	PIN6	PIN7	PIN8	PIN9	PIN10
MIY10	MIY11	MIY12	MIY13	MIY14	MIY15	MIY16
8	9	10	11	12	13	14
A25	A26	A27	A28	A29	A30	MA1
PIN11	PIN12	PIN13	PIN14	PIN15	PIN16	PIN17
MIY17	MIY18	MIY19	MIY20	MIY21	MIY22	MIY23
15	16	17	18	19	20	21
MA2	MA3	MA4	MA5	MA6	MA7	MA8
PIN18	PIN19	PIN20	PIN21	PIN22	PIN23	PIN24
MIY24	MIY25	MIY26	MIY27	MIY28	MIY29	MIY30
22	23	24	25	26	27	28
MA9	MA10	MA11	MA12	MA13	MA14	MA15
PIN25	PIN26	PIN27	PIN28	PIN29	PIN30	APP1
GIN1	GIN2	GIN3	GIN4	GIN5	GIN6	GIN7

UHURU 476/1992-93/6228-29/1984-85

Ashe	Ojobo	Iwa	Araorun	Ile	Itutu	Erinle
1	2	3	4	5	6	7
M21	M22	M23	M24	M25	M26	M27
KHN6	KHN7	KHN8	KHN9	KHN10	KHN11	KHN12
MAG12	MAG13	MAG14	MAG15	MAG16	MAG17	MAG18
8	9	10	11	12	13	14
M28	M29	M30	M31	A1	A2	A3
KHN13	KHN14	KHN15	KHN16	KHN17	KHN18	KHN19
MAG19	MAG20	MAG21	MAG22	MAG23	MAG24	MAG25
15	16	17	18	19	20	21
A4	A5	A6	A7	A8	A9	A10
KHN20	KHN21	KHN222	KHN23	KHN24	KHN25	KHN26
MAG26	MAG27	MAG28	MAG29	MAG30	MIY1	MIY2
22	23	24	25	26	27	28
A11	A12	A13	A14	A15	A16	A17
KHN27	KHN28	KHN29	KHN30	PIN1	PIN2	PIN3
MIY3	MIY4	MIY5	MIY6	MIY7	MIY8	MIY9

MABABA WEUSI 476/1992-93/6228-29/1984-85

Ashe	Ojobo	Iwa	Araorun	Ile	Itutu	Erinle
1	2	3	4	5	6	7
MA16	MA17	MA18	MA19	MA20	MA21	MA22
APP2	APP3	APP4	APP5	APP6	APP7	APP8
GIN8	GIN9	GIN10	GIN11	GIN12	GIN13	GIN14
8	9	10	11	12	13	14
MA23	MA24	MA25	MA26	MA27	MA28	MA29
APP9	APP10	APP11	APP12	APP13	APP14	APP15
GIN15	GIN16	GIN17	GIN18	GIN19	GIN20	GIN21
15	16	17	18	19	20	21
MA30	MA31	J1	J2	J3	J4	J5
APP16	APP17	APP18	APP19	APP20	APP21	APP22
GIN22	GIN23	GIN24	GIN25	GIN26	GIN27	GIN28
22	23	24	25	26	27	28
J6	J7	J8	J9	J10	J11	J12
APP23	APP24	APP25	APP26	APP27	APP28	APP29
GIN29	GIN30	SAN1	SAN2	SAN3	SAN4	SAN5

MWANAMKE WEUSI 476/1992-93/6228-29/1984-85

Ashe	Ojobo	Iwa	Araorun	Ile	Itutu	Erinle
1	2	3	4	5	6	7
J13	J14	J15	J16	J17	J18	J19
APP30	MSR1	MSR2	MSR3	MSR4	MSR5	MSR6
SAN6	SAN7	SAN8	SAN9	SAN10	SAN11	SAN12
8	9	10	11	12	13	14
J20	J21	J22	J23	J24	J25	J26
MSR7	MSR8	MSR9	MSR10	MSR11	MSR12	MSR13
SAN13	SAN14	SAN15	SAN16	SAN17	SAN18	SAN19
15	16	17	18	19	20	21
J27	J28	J29	J30	JU1	JU2	JU3
MSR14	MSR15	MSR16	MSR17	MSR18	MSR19	MSR20
SAN20	SAN21	SAN22	SAN23	SAN24	SAN25	SAN26
22	23	24	25	26	27	28
JU4	JU5	JU6	JU7	JU8	JU9	JU10
MSR21	MRS22	MSR23	MSR24	MSR25	MSR26	MSR27
SAN27	SAN28	SAN29	SAN30	HAM1	HAM2	HAM3

TAIFA WEUSI 476/1992-93/6228-29/1984-85

Ashe	Ojobo	Iwa	Araorun	Ile	Itutu	Erinle
1	2	3	4	5	6	7
AU8	AU9	AU10	AU11	AU12	AU13	AU14
THT20	THT21	THT22	THT23	THT24	THT25	THT26
NAH2	NAH3	NAH4	NAH5	NAH6	NAH7	NAH8
8	9	10	11	12	13	14
AU15	AU16	AU17	AU18	AU19	AU20	AU21
THT27	THT28	THT29	THT30	PAP1	PAP1	PAP2
NAH9	NAH10	NAH11	NAH12	NAH13	NAH14	NAH15
15	16	17	18	19	20	21
AU22	AU23	AU24	AU25	AU26	AU27	AU28
PAP3	PAP4	PAP5	PAP6	PAP7	PAP8	PAP9
NAH16	NAH17	NAH18	NAH19	NAH20	NAH21	NAH22
22	23	24	25	26	27	28
AU29	AU30	AU31	S1	S2	S3	S4
PAP10	PAP11	PAP12	PAP13	PAP14	PAP15	PAP16
NAH23	NAH24	NAH25	NAH26	NAH27	NAH28	NAH29

MTU WEUSI 476/1992-93/6228-29/1984-85

Ashe	Ojobo	Iwa	Araorun	Ile	Itutu	Erinle
1	2	3	4	5	6	7
JU11	JU12	JU13	JU14	JU15	JU16	JU17
MSR28	MSR29	MSR30	*ASR1*	*HRU2*	*STH3*	*AST4*
HAM4	HAM5	HAM6	HAM7	HAM8	HAM9	HAM10
8	9	10	11	12	13	14
JU18	JU19	JU20	JU21	JU22	JU23	JU24
NBH5	*ANU6*	THT1	THT2	THT3	THT4	THT5
HAM11	HAM12	HAM13	HAM14	HAM15	HAM16	HAM17
15	16	17	18	19	20	21
JU25	JU26	JU27	JU28	JU29	JU30	JU31
THT6	THT7	THT8	THT9	THT10	THT11	THT12
HAM18	HAM19	HAM20	HAM21	HAM22	HAM23	HAM24
22	23	24	25	26	27	28
AU1	AU2	AU3	AU4	AU5	AU6	AU7
THT13	THT14	THT15	THT16	THT17	THT18	THT19
HAM25	HAM26	HAM27	HAM28	HAM29	HAM30	NAH1

UWEZO 476/1992-93/6228-29/1984-85

Ashe	Ojobo	Iwa	Araorun	Ile	Itutu	Erinle
1	2	3	4	5	6	7
S5	S6	S7	S8	S9	S10	S11
PAP17	PAP18	PAP19	PAP20	PAP21	PAP22	PAP23
NAH30	*PAG1*	*PAG2*	*PAG3*	*PAG4*	*PAG5*	MAS1
8	9	10	11	12	13	14
S12	S13	S14	S15	S16	S17	S18
PAP24	PAP25	PAP26	PAP27	PAP28	PAP29	PAP30
MAS2	MAS3	MAS4	MAS5	MAS6	MAS7	MAS8
15	16	17	18	19	20	21
S19	S20	S21	S22	S23	S24	S25
HTH1	HTH2	HTH3	HTH4	HTH5	HTH6	HTH7
MAS9	MAS10	MAS11	MAS12	MAS13	MAS14	MAS15
22	23	24	25	26	27	28
S26	S27	S28	S29	S30	O1	O2
HTH8	HTH9	HTH10	HTH11	HTH12	HTH13	HTH14
MAS16	MAS17	MAS18	MAS19	MAS20	MAS21	MAS22

EKUNDU 476/1992-93/6228-29/1984-85						
Ashe	Ojobo	Iwa	Araorun	Ile	Itutu	Erinle
1	2	3	4	5	6	7
O3	O4	O5	O6	O7	O8	O9
HTH15	HTH16	HTH17	HTH18	HTH19	HTH20	HTH21
MAS23	MAS24	MAS25	MAS26	MAS27	MAS28	MAS29
8	9	10	11	12	13	14
O10	O11	O12	O13	O14	O15	O16
HTH22	HTH23	HTH24	HTH25	HTH26	HTH27	HTH28
MAS30	THi1	THi2	THi3	THi4	THi5	THi6
15	16	17	18	19	20	21
O17	O18	O19	O20	O21	O22	O23
HTH29	HTH30	KHK1	KHK2	KHK3	KHK4	KHK5
THi7	THi8	This9	THi10	THi11	THi12	THi13
22	23	24	25	26	27	28
O24	O25	O26	O27	O28	O29	O30
KHK6	KHK7	KHK8	KHK9	KHK10	KHK11	KHK12
Thi14	THi15	THi16	THi17	THi18	THi19	THi20

KIJANI 476/1992-93/6228-29/1984-85						
Ashe	Ojobo	Iwa	Araorun	Ile	Itutu	Erinle
1	2	3	4	5	6	7
N28	N29	N30	D1	D2	D3	D4
TBT11	TBT12	TBT13	TBT14	TBT15	TBT16	TBT17
HID19	HID20	HID21	HID22	HID23	HID24	HID25
8	9	10	11	12	13	14
D5	D6	D7	D8	D9	D10	D11
TBT18	TBT19	TBT20	TBT21	TBT22	TBT23	TBT24
HID26	HID27	HID28	HID29	HID30	TAH1	TAH2
15	16	17	18	19	20	21
D12	D13	D14	D15	D16	D17	D18
TBT25	TBT26	TBT27	TBT28	TBT29	TBT30	MKH1
TAH3	TAH4	TAH5	TAH6	TAH7	TAH8	TAH9
22	23	24	25	26	27	28
D19	D20	D21	D22	D23	D24	D25
MKH2	MKH3	MKH4	MKH5	MKH6	MKH7	MKH8
TAH10	TAH11	TAH12	TAH13	TAH14	TAH15	TAH16

EUSI 476/1992-93/6228-29/1984-85						
Ashe	Ojobo	Iwa	Araorun	Ile	Itutu	Erinle
1	2	3	4	5	6	7
O31	N1	N2	N3	N4	N5	N6
KHK13	KHK14	KHK15	KHK16	KHK17	KHK18	KHK19
THi21	THi22	THi23	THi24	THi25	THi26	THi27
8	9	10	11	12	13	14
N7	N8	N9	N10	N11	N12	N13
KHK20	KHK21	KHK22	KHK23	KHK24	KHK25	KHK26
THi28	THi29	THi30	HID1	HID2	HID3	HID4
15	16	17	18	19	20	21
N14	N15	N16	N17	N18	N19	N20
KHK27	KHK28	KHK29	KHK30	TBT1	TBT2	TBT3
HID5	HID6	HID7	HID8	HID9	HID10	HID11
22	23	24	25	26	27	28
N21	N22	N23	N24	N25	N26	N27
TBT4	TBT5	TBT6	TBT7	TBT8	TBT9	TBT10
HID12	HID13	HID14	HID15	HID16	HID17	HID18

KWANZA 476/1992-93/6228-29/1984-85						
Umoja	Kujichagulia	Ujima	Ujamaa	Nia	Kuumba	Imani
1	2	3	4	5	6	7
D26	D27	D28	D29	D30	D31	JA1
MKH9	MKH10	MKH11	MKH12	MKH13	MKH14	MKH1
TAH17	TAH18	TAH19	TAH20	TAH21	TAH22	TAH2
8	9	10	11	12	13	14
JA2	JA3	JA4	JA5	JA6	JA7	JA
MKH16	MKH17	MKH18	MKH19	MKH20	MKH21	MKH2
TAH24	TAH25	TAH26	TAH27	TAH28	TAH29	TAH3
15	16	17	18	19	20	21
JA9	JA10	JA11	JA12	JA13	JA14	JA1
MKH23	MKH24	MKH25	KMH26	MKH27	MKH28	MKH2
TIR1	TIR2	TIR3	TIR4	TIR5	TIR6	TIR7
22	23	24	25	26	27	28
JA16	JA17	JA18	JA19	JA20	JA21	JA2
MKH30	AMN1	AMN2	AMN3	AMN4	AMN5	AMN
TIR8	TIR9	TIR10	TIR11	TIR12	TIR13	TIR1

SHAHIDI 476/1992-93/6228-29/1984-85						
Ashe	Ojobo	Iwa	Araorun	Ile	Itutu	Erinle
1	2	3	4	5	6	7
JA23	JA24	JA25	JA26	JA27	JA28	JA29
AMN7	AMN8	AMN9	AMN10	AMN11	AMN12	AMN13
TIR15	TIR16	TIR17	TIR18	TIR19	TIR20	TIR21
8	9	10	11	12	13	14
JA30	JA31	F1	F2	F3	F4	F5
AMN14	AMN15	AMN16	AMN17	AMN18	AMN19	AMN20
TIR22	TIR23	TIR24	TIR25	TIR26	TIR27	TIR28
15	16	17	18	19	20	21
F6	F7	F8	F9	F10	F 11	F12
AMN21	AMN22	AMN23	AMN24	AMN25	AMN26	AMN28
TIR29	TIR30	YEK1	YEK2	YEK3	YEK4	YEK5
22	23	24	25	26	27	28
F13	F14	F15	F16	F17	F18	F19
AMN29	AMN30	RNT1	RNT2	RNT3	RNT4	RNT5
YEK6	YEK7	YEK8	YEK9	YEK10	YEK11	YEK12
SHAHIDI 27						
F20						
RNT6						
YEK13						

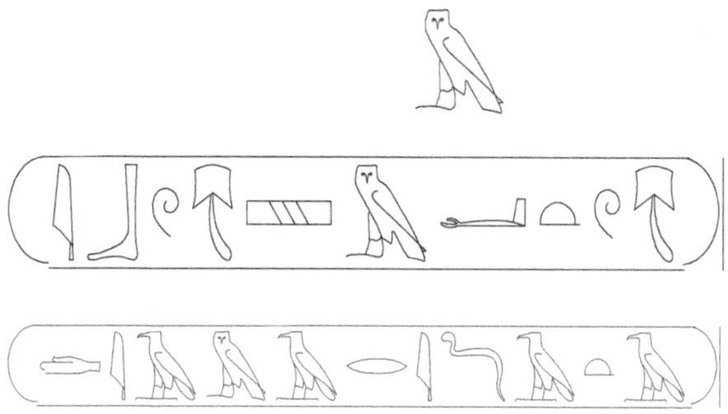

MAAFA 477 B.S.E./ASIENTO 1993-94 C.E./
KAM 6229-30 S.M./EC 1985-86 A.M. (ዓ.ም.)

AFU 477/1993-94/6229-30/1985-86

Ashe	Ojobo	Iwa	Araorun	Ile	Itutu	Erinle
1	2	3	4	5	6	7
21	F22	F23	F24	F25	F26	F27
NT7	RNT8	RNT9	RNT10	RNT11	RNT12	RNT13
EK14	YEK15	YEK16	YEK17	YEK18	YEK19	YEK20
	9	10	11	12	13	14
F28	M1	M2	M3	M4	M5	M6
NT14	RNT15	RNT16	RNT17	RNT18	RNT19	RNT20
EK21	YEK22	YEK23	YEK24	YEK25	YEK26	YEK27
5	16	17	18	19	20	21
M7	M8	M9	M10	M11	M12	M13
NT21	RNT22	RNT23	RNT24	RNT25	RNT26	RNT27
EK28	YEK29	YEK30	MAG1	MAG2	MAG3	MAG4
22	23	24	25	26	27	28
M14	M15	M16	M17	M18	M19	M20
NT28	RNT29	RNT30	KHN1	KHN2	KHN3	KHN4
AG5	MAG6	MAG7	MAG8	MAG9	MAG10	MAG11

KAWAIDA 477/1993-94/6229-30/1985-86

Ashe	Ojobo	Iwa	Araorun	Ile	Itutu	Erinle
1	2	3	4	5	6	7
A18	A19	A20	A21	A22	A23	A24
PIN3	PIN4	PIN5	PIN6	PIN7	PIN8	PIN9
MIY10	MIY11	MIY12	MIY13	MIY14	MIY15	MIY16
8	9	10	11	12	13	14
A25	A26	A27	A28	A29	A30	MA1
PIN10	PIN11	PIN12	PIN13	PIN14	PIN15	PIN16
MIY17	MIY18	MIY19	MIY20	MIY21	MIY22	MIY23
15	16	17	18	19	20	21
MA2	MA3	MA4	MA5	MA6	MA7	MA8
PIN17	PIN18	PIN19	PIN20	PIN21	PIN22	PIN23
MIY24	MIY25	MIY26	MIY27	MIY28	MIY29	MIY30
22	23	24	25	26	27	28
MA9	MA10	MA11	MA12	MA13	MA14	MA15
PIN24	PIN25	PIN26	PIN27	PIN28	PIN29	PIN30
GIN1	GIN2	GIN3	GIN4	GIN5	GIN6	GIN7

UHURU 477/1993-94/6229-30/1985-86

Ashe	Ojobo	Iwa	Araorun	Ile	Itutu	Erinle
1	2	3	4	5	6	7
M21	M22	M23	M24	M25	M26	M27
HN5	KHN6	KHN7	KHN8	KHN9	KHN10	KHN11
AG12	MAG13	MAG14	MAG15	MAG16	MAG17	MAG18
8	9	10	11	12	13	14
M28	M29	M30	M31	A1	A2	A3
HN12	KHN13	KHN14	KHN15	KHN16	KHN17	KHN18
AG19	MAG20	MAG21	MAG22	MAG23	MAG24	MAG25
15	16	17	18	19	20	21
A4	A5	A6	A7	A8	A9	A10
HN19	KHN20	KHN21	KHN22	KHN23	KHN24	KHN25
AG26	MAG27	MAG28	MAG29	MAG30	MIY1	MIY2
22	23	24	25	26	27	28
A11	A12	A13	A14	A15	A16	A17
HN26	KHN27	KHN28	KHN29	KHN30	PIN1	PIN2
IY3	MIY4	MIY5	MIY6	MIY7	MIY8	MIY9

MABABA WEUSI 477/1993-94/6229-30/1985-86

Ashe	Ojobo	Iwa	Araorun	Ile	Itutu	Erinle
1	2	3	4	5	6	7
MA16	MA17	MA18	MA19	MA20	MA21	MA22
APP1	APP2	APP3	APP4	APP5	APP6	APP7
GIN8	GIN9	GIN10	GIN11	GIN12	GIN13	GIN14
8	9	10	11	12	13	14
MA23	MA24	MA25	MA26	MA27	MA28	MA29
APP8	APP9	APP10	APP11	APP12	APP13	APP14
GIN15	GIN16	GIN17	GIN18	GIN19	GIN20	GIN21
15	16	17	18	19	20	21
MA30	MA31	J1	J2	J3	J4	J5
APP15	APP16	APP17	APP18	APP19	APP20	APP21
GIN22	GIN23	GIN24	GIN25	GIN26	GIN27	GIN28
22	23	24	25	26	27	28
J6	J7	J8	J9	J10	J11	J12
APP22	APP23	APP24	APP25	APP26	APP27	APP28
GIN29	GIN30	SAN1	SAN2	SAN3	SAN4	SAN5

MWANAMKE WEUSI 477/1993-94/6229-30/1985-86						
Ashe	Ojobo	Iwa	Araorun	Ile	Itutu	Erinle
1	2	3	4	5	6	7
J13	J14	J15	J16	J17	J18	J19
APP29	APP30	MSR1	MSR2	MSR3	MSR4	MSR5
SAN6	SAN7	SAN8	SAN9	SAN10	SAN11	SAN12
8	9	10	11	12	13	14
J20	J21	J22	J23	J24	J25	J26
MSR6	MSR7	MSR8	MSR9	MSR10	MSR11	MSR12
SAN13	SAN14	SAN15	SAN16	SAN17	SAN18	SAN19
15	16	17	18	19	20	21
J27	J28	J29	J30	JU1	JU2	JU3
MSR13	MSR14	MSR15	MSR16	MSR17	MSR18	MSR19
SAN20	SAN21	SAN22	SAN23	SAN24	SAN25	SAN26
22	23	24	25	26	27	28
JU4	JU5	JU6	JU7	JU8	JU9	JU10
MSR20	MRS21	MSR22	MSR23	MSR24	MSR25	MSR26
SAN27	SAN28	SAN29	SAN30	HAM1	HAM2	HAM3

MTU WEUSI 477/1993-94/6229-30/1985-86						
Ashe	Ojobo	Iwa	Araorun	Ile	Itutu	Erinle
1	2	3	4	5	6	7
JU11	JU12	JU13	JU14	JU15	JU16	JU17
MSR27	MSR28	MSR29	MSR30	*ASR1*	*HRU2*	*STH3*
HAM4	HAM5	HAM6	HAM7	HAM8	HAM9	HAM10
8	9	10	11	12	13	14
JU18	JU19	JU20	JU21	JU22	JU23	JU24
AST4	*NBH5*	THT1	THT2	THT3	THT4	THT5
HAM11	HAM12	HAM13	AM14	HAM15	HAM16	HAM17
15	16	17	18	19	20	21
JU25	JU26	JU27	JU28	JU29	JU30	JU31
THT6	THT7	THT8	THT9	THT10	THT11	THT12
HAM18	HAM19	HAM20	HAM21	HAM22	HAM23	HAM24
22	23	24	25	26	27	28
AU1	AU2	AU3	AU4	AU5	AU6	AU7
THT13	THT14	THT15	THT16	THT17	THT18	THT19
HAM25	HAM26	HAM27	HAM28	HAM29	HAM30	NAH1

TAIFA WEUSI 477/1993-94/6229-30/1985-86						
Ashe	Ojobo	Iwa	Araorun	Ile	Itutu	Erinle
1	2	3	4	5	6	7
AU8	AU9	AU10	AU11	AU12	AU13	AU14
THT20	THT21	THT22	THT23	THT24	THT25	THT26
NAH2	NAH3	NAH4	NAH5	NAH6	NAH7	NAH8
8	9	10	11	12	13	14
AU15	AU16	AU17	AU18	AU19	AU20	AU21
THT27	THT28	THT29	THT30	PAP1	PAP1	PAP2
NAH9	NAH10	NAH11	NAH12	NAH13	NAH14	NAH15
15	16	17	18	19	20	21
AU22	AU23	AU24	AU25	AU26	AU27	AU28
PAP3	PAP4	PAP5	PAP6	PAP7	PAP8	PAP9
NAH16	NAH17	NAH18	NAH19	NAH20	NAH21	NAH22
22	23	24	25	26	27	28
AU29	AU30	AU31	S1	S2	S3	S4
PAP10	PAP11	PAP12	PAP13	PAP14	PAP15	PAP16
NAH23	NAH24	NAH25	NAH26	NAH27	NAH28	NAH29

UWEZO 477/1993-94/6229-30/1985-86						
Ashe	Ojobo	Iwa	Araorun	Ile	Itutu	Erinle
1	2	3	4	5	6	7
S5	S6	S7	S8	S9	S10	S11
PAP17	PAP18	PAP19	PAP20	PAP21	PAP22	PAP23
NAH30	*PAG1*	*PAG2*	*PAG3*	*PAG4*	*PAG5*	MAS1
8	9	10	11	12	13	14
S12	S13	S14	S15	S16	S17	S18
PAP24	PAP25	PAP26	PAP27	PAP28	PAP29	PAP30
MAS2	MAS3	MAS4	MAS5	MAS6	MAS7	MAS8
15	16	17	18	19	20	21
S19	S20	S21	S22	S23	S24	S25
HTH1	HTH2	HTH3	HTH4	HTH5	HTH6	HTH7
MAS9	MAS10	MAS11	MAS12	MAS13	MAS14	MAS15
22	23	24	25	26	27	28
S26	S27	S28	S29	S30	O1	O2
HTH8	HTH9	HTH10	HTH11	HTH12	HTH13	HTH14
MAS16	MAS17	MAS18	MAS19	MAS20	MAS21	MAS22

EKUNDU 477/1993-94/6229-30/1985-86						
Ashe	Ojobo	Iwa	Araorun	Ile	Itutu	Erinle
1	2	3	4	5	6	7
O3	O4	O5	O6	O7	O8	O9
HTH15	HTH16	HTH17	HTH18	HTH19	HTH20	HTH21
MAS23	MAS24	MAS25	MAS26	MAS27	MAS28	MAS29
8	9	10	11	12	13	14
O10	O11	O12	O13	O14	O15	O16
HTH22	HTH23	HTH24	HTH25	HTH26	HTH27	HTH28
MAS30	THi1	THi2	THi3	THi4	THi5	THi6
15	16	17	18	19	20	21
O17	O18	O19	O20	O21	O22	O23
HTH29	HTH30	KHK1	KHK2	KHK3	KHK4	KHK5
THi7	THi8	THi9	THi10	THi11	THi12	THi13
22	23	24	25	26	27	28
O24	O25	O26	O27	O28	O29	O30
KHK6	KHK7	KHK8	KHK9	KHK10	KHK11	KHK12
THi14	THi15	THi16	THi17	THi18	THi19	THi20

KIJANI 477/1993-94/6229-30/1985-86						
Ashe	Ojobo	Iwa	Araorun	Ile	Itutu	Erinle
1	2	3	4	5	6	7
N28	N29	N30	D1	D2	D3	D4
TBT11	TBT12	TBT13	TBT14	TBT15	TBT16	TBT17
HID19	HID20	HID21	HID22	HID23	HID24	HID25
8	9	10	11	12	13	14
D5	D6	D7	D8	D9	D10	D11
TBT18	TBT19	TBT20	TBT21	TBT22	TBT23	TBT24
HID26	HID27	HID28	HID29	HID30	TAH1	TAH2
15	16	17	18	19	20	21
D12	D13	D14	D15	D16	D17	D18
TBT25	TBT26	TBT27	TBT28	TBT29	TBT30	MKH1
TAH3	TAH4	TAH5	TAH6	TAH7	TAH8	TAH9
22	23	24	25	26	27	28
D19	D20	D21	D22	D23	D24	D25
MKH2	MKH3	MKH4	MKH5	MKH6	MKH7	MKH8
TAH10	TAH11	TAH12	TAH13	TAH14	TAH15	TAH16

EUSI 477/1993-94/6229-30/1985-86						
Ashe	Ojobo	Iwa	Araorun	Ile	Itutu	Erinle
1	2	3	4	5	6	7
O31	N1	N2	N3	N4	N5	N6
KHK13	KHK14	KHK15	KHK16	KHK17	KHK18	KHK19
THi21	THi22	THi23	THi24	THi25	THi26	THi27
8	9	10	11	12	13	14
N7	N8	N9	N10	N11	N12	N13
KHK20	KHK21	KHK22	KHK23	KHK24	KHK25	KHK26
THi28	THi29	THi30	HID1	HID2	HID3	HID4
15	16	17	18	19	20	21
N14	N15	N16	N17	N18	N19	N20
KHK27	KHK28	KHK29	KHK30	TBT1	TBT2	TBT3
HID5	HID6	HID7	HID8	HID9	HID10	HID11
22	23	24	25	26	27	28
N21	N22	N23	N24	N25	N26	N27
TBT4	TBT5	TBT6	TBT7	TBT8	TBT9	TBT10
HID12	HID13	HID14	HID15	HID16	HID17	HID18

KWANZA 477/1993-94/6229-30/1985-86						
Umoja	Kujichagulia	Ujima	Ujamaa	Nia	Kuumba	Imani
1	2	3	4	5	6	7
D26	D27	D28	D29	D30	D31	JA1
MKH9	MKH10	MKH11	MKH12	MKH13	MKH14	MKH15
TAH17	TAH18	TAH19	TAH20	TAH21	TAH22	TAH23
8	9	10	11	12	13	14
JA2	JA3	JA4	JA5	JA6	JA7	JA8
MKH16	MKH17	MKH18	MKH19	MKH20	MKH21	MKH22
TAH24	TAH25	TAH26	TAH27	TAH28	TAH29	TAH30
15	16	17	18	19	20	21
JA9	JA10	JA11	JA12	JA13	JA14	JA15
MKH23	MKH24	MKH25	KMH26	MKH27	MKH28	MKH29
TIR1	TIR2	TIR3	TIR4	TIR5	TIR6	TIR7
22	23	24	25	26	27	28
JA16	JA17	JA18	JA19	JA20	JA21	JA22
MKH30	AMN1	AMN2	AMN3	AMN4	AMN5	AMN6
TIR8	TIR9	TIR10	TIR11	TIR12	TIR13	TIR14

SHAHIDI 477/1993-94/6229-30/1985-86						
Ashe	**Ojobo**	**Iwa**	**Araorun**	**Ile**	**Itutu**	**Erinle**
1	2	3	4	5	6	7
JA23	JA24	JA25	JA26	JA27	JA28	JA29
AMN7	AMN8	AMN9	AMN10	AMN11	AMN12	AMN13
TIR15	TIR16	TIR17	TIR18	TIR19	TIR20	TIR21
8	9	10	11	12	13	14
JA30	JA31	F1	F2	F3	F4	F5
AMN14	AMN15	AMN16	AMN17	AMN18	AMN19	AMN20
TIR22	TIR23	TIR24	TIR25	TIR26	TIR27	TIR28
15	16	17	18	19	20	21
F6	F7	F8	F9	F10	F 11	F12
AMN21	AMN22	AMN23	AMN24	AMN25	AMN26	AMN28
TIR29	TIR30	YEK1	YEK2	YEK3	YEK4	YEK5
22	23	24	25	26	27	28
F13	F14	F15	F16	F17	F18	F19
AMN29	AMN30	RNT1	RNT2	RNT3	RNT4	RNT5
YEK6	YEK7	YEK8	YEK9	YEK10	YEK11	YEK12
SHAHIDI 28						
F20						
RNT6						
YEK13						

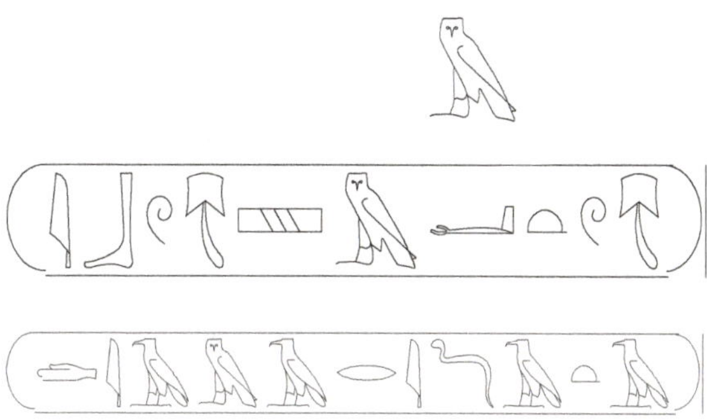

MAAFA 478 B.S.E./ASIENTO 1994-95 C.E./
KAM 6230-31 S.M./EC 1986-87 A.M. (ዓ.ም.)

AFU 478/1994-95/6230-31/1986-87

Ashe	Ojobo	Iwa	Araorun	Ile	Itutu	Erinle
1	2	3	4	5	6	7
F21	F22	F23	F24	F25	F26	F27
RNT7	RNT8	RNT9	RNT10	RNT11	RNT12	RNT13
YEK14	YEK15	YEK16	YEK17	YEK18	YEK19	YEK20
8	9	10	11	12	13	14
F28	M1	M2	M3	M4	M5	M6
RNT14	RNT15	RNT16	RNT17	RNT18	RNT19	RNT20
YEK21	YEK22	YEK23	YEK24	YEK25	YEK26	YEK27
15	16	17	18	19	20	21
M7	M8	M9	M10	M11	M12	M13
RNT21	RNT22	RNT23	RNT24	RNT25	RNT26	RNT27
YEK28	YEK29	YEK30	MAG1	MAG2	MAG3	MAG4
22	23	24	25	26	27	28
M14	M15	M16	M17	M18	M19	M20
RNT28	RNT29	RNT30	KHN1	KHN2	KHN3	KHN4
MAG5	MAG6	MAG7	MAG8	MAG9	MAG10	MAG11

KAWAIDA 478/1994-95/6230-31/1986-87

Ashe	Ojobo	Iwa	Araorun	Ile	Itutu	Erinle
1	2	3	4	5	6	7
A18	A19	A20	A21	A22	A23	A24
PIN3	PIN4	PIN5	PIN6	PIN7	PIN8	PIN9
MIY10	MIY11	MIY12	MIY13	MIY14	MIY15	MIY16
8	9	10	11	12	13	14
A25	A26	A27	A28	A29	A30	MA1
PIN10	PIN11	PIN12	PIN13	PIN14	PIN15	PIN16
MIY17	MIY18	MIY19	MIY20	MIY21	MIY22	MIY23
15	16	17	18	19	20	21
MA2	MA3	MA4	MA5	MA6	MA7	MA8
PIN17	PIN18	PIN19	PIN20	PIN21	PIN22	PIN23
MIY24	MIY25	MIY26	MIY27	MIY28	MIY29	MIY30
22	23	24	25	26	27	28
MA9	MA10	MA11	MA12	MA13	MA14	MA15
PIN24	PIN25	PIN26	PIN27	PIN28	PIN29	PIN30
GIN1	GIN2	GIN3	GIN4	GIN5	GIN6	GIN7

UHURU 478/1994-95/6230-31/1986-87

Ashe	Ojobo	Iwa	Araorun	Ile	Itutu	Erinle
1	2	3	4	5	6	7
M21	M22	M23	M24	M25	M26	M27
KHN5	KHN6	KHN7	KHN8	KHN9	KHN10	KHN11
MAG12	MAG13	MAG14	MAG15	MAG16	MAG17	MAG18
8	9	10	11	12	13	14
M28	M29	M30	M31	A1	A2	A3
KHN12	KHN13	KHN14	KHN15	KHN16	KHN17	KHN18
MAG19	MAG20	MAG21	MAG22	MAG23	MAG24	MAG25
15	16	17	18	19	20	21
A4	A5	A6	A7	A8	A9	A10
KHN19	KHN20	KHN21	KHN22	KHN23	KHN24	KHN25
MAG26	MAG27	MAG28	MAG29	MAG30	MIY1	MIY2
22	23	24	25	26	27	28
A11	A12	A13	A14	A15	A16	A17
KHN26	KHN27	KHN28	KHN29	KHN30	PIN1	PIN2
MIY3	MIY4	MIY5	MIY6	MIY7	MIY8	MIY9

MABABA WEUSI 478/1994-95/6230-31/1986-87

Ashe	Ojobo	Iwa	Araorun	Ile	Itutu	Erinle
1	2	3	4	5	6	7
MA16	MA17	MA18	MA19	MA20	MA21	MA22
APP1	APP2	APP3	APP4	APP5	APP6	APP7
GIN8	GIN9	GIN10	GIN11	GIN12	GIN13	GIN14
8	9	10	11	12	13	14
MA23	MA24	MA25	MA26	MA27	MA28	MA29
APP8	APP9	APP10	APP11	APP12	APP13	APP14
GIN15	GIN16	GIN17	GIN18	GIN19	GIN20	GIN21
15	16	17	18	19	20	21
MA30	MA31	J1	J2	J3	J4	J5
APP15	APP16	APP17	APP18	APP19	APP20	APP21
GIN22	GIN23	GIN24	GIN25	GIN26	GIN27	GIN28
22	23	24	25	26	27	28
J6	J7	J8	J9	J10	J11	J12
APP22	APP23	APP24	APP25	APP26	APP27	APP28
GIN29	GIN30	SAN1	SAN2	SAN3	SAN4	SAN5

MWANAMKE WEUSI 478/1994-95/6230-31/1986-87

Ashe	Ojobo	Iwa	Araorun	Ile	Itutu	Erinle
1	2	3	4	5	6	7
J13	J14	J15	J16	J17	J18	J19
APP29	APP30	MSR1	MSR2	MSR3	MSR4	MSR5
SAN6	SAN7	SAN8	SAN9	SAN10	SAN11	SAN12
8	9	10	11	12	13	14
J20	J21	J22	J23	J24	J25	J26
MSR6	MSR7	MSR8	MSR9	MSR10	MSR11	MSR12
SAN13	SAN14	SAN15	SAN16	SAN17	SAN18	SAN19
15	16	17	18	19	20	21
J27	J28	J29	J30	JU1	JU2	JU3
MSR13	MSR14	MSR15	MSR16	MSR17	MSR18	MSR19
SAN20	SAN21	SAN22	SAN23	SAN24	SAN25	SAN26
22	23	24	25	26	27	28
JU4	JU5	JU6	JU7	JU8	JU9	JU10
MSR20	MRS21	MSR22	MSR23	MSR24	MSR25	MSR26
SAN27	SAN28	SAN29	SAN30	HAM1	HAM2	HAM3

TAIFA WEUSI 478/1994-95/6230-31/1986-87

Ashe	Ojobo	Iwa	Araorun	Ile	Itutu	Erinle
1	2	3	4	5	6	7
AU8	AU9	AU10	AU11	AU12	AU13	AU14
THT20	THT21	THT22	THT23	THT24	THT25	THT26
NAH2	NAH3	NAH4	NAH5	NAH6	NAH7	NAH8
8	9	10	11	12	13	14
AU15	AU16	AU17	AU18	AU19	AU20	AU21
THT27	THT28	THT29	THT30	PAP1	PAP1	PAP2
NAH9	NAH10	NAH11	NAH12	NAH13	NAH14	NAH15
15	16	17	18	19	20	21
AU22	AU23	AU24	AU25	AU26	AU27	AU28
PAP3	PAP4	PAP5	PAP6	PAP7	PAP8	PAP9
NAH16	NAH17	NAH18	NAH19	NAH20	NAH21	NAH22
22	23	24	25	26	27	28
AU29	AU30	AU31	S1	S2	S3	S4
PAP10	PAP11	PAP12	PAP13	PAP14	PAP15	PAP16
NAH23	NAH24	NAH25	NAH26	NAH27	NAH28	NAH29

MTU WEUSI 478/1994-95/6230-31/1986-87

Ashe	Ojobo	Iwa	Araorun	Ile	Itutu	Erinle
1	2	3	4	5	6	7
JU11	JU12	JU13	JU14	JU15	JU16	JU17
MSR27	MSR28	MSR29	MSR30	*ASR1*	*HRU2*	*STH3*
HAM4	HAM5	HAM6	HAM7	HAM8	HAM9	HAM10
8	9	10	11	12	13	14
JU18	JU19	JU20	JU21	JU22	JU23	JU24
AST4	*NBH5*	THT1	THT2	THT3	THT4	THT5
HAM11	HAM12	HAM13	AM14	HAM15	HAM16	HAM17
15	16	17	18	19	20	21
JU25	JU26	JU27	JU28	JU29	JU30	JU31
THT6	THT7	THT8	THT9	THT10	THT11	THT12
HAM18	HAM19	HAM20	HAM21	HAM22	HAM23	HAM24
22	23	24	25	26	27	28
AU1	AU2	AU3	AU4	AU5	AU6	AU7
THT13	THT14	THT15	THT16	THT17	THT18	THT19
HAM25	HAM26	HAM27	HAM28	HAM29	HAM30	NAH1

UWEZO 478/1994-95/6230-31/1986-87

Ashe	Ojobo	Iwa	Araorun	Ile	Itutu	Erinle
1	2	3	4	5	6	7
S5	S6	S7	S8	S9	S10	S11
PAP17	PAP18	PAP19	PAP20	PAP21	PAP22	PAP23
NAH30	*PAG1*	*PAG2*	*PAG3*	*PAG4*	*PAG5*	MAS1
8	9	10	11	12	13	14
S12	S13	S14	S15	S16	S17	S18
PAP24	PAP25	PAP26	PAP27	PAP28	PAP29	PAP30
MAS2	MAS3	MAS4	MAS5	MAS6	MAS7	MAS8
15	16	17	18	19	20	21
S19	S20	S21	S22	S23	S24	S25
HTH1	HTH2	HTH3	HTH4	HTH5	HTH6	HTH7
MAS9	MAS10	MAS11	MAS12	MAS13	MAS14	MAS15
22	23	24	25	26	27	28
S26	S27	S28	S29	S30	O1	O2
HTH8	HTH9	HTH10	HTH11	HTH12	HTH13	HTH14
MAS16	MAS17	MAS18	MAS19	MAS20	MAS21	MAS22

EKUNDU 478/1994-95/6230-31/1986-87						
Ashe	Ojobo	Iwa	Araorun	Ile	Itutu	Erinle
1	2	3	4	5	6	7
O3	O4	O5	O6	O7	O8	O9
HTH15	HTH16	HTH17	HTH18	HTH19	HTH20	HTH21
MAS23	MAS24	MAS25	MAS26	MAS27	MAS28	MAS29
8	9	10	11	12	13	14
O10	O11	O12	O13	O14	O15	O16
HTH22	HTH23	HTH24	HTH25	HTH26	HTH27	HTH28
MAS30	THi1	THi2	THi3	THi4	THi5	THi6
15	16	17	18	19	20	21
O17	O18	O19	O20	O21	O22	O23
HTH29	HTH30	KHK1	KHK2	KHK3	KHK4	KHK5
THi7	THi8	THi9	THi10	THi11	THi12	THi13
22	23	24	25	26	27	28
O24	O25	O26	O27	O28	O29	O30
KHK6	KHK7	KHK8	KHK9	KHK10	KHK11	KHK12
THi14	THi15	THi16	THi17	THi18	THi19	THi20

KIJANI 478/1994-95/6230-31/1986-87						
Ashe	Ojobo	Iwa	Araorun	Ile	Itutu	Erinle
1	2	3	4	5	6	7
N28	N29	N30	D1	D2	D3	D4
TBT11	TBT12	TBT13	TBT14	TBT15	TBT16	TBT17
HID19	HID20	HID21	HID22	HID23	HID24	HID25
8	9	10	11	12	13	14
D5	D6	D7	D8	D9	D10	D11
TBT18	TBT19	TBT20	TBT21	TBT22	TBT23	TBT24
HID26	HID27	HID28	HID29	HID30	TAH1	TAH2
15	16	17	18	19	20	21
D12	D13	D14	D15	D16	D17	D18
TBT25	TBT26	TBT27	TBT28	TBT29	TBT30	MKH1
TAH3	TAH4	TAH5	TAH6	TAH7	TAH8	TAH9
22	23	24	25	26	27	28
D19	D20	D21	D22	D23	D24	D25
MKH2	MKH3	MKH4	MKH5	MKH6	MKH7	MKH8
TAH10	TAH11	TAH12	TAH13	TAH14	TAH15	TAH16

EUSI 478/1994-95/6230-31/1986-87						
Ashe	Ojobo	Iwa	Araorun	Ile	Itutu	Erinle
1	2	3	4	5	6	7
O31	N1	N2	N3	N4	N5	N6
KHK13	KHK14	KHK15	KHK16	KHK17	KHK18	KHK19
THi21	THi22	THi23	THi24	THi25	THi26	THi27
8	9	10	11	12	13	14
N7	N8	N9	N10	N11	N12	N13
KHK20	KHK21	KHK22	KHK23	KHK24	KHK25	KHK26
THi28	THi29	THi30	HID1	HID2	HID3	HID4
15	16	17	18	19	20	21
N14	N15	N16	N17	N18	N19	N20
KHK27	KHK28	KHK29	KHK30	TBT1	TBT2	TBT3
HID5	HID6	HID7	HID8	HID9	HID10	HID11
22	23	24	25	26	27	28
N21	N22	N23	N24	N25	N26	N27
TBT4	TBT5	TBT6	TBT7	TBT8	TBT9	TBT10
HID12	HID13	HID14	HID15	HID16	HID17	HID18

KWANZA 478/1994-95/6230-31/1986-87						
Umoja	Kujichagulia	Ujima	Ujamaa	Nia	Kuumba	Imani
1	2	3	4	5	6	7
D26	D27	D28	D29	D30	D31	JA1
MKH9	MKH10	MKH11	MKH12	MKH13	MKH14	MKH15
TAH17	TAH18	TAH19	TAH20	TAH21	TAH22	TAH23
8	9	10	11	12	13	14
JA2	JA3	JA4	JA5	JA6	JA7	JA8
MKH16	MKH17	MKH18	MKH19	MKH20	MKH21	MKH22
TAH24	TAH25	TAH26	TAH27	TAH28	TAH29	TAH30
15	16	17	18	19	20	21
JA9	JA10	JA11	JA12	JA13	JA14	JA15
MKH23	MKH24	MKH25	KMH26	MKH27	MKH28	MKH29
TIR1	TIR2	TIR3	TIR4	TIR5	TIR6	TIR7
22	23	24	25	26	27	28
JA16	JA17	JA18	JA19	JA20	JA21	JA22
MKH30	AMN1	AMN2	AMN3	AMN4	AMN5	AMN6
TIR8	TIR9	TIR10	TIR11	TIR12	TIR13	TIR14

SHAHIDI 478/1994-95/6230-31/1986-87						
Ashe	Ojobo	Iwa	Araorun	Ile	Itutu	Erinle
1	2	3	4	5	6	7
JA23	JA24	JA25	JA26	JA27	JA28	JA29
AMN7	AMN8	AMN9	AMN10	AMN11	AMN12	AMN13
TIR15	TIR16	TIR17	TIR18	TIR19	TIR20	TIR21
8	9	10	11	12	13	14
JA30	JA31	F1	F2	F3	F4	F5
AMN14	AMN15	AMN16	AMN17	AMN18	AMN19	AMN20
TIR22	TIR23	TIR24	TIR25	TIR26	TIR27	TIR28
15	16	17	18	19	20	21
F6	F7	F8	F9	F10	F 11	F12
AMN21	AMN22	AMN23	AMN24	AMN25	AMN26	AMN28
TIR29	TIR30	YEK1	YEK2	YEK3	YEK4	YEK5
22	23	24	25	26	27	28
F13	F14	F15	F16	F17	F18	F19
AMN29	AMN30	RNT1	RNT2	RNT3	RNT4	RNT5
YEK6	YEK7	YEK8	YEK9	YEK10	YEK11	YEK12
SHAHIDI 29						
F20						
RNT6						
YEK13						

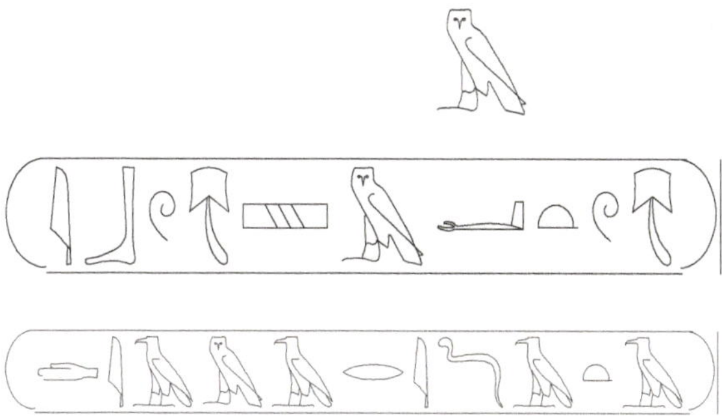

MAAFA 479 B.S.E./ASIENTO 1995-96 C.E./
KAM 6231-32 S.M./EC 1987-88 A.M. (ዓ.ም.)

AFU 479/1995-96/6231-32/1987-88

Ashe	Ojobo	Iwa	Araorun	Ile	Itutu	Erinle
1	2	3	4	5	6	7
F21	F22	F23	F24	F25	F26	F27
RNT7	RNT8	RNT9	RNT10	RNT11	RNT12	RNT13
YEK14	YEK15	YEK16	YEK17	YEK18	YEK19	YEK20
8	9	10	11	12	13	14
F28	M1	M2	M3	M4	M5	M6
RNT14	RNT15	RNT16	RNT17	RNT18	RNT19	RNT20
YEK21	YEK22	YEK23	YEK24	YEK25	YEK26	YEK27
15	16	17	18	19	20	21
M7	M8	M9	M10	M11	M12	M13
RNT21	RNT22	RNT23	RNT24	RNT25	RNT26	RNT27
YEK28	YEK 29	YEK30	MAG1	MAG2	MAG3	MAG4
22	23	24	25	26	27	28
M14	M15	M16	M17	M18	M19	M20
RNT28	RNT29	RNT30	KHN1	KHN2	KHN3	KHN4
MAG5	MAG6	MAG7	MAG8	MAG9	MAG10	MAG11

UHURU 479/1995-96/6231-32/1987-88

Ashe	Ojobo	Iwa	Araorun	Ile	Itutu	Erinle
1	2	3	4	5	6	7
M21	M22	M23	M24	M25	M26	M27
KHN5	KHN6	KHN7	KHN8	KHN9	KHN10	KHN11
MAG12	MAG13	MAG14	MAG15	MAG16	MAG17	MAG18
8	9	10	11	12	13	14
M28	M29	M30	M31	A1	A2	A3
KHN12	KHN13	KHN14	KHN15	KHN16	KHN17	KHN18
MAG19	MAG20	MAG21	MAG22	MAG23	MAG24	MAG25
15	16	17	18	19	20	21
A4	A5	A6	A7	A8	A9	A10
KHN19	KHN20	KHN21	KHN22	KHN23	KHN24	KHN25
MAG26	MAG27	MAG28	MAG29	MAG30	MIY1	MIY2
22	23	24	25	26	27	28
A11	A12	A13	A14	A15	A16	A17
KHN26	KHN27	KHN28	KHN29	KHN30	PIN1	PIN2
MIY3	MIY4	MIY5	MIY6	MIY7	MIY8	MIY9

KAWAIDA 479/1995-96/6231-32/1987-88

Ashe	Ojobo	Iwa	Araorun	Ile	Itutu	Erinle
1	2	3	4	5	6	7
A18	A19	A20	A21	A22	A23	A24
PIN3	PIN4	PIN5	PIN6	PIN7	PIN8	PIN9
MIY10	MIY11	MIY12	MIY13	MIY14	MIY15	MIY16
8	9	10	11	12	13	14
A25	A26	A27	A28	A29	A30	MA1
PIN10	PIN11	PIN12	PIN13	PIN14	PIN15	PIN16
MIY17	MIY18	MIY19	MIY20	MIY21	MIY22	MIY23
15	16	17	18	19	20	21
MA2	MA3	MA4	MA5	MA6	MA7	MA8
PIN17	PIN18	PIN19	PIN20	PIN21	PIN22	PIN23
MIY24	MIY25	MIY26	MIY27	MIY28	MIY29	MIY30
22	23	24	25	26	27	28
MA9	MA10	MA11	MA12	MA13	MA14	MA15
PIN24	PIN25	PIN26	PIN27	PIN28	PIN29	PIN30
GIN1	GIN2	GIN3	GIN4	GIN5	GIN6	GIN7

MABABA WEUSI 479/1995-96/6231-32/1987-88

Ashe	Ojobo	Iwa	Araorun	Ile	Itutu	Erinle
1	2	3	4	5	6	7
MA16	MA17	MA18	MA19	MA20	MA21	MA22
APP1	APP2	APP3	APP4	APP5	APP6	APP7
GIN8	GIN9	GIN10	GIN11	GIN12	GIN13	GIN14
8	9	10	11	12	13	14
MA23	MA24	MA25	MA26	MA27	MA28	MA29
APP8	APP9	APP10	APP11	APP12	APP13	APP14
GIN15	GIN16	GIN17	GIN18	GIN19	GIN20	GIN21
15	16	17	18	19	20	21
MA30	MA31	J1	J2	J3	J4	J5
APP15	APP16	APP17	APP18	APP19	APP20	APP21
GIN22	GIN23	GIN24	GIN25	GIN26	GIN27	GIN28
22	23	24	25	26	27	28
J6	J7	J8	J9	J10	J11	J12
APP22	APP23	APP24	APP25	APP26	APP27	APP28
GIN29	GIN30	SAN1	SAN2	SAN3	SAN4	SAN5

MWANAMKE WEUSI 479/1995-96/6231-32/1987-88

Ashe	Ojobo	Iwa	Araorun	Ile	Itutu	Erinle
1	2	3	4	5	6	7
J13	J14	J15	J16	J17	J18	J19
APP29	APP30	MSR1	MSR2	MSR3	MSR4	MSR5
SAN6	SAN7	SAN8	SAN9	SAN10	SAN11	SAN12
8	9	10	11	12	13	14
J20	J21	J22	J23	J24	J25	J26
MSR6	MSR7	MSR8	MSR9	MSR10	MSR11	MSR12
SAN13	SAN14	SAN15	SAN16	SAN17	SAN18	SAN19
15	16	17	18	19	20	21
J27	J28	J29	J30	JU1	JU2	JU3
MSR13	MSR14	MSR15	MSR16	MSR17	MSR18	MSR19
SAN20	SAN21	SAN22	SAN23	SAN24	SAN25	SAN26
22	23	24	25	26	27	28
JU4	JU5	JU6	JU7	JU8	JU9	JU10
MSR20	MRS21	MSR22	MSR23	MSR24	MSR25	MSR26
SAN27	SAN28	SAN29	SAN30	HAM1	HAM2	HAM3

TAIFA WEUSI 479/1995-96/6231-32/1987-88

Ashe	Ojobo	Iwa	Araorun	Ile	Itutu	Erinle
1	2	3	4	5	6	7
AU8	AU9	AU10	AU11	AU12	AU13	AU14
THT20	THT21	THT22	THT23	THT24	THT25	THT26
NAH2	NAH3	NAH4	NAH5	NAH6	NAH7	NAH8
8	9	10	11	12	13	14
AU15	AU16	AU17	AU18	AU19	AU20	AU21
THT27	THT28	THT29	THT30	PAP1	PAP1	PAP2
NAH9	NAH10	NAH11	NAH12	NAH13	NAH14	NAH15
15	16	17	18	19	20	21
AU22	AU23	AU24	AU25	AU26	AU27	AU28
PAP3	PAP4	PAP5	PAP6	PAP7	PAP8	PAP9
NAH16	NAH17	NAH18	NAH19	NAH20	NAH21	NAH22
22	23	24	25	26	27	28
AU29	AU30	AU31	S1	S2	S3	S4
PAP10	PAP11	PAP12	PAP13	PAP14	PAP15	PAP16
NAH23	NAH24	NAH25	NAH26	NAH27	NAH28	NAH29

MTU WEUSI 479/1995-96/6231-32/1987-88

Ashe	Ojobo	Iwa	Araorun	Ile	Itutu	Erinle
1	2	3	4	5	6	7
JU11	JU12	JU13	JU14	JU15	JU16	JU17
MSR27	MSR28	MSR29	MSR30	*ASR1*	*HRU2*	*STH3*
HAM4	HAM5	HAM6	HAM7	HAM8	HAM9	HAM10
8	9	10	11	12	13	14
JU18	JU19	JU20	JU21	JU22	JU23	JU24
AST4	*NBH5*	THT1	THT2	THT3	THT4	THT5
HAM11	HAM12	HAM13	HAM14	HAM15	HAM16	HAM17
15	16	17	18	19	20	21
JU25	JU26	JU27	JU28	JU29	JU30	JU31
THT6	THT7	THT8	THT9	THT10	THT11	THT12
HAM18	HAM19	HAM20	HAM21	HAM22	HAM23	HAM24
22	23	24	25	26	27	28
AU1	AU2	AU3	AU4	AU5	AU6	AU7
THT13	THT14	THT15	THT16	THT17	THT18	THT19
HAM25	HAM26	HAM27	HAM28	HAM29	HAM30	NAH1

UWEZO 479/1995-96/6231-32/1987-88

Ashe	Ojobo	Iwa	Araorun	Ile	Itutu	Erinle
1	2	3	4	5	6	7
S5	S6	S7	S8	S9	S10	S11
PAP17	PAP18	PAP19	PAP20	PAP21	PAP22	PAP23
NAH30	*QUA1*	*QUA2*	*QUA3*	*QUA4*	*QUA5*	*QUA6*
8	9	10	11	12	13	14
S12	S13	S14	S15	S16	S17	S18
PAP24	PAP25	PAP26	PAP27	PAP28	PAP29	PAP30
MAS1	MAS2	MAS3	MAS4	MAS5	MAS6	MAS7
15	16	17	18	19	20	21
S19	S20	S21	S22	S23	S24	S25
HTH1	HTH2	HTH3	HTH4	HTH5	HTH6	HTH7
MAS8	MAS9	MAS10	MAS11	MAS12	MAS13	MAS14
22	23	24	25	26	27	28
S26	S27	S28	S29	S30	O1	O2
HTH8	HTH9	HTH10	HTH11	HTH12	HTH13	HTH14
MAS15	MAS16	MAS17	MAS18	MAS19	MAS20	MAS21

EKUNDU 479/1995-96/6231-32/1987-88						
Ashe	Ojobo	Iwa	Araorun	Ile	Itutu	Erinle
1	2	3	4	5	6	7
O3	O4	O5	O6	O7	O8	O9
HTH15	HTH16	HTH17	HTH18	HTH19	HTH20	HTH21
MAS22	MAS23	MAS 24	MAS25	MAS26	MAS27	MAS28
8	9	10	11	12	13	14
O10	O11	O12	O13	O14	O15	O16
HTH22	HTH23	HTH24	HTH25	HTH26	HTH27	HTH28
MAS29	MAS30	THi1	THi2	THi3	THi4	THi5
15	16	17	18	19	20	21
O17	O18	O19	O20	O21	O22	O23
HTH29	HTH30	KHK1	KHK2	KHK3	KHK4	KHK5
THi6	THi7	THi8	THi9	THi10	THi11	THi12
22	23	24	25	26	27	28
O24	O25	O26	O27	O28	O29	O30
KHK6	KHK7	KHK8	KHK9	KHK10	KHK11	KHK12
THi13	THi14	THi15	THi16	THi17	THi18	THi19

KIJANI 479/1995-96/6231-32/1987-88						
Ashe	Ojobo	Iwa	Araorun	Ile	Itutu	Erinle
1	2	3	4	5	6	7
N28	N29	N30	D1	D2	D3	D4
TBT11	TBT12	TBT13	TBT14	TBT15	TBT16	TBT17
HID18	HID19	HID20	HID21	HID22	HID23	HID24
8	9	10	11	12	13	14
D5	D6	D7	D8	D9	D10	D11
TBT18	TBT19	TBT20	TBT21	TBT22	TBT23	TBT24
HID25	HID26	HID27	HID28	HID29	HID30	TAH1
15	16	17	18	19	20	21
D12	D13	D14	D15	D16	D17	D18
TBT25	TBT26	TBT27	TBT28	TBT29	TBT30	MKH1
TAH2	TAH3	TAH4	TAH5	TAH6	TAH7	TAH8
22	23	24	25	26	27	28
D19	D20	D21	D22	D23	D24	D25
MKH2	MKH3	MKH4	MKH5	MKH6	MKH7	MKH8
TAH9	TAH10	TAH11	TAH12	TAH13	TAH14	TAH15

EUSI 479/1995-96/6231-32/1987-88						
Ashe	Ojobo	Iwa	Araorun	Ile	Itutu	Erinle
1	2	3	4	5	6	7
O31	N1	N2	N3	N4	N5	N6
KHK13	KHK14	KHK15	KHK16	KHK17	KHK18	KHK19
THi20	THi21	THi22	THi23	THi24	THi25	THi26
8	9	10	11	12	13	14
N7	N8	N9	N10	N11	N12	N13
KHK20	KHK21	KHK22	KHK23	KHK24	KHK25	KHK26
THi27	THi28	THi29	THi30	HID1	HID2	HID3
15	16	17	18	19	20	21
N14	N15	N16	N17	N18	N19	N20
KHK27	KHK28	KHK29	KHK30	TBT1	TBT2	TBT3
HID4	HID5	HID6	HID7	HID8	HID9	HID10
22	23	24	25	26	27	28
N21	N22	N23	N24	N25	N26	N27
TBT4	TBT5	TBT6	TBT7	TBT8	TBT9	TBT10
HID11	HID12	HID13	HID14	HID15	HID16	HID17

KWANZA 479/1995-96/6231-32/1987-88						
Umoja	Kujichagulia	Ujima	Ujamaa	Nia	Kuumba	Imani
1	2	3	4	5	6	7
D26	D27	D28	D29	D30	D31	JA1
MKH9	MKH10	MKH11	MKH12	MKH13	MKH14	MKH15
TAH16	TAH17	TAH18	TAH19	TAH20	TAH21	TAH22
8	9	10	11	12	13	14
JA2	JA3	JA4	JA5	JA6	JA7	JA8
MKH16	MKH17	MKH18	MKH19	MKH20	MKH21	MKH22
TAH23	TAH24	TAH25	TAH26	TAH27	TAH28	TAH29
15	16	17	18	19	20	21
JA9	JA10	JA11	JA12	JA13	JA14	JA15
MKH23	MKH24	MKH25	KMH26	MKH27	MKH28	MKH29
TAH30	TIR1	TIR2	TIR3	TIR4	TIR5	TIR6
22	23	24	25	26	27	28
JA16	JA17	JA18	JA19	JA20	JA21	JA22
MKH30	AMN1	AMN2	AMN3	AMN4	AMN5	AMN6
TIR7	TIR8	TIR9	TIR10	TIR11	TIR12	TIR13

SHAHIDI 479/1995-96/6231-32/1987-88						
Ashe	Ojobo	Iwa	Araorun	Ile	Itutu	Erinle
1	2	3	4	5	6	7
JA23	JA24	JA25	JA26	JA27	JA28	JA29
AMN7	AMN8	AMN9	AMN10	AMN11	AMN12	AMN13
TIR14	TIR15	TIR16	TIR17	TIR18	TIR19	TIR20
8	9	10	11	12	13	14
JA30	JA31	F1	F2	F3	F4	F5
AMN14	AMN15	AMN16	AMN17	AMN18	AMN19	AMN20
TIR21	TIR22	TIR23	TIR24	TIR25	TIR26	TIR27
15	16	17	18	19	20	21
F6	F7	F8	F9	F10	F 11	F12
AMN21	AMN22	AMN23	AMN24	AMN25	AMN26	AMN28
TIR28	TIR29	TIR30	YEK1	YEK2	YEK3	YEK4
22	23	24	25	26	27	28
F13	F14	F15	F16	F17	F18	F19
AMN29	AMN30	RNT1	RNT2	RNT3	RNT4	RNT5
YEK5	YEK6	YEK7	YEK8	YEK9	YEK10	YEK11
SHAHIDI 30						
F20						
RNT6						
YEK12						

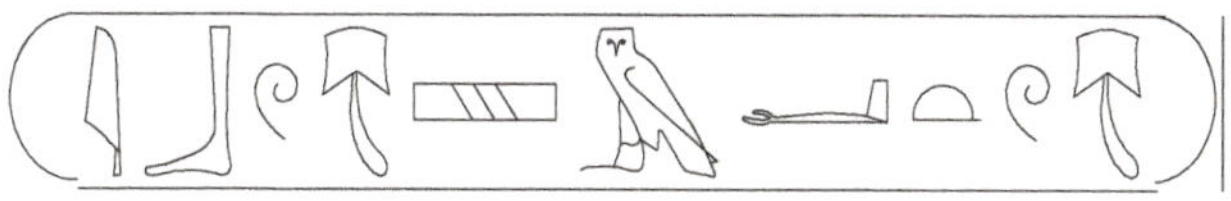

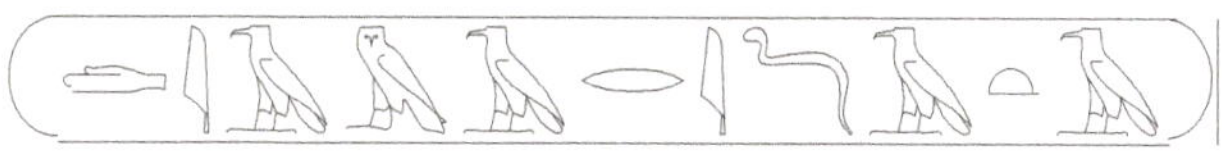

MAAFA 480 B.S.E./ASIENTO 1996-97 C.E./
KAM 6232-33 S.M./EC 1988-89 A.M. (ዓ.ም.)

AFU 480/1996-97/6232-33/1988-89						
Ashe	Ojobo	Iwa	Araorun	Ile	Itutu	Erinle
1	2	3	4	5	6	7
F21	F22	F23	F24	F25	F26	F27
RNT7	RNT8	RNT9	RNT10	RNT11	RNT12	RNT13
YEK13	YEK14	YEK15	YEK16	YEK17	YEK18	YEK19
8	9	10	11	12	13	14
F28	F29	M1	M2	M3	M4	M5
RNT14	RNT15	RNT16	RNT17	RNT18	RNT19	RNT20
YEK 20	YEK21	YEK22	YEK23	YEK24	YEK25	YEK26
15	16	17	18	19	20	21
M6	M7	M8	M9	M10	M11	M12
RNT21	RNT22	RNT23	RNT24	RNT25	RNT26	RNT27
YEK 27	YEK28	YEK29	YEK30	MAG1	MAG2	MAG3
22	23	24	25	26	27	28
M13	M14	M15	M16	M17	M18	M19
RNT28	RNT29	RNT30	KHN1	KHN2	KHN3	KHN4
MAG4	MAG5	MAG6	MAG7	MAG8	MAG9	MAG10
SHARIFU 117						
M20						
KHN5						
MAG11						

KAWAIDA 480/1996-97/6232-33/1988-89						
Ashe	Ojobo	Iwa	Araorun	Ile	Itutu	Erinle
1	2	3	4	5	6	7
A18	A19	A20	A21	A22	A23	A24
PIN4	PIN5	PIN6	PIN7	PIN8	PIN9	PIN10
MIY10	MIY11	MIY12	MIY13	MIY14	MIY15	MIY16
8	9	10	11	12	13	14
A25	A26	A27	A28	A29	A30	MA1
PIN11	PIN12	PIN13	PIN14	PIN15	PIN16	PIN17
MIY17	MIY18	MIY19	MIY20	MIY21	MIY22	MIY23
15	16	17	18	19	20	21
MA2	MA3	MA4	MA5	MA6	MA7	MA8
PIN18	PIN19	PIN20	PIN21	PIN22	PIN23	PIN24
MIY24	MIY25	MIY26	MIY27	MIY28	MIY29	MIY30
22	23	24	25	26	27	28
MA9	MA10	MA11	MA12	MA13	MA14	MA15
PIN25	PIN26	PIN27	PIN28	PIN29	PIN30	APP1
GIN1	GIN2	GIN3	GIN4	GIN5	GIN6	GIN7

UHURU 480/1996-97/6232-33/1988-89						
Ashe	Ojobo	Iwa	Araorun	Ile	Itutu	Erinle
1	2	3	4	5	6	7
M21	M22	M23	M24	M25	M26	M27
KHN6	KHN7	KHN8	KHN9	KHN10	KHN11	KHN12
MAG12	MAG13	MAG14	MAG15	MAG16	MAG17	MAG18
8	9	10	11	12	13	14
M28	M29	M30	M31	A1	A2	A3
KHN13	KHN14	KHN15	KHN16	KHN17	KHN18	KHN19
MAG19	MAG20	MAG21	MAG22	MAG23	MAG24	MAG25
15	16	17	18	19	20	21
A4	A5	A6	A7	A8	A9	A10
KHN20	KHN21	KHN222	KHN23	KHN24	KHN25	KHN26
MAG26	MAG27	MAG28	MAG29	MAG30	MIY1	MIY2
22	23	24	25	26	27	28
A11	A12	A13	A14	A15	A16	A17
KHN27	KHN28	KHN29	KHN30	PIN1	PIN2	PIN3
MIY3	MIY4	MIY5	MIY6	MIY7	MIY8	MIY9

MABABA WEUSI 480/1996-97/6232-33/1988-89						
Ashe	Ojobo	Iwa	Araorun	Ile	Itutu	Erinle
1	2	3	4	5	6	7
MA16	MA17	MA18	MA19	MA20	MA21	MA22
APP2	APP3	APP4	APP5	APP6	APP7	APP8
GIN8	GIN9	GIN10	GIN11	GIN12	GIN13	GIN14
8	9	10	11	12	13	14
MA23	MA24	MA25	MA26	MA27	MA28	MA29
APP9	APP10	APP11	APP12	APP13	APP14	APP15
GIN15	GIN16	GIN17	GIN18	GIN19	GIN20	GIN21
15	16	17	18	19	20	21
MA30	MA31	J1	J2	J3	J4	J5
APP16	APP17	APP18	APP19	APP20	APP21	APP22
GIN22	GIN23	GIN24	GIN25	GIN26	GIN27	GIN28
22	23	24	25	26	27	28
J6	J7	J8	J9	J10	J11	J12
APP23	APP24	APP25	APP26	APP27	APP28	APP29
GIN29	GIN30	SAN1	SAN2	SAN3	SAN4	SAN5

MWANAMKE WEUSI 480/1996-97/6232-33/1988-89

Ashe	Ojobo	Iwa	Araorun	Ile	Itutu	Erinle
1	2	3	4	5	6	7
J13	J14	J15	J16	J17	J18	J19
APP30	MSR1	MSR2	MSR3	MSR4	MSR5	MSR6
SAN6	SAN7	SAN8	SAN9	SAN10	SAN11	SAN12
8	9	10	11	12	13	14
J20	J21	J22	J23	J24	J25	J26
MSR7	MSR8	MSR9	MSR10	MSR11	MSR12	MSR13
SAN13	SAN14	SAN15	SAN16	SAN17	SAN18	SAN19
15	16	17	18	19	20	21
J27	J28	J29	J30	JU1	JU2	JU3
MSR14	MSR15	MSR16	MSR17	MSR18	MSR19	MSR20
SAN20	SAN21	SAN22	SAN23	SAN24	SAN25	SAN26
22	23	24	25	26	27	28
JU4	JU5	JU6	JU7	JU8	JU9	JU10
MSR21	MRS22	MSR23	MSR24	MSR25	MSR26	MSR27
SAN27	SAN28	SAN29	SAN30	HAM1	HAM2	HAM3

TAIFA WEUSI 480/1996-97/6232-33/1988-89

Ashe	Ojobo	Iwa	Araorun	Ile	Itutu	Erinle
1	2	3	4	5	6	7
AU8	AU9	AU10	AU11	AU12	AU13	AU14
THT20	THT21	THT22	THT23	THT24	THT25	THT26
NAH2	NAH3	NAH4	NAH5	NAH6	NAH7	NAH8
8	9	10	11	12	13	14
AU15	AU16	AU17	AU18	AU19	AU20	AU21
THT27	THT28	THT29	THT30	PAP1	PAP1	PAP2
NAH9	NAH10	NAH11	NAH12	NAH13	NAH14	NAH15
15	16	17	18	19	20	21
AU22	AU23	AU24	AU25	AU26	AU27	AU28
PAP3	PAP4	PAP5	PAP6	PAP7	PAP8	PAP9
NAH16	NAH17	NAH18	NAH19	NAH20	NAH21	NAH22
22	23	24	25	26	27	28
AU29	AU30	AU31	S1	S2	S3	S4
PAP10	PAP11	PAP12	PAP13	PAP14	PAP15	PAP16
NAH23	NAH24	NAH25	NAH26	NAH27	NAH28	NAH29

MTU WEUSI 480/1996-97/6232-33/1988-89

Ashe	Ojobo	Iwa	Araorun	Ile	Itutu	Erinle
1	2	3	4	5	6	7
JU11	JU12	JU13	JU14	JU15	JU16	JU17
MSR28	MSR29	MSR30	*ASR1*	*HRU2*	*STH3*	*AST4*
HAM4	HAM5	HAM6	HAM7	HAM8	HAM9	HAM10
8	9	10	11	12	13	14
JU18	JU19	JU20	JU21	JU22	JU23	JU24
NBH5	*ANU6*	THT1	THT2	THT3	THT4	THT5
HAM11	HAM12	HAM13	HAM14	HAM15	HAM16	HAM17
15	16	17	18	19	20	21
JU25	JU26	JU27	JU28	JU29	JU30	JU31
THT6	THT7	THT8	THT9	THT10	THT11	THT12
HAM18	HAM19	HAM20	HAM21	HAM22	HAM23	HAM24
22	23	24	25	26	27	28
AU1	AU2	AU3	AU4	AU5	AU6	AU7
THT13	THT14	THT15	THT16	THT17	THT18	THT19
HAM25	HAM26	HAM27	HAM28	HAM29	HAM30	NAH1

UWEZO 480/1996-97/6232-33/1988-89

Ashe	Ojobo	Iwa	Araorun	Ile	Itutu	Erinle
1	2	3	4	5	6	7
S5	S6	S7	S8	S9	S10	S11
PAP17	PAP18	PAP19	PAP20	PAP21	PAP22	PAP23
NAH30	*PAG1*	*PAG2*	*PAG3*	*PAG4*	*PAG5*	MAS1
8	9	10	11	12	13	14
S12	S13	S14	S15	S16	S17	S18
PAP24	PAP25	PAP26	PAP27	PAP28	PAP29	PAP30
MAS2	MAS3	MAS4	MAS5	MAS6	MAS7	MAS8
15	16	17	18	19	20	21
S19	S20	S21	S22	S23	S24	S25
HTH1	HTH2	HTH3	HTH4	HTH5	HTH6	HTH7
MAS9	MAS10	MAS11	MAS12	MAS13	MAS14	MAS15
22	23	24	25	26	27	28
S26	S27	S28	S29	S30	O1	O2
HTH8	HTH9	HTH10	HTH11	HTH12	HTH13	HTH14
MAS16	MAS17	MAS18	MAS19	MAS20	MAS21	MAS22

EKUNDU 480/1996-97/6232-33/1988-89						
Ashe	Ojobo	Iwa	Araorun	Ile	Itutu	Erinle
1	2	3	4	5	6	7
O3	O4	O5	O6	O7	O8	O9
HTH15	HTH16	HTH17	HTH18	HTH19	HTH20	HTH21
MAS23	MAS24	MAS25	MAS26	MAS27	MAS28	MAS29
8	9	10	11	12	13	14
O10	O11	O12	O13	O14	O15	O16
HTH22	HTH23	HTH24	HTH25	HTH26	HTH27	HTH28
MAS30	THi1	THi2	THi3	THi4	THi5	THi6
15	16	17	18	19	20	21
O17	O18	O19	O20	O21	O22	O23
HTH29	HTH30	KHK1	KHK2	KHK3	KHK4	KHK5
THi7	THi8	This9	THi10	THi11	THi12	THi13
22	23	24	25	26	27	28
O24	O25	O26	O27	O28	O29	O30
KHK6	KHK7	KHK8	KHK9	KHK10	KHK11	KHK12
Thi14	THi15	THi16	THi17	THi18	THi19	THi20

KIJANI 480/1996-97/6232-33/1988-89						
Ashe	Ojobo	Iwa	Araorun	Ile	Itutu	Erinle
1	2	3	4	5	6	7
N28	N29	N30	D1	D2	D3	D4
TBT11	TBT12	TBT13	TBT14	TBT15	TBT16	TBT17
HID19	HID20	HID21	HID22	HID23	HID24	HID25
8	9	10	11	12	13	14
D5	D6	D7	D8	D9	D10	D11
TBT18	TBT19	TBT20	TBT21	TBT22	TBT23	TBT24
HID26	HID27	HID28	HID29	HID30	TAH1	TAH2
15	16	17	18	19	20	21
D12	D13	D14	D15	D16	D17	D18
TBT25	TBT26	TBT27	TBT28	TBT29	TBT30	MKH1
TAH3	TAH4	TAH5	TAH6	TAH7	TAH8	TAH9
22	23	24	25	26	27	28
D19	D20	D21	D22	D23	D24	D25
MKH2	MKH3	MKH4	MKH5	MKH6	MKH7	MKH8
TAH10	TAH11	TAH12	TAH13	TAH14	TAH15	TAH16

EUSI 480/1996-97/6232-33/1988-89						
Ashe	Ojobo	Iwa	Araorun	Ile	Itutu	Erinle
1	2	3	4	5	6	7
O31	N1	N2	N3	N4	N5	N6
KHK13	KHK14	KHK15	KHK16	KHK17	KHK18	KHK19
THi21	THi22	THi23	THi24	THi25	THi26	THi27
8	9	10	11	12	13	14
N7	N8	N9	N10	N11	N12	N13
KHK20	KHK21	KHK22	KHK23	KHK24	KHK25	KHK26
THi28	THi29	THi30	HID1	HID2	HID3	HID4
15	16	17	18	19	20	21
N14	N15	N16	N17	N18	N19	N20
KHK27	KHK28	KHK29	KHK30	TBT1	TBT2	TBT3
HID5	HID6	HID7	HID8	HID9	HID10	HID11
22	23	24	25	26	27	28
N21	N22	N23	N24	N25	N26	N27
TBT4	TBT5	TBT6	TBT7	TBT8	TBT9	TBT10
HID12	HID13	HID14	HID15	HID16	HID17	HID18

KWANZA 480/1996-97/6232-33/1988-89						
Umoja	Kujichagulia	Ujima	Ujamaa	Nia	Kuumba	Imani
1	2	3	4	5	6	7
D26	D27	D28	D29	D30	D31	JA1
MKH9	MKH10	MKH11	MKH12	MKH13	MKH14	MKH15
TAH17	TAH18	TAH19	TAH20	TAH21	TAH22	TAH23
8	9	10	11	12	13	14
JA2	JA3	JA4	JA5	JA6	JA7	JA8
MKH16	MKH17	MKH18	MKH19	MKH20	MKH21	MKH22
TAH24	TAH25	TAH26	TAH27	TAH28	TAH29	TAH30
15	16	17	18	19	20	21
JA9	JA10	JA11	JA12	JA13	JA14	JA15
MKH23	MKH24	MKH25	KMH26	MKH27	MKH28	MKH29
TIR1	TIR2	TIR3	TIR4	TIR5	TIR6	TIR7
22	23	24	25	26	27	28
JA16	JA17	JA18	JA19	JA20	JA21	JA22
MKH30	AMN1	AMN2	AMN3	AMN4	AMN5	AMN6
TIR8	TIR9	TIR10	TIR11	TIR12	TIR13	TIR14

SHAHIDI 480/1996-97/6232-33/1988-89						
Ashe	Ojobo	Iwa	Araorun	Ile	Itutu	Erinle
1	2	3	4	5	6	7
JA23	JA24	JA25	JA26	JA27	JA28	JA29
AMN7	AMN8	AMN9	AMN10	AMN11	AMN12	AMN13
TIR15	TIR16	TIR17	TIR18	TIR19	TIR20	TIR21
8	9	10	11	12	13	14
JA30	JA31	F1	F2	F3	F4	F5
AMN14	AMN15	AMN16	AMN17	AMN18	AMN19	AMN20
TIR22	TIR23	TIR24	TIR25	TIR26	TIR27	TIR28
15	16	17	18	19	20	21
F6	F7	F8	F9	F10	F 11	F12
AMN21	AMN22	AMN23	AMN24	AMN25	AMN26	AMN28
TIR29	TIR30	YEK1	YEK2	YEK3	YEK4	YEK5
22	23	24	25	26	27	28
F13	F14	F15	F16	F17	F18	F19
AMN29	AMN30	RNT1	RNT2	RNT3	RNT4	RNT5
YEK6	YEK7	YEK8	YEK9	YEK10	YEK11	YEK12
SHAHIDI 31						
F20						
RNT6						
YEK13						

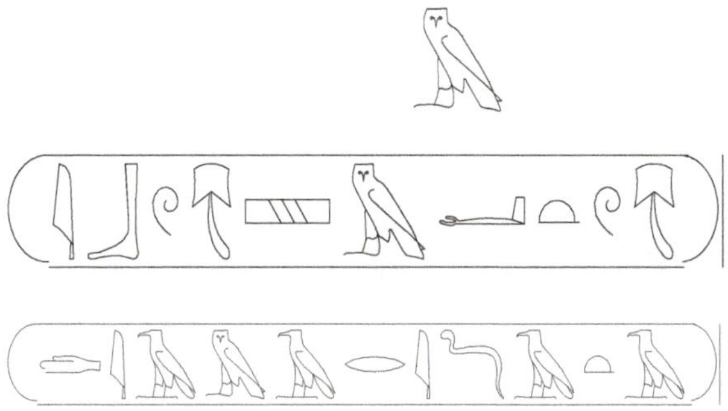

MAAFA 481 B.S.E./ASIENTO 1997-98 C.E./
KAM 6233-34 S.M./EC 1989-90 A.M. (ዓ.ም.)

AFU 481/1997-98/6233-34/1989-90

Ashe	Ojobo	Iwa	Araorun	Ile	Itutu	Erinle
1	2	3	4	5	6	7
F21	F22	F23	F24	F25	F26	F27
RNT7	RNT8	RNT9	RNT10	RNT11	RNT12	RNT13
YEK14	YEK15	YEK16	YEK17	YEK18	YEK19	YEK20
8	9	10	11	12	13	14
F28	M1	M2	M3	M4	M5	M6
RNT14	RNT15	RNT16	RNT17	RNT18	RNT19	RNT20
YEK21	YEK22	YEK23	YEK24	YEK25	YEK26	YEK27
15	16	17	18	19	20	21
M7	M8	M9	M10	M11	M12	M13
RNT21	RNT22	RNT23	RNT24	RNT25	RNT26	RNT27
YEK28	YEK29	YEK30	MAG1	MAG2	MAG3	MAG4
22	23	24	25	26	27	28
M14	M15	M16	M17	M18	M19	M20
RNT28	RNT29	RNT30	KHN1	KHN2	KHN3	KHN4
MAG5	MAG6	MAG7	MAG8	MAG9	MAG10	MAG11

KAWAIDA 481/1997-98/6233-34/1989-90

Ashe	Ojobo	Iwa	Araorun	Ile	Itutu	Erinle
1	2	3	4	5	6	7
A18	A19	A20	A21	A22	A23	A24
PIN3	PIN4	PIN5	PIN6	PIN7	PIN8	PIN9
MIY10	MIY11	MIY12	MIY13	MIY14	MIY15	MIY16
8	9	10	11	12	13	14
A25	A26	A27	A28	A29	A30	MA1
PIN10	PIN11	PIN12	PIN13	PIN14	PIN15	PIN16
MIY17	MIY18	MIY19	MIY20	MIY21	MIY22	MIY23
15	16	17	18	19	20	21
MA2	MA3	MA4	MA5	MA6	MA7	MA8
PIN17	PIN18	PIN19	PIN20	PIN21	PIN22	PIN23
MIY24	MIY25	MIY26	MIY27	MIY28	MIY29	MIY30
22	23	24	25	26	27	28
MA9	MA10	MA11	MA12	MA13	MA14	MA15
PIN24	PIN25	PIN26	PIN27	PIN28	PIN29	PIN30
GIN1	GIN2	GIN3	GIN4	GIN5	GIN6	GIN7

UHURU 481/1997-98/6233-34/1989-90

Ashe	Ojobo	Iwa	Araorun	Ile	Itutu	Erinle
1	2	3	4	5	6	7
M21	M22	M23	M24	M25	M26	M27
KHN5	KHN6	KHN7	KHN8	KHN9	KHN10	KHN11
MAG12	MAG13	MAG14	MAG15	MAG16	MAG17	MAG18
8	9	10	11	12	13	14
M28	M29	M30	M31	A1	A2	A3
KHN12	KHN13	KHN14	KHN15	KHN16	KHN17	KHN18
MAG19	MAG20	MAG21	MAG22	MAG23	MAG24	MAG25
15	16	17	18	19	20	21
A4	A5	A6	A7	A8	A9	A10
KHN19	KHN20	KHN21	KHN22	KHN23	KHN24	KHN25
MAG26	MAG27	MAG28	MAG29	MAG30	MIY1	MIY2
22	23	24	25	26	27	28
A11	A12	A13	A14	A15	A16	A17
KHN26	KHN27	KHN28	KHN29	KHN30	PIN1	PIN2
MIY3	MIY4	MIY5	MIY6	MIY7	MIY8	MIY9

MABABA WEUSI 481/1997-98/6233-34/1989-90

Ashe	Ojobo	Iwa	Araorun	Ile	Itutu	Erinle
1	2	3	4	5	6	7
MA16	MA17	MA18	MA19	MA20	MA21	MA22
APP1	APP2	APP3	APP4	APP5	APP6	APP7
GIN8	GIN9	GIN10	GIN11	GIN12	GIN13	GIN14
8	9	10	11	12	13	14
MA23	MA24	MA25	MA26	MA27	MA28	MA29
APP8	APP9	APP10	APP11	APP12	APP13	APP14
GIN15	GIN16	GIN17	GIN18	GIN19	GIN20	GIN21
15	16	17	18	19	20	21
MA30	MA31	J1	J2	J3	J4	J5
APP15	APP16	APP17	APP18	APP19	APP20	APP21
GIN22	GIN23	GIN24	GIN25	GIN26	GIN27	GIN28
22	23	24	25	26	27	28
J6	J7	J8	J9	J10	J11	J12
APP22	APP23	APP24	APP25	APP26	APP27	APP28
GIN29	GIN30	SAN1	SAN2	SAN3	SAN4	SAN5

MWANAMKE WEUSI 481/1997-98/6233-34/1989-90						
Ashe	Ojobo	Iwa	Araorun	Ile	Itutu	Erinle
1	2	3	4	5	6	7
J13	J14	J15	J16	J17	J18	J19
APP29	APP30	MSR1	MSR2	MSR3	MSR4	MSR5
SAN6	SAN7	SAN8	SAN9	SAN10	SAN11	SAN12
8	9	10	11	12	13	14
J20	J21	J22	J23	J24	J25	J26
MSR6	MSR7	MSR8	MSR9	MSR10	MSR11	MSR12
SAN13	SAN14	SAN15	SAN16	SAN17	SAN18	SAN19
15	16	17	18	19	20	21
J27	J28	J29	J30	JU1	JU2	JU3
MSR13	MSR14	MSR15	MSR16	MSR17	MSR18	MSR19
SAN20	SAN21	SAN22	SAN23	SAN24	SAN25	SAN26
22	23	24	25	26	27	28
JU4	JU5	JU6	JU7	JU8	JU9	JU10
MSR20	MRS21	MSR22	MSR23	MSR24	MSR25	MSR26
SAN27	SAN28	SAN29	SAN30	HAM1	HAM2	HAM3

TAIFA WEUSI 481/1997-98/6233-34/1989-90						
Ashe	Ojobo	Iwa	Araorun	Ile	Itutu	Erinle
1	2	3	4	5	6	7
AU8	AU9	AU10	AU11	AU12	AU13	AU14
THT20	THT21	THT22	THT23	THT24	THT25	THT26
NAH2	NAH3	NAH4	NAH5	NAH6	NAH7	NAH8
8	9	10	11	12	13	14
AU15	AU16	AU17	AU18	AU19	AU20	AU21
THT27	THT28	THT29	THT30	PAP1	PAP1	PAP2
NAH9	NAH10	NAH11	NAH12	NAH13	NAH14	NAH15
15	16	17	18	19	20	21
AU22	AU23	AU24	AU25	AU26	AU27	AU28
PAP3	PAP4	PAP5	PAP6	PAP7	PAP8	PAP9
NAH16	NAH17	NAH18	NAH19	NAH20	NAH21	NAH22
22	23	24	25	26	27	28
AU29	AU30	AU31	S1	S2	S3	S4
PAP10	PAP11	PAP12	PAP13	PAP14	PAP15	PAP16
NAH23	NAH24	NAH25	NAH26	NAH27	NAH28	NAH29

MTU WEUSI 481/1997-98/6233-34/1989-90						
Ashe	Ojobo	Iwa	Araorun	Ile	Itutu	Erinle
1	2	3	4	5	6	7
JU11	JU12	JU13	JU14	JU15	JU16	JU17
MSR27	MSR28	MSR29	MSR30	*ASR1*	*HRU2*	*STH3*
HAM4	HAM5	HAM6	HAM7	HAM8	HAM9	HAM10
8	9	10	11	12	13	14
JU18	JU19	JU20	JU21	JU22	JU23	JU24
AST4	*NBH5*	THT1	THT2	THT3	THT4	THT5
HAM11	HAM12	HAM13	AM14	HAM15	HAM16	HAM17
15	16	17	18	19	20	21
JU25	JU26	JU27	JU28	JU29	JU30	JU31
THT6	THT7	THT8	THT9	THT10	THT11	THT12
HAM18	HAM19	HAM20	HAM21	HAM22	HAM23	HAM24
22	23	24	25	26	27	28
AU1	AU2	AU3	AU4	AU5	AU6	AU7
THT13	THT14	THT15	THT16	THT17	THT18	THT19
HAM25	HAM26	HAM27	HAM28	HAM29	HAM30	NAH1

UWEZO 481/1997-98/6233-34/1989-90						
Ashe	Ojobo	Iwa	Araorun	Ile	Itutu	Erinle
1	2	3	4	5	6	7
S5	S6	S7	S8	S9	S10	S11
PAP17	PAP18	PAP19	PAP20	PAP21	PAP22	PAP23
NAH30	*PAG1*	*PAG2*	*PAG3*	*PAG4*	*PAG5*	MAS1
8	9	10	11	12	13	14
S12	S13	S14	S15	S16	S17	S18
PAP24	PAP25	PAP26	PAP27	PAP28	PAP29	PAP30
MAS2	MAS3	MAS4	MAS5	MAS6	MAS7	MAS8
15	16	17	18	19	20	21
S19	S20	S21	S22	S23	S24	S25
HTH1	HTH2	HTH3	HTH4	HTH5	HTH6	HTH7
MAS9	MAS10	MAS11	MAS12	MAS13	MAS14	MAS15
22	23	24	25	26	27	28
S26	S27	S28	S29	S30	O1	O2
HTH8	HTH9	HTH10	HTH11	HTH12	HTH13	HTH14
MAS16	MAS17	MAS18	MAS19	MAS20	MAS21	MAS22

EKUNDU 481/1997-98/6233-34/1989-90						
Ashe	Ojobo	Iwa	Araorun	Ile	Itutu	Erinle
1	2	3	4	5	6	7
O3	O4	O5	O6	O7	O8	O9
HTH15	HTH16	HTH17	HTH18	HTH19	HTH20	HTH21
MAS23	MAS24	MAS25	MAS26	MAS27	MAS28	MAS29
8	9	10	11	12	13	14
O10	O11	O12	O13	O14	O15	O16
HTH22	HTH23	HTH24	HTH25	HTH26	HTH27	HTH28
MAS30	THi1	THi2	THi3	THi4	THi5	THi6
15	16	17	18	19	20	21
O17	O18	O19	O20	O21	O22	O23
HTH29	HTH30	KHK1	KHK2	KHK3	KHK4	KHK5
THi7	THi8	THi9	THi10	THi11	THi12	THi13
22	23	24	25	26	27	28
O24	O25	O26	O27	O28	O29	O30
KHK6	KHK7	KHK8	KHK9	KHK10	KHK11	KHK12
THi14	THi15	THi16	THi17	THi18	THi19	THi20

KIJANI 481/1997-98/6233-34/1989-90						
Ashe	Ojobo	Iwa	Araorun	Ile	Itutu	Erinle
1	2	3	4	5	6	7
N28	N29	N30	D1	D2	D3	D4
TBT11	TBT12	TBT13	TBT14	TBT15	TBT16	TBT17
HID19	HID20	HID21	HID22	HID23	HID24	HID25
8	9	10	11	12	13	14
D5	D6	D7	D8	D9	D10	D11
TBT18	TBT19	TBT20	TBT21	TBT22	TBT23	TBT24
HID26	HID27	HID28	HID29	HID30	TAH1	TAH2
15	16	17	18	19	20	21
D12	D13	D14	D15	D16	D17	D18
TBT25	TBT26	TBT27	TBT28	TBT29	TBT30	MKH1
TAH3	TAH4	TAH5	TAH6	TAH7	TAH8	TAH9
22	23	24	25	26	27	28
D19	D20	D21	D22	D23	D24	D25
MKH2	MKH3	MKH4	MKH5	MKH6	MKH7	MKH8
TAH10	TAH11	TAH12	TAH13	TAH14	TAH15	TAH16

EUSI 481/1997-98/6233-34/1989-90						
Ashe	Ojobo	Iwa	Araorun	Ile	Itutu	Erinle
1	2	3	4	5	6	7
O31	N1	N2	N3	N4	N5	N6
KHK13	KHK14	KHK15	KHK16	KHK17	KHK18	KHK19
THi21	THi22	THi23	THi24	THi25	THi26	THi27
8	9	10	11	12	13	14
N7	N8	N9	N10	N11	N12	N13
KHK20	KHK21	KHK22	KHK23	KHK24	KHK25	KHK26
THi28	THi29	THi30	HID1	HID2	HID3	HID4
15	16	17	18	19	20	21
N14	N15	N16	N17	N18	N19	N20
KHK27	KHK28	KHK29	KHK30	TBT1	TBT2	TBT3
HID5	HID6	HID7	HID8	HID9	HID10	HID11
22	23	24	25	26	27	28
N21	N22	N23	N24	N25	N26	N27
TBT4	TBT5	TBT6	TBT7	TBT8	TBT9	TBT10
HID12	HID13	HID14	HID15	HID16	HID17	HID18

KWANZA 481/1997-98/6233-34/1989-90						
Umoja	Kujichagulia	Ujima	Ujamaa	Nia	Kuumba	Imani
1	2	3	4	5	6	7
D26	D27	D28	D29	D30	D31	JA1
MKH9	MKH10	MKH11	MKH12	MKH13	MKH14	MKH15
TAH17	TAH18	TAH19	TAH20	TAH21	TAH22	TAH23
8	9	10	11	12	13	14
JA2	JA3	JA4	JA5	JA6	JA7	JA8
MKH16	MKH17	MKH18	MKH19	MKH20	MKH21	MKH22
TAH24	TAH25	TAH26	TAH27	TAH28	TAH29	TAH30
15	16	17	18	19	20	21
JA9	JA10	JA11	JA12	JA13	JA14	JA15
MKH23	MKH24	MKH25	KMH26	MKH27	MKH28	MKH29
TIR1	TIR2	TIR3	TIR4	TIR5	TIR6	TIR7
22	23	24	25	26	27	28
JA16	JA17	JA18	JA19	JA20	JA21	JA22
MKH30	AMN1	AMN2	AMN3	AMN4	AMN5	AMN6
TIR8	TIR9	TIR10	TIR11	TIR12	TIR13	TIR14

SHAHIDI 481/1997-98/6233-34/1989-90						
Ashe	Ojobo	Iwa	Araorun	Ile	Itutu	Erinle
1	2	3	4	5	6	7
JA23	JA24	JA25	JA26	JA27	JA28	JA29
AMN7	AMN8	AMN9	AMN10	AMN11	AMN12	AMN13
TIR15	TIR16	TIR17	TIR18	TIR19	TIR20	TIR21
8	9	10	11	12	13	14
JA30	JA31	F1	F2	F3	F4	F5
AMN14	AMN15	AMN16	AMN17	AMN18	AMN19	AMN20
TIR22	TIR23	TIR24	TIR25	TIR26	TIR27	TIR28
15	16	17	18	19	20	21
F6	F7	F8	F9	F10	F 11	F12
AMN21	AMN22	AMN23	AMN24	AMN25	AMN26	AMN28
TIR29	TIR30	YEK1	YEK2	YEK3	YEK4	YEK5
22	23	24	25	26	27	28
F13	F14	F15	F16	F17	F18	F19
AMN29	AMN30	RNT1	RNT2	RNT3	RNT4	RNT5
YEK6	YEK7	YEK8	YEK9	YEK10	YEK11	YEK12
SHAHIDI 32						
F20						
RNT6						
YEK13						

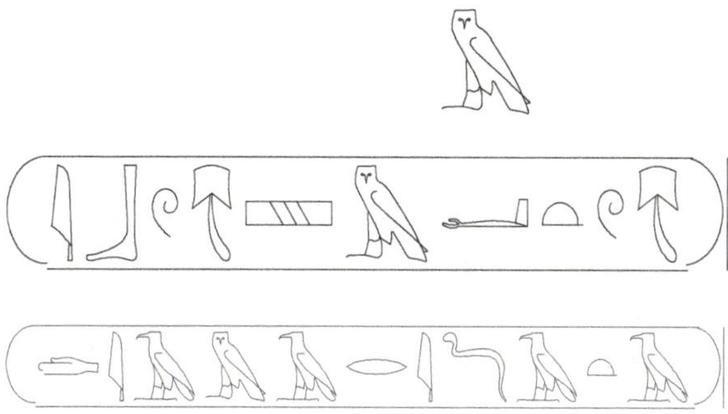

MAAFA 482 B.S.E./ASIENTO 1998-99 C.E./
KAM 6234-35 S.M./EC 1990-91 A.M. (ዓ.ም.)

AFU 482/1998-99/6234-35/1990-91

Ashe	Ojobo	Iwa	Araorun	Ile	Itutu	Erinle
1	2	3	4	5	6	7
F21	F22	F23	F24	F25	F26	F27
RNT7	RNT8	RNT9	RNT10	RNT11	RNT12	RNT13
YEK14	YEK15	YEK16	YEK17	YEK18	YEK19	YEK20
8	9	10	11	12	13	14
F28	M1	M2	M3	M4	M5	M6
RNT14	RNT15	RNT16	RNT17	RNT18	RNT19	RNT20
YEK21	YEK22	YEK23	YEK24	YEK25	YEK26	YEK27
15	16	17	18	19	20	21
M7	M8	M9	M10	M11	M12	M13
RNT21	RNT22	RNT23	RNT24	RNT25	RNT26	RNT27
YEK28	YEK29	YEK30	MAG1	MAG2	MAG3	MAG4
22	23	24	25	26	27	28
M14	M15	M16	M17	M18	M19	M20
RNT28	RNT29	RNT30	KHN1	KHN2	KHN3	KHN4
MAG5	MAG6	MAG7	MAG8	MAG9	MAG10	MAG11

KAWAIDA 482/1998-99/6234-35/1990-91

Ashe	Ojobo	Iwa	Araorun	Ile	Itutu	Erinle
1	2	3	4	5	6	7
A18	A19	A20	A21	A22	A23	A24
PIN3	PIN4	PIN5	PIN6	PIN7	PIN8	PIN9
MIY10	MIY11	MIY12	MIY13	MIY14	MIY15	MIY16
8	9	10	11	12	13	14
A25	A26	A27	A28	A29	A30	MA1
PIN10	PIN11	PIN12	PIN13	PIN14	PIN15	PIN16
MIY17	MIY18	MIY19	MIY20	MIY21	MIY22	MIY23
15	16	17	18	19	20	21
MA2	MA3	MA4	MA5	MA6	MA7	MA8
PIN17	PIN18	PIN19	PIN20	PIN21	PIN22	PIN23
MIY24	MIY25	MIY26	MIY27	MIY28	MIY29	MIY30
22	23	24	25	26	27	28
MA9	MA10	MA11	MA12	MA13	MA14	MA15
PIN24	PIN25	PIN26	PIN27	PIN28	PIN29	PIN30
GIN1	GIN2	GIN3	GIN4	GIN5	GIN6	GIN7

UHURU 482/1998-99/6234-35/1990-91

Ashe	Ojobo	Iwa	Araorun	Ile	Itutu	Erinle
1	2	3	4	5	6	7
M21	M22	M23	M24	M25	M26	M27
KHN5	KHN6	KHN7	KHN8	KHN9	KHN10	KHN11
MAG12	MAG13	MAG14	MAG15	MAG16	MAG17	MAG18
8	9	10	11	12	13	14
M28	M29	M30	M31	A1	A2	A3
KHN12	KHN13	KHN14	KHN15	KHN16	KHN17	KHN18
MAG19	MAG20	MAG21	MAG22	MAG23	MAG24	MAG25
15	16	17	18	19	20	21
A4	A5	A6	A7	A8	A9	A10
KHN19	KHN20	KHN21	KHN22	KHN23	KHN24	KHN25
MAG26	MAG27	MAG28	MAG29	MAG30	MIY1	MIY2
22	23	24	25	26	27	28
A11	A12	A13	A14	A15	A16	A17
KHN26	KHN27	KHN28	KHN29	KHN30	PIN1	PIN2
MIY3	MIY4	MIY5	MIY6	MIY7	MIY8	MIY9

MABABA WEUSI 482/1998-99/6234-35/1990-91

Ashe	Ojobo	Iwa	Araorun	Ile	Itutu	Erinle
1	2	3	4	5	6	7
MA16	MA17	MA18	MA19	MA20	MA21	MA22
APP1	APP2	APP3	APP4	APP5	APP6	APP7
GIN8	GIN9	GIN10	GIN11	GIN12	GIN13	GIN14
8	9	10	11	12	13	14
MA23	MA24	MA25	MA26	MA27	MA28	MA29
APP8	APP9	APP10	APP11	APP12	APP13	APP14
GIN15	GIN16	GIN17	GIN18	GIN19	GIN20	GIN21
15	16	17	18	19	20	21
MA30	MA31	J1	J2	J3	J4	J5
APP15	APP16	APP17	APP18	APP19	APP20	APP21
GIN22	GIN23	GIN24	GIN25	GIN26	GIN27	GIN28
22	23	24	25	26	27	28
J6	J7	J8	J9	J10	J11	J12
APP22	APP23	APP24	APP25	APP26	APP27	APP28
GIN29	GIN30	SAN1	SAN2	SAN3	SAN4	SAN5

MWANAMKE WEUSI 482/1998-99/6234-35/1990-91						
Ashe	Ojobo	Iwa	Araorun	Ile	Itutu	Erinle
1	2	3	4	5	6	7
J13	J14	J15	J16	J17	J18	J19
APP29	APP30	MSR1	MSR2	MSR3	MSR4	MSR5
SAN6	SAN7	SAN8	SAN9	SAN10	SAN11	SAN12
8	9	10	11	12	13	14
J20	J21	J22	J23	J24	J25	J26
MSR6	MSR7	MSR8	MSR9	MSR10	MSR11	MSR12
SAN13	SAN14	SAN15	SAN16	SAN17	SAN18	SAN19
15	16	17	18	19	20	21
J27	J28	J29	J30	JU1	JU2	JU3
MSR13	MSR14	MSR15	MSR16	MSR17	MSR18	MSR19
SAN20	SAN21	SAN22	SAN23	SAN24	SAN25	SAN26
22	23	24	25	26	27	28
JU4	JU5	JU6	JU7	JU8	JU9	JU10
MSR20	MRS21	MSR22	MSR23	MSR24	MSR25	MSR26
SAN27	SAN28	SAN29	SAN30	HAM1	HAM2	HAM3

TAIFA WEUSI 482/1998-99/6234-35/1990-91						
Ashe	Ojobo	Iwa	Araorun	Ile	Itutu	Erinle
1	2	3	4	5	6	7
AU8	AU9	AU10	AU11	AU12	AU13	AU14
THT20	THT21	THT22	THT23	THT24	THT25	THT26
NAH2	NAH3	NAH4	NAH5	NAH6	NAH7	NAH8
8	9	10	11	12	13	14
AU15	AU16	AU17	AU18	AU19	AU20	AU21
THT27	THT28	THT29	THT30	PAP1	PAP1	PAP2
NAH9	NAH10	NAH11	NAH12	NAH13	NAH14	NAH15
15	16	17	18	19	20	21
AU22	AU23	AU24	AU25	AU26	AU27	AU28
PAP3	PAP4	PAP5	PAP6	PAP7	PAP8	PAP9
NAH16	NAH17	NAH18	NAH19	NAH20	NAH21	NAH22
22	23	24	25	26	27	28
AU29	AU30	AU31	S1	S2	S3	S4
PAP10	PAP11	PAP12	PAP13	PAP14	PAP15	PAP16
NAH23	NAH24	NAH25	NAH26	NAH27	NAH28	NAH29

MTU WEUSI 482/1998-99/6234-35/1990-91						
Ashe	Ojobo	Iwa	Araorun	Ile	Itutu	Erinle
1	2	3	4	5	6	7
JU11	JU12	JU13	JU14	JU15	JU16	JU17
MSR27	MSR28	MSR29	MSR30	*ASR1*	*HRU2*	*STH3*
HAM4	HAM5	HAM6	HAM7	HAM8	HAM9	HAM10
8	9	10	11	12	13	14
JU18	JU19	JU20	JU21	JU22	JU23	JU24
AST4	*NBH5*	THT1	THT2	THT3	THT4	THT5
HAM11	HAM12	HAM13	AM14	HAM15	HAM16	HAM17
15	16	17	18	19	20	21
JU25	JU26	JU27	JU28	JU29	JU30	JU31
THT6	THT7	THT8	THT9	THT10	THT11	THT12
HAM18	HAM19	HAM20	HAM21	HAM22	HAM23	HAM24
22	23	24	25	26	27	28
AU1	AU2	AU3	AU4	AU5	AU6	AU7
THT13	THT14	THT15	THT16	THT17	THT18	THT19
HAM25	HAM26	HAM27	HAM28	HAM29	HAM30	NAH1

UWEZO 482/1998-99/6234-35/1990-91						
Ashe	Ojobo	Iwa	Araorun	Ile	Itutu	Erinle
1	2	3	4	5	6	7
S5	S6	S7	S8	S9	S10	S11
PAP17	PAP18	PAP19	PAP20	PAP21	PAP22	PAP23
NAH30	*PAG1*	*PAG2*	*PAG3*	*PAG4*	*PAG5*	MAS1
8	9	10	11	12	13	14
S12	S13	S14	S15	S16	S17	S18
PAP24	PAP25	PAP26	PAP27	PAP28	PAP29	PAP30
MAS2	MAS3	MAS4	MAS5	MAS6	MAS7	MAS8
15	16	17	18	19	20	21
S19	S20	S21	S22	S23	S24	S25
HTH1	HTH2	HTH3	HTH4	HTH5	HTH6	HTH7
MAS9	MAS10	MAS11	MAS12	MAS13	MAS14	MAS15
22	23	24	25	26	27	28
S26	S27	S28	S29	S30	O1	O2
HTH8	HTH9	HTH10	HTH11	HTH12	HTH13	HTH14
MAS16	MAS17	MAS18	MAS19	MAS20	MAS21	MAS22

EKUNDU 482/1998-99/6234-35/1990-91						
Ashe	Ojobo	Iwa	Araorun	Ile	Itutu	Erinle
1	2	3	4	5	6	7
O3	O4	O5	O6	O7	O8	O9
HTH15	HTH16	HTH17	HTH18	HTH19	HTH20	HTH21
MAS23	MAS24	MAS25	MAS26	MAS27	MAS28	MAS29
8	9	10	11	12	13	14
O10	O11	O12	O13	O14	O15	O16
HTH22	HTH23	HTH24	HTH25	HTH26	HTH27	HTH28
MAS30	THi1	THi2	THi3	THi4	THi5	THi6
15	16	17	18	19	20	21
O17	O18	O19	O20	O21	O22	O23
HTH29	HTH30	KHK1	KHK2	KHK3	KHK4	KHK5
THi7	THi8	THi9	THi10	THi11	THi12	THi13
22	23	24	25	26	27	28
O24	O25	O26	O27	O28	O29	O30
KHK6	KHK7	KHK8	KHK9	KHK10	KHK11	KHK12
THi14	THi15	THi16	THi17	THi18	THi19	THi20

KIJANI 482/1998-99/6234-35/1990-91						
Ashe	Ojobo	Iwa	Araorun	Ile	Itutu	Erinle
1	2	3	4	5	6	7
N28	N29	N30	D1	D2	D3	D4
TBT11	TBT12	TBT13	TBT14	TBT15	TBT16	TBT17
HID19	HID20	HID21	HID22	HID23	HID24	HID25
8	9	10	11	12	13	14
D5	D6	D7	D8	D9	D10	D11
TBT18	TBT19	TBT20	TBT21	TBT22	TBT23	TBT24
HID26	HID27	HID28	HID29	HID30	TAH1	TAH2
15	16	17	18	19	20	21
D12	D13	D14	D15	D16	D17	D18
TBT25	TBT26	TBT27	TBT28	TBT29	TBT30	MKH1
TAH3	TAH4	TAH5	TAH6	TAH7	TAH8	TAH9
22	23	24	25	26	27	28
D19	D20	D21	D22	D23	D24	D25
MKH2	MKH3	MKH4	MKH5	MKH6	MKH7	MKH8
TAH10	TAH11	TAH12	TAH13	TAH14	TAH15	TAH16

EUSI 482/1998-99/6234-35/1990-91						
Ashe	Ojobo	Iwa	Araorun	Ile	Itutu	Erinle
1	2	3	4	5	6	7
O31	N1	N2	N3	N4	N5	N6
KHK13	KHK14	KHK15	KHK16	KHK17	KHK18	KHK19
THi21	THi22	THi23	THi24	THi25	THi26	THi27
8	9	10	11	12	13	14
N7	N8	N9	N10	N11	N12	N13
KHK20	KHK21	KHK22	KHK23	KHK24	KHK25	KHK26
THi28	THi29	THi30	HID1	HID2	HID3	HID4
15	16	17	18	19	20	21
N14	N15	N16	N17	N18	N19	N20
KHK27	KHK28	KHK29	KHK30	TBT1	TBT2	TBT3
HID5	HID6	HID7	HID8	HID9	HID10	HID11
22	23	24	25	26	27	28
N21	N22	N23	N24	N25	N26	N27
TBT4	TBT5	TBT6	TBT7	TBT8	TBT9	TBT10
HID12	HID13	HID14	HID15	HID16	HID17	HID18

KWANZA 482/1998-99/6234-35/1990-91						
Umoja	Kujichagulia	Ujima	Ujamaa	Nia	Kuumba	Imani
1	2	3	4	5	6	7
D26	D27	D28	D29	D30	D31	JA1
MKH9	MKH10	MKH11	MKH12	MKH13	MKH14	MKH15
TAH17	TAH18	TAH19	TAH20	TAH21	TAH22	TAH23
8	9	10	11	12	13	14
JA2	JA3	JA4	JA5	JA6	JA7	JA8
MKH16	MKH17	MKH18	MKH19	MKH20	MKH21	MKH22
TAH24	TAH25	TAH26	TAH27	TAH28	TAH29	TAH30
15	16	17	18	19	20	21
JA9	JA10	JA11	JA12	JA13	JA14	JA15
MKH23	MKH24	MKH25	KMH26	MKH27	MKH28	MKH29
TIR1	TIR2	TIR3	TIR4	TIR5	TIR6	TIR7
22	23	24	25	26	27	28
JA16	JA17	JA18	JA19	JA20	JA21	JA22
MKH30	AMN1	AMN2	AMN3	AMN4	AMN5	AMN6
TIR8	TIR9	TIR10	TIR11	TIR12	TIR13	TIR14

SHAHIDI 482/1998-99/6234-35/1990-91						
Ashe	Ojobo	Iwa	Araorun	Ile	Itutu	Erinle
1	2	3	4	5	6	7
JA23	JA24	JA25	JA26	JA27	JA28	JA29
AMN7	AMN8	AMN9	AMN10	AMN11	AMN12	AMN13
TIR15	TIR16	TIR17	TIR18	TIR19	TIR20	TIR21
8	9	10	11	12	13	14
JA30	JA31	F1	F2	F3	F4	F5
AMN14	AMN15	AMN16	AMN17	AMN18	AMN19	AMN20
TIR22	TIR23	TIR24	TIR25	TIR26	TIR27	TIR28
15	16	17	18	19	20	21
F6	F7	F8	F9	F10	F 11	F12
AMN21	AMN22	AMN23	AMN24	AMN25	AMN26	AMN28
TIR29	TIR30	YEK1	YEK2	YEK3	YEK4	YEK5
22	23	24	25	26	27	28
F13	F14	F15	F16	F17	F18	F19
AMN29	AMN30	RNT1	RNT2	RNT3	RNT4	RNT5
YEK6	YEK7	YEK8	YEK9	YEK10	YEK11	YEK12
SHAHIDI 33						
F20						
RNT6						
YEK13						

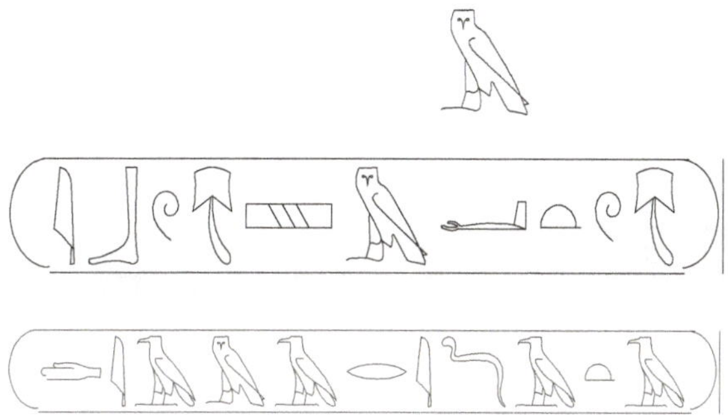

MAAFA 483 B.S.E./ASIENTO 1999-2000 C.E./ KAM 6235-36 S.M./EC 1991-92 A.M. (ዓ.ም.)

AFU 483/1999-00/6235-36/1991-92

Ashe	Ojobo	Iwa	Araorun	Ile	Itutu	Erinle
1	2	3	4	5	6	7
F21	F22	F23	F24	F25	F26	F27
RNT7	RNT8	RNT9	RNT10	RNT11	RNT12	RNT13
YEK14	YEK15	YEK16	YEK17	YEK18	YEK19	YEK20
8	9	10	11	12	13	14
F28	M1	M2	M3	M4	M5	M6
RNT14	RNT15	RNT16	RNT17	RNT18	RNT19	RNT20
YEK21	YEK22	YEK23	YEK24	YEK25	YEK26	YEK27
15	16	17	18	19	20	21
M7	M8	M9	M10	M11	M12	M13
RNT21	RNT22	RNT23	RNT24	RNT25	RNT26	RNT27
YEK28	YEK 29	YEK30	MAG1	MAG2	MAG3	MAG4
22	23	24	25	26	27	28
M14	M15	M16	M17	M18	M19	M20
RNT28	RNT29	RNT30	KHN1	KHN2	KHN3	KHN4
MAG5	MAG6	MAG7	MAG8	MAG9	MAG10	MAG11

UHURU 483/1999-00/6235-36/1991-92

Ashe	Ojobo	Iwa	Araorun	Ile	Itutu	Erinle
1	2	3	4	5	6	7
M21	M22	M23	M24	M25	M26	M27
KHN5	KHN6	KHN7	KHN8	KHN9	KHN10	KHN11
MAG12	MAG13	MAG14	MAG15	MAG16	MAG17	MAG18
8	9	10	11	12	13	14
M28	M29	M30	M31	A1	A2	A3
KHN12	KHN13	KHN14	KHN15	KHN16	KHN17	KHN18
MAG19	MAG20	MAG21	MAG22	MAG23	MAG24	MAG25
15	16	17	18	19	20	21
A4	A5	A6	A7	A8	A9	A10
KHN19	KHN20	KHN21	KHN22	KHN23	KHN24	KHN25
MAG26	MAG27	MAG28	MAG29	MAG30	MIY1	MIY2
22	23	24	25	26	27	28
A11	A12	A13	A14	A15	A16	A17
KHN26	KHN27	KHN28	KHN29	KHN30	PIN1	PIN2
MIY3	MIY4	MIY5	MIY6	MIY7	MIY8	MIY9

KAWAIDA 483/1999-00/6235-36/1991-92

Ashe	Ojobo	Iwa	Araorun	Ile	Itutu	Erinle
1	2	3	4	5	6	7
A18	A19	A20	A21	A22	A23	A24
PIN3	PIN4	PIN5	PIN6	PIN7	PIN8	PIN9
MIY10	MIY11	MIY12	MIY13	MIY14	MIY15	MIY16
8	9	10	11	12	13	14
A25	A26	A27	A28	A29	A30	MA1
PIN10	PIN11	PIN12	PIN13	PIN14	PIN15	PIN16
MIY17	MIY18	MIY19	MIY20	MIY21	MIY22	MIY23
15	16	17	18	19	20	21
MA2	MA3	MA4	MA5	MA6	MA7	MA8
PIN17	PIN18	PIN19	PIN20	PIN21	PIN22	PIN23
MIY24	MIY25	MIY26	MIY27	MIY28	MIY29	MIY30
22	23	24	25	26	27	28
MA9	MA10	MA11	MA12	MA13	MA14	MA15
PIN24	PIN25	PIN26	PIN27	PIN28	PIN29	PIN30
GIN1	GIN2	GIN3	GIN4	GIN5	GIN6	GIN7

MABABA WEUSI 483/1999-00/6235-36/1991-92

Ashe	Ojobo	Iwa	Araorun	Ile	Itutu	Erinle
1	2	3	4	5	6	7
MA16	MA17	MA18	MA19	MA20	MA21	MA22
APP1	APP2	APP3	APP4	APP5	APP6	APP7
GIN8	GIN9	GIN10	GIN11	GIN12	GIN13	GIN14
8	9	10	11	12	13	14
MA23	MA24	MA25	MA26	MA27	MA28	MA29
APP8	APP9	APP10	APP11	APP12	APP13	APP14
GIN15	GIN16	GIN17	GIN18	GIN19	GIN20	GIN21
15	16	17	18	19	20	21
MA30	MA31	J1	J2	J3	J4	J5
APP15	APP16	APP17	APP18	APP19	APP20	APP21
GIN22	GIN23	GIN24	GIN25	GIN26	GIN27	GIN28
22	23	24	25	26	27	28
J6	J7	J8	J9	J10	J11	J12
APP22	APP23	APP24	APP25	APP26	APP27	APP28
GIN29	GIN30	SAN1	SAN2	SAN3	SAN4	SAN5

MWANAMKE WEUSI 483/1999-00/6235-36/1991-92

Ashe	Ojobo	Iwa	Araorun	Ile	Itutu	Erinle
1	2	3	4	5	6	7
J13	J14	J15	J16	J17	J18	J19
APP29	APP30	MSR1	MSR2	MSR3	MSR4	MSR5
SAN6	SAN7	SAN8	SAN9	SAN10	SAN11	SAN12
8	9	10	11	12	13	14
J20	J21	J22	J23	J24	J25	J26
MSR6	MSR7	MSR8	MSR9	MSR10	MSR11	MSR12
SAN13	SAN14	SAN15	SAN16	SAN17	SAN18	SAN19
15	16	17	18	19	20	21
J27	J28	J29	J30	JU1	JU2	JU3
MSR13	MSR14	MSR15	MSR16	MSR17	MSR18	MSR19
SAN20	SAN21	SAN22	SAN23	SAN24	SAN25	SAN26
22	23	24	25	26	27	28
JU4	JU5	JU6	JU7	JU8	JU9	JU10
MSR20	MRS21	MSR22	MSR23	MSR24	MSR25	MSR26
SAN27	SAN28	SAN29	SAN30	HAM1	HAM2	HAM3

MTU WEUSI 483/1999-00/6235-36/1991-92

Ashe	Ojobo	Iwa	Araorun	Ile	Itutu	Erinle
1	2	3	4	5	6	7
JU11	JU12	JU13	JU14	JU15	JU16	JU17
MSR27	MSR28	MSR29	MSR30	*ASR1*	*HRU2*	*STH3*
HAM4	HAM5	HAM6	HAM7	HAM8	HAM9	HAM10
8	9	10	11	12	13	14
JU18	JU19	JU20	JU21	JU22	JU23	JU24
AST4	*NBH5*	THT1	THT2	THT3	THT4	THT5
HAM11	HAM12	HAM13	HAM14	HAM15	HAM16	HAM17
15	16	17	18	19	20	21
JU25	JU26	JU27	JU28	JU29	JU30	JU31
THT6	THT7	THT8	THT9	THT10	THT11	THT12
HAM18	HAM19	HAM20	HAM21	HAM22	HAM23	HAM24
22	23	24	25	26	27	28
AU1	AU2	AU3	AU4	AU5	AU6	AU7
THT13	THT14	THT15	THT16	THT17	THT18	THT19
HAM25	HAM26	HAM27	HAM28	HAM29	HAM30	NAH1

TAIFA WEUSI 483/1999-00/6235-36/1991-92

Ashe	Ojobo	Iwa	Araorun	Ile	Itutu	Erinle
1	2	3	4	5	6	7
AU8	AU9	AU10	AU11	AU12	AU13	AU14
THT20	THT21	THT22	THT23	THT24	THT25	THT26
NAH2	NAH3	NAH4	NAH5	NAH6	NAH7	NAH8
8	9	10	11	12	13	14
AU15	AU16	AU17	AU18	AU19	AU20	AU21
THT27	THT28	THT29	THT30	PAP1	PAP1	PAP2
NAH9	NAH10	NAH11	NAH12	NAH13	NAH14	NAH15
15	16	17	18	19	20	21
AU22	AU23	AU24	AU25	AU26	AU27	AU28
PAP3	PAP4	PAP5	PAP6	PAP7	PAP8	PAP9
NAH16	NAH17	NAH18	NAH19	NAH20	NAH21	NAH22
22	23	24	25	26	27	28
AU29	AU30	AU31	S1	S2	S3	S4
PAP10	PAP11	PAP12	PAP13	PAP14	PAP15	PAP16
NAH23	NAH24	NAH25	NAH26	NAH27	NAH28	NAH29

UWEZO 483/1999-00/6235-36/1991-92

Ashe	Ojobo	Iwa	Araorun	Ile	Itutu	Erinle
1	2	3	4	5	6	7
S5	S6	S7	S8	S9	S10	S11
PAP17	PAP18	PAP19	PAP20	PAP21	PAP22	PAP23
NAH30	*QUA1*	*QUA2*	*QUA3*	*QUA4*	*QUA5*	*QUA6*
8	9	10	11	12	13	14
S12	S13	S14	S15	S16	S17	S18
PAP24	PAP25	PAP26	PAP27	PAP28	PAP29	PAP30
MAS1	MAS2	MAS3	MAS4	MAS5	MAS6	MAS7
15	16	17	18	19	20	21
S19	S20	S21	S22	S23	S24	S25
HTH1	HTH2	HTH3	HTH4	HTH5	HTH6	HTH7
MAS8	MAS9	MAS10	MAS11	MAS12	MAS13	MAS14
22	23	24	25	26	27	28
S26	S27	S28	S29	S30	O1	O2
HTH8	HTH9	HTH10	HTH11	HTH12	HTH13	HTH14
MAS15	MAS16	MAS17	MAS18	MAS19	MAS20	MAS21

EKUNDU 483/1999-00/6235-36/1991-92						
Ashe	Ojobo	Iwa	Araorun	Ile	Itutu	Erinle
1	2	3	4	5	6	7
O3	O4	O5	O6	O7	O8	O9
HTH15	HTH16	HTH17	HTH18	HTH19	HTH20	HTH21
MAS22	MAS23	MAS 24	MAS25	MAS26	MAS27	MAS28
8	9	10	11	12	13	14
O10	O11	O12	O13	O14	O15	O16
HTH22	HTH23	HTH24	HTH25	HTH26	HTH27	HTH28
MAS29	MAS30	THi1	THi2	THi3	THi4	THi5
15	16	17	18	19	20	21
O17	O18	O19	O20	O21	O22	O23
HTH29	HTH30	KHK1	KHK2	KHK3	KHK4	KHK5
THi6	THi7	THi8	THi9	THi10	THi11	THi12
22	23	24	25	26	27	28
O24	O25	O26	O27	O28	O29	O30
KHK6	KHK7	KHK8	KHK9	KHK10	KHK11	KHK12
THi13	THi14	THi15	THi16	THi17	THi18	THi19

KIJANI 483/1999-00/6235-36/1991-92						
Ashe	Ojobo	Iwa	Araorun	Ile	Itutu	Erinle
1	2	3	4	5	6	7
N28	N29	N30	D1	D2	D3	D4
TBT11	TBT12	TBT13	TBT14	TBT15	TBT16	TBT17
HID18	HID19	HID20	HID21	HID22	HID23	HID24
8	9	10	11	12	13	14
D5	D6	D7	D8	D9	D10	D11
TBT18	TBT19	TBT20	TBT21	TBT22	TBT23	TBT24
HID25	HID26	HID27	HID28	HID29	HID30	TAH1
15	16	17	18	19	20	21
D12	D13	D14	D15	D16	D17	D18
TBT25	TBT26	TBT27	TBT28	TBT29	TBT30	MKH1
TAH2	TAH3	TAH4	TAH5	TAH6	TAH7	TAH8
22	23	24	25	26	27	28
D19	D20	D21	D22	D23	D24	D25
MKH2	MKH3	MKH4	MKH5	MKH6	MKH7	MKH8
TAH9	TAH10	TAH11	TAH12	TAH13	TAH14	TAH15

EUSI 483/1999-00/6235-36/1991-92						
Ashe	Ojobo	Iwa	Araorun	Ile	Itutu	Erinle
1	2	3	4	5	6	7
O31	N1	N2	N3	N4	N5	N6
KHK13	KHK14	KHK15	KHK16	KHK17	KHK18	KHK19
THi20	THi21	THi22	THi23	THi24	THi25	THi26
8	9	10	11	12	13	14
N7	N8	N9	N10	N11	N12	N13
KHK20	KHK21	KHK22	KHK23	KHK24	KHK25	KHK26
THi27	THi28	THi29	THi30	HID1	HID2	HID3
15	16	17	18	19	20	21
N14	N15	N16	N17	N18	N19	N20
KHK27	KHK28	KHK29	KHK30	TBT1	TBT2	TBT3
HID4	HID5	HID6	HID7	HID8	HID9	HID10
22	23	24	25	26	27	28
N21	N22	N23	N24	N25	N26	N27
TBT4	TBT5	TBT6	TBT7	TBT8	TBT9	TBT10
HID11	HID12	HID13	HID14	HID15	HID16	HID17

KWANZA 483/1999-00/6235-36/1991-92						
Umoja	Kujichagulia	Ujima	Ujamaa	Nia	Kuumba	Imani
1	2	3	4	5	6	7
D26	D27	D28	D29	D30	D31	JA1
MKH9	MKH10	MKH11	MKH12	MKH13	MKH14	MKH1
TAH16	TAH17	TAH18	TAH19	TAH20	TAH21	TAH2
8	9	10	11	12	13	14
JA2	JA3	JA4	JA5	JA6	JA7	JA
MKH16	MKH17	MKH18	MKH19	MKH20	MKH21	MKH2
TAH23	TAH24	TAH25	TAH26	TAH27	TAH28	TAH29
15	16	17	18	19	20	21
JA9	JA10	JA11	JA12	JA13	JA14	JA1
MKH23	MKH24	MKH25	KMH26	MKH27	MKH28	MKH2
TAH30	TIR1	TIR2	TIR3	TIR4	TIR5	TIR6
22	23	24	25	26	27	28
JA16	JA17	JA18	JA19	JA20	JA21	JA2
MKH30	AMN1	AMN2	AMN3	AMN4	AMN5	AMN
TIR7	TIR8	TIR9	TIR10	TIR11	TIR12	TIR13

SHAHIDI 483/1999-00/6235-36/1991-92						
Ashe	Ojobo	Iwa	Araorun	Ile	Itutu	Erinle
1	2	3	4	5	6	7
JA23	JA24	JA25	JA26	JA27	JA28	JA29
AMN7	AMN8	AMN9	AMN10	AMN11	AMN12	AMN13
TIR14	TIR15	TIR16	TIR17	TIR18	TIR19	TIR20
8	9	10	11	12	13	14
JA30	JA31	F1	F2	F3	F4	F5
AMN14	AMN15	AMN16	AMN17	AMN18	AMN19	AMN20
TIR21	TIR22	TIR23	TIR24	TIR25	TIR26	TIR27
15	16	17	18	19	20	21
F6	F7	F8	F9	F10	F 11	F12
AMN21	AMN22	AMN23	AMN24	AMN25	AMN26	AMN28
TIR28	TIR29	TIR30	YEK1	YEK2	YEK3	YEK4
22	23	24	25	26	27	28
F13	F14	F15	F16	F17	F18	F19
AMN29	AMN30	RNT1	RNT2	RNT3	RNT4	RNT5
YEK5	YEK6	YEK7	YEK8	YEK9	YEK10	YEK11
SHAHIDI 34						
F20						
RNT6						
YEK12						

THE NU AFRIKAN CALENDAR

MAAFA 484 B.S.E./ASIENTO 2000-01 C.E./
KAM 6236-37 S.M/EC 1992-93 A.M. (ዓ.ም.)

AFU 484/2000-01/6236-37/1992-93

Ashe	Ojobo	Iwa	Araorun	Ile	Itutu	Erinle
1	2	3	4	5	6	7
F21	F22	F23	F24	F25	F26	F27
RNT7	RNT8	RNT9	RNT10	RNT11	RNT12	RNT13
YEK13	YEK14	YEK15	YEK16	YEK17	YEK18	YEK19
8	9	10	11	12	13	14
F28	F29	M1	M2	M3	M4	M5
RNT14	RNT15	RNT16	RNT17	RNT18	RNT19	RNT20
YEK 20	YEK21	YEK22	YEK23	YEK24	YEK25	YEK26
15	16	17	18	19	20	21
M6	M7	M8	M9	M10	M11	M12
RNT21	RNT22	RNT23	RNT24	RNT25	RNT26	RNT27
YEK 27	YEK28	YEK29	YEK30	MAG1	MAG2	MAG3
22	23	24	25	26	27	28
M13	M14	M15	M16	M17	M18	M19
RNT28	RNT29	RNT30	KHN1	KHN2	KHN3	KHN4
MAG4	MAG5	MAG6	MAG7	MAG8	MAG9	MAG10
SHARIFU 118						
M20						
KHN5						
MAG11						

UHURU 484/2000-01/6236-37/1992-93

Ashe	Ojobo	Iwa	Araorun	Ile	Itutu	Erinle
1	2	3	4	5	6	7
M21	M22	M23	M24	M25	M26	M27
KHN6	KHN7	KHN8	KHN9	KHN10	KHN11	KHN12
MAG12	MAG13	MAG14	MAG15	MAG16	MAG17	MAG18
8	9	10	11	12	13	14
M28	M29	M30	M31	A1	A2	A3
KHN13	KHN14	KHN15	KHN16	KHN17	KHN18	KHN19
MAG19	MAG20	MAG21	MAG22	MAG23	MAG24	MAG25
15	16	17	18	19	20	21
A4	A5	A6	A7	A8	A9	A10
KHN20	KHN21	KHN222	KHN23	KHN24	KHN25	KHN26
MAG26	MAG27	MAG28	MAG29	MAG30	MIY1	MIY2
22	23	24	25	26	27	28
A11	A12	A13	A14	A15	A16	A17
KHN27	KHN28	KHN29	KHN30	PIN1	PIN2	PIN3
MIY3	MIY4	MIY5	MIY6	MIY7	MIY8	MIY9

KAWAIDA 484/2000-01/6236-37/1992-93

Ashe	Ojobo	Iwa	Araorun	Ile	Itutu	Erinl
1	2	3	4	5	6	7
A18	A19	A20	A21	A22	A23	A24
PIN4	PIN5	PIN6	PIN7	PIN8	PIN9	PIN10
MIY10	MIY11	MIY12	MIY13	MIY14	MIY15	MIY1
8	9	10	11	12	13	14
A25	A26	A27	A28	A29	A30	MA1
PIN11	PIN12	PIN13	PIN14	PIN15	PIN16	PIN17
MIY17	MIY18	MIY19	MIY20	MIY21	MIY22	MIY2
15	16	17	18	19	20	21
MA2	MA3	MA4	MA5	MA6	MA7	MA
PIN18	PIN19	PIN20	PIN21	PIN22	PIN23	PIN
MIY24	MIY25	MIY26	MIY27	MIY28	MIY29	MI
22	23	24	25	26	27	28
MA9	MA10	MA11	MA12	MA13	MA14	MA
PIN25	PIN26	PIN27	PIN28	PIN29	PIN30	AP
GIN1	GIN2	GIN3	GIN4	GIN5	GIN6	GI

MABABA WEUSI 484/2000-01/6236-37/1992-93

Ashe	Ojobo	Iwa	Araorun	Ile	Itutu	Erinl
1	2	3	4	5	6	7
MA16	MA17	MA18	MA19	MA20	MA21	MA
APP2	APP3	APP4	APP5	APP6	APP7	APP
GIN8	GIN9	GIN10	GIN11	GIN12	GIN13	GIN
8	9	10	11	12	13	14
MA23	MA24	MA25	MA26	MA27	MA28	MA
APP9	APP10	APP11	APP12	APP13	APP14	APP
GIN15	GIN16	GIN17	GIN18	GIN19	GIN20	GIN
15	16	17	18	19	20	21
MA30	MA31	J1	J2	J3	J4	J5
APP16	APP17	APP18	APP19	APP20	APP21	AP
GIN22	GIN23	GIN24	GIN25	GIN26	GIN27	GI
22	23	24	25	26	27	28
J6	J7	J8	J9	J10	J11	J1
APP23	APP24	APP25	APP26	APP27	APP28	AP
GIN29	GIN30	SAN1	SAN2	SAN3	SAN4	SA

THE NU AFRIKAN CALENDAR

MWANAMKE WEUSI 484/2000-01/6236-37/1992-93						
Ashe	Ojobo	Iwa	Araorun	Ile	Itutu	Erinle
1	2	3	4	5	6	7
J13	J14	J15	J16	J17	J18	J19
APP30	MSR1	MSR2	MSR3	MSR4	MSR5	MSR6
SAN6	SAN7	SAN8	SAN9	SAN10	SAN11	SAN12
8	9	10	11	12	13	14
J20	J21	J22	J23	J24	J25	J26
MSR7	MSR8	MSR9	MSR10	MSR11	MSR12	MSR13
SAN13	SAN14	SAN15	SAN16	SAN17	SAN18	SAN19
15	16	17	18	19	20	21
J27	J28	J29	J30	JU1	JU2	JU3
MSR14	MSR15	MSR16	MSR17	MSR18	MSR19	MSR20
SAN20	SAN21	SAN22	SAN23	SAN24	SAN25	SAN26
22	23	24	25	26	27	28
JU4	JU5	JU6	JU7	JU8	JU9	JU10
MSR21	MRS22	MSR23	MSR24	MSR25	MSR26	MSR27
SAN27	SAN28	SAN29	SAN30	HAM1	HAM2	HAM3

TAIFA WEUSI 484/2000-01/6236-37/1992-93						
Ashe	Ojobo	Iwa	Araorun	Ile	Itutu	Erinle
1	2	3	4	5	6	7
AU8	AU9	AU10	AU11	AU12	AU13	AU14
THT20	THT21	THT22	THT23	THT24	THT25	THT26
NAH2	NAH3	NAH4	NAH5	NAH6	NAH7	NAH8
8	9	10	11	12	13	14
AU15	AU16	AU17	AU18	AU19	AU20	AU21
THT27	THT28	THT29	THT30	PAP1	PAP1	PAP2
NAH9	NAH10	NAH11	NAH12	NAH13	NAH14	NAH15
15	16	17	18	19	20	21
AU22	AU23	AU24	AU25	AU26	AU27	AU28
PAP3	PAP4	PAP5	PAP6	PAP7	PAP8	PAP9
NAH16	NAH17	NAH18	NAH19	NAH20	NAH21	NAH22
22	23	24	25	26	27	28
AU29	AU30	AU31	S1	S2	S3	S4
PAP10	PAP11	PAP12	PAP13	PAP14	PAP15	PAP16
NAH23	NAH24	NAH25	NAH26	NAH27	NAH28	NAH29

MTU WEUSI 484/2000-01/6236-37/1992-93						
Ashe	Ojobo	Iwa	Araorun	Ile	Itutu	Erinle
1	2	3	4	5	6	7
JU11	JU12	JU13	JU14	JU15	JU16	JU17
MSR28	MSR29	MSR30	*ASR1*	*HRU2*	*STH3*	*AST4*
HAM4	HAM5	HAM6	HAM7	HAM8	HAM9	HAM10
8	9	10	11	12	13	14
JU18	JU19	JU20	JU21	JU22	JU23	JU24
NBH5	*ANU6*	THT1	THT2	THT3	THT4	THT5
HAM11	HAM12	HAM13	HAM14	HAM15	HAM16	HAM17
15	16	17	18	19	20	21
JU25	JU26	JU27	JU28	JU29	JU30	JU31
THT6	THT7	THT8	THT9	THT10	THT11	THT12
HAM18	HAM19	HAM20	HAM21	HAM22	HAM23	HAM24
22	23	24	25	26	27	28
AU1	AU2	AU3	AU4	AU5	AU6	AU7
THT13	THT14	THT15	THT16	THT17	THT18	THT19
HAM25	HAM26	HAM27	HAM28	HAM29	HAM30	NAH1

UWEZO 484/2000-01/6236-37/1992-93						
Ashe	Ojobo	Iwa	Araorun	Ile	Itutu	Erinle
1	2	3	4	5	6	7
S5	S6	S7	S8	S9	S10	S11
PAP17	PAP18	PAP19	PAP20	PAP21	PAP22	PAP23
NAH30	*PAG1*	*PAG2*	*PAG3*	*PAG4*	*PAG5*	MAS1
8	9	10	11	12	13	14
S12	S13	S14	S15	S16	S17	S18
PAP24	PAP25	PAP26	PAP27	PAP28	PAP29	PAP30
MAS2	MAS3	MAS4	MAS5	MAS6	MAS7	MAS8
15	16	17	18	19	20	21
S19	S20	S21	S22	S23	S24	S25
HTH1	HTH2	HTH3	HTH4	HTH5	HTH6	HTH7
MAS9	MAS10	MAS11	MAS12	MAS13	MAS14	MAS15
22	23	24	25	26	27	28
S26	S27	S28	S29	S30	O1	O2
HTH8	HTH9	HTH10	HTH11	HTH12	HTH13	HTH14
MAS16	MAS17	MAS18	MAS19	MAS20	MAS21	MAS22

THE NU AFRIKAN CALENDAR

EKUNDU 484/2000-01/6236-37/1992-93						
Ashe	Ojobo	Iwa	Araorun	Ile	Itutu	Erinle
1	2	3	4	5	6	7
O3	O4	O5	O6	O7	O8	O9
HTH15	HTH16	HTH17	HTH18	HTH19	HTH20	HTH21
MAS23	MAS24	MAS25	MAS26	MAS27	MAS28	MAS29
8	9	10	11	12	13	14
O10	O11	O12	O13	O14	O15	O16
HTH22	HTH23	HTH24	HTH25	HTH26	HTH27	HTH28
MAS30	THi1	THi2	THi3	THi4	THi5	THi6
15	16	17	18	19	20	21
O17	O18	O19	O20	O21	O22	O23
HTH29	HTH30	KHK1	KHK2	KHK3	KHK4	KHK5
THi7	THi8	This9	THi10	THi11	THi12	THi13
22	23	24	25	26	27	28
O24	O25	O26	O27	O28	O29	O30
KHK6	KHK7	KHK8	KHK9	KHK10	KHK11	KHK12
Thi14	THi15	THi16	THi17	THi18	THi19	THi20

KIJANI 484/2000-01/6236-37/1992-93						
Ashe	Ojobo	Iwa	Araorun	Ile	Itutu	Erinle
1	2	3	4	5	6	7
N28	N29	N30	D1	D2	D3	D4
TBT11	TBT12	TBT13	TBT14	TBT15	TBT16	TBT17
HID19	HID20	HID21	HID22	HID23	HID24	HID25
8	9	10	11	12	13	14
D5	D6	D7	D8	D9	D10	D11
TBT18	TBT19	TBT20	TBT21	TBT22	TBT23	TBT24
HID26	HID27	HID28	HID29	HID30	TAH1	TAH2
15	16	17	18	19	20	21
D12	D13	D14	D15	D16	D17	D18
TBT25	TBT26	TBT27	TBT28	TBT29	TBT30	MKH1
TAH3	TAH4	TAH5	TAH6	TAH7	TAH8	TAH9
22	23	24	25	26	27	28
D19	D20	D21	D22	D23	D24	D25
MKH2	MKH3	MKH4	MKH5	MKH6	MKH7	MKH8
TAH10	TAH11	TAH12	TAH13	TAH14	TAH15	TAH16

EUSI 484/2000-01/6236-37/1992-93						
Ashe	Ojobo	Iwa	Araorun	Ile	Itutu	Erinle
1	2	3	4	5	6	7
O31	N1	N2	N3	N4	N5	N6
KHK13	KHK14	KHK15	KHK16	KHK17	KHK18	KHK19
THi21	THi22	THi23	THi24	THi25	THi26	THi27
8	9	10	11	12	13	14
N7	N8	N9	N10	N11	N12	N13
KHK20	KHK21	KHK22	KHK23	KHK24	KHK25	KHK26
THi28	THi29	THi30	HID1	HID2	HID3	HID4
15	16	17	18	19	20	21
N14	N15	N16	N17	N18	N19	N20
KHK27	KHK28	KHK29	KHK30	TBT1	TBT2	TBT3
HID5	HID6	HID7	HID8	HID9	HID10	HID11
22	23	24	25	26	27	28
N21	N22	N23	N24	N25	N26	N27
TBT4	TBT5	TBT6	TBT7	TBT8	TBT9	TBT10
HID12	HID13	HID14	HID15	HID16	HID17	HID18

KWANZA 484/2000-01/6236-37/1992-93						
Umoja	Kujichagulia	Ujima	Ujamaa	Nia	Kuumba	Imani
1	2	3	4	5	6	7
D26	D27	D28	D29	D30	D31	JA1
MKH9	MKH10	MKH11	MKH12	MKH13	MKH14	MKH15
TAH17	TAH18	TAH19	TAH20	TAH21	TAH22	TAH23
8	9	10	11	12	13	14
JA2	JA3	JA4	JA5	JA6	JA7	JA8
MKH16	MKH17	MKH18	MKH19	MKH20	MKH21	MKH22
TAH24	TAH25	TAH26	TAH27	TAH28	TAH29	TAH30
15	16	17	18	19	20	21
JA9	JA10	JA11	JA12	JA13	JA14	JA15
MKH23	MKH24	MKH25	KMH26	MKH27	MKH28	MKH29
TIR1	TIR2	TIR3	TIR4	TIR5	TIR6	TIR7
22	23	24	25	26	27	28
JA16	JA17	JA18	JA19	JA20	JA21	JA22
MKH30	AMN1	AMN2	AMN3	AMN4	AMN5	AMN6
TIR8	TIR9	TIR10	TIR11	TIR12	TIR13	TIR14

THE NU AFRIKAN CALENDAR

SHAHIDI 484/2000-01/6236-37/1992-93						
Ashe	Ojobo	Iwa	Araorun	Ile	Itutu	Erinle
1	2	3	4	5	6	7
JA23	JA24	JA25	JA26	JA27	JA28	JA29
AMN7	AMN8	AMN9	AMN10	AMN11	AMN12	AMN13
TIR15	TIR16	TIR17	TIR18	TIR19	TIR20	TIR21
8	9	10	11	12	13	14
JA30	JA31	F1	F2	F3	F4	F5
AMN14	AMN15	AMN16	AMN17	AMN18	AMN19	AMN20
TIR22	TIR23	TIR24	TIR25	TIR26	TIR27	TIR28
15	16	17	18	19	20	21
F6	F7	F8	F9	F10	F 11	F12
AMN21	AMN22	AMN23	AMN24	AMN25	AMN26	AMN28
TIR29	TIR30	YEK1	YEK2	YEK3	YEK4	YEK5
22	23	24	25	26	27	28
F13	F14	F15	F16	F17	F18	F19
AMN29	AMN30	RNT1	RNT2	RNT3	RNT4	RNT5
YEK6	YEK7	YEK8	YEK9	YEK10	YEK11	YEK12
SHAHIDI 35						
F20						
RNT6						
YEK13						

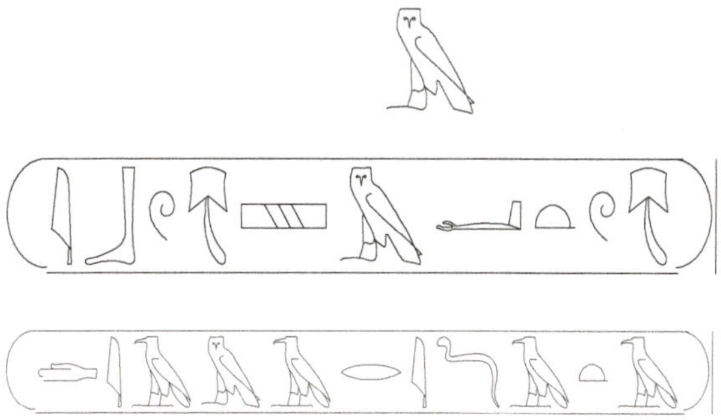

THE NU AFRIKAN CALENDAR

MAAFA 485 B.S.E./ASIENTO 2001-02 C.E.
KAM 6237-38S.M/EC 1993-94 A.M. (ዓ.ም.)

AFU 485/2001-02/6237-38/1993-94						
Ashe	Ojobo	Iwa	Araorun	Ile	Itutu	Erinle
1	2	3	4	5	6	7
F21	F22	F23	F24	F25	F26	F27
RNT7	RNT8	RNT9	RNT10	RNT11	RNT12	RNT13
YEK14	YEK15	YEK16	YEK17	YEK18	YEK19	YEK20
8	9	10	11	12	13	14
F28	M1	M2	M3	M4	M5	M6
RNT14	RNT15	RNT16	RNT17	RNT18	RNT19	RNT20
YEK21	YEK22	YEK23	YEK24	YEK25	YEK26	YEK27
15	16	17	18	19	20	21
M7	M8	M9	M10	M11	M12	M13
RNT21	RNT22	RNT23	RNT24	RNT25	RNT26	RNT27
YEK28	YEK29	YEK30	MAG1	MAG2	MAG3	MAG4
22	23	24	25	26	27	28
M14	M15	M16	M17	M18	M19	M20
RNT28	RNT29	RNT30	KHN1	KHN2	KHN3	KHN4
MAG5	MAG6	MAG7	MAG8	MAG9	MAG10	MAG11

KAWAIDA 485/2001-02/6237-38/1993-94						
Ashe	Ojobo	Iwa	Araorun	Ile	Itutu	Erinle
1	2	3	4	5	6	7
A18	A19	A20	A21	A22	A23	A24
PIN3	PIN4	PIN5	PIN6	PIN7	PIN8	PIN9
MIY10	MIY11	MIY12	MIY13	MIY14	MIY15	MIY16
8	9	10	11	12	13	14
A25	A26	A27	A28	A29	A30	MA1
PIN10	PIN11	PIN12	PIN13	PIN14	PIN15	PIN16
MIY17	MIY18	MIY19	MIY20	MIY21	MIY22	MIY23
15	16	17	18	19	20	21
MA2	MA3	MA4	MA5	MA6	MA7	MA8
PIN17	PIN18	PIN19	PIN20	PIN21	PIN22	PIN23
MIY24	MIY25	MIY26	MIY27	MIY28	MIY29	MIY30
22	23	24	25	26	27	28
MA9	MA10	MA11	MA12	MA13	MA14	MA15
PIN24	PIN25	PIN26	PIN27	PIN28	PIN29	PIN30
GIN1	GIN2	GIN3	GIN4	GIN5	GIN6	GIN7

UHURU 485/2001-02/6237-38/1993-94						
Ashe	Ojobo	Iwa	Araorun	Ile	Itutu	Erinle
1	2	3	4	5	6	7
M21	M22	M23	M24	M25	M26	M27
KHN5	KHN6	KHN7	KHN8	KHN9	KHN10	KHN11
MAG12	MAG13	MAG14	MAG15	MAG16	MAG17	MAG18
8	9	10	11	12	13	14
M28	M29	M30	M31	A1	A2	A3
KHN12	KHN13	KHN14	KHN15	KHN16	KHN17	KHN18
MAG19	MAG20	MAG21	MAG22	MAG23	MAG24	MAG25
15	16	17	18	19	20	21
A4	A5	A6	A7	A8	A9	A10
KHN19	KHN20	KHN21	KHN22	KHN23	KHN24	KHN25
MAG26	MAG27	MAG28	MAG29	MAG30	MIY1	MIY2
22	23	24	25	26	27	28
A11	A12	A13	A14	A15	A16	A17
KHN26	KHN27	KHN28	KHN29	KHN30	PIN1	PIN2
MIY3	MIY4	MIY5	MIY6	MIY7	MIY8	MIY9

MABABA WEUSI 485/2001-02/6237-38/1993-94						
Ashe	Ojobo	Iwa	Araorun	Ile	Itutu	Erinle
1	2	3	4	5	6	7
MA16	MA17	MA18	MA19	MA20	MA21	MA22
APP1	APP2	APP3	APP4	APP5	APP6	APP7
GIN8	GIN9	GIN10	GIN11	GIN12	GIN13	GIN14
8	9	10	11	12	13	14
MA23	MA24	MA25	MA26	MA27	MA28	MA29
APP8	APP9	APP10	APP11	APP12	APP13	APP14
GIN15	GIN16	GIN17	GIN18	GIN19	GIN20	GIN21
15	16	17	18	19	20	21
MA30	MA31	J1	J2	J3	J4	J5
APP15	APP16	APP17	APP18	APP19	APP20	APP21
GIN22	GIN23	GIN24	GIN25	GIN26	GIN27	GIN28
22	23	24	25	26	27	28
J6	J7	J8	J9	J10	J11	J12
APP22	APP23	APP24	APP25	APP26	APP27	APP28
GIN29	GIN30	SAN1	SAN2	SAN3	SAN4	SAN5

THE NU AFRIKAN CALENDAR

MWANAMKE WEUSI 485/2001-02/6237-38/1993-94

Ashe	Ojobo	Iwa	Araorun	Ile	Itutu	Erinle
1	2	3	4	5	6	7
J13	J14	J15	J16	J17	J18	J19
APP29	APP30	MSR1	MSR2	MSR3	MSR4	MSR5
SAN6	SAN7	SAN8	SAN9	SAN10	SAN11	SAN12
8	9	10	11	12	13	14
J20	J21	J22	J23	J24	J25	J26
MSR6	MSR7	MSR8	MSR9	MSR10	MSR11	MSR12
SAN13	SAN14	SAN15	SAN16	SAN17	SAN18	SAN19
15	16	17	18	19	20	21
J27	J28	J29	J30	JU1	JU2	JU3
MSR13	MSR14	MSR15	MSR16	MSR17	MSR18	MSR19
SAN20	SAN21	SAN22	SAN23	SAN24	SAN25	SAN26
22	23	24	25	26	27	28
JU4	JU5	JU6	JU7	JU8	JU9	JU10
MSR20	MRS21	MSR22	MSR23	MSR24	MSR25	MSR26
SAN27	SAN28	SAN29	SAN30	HAM1	HAM2	HAM3

TAIFA WEUSI 485/2001-02/6237-38/1993-94

Ashe	Ojobo	Iwa	Araorun	Ile	Itutu	Erinle
1	2	3	4	5	6	7
AU8	AU9	AU10	AU11	AU12	AU13	AU14
THT20	THT21	THT22	THT23	THT24	THT25	THT26
NAH2	NAH3	NAH4	NAH5	NAH6	NAH7	NAH8
8	9	10	11	12	13	14
AU15	AU16	AU17	AU18	AU19	AU20	AU21
THT27	THT28	THT29	THT30	PAP1	PAP1	PAP2
NAH9	NAH10	NAH11	NAH12	NAH13	NAH14	NAH15
15	16	17	18	19	20	21
AU22	AU23	AU24	AU25	AU26	AU27	AU28
PAP3	PAP4	PAP5	PAP6	PAP7	PAP8	PAP9
NAH16	NAH17	NAH18	NAH19	NAH20	NAH21	NAH22
22	23	24	25	26	27	28
AU29	AU30	AU31	S1	S2	S3	S4
PAP10	PAP11	PAP12	PAP13	PAP14	PAP15	PAP16
NAH23	NAH24	NAH25	NAH26	NAH27	NAH28	NAH29

MTU WEUSI 485/2001-02/6237-38/1993-94

Ashe	Ojobo	Iwa	Araorun	Ile	Itutu	Erinle
1	2	3	4	5	6	7
JU11	JU12	JU13	JU14	JU15	JU16	JU17
MSR27	MSR28	MSR29	MSR30	*ASR1*	*HRU2*	*STH3*
HAM4	HAM5	HAM6	HAM7	HAM8	HAM9	HAM10
8	9	10	11	12	13	14
JU18	JU19	JU20	JU21	JU22	JU23	JU24
AST4	*NBH5*	THT1	THT2	THT3	THT4	THT5
HAM11	HAM12	HAM13	AM14	HAM15	HAM16	HAM17
15	16	17	18	19	20	21
JU25	JU26	JU27	JU28	JU29	JU30	JU31
THT6	THT7	THT8	THT9	THT10	THT11	THT12
HAM18	HAM19	HAM20	HAM21	HAM22	HAM23	HAM24
22	23	24	25	26	27	28
AU1	AU2	AU3	AU4	AU5	AU6	AU7
THT13	THT14	THT15	THT16	THT17	THT18	THT19
HAM25	HAM26	HAM27	HAM28	HAM29	HAM30	NAH1

UWEZO 485/2001-02/6237-38/1993-94

Ashe	Ojobo	Iwa	Araorun	Ile	Itutu	Erinle
1	2	3	4	5	6	7
S5	S6	S7	S8	S9	S10	S11
PAP17	PAP18	PAP19	PAP20	PAP21	PAP22	PAP23
NAH30	*PAG1*	*PAG2*	*PAG3*	*PAG4*	*PAG5*	MAS1
8	9	10	11	12	13	14
S12	S13	S14	S15	S16	S17	S18
PAP24	PAP25	PAP26	PAP27	PAP28	PAP29	PAP30
MAS2	MAS3	MAS4	MAS5	MAS6	MAS7	MAS8
15	16	17	18	19	20	21
S19	S20	S21	S22	S23	S24	S25
HTH1	HTH2	HTH3	HTH4	HTH5	HTH6	HTH7
MAS9	MAS10	MAS11	MAS12	MAS13	MAS14	MAS15
22	23	24	25	26	27	28
S26	S27	S28	S29	S30	O1	O2
HTH8	HTH9	HTH10	HTH11	HTH12	HTH13	HTH14
MAS16	MAS17	MAS18	MAS19	MAS20	MAS21	MAS22

THE NU AFRIKAN CALENDAR

EKUNDU 485/2001-02/6237-38/1993-94						
Ashe	Ojobo	Iwa	Araorun	Ile	Itutu	Erinle
1	2	3	4	5	6	7
O3	O4	O5	O6	O7	O8	O9
HTH15	HTH16	HTH17	HTH18	HTH19	HTH20	HTH21
MAS23	MAS24	MAS25	MAS26	MAS27	MAS28	MAS29
8	9	10	11	12	13	14
O10	O11	O12	O13	O14	O15	O16
HTH22	HTH23	HTH24	HTH25	HTH26	HTH27	HTH28
MAS30	THi1	THi2	THi3	THi4	THi5	THi6
15	16	17	18	19	20	21
O17	O18	O19	O20	O21	O22	O23
HTH29	HTH30	KHK1	KHK2	KHK3	KHK4	KHK5
THi7	THi8	THi9	THi10	THi11	THi12	THi13
22	23	24	25	26	27	28
O24	O25	O26	O27	O28	O29	O30
KHK6	KHK7	KHK8	KHK9	KHK10	KHK11	KHK12
THi14	THi15	THi16	THi17	THi18	THi19	THi20

KIJANI 485/2001-02/6237-38/1993-94						
Ashe	Ojobo	Iwa	Araorun	Ile	Itutu	Erinle
1	2	3	4	5	6	7
N28	N29	N30	D1	D2	D3	D4
TBT11	TBT12	TBT13	TBT14	TBT15	TBT16	TBT17
HID19	HID20	HID21	HID22	HID23	HID24	HID25
8	9	10	11	12	13	14
D5	D6	D7	D8	D9	D10	D11
TBT18	TBT19	TBT20	TBT21	TBT22	TBT23	TBT24
HID26	HID27	HID28	HID29	HID30	TAH1	TAH2
15	16	17	18	19	20	21
D12	D13	D14	D15	D16	D17	D18
TBT25	TBT26	TBT27	TBT28	TBT29	TBT30	MKH1
TAH3	TAH4	TAH5	TAH6	TAH7	TAH8	TAH9
22	23	24	25	26	27	28
D19	D20	D21	D22	D23	D24	D25
MKH2	MKH3	MKH4	MKH5	MKH6	MKH7	MKH8
TAH10	TAH11	TAH12	TAH13	TAH14	TAH15	TAH16

EUSI 485/2001-02/6237-38/1993-94						
Ashe	Ojobo	Iwa	Araorun	Ile	Itutu	Erinle
1	2	3	4	5	6	7
O31	N1	N2	N3	N4	N5	N6
KHK13	KHK14	KHK15	KHK16	KHK17	KHK18	KHK19
THi21	THi22	THi23	THi24	THi25	THi26	THi27
8	9	10	11	12	13	14
N7	N8	N9	N10	N11	N12	N13
KHK20	KHK21	KHK22	KHK23	KHK24	KHK25	KHK26
THi28	THi29	THi30	HID1	HID2	HID3	HID4
15	16	17	18	19	20	21
N14	N15	N16	N17	N18	N19	N20
KHK27	KHK28	KHK29	KHK30	TBT1	TBT2	TBT3
HID5	HID6	HID7	HID8	HID9	HID10	HID11
22	23	24	25	26	27	28
N21	N22	N23	N24	N25	N26	N27
TBT4	TBT5	TBT6	TBT7	TBT8	TBT9	TBT10
HID12	HID13	HID14	HID15	HID16	HID17	HID18

KWANZA 485/2001-02/6237-38/1993-94						
Umoja	Kujichagulia	Ujima	Ujamaa	Nia	Kuumba	Imani
1	2	3	4	5	6	7
D26	D27	D28	D29	D30	D31	JA1
MKH9	MKH10	MKH11	MKH12	MKH13	MKH14	MKH15
TAH17	TAH18	TAH19	TAH20	TAH21	TAH22	TAH23
8	9	10	11	12	13	14
JA2	JA3	JA4	JA5	JA6	JA7	JA8
MKH16	MKH17	MKH18	MKH19	MKH20	MKH21	MKH22
TAH24	TAH25	TAH26	TAH27	TAH28	TAH29	TAH30
15	16	17	18	19	20	21
JA9	JA10	JA11	JA12	JA13	JA14	JA15
MKH23	MKH24	MKH25	KMH26	MKH27	MKH28	MKH29
TIR1	TIR2	TIR3	TIR4	TIR5	TIR6	TIR7
22	23	24	25	26	27	28
JA16	JA17	JA18	JA19	JA20	JA21	JA22
MKH30	AMN1	AMN2	AMN3	AMN4	AMN5	AMN6
TIR8	TIR9	TIR10	TIR11	TIR12	TIR13	TIR14

THE NU AFRIKAN CALENDAR

SHAHIDI 485/2001-02/6237-38/1993-94						
Ashe	Ojobo	Iwa	Araorun	Ile	Itutu	Erinle
1	2	3	4	5	6	7
JA23	JA24	JA25	JA26	JA27	JA28	JA29
AMN7	AMN8	AMN9	AMN10	AMN11	AMN12	AMN13
TIR15	TIR16	TIR17	TIR18	TIR19	TIR20	TIR21
8	9	10	11	12	13	14
JA30	JA31	F1	F2	F3	F4	F5
AMN14	AMN15	AMN16	AMN17	AMN18	AMN19	AMN20
TIR22	TIR23	TIR24	TIR25	TIR26	TIR27	TIR28
15	16	17	18	19	20	21
F6	F7	F8	F9	F10	F 11	F12
AMN21	AMN22	AMN23	AMN24	AMN25	AMN26	AMN28
TIR29	TIR30	YEK1	YEK2	YEK3	YEK4	YEK5
22	23	24	25	26	27	28
F13	F14	F15	F16	F17	F18	F19
AMN29	AMN30	RNT1	RNT2	RNT3	RNT4	RNT5
YEK6	YEK7	YEK8	YEK9	YEK10	YEK11	YEK12
SHAHIDI 36						
F20						
RNT6						
YEK13						

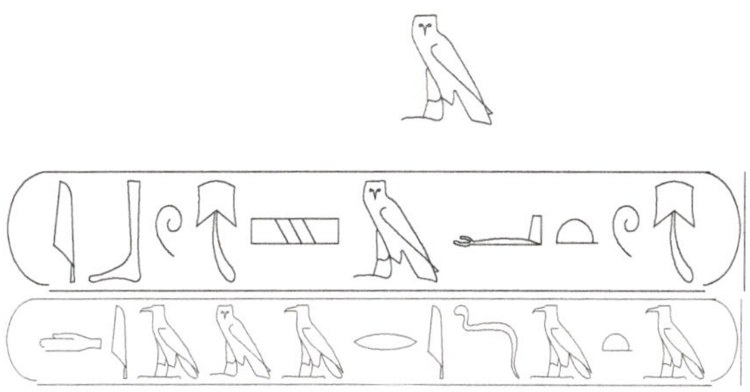

THE NU AFRIKAN CALENDAR

MAAFA 486 B.S.E./ASIENTO 2002-03 C.E.
KAM 6238-39 S.M/EC 1994-95 A.M. (ዓ.ም.)

AFU 486/2002-03/6238-39/1994-95						
e	Ojobo	Iwa	Araorun	Ile	Itutu	Erinle
	2	3	4	5	6	7
	F22	F23	F24	F25	F26	F27
7	RNT8	RNT9	RNT10	RNT11	RNT12	RNT13
14	YEK15	YEK16	YEK17	YEK18	YEK19	YEK20
	9	10	11	12	13	14
	M1	M2	M3	M4	M5	M6
14	RNT15	RNT16	RNT17	RNT18	RNT19	RNT20
21	YEK22	YEK23	YEK24	YEK25	YEK26	YEK27
	16	17	18	19	20	21
	M8	M9	M10	M11	M12	M13
21	RNT22	RNT23	RNT24	RNT25	RNT26	RNT27
28	YEK29	YEK30	MAG1	MAG2	MAG3	MAG4
	23	24	25	26	27	28
	M15	M16	M17	M18	M19	M20
28	RNT29	RNT30	KHN1	KHN2	KHN3	KHN4
5	MAG6	MAG7	MAG8	MAG9	MAG10	MAG11

KAWAIDA 486/2002-03/6238-39/1994-95						
Ashe	Ojobo	Iwa	Araorun	Ile	Itutu	Erinle
1	2	3	4	5	6	7
A18	A19	A20	A21	A22	A23	A24
PIN3	PIN4	PIN5	PIN6	PIN7	PIN8	PIN9
MIY10	MIY11	MIY12	MIY13	MIY14	MIY15	MIY16
8	9	10	11	12	13	14
A25	A26	A27	A28	A29	A30	MA1
PIN10	PIN11	PIN12	PIN13	PIN14	PIN15	PIN16
MIY17	MIY18	MIY19	MIY20	MIY21	MIY22	MIY23
15	16	17	18	19	20	21
MA2	MA3	MA4	MA5	MA6	MA7	MA8
PIN17	PIN18	PIN19	PIN20	PIN21	PIN22	PIN23
MIY24	MIY25	MIY26	MIY27	MIY28	MIY29	MIY30
22	23	24	25	26	27	28
MA9	MA10	MA11	MA12	MA13	MA14	MA15
PIN24	PIN25	PIN26	PIN27	PIN28	PIN29	PIN30
GIN1	GIN2	GIN3	GIN4	GIN5	GIN6	GIN7

UHURU 486/2002-03/6238-39/1994-95						
e	Ojobo	Iwa	Araorun	Ile	Itutu	Erinle
	2	3	4	5	6	7
	M22	M23	M24	M25	M26	M27
5	KHN6	KHN7	KHN8	KHN9	KHN10	KHN11
12	MAG13	MAG14	MAG15	MAG16	MAG17	MAG18
	9	10	11	12	13	14
	M29	M30	M31	A1	A2	A3
12	KHN13	KHN14	KHN15	KHN16	KHN17	KHN18
19	MAG20	MAG21	MAG22	MAG23	MAG24	MAG25
	16	17	18	19	20	21
	A5	A6	A7	A8	A9	A10
19	KHN20	KHN21	KHN22	KHN23	KHN24	KHN25
26	MAG27	MAG28	MAG29	MAG30	MIY1	MIY2
	23	24	25	26	27	28
	A12	A13	A14	A15	A16	A17
26	KHN27	KHN28	KHN29	KHN30	PIN1	PIN2
3	MIY4	MIY5	MIY6	MIY7	MIY8	MIY9

MABABA WEUSI 486/2002-03/6238-39/1994-95						
Ashe	Ojobo	Iwa	Araorun	Ile	Itutu	Erinle
1	2	3	4	5	6	7
MA16	MA17	MA18	MA19	MA20	MA21	MA22
APP1	APP2	APP3	APP4	APP5	APP6	APP7
GIN8	GIN9	GIN10	GIN11	GIN12	GIN13	GIN14
8	9	10	11	12	13	14
MA23	MA24	MA25	MA26	MA27	MA28	MA29
APP8	APP9	APP10	APP11	APP12	APP13	APP14
GIN15	GIN16	GIN17	GIN18	GIN19	GIN20	GIN21
15	16	17	18	19	20	21
MA30	MA31	J1	J2	J3	J4	J5
APP15	APP16	APP17	APP18	APP19	APP20	APP21
GIN22	GIN23	GIN24	GIN25	GIN26	GIN27	GIN28
22	23	24	25	26	27	28
J6	J7	J8	J9	J10	J11	J12
APP22	APP23	APP24	APP25	APP26	APP27	APP28
GIN29	GIN30	SAN1	SAN2	SAN3	SAN4	SAN5

THE NU AFRIKAN CALENDAR

MWANAMKE WEUSI 486/2002-03/6238-39/1994-95						
Ashe	Ojobo	Iwa	Araorun	Ile	Itutu	Erinle
1	2	3	4	5	6	7
J13	J14	J15	J16	J17	J18	J19
APP29	APP30	MSR1	MSR2	MSR3	MSR4	MSR5
SAN6	SAN7	SAN8	SAN9	SAN10	SAN11	SAN12
8	9	10	11	12	13	14
J20	J21	J22	J23	J24	J25	J26
MSR6	MSR7	MSR8	MSR9	MSR10	MSR11	MSR12
SAN13	SAN14	SAN15	SAN16	SAN17	SAN18	SAN19
15	16	17	18	19	20	21
J27	J28	J29	J30	JU1	JU2	JU3
MSR13	MSR14	MSR15	MSR16	MSR17	MSR18	MSR19
SAN20	SAN21	SAN22	SAN23	SAN24	SAN25	SAN26
22	23	24	25	26	27	28
JU4	JU5	JU6	JU7	JU8	JU9	JU10
MSR20	MRS21	MSR22	MSR23	MSR24	MSR25	MSR26
SAN27	SAN28	SAN29	SAN30	HAM1	HAM2	HAM3

TAIFA WEUSI 486/2002-03/6238-39/1994-95						
Ashe	Ojobo	Iwa	Araorun	Ile	Itutu	Erinle
1	2	3	4	5	6	7
AU8	AU9	AU10	AU11	AU12	AU13	AU14
THT20	THT21	THT22	THT23	THT24	THT25	THT26
NAH2	NAH3	NAH4	NAH5	NAH6	NAH7	NAH8
8	9	10	11	12	13	14
AU15	AU16	AU17	AU18	AU19	AU20	AU21
THT27	THT28	THT29	THT30	PAP1	PAP1	PAP2
NAH9	NAH10	NAH11	NAH12	NAH13	NAH14	NAH15
15	16	17	18	19	20	21
AU22	AU23	AU24	AU25	AU26	AU27	AU28
PAP3	PAP4	PAP5	PAP6	PAP7	PAP8	PAP9
NAH16	NAH17	NAH18	NAH19	NAH20	NAH21	NAH22
22	23	24	25	26	27	28
AU29	AU30	AU31	S1	S2	S3	S4
PAP10	PAP11	PAP12	PAP13	PAP14	PAP15	PAP16
NAH23	NAH24	NAH25	NAH26	NAH27	NAH28	NAH29

MTU WEUSI 486/2002-03/6238-39/1994-95						
Ashe	Ojobo	Iwa	Araorun	Ile	Itutu	Erinle
1	2	3	4	5	6	7
JU11	JU12	JU13	JU14	JU15	JU16	JU17
MSR27	MSR28	MSR29	MSR30	*ASR1*	*HRU2*	*STH3*
HAM4	HAM5	HAM6	HAM7	HAM8	HAM9	HAM10
8	9	10	11	12	13	14
JU18	JU19	JU20	JU21	JU22	JU23	JU24
AST4	*NBH5*	THT1	THT2	THT3	THT4	THT5
HAM11	HAM12	HAM13	AM14	HAM15	HAM16	HAM17
15	16	17	18	19	20	21
JU25	JU26	JU27	JU28	JU29	JU30	JU31
THT6	THT7	THT8	THT9	THT10	THT11	THT12
HAM18	HAM19	HAM20	HAM21	HAM22	HAM23	HAM24
22	23	24	25	26	27	28
AU1	AU2	AU3	AU4	AU5	AU6	AU7
THT13	THT14	THT15	THT16	THT17	THT18	THT19
HAM25	HAM26	HAM27	HAM28	HAM29	HAM30	NAH1

UWEZO 486/2002-03/6238-39/1994-95						
Ashe	Ojobo	Iwa	Araorun	Ile	Itutu	Erinle
1	2	3	4	5	6	7
S5	S6	S7	S8	S9	S10	S11
PAP17	PAP18	PAP19	PAP20	PAP21	PAP22	PAP23
NAH30	*PAG1*	*PAG2*	*PAG3*	*PAG4*	*PAG5*	MAS1
8	9	10	11	12	13	14
S12	S13	S14	S15	S16	S17	S18
PAP24	PAP25	PAP26	PAP27	PAP28	PAP29	PAP30
MAS2	MAS3	MAS4	MAS5	MAS6	MAS7	MAS8
15	16	17	18	19	20	21
S19	S20	S21	S22	S23	S24	S25
HTH1	HTH2	HTH3	HTH4	HTH5	HTH6	HTH7
MAS9	MAS10	MAS11	MAS12	MAS13	MAS14	MAS15
22	23	24	25	26	27	28
S26	S27	S28	S29	S30	O1	O2
HTH8	HTH9	HTH10	HTH11	HTH12	HTH13	HTH14
MAS16	MAS17	MAS18	MAS19	MAS20	MAS21	MAS22

THE NU AFRIKAN CALENDAR

EKUNDU 486/2002-03/6238-39/1994-95						
Ashe	Ojobo	Iwa	Araorun	Ile	Itutu	Erinle
1	2	3	4	5	6	7
O3	O4	O5	O6	O7	O8	O9
HTH15	HTH16	HTH17	HTH18	HTH19	HTH20	HTH21
MAS23	MAS24	MAS25	MAS26	MAS27	MAS28	MAS29
8	9	10	11	12	13	14
O10	O11	O12	O13	O14	O15	O16
HTH22	HTH23	HTH24	HTH25	HTH26	HTH27	HTH28
MAS30	THi1	THi2	THi3	THi4	THi5	THi6
15	16	17	18	19	20	21
O17	O18	O19	O20	O21	O22	O23
HTH29	HTH30	KHK1	KHK2	KHK3	KHK4	KHK5
THi7	THi8	THi9	THi10	THi11	THi12	THi13
22	23	24	25	26	27	28
O24	O25	O26	O27	O28	O29	O30
KHK6	KHK7	KHK8	KHK9	KHK10	KHK11	KHK12
THi14	THi15	THi16	THi17	THi18	THi19	THi20

KIJANI 486/2002-03/6238-39/1994-95						
Ashe	Ojobo	Iwa	Araorun	Ile	Itutu	Erinle
1	2	3	4	5	6	7
N28	N29	N30	D1	D2	D3	D4
TBT11	TBT12	TBT13	TBT14	TBT15	TBT16	TBT17
HID19	HID20	HID21	HID22	HID23	HID24	HID25
8	9	10	11	12	13	14
D5	D6	D7	D8	D9	D10	D11
TBT18	TBT19	TBT20	TBT21	TBT22	TBT23	TBT24
HID26	HID27	HID28	HID29	HID30	TAH1	TAH2
15	16	17	18	19	20	21
D12	D13	D14	D15	D16	D17	D18
TBT25	TBT26	TBT27	TBT28	TBT29	TBT30	MKH1
TAH3	TAH4	TAH5	TAH6	TAH7	TAH8	TAH9
22	23	24	25	26	27	28
D19	D20	D21	D22	D23	D24	D25
MKH2	MKH3	MKH4	MKH5	MKH6	MKH7	MKH8
TAH10	TAH11	TAH12	TAH13	TAH14	TAH15	TAH16

EUSI 486/2002-03/6238-39/1994-95						
Ashe	Ojobo	Iwa	Araorun	Ile	Itutu	Erinle
1	2	3	4	5	6	7
O31	N1	N2	N3	N4	N5	N6
KHK13	KHK14	KHK15	KHK16	KHK17	KHK18	KHK19
THi21	THi22	THi23	THi24	THi25	THi26	THi27
8	9	10	11	12	13	14
N7	N8	N9	N10	N11	N12	N13
KHK20	KHK21	KHK22	KHK23	KHK24	KHK25	KHK26
THi28	THi29	THi30	HID1	HID2	HID3	HID4
15	16	17	18	19	20	21
N14	N15	N16	N17	N18	N19	N20
KHK27	KHK28	KHK29	KHK30	TBT1	TBT2	TBT3
HID5	HID6	HID7	HID8	HID9	HID10	HID11
22	23	24	25	26	27	28
N21	N22	N23	N24	N25	N26	N27
TBT4	TBT5	TBT6	TBT7	TBT8	TBT9	TBT10
HID12	HID13	HID14	HID15	HID16	HID17	HID18

KWANZA 486/2002-03/6238-39/1994-95						
Umoja	Kujichagulia	Ujima	Ujamaa	Nia	Kuumba	Imani
1	2	3	4	5	6	7
D26	D27	D28	D29	D30	D31	JA1
MKH9	MKH10	MKH11	MKH12	MKH13	MKH14	MKH15
TAH17	TAH18	TAH19	TAH20	TAH21	TAH22	TAH23
8	9	10	11	12	13	14
JA2	JA3	JA4	JA5	JA6	JA7	JA8
MKH16	MKH17	MKH18	MKH19	MKH20	MKH21	MKH22
TAH24	TAH25	TAH26	TAH27	TAH28	TAH29	TAH30
15	16	17	18	19	20	21
JA9	JA10	JA11	JA12	JA13	JA14	JA15
MKH23	MKH24	MKH25	KMH26	MKH27	MKH28	MKH29
TIR1	TIR2	TIR3	TIR4	TIR5	TIR6	TIR7
22	23	24	25	26	27	28
JA16	JA17	JA18	JA19	JA20	JA21	JA22
MKH30	AMN1	AMN2	AMN3	AMN4	AMN5	AMN6
TIR8	TIR9	TIR10	TIR11	TIR12	TIR13	TIR14

THE NU AFRIKAN CALENDAR

SHAHIDI 486/2002-03/6238-39/1994-95						
Ashe	Ojobo	Iwa	Araorun	Ile	Itutu	Erinle
1	2	3	4	5	6	7
JA23	JA24	JA25	JA26	JA27	JA28	JA29
AMN7	AMN8	AMN9	AMN10	AMN11	AMN12	AMN13
TIR15	TIR16	TIR17	TIR18	TIR19	TIR20	TIR21
8	9	10	11	12	13	14
JA30	JA31	F1	F2	F3	F4	F5
AMN14	AMN15	AMN16	AMN17	AMN18	AMN19	AMN20
TIR22	TIR23	TIR24	TIR25	TIR26	TIR27	TIR28
15	16	17	18	19	20	21
F6	F7	F8	F9	F10	F 11	F12
AMN21	AMN22	AMN23	AMN24	AMN25	AMN26	AMN28
TIR29	TIR30	YEK1	YEK2	YEK3	YEK4	YEK5
22	23	24	25	26	27	28
F13	F14	F15	F16	F17	F18	F19
AMN29	AMN30	RNT1	RNT2	RNT3	RNT4	RNT5
YEK6	YEK7	YEK8	YEK9	YEK10	YEK11	YEK12
SHAHIDI 37						
F20						
RNT6						
YEK13						

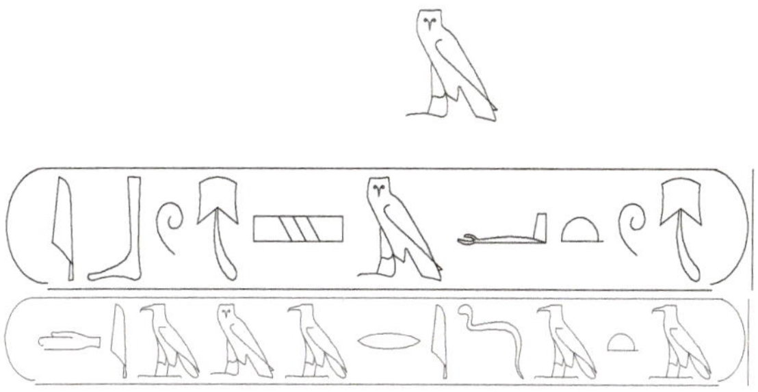

MAAFA 487 B.S.E./ASIENTO 2003-04 C.E./
KAM 6239-40 S.M/EC 1995-96 A.M. (ዓ.ም.)

AFU 487/2003-04/6239-40/1995-96

Ashe	Ojobo	Iwa	Araorun	Ile	Itutu	Erinle
1	2	3	4	5	6	7
F21	F22	F23	F24	F25	F26	F27
RNT7	RNT8	RNT9	RNT10	RNT11	RNT12	RNT13
YEK14	YEK15	YEK16	YEK17	YEK18	YEK19	YEK20
8	9	10	11	12	13	14
F28	M1	M2	M3	M4	M5	M6
RNT14	RNT15	RNT16	RNT17	RNT18	RNT19	RNT20
YEK21	YEK22	YEK23	YEK24	YEK25	YEK26	YEK27
15	16	17	18	19	20	21
M7	M8	M9	M10	M11	M12	M13
RNT21	RNT22	RNT23	RNT24	RNT25	RNT26	RNT27
YEK28	YEK 29	YEK30	MAG1	MAG2	MAG3	MAG4
22	23	24	25	26	27	28
M14	M15	M16	M17	M18	M19	M20
RNT28	RNT29	RNT30	KHN1	KHN2	KHN3	KHN4
MAG5	MAG6	MAG7	MAG8	MAG9	MAG10	MAG11

UHURU 487/2003-04/6239-40/1995-96

Ashe	Ojobo	Iwa	Araorun	Ile	Itutu	Erinle
1	2	3	4	5	6	7
M21	M22	M23	M24	M25	M26	M27
KHN5	KHN6	KHN7	KHN8	KHN9	KHN10	KHN11
MAG12	MAG13	MAG14	MAG15	MAG16	MAG17	MAG18
8	9	10	11	12	13	14
M28	M29	M30	M31	A1	A2	A3
KHN12	KHN13	KHN14	KHN15	KHN16	KHN17	KHN18
MAG19	MAG20	MAG21	MAG22	MAG23	MAG24	MAG25
15	16	17	18	19	20	21
A4	A5	A6	A7	A8	A9	A10
KHN19	KHN20	KHN21	KHN22	KHN23	KHN24	KHN25
MAG26	MAG27	MAG28	MAG29	MAG30	MIY1	MIY2
22	23	24	25	26	27	28
A11	A12	A13	A14	A15	A16	A17
KHN26	KHN27	KHN28	KHN29	KHN30	PIN1	PIN2
MIY3	MIY4	MIY5	MIY6	MIY7	MIY8	MIY9

KAWAIDA 487/2003-04/6239-40/1995-96

Ashe	Ojobo	Iwa	Araorun	Ile	Itutu	Erinle
1	2	3	4	5	6	7
A18	A19	A20	A21	A22	A23	A24
PIN3	PIN4	PIN5	PIN6	PIN7	PIN8	PIN9
MIY10	MIY11	MIY12	MIY13	MIY14	MIY15	MIY16
8	9	10	11	12	13	14
A25	A26	A27	A28	A29	A30	MA1
PIN10	PIN11	PIN12	PIN13	PIN14	PIN15	PIN16
MIY17	MIY18	MIY19	MIY20	MIY21	MIY22	MIY23
15	16	17	18	19	20	21
MA2	MA3	MA4	MA5	MA6	MA7	MA8
PIN17	PIN18	PIN19	PIN20	PIN21	PIN22	PIN23
MIY24	MIY25	MIY26	MIY27	MIY28	MIY29	MIY30
22	23	24	25	26	27	28
MA9	MA10	MA11	MA12	MA13	MA14	MA15
PIN24	PIN25	PIN26	PIN27	PIN28	PIN29	PIN30
GIN1	GIN2	GIN3	GIN4	GIN5	GIN6	GIN7

MABABA WEUSI 487/2003-04/6239-40/1995-96

Ashe	Ojobo	Iwa	Araorun	Ile	Itutu	Erinle
1	2	3	4	5	6	7
MA16	MA17	MA18	MA19	MA20	MA21	MA22
APP1	APP2	APP3	APP4	APP5	APP6	APP7
GIN8	GIN9	GIN10	GIN11	GIN12	GIN13	GIN14
8	9	10	11	12	13	14
MA23	MA24	MA25	MA26	MA27	MA28	MA29
APP8	APP9	APP10	APP11	APP12	APP13	APP14
GIN15	GIN16	GIN17	GIN18	GIN19	GIN20	GIN21
15	16	17	18	19	20	21
MA30	MA31	J1	J2	J3	J4	J5
APP15	APP16	APP17	APP18	APP19	APP20	APP21
GIN22	GIN23	GIN24	GIN25	GIN26	GIN27	GIN28
22	23	24	25	26	27	28
J6	J7	J8	J9	J10	J11	J12
APP22	APP23	APP24	APP25	APP26	APP27	APP28
GIN29	GIN30	SAN1	SAN2	SAN3	SAN4	SAN5

THE NU AFRIKAN CALENDAR

MWANAMKE WEUSI 487/2003-04/6239-40/1995-96						
Ashe	Ojobo	Iwa	Araorun	Ile	Itutu	Erinle
1	2	3	4	5	6	7
J13	J14	J15	J16	J17	J18	J19
APP29	APP30	MSR1	MSR2	MSR3	MSR4	MSR5
SAN6	SAN7	SAN8	SAN9	SAN10	SAN11	SAN12
8	9	10	11	12	13	14
J20	J21	J22	J23	J24	J25	J26
MSR6	MSR7	MSR8	MSR9	MSR10	MSR11	MSR12
SAN13	SAN14	SAN15	SAN16	SAN17	SAN18	SAN19
15	16	17	18	19	20	21
J27	J28	J29	J30	JU1	JU2	JU3
MSR13	MSR14	MSR15	MSR16	MSR17	MSR18	MSR19
SAN20	SAN21	SAN22	SAN23	SAN24	SAN25	SAN26
22	23	24	25	26	27	28
JU4	JU5	JU6	JU7	JU8	JU9	JU10
MSR20	MRS21	MSR22	MSR23	MSR24	MSR25	MSR26
SAN27	SAN28	SAN29	SAN30	HAM1	HAM2	HAM3

TAIFA WEUSI 487/2003-04/6239-40/1995-96						
Ashe	Ojobo	Iwa	Araorun	Ile	Itutu	Erinle
1	2	3	4	5	6	7
AU8	AU9	AU10	AU11	AU12	AU13	AU14
THT20	THT21	THT22	THT23	THT24	THT25	THT26
NAH2	NAH3	NAH4	NAH5	NAH6	NAH7	NAH8
8	9	10	11	12	13	14
AU15	AU16	AU17	AU18	AU19	AU20	AU21
THT27	THT28	THT29	THT30	PAP1	PAP1	PAP2
NAH9	NAH10	NAH11	NAH12	NAH13	NAH14	NAH15
15	16	17	18	19	20	21
AU22	AU23	AU24	AU25	AU26	AU27	AU28
PAP3	PAP4	PAP5	PAP6	PAP7	PAP8	PAP9
NAH16	NAH17	NAH18	NAH19	NAH20	NAH21	NAH22
22	23	24	25	26	27	28
AU29	AU30	AU31	S1	S2	S3	S4
PAP10	PAP11	PAP12	PAP13	PAP14	PAP15	PAP16
NAH23	NAH24	NAH25	NAH26	NAH27	NAH28	NAH29

MTU WEUSI 487/2003-04/6239-40/1995-96						
Ashe	Ojobo	Iwa	Araorun	Ile	Itutu	Erinle
1	2	3	4	5	6	7
JU11	JU12	JU13	JU14	JU15	JU16	JU17
MSR27	MSR28	MSR29	MSR30	*ASR1*	*HRU2*	*STH3*
HAM4	HAM5	HAM6	HAM7	HAM8	HAM9	HAM10
8	9	10	11	12	13	14
JU18	JU19	JU20	JU21	JU22	JU23	JU24
AST4	*NBH5*	THT1	THT2	THT3	THT4	THT5
HAM11	HAM12	HAM13	HAM14	HAM15	HAM16	HAM17
15	16	17	18	19	20	21
JU25	JU26	JU27	JU28	JU29	JU30	JU31
THT6	THT7	THT8	THT9	THT10	THT11	THT12
HAM18	HAM19	HAM20	HAM21	HAM22	HAM23	HAM24
22	23	24	25	26	27	28
AU1	AU2	AU3	AU4	AU5	AU6	AU7
THT13	THT14	THT15	THT16	THT17	THT18	THT19
HAM25	HAM26	HAM27	HAM28	HAM29	HAM30	NAH1

UWEZO 487/2003-04/6239-40/1995-96						
Ashe	Ojobo	Iwa	Araorun	Ile	Itutu	Erinle
1	2	3	4	5	6	7
S5	S6	S7	S8	S9	S10	S11
PAP17	PAP18	PAP19	PAP20	PAP21	PAP22	PAP23
NAH30	*QUA1*	*QUA2*	*QUA3*	*QUA4*	*QUA5*	*QUA6*
8	9	10	11	12	13	14
S12	S13	S14	S15	S16	S17	S18
PAP24	PAP25	PAP26	PAP27	PAP28	PAP29	PAP30
MAS1	MAS2	MAS3	MAS4	MAS5	MAS6	MAS7
15	16	17	18	19	20	21
S19	S20	S21	S22	S23	S24	S25
HTH1	HTH2	HTH3	HTH4	HTH5	HTH6	HTH7
MAS8	MAS9	MAS10	MAS11	MAS12	MAS13	MAS14
22	23	24	25	26	27	28
S26	S27	S28	S29	S30	O1	O2
HTH8	HTH9	HTH10	HTH11	HTH12	HTH13	HTH14
MAS15	MAS16	MAS17	MAS18	MAS19	MAS20	MAS21

THE NU AFRIKAN CALENDAR

EKUNDU 487/2003-04/6239-40/1995-96

Ashe	Ojobo	Iwa	Araorun	Ile	Itutu	Erinle
1	2	3	4	5	6	7
O3	O4	O5	O6	O7	O8	O9
HTH15	HTH16	HTH17	HTH18	HTH19	HTH20	HTH21
MAS22	MAS23	MAS 24	MAS25	MAS26	MAS27	MAS28
8	9	10	11	12	13	14
O10	O11	O12	O13	O14	O15	O16
HTH22	HTH23	HTH24	HTH25	HTH26	HTH27	HTH28
MAS29	MAS30	THi1	THi2	THi3	THi4	THi5
15	16	17	18	19	20	21
O17	O18	O19	O20	O21	O22	O23
HTH29	HTH30	KHK1	KHK2	KHK3	KHK4	KHK5
THi6	THi7	THi8	THi9	THi10	THi11	THi12
22	23	24	25	26	27	28
O24	O25	O26	O27	O28	O29	O30
KHK6	KHK7	KHK8	KHK9	KHK10	KHK11	KHK12
THi13	THi14	THi15	THi16	THi17	THi18	THi19

KIJANI 487/2003-04/6239-40/1995-96

Ashe	Ojobo	Iwa	Araorun	Ile	Itutu	Erinle
1	2	3	4	5	6	7
N28	N29	N30	D1	D2	D3	D4
TBT11	TBT12	TBT13	TBT14	TBT15	TBT16	TBT17
HID18	HID19	HID20	HID21	HID22	HID23	HID24
8	9	10	11	12	13	14
D5	D6	D7	D8	D9	D10	D11
TBT18	TBT19	TBT20	TBT21	TBT22	TBT23	TBT24
HID25	HID26	HID27	HID28	HID29	HID30	TAH1
15	16	17	18	19	20	21
D12	D13	D14	D15	D16	D17	D18
TBT25	TBT26	TBT27	TBT28	TBT29	TBT30	MKH1
TAH2	TAH3	TAH4	TAH5	TAH6	TAH7	TAH8
22	23	24	25	26	27	28
D19	D20	D21	D22	D23	D24	D25
MKH2	MKH3	MKH4	MKH5	MKH6	MKH7	MKH8
TAH9	TAH10	TAH11	TAH12	TAH13	TAH14	TAH15

EUSI 487/2003-04/6239-40/1995-96

Ashe	Ojobo	Iwa	Araorun	Ile	Itutu	Erinle
1	2	3	4	5	6	7
O31	N1	N2	N3	N4	N5	N6
KHK13	KHK14	KHK15	KHK16	KHK17	KHK18	KHK19
THi20	THi21	THi22	THi23	THi24	THi25	THi26
8	9	10	11	12	13	14
N7	N8	N9	N10	N11	N12	N13
KHK20	KHK21	KHK22	KHK23	KHK24	KHK25	KHK26
THi27	THi28	THi29	THi30	HID1	HID2	HID3
15	16	17	18	19	20	21
N14	N15	N16	N17	N18	N19	N20
KHK27	KHK28	KHK29	KHK30	TBT1	TBT2	TBT3
HID4	HID5	HID6	HID7	HID8	HID9	HID10
22	23	24	25	26	27	28
N21	N22	N23	N24	N25	N26	N27
TBT4	TBT5	TBT6	TBT7	TBT8	TBT9	TBT10
HID11	HID12	HID13	HID14	HID15	HID16	HID17

KWANZA 487/2003-04/6239-40/1995-96

Umoja	Kujichagulia	Ujima	Ujamaa	Nia	Kuumba	Imani
1	2	3	4	5	6	7
D26	D27	D28	D29	D30	D31	JA1
MKH9	MKH10	MKH11	MKH12	MKH13	MKH14	MKH15
TAH16	TAH17	TAH18	TAH19	TAH20	TAH21	TAH22
8	9	10	11	12	13	14
JA2	JA3	JA4	JA5	JA6	JA7	JA8
MKH16	MKH17	MKH18	MKH19	MKH20	MKH21	MKH22
TAH23	TAH24	TAH25	TAH26	TAH27	TAH28	TAH29
15	16	17	18	19	20	21
JA9	JA10	JA11	JA12	JA13	JA14	JA15
MKH23	MKH24	MKH25	KMH26	MKH27	MKH28	MKH29
TAH30	TIR1	TIR2	TIR3	TIR4	TIR5	TIR6
22	23	24	25	26	27	28
JA16	JA17	JA18	JA19	JA20	JA21	JA22
MKH30	AMN1	AMN2	AMN3	AMN4	AMN5	AMN6
TIR7	TIR8	TIR9	TIR10	TIR11	TIR12	TIR13

THE NU AFRIKAN CALENDAR

SHAHIDI 487/2003-04/6239-40/1995-96						
Ashe	Ojobo	Iwa	Araorun	Ile	Itutu	Erinle
1	2	3	4	5	6	7
JA23	JA24	JA25	JA26	JA27	JA28	JA29
AMN7	AMN8	AMN9	AMN10	AMN11	AMN12	AMN13
TIR14	TIR15	TIR16	TIR17	TIR18	TIR19	TIR20
8	9	10	11	12	13	14
JA30	JA31	F1	F2	F3	F4	F5
AMN14	AMN15	AMN16	AMN17	AMN18	AMN19	AMN20
TIR21	TIR22	TIR23	TIR24	TIR25	TIR26	TIR27
15	16	17	18	19	20	21
F6	F7	F8	F9	F10	F 11	F12
AMN21	AMN22	AMN23	AMN24	AMN25	AMN26	AMN28
TIR28	TIR29	TIR30	YEK1	YEK2	YEK3	YEK4
22	23	24	25	26	27	28
F13	F14	F15	F16	F17	F18	F19
AMN29	AMN30	RNT1	RNT2	RNT3	RNT4	RNT5
YEK5	YEK6	YEK7	YEK8	YEK9	YEK10	YEK11
SHAHIDI 38						
F20						
RNT6						
YEK12						

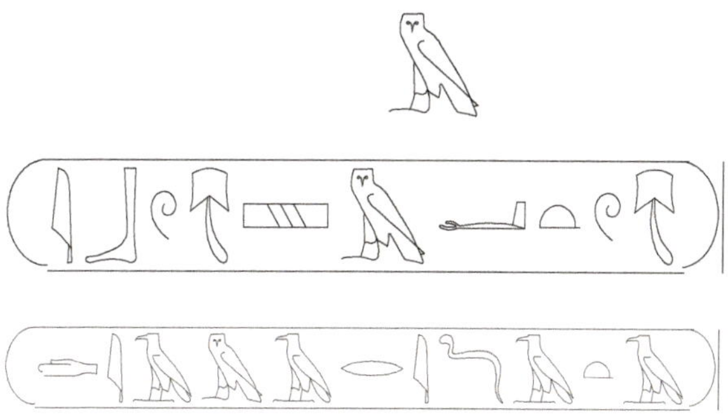

THE NU AFRIKAN CALENDAR

MAAFA 488 B.S.E./ASIENTO 2004-05 C.E./
KAM 6240-41 S.M/EC 1996-97 A.M. (ዓ.ም.)

AFU 488/2004-05/6240-41/1996-97

Ashe	Ojobo	Iwa	Araorun	Ile	Itutu	Erinle
1	2	3	4	5	6	7
F21	F22	F23	F24	F25	F26	F27
RNT7	RNT8	RNT9	RNT10	RNT11	RNT12	RNT13
YEK13	YEK14	YEK15	YEK16	YEK17	YEK18	YEK19
8	9	10	11	12	13	14
F28	F29	M1	M2	M3	M4	M5
RNT14	RNT15	RNT16	RNT17	RNT18	RNT19	RNT20
YEK 20	YEK21	YEK22	YEK23	YEK24	YEK25	YEK26
15	16	17	18	19	20	21
M6	M7	M8	M9	M10	M11	M12
RNT21	RNT22	RNT23	RNT24	RNT25	RNT26	RNT27
YEK 27	YEK28	YEK29	YEK30	MAG1	MAG2	MAG3
22	23	24	25	26	27	28
M13	M14	M15	M16	M17	M18	M19
RNT28	RNT29	RNT30	KHN1	KHN2	KHN3	KHN4
MAG4	MAG5	MAG6	MAG7	MAG8	MAG9	MAG10
SHARIFU 119						
M20						
KHN5						
MAG11						

KAWAIDA 488/2004-05/6240-41/1996-97

Ashe	Ojobo	Iwa	Araorun	Ile	Itutu	Erinle
1	2	3	4	5	6	7
A18	A19	A20	A21	A22	A23	A24
PIN4	PIN5	PIN6	PIN7	PIN8	PIN9	PIN10
MIY10	MIY11	MIY12	MIY13	MIY14	MIY15	MIY16
8	9	10	11	12	13	14
A25	A26	A27	A28	A29	A30	MA1
PIN11	PIN12	PIN13	PIN14	PIN15	PIN16	PIN17
MIY17	MIY18	MIY19	MIY20	MIY21	MIY22	MIY23
15	16	17	18	19	20	21
MA2	MA3	MA4	MA5	MA6	MA7	MA8
PIN18	PIN19	PIN20	PIN21	PIN22	PIN23	PIN24
MIY24	MIY25	MIY26	MIY27	MIY28	MIY29	MIY30
22	23	24	25	26	27	28
MA9	MA10	MA11	MA12	MA13	MA14	MA15
PIN25	PIN26	PIN27	PIN28	PIN29	PIN30	APP1
GIN1	GIN2	GIN3	GIN4	GIN5	GIN6	GIN7

UHURU 488/2004-05/6240-41/1996-97

Ashe	Ojobo	Iwa	Araorun	Ile	Itutu	Erinle
1	2	3	4	5	6	7
M21	M22	M23	M24	M25	M26	M27
KHN6	KHN7	KHN8	KHN9	KHN10	KHN11	KHN12
MAG12	MAG13	MAG14	MAG15	MAG16	MAG17	MAG18
8	9	10	11	12	13	14
M28	M29	M30	M31	A1	A2	A3
KHN13	KHN14	KHN15	KHN16	KHN17	KHN18	KHN19
MAG19	MAG20	MAG21	MAG22	MAG23	MAG24	MAG25
15	16	17	18	19	20	21
A4	A5	A6	A7	A8	A9	A10
KHN20	KHN21	KHN222	KHN23	KHN24	KHN25	KHN26
MAG26	MAG27	MAG28	MAG29	MAG30	MIY1	MIY2
22	23	24	25	26	27	28
A11	A12	A13	A14	A15	A16	A17
KHN27	KHN28	KHN29	KHN30	PIN1	PIN2	PIN3
MIY3	MIY4	MIY5	MIY6	MIY7	MIY8	MIY9

MABABA WEUSI 488/2004-05/6240-41/1996-97

Ashe	Ojobo	Iwa	Araorun	Ile	Itutu	Erinle
1	2	3	4	5	6	7
MA16	MA17	MA18	MA19	MA20	MA21	MA22
APP2	APP3	APP4	APP5	APP6	APP7	APP8
GIN8	GIN9	GIN10	GIN11	GIN12	GIN13	GIN14
8	9	10	11	12	13	14
MA23	MA24	MA25	MA26	MA27	MA28	MA29
APP9	APP10	APP11	APP12	APP13	APP14	APP15
GIN15	GIN16	GIN17	GIN18	GIN19	GIN20	GIN21
15	16	17	18	19	20	21
MA30	MA31	J1	J2	J3	J4	J5
APP16	APP17	APP18	APP19	APP20	APP21	APP22
GIN22	GIN23	GIN24	GIN25	GIN26	GIN27	GIN28
22	23	24	25	26	27	28
J6	J7	J8	J9	J10	J11	J12
APP23	APP24	APP25	APP26	APP27	APP28	APP29
GIN29	GIN30	SAN1	SAN2	SAN3	SAN4	SAN5

THE NU AFRIKAN CALENDAR

MWANAMKE WEUSI 488/2004-05/6240-41/1996-97						
Ashe	Ojobo	Iwa	Araorun	Ile	Itutu	Erinle
1	2	3	4	5	6	7
J13	J14	J15	J16	J17	J18	J19
APP30	MSR1	MSR2	MSR3	MSR4	MSR5	MSR6
SAN6	SAN7	SAN8	SAN9	SAN10	SAN11	SAN12
8	9	10	11	12	13	14
J20	J21	J22	J23	J24	J25	J26
MSR7	MSR8	MSR9	MSR10	MSR11	MSR12	MSR13
SAN13	SAN14	SAN15	SAN16	SAN17	SAN18	SAN19
15	16	17	18	19	20	21
J27	J28	J29	J30	JU1	JU2	JU3
MSR14	MSR15	MSR16	MSR17	MSR18	MSR19	MSR20
SAN20	SAN21	SAN22	SAN23	SAN24	SAN25	SAN26
22	23	24	25	26	27	28
JU4	JU5	JU6	JU7	JU8	JU9	JU10
MSR21	MRS22	MSR23	MSR24	MSR25	MSR26	MSR27
SAN27	SAN28	SAN29	SAN30	HAM1	HAM2	HAM3

TAIFA WEUSI 488/2004-05/6240-41/1996-97						
Ashe	Ojobo	Iwa	Araorun	Ile	Itutu	Erinle
1	2	3	4	5	6	7
AU8	AU9	AU10	AU11	AU12	AU13	AU14
THT20	THT21	THT22	THT23	THT24	THT25	THT26
NAH2	NAH3	NAH4	NAH5	NAH6	NAH7	NAH8
8	9	10	11	12	13	14
AU15	AU16	AU17	AU18	AU19	AU20	AU21
THT27	THT28	THT29	THT30	PAP1	PAP1	PAP2
NAH9	NAH10	NAH11	NAH12	NAH13	NAH14	NAH15
15	16	17	18	19	20	21
AU22	AU23	AU24	AU25	AU26	AU27	AU28
PAP3	PAP4	PAP5	PAP6	PAP7	PAP8	PAP9
NAH16	NAH17	NAH18	NAH19	NAH20	NAH21	NAH22
22	23	24	25	26	27	28
AU29	AU30	AU31	S1	S2	S3	S4
PAP10	PAP11	PAP12	PAP13	PAP14	PAP15	PAP16
NAH23	NAH24	NAH25	NAH26	NAH27	NAH28	NAH29

MTU WEUSI 488/2004-05/6240-41/1996-97						
Ashe	Ojobo	Iwa	Araorun	Ile	Itutu	Erinle
1	2	3	4	5	6	7
JU11	JU12	JU13	JU14	JU15	JU16	JU17
MSR28	MSR29	MSR30	*ASR1*	*HRU2*	*STH3*	*AST4*
HAM4	HAM5	HAM6	HAM7	HAM8	HAM9	HAM10
8	9	10	11	12	13	14
JU18	JU19	JU20	JU21	JU22	JU23	JU24
NBH5	*ANU6*	THT1	THT2	THT3	THT4	THT5
HAM11	HAM12	HAM13	HAM14	HAM15	HAM16	HAM17
15	16	17	18	19	20	21
JU25	JU26	JU27	JU28	JU29	JU30	JU31
THT6	THT7	THT8	THT9	THT10	THT11	THT12
HAM18	HAM19	HAM20	HAM21	HAM22	HAM23	HAM24
22	23	24	25	26	27	28
AU1	AU2	AU3	AU4	AU5	AU6	AU7
THT13	THT14	THT15	THT16	THT17	THT18	THT19
HAM25	HAM26	HAM27	HAM28	HAM29	HAM30	NAH1

UWEZO 488/2004-05/6240-41/1996-97						
Ashe	Ojobo	Iwa	Araorun	Ile	Itutu	Erinle
1	2	3	4	5	6	7
S5	S6	S7	S8	S9	S10	S11
PAP17	PAP18	PAP19	PAP20	PAP21	PAP22	PAP23
NAH30	*PAG1*	*PAG2*	*PAG3*	*PAG4*	*PAG5*	MAS1
8	9	10	11	12	13	14
S12	S13	S14	S15	S16	S17	S18
PAP24	PAP25	PAP26	PAP27	PAP28	PAP29	PAP30
MAS2	MAS3	MAS4	MAS5	MAS6	MAS7	MAS8
15	16	17	18	19	20	21
S19	S20	S21	S22	S23	S24	S25
HTH1	HTH2	HTH3	HTH4	HTH5	HTH6	HTH7
MAS9	MAS10	MAS11	MAS12	MAS13	MAS14	MAS15
22	23	24	25	26	27	28
S26	S27	S28	S29	S30	O1	O2
HTH8	HTH9	HTH10	HTH11	HTH12	HTH13	HTH14
MAS16	MAS17	MAS18	MAS19	MAS20	MAS21	MAS22

THE NU AFRIKAN CALENDAR

EKUNDU 488/2004-05/6240-41/1996-97

Ashe	Ojobo	Iwa	Araorun	Ile	Itutu	Erinle
1	2	3	4	5	6	7
O3	O4	O5	O6	O7	O8	O9
HTH15	HTH16	HTH17	HTH18	HTH19	HTH20	HTH21
MAS23	MAS24	MAS25	MAS26	MAS27	MAS28	MAS29
8	9	10	11	12	13	14
O10	O11	O12	O13	O14	O15	O16
HTH22	HTH23	HTH24	HTH25	HTH26	HTH27	HTH28
MAS30	THi1	THi2	THi3	THi4	THi5	THi6
15	16	17	18	19	20	21
O17	O18	O19	O20	O21	O22	O23
HTH29	HTH30	KHK1	KHK2	KHK3	KHK4	KHK5
THi7	THi8	This9	THi10	THi11	THi12	THi13
22	23	24	25	26	27	28
O24	O25	O26	O27	O28	O29	O30
KHK6	KHK7	KHK8	KHK9	KHK10	KHK11	KHK12
Thi14	THi15	THi16	THi17	THi18	THi19	THi20

KIJANI 488/2004-05/6240-41/1996-97

Ashe	Ojobo	Iwa	Araorun	Ile	Itutu	Erinle
1	2	3	4	5	6	7
N28	N29	N30	D1	D2	D3	D4
TBT11	TBT12	TBT13	TBT14	TBT15	TBT16	TBT17
HID19	HID20	HID21	HID22	HID23	HID24	HID25
8	9	10	11	12	13	14
D5	D6	D7	D8	D9	D10	D11
TBT18	TBT19	TBT20	TBT21	TBT22	TBT23	TBT24
HID26	HID27	HID28	HID29	HID30	TAH1	TAH2
15	16	17	18	19	20	21
D12	D13	D14	D15	D16	D17	D18
TBT25	TBT26	TBT27	TBT28	TBT29	TBT30	MKH1
TAH3	TAH4	TAH5	TAH6	TAH7	TAH8	TAH9
22	23	24	25	26	27	28
D19	D20	D21	D22	D23	D24	D25
MKH2	MKH3	MKH4	MKH5	MKH6	MKH7	MKH8
TAH10	TAH11	TAH12	TAH13	TAH14	TAH15	TAH16

EUSI 488/2004-05/6240-41/1996-97

Ashe	Ojobo	Iwa	Araorun	Ile	Itutu	Erinle
1	2	3	4	5	6	7
O31	N1	N2	N3	N4	N5	N6
KHK13	KHK14	KHK15	KHK16	KHK17	KHK18	KHK19
THi21	THi22	THi23	THi24	THi25	THi26	THi27
8	9	10	11	12	13	14
N7	N8	N9	N10	N11	N12	N13
KHK20	KHK21	KHK22	KHK23	KHK24	KHK25	KHK26
THi28	THi29	THi30	HID1	HID2	HID3	HID4
15	16	17	18	19	20	21
N14	N15	N16	N17	N18	N19	N20
KHK27	KHK28	KHK29	KHK30	TBT1	TBT2	TBT3
HID5	HID6	HID7	HID8	HID9	HID10	HID11
22	23	24	25	26	27	28
N21	N22	N23	N24	N25	N26	N27
TBT4	TBT5	TBT6	TBT7	TBT8	TBT9	TBT10
HID12	HID13	HID14	HID15	HID16	HID17	HID18

KWANZA 488/2004-05/6240-41/1996-97

Umoja	Kujichagulia	Ujima	Ujamaa	Nia	Kuumba	Imani
1	2	3	4	5	6	7
D26	D27	D28	D29	D30	D31	JA1
MKH9	MKH10	MKH11	MKH12	MKH13	MKH14	MKH15
TAH17	TAH18	TAH19	TAH20	TAH21	TAH22	TAH23
8	9	10	11	12	13	14
JA2	JA3	JA4	JA5	JA6	JA7	JA8
MKH16	MKH17	MKH18	MKH19	MKH20	MKH21	MKH22
TAH24	TAH25	TAH26	TAH27	TAH28	TAH29	TAH30
15	16	17	18	19	20	21
JA9	JA10	JA11	JA12	JA13	JA14	JA15
MKH23	MKH24	MKH25	KMH26	MKH27	MKH28	MKH29
TIR1	TIR2	TIR3	TIR4	TIR5	TIR6	TIR7
22	23	24	25	26	27	28
JA16	JA17	JA18	JA19	JA20	JA21	JA22
MKH30	AMN1	AMN2	AMN3	AMN4	AMN5	AMN6
TIR8	TIR9	TIR10	TIR11	TIR12	TIR13	TIR14

THE NU AFRIKAN CALENDAR

SHAHIDI 488/2004-05/6240-41/1996-97						
Ashe	Ojobo	Iwa	Araorun	Ile	Itutu	Erinle
1	2	3	4	5	6	7
JA23	JA24	JA25	JA26	JA27	JA28	JA29
AMN7	AMN8	AMN9	AMN10	AMN11	AMN12	AMN13
TIR15	TIR16	TIR17	TIR18	TIR19	TIR20	TIR21
8	9	10	11	12	13	14
JA30	JA31	F1	F2	F3	F4	F5
AMN14	AMN15	AMN16	AMN17	AMN18	AMN19	AMN20
TIR22	TIR23	TIR24	TIR25	TIR26	TIR27	TIR28
15	16	17	18	19	20	21
F6	F7	F8	F9	F10	F 11	F12
AMN21	AMN22	AMN23	AMN24	AMN25	AMN26	AMN28
TIR29	TIR30	YEK1	YEK2	YEK3	YEK4	YEK5
22	23	24	25	26	27	28
F13	F14	F15	F16	F17	F18	F19
AMN29	AMN30	RNT1	RNT2	RNT3	RNT4	RNT5
YEK6	YEK7	YEK8	YEK9	YEK10	YEK11	YEK12
SHAHIDI 39						
F20						
RNT6						
YEK13						

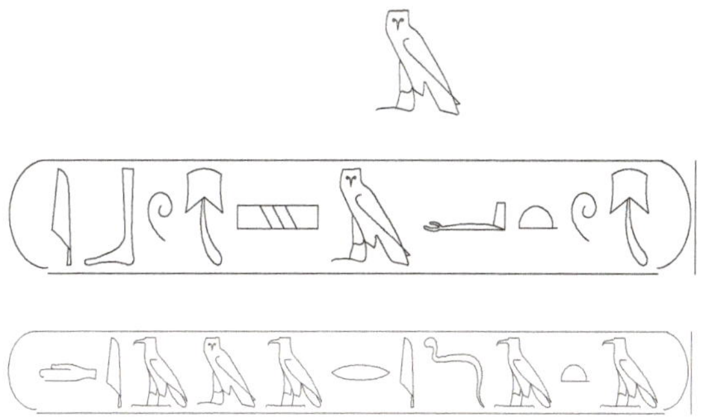

THE NU AFRIKAN CALENDAR

MAAFA 489 B.S.E./ASIENTO 2005-06 C.E.
KAM 6241-42 S.M/EC 1997-98 A.M. (ዓ.ም.)

AFU 489/2005-06/6241-42/1997-98

Ashe	Ojobo	Iwa	Araorun	Ile	Itutu	Erinle
1	2	3	4	5	6	7
F21	F22	F23	F24	F25	F26	F27
RNT7	RNT8	RNT9	RNT10	RNT11	RNT12	RNT13
YEK14	YEK15	YEK16	YEK17	YEK18	YEK19	YEK20
8	9	10	11	12	13	14
F28	M1	M2	M3	M4	M5	M6
RNT14	RNT15	RNT16	RNT17	RNT18	RNT19	RNT20
YEK21	YEK22	YEK23	YEK24	YEK25	YEK26	YEK27
15	16	17	18	19	20	21
M7	M8	M9	M10	M11	M12	M13
RNT21	RNT22	RNT23	RNT24	RNT25	RNT26	RNT27
YEK28	YEK29	YEK30	MAG1	MAG2	MAG3	MAG4
22	23	24	25	26	27	28
M14	M15	M16	M17	M18	M19	M20
RNT28	RNT29	RNT30	KHN1	KHN2	KHN3	KHN4
MAG5	MAG6	MAG7	MAG8	MAG9	MAG10	MAG11

KAWAIDA 489/2005-06/6241-42/1997-98

Ashe	Ojobo	Iwa	Araorun	Ile	Itutu	Erinle
1	2	3	4	5	6	7
A18	A19	A20	A21	A22	A23	A24
PIN3	PIN4	PIN5	PIN6	PIN7	PIN8	PIN9
MIY10	MIY11	MIY12	MIY13	MIY14	MIY15	MIY16
8	9	10	11	12	13	14
A25	A26	A27	A28	A29	A30	MA1
PIN10	PIN11	PIN12	PIN13	PIN14	PIN15	PIN16
MIY17	MIY18	MIY19	MIY20	MIY21	MIY22	MIY23
15	16	17	18	19	20	21
MA2	MA3	MA4	MA5	MA6	MA7	MA8
PIN17	PIN18	PIN19	PIN20	PIN21	PIN22	PIN23
MIY24	MIY25	MIY26	MIY27	MIY28	MIY29	MIY3
22	23	24	25	26	27	28
MA9	MA10	MA11	MA12	MA13	MA14	MA1
PIN24	PIN25	PIN26	PIN27	PIN28	PIN29	PIN3
GIN1	GIN2	GIN3	GIN4	GIN5	GIN6	GIN7

UHURU 489/2005-06/6241-42/1997-98

Ashe	Ojobo	Iwa	Araorun	Ile	Itutu	Erinle
1	2	3	4	5	6	7
M21	M22	M23	M24	M25	M26	M27
KHN5	KHN6	KHN7	KHN8	KHN9	KHN10	KHN11
MAG12	MAG13	MAG14	MAG15	MAG16	MAG17	MAG18
8	9	10	11	12	13	14
M28	M29	M30	M31	A1	A2	A3
KHN12	KHN13	KHN14	KHN15	KHN16	KHN17	KHN18
MAG19	MAG20	MAG21	MAG22	MAG23	MAG24	MAG25
15	16	17	18	19	20	21
A4	A5	A6	A7	A8	A9	A10
KHN19	KHN20	KHN21	KHN22	KHN23	KHN24	KHN25
MAG26	MAG27	MAG28	MAG29	MAG30	MIY1	MIY2
22	23	24	25	26	27	28
A11	A12	A13	A14	A15	A16	A17
KHN26	KHN27	KHN28	KHN29	KHN30	PIN1	PIN2
MIY3	MIY4	MIY5	MIY6	MIY7	MIY8	MIY9

MABABA WEUSI 489/2005-06/6241-42/1997-98

Ashe	Ojobo	Iwa	Araorun	Ile	Itutu	Erinle
1	2	3	4	5	6	7
MA16	MA17	MA18	MA19	MA20	MA21	MA22
APP1	APP2	APP3	APP4	APP5	APP6	APP7
GIN8	GIN9	GIN10	GIN11	GIN12	GIN13	GIN14
8	9	10	11	12	13	14
MA23	MA24	MA25	MA26	MA27	MA28	MA29
APP8	APP9	APP10	APP11	APP12	APP13	APP14
GIN15	GIN16	GIN17	GIN18	GIN19	GIN20	GIN21
15	16	17	18	19	20	21
MA30	MA31	J1	J2	J3	J4	J5
APP15	APP16	APP17	APP18	APP19	APP20	APP2
GIN22	GIN23	GIN24	GIN25	GIN26	GIN27	GIN2
22	23	24	25	26	27	28
J6	J7	J8	J9	J10	J11	J12
APP22	APP23	APP24	APP25	APP26	APP27	APP2
GIN29	GIN30	SAN1	SAN2	SAN3	SAN4	SAN5

THE NU AFRIKAN CALENDAR

MWANAMKE WEUSI 489/2005-06/6241-42/1997-98						
Ashe	Ojobo	Iwa	Araorun	Ile	Itutu	Erinle
1	2	3	4	5	6	7
J13	J14	J15	J16	J17	J18	J19
APP29	APP30	MSR1	MSR2	MSR3	MSR4	MSR5
SAN6	SAN7	SAN8	SAN9	SAN10	SAN11	SAN12
8	9	10	11	12	13	14
J20	J21	J22	J23	J24	J25	J26
MSR6	MSR7	MSR8	MSR9	MSR10	MSR11	MSR12
SAN13	SAN14	SAN15	SAN16	SAN17	SAN18	SAN19
15	16	17	18	19	20	21
J27	J28	J29	J30	JU1	JU2	JU3
MSR13	MSR14	MSR15	MSR16	MSR17	MSR18	MSR19
SAN20	SAN21	SAN22	SAN23	SAN24	SAN25	SAN26
22	23	24	25	26	27	28
JU4	JU5	JU6	JU7	JU8	JU9	JU10
MSR20	MRS21	MSR22	MSR23	MSR24	MSR25	MSR26
SAN27	SAN28	SAN29	SAN30	HAM1	HAM2	HAM3

TAIFA WEUSI 489/2005-06/6241-42/1997-98						
Ashe	Ojobo	Iwa	Araorun	Ile	Itutu	Erinle
1	2	3	4	5	6	7
AU8	AU9	AU10	AU11	AU12	AU13	AU14
THT20	THT21	THT22	THT23	THT24	THT25	THT26
NAH2	NAH3	NAH4	NAH5	NAH6	NAH7	NAH8
8	9	10	11	12	13	14
AU15	AU16	AU17	AU18	AU19	AU20	AU21
THT27	THT28	THT29	THT30	PAP1	PAP1	PAP2
NAH9	NAH10	NAH11	NAH12	NAH13	NAH14	NAH15
15	16	17	18	19	20	21
AU22	AU23	AU24	AU25	AU26	AU27	AU28
PAP3	PAP4	PAP5	PAP6	PAP7	PAP8	PAP9
NAH16	NAH17	NAH18	NAH19	NAH20	NAH21	NAH22
22	23	24	25	26	27	28
AU29	AU30	AU31	S1	S2	S3	S4
PAP10	PAP11	PAP12	PAP13	PAP14	PAP15	PAP16
NAH23	NAH24	NAH25	NAH26	NAH27	NAH28	NAH29

MTU WEUSI 489/2005-06/6241-42/1997-98						
Ashe	Ojobo	Iwa	Araorun	Ile	Itutu	Erinle
1	2	3	4	5	6	7
JU11	JU12	JU13	JU14	JU15	JU16	JU17
MSR27	MSR28	MSR29	MSR30	*ASR1*	*HRU2*	*STH3*
HAM4	HAM5	HAM6	HAM7	HAM8	HAM9	HAM10
8	9	10	11	12	13	14
JU18	JU19	JU20	JU21	JU22	JU23	JU24
AST4	*NBH5*	THT1	THT2	THT3	THT4	THT5
HAM11	HAM12	HAM13	AM14	HAM15	HAM16	HAM17
15	16	17	18	19	20	21
JU25	JU26	JU27	JU28	JU29	JU30	JU31
THT6	THT7	THT8	THT9	THT10	THT11	THT12
HAM18	HAM19	HAM20	HAM21	HAM22	HAM23	HAM24
22	23	24	25	26	27	28
AU1	AU2	AU3	AU4	AU5	AU6	AU7
THT13	THT14	THT15	THT16	THT17	THT18	THT19
HAM25	HAM26	HAM27	HAM28	HAM29	HAM30	NAH1

UWEZO 489/2005-06/6241-42/1997-98						
Ashe	Ojobo	Iwa	Araorun	Ile	Itutu	Erinle
1	2	3	4	5	6	7
S5	S6	S7	S8	S9	S10	S11
PAP17	PAP18	PAP19	PAP20	PAP21	PAP22	PAP23
NAH30	*PAG1*	*PAG2*	*PAG3*	*PAG4*	*PAG5*	MAS1
8	9	10	11	12	13	14
S12	S13	S14	S15	S16	S17	S18
PAP24	PAP25	PAP26	PAP27	PAP28	PAP29	PAP30
MAS2	MAS3	MAS4	MAS5	MAS6	MAS7	MAS8
15	16	17	18	19	20	21
S19	S20	S21	S22	S23	S24	S25
HTH1	HTH2	HTH3	HTH4	HTH5	HTH6	HTH7
MAS9	MAS10	MAS11	MAS12	MAS13	MAS14	MAS15
22	23	24	25	26	27	28
S26	S27	S28	S29	S30	O1	O2
HTH8	HTH9	HTH10	HTH11	HTH12	HTH13	HTH14
MAS16	MAS17	MAS18	MAS19	MAS20	MAS21	MAS22

THE NU AFRIKAN CALENDAR

EKUNDU 489/2005-06/6241-42/1997-98						
he	Ojobo	Iwa	Araorun	Ile	Itutu	Erinle
	2	3	4	5	6	7
	O4	O5	O6	O7	O8	O9
H15	HTH16	HTH17	HTH18	HTH19	HTH20	HTH21
AS23	MAS24	MAS25	MAS26	MAS27	MAS28	MAS29
	9	10	11	12	13	14
10	O11	O12	O13	O14	O15	O16
H22	HTH23	HTH24	HTH25	HTH26	HTH27	HTH28
AS30	THi1	THi2	THi3	THi4	THi5	THi6
	16	17	18	19	20	21
17	O18	O19	O20	O21	O22	O23
H29	HTH30	KHK1	KHK2	KHK3	KHK4	KHK5
i7	THi8	THi9	THi10	THi11	THi12	THi13
	23	24	25	26	27	28
4	O25	O26	O27	O28	O29	O30
IK6	KHK7	KHK8	KHK9	KHK10	KHK11	KHK12
i14	THi15	THi16	THi17	THi18	THi19	THi20

KIJANI 489/2005-06/6241-42/1997-98						
Ashe	Ojobo	Iwa	Araorun	Ile	Itutu	Erinle
1	2	3	4	5	6	7
N28	N29	N30	D1	D2	D3	D4
TBT11	TBT12	TBT13	TBT14	TBT15	TBT16	TBT17
HID19	HID20	HID21	HID22	HID23	HID24	HID25
8	9	10	11	12	13	14
D5	D6	D7	D8	D9	D10	D11
TBT18	TBT19	TBT20	TBT21	TBT22	TBT23	TBT24
HID26	HID27	HID28	HID29	HID30	TAH1	TAH2
15	16	17	18	19	20	21
D12	D13	D14	D15	D16	D17	D18
TBT25	TBT26	TBT27	TBT28	TBT29	TBT30	MKH1
TAH3	TAH4	TAH5	TAH6	TAH7	TAH8	TAH9
22	23	24	25	26	27	28
D19	D20	D21	D22	D23	D24	D25
MKH2	MKH3	MKH4	MKH5	MKH6	MKH7	MKH8
TAH10	TAH11	TAH12	TAH13	TAH14	TAH15	TAH16

EUSI 489/2005-06/6241-42/1997-98						
he	Ojobo	Iwa	Araorun	Ile	Itutu	Erinle
	2	3	4	5	6	7
1	N1	N2	N3	N4	N5	N6
IK13	KHK14	KHK15	KHK16	KHK17	KHK18	KHK19
i21	THi22	THi23	THi24	THi25	THi26	THi27
	9	10	11	12	13	14
7	N8	N9	N10	N11	N12	N13
IK20	KHK21	KHK22	KHK23	KHK24	KHK25	KHK26
i28	THi29	THi30	HID1	HID2	HID3	HID4
	16	17	18	19	20	21
4	N15	N16	N17	N18	N19	N20
IK27	KHK28	KHK29	KHK30	TBT1	TBT2	TBT3
D5	HID6	HID7	HID8	HID9	HID10	HID11
2	23	24	25	26	27	28
1	N22	N23	N24	N25	N26	N27
T4	TBT5	TBT6	TBT7	TBT8	TBT9	TBT10
D12	HID13	HID14	HID15	HID16	HID17	HID18

KWANZA 489/2005-06/6241-42/1997-98						
Umoja	Kujichagulia	Ujima	Ujamaa	Nia	Kuumba	Imani
1	2	3	4	5	6	7
D26	D27	D28	D29	D30	D31	JA1
MKH9	MKH10	MKH11	MKH12	MKH13	MKH14	MKH15
TAH17	TAH18	TAH19	TAH20	TAH21	TAH22	TAH23
8	9	10	11	12	13	14
JA2	JA3	JA4	JA5	JA6	JA7	JA8
MKH16	MKH17	MKH18	MKH19	MKH20	MKH21	MKH22
TAH24	TAH25	TAH26	TAH27	TAH28	TAH29	TAH30
15	16	17	18	19	20	21
JA9	JA10	JA11	JA12	JA13	JA14	JA15
MKH23	MKH24	MKH25	KMH26	MKH27	MKH28	MKH29
TIR1	TIR2	TIR3	TIR4	TIR5	TIR6	TIR7
22	23	24	25	26	27	28
JA16	JA17	JA18	JA19	JA20	JA21	JA22
MKH30	AMN1	AMN2	AMN3	AMN4	AMN5	AMN6
TIR8	TIR9	TIR10	TIR11	TIR12	TIR13	TIR14

THE NU AFRIKAN CALENDAR

SHAHIDI 489/2005-06/6241-42/1997-98						
Ashe	Ojobo	Iwa	Araorun	Ile	Itutu	Erinle
1	2	3	4	5	6	7
JA23	JA24	JA25	JA26	JA27	JA28	JA29
AMN7	AMN8	AMN9	AMN10	AMN11	AMN12	AMN13
TIR15	TIR16	TIR17	TIR18	TIR19	TIR20	TIR21
8	9	10	11	12	13	14
JA30	JA31	F1	F2	F3	F4	F5
AMN14	AMN15	AMN16	AMN17	AMN18	AMN19	AMN20
TIR22	TIR23	TIR24	TIR25	TIR26	TIR27	TIR28
15	16	17	18	19	20	21
F6	F7	F8	F9	F10	F 11	F12
AMN21	AMN22	AMN23	AMN24	AMN25	AMN26	AMN28
TIR29	TIR30	YEK1	YEK2	YEK3	YEK4	YEK5
22	23	24	25	26	27	28
F13	F14	F15	F16	F17	F18	F19
AMN29	AMN30	RNT1	RNT2	RNT3	RNT4	RNT5
YEK6	YEK7	YEK8	YEK9	YEK10	YEK11	YEK12
SHAHIDI 40						
F20						
RNT6						
YEK13						

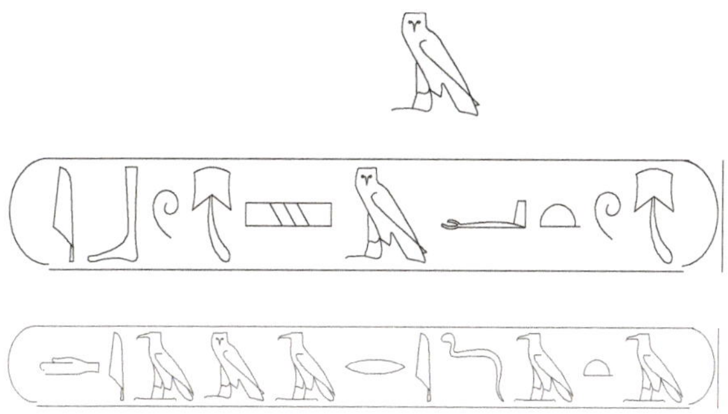

THE NU AFRIKAN CALENDAR

MAAFA 490 B.S.E./ASIENTO 2006-07 C.E.
KAM 6242-43 S.M/EC 1998-99 A.M. (ዓ.ም.)

AFU 490/2006-07/6242-41/1998-99						
Ashe	Ojobo	Iwa	Araorun	Ile	Itutu	Erinle
1	2	3	4	5	6	7
F21	F22	F23	F24	F25	F26	F27
RNT7	RNT8	RNT9	RNT10	RNT11	RNT12	RNT13
YEK14	YEK15	YEK16	YEK17	YEK18	YEK19	YEK20
8	9	10	11	12	13	14
F28	M1	M2	M3	M4	M5	M6
RNT14	RNT15	RNT16	RNT17	RNT18	RNT19	RNT20
YEK21	YEK22	YEK23	YEK24	YEK25	YEK26	YEK27
15	16	17	18	19	20	21
M7	M8	M9	M10	M11	M12	M13
RNT21	RNT22	RNT23	RNT24	RNT25	RNT26	RNT27
YEK28	YEK29	YEK30	MAG1	MAG2	MAG3	MAG4
22	23	24	25	26	27	28
M14	M15	M16	M17	M18	M19	M20
RNT28	RNT29	RNT30	KHN1	KHN2	KHN3	KHN4
MAG5	MAG6	MAG7	MAG8	MAG9	MAG10	MAG11

UHURU 490/2006-07/6242-41/1998-99						
Ashe	Ojobo	Iwa	Araorun	Ile	Itutu	Erinle
1	2	3	4	5	6	7
M21	M22	M23	M24	M25	M26	M27
KHN5	KHN6	KHN7	KHN8	KHN9	KHN10	KHN11
MAG12	MAG13	MAG14	MAG15	MAG16	MAG17	MAG18
8	9	10	11	12	13	14
M28	M29	M30	M31	A1	A2	A3
KHN12	KHN13	KHN14	KHN15	KHN16	KHN17	KHN18
MAG19	MAG20	MAG21	MAG22	MAG23	MAG24	MAG25
15	16	17	18	19	20	21
A4	A5	A6	A7	A8	A9	A10
KHN19	KHN20	KHN21	KHN22	KHN23	KHN24	KHN25
MAG26	MAG27	MAG28	MAG29	MAG30	MIY1	MIY2
22	23	24	25	26	27	28
A11	A12	A13	A14	A15	A16	A17
KHN26	KHN27	KHN28	KHN29	KHN30	PIN1	PIN2
MIY3	MIY4	MIY5	MIY6	MIY7	MIY8	MIY9

KAWAIDA 490/2006-07/6242-41/1998-99						
Ashe	Ojobo	Iwa	Araorun	Ile	Itutu	Erinle
1	2	3	4	5	6	7
A18	A19	A20	A21	A22	A23	A24
PIN3	PIN4	PIN5	PIN6	PIN7	PIN8	PIN9
MIY10	MIY11	MIY12	MIY13	MIY14	MIY15	MIY16
8	9	10	11	12	13	14
A25	A26	A27	A28	A29	A30	MA1
PIN10	PIN11	PIN12	PIN13	PIN14	PIN15	PIN16
MIY17	MIY18	MIY19	MIY20	MIY21	MIY22	MIY23
15	16	17	18	19	20	21
MA2	MA3	MA4	MA5	MA6	MA7	MA8
PIN17	PIN18	PIN19	PIN20	PIN21	PIN22	PIN23
MIY24	MIY25	MIY26	MIY27	MIY28	MIY29	MIY30
22	23	24	25	26	27	28
MA9	MA10	MA11	MA12	MA13	MA14	MA15
PIN24	PIN25	PIN26	PIN27	PIN28	PIN29	PIN30
GIN1	GIN2	GIN3	GIN4	GIN5	GIN6	GIN7

MABABA WEUSI 490/2006-07/6242-41/1998-99						
Ashe	Ojobo	Iwa	Araorun	Ile	Itutu	Erinle
1	2	3	4	5	6	7
MA16	MA17	MA18	MA19	MA20	MA21	MA22
APP1	APP2	APP3	APP4	APP5	APP6	APP7
GIN8	GIN9	GIN10	GIN11	GIN12	GIN13	GIN14
8	9	10	11	12	13	14
MA23	MA24	MA25	MA26	MA27	MA28	MA29
APP8	APP9	APP10	APP11	APP12	APP13	APP14
GIN15	GIN16	GIN17	GIN18	GIN19	GIN20	GIN21
15	16	17	18	19	20	21
MA30	MA31	J1	J2	J3	J4	J5
APP15	APP16	APP17	APP18	APP19	APP20	APP21
GIN22	GIN23	GIN24	GIN25	GIN26	GIN27	GIN28
22	23	24	25	26	27	28
J6	J7	J8	J9	J10	J11	J12
APP22	APP23	APP24	APP25	APP26	APP27	APP28
GIN29	GIN30	SAN1	SAN2	SAN3	SAN4	SAN5

THE NU AFRIKAN CALENDAR

MWANAMKE WEUSI 490/2006-07/6242-41/1998-99

Ashe	Ojobo	Iwa	Araorun	Ile	Itutu	Erinle
1	2	3	4	5	6	7
J13	J14	J15	J16	J17	J18	J19
APP29	APP30	MSR1	MSR2	MSR3	MSR4	MSR5
SAN6	SAN7	SAN8	SAN9	SAN10	SAN11	SAN12
8	9	10	11	12	13	14
J20	J21	J22	J23	J24	J25	J26
MSR6	MSR7	MSR8	MSR9	MSR10	MSR11	MSR12
SAN13	SAN14	SAN15	SAN16	SAN17	SAN18	SAN19
15	16	17	18	19	20	21
J27	J28	J29	J30	JU1	JU2	JU3
MSR13	MSR14	MSR15	MSR16	MSR17	MSR18	MSR19
SAN20	SAN21	SAN22	SAN23	SAN24	SAN25	SAN26
22	23	24	25	26	27	28
JU4	JU5	JU6	JU7	JU8	JU9	JU10
MSR20	MRS21	MSR22	MSR23	MSR24	MSR25	MSR26
SAN27	SAN28	SAN29	SAN30	HAM1	HAM2	HAM3

TAIFA WEUSI 490/2006-07/6242-41/1998-99

Ashe	Ojobo	Iwa	Araorun	Ile	Itutu	Erinle
1	2	3	4	5	6	7
AU8	AU9	AU10	AU11	AU12	AU13	AU14
THT20	THT21	THT22	THT23	THT24	THT25	THT26
NAH2	NAH3	NAH4	NAH5	NAH6	NAH7	NAH8
8	9	10	11	12	13	14
AU15	AU16	AU17	AU18	AU19	AU20	AU21
THT27	THT28	THT29	THT30	PAP1	PAP1	PAP2
NAH9	NAH10	NAH11	NAH12	NAH13	NAH14	NAH15
15	16	17	18	19	20	21
AU22	AU23	AU24	AU25	AU26	AU27	AU28
PAP3	PAP4	PAP5	PAP6	PAP7	PAP8	PAP9
NAH16	NAH17	NAH18	NAH19	NAH20	NAH21	NAH22
22	23	24	25	26	27	28
AU29	AU30	AU31	S1	S2	S3	S4
PAP10	PAP11	PAP12	PAP13	PAP14	PAP15	PAP16
NAH23	NAH24	NAH25	NAH26	NAH27	NAH28	NAH29

MTU WEUSI 490/2006-07/6242-41/1998-99

Ashe	Ojobo	Iwa	Araorun	Ile	Itutu	Erinle
1	2	3	4	5	6	7
JU11	JU12	JU13	JU14	JU15	JU16	JU17
MSR27	MSR28	MSR29	MSR30	*ASR1*	*HRU2*	*STH3*
HAM4	HAM5	HAM6	HAM7	HAM8	HAM9	HAM10
8	9	10	11	12	13	14
JU18	JU19	JU20	JU21	JU22	JU23	JU24
AST4	*NBH5*	THT1	THT2	THT3	THT4	THT5
HAM11	HAM12	HAM13	AM14	HAM15	HAM16	HAM17
15	16	17	18	19	20	21
JU25	JU26	JU27	JU28	JU29	JU30	JU31
THT6	THT7	THT8	THT9	THT10	THT11	THT12
HAM18	HAM19	HAM20	HAM21	HAM22	HAM23	HAM24
22	23	24	25	26	27	28
AU1	AU2	AU3	AU4	AU5	AU6	AU7
THT13	THT14	THT15	THT16	THT17	THT18	THT19
HAM25	HAM26	HAM27	HAM28	HAM29	HAM30	NAH1

UWEZO 490/2006-07/6242-41/1998-99

Ashe	Ojobo	Iwa	Araorun	Ile	Itutu	Erinle
1	2	3	4	5	6	7
S5	S6	S7	S8	S9	S10	S11
PAP17	PAP18	PAP19	PAP20	PAP21	PAP22	PAP23
NAH30	*PAG1*	*PAG2*	*PAG3*	*PAG4*	*PAG5*	MAS1
8	9	10	11	12	13	14
S12	S13	S14	S15	S16	S17	S18
PAP24	PAP25	PAP26	PAP27	PAP28	PAP29	PAP30
MAS2	MAS3	MAS4	MAS5	MAS6	MAS7	MAS8
15	16	17	18	19	20	21
S19	S20	S21	S22	S23	S24	S25
HTH1	HTH2	HTH3	HTH4	HTH5	HTH6	HTH7
MAS9	MAS10	MAS11	MAS12	MAS13	MAS14	MAS15
22	23	24	25	26	27	28
S26	S27	S28	S29	S30	O1	O2
HTH8	HTH9	HTH10	HTH11	HTH12	HTH13	HTH14
MAS16	MAS17	MAS18	MAS19	MAS20	MAS21	MAS22

THE NU AFRIKAN CALENDAR

EKUNDU 490/2006-07/6242-41/1998-99						
Ashe	Ojobo	Iwa	Araorun	Ile	Itutu	Erinle
1	2	3	4	5	6	7
O3	O4	O5	O6	O7	O8	O9
HTH15	HTH16	HTH17	HTH18	HTH19	HTH20	HTH21
MAS23	MAS24	MAS25	MAS26	MAS27	MAS28	MAS29
8	9	10	11	12	13	14
O10	O11	O12	O13	O14	O15	O16
HTH22	HTH23	HTH24	HTH25	HTH26	HTH27	HTH28
MAS30	THi1	THi2	THi3	THi4	THi5	THi6
15	16	17	18	19	20	21
O17	O18	O19	O20	O21	O22	O23
HTH29	HTH30	KHK1	KHK2	KHK3	KHK4	KHK5
THi7	THi8	THi9	THi10	THi11	THi12	THi13
22	23	24	25	26	27	28
O24	O25	O26	O27	O28	O29	O30
KHK6	KHK7	KHK8	KHK9	KHK10	KHK11	KHK12
THi14	THi15	THi16	THi17	THi18	THi19	THi20

KIJANI 490/2006-07/6242-41/1998-99						
Ashe	Ojobo	Iwa	Araorun	Ile	Itutu	Erinle
1	2	3	4	5	6	7
N28	N29	N30	D1	D2	D3	D4
TBT11	TBT12	TBT13	TBT14	TBT15	TBT16	TBT17
HID19	HID20	HID21	HID22	HID23	HID24	HID25
8	9	10	11	12	13	14
D5	D6	D7	D8	D9	D10	D11
TBT18	TBT19	TBT20	TBT21	TBT22	TBT23	TBT24
HID26	HID27	HID28	HID29	HID30	TAH1	TAH2
15	16	17	18	19	20	21
D12	D13	D14	D15	D16	D17	D18
TBT25	TBT26	TBT27	TBT28	TBT29	TBT30	MKH1
TAH3	TAH4	TAH5	TAH6	TAH7	TAH8	TAH9
22	23	24	25	26	27	28
D19	D20	D21	D22	D23	D24	D25
MKH2	MKH3	MKH4	MKH5	MKH6	MKH7	MKH8
TAH10	TAH11	TAH12	TAH13	TAH14	TAH15	TAH16

EUSI 490/2006-07/6242-41/1998-99						
Ashe	Ojobo	Iwa	Araorun	Ile	Itutu	Erinle
1	2	3	4	5	6	7
O31	N1	N2	N3	N4	N5	N6
KHK13	KHK14	KHK15	KHK16	KHK17	KHK18	KHK19
THi21	THi22	THi23	THi24	THi25	THi26	THi27
8	9	10	11	12	13	14
N7	N8	N9	N10	N11	N12	N13
KHK20	KHK21	KHK22	KHK23	KHK24	KHK25	KHK26
THi28	THi29	THi30	HID1	HID2	HID3	HID4
15	16	17	18	19	20	21
N14	N15	N16	N17	N18	N19	N20
KHK27	KHK28	KHK29	KHK30	TBT1	TBT2	TBT3
HID5	HID6	HID7	HID8	HID9	HID10	HID11
22	23	24	25	26	27	28
N21	N22	N23	N24	N25	N26	N27
TBT4	TBT5	TBT6	TBT7	TBT8	TBT9	TBT10
HID12	HID13	HID14	HID15	HID16	HID17	HID18

KWANZA 490/2006-07/6242-41/1998-99						
Umoja	Kujichagulia	Ujima	Ujamaa	Nia	Kuumba	Imani
1	2	3	4	5	6	7
D26	D27	D28	D29	D30	D31	JA1
MKH9	MKH10	MKH11	MKH12	MKH13	MKH14	MKH15
TAH17	TAH18	TAH19	TAH20	TAH21	TAH22	TAH23
8	9	10	11	12	13	14
JA2	JA3	JA4	JA5	JA6	JA7	JA8
MKH16	MKH17	MKH18	MKH19	MKH20	MKH21	MKH22
TAH24	TAH25	TAH26	TAH27	TAH28	TAH29	TAH30
15	16	17	18	19	20	21
JA9	JA10	JA11	JA12	JA13	JA14	JA15
MKH23	MKH24	MKH25	KMH26	MKH27	MKH28	MKH29
TIR1	TIR2	TIR3	TIR4	TIR5	TIR6	TIR7
22	23	24	25	26	27	28
JA16	JA17	JA18	JA19	JA20	JA21	JA22
MKH30	AMN1	AMN2	AMN3	AMN4	AMN5	AMN6
TIR8	TIR9	TIR10	TIR11	TIR12	TIR13	TIR14

THE NU AFRIKAN CALENDAR

SHAHIDI 490/2006-07/6242-41/1998-99						
Ashe	Ojobo	Iwa	Araorun	Ile	Itutu	Erinle
1	2	3	4	5	6	7
JA23	JA24	JA25	JA26	JA27	JA28	JA29
AMN7	AMN8	AMN9	AMN10	AMN11	AMN12	AMN13
TIR15	TIR16	TIR17	TIR18	TIR19	TIR20	TIR21
8	9	10	11	12	13	14
JA30	JA31	F1	F2	F3	F4	F5
AMN14	AMN15	AMN16	AMN17	AMN18	AMN19	AMN20
TIR22	TIR23	TIR24	TIR25	TIR26	TIR27	TIR28
15	16	17	18	19	20	21
F6	F7	F8	F9	F10	F 11	F12
AMN21	AMN22	AMN23	AMN24	AMN25	AMN26	AMN28
TIR29	TIR30	YEK1	YEK2	YEK3	YEK4	YEK5
22	23	24	25	26	27	28
F13	F14	F15	F16	F17	F18	F19
AMN29	AMN30	RNT1	RNT2	RNT3	RNT4	RNT5
YEK6	YEK7	YEK8	YEK9	YEK10	YEK11	YEK12
SHAHIDI 41						
F20						
RNT6						
YEK13						

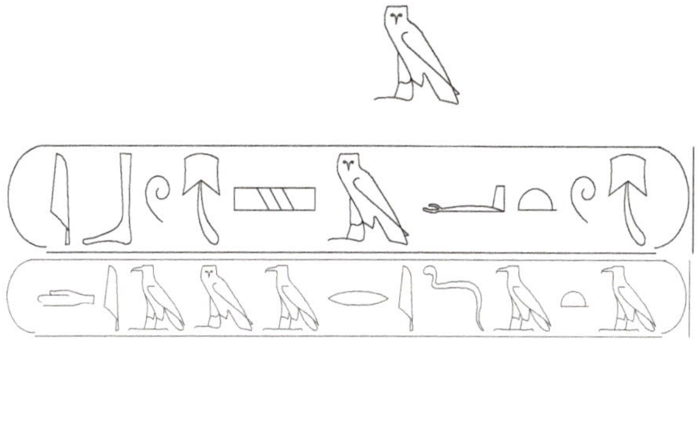

MAAFA 491 B.S.E./ASIENTO 2007-08 C.E./
KAM 6243-44 S.M/EC 1999-2000 A.M. (ዓ.ም.)

AFU 491/2007-08/6243-44/1999-2000						
Ashe	Ojobo	Iwa	Araorun	Ile	Itutu	Erinle
1	2	3	4	5	6	7
F21	F22	F23	F24	F25	F26	F27
RNT7	RNT8	RNT9	RNT10	RNT11	RNT12	RNT13
YEK14	YEK15	YEK16	YEK17	YEK18	YEK19	YEK20
8	9	10	11	12	13	14
F28	M1	M2	M3	M4	M5	M6
RNT14	RNT15	RNT16	RNT17	RNT18	RNT19	RNT20
YEK21	YEK22	YEK23	YEK24	YEK25	YEK26	YEK27
15	16	17	18	19	20	21
M7	M8	M9	M10	M11	M12	M13
RNT21	RNT22	RNT23	RNT24	RNT25	RNT26	RNT27
YEK28	YEK 29	YEK30	MAG1	MAG2	MAG3	MAG4
22	23	24	25	26	27	28
M14	M15	M16	M17	M18	M19	M20
RNT28	RNT29	RNT30	KHN1	KHN2	KHN3	KHN4
MAG5	MAG6	MAG7	MAG8	MAG9	MAG10	MAG11

UHURU 491/2007-08/6243-44/1999-2000						
Ashe	Ojobo	Iwa	Araorun	Ile	Itutu	Erinle
1	2	3	4	5	6	7
M21	M22	M23	M24	M25	M26	M27
KHN5	KHN6	KHN7	KHN8	KHN9	KHN10	KHN11
MAG12	MAG13	MAG14	MAG15	MAG16	MAG17	MAG18
8	9	10	11	12	13	14
M28	M29	M30	M31	A1	A2	A3
KHN12	KHN13	KHN14	KHN15	KHN16	KHN17	KHN18
MAG19	MAG20	MAG21	MAG22	MAG23	MAG24	MAG25
15	16	17	18	19	20	21
A4	A5	A6	A7	A8	A9	A10
KHN19	KHN20	KHN21	KHN22	KHN23	KHN24	KHN25
MAG26	MAG27	MAG28	MAG29	MAG30	MIY1	MIY2
22	23	24	25	26	27	28
A11	A12	A13	A14	A15	A16	A17
KHN26	KHN27	KHN28	KHN29	KHN30	PIN1	PIN2
MIY3	MIY4	MIY5	MIY6	MIY7	MIY8	MIY9

KAWAIDA 491/2007-08/6243-44/1999-2000						
Ashe	Ojobo	Iwa	Araorun	Ile	Itutu	Erinle
1	2	3	4	5	6	7
A18	A19	A20	A21	A22	A23	A24
PIN3	PIN4	PIN5	PIN6	PIN7	PIN8	PIN9
MIY10	MIY11	MIY12	MIY13	MIY14	MIY15	MIY16
8	9	10	11	12	13	14
A25	A26	A27	A28	A29	A30	MA1
PIN10	PIN11	PIN12	PIN13	PIN14	PIN15	PIN16
MIY17	MIY18	MIY19	MIY20	MIY21	MIY22	MIY23
15	16	17	18	19	20	21
MA2	MA3	MA4	MA5	MA6	MA7	MA8
PIN17	PIN18	PIN19	PIN20	PIN21	PIN22	PIN23
MIY24	MIY25	MIY26	MIY27	MIY28	MIY29	MIY30
22	23	24	25	26	27	28
MA9	MA10	MA11	MA12	MA13	MA14	MA15
PIN24	PIN25	PIN26	PIN27	PIN28	PIN29	PIN30
GIN1	GIN2	GIN3	GIN4	GIN5	GIN6	GIN7

MABABA WEUSI 491/2007-08/6243-44/1999-2000						
Ashe	Ojobo	Iwa	Araorun	Ile	Itutu	Erinle
1	2	3	4	5	6	7
MA16	MA17	MA18	MA19	MA20	MA21	MA22
APP1	APP2	APP3	APP4	APP5	APP6	APP7
GIN8	GIN9	GIN10	GIN11	GIN12	GIN13	GIN14
8	9	10	11	12	13	14
MA23	MA24	MA25	MA26	MA27	MA28	MA29
APP8	APP9	APP10	APP11	APP12	APP13	APP14
GIN15	GIN16	GIN17	GIN18	GIN19	GIN20	GIN21
15	16	17	18	19	20	21
MA30	MA31	J1	J2	J3	J4	J5
APP15	APP16	APP17	APP18	APP19	APP20	APP21
GIN22	GIN23	GIN24	GIN25	GIN26	GIN27	GIN28
22	23	24	25	26	27	28
J6	J7	J8	J9	J10	J11	J12
APP22	APP23	APP24	APP25	APP26	APP27	APP28
GIN29	GIN30	SAN1	SAN2	SAN3	SAN4	SAN5

THE NU AFRIKAN CALENDAR

MWANAMKE WEUSI 491/2007-08/6243-44/1999-2000

Ashe	Ojobo	Iwa	Araorun	Ile	Itutu	Erinle
1	2	3	4	5	6	7
J13	J14	J15	J16	J17	J18	J19
APP29	APP30	MSR1	MSR2	MSR3	MSR4	MSR5
SAN6	SAN7	SAN8	SAN9	SAN10	SAN11	SAN12
8	9	10	11	12	13	14
J20	J21	J22	J23	J24	J25	J26
MSR6	MSR7	MSR8	MSR9	MSR10	MSR11	MSR12
SAN13	SAN14	SAN15	SAN16	SAN17	SAN18	SAN19
15	16	17	18	19	20	21
J27	J28	J29	J30	JU1	JU2	JU3
MSR13	MSR14	MSR15	MSR16	MSR17	MSR18	MSR19
SAN20	SAN21	SAN22	SAN23	SAN24	SAN25	SAN26
22	23	24	25	26	27	28
JU4	JU5	JU6	JU7	JU8	JU9	JU10
MSR20	MRS21	MSR22	MSR23	MSR24	MSR25	MSR26
SAN27	SAN28	SAN29	SAN30	HAM1	HAM2	HAM3

TAIFA WEUSI 491/2007-08/6243-44/1999-2000

Ashe	Ojobo	Iwa	Araorun	Ile	Itutu	Erinle
1	2	3	4	5	6	7
AU8	AU9	AU10	AU11	AU12	AU13	AU14
THT20	THT21	THT22	THT23	THT24	THT25	THT26
NAH2	NAH3	NAH4	NAH5	NAH6	NAH7	NAH8
8	9	10	11	12	13	14
AU15	AU16	AU17	AU18	AU19	AU20	AU21
THT27	THT28	THT29	THT30	PAP1	PAP1	PAP2
NAH9	NAH10	NAH11	NAH12	NAH13	NAH14	NAH15
15	16	17	18	19	20	21
AU22	AU23	AU24	AU25	AU26	AU27	AU28
PAP3	PAP4	PAP5	PAP6	PAP7	PAP8	PAP9
NAH16	NAH17	NAH18	NAH19	NAH20	NAH21	NAH22
22	23	24	25	26	27	28
AU29	AU30	AU31	S1	S2	S3	S4
PAP10	PAP11	PAP12	PAP13	PAP14	PAP15	PAP16
NAH23	NAH24	NAH25	NAH26	NAH27	NAH28	NAH29

MTU WEUSI 491/2007-08/6243-44/1999-2000

Ashe	Ojobo	Iwa	Araorun	Ile	Itutu	Erinle
1	2	3	4	5	6	7
JU11	JU12	JU13	JU14	JU15	JU16	JU17
MSR27	MSR28	MSR29	MSR30	*ASR1*	*HRU2*	*STH3*
HAM4	HAM5	HAM6	HAM7	HAM8	HAM9	HAM10
8	9	10	11	12	13	14
JU18	JU19	JU20	JU21	JU22	JU23	JU24
AST4	*NBH5*	THT1	THT2	THT3	THT4	THT5
HAM11	HAM12	HAM13	HAM14	HAM15	HAM16	HAM17
15	16	17	18	19	20	21
JU25	JU26	JU27	JU28	JU29	JU30	JU31
THT6	THT7	THT8	THT9	THT10	THT11	THT12
HAM18	HAM19	HAM20	HAM21	HAM22	HAM23	HAM24
22	23	24	25	26	27	28
AU1	AU2	AU3	AU4	AU5	AU6	AU7
THT13	THT14	THT15	THT16	THT17	THT18	THT19
HAM25	HAM26	HAM27	HAM28	HAM29	HAM30	NAH1

UWEZO 491/2007-08/6243-44/1999-2000

Ashe	Ojobo	Iwa	Araorun	Ile	Itutu	Erinle
1	2	3	4	5	6	7
S5	S6	S7	S8	S9	S10	S11
PAP17	PAP18	PAP19	PAP20	PAP21	PAP22	PAP23
NAH30	*QUA1*	*QUA2*	*QUA3*	*QUA4*	*QUA5*	*QUA6*
8	9	10	11	12	13	14
S12	S13	S14	S15	S16	S17	S18
PAP24	PAP25	PAP26	PAP27	PAP28	PAP29	PAP30
MAS1	MAS2	MAS3	MAS4	MAS5	MAS6	MAS7
15	16	17	18	19	20	21
S19	S20	S21	S22	S23	S24	S25
HTH1	HTH2	HTH3	HTH4	HTH5	HTH6	HTH7
MAS8	MAS9	MAS10	MAS11	MAS12	MAS13	MAS14
22	23	24	25	26	27	28
S26	S27	S28	S29	S30	O1	O2
HTH8	HTH9	HTH10	HTH11	HTH12	HTH13	HTH14
MAS15	MAS16	MAS17	MAS18	MAS19	MAS20	MAS21

THE NU AFRIKAN CALENDAR

EKUNDU 491/2007-08/6243-44/1999-2000

Ashe	Ojobo	Iwa	Araorun	Ile	Itutu	Erinle
1	2	3	4	5	6	7
O3	O4	O5	O6	O7	O8	O9
HTH15	HTH16	HTH17	HTH18	HTH19	HTH20	HTH21
MAS22	MAS23	MAS 24	MAS25	MAS26	MAS27	MAS28
8	9	10	11	12	13	14
O10	O11	O12	O13	O14	O15	O16
HTH22	HTH23	HTH24	HTH25	HTH26	HTH27	HTH28
MAS29	MAS30	THi1	THi2	THi3	THi4	THi5
15	16	17	18	19	20	21
O17	O18	O19	O20	O21	O22	O23
HTH29	HTH30	KHK1	KHK2	KHK3	KHK4	KHK5
THi6	THi7	THi8	THi9	THi10	THi11	THi12
22	23	24	25	26	27	28
O24	O25	O26	O27	O28	O29	O30
KHK6	KHK7	KHK8	KHK9	KHK10	KHK11	KHK12
THi13	THi14	THi15	THi16	THi17	THi18	THi19

KIJANI 491/2007-08/6243-44/1999-2000

Ashe	Ojobo	Iwa	Araorun	Ile	Itutu	Erinle
1	2	3	4	5	6	7
N28	N29	N30	D1	D2	D3	D4
TBT11	TBT12	TBT13	TBT14	TBT15	TBT16	TBT17
HID18	HID19	HID20	HID21	HID22	HID23	HID24
8	9	10	11	12	13	14
D5	D6	D7	D8	D9	D10	D11
TBT18	TBT19	TBT20	TBT21	TBT22	TBT23	TBT24
HID25	HID26	HID27	HID28	HID29	HID30	TAH1
15	16	17	18	19	20	21
D12	D13	D14	D15	D16	D17	D18
TBT25	TBT26	TBT27	TBT28	TBT29	TBT30	MKH1
TAH2	TAH3	TAH4	TAH5	TAH6	TAH7	TAH8
22	23	24	25	26	27	28
D19	D20	D21	D22	D23	D24	D25
MKH2	MKH3	MKH4	MKH5	MKH6	MKH7	MKH8
TAH9	TAH10	TAH11	TAH12	TAH13	TAH14	TAH15

EUSI 491/2007-08/6243-44/1999-2000

Ashe	Ojobo	Iwa	Araorun	Ile	Itutu	Erinle
1	2	3	4	5	6	7
O31	N1	N2	N3	N4	N5	N6
KHK13	KHK14	KHK15	KHK16	KHK17	KHK18	KHK19
THi20	THi21	THi22	THi23	THi24	THi25	THi26
8	9	10	11	12	13	14
N7	N8	N9	N10	N11	N12	N13
KHK20	KHK21	KHK22	KHK23	KHK24	KHK25	KHK26
THi27	THi28	THi29	THi30	HID1	HID2	HID3
15	16	17	18	19	20	21
N14	N15	N16	N17	N18	N19	N20
KHK27	KHK28	KHK29	KHK30	TBT1	TBT2	TBT3
HID4	HID5	HID6	HID7	HID8	HID9	HID10
22	23	24	25	26	27	28
N21	N22	N23	N24	N25	N26	N27
TBT4	TBT5	TBT6	TBT7	TBT8	TBT9	TBT10
HID11	HID12	HID13	HID14	HID15	HID16	HID17

KWANZA 491/2007-08/6243-44/1999-2000

Umoja	Kujichagulia	Ujima	Ujamaa	Nia	Kuumba	Imani
1	2	3	4	5	6	7
D26	D27	D28	D29	D30	D31	JA1
MKH9	MKH10	MKH11	MKH12	MKH13	MKH14	MKH15
TAH16	TAH17	TAH18	TAH19	TAH20	TAH21	TAH22
8	9	10	11	12	13	14
JA2	JA3	JA4	JA5	JA6	JA7	JA8
MKH16	MKH17	MKH18	MKH19	MKH20	MKH21	MKH22
TAH23	TAH24	TAH25	TAH26	TAH27	TAH28	TAH29
15	16	17	18	19	20	21
JA9	JA10	JA11	JA12	JA13	JA14	JA15
MKH23	MKH24	MKH25	KMH26	MKH27	MKH28	MKH29
TAH30	TIR1	TIR2	TIR3	TIR4	TIR5	TIR6
22	23	24	25	26	27	28
JA16	JA17	JA18	JA19	JA20	JA21	JA22
MKH30	AMN1	AMN2	AMN3	AMN4	AMN5	AMN6
TIR7	TIR8	TIR9	TIR10	TIR11	TIR12	TIR13

THE NU AFRIKAN CALENDAR

SHAHIDI 491/2007-08/6243-44/1999-2000						
Ashe	Ojobo	Iwa	Araorun	Ile	Itutu	Erinle
1	2	3	4	5	6	7
JA23	JA24	JA25	JA26	JA27	JA28	JA29
AMN7	AMN8	AMN9	AMN10	AMN11	AMN12	AMN13
TIR14	TIR15	TIR16	TIR17	TIR18	TIR19	TIR20
8	9	10	11	12	13	14
JA30	JA31	F1	F2	F3	F4	F5
AMN14	AMN15	AMN16	AMN17	AMN18	AMN19	AMN20
TIR21	TIR22	TIR23	TIR24	TIR25	TIR26	TIR27
15	16	17	18	19	20	21
F6	F7	F8	F9	F10	F 11	F12
AMN21	AMN22	AMN23	AMN24	AMN25	AMN26	AMN28
TIR28	TIR29	TIR30	YEK1	YEK2	YEK3	YEK4
22	23	24	25	26	27	28
F13	F14	F15	F16	F17	F18	F19
AMN29	AMN30	RNT1	RNT2	RNT3	RNT4	RNT5
YEK5	YEK6	YEK7	YEK8	YEK9	YEK10	YEK11
SHAHIDI 42						
F20						
RNT6						
YEK12						

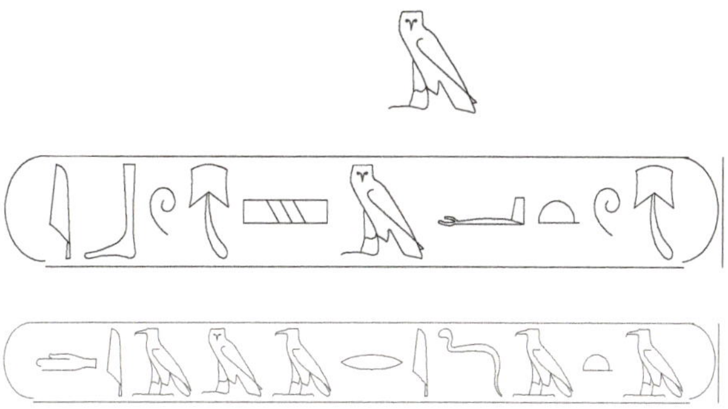

THE NU AFRIKAN CALENDAR

MAAFA 492 B.S.E./ASIENTO 2008-09 C.E./
KAM 6244-45 S.M/EC 2000-01 A.M. (ዓ.ም.)

AFU 492/2008-09/6244-45/2000-01

Ashe	Ojobo	Iwa	Araorun	Ile	Itutu	Erinle
1	2	3	4	5	6	7
F21	F22	F23	F24	F25	F26	F27
RNT7	RNT8	RNT9	RNT10	RNT11	RNT12	RNT13
YEK13	YEK14	YEK15	YEK16	YEK17	YEK18	YEK19
8	9	10	11	12	13	14
F28	F29	M1	M2	M3	M4	M5
RNT14	RNT15	RNT16	RNT17	RNT18	RNT19	RNT20
YEK 20	YEK21	YEK22	YEK23	YEK24	YEK25	YEK26
15	16	17	18	19	20	21
M6	M7	M8	M9	M10	M11	M12
RNT21	RNT22	RNT23	RNT24	RNT25	RNT26	RNT27
YEK 27	YEK28	YEK29	YEK30	MAG1	MAG2	MAG3
22	23	24	25	26	27	28
M13	M14	M15	M16	M17	M18	M19
RNT28	RNT29	RNT30	KHN1	KHN2	KHN3	KHN4
MAG4	MAG5	MAG6	MAG7	MAG8	MAG9	MAG10
SHARIFU 120						
M20						
KHN5						
MAG11						

KAWAIDA 492/2008-09/6244-45/2000-01

Ashe	Ojobo	Iwa	Araorun	Ile	Itutu	Erinle
1	2	3	4	5	6	7
A18	A19	A20	A21	A22	A23	A24
PIN4	PIN5	PIN6	PIN7	PIN8	PIN9	PIN10
MIY10	MIY11	MIY12	MIY13	MIY14	MIY15	MIY16
8	9	10	11	12	13	14
A25	A26	A27	A28	A29	A30	MA1
PIN11	PIN12	PIN13	PIN14	PIN15	PIN16	PIN17
MIY17	MIY18	MIY19	MIY20	MIY21	MIY22	MIY23
15	16	17	18	19	20	21
MA2	MA3	MA4	MA5	MA6	MA7	MA8
PIN18	PIN19	PIN20	PIN21	PIN22	PIN23	PIN24
MIY24	MIY25	MIY26	MIY27	MIY28	MIY29	MIY30
22	23	24	25	26	27	28
MA9	MA10	MA11	MA12	MA13	MA14	MA15
PIN25	PIN26	PIN27	PIN28	PIN29	PIN30	APP1
GIN1	GIN2	GIN3	GIN4	GIN5	GIN6	GIN7

UHURU 492/2008-09/6244-45/2000-01

Ashe	Ojobo	Iwa	Araorun	Ile	Itutu	Erinle
1	2	3	4	5	6	7
M21	M22	M23	M24	M25	M26	M27
KHN6	KHN7	KHN8	KHN9	KHN10	KHN11	KHN12
MAG12	MAG13	MAG14	MAG15	MAG16	MAG17	MAG18
8	9	10	11	12	13	14
M28	M29	M30	M31	A1	A2	A3
KHN13	KHN14	KHN15	KHN16	KHN17	KHN18	KHN19
MAG19	MAG20	MAG21	MAG22	MAG23	MAG24	MAG25
15	16	17	18	19	20	21
A4	A5	A6	A7	A8	A9	A10
KHN20	KHN21	KHN222	KHN23	KHN24	KHN25	KHN26
MAG26	MAG27	MAG28	MAG29	MAG30	MIY1	MIY2
22	23	24	25	26	27	28
A11	A12	A13	A14	A15	A16	A17
KHN27	KHN28	KHN29	KHN30	PIN1	PIN2	PIN3
MIY3	MIY4	MIY5	MIY6	MIY7	MIY8	MIY9

MABABA WEUSI 492/2008-09/6244-45/2000-01

Ashe	Ojobo	Iwa	Araorun	Ile	Itutu	Erinle
1	2	3	4	5	6	7
MA16	MA17	MA18	MA19	MA20	MA21	MA22
APP2	APP3	APP4	APP5	APP6	APP7	APP8
GIN8	GIN9	GIN10	GIN11	GIN12	GIN13	GIN14
8	9	10	11	12	13	14
MA23	MA24	MA25	MA26	MA27	MA28	MA29
APP9	APP10	APP11	APP12	APP13	APP14	APP15
GIN15	GIN16	GIN17	GIN18	GIN19	GIN20	GIN21
15	16	17	18	19	20	21
MA30	MA31	J1	J2	J3	J4	J5
APP16	APP17	APP18	APP19	APP20	APP21	APP22
GIN22	GIN23	GIN24	GIN25	GIN26	GIN27	GIN28
22	23	24	25	26	27	28
J6	J7	J8	J9	J10	J11	J12
APP23	APP24	APP25	APP26	APP27	APP28	APP29
GIN29	GIN30	SAN1	SAN2	SAN3	SAN4	SAN5

THE NU AFRIKAN CALENDAR

MWANAMKE WEUSI 492/2008-09/6244-45/2000-01						
Ashe	Ojobo	Iwa	Araorun	Ile	Itutu	Erinle
1	2	3	4	5	6	7
J13	J14	J15	J16	J17	J18	J19
APP30	MSR1	MSR2	MSR3	MSR4	MSR5	MSR6
SAN6	SAN7	SAN8	SAN9	SAN10	SAN11	SAN12
8	9	10	11	12	13	14
J20	J21	J22	J23	J24	J25	J26
MSR7	MSR8	MSR9	MSR10	MSR11	MSR12	MSR13
SAN13	SAN14	SAN15	SAN16	SAN17	SAN18	SAN19
15	16	17	18	19	20	21
J27	J28	J29	J30	JU1	JU2	JU3
MSR14	MSR15	MSR16	MSR17	MSR18	MSR19	MSR20
SAN20	SAN21	SAN22	SAN23	SAN24	SAN25	SAN26
22	23	24	25	26	27	28
JU4	JU5	JU6	JU7	JU8	JU9	JU10
MSR21	MRS22	MSR23	MSR24	MSR25	MSR26	MSR27
SAN27	SAN28	SAN29	SAN30	HAM1	HAM2	HAM3

TAIFA WEUSI 492/2008-09/6244-45/2000-01						
Ashe	Ojobo	Iwa	Araorun	Ile	Itutu	Erinle
1	2	3	4	5	6	7
AU8	AU9	AU10	AU11	AU12	AU13	AU14
THT20	THT21	THT22	THT23	THT24	THT25	THT26
NAH2	NAH3	NAH4	NAH5	NAH6	NAH7	NAH8
8	9	10	11	12	13	14
AU15	AU16	AU17	AU18	AU19	AU20	AU21
THT27	THT28	THT29	THT30	PAP1	PAP1	PAP2
NAH9	NAH10	NAH11	NAH12	NAH13	NAH14	NAH15
15	16	17	18	19	20	21
AU22	AU23	AU24	AU25	AU26	AU27	AU28
PAP3	PAP4	PAP5	PAP6	PAP7	PAP8	PAP9
NAH16	NAH17	NAH18	NAH19	NAH20	NAH21	NAH22
22	23	24	25	26	27	28
AU29	AU30	AU31	S1	S2	S3	S4
PAP10	PAP11	PAP12	PAP13	PAP14	PAP15	PAP16
NAH23	NAH24	NAH25	NAH26	NAH27	NAH28	NAH29

MTU WEUSI 492/2008-09/6244-45/2000-01						
Ashe	Ojobo	Iwa	Araorun	Ile	Itutu	Erinle
1	2	3	4	5	6	7
JU11	JU12	JU13	JU14	JU15	JU16	JU17
MSR28	MSR29	MSR30	*ASR1*	*HRU2*	*STH3*	*AST4*
HAM4	HAM5	HAM6	HAM7	HAM8	HAM9	HAM10
8	9	10	11	12	13	14
JU18	JU19	JU20	JU21	JU22	JU23	JU24
NBH5	*ANU6*	THT1	THT2	THT3	THT4	THT5
HAM11	HAM12	HAM13	HAM14	HAM15	HAM16	HAM17
15	16	17	18	19	20	21
JU25	JU26	JU27	JU28	JU29	JU30	JU31
THT6	THT7	THT8	THT9	THT10	THT11	THT12
HAM18	HAM19	HAM20	HAM21	HAM22	HAM23	HAM24
22	23	24	25	26	27	28
AU1	AU2	AU3	AU4	AU5	AU6	AU7
THT13	THT14	THT15	THT16	THT17	THT18	THT19
HAM25	HAM26	HAM27	HAM28	HAM29	HAM30	NAH1

UWEZO 492/2008-09/6244-45/2000-01						
Ashe	Ojobo	Iwa	Araorun	Ile	Itutu	Erinle
1	2	3	4	5	6	7
S5	S6	S7	S8	S9	S10	S11
PAP17	PAP18	PAP19	PAP20	PAP21	PAP22	PAP23
NAH30	*PAG1*	*PAG2*	*PAG3*	*PAG4*	*PAG5*	MAS1
8	9	10	11	12	13	14
S12	S13	S14	S15	S16	S17	S18
PAP24	PAP25	PAP26	PAP27	PAP28	PAP29	PAP30
MAS2	MAS3	MAS4	MAS5	MAS6	MAS7	MAS8
15	16	17	18	19	20	21
S19	S20	S21	S22	S23	S24	S25
HTH1	HTH2	HTH3	HTH4	HTH5	HTH6	HTH7
MAS9	MAS10	MAS11	MAS12	MAS13	MAS14	MAS15
22	23	24	25	26	27	28
S26	S27	S28	S29	S30	O1	O2
HTH8	HTH9	HTH10	HTH11	HTH12	HTH13	HTH14
MAS16	MAS17	MAS18	MAS19	MAS20	MAS21	MAS22

THE NU AFRIKAN CALENDAR

EKUNDU 492/2008-09/6244-45/2000-01

Ashe	Ojobo	Iwa	Araorun	Ile	Itutu	Erinle
1	2	3	4	5	6	7
O3	O4	O5	O6	O7	O8	O9
HTH15	HTH16	HTH17	HTH18	HTH19	HTH20	HTH21
MAS23	MAS24	MAS25	MAS26	MAS27	MAS28	MAS29
8	9	10	11	12	13	14
O10	O11	O12	O13	O14	O15	O16
HTH22	HTH23	HTH24	HTH25	HTH26	HTH27	HTH28
MAS30	THi1	THi2	THi3	THi4	THi5	THi6
15	16	17	18	19	20	21
O17	O18	O19	O20	O21	O22	O23
HTH29	HTH30	KHK1	KHK2	KHK3	KHK4	KHK5
THi7	THi8	This9	THi10	THi11	THi12	THi13
22	23	24	25	26	27	28
O24	O25	O26	O27	O28	O29	O30
KHK6	KHK7	KHK8	KHK9	KHK10	KHK11	KHK12
Thi14	THi15	THi16	THi17	THi18	THi19	THi20

KIJANI 492/2008-09/6244-45/2000-01

Ashe	Ojobo	Iwa	Araorun	Ile	Itutu	Erinle
1	2	3	4	5	6	7
N28	N29	N30	D1	D2	D3	D4
TBT11	TBT12	TBT13	TBT14	TBT15	TBT16	TBT17
HID19	HID20	HID21	HID22	HID23	HID24	HID25
8	9	10	11	12	13	14
D5	D6	D7	D8	D9	D10	D11
TBT18	TBT19	TBT20	TBT21	TBT22	TBT23	TBT24
HID26	HID27	HID28	HID29	HID30	TAH1	TAH2
15	16	17	18	19	20	21
D12	D13	D14	D15	D16	D17	D18
TBT25	TBT26	TBT27	TBT28	TBT29	TBT30	MKH1
TAH3	TAH4	TAH5	TAH6	TAH7	TAH8	TAH9
22	23	24	25	26	27	28
D19	D20	D21	D22	D23	D24	D25
MKH2	MKH3	MKH4	MKH5	MKH6	MKH7	MKH8
TAH10	TAH11	TAH12	TAH13	TAH14	TAH15	TAH16

EUSI 492/2008-09/6244-45/2000-01

Ashe	Ojobo	Iwa	Araorun	Ile	Itutu	Erinle
1	2	3	4	5	6	7
O31	N1	N2	N3	N4	N5	N6
KHK13	KHK14	KHK15	KHK16	KHK17	KHK18	KHK19
THi21	THi22	THi23	THi24	THi25	THi26	THi27
8	9	10	11	12	13	14
N7	N8	N9	N10	N11	N12	N13
KHK20	KHK21	KHK22	KHK23	KHK24	KHK25	KHK26
THi28	THi29	THi30	HID1	HID2	HID3	HID4
15	16	17	18	19	20	21
N14	N15	N16	N17	N18	N19	N20
KHK27	KHK28	KHK29	KHK30	TBT1	TBT2	TBT3
HID5	HID6	HID7	HID8	HID9	HID10	HID11
22	23	24	25	26	27	28
N21	N22	N23	N24	N25	N26	N27
TBT4	TBT5	TBT6	TBT7	TBT8	TBT9	TBT10
HID12	HID13	HID14	HID15	HID16	HID17	HID18

KWANZA 492/2008-09/6244-45/2000-01

Umoja	Kujichagulia	Ujima	Ujamaa	Nia	Kuumba	Imani
1	2	3	4	5	6	7
D26	D27	D28	D29	D30	D31	JA1
MKH9	MKH10	MKH11	MKH12	MKH13	MKH14	MKH15
TAH17	TAH18	TAH19	TAH20	TAH21	TAH22	TAH23
8	9	10	11	12	13	14
JA2	JA3	JA4	JA5	JA6	JA7	JA8
MKH16	MKH17	MKH18	MKH19	MKH20	MKH21	MKH22
TAH24	TAH25	TAH26	TAH27	TAH28	TAH29	TAH30
15	16	17	18	19	20	21
JA9	JA10	JA11	JA12	JA13	JA14	JA15
MKH23	MKH24	MKH25	KMH26	MKH27	MKH28	MKH29
TIR1	TIR2	TIR3	TIR4	TIR5	TIR6	TIR7
22	23	24	25	26	27	28
JA16	JA17	JA18	JA19	JA20	JA21	JA22
MKH30	AMN1	AMN2	AMN3	AMN4	AMN5	AMN6
TIR8	TIR9	TIR10	TIR11	TIR12	TIR13	TIR14

THE NU AFRIKAN CALENDAR

SHAHIDI 492/2008-09/6244-45/2000-01						
Ashe	Ojobo	Iwa	Araorun	Ile	Itutu	Erinle
1	2	3	4	5	6	7
JA23	JA24	JA25	JA26	JA27	JA28	JA29
AMN7	AMN8	AMN9	AMN10	AMN11	AMN12	AMN13
TIR15	TIR16	TIR17	TIR18	TIR19	TIR20	TIR21
8	9	10	11	12	13	14
JA30	JA31	F1	F2	F3	F4	F5
AMN14	AMN15	AMN16	AMN17	AMN18	AMN19	AMN20
TIR22	TIR23	TIR24	TIR25	TIR26	TIR27	TIR28
15	16	17	18	19	20	21
F6	F7	F8	F9	F10	F 11	F12
AMN21	AMN22	AMN23	AMN24	AMN25	AMN26	AMN28
TIR29	TIR30	YEK1	YEK2	YEK3	YEK4	YEK5
22	23	24	25	26	27	28
F13	F14	F15	F16	F17	F18	F19
AMN29	AMN30	RNT1	RNT2	RNT3	RNT4	RNT5
YEK6	YEK7	YEK8	YEK9	YEK10	YEK11	YEK12
SHAHIDI 43						
F20						
RNT6						
YEK13						

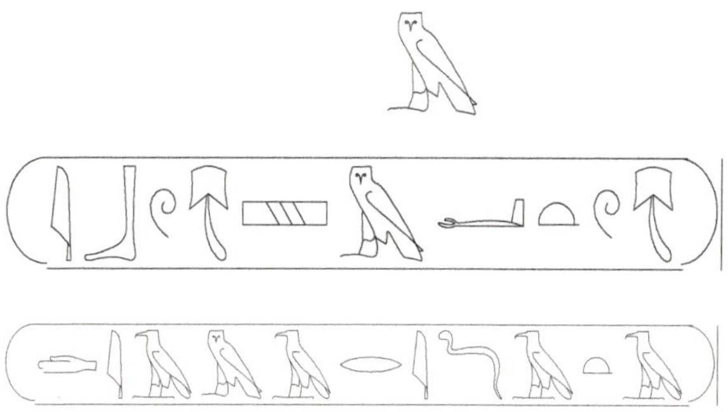

THE NU AFRIKAN CALENDAR

MAAFA 493 B.S.E./ASIENTO 2009-10 C.E.
KAM 6245-46 S.M/EC 2001-02 A.M. (ዓ.ም.)

AFU 493/2009-10/6245-46/2001-02

Ashe	Ojobo	Iwa	Araorun	Ile	Itutu	Erinle
1	2	3	4	5	6	7
F21	F22	F23	F24	F25	F26	F27
RNT7	RNT8	RNT9	RNT10	RNT11	RNT12	RNT13
YEK14	YEK15	YEK16	YEK17	YEK18	YEK19	YEK20
8	9	10	11	12	13	14
F28	M1	M2	M3	M4	M5	M6
RNT14	RNT15	RNT16	RNT17	RNT18	RNT19	RNT20
YEK21	YEK22	YEK23	YEK24	YEK25	YEK26	YEK27
15	16	17	18	19	20	21
M7	M8	M9	M10	M11	M12	M13
RNT21	RNT22	RNT23	RNT24	RNT25	RNT26	RNT27
YEK28	YEK29	YEK30	MAG1	MAG2	MAG3	MAG4
22	23	24	25	26	27	28
M14	M15	M16	M17	M18	M19	M20
RNT28	RNT29	RNT30	KHN1	KHN2	KHN3	KHN4
MAG5	MAG6	MAG7	MAG8	MAG9	MAG10	MAG11

KAWAIDA 493/2009-10/6245-46/2001-02

Ashe	Ojobo	Iwa	Araorun	Ile	Itutu	Erinle
1	2	3	4	5	6	7
A18	A19	A20	A21	A22	A23	A24
PIN3	PIN4	PIN5	PIN6	PIN7	PIN8	PIN9
MIY10	MIY11	MIY12	MIY13	MIY14	MIY15	MIY16
8	9	10	11	12	13	14
A25	A26	A27	A28	A29	A30	MA1
PIN10	PIN11	PIN12	PIN13	PIN14	PIN15	PIN16
MIY17	MIY18	MIY19	MIY20	MIY21	MIY22	MIY23
15	16	17	18	19	20	21
MA2	MA3	MA4	MA5	MA6	MA7	MA8
PIN17	PIN18	PIN19	PIN20	PIN21	PIN22	PIN23
MIY24	MIY25	MIY26	MIY27	MIY28	MIY29	MIY30
22	23	24	25	26	27	28
MA9	MA10	MA11	MA12	MA13	MA14	MA15
PIN24	PIN25	PIN26	PIN27	PIN28	PIN29	PIN30
GIN1	GIN2	GIN3	GIN4	GIN5	GIN6	GIN7

UHURU 493/2009-10/6245-46/2001-02

Ashe	Ojobo	Iwa	Araorun	Ile	Itutu	Erinle
1	2	3	4	5	6	7
M21	M22	M23	M24	M25	M26	M27
KHN5	KHN6	KHN7	KHN8	KHN9	KHN10	KHN11
MAG12	MAG13	MAG14	MAG15	MAG16	MAG17	MAG18
8	9	10	11	12	13	14
M28	M29	M30	M31	A1	A2	A3
KHN12	KHN13	KHN14	KHN15	KHN16	KHN17	KHN18
MAG19	MAG20	MAG21	MAG22	MAG23	MAG24	MAG25
15	16	17	18	19	20	21
A4	A5	A6	A7	A8	A9	A10
KHN19	KHN20	KHN21	KHN22	KHN23	KHN24	KHN25
MAG26	MAG27	MAG28	MAG29	MAG30	MIY1	MIY2
22	23	24	25	26	27	28
A11	A12	A13	A14	A15	A16	A17
KHN26	KHN27	KHN28	KHN29	KHN30	PIN1	PIN2
MIY3	MIY4	MIY5	MIY6	MIY7	MIY8	MIY9

MABABA WEUSI 493/2009-10/6245-46/2001-02

Ashe	Ojobo	Iwa	Araorun	Ile	Itutu	Erinle
1	2	3	4	5	6	7
MA16	MA17	MA18	MA19	MA20	MA21	MA22
APP1	APP2	APP3	APP4	APP5	APP6	APP7
GIN8	GIN9	GIN10	GIN11	GIN12	GIN13	GIN14
8	9	10	11	12	13	14
MA23	MA24	MA25	MA26	MA27	MA28	MA29
APP8	APP9	APP10	APP11	APP12	APP13	APP14
GIN15	GIN16	GIN17	GIN18	GIN19	GIN20	GIN21
15	16	17	18	19	20	21
MA30	MA31	J1	J2	J3	J4	J5
APP15	APP16	APP17	APP18	APP19	APP20	APP21
GIN22	GIN23	GIN24	GIN25	GIN26	GIN27	GIN28
22	23	24	25	26	27	28
J6	J7	J8	J9	J10	J11	J12
APP22	APP23	APP24	APP25	APP26	APP27	APP28
GIN29	GIN30	SAN1	SAN2	SAN3	SAN4	SAN5

THE NU AFRIKAN CALENDAR

MWANAMKE WEUSI 493/2009-10/6245-46/2001-02						
Ashe	Ojobo	Iwa	Araorun	Ile	Itutu	Erinle
1	2	3	4	5	6	7
J13	J14	J15	J16	J17	J18	J19
APP29	APP30	MSR1	MSR2	MSR3	MSR4	MSR5
SAN6	SAN7	SAN8	SAN9	SAN10	SAN11	SAN12
8	9	10	11	12	13	14
J20	J21	J22	J23	J24	J25	J26
MSR6	MSR7	MSR8	MSR9	MSR10	MSR11	MSR12
SAN13	SAN14	SAN15	SAN16	SAN17	SAN18	SAN19
15	16	17	18	19	20	21
J27	J28	J29	J30	JU1	JU2	JU3
MSR13	MSR14	MSR15	MSR16	MSR17	MSR18	MSR19
SAN20	SAN21	SAN22	SAN23	SAN24	SAN25	SAN26
22	23	24	25	26	27	28
JU4	JU5	JU6	JU7	JU8	JU9	JU10
MSR20	MRS21	MSR22	MSR23	MSR24	MSR25	MSR26
SAN27	SAN28	SAN29	SAN30	HAM1	HAM2	HAM3

TAIFA WEUSI 493/2009-10/6245-46/2001-02						
Ashe	Ojobo	Iwa	Araorun	Ile	Itutu	Erinle
1	2	3	4	5	6	7
AU8	AU9	AU10	AU11	AU12	AU13	AU14
THT20	THT21	THT22	THT23	THT24	THT25	THT26
NAH2	NAH3	NAH4	NAH5	NAH6	NAH7	NAH8
8	9	10	11	12	13	14
AU15	AU16	AU17	AU18	AU19	AU20	AU21
THT27	THT28	THT29	THT30	PAP1	PAP1	PAP2
NAH9	NAH10	NAH11	NAH12	NAH13	NAH14	NAH15
15	16	17	18	19	20	21
AU22	AU23	AU24	AU25	AU26	AU27	AU28
PAP3	PAP4	PAP5	PAP6	PAP7	PAP8	PAP9
NAH16	NAH17	NAH18	NAH19	NAH20	NAH21	NAH22
22	23	24	25	26	27	28
AU29	AU30	AU31	S1	S2	S3	S4
PAP10	PAP11	PAP12	PAP13	PAP14	PAP15	PAP16
NAH23	NAH24	NAH25	NAH26	NAH27	NAH28	NAH29

MTU WEUSI 493/2009-10/6245-46/2001-02						
Ashe	Ojobo	Iwa	Araorun	Ile	Itutu	Erinle
1	2	3	4	5	6	7
JU11	JU12	JU13	JU14	JU15	JU16	JU17
MSR27	MSR28	MSR29	MSR30	*ASR1*	*HRU2*	*STH3*
HAM4	HAM5	HAM6	HAM7	HAM8	HAM9	HAM10
8	9	10	11	12	13	14
JU18	JU19	JU20	JU21	JU22	JU23	JU24
AST4	*NBH5*	THT1	THT2	THT3	THT4	THT5
HAM11	HAM12	HAM13	AM14	HAM15	HAM16	HAM17
15	16	17	18	19	20	21
JU25	JU26	JU27	JU28	JU29	JU30	JU31
THT6	THT7	THT8	THT9	THT10	THT11	THT12
HAM18	HAM19	HAM20	HAM21	HAM22	HAM23	HAM24
22	23	24	25	26	27	28
AU1	AU2	AU3	AU4	AU5	AU6	AU7
THT13	THT14	THT15	THT16	THT17	THT18	THT19
HAM25	HAM26	HAM27	HAM28	HAM29	HAM30	NAH1

UWEZO 493/2009-10/6245-46/2001-02						
Ashe	Ojobo	Iwa	Araorun	Ile	Itutu	Erinle
1	2	3	4	5	6	7
S5	S6	S7	S8	S9	S10	S11
PAP17	PAP18	PAP19	PAP20	PAP21	PAP22	PAP23
NAH30	*PAG1*	*PAG2*	*PAG3*	*PAG4*	*PAG5*	MAS1
8	9	10	11	12	13	14
S12	S13	S14	S15	S16	S17	S18
PAP24	PAP25	PAP26	PAP27	PAP28	PAP29	PAP30
MAS2	MAS3	MAS4	MAS5	MAS6	MAS7	MAS8
15	16	17	18	19	20	21
S19	S20	S21	S22	S23	S24	S25
HTH1	HTH2	HTH3	HTH4	HTH5	HTH6	HTH7
MAS9	MAS10	MAS11	MAS12	MAS13	MAS14	MAS15
22	23	24	25	26	27	28
S26	S27	S28	S29	S30	O1	O2
HTH8	HTH9	HTH10	HTH11	HTH12	HTH13	HTH14
MAS16	MAS17	MAS18	MAS19	MAS20	MAS21	MAS22

THE NU AFRIKAN CALENDAR

EKUNDU 493/2009-10/6245-46/2001-02						
Ashe	Ojobo	Iwa	Araorun	Ile	Itutu	Erinle
1	2	3	4	5	6	7
O3	O4	O5	O6	O7	O8	O9
HTH15	HTH16	HTH17	HTH18	HTH19	HTH20	HTH21
MAS23	MAS24	MAS25	MAS26	MAS27	MAS28	MAS29
8	9	10	11	12	13	14
O10	O11	O12	O13	O14	O15	O16
HTH22	HTH23	HTH24	HTH25	HTH26	HTH27	HTH28
MAS30	THi1	THi2	THi3	THi4	THi5	THi6
15	16	17	18	19	20	21
O17	O18	O19	O20	O21	O22	O23
HTH29	HTH30	KHK1	KHK2	KHK3	KHK4	KHK5
THi7	THi8	THi9	THi10	THi11	THi12	THi13
22	23	24	25	26	27	28
O24	O25	O26	O27	O28	O29	O30
KHK6	KHK7	KHK8	KHK9	KHK10	KHK11	KHK12
THi14	THi15	THi16	THi17	THi18	THi19	THi20

KIJANI 493/2009-10/6245-46/2001-02						
Ashe	Ojobo	Iwa	Araorun	Ile	Itutu	Erinle
1	2	3	4	5	6	7
N28	N29	N30	D1	D2	D3	D4
TBT11	TBT12	TBT13	TBT14	TBT15	TBT16	TBT17
HID19	HID20	HID21	HID22	HID23	HID24	HID25
8	9	10	11	12	13	14
D5	D6	D7	D8	D9	D10	D11
TBT18	TBT19	TBT20	TBT21	TBT22	TBT23	TBT24
HID26	HID27	HID28	HID29	HID30	TAH1	TAH2
15	16	17	18	19	20	21
D12	D13	D14	D15	D16	D17	D18
TBT25	TBT26	TBT27	TBT28	TBT29	TBT30	MKH1
TAH3	TAH4	TAH5	TAH6	TAH7	TAH8	TAH9
22	23	24	25	26	27	28
D19	D20	D21	D22	D23	D24	D25
MKH2	MKH3	MKH4	MKH5	MKH6	MKH7	MKH8
TAH10	TAH11	TAH12	TAH13	TAH14	TAH15	TAH16

EUSI 493/2009-10/6245-46/2001-02						
Ashe	Ojobo	Iwa	Araorun	Ile	Itutu	Erinle
1	2	3	4	5	6	7
O31	N1	N2	N3	N4	N5	N6
KHK13	KHK14	KHK15	KHK16	KHK17	KHK18	KHK19
THi21	THi22	THi23	THi24	THi25	THi26	THi27
8	9	10	11	12	13	14
N7	N8	N9	N10	N11	N12	N13
KHK20	KHK21	KHK22	KHK23	KHK24	KHK25	KHK26
THi28	THi29	THi30	HID1	HID2	HID3	HID4
15	16	17	18	19	20	21
N14	N15	N16	N17	N18	N19	N20
KHK27	KHK28	KHK29	KHK30	TBT1	TBT2	TBT3
HID5	HID6	HID7	HID8	HID9	HID10	HID11
22	23	24	25	26	27	28
N21	N22	N23	N24	N25	N26	N27
TBT4	TBT5	TBT6	TBT7	TBT8	TBT9	TBT10
HID12	HID13	HID14	HID15	HID16	HID17	HID18

KWANZA 493/2009-10/6245-46/2001-02						
Umoja	Kujichagulia	Ujima	Ujamaa	Nia	Kuumba	Imani
1	2	3	4	5	6	7
D26	D27	D28	D29	D30	D31	JA1
MKH9	MKH10	MKH11	MKH12	MKH13	MKH14	MKH15
TAH17	TAH18	TAH19	TAH20	TAH21	TAH22	TAH23
8	9	10	11	12	13	14
JA2	JA3	JA4	JA5	JA6	JA7	JA8
MKH16	MKH17	MKH18	MKH19	MKH20	MKH21	MKH22
TAH24	TAH25	TAH26	TAH27	TAH28	TAH29	TAH30
15	16	17	18	19	20	21
JA9	JA10	JA11	JA12	JA13	JA14	JA15
MKH23	MKH24	MKH25	KMH26	MKH27	MKH28	MKH29
TIR1	TIR2	TIR3	TIR4	TIR5	TIR6	TIR7
22	23	24	25	26	27	28
JA16	JA17	JA18	JA19	JA20	JA21	JA22
MKH30	AMN1	AMN2	AMN3	AMN4	AMN5	AMN6
TIR8	TIR9	TIR10	TIR11	TIR12	TIR13	TIR14

THE NU AFRIKAN CALENDAR

SHAHIDI 493/2009-10/6245-46/2001-02						
Ashe	Ojobo	Iwa	Araorun	Ile	Itutu	Erinle
1	2	3	4	5	6	7
JA23	JA24	JA25	JA26	JA27	JA28	JA29
AMN7	AMN8	AMN9	AMN10	AMN11	AMN12	AMN13
TIR15	TIR16	TIR17	TIR18	TIR19	TIR20	TIR21
8	9	10	11	12	13	14
JA30	JA31	F1	F2	F3	F4	F5
AMN14	AMN15	AMN16	AMN17	AMN18	AMN19	AMN20
TIR22	TIR23	TIR24	TIR25	TIR26	TIR27	TIR28
15	16	17	18	19	20	21
F6	F7	F8	F9	F10	F 11	F12
AMN21	AMN22	AMN23	AMN24	AMN25	AMN26	AMN28
TIR29	TIR30	YEK1	YEK2	YEK3	YEK4	YEK5
22	23	24	25	26	27	28
F13	F14	F15	F16	F17	F18	F19
AMN29	AMN30	RNT1	RNT2	RNT3	RNT4	RNT5
YEK6	YEK7	YEK8	YEK9	YEK10	YEK11	YEK12
SHAHIDI 44						
F20						
RNT6						
YEK13						

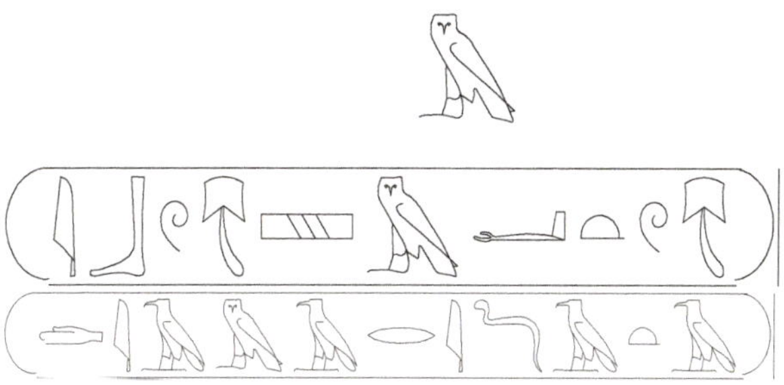

THE NU AFRIKAN CALENDAR

MAAFA 494 B.S.E./ASIENTO 2010-11 C.E.
KAM 6246-47 S.M/EC 2002-03 A.M. (ዓ.ም.)

AFU 494/2010-11/6246-47/2002-03

Ashe	Ojobo	Iwa	Araorun	Ile	Itutu	Erinle
1	2	3	4	5	6	7
F21	F22	F23	F24	F25	F26	F27
RNT7	RNT8	RNT9	RNT10	RNT11	RNT12	RNT13
YEK14	YEK15	YEK16	YEK17	YEK18	YEK19	YEK20
8	9	10	11	12	13	14
F28	M1	M2	M3	M4	M5	M6
RNT14	RNT15	RNT16	RNT17	RNT18	RNT19	RNT20
YEK21	YEK22	YEK23	YEK24	YEK25	YEK26	YEK27
15	16	17	18	19	20	21
M7	M8	M9	M10	M11	M12	M13
RNT21	RNT22	RNT23	RNT24	RNT25	RNT26	RNT27
YEK28	YEK29	YEK30	MAG1	MAG2	MAG3	MAG4
22	23	24	25	26	27	28
M14	M15	M16	M17	M18	M19	M20
RNT28	RNT29	RNT30	KHN1	KHN2	KHN3	KHN4
MAG5	MAG6	MAG7	MAG8	MAG9	MAG10	MAG11

KAWAIDA 494/2010-11/6246-47/2002-03

Ashe	Ojobo	Iwa	Araorun	Ile	Itutu	Erinle
1	2	3	4	5	6	7
A18	A19	A20	A21	A22	A23	A24
PIN3	PIN4	PIN5	PIN6	PIN7	PIN8	PIN9
MIY10	MIY11	MIY12	MIY13	MIY14	MIY15	MIY16
8	9	10	11	12	13	14
A25	A26	A27	A28	A29	A30	MA1
PIN10	PIN11	PIN12	PIN13	PIN14	PIN15	PIN16
MIY17	MIY18	MIY19	MIY20	MIY21	MIY22	MIY23
15	16	17	18	19	20	21
MA2	MA3	MA4	MA5	MA6	MA7	MA8
PIN17	PIN18	PIN19	PIN20	PIN21	PIN22	PIN23
MIY24	MIY25	MIY26	MIY27	MIY28	MIY29	MIY30
22	23	24	25	26	27	28
MA9	MA10	MA11	MA12	MA13	MA14	MA15
PIN24	PIN25	PIN26	PIN27	PIN28	PIN29	PIN30
GIN1	GIN2	GIN3	GIN4	GIN5	GIN6	GIN7

UHURU 494/2010-11/6246-47/2002-03

Ashe	Ojobo	Iwa	Araorun	Ile	Itutu	Erinle
1	2	3	4	5	6	7
M21	M22	M23	M24	M25	M26	M27
KHN5	KHN6	KHN7	KHN8	KHN9	KHN10	KHN11
MAG12	MAG13	MAG14	MAG15	MAG16	MAG17	MAG18
8	9	10	11	12	13	14
M28	M29	M30	M31	A1	A2	A3
KHN12	KHN13	KHN14	KHN15	KHN16	KHN17	KHN18
MAG19	MAG20	MAG21	MAG22	MAG23	MAG24	MAG25
15	16	17	18	19	20	21
A4	A5	A6	A7	A8	A9	A10
KHN19	KHN20	KHN21	KHN22	KHN23	KHN24	KHN25
MAG26	MAG27	MAG28	MAG29	MAG30	MIY1	MIY2
22	23	24	25	26	27	28
A11	A12	A13	A14	A15	A16	A17
KHN26	KHN27	KHN28	KHN29	KHN30	PIN1	PIN2
MIY3	MIY4	MIY5	MIY6	MIY7	MIY8	MIY9

MABABA WEUSI 494/2010-11/6246-47/2002-03

Ashe	Ojobo	Iwa	Araorun	Ile	Itutu	Erinle
1	2	3	4	5	6	7
MA16	MA17	MA18	MA19	MA20	MA21	MA22
APP1	APP2	APP3	APP4	APP5	APP6	APP7
GIN8	GIN9	GIN10	GIN11	GIN12	GIN13	GIN14
8	9	10	11	12	13	14
MA23	MA24	MA25	MA26	MA27	MA28	MA29
APP8	APP9	APP10	APP11	APP12	APP13	APP14
GIN15	GIN16	GIN17	GIN18	GIN19	GIN20	GIN21
15	16	17	18	19	20	21
MA30	MA31	J1	J2	J3	J4	J5
APP15	APP16	APP17	APP18	APP19	APP20	APP21
GIN22	GIN23	GIN24	GIN25	GIN26	GIN27	GIN28
22	23	24	25	26	27	28
J6	J7	J8	J9	J10	J11	J12
APP22	APP23	APP24	APP25	APP26	APP27	APP28
GIN29	GIN30	SAN1	SAN2	SAN3	SAN4	SAN5

THE NU AFRIKAN CALENDAR

MWANAMKE WEUSI 494/2010-11/6246-47/2002-03						
Ashe	Ojobo	Iwa	Araorun	Ile	Itutu	Erinle
1	2	3	4	5	6	7
J13	J14	J15	J16	J17	J18	J19
APP29	APP30	MSR1	MSR2	MSR3	MSR4	MSR5
SAN6	SAN7	SAN8	SAN9	SAN10	SAN11	SAN12
8	9	10	11	12	13	14
J20	J21	J22	J23	J24	J25	J26
MSR6	MSR7	MSR8	MSR9	MSR10	MSR11	MSR12
SAN13	SAN14	SAN15	SAN16	SAN17	SAN18	SAN19
15	16	17	18	19	20	21
J27	J28	J29	J30	JU1	JU2	JU3
MSR13	MSR14	MSR15	MSR16	MSR17	MSR18	MSR19
SAN20	SAN21	SAN22	SAN23	SAN24	SAN25	SAN26
22	23	24	25	26	27	28
JU4	JU5	JU6	JU7	JU8	JU9	JU10
MSR20	MRS21	MSR22	MSR23	MSR24	MSR25	MSR26
SAN27	SAN28	SAN29	SAN30	HAM1	HAM2	HAM3

TAIFA WEUSI 494/2010-11/6246-47/2002-03						
Ashe	Ojobo	Iwa	Araorun	Ile	Itutu	Erinle
1	2	3	4	5	6	7
AU8	AU9	AU10	AU11	AU12	AU13	AU14
THT20	THT21	THT22	THT23	THT24	THT25	THT26
NAH2	NAH3	NAH4	NAH5	NAH6	NAH7	NAH8
8	9	10	11	12	13	14
AU15	AU16	AU17	AU18	AU19	AU20	AU21
THT27	THT28	THT29	THT30	PAP1	PAP1	PAP2
NAH9	NAH10	NAH11	NAH12	NAH13	NAH14	NAH15
15	16	17	18	19	20	21
AU22	AU23	AU24	AU25	AU26	AU27	AU28
PAP3	PAP4	PAP5	PAP6	PAP7	PAP8	PAP9
NAH16	NAH17	NAH18	NAH19	NAH20	NAH21	NAH22
22	23	24	25	26	27	28
AU29	AU30	AU31	S1	S2	S3	S4
PAP10	PAP11	PAP12	PAP13	PAP14	PAP15	PAP16
NAH23	NAH24	NAH25	NAH26	NAH27	NAH28	NAH29

MTU WEUSI 494/2010-11/6246-47/2002-03						
Ashe	Ojobo	Iwa	Araorun	Ile	Itutu	Erinle
1	2	3	4	5	6	7
JU11	JU12	JU13	JU14	JU15	JU16	JU17
MSR27	MSR28	MSR29	MSR30	*ASR1*	*HRU2*	*STH3*
HAM4	HAM5	HAM6	HAM7	HAM8	HAM9	HAM10
8	9	10	11	12	13	14
JU18	JU19	JU20	JU21	JU22	JU23	JU24
AST4	*NBH5*	THT1	THT2	THT3	THT4	THT5
HAM11	HAM12	HAM13	AM14	HAM15	HAM16	HAM17
15	16	17	18	19	20	21
JU25	JU26	JU27	JU28	JU29	JU30	JU31
THT6	THT7	THT8	THT9	THT10	THT11	THT12
HAM18	HAM19	HAM20	HAM21	HAM22	HAM23	HAM24
22	23	24	25	26	27	28
AU1	AU2	AU3	AU4	AU5	AU6	AU7
THT13	THT14	THT15	THT16	THT17	THT18	THT19
HAM25	HAM26	HAM27	HAM28	HAM29	HAM30	NAH1

UWEZO 494/2010-11/6246-47/2002-03						
Ashe	Ojobo	Iwa	Araorun	Ile	Itutu	Erinle
1	2	3	4	5	6	7
S5	S6	S7	S8	S9	S10	S11
PAP17	PAP18	PAP19	PAP20	PAP21	PAP22	PAP23
NAH30	*PAG1*	*PAG2*	*PAG3*	*PAG4*	*PAG5*	MAS1
8	9	10	11	12	13	14
S12	S13	S14	S15	S16	S17	S18
PAP24	PAP25	PAP26	PAP27	PAP28	PAP29	PAP30
MAS2	MAS3	MAS4	MAS5	MAS6	MAS7	MAS8
15	16	17	18	19	20	21
S19	S20	S21	S22	S23	S24	S25
HTH1	HTH2	HTH3	HTH4	HTH5	HTH6	HTH7
MAS9	MAS10	MAS11	MAS12	MAS13	MAS14	MAS15
22	23	24	25	26	27	28
S26	S27	S28	S29	S30	O1	O2
HTH8	HTH9	HTH10	HTH11	HTH12	HTH13	HTH14
MAS16	MAS17	MAS18	MAS19	MAS20	MAS21	MAS22

THE NU AFRIKAN CALENDAR

EKUNDU 494/2010-11/6246-47/2002-03						
Ashe	Ojobo	Iwa	Araorun	Ile	Itutu	Erinle
1	2	3	4	5	6	7
O3	O4	O5	O6	O7	O8	O9
HTH15	HTH16	HTH17	HTH18	HTH19	HTH20	HTH21
MAS23	MAS24	MAS25	MAS26	MAS27	MAS28	MAS29
8	9	10	11	12	13	14
O10	O11	O12	O13	O14	O15	O16
HTH22	HTH23	HTH24	HTH25	HTH26	HTH27	HTH28
MAS30	THi1	THi2	THi3	THi4	THi5	THi6
15	16	17	18	19	20	21
O17	O18	O19	O20	O21	O22	O23
HTH29	HTH30	KHK1	KHK2	KHK3	KHK4	KHK5
THi7	THi8	THi9	THi10	THi11	THi12	THi13
22	23	24	25	26	27	28
O24	O25	O26	O27	O28	O29	O30
KHK6	KHK7	KHK8	KHK9	KHK10	KHK11	KHK12
THi14	THi15	THi16	THi17	THi18	THi19	THi20

KIJANI 494/2010-11/6246-47/2002-03						
Ashe	Ojobo	Iwa	Araorun	Ile	Itutu	Erinle
1	2	3	4	5	6	7
N28	N29	N30	D1	D2	D3	D4
TBT11	TBT12	TBT13	TBT14	TBT15	TBT16	TBT17
HID19	HID20	HID21	HID22	HID23	HID24	HID25
8	9	10	11	12	13	14
D5	D6	D7	D8	D9	D10	D11
TBT18	TBT19	TBT20	TBT21	TBT22	TBT23	TBT24
HID26	HID27	HID28	HID29	HID30	TAH1	TAH2
15	16	17	18	19	20	21
D12	D13	D14	D15	D16	D17	D18
TBT25	TBT26	TBT27	TBT28	TBT29	TBT30	MKH1
TAH3	TAH4	TAH5	TAH6	TAH7	TAH8	TAH9
22	23	24	25	26	27	28
D19	D20	D21	D22	D23	D24	D25
MKH2	MKH3	MKH4	MKH5	MKH6	MKH7	MKH8
TAH10	TAH11	TAH12	TAH13	TAH14	TAH15	TAH16

EUSI 494/2010-11/6246-47/2002-03						
Ashe	Ojobo	Iwa	Araorun	Ile	Itutu	Erinle
1	2	3	4	5	6	7
O31	N1	N2	N3	N4	N5	N6
KHK13	KHK14	KHK15	KHK16	KHK17	KHK18	KHK19
THi21	THi22	THi23	THi24	THi25	THi26	THi27
8	9	10	11	12	13	14
N7	N8	N9	N10	N11	N12	N13
KHK20	KHK21	KHK22	KHK23	KHK24	KHK25	KHK26
THi28	THi29	THi30	HID1	HID2	HID3	HID4
15	16	17	18	19	20	21
N14	N15	N16	N17	N18	N19	N20
KHK27	KHK28	KHK29	KHK30	TBT1	TBT2	TBT3
HID5	HID6	HID7	HID8	HID9	HID10	HID11
22	23	24	25	26	27	28
N21	N22	N23	N24	N25	N26	N27
TBT4	TBT5	TBT6	TBT7	TBT8	TBT9	TBT10
HID12	HID13	HID14	HID15	HID16	HID17	HID18

KWANZA 494/2010-11/6246-47/2002-03						
Umoja	Kujichagulia	Ujima	Ujamaa	Nia	Kuumba	Imani
1	2	3	4	5	6	7
D26	D27	D28	D29	D30	D31	JA1
MKH9	MKH10	MKH11	MKH12	MKH13	MKH14	MKH15
TAH17	TAH18	TAH19	TAH20	TAH21	TAH22	TAH23
8	9	10	11	12	13	14
JA2	JA3	JA4	JA5	JA6	JA7	JA8
MKH16	MKH17	MKH18	MKH19	MKH20	MKH21	MKH22
TAH24	TAH25	TAH26	TAH27	TAH28	TAH29	TAH30
15	16	17	18	19	20	21
JA9	JA10	JA11	JA12	JA13	JA14	JA15
MKH23	MKH24	MKH25	KMH26	MKH27	MKH28	MKH29
TIR1	TIR2	TIR3	TIR4	TIR5	TIR6	TIR7
22	23	24	25	26	27	28
JA16	JA17	JA18	JA19	JA20	JA21	JA22
MKH30	AMN1	AMN2	AMN3	AMN4	AMN5	AMN6
TIR8	TIR9	TIR10	TIR11	TIR12	TIR13	TIR14

THE NU AFRIKAN CALENDAR

SHAHIDI 494/2010-11/6246-47/2002-03						
Ashe	Ojobo	Iwa	Araorun	Ile	Itutu	Erinle
1	2	3	4	5	6	7
JA23	JA24	JA25	JA26	JA27	JA28	JA29
AMN7	AMN8	AMN9	AMN10	AMN11	AMN12	AMN13
TIR15	TIR16	TIR17	TIR18	TIR19	TIR20	TIR21
8	9	10	11	12	13	14
JA30	JA31	F1	F2	F3	F4	F5
AMN14	AMN15	AMN16	AMN17	AMN18	AMN19	AMN20
TIR22	TIR23	TIR24	TIR25	TIR26	TIR27	TIR28
15	16	17	18	19	20	21
F6	F7	F8	F9	F10	F 11	F12
AMN21	AMN22	AMN23	AMN24	AMN25	AMN26	AMN28
TIR29	TIR30	YEK1	YEK2	YEK3	YEK4	YEK5
22	23	24	25	26	27	28
F13	F14	F15	F16	F17	F18	F19
AMN29	AMN30	RNT1	RNT2	RNT3	RNT4	RNT5
YEK6	YEK7	YEK8	YEK9	YEK10	YEK11	YEK12
SHAHIDI 45						
F20						
RNT6						
YEK13						

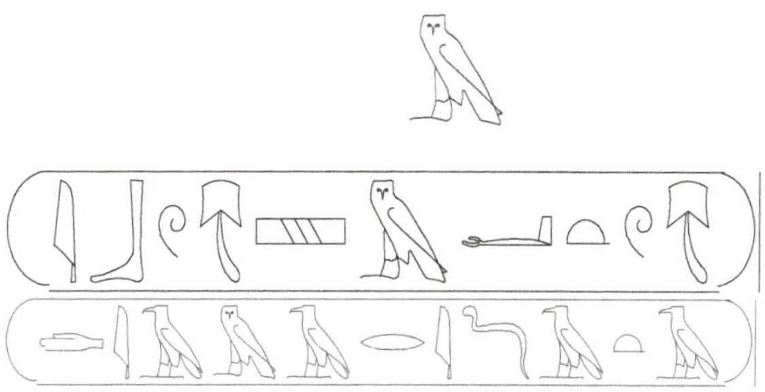

MAAFA 495 B.S.E./ASIENTO 2011-12 C.E./ KAM 6247-48 S.M/EC 2003-04 A.M. (ዓ.ም.)

AFU 495/2011-12/6247-48/2003-04

Ashe	Ojobo	Iwa	Araorun	Ile	Itutu	Erinle
1	2	3	4	5	6	7
F21	F22	F23	F24	F25	F26	F27
RNT7	RNT8	RNT9	RNT10	RNT11	RNT12	RNT13
YEK14	YEK15	YEK16	YEK17	YEK18	YEK19	YEK20
8	9	10	11	12	13	14
F28	M1	M2	M3	M4	M5	M6
RNT14	RNT15	RNT16	RNT17	RNT18	RNT19	RNT20
YEK21	YEK22	YEK23	YEK24	YEK25	YEK26	YEK27
15	16	17	18	19	20	21
M7	M8	M9	M10	M11	M12	M13
RNT21	RNT22	RNT23	RNT24	RNT25	RNT26	RNT27
YEK28	YEK 29	YEK30	MAG1	MAG2	MAG3	MAG4
22	23	24	25	26	27	28
M14	M15	M16	M17	M18	M19	M20
RNT28	RNT29	RNT30	KHN1	KHN2	KHN3	KHN4
MAG5	MAG6	MAG7	MAG8	MAG9	MAG10	MAG11

UHURU 495/2011-12/6247-48/2003-04

Ashe	Ojobo	Iwa	Araorun	Ile	Itutu	Erinle
1	2	3	4	5	6	7
M21	M22	M23	M24	M25	M26	M27
KHN5	KHN6	KHN7	KHN8	KHN9	KHN10	KHN11
MAG12	MAG13	MAG14	MAG15	MAG16	MAG17	MAG18
8	9	10	11	12	13	14
M28	M29	M30	M31	A1	A2	A3
KHN12	KHN13	KHN14	KHN15	KHN16	KHN17	KHN18
MAG19	MAG20	MAG21	MAG22	MAG23	MAG24	MAG25
15	16	17	18	19	20	21
A4	A5	A6	A7	A8	A9	A10
KHN19	KHN20	KHN21	KHN22	KHN23	KHN24	KHN25
MAG26	MAG27	MAG28	MAG29	MAG30	MIY1	MIY2
22	23	24	25	26	27	28
A11	A12	A13	A14	A15	A16	A17
KHN26	KHN27	KHN28	KHN29	KHN30	PIN1	PIN2
MIY3	MIY4	MIY5	MIY6	MIY7	MIY8	MIY9

KAWAIDA 495/2011-12/6247-48/2003-04

Ashe	Ojobo	Iwa	Araorun	Ile	Itutu	Erinle
1	2	3	4	5	6	7
A18	A19	A20	A21	A22	A23	A24
PIN3	PIN4	PIN5	PIN6	PIN7	PIN8	PIN9
MIY10	MIY11	MIY12	MIY13	MIY14	MIY15	MIY16
8	9	10	11	12	13	14
A25	A26	A27	A28	A29	A30	MA1
PIN10	PIN11	PIN12	PIN13	PIN14	PIN15	PIN16
MIY17	MIY18	MIY19	MIY20	MIY21	MIY22	MIY23
15	16	17	18	19	20	21
MA2	MA3	MA4	MA5	MA6	MA7	MA8
PIN17	PIN18	PIN19	PIN20	PIN21	PIN22	PIN23
MIY24	MIY25	MIY26	MIY27	MIY28	MIY29	MIY30
22	23	24	25	26	27	28
MA9	MA10	MA11	MA12	MA13	MA14	MA15
PIN24	PIN25	PIN26	PIN27	PIN28	PIN29	PIN30
GIN1	GIN2	GIN3	GIN4	GIN5	GIN6	GIN7

MABABA WEUSI 495/2011-12/6247-48/2003-04

Ashe	Ojobo	Iwa	Araorun	Ile	Itutu	Erinle
1	2	3	4	5	6	7
MA16	MA17	MA18	MA19	MA20	MA21	MA22
APP1	APP2	APP3	APP4	APP5	APP6	APP7
GIN8	GIN9	GIN10	GIN11	GIN12	GIN13	GIN14
8	9	10	11	12	13	14
MA23	MA24	MA25	MA26	MA27	MA28	MA29
APP8	APP9	APP10	APP11	APP12	APP13	APP14
GIN15	GIN16	GIN17	GIN18	GIN19	GIN20	GIN21
15	16	17	18	19	20	21
MA30	MA31	J1	J2	J3	J4	J5
APP15	APP16	APP17	APP18	APP19	APP20	APP21
GIN22	GIN23	GIN24	GIN25	GIN26	GIN27	GIN28
22	23	24	25	26	27	28
J6	J7	J8	J9	J10	J11	J12
APP22	APP23	APP24	APP25	APP26	APP27	APP28
GIN29	GIN30	SAN1	SAN2	SAN3	SAN4	SAN5

THE NU AFRIKAN CALENDAR

MWANAMKE WEUSI 495/2011-12/6247-48/2003-04						
Ashe	Ojobo	Iwa	Araorun	Ile	Itutu	Erinle
1	2	3	4	5	6	7
J13	J14	J15	J16	J17	J18	J19
APP29	APP30	MSR1	MSR2	MSR3	MSR4	MSR5
SAN6	SAN7	SAN8	SAN9	SAN10	SAN11	SAN12
8	9	10	11	12	13	14
J20	J21	J22	J23	J24	J25	J26
MSR6	MSR7	MSR8	MSR9	MSR10	MSR11	MSR12
SAN13	SAN14	SAN15	SAN16	SAN17	SAN18	SAN19
15	16	17	18	19	20	21
J27	J28	J29	J30	JU1	JU2	JU3
MSR13	MSR14	MSR15	MSR16	MSR17	MSR18	MSR19
SAN20	SAN21	SAN22	SAN23	SAN24	SAN25	SAN26
22	23	24	25	26	27	28
JU4	JU5	JU6	JU7	JU8	JU9	JU10
MSR20	MRS21	MSR22	MSR23	MSR24	MSR25	MSR26
SAN27	SAN28	SAN29	SAN30	HAM1	HAM2	HAM3

TAIFA WEUSI 495/2011-12/6247-48/2003-04						
Ashe	Ojobo	Iwa	Araorun	Ile	Itutu	Erinle
1	2	3	4	5	6	7
AU8	AU9	AU10	AU11	AU12	AU13	AU14
THT20	THT21	THT22	THT23	THT24	THT25	THT26
NAH2	NAH3	NAH4	NAH5	NAH6	NAH7	NAH8
8	9	10	11	12	13	14
AU15	AU16	AU17	AU18	AU19	AU20	AU21
THT27	THT28	THT29	THT30	PAP1	PAP1	PAP2
NAH9	NAH10	NAH11	NAH12	NAH13	NAH14	NAH15
15	16	17	18	19	20	21
AU22	AU23	AU24	AU25	AU26	AU27	AU28
PAP3	PAP4	PAP5	PAP6	PAP7	PAP8	PAP9
NAH16	NAH17	NAH18	NAH19	NAH20	NAH21	NAH22
22	23	24	25	26	27	28
AU29	AU30	AU31	S1	S2	S3	S4
PAP10	PAP11	PAP12	PAP13	PAP14	PAP15	PAP16
NAH23	NAH24	NAH25	NAH26	NAH27	NAH28	NAH29

MTU WEUSI 495/2011-12/6247-48/2003-04						
Ashe	Ojobo	Iwa	Araorun	Ile	Itutu	Erinle
1	2	3	4	5	6	7
JU11	JU12	JU13	JU14	JU15	JU16	JU17
MSR27	MSR28	MSR29	MSR30	*ASR1*	*HRU2*	*STH3*
HAM4	HAM5	HAM6	HAM7	HAM8	HAM9	HAM10
8	9	10	11	12	13	14
JU18	JU19	JU20	JU21	JU22	JU23	JU24
AST4	*NBH5*	THT1	THT2	THT3	THT4	THT5
HAM11	HAM12	HAM13	HAM14	HAM15	HAM16	HAM17
15	16	17	18	19	20	21
JU25	JU26	JU27	JU28	JU29	JU30	JU31
THT6	THT7	THT8	THT9	THT10	THT11	THT12
HAM18	HAM19	HAM20	HAM21	HAM22	HAM23	HAM24
22	23	24	25	26	27	28
AU1	AU2	AU3	AU4	AU5	AU6	AU7
THT13	THT14	THT15	THT16	THT17	THT18	THT19
HAM25	HAM26	HAM27	HAM28	HAM29	HAM30	NAH1

UWEZO 495/2011-12/6247-48/2003-04						
Ashe	Ojobo	Iwa	Araorun	Ile	Itutu	Erinle
1	2	3	4	5	6	7
S5	S6	S7	S8	S9	S10	S11
PAP17	PAP18	PAP19	PAP20	PAP21	PAP22	PAP23
NAH30	*QUA1*	*QUA2*	*QUA3*	*QUA4*	*QUA5*	*QUA6*
8	9	10	11	12	13	14
S12	S13	S14	S15	S16	S17	S18
PAP24	PAP25	PAP26	PAP27	PAP28	PAP29	PAP30
MAS1	MAS2	MAS3	MAS4	MAS5	MAS6	MAS7
15	16	17	18	19	20	21
S19	S20	S21	S22	S23	S24	S25
HTH1	HTH2	HTH3	HTH4	HTH5	HTH6	HTH7
MAS8	MAS9	MAS10	MAS11	MAS12	MAS13	MAS14
22	23	24	25	26	27	28
S26	S27	S28	S29	S30	O1	O2
HTH8	HTH9	HTH10	HTH11	HTH12	HTH13	HTH14
MAS15	MAS16	MAS17	MAS18	MAS19	MAS20	MAS21

THE NU AFRIKAN CALENDAR

EKUNDU 495/2011-12/6247-48/2003-04						
Ashe	Ojobo	Iwa	Araorun	Ile	Itutu	Erinle
1	2	3	4	5	6	7
O3	O4	O5	O6	O7	O8	O9
HTH15	HTH16	HTH17	HTH18	HTH19	HTH20	HTH21
MAS22	MAS23	MAS 24	MAS25	MAS26	MAS27	MAS28
8	9	10	11	12	13	14
O10	O11	O12	O13	O14	O15	O16
HTH22	HTH23	HTH24	HTH25	HTH26	HTH27	HTH28
MAS29	MAS30	THi1	THi2	THi3	THi4	THi5
15	16	17	18	19	20	21
O17	O18	O19	O20	O21	O22	O23
HTH29	HTH30	KHK1	KHK2	KHK3	KHK4	KHK5
THi6	THi7	THi8	THi9	THi10	THi11	THi12
22	23	24	25	26	27	28
O24	O25	O26	O27	O28	O29	O30
KHK6	KHK7	KHK8	KHK9	KHK10	KHK11	KHK12
THi13	THi14	THi15	THi16	THi17	THi18	THi19

KIJANI 495/2011-12/6247-48/2003-04						
Ashe	Ojobo	Iwa	Araorun	Ile	Itutu	Erinle
1	2	3	4	5	6	7
N28	N29	N30	D1	D2	D3	D4
TBT11	TBT12	TBT13	TBT14	TBT15	TBT16	TBT17
HID18	HID19	HID20	HID21	HID22	HID23	HID24
8	9	10	11	12	13	14
D5	D6	D7	D8	D9	D10	D11
TBT18	TBT19	TBT20	TBT21	TBT22	TBT23	TBT24
HID25	HID26	HID27	HID28	HID29	HID30	TAH1
15	16	17	18	19	20	21
D12	D13	D14	D15	D16	D17	D18
TBT25	TBT26	TBT27	TBT28	TBT29	TBT30	MKH1
TAH2	TAH3	TAH4	TAH5	TAH6	TAH7	TAH8
22	23	24	25	26	27	28
D19	D20	D21	D22	D23	D24	D25
MKH2	MKH3	MKH4	MKH5	MKH6	MKH7	MKH8
TAH9	TAH10	TAH11	TAH12	TAH13	TAH14	TAH15

EUSI 495/2011-12/6247-48/2003-04						
Ashe	Ojobo	Iwa	Araorun	Ile	Itutu	Erinle
1	2	3	4	5	6	7
O31	N1	N2	N3	N4	N5	N6
KHK13	KHK14	KHK15	KHK16	KHK17	KHK18	KHK19
THi20	THi21	THi22	THi23	THi24	THi25	THi26
8	9	10	11	12	13	14
N7	N8	N9	N10	N11	N12	N13
KHK20	KHK21	KHK22	KHK23	KHK24	KHK25	KHK26
THi27	THi28	THi29	THi30	HID1	HID2	HID3
15	16	17	18	19	20	21
N14	N15	N16	N17	N18	N19	N20
KHK27	KHK28	KHK29	KHK30	TBT1	TBT2	TBT3
HID4	HID5	HID6	HID7	HID8	HID9	HID10
22	23	24	25	26	27	28
N21	N22	N23	N24	N25	N26	N27
TBT4	TBT5	TBT6	TBT7	TBT8	TBT9	TBT10
HID11	HID12	HID13	HID14	HID15	HID16	HID17

KWANZA 495/2011-12/6247-48/2003-04						
Umoja	Kujichagulia	Ujima	Ujamaa	Nia	Kuumba	Imani
1	2	3	4	5	6	7
D26	D27	D28	D29	D30	D31	JA1
MKH9	MKH10	MKH11	MKH12	MKH13	MKH14	MKH15
TAH16	TAH17	TAH18	TAH19	TAH20	TAH21	TAH22
8	9	10	11	12	13	14
JA2	JA3	JA4	JA5	JA6	JA7	JA8
MKH16	MKH17	MKH18	MKH19	MKH20	MKH21	MKH22
TAH23	TAH24	TAH25	TAH26	TAH27	TAH28	TAH29
15	16	17	18	19	20	21
JA9	JA10	JA11	JA12	JA13	JA14	JA15
MKH23	MKH24	MKH25	KMH26	MKH27	MKH28	MKH29
TAH30	TIR1	TIR2	TIR3	TIR4	TIR5	TIR6
22	23	24	25	26	27	28
JA16	JA17	JA18	JA19	JA20	JA21	JA22
MKH30	AMN1	AMN2	AMN3	AMN4	AMN5	AMN6
TIR7	TIR8	TIR9	TIR10	TIR11	TIR12	TIR13

THE NU AFRIKAN CALENDAR

SHAHIDI 495/2011-12/6247-48/2003-04						
Ashe	Ojobo	Iwa	Araorun	Ile	Itutu	Erinle
1	2	3	4	5	6	7
JA23	JA24	JA25	JA26	JA27	JA28	JA29
AMN7	AMN8	AMN9	AMN10	AMN11	AMN12	AMN13
TIR14	TIR15	TIR16	TIR17	TIR18	TIR19	TIR20
8	9	10	11	12	13	14
JA30	JA31	F1	F2	F3	F4	F5
AMN14	AMN15	AMN16	AMN17	AMN18	AMN19	AMN20
TIR21	TIR22	TIR23	TIR24	TIR25	TIR26	TIR27
15	16	17	18	19	20	21
F6	F7	F8	F9	F10	F 11	F12
AMN21	AMN22	AMN23	AMN24	AMN25	AMN26	AMN28
TIR28	TIR29	TIR30	YEK1	YEK2	YEK3	YEK4
22	23	24	25	26	27	28
F13	F14	F15	F16	F17	F18	F19
AMN29	AMN30	RNT1	RNT2	RNT3	RNT4	RNT5
YEK5	YEK6	YEK7	YEK8	YEK9	YEK10	YEK11
SHAHIDI 46						
F20						
RNT6						
YEK12						

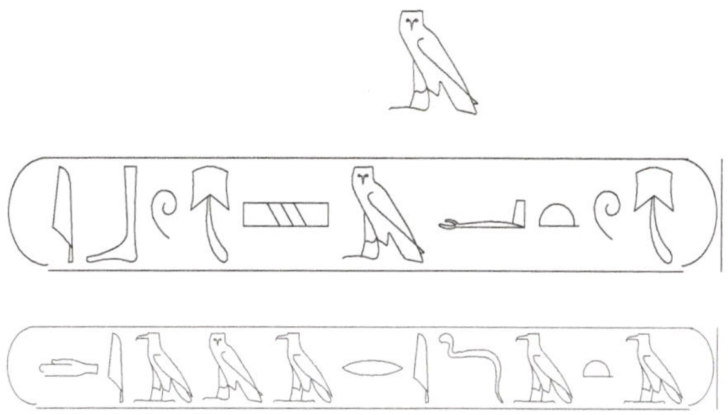

THE NU AFRIKAN CALENDAR

MAAFA 496 B.S.E./ASIENTO 2012-13 C.E./
KAM 6248-49 S.M/EC 2004-05 A.M. (ዓ.ም.)

AFU 496/2012-13/6248-49/2004-05

Ashe	Ojobo	Iwa	Araorun	Ile	Itutu	Erinle
1	2	3	4	5	6	7
F21	F22	F23	F24	F25	F26	F27
RNT7	RNT8	RNT9	RNT10	RNT11	RNT12	RNT13
YEK13	YEK14	YEK15	YEK16	YEK17	YEK18	YEK19
8	9	10	11	12	13	14
F28	F29	M1	M2	M3	M4	M5
RNT14	RNT15	RNT16	RNT17	RNT18	RNT19	RNT20
YEK 20	YEK21	YEK22	YEK23	YEK24	YEK25	YEK26
15	16	17	18	19	20	21
M6	M7	M8	M9	M10	M11	M12
RNT21	RNT22	RNT23	RNT24	RNT25	RNT26	RNT27
YEK 27	YEK28	YEK29	YEK30	MAG1	MAG2	MAG3
22	23	24	25	26	27	28
M13	M14	M15	M16	M17	M18	M19
RNT28	RNT29	RNT30	KHN1	KHN2	KHN3	KHN4
MAG4	MAG5	MAG6	MAG7	MAG8	MAG9	MAG10
SHARIFU 121						
M20						
KHN5						
MAG11						

UHURU 496/2012-13/6248-49/2004-05

Ashe	Ojobo	Iwa	Araorun	Ile	Itutu	Erinle
1	2	3	4	5	6	7
M21	M22	M23	M24	M25	M26	M27
KHN6	KHN7	KHN8	KHN9	KHN10	KHN11	KHN12
MAG12	MAG13	MAG14	MAG15	MAG16	MAG17	MAG18
8	9	10	11	12	13	14
M28	M29	M30	M31	A1	A2	A3
KHN13	KHN14	KHN15	KHN16	KHN17	KHN18	KHN19
MAG19	MAG20	MAG21	MAG22	MAG23	MAG24	MAG25
15	16	17	18	19	20	21
A4	A5	A6	A7	A8	A9	A10
KHN20	KHN21	KHN222	KHN23	KHN24	KHN25	KHN26
MAG26	MAG27	MAG28	MAG29	MAG30	MIY1	MIY2
22	23	24	25	26	27	28
A11	A12	A13	A14	A15	A16	A17
KHN27	KHN28	KHN29	KHN30	PIN1	PIN2	PIN3
MIY3	MIY4	MIY5	MIY6	MIY7	MIY8	MIY9

KAWAIDA 496/2012-13/6248-49/2004-05

Ashe	Ojobo	Iwa	Araorun	Ile	Itutu	Erinle
1	2	3	4	5	6	7
A18	A19	A20	A21	A22	A23	A24
PIN4	PIN5	PIN6	PIN7	PIN8	PIN9	PIN10
MIY10	MIY11	MIY12	MIY13	MIY14	MIY15	MIY16
8	9	10	11	12	13	14
A25	A26	A27	A28	A29	A30	MA1
PIN11	PIN12	PIN13	PIN14	PIN15	PIN16	PIN17
MIY17	MIY18	MIY19	MIY20	MIY21	MIY22	MIY23
15	16	17	18	19	20	21
MA2	MA3	MA4	MA5	MA6	MA7	MA8
PIN18	PIN19	PIN20	PIN21	PIN22	PIN23	PIN24
MIY24	MIY25	MIY26	MIY27	MIY28	MIY29	MIY30
22	23	24	25	26	27	28
MA9	MA10	MA11	MA12	MA13	MA14	MA15
PIN25	PIN26	PIN27	PIN28	PIN29	PIN30	APP1
GIN1	GIN2	GIN3	GIN4	GIN5	GIN6	GIN7

MABABA WEUSI 496/2012-13/6248-49/2004-05

Ashe	Ojobo	Iwa	Araorun	Ile	Itutu	Erinle
1	2	3	4	5	6	7
MA16	MA17	MA18	MA19	MA20	MA21	MA22
APP2	APP3	APP4	APP5	APP6	APP7	APP8
GIN8	GIN9	GIN10	GIN11	GIN12	GIN13	GIN14
8	9	10	11	12	13	14
MA23	MA24	MA25	MA26	MA27	MA28	MA29
APP9	APP10	APP11	APP12	APP13	APP14	APP15
GIN15	GIN16	GIN17	GIN18	GIN19	GIN20	GIN21
15	16	17	18	19	20	21
MA30	MA31	J1	J2	J3	J4	J5
APP16	APP17	APP18	APP19	APP20	APP21	APP22
GIN22	GIN23	GIN24	GIN25	GIN26	GIN27	GIN28
22	23	24	25	26	27	28
J6	J7	J8	J9	J10	J11	J12
APP23	APP24	APP25	APP26	APP27	APP28	APP29
GIN29	GIN30	SAN1	SAN2	SAN3	SAN4	SAN5

THE NU AFRIKAN CALENDAR

MWANAMKE WEUSI 496/2012-13/6248-49/2004-05

Ashe	Ojobo	Iwa	Araorun	Ile	Itutu	Erinle
1	2	3	4	5	6	7
J13	J14	J15	J16	J17	J18	J19
APP30	MSR1	MSR2	MSR3	MSR4	MSR5	MSR6
SAN6	SAN7	SAN8	SAN9	SAN10	SAN11	SAN12
8	9	10	11	12	13	14
J20	J21	J22	J23	J24	J25	J26
MSR7	MSR8	MSR9	MSR10	MSR11	MSR12	MSR13
SAN13	SAN14	SAN15	SAN16	SAN17	SAN18	SAN19
15	16	17	18	19	20	21
J27	J28	J29	J30	JU1	JU2	JU3
MSR14	MSR15	MSR16	MSR17	MSR18	MSR19	MSR20
SAN20	SAN21	SAN22	SAN23	SAN24	SAN25	SAN26
22	23	24	25	26	27	28
JU4	JU5	JU6	JU7	JU8	JU9	JU10
MSR21	MRS22	MSR23	MSR24	MSR25	MSR26	MSR27
SAN27	SAN28	SAN29	SAN30	HAM1	HAM2	HAM3

TAIFA WEUSI 496/2012-13/6248-49/2004-05

Ashe	Ojobo	Iwa	Araorun	Ile	Itutu	Erinle
1	2	3	4	5	6	7
AU8	AU9	AU10	AU11	AU12	AU13	AU14
THT20	THT21	THT22	THT23	THT24	THT25	THT26
NAH2	NAH3	NAH4	NAH5	NAH6	NAH7	NAH8
8	9	10	11	12	13	14
AU15	AU16	AU17	AU18	AU19	AU20	AU21
THT27	THT28	THT29	THT30	PAP1	PAP1	PAP2
NAH9	NAH10	NAH11	NAH12	NAH13	NAH14	NAH15
15	16	17	18	19	20	21
AU22	AU23	AU24	AU25	AU26	AU27	AU28
PAP3	PAP4	PAP5	PAP6	PAP7	PAP8	PAP9
NAH16	NAH17	NAH18	NAH19	NAH20	NAH21	NAH22
22	23	24	25	26	27	28
AU29	AU30	AU31	S1	S2	S3	S4
PAP10	PAP11	PAP12	PAP13	PAP14	PAP15	PAP16
NAH23	NAH24	NAH25	NAH26	NAH27	NAH28	NAH29

MTU WEUSI 496/2012-13/6248-49/2004-05

Ashe	Ojobo	Iwa	Araorun	Ile	Itutu	Erinle
1	2	3	4	5	6	7
JU11	JU12	JU13	JU14	JU15	JU16	JU17
MSR28	MSR29	MSR30	*ASR1*	*HRU2*	*STH3*	*AST4*
HAM4	HAM5	HAM6	HAM7	HAM8	HAM9	HAM10
8	9	10	11	12	13	14
JU18	JU19	JU20	JU21	JU22	JU23	JU24
NBH5	*ANU6*	THT1	THT2	THT3	THT4	THT5
HAM11	HAM12	HAM13	HAM14	HAM15	HAM16	HAM17
15	16	17	18	19	20	21
JU25	JU26	JU27	JU28	JU29	JU30	JU31
THT6	THT7	THT8	THT9	THT10	THT11	THT12
HAM18	HAM19	HAM20	HAM21	HAM22	HAM23	HAM24
22	23	24	25	26	27	28
AU1	AU2	AU3	AU4	AU5	AU6	AU7
THT13	THT14	THT15	THT16	THT17	THT18	THT19
HAM25	HAM26	HAM27	HAM28	HAM29	HAM30	NAH1

UWEZO 496/2012-13/6248-49/2004-05

Ashe	Ojobo	Iwa	Araorun	Ile	Itutu	Erinle
1	2	3	4	5	6	7
S5	S6	S7	S8	S9	S10	S11
PAP17	PAP18	PAP19	PAP20	PAP21	PAP22	PAP23
NAH30	*PAG1*	*PAG2*	*PAG3*	*PAG4*	*PAG5*	MAS1
8	9	10	11	12	13	14
S12	S13	S14	S15	S16	S17	S18
PAP24	PAP25	PAP26	PAP27	PAP28	PAP29	PAP30
MAS2	MAS3	MAS4	MAS5	MAS6	MAS7	MAS8
15	16	17	18	19	20	21
S19	S20	S21	S22	S23	S24	S25
HTH1	HTH2	HTH3	HTH4	HTH5	HTH6	HTH7
MAS9	MAS10	MAS11	MAS12	MAS13	MAS14	MAS15
22	23	24	25	26	27	28
S26	S27	S28	S29	S30	O1	O2
HTH8	HTH9	HTH10	HTH11	HTH12	HTH13	HTH14
MAS16	MAS17	MAS18	MAS19	MAS20	MAS21	MAS22

THE NU AFRIKAN CALENDAR

EKUNDU 496/2012-13/6248-49/2004-05						
Ashe	Ojobo	Iwa	Araorun	Ile	Itutu	Erinle
1	2	3	4	5	6	7
O3	O4	O5	O6	O7	O8	O9
HTH15	HTH16	HTH17	HTH18	HTH19	HTH20	HTH21
MAS23	MAS24	MAS25	MAS26	MAS27	MAS28	MAS29
8	9	10	11	12	13	14
O10	O11	O12	O13	O14	O15	O16
HTH22	HTH23	HTH24	HTH25	HTH26	HTH27	HTH28
MAS30	THi1	THi2	THi3	THi4	THi5	THi6
15	16	17	18	19	20	21
O17	O18	O19	O20	O21	O22	O23
HTH29	HTH30	KHK1	KHK2	KHK3	KHK4	KHK5
THi7	THi8	This9	THi10	THi11	THi12	THi13
22	23	24	25	26	27	28
O24	O25	O26	O27	O28	O29	O30
KHK6	KHK7	KHK8	KHK9	KHK10	KHK11	KHK12
Thi14	THi15	THi16	THi17	THi18	THi19	THi20

KIJANI 496/2012-13/6248-49/2004-05						
Ashe	Ojobo	Iwa	Araorun	Ile	Itutu	Erinle
1	2	3	4	5	6	7
N28	N29	N30	D1	D2	D3	D4
TBT11	TBT12	TBT13	TBT14	TBT15	TBT16	TBT17
HID19	HID20	HID21	HID22	HID23	HID24	HID25
8	9	10	11	12	13	14
D5	D6	D7	D8	D9	D10	D11
TBT18	TBT19	TBT20	TBT21	TBT22	TBT23	TBT24
HID26	HID27	HID28	HID29	HID30	TAH1	TAH2
15	16	17	18	19	20	21
D12	D13	D14	D15	D16	D17	D18
TBT25	TBT26	TBT27	TBT28	TBT29	TBT30	MKH1
TAH3	TAH4	TAH5	TAH6	TAH7	TAH8	TAH9
22	23	24	25	26	27	28
D19	D20	D21	D22	D23	D24	D25
MKH2	MKH3	MKH4	MKH5	MKH6	MKH7	MKH8
TAH10	TAH11	TAH12	TAH13	TAH14	TAH15	TAH16

EUSI 496/2012-13/6248-49/2004-05						
Ashe	Ojobo	Iwa	Araorun	Ile	Itutu	Erinle
1	2	3	4	5	6	7
O31	N1	N2	N3	N4	N5	N6
KHK13	KHK14	KHK15	KHK16	KHK17	KHK18	KHK19
THi21	THi22	THi23	THi24	THi25	THi26	THi27
8	9	10	11	12	13	14
N7	N8	N9	N10	N11	N12	N13
KHK20	KHK21	KHK22	KHK23	KHK24	KHK25	KHK26
THi28	THi29	THi30	HID1	HID2	HID3	HID4
15	16	17	18	19	20	21
N14	N15	N16	N17	N18	N19	N20
KHK27	KHK28	KHK29	KHK30	TBT1	TBT2	TBT3
HID5	HID6	HID7	HID8	HID9	HID10	HID11
22	23	24	25	26	27	28
N21	N22	N23	N24	N25	N26	N27
TBT4	TBT5	TBT6	TBT7	TBT8	TBT9	TBT10
HID12	HID13	HID14	HID15	HID16	HID17	HID18

KWANZA 496/2012-13/6248-49/2004-05						
Umoja	Kujichagulia	Ujima	Ujamaa	Nia	Kuumba	Imani
1	2	3	4	5	6	7
D26	D27	D28	D29	D30	D31	JA1
MKH9	MKH10	MKH11	MKH12	MKH13	MKH14	MKH15
TAH17	TAH18	TAH19	TAH20	TAH21	TAH22	TAH23
8	9	10	11	12	13	14
JA2	JA3	JA4	JA5	JA6	JA7	JA8
MKH16	MKH17	MKH18	MKH19	MKH20	MKH21	MKH22
TAH24	TAH25	TAH26	TAH27	TAH28	TAH29	TAH30
15	16	17	18	19	20	21
JA9	JA10	JA11	JA12	JA13	JA14	JA15
MKH23	MKH24	MKH25	KMH26	MKH27	MKH28	MKH29
TIR1	TIR2	TIR3	TIR4	TIR5	TIR6	TIR7
22	23	24	25	26	27	28
JA16	JA17	JA18	JA19	JA20	JA21	JA22
MKH30	AMN1	AMN2	AMN3	AMN4	AMN5	AMN6
TIR8	TIR9	TIR10	TIR11	TIR12	TIR13	TIR14

THE NU AFRIKAN CALENDAR

SHAHIDI 496/2012-13/6248-49/2004-05						
Ashe	Ojobo	Iwa	Araorun	Ile	Itutu	Erinle
1	2	3	4	5	6	7
JA23	JA24	JA25	JA26	JA27	JA28	JA29
AMN7	AMN8	AMN9	AMN10	AMN11	AMN12	AMN13
TIR15	TIR16	TIR17	TIR18	TIR19	TIR20	TIR21
8	9	10	11	12	13	14
JA30	JA31	F1	F2	F3	F4	F5
AMN14	AMN15	AMN16	AMN17	AMN18	AMN19	AMN20
TIR22	TIR23	TIR24	TIR25	TIR26	TIR27	TIR28
15	16	17	18	19	20	21
F6	F7	F8	F9	F10	F 11	F12
AMN21	AMN22	AMN23	AMN24	AMN25	AMN26	AMN28
TIR29	TIR30	YEK1	YEK2	YEK3	YEK4	YEK5
22	23	24	25	26	27	28
F13	F14	F15	F16	F17	F18	F19
AMN29	AMN30	RNT1	RNT2	RNT3	RNT4	RNT5
YEK6	YEK7	YEK8	YEK9	YEK10	YEK11	YEK12
SHAHIDI 47						
F20						
RNT6						
YEK13						

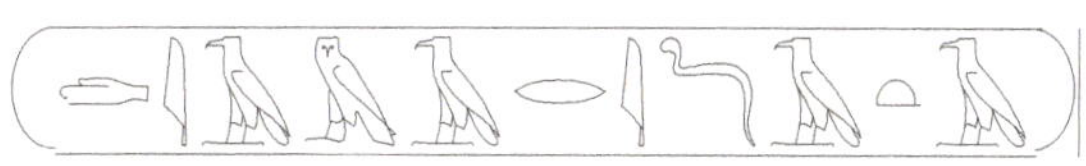

THE NU AFRIKAN CALENDAR

MAAFA 497 B.S.E./ASIENTO 2013-14 C.E.
KAM 6249-50 S.M/EC 2005-06 A.M. (ዓ.ም.)

AFU 497/2013-14/6249-50/2005-06

Ashe	Ojobo	Iwa	Araorun	Ile	Itutu	Erinle
1	2	3	4	5	6	7
F21	F22	F23	F24	F25	F26	F27
RNT7	RNT8	RNT9	RNT10	RNT11	RNT12	RNT13
YEK14	YEK15	YEK16	YEK17	YEK18	YEK19	YEK20
8	9	10	11	12	13	14
F28	M1	M2	M3	M4	M5	M6
RNT14	RNT15	RNT16	RNT17	RNT18	RNT19	RNT20
YEK21	YEK22	YEK23	YEK24	YEK25	YEK26	YEK27
15	16	17	18	19	20	21
M7	M8	M9	M10	M11	M12	M13
RNT21	RNT22	RNT23	RNT24	RNT25	RNT26	RNT27
YEK28	YEK29	YEK30	MAG1	MAG2	MAG3	MAG4
22	23	24	25	26	27	28
M14	M15	M16	M17	M18	M19	M20
RNT28	RNT29	RNT30	KHN1	KHN2	KHN3	KHN4
MAG5	MAG6	MAG7	MAG8	MAG9	MAG10	MAG11

KAWAIDA 497/2013-14/6249-50/2005-06

Ashe	Ojobo	Iwa	Araorun	Ile	Itutu	Erinle
1	2	3	4	5	6	7
A18	A19	A20	A21	A22	A23	A24
PIN3	PIN4	PIN5	PIN6	PIN7	PIN8	PIN9
MIY10	MIY11	MIY12	MIY13	MIY14	MIY15	MIY16
8	9	10	11	12	13	14
A25	A26	A27	A28	A29	A30	MA1
PIN10	PIN11	PIN12	PIN13	PIN14	PIN15	PIN16
MIY17	MIY18	MIY19	MIY20	MIY21	MIY22	MIY23
15	16	17	18	19	20	21
MA2	MA3	MA4	MA5	MA6	MA7	MA8
PIN17	PIN18	PIN19	PIN20	PIN21	PIN22	PIN23
MIY24	MIY25	MIY26	MIY27	MIY28	MIY29	MIY30
22	23	24	25	26	27	28
MA9	MA10	MA11	MA12	MA13	MA14	MA15
PIN24	PIN25	PIN26	PIN27	PIN28	PIN29	PIN30
GIN1	GIN2	GIN3	GIN4	GIN5	GIN6	GIN7

UHURU 497/2013-14/6249-50/2005-06

Ashe	Ojobo	Iwa	Araorun	Ile	Itutu	Erinle
1	2	3	4	5	6	7
M21	M22	M23	M24	M25	M26	M27
KHN5	KHN6	KHN7	KHN8	KHN9	KHN10	KHN11
MAG12	MAG13	MAG14	MAG15	MAG16	MAG17	MAG18
8	9	10	11	12	13	14
M28	M29	M30	M31	A1	A2	A3
KHN12	KHN13	KHN14	KHN15	KHN16	KHN17	KHN18
MAG19	MAG20	MAG21	MAG22	MAG23	MAG24	MAG25
15	16	17	18	19	20	21
A4	A5	A6	A7	A8	A9	A10
KHN19	KHN20	KHN21	KHN22	KHN23	KHN24	KHN25
MAG26	MAG27	MAG28	MAG29	MAG30	MIY1	MIY2
22	23	24	25	26	27	28
A11	A12	A13	A14	A15	A16	A17
KHN26	KHN27	KHN28	KHN29	KHN30	PIN1	PIN2
MIY3	MIY4	MIY5	MIY6	MIY7	MIY8	MIY9

MABABA WEUSI 497/2013-14/6249-50/2005-06

Ashe	Ojobo	Iwa	Araorun	Ile	Itutu	Erinle
1	2	3	4	5	6	7
MA16	MA17	MA18	MA19	MA20	MA21	MA22
APP1	APP2	APP3	APP4	APP5	APP6	APP7
GIN8	GIN9	GIN10	GIN11	GIN12	GIN13	GIN14
8	9	10	11	12	13	14
MA23	MA24	MA25	MA26	MA27	MA28	MA29
APP8	APP9	APP10	APP11	APP12	APP13	APP14
GIN15	GIN16	GIN17	GIN18	GIN19	GIN20	GIN21
15	16	17	18	19	20	21
MA30	MA31	J1	J2	J3	J4	J5
APP15	APP16	APP17	APP18	APP19	APP20	APP21
GIN22	GIN23	GIN24	GIN25	GIN26	GIN27	GIN28
22	23	24	25	26	27	28
J6	J7	J8	J9	J10	J11	J12
APP22	APP23	APP24	APP25	APP26	APP27	APP28
GIN29	GIN30	SAN1	SAN2	SAN3	SAN4	SAN5

THE NU AFRIKAN CALENDAR

MWANAMKE WEUSI 497/2013-14/6249-50/2005-06						
Ashe	Ojobo	Iwa	Araorun	Ile	Itutu	Erinle
1	2	3	4	5	6	7
J13	J14	J15	J16	J17	J18	J19
APP29	APP30	MSR1	MSR2	MSR3	MSR4	MSR5
SAN6	SAN7	SAN8	SAN9	SAN10	SAN11	SAN12
8	9	10	11	12	13	14
J20	J21	J22	J23	J24	J25	J26
MSR6	MSR7	MSR8	MSR9	MSR10	MSR11	MSR12
SAN13	SAN14	SAN15	SAN16	SAN17	SAN18	SAN19
15	16	17	18	19	20	21
J27	J28	J29	J30	JU1	JU2	JU3
MSR13	MSR14	MSR15	MSR16	MSR17	MSR18	MSR19
SAN20	SAN21	SAN22	SAN23	SAN24	SAN25	SAN26
22	23	24	25	26	27	28
JU4	JU5	JU6	JU7	JU8	JU9	JU10
MSR20	MRS21	MSR22	MSR23	MSR24	MSR25	MSR26
SAN27	SAN28	SAN29	SAN30	HAM1	HAM2	HAM3

TAIFA WEUSI 497/2013-14/6249-50/2005-06						
Ashe	Ojobo	Iwa	Araorun	Ile	Itutu	Erinle
1	2	3	4	5	6	7
AU8	AU9	AU10	AU11	AU12	AU13	AU14
THT20	THT21	THT22	THT23	THT24	THT25	THT26
NAH2	NAH3	NAH4	NAH5	NAH6	NAH7	NAH8
8	9	10	11	12	13	14
AU15	AU16	AU17	AU18	AU19	AU20	AU21
THT27	THT28	THT29	THT30	PAP1	PAP1	PAP2
NAH9	NAH10	NAH11	NAH12	NAH13	NAH14	NAH15
15	16	17	18	19	20	21
AU22	AU23	AU24	AU25	AU26	AU27	AU28
PAP3	PAP4	PAP5	PAP6	PAP7	PAP8	PAP9
NAH16	NAH17	NAH18	NAH19	NAH20	NAH21	NAH22
22	23	24	25	26	27	28
AU29	AU30	AU31	S1	S2	S3	S4
PAP10	PAP11	PAP12	PAP13	PAP14	PAP15	PAP16
NAH23	NAH24	NAH25	NAH26	NAH27	NAH28	NAH29

MTU WEUSI 497/2013-14/6249-50/2005-06						
Ashe	Ojobo	Iwa	Araorun	Ile	Itutu	Erinle
1	2	3	4	5	6	7
JU11	JU12	JU13	JU14	JU15	JU16	JU17
MSR27	MSR28	MSR29	MSR30	*ASR1*	*HRU2*	*STH3*
HAM4	HAM5	HAM6	HAM7	HAM8	HAM9	HAM10
8	9	10	11	12	13	14
JU18	JU19	JU20	JU21	JU22	JU23	JU24
AST4	*NBH5*	THT1	THT2	THT3	THT4	THT5
HAM11	HAM12	HAM13	AM14	HAM15	HAM16	HAM17
15	16	17	18	19	20	21
JU25	JU26	JU27	JU28	JU29	JU30	JU31
THT6	THT7	THT8	THT9	THT10	THT11	THT12
HAM18	HAM19	HAM20	HAM21	HAM22	HAM23	HAM24
22	23	24	25	26	27	28
AU1	AU2	AU3	AU4	AU5	AU6	AU7
THT13	THT14	THT15	THT16	THT17	THT18	THT19
HAM25	HAM26	HAM27	HAM28	HAM29	HAM30	NAH1

UWEZO 497/2013-14/6249-50/2005-06						
Ashe	Ojobo	Iwa	Araorun	Ile	Itutu	Erinle
1	2	3	4	5	6	7
S5	S6	S7	S8	S9	S10	S11
PAP17	PAP18	PAP19	PAP20	PAP21	PAP22	PAP23
NAH30	*PAG1*	*PAG2*	*PAG3*	*PAG4*	*PAG5*	MAS1
8	9	10	11	12	13	14
S12	S13	S14	S15	S16	S17	S18
PAP24	PAP25	PAP26	PAP27	PAP28	PAP29	PAP30
MAS2	MAS3	MAS4	MAS5	MAS6	MAS7	MAS8
15	16	17	18	19	20	21
S19	S20	S21	S22	S23	S24	S25
HTH1	HTH2	HTH3	HTH4	HTH5	HTH6	HTH7
MAS9	MAS10	MAS11	MAS12	MAS13	MAS14	MAS15
22	23	24	25	26	27	28
S26	S27	S28	S29	S30	O1	O2
HTH8	HTH9	HTH10	HTH11	HTH12	HTH13	HTH14
MAS16	MAS17	MAS18	MAS19	MAS20	MAS21	MAS22

THE NU AFRIKAN CALENDAR

EKUNDU 497/2013-14/6249-50/2005-06						
Ashe	Ojobo	Iwa	Araorun	Ile	Itutu	Erinle
1	2	3	4	5	6	7
O3	O4	O5	O6	O7	O8	O9
HTH15	HTH16	HTH17	HTH18	HTH19	HTH20	HTH21
MAS23	MAS24	MAS25	MAS26	MAS27	MAS28	MAS29
8	9	10	11	12	13	14
O10	O11	O12	O13	O14	O15	O16
HTH22	HTH23	HTH24	HTH25	HTH26	HTH27	HTH28
MAS30	THi1	THi2	THi3	THi4	THi5	THi6
15	16	17	18	19	20	21
O17	O18	O19	O20	O21	O22	O23
HTH29	HTH30	KHK1	KHK2	KHK3	KHK4	KHK5
THi7	THi8	THi9	THi10	THi11	THi12	THi13
22	23	24	25	26	27	28
O24	O25	O26	O27	O28	O29	O30
KHK6	KHK7	KHK8	KHK9	KHK10	KHK11	KHK12
THi14	THi15	THi16	THi17	THi18	THi19	THi20

KIJANI 497/2013-14/6249-50/2005-06						
Ashe	Ojobo	Iwa	Araorun	Ile	Itutu	Erinle
1	2	3	4	5	6	7
N28	N29	N30	D1	D2	D3	D4
TBT11	TBT12	TBT13	TBT14	TBT15	TBT16	TBT17
HID19	HID20	HID21	HID22	HID23	HID24	HID25
8	9	10	11	12	13	14
D5	D6	D7	D8	D9	D10	D11
TBT18	TBT19	TBT20	TBT21	TBT22	TBT23	TBT24
HID26	HID27	HID28	HID29	HID30	TAH1	TAH2
15	16	17	18	19	20	21
D12	D13	D14	D15	D16	D17	D18
TBT25	TBT26	TBT27	TBT28	TBT29	TBT30	MKH1
TAH3	TAH4	TAH5	TAH6	TAH7	TAH8	TAH9
22	23	24	25	26	27	28
D19	D20	D21	D22	D23	D24	D25
MKH2	MKH3	MKH4	MKH5	MKH6	MKH7	MKH8
TAH10	TAH11	TAH12	TAH13	TAH14	TAH15	TAH16

EUSI 497/2013-14/6249-50/2005-06						
Ashe	Ojobo	Iwa	Araorun	Ile	Itutu	Erinle
1	2	3	4	5	6	7
O31	N1	N2	N3	N4	N5	N6
KHK13	KHK14	KHK15	KHK16	KHK17	KHK18	KHK19
THi21	THi22	THi23	THi24	THi25	THi26	THi27
8	9	10	11	12	13	14
N7	N8	N9	N10	N11	N12	N13
KHK20	KHK21	KHK22	KHK23	KHK24	KHK25	KHK26
THi28	THi29	THi30	HID1	HID2	HID3	HID4
15	16	17	18	19	20	21
N14	N15	N16	N17	N18	N19	N20
KHK27	KHK28	KHK29	KHK30	TBT1	TBT2	TBT3
HID5	HID6	HID7	HID8	HID9	HID10	HID11
22	23	24	25	26	27	28
N21	N22	N23	N24	N25	N26	N27
TBT4	TBT5	TBT6	TBT7	TBT8	TBT9	TBT10
HID12	HID13	HID14	HID15	HID16	HID17	HID18

KWANZA 497/2013-14/6249-50/2005-06						
Umoja	Kujichagulia	Ujima	Ujamaa	Nia	Kuumba	Imani
1	2	3	4	5	6	7
D26	D27	D28	D29	D30	D31	JA1
MKH9	MKH10	MKH11	MKH12	MKH13	MKH14	MKH15
TAH17	TAH18	TAH19	TAH20	TAH21	TAH22	TAH23
8	9	10	11	12	13	14
JA2	JA3	JA4	JA5	JA6	JA7	JA8
MKH16	MKH17	MKH18	MKH19	MKH20	MKH21	MKH22
TAH24	TAH25	TAH26	TAH27	TAH28	TAH29	TAH30
15	16	17	18	19	20	21
JA9	JA10	JA11	JA12	JA13	JA14	JA15
MKH23	MKH24	MKH25	KMH26	MKH27	MKH28	MKH29
TIR1	TIR2	TIR3	TIR4	TIR5	TIR6	TIR7
22	23	24	25	26	27	28
JA16	JA17	JA18	JA19	JA20	JA21	JA22
MKH30	AMN1	AMN2	AMN3	AMN4	AMN5	AMN6
TIR8	TIR9	TIR10	TIR11	TIR12	TIR13	TIR14

THE NU AFRIKAN CALENDAR

SHAHIDI 497/2013-14/6249-50/2005-06						
Ashe	Ojobo	Iwa	Araorun	Ile	Itutu	Erinle
1	2	3	4	5	6	7
JA23	JA24	JA25	JA26	JA27	JA28	JA29
AMN7	AMN8	AMN9	AMN10	AMN11	AMN12	AMN13
TIR15	TIR16	TIR17	TIR18	TIR19	TIR20	TIR21
8	9	10	11	12	13	14
JA30	JA31	F1	F2	F3	F4	F5
AMN14	AMN15	AMN16	AMN17	AMN18	AMN19	AMN20
TIR22	TIR23	TIR24	TIR25	TIR26	TIR27	TIR28
15	16	17	18	19	20	21
F6	F7	F8	F9	F10	F 11	F12
AMN21	AMN22	AMN23	AMN24	AMN25	AMN26	AMN28
TIR29	TIR30	YEK1	YEK2	YEK3	YEK4	YEK5
22	23	24	25	26	27	28
F13	F14	F15	F16	F17	F18	F19
AMN29	AMN30	RNT1	RNT2	RNT3	RNT4	RNT5
YEK6	YEK7	YEK8	YEK9	YEK10	YEK11	YEK12
SHAHIDI 48						
F20						
RNT6						
YEK13						

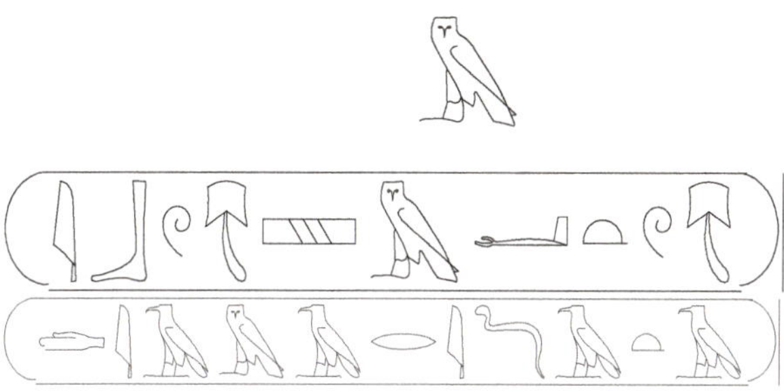

THE NU AFRIKAN CALENDAR

MAAFA 498 B.S.E./ASIENTO 2014-15 C.E.
KAM 6250-51 S.M/EC 2006-07 A.M. (ዓ.ም.)

AFU 498/2014-15/6250-51/2006-07						
Ashe	Ojobo	Iwa	Araorun	Ile	Itutu	Erinle
1	2	3	4	5	6	7
F21	F22	F23	F24	F25	F26	F27
RNT7	RNT8	RNT9	RNT10	RNT11	RNT12	RNT13
YEK14	YEK15	YEK16	YEK17	YEK18	YEK19	YEK20
8	9	10	11	12	13	14
F28	M1	M2	M3	M4	M5	M6
RNT14	RNT15	RNT16	RNT17	RNT18	RNT19	RNT20
YEK21	YEK22	YEK23	YEK24	YEK25	YEK26	YEK27
15	16	17	18	19	20	21
M7	M8	M9	M10	M11	M12	M13
RNT21	RNT22	RNT23	RNT24	RNT25	RNT26	RNT27
YEK28	YEK29	YEK30	MAG1	MAG2	MAG3	MAG4
22	23	24	25	26	27	28
M14	M15	M16	M17	M18	M19	M20
RNT28	RNT29	RNT30	KHN1	KHN2	KHN3	KHN4
MAG5	MAG6	MAG7	MAG8	MAG9	MAG10	MAG11

KAWAIDA 498/2014-15/6250-51/2006-07						
Ashe	Ojobo	Iwa	Araorun	Ile	Itutu	Erinle
1	2	3	4	5	6	7
A18	A19	A20	A21	A22	A23	A24
PIN3	PIN4	PIN5	PIN6	PIN7	PIN8	PIN9
MIY10	MIY11	MIY12	MIY13	MIY14	MIY15	MIY16
8	9	10	11	12	13	14
A25	A26	A27	A28	A29	A30	MA1
PIN10	PIN11	PIN12	PIN13	PIN14	PIN15	PIN16
MIY17	MIY18	MIY19	MIY20	MIY21	MIY22	MIY23
15	16	17	18	19	20	21
MA2	MA3	MA4	MA5	MA6	MA7	MA8
PIN17	PIN18	PIN19	PIN20	PIN21	PIN22	PIN23
MIY24	MIY25	MIY26	MIY27	MIY28	MIY29	MIY30
22	23	24	25	26	27	28
MA9	MA10	MA11	MA12	MA13	MA14	MA15
PIN24	PIN25	PIN26	PIN27	PIN28	PIN29	PIN30
GIN1	GIN2	GIN3	GIN4	GIN5	GIN6	GIN7

UHURU 498/2014-15/6250-51/2006-07						
Ashe	Ojobo	Iwa	Araorun	Ile	Itutu	Erinle
1	2	3	4	5	6	7
M21	M22	M23	M24	M25	M26	M27
KHN5	KHN6	KHN7	KHN8	KHN9	KHN10	KHN11
MAG12	MAG13	MAG14	MAG15	MAG16	MAG17	MAG18
8	9	10	11	12	13	14
M28	M29	M30	M31	A1	A2	A3
KHN12	KHN13	KHN14	KHN15	KHN16	KHN17	KHN18
MAG19	MAG20	MAG21	MAG22	MAG23	MAG24	MAG25
15	16	17	18	19	20	21
A4	A5	A6	A7	A8	A9	A10
KHN19	KHN20	KHN21	KHN22	KHN23	KHN24	KHN25
MAG26	MAG27	MAG28	MAG29	MAG30	MIY1	MIY2
22	23	24	25	26	27	28
A11	A12	A13	A14	A15	A16	A17
KHN26	KHN27	KHN28	KHN29	KHN30	PIN1	PIN2
MIY3	MIY4	MIY5	MIY6	MIY7	MIY8	MIY9

MABABA WEUSI 498/2014-15/6250-51/2006-07						
Ashe	Ojobo	Iwa	Araorun	Ile	Itutu	Erinle
1	2	3	4	5	6	7
MA16	MA17	MA18	MA19	MA20	MA21	MA22
APP1	APP2	APP3	APP4	APP5	APP6	APP7
GIN8	GIN9	GIN10	GIN11	GIN12	GIN13	GIN14
8	9	10	11	12	13	14
MA23	MA24	MA25	MA26	MA27	MA28	MA29
APP8	APP9	APP10	APP11	APP12	APP13	APP14
GIN15	GIN16	GIN17	GIN18	GIN19	GIN20	GIN21
15	16	17	18	19	20	21
MA30	MA31	J1	J2	J3	J4	J5
APP15	APP16	APP17	APP18	APP19	APP20	APP21
GIN22	GIN23	GIN24	GIN25	GIN26	GIN27	GIN28
22	23	24	25	26	27	28
J6	J7	J8	J9	J10	J11	J12
APP22	APP23	APP24	APP25	APP26	APP27	APP28
GIN29	GIN30	SAN1	SAN2	SAN3	SAN4	SAN5

THE NU AFRIKAN CALENDAR

MWANAMKE WEUSI 498/2014-15/6250-51/2006-07

Ashe	Ojobo	Iwa	Araorun	Ile	Itutu	Erinle
1	2	3	4	5	6	7
J13	J14	J15	J16	J17	J18	J19
APP29	APP30	MSR1	MSR2	MSR3	MSR4	MSR5
SAN6	SAN7	SAN8	SAN9	SAN10	SAN11	SAN12
8	9	10	11	12	13	14
J20	J21	J22	J23	J24	J25	J26
MSR6	MSR7	MSR8	MSR9	MSR10	MSR11	MSR12
SAN13	SAN14	SAN15	SAN16	SAN17	SAN18	SAN19
15	16	17	18	19	20	21
J27	J28	J29	J30	JU1	JU2	JU3
MSR13	MSR14	MSR15	MSR16	MSR17	MSR18	MSR19
SAN20	SAN21	SAN22	SAN23	SAN24	SAN25	SAN26
22	23	24	25	26	27	28
JU4	JU5	JU6	JU7	JU8	JU9	JU10
MSR20	MRS21	MSR22	MSR23	MSR24	MSR25	MSR26
SAN27	SAN28	SAN29	SAN30	HAM1	HAM2	HAM3

TAIFA WEUSI 498/2014-15/6250-51/2006-07

Ashe	Ojobo	Iwa	Araorun	Ile	Itutu	Erinle
1	2	3	4	5	6	7
AU8	AU9	AU10	AU11	AU12	AU13	AU14
THT20	THT21	THT22	THT23	THT24	THT25	THT26
NAH2	NAH3	NAH4	NAH5	NAH6	NAH7	NAH8
8	9	10	11	12	13	14
AU15	AU16	AU17	AU18	AU19	AU20	AU21
THT27	THT28	THT29	THT30	PAP1	PAP1	PAP2
NAH9	NAH10	NAH11	NAH12	NAH13	NAH14	NAH15
15	16	17	18	19	20	21
AU22	AU23	AU24	AU25	AU26	AU27	AU28
PAP3	PAP4	PAP5	PAP6	PAP7	PAP8	PAP9
NAH16	NAH17	NAH18	NAH19	NAH20	NAH21	NAH22
22	23	24	25	26	27	28
AU29	AU30	AU31	S1	S2	S3	S4
PAP10	PAP11	PAP12	PAP13	PAP14	PAP15	PAP16
NAH23	NAH24	NAH25	NAH26	NAH27	NAH28	NAH29

MTU WEUSI 498/2014-15/6250-51/2006-07

Ashe	Ojobo	Iwa	Araorun	Ile	Itutu	Erinle
1	2	3	4	5	6	7
JU11	JU12	JU13	JU14	JU15	JU16	JU17
MSR27	MSR28	MSR29	MSR30	*ASR1*	*HRU2*	*STH3*
HAM4	HAM5	HAM6	HAM7	HAM8	HAM9	HAM10
8	9	10	11	12	13	14
JU18	JU19	JU20	JU21	JU22	JU23	JU24
AST4	*NBH5*	THT1	THT2	THT3	THT4	THT5
HAM11	HAM12	HAM13	AM14	HAM15	HAM16	HAM17
15	16	17	18	19	20	21
JU25	JU26	JU27	JU28	JU29	JU30	JU31
THT6	THT7	THT8	THT9	THT10	THT11	THT12
HAM18	HAM19	HAM20	HAM21	HAM22	HAM23	HAM24
22	23	24	25	26	27	28
AU1	AU2	AU3	AU4	AU5	AU6	AU7
THT13	THT14	THT15	THT16	THT17	THT18	THT19
HAM25	HAM26	HAM27	HAM28	HAM29	HAM30	NAH1

UWEZO 498/2014-15/6250-51/2006-07

Ashe	Ojobo	Iwa	Araorun	Ile	Itutu	Erinle
1	2	3	4	5	6	7
S5	S6	S7	S8	S9	S10	S11
PAP17	PAP18	PAP19	PAP20	PAP21	PAP22	PAP23
NAH30	*PAG1*	*PAG2*	*PAG3*	*PAG4*	*PAG5*	MAS1
8	9	10	11	12	13	14
S12	S13	S14	S15	S16	S17	S18
PAP24	PAP25	PAP26	PAP27	PAP28	PAP29	PAP30
MAS2	MAS3	MAS4	MAS5	MAS6	MAS7	MAS8
15	16	17	18	19	20	21
S19	S20	S21	S22	S23	S24	S25
HTH1	HTH2	HTH3	HTH4	HTH5	HTH6	HTH7
MAS9	MAS10	MAS11	MAS12	MAS13	MAS14	MAS15
22	23	24	25	26	27	28
S26	S27	S28	S29	S30	O1	O2
HTH8	HTH9	HTH10	HTH11	HTH12	HTH13	HTH14
MAS16	MAS17	MAS18	MAS19	MAS20	MAS21	MAS22

THE NU AFRIKAN CALENDAR

EKUNDU 498/2014-15/6250-51/2006-07						
Ashe	Ojobo	Iwa	Araorun	Ile	Itutu	Erinle
1	2	3	4	5	6	7
O3	O4	O5	O6	O7	O8	O9
HTH15	HTH16	HTH17	HTH18	HTH19	HTH20	HTH21
MAS23	MAS24	MAS25	MAS26	MAS27	MAS28	MAS29
8	9	10	11	12	13	14
O10	O11	O12	O13	O14	O15	O16
HTH22	HTH23	HTH24	HTH25	HTH26	HTH27	HTH28
MAS30	THi1	THi2	THi3	THi4	THi5	THi6
15	16	17	18	19	20	21
O17	O18	O19	O20	O21	O22	O23
HTH29	HTH30	KHK1	KHK2	KHK3	KHK4	KHK5
THi7	THi8	THi9	THi10	THi11	THi12	THi13
22	23	24	25	26	27	28
O24	O25	O26	O27	O28	O29	O30
KHK6	KHK7	KHK8	KHK9	KHK10	KHK11	KHK12
THi14	THi15	THi16	THi17	THi18	THi19	THi20

KIJANI 498/2014-15/6250-51/2006-07						
Ashe	Ojobo	Iwa	Araorun	Ile	Itutu	Erinle
1	2	3	4	5	6	7
N28	N29	N30	D1	D2	D3	D4
TBT11	TBT12	TBT13	TBT14	TBT15	TBT16	TBT17
HID19	HID20	HID21	HID22	HID23	HID24	HID25
8	9	10	11	12	13	14
D5	D6	D7	D8	D9	D10	D11
TBT18	TBT19	TBT20	TBT21	TBT22	TBT23	TBT24
HID26	HID27	HID28	HID29	HID30	TAH1	TAH2
15	16	17	18	19	20	21
D12	D13	D14	D15	D16	D17	D18
TBT25	TBT26	TBT27	TBT28	TBT29	TBT30	MKH1
TAH3	TAH4	TAH5	TAH6	TAH7	TAH8	TAH9
22	23	24	25	26	27	28
D19	D20	D21	D22	D23	D24	D25
MKH2	MKH3	MKH4	MKH5	MKH6	MKH7	MKH8
TAH10	TAH11	TAH12	TAH13	TAH14	TAH15	TAH16

EUSI 498/2014-15/6250-51/2006-07						
Ashe	Ojobo	Iwa	Araorun	Ile	Itutu	Erinle
1	2	3	4	5	6	7
O31	N1	N2	N3	N4	N5	N6
KHK13	KHK14	KHK15	KHK16	KHK17	KHK18	KHK19
THi21	THi22	THi23	THi24	THi25	THi26	THi27
8	9	10	11	12	13	14
N7	N8	N9	N10	N11	N12	N13
KHK20	KHK21	KHK22	KHK23	KHK24	KHK25	KHK26
THi28	THi29	THi30	HID1	HID2	HID3	HID4
15	16	17	18	19	20	21
N14	N15	N16	N17	N18	N19	N20
KHK27	KHK28	KHK29	KHK30	TBT1	TBT2	TBT3
HID5	HID6	HID7	HID8	HID9	HID10	HID11
22	23	24	25	26	27	28
N21	N22	N23	N24	N25	N26	N27
TBT4	TBT5	TBT6	TBT7	TBT8	TBT9	TBT10
HID12	HID13	HID14	HID15	HID16	HID17	HID18

KWANZA 498/2014-15/6250-51/2006-07						
Umoja	Kujichagulia	Ujima	Ujamaa	Nia	Kuumba	Imani
1	2	3	4	5	6	7
D26	D27	D28	D29	D30	D31	JA1
MKH9	MKH10	MKH11	MKH12	MKH13	MKH14	MKH15
TAH17	TAH18	TAH19	TAH20	TAH21	TAH22	TAH23
8	9	10	11	12	13	14
JA2	JA3	JA4	JA5	JA6	JA7	JA8
MKH16	MKH17	MKH18	MKH19	MKH20	MKH21	MKH22
TAH24	TAH25	TAH26	TAH27	TAH28	TAH29	TAH30
15	16	17	18	19	20	21
JA9	JA10	JA11	JA12	JA13	JA14	JA15
MKH23	MKH24	MKH25	KMH26	MKH27	MKH28	MKH29
TIR1	TIR2	TIR3	TIR4	TIR5	TIR6	TIR7
22	23	24	25	26	27	28
JA16	JA17	JA18	JA19	JA20	JA21	JA22
MKH30	AMN1	AMN2	AMN3	AMN4	AMN5	AMN6
TIR8	TIR9	TIR10	TIR11	TIR12	TIR13	TIR14

THE NU AFRIKAN CALENDAR

SHAHIDI 498/2014-15/6250-51/2006-07						
Ashe	Ojobo	Iwa	Araorun	Ile	Itutu	Erinle
1	2	3	4	5	6	7
JA23	JA24	JA25	JA26	JA27	JA28	JA29
AMN7	AMN8	AMN9	AMN10	AMN11	AMN12	AMN13
TIR15	TIR16	TIR17	TIR18	TIR19	TIR20	TIR21
8	9	10	11	12	13	14
JA30	JA31	F1	F2	F3	F4	F5
AMN14	AMN15	AMN16	AMN17	AMN18	AMN19	AMN20
TIR22	TIR23	TIR24	TIR25	TIR26	TIR27	TIR28
15	16	17	18	19	20	21
F6	F7	F8	F9	F10	F 11	F12
AMN21	AMN22	AMN23	AMN24	AMN25	AMN26	AMN28
TIR29	TIR30	YEK1	YEK2	YEK3	YEK4	YEK5
22	23	24	25	26	27	28
F13	F14	F15	F16	F17	F18	F19
AMN29	AMN30	RNT1	RNT2	RNT3	RNT4	RNT5
YEK6	YEK7	YEK8	YEK9	YEK10	YEK11	YEK12
SHAHIDI 49						
F20						
RNT6						
YEK13						

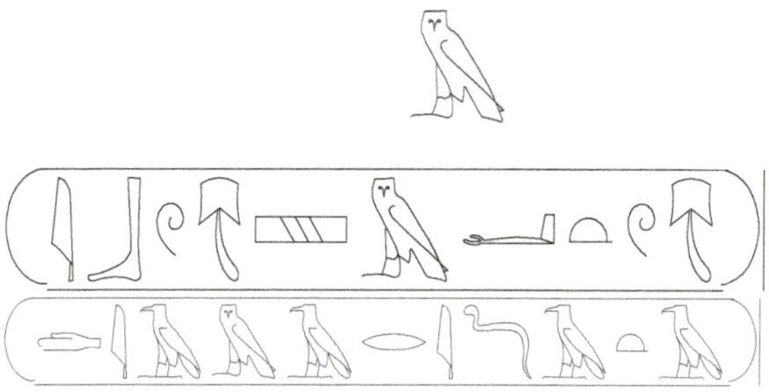

MAAFA 499 B.S.E./ASIENTO 2015-16 C.E./
KAM 6251-52 S.M/EC 2007-08 A.M. (ዓ.ም.)

AFU 499/2015-16/6251-52/2007-08

Ashe	Ojobo	Iwa	Araorun	Ile	Itutu	Erinle
1	2	3	4	5	6	7
F21	F22	F23	F24	F25	F26	F27
RNT7	RNT8	RNT9	RNT10	RNT11	RNT12	RNT13
YEK14	YEK15	YEK16	YEK17	YEK18	YEK19	YEK20
8	9	10	11	12	13	14
F28	M1	M2	M3	M4	M5	M6
RNT14	RNT15	RNT16	RNT17	RNT18	RNT19	RNT20
YEK21	YEK22	YEK23	YEK24	YEK25	YEK26	YEK27
15	16	17	18	19	20	21
M7	M8	M9	M10	M11	M12	M13
RNT21	RNT22	RNT23	RNT24	RNT25	RNT26	RNT27
YEK28	YEK 29	YEK30	MAG1	MAG2	MAG3	MAG4
22	23	24	25	26	27	28
M14	M15	M16	M17	M18	M19	M20
RNT28	RNT29	RNT30	KHN1	KHN2	KHN3	KHN4
MAG5	MAG6	MAG7	MAG8	MAG9	MAG10	MAG11

UHURU 499/2015-16/6251-52/2007-08

Ashe	Ojobo	Iwa	Araorun	Ile	Itutu	Erinle
1	2	3	4	5	6	7
M21	M22	M23	M24	M25	M26	M27
KHN5	KHN6	KHN7	KHN8	KHN9	KHN10	KHN11
MAG12	MAG13	MAG14	MAG15	MAG16	MAG17	MAG18
8	9	10	11	12	13	14
M28	M29	M30	M31	A1	A2	A3
KHN12	KHN13	KHN14	KHN15	KHN16	KHN17	KHN18
MAG19	MAG20	MAG21	MAG22	MAG23	MAG24	MAG25
15	16	17	18	19	20	21
A4	A5	A6	A7	A8	A9	A10
KHN19	KHN20	KHN21	KHN22	KHN23	KHN24	KHN25
MAG26	MAG27	MAG28	MAG29	MAG30	MIY1	MIY2
22	23	24	25	26	27	28
A11	A12	A13	A14	A15	A16	A17
KHN26	KHN27	KHN28	KHN29	KHN30	PIN1	PIN2
MIY3	MIY4	MIY5	MIY6	MIY7	MIY8	MIY9

KAWAIDA 499/2015-16/6251-52/2007-08

Ashe	Ojobo	Iwa	Araorun	Ile	Itutu	Erinle
1	2	3	4	5	6	7
A18	A19	A20	A21	A22	A23	A24
PIN3	PIN4	PIN5	PIN6	PIN7	PIN8	PIN9
MIY10	MIY11	MIY12	MIY13	MIY14	MIY15	MIY16
8	9	10	11	12	13	14
A25	A26	A27	A28	A29	A30	MA1
PIN10	PIN11	PIN12	PIN13	PIN14	PIN15	PIN16
MIY17	MIY18	MIY19	MIY20	MIY21	MIY22	MIY23
15	16	17	18	19	20	21
MA2	MA3	MA4	MA5	MA6	MA7	MA8
PIN17	PIN18	PIN19	PIN20	PIN21	PIN22	PIN23
MIY24	MIY25	MIY26	MIY27	MIY28	MIY29	MIY30
22	23	24	25	26	27	28
MA9	MA10	MA11	MA12	MA13	MA14	MA15
PIN24	PIN25	PIN26	PIN27	PIN28	PIN29	PIN30
GIN1	GIN2	GIN3	GIN4	GIN5	GIN6	GIN7

MABABA WEUSI 499/2015-16/6251-52/2007-08

Ashe	Ojobo	Iwa	Araorun	Ile	Itutu	Erinle
1	2	3	4	5	6	7
MA16	MA17	MA18	MA19	MA20	MA21	MA22
APP1	APP2	APP3	APP4	APP5	APP6	APP7
GIN8	GIN9	GIN10	GIN11	GIN12	GIN13	GIN14
8	9	10	11	12	13	14
MA23	MA24	MA25	MA26	MA27	MA28	MA29
APP8	APP9	APP10	APP11	APP12	APP13	APP14
GIN15	GIN16	GIN17	GIN18	GIN19	GIN20	GIN21
15	16	17	18	19	20	21
MA30	MA31	J1	J2	J3	J4	J5
APP15	APP16	APP17	APP18	APP19	APP20	APP21
GIN22	GIN23	GIN24	GIN25	GIN26	GIN27	GIN28
22	23	24	25	26	27	28
J6	J7	J8	J9	J10	J11	J12
APP22	APP23	APP24	APP25	APP26	APP27	APP28
GIN29	GIN30	SAN1	SAN2	SAN3	SAN4	SAN5

THE NU AFRIKAN CALENDAR

MWANAMKE 499/2015-16/6251-52/2007-08

Ashe	Ojobo	Iwa	Araorun	Ile	Itutu	Erinle
1	2	3	4	5	6	7
J13	J14	J15	J16	J17	J18	J19
APP29	APP30	MSR1	MSR2	MSR3	MSR4	MSR5
SAN6	SAN7	SAN8	SAN9	SAN10	SAN11	SAN12
8	9	10	11	12	13	14
J20	J21	J22	J23	J24	J25	J26
MSR6	MSR7	MSR8	MSR9	MSR10	MSR11	MSR12
SAN13	SAN14	SAN15	SAN16	SAN17	SAN18	SAN19
15	16	17	18	19	20	21
J27	J28	J29	J30	JU1	JU2	JU3
MSR13	MSR14	MSR15	MSR16	MSR17	MSR18	MSR19
SAN20	SAN21	SAN22	SAN23	SAN24	SAN25	SAN26
22	23	24	25	26	27	28
JU4	JU5	JU6	JU7	JU8	JU9	JU10
MSR20	MRS21	MSR22	MSR23	MSR24	MSR25	MSR26
SAN27	SAN28	SAN29	SAN30	HAM1	HAM2	HAM3

TAIFA WEUSI 499/2015-16/6251-52/2007-08

Ashe	Ojobo	Iwa	Araorun	Ile	Itutu	Erinle
1	2	3	4	5	6	7
AU8	AU9	AU10	AU11	AU12	AU13	AU14
THT20	THT21	THT22	THT23	THT24	THT25	THT26
NAH2	NAH3	NAH4	NAH5	NAH6	NAH7	NAH8
8	9	10	11	12	13	14
AU15	AU16	AU17	AU18	AU19	AU20	AU21
THT27	THT28	THT29	THT30	PAP1	PAP1	PAP2
NAH9	NAH10	NAH11	NAH12	NAH13	NAH14	NAH15
15	16	17	18	19	20	21
AU22	AU23	AU24	AU25	AU26	AU27	AU28
PAP3	PAP4	PAP5	PAP6	PAP7	PAP8	PAP9
NAH16	NAH17	NAH18	NAH19	NAH20	NAH21	NAH22
22	23	24	25	26	27	28
AU29	AU30	AU31	S1	S2	S3	S4
PAP10	PAP11	PAP12	PAP13	PAP14	PAP15	PAP16
NAH23	NAH24	NAH25	NAH26	NAH27	NAH28	NAH29

MTU WEUSI 499/2015-16/6251-52/2007-08

Ashe	Ojobo	Iwa	Araorun	Ile	Itutu	Erinle
1	2	3	4	5	6	7
JU11	JU12	JU13	JU14	JU15	JU16	JU17
MSR27	MSR28	MSR29	MSR30	*ASR1*	*HRU2*	*STH3*
HAM4	HAM5	HAM6	HAM7	HAM8	HAM9	HAM10
8	9	10	11	12	13	14
JU18	JU19	JU20	JU21	JU22	JU23	JU24
AST4	*NBH5*	THT1	THT2	THT3	THT4	THT5
HAM11	HAM12	HAM13	HAM14	HAM15	HAM16	HAM17
15	16	17	18	19	20	21
JU25	JU26	JU27	JU28	JU29	JU30	JU31
THT6	THT7	THT8	THT9	THT10	THT11	THT12
HAM18	HAM19	HAM20	HAM21	HAM22	HAM23	HAM24
22	23	24	25	26	27	28
AU1	AU2	AU3	AU4	AU5	AU6	AU7
THT13	THT14	THT15	THT16	THT17	THT18	THT19
HAM25	HAM26	HAM27	HAM28	HAM29	HAM30	NAH1

UWEZO 499/2015-16/6251-52/2007-08

Ashe	Ojobo	Iwa	Araorun	Ile	Itutu	Erinle
1	2	3	4	5	6	7
S5	S6	S7	S8	S9	S10	S11
PAP17	PAP18	PAP19	PAP20	PAP21	PAP22	PAP23
NAH30	*QUA1*	*QUA2*	*QUA3*	*QUA4*	*QUA5*	*QUA6*
8	9	10	11	12	13	14
S12	S13	S14	S15	S16	S17	S18
PAP24	PAP25	PAP26	PAP27	PAP28	PAP29	PAP30
MAS1	MAS2	MAS3	MAS4	MAS5	MAS6	MAS7
15	16	17	18	19	20	21
S19	S20	S21	S22	S23	S24	S25
HTH1	HTH2	HTH3	HTH4	HTH5	HTH6	HTH7
MAS8	MAS9	MAS10	MAS11	MAS12	MAS13	MAS14
22	23	24	25	26	27	28
S26	S27	S28	S29	S30	O1	O2
HTH8	HTH9	HTH10	HTH11	HTH12	HTH13	HTH14
MAS15	MAS16	MAS17	MAS18	MAS19	MAS20	MAS21

THE NU AFRIKAN CALENDAR

EKUNDU 499/2015-16/6251-52/2007-08						
Ashe	Ojobo	Iwa	Araorun	Ile	Itutu	Erinle
1	2	3	4	5	6	7
O3	O4	O5	O6	O7	O8	O9
HTH15	HTH16	HTH17	HTH18	HTH19	HTH20	HTH21
MAS22	MAS23	MAS 24	MAS25	MAS26	MAS27	MAS28
8	9	10	11	12	13	14
O10	O11	O12	O13	O14	O15	O16
HTH22	HTH23	HTH24	HTH25	HTH26	HTH27	HTH28
MAS29	MAS30	THi1	THi2	THi3	THi4	THi5
15	16	17	18	19	20	21
O17	O18	O19	O20	O21	O22	O23
HTH29	HTH30	KHK1	KHK2	KHK3	KHK4	KHK5
THi6	THi7	THi8	THi9	THi10	THi11	THi12
22	23	24	25	26	27	28
O24	O25	O26	O27	O28	O29	O30
KHK6	KHK7	KHK8	KHK9	KHK10	KHK11	KHK12
THi13	THi14	THi15	THi16	THi17	THi18	THi19

KIJANI 499/2015-16/6251-52/2007-08						
Ashe	Ojobo	Iwa	Araorun	Ile	Itutu	Erin
1	2	3	4	5	6	7
N28	N29	N30	D1	D2	D3	D4
TBT11	TBT12	TBT13	TBT14	TBT15	TBT16	TBT
HID18	HID19	HID20	HID21	HID22	HID23	HID2
8	9	10	11	12	13	14
D5	D6	D7	D8	D9	D10	D11
TBT18	TBT19	TBT20	TBT21	TBT22	TBT23	TBT
HID25	HID26	HID27	HID28	HID29	HID30	TAH
15	16	17	18	19	20	21
D12	D13	D14	D15	D16	D17	D18
TBT25	TBT26	TBT27	TBT28	TBT29	TBT30	MK
TAH2	TAH3	TAH4	TAH5	TAH6	TAH7	TA
22	23	24	25	26	27	28
D19	D20	D21	D22	D23	D24	D25
MKH2	MKH3	MKH4	MKH5	MKH6	MKH7	MK
TAH9	TAH10	TAH11	TAH12	TAH13	TAH14	TAH

EUSI 499/2015-16/6251-52/2007-08						
Ashe	Ojobo	Iwa	Araorun	Ile	Itutu	Erinle
1	2	3	4	5	6	7
O31	N1	N2	N3	N4	N5	N6
KHK13	KHK14	KHK15	KHK16	KHK17	KHK18	KHK19
THi20	THi21	THi22	THi23	THi24	THi25	THi26
8	9	10	11	12	13	14
N7	N8	N9	N10	N11	N12	N13
KHK20	KHK21	KHK22	KHK23	KHK24	KHK25	KHK26
THi27	THi28	THi29	THi30	HID1	HID2	HID3
15	16	17	18	19	20	21
N14	N15	N16	N17	N18	N19	N20
KHK27	KHK28	KHK29	KHK30	TBT1	TBT2	TBT3
HID4	HID5	HID6	HID7	HID8	HID9	HID10
22	23	24	25	26	27	28
N21	N22	N23	N24	N25	N26	N27
TBT4	TBT5	TBT6	TBT7	TBT8	TBT9	TBT10
HID11	HID12	HID13	HID14	HID15	HID16	HID17

KWANZA 499/2015-16/6251-52/2007-08						
Umoja	Kujichagulia	Ujima	Ujamaa	Nia	Kuumba	In
1	2	3	4	5	6	
D26	D27	D28	D29	D30	D31	JA
MKH9	MKH10	MKH11	MKH12	MKH13	MKH14	MK
TAH16	TAH17	TAH18	TAH19	TAH20	TAH21	TA
8	9	10	11	12	13	1
JA2	JA3	JA4	JA5	JA6	JA7	
MKH16	MKH17	MKH18	MKH19	MKH20	MKH21	MK
TAH23	TAH24	TAH25	TAH26	TAH27	TAH28	TA
15	16	17	18	19	20	2
JA9	JA10	JA11	JA12	JA13	JA14	J
MKH23	MKH24	MKH25	KMH26	MKH27	MKH28	MK
TAH30	TIR1	TIR2	TIR3	TIR4	TIR5	TIR
22	23	24	25	26	27	2
JA16	JA17	JA18	JA19	JA20	JA21	J
MKH30	AMN1	AMN2	AMN3	AMN4	AMN5	A
TIR7	TIR8	TIR9	TIR10	TIR11	TIR12	TI

THE NU AFRIKAN CALENDAR

SHAHIDI 499/2015-16/6251-52/2007-08						
Ashe	Ojobo	Iwa	Araorun	Ile	Itutu	Erinle
1	2	3	4	5	6	7
JA23	JA24	JA25	JA26	JA27	JA28	JA29
AMN7	AMN8	AMN9	AMN10	AMN11	AMN12	AMN13
TIR14	TIR15	TIR16	TIR17	TIR18	TIR19	TIR20
8	9	10	11	12	13	14
JA30	JA31	F1	F2	F3	F4	F5
AMN14	AMN15	AMN16	AMN17	AMN18	AMN19	AMN20
TIR21	TIR22	TIR23	TIR24	TIR25	TIR26	TIR27
15	16	17	18	19	20	21
F6	F7	F8	F9	F10	F 11	F12
AMN21	AMN22	AMN23	AMN24	AMN25	AMN26	AMN28
TIR28	TIR29	TIR30	YEK1	YEK2	YEK3	YEK4
22	23	24	25	26	27	28
F13	F14	F15	F16	F17	F18	F19
AMN29	AMN30	RNT1	RNT2	RNT3	RNT4	RNT5
YEK5	YEK6	YEK7	YEK8	YEK9	YEK10	YEK11
SHAHIDI 50						
F20						
RNT6						
YEK12						

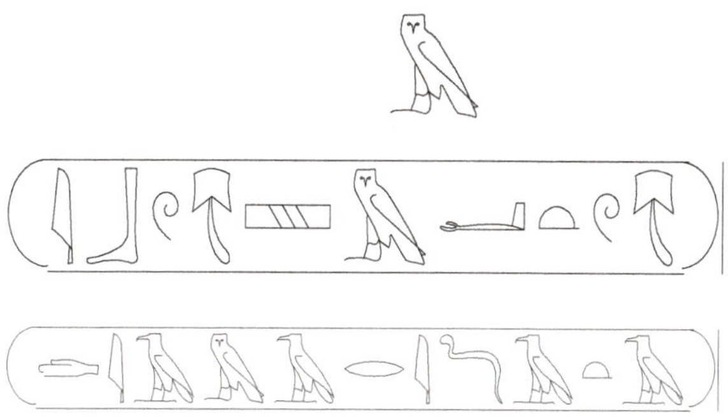

THE NU AFRIKAN CALENDAR

MAAFA 500 B.S.E./ASIENTO 2016-17 C.E./
KAM 6252-53 S.M/EC 2008-09 A.M. (ዓ.ም.)

AFU 500/2016-17/6252-53/2008-09

Ashe	Ojobo	Iwa	Araorun	Ile	Itutu	Erinle
1	2	3	4	5	6	7
F21	F22	F23	F24	F25	F26	F27
RNT7	RNT8	RNT9	RNT10	RNT11	RNT12	RNT13
YEK13	YEK14	YEK15	YEK16	YEK17	YEK18	YEK19
8	9	10	11	12	13	14
F28	F29	M1	M2	M3	M4	M5
RNT14	RNT15	RNT16	RNT17	RNT18	RNT19	RNT20
YEK 20	YEK21	YEK22	YEK23	YEK24	YEK25	YEK26
15	16	17	18	19	20	21
M6	M7	M8	M9	M10	M11	M12
RNT21	RNT22	RNT23	RNT24	RNT25	RNT26	RNT27
YEK 27	YEK28	YEK29	YEK30	MAG1	MAG2	MAG3
22	23	24	25	26	27	28
M13	M14	M15	M16	M17	M18	M19
RNT28	RNT29	RNT30	KHN1	KHN2	KHN3	KHN4
MAG4	MAG5	MAG6	MAG7	MAG8	MAG9	MAG10
SHARIFU 122						
M20						
KHN5						
MAG11						

UHURU 500/2016-17/6252-53/2008-09

Ashe	Ojobo	Iwa	Araorun	Ile	Itutu	Erinle
1	2	3	4	5	6	7
M21	M22	M23	M24	M25	M26	M27
KHN6	KHN7	KHN8	KHN9	KHN10	KHN11	KHN12
MAG12	MAG13	MAG14	MAG15	MAG16	MAG17	MAG18
8	9	10	11	12	13	14
M28	M29	M30	M31	A1	A2	A3
KHN13	KHN14	KHN15	KHN16	KHN17	KHN18	KHN19
MAG19	MAG20	MAG21	MAG22	MAG23	MAG24	MAG25
15	16	17	18	19	20	21
A4	A5	A6	A7	A8	A9	A10
KHN20	KHN21	KHN222	KHN23	KHN24	KHN25	KHN26
MAG26	MAG27	MAG28	MAG29	MAG30	MIY1	MIY2
22	23	24	25	26	27	28
A11	A12	A13	A14	A15	A16	A17
KHN27	KHN28	KHN29	KHN30	PIN1	PIN2	PIN3
MIY3	MIY4	MIY5	MIY6	MIY7	MIY8	MIY9

KAWAIDA 500/2016-17/6252-53/2008-09

Ashe	Ojobo	Iwa	Araorun	Ile	Itutu	Erinle
1	2	3	4	5	6	7
A18	A19	A20	A21	A22	A23	A24
PIN4	PIN5	PIN6	PIN7	PIN8	PIN9	PIN10
MIY10	MIY11	MIY12	MIY13	MIY14	MIY15	MIY16
8	9	10	11	12	13	14
A25	A26	A27	A28	A29	A30	MA1
PIN11	PIN12	PIN13	PIN14	PIN15	PIN16	PIN17
MIY17	MIY18	MIY19	MIY20	MIY21	MIY22	MIY23
15	16	17	18	19	20	21
MA2	MA3	MA4	MA5	MA6	MA7	MA8
PIN18	PIN19	PIN20	PIN21	PIN22	PIN23	PIN24
MIY24	MIY25	MIY26	MIY27	MIY28	MIY29	MIY30
22	23	24	25	26	27	28
MA9	MA10	MA11	MA12	MA13	MA14	MA15
PIN25	PIN26	PIN27	PIN28	PIN29	PIN30	APP1
GIN1	GIN2	GIN3	GIN4	GIN5	GIN6	GIN7

MABABA WEUSI 500/2016-17/6252-53/2008-09

Ashe	Ojobo	Iwa	Araorun	Ile	Itutu	Erinle
1	2	3	4	5	6	7
MA16	MA17	MA18	MA19	MA20	MA21	MA22
APP2	APP3	APP4	APP5	APP6	APP7	APP8
GIN8	GIN9	GIN10	GIN11	GIN12	GIN13	GIN14
8	9	10	11	12	13	14
MA23	MA24	MA25	MA26	MA27	MA28	MA29
APP9	APP10	APP11	APP12	APP13	APP14	APP15
GIN15	GIN16	GIN17	GIN18	GIN19	GIN20	GIN21
15	16	17	18	19	20	21
MA30	MA31	J1	J2	J3	J4	J5
APP16	APP17	APP18	APP19	APP20	APP21	APP22
GIN22	GIN23	GIN24	GIN25	GIN26	GIN27	GIN28
22	23	24	25	26	27	28
J6	J7	J8	J9	J10	J11	J12
APP23	APP24	APP25	APP26	APP27	APP28	APP29
GIN29	GIN30	SAN1	SAN2	SAN3	SAN4	SAN5

THE NU AFRIKAN CALENDAR

MWANAMKE WEUSI 500/2016-17/6252-53/2008-09

Ashe	Ojobo	Iwa	Araorun	Ile	Itutu	Erinle
1	2	3	4	5	6	7
J13	J14	J15	J16	J17	J18	J19
APP30	MSR1	MSR2	MSR3	MSR4	MSR5	MSR6
SAN6	SAN7	SAN8	SAN9	SAN10	SAN11	SAN12
8	9	10	11	12	13	14
J20	J21	J22	J23	J24	J25	J26
MSR7	MSR8	MSR9	MSR10	MSR11	MSR12	MSR13
SAN13	SAN14	SAN15	SAN16	SAN17	SAN18	SAN19
15	16	17	18	19	20	21
J27	J28	J29	J30	JU1	JU2	JU3
MSR14	MSR15	MSR16	MSR17	MSR18	MSR19	MSR20
SAN20	SAN21	SAN22	SAN23	SAN24	SAN25	SAN26
22	23	24	25	26	27	28
JU4	JU5	JU6	JU7	JU8	JU9	JU10
MSR21	MRS22	MSR23	MSR24	MSR25	MSR26	MSR27
SAN27	SAN28	SAN29	SAN30	HAM1	HAM2	HAM3

TAIFA WEUSI 500/2016-17/6252-53/2008-09

Ashe	Ojobo	Iwa	Araorun	Ile	Itutu	Erinle
1	2	3	4	5	6	7
AU8	AU9	AU10	AU11	AU12	AU13	AU14
THT20	THT21	THT22	THT23	THT24	THT25	THT26
NAH2	NAH3	NAH4	NAH5	NAH6	NAH7	NAH8
8	9	10	11	12	13	14
AU15	AU16	AU17	AU18	AU19	AU20	AU21
THT27	THT28	THT29	THT30	PAP1	PAP1	PAP2
NAH9	NAH10	NAH11	NAH12	NAH13	NAH14	NAH15
15	16	17	18	19	20	21
AU22	AU23	AU24	AU25	AU26	AU27	AU28
PAP3	PAP4	PAP5	PAP6	PAP7	PAP8	PAP9
NAH16	NAH17	NAH18	NAH19	NAH20	NAH21	NAH22
22	23	24	25	26	27	28
AU29	AU30	AU31	S1	S2	S3	S4
PAP10	PAP11	PAP12	PAP13	PAP14	PAP15	PAP16
NAH23	NAH24	NAH25	NAH26	NAH27	NAH28	NAH29

MTU WEUSI 500/2016-17/6252-53/2008-09

Ashe	Ojobo	Iwa	Araorun	Ile	Itutu	Erinle
1	2	3	4	5	6	7
JU11	JU12	JU13	JU14	JU15	JU16	JU17
MSR28	MSR29	MSR30	*ASR1*	*HRU2*	*STH3*	*AST4*
HAM4	HAM5	HAM6	HAM7	HAM8	HAM9	HAM10
8	9	10	11	12	13	14
JU18	JU19	JU20	JU21	JU22	JU23	JU24
NBH5	*ANU6*	THT1	THT2	THT3	THT4	THT5
HAM11	HAM12	HAM13	HAM14	HAM15	HAM16	HAM17
15	16	17	18	19	20	21
JU25	JU26	JU27	JU28	JU29	JU30	JU31
THT6	THT7	THT8	THT9	THT10	THT11	THT12
HAM18	HAM19	HAM20	HAM21	HAM22	HAM23	HAM24
22	23	24	25	26	27	28
AU1	AU2	AU3	AU4	AU5	AU6	AU7
THT13	THT14	THT15	THT16	THT17	THT18	THT19
HAM25	HAM26	HAM27	HAM28	HAM29	HAM30	NAH1

UWEZO 500/2016-17/6252-53/2008-09

Ashe	Ojobo	Iwa	Araorun	Ile	Itutu	Erinle
1	2	3	4	5	6	7
S5	S6	S7	S8	S9	S10	S11
PAP17	PAP18	PAP19	PAP20	PAP21	PAP22	PAP23
NAH30	*PAG1*	*PAG2*	*PAG3*	*PAG4*	*PAG5*	MAS1
8	9	10	11	12	13	14
S12	S13	S14	S15	S16	S17	S18
PAP24	PAP25	PAP26	PAP27	PAP28	PAP29	PAP30
MAS2	MAS3	MAS4	MAS5	MAS6	MAS7	MAS8
15	16	17	18	19	20	21
S19	S20	S21	S22	S23	S24	S25
HTH1	HTH2	HTH3	HTH4	HTH5	HTH6	HTH7
MAS9	MAS10	MAS11	MAS12	MAS13	MAS14	MAS15
22	23	24	25	26	27	28
S26	S27	S28	S29	S30	O1	O2
HTH8	HTH9	HTH10	HTH11	HTH12	HTH13	HTH14
MAS16	MAS17	MAS18	MAS19	MAS20	MAS21	MAS22

THE NU AFRIKAN CALENDAR

EKUNDU 500/2016-17/6252-53/2008-09						
Ashe	Ojobo	Iwa	Araorun	Ile	Itutu	Erinle
1	2	3	4	5	6	7
O3	O4	O5	O6	O7	O8	O9
HTH15	HTH16	HTH17	HTH18	HTH19	HTH20	HTH21
MAS23	MAS24	MAS25	MAS26	MAS27	MAS28	MAS29
8	9	10	11	12	13	14
O10	O11	O12	O13	O14	O15	O16
HTH22	HTH23	HTH24	HTH25	HTH26	HTH27	HTH28
MAS30	THi1	THi2	THi3	THi4	THi5	THi6
15	16	17	18	19	20	21
O17	O18	O19	O20	O21	O22	O23
HTH29	HTH30	KHK1	KHK2	KHK3	KHK4	KHK5
THi7	THi8	This9	THi10	THi11	THi12	THi13
22	23	24	25	26	27	28
O24	O25	O26	O27	O28	O29	O30
KHK6	KHK7	KHK8	KHK9	KHK10	KHK11	KHK12
Thi14	THi15	THi16	THi17	THi18	THi19	THi20

KIJANI 500/2016-17/6252-53/2008-09						
Ashe	Ojobo	Iwa	Araorun	Ile	Itutu	Erinle
1	2	3	4	5	6	7
N28	N29	N30	D1	D2	D3	D4
TBT11	TBT12	TBT13	TBT14	TBT15	TBT16	TBT17
HID19	HID20	HID21	HID22	HID23	HID24	HID25
8	9	10	11	12	13	14
D5	D6	D7	D8	D9	D10	D11
TBT18	TBT19	TBT20	TBT21	TBT22	TBT23	TBT24
HID26	HID27	HID28	HID29	HID30	TAH1	TAH2
15	16	17	18	19	20	21
D12	D13	D14	D15	D16	D17	D18
TBT25	TBT26	TBT27	TBT28	TBT29	TBT30	MKH1
TAH3	TAH4	TAH5	TAH6	TAH7	TAH8	TAH9
22	23	24	25	26	27	28
D19	D20	D21	D22	D23	D24	D25
MKH2	MKH3	MKH4	MKH5	MKH6	MKH7	MKH8
TAH10	TAH11	TAH12	TAH13	TAH14	TAH15	TAH16

EUSI 500/2016-17/6252-53/2008-09						
Ashe	Ojobo	Iwa	Araorun	Ile	Itutu	Erinle
1	2	3	4	5	6	7
O31	N1	N2	N3	N4	N5	N6
KHK13	KHK14	KHK15	KHK16	KHK17	KHK18	KHK19
THi21	THi22	THi23	THi24	THi25	THi26	THi27
8	9	10	11	12	13	14
N7	N8	N9	N10	N11	N12	N13
KHK20	KHK21	KHK22	KHK23	KHK24	KHK25	KHK26
THi28	THi29	THi30	HID1	HID2	HID3	HID4
15	16	17	18	19	20	21
N14	N15	N16	N17	N18	N19	N20
KHK27	KHK28	KHK29	KHK30	TBT1	TBT2	TBT3
HID5	HID6	HID7	HID8	HID9	HID10	HID11
22	23	24	25	26	27	28
N21	N22	N23	N24	N25	N26	N27
TBT4	TBT5	TBT6	TBT7	TBT8	TBT9	TBT10
HID12	HID13	HID14	HID15	HID16	HID17	HID18

KWANZA 500/2016-17/6252-53/2008-09						
Umoja	Kujichagulia	Ujima	Ujamaa	Nia	Kuumba	Imani
1	2	3	4	5	6	7
D26	D27	D28	D29	D30	D31	JA1
MKH9	MKH10	MKH11	MKH12	MKH13	MKH14	MKH15
TAH17	TAH18	TAH19	TAH20	TAH21	TAH22	TAH23
8	9	10	11	12	13	14
JA2	JA3	JA4	JA5	JA6	JA7	JA8
MKH16	MKH17	MKH18	MKH19	MKH20	MKH21	MKH22
TAH24	TAH25	TAH26	TAH27	TAH28	TAH29	TAH30
15	16	17	18	19	20	21
JA9	JA10	JA11	JA12	JA13	JA14	JA15
MKH23	MKH24	MKH25	KMH26	MKH27	MKH28	MKH29
TIR1	TIR2	TIR3	TIR4	TIR5	TIR6	TIR7
22	23	24	25	26	27	28
JA16	JA17	JA18	JA19	JA20	JA21	JA22
MKH30	AMN1	AMN2	AMN3	AMN4	AMN5	AMN6
TIR8	TIR9	TIR10	TIR11	TIR12	TIR13	TIR14

THE NU AFRIKAN CALENDAR

SHAHIDI 500/2016-17/6252-53/2008-09						
Ashe	Ojobo	Iwa	Araorun	Ile	Itutu	Erinle
1	2	3	4	5	6	7
JA23	JA24	JA25	JA26	JA27	JA28	JA29
AMN7	AMN8	AMN9	AMN10	AMN11	AMN12	AMN13
TIR15	TIR16	TIR17	TIR18	TIR19	TIR20	TIR21
8	9	10	11	12	13	14
JA30	JA31	F1	F2	F3	F4	F5
AMN14	AMN15	AMN16	AMN17	AMN18	AMN19	AMN20
TIR22	TIR23	TIR24	TIR25	TIR26	TIR27	TIR28
15	16	17	18	19	20	21
F6	F7	F8	F9	F10	F 11	F12
AMN21	AMN22	AMN23	AMN24	AMN25	AMN26	AMN28
TIR29	TIR30	YEK1	YEK2	YEK3	YEK4	YEK5
22	23	24	25	26	27	28
F13	F14	F15	F16	F17	F18	F19
AMN29	AMN30	RNT1	RNT2	RNT3	RNT4	RNT5
YEK6	YEK7	YEK8	YEK9	YEK10	YEK11	YEK12
SHAHIDI 51						
F20						
RNT6						
YEK13						

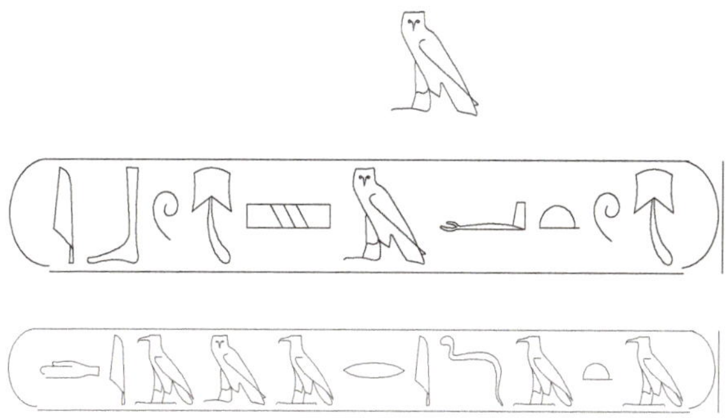

THE NU AFRIKAN CALENDAR

MAAFA 501 B.S.E./ASIENTO 2017-18 C.E.
KAM 6253-54 S.M/EC 2009-10 A.M. (ዓ.ም.)

AFU 501/2017-18/6253-54/2009-10

Ashe	Ojobo	Iwa	Araorun	Ile	Itutu	Erinle
1	2	3	4	5	6	7
F21	F22	F23	F24	F25	F26	F27
RNT7	RNT8	RNT9	RNT10	RNT11	RNT12	RNT13
YEK14	YEK15	YEK16	YEK17	YEK18	YEK19	YEK20
8	9	10	11	12	13	14
F28	M1	M2	M3	M4	M5	M6
RNT14	RNT15	RNT16	RNT17	RNT18	RNT19	RNT20
YEK21	YEK22	YEK23	YEK24	YEK25	YEK26	YEK27
15	16	17	18	19	20	21
M7	M8	M9	M10	M11	M12	M13
RNT21	RNT22	RNT23	RNT24	RNT25	RNT26	RNT27
YEK28	YEK29	YEK30	MAG1	MAG2	MAG3	MAG4
22	23	24	25	26	27	28
M14	M15	M16	M17	M18	M19	M20
RNT28	RNT29	RNT30	KHN1	KHN2	KHN3	KHN4
MAG5	MAG6	MAG7	MAG8	MAG9	MAG10	MAG11

KAWAIDA 501/2017-18/6253-54/2009-10

Ashe	Ojobo	Iwa	Araorun	Ile	Itutu	Erinle
1	2	3	4	5	6	7
A18	A19	A20	A21	A22	A23	A24
PIN3	PIN4	PIN5	PIN6	PIN7	PIN8	PIN9
MIY10	MIY11	MIY12	MIY13	MIY14	MIY15	MIY16
8	9	10	11	12	13	14
A25	A26	A27	A28	A29	A30	MA1
PIN10	PIN11	PIN12	PIN13	PIN14	PIN15	PIN16
MIY17	MIY18	MIY19	MIY20	MIY21	MIY22	MIY23
15	16	17	18	19	20	21
MA2	MA3	MA4	MA5	MA6	MA7	MA8
PIN17	PIN18	PIN19	PIN20	PIN21	PIN22	PIN23
MIY24	MIY25	MIY26	MIY27	MIY28	MIY29	MIY30
22	23	24	25	26	27	28
MA9	MA10	MA11	MA12	MA13	MA14	MA15
PIN24	PIN25	PIN26	PIN27	PIN28	PIN29	PIN30
GIN1	GIN2	GIN3	GIN4	GIN5	GIN6	GIN7

UHURU 501/2017-18/6253-54/2009-10

Ashe	Ojobo	Iwa	Araorun	Ile	Itutu	Erinle
1	2	3	4	5	6	7
M21	M22	M23	M24	M25	M26	M27
KHN5	KHN6	KHN7	KHN8	KHN9	KHN10	KHN11
MAG12	MAG13	MAG14	MAG15	MAG16	MAG17	MAG18
8	9	10	11	12	13	14
M28	M29	M30	M31	A1	A2	A3
KHN12	KHN13	KHN14	KHN15	KHN16	KHN17	KHN18
MAG19	MAG20	MAG21	MAG22	MAG23	MAG24	MAG25
15	16	17	18	19	20	21
A4	A5	A6	A7	A8	A9	A10
KHN19	KHN20	KHN21	KHN22	KHN23	KHN24	KHN25
MAG26	MAG27	MAG28	MAG29	MAG30	MIY1	MIY2
22	23	24	25	26	27	28
A11	A12	A13	A14	A15	A16	A17
KHN26	KHN27	KHN28	KHN29	KHN30	PIN1	PIN2
MIY3	MIY4	MIY5	MIY6	MIY7	MIY8	MIY9

MABABA WEUSI 501/2017-18/6253-54/2009-10

Ashe	Ojobo	Iwa	Araorun	Ile	Itutu	Erinle
1	2	3	4	5	6	7
MA16	MA17	MA18	MA19	MA20	MA21	MA22
APP1	APP2	APP3	APP4	APP5	APP6	APP7
GIN8	GIN9	GIN10	GIN11	GIN12	GIN13	GIN14
8	9	10	11	12	13	14
MA23	MA24	MA25	MA26	MA27	MA28	MA29
APP8	APP9	APP10	APP11	APP12	APP13	APP14
GIN15	GIN16	GIN17	GIN18	GIN19	GIN20	GIN21
15	16	17	18	19	20	21
MA30	MA31	J1	J2	J3	J4	J5
APP15	APP16	APP17	APP18	APP19	APP20	APP21
GIN22	GIN23	GIN24	GIN25	GIN26	GIN27	GIN28
22	23	24	25	26	27	28
J6	J7	J8	J9	J10	J11	J12
APP22	APP23	APP24	APP25	APP26	APP27	APP28
GIN29	GIN30	SAN1	SAN2	SAN3	SAN4	SAN5

THE NU AFRIKAN CALENDAR

MWANAMKE WEUSI 501/2017-18/6253-54/2009-10						
Ashe	Ojobo	Iwa	Araorun	Ile	Itutu	Erinle
1	2	3	4	5	6	7
J13	J14	J15	J16	J17	J18	J19
APP29	APP30	MSR1	MSR2	MSR3	MSR4	MSR5
SAN6	SAN7	SAN8	SAN9	SAN10	SAN11	SAN12
8	9	10	11	12	13	14
J20	J21	J22	J23	J24	J25	J26
MSR6	MSR7	MSR8	MSR9	MSR10	MSR11	MSR12
SAN13	SAN14	SAN15	SAN16	SAN17	SAN18	SAN19
15	16	17	18	19	20	21
J27	J28	J29	J30	JU1	JU2	JU3
MSR13	MSR14	MSR15	MSR16	MSR17	MSR18	MSR19
SAN20	SAN21	SAN22	SAN23	SAN24	SAN25	SAN26
22	23	24	25	26	27	28
JU4	JU5	JU6	JU7	JU8	JU9	JU10
MSR20	MRS21	MSR22	MSR23	MSR24	MSR25	MSR26
SAN27	SAN28	SAN29	SAN30	HAM1	HAM2	HAM3

TAIFA WEUSI 501/2017-18/6253-54/2009-10						
Ashe	Ojobo	Iwa	Araorun	Ile	Itutu	Erinle
1	2	3	4	5	6	7
AU8	AU9	AU10	AU11	AU12	AU13	AU14
THT20	THT21	THT22	THT23	THT24	THT25	THT26
NAH2	NAH3	NAH4	NAH5	NAH6	NAH7	NAH8
8	9	10	11	12	13	14
AU15	AU16	AU17	AU18	AU19	AU20	AU21
THT27	THT28	THT29	THT30	PAP1	PAP1	PAP2
NAH9	NAH10	NAH11	NAH12	NAH13	NAH14	NAH15
15	16	17	18	19	20	21
AU22	AU23	AU24	AU25	AU26	AU27	AU28
PAP3	PAP4	PAP5	PAP6	PAP7	PAP8	PAP9
NAH16	NAH17	NAH18	NAH19	NAH20	NAH21	NAH22
22	23	24	25	26	27	28
AU29	AU30	AU31	S1	S2	S3	S4
PAP10	PAP11	PAP12	PAP13	PAP14	PAP15	PAP16
NAH23	NAH24	NAH25	NAH26	NAH27	NAH28	NAH29

MTU WEUSI 501/2017-18/6253-54/2009-10						
Ashe	Ojobo	Iwa	Araorun	Ile	Itutu	Erinle
1	2	3	4	5	6	7
JU11	JU12	JU13	JU14	JU15	JU16	JU17
MSR27	MSR28	MSR29	MSR30	*ASR1*	*HRU2*	*STH3*
HAM4	HAM5	HAM6	HAM7	HAM8	HAM9	HAM10
8	9	10	11	12	13	14
JU18	JU19	JU20	JU21	JU22	JU23	JU24
AST4	*NBH5*	THT1	THT2	THT3	THT4	THT5
HAM11	HAM12	HAM13	AM14	HAM15	HAM16	HAM17
15	16	17	18	19	20	21
JU25	JU26	JU27	JU28	JU29	JU30	JU31
THT6	THT7	THT8	THT9	THT10	THT11	THT12
HAM18	HAM19	HAM20	HAM21	HAM22	HAM23	HAM24
22	23	24	25	26	27	28
AU1	AU2	AU3	AU4	AU5	AU6	AU7
THT13	THT14	THT15	THT16	THT17	THT18	THT19
HAM25	HAM26	HAM27	HAM28	HAM29	HAM30	NAH1

UWEZO 501/2017-18/6253-54/2009-10						
Ashe	Ojobo	Iwa	Araorun	Ile	Itutu	Erinle
1	2	3	4	5	6	7
S5	S6	S7	S8	S9	S10	S11
PAP17	PAP18	PAP19	PAP20	PAP21	PAP22	PAP23
NAH30	*PAG1*	*PAG2*	*PAG3*	*PAG4*	*PAG5*	MAS1
8	9	10	11	12	13	14
S12	S13	S14	S15	S16	S17	S18
PAP24	PAP25	PAP26	PAP27	PAP28	PAP29	PAP30
MAS2	MAS3	MAS4	MAS5	MAS6	MAS7	MAS8
15	16	17	18	19	20	21
S19	S20	S21	S22	S23	S24	S25
HTH1	HTH2	HTH3	HTH4	HTH5	HTH6	HTH7
MAS9	MAS10	MAS11	MAS12	MAS13	MAS14	MAS15
22	23	24	25	26	27	28
S26	S27	S28	S29	S30	O1	O2
HTH8	HTH9	HTH10	HTH11	HTH12	HTH13	HTH14
MAS16	MAS17	MAS18	MAS19	MAS20	MAS21	MAS22

THE NU AFRIKAN CALENDAR

EKUNDU 501/2017-18/6253-54/2009-10						
Ashe	Ojobo	Iwa	Araorun	Ile	Itutu	Erinle
1	2	3	4	5	6	7
O3	O4	O5	O6	O7	O8	O9
HTH15	HTH16	HTH17	HTH18	HTH19	HTH20	HTH21
MAS23	MAS24	MAS25	MAS26	MAS27	MAS28	MAS29
8	9	10	11	12	13	14
O10	O11	O12	O13	O14	O15	O16
HTH22	HTH23	HTH24	HTH25	HTH26	HTH27	HTH28
MAS30	THi1	THi2	THi3	THi4	THi5	THi6
15	16	17	18	19	20	21
O17	O18	O19	O20	O21	O22	O23
HTH29	HTH30	KHK1	KHK2	KHK3	KHK4	KHK5
THi7	THi8	THi9	THi10	THi11	THi12	THi13
22	23	24	25	26	27	28
O24	O25	O26	O27	O28	O29	O30
KHK6	KHK7	KHK8	KHK9	KHK10	KHK11	KHK12
THi14	THi15	THi16	THi17	THi18	THi19	THi20

KIJANI 501/2017-18/6253-54/2009-10						
Ashe	Ojobo	Iwa	Araorun	Ile	Itutu	Erinle
1	2	3	4	5	6	7
N28	N29	N30	D1	D2	D3	D4
TBT11	TBT12	TBT13	TBT14	TBT15	TBT16	TBT17
HID19	HID20	HID21	HID22	HID23	HID24	HID25
8	9	10	11	12	13	14
D5	D6	D7	D8	D9	D10	D11
TBT18	TBT19	TBT20	TBT21	TBT22	TBT23	TBT24
HID26	HID27	HID28	HID29	HID30	TAH1	TAH2
15	16	17	18	19	20	21
D12	D13	D14	D15	D16	D17	D18
TBT25	TBT26	TBT27	TBT28	TBT29	TBT30	MKH1
TAH3	TAH4	TAH5	TAH6	TAH7	TAH8	TAH9
22	23	24	25	26	27	28
D19	D20	D21	D22	D23	D24	D25
MKH2	MKH3	MKH4	MKH5	MKH6	MKH7	MKH8
TAH10	TAH11	TAH12	TAH13	TAH14	TAH15	TAH16

EUSI 501/2017-18/6253-54/2009-10						
Ashe	Ojobo	Iwa	Araorun	Ile	Itutu	Erinle
1	2	3	4	5	6	7
O31	N1	N2	N3	N4	N5	N6
KHK13	KHK14	KHK15	KHK16	KHK17	KHK18	KHK19
THi21	THi22	THi23	THi24	THi25	THi26	THi27
8	9	10	11	12	13	14
N7	N8	N9	N10	N11	N12	N13
KHK20	KHK21	KHK22	KHK23	KHK24	KHK25	KHK26
THi28	THi29	THi30	HID1	HID2	HID3	HID4
15	16	17	18	19	20	21
N14	N15	N16	N17	N18	N19	N20
KHK27	KHK28	KHK29	KHK30	TBT1	TBT2	TBT3
HID5	HID6	HID7	HID8	HID9	HID10	HID11
22	23	24	25	26	27	28
N21	N22	N23	N24	N25	N26	N27
TBT4	TBT5	TBT6	TBT7	TBT8	TBT9	TBT10
HID12	HID13	HID14	HID15	HID16	HID17	HID18

KWANZA 501/2017-18/6253-54/2009-10						
Umoja	Kujichagulia	Ujima	Ujamaa	Nia	Kuumba	Imani
1	2	3	4	5	6	7
D26	D27	D28	D29	D30	D31	JA1
MKH9	MKH10	MKH11	MKH12	MKH13	MKH14	MKH15
TAH17	TAH18	TAH19	TAH20	TAH21	TAH22	TAH23
8	9	10	11	12	13	14
JA2	JA3	JA4	JA5	JA6	JA7	JA8
MKH16	MKH17	MKH18	MKH19	MKH20	MKH21	MKH22
TAH24	TAH25	TAH26	TAH27	TAH28	TAH29	TAH30
15	16	17	18	19	20	21
JA9	JA10	JA11	JA12	JA13	JA14	JA15
MKH23	MKH24	MKH25	KMH26	MKH27	MKH28	MKH29
TIR1	TIR2	TIR3	TIR4	TIR5	TIR6	TIR7
22	23	24	25	26	27	28
JA16	JA17	JA18	JA19	JA20	JA21	JA22
MKH30	AMN1	AMN2	AMN3	AMN4	AMN5	AMN6
TIR8	TIR9	TIR10	TIR11	TIR12	TIR13	TIR14

THE NU AFRIKAN CALENDAR

SHAHIDI 501/2017-18/6253-54/2009-10						
Ashe	Ojobo	Iwa	Araorun	Ile	Itutu	Erinle
1	2	3	4	5	6	7
JA23	JA24	JA25	JA26	JA27	JA28	JA29
AMN7	AMN8	AMN9	AMN10	AMN11	AMN12	AMN13
TIR15	TIR16	TIR17	TIR18	TIR19	TIR20	TIR21
8	9	10	11	12	13	14
JA30	JA31	F1	F2	F3	F4	F5
AMN14	AMN15	AMN16	AMN17	AMN18	AMN19	AMN20
TIR22	TIR23	TIR24	TIR25	TIR26	TIR27	TIR28
15	16	17	18	19	20	21
F6	F7	F8	F9	F10	F 11	F12
AMN21	AMN22	AMN23	AMN24	AMN25	AMN26	AMN28
TIR29	TIR30	YEK1	YEK2	YEK3	YEK4	YEK5
22	23	24	25	26	27	28
F13	F14	F15	F16	F17	F18	F19
AMN29	AMN30	RNT1	RNT2	RNT3	RNT4	RNT5
YEK6	YEK7	YEK8	YEK9	YEK10	YEK11	YEK12
SHAHIDI 52						
F20						
RNT6						
YEK13						

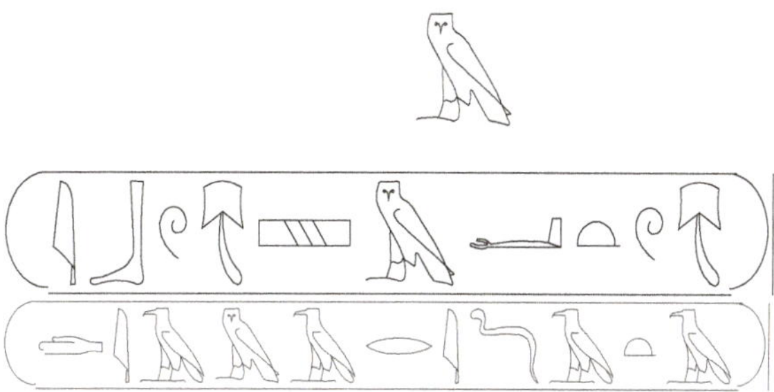

THE NU AFRIKAN CALENDAR

MAAFA 502 B.S.E./ASIENTO 2018-19 C.E.
KAM 6254-55 S.M/EC 2010-11 A.M. (ዓ.ም.)

AFU 502/2018-19/6254-55/2010-11						
Ashe	Ojobo	Iwa	Araorun	Ile	Itutu	Erinle
1	2	3	4	5	6	7
F21	F22	F23	F24	F25	F26	F27
RNT7	RNT8	RNT9	RNT10	RNT11	RNT12	RNT13
YEK14	YEK15	YEK16	YEK17	YEK18	YEK19	YEK20
8	9	10	11	12	13	14
F28	M1	M2	M3	M4	M5	M6
RNT14	RNT15	RNT16	RNT17	RNT18	RNT19	RNT20
YEK21	YEK22	YEK23	YEK24	YEK25	YEK26	YEK27
15	16	17	18	19	20	21
M7	M8	M9	M10	M11	M12	M13
RNT21	RNT22	RNT23	RNT24	RNT25	RNT26	RNT27
YEK28	YEK29	YEK30	MAG1	MAG2	MAG3	MAG4
22	23	24	25	26	27	28
M14	M15	M16	M17	M18	M19	M20
RNT28	RNT29	RNT30	KHN1	KHN2	KHN3	KHN4
MAG5	MAG6	MAG7	MAG8	MAG9	MAG10	MAG11

KAWAIDA 502/2018-19/6254-55/2010-11						
Ashe	Ojobo	Iwa	Araorun	Ile	Itutu	Erinle
1	2	3	4	5	6	7
A18	A19	A20	A21	A22	A23	A24
PIN3	PIN4	PIN5	PIN6	PIN7	PIN8	PIN9
MIY10	MIY11	MIY12	MIY13	MIY14	MIY15	MIY16
8	9	10	11	12	13	14
A25	A26	A27	A28	A29	A30	MA1
PIN10	PIN11	PIN12	PIN13	PIN14	PIN15	PIN16
MIY17	MIY18	MIY19	MIY20	MIY21	MIY22	MIY23
15	16	17	18	19	20	21
MA2	MA3	MA4	MA5	MA6	MA7	MA8
PIN17	PIN18	PIN19	PIN20	PIN21	PIN22	PIN23
MIY24	MIY25	MIY26	MIY27	MIY28	MIY29	MIY30
22	23	24	25	26	27	28
MA9	MA10	MA11	MA12	MA13	MA14	MA15
PIN24	PIN25	PIN26	PIN27	PIN28	PIN29	PIN30
GIN1	GIN2	GIN3	GIN4	GIN5	GIN6	GIN7

UHURU 502/2018-19/6254-55/2010-11						
Ashe	Ojobo	Iwa	Araorun	Ile	Itutu	Erinle
1	2	3	4	5	6	7
M21	M22	M23	M24	M25	M26	M27
KHN5	KHN6	KHN7	KHN8	KHN9	KHN10	KHN11
MAG12	MAG13	MAG14	MAG15	MAG16	MAG17	MAG18
8	9	10	11	12	13	14
M28	M29	M30	M31	A1	A2	A3
KHN12	KHN13	KHN14	KHN15	KHN16	KHN17	KHN18
MAG19	MAG20	MAG21	MAG22	MAG23	MAG24	MAG25
15	16	17	18	19	20	21
A4	A5	A6	A7	A8	A9	A10
KHN19	KHN20	KHN21	KHN22	KHN23	KHN24	KHN25
MAG26	MAG27	MAG28	MAG29	MAG30	MIY1	MIY2
22	23	24	25	26	27	28
A11	A12	A13	A14	A15	A16	A17
KHN26	KHN27	KHN28	KHN29	KHN30	PIN1	PIN2
MIY3	MIY4	MIY5	MIY6	MIY7	MIY8	MIY9

MABABA WEUSI 502/2018-19/6254-55/2010-11						
Ashe	Ojobo	Iwa	Araorun	Ile	Itutu	Erinle
1	2	3	4	5	6	7
MA16	MA17	MA18	MA19	MA20	MA21	MA22
APP1	APP2	APP3	APP4	APP5	APP6	APP7
GIN8	GIN9	GIN10	GIN11	GIN12	GIN13	GIN14
8	9	10	11	12	13	14
MA23	MA24	MA25	MA26	MA27	MA28	MA29
APP8	APP9	APP10	APP11	APP12	APP13	APP14
GIN15	GIN16	GIN17	GIN18	GIN19	GIN20	GIN21
15	16	17	18	19	20	21
MA30	MA31	J1	J2	J3	J4	J5
APP15	APP16	APP17	APP18	APP19	APP20	APP21
GIN22	GIN23	GIN24	GIN25	GIN26	GIN27	GIN28
22	23	24	25	26	27	28
J6	J7	J8	J9	J10	J11	J12
APP22	APP23	APP24	APP25	APP26	APP27	APP28
GIN29	GIN30	SAN1	SAN2	SAN3	SAN4	SAN5

THE NU AFRIKAN CALENDAR

MWANAMKE WEUSI 502/2018-19/6254-55/2010-11						
Ashe	Ojobo	Iwa	Araorun	Ile	Itutu	Erinle
1	2	3	4	5	6	7
J13	J14	J15	J16	J17	J18	J19
APP29	APP30	MSR1	MSR2	MSR3	MSR4	MSR5
SAN6	SAN7	SAN8	SAN9	SAN10	SAN11	SAN12
8	9	10	11	12	13	14
J20	J21	J22	J23	J24	J25	J26
MSR6	MSR7	MSR8	MSR9	MSR10	MSR11	MSR12
SAN13	SAN14	SAN15	SAN16	SAN17	SAN18	SAN19
15	16	17	18	19	20	21
J27	J28	J29	J30	JU1	JU2	JU3
MSR13	MSR14	MSR15	MSR16	MSR17	MSR18	MSR19
SAN20	SAN21	SAN22	SAN23	SAN24	SAN25	SAN26
22	23	24	25	26	27	28
JU4	JU5	JU6	JU7	JU8	JU9	JU10
MSR20	MRS21	MSR22	MSR23	MSR24	MSR25	MSR26
SAN27	SAN28	SAN29	SAN30	HAM1	HAM2	HAM3

TAIFA WEUSI 502/2018-19/6254-55/2010-11						
Ashe	Ojobo	Iwa	Araorun	Ile	Itutu	Erinle
1	2	3	4	5	6	7
AU8	AU9	AU10	AU11	AU12	AU13	AU14
THT20	THT21	THT22	THT23	THT24	THT25	THT26
NAH2	NAH3	NAH4	NAH5	NAH6	NAH7	NAH8
8	9	10	11	12	13	14
AU15	AU16	AU17	AU18	AU19	AU20	AU21
THT27	THT28	THT29	THT30	PAP1	PAP1	PAP2
NAH9	NAH10	NAH11	NAH12	NAH13	NAH14	NAH15
15	16	17	18	19	20	21
AU22	AU23	AU24	AU25	AU26	AU27	AU28
PAP3	PAP4	PAP5	PAP6	PAP7	PAP8	PAP9
NAH16	NAH17	NAH18	NAH19	NAH20	NAH21	NAH22
22	23	24	25	26	27	28
AU29	AU30	AU31	S1	S2	S3	S4
PAP10	PAP11	PAP12	PAP13	PAP14	PAP15	PAP16
NAH23	NAH24	NAH25	NAH26	NAH27	NAH28	NAH29

MTU WEUSI 502/2018-19/6254-55/2010-11						
Ashe	Ojobo	Iwa	Araorun	Ile	Itutu	Erinle
1	2	3	4	5	6	7
JU11	JU12	JU13	JU14	JU15	JU16	JU17
MSR27	MSR28	MSR29	MSR30	*ASR1*	*HRU2*	*STH3*
HAM4	HAM5	HAM6	HAM7	HAM8	HAM9	HAM10
8	9	10	11	12	13	14
JU18	JU19	JU20	JU21	JU22	JU23	JU24
AST4	*NBH5*	THT1	THT2	THT3	THT4	THT5
HAM11	HAM12	HAM13	AM14	HAM15	HAM16	HAM17
15	16	17	18	19	20	21
JU25	JU26	JU27	JU28	JU29	JU30	JU31
THT6	THT7	THT8	THT9	THT10	THT11	THT12
HAM18	HAM19	HAM20	HAM21	HAM22	HAM23	HAM24
22	23	24	25	26	27	28
AU1	AU2	AU3	AU4	AU5	AU6	AU7
THT13	THT14	THT15	THT16	THT17	THT18	THT19
HAM25	HAM26	HAM27	HAM28	HAM29	HAM30	NAH1

UWEZO 502/2018-19/6254-55/2010-11						
Ashe	Ojobo	Iwa	Araorun	Ile	Itutu	Erinle
1	2	3	4	5	6	7
S5	S6	S7	S8	S9	S10	S11
PAP17	PAP18	PAP19	PAP20	PAP21	PAP22	PAP23
NAH30	*PAG1*	*PAG2*	*PAG3*	*PAG4*	*PAG5*	MAS1
8	9	10	11	12	13	14
S12	S13	S14	S15	S16	S17	S18
PAP24	PAP25	PAP26	PAP27	PAP28	PAP29	PAP30
MAS2	MAS3	MAS4	MAS5	MAS6	MAS7	MAS8
15	16	17	18	19	20	21
S19	S20	S21	S22	S23	S24	S25
HTH1	HTH2	HTH3	HTH4	HTH5	HTH6	HTH7
MAS9	MAS10	MAS11	MAS12	MAS13	MAS14	MAS15
22	23	24	25	26	27	28
S26	S27	S28	S29	S30	O1	O2
HTH8	HTH9	HTH10	HTH11	HTH12	HTH13	HTH14
MAS16	MAS17	MAS18	MAS19	MAS20	MAS21	MAS22

THE NU AFRIKAN CALENDAR

EKUNDU 502/2018-19/6254-55/2010-11						
Ashe	Ojobo	Iwa	Araorun	Ile	Itutu	Erinle
1	2	3	4	5	6	7
O3	O4	O5	O6	O7	O8	O9
HTH15	HTH16	HTH17	HTH18	HTH19	HTH20	HTH21
MAS23	MAS24	MAS25	MAS26	MAS27	MAS28	MAS29
8	9	10	11	12	13	14
O10	O11	O12	O13	O14	O15	O16
HTH22	HTH23	HTH24	HTH25	HTH26	HTH27	HTH28
MAS30	THi1	THi2	THi3	THi4	THi5	THi6
15	16	17	18	19	20	21
O17	O18	O19	O20	O21	O22	O23
HTH29	HTH30	KHK1	KHK2	KHK3	KHK4	KHK5
THi7	THi8	THi9	THi10	THi11	THi12	THi13
22	23	24	25	26	27	28
O24	O25	O26	O27	O28	O29	O30
KHK6	KHK7	KHK8	KHK9	KHK10	KHK11	KHK12
THi14	THi15	THi16	THi17	THi18	THi19	THi20

KIJANI 502/2018-19/6254-55/2010-11						
Ashe	Ojobo	Iwa	Araorun	Ile	Itutu	Erinle
1	2	3	4	5	6	7
N28	N29	N30	D1	D2	D3	D4
TBT11	TBT12	TBT13	TBT14	TBT15	TBT16	TBT17
HID19	HID20	HID21	HID22	HID23	HID24	HID25
8	9	10	11	12	13	14
D5	D6	D7	D8	D9	D10	D11
TBT18	TBT19	TBT20	TBT21	TBT22	TBT23	TBT24
HID26	HID27	HID28	HID29	HID30	TAH1	TAH2
15	16	17	18	19	20	21
D12	D13	D14	D15	D16	D17	D18
TBT25	TBT26	TBT27	TBT28	TBT29	TBT30	MKH1
TAH3	TAH4	TAH5	TAH6	TAH7	TAH8	TAH9
22	23	24	25	26	27	28
D19	D20	D21	D22	D23	D24	D25
MKH2	MKH3	MKH4	MKH5	MKH6	MKH7	MKH8
TAH10	TAH11	TAH12	TAH13	TAH14	TAH15	TAH16

EUSI 502/2018-19/6254-55/2010-11						
Ashe	Ojobo	Iwa	Araorun	Ile	Itutu	Erinle
1	2	3	4	5	6	7
O31	N1	N2	N3	N4	N5	N6
KHK13	KHK14	KHK15	KHK16	KHK17	KHK18	KHK19
THi21	THi22	THi23	THi24	THi25	THi26	THi27
8	9	10	11	12	13	14
N7	N8	N9	N10	N11	N12	N13
KHK20	KHK21	KHK22	KHK23	KHK24	KHK25	KHK26
THi28	THi29	THi30	HID1	HID2	HID3	HID4
15	16	17	18	19	20	21
N14	N15	N16	N17	N18	N19	N20
KHK27	KHK28	KHK29	KHK30	TBT1	TBT2	TBT3
HID5	HID6	HID7	HID8	HID9	HID10	HID11
22	23	24	25	26	27	28
N21	N22	N23	N24	N25	N26	N27
TBT4	TBT5	TBT6	TBT7	TBT8	TBT9	TBT10
HID12	HID13	HID14	HID15	HID16	HID17	HID18

KWANZA 502/2018-19/6254-55/2010-11						
Umoja	Kujichagulia	Ujima	Ujamaa	Nia	Kuumba	Imani
1	2	3	4	5	6	7
D26	D27	D28	D29	D30	D31	JA1
MKH9	MKH10	MKH11	MKH12	MKH13	MKH14	MKH15
TAH17	TAH18	TAH19	TAH20	TAH21	TAH22	TAH23
8	9	10	11	12	13	14
JA2	JA3	JA4	JA5	JA6	JA7	JA8
MKH16	MKH17	MKH18	MKH19	MKH20	MKH21	MKH22
TAH24	TAH25	TAH26	TAH27	TAH28	TAH29	TAH30
15	16	17	18	19	20	21
JA9	JA10	JA11	JA12	JA13	JA14	JA15
MKH23	MKH24	MKH25	KMH26	MKH27	MKH28	MKH29
TIR1	TIR2	TIR3	TIR4	TIR5	TIR6	TIR7
22	23	24	25	26	27	28
JA16	JA17	JA18	JA19	JA20	JA21	JA22
MKH30	AMN1	AMN2	AMN3	AMN4	AMN5	AMN6
TIR8	TIR9	TIR10	TIR11	TIR12	TIR13	TIR14

THE NU AFRIKAN CALENDAR

SHAHIDI 502/2018-19/6254-55/2010-11						
Ashe	Ojobo	Iwa	Araorun	Ile	Itutu	Erinle
1	2	3	4	5	6	7
JA23	JA24	JA25	JA26	JA27	JA28	JA29
AMN7	AMN8	AMN9	AMN10	AMN11	AMN12	AMN13
TIR15	TIR16	TIR17	TIR18	TIR19	TIR20	TIR21
8	9	10	11	12	13	14
JA30	JA31	F1	F2	F3	F4	F5
AMN14	AMN15	AMN16	AMN17	AMN18	AMN19	AMN20
TIR22	TIR23	TIR24	TIR25	TIR26	TIR27	TIR28
15	16	17	18	19	20	21
F6	F7	F8	F9	F10	F 11	F12
AMN21	AMN22	AMN23	AMN24	AMN25	AMN26	AMN28
TIR29	TIR30	YEK1	YEK2	YEK3	YEK4	YEK5
22	23	24	25	26	27	28
F13	F14	F15	F16	F17	F18	F19
AMN29	AMN30	RNT1	RNT2	RNT3	RNT4	RNT5
YEK6	YEK7	YEK8	YEK9	YEK10	YEK11	YEK12
SHAHIDI 53						
F20						
RNT6						
YEK13						

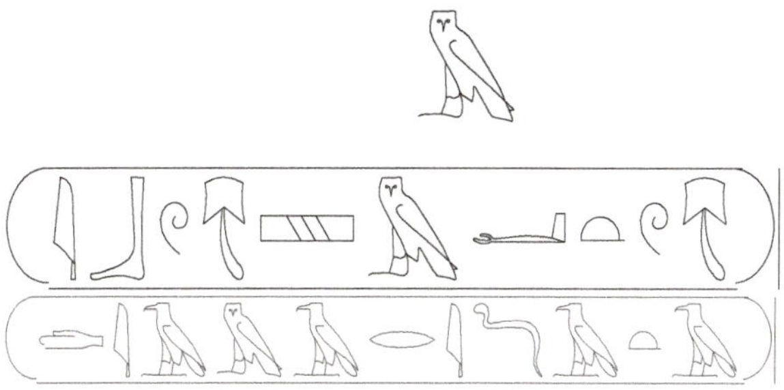

MAAFA 503 B.S.E./ASIENTO 2019-20 C.E./
KAM 6255-56 S.M/EC 2011-12 A.M. (ዓ.ም.)

AFU 503/2019-20/6255-56/2011-12

Ashe	Ojobo	Iwa	Araorun	Ile	Itutu	Erinle
1	2	3	4	5	6	7
F21	F22	F23	F24	F25	F26	F27
RNT7	RNT8	RNT9	RNT10	RNT11	RNT12	RNT13
YEK14	YEK15	YEK16	YEK17	YEK18	YEK19	YEK20
8	9	10	11	12	13	14
F28	M1	M2	M3	M4	M5	M6
RNT14	RNT15	RNT16	RNT17	RNT18	RNT19	RNT20
YEK21	YEK22	YEK23	YEK24	YEK25	YEK26	YEK27
15	16	17	18	19	20	21
M7	M8	M9	M10	M11	M12	M13
RNT21	RNT22	RNT23	RNT24	RNT25	RNT26	RNT27
YEK28	YEK 29	YEK30	MAG1	MAG2	MAG3	MAG4
22	23	24	25	26	27	28
M14	M15	M16	M17	M18	M19	M20
RNT28	RNT29	RNT30	KHN1	KHN2	KHN3	KHN4
MAG5	MAG6	MAG7	MAG8	MAG9	MAG10	MAG11

UHURU 503/2019-20/6255-56/2011-12

Ashe	Ojobo	Iwa	Araorun	Ile	Itutu	Erinle
1	2	3	4	5	6	7
M21	M22	M23	M24	M25	M26	M27
KHN5	KHN6	KHN7	KHN8	KHN9	KHN10	KHN11
MAG12	MAG13	MAG14	MAG15	MAG16	MAG17	MAG18
8	9	10	11	12	13	14
M28	M29	M30	M31	A1	A2	A3
KHN12	KHN13	KHN14	KHN15	KHN16	KHN17	KHN18
MAG19	MAG20	MAG21	MAG22	MAG23	MAG24	MAG25
15	16	17	18	19	20	21
A4	A5	A6	A7	A8	A9	A10
KHN19	KHN20	KHN21	KHN22	KHN23	KHN24	KHN25
MAG26	MAG27	MAG28	MAG29	MAG30	MIY1	MIY2
22	23	24	25	26	27	28
A11	A12	A13	A14	A15	A16	A17
KHN26	KHN27	KHN28	KHN29	KHN30	PIN1	PIN2
MIY3	MIY4	MIY5	MIY6	MIY7	MIY8	MIY9

KAWAIDA 503/2019-20/6255-56/2011-12

Ashe	Ojobo	Iwa	Araorun	Ile	Itutu	Erinle
1	2	3	4	5	6	7
A18	A19	A20	A21	A22	A23	A24
PIN3	PIN4	PIN5	PIN6	PIN7	PIN8	PIN9
MIY10	MIY11	MIY12	MIY13	MIY14	MIY15	MIY16
8	9	10	11	12	13	14
A25	A26	A27	A28	A29	A30	MA1
PIN10	PIN11	PIN12	PIN13	PIN14	PIN15	PIN16
MIY17	MIY18	MIY19	MIY20	MIY21	MIY22	MIY23
15	16	17	18	19	20	21
MA2	MA3	MA4	MA5	MA6	MA7	MA8
PIN17	PIN18	PIN19	PIN20	PIN21	PIN22	PIN23
MIY24	MIY25	MIY26	MIY27	MIY28	MIY29	MIY30
22	23	24	25	26	27	28
MA9	MA10	MA11	MA12	MA13	MA14	MA15
PIN24	PIN25	PIN26	PIN27	PIN28	PIN29	PIN30
GIN1	GIN2	GIN3	GIN4	GIN5	GIN6	GIN7

MABABA WEUSI 503/2019-20/6255-56/2011-12

Ashe	Ojobo	Iwa	Araorun	Ile	Itutu	Erinle
1	2	3	4	5	6	7
MA16	MA17	MA18	MA19	MA20	MA21	MA22
APP1	APP2	APP3	APP4	APP5	APP6	APP7
GIN8	GIN9	GIN10	GIN11	GIN12	GIN13	GIN14
8	9	10	11	12	13	14
MA23	MA24	MA25	MA26	MA27	MA28	MA29
APP8	APP9	APP10	APP11	APP12	APP13	APP14
GIN15	GIN16	GIN17	GIN18	GIN19	GIN20	GIN21
15	16	17	18	19	20	21
MA30	MA31	J1	J2	J3	J4	J5
APP15	APP16	APP17	APP18	APP19	APP20	APP21
GIN22	GIN23	GIN24	GIN25	GIN26	GIN27	GIN28
22	23	24	25	26	27	28
J6	J7	J8	J9	J10	J11	J12
APP22	APP23	APP24	APP25	APP26	APP27	APP28
GIN29	GIN30	SAN1	SAN2	SAN3	SAN4	SAN5

THE NU AFRIKAN CALENDAR

MWANAMKE 503/2019-20/6255-56/2011-12

Ashe	Ojobo	Iwa	Araorun	Ile	Itutu	Erinle
1	2	3	4	5	6	7
J13	J14	J15	J16	J17	J18	J19
APP29	APP30	MSR1	MSR2	MSR3	MSR4	MSR5
SAN6	SAN7	SAN8	SAN9	SAN10	SAN11	SAN12
8	9	10	11	12	13	14
J20	J21	J22	J23	J24	J25	J26
MSR6	MSR7	MSR8	MSR9	MSR10	MSR11	MSR12
SAN13	SAN14	SAN15	SAN16	SAN17	SAN18	SAN19
15	16	17	18	19	20	21
J27	J28	J29	J30	JU1	JU2	JU3
MSR13	MSR14	MSR15	MSR16	MSR17	MSR18	MSR19
SAN20	SAN21	SAN22	SAN23	SAN24	SAN25	SAN26
22	23	24	25	26	27	28
JU4	JU5	JU6	JU7	JU8	JU9	JU10
MSR20	MRS21	MSR22	MSR23	MSR24	MSR25	MSR26
SAN27	SAN28	SAN29	SAN30	HAM1	HAM2	HAM3

TAIFA WEUSI 503/2019-20/6255-56/2011-12

Ashe	Ojobo	Iwa	Araorun	Ile	Itutu	Erinle
1	2	3	4	5	6	7
AU8	AU9	AU10	AU11	AU12	AU13	AU14
THT20	THT21	THT22	THT23	THT24	THT25	THT26
NAH2	NAH3	NAH4	NAH5	NAH6	NAH7	NAH8
8	9	10	11	12	13	14
AU15	AU16	AU17	AU18	AU19	AU20	AU21
THT27	THT28	THT29	THT30	PAP1	PAP1	PAP2
NAH9	NAH10	NAH11	NAH12	NAH13	NAH14	NAH15
15	16	17	18	19	20	21
AU22	AU23	AU24	AU25	AU26	AU27	AU28
PAP3	PAP4	PAP5	PAP6	PAP7	PAP8	PAP9
NAH16	NAH17	NAH18	NAH19	NAH20	NAH21	NAH22
22	23	24	25	26	27	28
AU29	AU30	AU31	S1	S2	S3	S4
PAP10	PAP11	PAP12	PAP13	PAP14	PAP15	PAP16
NAH23	NAH24	NAH25	NAH26	NAH27	NAH28	NAH29

MTU WEUSI 503/2019-20/6255-56/2011-12

Ashe	Ojobo	Iwa	Araorun	Ile	Itutu	Erinle
1	2	3	4	5	6	7
JU11	JU12	JU13	JU14	JU15	JU16	JU17
MSR27	MSR28	MSR29	MSR30	*ASR1*	*HRU2*	*STH3*
HAM4	HAM5	HAM6	HAM7	HAM8	HAM9	HAM10
8	9	10	11	12	13	14
JU18	JU19	JU20	JU21	JU22	JU23	JU24
AST4	*NBH5*	THT1	THT2	THT3	THT4	THT5
HAM11	HAM12	HAM13	HAM14	HAM15	HAM16	HAM17
15	16	17	18	19	20	21
JU25	JU26	JU27	JU28	JU29	JU30	JU31
THT6	THT7	THT8	THT9	THT10	THT11	THT12
HAM18	HAM19	HAM20	HAM21	HAM22	HAM23	HAM24
22	23	24	25	26	27	28
AU1	AU2	AU3	AU4	AU5	AU6	AU7
THT13	THT14	THT15	THT16	THT17	THT18	THT19
HAM25	HAM26	HAM27	HAM28	HAM29	HAM30	NAH1

UWEZO 503/2019-20/6255-56/2011-12

Ashe	Ojobo	Iwa	Araorun	Ile	Itutu	Erinle
1	2	3	4	5	6	7
S5	S6	S7	S8	S9	S10	S11
PAP17	PAP18	PAP19	PAP20	PAP21	PAP22	PAP23
NAH30	*QUA1*	*QUA2*	*QUA3*	*QUA4*	*QUA5*	*QUA6*
8	9	10	11	12	13	14
S12	S13	S14	S15	S16	S17	S18
PAP24	PAP25	PAP26	PAP27	PAP28	PAP29	PAP30
MAS1	MAS2	MAS3	MAS4	MAS5	MAS6	MAS7
15	16	17	18	19	20	21
S19	S20	S21	S22	S23	S24	S25
HTH1	HTH2	HTH3	HTH4	HTH5	HTH6	HTH7
MAS8	MAS9	MAS10	MAS11	MAS12	MAS13	MAS14
22	23	24	25	26	27	28
S26	S27	S28	S29	S30	O1	O2
HTH8	HTH9	HTH10	HTH11	HTH12	HTH13	HTH14
MAS15	MAS16	MAS17	MAS18	MAS19	MAS20	MAS21

THE NU AFRIKAN CALENDAR

EKUNDU 503/2019-20/6255-56/2011-12						
Ashe	Ojobo	Iwa	Araorun	Ile	Itutu	Erinle
1	2	3	4	5	6	7
O3	O4	O5	O6	O7	O8	O9
HTH15	HTH16	HTH17	HTH18	HTH19	HTH20	HTH21
MAS22	MAS23	MAS 24	MAS25	MAS26	MAS27	MAS28
8	9	10	11	12	13	14
O10	O11	O12	O13	O14	O15	O16
HTH22	HTH23	HTH24	HTH25	HTH26	HTH27	HTH28
MAS29	MAS30	THi1	THi2	THi3	THi4	THi5
15	16	17	18	19	20	21
O17	O18	O19	O20	O21	O22	O23
HTH29	HTH30	KHK1	KHK2	KHK3	KHK4	KHK5
THi6	THi7	THi8	THi9	THi10	THi11	THi12
22	23	24	25	26	27	28
O24	O25	O26	O27	O28	O29	O30
KHK6	KHK7	KHK8	KHK9	KHK10	KHK11	KHK12
THi13	THi14	THi15	THi16	THi17	THi18	THi19

KIJANI 503/2019-20/6255-56/2011-12						
Ashe	Ojobo	Iwa	Araorun	Ile	Itutu	Erinle
1	2	3	4	5	6	7
N28	N29	N30	D1	D2	D3	D4
TBT11	TBT12	TBT13	TBT14	TBT15	TBT16	TBT17
HID18	HID19	HID20	HID21	HID22	HID23	HID24
8	9	10	11	12	13	14
D5	D6	D7	D8	D9	D10	D11
TBT18	TBT19	TBT20	TBT21	TBT22	TBT23	TBT24
HID25	HID26	HID27	HID28	HID29	HID30	TAH1
15	16	17	18	19	20	21
D12	D13	D14	D15	D16	D17	D18
TBT25	TBT26	TBT27	TBT28	TBT29	TBT30	MKH1
TAH2	TAH3	TAH4	TAH5	TAH6	TAH7	TAH8
22	23	24	25	26	27	28
D19	D20	D21	D22	D23	D24	D25
MKH2	MKH3	MKH4	MKH5	MKH6	MKH7	MKH8
TAH9	TAH10	TAH11	TAH12	TAH13	TAH14	TAH15

EUSI 503/2019-20/6255-56/2011-12						
Ashe	Ojobo	Iwa	Araorun	Ile	Itutu	Erinle
1	2	3	4	5	6	7
O31	N1	N2	N3	N4	N5	N6
KHK13	KHK14	KHK15	KHK16	KHK17	KHK18	KHK19
THi20	THi21	THi22	THi23	THi24	THi25	THi26
8	9	10	11	12	13	14
N7	N8	N9	N10	N11	N12	N13
KHK20	KHK21	KHK22	KHK23	KHK24	KHK25	KHK26
THi27	THi28	THi29	THi30	HID1	HID2	HID3
15	16	17	18	19	20	21
N14	N15	N16	N17	N18	N19	N20
KHK27	KHK28	KHK29	KHK30	TBT1	TBT2	TBT3
HID4	HID5	HID6	HID7	HID8	HID9	HID10
22	23	24	25	26	27	28
N21	N22	N23	N24	N25	N26	N27
TBT4	TBT5	TBT6	TBT7	TBT8	TBT9	TBT10
HID11	HID12	HID13	HID14	HID15	HID16	HID17

KWANZA 503/2019-20/6255-56/2011-12						
Umoja	Kujichagulia	Ujima	Ujamaa	Nia	Kuumba	Imani
1	2	3	4	5	6	7
D26	D27	D28	D29	D30	D31	JA1
MKH9	MKH10	MKH11	MKH12	MKH13	MKH14	MKH15
TAH16	TAH17	TAH18	TAH19	TAH20	TAH21	TAH22
8	9	10	11	12	13	14
JA2	JA3	JA4	JA5	JA6	JA7	JA8
MKH16	MKH17	MKH18	MKH19	MKH20	MKH21	MKH22
TAH23	TAH24	TAH25	TAH26	TAH27	TAH28	TAH29
15	16	17	18	19	20	21
JA9	JA10	JA11	JA12	JA13	JA14	JA15
MKH23	MKH24	MKH25	KMH26	MKH27	MKH28	MKH29
TAH30	TIR1	TIR2	TIR3	TIR4	TIR5	TIR6
22	23	24	25	26	27	28
JA16	JA17	JA18	JA19	JA20	JA21	JA22
MKH30	AMN1	AMN2	AMN3	AMN4	AMN5	AMN6
TIR7	TIR8	TIR9	TIR10	TIR11	TIR12	TIR13

THE NU AFRIKAN CALENDAR

SHAHIDI 503/2019-20/6255-56/2011-12						
Ashe	Ojobo	Iwa	Araorun	Ile	Itutu	Erinle
1	2	3	4	5	6	7
JA23	JA24	JA25	JA26	JA27	JA28	JA29
AMN7	AMN8	AMN9	AMN10	AMN11	AMN12	AMN13
TIR14	TIR15	TIR16	TIR17	TIR18	TIR19	TIR20
8	9	10	11	12	13	14
JA30	JA31	F1	F2	F3	F4	F5
AMN14	AMN15	AMN16	AMN17	AMN18	AMN19	AMN20
TIR21	TIR22	TIR23	TIR24	TIR25	TIR26	TIR27
15	16	17	18	19	20	21
F6	F7	F8	F9	F10	F 11	F12
AMN21	AMN22	AMN23	AMN24	AMN25	AMN26	AMN28
TIR28	TIR29	TIR30	YEK1	YEK2	YEK3	YEK4
22	23	24	25	26	27	28
F13	F14	F15	F16	F17	F18	F19
AMN29	AMN30	RNT1	RNT2	RNT3	RNT4	RNT5
YEK5	YEK6	YEK7	YEK8	YEK9	YEK10	YEK11
SHAHIDI 54						
F20						
RNT6						
YEK12						

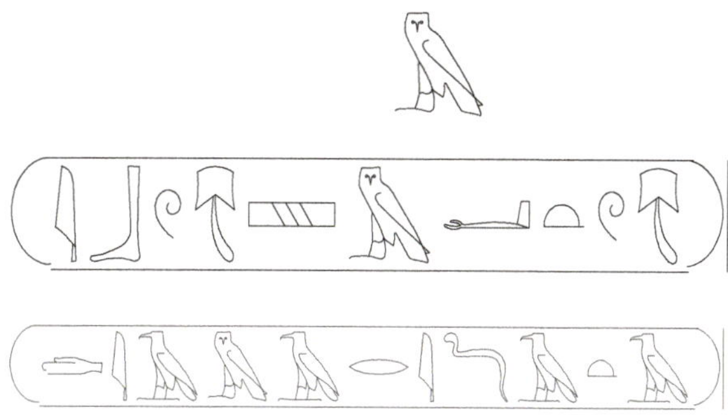

THE NU AFRIKAN CALENDAR

MAAFA 504 B.S.E./ASIENTO 2020-21 C.E./
KAM 6256-57 S.M/EC 2012-13 A.M. (ዓ.ም.)

AFU 504/2020-21/6256-57/2012-13

Ashe	Ojobo	Iwa	Araorun	Ile	Itutu	Erinle
1	2	3	4	5	6	7
F21	F22	F23	F24	F25	F26	F27
RNT7	RNT8	RNT9	RNT10	RNT11	RNT12	RNT13
YEK13	YEK14	YEK15	YEK16	YEK17	YEK18	YEK19
8	9	10	11	12	13	14
F28	F29	M1	M2	M3	M4	M5
RNT14	RNT15	RNT16	RNT17	RNT18	RNT19	RNT20
YEK 20	YEK21	YEK22	YEK23	YEK24	YEK25	YEK26
15	16	17	18	19	20	21
M6	M7	M8	M9	M10	M11	M12
RNT21	RNT22	RNT23	RNT24	RNT25	RNT26	RNT27
YEK 27	YEK28	YEK29	YEK30	MAG1	MAG2	MAG3
22	23	24	25	26	27	28
M13	M14	M15	M16	M17	M18	M19
RNT28	RNT29	RNT30	KHN1	KHN2	KHN3	KHN4
MAG4	MAG5	MAG6	MAG7	MAG8	MAG9	MAG10
SHARIFU 123						
M20						
KHN5						
MAG11						

UHURU 504/2020-21/6256-57/2012-13

Ashe	Ojobo	Iwa	Araorun	Ile	Itutu	Erinle
1	2	3	4	5	6	7
M21	M22	M23	M24	M25	M26	M27
KHN6	KHN7	KHN8	KHN9	KHN10	KHN11	KHN12
MAG12	MAG13	MAG14	MAG15	MAG16	MAG17	MAG18
8	9	10	11	12	13	14
M28	M29	M30	M31	A1	A2	A3
KHN13	KHN14	KHN15	KHN16	KHN17	KHN18	KHN19
MAG19	MAG20	MAG21	MAG22	MAG23	MAG24	MAG25
15	16	17	18	19	20	21
A4	A5	A6	A7	A8	A9	A10
KHN20	KHN21	KHN222	KHN23	KHN24	KHN25	KHN26
MAG26	MAG27	MAG28	MAG29	MAG30	MIY1	MIY2
22	23	24	25	26	27	28
A11	A12	A13	A14	A15	A16	A17
KHN27	KHN28	KHN29	KHN30	PIN1	PIN2	PIN3
MIY3	MIY4	MIY5	MIY6	MIY7	MIY8	MIY9

KAWAIDA 504/2020-21/6256-57/2012-13

Ashe	Ojobo	Iwa	Araorun	Ile	Itutu	Erinle
1	2	3	4	5	6	7
A18	A19	A20	A21	A22	A23	A24
PIN4	PIN5	PIN6	PIN7	PIN8	PIN9	PIN10
MIY10	MIY11	MIY12	MIY13	MIY14	MIY15	MIY16
8	9	10	11	12	13	14
A25	A26	A27	A28	A29	A30	MA1
PIN11	PIN12	PIN13	PIN14	PIN15	PIN16	PIN17
MIY17	MIY18	MIY19	MIY20	MIY21	MIY22	MIY23
15	16	17	18	19	20	21
MA2	MA3	MA4	MA5	MA6	MA7	MA8
PIN18	PIN19	PIN20	PIN21	PIN22	PIN23	PIN24
MIY24	MIY25	MIY26	MIY27	MIY28	MIY29	MIY30
22	23	24	25	26	27	28
MA9	MA10	MA11	MA12	MA13	MA14	MA15
PIN25	PIN26	PIN27	PIN28	PIN29	PIN30	APP1
GIN1	GIN2	GIN3	GIN4	GIN5	GIN6	GIN7

MABABA WEUSI 504/2020-21/6256-57/2012-13

Ashe	Ojobo	Iwa	Araorun	Ile	Itutu	Erinle
1	2	3	4	5	6	7
MA16	MA17	MA18	MA19	MA20	MA21	MA22
APP2	APP3	APP4	APP5	APP6	APP7	APP8
GIN8	GIN9	GIN10	GIN11	GIN12	GIN13	GIN14
8	9	10	11	12	13	14
MA23	MA24	MA25	MA26	MA27	MA28	MA29
APP9	APP10	APP11	APP12	APP13	APP14	APP15
GIN15	GIN16	GIN17	GIN18	GIN19	GIN20	GIN21
15	16	17	18	19	20	21
MA30	MA31	J1	J2	J3	J4	J5
APP16	APP17	APP18	APP19	APP20	APP21	APP22
GIN22	GIN23	GIN24	GIN25	GIN26	GIN27	GIN28
22	23	24	25	26	27	28
J6	J7	J8	J9	J10	J11	J12
APP23	APP24	APP25	APP26	APP27	APP28	APP29
GIN29	GIN30	SAN1	SAN2	SAN3	SAN4	SAN5

THE NU AFRIKAN CALENDAR

MWANAMKE WEUSI 504/2020-21/6256-57/2012-13						
Ashe	Ojobo	Iwa	Araorun	Ile	Itutu	Erinle
1	2	3	4	5	6	7
J13	J14	J15	J16	J17	J18	J19
APP30	MSR1	MSR2	MSR3	MSR4	MSR5	MSR6
SAN6	SAN7	SAN8	SAN9	SAN10	SAN11	SAN12
8	9	10	11	12	13	14
J20	J21	J22	J23	J24	J25	J26
MSR7	MSR8	MSR9	MSR10	MSR11	MSR12	MSR13
SAN13	SAN14	SAN15	SAN16	SAN17	SAN18	SAN19
15	16	17	18	19	20	21
J27	J28	J29	J30	JU1	JU2	JU3
MSR14	MSR15	MSR16	MSR17	MSR18	MSR19	MSR20
SAN20	SAN21	SAN22	SAN23	SAN24	SAN25	SAN26
22	23	24	25	26	27	28
JU4	JU5	JU6	JU7	JU8	JU9	JU10
MSR21	MRS22	MSR23	MSR24	MSR25	MSR26	MSR27
SAN27	SAN28	SAN29	SAN30	HAM1	HAM2	HAM3

TAIFA WEUSI 504/2020-21/6256-57/2012-13						
Ashe	Ojobo	Iwa	Araorun	Ile	Itutu	Erinle
1	2	3	4	5	6	7
AU8	AU9	AU10	AU11	AU12	AU13	AU14
THT20	THT21	THT22	THT23	THT24	THT25	THT26
NAH2	NAH3	NAH4	NAH5	NAH6	NAH7	NAH8
8	9	10	11	12	13	14
AU15	AU16	AU17	AU18	AU19	AU20	AU21
THT27	THT28	THT29	THT30	PAP1	PAP1	PAP2
NAH9	NAH10	NAH11	NAH12	NAH13	NAH14	NAH15
15	16	17	18	19	20	21
AU22	AU23	AU24	AU25	AU26	AU27	AU28
PAP3	PAP4	PAP5	PAP6	PAP7	PAP8	PAP9
NAH16	NAH17	NAH18	NAH19	NAH20	NAH21	NAH22
22	23	24	25	26	27	28
AU29	AU30	AU31	S1	S2	S3	S4
PAP10	PAP11	PAP12	PAP13	PAP14	PAP15	PAP16
NAH23	NAH24	NAH25	NAH26	NAH27	NAH28	NAH29

MTU WEUSI 504/2020-21/6256-57/2012-13						
Ashe	Ojobo	Iwa	Araorun	Ile	Itutu	Erinle
1	2	3	4	5	6	7
JU11	JU12	JU13	JU14	JU15	JU16	JU17
MSR28	MSR29	MSR30	*ASR1*	*HRU2*	*STH3*	*AST4*
HAM4	HAM5	HAM6	HAM7	HAM8	HAM9	HAM10
8	9	10	11	12	13	14
JU18	JU19	JU20	JU21	JU22	JU23	JU24
NBH5	*ANU6*	THT1	THT2	THT3	THT4	THT5
HAM11	HAM12	HAM13	HAM14	HAM15	HAM16	HAM17
15	16	17	18	19	20	21
JU25	JU26	JU27	JU28	JU29	JU30	JU31
THT6	THT7	THT8	THT9	THT10	THT11	THT12
HAM18	HAM19	HAM20	HAM21	HAM22	HAM23	HAM24
22	23	24	25	26	27	28
AU1	AU2	AU3	AU4	AU5	AU6	AU7
THT13	THT14	THT15	THT16	THT17	THT18	THT19
HAM25	HAM26	HAM27	HAM28	HAM29	HAM30	NAH1

UWEZO 504/2020-21/6256-57/2012-13						
Ashe	Ojobo	Iwa	Araorun	Ile	Itutu	Erinle
1	2	3	4	5	6	7
S5	S6	S7	S8	S9	S10	S11
PAP17	PAP18	PAP19	PAP20	PAP21	PAP22	PAP23
NAH30	*PAG1*	*PAG2*	*PAG3*	*PAG4*	*PAG5*	MAS1
8	9	10	11	12	13	14
S12	S13	S14	S15	S16	S17	S18
PAP24	PAP25	PAP26	PAP27	PAP28	PAP29	PAP30
MAS2	MAS3	MAS4	MAS5	MAS6	MAS7	MAS8
15	16	17	18	19	20	21
S19	S20	S21	S22	S23	S24	S25
HTH1	HTH2	HTH3	HTH4	HTH5	HTH6	HTH7
MAS9	MAS10	MAS11	MAS12	MAS13	MAS14	MAS15
22	23	24	25	26	27	28
S26	S27	S28	S29	S30	O1	O2
HTH8	HTH9	HTH10	HTH11	HTH12	HTH13	HTH14
MAS16	MAS17	MAS18	MAS19	MAS20	MAS21	MAS22

THE NU AFRIKAN CALENDAR

EKUNDU 504/2020-21/6256-57/2012-13						
Ashe	Ojobo	Iwa	Araorun	Ile	Itutu	Erinle
1	2	3	4	5	6	7
O3	O4	O5	O6	O7	O8	O9
HTH15	HTH16	HTH17	HTH18	HTH19	HTH20	HTH21
MAS23	MAS24	MAS25	MAS26	MAS27	MAS28	MAS29
8	9	10	11	12	13	14
O10	O11	O12	O13	O14	O15	O16
HTH22	HTH23	HTH24	HTH25	HTH26	HTH27	HTH28
MAS30	THi1	THi2	THi3	THi4	THi5	THi6
15	16	17	18	19	20	21
O17	O18	O19	O20	O21	O22	O23
HTH29	HTH30	KHK1	KHK2	KHK3	KHK4	KHK5
THi7	THi8	This9	THi10	THi11	THi12	THi13
22	23	24	25	26	27	28
O24	O25	O26	O27	O28	O29	O30
KHK6	KHK7	KHK8	KHK9	KHK10	KHK11	KHK12
Thi14	THi15	THi16	THi17	THi18	THi19	THi20

KIJANI 504/2020-21/6256-57/2012-13						
Ashe	Ojobo	Iwa	Araorun	Ile	Itutu	Erinle
1	2	3	4	5	6	7
N28	N29	N30	D1	D2	D3	D4
TBT11	TBT12	TBT13	TBT14	TBT15	TBT16	TBT17
HID19	HID20	HID21	HID22	HID23	HID24	HID25
8	9	10	11	12	13	14
D5	D6	D7	D8	D9	D10	D11
TBT18	TBT19	TBT20	TBT21	TBT22	TBT23	TBT24
HID26	HID27	HID28	HID29	HID30	TAH1	TAH2
15	16	17	18	19	20	21
D12	D13	D14	D15	D16	D17	D18
TBT25	TBT26	TBT27	TBT28	TBT29	TBT30	MKH1
TAH3	TAH4	TAH5	TAH6	TAH7	TAH8	TAH9
22	23	24	25	26	27	28
D19	D20	D21	D22	D23	D24	D25
MKH2	MKH3	MKH4	MKH5	MKH6	MKH7	MKH8
TAH10	TAH11	TAH12	TAH13	TAH14	TAH15	TAH16

EUSI 504/2020-21/6256-57/2012-13						
Ashe	Ojobo	Iwa	Araorun	Ile	Itutu	Erinle
1	2	3	4	5	6	7
O31	N1	N2	N3	N4	N5	N6
KHK13	KHK14	KHK15	KHK16	KHK17	KHK18	KHK19
THi21	THi22	THi23	THi24	THi25	THi26	THi27
8	9	10	11	12	13	14
N7	N8	N9	N10	N11	N12	N13
KHK20	KHK21	KHK22	KHK23	KHK24	KHK25	KHK26
THi28	THi29	THi30	HID1	HID2	HID3	HID4
15	16	17	18	19	20	21
N14	N15	N16	N17	N18	N19	N20
KHK27	KHK28	KHK29	KHK30	TBT1	TBT2	TBT3
HID5	HID6	HID7	HID8	HID9	HID10	HID11
22	23	24	25	26	27	28
N21	N22	N23	N24	N25	N26	N27
TBT4	TBT5	TBT6	TBT7	TBT8	TBT9	TBT10
HID12	HID13	HID14	HID15	HID16	HID17	HID18

KWANZA 504/2020-21/6256-57/2012-13						
Umoja	Kujichagulia	Ujima	Ujamaa	Nia	Kuumba	Imani
1	2	3	4	5	6	7
D26	D27	D28	D29	D30	D31	JA1
MKH9	MKH10	MKH11	MKH12	MKH13	MKH14	MKH15
TAH17	TAH18	TAH19	TAH20	TAH21	TAH22	TAH23
8	9	10	11	12	13	14
JA2	JA3	JA4	JA5	JA6	JA7	JA8
MKH16	MKH17	MKH18	MKH19	MKH20	MKH21	MKH22
TAH24	TAH25	TAH26	TAH27	TAH28	TAH29	TAH30
15	16	17	18	19	20	21
JA9	JA10	JA11	JA12	JA13	JA14	JA15
MKH23	MKH24	MKH25	KMH26	MKH27	MKH28	MKH29
TIR1	TIR2	TIR3	TIR4	TIR5	TIR6	TIR7
22	23	24	25	26	27	28
JA16	JA17	JA18	JA19	JA20	JA21	JA22
MKH30	AMN1	AMN2	AMN3	AMN4	AMN5	AMN6
TIR8	TIR9	TIR10	TIR11	TIR12	TIR13	TIR14

THE NU AFRIKAN CALENDAR

SHAHIDI 504/2020-21/6256-57/2012-13						
Ashe	Ojobo	Iwa	Araorun	Ile	Itutu	Erinle
1	2	3	4	5	6	7
JA23	JA24	JA25	JA26	JA27	JA28	JA29
AMN7	AMN8	AMN9	AMN10	AMN11	AMN12	AMN13
TIR15	TIR16	TIR17	TIR18	TIR19	TIR20	TIR21
8	9	10	11	12	13	14
JA30	JA31	F1	F2	F3	F4	F5
AMN14	AMN15	AMN16	AMN17	AMN18	AMN19	AMN20
TIR22	TIR23	TIR24	TIR25	TIR26	TIR27	TIR28
15	16	17	18	19	20	21
F6	F7	F8	F9	F10	F 11	F12
AMN21	AMN22	AMN23	AMN24	AMN25	AMN26	AMN28
TIR29	TIR30	YEK1	YEK2	YEK3	YEK4	YEK5
22	23	24	25	26	27	28
F13	F14	F15	F16	F17	F18	F19
AMN29	AMN30	RNT1	RNT2	RNT3	RNT4	RNT5
YEK6	YEK7	YEK8	YEK9	YEK10	YEK11	YEK12
SHAHIDI 55						
F20						
RNT6						
YEK13						

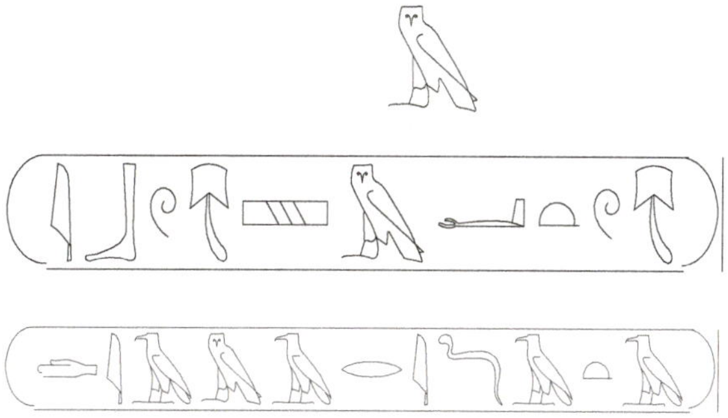

THE NU AFRIKAN CALENDAR

MAAFA 505 B.S.E./ASIENTO 2021-22 C.E.
KAM 6257-58 S.M./EC 2013-14 A.M. (ዓ.ም.)

AFU 505/2021-22/6257-58/2013-14

Ashe	Ojobo	Iwa	Araorun	Ile	Itutu	Erinle
1	2	3	4	5	6	7
F21	F22	F23	F24	F25	F26	F27
RNT7	RNT8	RNT9	RNT10	RNT11	RNT12	RNT13
YEK14	YEK15	YEK16	YEK17	YEK18	YEK19	YEK20
8	9	10	11	12	13	14
F28	M1	M2	M3	M4	M5	M6
RNT14	RNT15	RNT16	RNT17	RNT18	RNT19	RNT20
YEK21	YEK22	YEK23	YEK24	YEK25	YEK26	YEK27
15	16	17	18	19	20	21
M7	M8	M9	M10	M11	M12	M13
RNT21	RNT22	RNT23	RNT24	RNT25	RNT26	RNT27
YEK28	YEK29	YEK30	MAG1	MAG2	MAG3	MAG4
22	23	24	25	26	27	28
M14	M15	M16	M17	M18	M19	M20
RNT28	RNT29	RNT30	KHN1	KHN2	KHN3	KHN4
MAG5	MAG6	MAG7	MAG8	MAG9	MAG10	MAG11

KAWAIDA 505/2021-22/6257-58/2013-14

Ashe	Ojobo	Iwa	Araorun	Ile	Itutu	Erinle
1	2	3	4	5	6	7
A18	A19	A20	A21	A22	A23	A24
PIN3	PIN4	PIN5	PIN6	PIN7	PIN8	PIN9
MIY10	MIY11	MIY12	MIY13	MIY14	MIY15	MIY16
8	9	10	11	12	13	14
A25	A26	A27	A28	A29	A30	MA1
PIN10	PIN11	PIN12	PIN13	PIN14	PIN15	PIN16
MIY17	MIY18	MIY19	MIY20	MIY21	MIY22	MIY23
15	16	17	18	19	20	21
MA2	MA3	MA4	MA5	MA6	MA7	MA8
PIN17	PIN18	PIN19	PIN20	PIN21	PIN22	PIN23
MIY24	MIY25	MIY26	MIY27	MIY28	MIY29	MIY30
22	23	24	25	26	27	28
MA9	MA10	MA11	MA12	MA13	MA14	MA15
PIN24	PIN25	PIN26	PIN27	PIN28	PIN29	PIN30
GIN1	GIN2	GIN3	GIN4	GIN5	GIN6	GIN7

UHURU 505/2021-22/6257-58/2013-14

Ashe	Ojobo	Iwa	Araorun	Ile	Itutu	Erinle
1	2	3	4	5	6	7
M21	M22	M23	M24	M25	M26	M27
KHN5	KHN6	KHN7	KHN8	KHN9	KHN10	KHN11
MAG12	MAG13	MAG14	MAG15	MAG16	MAG17	MAG18
8	9	10	11	12	13	14
M28	M29	M30	M31	A1	A2	A3
KHN12	KHN13	KHN14	KHN15	KHN16	KHN17	KHN18
MAG19	MAG20	MAG21	MAG22	MAG23	MAG24	MAG25
15	16	17	18	19	20	21
A4	A5	A6	A7	A8	A9	A10
KHN19	KHN20	KHN21	KHN22	KHN23	KHN24	KHN25
MAG26	MAG27	MAG28	MAG29	MAG30	MIY1	MIY2
22	23	24	25	26	27	28
A11	A12	A13	A14	A15	A16	A17
KHN26	KHN27	KHN28	KHN29	KHN30	PIN1	PIN2
MIY3	MIY4	MIY5	MIY6	MIY7	MIY8	MIY9

MABABA WEUSI 505/2021-22/6257-58/2013-14

Ashe	Ojobo	Iwa	Araorun	Ile	Itutu	Erinle
1	2	3	4	5	6	7
MA16	MA17	MA18	MA19	MA20	MA21	MA22
APP1	APP2	APP3	APP4	APP5	APP6	APP7
GIN8	GIN9	GIN10	GIN11	GIN12	GIN13	GIN14
8	9	10	11	12	13	14
MA23	MA24	MA25	MA26	MA27	MA28	MA29
APP8	APP9	APP10	APP11	APP12	APP13	APP14
GIN15	GIN16	GIN17	GIN18	GIN19	GIN20	GIN21
15	16	17	18	19	20	21
MA30	MA31	J1	J2	J3	J4	J5
APP15	APP16	APP17	APP18	APP19	APP20	APP21
GIN22	GIN23	GIN24	GIN25	GIN26	GIN27	GIN28
22	23	24	25	26	27	28
J6	J7	J8	J9	J10	J11	J12
APP22	APP23	APP24	APP25	APP26	APP27	APP28
GIN29	GIN30	SAN1	SAN2	SAN3	SAN4	SAN5

THE NU AFRIKAN CALENDAR

MWANAMKE WEUSI 505/2021-22/6257-58/2013-14						
Ashe	Ojobo	Iwa	Araorun	Ile	Itutu	Erinle
1	2	3	4	5	6	7
J13	J14	J15	J16	J17	J18	J19
APP29	APP30	MSR1	MSR2	MSR3	MSR4	MSR5
SAN6	SAN7	SAN8	SAN9	SAN10	SAN11	SAN12
8	9	10	11	12	13	14
J20	J21	J22	J23	J24	J25	J26
MSR6	MSR7	MSR8	MSR9	MSR10	MSR11	MSR12
SAN13	SAN14	SAN15	SAN16	SAN17	SAN18	SAN19
15	16	17	18	19	20	21
J27	J28	J29	J30	JU1	JU2	JU3
MSR13	MSR14	MSR15	MSR16	MSR17	MSR18	MSR19
SAN20	SAN21	SAN22	SAN23	SAN24	SAN25	SAN26
22	23	24	25	26	27	28
JU4	JU5	JU6	JU7	JU8	JU9	JU10
MSR20	MRS21	MSR22	MSR23	MSR24	MSR25	MSR26
SAN27	SAN28	SAN29	SAN30	HAM1	HAM2	HAM3

TAIFA WEUSI 505/2021-22/6257-58/2013-14						
Ashe	Ojobo	Iwa	Araorun	Ile	Itutu	Erinle
1	2	3	4	5	6	7
AU8	AU9	AU10	AU11	AU12	AU13	AU14
THT20	THT21	THT22	THT23	THT24	THT25	THT26
NAH2	NAH3	NAH4	NAH5	NAH6	NAH7	NAH8
8	9	10	11	12	13	14
AU15	AU16	AU17	AU18	AU19	AU20	AU21
THT27	THT28	THT29	THT30	PAP1	PAP1	PAP2
NAH9	NAH10	NAH11	NAH12	NAH13	NAH14	NAH15
15	16	17	18	19	20	21
AU22	AU23	AU24	AU25	AU26	AU27	AU28
PAP3	PAP4	PAP5	PAP6	PAP7	PAP8	PAP9
NAH16	NAH17	NAH18	NAH19	NAH20	NAH21	NAH22
22	23	24	25	26	27	28
AU29	AU30	AU31	S1	S2	S3	S4
PAP10	PAP11	PAP12	PAP13	PAP14	PAP15	PAP16
NAH23	NAH24	NAH25	NAH26	NAH27	NAH28	NAH29

MTU WEUSI 505/2021-22/6257-58/2013-14						
Ashe	Ojobo	Iwa	Araorun	Ile	Itutu	Erinle
1	2	3	4	5	6	7
JU11	JU12	JU13	JU14	JU15	JU16	JU17
MSR27	MSR28	MSR29	MSR30	*ASR1*	*HRU2*	*STH3*
HAM4	HAM5	HAM6	HAM7	HAM8	HAM9	HAM10
8	9	10	11	12	13	14
JU18	JU19	JU20	JU21	JU22	JU23	JU24
AST4	*NBH5*	THT1	THT2	THT3	THT4	THT5
HAM11	HAM12	HAM13	AM14	HAM15	HAM16	HAM17
15	16	17	18	19	20	21
JU25	JU26	JU27	JU28	JU29	JU30	JU31
THT6	THT7	THT8	THT9	THT10	THT11	THT12
HAM18	HAM19	HAM20	HAM21	HAM22	HAM23	HAM24
22	23	24	25	26	27	28
AU1	AU2	AU3	AU4	AU5	AU6	AU7
THT13	THT14	THT15	THT16	THT17	THT18	THT19
HAM25	HAM26	HAM27	HAM28	HAM29	HAM30	NAH1

UWEZO 505/2021-22/6257-58/2013-14						
Ashe	Ojobo	Iwa	Araorun	Ile	Itutu	Erinle
1	2	3	4	5	6	7
S5	S6	S7	S8	S9	S10	S11
PAP17	PAP18	PAP19	PAP20	PAP21	PAP22	PAP23
NAH30	*PAG1*	*PAG2*	*PAG3*	*PAG4*	*PAG5*	MAS1
8	9	10	11	12	13	14
S12	S13	S14	S15	S16	S17	S18
PAP24	PAP25	PAP26	PAP27	PAP28	PAP29	PAP30
MAS2	MAS3	MAS4	MAS5	MAS6	MAS7	MAS8
15	16	17	18	19	20	21
S19	S20	S21	S22	S23	S24	S25
HTH1	HTH2	HTH3	HTH4	HTH5	HTH6	HTH7
MAS9	MAS10	MAS11	MAS12	MAS13	MAS14	MAS15
22	23	24	25	26	27	28
S26	S27	S28	S29	S30	O1	O2
HTH8	HTH9	HTH10	HTH11	HTH12	HTH13	HTH14
MAS16	MAS17	MAS18	MAS19	MAS20	MAS21	MAS22

THE NU AFRIKAN CALENDAR

EKUNDU 505/2021-22/6257-58/2013-14						
Ashe	Ojobo	Iwa	Araorun	Ile	Itutu	Erinle
1	2	3	4	5	6	7
O3	O4	O5	O6	O7	O8	O9
HTH15	HTH16	HTH17	HTH18	HTH19	HTH20	HTH21
MAS23	MAS24	MAS25	MAS26	MAS27	MAS28	MAS29
8	9	10	11	12	13	14
O10	O11	O12	O13	O14	O15	O16
HTH22	HTH23	HTH24	HTH25	HTH26	HTH27	HTH28
MAS30	THi1	THi2	THi3	THi4	THi5	THi6
15	16	17	18	19	20	21
O17	O18	O19	O20	O21	O22	O23
HTH29	HTH30	KHK1	KHK2	KHK3	KHK4	KHK5
THi7	THi8	THi9	THi10	THi11	THi12	THi13
22	23	24	25	26	27	28
O24	O25	O26	O27	O28	O29	O30
KHK6	KHK7	KHK8	KHK9	KHK10	KHK11	KHK12
THi14	THi15	THi16	THi17	THi18	THi19	THi20

KIJANI 505/2021-22/6257-58/2013-14						
Ashe	Ojobo	Iwa	Araorun	Ile	Itutu	Erinle
1	2	3	4	5	6	7
N28	N29	N30	D1	D2	D3	D4
TBT11	TBT12	TBT13	TBT14	TBT15	TBT16	TBT17
HID19	HID20	HID21	HID22	HID23	HID24	HID25
8	9	10	11	12	13	14
D5	D6	D7	D8	D9	D10	D11
TBT18	TBT19	TBT20	TBT21	TBT22	TBT23	TBT24
HID26	HID27	HID28	HID29	HID30	TAH1	TAH2
15	16	17	18	19	20	21
D12	D13	D14	D15	D16	D17	D18
TBT25	TBT26	TBT27	TBT28	TBT29	TBT30	MKH1
TAH3	TAH4	TAH5	TAH6	TAH7	TAH8	TAH9
22	23	24	25	26	27	28
D19	D20	D21	D22	D23	D24	D25
MKH2	MKH3	MKH4	MKH5	MKH6	MKH7	MKH8
TAH10	TAH11	TAH12	TAH13	TAH14	TAH15	TAH16

EUSI 505/2021-22/6257-58/2013-14						
Ashe	Ojobo	Iwa	Araorun	Ile	Itutu	Erinle
1	2	3	4	5	6	7
O31	N1	N2	N3	N4	N5	N6
KHK13	KHK14	KHK15	KHK16	KHK17	KHK18	KHK19
THi21	THi22	THi23	THi24	THi25	THi26	THi27
8	9	10	11	12	13	14
N7	N8	N9	N10	N11	N12	N13
KHK20	KHK21	KHK22	KHK23	KHK24	KHK25	KHK26
THi28	THi29	THi30	HID1	HID2	HID3	HID4
15	16	17	18	19	20	21
N14	N15	N16	N17	N18	N19	N20
KHK27	KHK28	KHK29	KHK30	TBT1	TBT2	TBT3
HID5	HID6	HID7	HID8	HID9	HID10	HID11
22	23	24	25	26	27	28
N21	N22	N23	N24	N25	N26	N27
TBT4	TBT5	TBT6	TBT7	TBT8	TBT9	TBT10
HID12	HID13	HID14	HID15	HID16	HID17	HID18

KWANZA 505/2021-22/6257-58/2013-14						
Umoja	Kujichagulia	Ujima	Ujamaa	Nia	Kuumba	Imani
1	2	3	4	5	6	7
D26	D27	D28	D29	D30	D31	JA1
MKH9	MKH10	MKH11	MKH12	MKH13	MKH14	MKH1
TAH17	TAH18	TAH19	TAH20	TAH21	TAH22	TAH23
8	9	10	11	12	13	14
JA2	JA3	JA4	JA5	JA6	JA7	JA8
MKH16	MKH17	MKH18	MKH19	MKH20	MKH21	MKH2
TAH24	TAH25	TAH26	TAH27	TAH28	TAH29	TAH3
15	16	17	18	19	20	21
JA9	JA10	JA11	JA12	JA13	JA14	JA1
MKH23	MKH24	MKH25	KMH26	MKH27	MKH28	MKH2
TIR1	TIR2	TIR3	TIR4	TIR5	TIR6	TIR7
22	23	24	25	26	27	28
JA16	JA17	JA18	JA19	JA20	JA21	JA2
MKH30	AMN1	AMN2	AMN3	AMN4	AMN5	AMN
TIR8	TIR9	TIR10	TIR11	TIR12	TIR13	TIR14

THE NU AFRIKAN CALENDAR

SHAHIDI 505/2021-22/6257-58/2013-14						
Ashe	Ojobo	Iwa	Araorun	Ile	Itutu	Erinle
1	2	3	4	5	6	7
JA23	JA24	JA25	JA26	JA27	JA28	JA29
AMN7	AMN8	AMN9	AMN10	AMN11	AMN12	AMN13
TIR15	TIR16	TIR17	TIR18	TIR19	TIR20	TIR21
8	9	10	11	12	13	14
JA30	JA31	F1	F2	F3	F4	F5
AMN14	AMN15	AMN16	AMN17	AMN18	AMN19	AMN20
TIR22	TIR23	TIR24	TIR25	TIR26	TIR27	TIR28
15	16	17	18	19	20	21
F6	F7	F8	F9	F10	F 11	F12
AMN21	AMN22	AMN23	AMN24	AMN25	AMN26	AMN28
TIR29	TIR30	YEK1	YEK2	YEK3	YEK4	YEK5
22	23	24	25	26	27	28
F13	F14	F15	F16	F17	F18	F19
AMN29	AMN30	RNT1	RNT2	RNT3	RNT4	RNT5
YEK6	YEK7	YEK8	YEK9	YEK10	YEK11	YEK12
SHAHIDI 56						
F20						
RNT6						
YEK13						

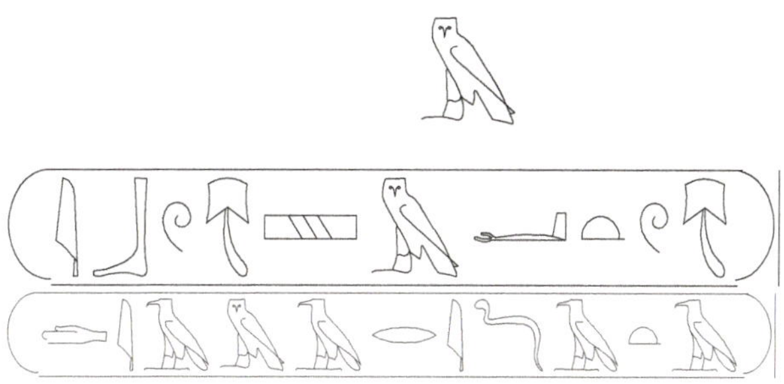

THE NU AFRIKAN CALENDAR

MAAFA 506 B.S.E./ASIENTO 2022-23 C.E.
KAM 6258-59 S.M./EC 2014-15 A.M. (ዓ.ም.)

AFU 506/2022-23/6258-59/2014-15

Ashe	Ojobo	Iwa	Araorun	Ile	Itutu	Erinle
1	2	3	4	5	6	7
F21	F22	F23	F24	F25	F26	F27
RNT7	RNT8	RNT9	RNT10	RNT11	RNT12	RNT13
YEK14	YEK15	YEK16	YEK17	YEK18	YEK19	YEK20
8	9	10	11	12	13	14
F28	M1	M2	M3	M4	M5	M6
RNT14	RNT15	RNT16	RNT17	RNT18	RNT19	RNT20
YEK21	YEK22	YEK23	YEK24	YEK25	YEK26	YEK27
15	16	17	18	19	20	21
M7	M8	M9	M10	M11	M12	M13
RNT21	RNT22	RNT23	RNT24	RNT25	RNT26	RNT27
YEK28	YEK29	YEK30	MAG1	MAG2	MAG3	MAG4
22	23	24	25	26	27	28
M14	M15	M16	M17	M18	M19	M20
RNT28	RNT29	RNT30	KHN1	KHN2	KHN3	KHN4
MAG5	MAG6	MAG7	MAG8	MAG9	MAG10	MAG11

KAWAIDA 506/2022-23/6258-59/2014-15

Ashe	Ojobo	Iwa	Araorun	Ile	Itutu	Erinle
1	2	3	4	5	6	7
A18	A19	A20	A21	A22	A23	A24
PIN3	PIN4	PIN5	PIN6	PIN7	PIN8	PIN9
MIY10	MIY11	MIY12	MIY13	MIY14	MIY15	MIY16
8	9	10	11	12	13	14
A25	A26	A27	A28	A29	A30	MA1
PIN10	PIN11	PIN12	PIN13	PIN14	PIN15	PIN16
MIY17	MIY18	MIY19	MIY20	MIY21	MIY22	MIY23
15	16	17	18	19	20	21
MA2	MA3	MA4	MA5	MA6	MA7	MA8
PIN17	PIN18	PIN19	PIN20	PIN21	PIN22	PIN23
MIY24	MIY25	MIY26	MIY27	MIY28	MIY29	MIY30
22	23	24	25	26	27	28
MA9	MA10	MA11	MA12	MA13	MA14	MA15
PIN24	PIN25	PIN26	PIN27	PIN28	PIN29	PIN30
GIN1	GIN2	GIN3	GIN4	GIN5	GIN6	GIN7

UHURU 506/2022-23/6258-59/2014-15

Ashe	Ojobo	Iwa	Araorun	Ile	Itutu	Erinle
1	2	3	4	5	6	7
M21	M22	M23	M24	M25	M26	M27
KHN5	KHN6	KHN7	KHN8	KHN9	KHN10	KHN11
MAG12	MAG13	MAG14	MAG15	MAG16	MAG17	MAG18
8	9	10	11	12	13	14
M28	M29	M30	M31	A1	A2	A3
KHN12	KHN13	KHN14	KHN15	KHN16	KHN17	KHN18
MAG19	MAG20	MAG21	MAG22	MAG23	MAG24	MAG25
15	16	17	18	19	20	21
A4	A5	A6	A7	A8	A9	A10
KHN19	KHN20	KHN21	KHN22	KHN23	KHN24	KHN25
MAG26	MAG27	MAG28	MAG29	MAG30	MIY1	MIY2
22	23	24	25	26	27	28
A11	A12	A13	A14	A15	A16	A17
KHN26	KHN27	KHN28	KHN29	KHN30	PIN1	PIN2
MIY3	MIY4	MIY5	MIY6	MIY7	MIY8	MIY9

MABABA WEUSI 506/2022-23/6258-59/2014-15

Ashe	Ojobo	Iwa	Araorun	Ile	Itutu	Erinle
1	2	3	4	5	6	7
MA16	MA17	MA18	MA19	MA20	MA21	MA22
APP1	APP2	APP3	APP4	APP5	APP6	APP7
GIN8	GIN9	GIN10	GIN11	GIN12	GIN13	GIN14
8	9	10	11	12	13	14
MA23	MA24	MA25	MA26	MA27	MA28	MA29
APP8	APP9	APP10	APP11	APP12	APP13	APP14
GIN15	GIN16	GIN17	GIN18	GIN19	GIN20	GIN21
15	16	17	18	19	20	21
MA30	MA31	J1	J2	J3	J4	J5
APP15	APP16	APP17	APP18	APP19	APP20	APP21
GIN22	GIN23	GIN24	GIN25	GIN26	GIN27	GIN28
22	23	24	25	26	27	28
J6	J7	J8	J9	J10	J11	J12
APP22	APP23	APP24	APP25	APP26	APP27	APP28
GIN29	GIN30	SAN1	SAN2	SAN3	SAN4	SAN5

THE NU AFRIKAN CALENDAR

MWANAMKE WEUSI 506/2022-23/6258-59/2014-15						
Ashe	Ojobo	Iwa	Araorun	Ile	Itutu	Erinle
1	2	3	4	5	6	7
J13	J14	J15	J16	J17	J18	J19
APP29	APP30	MSR1	MSR2	MSR3	MSR4	MSR5
SAN6	SAN7	SAN8	SAN9	SAN10	SAN11	SAN12
8	9	10	11	12	13	14
J20	J21	J22	J23	J24	J25	J26
MSR6	MSR7	MSR8	MSR9	MSR10	MSR11	MSR12
SAN13	SAN14	SAN15	SAN16	SAN17	SAN18	SAN19
15	16	17	18	19	20	21
J27	J28	J29	J30	JU1	JU2	JU3
MSR13	MSR14	MSR15	MSR16	MSR17	MSR18	MSR19
SAN20	SAN21	SAN22	SAN23	SAN24	SAN25	SAN26
22	23	24	25	26	27	28
JU4	JU5	JU6	JU7	JU8	JU9	JU10
MSR20	MRS21	MSR22	MSR23	MSR24	MSR25	MSR26
SAN27	SAN28	SAN29	SAN30	HAM1	HAM2	HAM3

TAIFA WEUSI 506/2022-23/6258-59/2014-15						
Ashe	Ojobo	Iwa	Araorun	Ile	Itutu	Erinle
1	2	3	4	5	6	7
AU8	AU9	AU10	AU11	AU12	AU13	AU14
THT20	THT21	THT22	THT23	THT24	THT25	THT26
NAH2	NAH3	NAH4	NAH5	NAH6	NAH7	NAH8
8	9	10	11	12	13	14
AU15	AU16	AU17	AU18	AU19	AU20	AU21
THT27	THT28	THT29	THT30	PAP1	PAP1	PAP2
NAH9	NAH10	NAH11	NAH12	NAH13	NAH14	NAH15
15	16	17	18	19	20	21
AU22	AU23	AU24	AU25	AU26	AU27	AU28
PAP3	PAP4	PAP5	PAP6	PAP7	PAP8	PAP9
NAH16	NAH17	NAH18	NAH19	NAH20	NAH21	NAH22
22	23	24	25	26	27	28
AU29	AU30	AU31	S1	S2	S3	S4
PAP10	PAP11	PAP12	PAP13	PAP14	PAP15	PAP16
NAH23	NAH24	NAH25	NAH26	NAH27	NAH28	NAH29

MTU WEUSI 506/2022-23/6258-59/2014-15						
Ashe	Ojobo	Iwa	Araorun	Ile	Itutu	Erinle
1	2	3	4	5	6	7
JU11	JU12	JU13	JU14	JU15	JU16	JU17
MSR27	MSR28	MSR29	MSR30	*ASR1*	*HRU2*	*STH3*
HAM4	HAM5	HAM6	HAM7	HAM8	HAM9	HAM10
8	9	10	11	12	13	14
JU18	JU19	JU20	JU21	JU22	JU23	JU24
AST4	*NBH5*	THT1	THT2	THT3	THT4	THT5
HAM11	HAM12	HAM13	AM14	HAM15	HAM16	HAM17
15	16	17	18	19	20	21
JU25	JU26	JU27	JU28	JU29	JU30	JU31
THT6	THT7	THT8	THT9	THT10	THT11	THT12
HAM18	HAM19	HAM20	HAM21	HAM22	HAM23	HAM24
22	23	24	25	26	27	28
AU1	AU2	AU3	AU4	AU5	AU6	AU7
THT13	THT14	THT15	THT16	THT17	THT18	THT19
HAM25	HAM26	HAM27	HAM28	HAM29	HAM30	NAH1

UWEZO 506/2022-23/6258-59/2014-15						
Ashe	Ojobo	Iwa	Araorun	Ile	Itutu	Erinle
1	2	3	4	5	6	7
S5	S6	S7	S8	S9	S10	S11
PAP17	PAP18	PAP19	PAP20	PAP21	PAP22	PAP23
NAH30	*PAG1*	*PAG2*	*PAG3*	*PAG4*	*PAG5*	MAS1
8	9	10	11	12	13	14
S12	S13	S14	S15	S16	S17	S18
PAP24	PAP25	PAP26	PAP27	PAP28	PAP29	PAP30
MAS2	MAS3	MAS4	MAS5	MAS6	MAS7	MAS8
15	16	17	18	19	20	21
S19	S20	S21	S22	S23	S24	S25
HTH1	HTH2	HTH3	HTH4	HTH5	HTH6	HTH7
MAS9	MAS10	MAS11	MAS12	MAS13	MAS14	MAS15
22	23	24	25	26	27	28
S26	S27	S28	S29	S30	O1	O2
HTH8	HTH9	HTH10	HTH11	HTH12	HTH13	HTH14
MAS16	MAS17	MAS18	MAS19	MAS20	MAS21	MAS22

THE NU AFRIKAN CALENDAR

EKUNDU 506/2022-23/6258-59/2014-15						
Ashe	Ojobo	Iwa	Araorun	Ile	Itutu	Erinle
1	2	3	4	5	6	7
O3	O4	O5	O6	O7	O8	O9
HTH15	HTH16	HTH17	HTH18	HTH19	HTH20	HTH21
MAS23	MAS24	MAS25	MAS26	MAS27	MAS28	MAS29
8	9	10	11	12	13	14
O10	O11	O12	O13	O14	O15	O16
HTH22	HTH23	HTH24	HTH25	HTH26	HTH27	HTH28
MAS30	THi1	THi2	THi3	THi4	THi5	THi6
15	16	17	18	19	20	21
O17	O18	O19	O20	O21	O22	O23
HTH29	HTH30	KHK1	KHK2	KHK3	KHK4	KHK5
THi7	THi8	THi9	THi10	THi11	THi12	THi13
22	23	24	25	26	27	28
O24	O25	O26	O27	O28	O29	O30
KHK6	KHK7	KHK8	KHK9	KHK10	KHK11	KHK12
THi14	THi15	THi16	THi17	THi18	THi19	THi20

KIJANI 506/2022-23/6258-59/2014-15						
Ashe	Ojobo	Iwa	Araorun	Ile	Itutu	Erinle
1	2	3	4	5	6	7
N28	N29	N30	D1	D2	D3	D4
TBT11	TBT12	TBT13	TBT14	TBT15	TBT16	TBT17
HID19	HID20	HID21	HID22	HID23	HID24	HID25
8	9	10	11	12	13	14
D5	D6	D7	D8	D9	D10	D11
TBT18	TBT19	TBT20	TBT21	TBT22	TBT23	TBT24
HID26	HID27	HID28	HID29	HID30	TAH1	TAH2
15	16	17	18	19	20	21
D12	D13	D14	D15	D16	D17	D18
TBT25	TBT26	TBT27	TBT28	TBT29	TBT30	MKH1
TAH3	TAH4	TAH5	TAH6	TAH7	TAH8	TAH9
22	23	24	25	26	27	28
D19	D20	D21	D22	D23	D24	D25
MKH2	MKH3	MKH4	MKH5	MKH6	MKH7	MKH8
TAH10	TAH11	TAH12	TAH13	TAH14	TAH15	TAH16

EUSI 506/2022-23/6258-59/2014-15						
Ashe	Ojobo	Iwa	Araorun	Ile	Itutu	Erinle
1	2	3	4	5	6	7
O31	N1	N2	N3	N4	N5	N6
KHK13	KHK14	KHK15	KHK16	KHK17	KHK18	KHK19
THi21	THi22	THi23	THi24	THi25	THi26	THi27
8	9	10	11	12	13	14
N7	N8	N9	N10	N11	N12	N13
KHK20	KHK21	KHK22	KHK23	KHK24	KHK25	KHK26
THi28	THi29	THi30	HID1	HID2	HID3	HID4
15	16	17	18	19	20	21
N14	N15	N16	N17	N18	N19	N20
KHK27	KHK28	KHK29	KHK30	TBT1	TBT2	TBT3
HID5	HID6	HID7	HID8	HID9	HID10	HID11
22	23	24	25	26	27	28
N21	N22	N23	N24	N25	N26	N27
TBT4	TBT5	TBT6	TBT7	TBT8	TBT9	TBT10
HID12	HID13	HID14	HID15	HID16	HID17	HID18

KWANZA 506/2022-23/6258-59/2014-15						
Umoja	Kujichagulia	Ujima	Ujamaa	Nia	Kuumba	Imani
1	2	3	4	5	6	7
D26	D27	D28	D29	D30	D31	JA1
MKH9	MKH10	MKH11	MKH12	MKH13	MKH14	MKH15
TAH17	TAH18	TAH19	TAH20	TAH21	TAH22	TAH23
8	9	10	11	12	13	14
JA2	JA3	JA4	JA5	JA6	JA7	JA8
MKH16	MKH17	MKH18	MKH19	MKH20	MKH21	MKH22
TAH24	TAH25	TAH26	TAH27	TAH28	TAH29	TAH30
15	16	17	18	19	20	21
JA9	JA10	JA11	JA12	JA13	JA14	JA15
MKH23	MKH24	MKH25	KMH26	MKH27	MKH28	MKH29
TIR1	TIR2	TIR3	TIR4	TIR5	TIR6	TIR7
22	23	24	25	26	27	28
JA16	JA17	JA18	JA19	JA20	JA21	JA22
MKH30	AMN1	AMN2	AMN3	AMN4	AMN5	AMN6
TIR8	TIR9	TIR10	TIR11	TIR12	TIR13	TIR14

SHAHIDI 506/2022-23/6258-59/2014-15						
ne	Ojobo	Iwa	Araorun	Ile	Itutu	Erinle
	2	3	4	5	6	7
23	JA24	JA25	JA26	JA27	JA28	JA29
IN7	AMN8	AMN9	AMN10	AMN11	AMN12	AMN13
R15	TIR16	TIR17	TIR18	TIR19	TIR20	TIR21
	9	10	11	12	13	14
.30	JA31	F1	F2	F3	F4	F5
N14	AMN15	AMN16	AMN17	AMN18	AMN19	AMN20
R22	TIR23	TIR24	TIR25	TIR26	TIR27	TIR28
	16	17	18	19	20	21
	F7	F8	F9	F10	F 11	F12
N21	AMN22	AMN23	AMN24	AMN25	AMN26	AMN28
29	TIR30	YEK1	YEK2	YEK3	YEK4	YEK5
	23	24	25	26	27	28
3	F14	F15	F16	F17	F18	F19
N29	AMN30	RNT1	RNT2	RNT3	RNT4	RNT5
K6	YEK7	YEK8	YEK9	YEK10	YEK11	YEK12
AHIDI 57						
T6						
K13						

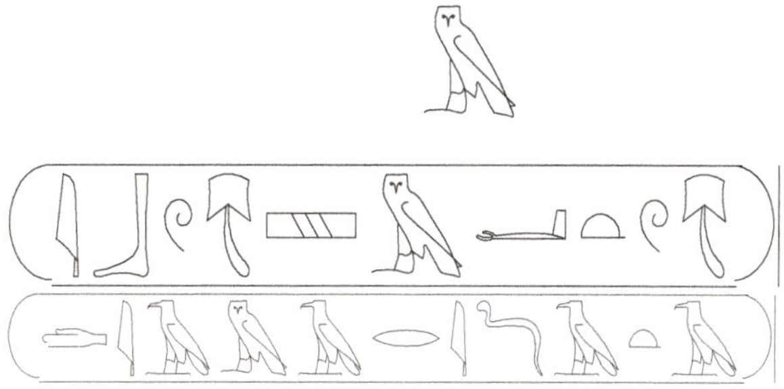

THE NU AFRIKAN CALENDAR

MAAFA 507 B.S.E./ASIENTO 2023-24 C.E./
KAM 6259-60 S.M/EC 2015-16 A.M. (ዓ.ም.)

AFU 507/2023-24/6259-60/2015-16						
Ashe	Ojobo	Iwa	Araorun	Ile	Itutu	Erinle
1	2	3	4	5	6	7
F21	F22	F23	F24	F25	F26	F27
RNT7	RNT8	RNT9	RNT10	RNT11	RNT12	RNT13
YEK14	YEK15	YEK16	YEK17	YEK18	YEK19	YEK20
8	9	10	11	12	13	14
F28	M1	M2	M3	M4	M5	M6
RNT14	RNT15	RNT16	RNT17	RNT18	RNT19	RNT20
YEK21	YEK22	YEK23	YEK24	YEK25	YEK26	YEK27
15	16	17	18	19	20	21
M7	M8	M9	M10	M11	M12	M13
RNT21	RNT22	RNT23	RNT24	RNT25	RNT26	RNT27
YEK28	YEK 29	YEK30	MAG1	MAG2	MAG3	MAG4
22	23	24	25	26	27	28
M14	M15	M16	M17	M18	M19	M20
RNT28	RNT29	RNT30	KHN1	KHN2	KHN3	KHN4
MAG5	MAG6	MAG7	MAG8	MAG9	MAG10	MAG11

KAWAIDA 507/2023-24/6259-60/2015-16						
Ashe	Ojobo	Iwa	Araorun	Ile	Itutu	Erinle
1	2	3	4	5	6	7
A18	A19	A20	A21	A22	A23	A24
PIN3	PIN4	PIN5	PIN6	PIN7	PIN8	PIN9
MIY10	MIY11	MIY12	MIY13	MIY14	MIY15	MIY16
8	9	10	11	12	13	14
A25	A26	A27	A28	A29	A30	MA1
PIN10	PIN11	PIN12	PIN13	PIN14	PIN15	PIN16
MIY17	MIY18	MIY19	MIY20	MIY21	MIY22	MIY23
15	16	17	18	19	20	21
MA2	MA3	MA4	MA5	MA6	MA7	MA8
PIN17	PIN18	PIN19	PIN20	PIN21	PIN22	PIN23
MIY24	MIY25	MIY26	MIY27	MIY28	MIY29	MIY30
22	23	24	25	26	27	28
MA9	MA10	MA11	MA12	MA13	MA14	MA15
PIN24	PIN25	PIN26	PIN27	PIN28	PIN29	PIN30
GIN1	GIN2	GIN3	GIN4	GIN5	GIN6	GIN7

UHURU 507/2023-24/6259-60/2015-16						
Ashe	Ojobo	Iwa	Araorun	Ile	Itutu	Erinle
1	2	3	4	5	6	7
M21	M22	M23	M24	M25	M26	M27
KHN5	KHN6	KHN7	KHN8	KHN9	KHN10	KHN11
MAG12	MAG13	MAG14	MAG15	MAG16	MAG17	MAG18
8	9	10	11	12	13	14
M28	M29	M30	M31	A1	A2	A3
KHN12	KHN13	KHN14	KHN15	KHN16	KHN17	KHN18
MAG19	MAG20	MAG21	MAG22	MAG23	MAG24	MAG25
15	16	17	18	19	20	21
A4	A5	A6	A7	A8	A9	A10
KHN19	KHN20	KHN21	KHN22	KHN23	KHN24	KHN25
MAG26	MAG27	MAG28	MAG29	MAG30	MIY1	MIY2
22	23	24	25	26	27	28
A11	A12	A13	A14	A15	A16	A17
KHN26	KHN27	KHN28	KHN29	KHN30	PIN1	PIN2
MIY3	MIY4	MIY5	MIY6	MIY7	MIY8	MIY9

MABABA WEUSI 507/2023-24/6259-60/2015-16						
Ashe	Ojobo	Iwa	Araorun	Ile	Itutu	Erinle
1	2	3	4	5	6	7
MA16	MA17	MA18	MA19	MA20	MA21	MA22
APP1	APP2	APP3	APP4	APP5	APP6	APP7
GIN8	GIN9	GIN10	GIN11	GIN12	GIN13	GIN14
8	9	10	11	12	13	14
MA23	MA24	MA25	MA26	MA27	MA28	MA29
APP8	APP9	APP10	APP11	APP12	APP13	APP14
GIN15	GIN16	GIN17	GIN18	GIN19	GIN20	GIN21
15	16	17	18	19	20	21
MA30	MA31	J1	J2	J3	J4	J5
APP15	APP16	APP17	APP18	APP19	APP20	APP21
GIN22	GIN23	GIN24	GIN25	GIN26	GIN27	GIN28
22	23	24	25	26	27	28
J6	J7	J8	J9	J10	J11	J12
APP22	APP23	APP24	APP25	APP26	APP27	APP28
GIN29	GIN30	SAN1	SAN2	SAN3	SAN4	SAN5

THE NU AFRIKAN CALENDAR

MWANAMKE 507/2023-24/6259-60/2015-16

Ashe	Ojobo	Iwa	Araorun	Ile	Itutu	Erinle
1	2	3	4	5	6	7
J13	J14	J15	J16	J17	J18	J19
APP29	APP30	MSR1	MSR2	MSR3	MSR4	MSR5
SAN6	SAN7	SAN8	SAN9	SAN10	SAN11	SAN12
8	9	10	11	12	13	14
J20	J21	J22	J23	J24	J25	J26
MSR6	MSR7	MSR8	MSR9	MSR10	MSR11	MSR12
SAN13	SAN14	SAN15	SAN16	SAN17	SAN18	SAN19
15	16	17	18	19	20	21
J27	J28	J29	J30	JU1	JU2	JU3
MSR13	MSR14	MSR15	MSR16	MSR17	MSR18	MSR19
SAN20	SAN21	SAN22	SAN23	SAN24	SAN25	SAN26
22	23	24	25	26	27	28
JU4	JU5	JU6	JU7	JU8	JU9	JU10
MSR20	MRS21	MSR22	MSR23	MSR24	MSR25	MSR26
SAN27	SAN28	SAN29	SAN30	HAM1	HAM2	HAM3

TAIFA WEUSI 507/2023-24/6259-60/2015-16

Ashe	Ojobo	Iwa	Araorun	Ile	Itutu	Erinle
1	2	3	4	5	6	7
AU8	AU9	AU10	AU11	AU12	AU13	AU14
THT20	THT21	THT22	THT23	THT24	THT25	THT26
NAH2	NAH3	NAH4	NAH5	NAH6	NAH7	NAH8
8	9	10	11	12	13	14
AU15	AU16	AU17	AU18	AU19	AU20	AU21
THT27	THT28	THT29	THT30	PAP1	PAP1	PAP2
NAH9	NAH10	NAH11	NAH12	NAH13	NAH14	NAH15
15	16	17	18	19	20	21
AU22	AU23	AU24	AU25	AU26	AU27	AU28
PAP3	PAP4	PAP5	PAP6	PAP7	PAP8	PAP9
NAH16	NAH17	NAH18	NAH19	NAH20	NAH21	NAH22
22	23	24	25	26	27	28
AU29	AU30	AU31	S1	S2	S3	S4
PAP10	PAP11	PAP12	PAP13	PAP14	PAP15	PAP16
NAH23	NAH24	NAH25	NAH26	NAH27	NAH28	NAH29

MTU WEUSI 507/2023-24/6259-60/2015-16

Ashe	Ojobo	Iwa	Araorun	Ile	Itutu	Erinle
1	2	3	4	5	6	7
JU11	JU12	JU13	JU14	JU15	JU16	JU17
MSR27	MSR28	MSR29	MSR30	*ASR1*	*HRU2*	*STH3*
HAM4	HAM5	HAM6	HAM7	HAM8	HAM9	HAM10
8	9	10	11	12	13	14
JU18	JU19	JU20	JU21	JU22	JU23	JU24
AST4	*NBH5*	THT1	THT2	THT3	THT4	THT5
HAM11	HAM12	HAM13	HAM14	HAM15	HAM16	HAM17
15	16	17	18	19	20	21
JU25	JU26	JU27	JU28	JU29	JU30	JU31
THT6	THT7	THT8	THT9	THT10	THT11	THT12
HAM18	HAM19	HAM20	HAM21	HAM22	HAM23	HAM24
22	23	24	25	26	27	28
AU1	AU2	AU3	AU4	AU5	AU6	AU7
THT13	THT14	THT15	THT16	THT17	THT18	THT19
HAM25	HAM26	HAM27	HAM28	HAM29	HAM30	NAH1

UWEZO 507/2023-24/6259-60/2015-16

Ashe	Ojobo	Iwa	Araorun	Ile	Itutu	Erinle
1	2	3	4	5	6	7
S5	S6	S7	S8	S9	S10	S11
PAP17	PAP18	PAP19	PAP20	PAP21	PAP22	PAP23
NAH30	*QUA1*	*QUA2*	*QUA3*	*QUA4*	*QUA5*	*QUA6*
8	9	10	11	12	13	14
S12	S13	S14	S15	S16	S17	S18
PAP24	PAP25	PAP26	PAP27	PAP28	PAP29	PAP30
MAS1	MAS2	MAS3	MAS4	MAS5	MAS6	MAS7
15	16	17	18	19	20	21
S19	S20	S21	S22	S23	S24	S25
HTH1	HTH2	HTH3	HTH4	HTH5	HTH6	HTH7
MAS8	MAS9	MAS10	MAS11	MAS12	MAS13	MAS14
22	23	24	25	26	27	28
S26	S27	S28	S29	S30	O1	O2
HTH8	HTH9	HTH10	HTH11	HTH12	HTH13	HTH14
MAS15	MAS16	MAS17	MAS18	MAS19	MAS20	MAS21

THE NU AFRIKAN CALENDAR

EKUNDU 507/2023-24/6259-60/2015-16

Ashe	Ojobo	Iwa	Araorun	Ile	Itutu	Erinle
1	2	3	4	5	6	7
O3	O4	O5	O6	O7	O8	O9
HTH15	HTH16	HTH17	HTH18	HTH19	HTH20	HTH21
MAS22	MAS23	MAS 24	MAS25	MAS26	MAS27	MAS28
8	9	10	11	12	13	14
O10	O11	O12	O13	O14	O15	O16
HTH22	HTH23	HTH24	HTH25	HTH26	HTH27	HTH28
MAS29	MAS30	THi1	THi2	THi3	THi4	THi5
15	16	17	18	19	20	21
O17	O18	O19	O20	O21	O22	O23
HTH29	HTH30	KHK1	KHK2	KHK3	KHK4	KHK5
THi6	THi7	THi8	THi9	THi10	THi11	THi12
22	23	24	25	26	27	28
O24	O25	O26	O27	O28	O29	O30
KHK6	KHK7	KHK8	KHK9	KHK10	KHK11	KHK12
THi13	THi14	THi15	THi16	THi17	THi18	THi19

KIJANI 507/2023-24/6259-60/2015-16

Ashe	Ojobo	Iwa	Araorun	Ile	Itutu	Erinle
1	2	3	4	5	6	7
N28	N29	N30	D1	D2	D3	D4
TBT11	TBT12	TBT13	TBT14	TBT15	TBT16	TBT17
HID18	HID19	HID20	HID21	HID22	HID23	HID24
8	9	10	11	12	13	14
D5	D6	D7	D8	D9	D10	D11
TBT18	TBT19	TBT20	TBT21	TBT22	TBT23	TBT24
HID25	HID26	HID27	HID28	HID29	HID30	TAH1
15	16	17	18	19	20	21
D12	D13	D14	D15	D16	D17	D18
TBT25	TBT26	TBT27	TBT28	TBT29	TBT30	MKH1
TAH2	TAH3	TAH4	TAH5	TAH6	TAH7	TAH8
22	23	24	25	26	27	28
D19	D20	D21	D22	D23	D24	D25
MKH2	MKH3	MKH4	MKH5	MKH6	MKH7	MKH8
TAH9	TAH10	TAH11	TAH12	TAH13	TAH14	TAH15

EUSI 507/2023-24/6259-60/2015-16

Ashe	Ojobo	Iwa	Araorun	Ile	Itutu	Erinle
1	2	3	4	5	6	7
O31	N1	N2	N3	N4	N5	N6
KHK13	KHK14	KHK15	KHK16	KHK17	KHK18	KHK19
THi20	THi21	THi22	THi23	THi24	THi25	THi26
8	9	10	11	12	13	14
N7	N8	N9	N10	N11	N12	N13
KHK20	KHK21	KHK22	KHK23	KHK24	KHK25	KHK26
THi27	THi28	THi29	THi30	HID1	HID2	HID3
15	16	17	18	19	20	21
N14	N15	N16	N17	N18	N19	N20
KHK27	KHK28	KHK29	KHK30	TBT1	TBT2	TBT3
HID4	HID5	HID6	HID7	HID8	HID9	HID10
22	23	24	25	26	27	28
N21	N22	N23	N24	N25	N26	N27
TBT4	TBT5	TBT6	TBT7	TBT8	TBT9	TBT10
HID11	HID12	HID13	HID14	HID15	HID16	HID17

KWANZA 507/2023-24/6259-60/2015-16

Umoja	Kujichagulia	Ujima	Ujamaa	Nia	Kuumba	Imani
1	2	3	4	5	6	7
D26	D27	D28	D29	D30	D31	JA1
MKH9	MKH10	MKH11	MKH12	MKH13	MKH14	MKH15
TAH16	TAH17	TAH18	TAH19	TAH20	TAH21	TAH22
8	9	10	11	12	13	14
JA2	JA3	JA4	JA5	JA6	JA7	JA8
MKH16	MKH17	MKH18	MKH19	MKH20	MKH21	MKH22
TAH23	TAH24	TAH25	TAH26	TAH27	TAH28	TAH29
15	16	17	18	19	20	21
JA9	JA10	JA11	JA12	JA13	JA14	JA15
MKH23	MKH24	MKH25	KMH26	MKH27	MKH28	MKH29
TAH30	TIR1	TIR2	TIR3	TIR4	TIR5	TIR6
22	23	24	25	26	27	28
JA16	JA17	JA18	JA19	JA20	JA21	JA22
MKH30	AMN1	AMN2	AMN3	AMN4	AMN5	AMN6
TIR7	TIR8	TIR9	TIR10	TIR11	TIR12	TIR13

THE NU AFRIKAN CALENDAR

SHAHIDI 507/2023-24/6259-60/2015-16						
Ashe	Ojobo	Iwa	Araorun	Ile	Itutu	Erinle
1	2	3	4	5	6	7
JA23	JA24	JA25	JA26	JA27	JA28	JA29
AMN7	AMN8	AMN9	AMN10	AMN11	AMN12	AMN13
TIR14	TIR15	TIR16	TIR17	TIR18	TIR19	TIR20
8	9	10	11	12	13	14
JA30	JA31	F1	F2	F3	F4	F5
AMN14	AMN15	AMN16	AMN17	AMN18	AMN19	AMN20
TIR21	TIR22	TIR23	TIR24	TIR25	TIR26	TIR27
15	16	17	18	19	20	21
F6	F7	F8	F9	F10	F 11	F12
AMN21	AMN22	AMN23	AMN24	AMN25	AMN26	AMN28
TIR28	TIR29	TIR30	YEK1	YEK2	YEK3	YEK4
22	23	24	25	26	27	28
F13	F14	F15	F16	F17	F18	F19
AMN29	AMN30	RNT1	RNT2	RNT3	RNT4	RNT5
YEK5	YEK6	YEK7	YEK8	YEK9	YEK10	YEK11
SHAHIDI 58						
F20						
RNT6						
YEK12						

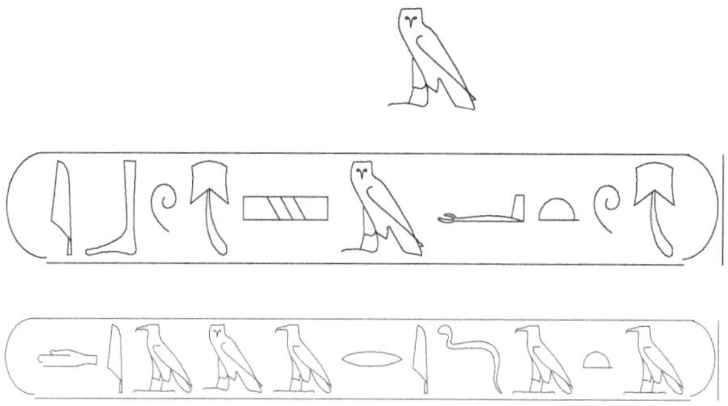

THE NU AFRIKAN CALENDAR

MAAFA 508 B.S.E./ASIENTO 2024-25 C.E./
KAM 6260-61 S.M./EC 2016-17 A.M. (ዓ.ም.)

AFU 508/2024-25/6260-61/2016-17

Ashe	Ojobo	Iwa	Araorun	Ile	Itutu	Erinle
1	2	3	4	5	6	7
F21	F22	F23	F24	F25	F26	F27
RNT7	RNT8	RNT9	RNT10	RNT11	RNT12	RNT13
YEK13	YEK14	YEK15	YEK16	YEK17	YEK18	YEK19
8	9	10	11	12	13	14
F28	F29	M1	M2	M3	M4	M5
RNT14	RNT15	RNT16	RNT17	RNT18	RNT19	RNT20
YEK 20	YEK21	YEK22	YEK23	YEK24	YEK25	YEK26
15	16	17	18	19	20	21
M6	M7	M8	M9	M10	M11	M12
RNT21	RNT22	RNT23	RNT24	RNT25	RNT26	RNT27
YEK 27	YEK28	YEK29	YEK30	MAG1	MAG2	MAG3
22	23	24	25	26	27	28
M13	M14	M15	M16	M17	M18	M19
RNT28	RNT29	RNT30	KHN1	KHN2	KHN3	KHN4
MAG4	MAG5	MAG6	MAG7	MAG8	MAG9	MAG10
SHARIFU 124						
M20						
KHN5						
MAG11						

UHURU 508/2024-25/6260-61/2016-17

Ashe	Ojobo	Iwa	Araorun	Ile	Itutu	Erinle
1	2	3	4	5	6	7
M21	M22	M23	M24	M25	M26	M27
KHN6	KHN7	KHN8	KHN9	KHN10	KHN11	KHN12
MAG12	MAG13	MAG14	MAG15	MAG16	MAG17	MAG18
8	9	10	11	12	13	14
M28	M29	M30	M31	A1	A2	A3
KHN13	KHN14	KHN15	KHN16	KHN17	KHN18	KHN19
MAG19	MAG20	MAG21	MAG22	MAG23	MAG24	MAG25
15	16	17	18	19	20	21
A4	A5	A6	A7	A8	A9	A10
KHN20	KHN21	KHN222	KHN23	KHN24	KHN25	KHN26
MAG26	MAG27	MAG28	MAG29	MAG30	MIY1	MIY2
22	23	24	25	26	27	28
A11	A12	A13	A14	A15	A16	A17
KHN27	KHN28	KHN29	KHN30	PIN1	PIN2	PIN3
MIY3	MIY4	MIY5	MIY6	MIY7	MIY8	MIY9

KAWAIDA 508/2024-25/6260-61/2016-17

Ashe	Ojobo	Iwa	Araorun	Ile	Itutu	Erinle
1	2	3	4	5	6	7
A18	A19	A20	A21	A22	A23	A24
PIN4	PIN5	PIN6	PIN7	PIN8	PIN9	PIN10
MIY10	MIY11	MIY12	MIY13	MIY14	MIY15	MIY16
8	9	10	11	12	13	14
A25	A26	A27	A28	A29	A30	MA1
PIN11	PIN12	PIN13	PIN14	PIN15	PIN16	PIN17
MIY17	MIY18	MIY19	MIY20	MIY21	MIY22	MIY23
15	16	17	18	19	20	21
MA2	MA3	MA4	MA5	MA6	MA7	MA8
PIN18	PIN19	PIN20	PIN21	PIN22	PIN23	PIN24
MIY24	MIY25	MIY26	MIY27	MIY28	MIY29	MIY30
22	23	24	25	26	27	28
MA9	MA10	MA11	MA12	MA13	MA14	MA15
PIN25	PIN26	PIN27	PIN28	PIN29	PIN30	APP1
GIN1	GIN2	GIN3	GIN4	GIN5	GIN6	GIN7

MABABA WEUSI 508/2024-25/6260-61/2016-17

Ashe	Ojobo	Iwa	Araorun	Ile	Itutu	Erinle
1	2	3	4	5	6	7
MA16	MA17	MA18	MA19	MA20	MA21	MA22
APP2	APP3	APP4	APP5	APP6	APP7	APP8
GIN8	GIN9	GIN10	GIN11	GIN12	GIN13	GIN14
8	9	10	11	12	13	14
MA23	MA24	MA25	MA26	MA27	MA28	MA29
APP9	APP10	APP11	APP12	APP13	APP14	APP15
GIN15	GIN16	GIN17	GIN18	GIN19	GIN20	GIN21
15	16	17	18	19	20	21
MA30	MA31	J1	J2	J3	J4	J5
APP16	APP17	APP18	APP19	APP20	APP21	APP22
GIN22	GIN23	GIN24	GIN25	GIN26	GIN27	GIN28
22	23	24	25	26	27	28
J6	J7	J8	J9	J10	J11	J12
APP23	APP24	APP25	APP26	APP27	APP28	APP29
GIN29	GIN30	SAN1	SAN2	SAN3	SAN4	SAN5

THE NU AFRIKAN CALENDAR

MWANAMKE WEUSI 508/2024-25/6260-61/2016-17

Ashe	Ojobo	Iwa	Araorun	Ile	Itutu	Erinle
1	2	3	4	5	6	7
J13	J14	J15	J16	J17	J18	J19
APP30	MSR1	MSR2	MSR3	MSR4	MSR5	MSR6
SAN6	SAN7	SAN8	SAN9	SAN10	SAN11	SAN12
8	9	10	11	12	13	14
J20	J21	J22	J23	J24	J25	J26
MSR7	MSR8	MSR9	MSR10	MSR11	MSR12	MSR13
SAN13	SAN14	SAN15	SAN16	SAN17	SAN18	SAN19
15	16	17	18	19	20	21
J27	J28	J29	J30	JU1	JU2	JU3
MSR14	MSR15	MSR16	MSR17	MSR18	MSR19	MSR20
SAN20	SAN21	SAN22	SAN23	SAN24	SAN25	SAN26
22	23	24	25	26	27	28
JU4	JU5	JU6	JU7	JU8	JU9	JU10
MSR21	MRS22	MSR23	MSR24	MSR25	MSR26	MSR27
SAN27	SAN28	SAN29	SAN30	HAM1	HAM2	HAM3

TAIFA WEUSI 508/2024-25/6260-61/2016-17

Ashe	Ojobo	Iwa	Araorun	Ile	Itutu	Erinle
1	2	3	4	5	6	7
AU8	AU9	AU10	AU11	AU12	AU13	AU14
THT20	THT21	THT22	THT23	THT24	THT25	THT26
NAH2	NAH3	NAH4	NAH5	NAH6	NAH7	NAH8
8	9	10	11	12	13	14
AU15	AU16	AU17	AU18	AU19	AU20	AU21
THT27	THT28	THT29	THT30	PAP1	PAP1	PAP2
NAH9	NAH10	NAH11	NAH12	NAH13	NAH14	NAH15
15	16	17	18	19	20	21
AU22	AU23	AU24	AU25	AU26	AU27	AU28
PAP3	PAP4	PAP5	PAP6	PAP7	PAP8	PAP9
NAH16	NAH17	NAH18	NAH19	NAH20	NAH21	NAH22
22	23	24	25	26	27	28
AU29	AU30	AU31	S1	S2	S3	S4
PAP10	PAP11	PAP12	PAP13	PAP14	PAP15	PAP16
NAH23	NAH24	NAH25	NAH26	NAH27	NAH28	NAH29

MTU WEUSI 508/2024-25/6260-61/2016-17

Ashe	Ojobo	Iwa	Araorun	Ile	Itutu	Erinle
1	2	3	4	5	6	7
JU11	JU12	JU13	JU14	JU15	JU16	JU17
MSR28	MSR29	MSR30	*ASR1*	*HRU2*	*STH3*	*AST4*
HAM4	HAM5	HAM6	HAM7	HAM8	HAM9	HAM10
8	9	10	11	12	13	14
JU18	JU19	JU20	JU21	JU22	JU23	JU24
NBH5	*ANU6*	THT1	THT2	THT3	THT4	THT5
HAM11	HAM12	HAM13	HAM14	HAM15	HAM16	HAM17
15	16	17	18	19	20	21
JU25	JU26	JU27	JU28	JU29	JU30	JU31
THT6	THT7	THT8	THT9	THT10	THT11	THT12
HAM18	HAM19	HAM20	HAM21	HAM22	HAM23	HAM24
22	23	24	25	26	27	28
AU1	AU2	AU3	AU4	AU5	AU6	AU7
THT13	THT14	THT15	THT16	THT17	THT18	THT19
HAM25	HAM26	HAM27	HAM28	HAM29	HAM30	NAH1

UWEZO 508/2024-25/6260-61/2016-17

Ashe	Ojobo	Iwa	Araorun	Ile	Itutu	Erinle
1	2	3	4	5	6	7
S5	S6	S7	S8	S9	S10	S11
PAP17	PAP18	PAP19	PAP20	PAP21	PAP22	PAP23
NAH30	*PAG1*	*PAG2*	*PAG3*	*PAG4*	*PAG5*	MAS1
8	9	10	11	12	13	14
S12	S13	S14	S15	S16	S17	S18
PAP24	PAP25	PAP26	PAP27	PAP28	PAP29	PAP30
MAS2	MAS3	MAS4	MAS5	MAS6	MAS7	MAS8
15	16	17	18	19	20	21
S19	S20	S21	S22	S23	S24	S25
HTH1	HTH2	HTH3	HTH4	HTH5	HTH6	HTH7
MAS9	MAS10	MAS11	MAS12	MAS13	MAS14	MAS15
22	23	24	25	26	27	28
S26	S27	S28	S29	S30	O1	O2
HTH8	HTH9	HTH10	HTH11	HTH12	HTH13	HTH14
MAS16	MAS17	MAS18	MAS19	MAS20	MAS21	MAS22

THE NU AFRIKAN CALENDAR

EKUNDU 508/2024-25/6260-61/2016-17						
Ashe	Ojobo	Iwa	Araorun	Ile	Itutu	Erinle
1	2	3	4	5	6	7
O3	O4	O5	O6	O7	O8	O9
HTH15	HTH16	HTH17	HTH18	HTH19	HTH20	HTH21
MAS23	MAS24	MAS25	MAS26	MAS27	MAS28	MAS29
8	9	10	11	12	13	14
O10	O11	O12	O13	O14	O15	O16
HTH22	HTH23	HTH24	HTH25	HTH26	HTH27	HTH28
MAS30	THi1	THi2	THi3	THi4	THi5	THi6
15	16	17	18	19	20	21
O17	O18	O19	O20	O21	O22	O23
HTH29	HTH30	KHK1	KHK2	KHK3	KHK4	KHK5
THi7	THi8	This9	THi10	THi11	THi12	THi13
22	23	24	25	26	27	28
O24	O25	O26	O27	O28	O29	O30
KHK6	KHK7	KHK8	KHK9	KHK10	KHK11	KHK12
Thi14	THi15	THi16	THi17	THi18	THi19	THi20

KIJANI 508/2024-25/6260-61/2016-17						
Ashe	Ojobo	Iwa	Araorun	Ile	Itutu	Erinle
1	2	3	4	5	6	7
N28	N29	N30	D1	D2	D3	D4
TBT11	TBT12	TBT13	TBT14	TBT15	TBT16	TBT17
HID19	HID20	HID21	HID22	HID23	HID24	HID25
8	9	10	11	12	13	14
D5	D6	D7	D8	D9	D10	D11
TBT18	TBT19	TBT20	TBT21	TBT22	TBT23	TBT24
HID26	HID27	HID28	HID29	HID30	TAH1	TAH2
15	16	17	18	19	20	21
D12	D13	D14	D15	D16	D17	D18
TBT25	TBT26	TBT27	TBT28	TBT29	TBT30	MKH1
TAH3	TAH4	TAH5	TAH6	TAH7	TAH8	TAH9
22	23	24	25	26	27	28
D19	D20	D21	D22	D23	D24	D25
MKH2	MKH3	MKH4	MKH5	MKH6	MKH7	MKH8
TAH10	TAH11	TAH12	TAH13	TAH14	TAH15	TAH16

EUSI 508/2024-25/6260-61/2016-17						
Ashe	Ojobo	Iwa	Araorun	Ile	Itutu	Erinle
1	2	3	4	5	6	7
O31	N1	N2	N3	N4	N5	N6
KHK13	KHK14	KHK15	KHK16	KHK17	KHK18	KHK19
THi21	THi22	THi23	THi24	THi25	THi26	THi27
8	9	10	11	12	13	14
N7	N8	N9	N10	N11	N12	N13
KHK20	KHK21	KHK22	KHK23	KHK24	KHK25	KHK26
THi28	THi29	THi30	HID1	HID2	HID3	HID4
15	16	17	18	19	20	21
N14	N15	N16	N17	N18	N19	N20
KHK27	KHK28	KHK29	KHK30	TBT1	TBT2	TBT3
HID5	HID6	HID7	HID8	HID9	HID10	HID11
22	23	24	25	26	27	28
N21	N22	N23	N24	N25	N26	N27
TBT4	TBT5	TBT6	TBT7	TBT8	TBT9	TBT10
HID12	HID13	HID14	HID15	HID16	HID17	HID18

KWANZA 508/2024-25/6260-61/2016-17						
Umoja	Kujichagulia	Ujima	Ujamaa	Nia	Kuumba	Imani
1	2	3	4	5	6	7
D26	D27	D28	D29	D30	D31	JA1
MKH9	MKH10	MKH11	MKH12	MKH13	MKH14	MKH15
TAH17	TAH18	TAH19	TAH20	TAH21	TAH22	TAH23
8	9	10	11	12	13	14
JA2	JA3	JA4	JA5	JA6	JA7	JA8
MKH16	MKH17	MKH18	MKH19	MKH20	MKH21	MKH22
TAH24	TAH25	TAH26	TAH27	TAH28	TAH29	TAH30
15	16	17	18	19	20	21
JA9	JA10	JA11	JA12	JA13	JA14	JA15
MKH23	MKH24	MKH25	KMH26	MKH27	MKH28	MKH29
TIR1	TIR2	TIR3	TIR4	TIR5	TIR6	TIR7
22	23	24	25	26	27	28
JA16	JA17	JA18	JA19	JA20	JA21	JA22
MKH30	AMN1	AMN2	AMN3	AMN4	AMN5	AMN6
TIR8	TIR9	TIR10	TIR11	TIR12	TIR13	TIR14

THE NU AFRIKAN CALENDAR

SHAHIDI 508/2024-25/6260-61/2016-17						
Ashe	Ojobo	Iwa	Araorun	Ile	Itutu	Erinle
1	2	3	4	5	6	7
JA23	JA24	JA25	JA26	JA27	JA28	JA29
AMN7	AMN8	AMN9	AMN10	AMN11	AMN12	AMN13
TIR15	TIR16	TIR17	TIR18	TIR19	TIR20	TIR21
8	9	10	11	12	13	14
JA30	JA31	F1	F2	F3	F4	F5
AMN14	AMN15	AMN16	AMN17	AMN18	AMN19	AMN20
TIR22	TIR23	TIR24	TIR25	TIR26	TIR27	TIR28
15	16	17	18	19	20	21
F6	F7	F8	F9	F10	F 11	F12
AMN21	AMN22	AMN23	AMN24	AMN25	AMN26	AMN28
TIR29	TIR30	YEK1	YEK2	YEK3	YEK4	YEK5
22	23	24	25	26	27	28
F13	F14	F15	F16	F17	F18	F19
AMN29	AMN30	RNT1	RNT2	RNT3	RNT4	RNT5
YEK6	YEK7	YEK8	YEK9	YEK10	YEK11	YEK12
SHAHIDI 59						
F20						
RNT6						
YEK13						

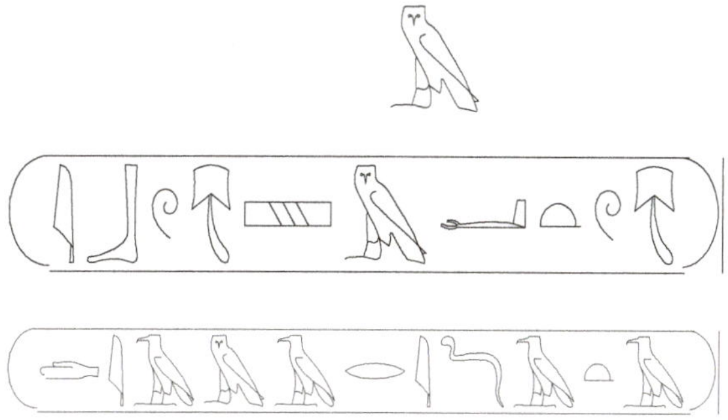

THE NU AFRIKAN CALENDAR

MAAFA 509 B.S.E./ASIENTO 2025-26 C.E.
KAM 6261-62 S.M./EC 2017-18 A.M. (ዓ.ም.)

AFU 509/2025-26/6261-62/2017-18						
Ashe	Ojobo	Iwa	Araorun	Ile	Itutu	Erinle
1	2	3	4	5	6	7
F21	F22	F23	F24	F25	F26	F27
RNT7	RNT8	RNT9	RNT10	RNT11	RNT12	RNT13
YEK14	YEK15	YEK16	YEK17	YEK18	YEK19	YEK20
8	9	10	11	12	13	14
F28	M1	M2	M3	M4	M5	M6
RNT14	RNT15	RNT16	RNT17	RNT18	RNT19	RNT20
YEK21	YEK22	YEK23	YEK24	YEK25	YEK26	YEK27
15	16	17	18	19	20	21
M7	M8	M9	M10	M11	M12	M13
RNT21	RNT22	RNT23	RNT24	RNT25	RNT26	RNT27
YEK28	YEK29	YEK30	MAG1	MAG2	MAG3	MAG4
22	23	24	25	26	27	28
M14	M15	M16	M17	M18	M19	M20
RNT28	RNT29	RNT30	KHN1	KHN2	KHN3	KHN4
MAG5	MAG6	MAG7	MAG8	MAG9	MAG10	MAG11

KAWAIDA 509/2025-26/6261-62/2017-18						
Ashe	Ojobo	Iwa	Araorun	Ile	Itutu	Erinle
1	2	3	4	5	6	7
A18	A19	A20	A21	A22	A23	A24
PIN3	PIN4	PIN5	PIN6	PIN7	PIN8	PIN9
MIY10	MIY11	MIY12	MIY13	MIY14	MIY15	MIY16
8	9	10	11	12	13	14
A25	A26	A27	A28	A29	A30	MA1
PIN10	PIN11	PIN12	PIN13	PIN14	PIN15	PIN16
MIY17	MIY18	MIY19	MIY20	MIY21	MIY22	MIY23
15	16	17	18	19	20	21
MA2	MA3	MA4	MA5	MA6	MA7	MA8
PIN17	PIN18	PIN19	PIN20	PIN21	PIN22	PIN23
MIY24	MIY25	MIY26	MIY27	MIY28	MIY29	MIY30
22	23	24	25	26	27	28
MA9	MA10	MA11	MA12	MA13	MA14	MA15
PIN24	PIN25	PIN26	PIN27	PIN28	PIN29	PIN30
GIN1	GIN2	GIN3	GIN4	GIN5	GIN6	GIN7

UHURU 509/2025-26/6261-62/2017-18						
Ashe	Ojobo	Iwa	Araorun	Ile	Itutu	Erinle
1	2	3	4	5	6	7
M21	M22	M23	M24	M25	M26	M27
KHN5	KHN6	KHN7	KHN8	KHN9	KHN10	KHN11
MAG12	MAG13	MAG14	MAG15	MAG16	MAG17	MAG18
8	9	10	11	12	13	14
M28	M29	M30	M31	A1	A2	A3
KHN12	KHN13	KHN14	KHN15	KHN16	KHN17	KHN18
MAG19	MAG20	MAG21	MAG22	MAG23	MAG24	MAG25
15	16	17	18	19	20	21
A4	A5	A6	A7	A8	A9	A10
KHN19	KHN20	KHN21	KHN22	KHN23	KHN24	KHN25
MAG26	MAG27	MAG28	MAG29	MAG30	MIY1	MIY2
22	23	24	25	26	27	28
A11	A12	A13	A14	A15	A16	A17
KHN26	KHN27	KHN28	KHN29	KHN30	PIN1	PIN2
MIY3	MIY4	MIY5	MIY6	MIY7	MIY8	MIY9

MABABA WEUSI 509/2025-26/6261-62/2017-18						
Ashe	Ojobo	Iwa	Araorun	Ile	Itutu	Erinle
1	2	3	4	5	6	7
MA16	MA17	MA18	MA19	MA20	MA21	MA22
APP1	APP2	APP3	APP4	APP5	APP6	APP7
GIN8	GIN9	GIN10	GIN11	GIN12	GIN13	GIN14
8	9	10	11	12	13	14
MA23	MA24	MA25	MA26	MA27	MA28	MA29
APP8	APP9	APP10	APP11	APP12	APP13	APP14
GIN15	GIN16	GIN17	GIN18	GIN19	GIN20	GIN21
15	16	17	18	19	20	21
MA30	MA31	J1	J2	J3	J4	J5
APP15	APP16	APP17	APP18	APP19	APP20	APP21
GIN22	GIN23	GIN24	GIN25	GIN26	GIN27	GIN28
22	23	24	25	26	27	28
J6	J7	J8	J9	J10	J11	J12
APP22	APP23	APP24	APP25	APP26	APP27	APP28
GIN29	GIN30	SAN1	SAN2	SAN3	SAN4	SAN5

THE NU AFRIKAN CALENDAR

MWANAMKE WEUSI 509/2025-26/6261-62/2017-18						
Ashe	Ojobo	Iwa	Araorun	Ile	Itutu	Erinle
1	2	3	4	5	6	7
J13	J14	J15	J16	J17	J18	J19
APP29	APP30	MSR1	MSR2	MSR3	MSR4	MSR5
SAN6	SAN7	SAN8	SAN9	SAN10	SAN11	SAN12
8	9	10	11	12	13	14
J20	J21	J22	J23	J24	J25	J26
MSR6	MSR7	MSR8	MSR9	MSR10	MSR11	MSR12
SAN13	SAN14	SAN15	SAN16	SAN17	SAN18	SAN19
15	16	17	18	19	20	21
J27	J28	J29	J30	JU1	JU2	JU3
MSR13	MSR14	MSR15	MSR16	MSR17	MSR18	MSR19
SAN20	SAN21	SAN22	SAN23	SAN24	SAN25	SAN26
22	23	24	25	26	27	28
JU4	JU5	JU6	JU7	JU8	JU9	JU10
MSR20	MRS21	MSR22	MSR23	MSR24	MSR25	MSR26
SAN27	SAN28	SAN29	SAN30	HAM1	HAM2	HAM3

TAIFA WEUSI 509/2025-26/6261-62/2017-18						
Ashe	Ojobo	Iwa	Araorun	Ile	Itutu	Erinle
1	2	3	4	5	6	7
AU8	AU9	AU10	AU11	AU12	AU13	AU14
THT20	THT21	THT22	THT23	THT24	THT25	THT26
NAH2	NAH3	NAH4	NAH5	NAH6	NAH7	NAH8
8	9	10	11	12	13	14
AU15	AU16	AU17	AU18	AU19	AU20	AU21
THT27	THT28	THT29	THT30	PAP1	PAP1	PAP2
NAH9	NAH10	NAH11	NAH12	NAH13	NAH14	NAH15
15	16	17	18	19	20	21
AU22	AU23	AU24	AU25	AU26	AU27	AU28
PAP3	PAP4	PAP5	PAP6	PAP7	PAP8	PAP9
NAH16	NAH17	NAH18	NAH19	NAH20	NAH21	NAH22
22	23	24	25	26	27	28
AU29	AU30	AU31	S1	S2	S3	S4
PAP10	PAP11	PAP12	PAP13	PAP14	PAP15	PAP16
NAH23	NAH24	NAH25	NAH26	NAH27	NAH28	NAH29

MTU WEUSI 509/2025-26/6261-62/2017-18						
Ashe	Ojobo	Iwa	Araorun	Ile	Itutu	Erinle
1	2	3	4	5	6	7
JU11	JU12	JU13	JU14	JU15	JU16	JU17
MSR27	MSR28	MSR29	MSR30	*ASR1*	*HRU2*	*STH3*
HAM4	HAM5	HAM6	HAM7	HAM8	HAM9	HAM10
8	9	10	11	12	13	14
JU18	JU19	JU20	JU21	JU22	JU23	JU24
AST4	*NBH5*	THT1	THT2	THT3	THT4	THT5
HAM11	HAM12	HAM13	AM14	HAM15	HAM16	HAM17
15	16	17	18	19	20	21
JU25	JU26	JU27	JU28	JU29	JU30	JU31
THT6	THT7	THT8	THT9	THT10	THT11	THT12
HAM18	HAM19	HAM20	HAM21	HAM22	HAM23	HAM24
22	23	24	25	26	27	28
AU1	AU2	AU3	AU4	AU5	AU6	AU7
THT13	THT14	THT15	THT16	THT17	THT18	THT19
HAM25	HAM26	HAM27	HAM28	HAM29	HAM30	NAH1

UWEZO 509/2025-26/6261-62/2017-18						
Ashe	Ojobo	Iwa	Araorun	Ile	Itutu	Erinle
1	2	3	4	5	6	7
S5	S6	S7	S8	S9	S10	S11
PAP17	PAP18	PAP19	PAP20	PAP21	PAP22	PAP23
NAH30	*PAG1*	*PAG2*	*PAG3*	*PAG4*	*PAG5*	MAS1
8	9	10	11	12	13	14
S12	S13	S14	S15	S16	S17	S18
PAP24	PAP25	PAP26	PAP27	PAP28	PAP29	PAP30
MAS2	MAS3	MAS4	MAS5	MAS6	MAS7	MAS8
15	16	17	18	19	20	21
S19	S20	S21	S22	S23	S24	S25
HTH1	HTH2	HTH3	HTH4	HTH5	HTH6	HTH7
MAS9	MAS10	MAS11	MAS12	MAS13	MAS14	MAS15
22	23	24	25	26	27	28
S26	S27	S28	S29	S30	O1	O2
HTH8	HTH9	HTH10	HTH11	HTH12	HTH13	HTH14
MAS16	MAS17	MAS18	MAS19	MAS20	MAS21	MAS22

THE NU AFRIKAN CALENDAR

EKUNDU 509/2025-26/6261-62/2017-18						
Ashe	Ojobo	Iwa	Araorun	Ile	Itutu	Erinle
1	2	3	4	5	6	7
O3	O4	O5	O6	O7	O8	O9
HTH15	HTH16	HTH17	HTH18	HTH19	HTH20	HTH21
MAS23	MAS24	MAS25	MAS26	MAS27	MAS28	MAS29
8	9	10	11	12	13	14
O10	O11	O12	O13	O14	O15	O16
HTH22	HTH23	HTH24	HTH25	HTH26	HTH27	HTH28
MAS30	THi1	THi2	THi3	THi4	THi5	THi6
15	16	17	18	19	20	21
O17	O18	O19	O20	O21	O22	O23
HTH29	HTH30	KHK1	KHK2	KHK3	KHK4	KHK5
THi7	THi8	THi9	THi10	THi11	THi12	THi13
22	23	24	25	26	27	28
O24	O25	O26	O27	O28	O29	O30
KHK6	KHK7	KHK8	KHK9	KHK10	KHK11	KHK12
THi14	THi15	THi16	THi17	THi18	THi19	THi20

KIJANI 509/2025-26/6261-62/2017-18						
Ashe	Ojobo	Iwa	Araorun	Ile	Itutu	Erinle
1	2	3	4	5	6	7
N28	N29	N30	D1	D2	D3	D4
TBT11	TBT12	TBT13	TBT14	TBT15	TBT16	TBT17
HID19	HID20	HID21	HID22	HID23	HID24	HID25
8	9	10	11	12	13	14
D5	D6	D7	D8	D9	D10	D11
TBT18	TBT19	TBT20	TBT21	TBT22	TBT23	TBT24
HID26	HID27	HID28	HID29	HID30	TAH1	TAH2
15	16	17	18	19	20	21
D12	D13	D14	D15	D16	D17	D18
TBT25	TBT26	TBT27	TBT28	TBT29	TBT30	MKH1
TAH3	TAH4	TAH5	TAH6	TAH7	TAH8	TAH9
22	23	24	25	26	27	28
D19	D20	D21	D22	D23	D24	D25
MKH2	MKH3	MKH4	MKH5	MKH6	MKH7	MKH8
TAH10	TAH11	TAH12	TAH13	TAH14	TAH15	TAH16

EUSI 509/2025-26/6261-62/2017-18						
Ashe	Ojobo	Iwa	Araorun	Ile	Itutu	Erinle
1	2	3	4	5	6	7
O31	N1	N2	N3	N4	N5	N6
KHK13	KHK14	KHK15	KHK16	KHK17	KHK18	KHK19
THi21	THi22	THi23	THi24	THi25	THi26	THi27
8	9	10	11	12	13	14
N7	N8	N9	N10	N11	N12	N13
KHK20	KHK21	KHK22	KHK23	KHK24	KHK25	KHK26
THi28	THi29	THi30	HID1	HID2	HID3	HID4
15	16	17	18	19	20	21
N14	N15	N16	N17	N18	N19	N20
KHK27	KHK28	KHK29	KHK30	TBT1	TBT2	TBT3
HID5	HID6	HID7	HID8	HID9	HID10	HID11
22	23	24	25	26	27	28
N21	N22	N23	N24	N25	N26	N27
TBT4	TBT5	TBT6	TBT7	TBT8	TBT9	TBT10
HID12	HID13	HID14	HID15	HID16	HID17	HID18

KWANZA 509/2025-26/6261-62/2017-18						
Umoja	Kujichagulia	Ujima	Ujamaa	Nia	Kuumba	Imani
1	2	3	4	5	6	7
D26	D27	D28	D29	D30	D31	JA1
MKH9	MKH10	MKH11	MKH12	MKH13	MKH14	MKH15
TAH17	TAH18	TAH19	TAH20	TAH21	TAH22	TAH23
8	9	10	11	12	13	14
JA2	JA3	JA4	JA5	JA6	JA7	JA8
MKH16	MKH17	MKH18	MKH19	MKH20	MKH21	MKH22
TAH24	TAH25	TAH26	TAH27	TAH28	TAH29	TAH30
15	16	17	18	19	20	21
JA9	JA10	JA11	JA12	JA13	JA14	JA15
MKH23	MKH24	MKH25	KMH26	MKH27	MKH28	MKH29
TIR1	TIR2	TIR3	TIR4	TIR5	TIR6	TIR7
22	23	24	25	26	27	28
JA16	JA17	JA18	JA19	JA20	JA21	JA22
MKH30	AMN1	AMN2	AMN3	AMN4	AMN5	AMN6
TIR8	TIR9	TIR10	TIR11	TIR12	TIR13	TIR14

THE NU AFRIKAN CALENDAR

SHAHIDI 509/2025-26/6261-62/2017-18						
Ashe	Ojobo	Iwa	Araorun	Ile	Itutu	Erinle
1	2	3	4	5	6	7
JA23	JA24	JA25	JA26	JA27	JA28	JA29
AMN7	AMN8	AMN9	AMN10	AMN11	AMN12	AMN13
TIR15	TIR16	TIR17	TIR18	TIR19	TIR20	TIR21
8	9	10	11	12	13	14
JA30	JA31	F1	F2	F3	F4	F5
AMN14	AMN15	AMN16	AMN17	AMN18	AMN19	AMN20
TIR22	TIR23	TIR24	TIR25	TIR26	TIR27	TIR28
15	16	17	18	19	20	21
F6	F7	F8	F9	F10	F 11	F12
AMN21	AMN22	AMN23	AMN24	AMN25	AMN26	AMN28
TIR29	TIR30	YEK1	YEK2	YEK3	YEK4	YEK5
22	23	24	25	26	27	28
F13	F14	F15	F16	F17	F18	F19
AMN29	AMN30	RNT1	RNT2	RNT3	RNT4	RNT5
YEK6	YEK7	YEK8	YEK9	YEK10	YEK11	YEK12
SHAHIDI 60						
F20						
RNT6						
YEK13						

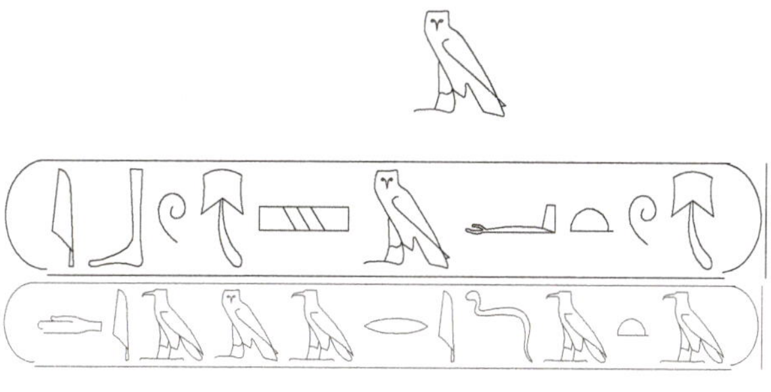

THE NU AFRIKAN CALENDAR

MAAFA 510 B.S.E./ASIENTO 2026-27 C.E.
KAM 6262-63 S.M./EC 2018-19 A.M. (ዓ.ም.)

AFU 510/2026-27/6262-63/2018-19

Ashe	Ojobo	Iwa	Araorun	Ile	Itutu	Erinle
1	2	3	4	5	6	7
F21	F22	F23	F24	F25	F26	F27
RNT7	RNT8	RNT9	RNT10	RNT11	RNT12	RNT13
YEK14	YEK15	YEK16	YEK17	YEK18	YEK19	YEK20
8	9	10	11	12	13	14
F28	M1	M2	M3	M4	M5	M6
RNT14	RNT15	RNT16	RNT17	RNT18	RNT19	RNT20
YEK21	YEK22	YEK23	YEK24	YEK25	YEK26	YEK27
15	16	17	18	19	20	21
M7	M8	M9	M10	M11	M12	M13
RNT21	RNT22	RNT23	RNT24	RNT25	RNT26	RNT27
YEK28	YEK29	YEK30	MAG1	MAG2	MAG3	MAG4
22	23	24	25	26	27	28
M14	M15	M16	M17	M18	M19	M20
RNT28	RNT29	RNT30	KHN1	KHN2	KHN3	KHN4
MAG5	MAG6	MAG7	MAG8	MAG9	MAG10	MAG11

KAWAIDA 510/2026-27/6262-63/2018-19

Ashe	Ojobo	Iwa	Araorun	Ile	Itutu	Erinle
1	2	3	4	5	6	7
A18	A19	A20	A21	A22	A23	A24
PIN3	PIN4	PIN5	PIN6	PIN7	PIN8	PIN9
MIY10	MIY11	MIY12	MIY13	MIY14	MIY15	MIY16
8	9	10	11	12	13	14
A25	A26	A27	A28	A29	A30	MA1
PIN10	PIN11	PIN12	PIN13	PIN14	PIN15	PIN16
MIY17	MIY18	MIY19	MIY20	MIY21	MIY22	MIY23
15	16	17	18	19	20	21
MA2	MA3	MA4	MA5	MA6	MA7	MA8
PIN17	PIN18	PIN19	PIN20	PIN21	PIN22	PIN23
MIY24	MIY25	MIY26	MIY27	MIY28	MIY29	MIY30
22	23	24	25	26	27	28
MA9	MA10	MA11	MA12	MA13	MA14	MA15
PIN24	PIN25	PIN26	PIN27	PIN28	PIN29	PIN30
GIN1	GIN2	GIN3	GIN4	GIN5	GIN6	GIN7

UHURU 510/2026-27/6262-63/2018-19

Ashe	Ojobo	Iwa	Araorun	Ile	Itutu	Erinle
1	2	3	4	5	6	7
M21	M22	M23	M24	M25	M26	M27
KHN5	KHN6	KHN7	KHN8	KHN9	KHN10	KHN11
MAG12	MAG13	MAG14	MAG15	MAG16	MAG17	MAG18
8	9	10	11	12	13	14
M28	M29	M30	M31	A1	A2	A3
KHN12	KHN13	KHN14	KHN15	KHN16	KHN17	KHN18
MAG19	MAG20	MAG21	MAG22	MAG23	MAG24	MAG25
15	16	17	18	19	20	21
A4	A5	A6	A7	A8	A9	A10
KHN19	KHN20	KHN21	KHN22	KHN23	KHN24	KHN25
MAG26	MAG27	MAG28	MAG29	MAG30	MIY1	MIY2
22	23	24	25	26	27	28
A11	A12	A13	A14	A15	A16	A17
KHN26	KHN27	KHN28	KHN29	KHN30	PIN1	PIN2
MIY3	MIY4	MIY5	MIY6	MIY7	MIY8	MIY9

MABABA WEUSI 510/2026-27/6262-63/2018-19

Ashe	Ojobo	Iwa	Araorun	Ile	Itutu	Erinle
1	2	3	4	5	6	7
MA16	MA17	MA18	MA19	MA20	MA21	MA22
APP1	APP2	APP3	APP4	APP5	APP6	APP7
GIN8	GIN9	GIN10	GIN11	GIN12	GIN13	GIN14
8	9	10	11	12	13	14
MA23	MA24	MA25	MA26	MA27	MA28	MA29
APP8	APP9	APP10	APP11	APP12	APP13	APP14
GIN15	GIN16	GIN17	GIN18	GIN19	GIN20	GIN21
15	16	17	18	19	20	21
MA30	MA31	J1	J2	J3	J4	J5
APP15	APP16	APP17	APP18	APP19	APP20	APP21
GIN22	GIN23	GIN24	GIN25	GIN26	GIN27	GIN28
22	23	24	25	26	27	28
J6	J7	J8	J9	J10	J11	J12
APP22	APP23	APP24	APP25	APP26	APP27	APP28
GIN29	GIN30	SAN1	SAN2	SAN3	SAN4	SAN5

THE NU AFRIKAN CALENDAR

MWANAMKE WEUSI 510/2026-27/6262-63/2018-19

Ashe	Ojobo	Iwa	Araorun	Ile	Itutu	Erinle
1	2	3	4	5	6	7
J13	J14	J15	J16	J17	J18	J19
APP29	APP30	MSR1	MSR2	MSR3	MSR4	MSR5
SAN6	SAN7	SAN8	SAN9	SAN10	SAN11	SAN12
8	9	10	11	12	13	14
J20	J21	J22	J23	J24	J25	J26
MSR6	MSR7	MSR8	MSR9	MSR10	MSR11	MSR12
SAN13	SAN14	SAN15	SAN16	SAN17	SAN18	SAN19
15	16	17	18	19	20	21
J27	J28	J29	J30	JU1	JU2	JU3
MSR13	MSR14	MSR15	MSR16	MSR17	MSR18	MSR19
SAN20	SAN21	SAN22	SAN23	SAN24	SAN25	SAN26
22	23	24	25	26	27	28
JU4	JU5	JU6	JU7	JU8	JU9	JU10
MSR20	MRS21	MSR22	MSR23	MSR24	MSR25	MSR26
SAN27	SAN28	SAN29	SAN30	HAM1	HAM2	HAM3

TAIFA WEUSI 510/2026-27/6262-63/2018-19

Ashe	Ojobo	Iwa	Araorun	Ile	Itutu	Erinle
1	2	3	4	5	6	7
AU8	AU9	AU10	AU11	AU12	AU13	AU14
THT20	THT21	THT22	THT23	THT24	THT25	THT26
NAH2	NAH3	NAH4	NAH5	NAH6	NAH7	NAH8
8	9	10	11	12	13	14
AU15	AU16	AU17	AU18	AU19	AU20	AU21
THT27	THT28	THT29	THT30	PAP1	PAP1	PAP2
NAH9	NAH10	NAH11	NAH12	NAH13	NAH14	NAH15
15	16	17	18	19	20	21
AU22	AU23	AU24	AU25	AU26	AU27	AU28
PAP3	PAP4	PAP5	PAP6	PAP7	PAP8	PAP9
NAH16	NAH17	NAH18	NAH19	NAH20	NAH21	NAH22
22	23	24	25	26	27	28
AU29	AU30	AU31	S1	S2	S3	S4
PAP10	PAP11	PAP12	PAP13	PAP14	PAP15	PAP16
NAH23	NAH24	NAH25	NAH26	NAH27	NAH28	NAH29

MTU WEUSI 510/2026-27/6262-63/2018-19

Ashe	Ojobo	Iwa	Araorun	Ile	Itutu	Erinle
1	2	3	4	5	6	7
JU11	JU12	JU13	JU14	JU15	JU16	JU17
MSR27	MSR28	MSR29	MSR30	*ASR1*	*HRU2*	*STH3*
HAM4	HAM5	HAM6	HAM7	HAM8	HAM9	HAM10
8	9	10	11	12	13	14
JU18	JU19	JU20	JU21	JU22	JU23	JU24
AST4	*NBH5*	THT1	THT2	THT3	THT4	THT5
HAM11	HAM12	HAM13	AM14	HAM15	HAM16	HAM17
15	16	17	18	19	20	21
JU25	JU26	JU27	JU28	JU29	JU30	JU31
THT6	THT7	THT8	THT9	THT10	THT11	THT12
HAM18	HAM19	HAM20	HAM21	HAM22	HAM23	HAM24
22	23	24	25	26	27	28
AU1	AU2	AU3	AU4	AU5	AU6	AU7
THT13	THT14	THT15	THT16	THT17	THT18	THT19
HAM25	HAM26	HAM27	HAM28	HAM29	HAM30	NAH1

UWEZO 510/2026-27/6262-63/2018-19

Ashe	Ojobo	Iwa	Araorun	Ile	Itutu	Erinle
1	2	3	4	5	6	7
S5	S6	S7	S8	S9	S10	S11
PAP17	PAP18	PAP19	PAP20	PAP21	PAP22	PAP23
NAH30	*PAG1*	*PAG2*	*PAG3*	*PAG4*	*PAG5*	MAS1
8	9	10	11	12	13	14
S12	S13	S14	S15	S16	S17	S18
PAP24	PAP25	PAP26	PAP27	PAP28	PAP29	PAP30
MAS2	MAS3	MAS4	MAS5	MAS6	MAS7	MAS8
15	16	17	18	19	20	21
S19	S20	S21	S22	S23	S24	S25
HTH1	HTH2	HTH3	HTH4	HTH5	HTH6	HTH7
MAS9	MAS10	MAS11	MAS12	MAS13	MAS14	MAS15
22	23	24	25	26	27	28
S26	S27	S28	S29	S30	O1	O2
HTH8	HTH9	HTH10	HTH11	HTH12	HTH13	HTH14
MAS16	MAS17	MAS18	MAS19	MAS20	MAS21	MAS22

THE NU AFRIKAN CALENDAR

EKUNDU 510/2026-27/6262-63/2018-19						
Ashe	Ojobo	Iwa	Araorun	Ile	Itutu	Erinle
1	2	3	4	5	6	7
O3	O4	O5	O6	O7	O8	O9
HTH15	HTH16	HTH17	HTH18	HTH19	HTH20	HTH21
MAS23	MAS24	MAS25	MAS26	MAS27	MAS28	MAS29
8	9	10	11	12	13	14
O10	O11	O12	O13	O14	O15	O16
HTH22	HTH23	HTH24	HTH25	HTH26	HTH27	HTH28
MAS30	THi1	THi2	THi3	THi4	THi5	THi6
15	16	17	18	19	20	21
O17	O18	O19	O20	O21	O22	O23
HTH29	HTH30	KHK1	KHK2	KHK3	KHK4	KHK5
THi7	THi8	THi9	THi10	THi11	THi12	THi13
22	23	24	25	26	27	28
O24	O25	O26	O27	O28	O29	O30
KHK6	KHK7	KHK8	KHK9	KHK10	KHK11	KHK12
THi14	THi15	THi16	THi17	THi18	THi19	THi20

KIJANI 510/2026-27/6262-63/2018-19						
Ashe	Ojobo	Iwa	Araorun	Ile	Itutu	Erinle
1	2	3	4	5	6	7
N28	N29	N30	D1	D2	D3	D4
TBT11	TBT12	TBT13	TBT14	TBT15	TBT16	TBT17
HID19	HID20	HID21	HID22	HID23	HID24	HID25
8	9	10	11	12	13	14
D5	D6	D7	D8	D9	D10	D11
TBT18	TBT19	TBT20	TBT21	TBT22	TBT23	TBT24
HID26	HID27	HID28	HID29	HID30	TAH1	TAH2
15	16	17	18	19	20	21
D12	D13	D14	D15	D16	D17	D18
TBT25	TBT26	TBT27	TBT28	TBT29	TBT30	MKH1
TAH3	TAH4	TAH5	TAH6	TAH7	TAH8	TAH9
22	23	24	25	26	27	28
D19	D20	D21	D22	D23	D24	D25
MKH2	MKH3	MKH4	MKH5	MKH6	MKH7	MKH8
TAH10	TAH11	TAH12	TAH13	TAH14	TAH15	TAH16

EUSI 510/2026-27/6262-63/2018-19						
Ashe	Ojobo	Iwa	Araorun	Ile	Itutu	Erinle
1	2	3	4	5	6	7
O31	N1	N2	N3	N4	N5	N6
KHK13	KHK14	KHK15	KHK16	KHK17	KHK18	KHK19
THi21	THi22	THi23	THi24	THi25	THi26	THi27
8	9	10	11	12	13	14
N7	N8	N9	N10	N11	N12	N13
KHK20	KHK21	KHK22	KHK23	KHK24	KHK25	KHK26
THi28	THi29	THi30	HID1	HID2	HID3	HID4
15	16	17	18	19	20	21
N14	N15	N16	N17	N18	N19	N20
KHK27	KHK28	KHK29	KHK30	TBT1	TBT2	TBT3
HID5	HID6	HID7	HID8	HID9	HID10	HID11
22	23	24	25	26	27	28
N21	N22	N23	N24	N25	N26	N27
TBT4	TBT5	TBT6	TBT7	TBT8	TBT9	TBT10
HID12	HID13	HID14	HID15	HID16	HID17	HID18

KWANZA 510/2026-27/6262-63/2018-19						
Umoja	Kujichagulia	Ujima	Ujamaa	Nia	Kuumba	Imani
1	2	3	4	5	6	7
D26	D27	D28	D29	D30	D31	JA1
MKH9	MKH10	MKH11	MKH12	MKH13	MKH14	MKH15
TAH17	TAH18	TAH19	TAH20	TAH21	TAH22	TAH23
8	9	10	11	12	13	14
JA2	JA3	JA4	JA5	JA6	JA7	JA8
MKH16	MKH17	MKH18	MKH19	MKH20	MKH21	MKH22
TAH24	TAH25	TAH26	TAH27	TAH28	TAH29	TAH30
15	16	17	18	19	20	21
JA9	JA10	JA11	JA12	JA13	JA14	JA15
MKH23	MKH24	MKH25	KMH26	MKH27	MKH28	MKH29
TIR1	TIR2	TIR3	TIR4	TIR5	TIR6	TIR7
22	23	24	25	26	27	28
JA16	JA17	JA18	JA19	JA20	JA21	JA22
MKH30	AMN1	AMN2	AMN3	AMN4	AMN5	AMN6
TIR8	TIR9	TIR10	TIR11	TIR12	TIR13	TIR14

THE NU AFRIKAN CALENDAR

SHAHIDI 510/2026-27/6262-63/2018-19						
Ashe	Ojobo	Iwa	Araorun	Ile	Itutu	Erinle
1	2	3	4	5	6	7
JA23	JA24	JA25	JA26	JA27	JA28	JA29
AMN7	AMN8	AMN9	AMN10	AMN11	AMN12	AMN13
TIR15	TIR16	TIR17	TIR18	TIR19	TIR20	TIR21
8	9	10	11	12	13	14
JA30	JA31	F1	F2	F3	F4	F5
AMN14	AMN15	AMN16	AMN17	AMN18	AMN19	AMN20
TIR22	TIR23	TIR24	TIR25	TIR26	TIR27	TIR28
15	16	17	18	19	20	21
F6	F7	F8	F9	F10	F 11	F12
AMN21	AMN22	AMN23	AMN24	AMN25	AMN26	AMN28
TIR29	TIR30	YEK1	YEK2	YEK3	YEK4	YEK5
22	23	24	25	26	27	28
F13	F14	F15	F16	F17	F18	F19
AMN29	AMN30	RNT1	RNT2	RNT3	RNT4	RNT5
YEK6	YEK7	YEK8	YEK9	YEK10	YEK11	YEK12
SHAHIDI 61						
F20						
RNT6						
YEK13						

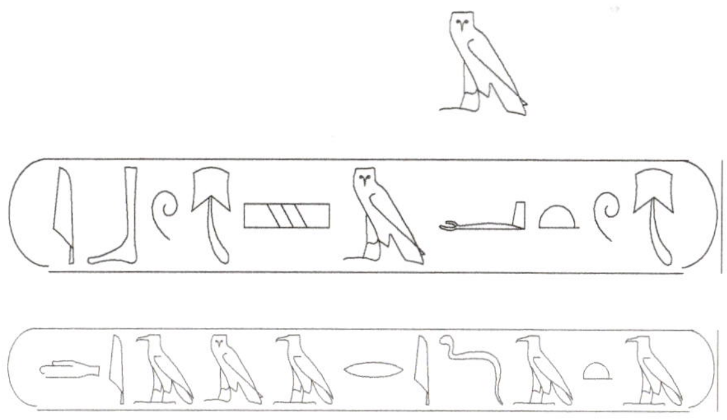

THE NU AFRIKAN CALENDAR

MAAFA 511 B.S.E./ASIENTO 2027-28 C.E./
KAM 6263-64 S.M./EC 2019-20 A.M. (ዓ.ም.)

AFU 511/2027-28/6263-64/2019-20

Ashe	Ojobo	Iwa	Araorun	Ile	Itutu	Erinle
1	2	3	4	5	6	7
F21	F22	F23	F24	F25	F26	F27
RNT7	RNT8	RNT9	RNT10	RNT11	RNT12	RNT13
YEK14	YEK15	YEK16	YEK17	YEK18	YEK19	YEK20
8	9	10	11	12	13	14
F28	M1	M2	M3	M4	M5	M6
RNT14	RNT15	RNT16	RNT17	RNT18	RNT19	RNT20
YEK21	YEK22	YEK23	YEK24	YEK25	YEK26	YEK27
15	16	17	18	19	20	21
M7	M8	M9	M10	M11	M12	M13
RNT21	RNT22	RNT23	RNT24	RNT25	RNT26	RNT27
YEK28	YEK 29	YEK30	MAG1	MAG2	MAG3	MAG4
22	23	24	25	26	27	28
M14	M15	M16	M17	M18	M19	M20
RNT28	RNT29	RNT30	KHN1	KHN2	KHN3	KHN4
MAG5	MAG6	MAG7	MAG8	MAG9	MAG10	MAG11

KAWAIDA 511/2027-28/6263-64/2019-20

Ashe	Ojobo	Iwa	Araorun	Ile	Itutu	Erinle
1	2	3	4	5	6	7
A18	A19	A20	A21	A22	A23	A24
PIN3	PIN4	PIN5	PIN6	PIN7	PIN8	PIN9
MIY10	MIY11	MIY12	MIY13	MIY14	MIY15	MIY16
8	9	10	11	12	13	14
A25	A26	A27	A28	A29	A30	MA1
PIN10	PIN11	PIN12	PIN13	PIN14	PIN15	PIN16
MIY17	MIY18	MIY19	MIY20	MIY21	MIY22	MIY23
15	16	17	18	19	20	21
MA2	MA3	MA4	MA5	MA6	MA7	MA8
PIN17	PIN18	PIN19	PIN20	PIN21	PIN22	PIN23
MIY24	MIY25	MIY26	MIY27	MIY28	MIY29	MIY30
22	23	24	25	26	27	28
MA9	MA10	MA11	MA12	MA13	MA14	MA15
PIN24	PIN25	PIN26	PIN27	PIN28	PIN29	PIN30
GIN1	GIN2	GIN3	GIN4	GIN5	GIN6	GIN7

UHURU 511/2027-28/6263-64/2019-20

Ashe	Ojobo	Iwa	Araorun	Ile	Itutu	Erinle
1	2	3	4	5	6	7
M21	M22	M23	M24	M25	M26	M27
KHN5	KHN6	KHN7	KHN8	KHN9	KHN10	KHN11
MAG12	MAG13	MAG14	MAG15	MAG16	MAG17	MAG18
8	9	10	11	12	13	14
M28	M29	M30	M31	A1	A2	A3
KHN12	KHN13	KHN14	KHN15	KHN16	KHN17	KHN18
MAG19	MAG20	MAG21	MAG22	MAG23	MAG24	MAG25
15	16	17	18	19	20	21
A4	A5	A6	A7	A8	A9	A10
KHN19	KHN20	KHN21	KHN22	KHN23	KHN24	KHN25
MAG26	MAG27	MAG28	MAG29	MAG30	MIY1	MIY2
22	23	24	25	26	27	28
A11	A12	A13	A14	A15	A16	A17
KHN26	KHN27	KHN28	KHN29	KHN30	PIN1	PIN2
MIY3	MIY4	MIY5	MIY6	MIY7	MIY8	MIY9

MABABA WEUSI 511/2027-28/6263-64/2019-20

Ashe	Ojobo	Iwa	Araorun	Ile	Itutu	Erinle
1	2	3	4	5	6	7
MA16	MA17	MA18	MA19	MA20	MA21	MA22
APP1	APP2	APP3	APP4	APP5	APP6	APP7
GIN8	GIN9	GIN10	GIN11	GIN12	GIN13	GIN14
8	9	10	11	12	13	14
MA23	MA24	MA25	MA26	MA27	MA28	MA29
APP8	APP9	APP10	APP11	APP12	APP13	APP14
GIN15	GIN16	GIN17	GIN18	GIN19	GIN20	GIN21
15	16	17	18	19	20	21
MA30	MA31	J1	J2	J3	J4	J5
APP15	APP16	APP17	APP18	APP19	APP20	APP21
GIN22	GIN23	GIN24	GIN25	GIN26	GIN27	GIN28
22	23	24	25	26	27	28
J6	J7	J8	J9	J10	J11	J12
APP22	APP23	APP24	APP25	APP26	APP27	APP28
GIN29	GIN30	SAN1	SAN2	SAN3	SAN4	SAN5

THE NU AFRIKAN CALENDAR

MWANAMKE WEUSI 511/2027-28/6263-64/2019-20						
Ashe	Ojobo	Iwa	Araorun	Ile	Itutu	Erinle
1	2	3	4	5	6	7
J13	J14	J15	J16	J17	J18	J19
APP29	APP30	MSR1	MSR2	MSR3	MSR4	MSR5
SAN6	SAN7	SAN8	SAN9	SAN10	SAN11	SAN12
8	9	10	11	12	13	14
J20	J21	J22	J23	J24	J25	J26
MSR6	MSR7	MSR8	MSR9	MSR10	MSR11	MSR12
SAN13	SAN14	SAN15	SAN16	SAN17	SAN18	SAN19
15	16	17	18	19	20	21
J27	J28	J29	J30	JU1	JU2	JU3
MSR13	MSR14	MSR15	MSR16	MSR17	MSR18	MSR19
SAN20	SAN21	SAN22	SAN23	SAN24	SAN25	SAN26
22	23	24	25	26	27	28
JU4	JU5	JU6	JU7	JU8	JU9	JU10
MSR20	MRS21	MSR22	MSR23	MSR24	MSR25	MSR26
SAN27	SAN28	SAN29	SAN30	HAM1	HAM2	HAM3

MTU WEUSI 511/2027-28/6263-64/2019-20						
Ashe	Ojobo	Iwa	Araorun	Ile	Itutu	Erinle
1	2	3	4	5	6	7
JU11	JU12	JU13	JU14	JU15	JU16	JU17
MSR27	MSR28	MSR29	MSR30	*ASR1*	*HRU2*	*STH3*
HAM4	HAM5	HAM6	HAM7	HAM8	HAM9	HAM10
8	9	10	11	12	13	14
JU18	JU19	JU20	JU21	JU22	JU23	JU24
AST4	*NBH5*	THT1	THT2	THT3	THT4	THT5
HAM11	HAM12	HAM13	HAM14	HAM15	HAM16	HAM17
15	16	17	18	19	20	21
JU25	JU26	JU27	JU28	JU29	JU30	JU31
THT6	THT7	THT8	THT9	THT10	THT11	THT12
HAM18	HAM19	HAM20	HAM21	HAM22	HAM23	HAM24
22	23	24	25	26	27	28
AU1	AU2	AU3	AU4	AU5	AU6	AU7
THT13	THT14	THT15	THT16	THT17	THT18	THT19
HAM25	HAM26	HAM27	HAM28	HAM29	HAM30	NAH1

TAIFA WEUSI 511/2027-28/6263-64/2019-20						
Ashe	Ojobo	Iwa	Araorun	Ile	Itutu	Erinle
1	2	3	4	5	6	7
AU8	AU9	AU10	AU11	AU12	AU13	AU14
THT20	THT21	THT22	THT23	THT24	THT25	THT26
NAH2	NAH3	NAH4	NAH5	NAH6	NAH7	NAH8
8	9	10	11	12	13	14
AU15	AU16	AU17	AU18	AU19	AU20	AU21
THT27	THT28	THT29	THT30	PAP1	PAP1	PAP2
NAH9	NAH10	NAH11	NAH12	NAH13	NAH14	NAH15
15	16	17	18	19	20	21
AU22	AU23	AU24	AU25	AU26	AU27	AU28
PAP3	PAP4	PAP5	PAP6	PAP7	PAP8	PAP9
NAH16	NAH17	NAH18	NAH19	NAH20	NAH21	NAH22
22	23	24	25	26	27	28
AU29	AU30	AU31	S1	S2	S3	S4
PAP10	PAP11	PAP12	PAP13	PAP14	PAP15	PAP16
NAH23	NAH24	NAH25	NAH26	NAH27	NAH28	NAH29

UWEZO 511/2027-28/6263-64/2019-20						
Ashe	Ojobo	Iwa	Araorun	Ile	Itutu	Erinle
1	2	3	4	5	6	7
S5	S6	S7	S8	S9	S10	S11
PAP17	PAP18	PAP19	PAP20	PAP21	PAP22	PAP23
NAH30	*QUA1*	*QUA2*	*QUA3*	*QUA4*	*QUA5*	*QUA6*
8	9	10	11	12	13	14
S12	S13	S14	S15	S16	S17	S18
PAP24	PAP25	PAP26	PAP27	PAP28	PAP29	PAP30
MAS1	MAS2	MAS3	MAS4	MAS5	MAS6	MAS7
15	16	17	18	19	20	21
S19	S20	S21	S22	S23	S24	S25
HTH1	HTH2	HTH3	HTH4	HTH5	HTH6	HTH7
MAS8	MAS9	MAS10	MAS11	MAS12	MAS13	MAS14
22	23	24	25	26	27	28
S26	S27	S28	S29	S30	O1	O2
HTH8	HTH9	HTH10	HTH11	HTH12	HTH13	HTH14
MAS15	MAS16	MAS17	MAS18	MAS19	MAS20	MAS21

THE NU AFRIKAN CALENDAR

EKUNDU 511/2027-28/6263-64/2019-20

Ashe	Ojobo	Iwa	Araorun	Ile	Itutu	Erinle
1	2	3	4	5	6	7
O3	O4	O5	O6	O7	O8	O9
HTH15	HTH16	HTH17	HTH18	HTH19	HTH20	HTH21
MAS22	MAS23	MAS 24	MAS25	MAS26	MAS27	MAS28
8	9	10	11	12	13	14
O10	O11	O12	O13	O14	O15	O16
HTH22	HTH23	HTH24	HTH25	HTH26	HTH27	HTH28
MAS29	MAS30	THi1	THi2	THi3	THi4	THi5
15	16	17	18	19	20	21
O17	O18	O19	O20	O21	O22	O23
HTH29	HTH30	KHK1	KHK2	KHK3	KHK4	KHK5
THi6	THi7	THi8	THi9	THi10	THi11	THi12
22	23	24	25	26	27	28
O24	O25	O26	O27	O28	O29	O30
KHK6	KHK7	KHK8	KHK9	KHK10	KHK11	KHK12
THi13	THi14	THi15	THi16	THi17	THi18	THi19

KIJANI 511/2027-28/6263-64/2019-20

Ashe	Ojobo	Iwa	Araorun	Ile	Itutu	Erinle
1	2	3	4	5	6	7
N28	N29	N30	D1	D2	D3	D4
TBT11	TBT12	TBT13	TBT14	TBT15	TBT16	TBT17
HID18	HID19	HID20	HID21	HID22	HID23	HID24
8	9	10	11	12	13	14
D5	D6	D7	D8	D9	D10	D11
TBT18	TBT19	TBT20	TBT21	TBT22	TBT23	TBT24
HID25	HID26	HID27	HID28	HID29	HID30	TAH1
15	16	17	18	19	20	21
D12	D13	D14	D15	D16	D17	D18
TBT25	TBT26	TBT27	TBT28	TBT29	TBT30	MKH1
TAH2	TAH3	TAH4	TAH5	TAH6	TAH7	TAH8
22	23	24	25	26	27	28
D19	D20	D21	D22	D23	D24	D25
MKH2	MKH3	MKH4	MKH5	MKH6	MKH7	MKH8
TAH9	TAH10	TAH11	TAH12	TAH13	TAH14	TAH15

EUSI 511/2027-28/6263-64/2019-20

Ashe	Ojobo	Iwa	Araorun	Ile	Itutu	Erinle
1	2	3	4	5	6	7
O31	N1	N2	N3	N4	N5	N6
KHK13	KHK14	KHK15	KHK16	KHK17	KHK18	KHK19
THi20	THi21	THi22	THi23	THi24	THi25	THi26
8	9	10	11	12	13	14
N7	N8	N9	N10	N11	N12	N13
KHK20	KHK21	KHK22	KHK23	KHK24	KHK25	KHK26
THi27	THi28	THi29	THi30	HID1	HID2	HID3
15	16	17	18	19	20	21
N14	N15	N16	N17	N18	N19	N20
KHK27	KHK28	KHK29	KHK30	TBT1	TBT2	TBT3
HID4	HID5	HID6	HID7	HID8	HID9	HID10
22	23	24	25	26	27	28
N21	N22	N23	N24	N25	N26	N27
TBT4	TBT5	TBT6	TBT7	TBT8	TBT9	TBT10
HID11	HID12	HID13	HID14	HID15	HID16	HID17

KWANZA 511/2027-28/6263-64/2019-20

Umoja	Kujichagulia	Ujima	Ujamaa	Nia	Kuumba	Imani
1	2	3	4	5	6	7
D26	D27	D28	D29	D30	D31	JA1
MKH9	MKH10	MKH11	MKH12	MKH13	MKH14	MKH15
TAH16	TAH17	TAH18	TAH19	TAH20	TAH21	TAH22
8	9	10	11	12	13	14
JA2	JA3	JA4	JA5	JA6	JA7	JA8
MKH16	MKH17	MKH18	MKH19	MKH20	MKH21	MKH22
TAH23	TAH24	TAH25	TAH26	TAH27	TAH28	TAH29
15	16	17	18	19	20	21
JA9	JA10	JA11	JA12	JA13	JA14	JA15
MKH23	MKH24	MKH25	KMH26	MKH27	MKH28	MKH29
TAH30	TIR1	TIR2	TIR3	TIR4	TIR5	TIR6
22	23	24	25	26	27	28
JA16	JA17	JA18	JA19	JA20	JA21	JA22
MKH30	AMN1	AMN2	AMN3	AMN4	AMN5	AMN6
TIR7	TIR8	TIR9	TIR10	TIR11	TIR12	TIR13

THE NU AFRIKAN CALENDAR

SHAHIDI 511/2027-28/6263-64/2019-20						
Ashe	Ojobo	Iwa	Araorun	Ile	Itutu	Erinle
1	2	3	4	5	6	7
JA23	JA24	JA25	JA26	JA27	JA28	JA29
AMN7	AMN8	AMN9	AMN10	AMN11	AMN12	AMN13
TIR14	TIR15	TIR16	TIR17	TIR18	TIR19	TIR20
8	9	10	11	12	13	14
JA30	JA31	F1	F2	F3	F4	F5
AMN14	AMN15	AMN16	AMN17	AMN18	AMN19	AMN20
TIR21	TIR22	TIR23	TIR24	TIR25	TIR26	TIR27
15	16	17	18	19	20	21
F6	F7	F8	F9	F10	F 11	F12
AMN21	AMN22	AMN23	AMN24	AMN25	AMN26	AMN28
TIR28	TIR29	TIR30	YEK1	YEK2	YEK3	YEK4
22	23	24	25	26	27	28
F13	F14	F15	F16	F17	F18	F19
AMN29	AMN30	RNT1	RNT2	RNT3	RNT4	RNT5
YEK5	YEK6	YEK7	YEK8	YEK9	YEK10	YEK11
SHAHIDI 62						
F20						
RNT6						
YEK12						

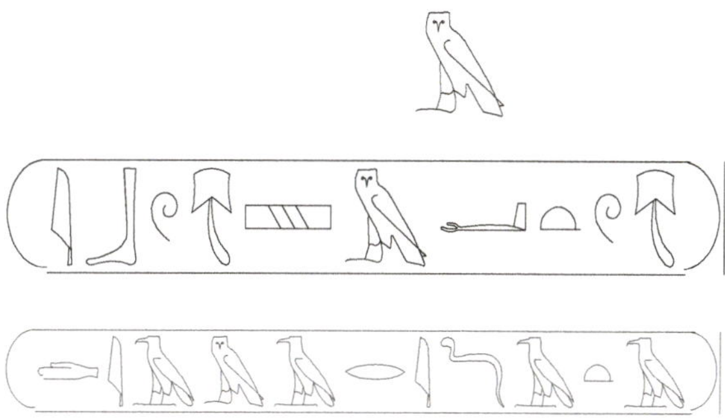

THE NU AFRIKAN CALENDAR

MAAFA 512 B.S.E./ASIENTO 2028-29 C.E./
KAM 6264-65 S.M./EC 2020-21 A.M. (ዓ.ም.)

AFU 512/2028-29/6264-65/2020-21						
Ashe	Ojobo	Iwa	Araorun	Ile	Itutu	Erinle
1	2	3	4	5	6	7
F21	F22	F23	F24	F25	F26	F27
RNT7	RNT8	RNT9	RNT10	RNT11	RNT12	RNT13
YEK13	YEK14	YEK15	YEK16	YEK17	YEK18	YEK19
8	9	10	11	12	13	14
F28	F29	M1	M2	M3	M4	M5
RNT14	RNT15	RNT16	RNT17	RNT18	RNT19	RNT20
YEK 20	YEK21	YEK22	YEK23	YEK24	YEK25	YEK26
15	16	17	18	19	20	21
M6	M7	M8	M9	M10	M11	M12
RNT21	RNT22	RNT23	RNT24	RNT25	RNT26	RNT27
YEK 27	YEK28	YEK29	YEK30	MAG1	MAG2	MAG3
22	23	24	25	26	27	28
M13	M14	M15	M16	M17	M18	M19
RNT28	RNT29	RNT30	KHN1	KHN2	KHN3	KHN4
MAG4	MAG5	MAG6	MAG7	MAG8	MAG9	MAG10
SHARIFU 125						
M20						
KHN5						
MAG11						

KAWAIDA 512/2028-29/6264-65/2020-21						
Ashe	Ojobo	Iwa	Araorun	Ile	Itutu	Erinle
1	2	3	4	5	6	7
A18	A19	A20	A21	A22	A23	A24
PIN4	PIN5	PIN6	PIN7	PIN8	PIN9	PIN10
MIY10	MIY11	MIY12	MIY13	MIY14	MIY15	MIY16
8	9	10	11	12	13	14
A25	A26	A27	A28	A29	A30	MA1
PIN11	PIN12	PIN13	PIN14	PIN15	PIN16	PIN17
MIY17	MIY18	MIY19	MIY20	MIY21	MIY22	MIY23
15	16	17	18	19	20	21
MA2	MA3	MA4	MA5	MA6	MA7	MA8
PIN18	PIN19	PIN20	PIN21	PIN22	PIN23	PIN24
MIY24	MIY25	MIY26	MIY27	MIY28	MIY29	MIY30
22	23	24	25	26	27	28
MA9	MA10	MA11	MA12	MA13	MA14	MA15
PIN25	PIN26	PIN27	PIN28	PIN29	PIN30	APP1
GIN1	GIN2	GIN3	GIN4	GIN5	GIN6	GIN7

UHURU 512/2028-29/6264-65/2020-21						
Ashe	Ojobo	Iwa	Araorun	Ile	Itutu	Erinle
1	2	3	4	5	6	7
M21	M22	M23	M24	M25	M26	M27
KHN6	KHN7	KHN8	KHN9	KHN10	KHN11	KHN12
MAG12	MAG13	MAG14	MAG15	MAG16	MAG17	MAG18
8	9	10	11	12	13	14
M28	M29	M30	M31	A1	A2	A3
KHN13	KHN14	KHN15	KHN16	KHN17	KHN18	KHN19
MAG19	MAG20	MAG21	MAG22	MAG23	MAG24	MAG25
15	16	17	18	19	20	21
A4	A5	A6	A7	A8	A9	A10
KHN20	KHN21	KHN222	KHN23	KHN24	KHN25	KHN26
MAG26	MAG27	MAG28	MAG29	MAG30	MIY1	MIY2
22	23	24	25	26	27	28
A11	A12	A13	A14	A15	A16	A17
KHN27	KHN28	KHN29	KHN30	PIN1	PIN2	PIN3
MIY3	MIY4	MIY5	MIY6	MIY7	MIY8	MIY9

MABABA WEUSI 512/2028-29/6264-65/2020-21						
Ashe	Ojobo	Iwa	Araorun	Ile	Itutu	Erinle
1	2	3	4	5	6	7
MA16	MA17	MA18	MA19	MA20	MA21	MA22
APP2	APP3	APP4	APP5	APP6	APP7	APP8
GIN8	GIN9	GIN10	GIN11	GIN12	GIN13	GIN14
8	9	10	11	12	13	14
MA23	MA24	MA25	MA26	MA27	MA28	MA29
APP9	APP10	APP11	APP12	APP13	APP14	APP15
GIN15	GIN16	GIN17	GIN18	GIN19	GIN20	GIN21
15	16	17	18	19	20	21
MA30	MA31	J1	J2	J3	J4	J5
APP16	APP17	APP18	APP19	APP20	APP21	APP22
GIN22	GIN23	GIN24	GIN25	GIN26	GIN27	GIN28
22	23	24	25	26	27	28
J6	J7	J8	J9	J10	J11	J12
APP23	APP24	APP25	APP26	APP27	APP28	APP29
GIN29	GIN30	SAN1	SAN2	SAN3	SAN4	SAN5

THE NU AFRIKAN CALENDAR

MWANAMKE WEUSI 512/2028-29/6264-65/2020-21						
Ashe	Ojobo	Iwa	Araorun	Ile	Itutu	Erinle
1	2	3	4	5	6	7
J13	J14	J15	J16	J17	J18	J19
APP30	MSR1	MSR2	MSR3	MSR4	MSR5	MSR6
SAN6	SAN7	SAN8	SAN9	SAN10	SAN11	SAN12
8	9	10	11	12	13	14
J20	J21	J22	J23	J24	J25	J26
MSR7	MSR8	MSR9	MSR10	MSR11	MSR12	MSR13
SAN13	SAN14	SAN15	SAN16	SAN17	SAN18	SAN19
15	16	17	18	19	20	21
J27	J28	J29	J30	JU1	JU2	JU3
MSR14	MSR15	MSR16	MSR17	MSR18	MSR19	MSR20
SAN20	SAN21	SAN22	SAN23	SAN24	SAN25	SAN26
22	23	24	25	26	27	28
JU4	JU5	JU6	JU7	JU8	JU9	JU10
MSR21	MRS22	MSR23	MSR24	MSR25	MSR26	MSR27
SAN27	SAN28	SAN29	SAN30	HAM1	HAM2	HAM3

TAIFA WEUSI 512/2028-29/6264-65/2020-21						
Ashe	Ojobo	Iwa	Araorun	Ile	Itutu	Erinle
1	2	3	4	5	6	7
AU8	AU9	AU10	AU11	AU12	AU13	AU14
THT20	THT21	THT22	THT23	THT24	THT25	THT26
NAH2	NAH3	NAH4	NAH5	NAH6	NAH7	NAH8
8	9	10	11	12	13	14
AU15	AU16	AU17	AU18	AU19	AU20	AU21
THT27	THT28	THT29	THT30	PAP1	PAP1	PAP2
NAH9	NAH10	NAH11	NAH12	NAH13	NAH14	NAH15
15	16	17	18	19	20	21
AU22	AU23	AU24	AU25	AU26	AU27	AU28
PAP3	PAP4	PAP5	PAP6	PAP7	PAP8	PAP9
NAH16	NAH17	NAH18	NAH19	NAH20	NAH21	NAH22
22	23	24	25	26	27	28
AU29	AU30	AU31	S1	S2	S3	S4
PAP10	PAP11	PAP12	PAP13	PAP14	PAP15	PAP16
NAH23	NAH24	NAH25	NAH26	NAH27	NAH28	NAH29

MTU WEUSI 512/2028-29/6264-65/2020-21						
Ashe	Ojobo	Iwa	Araorun	Ile	Itutu	Erinle
1	2	3	4	5	6	7
JU11	JU12	JU13	JU14	JU15	JU16	JU17
MSR28	MSR29	MSR30	*ASR1*	*HRU2*	*STH3*	*AST4*
HAM4	HAM5	HAM6	HAM7	HAM8	HAM9	HAM10
8	9	10	11	12	13	14
JU18	JU19	JU20	JU21	JU22	JU23	JU24
NBH5	*ANU6*	THT1	THT2	THT3	THT4	THT5
HAM11	HAM12	HAM13	HAM14	HAM15	HAM16	HAM17
15	16	17	18	19	20	21
JU25	JU26	JU27	JU28	JU29	JU30	JU31
THT6	THT7	THT8	THT9	THT10	THT11	THT12
HAM18	HAM19	HAM20	HAM21	HAM22	HAM23	HAM24
22	23	24	25	26	27	28
AU1	AU2	AU3	AU4	AU5	AU6	AU7
THT13	THT14	THT15	THT16	THT17	THT18	THT19
HAM25	HAM26	HAM27	HAM28	HAM29	HAM30	NAH1

UWEZO 512/2028-29/6264-65/2020-21						
Ashe	Ojobo	Iwa	Araorun	Ile	Itutu	Erinle
1	2	3	4	5	6	7
S5	S6	S7	S8	S9	S10	S11
PAP17	PAP18	PAP19	PAP20	PAP21	PAP22	PAP23
NAH30	*PAG1*	*PAG2*	*PAG3*	*PAG4*	*PAG5*	MAS1
8	9	10	11	12	13	14
S12	S13	S14	S15	S16	S17	S18
PAP24	PAP25	PAP26	PAP27	PAP28	PAP29	PAP30
MAS2	MAS3	MAS4	MAS5	MAS6	MAS7	MAS8
15	16	17	18	19	20	21
S19	S20	S21	S22	S23	S24	S25
HTH1	HTH2	HTH3	HTH4	HTH5	HTH6	HTH7
MAS9	MAS10	MAS11	MAS12	MAS13	MAS14	MAS15
22	23	24	25	26	27	28
S26	S27	S28	S29	S30	O1	O2
HTH8	HTH9	HTH10	HTH11	HTH12	HTH13	HTH14
MAS16	MAS17	MAS18	MAS19	MAS20	MAS21	MAS22

THE NU AFRIKAN CALENDAR

EKUNDU 512/2028-29/6264-65/2020-21						
Ashe	Ojobo	Iwa	Araorun	Ile	Itutu	Erinle
1	2	3	4	5	6	7
O3	O4	O5	O6	O7	O8	O9
HTH15	HTH16	HTH17	HTH18	HTH19	HTH20	HTH21
MAS23	MAS24	MAS25	MAS26	MAS27	MAS28	MAS29
8	9	10	11	12	13	14
O10	O11	O12	O13	O14	O15	O16
HTH22	HTH23	HTH24	HTH25	HTH26	HTH27	HTH28
MAS30	THi1	THi2	THi3	THi4	THi5	THi6
15	16	17	18	19	20	21
O17	O18	O19	O20	O21	O22	O23
HTH29	HTH30	KHK1	KHK2	KHK3	KHK4	KHK5
THi7	THi8	This9	THi10	THi11	THi12	THi13
22	23	24	25	26	27	28
O24	O25	O26	O27	O28	O29	O30
KHK6	KHK7	KHK8	KHK9	KHK10	KHK11	KHK12
Thi14	THi15	THi16	THi17	THi18	THi19	THi20

KIJANI 512/2028-29/6264-65/2020-21						
Ashe	Ojobo	Iwa	Araorun	Ile	Itutu	Erinle
1	2	3	4	5	6	7
N28	N29	N30	D1	D2	D3	D4
TBT11	TBT12	TBT13	TBT14	TBT15	TBT16	TBT17
HID19	HID20	HID21	HID22	HID23	HID24	HID25
8	9	10	11	12	13	14
D5	D6	D7	D8	D9	D10	D11
TBT18	TBT19	TBT20	TBT21	TBT22	TBT23	TBT24
HID26	HID27	HID28	HID29	HID30	TAH1	TAH2
15	16	17	18	19	20	21
D12	D13	D14	D15	D16	D17	D18
TBT25	TBT26	TBT27	TBT28	TBT29	TBT30	MKH1
TAH3	TAH4	TAH5	TAH6	TAH7	TAH8	TAH9
22	23	24	25	26	27	28
D19	D20	D21	D22	D23	D24	D25
MKH2	MKH3	MKH4	MKH5	MKH6	MKH7	MKH8
TAH10	TAH11	TAH12	TAH13	TAH14	TAH15	TAH16

EUSI 512/2028-29/6264-65/2020-21						
Ashe	Ojobo	Iwa	Araorun	Ile	Itutu	Erinle
1	2	3	4	5	6	7
O31	N1	N2	N3	N4	N5	N6
KHK13	KHK14	KHK15	KHK16	KHK17	KHK18	KHK19
THi21	THi22	THi23	THi24	THi25	THi26	THi27
8	9	10	11	12	13	14
N7	N8	N9	N10	N11	N12	N13
KHK20	KHK21	KHK22	KHK23	KHK24	KHK25	KHK26
THi28	THi29	THi30	HID1	HID2	HID3	HID4
15	16	17	18	19	20	21
N14	N15	N16	N17	N18	N19	N20
KHK27	KHK28	KHK29	KHK30	TBT1	TBT2	TBT3
HID5	HID6	HID7	HID8	HID9	HID10	HID11
22	23	24	25	26	27	28
N21	N22	N23	N24	N25	N26	N27
TBT4	TBT5	TBT6	TBT7	TBT8	TBT9	TBT10
HID12	HID13	HID14	HID15	HID16	HID17	HID18

KWANZA 512/2028-29/6264-65/2020-21						
Umoja	Kujichagulia	Ujima	Ujamaa	Nia	Kuumba	Imani
1	2	3	4	5	6	7
D26	D27	D28	D29	D30	D31	JA1
MKH9	MKH10	MKH11	MKH12	MKH13	MKH14	MKH15
TAH17	TAH18	TAH19	TAH20	TAH21	TAH22	TAH23
8	9	10	11	12	13	14
JA2	JA3	JA4	JA5	JA6	JA7	JA8
MKH16	MKH17	MKH18	MKH19	MKH20	MKH21	MKH22
TAH24	TAH25	TAH26	TAH27	TAH28	TAH29	TAH30
15	16	17	18	19	20	21
JA9	JA10	JA11	JA12	JA13	JA14	JA15
MKH23	MKH24	MKH25	KMH26	MKH27	MKH28	MKH29
TIR1	TIR2	TIR3	TIR4	TIR5	TIR6	TIR7
22	23	24	25	26	27	28
JA16	JA17	JA18	JA19	JA20	JA21	JA22
MKH30	AMN1	AMN2	AMN3	AMN4	AMN5	AMN6
TIR8	TIR9	TIR10	TIR11	TIR12	TIR13	TIR14

THE NU AFRIKAN CALENDAR

SHAHIDI 512/2028-29/6264-65/2020-21						
Ashe	Ojobo	Iwa	Araorun	Ile	Itutu	Erinle
1	2	3	4	5	6	7
JA23	JA24	JA25	JA26	JA27	JA28	JA29
AMN7	AMN8	AMN9	AMN10	AMN11	AMN12	AMN13
TIR15	TIR16	TIR17	TIR18	TIR19	TIR20	TIR21
8	9	10	11	12	13	14
JA30	JA31	F1	F2	F3	F4	F5
AMN14	AMN15	AMN16	AMN17	AMN18	AMN19	AMN20
TIR22	TIR23	TIR24	TIR25	TIR26	TIR27	TIR28
15	16	17	18	19	20	21
F6	F7	F8	F9	F10	F 11	F12
AMN21	AMN22	AMN23	AMN24	AMN25	AMN26	AMN28
TIR29	TIR30	YEK1	YEK2	YEK3	YEK4	YEK5
22	23	24	25	26	27	28
F13	F14	F15	F16	F17	F18	F19
AMN29	AMN30	RNT1	RNT2	RNT3	RNT4	RNT5
YEK6	YEK7	YEK8	YEK9	YEK10	YEK11	YEK12
SHAHIDI 63						
F20						
RNT6						
YEK13						

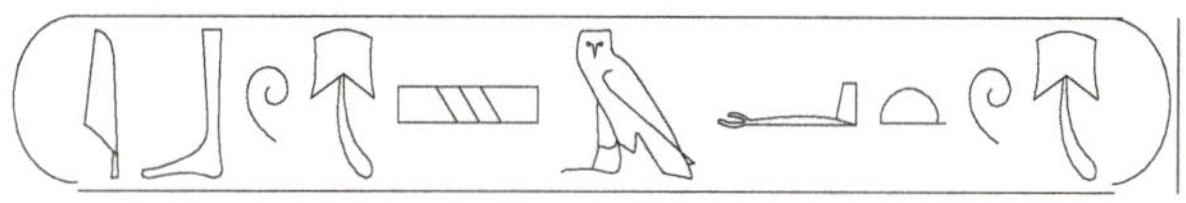

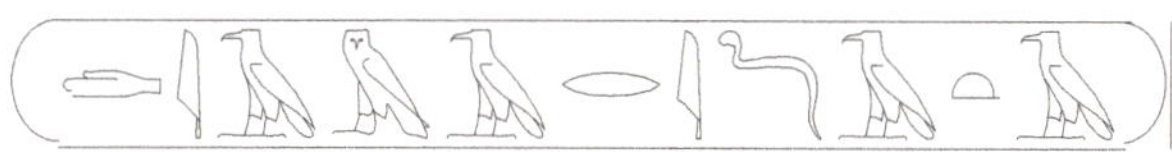

THE NU AFRIKAN CALENDAR

MAAFA 513 B.S.E./ASIENTO 2029-30 C.E.
KAM 6265-66 S.M. /EC 2021-22 A.M. (ዓ.ም.)

AFU 513/2029-30/6265-66/2021-22

Ashe	Ojobo	Iwa	Araorun	Ile	Itutu	Erinle
1	2	3	4	5	6	7
F21	F22	F23	F24	F25	F26	F27
RNT7	RNT8	RNT9	RNT10	RNT11	RNT12	RNT13
YEK14	YEK15	YEK16	YEK17	YEK18	YEK19	YEK20
8	9	10	11	12	13	14
F28	M1	M2	M3	M4	M5	M6
RNT14	RNT15	RNT16	RNT17	RNT18	RNT19	RNT20
YEK21	YEK22	YEK23	YEK24	YEK25	YEK26	YEK27
15	16	17	18	19	20	21
M7	M8	M9	M10	M11	M12	M13
RNT21	RNT22	RNT23	RNT24	RNT25	RNT26	RNT27
YEK28	YEK29	YEK30	MAG1	MAG2	MAG3	MAG4
22	23	24	25	26	27	28
M14	M15	M16	M17	M18	M19	M20
RNT28	RNT29	RNT30	KHN1	KHN2	KHN3	KHN4
MAG5	MAG6	MAG7	MAG8	MAG9	MAG10	MAG11

UHURU 513/2029-30/6265-66/2021-22

Ashe	Ojobo	Iwa	Araorun	Ile	Itutu	Erinle
1	2	3	4	5	6	7
M21	M22	M23	M24	M25	M26	M27
KHN5	KHN6	KHN7	KHN8	KHN9	KHN10	KHN11
MAG12	MAG13	MAG14	MAG15	MAG16	MAG17	MAG18
8	9	10	11	12	13	14
M28	M29	M30	M31	A1	A2	A3
KHN12	KHN13	KHN14	KHN15	KHN16	KHN17	KHN18
MAG19	MAG20	MAG21	MAG22	MAG23	MAG24	MAG25
15	16	17	18	19	20	21
A4	A5	A6	A7	A8	A9	A10
KHN19	KHN20	KHN21	KHN22	KHN23	KHN24	KHN25
MAG26	MAG27	MAG28	MAG29	MAG30	MIY1	MIY2
22	23	24	25	26	27	28
A11	A12	A13	A14	A15	A16	A17
KHN26	KHN27	KHN28	KHN29	KHN30	PIN1	PIN2
MIY3	MIY4	MIY5	MIY6	MIY7	MIY8	MIY9

KAWAIDA 513/2029-30/6265-66/2021-22

Ashe	Ojobo	Iwa	Araorun	Ile	Itutu	Erinle
1	2	3	4	5	6	7
A18	A19	A20	A21	A22	A23	A24
PIN3	PIN4	PIN5	PIN6	PIN7	PIN8	PIN9
MIY10	MIY11	MIY12	MIY13	MIY14	MIY15	MIY16
8	9	10	11	12	13	14
A25	A26	A27	A28	A29	A30	MA1
PIN10	PIN11	PIN12	PIN13	PIN14	PIN15	PIN16
MIY17	MIY18	MIY19	MIY20	MIY21	MIY22	MIY23
15	16	17	18	19	20	21
MA2	MA3	MA4	MA5	MA6	MA7	MA8
PIN17	PIN18	PIN19	PIN20	PIN21	PIN22	PIN23
MIY24	MIY25	MIY26	MIY27	MIY28	MIY29	MIY30
22	23	24	25	26	27	28
MA9	MA10	MA11	MA12	MA13	MA14	MA15
PIN24	PIN25	PIN26	PIN27	PIN28	PIN29	PIN30
GIN1	GIN2	GIN3	GIN4	GIN5	GIN6	GIN7

MABABA WEUSI 513/2029-30/6265-66/2021-22

Ashe	Ojobo	Iwa	Araorun	Ile	Itutu	Erinle
1	2	3	4	5	6	7
MA16	MA17	MA18	MA19	MA20	MA21	MA22
APP1	APP2	APP3	APP4	APP5	APP6	APP7
GIN8	GIN9	GIN10	GIN11	GIN12	GIN13	GIN14
8	9	10	11	12	13	14
MA23	MA24	MA25	MA26	MA27	MA28	MA29
APP8	APP9	APP10	APP11	APP12	APP13	APP14
GIN15	GIN16	GIN17	GIN18	GIN19	GIN20	GIN21
15	16	17	18	19	20	21
MA30	MA31	J1	J2	J3	J4	J5
APP15	APP16	APP17	APP18	APP19	APP20	APP21
GIN22	GIN23	GIN24	GIN25	GIN26	GIN27	GIN28
22	23	24	25	26	27	28
J6	J7	J8	J9	J10	J11	J12
APP22	APP23	APP24	APP25	APP26	APP27	APP28
GIN29	GIN30	SAN1	SAN2	SAN3	SAN4	SAN5

THE NU AFRIKAN CALENDAR

MWANAMKE WEUSI 513/2029-30/6265-66/2021-22						
Ashe	Ojobo	Iwa	Araorun	Ile	Itutu	Erinle
1	2	3	4	5	6	7
J13	J14	J15	J16	J17	J18	J19
APP29	APP30	MSR1	MSR2	MSR3	MSR4	MSR5
SAN6	SAN7	SAN8	SAN9	SAN10	SAN11	SAN12
8	9	10	11	12	13	14
J20	J21	J22	J23	J24	J25	J26
MSR6	MSR7	MSR8	MSR9	MSR10	MSR11	MSR12
SAN13	SAN14	SAN15	SAN16	SAN17	SAN18	SAN19
15	16	17	18	19	20	21
J27	J28	J29	J30	JU1	JU2	JU3
MSR13	MSR14	MSR15	MSR16	MSR17	MSR18	MSR19
SAN20	SAN21	SAN22	SAN23	SAN24	SAN25	SAN26
22	23	24	25	26	27	28
JU4	JU5	JU6	JU7	JU8	JU9	JU10
MSR20	MRS21	MSR22	MSR23	MSR24	MSR25	MSR26
SAN27	SAN28	SAN29	SAN30	HAM1	HAM2	HAM3

TAIFA WEUSI 513/2029-30/6265-66/2021-22						
Ashe	Ojobo	Iwa	Araorun	Ile	Itutu	Erinle
1	2	3	4	5	6	7
AU8	AU9	AU10	AU11	AU12	AU13	AU14
THT20	THT21	THT22	THT23	THT24	THT25	THT26
NAH2	NAH3	NAH4	NAH5	NAH6	NAH7	NAH8
8	9	10	11	12	13	14
AU15	AU16	AU17	AU18	AU19	AU20	AU21
THT27	THT28	THT29	THT30	PAP1	PAP1	PAP2
NAH9	NAH10	NAH11	NAH12	NAH13	NAH14	NAH15
15	16	17	18	19	20	21
AU22	AU23	AU24	AU25	AU26	AU27	AU28
PAP3	PAP4	PAP5	PAP6	PAP7	PAP8	PAP9
NAH16	NAH17	NAH18	NAH19	NAH20	NAH21	NAH22
22	23	24	25	26	27	28
AU29	AU30	AU31	S1	S2	S3	S4
PAP10	PAP11	PAP12	PAP13	PAP14	PAP15	PAP16
NAH23	NAH24	NAH25	NAH26	NAH27	NAH28	NAH29

MTU WEUSI 513/2029-30/6265-66/2021-22						
Ashe	Ojobo	Iwa	Araorun	Ile	Itutu	Erinle
1	2	3	4	5	6	7
JU11	JU12	JU13	JU14	JU15	JU16	JU17
MSR27	MSR28	MSR29	MSR30	*ASR1*	*HRU2*	*STH3*
HAM4	HAM5	HAM6	HAM7	HAM8	HAM9	HAM10
8	9	10	11	12	13	14
JU18	JU19	JU20	JU21	JU22	JU23	JU24
AST4	*NBH5*	THT1	THT2	THT3	THT4	THT5
HAM11	HAM12	HAM13	AM14	HAM15	HAM16	HAM17
15	16	17	18	19	20	21
JU25	JU26	JU27	JU28	JU29	JU30	JU31
THT6	THT7	THT8	THT9	THT10	THT11	THT12
HAM18	HAM19	HAM20	HAM21	HAM22	HAM23	HAM24
22	23	24	25	26	27	28
AU1	AU2	AU3	AU4	AU5	AU6	AU7
THT13	THT14	THT15	THT16	THT17	THT18	THT19
HAM25	HAM26	HAM27	HAM28	HAM29	HAM30	NAH1

UWEZO 513/2029-30/6265-66/2021-22						
Ashe	Ojobo	Iwa	Araorun	Ile	Itutu	Erinle
1	2	3	4	5	6	7
S5	S6	S7	S8	S9	S10	S11
PAP17	PAP18	PAP19	PAP20	PAP21	PAP22	PAP23
NAH30	*PAG1*	*PAG2*	*PAG3*	*PAG4*	*PAG5*	MAS1
8	9	10	11	12	13	14
S12	S13	S14	S15	S16	S17	S18
PAP24	PAP25	PAP26	PAP27	PAP28	PAP29	PAP30
MAS2	MAS3	MAS4	MAS5	MAS6	MAS7	MAS8
15	16	17	18	19	20	21
S19	S20	S21	S22	S23	S24	S25
HTH1	HTH2	HTH3	HTH4	HTH5	HTH6	HTH7
MAS9	MAS10	MAS11	MAS12	MAS13	MAS14	MAS15
22	23	24	25	26	27	28
S26	S27	S28	S29	S30	O1	O2
HTH8	HTH9	HTH10	HTH11	HTH12	HTH13	HTH14
MAS16	MAS17	MAS18	MAS19	MAS20	MAS21	MAS22

THE NU AFRIKAN CALENDAR

EKUNDU 513/2029-30/6265-66/2021-22						
Ashe	Ojobo	Iwa	Araorun	Ile	Itutu	Erinle
1	2	3	4	5	6	7
O3	O4	O5	O6	O7	O8	O9
HTH15	HTH16	HTH17	HTH18	HTH19	HTH20	HTH21
MAS23	MAS24	MAS25	MAS26	MAS27	MAS28	MAS29
8	9	10	11	12	13	14
O10	O11	O12	O13	O14	O15	O16
HTH22	HTH23	HTH24	HTH25	HTH26	HTH27	HTH28
MAS30	THi1	THi2	THi3	THi4	THi5	THi6
15	16	17	18	19	20	21
O17	O18	O19	O20	O21	O22	O23
HTH29	HTH30	KHK1	KHK2	KHK3	KHK4	KHK5
THi7	THi8	THi9	THi10	THi11	THi12	THi13
22	23	24	25	26	27	28
O24	O25	O26	O27	O28	O29	O30
KHK6	KHK7	KHK8	KHK9	KHK10	KHK11	KHK12
THi14	THi15	THi16	THi17	THi18	THi19	THi20

KIJANI 513/2029-30/6265-66/2021-22						
Ashe	Ojobo	Iwa	Araorun	Ile	Itutu	Erinle
1	2	3	4	5	6	7
N28	N29	N30	D1	D2	D3	D4
TBT11	TBT12	TBT13	TBT14	TBT15	TBT16	TBT17
HID19	HID20	HID21	HID22	HID23	HID24	HID25
8	9	10	11	12	13	14
D5	D6	D7	D8	D9	D10	D11
TBT18	TBT19	TBT20	TBT21	TBT22	TBT23	TBT24
HID26	HID27	HID28	HID29	HID30	TAH1	TAH2
15	16	17	18	19	20	21
D12	D13	D14	D15	D16	D17	D18
TBT25	TBT26	TBT27	TBT28	TBT29	TBT30	MKH1
TAH3	TAH4	TAH5	TAH6	TAH7	TAH8	TAH9
22	23	24	25	26	27	28
D19	D20	D21	D22	D23	D24	D25
MKH2	MKH3	MKH4	MKH5	MKH6	MKH7	MKH8
TAH10	TAH11	TAH12	TAH13	TAH14	TAH15	TAH16

EUSI 513/2029-30/6265-66/2021-22						
Ashe	Ojobo	Iwa	Araorun	Ile	Itutu	Erinle
1	2	3	4	5	6	7
O31	N1	N2	N3	N4	N5	N6
KHK13	KHK14	KHK15	KHK16	KHK17	KHK18	KHK19
THi21	THi22	THi23	THi24	THi25	THi26	THi27
8	9	10	11	12	13	14
N7	N8	N9	N10	N11	N12	N13
KHK20	KHK21	KHK22	KHK23	KHK24	KHK25	KHK26
THi28	THi29	THi30	HID1	HID2	HID3	HID4
15	16	17	18	19	20	21
N14	N15	N16	N17	N18	N19	N20
KHK27	KHK28	KHK29	KHK30	TBT1	TBT2	TBT3
HID5	HID6	HID7	HID8	HID9	HID10	HID11
22	23	24	25	26	27	28
N21	N22	N23	N24	N25	N26	N27
TBT4	TBT5	TBT6	TBT7	TBT8	TBT9	TBT10
HID12	HID13	HID14	HID15	HID16	HID17	HID18

KWANZA 513/2029-30/6265-66/2021-22						
Umoja	Kujichagulia	Ujima	Ujamaa	Nia	Kuumba	Imani
1	2	3	4	5	6	7
D26	D27	D28	D29	D30	D31	JA1
MKH9	MKH10	MKH11	MKH12	MKH13	MKH14	MKH15
TAH17	TAH18	TAH19	TAH20	TAH21	TAH22	TAH23
8	9	10	11	12	13	14
JA2	JA3	JA4	JA5	JA6	JA7	JA8
MKH16	MKH17	MKH18	MKH19	MKH20	MKH21	MKH22
TAH24	TAH25	TAH26	TAH27	TAH28	TAH29	TAH30
15	16	17	18	19	20	21
JA9	JA10	JA11	JA12	JA13	JA14	JA15
MKH23	MKH24	MKH25	KMH26	MKH27	MKH28	MKH29
TIR1	TIR2	TIR3	TIR4	TIR5	TIR6	TIR7
22	23	24	25	26	27	28
JA16	JA17	JA18	JA19	JA20	JA21	JA22
MKH30	AMN1	AMN2	AMN3	AMN4	AMN5	AMN6
TIR8	TIR9	TIR10	TIR11	TIR12	TIR13	TIR14

THE NU AFRIKAN CALENDAR

SHAHIDI 513/2029-30/6265-66/2021-22						
Ashe	Ojobo	Iwa	Araorun	Ile	Itutu	Erinle
1	2	3	4	5	6	7
JA23	JA24	JA25	JA26	JA27	JA28	JA29
AMN7	AMN8	AMN9	AMN10	AMN11	AMN12	AMN13
TIR15	TIR16	TIR17	TIR18	TIR19	TIR20	TIR21
8	9	10	11	12	13	14
JA30	JA31	F1	F2	F3	F4	F5
AMN14	AMN15	AMN16	AMN17	AMN18	AMN19	AMN20
TIR22	TIR23	TIR24	TIR25	TIR26	TIR27	TIR28
15	16	17	18	19	20	21
F6	F7	F8	F9	F10	F 11	F12
AMN21	AMN22	AMN23	AMN24	AMN25	AMN26	AMN28
TIR29	TIR30	YEK1	YEK2	YEK3	YEK4	YEK5
22	23	24	25	26	27	28
F13	F14	F15	F16	F17	F18	F19
AMN29	AMN30	RNT1	RNT2	RNT3	RNT4	RNT5
YEK6	YEK7	YEK8	YEK9	YEK10	YEK11	YEK12
SHAHIDI 64						
F20						
RNT6						
YEK13						

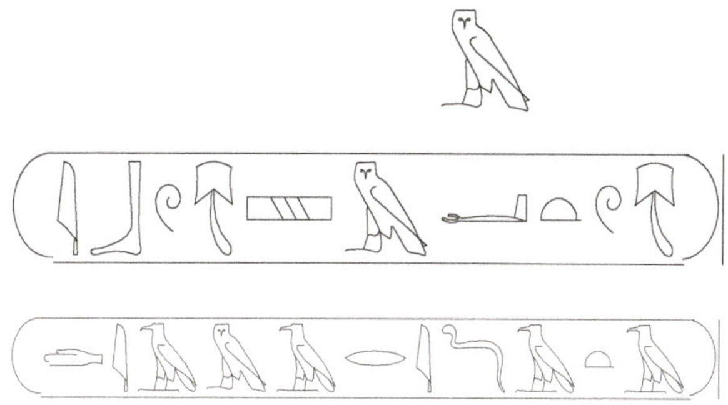

THE NU AFRIKAN CALENDAR

MAAFA 514 B.S.E./ASIENTO 2030-31 C.E.
KAM 6266-67 S.M./EC 2022-23 A.M. (ዓ.ም.)

AFU 514/2030-31/6266-67/2022-23						
Ashe	Ojobo	Iwa	Araorun	Ile	Itutu	Erinle
1	2	3	4	5	6	7
F21	F22	F23	F24	F25	F26	F27
RNT7	RNT8	RNT9	RNT10	RNT11	RNT12	RNT13
YEK14	YEK15	YEK16	YEK17	YEK18	YEK19	YEK20
8	9	10	11	12	13	14
F28	M1	M2	M3	M4	M5	M6
RNT14	RNT15	RNT16	RNT17	RNT18	RNT19	RNT20
YEK21	YEK22	YEK23	YEK24	YEK25	YEK26	YEK27
15	16	17	18	19	20	21
M7	M8	M9	M10	M11	M12	M13
RNT21	RNT22	RNT23	RNT24	RNT25	RNT26	RNT27
YEK28	YEK29	YEK30	MAG1	MAG2	MAG3	MAG4
22	23	24	25	26	27	28
M14	M15	M16	M17	M18	M19	M20
RNT28	RNT29	RNT30	KHN1	KHN2	KHN3	KHN4
MAG5	MAG6	MAG7	MAG8	MAG9	MAG10	MAG11

KAWAIDA 514/2030-31/6266-67/2022-23						
Ashe	Ojobo	Iwa	Araorun	Ile	Itutu	Erinle
1	2	3	4	5	6	7
A18	A19	A20	A21	A22	A23	A24
PIN3	PIN4	PIN5	PIN6	PIN7	PIN8	PIN9
MIY10	MIY11	MIY12	MIY13	MIY14	MIY15	MIY16
8	9	10	11	12	13	14
A25	A26	A27	A28	A29	A30	MA1
PIN10	PIN11	PIN12	PIN13	PIN14	PIN15	PIN16
MIY17	MIY18	MIY19	MIY20	MIY21	MIY22	MIY23
15	16	17	18	19	20	21
MA2	MA3	MA4	MA5	MA6	MA7	MA8
PIN17	PIN18	PIN19	PIN20	PIN21	PIN22	PIN23
MIY24	MIY25	MIY26	MIY27	MIY28	MIY29	MIY30
22	23	24	25	26	27	28
MA9	MA10	MA11	MA12	MA13	MA14	MA15
PIN24	PIN25	PIN26	PIN27	PIN28	PIN29	PIN30
GIN1	GIN2	GIN3	GIN4	GIN5	GIN6	GIN7

UHURU 514/2030-31/6266-67/2022-23						
Ashe	Ojobo	Iwa	Araorun	Ile	Itutu	Erinle
1	2	3	4	5	6	7
M21	M22	M23	M24	M25	M26	M27
KHN5	KHN6	KHN7	KHN8	KHN9	KHN10	KHN11
MAG12	MAG13	MAG14	MAG15	MAG16	MAG17	MAG18
8	9	10	11	12	13	14
M28	M29	M30	M31	A1	A2	A3
KHN12	KHN13	KHN14	KHN15	KHN16	KHN17	KHN18
MAG19	MAG20	MAG21	MAG22	MAG23	MAG24	MAG25
15	16	17	18	19	20	21
A4	A5	A6	A7	A8	A9	A10
KHN19	KHN20	KHN21	KHN22	KHN23	KHN24	KHN25
MAG26	MAG27	MAG28	MAG29	MAG30	MIY1	MIY2
22	23	24	25	26	27	28
A11	A12	A13	A14	A15	A16	A17
KHN26	KHN27	KHN28	KHN29	KHN30	PIN1	PIN2
MIY3	MIY4	MIY5	MIY6	MIY7	MIY8	MIY9

MABABA WEUSI 514/2030-31/6266-67/2022-23						
Ashe	Ojobo	Iwa	Araorun	Ile	Itutu	Erinle
1	2	3	4	5	6	7
MA16	MA17	MA18	MA19	MA20	MA21	MA22
APP1	APP2	APP3	APP4	APP5	APP6	APP7
GIN8	GIN9	GIN10	GIN11	GIN12	GIN13	GIN14
8	9	10	11	12	13	14
MA23	MA24	MA25	MA26	MA27	MA28	MA29
APP8	APP9	APP10	APP11	APP12	APP13	APP14
GIN15	GIN16	GIN17	GIN18	GIN19	GIN20	GIN21
15	16	17	18	19	20	21
MA30	MA31	J1	J2	J3	J4	J5
APP15	APP16	APP17	APP18	APP19	APP20	APP21
GIN22	GIN23	GIN24	GIN25	GIN26	GIN27	GIN28
22	23	24	25	26	27	28
J6	J7	J8	J9	J10	J11	J12
APP22	APP23	APP24	APP25	APP26	APP27	APP28
GIN29	GIN30	SAN1	SAN2	SAN3	SAN4	SAN5

THE NU AFRIKAN CALENDAR

MWANAMKE WEUSI 514/2030-31/6266-67/2022-23

Ashe	Ojobo	Iwa	Araorun	Ile	Itutu	Erinle
1	2	3	4	5	6	7
J13	J14	J15	J16	J17	J18	J19
APP29	APP30	MSR1	MSR2	MSR3	MSR4	MSR5
SAN6	SAN7	SAN8	SAN9	SAN10	SAN11	SAN12
8	9	10	11	12	13	14
J20	J21	J22	J23	J24	J25	J26
MSR6	MSR7	MSR8	MSR9	MSR10	MSR11	MSR12
SAN13	SAN14	SAN15	SAN16	SAN17	SAN18	SAN19
15	16	17	18	19	20	21
J27	J28	J29	J30	JU1	JU2	JU3
MSR13	MSR14	MSR15	MSR16	MSR17	MSR18	MSR19
SAN20	SAN21	SAN22	SAN23	SAN24	SAN25	SAN26
22	23	24	25	26	27	28
JU4	JU5	JU6	JU7	JU8	JU9	JU10
MSR20	MRS21	MSR22	MSR23	MSR24	MSR25	MSR26
SAN27	SAN28	SAN29	SAN30	HAM1	HAM2	HAM3

TAIFA WEUSI 514/2030-31/6266-67/2022-23

Ashe	Ojobo	Iwa	Araorun	Ile	Itutu	Erinle
1	2	3	4	5	6	7
AU8	AU9	AU10	AU11	AU12	AU13	AU14
THT20	THT21	THT22	THT23	THT24	THT25	THT26
NAH2	NAH3	NAH4	NAH5	NAH6	NAH7	NAH8
8	9	10	11	12	13	14
AU15	AU16	AU17	AU18	AU19	AU20	AU21
THT27	THT28	THT29	THT30	PAP1	PAP1	PAP2
NAH9	NAH10	NAH11	NAH12	NAH13	NAH14	NAH15
15	16	17	18	19	20	21
AU22	AU23	AU24	AU25	AU26	AU27	AU28
PAP3	PAP4	PAP5	PAP6	PAP7	PAP8	PAP9
NAH16	NAH17	NAH18	NAH19	NAH20	NAH21	NAH22
22	23	24	25	26	27	28
AU29	AU30	AU31	S1	S2	S3	S4
PAP10	PAP11	PAP12	PAP13	PAP14	PAP15	PAP16
NAH23	NAH24	NAH25	NAH26	NAH27	NAH28	NAH29

MTU WEUSI 514/2030-31/6266-67/2022-23

Ashe	Ojobo	Iwa	Araorun	Ile	Itutu	Erinle
1	2	3	4	5	6	7
JU11	JU12	JU13	JU14	JU15	JU16	JU17
MSR27	MSR28	MSR29	MSR30	*ASR1*	*HRU2*	*STH3*
HAM4	HAM5	HAM6	HAM7	HAM8	HAM9	HAM10
8	9	10	11	12	13	14
JU18	JU19	JU20	JU21	JU22	JU23	JU24
AST4	*NBH5*	THT1	THT2	THT3	THT4	THT5
HAM11	HAM12	HAM13	AM14	HAM15	HAM16	HAM17
15	16	17	18	19	20	21
JU25	JU26	JU27	JU28	JU29	JU30	JU31
THT6	THT7	THT8	THT9	THT10	THT11	THT12
HAM18	HAM19	HAM20	HAM21	HAM22	HAM23	HAM24
22	23	24	25	26	27	28
AU1	AU2	AU3	AU4	AU5	AU6	AU7
THT13	THT14	THT15	THT16	THT17	THT18	THT19
HAM25	HAM26	HAM27	HAM28	HAM29	HAM30	NAH1

UWEZO 514/2030-31/6266-67/2022-23

Ashe	Ojobo	Iwa	Araorun	Ile	Itutu	Erinle
1	2	3	4	5	6	7
S5	S6	S7	S8	S9	S10	S11
PAP17	PAP18	PAP19	PAP20	PAP21	PAP22	PAP23
NAH30	*PAG1*	*PAG2*	*PAG3*	*PAG4*	*PAG5*	MAS1
8	9	10	11	12	13	14
S12	S13	S14	S15	S16	S17	S18
PAP24	PAP25	PAP26	PAP27	PAP28	PAP29	PAP30
MAS2	MAS3	MAS4	MAS5	MAS6	MAS7	MAS8
15	16	17	18	19	20	21
S19	S20	S21	S22	S23	S24	S25
HTH1	HTH2	HTH3	HTH4	HTH5	HTH6	HTH7
MAS9	MAS10	MAS11	MAS12	MAS13	MAS14	MAS15
22	23	24	25	26	27	28
S26	S27	S28	S29	S30	O1	O2
HTH8	HTH9	HTH10	HTH11	HTH12	HTH13	HTH14
MAS16	MAS17	MAS18	MAS19	MAS20	MAS21	MAS22

THE NU AFRIKAN CALENDAR

EKUNDU 514/2030-31/6266-67/2022-23						
Ashe	Ojobo	Iwa	Araorun	Ile	Itutu	Erinle
1	2	3	4	5	6	7
O3	O4	O5	O6	O7	O8	O9
HTH15	HTH16	HTH17	HTH18	HTH19	HTH20	HTH21
MAS23	MAS24	MAS25	MAS26	MAS27	MAS28	MAS29
8	9	10	11	12	13	14
O10	O11	O12	O13	O14	O15	O16
HTH22	HTH23	HTH24	HTH25	HTH26	HTH27	HTH28
MAS30	THi1	THi2	THi3	THi4	THi5	THi6
15	16	17	18	19	20	21
O17	O18	O19	O20	O21	O22	O23
HTH29	HTH30	KHK1	KHK2	KHK3	KHK4	KHK5
THi7	THi8	THi9	THi10	THi11	THi12	THi13
22	23	24	25	26	27	28
O24	O25	O26	O27	O28	O29	O30
KHK6	KHK7	KHK8	KHK9	KHK10	KHK11	KHK12
THi14	THi15	THi16	THi17	THi18	THi19	THi20

KIJANI 514/2030-31/6266-67/2022-23						
Ashe	Ojobo	Iwa	Araorun	Ile	Itutu	Erinle
1	2	3	4	5	6	7
N28	N29	N30	D1	D2	D3	D4
TBT11	TBT12	TBT13	TBT14	TBT15	TBT16	TBT17
HID19	HID20	HID21	HID22	HID23	HID24	HID25
8	9	10	11	12	13	14
D5	D6	D7	D8	D9	D10	D11
TBT18	TBT19	TBT20	TBT21	TBT22	TBT23	TBT24
HID26	HID27	HID28	HID29	HID30	TAH1	TAH2
15	16	17	18	19	20	21
D12	D13	D14	D15	D16	D17	D18
TBT25	TBT26	TBT27	TBT28	TBT29	TBT30	MKH1
TAH3	TAH4	TAH5	TAH6	TAH7	TAH8	TAH9
22	23	24	25	26	27	28
D19	D20	D21	D22	D23	D24	D25
MKH2	MKH3	MKH4	MKH5	MKH6	MKH7	MKH8
TAH10	TAH11	TAH12	TAH13	TAH14	TAH15	TAH16

EUSI 514/2030-31/6266-67/2022-23						
Ashe	Ojobo	Iwa	Araorun	Ile	Itutu	Erinle
1	2	3	4	5	6	7
O31	N1	N2	N3	N4	N5	N6
KHK13	KHK14	KHK15	KHK16	KHK17	KHK18	KHK19
THi21	THi22	THi23	THi24	THi25	THi26	THi27
8	9	10	11	12	13	14
N7	N8	N9	N10	N11	N12	N13
KHK20	KHK21	KHK22	KHK23	KHK24	KHK25	KHK26
THi28	THi29	THi30	HID1	HID2	HID3	HID4
15	16	17	18	19	20	21
N14	N15	N16	N17	N18	N19	N20
KHK27	KHK28	KHK29	KHK30	TBT1	TBT2	TBT3
HID5	HID6	HID7	HID8	HID9	HID10	HID11
22	23	24	25	26	27	28
N21	N22	N23	N24	N25	N26	N27
TBT4	TBT5	TBT6	TBT7	TBT8	TBT9	TBT10
HID12	HID13	HID14	HID15	HID16	HID17	HID18

KWANZA 514/2030-31/6266-67/2022-23						
Umoja	Kujichagulia	Ujima	Ujamaa	Nia	Kuumba	Imani
1	2	3	4	5	6	7
D26	D27	D28	D29	D30	D31	JA1
MKH9	MKH10	MKH11	MKH12	MKH13	MKH14	MKH15
TAH17	TAH18	TAH19	TAH20	TAH21	TAH22	TAH23
8	9	10	11	12	13	14
JA2	JA3	JA4	JA5	JA6	JA7	JA8
MKH16	MKH17	MKH18	MKH19	MKH20	MKH21	MKH22
TAH24	TAH25	TAH26	TAH27	TAH28	TAH29	TAH30
15	16	17	18	19	20	21
JA9	JA10	JA11	JA12	JA13	JA14	JA15
MKH23	MKH24	MKH25	KMH26	MKH27	MKH28	MKH29
TIR1	TIR2	TIR3	TIR4	TIR5	TIR6	TIR7
22	23	24	25	26	27	28
JA16	JA17	JA18	JA19	JA20	JA21	JA22
MKH30	AMN1	AMN2	AMN3	AMN4	AMN5	AMN6
TIR8	TIR9	TIR10	TIR11	TIR12	TIR13	TIR14

THE NU AFRIKAN CALENDAR

SHAHIDI 514/2030-31/6266-67/2022-23						
Ashe	Ojobo	Iwa	Araorun	Ile	Itutu	Erinle
1	2	3	4	5	6	7
JA23	JA24	JA25	JA26	JA27	JA28	JA29
AMN7	AMN8	AMN9	AMN10	AMN11	AMN12	AMN13
TIR15	TIR16	TIR17	TIR18	TIR19	TIR20	TIR21
8	9	10	11	12	13	14
JA30	JA31	F1	F2	F3	F4	F5
AMN14	AMN15	AMN16	AMN17	AMN18	AMN19	AMN20
TIR22	TIR23	TIR24	TIR25	TIR26	TIR27	TIR28
15	16	17	18	19	20	21
F6	F7	F8	F9	F10	F 11	F12
AMN21	AMN22	AMN23	AMN24	AMN25	AMN26	AMN28
TIR29	TIR30	YEK1	YEK2	YEK3	YEK4	YEK5
22	23	24	25	26	27	28
F13	F14	F15	F16	F17	F18	F19
AMN29	AMN30	RNT1	RNT2	RNT3	RNT4	RNT5
YEK6	YEK7	YEK8	YEK9	YEK10	YEK11	YEK12
SHAHIDI 65						
F20						
RNT6						
YEK13						

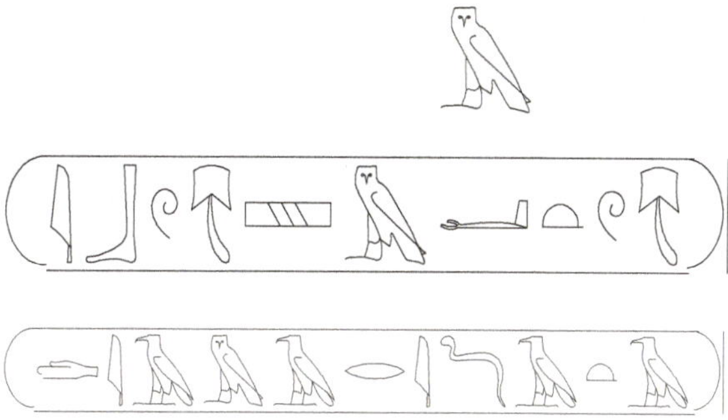

MAAFA 515 B.S.E./ASIENTO 2031-32 C.E./
KAM 6267-68 S.M./EC 2023-24 A.M. (ዓ.ም.)

AFU 515/2031-32/6267-68/2023-24

Ashe	Ojobo	Iwa	Araorun	Ile	Itutu	Erinle
1	2	3	4	5	6	7
F21	F22	F23	F24	F25	F26	F27
RNT7	RNT8	RNT9	RNT10	RNT11	RNT12	RNT13
YEK14	YEK15	YEK16	YEK17	YEK18	YEK19	YEK20
8	9	10	11	12	13	14
F28	M1	M2	M3	M4	M5	M6
RNT14	RNT15	RNT16	RNT17	RNT18	RNT19	RNT20
YEK21	YEK22	YEK23	YEK24	YEK25	YEK26	YEK27
15	16	17	18	19	20	21
M7	M8	M9	M10	M11	M12	M13
RNT21	RNT22	RNT23	RNT24	RNT25	RNT26	RNT27
YEK28	YEK 29	YEK30	MAG1	MAG2	MAG3	MAG4
22	23	24	25	26	27	28
M14	M15	M16	M17	M18	M19	M20
RNT28	RNT29	RNT30	KHN1	KHN2	KHN3	KHN4
MAG5	MAG6	MAG7	MAG8	MAG9	MAG10	MAG11

UHURU 515/2031-32/6267-68/2023-24

Ashe	Ojobo	Iwa	Araorun	Ile	Itutu	Erinle
1	2	3	4	5	6	7
M21	M22	M23	M24	M25	M26	M27
KHN5	KHN6	KHN7	KHN8	KHN9	KHN10	KHN11
MAG12	MAG13	MAG14	MAG15	MAG16	MAG17	MAG18
8	9	10	11	12	13	14
M28	M29	M30	M31	A1	A2	A3
KHN12	KHN13	KHN14	KHN15	KHN16	KHN17	KHN18
MAG19	MAG20	MAG21	MAG22	MAG23	MAG24	MAG25
15	16	17	18	19	20	21
A4	A5	A6	A7	A8	A9	A10
KHN19	KHN20	KHN21	KHN22	KHN23	KHN24	KHN25
MAG26	MAG27	MAG28	MAG29	MAG30	MIY1	MIY2
22	23	24	25	26	27	28
A11	A12	A13	A14	A15	A16	A17
KHN26	KHN27	KHN28	KHN29	KHN30	PIN1	PIN2
MIY3	MIY4	MIY5	MIY6	MIY7	MIY8	MIY9

KAWAIDA 515/2031-32/6267-68/2023-24

Ashe	Ojobo	Iwa	Araorun	Ile	Itutu	Erinle
1	2	3	4	5	6	7
A18	A19	A20	A21	A22	A23	A24
PIN3	PIN4	PIN5	PIN6	PIN7	PIN8	PIN9
MIY10	MIY11	MIY12	MIY13	MIY14	MIY15	MIY16
8	9	10	11	12	13	14
A25	A26	A27	A28	A29	A30	MA1
PIN10	PIN11	PIN12	PIN13	PIN14	PIN15	PIN16
MIY17	MIY18	MIY19	MIY20	MIY21	MIY22	MIY23
15	16	17	18	19	20	21
MA2	MA3	MA4	MA5	MA6	MA7	MA8
PIN17	PIN18	PIN19	PIN20	PIN21	PIN22	PIN23
MIY24	MIY25	MIY26	MIY27	MIY28	MIY29	MIY30
22	23	24	25	26	27	28
MA9	MA10	MA11	MA12	MA13	MA14	MA15
PIN24	PIN25	PIN26	PIN27	PIN28	PIN29	PIN30
GIN1	GIN2	GIN3	GIN4	GIN5	GIN6	GIN7

MABABA WEUSI 515/2031-32/6267-68/2023-24

Ashe	Ojobo	Iwa	Araorun	Ile	Itutu	Erinle
1	2	3	4	5	6	7
MA16	MA17	MA18	MA19	MA20	MA21	MA22
APP1	APP2	APP3	APP4	APP5	APP6	APP7
GIN8	GIN9	GIN10	GIN11	GIN12	GIN13	GIN14
8	9	10	11	12	13	14
MA23	MA24	MA25	MA26	MA27	MA28	MA29
APP8	APP9	APP10	APP11	APP12	APP13	APP14
GIN15	GIN16	GIN17	GIN18	GIN19	GIN20	GIN21
15	16	17	18	19	20	21
MA30	MA31	J1	J2	J3	J4	J5
APP15	APP16	APP17	APP18	APP19	APP20	APP21
GIN22	GIN23	GIN24	GIN25	GIN26	GIN27	GIN28
22	23	24	25	26	27	28
J6	J7	J8	J9	J10	J11	J12
APP22	APP23	APP24	APP25	APP26	APP27	APP28
GIN29	GIN30	SAN1	SAN2	SAN3	SAN4	SAN5

THE NU AFRIKAN CALENDAR

MWANAMKE WEUSI 515/2031-32/6267-68/2023-24						
Ashe	Ojobo	Iwa	Araorun	Ile	Itutu	Erinle
1	2	3	4	5	6	7
J13	J14	J15	J16	J17	J18	J19
APP29	APP30	MSR1	MSR2	MSR3	MSR4	MSR5
SAN6	SAN7	SAN8	SAN9	SAN10	SAN11	SAN12
8	9	10	11	12	13	14
J20	J21	J22	J23	J24	J25	J26
MSR6	MSR7	MSR8	MSR9	MSR10	MSR11	MSR12
SAN13	SAN14	SAN15	SAN16	SAN17	SAN18	SAN19
15	16	17	18	19	20	21
J27	J28	J29	J30	JU1	JU2	JU3
MSR13	MSR14	MSR15	MSR16	MSR17	MSR18	MSR19
SAN20	SAN21	SAN22	SAN23	SAN24	SAN25	SAN26
22	23	24	25	26	27	28
JU4	JU5	JU6	JU7	JU8	JU9	JU10
MSR20	MRS21	MSR22	MSR23	MSR24	MSR25	MSR26
SAN27	SAN28	SAN29	SAN30	HAM1	HAM2	HAM3

TAIFA WEUSI 515/2031-32/6267-68/2023-24						
Ashe	Ojobo	Iwa	Araorun	Ile	Itutu	Erinle
1	2	3	4	5	6	7
AU8	AU9	AU10	AU11	AU12	AU13	AU14
THT20	THT21	THT22	THT23	THT24	THT25	THT26
NAH2	NAH3	NAH4	NAH5	NAH6	NAH7	NAH8
8	9	10	11	12	13	14
AU15	AU16	AU17	AU18	AU19	AU20	AU21
THT27	THT28	THT29	THT30	PAP1	PAP1	PAP2
NAH9	NAH10	NAH11	NAH12	NAH13	NAH14	NAH15
15	16	17	18	19	20	21
AU22	AU23	AU24	AU25	AU26	AU27	AU28
PAP3	PAP4	PAP5	PAP6	PAP7	PAP8	PAP9
NAH16	NAH17	NAH18	NAH19	NAH20	NAH21	NAH22
22	23	24	25	26	27	28
AU29	AU30	AU31	S1	S2	S3	S4
PAP10	PAP11	PAP12	PAP13	PAP14	PAP15	PAP16
NAH23	NAH24	NAH25	NAH26	NAH27	NAH28	NAH29

MTU WEUSI 515/2031-32/6267-68/2023-24						
Ashe	Ojobo	Iwa	Araorun	Ile	Itutu	Erinle
1	2	3	4	5	6	7
JU11	JU12	JU13	JU14	JU15	JU16	JU17
MSR27	MSR28	MSR29	MSR30	*ASR1*	*HRU2*	*STH3*
HAM4	HAM5	HAM6	HAM7	HAM8	HAM9	HAM10
8	9	10	11	12	13	14
JU18	JU19	JU20	JU21	JU22	JU23	JU24
AST4	*NBH5*	THT1	THT2	THT3	THT4	THT5
HAM11	HAM12	HAM13	HAM14	HAM15	HAM16	HAM17
15	16	17	18	19	20	21
JU25	JU26	JU27	JU28	JU29	JU30	JU31
THT6	THT7	THT8	THT9	THT10	THT11	THT12
HAM18	HAM19	HAM20	HAM21	HAM22	HAM23	HAM24
22	23	24	25	26	27	28
AU1	AU2	AU3	AU4	AU5	AU6	AU7
THT13	THT14	THT15	THT16	THT17	THT18	THT19
HAM25	HAM26	HAM27	HAM28	HAM29	HAM30	NAH1

UWEZO 515/2031-32/6267-68/2023-24						
Ashe	Ojobo	Iwa	Araorun	Ile	Itutu	Erinle
1	2	3	4	5	6	7
S5	S6	S7	S8	S9	S10	S11
PAP17	PAP18	PAP19	PAP20	PAP21	PAP22	PAP23
NAH30	*QUA1*	*QUA2*	*QUA3*	*QUA4*	*QUA5*	*QUA6*
8	9	10	11	12	13	14
S12	S13	S14	S15	S16	S17	S18
PAP24	PAP25	PAP26	PAP27	PAP28	PAP29	PAP30
MAS1	MAS2	MAS3	MAS4	MAS5	MAS6	MAS7
15	16	17	18	19	20	21
S19	S20	S21	S22	S23	S24	S25
HTH1	HTH2	HTH3	HTH4	HTH5	HTH6	HTH7
MAS8	MAS9	MAS10	MAS11	MAS12	MAS13	MAS14
22	23	24	25	26	27	28
S26	S27	S28	S29	S30	O1	O2
HTH8	HTH9	HTH10	HTH11	HTH12	HTH13	HTH14
MAS15	MAS16	MAS17	MAS18	MAS19	MAS20	MAS21

THE NU AFRIKAN CALENDAR

EKUNDU 515/2031-32/6267-68/2023-24						
Ashe	Ojobo	Iwa	Araorun	Ile	Itutu	Erinle
1	2	3	4	5	6	7
O3	O4	O5	O6	O7	O8	O9
HTH15	HTH16	HTH17	HTH18	HTH19	HTH20	HTH21
MAS22	MAS23	MAS 24	MAS25	MAS26	MAS27	MAS28
8	9	10	11	12	13	14
O10	O11	O12	O13	O14	O15	O16
HTH22	HTH23	HTH24	HTH25	HTH26	HTH27	HTH28
MAS29	MAS30	THi1	THi2	THi3	THi4	THi5
15	16	17	18	19	20	21
O17	O18	O19	O20	O21	O22	O23
HTH29	HTH30	KHK1	KHK2	KHK3	KHK4	KHK5
THi6	THi7	THi8	THi9	THi10	THi11	THi12
22	23	24	25	26	27	28
O24	O25	O26	O27	O28	O29	O30
KHK6	KHK7	KHK8	KHK9	KHK10	KHK11	KHK12
THi13	THi14	THi15	THi16	THi17	THi18	THi19

KIJANI 515/2031-32/6267-68/2023-24						
Ashe	Ojobo	Iwa	Araorun	Ile	Itutu	Erinle
1	2	3	4	5	6	7
N28	N29	N30	D1	D2	D3	D4
TBT11	TBT12	TBT13	TBT14	TBT15	TBT16	TBT17
HID18	HID19	HID20	HID21	HID22	HID23	HID24
8	9	10	11	12	13	14
D5	D6	D7	D8	D9	D10	D11
TBT18	TBT19	TBT20	TBT21	TBT22	TBT23	TBT24
HID25	HID26	HID27	HID28	HID29	HID30	TAH1
15	16	17	18	19	20	21
D12	D13	D14	D15	D16	D17	D18
TBT25	TBT26	TBT27	TBT28	TBT29	TBT30	MKH1
TAH2	TAH3	TAH4	TAH5	TAH6	TAH7	TAH8
22	23	24	25	26	27	28
D19	D20	D21	D22	D23	D24	D25
MKH2	MKH3	MKH4	MKH5	MKH6	MKH7	MKH8
TAH9	TAH10	TAH11	TAH12	TAH13	TAH14	TAH15

EUSI 515/2031-32/6267-68/2023-24						
Ashe	Ojobo	Iwa	Araorun	Ile	Itutu	Erinle
1	2	3	4	5	6	7
O31	N1	N2	N3	N4	N5	N6
KHK13	KHK14	KHK15	KHK16	KHK17	KHK18	KHK19
THi20	THi21	THi22	THi23	THi24	THi25	THi26
8	9	10	11	12	13	14
N7	N8	N9	N10	N11	N12	N13
KHK20	KHK21	KHK22	KHK23	KHK24	KHK25	KHK26
THi27	THi28	THi29	THi30	HID1	HID2	HID3
15	16	17	18	19	20	21
N14	N15	N16	N17	N18	N19	N20
KHK27	KHK28	KHK29	KHK30	TBT1	TBT2	TBT3
HID4	HID5	HID6	HID7	HID8	HID9	HID10
22	23	24	25	26	27	28
N21	N22	N23	N24	N25	N26	N27
TBT4	TBT5	TBT6	TBT7	TBT8	TBT9	TBT10
HID11	HID12	HID13	HID14	HID15	HID16	HID17

KWANZA 515/2031-32/6267-68/2023-24						
Umoja	Kujichagulia	Ujima	Ujamaa	Nia	Kuumba	Imani
1	2	3	4	5	6	7
D26	D27	D28	D29	D30	D31	JA1
MKH9	MKH10	MKH11	MKH12	MKH13	MKH14	MKH15
TAH16	TAH17	TAH18	TAH19	TAH20	TAH21	TAH22
8	9	10	11	12	13	14
JA2	JA3	JA4	JA5	JA6	JA7	JA8
MKH16	MKH17	MKH18	MKH19	MKH20	MKH21	MKH22
TAH23	TAH24	TAH25	TAH26	TAH27	TAH28	TAH29
15	16	17	18	19	20	21
JA9	JA10	JA11	JA12	JA13	JA14	JA15
MKH23	MKH24	MKH25	KMH26	MKH27	MKH28	MKH29
TAH30	TIR1	TIR2	TIR3	TIR4	TIR5	TIR6
22	23	24	25	26	27	28
JA16	JA17	JA18	JA19	JA20	JA21	JA22
MKH30	AMN1	AMN2	AMN3	AMN4	AMN5	AMN6
TIR7	TIR8	TIR9	TIR10	TIR11	TIR12	TIR13

THE NU AFRIKAN CALENDAR

SHAHIDI 515/2031-32/6267-68/2023-24						
Ashe	Ojobo	Iwa	Araorun	Ile	Itutu	Erinle
1	2	3	4	5	6	7
JA23	JA24	JA25	JA26	JA27	JA28	JA29
AMN7	AMN8	AMN9	AMN10	AMN11	AMN12	AMN13
TIR14	TIR15	TIR16	TIR17	TIR18	TIR19	TIR20
8	9	10	11	12	13	14
JA30	JA31	F1	F2	F3	F4	F5
AMN14	AMN15	AMN16	AMN17	AMN18	AMN19	AMN20
TIR21	TIR22	TIR23	TIR24	TIR25	TIR26	TIR27
15	16	17	18	19	20	21
F6	F7	F8	F9	F10	F 11	F12
AMN21	AMN22	AMN23	AMN24	AMN25	AMN26	AMN28
TIR28	TIR29	TIR30	YEK1	YEK2	YEK3	YEK4
22	23	24	25	26	27	28
F13	F14	F15	F16	F17	F18	F19
AMN29	AMN30	RNT1	RNT2	RNT3	RNT4	RNT5
YEK5	YEK6	YEK7	YEK8	YEK9	YEK10	YEK11
SHAHIDI 66						
F20						
RNT6						
YEK12						

THE NU AFRIKAN CALENDAR

MAAFA 516 B.S.E./ASIENTO 2032-33 C.E./
KAM 6268-69 S.M./EC 2024-25 A.M. (ዓ.ም.)

AFU 516/2032-33/6268-69/2024-25

Ashe	Ojobo	Iwa	Araorun	Ile	Itutu	Erinle
1	2	3	4	5	6	7
F21	F22	F23	F24	F25	F26	F27
RNT7	RNT8	RNT9	RNT10	RNT11	RNT12	RNT13
YEK13	YEK14	YEK15	YEK16	YEK17	YEK18	YEK19
8	9	10	11	12	13	14
F28	F29	M1	M2	M3	M4	M5
RNT14	RNT15	RNT16	RNT17	RNT18	RNT19	RNT20
YEK 20	YEK21	YEK22	YEK23	YEK24	YEK25	YEK26
15	16	17	18	19	20	21
M6	M7	M8	M9	M10	M11	M12
RNT21	RNT22	RNT23	RNT24	RNT25	RNT26	RNT27
YEK 27	YEK28	YEK29	YEK30	MAG1	MAG2	MAG3
22	23	24	25	26	27	28
M13	M14	M15	M16	M17	M18	M19
RNT28	RNT29	RNT30	KHN1	KHN2	KHN3	KHN4
MAG4	MAG5	MAG6	MAG7	MAG8	MAG9	MAG10
SHARIFU 126						
M20						
KHN5						
MAG11						

UHURU 516/2032-33/6268-69/2024-25

Ashe	Ojobo	Iwa	Araorun	Ile	Itutu	Erinle
1	2	3	4	5	6	7
M21	M22	M23	M24	M25	M26	M27
KHN6	KHN7	KHN8	KHN9	KHN10	KHN11	KHN12
MAG12	MAG13	MAG14	MAG15	MAG16	MAG17	MAG18
8	9	10	11	12	13	14
M28	M29	M30	M31	A1	A2	A3
KHN13	KHN14	KHN15	KHN16	KHN17	KHN18	KHN19
MAG19	MAG20	MAG21	MAG22	MAG23	MAG24	MAG25
15	16	17	18	19	20	21
A4	A5	A6	A7	A8	A9	A10
KHN20	KHN21	KHN222	KHN23	KHN24	KHN25	KHN26
MAG26	MAG27	MAG28	MAG29	MAG30	MIY1	MIY2
22	23	24	25	26	27	28
A11	A12	A13	A14	A15	A16	A17
KHN27	KHN28	KHN29	KHN30	PIN1	PIN2	PIN3
MIY3	MIY4	MIY5	MIY6	MIY7	MIY8	MIY9

KAWAIDA 516/2032-33/6268-69/2024-25

Ashe	Ojobo	Iwa	Araorun	Ile	Itutu	Erinle
1	2	3	4	5	6	7
A18	A19	A20	A21	A22	A23	A24
PIN4	PIN5	PIN6	PIN7	PIN8	PIN9	PIN10
MIY10	MIY11	MIY12	MIY13	MIY14	MIY15	MIY16
8	9	10	11	12	13	14
A25	A26	A27	A28	A29	A30	MA1
PIN11	PIN12	PIN13	PIN14	PIN15	PIN16	PIN17
MIY17	MIY18	MIY19	MIY20	MIY21	MIY22	MIY23
15	16	17	18	19	20	21
MA2	MA3	MA4	MA5	MA6	MA7	MA8
PIN18	PIN19	PIN20	PIN21	PIN22	PIN23	PIN24
MIY24	MIY25	MIY26	MIY27	MIY28	MIY29	MIY30
22	23	24	25	26	27	28
MA9	MA10	MA11	MA12	MA13	MA14	MA15
PIN25	PIN26	PIN27	PIN28	PIN29	PIN30	APP1
GIN1	GIN2	GIN3	GIN4	GIN5	GIN6	GIN7

MABABA WEUSI 516/2032-33/6268-69/2024-25

Ashe	Ojobo	Iwa	Araorun	Ile	Itutu	Erinle
1	2	3	4	5	6	7
MA16	MA17	MA18	MA19	MA20	MA21	MA22
APP2	APP3	APP4	APP5	APP6	APP7	APP8
GIN8	GIN9	GIN10	GIN11	GIN12	GIN13	GIN14
8	9	10	11	12	13	14
MA23	MA24	MA25	MA26	MA27	MA28	MA29
APP9	APP10	APP11	APP12	APP13	APP14	APP15
GIN15	GIN16	GIN17	GIN18	GIN19	GIN20	GIN21
15	16	17	18	19	20	21
MA30	MA31	J1	J2	J3	J4	J5
APP16	APP17	APP18	APP19	APP20	APP21	APP22
GIN22	GIN23	GIN24	GIN25	GIN26	GIN27	GIN28
22	23	24	25	26	27	28
J6	J7	J8	J9	J10	J11	J12
APP23	APP24	APP25	APP26	APP27	APP28	APP29
GIN29	GIN30	SAN1	SAN2	SAN3	SAN4	SAN5

THE NU AFRIKAN CALENDAR

MWANAMKE WEUSI 516/2032-33/6268-69/2024-25

Ashe	Ojobo	Iwa	Araorun	Ile	Itutu	Erinle
1	2	3	4	5	6	7
J13	J14	J15	J16	J17	J18	J19
APP30	MSR1	MSR2	MSR3	MSR4	MSR5	MSR6
SAN6	SAN7	SAN8	SAN9	SAN10	SAN11	SAN12
8	9	10	11	12	13	14
J20	J21	J22	J23	J24	J25	J26
MSR7	MSR8	MSR9	MSR10	MSR11	MSR12	MSR13
SAN13	SAN14	SAN15	SAN16	SAN17	SAN18	SAN19
15	16	17	18	19	20	21
J27	J28	J29	J30	JU1	JU2	JU3
MSR14	MSR15	MSR16	MSR17	MSR18	MSR19	MSR20
SAN20	SAN21	SAN22	SAN23	SAN24	SAN25	SAN26
22	23	24	25	26	27	28
JU4	JU5	JU6	JU7	JU8	JU9	JU10
MSR21	MRS22	MSR23	MSR24	MSR25	MSR26	MSR27
SAN27	SAN28	SAN29	SAN30	HAM1	HAM2	HAM3

TAIFA WEUSI 516/2032-33/6268-69/2024-25

Ashe	Ojobo	Iwa	Araorun	Ile	Itutu	Erinle
1	2	3	4	5	6	7
AU8	AU9	AU10	AU11	AU12	AU13	AU14
THT20	THT21	THT22	THT23	THT24	THT25	THT26
NAH2	NAH3	NAH4	NAH5	NAH6	NAH7	NAH8
8	9	10	11	12	13	14
AU15	AU16	AU17	AU18	AU19	AU20	AU21
THT27	THT28	THT29	THT30	PAP1	PAP1	PAP2
NAH9	NAH10	NAH11	NAH12	NAH13	NAH14	NAH15
15	16	17	18	19	20	21
AU22	AU23	AU24	AU25	AU26	AU27	AU28
PAP3	PAP4	PAP5	PAP6	PAP7	PAP8	PAP9
NAH16	NAH17	NAH18	NAH19	NAH20	NAH21	NAH22
22	23	24	25	26	27	28
AU29	AU30	AU31	S1	S2	S3	S4
PAP10	PAP11	PAP12	PAP13	PAP14	PAP15	PAP16
NAH23	NAH24	NAH25	NAH26	NAH27	NAH28	NAH29

MTU WEUSI 516/2032-33/6268-69/2024-25

Ashe	Ojobo	Iwa	Araorun	Ile	Itutu	Erinle
1	2	3	4	5	6	7
JU11	JU12	JU13	JU14	JU15	JU16	JU17
MSR28	MSR29	MSR30	*ASR1*	*HRU2*	*STH3*	*AST4*
HAM4	HAM5	HAM6	HAM7	HAM8	HAM9	HAM10
8	9	10	11	12	13	14
JU18	JU19	JU20	JU21	JU22	JU23	JU24
NBH5	*ANU6*	THT1	THT2	THT3	THT4	THT5
HAM11	HAM12	HAM13	HAM14	HAM15	HAM16	HAM17
15	16	17	18	19	20	21
JU25	JU26	JU27	JU28	JU29	JU30	JU31
THT6	THT7	THT8	THT9	THT10	THT11	THT12
HAM18	HAM19	HAM20	HAM21	HAM22	HAM23	HAM24
22	23	24	25	26	27	28
AU1	AU2	AU3	AU4	AU5	AU6	AU7
THT13	THT14	THT15	THT16	THT17	THT18	THT19
HAM25	HAM26	HAM27	HAM28	HAM29	HAM30	NAH1

UWEZO 516/2032-33/6268-69/2024-25

Ashe	Ojobo	Iwa	Araorun	Ile	Itutu	Erinle
1	2	3	4	5	6	7
S5	S6	S7	S8	S9	S10	S11
PAP17	PAP18	PAP19	PAP20	PAP21	PAP22	PAP23
NAH30	*PAG1*	*PAG2*	*PAG3*	*PAG4*	*PAG5*	MAS1
8	9	10	11	12	13	14
S12	S13	S14	S15	S16	S17	S18
PAP24	PAP25	PAP26	PAP27	PAP28	PAP29	PAP30
MAS2	MAS3	MAS4	MAS5	MAS6	MAS7	MAS8
15	16	17	18	19	20	21
S19	S20	S21	S22	S23	S24	S25
HTH1	HTH2	HTH3	HTH4	HTH5	HTH6	HTH7
MAS9	MAS10	MAS11	MAS12	MAS13	MAS14	MAS15
22	23	24	25	26	27	28
S26	S27	S28	S29	S30	O1	O2
HTH8	HTH9	HTH10	HTH11	HTH12	HTH13	HTH14
MAS16	MAS17	MAS18	MAS19	MAS20	MAS21	MAS22

THE NU AFRIKAN CALENDAR

EKUNDU 516/2032-33/6268-69/2024-25						
Ashe	Ojobo	Iwa	Araorun	Ile	Itutu	Erinle
1	2	3	4	5	6	7
O3	O4	O5	O6	O7	O8	O9
HTH15	HTH16	HTH17	HTH18	HTH19	HTH20	HTH21
MAS23	MAS24	MAS25	MAS26	MAS27	MAS28	MAS29
8	9	10	11	12	13	14
O10	O11	O12	O13	O14	O15	O16
HTH22	HTH23	HTH24	HTH25	HTH26	HTH27	HTH28
MAS30	THi1	THi2	THi3	THi4	THi5	THi6
15	16	17	18	19	20	21
O17	O18	O19	O20	O21	O22	O23
HTH29	HTH30	KHK1	KHK2	KHK3	KHK4	KHK5
THi7	THi8	This9	THi10	THi11	THi12	THi13
22	23	24	25	26	27	28
O24	O25	O26	O27	O28	O29	O30
KHK6	KHK7	KHK8	KHK9	KHK10	KHK11	KHK12
Thi14	THi15	THi16	THi17	THi18	THi19	THi20

KIJANI 516/2032-33/6268-69/2024-25						
Ashe	Ojobo	Iwa	Araorun	Ile	Itutu	Erinle
1	2	3	4	5	6	7
N28	N29	N30	D1	D2	D3	D4
TBT11	TBT12	TBT13	TBT14	TBT15	TBT16	TBT17
HID19	HID20	HID21	HID22	HID23	HID24	HID25
8	9	10	11	12	13	14
D5	D6	D7	D8	D9	D10	D11
TBT18	TBT19	TBT20	TBT21	TBT22	TBT23	TBT24
HID26	HID27	HID28	HID29	HID30	TAH1	TAH2
15	16	17	18	19	20	21
D12	D13	D14	D15	D16	D17	D18
TBT25	TBT26	TBT27	TBT28	TBT29	TBT30	MKH1
TAH3	TAH4	TAH5	TAH6	TAH7	TAH8	TAH9
22	23	24	25	26	27	28
D19	D20	D21	D22	D23	D24	D25
MKH2	MKH3	MKH4	MKH5	MKH6	MKH7	MKH8
TAH10	TAH11	TAH12	TAH13	TAH14	TAH15	TAH16

EUSI 516/2032-33/6268-69/2024-25						
Ashe	Ojobo	Iwa	Araorun	Ile	Itutu	Erinle
1	2	3	4	5	6	7
O31	N1	N2	N3	N4	N5	N6
KHK13	KHK14	KHK15	KHK16	KHK17	KHK18	KHK19
THi21	THi22	THi23	THi24	THi25	THi26	THi27
8	9	10	11	12	13	14
N7	N8	N9	N10	N11	N12	N13
KHK20	KHK21	KHK22	KHK23	KHK24	KHK25	KHK26
THi28	THi29	THi30	HID1	HID2	HID3	HID4
15	16	17	18	19	20	21
N14	N15	N16	N17	N18	N19	N20
KHK27	KHK28	KHK29	KHK30	TBT1	TBT2	TBT3
HID5	HID6	HID7	HID8	HID9	HID10	HID11
22	23	24	25	26	27	28
N21	N22	N23	N24	N25	N26	N27
TBT4	TBT5	TBT6	TBT7	TBT8	TBT9	TBT10
HID12	HID13	HID14	HID15	HID16	HID17	HID18

KWANZA 516/2032-33/6268-69/2024-25						
Umoja	Kujichagulia	Ujima	Ujamaa	Nia	Kuumba	Imani
1	2	3	4	5	6	7
D26	D27	D28	D29	D30	D31	JA1
MKH9	MKH10	MKH11	MKH12	MKH13	MKH14	MKH15
TAH17	TAH18	TAH19	TAH20	TAH21	TAH22	TAH23
8	9	10	11	12	13	14
JA2	JA3	JA4	JA5	JA6	JA7	JA8
MKH16	MKH17	MKH18	MKH19	MKH20	MKH21	MKH22
TAH24	TAH25	TAH26	TAH27	TAH28	TAH29	TAH30
15	16	17	18	19	20	21
JA9	JA10	JA11	JA12	JA13	JA14	JA15
MKH23	MKH24	MKH25	KMH26	MKH27	MKH28	MKH29
TIR1	TIR2	TIR3	TIR4	TIR5	TIR6	TIR7
22	23	24	25	26	27	28
JA16	JA17	JA18	JA19	JA20	JA21	JA22
MKH30	AMN1	AMN2	AMN3	AMN4	AMN5	AMN6
TIR8	TIR9	TIR10	TIR11	TIR12	TIR13	TIR14

THE NU AFRIKAN CALENDAR

SHAHIDI 516/2032-33/6268-69/2024-25						
Ashe	Ojobo	Iwa	Araorun	Ile	Itutu	Erinle
1	2	3	4	5	6	7
JA23	JA24	JA25	JA26	JA27	JA28	JA29
AMN7	AMN8	AMN9	AMN10	AMN11	AMN12	AMN13
TIR15	TIR16	TIR17	TIR18	TIR19	TIR20	TIR21
8	9	10	11	12	13	14
JA30	JA31	F1	F2	F3	F4	F5
AMN14	AMN15	AMN16	AMN17	AMN18	AMN19	AMN20
TIR22	TIR23	TIR24	TIR25	TIR26	TIR27	TIR28
15	16	17	18	19	20	21
F6	F7	F8	F9	F10	F 11	F12
AMN21	AMN22	AMN23	AMN24	AMN25	AMN26	AMN28
TIR29	TIR30	YEK1	YEK2	YEK3	YEK4	YEK5
22	23	24	25	26	27	28
F13	F14	F15	F16	F17	F18	F19
AMN29	AMN30	RNT1	RNT2	RNT3	RNT4	RNT5
YEK6	YEK7	YEK8	YEK9	YEK10	YEK11	YEK12
SHAHIDI 67						
F20						
RNT6						
YEK13						

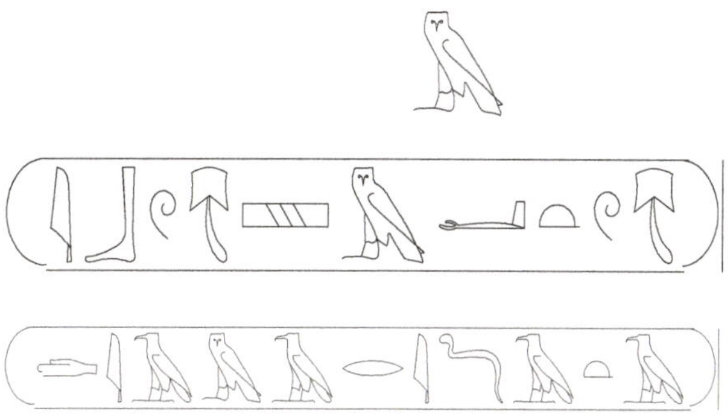

THE NU AFRIKAN CALENDAR

MAAFA 517 B.S.E./ASIENTO 2033-34 C.E.
KAM 6269-70 S.M./EC 2025-26 A.M. (ዓ.ም.)

AFU 517/2033-34/6269-70/2025-26

Ashe	Ojobo	Iwa	Araorun	Ile	Itutu	Erinle
1	2	3	4	5	6	7
F21	F22	F23	F24	F25	F26	F27
RNT7	RNT8	RNT9	RNT10	RNT11	RNT12	RNT13
YEK14	YEK15	YEK16	YEK17	YEK18	YEK19	YEK20
8	9	10	11	12	13	14
F28	M1	M2	M3	M4	M5	M6
RNT14	RNT15	RNT16	RNT17	RNT18	RNT19	RNT20
YEK21	YEK22	YEK23	YEK24	YEK25	YEK26	YEK27
15	16	17	18	19	20	21
M7	M8	M9	M10	M11	M12	M13
RNT21	RNT22	RNT23	RNT24	RNT25	RNT26	RNT27
YEK28	YEK29	YEK30	MAG1	MAG2	MAG3	MAG4
22	23	24	25	26	27	28
M14	M15	M16	M17	M18	M19	M20
RNT28	RNT29	RNT30	KHN1	KHN2	KHN3	KHN4
MAG5	MAG6	MAG7	MAG8	MAG9	MAG10	MAG11

KAWAIDA 517/2033-34/6269-70/2025-26

Ashe	Ojobo	Iwa	Araorun	Ile	Itutu	Erinle
1	2	3	4	5	6	7
A18	A19	A20	A21	A22	A23	A24
PIN3	PIN4	PIN5	PIN6	PIN7	PIN8	PIN9
MIY10	MIY11	MIY12	MIY13	MIY14	MIY15	MIY16
8	9	10	11	12	13	14
A25	A26	A27	A28	A29	A30	MA1
PIN10	PIN11	PIN12	PIN13	PIN14	PIN15	PIN16
MIY17	MIY18	MIY19	MIY20	MIY21	MIY22	MIY23
15	16	17	18	19	20	21
MA2	MA3	MA4	MA5	MA6	MA7	MA8
PIN17	PIN18	PIN19	PIN20	PIN21	PIN22	PIN23
MIY24	MIY25	MIY26	MIY27	MIY28	MIY29	MIY30
22	23	24	25	26	27	28
MA9	MA10	MA11	MA12	MA13	MA14	MA15
PIN24	PIN25	PIN26	PIN27	PIN28	PIN29	PIN30
GIN1	GIN2	GIN3	GIN4	GIN5	GIN6	GIN7

UHURU 517/2033-34/6269-70/2025-26

Ashe	Ojobo	Iwa	Araorun	Ile	Itutu	Erinle
1	2	3	4	5	6	7
M21	M22	M23	M24	M25	M26	M27
KHN5	KHN6	KHN7	KHN8	KHN9	KHN10	KHN11
MAG12	MAG13	MAG14	MAG15	MAG16	MAG17	MAG18
8	9	10	11	12	13	14
M28	M29	M30	M31	A1	A2	A3
KHN12	KHN13	KHN14	KHN15	KHN16	KHN17	KHN18
MAG19	MAG20	MAG21	MAG22	MAG23	MAG24	MAG25
15	16	17	18	19	20	21
A4	A5	A6	A7	A8	A9	A10
KHN19	KHN20	KHN21	KHN22	KHN23	KHN24	KHN25
MAG26	MAG27	MAG28	MAG29	MAG30	MIY1	MIY2
22	23	24	25	26	27	28
A11	A12	A13	A14	A15	A16	A17
KHN26	KHN27	KHN28	KHN29	KHN30	PIN1	PIN2
MIY3	MIY4	MIY5	MIY6	MIY7	MIY8	MIY9

MABABA WEUSI 517/2033-34/6269-70/2025-26

Ashe	Ojobo	Iwa	Araorun	Ile	Itutu	Erinle
1	2	3	4	5	6	7
MA16	MA17	MA18	MA19	MA20	MA21	MA22
APP1	APP2	APP3	APP4	APP5	APP6	APP7
GIN8	GIN9	GIN10	GIN11	GIN12	GIN13	GIN14
8	9	10	11	12	13	14
MA23	MA24	MA25	MA26	MA27	MA28	MA29
APP8	APP9	APP10	APP11	APP12	APP13	APP14
GIN15	GIN16	GIN17	GIN18	GIN19	GIN20	GIN21
15	16	17	18	19	20	21
MA30	MA31	J1	J2	J3	J4	J5
APP15	APP16	APP17	APP18	APP19	APP20	APP21
GIN22	GIN23	GIN24	GIN25	GIN26	GIN27	GIN28
22	23	24	25	26	27	28
J6	J7	J8	J9	J10	J11	J12
APP22	APP23	APP24	APP25	APP26	APP27	APP28
GIN29	GIN30	SAN1	SAN2	SAN3	SAN4	SAN5

THE NU AFRIKAN CALENDAR

MWANAMKE WEUSI 517/2033-34/6269-70/2025-26

Ashe	Ojobo	Iwa	Araorun	Ile	Itutu	Erinle
1	2	3	4	5	6	7
J13	J14	J15	J16	J17	J18	J19
APP29	APP30	MSR1	MSR2	MSR3	MSR4	MSR5
SAN6	SAN7	SAN8	SAN9	SAN10	SAN11	SAN12
8	9	10	11	12	13	14
J20	J21	J22	J23	J24	J25	J26
MSR6	MSR7	MSR8	MSR9	MSR10	MSR11	MSR12
SAN13	SAN14	SAN15	SAN16	SAN17	SAN18	SAN19
15	16	17	18	19	20	21
J27	J28	J29	J30	JU1	JU2	JU3
MSR13	MSR14	MSR15	MSR16	MSR17	MSR18	MSR19
SAN20	SAN21	SAN22	SAN23	SAN24	SAN25	SAN26
22	23	24	25	26	27	28
JU4	JU5	JU6	JU7	JU8	JU9	JU10
MSR20	MRS21	MSR22	MSR23	MSR24	MSR25	MSR26
SAN27	SAN28	SAN29	SAN30	HAM1	HAM2	HAM3

TAIFA WEUSI 517/2033-34/6269-70/2025-26

Ashe	Ojobo	Iwa	Araorun	Ile	Itutu	Erinle
1	2	3	4	5	6	7
AU8	AU9	AU10	AU11	AU12	AU13	AU14
THT20	THT21	THT22	THT23	THT24	THT25	THT26
NAH2	NAH3	NAH4	NAH5	NAH6	NAH7	NAH8
8	9	10	11	12	13	14
AU15	AU16	AU17	AU18	AU19	AU20	AU21
THT27	THT28	THT29	THT30	PAP1	PAP1	PAP2
NAH9	NAH10	NAH11	NAH12	NAH13	NAH14	NAH15
15	16	17	18	19	20	21
AU22	AU23	AU24	AU25	AU26	AU27	AU28
PAP3	PAP4	PAP5	PAP6	PAP7	PAP8	PAP9
NAH16	NAH17	NAH18	NAH19	NAH20	NAH21	NAH22
22	23	24	25	26	27	28
AU29	AU30	AU31	S1	S2	S3	S4
PAP10	PAP11	PAP12	PAP13	PAP14	PAP15	PAP16
NAH23	NAH24	NAH25	NAH26	NAH27	NAH28	NAH29

MTU WEUSI 517/2033-34/6269-70/2025-26

Ashe	Ojobo	Iwa	Araorun	Ile	Itutu	Erinle
1	2	3	4	5	6	7
JU11	JU12	JU13	JU14	JU15	JU16	JU17
MSR27	MSR28	MSR29	MSR30	*ASR1*	*HRU2*	*STH3*
HAM4	HAM5	HAM6	HAM7	HAM8	HAM9	HAM10
8	9	10	11	12	13	14
JU18	JU19	JU20	JU21	JU22	JU23	JU24
AST4	*NBH5*	THT1	THT2	THT3	THT4	THT5
HAM11	HAM12	HAM13	AM14	HAM15	HAM16	HAM17
15	16	17	18	19	20	21
JU25	JU26	JU27	JU28	JU29	JU30	JU31
THT6	THT7	THT8	THT9	THT10	THT11	THT12
HAM18	HAM19	HAM20	HAM21	HAM22	HAM23	HAM24
22	23	24	25	26	27	28
AU1	AU2	AU3	AU4	AU5	AU6	AU7
THT13	THT14	THT15	THT16	THT17	THT18	THT19
HAM25	HAM26	HAM27	HAM28	HAM29	HAM30	NAH1

UWEZO 517/2033-34/6269-70/2025-26

Ashe	Ojobo	Iwa	Araorun	Ile	Itutu	Erinle
1	2	3	4	5	6	7
S5	S6	S7	S8	S9	S10	S11
PAP17	PAP18	PAP19	PAP20	PAP21	PAP22	PAP23
NAH30	*PAG1*	*PAG2*	*PAG3*	*PAG4*	*PAG5*	MAS1
8	9	10	11	12	13	14
S12	S13	S14	S15	S16	S17	S18
PAP24	PAP25	PAP26	PAP27	PAP28	PAP29	PAP30
MAS2	MAS3	MAS4	MAS5	MAS6	MAS7	MAS8
15	16	17	18	19	20	21
S19	S20	S21	S22	S23	S24	S25
HTH1	HTH2	HTH3	HTH4	HTH5	HTH6	HTH7
MAS9	MAS10	MAS11	MAS12	MAS13	MAS14	MAS15
22	23	24	25	26	27	28
S26	S27	S28	S29	S30	O1	O2
HTH8	HTH9	HTH10	HTH11	HTH12	HTH13	HTH14
MAS16	MAS17	MAS18	MAS19	MAS20	MAS21	MAS22

THE NU AFRIKAN CALENDAR

EKUNDU 517/2033-34/6269-70/2025-26						
Ashe	Ojobo	Iwa	Araorun	Ile	Itutu	Erinle
1	2	3	4	5	6	7
O3	O4	O5	O6	O7	O8	O9
HTH15	HTH16	HTH17	HTH18	HTH19	HTH20	HTH21
MAS23	MAS24	MAS25	MAS26	MAS27	MAS28	MAS29
8	9	10	11	12	13	14
O10	O11	O12	O13	O14	O15	O16
HTH22	HTH23	HTH24	HTH25	HTH26	HTH27	HTH28
MAS30	THi1	THi2	THi3	THi4	THi5	THi6
15	16	17	18	19	20	21
O17	O18	O19	O20	O21	O22	O23
HTH29	HTH30	KHK1	KHK2	KHK3	KHK4	KHK5
THi7	THi8	THi9	THi10	THi11	THi12	THi13
22	23	24	25	26	27	28
O24	O25	O26	O27	O28	O29	O30
KHK6	KHK7	KHK8	KHK9	KHK10	KHK11	KHK12
THi14	THi15	THi16	THi17	THi18	THi19	THi20

KIJANI 517/2033-34/6269-70/2025-26						
Ashe	Ojobo	Iwa	Araorun	Ile	Itutu	Erinle
1	2	3	4	5	6	7
N28	N29	N30	D1	D2	D3	D4
TBT11	TBT12	TBT13	TBT14	TBT15	TBT16	TBT17
HID19	HID20	HID21	HID22	HID23	HID24	HID25
8	9	10	11	12	13	14
D5	D6	D7	D8	D9	D10	D11
TBT18	TBT19	TBT20	TBT21	TBT22	TBT23	TBT24
HID26	HID27	HID28	HID29	HID30	TAH1	TAH2
15	16	17	18	19	20	21
D12	D13	D14	D15	D16	D17	D18
TBT25	TBT26	TBT27	TBT28	TBT29	TBT30	MKH1
TAH3	TAH4	TAH5	TAH6	TAH7	TAH8	TAH9
22	23	24	25	26	27	28
D19	D20	D21	D22	D23	D24	D25
MKH2	MKH3	MKH4	MKH5	MKH6	MKH7	MKH8
TAH10	TAH11	TAH12	TAH13	TAH14	TAH15	TAH16

EUSI 517/2033-34/6269-70/2025-26						
Ashe	Ojobo	Iwa	Araorun	Ile	Itutu	Erinle
1	2	3	4	5	6	7
O31	N1	N2	N3	N4	N5	N6
KHK13	KHK14	KHK15	KHK16	KHK17	KHK18	KHK19
THi21	THi22	THi23	THi24	THi25	THi26	THi27
8	9	10	11	12	13	14
N7	N8	N9	N10	N11	N12	N13
KHK20	KHK21	KHK22	KHK23	KHK24	KHK25	KHK26
THi28	THi29	THi30	HID1	HID2	HID3	HID4
15	16	17	18	19	20	21
N14	N15	N16	N17	N18	N19	N20
KHK27	KHK28	KHK29	KHK30	TBT1	TBT2	TBT3
HID5	HID6	HID7	HID8	HID9	HID10	HID11
22	23	24	25	26	27	28
N21	N22	N23	N24	N25	N26	N27
TBT4	TBT5	TBT6	TBT7	TBT8	TBT9	TBT10
HID12	HID13	HID14	HID15	HID16	HID17	HID18

KWANZA 517/2033-34/6269-70/2025-26						
Umoja	Kujichagulia	Ujima	Ujamaa	Nia	Kuumba	Imani
1	2	3	4	5	6	7
D26	D27	D28	D29	D30	D31	JA1
MKH9	MKH10	MKH11	MKH12	MKH13	MKH14	MKH15
TAH17	TAH18	TAH19	TAH20	TAH21	TAH22	TAH23
8	9	10	11	12	13	14
JA2	JA3	JA4	JA5	JA6	JA7	JA8
MKH16	MKH17	MKH18	MKH19	MKH20	MKH21	MKH22
TAH24	TAH25	TAH26	TAH27	TAH28	TAH29	TAH30
15	16	17	18	19	20	21
JA9	JA10	JA11	JA12	JA13	JA14	JA15
MKH23	MKH24	MKH25	KMH26	MKH27	MKH28	MKH29
TIR1	TIR2	TIR3	TIR4	TIR5	TIR6	TIR7
22	23	24	25	26	27	28
JA16	JA17	JA18	JA19	JA20	JA21	JA22
MKH30	AMN1	AMN2	AMN3	AMN4	AMN5	AMN6
TIR8	TIR9	TIR10	TIR11	TIR12	TIR13	TIR14

THE NU AFRIKAN CALENDAR

SHAHIDI 517/2033-34/6269-70/2025-26						
Ashe	Ojobo	Iwa	Araorun	Ile	Itutu	Erinle
1	2	3	4	5	6	7
JA23	JA24	JA25	JA26	JA27	JA28	JA29
AMN7	AMN8	AMN9	AMN10	AMN11	AMN12	AMN13
TIR15	TIR16	TIR17	TIR18	TIR19	TIR20	TIR21
8	9	10	11	12	13	14
JA30	JA31	F1	F2	F3	F4	F5
AMN14	AMN15	AMN16	AMN17	AMN18	AMN19	AMN20
TIR22	TIR23	TIR24	TIR25	TIR26	TIR27	TIR28
15	16	17	18	19	20	21
F6	F7	F8	F9	F10	F 11	F12
AMN21	AMN22	AMN23	AMN24	AMN25	AMN26	AMN28
TIR29	TIR30	YEK1	YEK2	YEK3	YEK4	YEK5
22	23	24	25	26	27	28
F13	F14	F15	F16	F17	F18	F19
AMN29	AMN30	RNT1	RNT2	RNT3	RNT4	RNT5
YEK6	YEK7	YEK8	YEK9	YEK10	YEK11	YEK12
SHAHIDI 68						
F20						
RNT6						
YEK13						

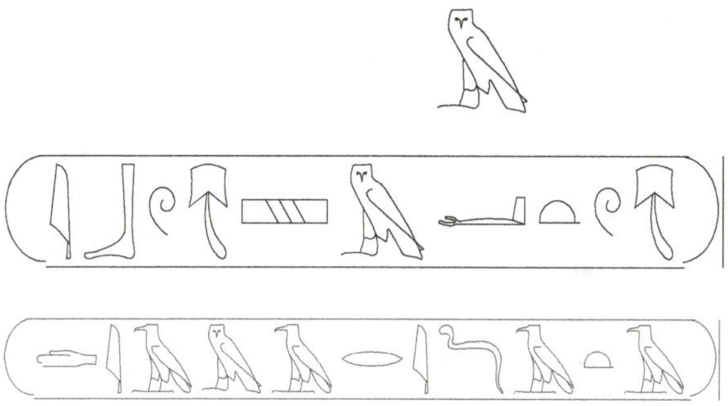

THE NU AFRIKAN CALENDAR

MAAFA 518 B.S.E./ASIENTO 2034-35 C.E.
KAM 6270-71 S.M./EC 2026-27 A.M. (ዓ.ም.)

AFU 518/2034-35/6270-71/2026-27

Ashe	Ojobo	Iwa	Araorun	Ile	Itutu	Erinle
1	2	3	4	5	6	7
F21	F22	F23	F24	F25	F26	F27
RNT7	RNT8	RNT9	RNT10	RNT11	RNT12	RNT13
YEK14	YEK15	YEK16	YEK17	YEK18	YEK19	YEK20
8	9	10	11	12	13	14
F28	M1	M2	M3	M4	M5	M6
RNT14	RNT15	RNT16	RNT17	RNT18	RNT19	RNT20
YEK21	YEK22	YEK23	YEK24	YEK25	YEK26	YEK27
15	16	17	18	19	20	21
M7	M8	M9	M10	M11	M12	M13
RNT21	RNT22	RNT23	RNT24	RNT25	RNT26	RNT27
YEK28	YEK29	YEK30	MAG1	MAG2	MAG3	MAG4
22	23	24	25	26	27	28
M14	M15	M16	M17	M18	M19	M20
RNT28	RNT29	RNT30	KHN1	KHN2	KHN3	KHN4
MAG5	MAG6	MAG7	MAG8	MAG9	MAG10	MAG11

KAWAIDA 518/2034-35/6270-71/2026-27

Ashe	Ojobo	Iwa	Araorun	Ile	Itutu	Erinle
1	2	3	4	5	6	7
A18	A19	A20	A21	A22	A23	A24
PIN3	PIN4	PIN5	PIN6	PIN7	PIN8	PIN9
MIY10	MIY11	MIY12	MIY13	MIY14	MIY15	MIY16
8	9	10	11	12	13	14
A25	A26	A27	A28	A29	A30	MA1
PIN10	PIN11	PIN12	PIN13	PIN14	PIN15	PIN16
MIY17	MIY18	MIY19	MIY20	MIY21	MIY22	MIY23
15	16	17	18	19	20	21
MA2	MA3	MA4	MA5	MA6	MA7	MA8
PIN17	PIN18	PIN19	PIN20	PIN21	PIN22	PIN23
MIY24	MIY25	MIY26	MIY27	MIY28	MIY29	MIY30
22	23	24	25	26	27	28
MA9	MA10	MA11	MA12	MA13	MA14	MA15
PIN24	PIN25	PIN26	PIN27	PIN28	PIN29	PIN30
GIN1	GIN2	GIN3	GIN4	GIN5	GIN6	GIN7

UHURU 518/2034-35/6270-71/2026-27

Ashe	Ojobo	Iwa	Araorun	Ile	Itutu	Erinle
1	2	3	4	5	6	7
M21	M22	M23	M24	M25	M26	M27
KHN5	KHN6	KHN7	KHN8	KHN9	KHN10	KHN11
MAG12	MAG13	MAG14	MAG15	MAG16	MAG17	MAG18
8	9	10	11	12	13	14
M28	M29	M30	M31	A1	A2	A3
KHN12	KHN13	KHN14	KHN15	KHN16	KHN17	KHN18
MAG19	MAG20	MAG21	MAG22	MAG23	MAG24	MAG25
15	16	17	18	19	20	21
A4	A5	A6	A7	A8	A9	A10
KHN19	KHN20	KHN21	KHN22	KHN23	KHN24	KHN25
MAG26	MAG27	MAG28	MAG29	MAG30	MIY1	MIY2
22	23	24	25	26	27	28
A11	A12	A13	A14	A15	A16	A17
KHN26	KHN27	KHN28	KHN29	KHN30	PIN1	PIN2
MIY3	MIY4	MIY5	MIY6	MIY7	MIY8	MIY9

MABABA WEUSI 518/2034-35/6270-71/2026-27

Ashe	Ojobo	Iwa	Araorun	Ile	Itutu	Erinle
1	2	3	4	5	6	7
MA16	MA17	MA18	MA19	MA20	MA21	MA22
APP1	APP2	APP3	APP4	APP5	APP6	APP7
GIN8	GIN9	GIN10	GIN11	GIN12	GIN13	GIN14
8	9	10	11	12	13	14
MA23	MA24	MA25	MA26	MA27	MA28	MA29
APP8	APP9	APP10	APP11	APP12	APP13	APP14
GIN15	GIN16	GIN17	GIN18	GIN19	GIN20	GIN21
15	16	17	18	19	20	21
MA30	MA31	J1	J2	J3	J4	J5
APP15	APP16	APP17	APP18	APP19	APP20	APP21
GIN22	GIN23	GIN24	GIN25	GIN26	GIN27	GIN28
22	23	24	25	26	27	28
J6	J7	J8	J9	J10	J11	J12
APP22	APP23	APP24	APP25	APP26	APP27	APP28
GIN29	GIN30	SAN1	SAN2	SAN3	SAN4	SAN5

THE NU AFRIKAN CALENDAR

MWANAMKE WEUSI 518/2034-35/6270-71/2026-27						
Ashe	Ojobo	Iwa	Araorun	Ile	Itutu	Erinle
1	2	3	4	5	6	7
J13	J14	J15	J16	J17	J18	J19
APP29	APP30	MSR1	MSR2	MSR3	MSR4	MSR5
SAN6	SAN7	SAN8	SAN9	SAN10	SAN11	SAN12
8	9	10	11	12	13	14
J20	J21	J22	J23	J24	J25	J26
MSR6	MSR7	MSR8	MSR9	MSR10	MSR11	MSR12
SAN13	SAN14	SAN15	SAN16	SAN17	SAN18	SAN19
15	16	17	18	19	20	21
J27	J28	J29	J30	JU1	JU2	JU3
MSR13	MSR14	MSR15	MSR16	MSR17	MSR18	MSR19
SAN20	SAN21	SAN22	SAN23	SAN24	SAN25	SAN26
22	23	24	25	26	27	28
JU4	JU5	JU6	JU7	JU8	JU9	JU10
MSR20	MRS21	MSR22	MSR23	MSR24	MSR25	MSR26
SAN27	SAN28	SAN29	SAN30	HAM1	HAM2	HAM3

TAIFA WEUSI 518/2034-35/6270-71/2026-27						
Ashe	Ojobo	Iwa	Araorun	Ile	Itutu	Erinle
1	2	3	4	5	6	7
AU8	AU9	AU10	AU11	AU12	AU13	AU14
THT20	THT21	THT22	THT23	THT24	THT25	THT26
NAH2	NAH3	NAH4	NAH5	NAH6	NAH7	NAH8
8	9	10	11	12	13	14
AU15	AU16	AU17	AU18	AU19	AU20	AU21
THT27	THT28	THT29	THT30	PAP1	PAP1	PAP2
NAH9	NAH10	NAH11	NAH12	NAH13	NAH14	NAH15
15	16	17	18	19	20	21
AU22	AU23	AU24	AU25	AU26	AU27	AU28
PAP3	PAP4	PAP5	PAP6	PAP7	PAP8	PAP9
NAH16	NAH17	NAH18	NAH19	NAH20	NAH21	NAH22
22	23	24	25	26	27	28
AU29	AU30	AU31	S1	S2	S3	S4
PAP10	PAP11	PAP12	PAP13	PAP14	PAP15	PAP16
NAH23	NAH24	NAH25	NAH26	NAH27	NAH28	NAH29

MTU WEUSI 518/2034-35/6270-71/2026-27						
Ashe	Ojobo	Iwa	Araorun	Ile	Itutu	Erinle
1	2	3	4	5	6	7
JU11	JU12	JU13	JU14	JU15	JU16	JU17
MSR27	MSR28	MSR29	MSR30	*ASR1*	*HRU2*	*STH3*
HAM4	HAM5	HAM6	HAM7	HAM8	HAM9	HAM10
8	9	10	11	12	13	14
JU18	JU19	JU20	JU21	JU22	JU23	JU24
AST4	*NBH5*	THT1	THT2	THT3	THT4	THT5
HAM11	HAM12	HAM13	AM14	HAM15	HAM16	HAM17
15	16	17	18	19	20	21
JU25	JU26	JU27	JU28	JU29	JU30	JU31
THT6	THT7	THT8	THT9	THT10	THT11	THT12
HAM18	HAM19	HAM20	HAM21	HAM22	HAM23	HAM24
22	23	24	25	26	27	28
AU1	AU2	AU3	AU4	AU5	AU6	AU7
THT13	THT14	THT15	THT16	THT17	THT18	THT19
HAM25	HAM26	HAM27	HAM28	HAM29	HAM30	NAH1

UWEZO 518/2034-35/6270-71/2026-27						
Ashe	Ojobo	Iwa	Araorun	Ile	Itutu	Erinle
1	2	3	4	5	6	7
S5	S6	S7	S8	S9	S10	S11
PAP17	PAP18	PAP19	PAP20	PAP21	PAP22	PAP23
NAH30	*PAG1*	*PAG2*	*PAG3*	*PAG4*	*PAG5*	MAS1
8	9	10	11	12	13	14
S12	S13	S14	S15	S16	S17	S18
PAP24	PAP25	PAP26	PAP27	PAP28	PAP29	PAP30
MAS2	MAS3	MAS4	MAS5	MAS6	MAS7	MAS8
15	16	17	18	19	20	21
S19	S20	S21	S22	S23	S24	S25
HTH1	HTH2	HTH3	HTH4	HTH5	HTH6	HTH7
MAS9	MAS10	MAS11	MAS12	MAS13	MAS14	MAS15
22	23	24	25	26	27	28
S26	S27	S28	S29	S30	O1	O2
HTH8	HTH9	HTH10	HTH11	HTH12	HTH13	HTH14
MAS16	MAS17	MAS18	MAS19	MAS20	MAS21	MAS22

THE NU AFRIKAN CALENDAR

EKUNDU 518/2034-35/6270-71/2026-27						
Ashe	Ojobo	Iwa	Araorun	Ile	Itutu	Erinle
1	2	3	4	5	6	7
O3	O4	O5	O6	O7	O8	O9
HTH15	HTH16	HTH17	HTH18	HTH19	HTH20	HTH21
MAS23	MAS24	MAS25	MAS26	MAS27	MAS28	MAS29
8	9	10	11	12	13	14
O10	O11	O12	O13	O14	O15	O16
HTH22	HTH23	HTH24	HTH25	HTH26	HTH27	HTH28
MAS30	THi1	THi2	THi3	THi4	THi5	THi6
15	16	17	18	19	20	21
O17	O18	O19	O20	O21	O22	O23
HTH29	HTH30	KHK1	KHK2	KHK3	KHK4	KHK5
THi7	THi8	THi9	THi10	THi11	THi12	THi13
22	23	24	25	26	27	28
O24	O25	O26	O27	O28	O29	O30
KHK6	KHK7	KHK8	KHK9	KHK10	KHK11	KHK12
THi14	THi15	THi16	THi17	THi18	THi19	THi20

KIJANI 518/2034-35/6270-71/2026-27						
Ashe	Ojobo	Iwa	Araorun	Ile	Itutu	Erinle
1	2	3	4	5	6	7
N28	N29	N30	D1	D2	D3	D4
TBT11	TBT12	TBT13	TBT14	TBT15	TBT16	TBT17
HID19	HID20	HID21	HID22	HID23	HID24	HID25
8	9	10	11	12	13	14
D5	D6	D7	D8	D9	D10	D11
TBT18	TBT19	TBT20	TBT21	TBT22	TBT23	TBT24
HID26	HID27	HID28	HID29	HID30	TAH1	TAH2
15	16	17	18	19	20	21
D12	D13	D14	D15	D16	D17	D18
TBT25	TBT26	TBT27	TBT28	TBT29	TBT30	MKH1
TAH3	TAH4	TAH5	TAH6	TAH7	TAH8	TAH9
22	23	24	25	26	27	28
D19	D20	D21	D22	D23	D24	D25
MKH2	MKH3	MKH4	MKH5	MKH6	MKH7	MKH8
TAH10	TAH11	TAH12	TAH13	TAH14	TAH15	TAH16

EUSI 518/2034-35/6270-71/2026-27						
Ashe	Ojobo	Iwa	Araorun	Ile	Itutu	Erinle
1	2	3	4	5	6	7
O31	N1	N2	N3	N4	N5	N6
KHK13	KHK14	KHK15	KHK16	KHK17	KHK18	KHK19
THi21	THi22	THi23	THi24	THi25	THi26	THi27
8	9	10	11	12	13	14
N7	N8	N9	N10	N11	N12	N13
KHK20	KHK21	KHK22	KHK23	KHK24	KHK25	KHK26
THi28	THi29	THi30	HID1	HID2	HID3	HID4
15	16	17	18	19	20	21
N14	N15	N16	N17	N18	N19	N20
KHK27	KHK28	KHK29	KHK30	TBT1	TBT2	TBT3
HID5	HID6	HID7	HID8	HID9	HID10	HID11
22	23	24	25	26	27	28
N21	N22	N23	N24	N25	N26	N27
TBT4	TBT5	TBT6	TBT7	TBT8	TBT9	TBT10
HID12	HID13	HID14	HID15	HID16	HID17	HID18

KWANZA 518/2034-35/6270-71/2026-27						
Umoja	Kujichagulia	Ujima	Ujamaa	Nia	Kuumba	Imani
1	2	3	4	5	6	7
D26	D27	D28	D29	D30	D31	JA1
MKH9	MKH10	MKH11	MKH12	MKH13	MKH14	MKH15
TAH17	TAH18	TAH19	TAH20	TAH21	TAH22	TAH23
8	9	10	11	12	13	14
JA2	JA3	JA4	JA5	JA6	JA7	JA8
MKH16	MKH17	MKH18	MKH19	MKH20	MKH21	MKH22
TAH24	TAH25	TAH26	TAH27	TAH28	TAH29	TAH30
15	16	17	18	19	20	21
JA9	JA10	JA11	JA12	JA13	JA14	JA15
MKH23	MKH24	MKH25	KMH26	MKH27	MKH28	MKH29
TIR1	TIR2	TIR3	TIR4	TIR5	TIR6	TIR7
22	23	24	25	26	27	28
JA16	JA17	JA18	JA19	JA20	JA21	JA22
MKH30	AMN1	AMN2	AMN3	AMN4	AMN5	AMN6
TIR8	TIR9	TIR10	TIR11	TIR12	TIR13	TIR14

THE NU AFRIKAN CALENDAR

SHAHIDI 518/2034-35/6270-71/2026-27						
Ashe	Ojobo	Iwa	Araorun	Ile	Itutu	Erinle
1	2	3	4	5	6	7
JA23	JA24	JA25	JA26	JA27	JA28	JA29
AMN7	AMN8	AMN9	AMN10	AMN11	AMN12	AMN13
TIR15	TIR16	TIR17	TIR18	TIR19	TIR20	TIR21
8	9	10	11	12	13	14
JA30	JA31	F1	F2	F3	F4	F5
AMN14	AMN15	AMN16	AMN17	AMN18	AMN19	AMN20
TIR22	TIR23	TIR24	TIR25	TIR26	TIR27	TIR28
15	16	17	18	19	20	21
F6	F7	F8	F9	F10	F 11	F12
AMN21	AMN22	AMN23	AMN24	AMN25	AMN26	AMN28
TIR29	TIR30	YEK1	YEK2	YEK3	YEK4	YEK5
22	23	24	25	26	27	28
F13	F14	F15	F16	F17	F18	F19
AMN29	AMN30	RNT1	RNT2	RNT3	RNT4	RNT5
YEK6	YEK7	YEK8	YEK9	YEK10	YEK11	YEK12
SHAHIDI 69						
F20						
RNT6						
YEK13						

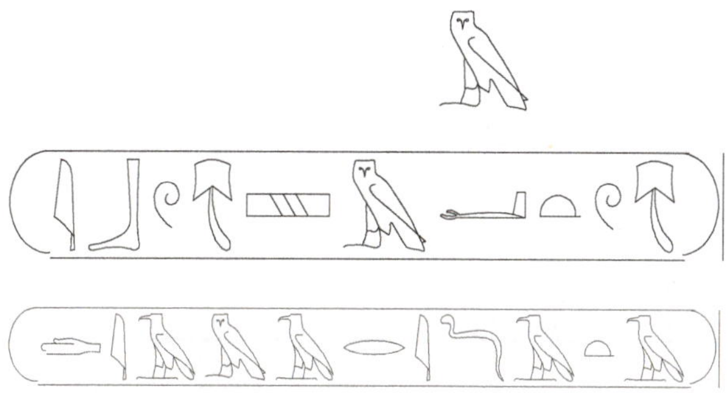

MAAFA 519 B.S.E./ASIENTO 2035-36 C.E./
KAM 6271-72 S.M./EC 2027-28 A.M. (ዓ.ም.)

AFU 519/2035-36/6271-72/2027-28						
Ashe	Ojobo	Iwa	Araorun	Ile	Itutu	Erinle
1	2	3	4	5	6	7
F21	F22	F23	F24	F25	F26	F27
RNT7	RNT8	RNT9	RNT10	RNT11	RNT12	RNT13
YEK14	YEK15	YEK16	YEK17	YEK18	YEK19	YEK20
8	9	10	11	12	13	14
F28	M1	M2	M3	M4	M5	M6
RNT14	RNT15	RNT16	RNT17	RNT18	RNT19	RNT20
YEK21	YEK22	YEK23	YEK24	YEK25	YEK26	YEK27
15	16	17	18	19	20	21
M7	M8	M9	M10	M11	M12	M13
RNT21	RNT22	RNT23	RNT24	RNT25	RNT26	RNT27
YEK28	YEK 29	YEK30	MAG1	MAG2	MAG3	MAG4
22	23	24	25	26	27	28
M14	M15	M16	M17	M18	M19	M20
RNT28	RNT29	RNT30	KHN1	KHN2	KHN3	KHN4
MAG5	MAG6	MAG7	MAG8	MAG9	MAG10	MAG11

KAWAIDA 519/2035-36/6271-72/2027-28						
Ashe	Ojobo	Iwa	Araorun	Ile	Itutu	Erinle
1	2	3	4	5	6	7
A18	A19	A20	A21	A22	A23	A24
PIN3	PIN4	PIN5	PIN6	PIN7	PIN8	PIN9
MIY10	MIY11	MIY12	MIY13	MIY14	MIY15	MIY16
8	9	10	11	12	13	14
A25	A26	A27	A28	A29	A30	MA1
PIN10	PIN11	PIN12	PIN13	PIN14	PIN15	PIN16
MIY17	MIY18	MIY19	MIY20	MIY21	MIY22	MIY23
15	16	17	18	19	20	21
MA2	MA3	MA4	MA5	MA6	MA7	MA8
PIN17	PIN18	PIN19	PIN20	PIN21	PIN22	PIN23
MIY24	MIY25	MIY26	MIY27	MIY28	MIY29	MIY30
22	23	24	25	26	27	28
MA9	MA10	MA11	MA12	MA13	MA14	MA15
PIN24	PIN25	PIN26	PIN27	PIN28	PIN29	PIN30
GIN1	GIN2	GIN3	GIN4	GIN5	GIN6	GIN7

UHURU 519/2035-36/6271-72/2027-28						
Ashe	Ojobo	Iwa	Araorun	Ile	Itutu	Erinle
1	2	3	4	5	6	7
M21	M22	M23	M24	M25	M26	M27
KHN5	KHN6	KHN7	KHN8	KHN9	KHN10	KHN11
MAG12	MAG13	MAG14	MAG15	MAG16	MAG17	MAG18
8	9	10	11	12	13	14
M28	M29	M30	M31	A1	A2	A3
KHN12	KHN13	KHN14	KHN15	KHN16	KHN17	KHN18
MAG19	MAG20	MAG21	MAG22	MAG23	MAG24	MAG25
15	16	17	18	19	20	21
A4	A5	A6	A7	A8	A9	A10
KHN19	KHN20	KHN21	KHN22	KHN23	KHN24	KHN25
MAG26	MAG27	MAG28	MAG29	MAG30	MIY1	MIY2
22	23	24	25	26	27	28
A11	A12	A13	A14	A15	A16	A17
KHN26	KHN27	KHN28	KHN29	KHN30	PIN1	PIN2
MIY3	MIY4	MIY5	MIY6	MIY7	MIY8	MIY9

MABABA WEUSI 519/2035-36/6271-72/2027-28						
Ashe	Ojobo	Iwa	Araorun	Ile	Itutu	Erinle
1	2	3	4	5	6	7
MA16	MA17	MA18	MA19	MA20	MA21	MA22
APP1	APP2	APP3	APP4	APP5	APP6	APP7
GIN8	GIN9	GIN10	GIN11	GIN12	GIN13	GIN14
8	9	10	11	12	13	14
MA23	MA24	MA25	MA26	MA27	MA28	MA29
APP8	APP9	APP10	APP11	APP12	APP13	APP14
GIN15	GIN16	GIN17	GIN18	GIN19	GIN20	GIN21
15	16	17	18	19	20	21
MA30	MA31	J1	J2	J3	J4	J5
APP15	APP16	APP17	APP18	APP19	APP20	APP21
GIN22	GIN23	GIN24	GIN25	GIN26	GIN27	GIN28
22	23	24	25	26	27	28
J6	J7	J8	J9	J10	J11	J12
APP22	APP23	APP24	APP25	APP26	APP27	APP28
GIN29	GIN30	SAN1	SAN2	SAN3	SAN4	SAN5

THE NU AFRIKAN CALENDAR

MWANAMKE WEUSI 519/2035-36/6271-72/2027-28						
Ashe	Ojobo	Iwa	Araorun	Ile	Itutu	Erinle
1	2	3	4	5	6	7
J13	J14	J15	J16	J17	J18	J19
APP29	APP30	MSR1	MSR2	MSR3	MSR4	MSR5
SAN6	SAN7	SAN8	SAN9	SAN10	SAN11	SAN12
8	9	10	11	12	13	14
J20	J21	J22	J23	J24	J25	J26
MSR6	MSR7	MSR8	MSR9	MSR10	MSR11	MSR12
SAN13	SAN14	SAN15	SAN16	SAN17	SAN18	SAN19
15	16	17	18	19	20	21
J27	J28	J29	J30	JU1	JU2	JU3
MSR13	MSR14	MSR15	MSR16	MSR17	MSR18	MSR19
SAN20	SAN21	SAN22	SAN23	SAN24	SAN25	SAN26
22	23	24	25	26	27	28
JU4	JU5	JU6	JU7	JU8	JU9	JU10
MSR20	MRS21	MSR22	MSR23	MSR24	MSR25	MSR26
SAN27	SAN28	SAN29	SAN30	HAM1	HAM2	HAM3

TAIFA WEUSI 519/2035-36/6271-72/2027-28						
Ashe	Ojobo	Iwa	Araorun	Ile	Itutu	Erinle
1	2	3	4	5	6	7
AU8	AU9	AU10	AU11	AU12	AU13	AU14
THT20	THT21	THT22	THT23	THT24	THT25	THT26
NAH2	NAH3	NAH4	NAH5	NAH6	NAH7	NAH8
8	9	10	11	12	13	14
AU15	AU16	AU17	AU18	AU19	AU20	AU21
THT27	THT28	THT29	THT30	PAP1	PAP1	PAP2
NAH9	NAH10	NAH11	NAH12	NAH13	NAH14	NAH15
15	16	17	18	19	20	21
AU22	AU23	AU24	AU25	AU26	AU27	AU28
PAP3	PAP4	PAP5	PAP6	PAP7	PAP8	PAP9
NAH16	NAH17	NAH18	NAH19	NAH20	NAH21	NAH22
22	23	24	25	26	27	28
AU29	AU30	AU31	S1	S2	S3	S4
PAP10	PAP11	PAP12	PAP13	PAP14	PAP15	PAP16
NAH23	NAH24	NAH25	NAH26	NAH27	NAH28	NAH29

MTU WEUSI 519/2035-36/6271-72/2027-28						
Ashe	Ojobo	Iwa	Araorun	Ile	Itutu	Erinle
1	2	3	4	5	6	7
JU11	JU12	JU13	JU14	JU15	JU16	JU17
MSR27	MSR28	MSR29	MSR30	*ASR1*	*HRU2*	*STH3*
HAM4	HAM5	HAM6	HAM7	HAM8	HAM9	HAM10
8	9	10	11	12	13	14
JU18	JU19	JU20	JU21	JU22	JU23	JU24
AST4	*NBH5*	THT1	THT2	THT3	THT4	THT5
HAM11	HAM12	HAM13	HAM14	HAM15	HAM16	HAM17
15	16	17	18	19	20	21
JU25	JU26	JU27	JU28	JU29	JU30	JU31
THT6	THT7	THT8	THT9	THT10	THT11	THT12
HAM18	HAM19	HAM20	HAM21	HAM22	HAM23	HAM24
22	23	24	25	26	27	28
AU1	AU2	AU3	AU4	AU5	AU6	AU7
THT13	THT14	THT15	THT16	THT17	THT18	THT19
HAM25	HAM26	HAM27	HAM28	HAM29	HAM30	NAH1

UWEZO 519/2035-36/6271-72/2027-28						
Ashe	Ojobo	Iwa	Araorun	Ile	Itutu	Erinle
1	2	3	4	5	6	7
S5	S6	S7	S8	S9	S10	S11
PAP17	PAP18	PAP19	PAP20	PAP21	PAP22	PAP23
NAH30	*QUA1*	*QUA2*	*QUA3*	*QUA4*	*QUA5*	*QUA6*
8	9	10	11	12	13	14
S12	S13	S14	S15	S16	S17	S18
PAP24	PAP25	PAP26	PAP27	PAP28	PAP29	PAP30
MAS1	MAS2	MAS3	MAS4	MAS5	MAS6	MAS7
15	16	17	18	19	20	21
S19	S20	S21	S22	S23	S24	S25
HTH1	HTH2	HTH3	HTH4	HTH5	HTH6	HTH7
MAS8	MAS9	MAS10	MAS11	MAS12	MAS13	MAS14
22	23	24	25	26	27	28
S26	S27	S28	S29	S30	O1	O2
HTH8	HTH9	HTH10	HTH11	HTH12	HTH13	HTH14
MAS15	MAS16	MAS17	MAS18	MAS19	MAS20	MAS21

THE NU AFRIKAN CALENDAR

EKUNDU 519/2035-36/6271-72/2027-28

Ashe	Ojobo	Iwa	Araorun	Ile	Itutu	Erinle
1	2	3	4	5	6	7
O3	O4	O5	O6	O7	O8	O9
HTH15	HTH16	HTH17	HTH18	HTH19	HTH20	HTH21
MAS22	MAS23	MAS 24	MAS25	MAS26	MAS27	MAS28
8	9	10	11	12	13	14
O10	O11	O12	O13	O14	O15	O16
HTH22	HTH23	HTH24	HTH25	HTH26	HTH27	HTH28
MAS29	MAS30	THi1	THi2	THi3	THi4	THi5
15	16	17	18	19	20	21
O17	O18	O19	O20	O21	O22	O23
HTH29	HTH30	KHK1	KHK2	KHK3	KHK4	KHK5
THi6	THi7	THi8	THi9	THi10	THi11	THi12
22	23	24	25	26	27	28
O24	O25	O26	O27	O28	O29	O30
KHK6	KHK7	KHK8	KHK9	KHK10	KHK11	KHK12
THi13	THi14	THi15	THi16	THi17	THi18	THi19

KIJANI 519/2035-36/6271-72/2027-28

Ashe	Ojobo	Iwa	Araorun	Ile	Itutu	Erinle
1	2	3	4	5	6	7
N28	N29	N30	D1	D2	D3	D4
TBT11	TBT12	TBT13	TBT14	TBT15	TBT16	TBT17
HID18	HID19	HID20	HID21	HID22	HID23	HID24
8	9	10	11	12	13	14
D5	D6	D7	D8	D9	D10	D11
TBT18	TBT19	TBT20	TBT21	TBT22	TBT23	TBT24
HID25	HID26	HID27	HID28	HID29	HID30	TAH1
15	16	17	18	19	20	21
D12	D13	D14	D15	D16	D17	D18
TBT25	TBT26	TBT27	TBT28	TBT29	TBT30	MKH1
TAH2	TAH3	TAH4	TAH5	TAH6	TAH7	TAH8
22	23	24	25	26	27	28
D19	D20	D21	D22	D23	D24	D25
MKH2	MKH3	MKH4	MKH5	MKH6	MKH7	MKH8
TAH9	TAH10	TAH11	TAH12	TAH13	TAH14	TAH15

EUSI 519/2035-36/6271-72/2027-28

Ashe	Ojobo	Iwa	Araorun	Ile	Itutu	Erinle
1	2	3	4	5	6	7
O31	N1	N2	N3	N4	N5	N6
KHK13	KHK14	KHK15	KHK16	KHK17	KHK18	KHK19
THi20	THi21	THi22	THi23	THi24	THi25	THi26
8	9	10	11	12	13	14
N7	N8	N9	N10	N11	N12	N13
KHK20	KHK21	KHK22	KHK23	KHK24	KHK25	KHK26
THi27	THi28	THi29	THi30	HID1	HID2	HID3
15	16	17	18	19	20	21
N14	N15	N16	N17	N18	N19	N20
KHK27	KHK28	KHK29	KHK30	TBT1	TBT2	TBT3
HID4	HID5	HID6	HID7	HID8	HID9	HID10
22	23	24	25	26	27	28
N21	N22	N23	N24	N25	N26	N27
TBT4	TBT5	TBT6	TBT7	TBT8	TBT9	TBT10
HID11	HID12	HID13	HID14	HID15	HID16	HID17

KWANZA 519/2035-36/6271-72/2027-28

Umoja	Kujichagulia	Ujima	Ujamaa	Nia	Kuumba	Imani
1	2	3	4	5	6	7
D26	D27	D28	D29	D30	D31	JA1
MKH9	MKH10	MKH11	MKH12	MKH13	MKH14	MKH15
TAH16	TAH17	TAH18	TAH19	TAH20	TAH21	TAH22
8	9	10	11	12	13	14
JA2	JA3	JA4	JA5	JA6	JA7	JA8
MKH16	MKH17	MKH18	MKH19	MKH20	MKH21	MKH22
TAH23	TAH24	TAH25	TAH26	TAH27	TAH28	TAH29
15	16	17	18	19	20	21
JA9	JA10	JA11	JA12	JA13	JA14	JA15
MKH23	MKH24	MKH25	KMH26	MKH27	MKH28	MKH29
TAH30	TIR1	TIR2	TIR3	TIR4	TIR5	TIR6
22	23	24	25	26	27	28
JA16	JA17	JA18	JA19	JA20	JA21	JA22
MKH30	AMN1	AMN2	AMN3	AMN4	AMN5	AMN6
TIR7	TIR8	TIR9	TIR10	TIR11	TIR12	TIR13

THE NU AFRIKAN CALENDAR

SHAHIDI 519/2035-36/6271-72/2027-28						
Ashe	**Ojobo**	**Iwa**	**Araorun**	**Ile**	**Itutu**	**Erinle**
1	2	3	4	5	6	7
JA23	JA24	JA25	JA26	JA27	JA28	JA29
AMN7	AMN8	AMN9	AMN10	AMN11	AMN12	AMN13
TIR14	TIR15	TIR16	TIR17	TIR18	TIR19	TIR20
8	9	10	11	12	13	14
JA30	JA31	F1	F2	F3	F4	F5
AMN14	AMN15	AMN16	AMN17	AMN18	AMN19	AMN20
TIR21	TIR22	TIR23	TIR24	TIR25	TIR26	TIR27
15	16	17	18	19	20	21
F6	F7	F8	F9	F10	F 11	F12
AMN21	AMN22	AMN23	AMN24	AMN25	AMN26	AMN28
TIR28	TIR29	TIR30	YEK1	YEK2	YEK3	YEK4
22	23	24	25	26	27	28
F13	F14	F15	F16	F17	F18	F19
AMN29	AMN30	RNT1	RNT2	RNT3	RNT4	RNT5
YEK5	YEK6	YEK7	YEK8	YEK9	YEK10	YEK11
SHAHIDI 70						
F20						
RNT6						
YEK12						

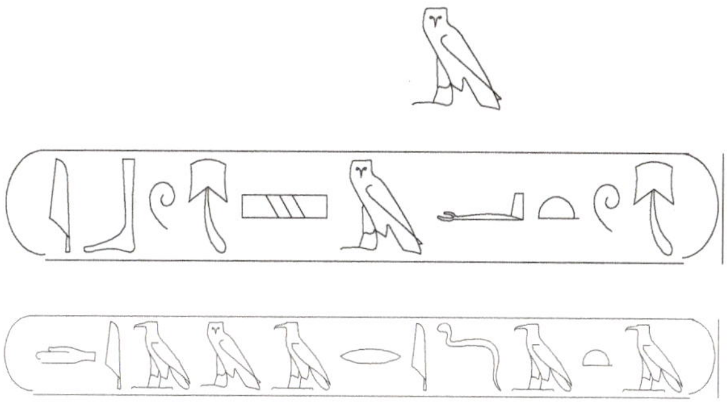

THE NU AFRIKAN CALENDAR

MAAFA 520 B.S.E./ASIENTO 2036-37 C.E./
KAM 6272-73 S.M./EC 2028-29 A.M. (ዓ.ም.)

AFU 520/2036-37/6272-73/2028-29

Ashe	Ojobo	Iwa	Araorun	Ile	Itutu	Erinle
1	2	3	4	5	6	7
F21	F22	F23	F24	F25	F26	F27
RNT7	RNT8	RNT9	RNT10	RNT11	RNT12	RNT13
YEK13	YEK14	YEK15	YEK16	YEK17	YEK18	YEK19
8	9	10	11	12	13	14
F28	F29	M1	M2	M3	M4	M5
RNT14	RNT15	RNT16	RNT17	RNT18	RNT19	RNT20
YEK 20	YEK21	YEK22	YEK23	YEK24	YEK25	YEK26
15	16	17	18	19	20	21
M6	M7	M8	M9	M10	M11	M12
RNT21	RNT22	RNT23	RNT24	RNT25	RNT26	RNT27
YEK 27	YEK28	YEK29	YEK30	MAG1	MAG2	MAG3
22	23	24	25	26	27	28
M13	M14	M15	M16	M17	M18	M19
RNT28	RNT29	RNT30	KHN1	KHN2	KHN3	KHN4
MAG4	MAG5	MAG6	MAG7	MAG8	MAG9	MAG10
SHARIFU 127						
M20						
KHN5						
MAG11						

KAWAIDA 520/2036-37/6272-73/2028-29

Ashe	Ojobo	Iwa	Araorun	Ile	Itutu	Erinle
1	2	3	4	5	6	7
A18	A19	A20	A21	A22	A23	A24
PIN4	PIN5	PIN6	PIN7	PIN8	PIN9	PIN10
MIY10	MIY11	MIY12	MIY13	MIY14	MIY15	MIY16
8	9	10	11	12	13	14
A25	A26	A27	A28	A29	A30	MA1
PIN11	PIN12	PIN13	PIN14	PIN15	PIN16	PIN17
MIY17	MIY18	MIY19	MIY20	MIY21	MIY22	MIY23
15	16	17	18	19	20	21
MA2	MA3	MA4	MA5	MA6	MA7	MA8
PIN18	PIN19	PIN20	PIN21	PIN22	PIN23	PIN24
MIY24	MIY25	MIY26	MIY27	MIY28	MIY29	MIY30
22	23	24	25	26	27	28
MA9	MA10	MA11	MA12	MA13	MA14	MA15
PIN25	PIN26	PIN27	PIN28	PIN29	PIN30	APP1
GIN1	GIN2	GIN3	GIN4	GIN5	GIN6	GIN7

UHURU 520/2036-37/6272-73/2028-29

Ashe	Ojobo	Iwa	Araorun	Ile	Itutu	Erinle
1	2	3	4	5	6	7
M21	M22	M23	M24	M25	M26	M27
KHN6	KHN7	KHN8	KHN9	KHN10	KHN11	KHN12
MAG12	MAG13	MAG14	MAG15	MAG16	MAG17	MAG18
8	9	10	11	12	13	14
M28	M29	M30	M31	A1	A2	A3
KHN13	KHN14	KHN15	KHN16	KHN17	KHN18	KHN19
MAG19	MAG20	MAG21	MAG22	MAG23	MAG24	MAG25
15	16	17	18	19	20	21
A4	A5	A6	A7	A8	A9	A10
KHN20	KHN21	KHN222	KHN23	KHN24	KHN25	KHN26
MAG26	MAG27	MAG28	MAG29	MAG30	MIY1	MIY2
22	23	24	25	26	27	28
A11	A12	A13	A14	A15	A16	A17
KHN27	KHN28	KHN29	KHN30	PIN1	PIN2	PIN3
MIY3	MIY4	MIY5	MIY6	MIY7	MIY8	MIY9

MABABA WEUSI 520/2036-37/6272-73/2028-29

Ashe	Ojobo	Iwa	Araorun	Ile	Itutu	Erinle
1	2	3	4	5	6	7
MA16	MA17	MA18	MA19	MA20	MA21	MA22
APP2	APP3	APP4	APP5	APP6	APP7	APP8
GIN8	GIN9	GIN10	GIN11	GIN12	GIN13	GIN14
8	9	10	11	12	13	14
MA23	MA24	MA25	MA26	MA27	MA28	MA29
APP9	APP10	APP11	APP12	APP13	APP14	APP15
GIN15	GIN16	GIN17	GIN18	GIN19	GIN20	GIN21
15	16	17	18	19	20	21
MA30	MA31	J1	J2	J3	J4	J5
APP16	APP17	APP18	APP19	APP20	APP21	APP22
GIN22	GIN23	GIN24	GIN25	GIN26	GIN27	GIN28
22	23	24	25	26	27	28
J6	J7	J8	J9	J10	J11	J12
APP23	APP24	APP25	APP26	APP27	APP28	APP29
GIN29	GIN30	SAN1	SAN2	SAN3	SAN4	SAN5

THE NU AFRIKAN CALENDAR

MWANAMKE WEUSI 520/2036-37/6272-73/2028-29						
Ashe	Ojobo	Iwa	Araorun	Ile	Itutu	Erinle
1	2	3	4	5	6	7
J13	J14	J15	J16	J17	J18	J19
APP30	MSR1	MSR2	MSR3	MSR4	MSR5	MSR6
SAN6	SAN7	SAN8	SAN9	SAN10	SAN11	SAN12
8	9	10	11	12	13	14
J20	J21	J22	J23	J24	J25	J26
MSR7	MSR8	MSR9	MSR10	MSR11	MSR12	MSR13
SAN13	SAN14	SAN15	SAN16	SAN17	SAN18	SAN19
15	16	17	18	19	20	21
J27	J28	J29	J30	JU1	JU2	JU3
MSR14	MSR15	MSR16	MSR17	MSR18	MSR19	MSR20
SAN20	SAN21	SAN22	SAN23	SAN24	SAN25	SAN26
22	23	24	25	26	27	28
JU4	JU5	JU6	JU7	JU8	JU9	JU10
MSR21	MRS22	MSR23	MSR24	MSR25	MSR26	MSR27
SAN27	SAN28	SAN29	SAN30	HAM1	HAM2	HAM3

TAIFA WEUSI 520/2036-37/6272-73/2028-29						
Ashe	Ojobo	Iwa	Araorun	Ile	Itutu	Erinle
1	2	3	4	5	6	7
AU8	AU9	AU10	AU11	AU12	AU13	AU14
THT20	THT21	THT22	THT23	THT24	THT25	THT26
NAH2	NAH3	NAH4	NAH5	NAH6	NAH7	NAH8
8	9	10	11	12	13	14
AU15	AU16	AU17	AU18	AU19	AU20	AU21
THT27	THT28	THT29	THT30	PAP1	PAP1	PAP2
NAH9	NAH10	NAH11	NAH12	NAH13	NAH14	NAH15
15	16	17	18	19	20	21
AU22	AU23	AU24	AU25	AU26	AU27	AU28
PAP3	PAP4	PAP5	PAP6	PAP7	PAP8	PAP9
NAH16	NAH17	NAH18	NAH19	NAH20	NAH21	NAH22
22	23	24	25	26	27	28
AU29	AU30	AU31	S1	S2	S3	S4
PAP10	PAP11	PAP12	PAP13	PAP14	PAP15	PAP16
NAH23	NAH24	NAH25	NAH26	NAH27	NAH28	NAH29

MTU WEUSI 520/2036-37/6272-73/2028-29						
Ashe	Ojobo	Iwa	Araorun	Ile	Itutu	Erinle
1	2	3	4	5	6	7
JU11	JU12	JU13	JU14	JU15	JU16	JU17
MSR28	MSR29	MSR30	*ASR1*	*HRU2*	*STH3*	*AST4*
HAM4	HAM5	HAM6	HAM7	HAM8	HAM9	HAM10
8	9	10	11	12	13	14
JU18	JU19	JU20	JU21	JU22	JU23	JU24
NBH5	*ANU6*	THT1	THT2	THT3	THT4	THT5
HAM11	HAM12	HAM13	HAM14	HAM15	HAM16	HAM17
15	16	17	18	19	20	21
JU25	JU26	JU27	JU28	JU29	JU30	JU31
THT6	THT7	THT8	THT9	THT10	THT11	THT12
HAM18	HAM19	HAM20	HAM21	HAM22	HAM23	HAM24
22	23	24	25	26	27	28
AU1	AU2	AU3	AU4	AU5	AU6	AU7
THT13	THT14	THT15	THT16	THT17	THT18	THT19
HAM25	HAM26	HAM27	HAM28	HAM29	HAM30	NAH1

UWEZO 520/2036-37/6272-73/2028-29						
Ashe	Ojobo	Iwa	Araorun	Ile	Itutu	Erinle
1	2	3	4	5	6	7
S5	S6	S7	S8	S9	S10	S11
PAP17	PAP18	PAP19	PAP20	PAP21	PAP22	PAP23
NAH30	*PAG1*	*PAG2*	*PAG3*	*PAG4*	*PAG5*	MAS1
8	9	10	11	12	13	14
S12	S13	S14	S15	S16	S17	S18
PAP24	PAP25	PAP26	PAP27	PAP28	PAP29	PAP30
MAS2	MAS3	MAS4	MAS5	MAS6	MAS7	MAS8
15	16	17	18	19	20	21
S19	S20	S21	S22	S23	S24	S25
HTH1	HTH2	HTH3	HTH4	HTH5	HTH6	HTH7
MAS9	MAS10	MAS11	MAS12	MAS13	MAS14	MAS15
22	23	24	25	26	27	28
S26	S27	S28	S29	S30	O1	O2
HTH8	HTH9	HTH10	HTH11	HTH12	HTH13	HTH14
MAS16	MAS17	MAS18	MAS19	MAS20	MAS21	MAS22

THE NU AFRIKAN CALENDAR

EKUNDU 520/2036-37/6272-73/2028-29						
Ashe	Ojobo	Iwa	Araorun	Ile	Itutu	Erinle
1	2	3	4	5	6	7
O3	O4	O5	O6	O7	O8	O9
HTH15	HTH16	HTH17	HTH18	HTH19	HTH20	HTH21
MAS23	MAS24	MAS25	MAS26	MAS27	MAS28	MAS29
8	9	10	11	12	13	14
O10	O11	O12	O13	O14	O15	O16
HTH22	HTH23	HTH24	HTH25	HTH26	HTH27	HTH28
MAS30	THi1	THi2	THi3	THi4	THi5	THi6
15	16	17	18	19	20	21
O17	O18	O19	O20	O21	O22	O23
HTH29	HTH30	KHK1	KHK2	KHK3	KHK4	KHK5
THi7	THi8	This9	THi10	THi11	THi12	THi13
22	23	24	25	26	27	28
O24	O25	O26	O27	O28	O29	O30
KHK6	KHK7	KHK8	KHK9	KHK10	KHK11	KHK12
Thi14	THi15	THi16	THi17	THi18	THi19	THi20

KIJANI 520/2036-37/6272-73/2028-29						
Ashe	Ojobo	Iwa	Araorun	Ile	Itutu	Erinle
1	2	3	4	5	6	7
N28	N29	N30	D1	D2	D3	D4
TBT11	TBT12	TBT13	TBT14	TBT15	TBT16	TBT17
HID19	HID20	HID21	HID22	HID23	HID24	HID25
8	9	10	11	12	13	14
D5	D6	D7	D8	D9	D10	D11
TBT18	TBT19	TBT20	TBT21	TBT22	TBT23	TBT24
HID26	HID27	HID28	HID29	HID30	TAH1	TAH2
15	16	17	18	19	20	21
D12	D13	D14	D15	D16	D17	D18
TBT25	TBT26	TBT27	TBT28	TBT29	TBT30	MKH1
TAH3	TAH4	TAH5	TAH6	TAH7	TAH8	TAH9
22	23	24	25	26	27	28
D19	D20	D21	D22	D23	D24	D25
MKH2	MKH3	MKH4	MKH5	MKH6	MKH7	MKH8
TAH10	TAH11	TAH12	TAH13	TAH14	TAH15	TAH16

EUSI 520/2036-37/6272-73/2028-29						
Ashe	Ojobo	Iwa	Araorun	Ile	Itutu	Erinle
1	2	3	4	5	6	7
O31	N1	N2	N3	N4	N5	N6
KHK13	KHK14	KHK15	KHK16	KHK17	KHK18	KHK19
THi21	THi22	THi23	THi24	THi25	THi26	THi27
8	9	10	11	12	13	14
N7	N8	N9	N10	N11	N12	N13
KHK20	KHK21	KHK22	KHK23	KHK24	KHK25	KHK26
THi28	THi29	THi30	HID1	HID2	HID3	HID4
15	16	17	18	19	20	21
N14	N15	N16	N17	N18	N19	N20
KHK27	KHK28	KHK29	KHK30	TBT1	TBT2	TBT3
HID5	HID6	HID7	HID8	HID9	HID10	HID11
22	23	24	25	26	27	28
N21	N22	N23	N24	N25	N26	N27
TBT4	TBT5	TBT6	TBT7	TBT8	TBT9	TBT10
HID12	HID13	HID14	HID15	HID16	HID17	HID18

KWANZA 520/2036-37/6272-73/2028-29						
Umoja	Kujichagulia	Ujima	Ujamaa	Nia	Kuumba	Imani
1	2	3	4	5	6	7
D26	D27	D28	D29	D30	D31	JA1
MKH9	MKH10	MKH11	MKH12	MKH13	MKH14	MKH15
TAH17	TAH18	TAH19	TAH20	TAH21	TAH22	TAH23
8	9	10	11	12	13	14
JA2	JA3	JA4	JA5	JA6	JA7	JA8
MKH16	MKH17	MKH18	MKH19	MKH20	MKH21	MKH22
TAH24	TAH25	TAH26	TAH27	TAH28	TAH29	TAH30
15	16	17	18	19	20	21
JA9	JA10	JA11	JA12	JA13	JA14	JA15
MKH23	MKH24	MKH25	KMH26	MKH27	MKH28	MKH29
TIR1	TIR2	TIR3	TIR4	TIR5	TIR6	TIR7
22	23	24	25	26	27	28
JA16	JA17	JA18	JA19	JA20	JA21	JA22
MKH30	AMN1	AMN2	AMN3	AMN4	AMN5	AMN6
TIR8	TIR9	TIR10	TIR11	TIR12	TIR13	TIR14

THE NU AFRIKAN CALENDAR

SHAHIDI 520/2036-37/6272-73/2028-29						
Ashe	Ojobo	Iwa	Araorun	Ile	Itutu	Erinle
1	2	3	4	5	6	7
JA23	JA24	JA25	JA26	JA27	JA28	JA29
AMN7	AMN8	AMN9	AMN10	AMN11	AMN12	AMN13
TIR15	TIR16	TIR17	TIR18	TIR19	TIR20	TIR21
8	9	10	11	12	13	14
JA30	JA31	F1	F2	F3	F4	F5
AMN14	AMN15	AMN16	AMN17	AMN18	AMN19	AMN20
TIR22	TIR23	TIR24	TIR25	TIR26	TIR27	TIR28
15	16	17	18	19	20	21
F6	F7	F8	F9	F10	F 11	F12
AMN21	AMN22	AMN23	AMN24	AMN25	AMN26	AMN28
TIR29	TIR30	YEK1	YEK2	YEK3	YEK4	YEK5
22	23	24	25	26	27	28
F13	F14	F15	F16	F17	F18	F19
AMN29	AMN30	RNT1	RNT2	RNT3	RNT4	RNT5
YEK6	YEK7	YEK8	YEK9	YEK10	YEK11	YEK12
SHAHIDI 71						
F20						
RNT6						
YEK13						

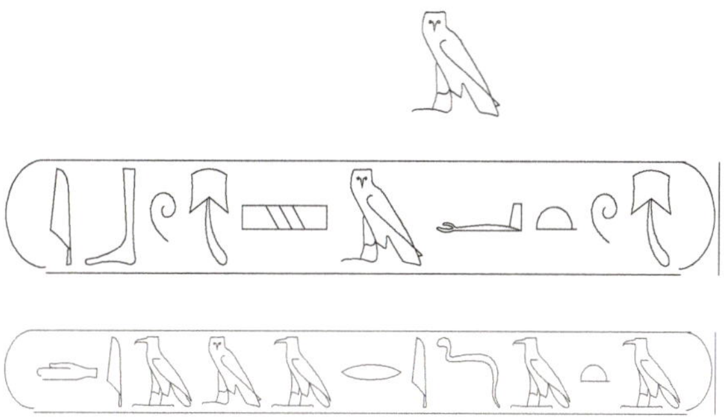

THE NU AFRIKAN CALENDAR

MAAFA 521 B.S.E./ASIENTO 2037-38 C.E.
KAM 6273-74 S.M./EC 2029-30 A.M. (ዓ.ም.)

AFU 521/2037-38/6273-74/2029-30

Ashe	Ojobo	Iwa	Araorun	Ile	Itutu	Erinle
1	2	3	4	5	6	7
F21	F22	F23	F24	F25	F26	F27
RNT7	RNT8	RNT9	RNT10	RNT11	RNT12	RNT13
YEK14	YEK15	YEK16	YEK17	YEK18	YEK19	YEK20
8	9	10	11	12	13	14
F28	M1	M2	M3	M4	M5	M6
RNT14	RNT15	RNT16	RNT17	RNT18	RNT19	RNT20
YEK21	YEK22	YEK23	YEK24	YEK25	YEK26	YEK27
15	16	17	18	19	20	21
M7	M8	M9	M10	M11	M12	M13
RNT21	RNT22	RNT23	RNT24	RNT25	RNT26	RNT27
YEK28	YEK29	YEK30	MAG1	MAG2	MAG3	MAG4
22	23	24	25	26	27	28
M14	M15	M16	M17	M18	M19	M20
RNT28	RNT29	RNT30	KHN1	KHN2	KHN3	KHN4
MAG5	MAG6	MAG7	MAG8	MAG9	MAG10	MAG11

KAWAIDA 521/2037-38/6273-74/2029-30

Ashe	Ojobo	Iwa	Araorun	Ile	Itutu	Erinle
1	2	3	4	5	6	7
A18	A19	A20	A21	A22	A23	A24
PIN3	PIN4	PIN5	PIN6	PIN7	PIN8	PIN9
MIY10	MIY11	MIY12	MIY13	MIY14	MIY15	MIY16
8	9	10	11	12	13	14
A25	A26	A27	A28	A29	A30	MA1
PIN10	PIN11	PIN12	PIN13	PIN14	PIN15	PIN16
MIY17	MIY18	MIY19	MIY20	MIY21	MIY22	MIY23
15	16	17	18	19	20	21
MA2	MA3	MA4	MA5	MA6	MA7	MA8
PIN17	PIN18	PIN19	PIN20	PIN21	PIN22	PIN23
MIY24	MIY25	MIY26	MIY27	MIY28	MIY29	MIY30
22	23	24	25	26	27	28
MA9	MA10	MA11	MA12	MA13	MA14	MA15
PIN24	PIN25	PIN26	PIN27	PIN28	PIN29	PIN30
GIN1	GIN2	GIN3	GIN4	GIN5	GIN6	GIN7

UHURU 521/2037-38/6273-74/2029-30

Ashe	Ojobo	Iwa	Araorun	Ile	Itutu	Erinle
1	2	3	4	5	6	7
M21	M22	M23	M24	M25	M26	M27
KHN5	KHN6	KHN7	KHN8	KHN9	KHN10	KHN11
MAG12	MAG13	MAG14	MAG15	MAG16	MAG17	MAG18
8	9	10	11	12	13	14
M28	M29	M30	M31	A1	A2	A3
KHN12	KHN13	KHN14	KHN15	KHN16	KHN17	KHN18
MAG19	MAG20	MAG21	MAG22	MAG23	MAG24	MAG25
15	16	17	18	19	20	21
A4	A5	A6	A7	A8	A9	A10
KHN19	KHN20	KHN21	KHN22	KHN23	KHN24	KHN25
MAG26	MAG27	MAG28	MAG29	MAG30	MIY1	MIY2
22	23	24	25	26	27	28
A11	A12	A13	A14	A15	A16	A17
KHN26	KHN27	KHN28	KHN29	KHN30	PIN1	PIN2
MIY3	MIY4	MIY5	MIY6	MIY7	MIY8	MIY9

MABABA WEUSI 521/2037-38/6273-74/2029-30

Ashe	Ojobo	Iwa	Araorun	Ile	Itutu	Erinle
1	2	3	4	5	6	7
MA16	MA17	MA18	MA19	MA20	MA21	MA22
APP1	APP2	APP3	APP4	APP5	APP6	APP7
GIN8	GIN9	GIN10	GIN11	GIN12	GIN13	GIN14
8	9	10	11	12	13	14
MA23	MA24	MA25	MA26	MA27	MA28	MA29
APP8	APP9	APP10	APP11	APP12	APP13	APP14
GIN15	GIN16	GIN17	GIN18	GIN19	GIN20	GIN21
15	16	17	18	19	20	21
MA30	MA31	J1	J2	J3	J4	J5
APP15	APP16	APP17	APP18	APP19	APP20	APP21
GIN22	GIN23	GIN24	GIN25	GIN26	GIN27	GIN28
22	23	24	25	26	27	28
J6	J7	J8	J9	J10	J11	J12
APP22	APP23	APP24	APP25	APP26	APP27	APP28
GIN29	GIN30	SAN1	SAN2	SAN3	SAN4	SAN5

THE NU AFRIKAN CALENDAR

MWANAMKE WEUSI 521/2037-38/6273-74/2029-30

Ashe	Ojobo	Iwa	Araorun	Ile	Itutu	Erinle
1	2	3	4	5	6	7
J13	J14	J15	J16	J17	J18	J19
APP29	APP30	MSR1	MSR2	MSR3	MSR4	MSR5
SAN6	SAN7	SAN8	SAN9	SAN10	SAN11	SAN12
8	9	10	11	12	13	14
J20	J21	J22	J23	J24	J25	J26
MSR6	MSR7	MSR8	MSR9	MSR10	MSR11	MSR12
SAN13	SAN14	SAN15	SAN16	SAN17	SAN18	SAN19
15	16	17	18	19	20	21
J27	J28	J29	J30	JU1	JU2	JU3
MSR13	MSR14	MSR15	MSR16	MSR17	MSR18	MSR19
SAN20	SAN21	SAN22	SAN23	SAN24	SAN25	SAN26
22	23	24	25	26	27	28
JU4	JU5	JU6	JU7	JU8	JU9	JU10
MSR20	MRS21	MSR22	MSR23	MSR24	MSR25	MSR26
SAN27	SAN28	SAN29	SAN30	HAM1	HAM2	HAM3

TAIFA WEUSI 521/2037-38/6273-74/2029-30

Ashe	Ojobo	Iwa	Araorun	Ile	Itutu	Erinle
1	2	3	4	5	6	7
AU8	AU9	AU10	AU11	AU12	AU13	AU14
THT20	THT21	THT22	THT23	THT24	THT25	THT26
NAH2	NAH3	NAH4	NAH5	NAH6	NAH7	NAH8
8	9	10	11	12	13	14
AU15	AU16	AU17	AU18	AU19	AU20	AU21
THT27	THT28	THT29	THT30	PAP1	PAP1	PAP2
NAH9	NAH10	NAH11	NAH12	NAH13	NAH14	NAH15
15	16	17	18	19	20	21
AU22	AU23	AU24	AU25	AU26	AU27	AU28
PAP3	PAP4	PAP5	PAP6	PAP7	PAP8	PAP9
NAH16	NAH17	NAH18	NAH19	NAH20	NAH21	NAH22
22	23	24	25	26	27	28
AU29	AU30	AU31	S1	S2	S3	S4
PAP10	PAP11	PAP12	PAP13	PAP14	PAP15	PAP16
NAH23	NAH24	NAH25	NAH26	NAH27	NAH28	NAH29

MTU WEUSI 521/2037-38/6273-74/2029-30

Ashe	Ojobo	Iwa	Araorun	Ile	Itutu	Erinle
1	2	3	4	5	6	7
JU11	JU12	JU13	JU14	JU15	JU16	JU17
MSR27	MSR28	MSR29	MSR30	*ASR1*	*HRU2*	*STH3*
HAM4	HAM5	HAM6	HAM7	HAM8	HAM9	HAM10
8	9	10	11	12	13	14
JU18	JU19	JU20	JU21	JU22	JU23	JU24
AST4	*NBH5*	THT1	THT2	THT3	THT4	THT5
HAM11	HAM12	HAM13	AM14	HAM15	HAM16	HAM17
15	16	17	18	19	20	21
JU25	JU26	JU27	JU28	JU29	JU30	JU31
THT6	THT7	THT8	THT9	THT10	THT11	THT12
HAM18	HAM19	HAM20	HAM21	HAM22	HAM23	HAM24
22	23	24	25	26	27	28
AU1	AU2	AU3	AU4	AU5	AU6	AU7
THT13	THT14	THT15	THT16	THT17	THT18	THT19
HAM25	HAM26	HAM27	HAM28	HAM29	HAM30	NAH1

UWEZO 521/2037-38/6273-74/2029-30

Ashe	Ojobo	Iwa	Araorun	Ile	Itutu	Erinle
1	2	3	4	5	6	7
S5	S6	S7	S8	S9	S10	S11
PAP17	PAP18	PAP19	PAP20	PAP21	PAP22	PAP23
NAH30	*PAG1*	*PAG2*	*PAG3*	*PAG4*	*PAG5*	MAS1
8	9	10	11	12	13	14
S12	S13	S14	S15	S16	S17	S18
PAP24	PAP25	PAP26	PAP27	PAP28	PAP29	PAP30
MAS2	MAS3	MAS4	MAS5	MAS6	MAS7	MAS8
15	16	17	18	19	20	21
S19	S20	S21	S22	S23	S24	S25
HTH1	HTH2	HTH3	HTH4	HTH5	HTH6	HTH7
MAS9	MAS10	MAS11	MAS12	MAS13	MAS14	MAS15
22	23	24	25	26	27	28
S26	S27	S28	S29	S30	O1	O2
HTH8	HTH9	HTH10	HTH11	HTH12	HTH13	HTH14
MAS16	MAS17	MAS18	MAS19	MAS20	MAS21	MAS22

THE NU AFRIKAN CALENDAR

EKUNDU 521/2037-38/6273-74/2029-30						
Ashe	Ojobo	Iwa	Araorun	Ile	Itutu	Erinle
1	2	3	4	5	6	7
O3	O4	O5	O6	O7	O8	O9
HTH15	HTH16	HTH17	HTH18	HTH19	HTH20	HTH21
MAS23	MAS24	MAS25	MAS26	MAS27	MAS28	MAS29
8	9	10	11	12	13	14
O10	O11	O12	O13	O14	O15	O16
HTH22	HTH23	HTH24	HTH25	HTH26	HTH27	HTH28
MAS30	THi1	THi2	THi3	THi4	THi5	THi6
15	16	17	18	19	20	21
O17	O18	O19	O20	O21	O22	O23
HTH29	HTH30	KHK1	KHK2	KHK3	KHK4	KHK5
THi7	THi8	THi9	THi10	THi11	THi12	THi13
22	23	24	25	26	27	28
O24	O25	O26	O27	O28	O29	O30
KHK6	KHK7	KHK8	KHK9	KHK10	KHK11	KHK12
THi14	THi15	THi16	THi17	THi18	THi19	THi20

KIJANI 521/2037-38/6273-74/2029-30						
Ashe	Ojobo	Iwa	Araorun	Ile	Itutu	Erinle
1	2	3	4	5	6	7
N28	N29	N30	D1	D2	D3	D4
TBT11	TBT12	TBT13	TBT14	TBT15	TBT16	TBT17
HID19	HID20	HID21	HID22	HID23	HID24	HID25
8	9	10	11	12	13	14
D5	D6	D7	D8	D9	D10	D11
TBT18	TBT19	TBT20	TBT21	TBT22	TBT23	TBT24
HID26	HID27	HID28	HID29	HID30	TAH1	TAH2
15	16	17	18	19	20	21
D12	D13	D14	D15	D16	D17	D18
TBT25	TBT26	TBT27	TBT28	TBT29	TBT30	MKH1
TAH3	TAH4	TAH5	TAH6	TAH7	TAH8	TAH9
22	23	24	25	26	27	28
D19	D20	D21	D22	D23	D24	D25
MKH2	MKH3	MKH4	MKH5	MKH6	MKH7	MKH8
TAH10	TAH11	TAH12	TAH13	TAH14	TAH15	TAH16

EUSI 521/2037-38/6273-74/2029-30						
Ashe	Ojobo	Iwa	Araorun	Ile	Itutu	Erinle
1	2	3	4	5	6	7
O31	N1	N2	N3	N4	N5	N6
KHK13	KHK14	KHK15	KHK16	KHK17	KHK18	KHK19
THi21	THi22	THi23	THi24	THi25	THi26	THi27
8	9	10	11	12	13	14
N7	N8	N9	N10	N11	N12	N13
KHK20	KHK21	KHK22	KHK23	KHK24	KHK25	KHK26
THi28	THi29	THi30	HID1	HID2	HID3	HID4
15	16	17	18	19	20	21
N14	N15	N16	N17	N18	N19	N20
KHK27	KHK28	KHK29	KHK30	TBT1	TBT2	TBT3
HID5	HID6	HID7	HID8	HID9	HID10	HID11
22	23	24	25	26	27	28
N21	N22	N23	N24	N25	N26	N27
TBT4	TBT5	TBT6	TBT7	TBT8	TBT9	TBT10
HID12	HID13	HID14	HID15	HID16	HID17	HID18

KWANZA 521/2037-38/6273-74/2029-30						
Umoja	Kujichagulia	Ujima	Ujamaa	Nia	Kuumba	Imani
1	2	3	4	5	6	7
D26	D27	D28	D29	D30	D31	JA1
MKH9	MKH10	MKH11	MKH12	MKH13	MKH14	MKH15
TAH17	TAH18	TAH19	TAH20	TAH21	TAH22	TAH23
8	9	10	11	12	13	14
JA2	JA3	JA4	JA5	JA6	JA7	JA8
MKH16	MKH17	MKH18	MKH19	MKH20	MKH21	MKH22
TAH24	TAH25	TAH26	TAH27	TAH28	TAH29	TAH30
15	16	17	18	19	20	21
JA9	JA10	JA11	JA12	JA13	JA14	JA15
MKH23	MKH24	MKH25	KMH26	MKH27	MKH28	MKH29
TIR1	TIR2	TIR3	TIR4	TIR5	TIR6	TIR7
22	23	24	25	26	27	28
JA16	JA17	JA18	JA19	JA20	JA21	JA22
MKH30	AMN1	AMN2	AMN3	AMN4	AMN5	AMN6
TIR8	TIR9	TIR10	TIR11	TIR12	TIR13	TIR14

THE NU AFRIKAN CALENDAR

SHAHIDI 521/2037-38/6273-74/2029-30						
Ashe	Ojobo	Iwa	Araorun	Ile	Itutu	Erinle
1	2	3	4	5	6	7
JA23	JA24	JA25	JA26	JA27	JA28	JA29
AMN7	AMN8	AMN9	AMN10	AMN11	AMN12	AMN13
TIR15	TIR16	TIR17	TIR18	TIR19	TIR20	TIR21
8	9	10	11	12	13	14
JA30	JA31	F1	F2	F3	F4	F5
AMN14	AMN15	AMN16	AMN17	AMN18	AMN19	AMN20
TIR22	TIR23	TIR24	TIR25	TIR26	TIR27	TIR28
15	16	17	18	19	20	21
F6	F7	F8	F9	F10	F 11	F12
AMN21	AMN22	AMN23	AMN24	AMN25	AMN26	AMN28
TIR29	TIR30	YEK1	YEK2	YEK3	YEK4	YEK5
22	23	24	25	26	27	28
F13	F14	F15	F16	F17	F18	F19
AMN29	AMN30	RNT1	RNT2	RNT3	RNT4	RNT5
YEK6	YEK7	YEK8	YEK9	YEK10	YEK11	YEK12
SHAHIDI 72						
F20						
RNT6						
YEK13						

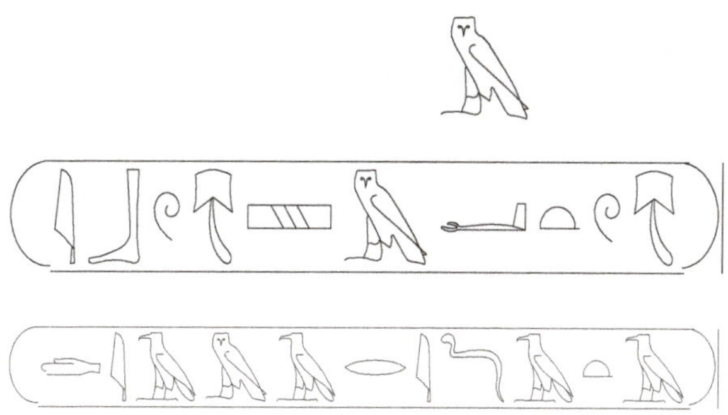

THE NU AFRIKAN CALENDAR

MAAFA 522 B.S.E./ASIENTO 2038-39 C.E.
KAM 6274-75 S.M./EC 2030-2031 A.M. (ዓ.ም.)

AFU 522/2038-39/6274-75/2030-31

Ashe	Ojobo	Iwa	Araorun	Ile	Itutu	Erinle
1	2	3	4	5	6	7
F21	F22	F23	F24	F25	F26	F27
RNT7	RNT8	RNT9	RNT10	RNT11	RNT12	RNT13
YEK14	YEK15	YEK16	YEK17	YEK18	YEK19	YEK20
8	9	10	11	12	13	14
F28	M1	M2	M3	M4	M5	M6
RNT14	RNT15	RNT16	RNT17	RNT18	RNT19	RNT20
YEK21	YEK22	YEK23	YEK24	YEK25	YEK26	YEK27
15	16	17	18	19	20	21
M7	M8	M9	M10	M11	M12	M13
RNT21	RNT22	RNT23	RNT24	RNT25	RNT26	RNT27
YEK28	YEK29	YEK30	MAG1	MAG2	MAG3	MAG4
22	23	24	25	26	27	28
M14	M15	M16	M17	M18	M19	M20
RNT28	RNT29	RNT30	KHN1	KHN2	KHN3	KHN4
MAG5	MAG6	MAG7	MAG8	MAG9	MAG10	MAG11

KAWAIDA 522/2038-39/6274-75/2030-31

Ashe	Ojobo	Iwa	Araorun	Ile	Itutu	Erinle
1	2	3	4	5	6	7
A18	A19	A20	A21	A22	A23	A24
PIN3	PIN4	PIN5	PIN6	PIN7	PIN8	PIN9
MIY10	MIY11	MIY12	MIY13	MIY14	MIY15	MIY16
8	9	10	11	12	13	14
A25	A26	A27	A28	A29	A30	MA1
PIN10	PIN11	PIN12	PIN13	PIN14	PIN15	PIN16
MIY17	MIY18	MIY19	MIY20	MIY21	MIY22	MIY23
15	16	17	18	19	20	21
MA2	MA3	MA4	MA5	MA6	MA7	MA8
PIN17	PIN18	PIN19	PIN20	PIN21	PIN22	PIN23
MIY24	MIY25	MIY26	MIY27	MIY28	MIY29	MIY30
22	23	24	25	26	27	28
MA9	MA10	MA11	MA12	MA13	MA14	MA15
PIN24	PIN25	PIN26	PIN27	PIN28	PIN29	PIN30
GIN1	GIN2	GIN3	GIN4	GIN5	GIN6	GIN7

UHURU 522/2038-39/6274-75/2030-31

Ashe	Ojobo	Iwa	Araorun	Ile	Itutu	Erinle
1	2	3	4	5	6	7
M21	M22	M23	M24	M25	M26	M27
KHN5	KHN6	KHN7	KHN8	KHN9	KHN10	KHN11
MAG12	MAG13	MAG14	MAG15	MAG16	MAG17	MAG18
8	9	10	11	12	13	14
M28	M29	M30	M31	A1	A2	A3
KHN12	KHN13	KHN14	KHN15	KHN16	KHN17	KHN18
MAG19	MAG20	MAG21	MAG22	MAG23	MAG24	MAG25
15	16	17	18	19	20	21
A4	A5	A6	A7	A8	A9	A10
KHN19	KHN20	KHN21	KHN22	KHN23	KHN24	KHN25
MAG26	MAG27	MAG28	MAG29	MAG30	MIY1	MIY2
22	23	24	25	26	27	28
A11	A12	A13	A14	A15	A16	A17
KHN26	KHN27	KHN28	KHN29	KHN30	PIN1	PIN2
MIY3	MIY4	MIY5	MIY6	MIY7	MIY8	MIY9

MABABA WEUSI 522/2038-39/6274-75/2030-31

Ashe	Ojobo	Iwa	Araorun	Ile	Itutu	Erinle
1	2	3	4	5	6	7
MA16	MA17	MA18	MA19	MA20	MA21	MA22
APP1	APP2	APP3	APP4	APP5	APP6	APP7
GIN8	GIN9	GIN10	GIN11	GIN12	GIN13	GIN14
8	9	10	11	12	13	14
MA23	MA24	MA25	MA26	MA27	MA28	MA29
APP8	APP9	APP10	APP11	APP12	APP13	APP14
GIN15	GIN16	GIN17	GIN18	GIN19	GIN20	GIN21
15	16	17	18	19	20	21
MA30	MA31	J1	J2	J3	J4	J5
APP15	APP16	APP17	APP18	APP19	APP20	APP21
GIN22	GIN23	GIN24	GIN25	GIN26	GIN27	GIN28
22	23	24	25	26	27	28
J6	J7	J8	J9	J10	J11	J12
APP22	APP23	APP24	APP25	APP26	APP27	APP28
GIN29	GIN30	SAN1	SAN2	SAN3	SAN4	SAN5

THE NU AFRIKAN CALENDAR

MWANAMKE WEUSI 522/2038-39/6274-75/2030-31

Ashe	Ojobo	Iwa	Araorun	Ile	Itutu	Erinle
1	2	3	4	5	6	7
J13	J14	J15	J16	J17	J18	J19
APP29	APP30	MSR1	MSR2	MSR3	MSR4	MSR5
SAN6	SAN7	SAN8	SAN9	SAN10	SAN11	SAN12
8	9	10	11	12	13	14
J20	J21	J22	J23	J24	J25	J26
MSR6	MSR7	MSR8	MSR9	MSR10	MSR11	MSR12
SAN13	SAN14	SAN15	SAN16	SAN17	SAN18	SAN19
15	16	17	18	19	20	21
J27	J28	J29	J30	JU1	JU2	JU3
MSR13	MSR14	MSR15	MSR16	MSR17	MSR18	MSR19
SAN20	SAN21	SAN22	SAN23	SAN24	SAN25	SAN26
22	23	24	25	26	27	28
JU4	JU5	JU6	JU7	JU8	JU9	JU10
MSR20	MRS21	MSR22	MSR23	MSR24	MSR25	MSR26
SAN27	SAN28	SAN29	SAN30	HAM1	HAM2	HAM3

TAIFA WEUSI 522/2038-39/6274-75/2030-31

Ashe	Ojobo	Iwa	Araorun	Ile	Itutu	Erinle
1	2	3	4	5	6	7
AU8	AU9	AU10	AU11	AU12	AU13	AU14
THT20	THT21	THT22	THT23	THT24	THT25	THT26
NAH2	NAH3	NAH4	NAH5	NAH6	NAH7	NAH8
8	9	10	11	12	13	14
AU15	AU16	AU17	AU18	AU19	AU20	AU21
THT27	THT28	THT29	THT30	PAP1	PAP1	PAP2
NAH9	NAH10	NAH11	NAH12	NAH13	NAH14	NAH15
15	16	17	18	19	20	21
AU22	AU23	AU24	AU25	AU26	AU27	AU28
PAP3	PAP4	PAP5	PAP6	PAP7	PAP8	PAP9
NAH16	NAH17	NAH18	NAH19	NAH20	NAH21	NAH22
22	23	24	25	26	27	28
AU29	AU30	AU31	S1	S2	S3	S4
PAP10	PAP11	PAP12	PAP13	PAP14	PAP15	PAP16
NAH23	NAH24	NAH25	NAH26	NAH27	NAH28	NAH29

MTU WEUSI 522/2038-39/6274-75/2030-31

Ashe	Ojobo	Iwa	Araorun	Ile	Itutu	Erinle
1	2	3	4	5	6	7
JU11	JU12	JU13	JU14	JU15	JU16	JU17
MSR27	MSR28	MSR29	MSR30	*ASR1*	*HRU2*	*STH3*
HAM4	HAM5	HAM6	HAM7	HAM8	HAM9	HAM10
8	9	10	11	12	13	14
JU18	JU19	JU20	JU21	JU22	JU23	JU24
AST4	*NBH5*	THT1	THT2	THT3	THT4	THT5
HAM11	HAM12	HAM13	AM14	HAM15	HAM16	HAM17
15	16	17	18	19	20	21
JU25	JU26	JU27	JU28	JU29	JU30	JU31
THT6	THT7	THT8	THT9	THT10	THT11	THT12
HAM18	HAM19	HAM20	HAM21	HAM22	HAM23	HAM24
22	23	24	25	26	27	28
AU1	AU2	AU3	AU4	AU5	AU6	AU7
THT13	THT14	THT15	THT16	THT17	THT18	THT19
HAM25	HAM26	HAM27	HAM28	HAM29	HAM30	NAH1

UWEZO 522/2038-39/6274-75/2030-31

Ashe	Ojobo	Iwa	Araorun	Ile	Itutu	Erinle
1	2	3	4	5	6	7
S5	S6	S7	S8	S9	S10	S11
PAP17	PAP18	PAP19	PAP20	PAP21	PAP22	PAP23
NAH30	*PAG1*	*PAG2*	*PAG3*	*PAG4*	*PAG5*	MAS1
8	9	10	11	12	13	14
S12	S13	S14	S15	S16	S17	S18
PAP24	PAP25	PAP26	PAP27	PAP28	PAP29	PAP30
MAS2	MAS3	MAS4	MAS5	MAS6	MAS7	MAS8
15	16	17	18	19	20	21
S19	S20	S21	S22	S23	S24	S25
HTH1	HTH2	HTH3	HTH4	HTH5	HTH6	HTH7
MAS9	MAS10	MAS11	MAS12	MAS13	MAS14	MAS15
22	23	24	25	26	27	28
S26	S27	S28	S29	S30	O1	O2
HTH8	HTH9	HTH10	HTH11	HTH12	HTH13	HTH14
MAS16	MAS17	MAS18	MAS19	MAS20	MAS21	MAS22

THE NU AFRIKAN CALENDAR

EKUNDU 522/2038-39/6274-75/2030-31						
Ashe	Ojobo	Iwa	Araorun	Ile	Itutu	Erinle
1	2	3	4	5	6	7
O3	O4	O5	O6	O7	O8	O9
HTH15	HTH16	HTH17	HTH18	HTH19	HTH20	HTH21
MAS23	MAS24	MAS25	MAS26	MAS27	MAS28	MAS29
8	9	10	11	12	13	14
O10	O11	O12	O13	O14	O15	O16
HTH22	HTH23	HTH24	HTH25	HTH26	HTH27	HTH28
MAS30	THi1	THi2	THi3	THi4	THi5	THi6
15	16	17	18	19	20	21
O17	O18	O19	O20	O21	O22	O23
HTH29	HTH30	KHK1	KHK2	KHK3	KHK4	KHK5
THi7	THi8	THi9	THi10	THi11	THi12	THi13
22	23	24	25	26	27	28
O24	O25	O26	O27	O28	O29	O30
KHK6	KHK7	KHK8	KHK9	KHK10	KHK11	KHK12
THi14	THi15	THi16	THi17	THi18	THi19	THi20

KIJANI 522/2038-39/6274-75/2030-31						
Ashe	Ojobo	Iwa	Araorun	Ile	Itutu	Erinle
1	2	3	4	5	6	7
N28	N29	N30	D1	D2	D3	D4
TBT11	TBT12	TBT13	TBT14	TBT15	TBT16	TBT17
HID19	HID20	HID21	HID22	HID23	HID24	HID25
8	9	10	11	12	13	14
D5	D6	D7	D8	D9	D10	D11
TBT18	TBT19	TBT20	TBT21	TBT22	TBT23	TBT24
HID26	HID27	HID28	HID29	HID30	TAH1	TAH2
15	16	17	18	19	20	21
D12	D13	D14	D15	D16	D17	D18
TBT25	TBT26	TBT27	TBT28	TBT29	TBT30	MKH1
TAH3	TAH4	TAH5	TAH6	TAH7	TAH8	TAH9
22	23	24	25	26	27	28
D19	D20	D21	D22	D23	D24	D25
MKH2	MKH3	MKH4	MKH5	MKH6	MKH7	MKH8
TAH10	TAH11	TAH12	TAH13	TAH14	TAH15	TAH16

EUSI 522/2038-39/6274-75/2030-31						
Ashe	Ojobo	Iwa	Araorun	Ile	Itutu	Erinle
1	2	3	4	5	6	7
O31	N1	N2	N3	N4	N5	N6
KHK13	KHK14	KHK15	KHK16	KHK17	KHK18	KHK19
THi21	THi22	THi23	THi24	THi25	THi26	THi27
8	9	10	11	12	13	14
N7	N8	N9	N10	N11	N12	N13
KHK20	KHK21	KHK22	KHK23	KHK24	KHK25	KHK26
THi28	THi29	THi30	HID1	HID2	HID3	HID4
15	16	17	18	19	20	21
N14	N15	N16	N17	N18	N19	N20
KHK27	KHK28	KHK29	KHK30	TBT1	TBT2	TBT3
HID5	HID6	HID7	HID8	HID9	HID10	HID11
22	23	24	25	26	27	28
N21	N22	N23	N24	N25	N26	N27
TBT4	TBT5	TBT6	TBT7	TBT8	TBT9	TBT10
HID12	HID13	HID14	HID15	HID16	HID17	HID18

KWANZA 522/2038-39/6274-75/2030-31						
Umoja	Kujichagulia	Ujima	Ujamaa	Nia	Kuumba	Imani
1	2	3	4	5	6	7
D26	D27	D28	D29	D30	D31	JA1
MKH9	MKH10	MKH11	MKH12	MKH13	MKH14	MKH15
TAH17	TAH18	TAH19	TAH20	TAH21	TAH22	TAH23
8	9	10	11	12	13	14
JA2	JA3	JA4	JA5	JA6	JA7	JA8
MKH16	MKH17	MKH18	MKH19	MKH20	MKH21	MKH22
TAH24	TAH25	TAH26	TAH27	TAH28	TAH29	TAH30
15	16	17	18	19	20	21
JA9	JA10	JA11	JA12	JA13	JA14	JA15
MKH23	MKH24	MKH25	KMH26	MKH27	MKH28	MKH29
TIR1	TIR2	TIR3	TIR4	TIR5	TIR6	TIR7
22	23	24	25	26	27	28
JA16	JA17	JA18	JA19	JA20	JA21	JA22
MKH30	AMN1	AMN2	AMN3	AMN4	AMN5	AMN6
TIR8	TIR9	TIR10	TIR11	TIR12	TIR13	TIR14

THE NU AFRIKAN CALENDAR

SHAHIDI 522/2038-39/6274-75/2030-31						
Ashe	Ojobo	Iwa	Araorun	Ile	Itutu	Erinle
1	2	3	4	5	6	7
JA23	JA24	JA25	JA26	JA27	JA28	JA29
AMN7	AMN8	AMN9	AMN10	AMN11	AMN12	AMN13
TIR15	TIR16	TIR17	TIR18	TIR19	TIR20	TIR21
8	9	10	11	12	13	14
JA30	JA31	F1	F2	F3	F4	F5
AMN14	AMN15	AMN16	AMN17	AMN18	AMN19	AMN20
TIR22	TIR23	TIR24	TIR25	TIR26	TIR27	TIR28
15	16	17	18	19	20	21
F6	F7	F8	F9	F10	F 11	F12
AMN21	AMN22	AMN23	AMN24	AMN25	AMN26	AMN28
TIR29	TIR30	YEK1	YEK2	YEK3	YEK4	YEK5
22	23	24	25	26	27	28
F13	F14	F15	F16	F17	F18	F19
AMN29	AMN30	RNT1	RNT2	RNT3	RNT4	RNT5
YEK6	YEK7	YEK8	YEK9	YEK10	YEK11	YEK12
SHAHIDI 73						
F20						
RNT6						
YEK13						

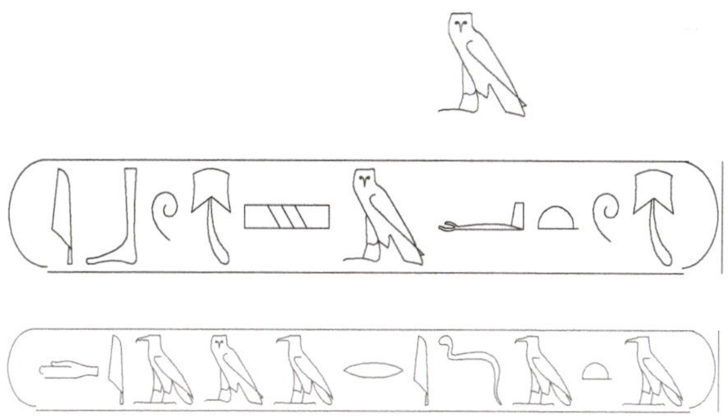

MAAFA 523 B.S.E./ASIENTO 2039-40 C.E./
KAM 6275-76 S.M./EC 2031-32 A.M. (ዓ.ም.)

AFU 523/2039-40/6275-76/2031-32						
Ashe	Ojobo	Iwa	Araorun	Ile	Itutu	Erinle
1	2	3	4	5	6	7
F21	F22	F23	F24	F25	F26	F27
RNT7	RNT8	RNT9	RNT10	RNT11	RNT12	RNT13
YEK14	YEK15	YEK16	YEK17	YEK18	YEK19	YEK20
8	9	10	11	12	13	14
F28	M1	M2	M3	M4	M5	M6
RNT14	RNT15	RNT16	RNT17	RNT18	RNT19	RNT20
YEK21	YEK22	YEK23	YEK24	YEK25	YEK26	YEK27
15	16	17	18	19	20	21
M7	M8	M9	M10	M11	M12	M13
RNT21	RNT22	RNT23	RNT24	RNT25	RNT26	RNT27
YEK28	YEK 29	YEK30	MAG1	MAG2	MAG3	MAG4
22	23	24	25	26	27	28
M14	M15	M16	M17	M18	M19	M20
RNT28	RNT29	RNT30	KHN1	KHN2	KHN3	KHN4
MAG5	MAG6	MAG7	MAG8	MAG9	MAG10	MAG11

UHURU 523/2039-40/6275-76/2031-32						
Ashe	Ojobo	Iwa	Araorun	Ile	Itutu	Erinle
1	2	3	4	5	6	7
M21	M22	M23	M24	M25	M26	M27
KHN5	KHN6	KHN7	KHN8	KHN9	KHN10	KHN11
MAG12	MAG13	MAG14	MAG15	MAG16	MAG17	MAG18
8	9	10	11	12	13	14
M28	M29	M30	M31	A1	A2	A3
KHN12	KHN13	KHN14	KHN15	KHN16	KHN17	KHN18
MAG19	MAG20	MAG21	MAG22	MAG23	MAG24	MAG25
15	16	17	18	19	20	21
A4	A5	A6	A7	A8	A9	A10
KHN19	KHN20	KHN21	KHN22	KHN23	KHN24	KHN25
MAG26	MAG27	MAG28	MAG29	MAG30	MIY1	MIY2
22	23	24	25	26	27	28
A11	A12	A13	A14	A15	A16	A17
KHN26	KHN27	KHN28	KHN29	KHN30	PIN1	PIN2
MIY3	MIY4	MIY5	MIY6	MIY7	MIY8	MIY9

KAWAIDA 523/2039-40/6275-76/2031-32						
Ashe	Ojobo	Iwa	Araorun	Ile	Itutu	Erinle
1	2	3	4	5	6	7
A18	A19	A20	A21	A22	A23	A24
PIN3	PIN4	PIN5	PIN6	PIN7	PIN8	PIN9
MIY10	MIY11	MIY12	MIY13	MIY14	MIY15	MIY16
8	9	10	11	12	13	14
A25	A26	A27	A28	A29	A30	MA1
PIN10	PIN11	PIN12	PIN13	PIN14	PIN15	PIN16
MIY17	MIY18	MIY19	MIY20	MIY21	MIY22	MIY23
15	16	17	18	19	20	21
MA2	MA3	MA4	MA5	MA6	MA7	MA8
PIN17	PIN18	PIN19	PIN20	PIN21	PIN22	PIN23
MIY24	MIY25	MIY26	MIY27	MIY28	MIY29	MIY30
22	23	24	25	26	27	28
MA9	MA10	MA11	MA12	MA13	MA14	MA15
PIN24	PIN25	PIN26	PIN27	PIN28	PIN29	PIN30
GIN1	GIN2	GIN3	GIN4	GIN5	GIN6	GIN7

MABABA WEUSI 523/2039-40/6275-76/2031-32						
Ashe	Ojobo	Iwa	Araorun	Ile	Itutu	Erinle
1	2	3	4	5	6	7
MA16	MA17	MA18	MA19	MA20	MA21	MA22
APP1	APP2	APP3	APP4	APP5	APP6	APP7
GIN8	GIN9	GIN10	GIN11	GIN12	GIN13	GIN14
8	9	10	11	12	13	14
MA23	MA24	MA25	MA26	MA27	MA28	MA29
APP8	APP9	APP10	APP11	APP12	APP13	APP14
GIN15	GIN16	GIN17	GIN18	GIN19	GIN20	GIN21
15	16	17	18	19	20	21
MA30	MA31	J1	J2	J3	J4	J5
APP15	APP16	APP17	APP18	APP19	APP20	APP21
GIN22	GIN23	GIN24	GIN25	GIN26	GIN27	GIN28
22	23	24	25	26	27	28
J6	J7	J8	J9	J10	J11	J12
APP22	APP23	APP24	APP25	APP26	APP27	APP28
GIN29	GIN30	SAN1	SAN2	SAN3	SAN4	SAN5

THE NU AFRIKAN CALENDAR

MWANAMKE WEUSI 523/2039-40/6275-76/2031-32

Ashe	Ojobo	Iwa	Araorun	Ile	Itutu	Erinle
1	2	3	4	5	6	7
J13	J14	J15	J16	J17	J18	J19
APP29	APP30	MSR1	MSR2	MSR3	MSR4	MSR5
SAN6	SAN7	SAN8	SAN9	SAN10	SAN11	SAN12
8	9	10	11	12	13	14
J20	J21	J22	J23	J24	J25	J26
MSR6	MSR7	MSR8	MSR9	MSR10	MSR11	MSR12
SAN13	SAN14	SAN15	SAN16	SAN17	SAN18	SAN19
15	16	17	18	19	20	21
J27	J28	J29	J30	JU1	JU2	JU3
MSR13	MSR14	MSR15	MSR16	MSR17	MSR18	MSR19
SAN20	SAN21	SAN22	SAN23	SAN24	SAN25	SAN26
22	23	24	25	26	27	28
JU4	JU5	JU6	JU7	JU8	JU9	JU10
MSR20	MRS21	MSR22	MSR23	MSR24	MSR25	MSR26
SAN27	SAN28	SAN29	SAN30	HAM1	HAM2	HAM3

TAIFA WEUSI 523/2039-40/6275-76/2031-32

Ashe	Ojobo	Iwa	Araorun	Ile	Itutu	Erinle
1	2	3	4	5	6	7
AU8	AU9	AU10	AU11	AU12	AU13	AU14
THT20	THT21	THT22	THT23	THT24	THT25	THT26
NAH2	NAH3	NAH4	NAH5	NAH6	NAH7	NAH8
8	9	10	11	12	13	14
AU15	AU16	AU17	AU18	AU19	AU20	AU21
THT27	THT28	THT29	THT30	PAP1	PAP1	PAP2
NAH9	NAH10	NAH11	NAH12	NAH13	NAH14	NAH15
15	16	17	18	19	20	21
AU22	AU23	AU24	AU25	AU26	AU27	AU28
PAP3	PAP4	PAP5	PAP6	PAP7	PAP8	PAP9
NAH16	NAH17	NAH18	NAH19	NAH20	NAH21	NAH22
22	23	24	25	26	27	28
AU29	AU30	AU31	S1	S2	S3	S4
PAP10	PAP11	PAP12	PAP13	PAP14	PAP15	PAP16
NAH23	NAH24	NAH25	NAH26	NAH27	NAH28	NAH29

MTU WEUSI 523/2039-40/6275-76/2031-32

Ashe	Ojobo	Iwa	Araorun	Ile	Itutu	Erinle
1	2	3	4	5	6	7
JU11	JU12	JU13	JU14	JU15	JU16	JU17
MSR27	MSR28	MSR29	MSR30	*ASR1*	*HRU2*	*STH3*
HAM4	HAM5	HAM6	HAM7	HAM8	HAM9	HAM10
8	9	10	11	12	13	14
JU18	JU19	JU20	JU21	JU22	JU23	JU24
AST4	*NBH5*	THT1	THT2	THT3	THT4	THT5
HAM11	HAM12	HAM13	HAM14	HAM15	HAM16	HAM17
15	16	17	18	19	20	21
JU25	JU26	JU27	JU28	JU29	JU30	JU31
THT6	THT7	THT8	THT9	THT10	THT11	THT12
HAM18	HAM19	HAM20	HAM21	HAM22	HAM23	HAM24
22	23	24	25	26	27	28
AU1	AU2	AU3	AU4	AU5	AU6	AU7
THT13	THT14	THT15	THT16	THT17	THT18	THT19
HAM25	HAM26	HAM27	HAM28	HAM29	HAM30	NAH1

UWEZO 523/2039-40/6275-76/2031-32

Ashe	Ojobo	Iwa	Araorun	Ile	Itutu	Erinle
1	2	3	4	5	6	7
S5	S6	S7	S8	S9	S10	S11
PAP17	PAP18	PAP19	PAP20	PAP21	PAP22	PAP23
NAH30	*QUA1*	*QUA2*	*QUA3*	*QUA4*	*QUA5*	*QUA6*
8	9	10	11	12	13	14
S12	S13	S14	S15	S16	S17	S18
PAP24	PAP25	PAP26	PAP27	PAP28	PAP29	PAP30
MAS1	MAS2	MAS3	MAS4	MAS5	MAS6	MAS7
15	16	17	18	19	20	21
S19	S20	S21	S22	S23	S24	S25
HTH1	HTH2	HTH3	HTH4	HTH5	HTH6	HTH7
MAS8	MAS9	MAS10	MAS11	MAS12	MAS13	MAS14
22	23	24	25	26	27	28
S26	S27	S28	S29	S30	O1	O2
HTH8	HTH9	HTH10	HTH11	HTH12	HTH13	HTH14
MAS15	MAS16	MAS17	MAS18	MAS19	MAS20	MAS21

THE NU AFRIKAN CALENDAR

EKUNDU 523/2039-40/6275-76/2031-32

Ashe	Ojobo	Iwa	Araorun	Ile	Itutu	Erinle
1	2	3	4	5	6	7
O3	O4	O5	O6	O7	O8	O9
HTH15	HTH16	HTH17	HTH18	HTH19	HTH20	HTH21
MAS22	MAS23	MAS 24	MAS25	MAS26	MAS27	MAS28
8	9	10	11	12	13	14
O10	O11	O12	O13	O14	O15	O16
HTH22	HTH23	HTH24	HTH25	HTH26	HTH27	HTH28
MAS29	MAS30	THi1	THi2	THi3	THi4	THi5
15	16	17	18	19	20	21
O17	O18	O19	O20	O21	O22	O23
HTH29	HTH30	KHK1	KHK2	KHK3	KHK4	KHK5
THi6	THi7	THi8	THi9	THi10	THi11	THi12
22	23	24	25	26	27	28
O24	O25	O26	O27	O28	O29	O30
KHK6	KHK7	KHK8	KHK9	KHK10	KHK11	KHK12
THi13	THi14	THi15	THi16	THi17	THi18	THi19

KIJANI 523/2039-40/6275-76/2031-32

Ashe	Ojobo	Iwa	Araorun	Ile	Itutu	Erinle
1	2	3	4	5	6	7
N28	N29	N30	D1	D2	D3	D4
TBT11	TBT12	TBT13	TBT14	TBT15	TBT16	TBT17
HID18	HID19	HID20	HID21	HID22	HID23	HID24
8	9	10	11	12	13	14
D5	D6	D7	D8	D9	D10	D11
TBT18	TBT19	TBT20	TBT21	TBT22	TBT23	TBT24
HID25	HID26	HID27	HID28	HID29	HID30	TAH1
15	16	17	18	19	20	21
D12	D13	D14	D15	D16	D17	D18
TBT25	TBT26	TBT27	TBT28	TBT29	TBT30	MKH1
TAH2	TAH3	TAH4	TAH5	TAH6	TAH7	TAH8
22	23	24	25	26	27	28
D19	D20	D21	D22	D23	D24	D25
MKH2	MKH3	MKH4	MKH5	MKH6	MKH7	MKH8
TAH9	TAH10	TAH11	TAH12	TAH13	TAH14	TAH15

EUSI 523/2039-40/6275-76/2031-32

Ashe	Ojobo	Iwa	Araorun	Ile	Itutu	Erinle
1	2	3	4	5	6	7
O31	N1	N2	N3	N4	N5	N6
KHK13	KHK14	KHK15	KHK16	KHK17	KHK18	KHK19
THi20	THi21	THi22	THi23	THi24	THi25	THi26
8	9	10	11	12	13	14
N7	N8	N9	N10	N11	N12	N13
KHK20	KHK21	KHK22	KHK23	KHK24	KHK25	KHK26
THi27	THi28	THi29	THi30	HID1	HID2	HID3
15	16	17	18	19	20	21
N14	N15	N16	N17	N18	N19	N20
KHK27	KHK28	KHK29	KHK30	TBT1	TBT2	TBT3
HID4	HID5	HID6	HID7	HID8	HID9	HID10
22	23	24	25	26	27	28
N21	N22	N23	N24	N25	N26	N27
TBT4	TBT5	TBT6	TBT7	TBT8	TBT9	TBT10
HID11	HID12	HID13	HID14	HID15	HID16	HID17

KWANZA 523/2039-40/6275-76/2031-32

Umoja	Kujichagulia	Ujima	Ujamaa	Nia	Kuumba	Imani
1	2	3	4	5	6	7
D26	D27	D28	D29	D30	D31	JA1
MKH9	MKH10	MKH11	MKH12	MKH13	MKH14	MKH15
TAH16	TAH17	TAH18	TAH19	TAH20	TAH21	TAH22
8	9	10	11	12	13	14
JA2	JA3	JA4	JA5	JA6	JA7	JA8
MKH16	MKH17	MKH18	MKH19	MKH20	MKH21	MKH22
TAH23	TAH24	TAH25	TAH26	TAH27	TAH28	TAH29
15	16	17	18	19	20	21
JA9	JA10	JA11	JA12	JA13	JA14	JA15
MKH23	MKH24	MKH25	KMH26	MKH27	MKH28	MKH29
TAH30	TIR1	TIR2	TIR3	TIR4	TIR5	TIR6
22	23	24	25	26	27	28
JA16	JA17	JA18	JA19	JA20	JA21	JA22
MKH30	AMN1	AMN2	AMN3	AMN4	AMN5	AMN6
TIR7	TIR8	TIR9	TIR10	TIR11	TIR12	TIR13

THE NU AFRIKAN CALENDAR

SHAHIDI 523/2039-40/6275-76/2031-32						
Ashe	Ojobo	Iwa	Araorun	Ile	Itutu	Erinle
1	2	3	4	5	6	7
JA23	JA24	JA25	JA26	JA27	JA28	JA29
AMN7	AMN8	AMN9	AMN10	AMN11	AMN12	AMN13
TIR14	TIR15	TIR16	TIR17	TIR18	TIR19	TIR20
8	9	10	11	12	13	14
JA30	JA31	F1	F2	F3	F4	F5
AMN14	AMN15	AMN16	AMN17	AMN18	AMN19	AMN20
TIR21	TIR22	TIR23	TIR24	TIR25	TIR26	TIR27
15	16	17	18	19	20	21
F6	F7	F8	F9	F10	F 11	F12
AMN21	AMN22	AMN23	AMN24	AMN25	AMN26	AMN28
TIR28	TIR29	TIR30	YEK1	YEK2	YEK3	YEK4
22	23	24	25	26	27	28
F13	F14	F15	F16	F17	F18	F19
AMN29	AMN30	RNT1	RNT2	RNT3	RNT4	RNT5
YEK5	YEK6	YEK7	YEK8	YEK9	YEK10	YEK11
SHAHIDI 74						
F20						
RNT6						
YEK12						

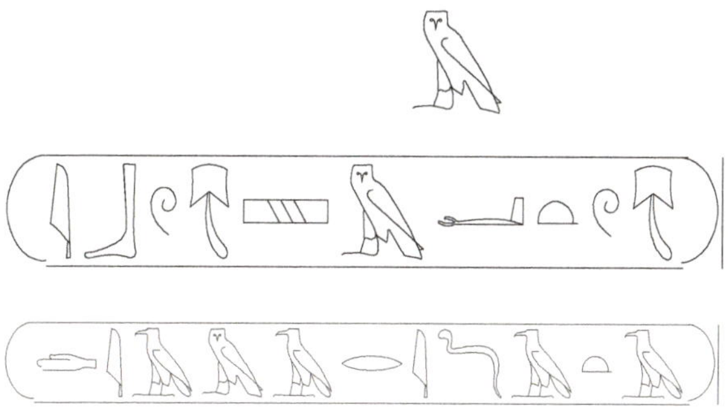

THE NU AFRIKAN CALENDAR

MAAFA 524 B.S.E./ASIENTO 2040-41 C.E./
KAM 6276-77 S.M./EC 2032-2033 A.M. (ዓ.ም.)

AFU 524/2040-41/6276-77/2032-33						
Ashe	Ojobo	Iwa	Araorun	Ile	Itutu	Erinle
1	2	3	4	5	6	7
F21	F22	F23	F24	F25	F26	F27
RNT7	RNT8	RNT9	RNT10	RNT11	RNT12	RNT13
YEK13	YEK14	YEK15	YEK16	YEK17	YEK18	YEK19
8	9	10	11	12	13	14
F28	F29	M1	M2	M3	M4	M5
RNT14	RNT15	RNT16	RNT17	RNT18	RNT19	RNT20
YEK 20	YEK21	YEK22	YEK23	YEK24	YEK25	YEK26
15	16	17	18	19	20	21
M6	M7	M8	M9	M10	M11	M12
RNT21	RNT22	RNT23	RNT24	RNT25	RNT26	RNT27
YEK 27	YEK28	YEK29	YEK30	MAG1	MAG2	MAG3
22	23	24	25	26	27	28
M13	M14	M15	M16	M17	M18	M19
RNT28	RNT29	RNT30	KHN1	KHN2	KHN3	KHN4
MAG4	MAG5	MAG6	MAG7	MAG8	MAG9	MAG10
SHARIFU 128						
M20						
KHN5						
MAG11						

KAWAIDA 524/2040-41/6276-77/2032-33						
Ashe	Ojobo	Iwa	Araorun	Ile	Itutu	Erinle
1	2	3	4	5	6	7
A18	A19	A20	A21	A22	A23	A24
PIN4	PIN5	PIN6	PIN7	PIN8	PIN9	PIN10
MIY10	MIY11	MIY12	MIY13	MIY14	MIY15	MIY16
8	9	10	11	12	13	14
A25	A26	A27	A28	A29	A30	MA1
PIN11	PIN12	PIN13	PIN14	PIN15	PIN16	PIN17
MIY17	MIY18	MIY19	MIY20	MIY21	MIY22	MIY23
15	16	17	18	19	20	21
MA2	MA3	MA4	MA5	MA6	MA7	MA8
PIN18	PIN19	PIN20	PIN21	PIN22	PIN23	PIN24
MIY24	MIY25	MIY26	MIY27	MIY28	MIY29	MIY30
22	23	24	25	26	27	28
MA9	MA10	MA11	MA12	MA13	MA14	MA15
PIN25	PIN26	PIN27	PIN28	PIN29	PIN30	APP1
GIN1	GIN2	GIN3	GIN4	GIN5	GIN6	GIN7

UHURU 524/2040-41/6276-77/2032-33						
Ashe	Ojobo	Iwa	Araorun	Ile	Itutu	Erinle
1	2	3	4	5	6	7
M21	M22	M23	M24	M25	M26	M27
KHN6	KHN7	KHN8	KHN9	KHN10	KHN11	KHN12
MAG12	MAG13	MAG14	MAG15	MAG16	MAG17	MAG18
8	9	10	11	12	13	14
M28	M29	M30	M31	A1	A2	A3
KHN13	KHN14	KHN15	KHN16	KHN17	KHN18	KHN19
MAG19	MAG20	MAG21	MAG22	MAG23	MAG24	MAG25
15	16	17	18	19	20	21
A4	A5	A6	A7	A8	A9	A10
KHN20	KHN21	KHN222	KHN23	KHN24	KHN25	KHN26
MAG26	MAG27	MAG28	MAG29	MAG30	MIY1	MIY2
22	23	24	25	26	27	28
A11	A12	A13	A14	A15	A16	A17
KHN27	KHN28	KHN29	KHN30	PIN1	PIN2	PIN3
MIY3	MIY4	MIY5	MIY6	MIY7	MIY8	MIY9

MABABA WEUSI 524/2040-41/6276-77/2032-33						
Ashe	Ojobo	Iwa	Araorun	Ile	Itutu	Erinle
1	2	3	4	5	6	7
MA16	MA17	MA18	MA19	MA20	MA21	MA22
APP2	APP3	APP4	APP5	APP6	APP7	APP8
GIN8	GIN9	GIN10	GIN11	GIN12	GIN13	GIN14
8	9	10	11	12	13	14
MA23	MA24	MA25	MA26	MA27	MA28	MA29
APP9	APP10	APP11	APP12	APP13	APP14	APP15
GIN15	GIN16	GIN17	GIN18	GIN19	GIN20	GIN21
15	16	17	18	19	20	21
MA30	MA31	J1	J2	J3	J4	J5
APP16	APP17	APP18	APP19	APP20	APP21	APP22
GIN22	GIN23	GIN24	GIN25	GIN26	GIN27	GIN28
22	23	24	25	26	27	28
J6	J7	J8	J9	J10	J11	J12
APP23	APP24	APP25	APP26	APP27	APP28	APP29
GIN29	GIN30	SAN1	SAN2	SAN3	SAN4	SAN5

THE NU AFRIKAN CALENDAR

MWANAMKE WEUSI 524/2040-41/6276-77/2032-33

Ashe	Ojobo	Iwa	Araorun	Ile	Itutu	Erinle
1	2	3	4	5	6	7
J13	J14	J15	J16	J17	J18	J19
APP30	MSR1	MSR2	MSR3	MSR4	MSR5	MSR6
SAN6	SAN7	SAN8	SAN9	SAN10	SAN11	SAN12
8	9	10	11	12	13	14
J20	J21	J22	J23	J24	J25	J26
MSR7	MSR8	MSR9	MSR10	MSR11	MSR12	MSR13
SAN13	SAN14	SAN15	SAN16	SAN17	SAN18	SAN19
15	16	17	18	19	20	21
J27	J28	J29	J30	JU1	JU2	JU3
MSR14	MSR15	MSR16	MSR17	MSR18	MSR19	MSR20
SAN20	SAN21	SAN22	SAN23	SAN24	SAN25	SAN26
22	23	24	25	26	27	28
JU4	JU5	JU6	JU7	JU8	JU9	JU10
MSR21	MRS22	MSR23	MSR24	MSR25	MSR26	MSR27
SAN27	SAN28	SAN29	SAN30	HAM1	HAM2	HAM3

TAIFA WEUSI 524/2040-41/6276-77/2032-33

Ashe	Ojobo	Iwa	Araorun	Ile	Itutu	Erinle
1	2	3	4	5	6	7
AU8	AU9	AU10	AU11	AU12	AU13	AU14
THT20	THT21	THT22	THT23	THT24	THT25	THT26
NAH2	NAH3	NAH4	NAH5	NAH6	NAH7	NAH8
8	9	10	11	12	13	14
AU15	AU16	AU17	AU18	AU19	AU20	AU21
THT27	THT28	THT29	THT30	PAP1	PAP1	PAP2
NAH9	NAH10	NAH11	NAH12	NAH13	NAH14	NAH15
15	16	17	18	19	20	21
AU22	AU23	AU24	AU25	AU26	AU27	AU28
PAP3	PAP4	PAP5	PAP6	PAP7	PAP8	PAP9
NAH16	NAH17	NAH18	NAH19	NAH20	NAH21	NAH22
22	23	24	25	26	27	28
AU29	AU30	AU31	S1	S2	S3	S4
PAP10	PAP11	PAP12	PAP13	PAP14	PAP15	PAP16
NAH23	NAH24	NAH25	NAH26	NAH27	NAH28	NAH29

MTU WEUSI 524/2040-41/6276-77/2032-33

Ashe	Ojobo	Iwa	Araorun	Ile	Itutu	Erinle
1	2	3	4	5	6	7
JU11	JU12	JU13	JU14	JU15	JU16	JU17
MSR28	MSR29	MSR30	*ASR1*	*HRU2*	*STH3*	*AST4*
HAM4	HAM5	HAM6	HAM7	HAM8	HAM9	HAM10
8	9	10	11	12	13	14
JU18	JU19	JU20	JU21	JU22	JU23	JU24
NBH5	*ANU6*	THT1	THT2	THT3	THT4	THT5
HAM11	HAM12	HAM13	HAM14	HAM15	HAM16	HAM17
15	16	17	18	19	20	21
JU25	JU26	JU27	JU28	JU29	JU30	JU31
THT6	THT7	THT8	THT9	THT10	THT11	THT12
HAM18	HAM19	HAM20	HAM21	HAM22	HAM23	HAM24
22	23	24	25	26	27	28
AU1	AU2	AU3	AU4	AU5	AU6	AU7
THT13	THT14	THT15	THT16	THT17	THT18	THT19
HAM25	HAM26	HAM27	HAM28	HAM29	HAM30	NAH1

UWEZO 524/2040-41/6276-77/2032-33

Ashe	Ojobo	Iwa	Araorun	Ile	Itutu	Erinle
1	2	3	4	5	6	7
S5	S6	S7	S8	S9	S10	S11
PAP17	PAP18	PAP19	PAP20	PAP21	PAP22	PAP23
NAH30	*PAG1*	*PAG2*	*PAG3*	*PAG4*	*PAG5*	MAS1
8	9	10	11	12	13	14
S12	S13	S14	S15	S16	S17	S18
PAP24	PAP25	PAP26	PAP27	PAP28	PAP29	PAP30
MAS2	MAS3	MAS4	MAS5	MAS6	MAS7	MAS8
15	16	17	18	19	20	21
S19	S20	S21	S22	S23	S24	S25
HTH1	HTH2	HTH3	HTH4	HTH5	HTH6	HTH7
MAS9	MAS10	MAS11	MAS12	MAS13	MAS14	MAS15
22	23	24	25	26	27	28
S26	S27	S28	S29	S30	O1	O2
HTH8	HTH9	HTH10	HTH11	HTH12	HTH13	HTH14
MAS16	MAS17	MAS18	MAS19	MAS20	MAS21	MAS22

THE NU AFRIKAN CALENDAR

EKUNDU 524/2040-41/6276-77/2032-33						
Ashe	Ojobo	Iwa	Araorun	Ile	Itutu	Erinle
1	2	3	4	5	6	7
O3	O4	O5	O6	O7	O8	O9
HTH15	HTH16	HTH17	HTH18	HTH19	HTH20	HTH21
MAS23	MAS24	MAS25	MAS26	MAS27	MAS28	MAS29
8	9	10	11	12	13	14
O10	O11	O12	O13	O14	O15	O16
HTH22	HTH23	HTH24	HTH25	HTH26	HTH27	HTH28
MAS30	THi1	THi2	THi3	THi4	THi5	THi6
15	16	17	18	19	20	21
O17	O18	O19	O20	O21	O22	O23
HTH29	HTH30	KHK1	KHK2	KHK3	KHK4	KHK5
THi7	THi8	This9	THi10	THi11	THi12	THi13
22	23	24	25	26	27	28
O24	O25	O26	O27	O28	O29	O30
KHK6	KHK7	KHK8	KHK9	KHK10	KHK11	KHK12
Thi14	THi15	THi16	THi17	THi18	THi19	THi20

EUSI 524/2040-41/6276-77/2032-33						
Ashe	Ojobo	Iwa	Araorun	Ile	Itutu	Erinle
1	2	3	4	5	6	7
O31	N1	N2	N3	N4	N5	N6
KHK13	KHK14	KHK15	KHK16	KHK17	KHK18	KHK19
THi21	THi22	THi23	THi24	THi25	THi26	THi27
8	9	10	11	12	13	14
N7	N8	N9	N10	N11	N12	N13
KHK20	KHK21	KHK22	KHK23	KHK24	KHK25	KHK26
THi28	THi29	THi30	HID1	HID2	HID3	HID4
15	16	17	18	19	20	21
N14	N15	N16	N17	N18	N19	N20
KHK27	KHK28	KHK29	KHK30	TBT1	TBT2	TBT3
HID5	HID6	HID7	HID8	HID9	HID10	HID11
22	23	24	25	26	27	28
N21	N22	N23	N24	N25	N26	N27
TBT4	TBT5	TBT6	TBT7	TBT8	TBT9	TBT10
HID12	HID13	HID14	HID15	HID16	HID17	HID18

KIJANI 524/2040-41/6276-77/2032-33						
Ashe	Ojobo	Iwa	Araorun	Ile	Itutu	Erinle
1	2	3	4	5	6	7
N28	N29	N30	D1	D2	D3	D4
TBT11	TBT12	TBT13	TBT14	TBT15	TBT16	TBT17
HID19	HID20	HID21	HID22	HID23	HID24	HID25
8	9	10	11	12	13	14
D5	D6	D7	D8	D9	D10	D11
TBT18	TBT19	TBT20	TBT21	TBT22	TBT23	TBT24
HID26	HID27	HID28	HID29	HID30	TAH1	TAH2
15	16	17	18	19	20	21
D12	D13	D14	D15	D16	D17	D18
TBT25	TBT26	TBT27	TBT28	TBT29	TBT30	MKH1
TAH3	TAH4	TAH5	TAH6	TAH7	TAH8	TAH9
22	23	24	25	26	27	28
D19	D20	D21	D22	D23	D24	D25
MKH2	MKH3	MKH4	MKH5	MKH6	MKH7	MKH8
TAH10	TAH11	TAH12	TAH13	TAH14	TAH15	TAH16

KWANZA 524/2040-41/6276-77/2032-33						
Umoja	Kujichagulia	Ujima	Ujamaa	Nia	Kuumba	Imani
1	2	3	4	5	6	7
D26	D27	D28	D29	D30	D31	JA1
MKH9	MKH10	MKH11	MKH12	MKH13	MKH14	MKH15
TAH17	TAH18	TAH19	TAH20	TAH21	TAH22	TAH23
8	9	10	11	12	13	14
JA2	JA3	JA4	JA5	JA6	JA7	JA8
MKH16	MKH17	MKH18	MKH19	MKH20	MKH21	MKH22
TAH24	TAH25	TAH26	TAH27	TAH28	TAH29	TAH30
15	16	17	18	19	20	21
JA9	JA10	JA11	JA12	JA13	JA14	JA15
MKH23	MKH24	MKH25	KMH26	MKH27	MKH28	MKH29
TIR1	TIR2	TIR3	TIR4	TIR5	TIR6	TIR7
22	23	24	25	26	27	28
JA16	JA17	JA18	JA19	JA20	JA21	JA22
MKH30	AMN1	AMN2	AMN3	AMN4	AMN5	AMN6
TIR8	TIR9	TIR10	TIR11	TIR12	TIR13	TIR14

THE NU AFRIKAN CALENDAR

SHAHIDI 524/2040-41/6276-77/2032-33						
Ashe	Ojobo	Iwa	Araorun	Ile	Itutu	Erinle
1	2	3	4	5	6	7
JA23	JA24	JA25	JA26	JA27	JA28	JA29
AMN7	AMN8	AMN9	AMN10	AMN11	AMN12	AMN13
TIR15	TIR16	TIR17	TIR18	TIR19	TIR20	TIR21
8	9	10	11	12	13	14
JA30	JA31	F1	F2	F3	F4	F5
AMN14	AMN15	AMN16	AMN17	AMN18	AMN19	AMN20
TIR22	TIR23	TIR24	TIR25	TIR26	TIR27	TIR28
15	16	17	18	19	20	21
F6	F7	F8	F9	F10	F 11	F12
AMN21	AMN22	AMN23	AMN24	AMN25	AMN26	AMN28
TIR29	TIR30	YEK1	YEK2	YEK3	YEK4	YEK5
22	23	24	25	26	27	28
F13	F14	F15	F16	F17	F18	F19
AMN29	AMN30	RNT1	RNT2	RNT3	RNT4	RNT5
YEK6	YEK7	YEK8	YEK9	YEK10	YEK11	YEK12
SHAHIDI 75						
F20						
RNT6						
YEK13						

THE NU AFRIKAN CALENDAR

MAAFA 525 B.S.E./ASIENTO 2041-42 C.E.
KAM 6277-78 S.M./EC 2033-34 A.M. (ዓ.ም.)

AFU 525/2041-42/6277-78/2033-34

Ashe	Ojobo	Iwa	Araorun	Ile	Itutu	Erinle
1	2	3	4	5	6	7
F21	F22	F23	F24	F25	F26	F27
RNT7	RNT8	RNT9	RNT10	RNT11	RNT12	RNT13
YEK14	YEK15	YEK16	YEK17	YEK18	YEK19	YEK20
8	9	10	11	12	13	14
F28	M1	M2	M3	M4	M5	M6
RNT14	RNT15	RNT16	RNT17	RNT18	RNT19	RNT20
YEK21	YEK22	YEK23	YEK24	YEK25	YEK26	YEK27
15	16	17	18	19	20	21
M7	M8	M9	M10	M11	M12	M13
RNT21	RNT22	RNT23	RNT24	RNT25	RNT26	RNT27
YEK28	YEK29	YEK30	MAG1	MAG2	MAG3	MAG4
22	23	24	25	26	27	28
M14	M15	M16	M17	M18	M19	M20
RNT28	RNT29	RNT30	KHN1	KHN2	KHN3	KHN4
MAG5	MAG6	MAG7	MAG8	MAG9	MAG10	MAG11

KAWAIDA 525/2041-42/6277-78/2033-34

Ashe	Ojobo	Iwa	Araorun	Ile	Itutu	Erinle
1	2	3	4	5	6	7
A18	A19	A20	A21	A22	A23	A24
PIN3	PIN4	PIN5	PIN6	PIN7	PIN8	PIN9
MIY10	MIY11	MIY12	MIY13	MIY14	MIY15	MIY16
8	9	10	11	12	13	14
A25	A26	A27	A28	A29	A30	MA1
PIN10	PIN11	PIN12	PIN13	PIN14	PIN15	PIN16
MIY17	MIY18	MIY19	MIY20	MIY21	MIY22	MIY23
15	16	17	18	19	20	21
MA2	MA3	MA4	MA5	MA6	MA7	MA8
PIN17	PIN18	PIN19	PIN20	PIN21	PIN22	PIN23
MIY24	MIY25	MIY26	MIY27	MIY28	MIY29	MIY30
22	23	24	25	26	27	28
MA9	MA10	MA11	MA12	MA13	MA14	MA15
PIN24	PIN25	PIN26	PIN27	PIN28	PIN29	PIN30
GIN1	GIN2	GIN3	GIN4	GIN5	GIN6	GIN7

UHURU 525/2041-42/6277-78/2033-34

Ashe	Ojobo	Iwa	Araorun	Ile	Itutu	Erinle
1	2	3	4	5	6	7
M21	M22	M23	M24	M25	M26	M27
KHN5	KHN6	KHN7	KHN8	KHN9	KHN10	KHN11
MAG12	MAG13	MAG14	MAG15	MAG16	MAG17	MAG18
8	9	10	11	12	13	14
M28	M29	M30	M31	A1	A2	A3
KHN12	KHN13	KHN14	KHN15	KHN16	KHN17	KHN18
MAG19	MAG20	MAG21	MAG22	MAG23	MAG24	MAG25
15	16	17	18	19	20	21
A4	A5	A6	A7	A8	A9	A10
KHN19	KHN20	KHN21	KHN22	KHN23	KHN24	KHN25
MAG26	MAG27	MAG28	MAG29	MAG30	MIY1	MIY2
22	23	24	25	26	27	28
A11	A12	A13	A14	A15	A16	A17
KHN26	KHN27	KHN28	KHN29	KHN30	PIN1	PIN2
MIY3	MIY4	MIY5	MIY6	MIY7	MIY8	MIY9

MABABA WEUSI 525/2041-42/6277-78/2033-34

Ashe	Ojobo	Iwa	Araorun	Ile	Itutu	Erinle
1	2	3	4	5	6	7
MA16	MA17	MA18	MA19	MA20	MA21	MA22
APP1	APP2	APP3	APP4	APP5	APP6	APP7
GIN8	GIN9	GIN10	GIN11	GIN12	GIN13	GIN14
8	9	10	11	12	13	14
MA23	MA24	MA25	MA26	MA27	MA28	MA29
APP8	APP9	APP10	APP11	APP12	APP13	APP14
GIN15	GIN16	GIN17	GIN18	GIN19	GIN20	GIN21
15	16	17	18	19	20	21
MA30	MA31	J1	J2	J3	J4	J5
APP15	APP16	APP17	APP18	APP19	APP20	APP21
GIN22	GIN23	GIN24	GIN25	GIN26	GIN27	GIN28
22	23	24	25	26	27	28
J6	J7	J8	J9	J10	J11	J12
APP22	APP23	APP24	APP25	APP26	APP27	APP28
GIN29	GIN30	SAN1	SAN2	SAN3	SAN4	SAN5

THE NU AFRIKAN CALENDAR

MWANAMKE WEUSI 525/2041-42/6277-78/2033-34

Ashe	Ojobo	Iwa	Araorun	Ile	Itutu	Erinle
1	2	3	4	5	6	7
J13	J14	J15	J16	J17	J18	J19
APP29	APP30	MSR1	MSR2	MSR3	MSR4	MSR5
SAN6	SAN7	SAN8	SAN9	SAN10	SAN11	SAN12
8	9	10	11	12	13	14
J20	J21	J22	J23	J24	J25	J26
MSR6	MSR7	MSR8	MSR9	MSR10	MSR11	MSR12
SAN13	SAN14	SAN15	SAN16	SAN17	SAN18	SAN19
15	16	17	18	19	20	21
J27	J28	J29	J30	JU1	JU2	JU3
MSR13	MSR14	MSR15	MSR16	MSR17	MSR18	MSR19
SAN20	SAN21	SAN22	SAN23	SAN24	SAN25	SAN26
22	23	24	25	26	27	28
JU4	JU5	JU6	JU7	JU8	JU9	JU10
MSR20	MRS21	MSR22	MSR23	MSR24	MSR25	MSR26
SAN27	SAN28	SAN29	SAN30	HAM1	HAM2	HAM3

TAIFA WEUSI 525/2041-42/6277-78/2033-34

Ashe	Ojobo	Iwa	Araorun	Ile	Itutu	Erinle
1	2	3	4	5	6	7
AU8	AU9	AU10	AU11	AU12	AU13	AU14
THT20	THT21	THT22	THT23	THT24	THT25	THT26
NAH2	NAH3	NAH4	NAH5	NAH6	NAH7	NAH8
8	9	10	11	12	13	14
AU15	AU16	AU17	AU18	AU19	AU20	AU21
THT27	THT28	THT29	THT30	PAP1	PAP1	PAP2
NAH9	NAH10	NAH11	NAH12	NAH13	NAH14	NAH15
15	16	17	18	19	20	21
AU22	AU23	AU24	AU25	AU26	AU27	AU28
PAP3	PAP4	PAP5	PAP6	PAP7	PAP8	PAP9
NAH16	NAH17	NAH18	NAH19	NAH20	NAH21	NAH22
22	23	24	25	26	27	28
AU29	AU30	AU31	S1	S2	S3	S4
PAP10	PAP11	PAP12	PAP13	PAP14	PAP15	PAP16
NAH23	NAH24	NAH25	NAH26	NAH27	NAH28	NAH29

MTU WEUSI 525/2041-42/6277-78/2033-34

Ashe	Ojobo	Iwa	Araorun	Ile	Itutu	Erinle
1	2	3	4	5	6	7
JU11	JU12	JU13	JU14	JU15	JU16	JU17
MSR27	MSR28	MSR29	MSR30	*ASR1*	*HRU2*	*STH3*
HAM4	HAM5	HAM6	HAM7	HAM8	HAM9	HAM10
8	9	10	11	12	13	14
JU18	JU19	JU20	JU21	JU22	JU23	JU24
AST4	*NBH5*	THT1	THT2	THT3	THT4	THT5
HAM11	HAM12	HAM13	AM14	HAM15	HAM16	HAM17
15	16	17	18	19	20	21
JU25	JU26	JU27	JU28	JU29	JU30	JU31
THT6	THT7	THT8	THT9	THT10	THT11	THT12
HAM18	HAM19	HAM20	HAM21	HAM22	HAM23	HAM24
22	23	24	25	26	27	28
AU1	AU2	AU3	AU4	AU5	AU6	AU7
THT13	THT14	THT15	THT16	THT17	THT18	THT19
HAM25	HAM26	HAM27	HAM28	HAM29	HAM30	NAH1

UWEZO 525/2041-42/6277-78/2033-34

Ashe	Ojobo	Iwa	Araorun	Ile	Itutu	Erinle
1	2	3	4	5	6	7
S5	S6	S7	S8	S9	S10	S11
PAP17	PAP18	PAP19	PAP20	PAP21	PAP22	PAP23
NAH30	*PAG1*	*PAG2*	*PAG3*	*PAG4*	*PAG5*	MAS1
8	9	10	11	12	13	14
S12	S13	S14	S15	S16	S17	S18
PAP24	PAP25	PAP26	PAP27	PAP28	PAP29	PAP30
MAS2	MAS3	MAS4	MAS5	MAS6	MAS7	MAS8
15	16	17	18	19	20	21
S19	S20	S21	S22	S23	S24	S25
HTH1	HTH2	HTH3	HTH4	HTH5	HTH6	HTH7
MAS9	MAS10	MAS11	MAS12	MAS13	MAS14	MAS15
22	23	24	25	26	27	28
S26	S27	S28	S29	S30	O1	O2
HTH8	HTH9	HTH10	HTH11	HTH12	HTH13	HTH14
MAS16	MAS17	MAS18	MAS19	MAS20	MAS21	MAS22

THE NU AFRIKAN CALENDAR

EKUNDU 525/2041-42/6277-78/2033-34						
Ashe	Ojobo	Iwa	Araorun	Ile	Itutu	Erinle
1	2	3	4	5	6	7
O3	O4	O5	O6	O7	O8	O9
HTH15	HTH16	HTH17	HTH18	HTH19	HTH20	HTH21
MAS23	MAS24	MAS25	MAS26	MAS27	MAS28	MAS29
8	9	10	11	12	13	14
O10	O11	O12	O13	O14	O15	O16
HTH22	HTH23	HTH24	HTH25	HTH26	HTH27	HTH28
MAS30	THi1	THi2	THi3	THi4	THi5	THi6
15	16	17	18	19	20	21
O17	O18	O19	O20	O21	O22	O23
HTH29	HTH30	KHK1	KHK2	KHK3	KHK4	KHK5
THi7	THi8	THi9	THi10	THi11	THi12	THi13
22	23	24	25	26	27	28
O24	O25	O26	O27	O28	O29	O30
KHK6	KHK7	KHK8	KHK9	KHK10	KHK11	KHK12
THi14	THi15	THi16	THi17	THi18	THi19	THi20

KIJANI 525/2041-42/6277-78/2033-34						
Ashe	Ojobo	Iwa	Araorun	Ile	Itutu	Erinle
1	2	3	4	5	6	7
N28	N29	N30	D1	D2	D3	D4
TBT11	TBT12	TBT13	TBT14	TBT15	TBT16	TBT17
HID19	HID20	HID21	HID22	HID23	HID24	HID25
8	9	10	11	12	13	14
D5	D6	D7	D8	D9	D10	D11
TBT18	TBT19	TBT20	TBT21	TBT22	TBT23	TBT24
HID26	HID27	HID28	HID29	HID30	TAH1	TAH2
15	16	17	18	19	20	21
D12	D13	D14	D15	D16	D17	D18
TBT25	TBT26	TBT27	TBT28	TBT29	TBT30	MKH1
TAH3	TAH4	TAH5	TAH6	TAH7	TAH8	TAH9
22	23	24	25	26	27	28
D19	D20	D21	D22	D23	D24	D25
MKH2	MKH3	MKH4	MKH5	MKH6	MKH7	MKH8
TAH10	TAH11	TAH12	TAH13	TAH14	TAH15	TAH16

EUSI 525/2041-42/6277-78/2033-34						
Ashe	Ojobo	Iwa	Araorun	Ile	Itutu	Erinle
1	2	3	4	5	6	7
O31	N1	N2	N3	N4	N5	N6
KHK13	KHK14	KHK15	KHK16	KHK17	KHK18	KHK19
THi21	THi22	THi23	THi24	THi25	THi26	THi27
8	9	10	11	12	13	14
N7	N8	N9	N10	N11	N12	N13
KHK20	KHK21	KHK22	KHK23	KHK24	KHK25	KHK26
THi28	THi29	THi30	HID1	HID2	HID3	HID4
15	16	17	18	19	20	21
N14	N15	N16	N17	N18	N19	N20
KHK27	KHK28	KHK29	KHK30	TBT1	TBT2	TBT3
HID5	HID6	HID7	HID8	HID9	HID10	HID11
22	23	24	25	26	27	28
N21	N22	N23	N24	N25	N26	N27
TBT4	TBT5	TBT6	TBT7	TBT8	TBT9	TBT10
HID12	HID13	HID14	HID15	HID16	HID17	HID18

KWANZA 525/2041-42/6277-78/2033-34						
Umoja	Kujichagulia	Ujima	Ujamaa	Nia	Kuumba	Imani
1	2	3	4	5	6	7
D26	D27	D28	D29	D30	D31	JA1
MKH9	MKH10	MKH11	MKH12	MKH13	MKH14	MKH15
TAH17	TAH18	TAH19	TAH20	TAH21	TAH22	TAH23
8	9	10	11	12	13	14
JA2	JA3	JA4	JA5	JA6	JA7	JA8
MKH16	MKH17	MKH18	MKH19	MKH20	MKH21	MKH22
TAH24	TAH25	TAH26	TAH27	TAH28	TAH29	TAH30
15	16	17	18	19	20	21
JA9	JA10	JA11	JA12	JA13	JA14	JA15
MKH23	MKH24	MKH25	KMH26	MKH27	MKH28	MKH29
TIR1	TIR2	TIR3	TIR4	TIR5	TIR6	TIR7
22	23	24	25	26	27	28
JA16	JA17	JA18	JA19	JA20	JA21	JA22
MKH30	AMN1	AMN2	AMN3	AMN4	AMN5	AMN6
TIR8	TIR9	TIR10	TIR11	TIR12	TIR13	TIR14

THE NU AFRIKAN CALENDAR

SHAHIDI 525/2041-42/6277-78/2033-34						
Ashe	Ojobo	Iwa	Araorun	Ile	Itutu	Erinle
1	2	3	4	5	6	7
JA23	JA24	JA25	JA26	JA27	JA28	JA29
AMN7	AMN8	AMN9	AMN10	AMN11	AMN12	AMN13
TIR15	TIR16	TIR17	TIR18	TIR19	TIR20	TIR21
8	9	10	11	12	13	14
JA30	JA31	F1	F2	F3	F4	F5
AMN14	AMN15	AMN16	AMN17	AMN18	AMN19	AMN20
TIR22	TIR23	TIR24	TIR25	TIR26	TIR27	TIR28
15	16	17	18	19	20	21
F6	F7	F8	F9	F10	F 11	F12
AMN21	AMN22	AMN23	AMN24	AMN25	AMN26	AMN28
TIR29	TIR30	YEK1	YEK2	YEK3	YEK4	YEK5
22	23	24	25	26	27	28
F13	F14	F15	F16	F17	F18	F19
AMN29	AMN30	RNT1	RNT2	RNT3	RNT4	RNT5
YEK6	YEK7	YEK8	YEK9	YEK10	YEK11	YEK12
SHAHIDI 76						
F20						
RNT6						
YEK13						

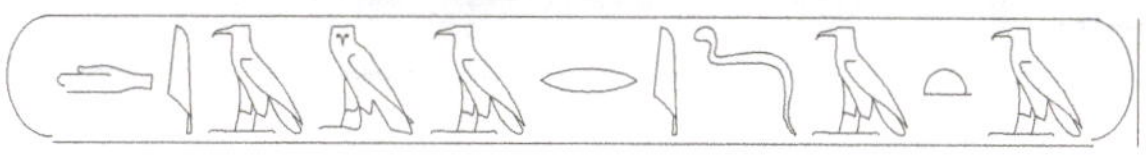

THE NU AFRIKAN CALENDAR

MAAFA 526 B.S.E./ASIENTO 2042-43 C.E.
KAM 6278-79 S.M./EC 2034-35 A.M. (ዓ.ም.)

AFU 526/2042-43/6278-79/2034-35

Ashe	Ojobo	Iwa	Araorun	Ile	Itutu	Erinle
1	2	3	4	5	6	7
F21	F22	F23	F24	F25	F26	F27
RNT7	RNT8	RNT9	RNT10	RNT11	RNT12	RNT13
YEK14	YEK15	YEK16	YEK17	YEK18	YEK19	YEK20
8	9	10	11	12	13	14
F28	M1	M2	M3	M4	M5	M6
RNT14	RNT15	RNT16	RNT17	RNT18	RNT19	RNT20
YEK21	YEK22	YEK23	YEK24	YEK25	YEK26	YEK27
15	16	17	18	19	20	21
M7	M8	M9	M10	M11	M12	M13
RNT21	RNT22	RNT23	RNT24	RNT25	RNT26	RNT27
YEK28	YEK29	YEK30	MAG1	MAG2	MAG3	MAG4
22	23	24	25	26	27	28
M14	M15	M16	M17	M18	M19	M20
RNT28	RNT29	RNT30	KHN1	KHN2	KHN3	KHN4
MAG5	MAG6	MAG7	MAG8	MAG9	MAG10	MAG11

UHURU 526/2042-43/6278-79/2034-35

Ashe	Ojobo	Iwa	Araorun	Ile	Itutu	Erinle
1	2	3	4	5	6	7
M21	M22	M23	M24	M25	M26	M27
KHN5	KHN6	KHN7	KHN8	KHN9	KHN10	KHN11
MAG12	MAG13	MAG14	MAG15	MAG16	MAG17	MAG18
8	9	10	11	12	13	14
M28	M29	M30	M31	A1	A2	A3
KHN12	KHN13	KHN14	KHN15	KHN16	KHN17	KHN18
MAG19	MAG20	MAG21	MAG22	MAG23	MAG24	MAG25
15	16	17	18	19	20	21
A4	A5	A6	A7	A8	A9	A10
KHN19	KHN20	KHN21	KHN22	KHN23	KHN24	KHN25
MAG26	MAG27	MAG28	MAG29	MAG30	MIY1	MIY2
22	23	24	25	26	27	28
A11	A12	A13	A14	A15	A16	A17
KHN26	KHN27	KHN28	KHN29	KHN30	PIN1	PIN2
MIY3	MIY4	MIY5	MIY6	MIY7	MIY8	MIY9

KAWAIDA 526/2042-43/6278-79/2034-35

Ashe	Ojobo	Iwa	Araorun	Ile	Itutu	Erinle
1	2	3	4	5	6	7
A18	A19	A20	A21	A22	A23	A24
PIN3	PIN4	PIN5	PIN6	PIN7	PIN8	PIN9
MIY10	MIY11	MIY12	MIY13	MIY14	MIY15	MIY16
8	9	10	11	12	13	14
A25	A26	A27	A28	A29	A30	MA1
PIN10	PIN11	PIN12	PIN13	PIN14	PIN15	PIN16
MIY17	MIY18	MIY19	MIY20	MIY21	MIY22	MIY23
15	16	17	18	19	20	21
MA2	MA3	MA4	MA5	MA6	MA7	MA8
PIN17	PIN18	PIN19	PIN20	PIN21	PIN22	PIN23
MIY24	MIY25	MIY26	MIY27	MIY28	MIY29	MIY30
22	23	24	25	26	27	28
MA9	MA10	MA11	MA12	MA13	MA14	MA15
PIN24	PIN25	PIN26	PIN27	PIN28	PIN29	PIN30
GIN1	GIN2	GIN3	GIN4	GIN5	GIN6	GIN7

MABABA WEUSI 526/2042-43/6278-79/2034-35

Ashe	Ojobo	Iwa	Araorun	Ile	Itutu	Erinle
1	2	3	4	5	6	7
MA16	MA17	MA18	MA19	MA20	MA21	MA22
APP1	APP2	APP3	APP4	APP5	APP6	APP7
GIN8	GIN9	GIN10	GIN11	GIN12	GIN13	GIN14
8	9	10	11	12	13	14
MA23	MA24	MA25	MA26	MA27	MA28	MA29
APP8	APP9	APP10	APP11	APP12	APP13	APP14
GIN15	GIN16	GIN17	GIN18	GIN19	GIN20	GIN21
15	16	17	18	19	20	21
MA30	MA31	J1	J2	J3	J4	J5
APP15	APP16	APP17	APP18	APP19	APP20	APP21
GIN22	GIN23	GIN24	GIN25	GIN26	GIN27	GIN28
22	23	24	25	26	27	28
J6	J7	J8	J9	J10	J11	J12
APP22	APP23	APP24	APP25	APP26	APP27	APP28
GIN29	GIN30	SAN1	SAN2	SAN3	SAN4	SAN5

THE NU AFRIKAN CALENDAR

MWANAMKE WEUSI 526/2042-43/6278-79/2034-35						
Ashe	Ojobo	Iwa	Araorun	Ile	Itutu	Erinle
1	2	3	4	5	6	7
J13	J14	J15	J16	J17	J18	J19
APP29	APP30	MSR1	MSR2	MSR3	MSR4	MSR5
SAN6	SAN7	SAN8	SAN9	SAN10	SAN11	SAN12
8	9	10	11	12	13	14
J20	J21	J22	J23	J24	J25	J26
MSR6	MSR7	MSR8	MSR9	MSR10	MSR11	MSR12
SAN13	SAN14	SAN15	SAN16	SAN17	SAN18	SAN19
15	16	17	18	19	20	21
J27	J28	J29	J30	JU1	JU2	JU3
MSR13	MSR14	MSR15	MSR16	MSR17	MSR18	MSR19
SAN20	SAN21	SAN22	SAN23	SAN24	SAN25	SAN26
22	23	24	25	26	27	28
JU4	JU5	JU6	JU7	JU8	JU9	JU10
MSR20	MRS21	MSR22	MSR23	MSR24	MSR25	MSR26
SAN27	SAN28	SAN29	SAN30	HAM1	HAM2	HAM3

TAIFA WEUSI 526/2042-43/6278-79/2034-35						
Ashe	Ojobo	Iwa	Araorun	Ile	Itutu	Erinle
1	2	3	4	5	6	7
AU8	AU9	AU10	AU11	AU12	AU13	AU14
THT20	THT21	THT22	THT23	THT24	THT25	THT26
NAH2	NAH3	NAH4	NAH5	NAH6	NAH7	NAH8
8	9	10	11	12	13	14
AU15	AU16	AU17	AU18	AU19	AU20	AU21
THT27	THT28	THT29	THT30	PAP1	PAP1	PAP2
NAH9	NAH10	NAH11	NAH12	NAH13	NAH14	NAH15
15	16	17	18	19	20	21
AU22	AU23	AU24	AU25	AU26	AU27	AU28
PAP3	PAP4	PAP5	PAP6	PAP7	PAP8	PAP9
NAH16	NAH17	NAH18	NAH19	NAH20	NAH21	NAH22
22	23	24	25	26	27	28
AU29	AU30	AU31	S1	S2	S3	S4
PAP10	PAP11	PAP12	PAP13	PAP14	PAP15	PAP16
NAH23	NAH24	NAH25	NAH26	NAH27	NAH28	NAH29

MTU WEUSI 526/2042-43/6278-79/2034-35						
Ashe	Ojobo	Iwa	Araorun	Ile	Itutu	Erinle
1	2	3	4	5	6	7
JU11	JU12	JU13	JU14	JU15	JU16	JU17
MSR27	MSR28	MSR29	MSR30	*ASR1*	*HRU2*	*STH3*
HAM4	HAM5	HAM6	HAM7	HAM8	HAM9	HAM10
8	9	10	11	12	13	14
JU18	JU19	JU20	JU21	JU22	JU23	JU24
AST4	*NBH5*	THT1	THT2	THT3	THT4	THT5
HAM11	HAM12	HAM13	AM14	HAM15	HAM16	HAM17
15	16	17	18	19	20	21
JU25	JU26	JU27	JU28	JU29	JU30	JU31
THT6	THT7	THT8	THT9	THT10	THT11	THT12
HAM18	HAM19	HAM20	HAM21	HAM22	HAM23	HAM24
22	23	24	25	26	27	28
AU1	AU2	AU3	AU4	AU5	AU6	AU7
THT13	THT14	THT15	THT16	THT17	THT18	THT19
HAM25	HAM26	HAM27	HAM28	HAM29	HAM30	NAH1

UWEZO 526/2042-43/6278-79/2034-35						
Ashe	Ojobo	Iwa	Araorun	Ile	Itutu	Erinle
1	2	3	4	5	6	7
S5	S6	S7	S8	S9	S10	S11
PAP17	PAP18	PAP19	PAP20	PAP21	PAP22	PAP23
NAH30	*PAG1*	*PAG2*	*PAG3*	*PAG4*	*PAG5*	MAS1
8	9	10	11	12	13	14
S12	S13	S14	S15	S16	S17	S18
PAP24	PAP25	PAP26	PAP27	PAP28	PAP29	PAP30
MAS2	MAS3	MAS4	MAS5	MAS6	MAS7	MAS8
15	16	17	18	19	20	21
S19	S20	S21	S22	S23	S24	S25
HTH1	HTH2	HTH3	HTH4	HTH5	HTH6	HTH7
MAS9	MAS10	MAS11	MAS12	MAS13	MAS14	MAS15
22	23	24	25	26	27	28
S26	S27	S28	S29	S30	O1	O2
HTH8	HTH9	HTH10	HTH11	HTH12	HTH13	HTH14
MAS16	MAS17	MAS18	MAS19	MAS20	MAS21	MAS22

THE NU AFRIKAN CALENDAR

EKUNDU 526/2042-43/6278-79/2034-35						
Ashe	**Ojobo**	**Iwa**	**Araorun**	**Ile**	**Itutu**	**Erinle**
1	2	3	4	5	6	7
O3	O4	O5	O6	O7	O8	O9
HTH15	HTH16	HTH17	HTH18	HTH19	HTH20	HTH21
MAS23	MAS24	MAS25	MAS26	MAS27	MAS28	MAS29
8	9	10	11	12	13	14
O10	O11	O12	O13	O14	O15	O16
HTH22	HTH23	HTH24	HTH25	HTH26	HTH27	HTH28
MAS30	THi1	THi2	THi3	THi4	THi5	THi6
15	16	17	18	19	20	21
O17	O18	O19	O20	O21	O22	O23
HTH29	HTH30	KHK1	KHK2	KHK3	KHK4	KHK5
THi7	THi8	THi9	THi10	THi11	THi12	THi13
22	23	24	25	26	27	28
O24	O25	O26	O27	O28	O29	O30
KHK6	KHK7	KHK8	KHK9	KHK10	KHK11	KHK12
THi14	THi15	THi16	THi17	THi18	THi19	THi20

KIJANI 526/2042-43/6278-79/2034-35						
Ashe	**Ojobo**	**Iwa**	**Araorun**	**Ile**	**Itutu**	**Erinle**
1	2	3	4	5	6	7
N28	N29	N30	D1	D2	D3	D4
TBT11	TBT12	TBT13	TBT14	TBT15	TBT16	TBT17
HID19	HID20	HID21	HID22	HID23	HID24	HID25
8	9	10	11	12	13	14
D5	D6	D7	D8	D9	D10	D11
TBT18	TBT19	TBT20	TBT21	TBT22	TBT23	TBT24
HID26	HID27	HID28	HID29	HID30	TAH1	TAH2
15	16	17	18	19	20	21
D12	D13	D14	D15	D16	D17	D18
TBT25	TBT26	TBT27	TBT28	TBT29	TBT30	MKH1
TAH3	TAH4	TAH5	TAH6	TAH7	TAH8	TAH9
22	23	24	25	26	27	28
D19	D20	D21	D22	D23	D24	D25
MKH2	MKH3	MKH4	MKH5	MKH6	MKH7	MKH8
TAH10	TAH11	TAH12	TAH13	TAH14	TAH15	TAH16

EUSI 526/2042-43/6278-79/2034-35						
Ashe	**Ojobo**	**Iwa**	**Araorun**	**Ile**	**Itutu**	**Erinle**
1	2	3	4	5	6	7
O31	N1	N2	N3	N4	N5	N6
KHK13	KHK14	KHK15	KHK16	KHK17	KHK18	KHK19
THi21	THi22	THi23	THi24	THi25	THi26	THi27
8	9	10	11	12	13	14
N7	N8	N9	N10	N11	N12	N13
KHK20	KHK21	KHK22	KHK23	KHK24	KHK25	KHK26
THi28	THi29	THi30	HID1	HID2	HID3	HID4
15	16	17	18	19	20	21
N14	N15	N16	N17	N18	N19	N20
KHK27	KHK28	KHK29	KHK30	TBT1	TBT2	TBT3
HID5	HID6	HID7	HID8	HID9	HID10	HID11
22	23	24	25	26	27	28
N21	N22	N23	N24	N25	N26	N27
TBT4	TBT5	TBT6	TBT7	TBT8	TBT9	TBT10
HID12	HID13	HID14	HID15	HID16	HID17	HID18

KWANZA 526/2042-43/6278-79/2034-35						
Umoja	**Kujichagulia**	**Ujima**	**Ujamaa**	**Nia**	**Kuumba**	**Imani**
1	2	3	4	5	6	7
D26	D27	D28	D29	D30	D31	JA1
MKH9	MKH10	MKH11	MKH12	MKH13	MKH14	MKH1
TAH17	TAH18	TAH19	TAH20	TAH21	TAH22	TAH23
8	9	10	11	12	13	14
JA2	JA3	JA4	JA5	JA6	JA7	JA8
MKH16	MKH17	MKH18	MKH19	MKH20	MKH21	MKH2
TAH24	TAH25	TAH26	TAH27	TAH28	TAH29	TAH30
15	16	17	18	19	20	21
JA9	JA10	JA11	JA12	JA13	JA14	JA15
MKH23	MKH24	MKH25	KMH26	MKH27	MKH28	MKH2
TIR1	TIR2	TIR3	TIR4	TIR5	TIR6	TIR7
22	23	24	25	26	27	28
JA16	JA17	JA18	JA19	JA20	JA21	JA22
MKH30	AMN1	AMN2	AMN3	AMN4	AMN5	AMN6
TIR8	TIR9	TIR10	TIR11	TIR12	TIR13	TIR14

THE NU AFRIKAN CALENDAR

SHAHIDI 526/2042-43/6278-79/2034-35						
Ashe	Ojobo	Iwa	Araorun	Ile	Itutu	Erinle
1	2	3	4	5	6	7
JA23	JA24	JA25	JA26	JA27	JA28	JA29
AMN7	AMN8	AMN9	AMN10	AMN11	AMN12	AMN13
TIR15	TIR16	TIR17	TIR18	TIR19	TIR20	TIR21
8	9	10	11	12	13	14
JA30	JA31	F1	F2	F3	F4	F5
AMN14	AMN15	AMN16	AMN17	AMN18	AMN19	AMN20
TIR22	TIR23	TIR24	TIR25	TIR26	TIR27	TIR28
15	16	17	18	19	20	21
F6	F7	F8	F9	F10	F 11	F12
AMN21	AMN22	AMN23	AMN24	AMN25	AMN26	AMN28
TIR29	TIR30	YEK1	YEK2	YEK3	YEK4	YEK5
22	23	24	25	26	27	28
F13	F14	F15	F16	F17	F18	F19
AMN29	AMN30	RNT1	RNT2	RNT3	RNT4	RNT5
YEK6	YEK7	YEK8	YEK9	YEK10	YEK11	YEK12
SHAHIDI 77						
F20						
RNT6						
YEK13						

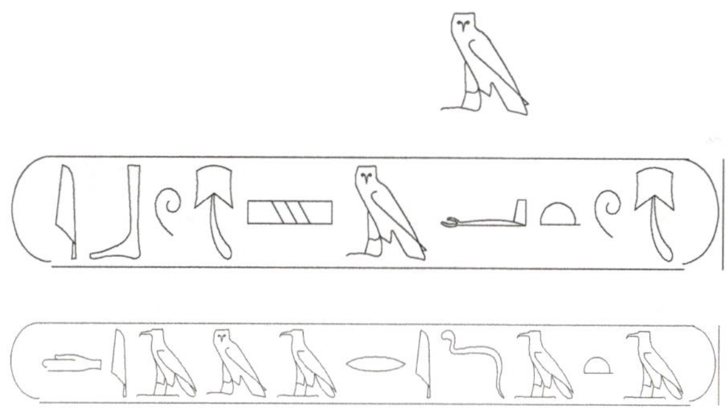

THE NU AFRIKAN CALENDAR

MAAFA 527 B.S.E./ASIENTO 2043-44 C.E./
KAM 6279-80 S.M./EC 2035-36 A.M. (ዓ.ም.)

AFU 527/2043-44/6279-80/2035-36						
Ashe	Ojobo	Iwa	Araorun	Ile	Itutu	Erinle
1	2	3	4	5	6	7
F21	F22	F23	F24	F25	F26	F27
RNT7	RNT8	RNT9	RNT10	RNT11	RNT12	RNT13
YEK14	YEK15	YEK16	YEK17	YEK18	YEK19	YEK20
8	9	10	11	12	13	14
F28	M1	M2	M3	M4	M5	M6
RNT14	RNT15	RNT16	RNT17	RNT18	RNT19	RNT20
YEK21	YEK22	YEK23	YEK24	YEK25	YEK26	YEK27
15	16	17	18	19	20	21
M7	M8	M9	M10	M11	M12	M13
RNT21	RNT22	RNT23	RNT24	RNT25	RNT26	RNT27
YEK28	YEK 29	YEK30	MAG1	MAG2	MAG3	MAG4
22	23	24	25	26	27	28
M14	M15	M16	M17	M18	M19	M20
RNT28	RNT29	RNT30	KHN1	KHN2	KHN3	KHN4
MAG5	MAG6	MAG7	MAG8	MAG9	MAG10	MAG11

KAWAIDA AFU 527/2043-44/6279-80/2035-36						
Ashe	Ojobo	Iwa	Araorun	Ile	Itutu	Erinle
1	2	3	4	5	6	7
A18	A19	A20	A21	A22	A23	A24
PIN3	PIN4	PIN5	PIN6	PIN7	PIN8	PIN9
MIY10	MIY11	MIY12	MIY13	MIY14	MIY15	MIY16
8	9	10	11	12	13	14
A25	A26	A27	A28	A29	A30	MA1
PIN10	PIN11	PIN12	PIN13	PIN14	PIN15	PIN16
MIY17	MIY18	MIY19	MIY20	MIY21	MIY22	MIY23
15	16	17	18	19	20	21
MA2	MA3	MA4	MA5	MA6	MA7	MA8
PIN17	PIN18	PIN19	PIN20	PIN21	PIN22	PIN23
MIY24	MIY25	MIY26	MIY27	MIY28	MIY29	MIY30
22	23	24	25	26	27	28
MA9	MA10	MA11	MA12	MA13	MA14	MA15
PIN24	PIN25	PIN26	PIN27	PIN28	PIN29	PIN30
GIN1	GIN2	GIN3	GIN4	GIN5	GIN6	GIN7

UHURU AFU 527/2043-44/6279-80/2035-36						
Ashe	Ojobo	Iwa	Araorun	Ile	Itutu	Erinle
1	2	3	4	5	6	7
M21	M22	M23	M24	M25	M26	M27
KHN5	KHN6	KHN7	KHN8	KHN9	KHN10	KHN11
MAG12	MAG13	MAG14	MAG15	MAG16	MAG17	MAG18
8	9	10	11	12	13	14
M28	M29	M30	M31	A1	A2	A3
KHN12	KHN13	KHN14	KHN15	KHN16	KHN17	KHN18
MAG19	MAG20	MAG21	MAG22	MAG23	MAG24	MAG25
15	16	17	18	19	20	21
A4	A5	A6	A7	A8	A9	A10
KHN19	KHN20	KHN21	KHN22	KHN23	KHN24	KHN25
MAG26	MAG27	MAG28	MAG29	MAG30	MIY1	MIY2
22	23	24	25	26	27	28
A11	A12	A13	A14	A15	A16	A17
KHN26	KHN27	KHN28	KHN29	KHN30	PIN1	PIN2
MIY3	MIY4	MIY5	MIY6	MIY7	MIY8	MIY9

MABABA WEUSI AFU 527/2043-44/6279-80/2035-36						
Ashe	Ojobo	Iwa	Araorun	Ile	Itutu	Erinle
1	2	3	4	5	6	7
MA16	MA17	MA18	MA19	MA20	MA21	MA22
APP1	APP2	APP3	APP4	APP5	APP6	APP7
GIN8	GIN9	GIN10	GIN11	GIN12	GIN13	GIN14
8	9	10	11	12	13	14
MA23	MA24	MA25	MA26	MA27	MA28	MA29
APP8	APP9	APP10	APP11	APP12	APP13	APP14
GIN15	GIN16	GIN17	GIN18	GIN19	GIN20	GIN21
15	16	17	18	19	20	21
MA30	MA31	J1	J2	J3	J4	J5
APP15	APP16	APP17	APP18	APP19	APP20	APP21
GIN22	GIN23	GIN24	GIN25	GIN26	GIN27	GIN28
22	23	24	25	26	27	28
J6	J7	J8	J9	J10	J11	J12
APP22	APP23	APP24	APP25	APP26	APP27	APP28
GIN29	GIN30	SAN1	SAN2	SAN3	SAN4	SAN5

THE NU AFRIKAN CALENDAR

MWANAMKE WEUSI AFU 527/2043-44/6279-80/2035-36

Ashe	Ojobo	Iwa	Araorun	Ile	Itutu	Erinle
1	2	3	4	5	6	7
J13	J14	J15	J16	J17	J18	J19
APP29	APP30	MSR1	MSR2	MSR3	MSR4	MSR5
SAN6	SAN7	SAN8	SAN9	SAN10	SAN11	SAN12
8	9	10	11	12	13	14
J20	J21	J22	J23	J24	J25	J26
MSR6	MSR7	MSR8	MSR9	MSR10	MSR11	MSR12
SAN13	SAN14	SAN15	SAN16	SAN17	SAN18	SAN19
15	16	17	18	19	20	21
J27	J28	J29	J30	JU1	JU2	JU3
MSR13	MSR14	MSR15	MSR16	MSR17	MSR18	MSR19
SAN20	SAN21	SAN22	SAN23	SAN24	SAN25	SAN26
22	23	24	25	26	27	28
JU4	JU5	JU6	JU7	JU8	JU9	JU10
MSR20	MRS21	MSR22	MSR23	MSR24	MSR25	MSR26
SAN27	SAN28	SAN29	SAN30	HAM1	HAM2	HAM3

TAIFA WEUSI AFU 527/2043-44/6279-80/2035-36

Ashe	Ojobo	Iwa	Araorun	Ile	Itutu	Erinle
1	2	3	4	5	6	7
AU8	AU9	AU10	AU11	AU12	AU13	AU14
THT20	THT21	THT22	THT23	THT24	THT25	THT26
NAH2	NAH3	NAH4	NAH5	NAH6	NAH7	NAH8
8	9	10	11	12	13	14
AU15	AU16	AU17	AU18	AU19	AU20	AU21
THT27	THT28	THT29	THT30	PAP1	PAP1	PAP2
NAH9	NAH10	NAH11	NAH12	NAH13	NAH14	NAH15
15	16	17	18	19	20	21
AU22	AU23	AU24	AU25	AU26	AU27	AU28
PAP3	PAP4	PAP5	PAP6	PAP7	PAP8	PAP9
NAH16	NAH17	NAH18	NAH19	NAH20	NAH21	NAH22
22	23	24	25	26	27	28
AU29	AU30	AU31	S1	S2	S3	S4
PAP10	PAP11	PAP12	PAP13	PAP14	PAP15	PAP16
NAH23	NAH24	NAH25	NAH26	NAH27	NAH28	NAH29

MTU WEUSI AFU 527/2043-44/6279-80/2035-36

Ashe	Ojobo	Iwa	Araorun	Ile	Itutu	Erinle
1	2	3	4	5	6	7
JU11	JU12	JU13	JU14	JU15	JU16	JU17
MSR27	MSR28	MSR29	MSR30	*ASR1*	*HRU2*	*STH3*
HAM4	HAM5	HAM6	HAM7	HAM8	HAM9	HAM10
8	9	10	11	12	13	14
JU18	JU19	JU20	JU21	JU22	JU23	JU24
AST4	*NBH5*	THT1	THT2	THT3	THT4	THT5
HAM11	HAM12	HAM13	HAM14	HAM15	HAM16	HAM17
15	16	17	18	19	20	21
JU25	JU26	JU27	JU28	JU29	JU30	JU31
THT6	THT7	THT8	THT9	THT10	THT11	THT12
HAM18	HAM19	HAM20	HAM21	HAM22	HAM23	HAM24
22	23	24	25	26	27	28
AU1	AU2	AU3	AU4	AU5	AU6	AU7
THT13	THT14	THT15	THT16	THT17	THT18	THT19
HAM25	HAM26	HAM27	HAM28	HAM29	HAM30	NAH1

UWEZO AFU 527/2043-44/6279-80/2035-36

Ashe	Ojobo	Iwa	Araorun	Ile	Itutu	Erinle
1	2	3	4	5	6	7
S5	S6	S7	S8	S9	S10	S11
PAP17	PAP18	PAP19	PAP20	PAP21	PAP22	PAP23
NAH30	*QUA1*	*QUA2*	*QUA3*	*QUA4*	*QUA5*	*QUA6*
8	9	10	11	12	13	14
S12	S13	S14	S15	S16	S17	S18
PAP24	PAP25	PAP26	PAP27	PAP28	PAP29	PAP30
MAS1	MAS2	MAS3	MAS4	MAS5	MAS6	MAS7
15	16	17	18	19	20	21
S19	S20	S21	S22	S23	S24	S25
HTH1	HTH2	HTH3	HTH4	HTH5	HTH6	HTH7
MAS8	MAS9	MAS10	MAS11	MAS12	MAS13	MAS14
22	23	24	25	26	27	28
S26	S27	S28	S29	S30	O1	O2
HTH8	HTH9	HTH10	HTH11	HTH12	HTH13	HTH14
MAS15	MAS16	MAS17	MAS18	MAS19	MAS20	MAS21

THE NU AFRIKAN CALENDAR

EKUNDU AFU 527/2043-44/6279-80/2035-36

Ashe	Ojobo	Iwa	Araorun	Ile	Itutu	Erinle
1	2	3	4	5	6	7
O3	O4	O5	O6	O7	O8	O9
HTH15	HTH16	HTH17	HTH18	HTH19	HTH20	HTH21
MAS22	MAS23	MAS 24	MAS25	MAS26	MAS27	MAS28
8	9	10	11	12	13	14
O10	O11	O12	O13	O14	O15	O16
HTH22	HTH23	HTH24	HTH25	HTH26	HTH27	HTH28
MAS29	MAS30	THi1	THi2	THi3	THi4	THi5
15	16	17	18	19	20	21
O17	O18	O19	O20	O21	O22	O23
HTH29	HTH30	KHK1	KHK2	KHK3	KHK4	KHK5
THi6	THi7	THi8	THi9	THi10	THi11	THi12
22	23	24	25	26	27	28
O24	O25	O26	O27	O28	O29	O30
KHK6	KHK7	KHK8	KHK9	KHK10	KHK11	KHK12
THi13	THi14	THi15	THi16	THi17	THi18	THi19

KIJANI AFU 527/2043-44/6279-80/2035-36

Ashe	Ojobo	Iwa	Araorun	Ile	Itutu	Erinle
1	2	3	4	5	6	7
N28	N29	N30	D1	D2	D3	D4
TBT11	TBT12	TBT13	TBT14	TBT15	TBT16	TBT17
HID18	HID19	HID20	HID21	HID22	HID23	HID24
8	9	10	11	12	13	14
D5	D6	D7	D8	D9	D10	D11
TBT18	TBT19	TBT20	TBT21	TBT22	TBT23	TBT24
HID25	HID26	HID27	HID28	HID29	HID30	TAH1
15	16	17	18	19	20	21
D12	D13	D14	D15	D16	D17	D18
TBT25	TBT26	TBT27	TBT28	TBT29	TBT30	MKH1
TAH2	TAH3	TAH4	TAH5	TAH6	TAH7	TAH8
22	23	24	25	26	27	28
D19	D20	D21	D22	D23	D24	D25
MKH2	MKH3	MKH4	MKH5	MKH6	MKH7	MKH8
TAH9	TAH10	TAH11	TAH12	TAH13	TAH14	TAH15

EUSI AFU 527/2043-44/6279-80/2035-36

Ashe	Ojobo	Iwa	Araorun	Ile	Itutu	Erinle
1	2	3	4	5	6	7
O31	N1	N2	N3	N4	N5	N6
KHK13	KHK14	KHK15	KHK16	KHK17	KHK18	KHK19
THi20	THi21	THi22	THi23	THi24	THi25	THi26
8	9	10	11	12	13	14
N7	N8	N9	N10	N11	N12	N13
KHK20	KHK21	KHK22	KHK23	KHK24	KHK25	KHK26
THi27	THi28	THi29	THi30	HID1	HID2	HID3
15	16	17	18	19	20	21
N14	N15	N16	N17	N18	N19	N20
KHK27	KHK28	KHK29	KHK30	TBT1	TBT2	TBT3
HID4	HID5	HID6	HID7	HID8	HID9	HID10
22	23	24	25	26	27	28
N21	N22	N23	N24	N25	N26	N27
TBT4	TBT5	TBT6	TBT7	TBT8	TBT9	TBT10
HID11	HID12	HID13	HID14	HID15	HID16	HID17

KWANZA AFU 527/2043-44/6279-80/2035-36

Umoja	Kujichagulia	Ujima	Ujamaa	Nia	Kuumba	Imani
1	2	3	4	5	6	7
D26	D27	D28	D29	D30	D31	JA1
MKH9	MKH10	MKH11	MKH12	MKH13	MKH14	MKH15
TAH16	TAH17	TAH18	TAH19	TAH20	TAH21	TAH22
8	9	10	11	12	13	14
JA2	JA3	JA4	JA5	JA6	JA7	JA8
MKH16	MKH17	MKH18	MKH19	MKH20	MKH21	MKH22
TAH23	TAH24	TAH25	TAH26	TAH27	TAH28	TAH29
15	16	17	18	19	20	21
JA9	JA10	JA11	JA12	JA13	JA14	JA15
MKH23	MKH24	MKH25	KMH26	MKH27	MKH28	MKH29
TAH30	TIR1	TIR2	TIR3	TIR4	TIR5	TIR6
22	23	24	25	26	27	28
JA16	JA17	JA18	JA19	JA20	JA21	JA22
MKH30	AMN1	AMN2	AMN3	AMN4	AMN5	AMN6
TIR7	TIR8	TIR9	TIR10	TIR11	TIR12	TIR13

THE NU AFRIKAN CALENDAR

SHAHIDI AFU 527/2043-44/6279-80/2035-36						
Ashe	**Ojobo**	**Iwa**	**Araorun**	**Ile**	**Itutu**	**Erinle**
1	2	3	4	5	6	7
JA23	JA24	JA25	JA26	JA27	JA28	JA29
AMN7	AMN8	AMN9	AMN10	AMN11	AMN12	AMN13
TIR14	TIR15	TIR16	TIR17	TIR18	TIR19	TIR20
8	9	10	11	12	13	14
JA30	JA31	F1	F2	F3	F4	F5
AMN14	AMN15	AMN16	AMN17	AMN18	AMN19	AMN20
TIR21	TIR22	TIR23	TIR24	TIR25	TIR26	TIR27
15	16	17	18	19	20	21
F6	F7	F8	F9	F10	F 11	F12
AMN21	AMN22	AMN23	AMN24	AMN25	AMN26	AMN28
TIR28	TIR29	TIR30	YEK1	YEK2	YEK3	YEK4
22	23	24	25	26	27	28
F13	F14	F15	F16	F17	F18	F19
AMN29	AMN30	RNT1	RNT2	RNT3	RNT4	RNT5
YEK5	YEK6	YEK7	YEK8	YEK9	YEK10	YEK11
SHAHIDI 78						
F20						
RNT6						
YEK12						

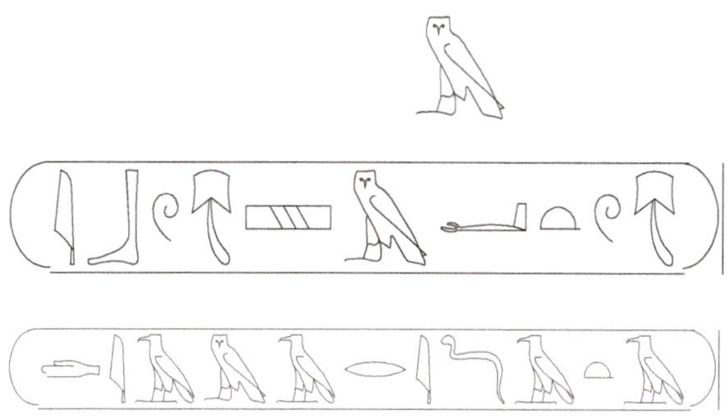

THE NU AFRIKAN CALENDAR

MAAFA 528 B.S.E./ASIENTO 2044-45 C.E./
KAM 6280-81 S.M./EC 2036-37 A.M. (ዓ.ም.)

AFU 528/2044-45/6280-81/2036-37

Ashe	Ojobo	Iwa	Araorun	Ile	Itutu	Erinle
1	2	3	4	5	6	7
F21	F22	F23	F24	F25	F26	F27
RNT7	RNT8	RNT9	RNT10	RNT11	RNT12	RNT13
YEK13	YEK14	YEK15	YEK16	YEK17	YEK18	YEK19
8	9	10	11	12	13	14
F28	F29	M1	M2	M3	M4	M5
RNT14	RNT15	RNT16	RNT17	RNT18	RNT19	RNT20
YEK 20	YEK21	YEK22	YEK23	YEK24	YEK25	YEK26
15	16	17	18	19	20	21
M6	M7	M8	M9	M10	M11	M12
RNT21	RNT22	RNT23	RNT24	RNT25	RNT26	RNT27
YEK 27	YEK28	YEK29	YEK30	MAG1	MAG2	MAG3
22	23	24	25	26	27	28
M13	M14	M15	M16	M17	M18	M19
RNT28	RNT29	RNT30	KHN1	KHN2	KHN3	KHN4
MAG4	MAG5	MAG6	MAG7	MAG8	MAG9	MAG10
SHARIFU 129						
M20						
KHN5						
MAG11						

KAWAIDA 528/2044-45/6280-81/2036-37

Ashe	Ojobo	Iwa	Araorun	Ile	Itutu	Erinle
1	2	3	4	5	6	7
A18	A19	A20	A21	A22	A23	A24
PIN4	PIN5	PIN6	PIN7	PIN8	PIN9	PIN10
MIY10	MIY11	MIY12	MIY13	MIY14	MIY15	MIY16
8	9	10	11	12	13	14
A25	A26	A27	A28	A29	A30	MA1
PIN11	PIN12	PIN13	PIN14	PIN15	PIN16	PIN17
MIY17	MIY18	MIY19	MIY20	MIY21	MIY22	MIY23
15	16	17	18	19	20	21
MA2	MA3	MA4	MA5	MA6	MA7	MA8
PIN18	PIN19	PIN20	PIN21	PIN22	PIN23	PIN24
MIY24	MIY25	MIY26	MIY27	MIY28	MIY29	MIY30
22	23	24	25	26	27	28
MA9	MA10	MA11	MA12	MA13	MA14	MA15
PIN25	PIN26	PIN27	PIN28	PIN29	PIN30	APP1
GIN1	GIN2	GIN3	GIN4	GIN5	GIN6	GIN7

UHURU 528/2044-45/6280-81/2036-37

Ashe	Ojobo	Iwa	Araorun	Ile	Itutu	Erinle
1	2	3	4	5	6	7
M21	M22	M23	M24	M25	M26	M27
KHN6	KHN7	KHN8	KHN9	KHN10	KHN11	KHN12
MAG12	MAG13	MAG14	MAG15	MAG16	MAG17	MAG18
8	9	10	11	12	13	14
M28	M29	M30	M31	A1	A2	A3
KHN13	KHN14	KHN15	KHN16	KHN17	KHN18	KHN19
MAG19	MAG20	MAG21	MAG22	MAG23	MAG24	MAG25
15	16	17	18	19	20	21
A4	A5	A6	A7	A8	A9	A10
KHN20	KHN21	KHN222	KHN23	KHN24	KHN25	KHN26
MAG26	MAG27	MAG28	MAG29	MAG30	MIY1	MIY2
22	23	24	25	26	27	28
A11	A12	A13	A14	A15	A16	A17
KHN27	KHN28	KHN29	KHN30	PIN1	PIN2	PIN3
MIY3	MIY4	MIY5	MIY6	MIY7	MIY8	MIY9

MABABA WEUSI 528/2044-45/6280-81/2036-37

Ashe	Ojobo	Iwa	Araorun	Ile	Itutu	Erinle
1	2	3	4	5	6	7
MA16	MA17	MA18	MA19	MA20	MA21	MA22
APP2	APP3	APP4	APP5	APP6	APP7	APP8
GIN8	GIN9	GIN10	GIN11	GIN12	GIN13	GIN14
8	9	10	11	12	13	14
MA23	MA24	MA25	MA26	MA27	MA28	MA29
APP9	APP10	APP11	APP12	APP13	APP14	APP15
GIN15	GIN16	GIN17	GIN18	GIN19	GIN20	GIN21
15	16	17	18	19	20	21
MA30	MA31	J1	J2	J3	J4	J5
APP16	APP17	APP18	APP19	APP20	APP21	APP22
GIN22	GIN23	GIN24	GIN25	GIN26	GIN27	GIN28
22	23	24	25	26	27	28
J6	J7	J8	J9	J10	J11	J12
APP23	APP24	APP25	APP26	APP27	APP28	APP29
GIN29	GIN30	SAN1	SAN2	SAN3	SAN4	SAN5

THE NU AFRIKAN CALENDAR

MWANAMKE WEUSI 528/2044-45/6280-81/2036-37						
Ashe	Ojobo	Iwa	Araorun	Ile	Itutu	Erinle
1	2	3	4	5	6	7
J13	J14	J15	J16	J17	J18	J19
APP30	MSR1	MSR2	MSR3	MSR4	MSR5	MSR6
SAN6	SAN7	SAN8	SAN9	SAN10	SAN11	SAN12
8	9	10	11	12	13	14
J20	J21	J22	J23	J24	J25	J26
MSR7	MSR8	MSR9	MSR10	MSR11	MSR12	MSR13
SAN13	SAN14	SAN15	SAN16	SAN17	SAN18	SAN19
15	16	17	18	19	20	21
J27	J28	J29	J30	JU1	JU2	JU3
MSR14	MSR15	MSR16	MSR17	MSR18	MSR19	MSR20
SAN20	SAN21	SAN22	SAN23	SAN24	SAN25	SAN26
22	23	24	25	26	27	28
JU4	JU5	JU6	JU7	JU8	JU9	JU10
MSR21	MRS22	MSR23	MSR24	MSR25	MSR26	MSR27
SAN27	SAN28	SAN29	SAN30	HAM1	HAM2	HAM3

TAIFA WEUSI 528/2044-45/6280-81/2036-37						
Ashe	Ojobo	Iwa	Araorun	Ile	Itutu	Erinle
1	2	3	4	5	6	7
AU8	AU9	AU10	AU11	AU12	AU13	AU14
THT20	THT21	THT22	THT23	THT24	THT25	THT26
NAH2	NAH3	NAH4	NAH5	NAH6	NAH7	NAH8
8	9	10	11	12	13	14
AU15	AU16	AU17	AU18	AU19	AU20	AU21
THT27	THT28	THT29	THT30	PAP1	PAP1	PAP2
NAH9	NAH10	NAH11	NAH12	NAH13	NAH14	NAH15
15	16	17	18	19	20	21
AU22	AU23	AU24	AU25	AU26	AU27	AU28
PAP3	PAP4	PAP5	PAP6	PAP7	PAP8	PAP9
NAH16	NAH17	NAH18	NAH19	NAH20	NAH21	NAH22
22	23	24	25	26	27	28
AU29	AU30	AU31	S1	S2	S3	S4
PAP10	PAP11	PAP12	PAP13	PAP14	PAP15	PAP16
NAH23	NAH24	NAH25	NAH26	NAH27	NAH28	NAH29

MTU WEUSI 528/2044-45/6280-81/2036-37						
Ashe	Ojobo	Iwa	Araorun	Ile	Itutu	Erinle
1	2	3	4	5	6	7
JU11	JU12	JU13	JU14	JU15	JU16	JU17
MSR28	MSR29	MSR30	*ASR1*	*HRU2*	*STH3*	*AST4*
HAM4	HAM5	HAM6	HAM7	HAM8	HAM9	HAM10
8	9	10	11	12	13	14
JU18	JU19	JU20	JU21	JU22	JU23	JU24
NBH5	*ANU6*	THT1	THT2	THT3	THT4	THT5
HAM11	HAM12	HAM13	HAM14	HAM15	HAM16	HAM17
15	16	17	18	19	20	21
JU25	JU26	JU27	JU28	JU29	JU30	JU31
THT6	THT7	THT8	THT9	THT10	THT11	THT12
HAM18	HAM19	HAM20	HAM21	HAM22	HAM23	HAM24
22	23	24	25	26	27	28
AU1	AU2	AU3	AU4	AU5	AU6	AU7
THT13	THT14	THT15	THT16	THT17	THT18	THT19
HAM25	HAM26	HAM27	HAM28	HAM29	HAM30	NAH1

UWEZO 528/2044-45/6280-81/2036-37						
Ashe	Ojobo	Iwa	Araorun	Ile	Itutu	Erinle
1	2	3	4	5	6	7
S5	S6	S7	S8	S9	S10	S11
PAP17	PAP18	PAP19	PAP20	PAP21	PAP22	PAP23
NAH30	*PAG1*	*PAG2*	*PAG3*	*PAG4*	*PAG5*	MAS1
8	9	10	11	12	13	14
S12	S13	S14	S15	S16	S17	S18
PAP24	PAP25	PAP26	PAP27	PAP28	PAP29	PAP30
MAS2	MAS3	MAS4	MAS5	MAS6	MAS7	MAS8
15	16	17	18	19	20	21
S19	S20	S21	S22	S23	S24	S25
HTH1	HTH2	HTH3	HTH4	HTH5	HTH6	HTH7
MAS9	MAS10	MAS11	MAS12	MAS13	MAS14	MAS15
22	23	24	25	26	27	28
S26	S27	S28	S29	S30	O1	O2
HTH8	HTH9	HTH10	HTH11	HTH12	HTH13	HTH14
MAS16	MAS17	MAS18	MAS19	MAS20	MAS21	MAS22

THE NU AFRIKAN CALENDAR

EKUNDU 528/2044-45/6280-81/2036-37						
Ashe	Ojobo	Iwa	Araorun	Ile	Itutu	Erinle
1	2	3	4	5	6	7
O3	O4	O5	O6	O7	O8	O9
HTH15	HTH16	HTH17	HTH18	HTH19	HTH20	HTH21
MAS23	MAS24	MAS25	MAS26	MAS27	MAS28	MAS29
8	9	10	11	12	13	14
O10	O11	O12	O13	O14	O15	O16
HTH22	HTH23	HTH24	HTH25	HTH26	HTH27	HTH28
MAS30	THi1	THi2	THi3	THi4	THi5	THi6
15	16	17	18	19	20	21
O17	O18	O19	O20	O21	O22	O23
HTH29	HTH30	KHK1	KHK2	KHK3	KHK4	KHK5
THi7	THi8	This9	THi10	THi11	THi12	THi13
22	23	24	25	26	27	28
O24	O25	O26	O27	O28	O29	O30
KHK6	KHK7	KHK8	KHK9	KHK10	KHK11	KHK12
Thi14	THi15	THi16	THi17	THi18	THi19	THi20

KIJANI 528/2044-45/6280-81/2036-37						
Ashe	Ojobo	Iwa	Araorun	Ile	Itutu	Erinle
1	2	3	4	5	6	7
N28	N29	N30	D1	D2	D3	D4
TBT11	TBT12	TBT13	TBT14	TBT15	TBT16	TBT17
HID19	HID20	HID21	HID22	HID23	HID24	HID25
8	9	10	11	12	13	14
D5	D6	D7	D8	D9	D10	D11
TBT18	TBT19	TBT20	TBT21	TBT22	TBT23	TBT24
HID26	HID27	HID28	HID29	HID30	TAH1	TAH2
15	16	17	18	19	20	21
D12	D13	D14	D15	D16	D17	D18
TBT25	TBT26	TBT27	TBT28	TBT29	TBT30	MKH1
TAH3	TAH4	TAH5	TAH6	TAH7	TAH8	TAH9
22	23	24	25	26	27	28
D19	D20	D21	D22	D23	D24	D25
MKH2	MKH3	MKH4	MKH5	MKH6	MKH7	MKH8
TAH10	TAH11	TAH12	TAH13	TAH14	TAH15	TAH16

EUSI 528/2044-45/6280-81/2036-37						
Ashe	Ojobo	Iwa	Araorun	Ile	Itutu	Erinle
1	2	3	4	5	6	7
O31	N1	N2	N3	N4	N5	N6
KHK13	KHK14	KHK15	KHK16	KHK17	KHK18	KHK19
THi21	THi22	THi23	THi24	THi25	THi26	THi27
8	9	10	11	12	13	14
N7	N8	N9	N10	N11	N12	N13
KHK20	KHK21	KHK22	KHK23	KHK24	KHK25	KHK26
THi28	THi29	THi30	HID1	HID2	HID3	HID4
15	16	17	18	19	20	21
N14	N15	N16	N17	N18	N19	N20
KHK27	KHK28	KHK29	KHK30	TBT1	TBT2	TBT3
HID5	HID6	HID7	HID8	HID9	HID10	HID11
22	23	24	25	26	27	28
N21	N22	N23	N24	N25	N26	N27
TBT4	TBT5	TBT6	TBT7	TBT8	TBT9	TBT10
HID12	HID13	HID14	HID15	HID16	HID17	HID18

KWANZA 528/2044-45/6280-81/2036-37						
Umoja	Kujichagulia	Ujima	Ujamaa	Nia	Kuumba	Imani
1	2	3	4	5	6	7
D26	D27	D28	D29	D30	D31	JA1
MKH9	MKH10	MKH11	MKH12	MKH13	MKH14	MKH15
TAH17	TAH18	TAH19	TAH20	TAH21	TAH22	TAH23
8	9	10	11	12	13	14
JA2	JA3	JA4	JA5	JA6	JA7	JA8
MKH16	MKH17	MKH18	MKH19	MKH20	MKH21	MKH22
TAH24	TAH25	TAH26	TAH27	TAH28	TAH29	TAH30
15	16	17	18	19	20	21
JA9	JA10	JA11	JA12	JA13	JA14	JA15
MKH23	MKH24	MKH25	KMH26	MKH27	MKH28	MKH29
TIR1	TIR2	TIR3	TIR4	TIR5	TIR6	TIR7
22	23	24	25	26	27	28
JA16	JA17	JA18	JA19	JA20	JA21	JA22
MKH30	AMN1	AMN2	AMN3	AMN4	AMN5	AMN6
TIR8	TIR9	TIR10	TIR11	TIR12	TIR13	TIR14

THE NU AFRIKAN CALENDAR

SHAHIDI 528/2044-45/6280-81/2036-37						
Ashe	Ojobo	Iwa	Araorun	Ile	Itutu	Erinle
1	2	3	4	5	6	7
JA23	JA24	JA25	JA26	JA27	JA28	JA29
AMN7	AMN8	AMN9	AMN10	AMN11	AMN12	AMN13
TIR15	TIR16	TIR17	TIR18	TIR19	TIR20	TIR21
8	9	10	11	12	13	14
JA30	JA31	F1	F2	F3	F4	F5
AMN14	AMN15	AMN16	AMN17	AMN18	AMN19	AMN20
TIR22	TIR23	TIR24	TIR25	TIR26	TIR27	TIR28
15	16	17	18	19	20	21
F6	F7	F8	F9	F10	F 11	F12
AMN21	AMN22	AMN23	AMN24	AMN25	AMN26	AMN28
TIR29	TIR30	YEK1	YEK2	YEK3	YEK4	YEK5
22	23	24	25	26	27	28
F13	F14	F15	F16	F17	F18	F19
AMN29	AMN30	RNT1	RNT2	RNT3	RNT4	RNT5
YEK6	YEK7	YEK8	YEK9	YEK10	YEK11	YEK12
SHAHIDI 79						
F20						
RNT6						
YEK13						

THE NU AFRIKAN CALENDAR

MAAFA 529 B.S.E./ASIENTO 2045-46 C.E.
KAM 6281-82 S.M./EC 2037-38 A.M. (ዓ.ም.)

AFU 529/2045-46/6281-82/2037-38						
Ashe	Ojobo	Iwa	Araorun	Ile	Itutu	Erinle
1	2	3	4	5	6	7
F21	F22	F23	F24	F25	F26	F27
RNT7	RNT8	RNT9	RNT10	RNT11	RNT12	RNT13
YEK14	YEK15	YEK16	YEK17	YEK18	YEK19	YEK20
8	9	10	11	12	13	14
F28	M1	M2	M3	M4	M5	M6
RNT14	RNT15	RNT16	RNT17	RNT18	RNT19	RNT20
YEK21	YEK22	YEK23	YEK24	YEK25	YEK26	YEK27
15	16	17	18	19	20	21
M7	M8	M9	M10	M11	M12	M13
RNT21	RNT22	RNT23	RNT24	RNT25	RNT26	RNT27
YEK28	YEK29	YEK30	MAG1	MAG2	MAG3	MAG4
22	23	24	25	26	27	28
M14	M15	M16	M17	M18	M19	M20
RNT28	RNT29	RNT30	KHN1	KHN2	KHN3	KHN4
MAG5	MAG6	MAG7	MAG8	MAG9	MAG10	MAG11

KAWAIDA 529/2045-46/6281-82/2037-38						
Ashe	Ojobo	Iwa	Araorun	Ile	Itutu	Erinle
1	2	3	4	5	6	7
A18	A19	A20	A21	A22	A23	A24
PIN3	PIN4	PIN5	PIN6	PIN7	PIN8	PIN9
MIY10	MIY11	MIY12	MIY13	MIY14	MIY15	MIY16
8	9	10	11	12	13	14
A25	A26	A27	A28	A29	A30	MA1
PIN10	PIN11	PIN12	PIN13	PIN14	PIN15	PIN16
MIY17	MIY18	MIY19	MIY20	MIY21	MIY22	MIY23
15	16	17	18	19	20	21
MA2	MA3	MA4	MA5	MA6	MA7	MA8
PIN17	PIN18	PIN19	PIN20	PIN21	PIN22	PIN23
MIY24	MIY25	MIY26	MIY27	MIY28	MIY29	MIY30
22	23	24	25	26	27	28
MA9	MA10	MA11	MA12	MA13	MA14	MA15
PIN24	PIN25	PIN26	PIN27	PIN28	PIN29	PIN30
GIN1	GIN2	GIN3	GIN4	GIN5	GIN6	GIN7

UHURU 529/2045-46/6281-82/2037-38						
Ashe	Ojobo	Iwa	Araorun	Ile	Itutu	Erinle
1	2	3	4	5	6	7
M21	M22	M23	M24	M25	M26	M27
KHN5	KHN6	KHN7	KHN8	KHN9	KHN10	KHN11
MAG12	MAG13	MAG14	MAG15	MAG16	MAG17	MAG18
8	9	10	11	12	13	14
M28	M29	M30	M31	A1	A2	A3
KHN12	KHN13	KHN14	KHN15	KHN16	KHN17	KHN18
MAG19	MAG20	MAG21	MAG22	MAG23	MAG24	MAG25
15	16	17	18	19	20	21
A4	A5	A6	A7	A8	A9	A10
KHN19	KHN20	KHN21	KHN22	KHN23	KHN24	KHN25
MAG26	MAG27	MAG28	MAG29	MAG30	MIY1	MIY2
22	23	24	25	26	27	28
A11	A12	A13	A14	A15	A16	A17
KHN26	KHN27	KHN28	KHN29	KHN30	PIN1	PIN2
MIY3	MIY4	MIY5	MIY6	MIY7	MIY8	MIY9

MABABA WEUSI 529/2045-46/6281-82/2037-38						
Ashe	Ojobo	Iwa	Araorun	Ile	Itutu	Erinle
1	2	3	4	5	6	7
MA16	MA17	MA18	MA19	MA20	MA21	MA22
APP1	APP2	APP3	APP4	APP5	APP6	APP7
GIN8	GIN9	GIN10	GIN11	GIN12	GIN13	GIN14
8	9	10	11	12	13	14
MA23	MA24	MA25	MA26	MA27	MA28	MA29
APP8	APP9	APP10	APP11	APP12	APP13	APP14
GIN15	GIN16	GIN17	GIN18	GIN19	GIN20	GIN21
15	16	17	18	19	20	21
MA30	MA31	J1	J2	J3	J4	J5
APP15	APP16	APP17	APP18	APP19	APP20	APP21
GIN22	GIN23	GIN24	GIN25	GIN26	GIN27	GIN28
22	23	24	25	26	27	28
J6	J7	J8	J9	J10	J11	J12
APP22	APP23	APP24	APP25	APP26	APP27	APP28
GIN29	GIN30	SAN1	SAN2	SAN3	SAN4	SAN5

THE NU AFRIKAN CALENDAR

EKUNDU 529/2045-46/6281-82/2037-38						
Ashe	Ojobo	Iwa	Araorun	Ile	Itutu	Erinle
1	2	3	4	5	6	7
O3	O4	O5	O6	O7	O8	O9
HTH15	HTH16	HTH17	HTH18	HTH19	HTH20	HTH21
MAS23	MAS24	MAS25	MAS26	MAS27	MAS28	MAS29
8	9	10	11	12	13	14
O10	O11	O12	O13	O14	O15	O16
HTH22	HTH23	HTH24	HTH25	HTH26	HTH27	HTH28
MAS30	THi1	THi2	THi3	THi4	THi5	THi6
15	16	17	18	19	20	21
O17	O18	O19	O20	O21	O22	O23
HTH29	HTH30	KHK1	KHK2	KHK3	KHK4	KHK5
THi7	THi8	THi9	THi10	THi11	THi12	THi13
22	23	24	25	26	27	28
O24	O25	O26	O27	O28	O29	O30
KHK6	KHK7	KHK8	KHK9	KHK10	KHK11	KHK12
THi14	THi15	THi16	THi17	THi18	THi19	THi20

KIJANI 529/2045-46/6281-82/2037-38						
Ashe	Ojobo	Iwa	Araorun	Ile	Itutu	Erinle
1	2	3	4	5	6	7
N28	N29	N30	D1	D2	D3	D4
TBT11	TBT12	TBT13	TBT14	TBT15	TBT16	TBT17
HID19	HID20	HID21	HID22	HID23	HID24	HID25
8	9	10	11	12	13	14
D5	D6	D7	D8	D9	D10	D11
TBT18	TBT19	TBT20	TBT21	TBT22	TBT23	TBT24
HID26	HID27	HID28	HID29	HID30	TAH1	TAH2
15	16	17	18	19	20	21
D12	D13	D14	D15	D16	D17	D18
TBT25	TBT26	TBT27	TBT28	TBT29	TBT30	MKH1
TAH3	TAH4	TAH5	TAH6	TAH7	TAH8	TAH9
22	23	24	25	26	27	28
D19	D20	D21	D22	D23	D24	D25
MKH2	MKH3	MKH4	MKH5	MKH6	MKH7	MKH8
TAH10	TAH11	TAH12	TAH13	TAH14	TAH15	TAH16

EUSI 529/2045-46/6281-82/2037-38						
Ashe	Ojobo	Iwa	Araorun	Ile	Itutu	Erinle
1	2	3	4	5	6	7
O31	N1	N2	N3	N4	N5	N6
KHK13	KHK14	KHK15	KHK16	KHK17	KHK18	KHK19
THi21	THi22	THi23	THi24	THi25	THi26	THi27
8	9	10	11	12	13	14
N7	N8	N9	N10	N11	N12	N13
KHK20	KHK21	KHK22	KHK23	KHK24	KHK25	KHK26
THi28	THi29	THi30	HID1	HID2	HID3	HID4
15	16	17	18	19	20	21
N14	N15	N16	N17	N18	N19	N20
KHK27	KHK28	KHK29	KHK30	TBT1	TBT2	TBT3
HID5	HID6	HID7	HID8	HID9	HID10	HID11
22	23	24	25	26	27	28
N21	N22	N23	N24	N25	N26	N27
TBT4	TBT5	TBT6	TBT7	TBT8	TBT9	TBT10
HID12	HID13	HID14	HID15	HID16	HID17	HID18

KWANZA 529/2045-46/6281-82/2037-38						
Umoja	Kujichagulia	Ujima	Ujamaa	Nia	Kuumba	Imani
1	2	3	4	5	6	7
D26	D27	D28	D29	D30	D31	JA1
MKH9	MKH10	MKH11	MKH12	MKH13	MKH14	MKH15
TAH17	TAH18	TAH19	TAH20	TAH21	TAH22	TAH23
8	9	10	11	12	13	14
JA2	JA3	JA4	JA5	JA6	JA7	JA8
MKH16	MKH17	MKH18	MKH19	MKH20	MKH21	MKH22
TAH24	TAH25	TAH26	TAH27	TAH28	TAH29	TAH30
15	16	17	18	19	20	21
JA9	JA10	JA11	JA12	JA13	JA14	JA15
MKH23	MKH24	MKH25	KMH26	MKH27	MKH28	MKH29
TIR1	TIR2	TIR3	TIR4	TIR5	TIR6	TIR7
22	23	24	25	26	27	28
JA16	JA17	JA18	JA19	JA20	JA21	JA22
MKH30	AMN1	AMN2	AMN3	AMN4	AMN5	AMN6
TIR8	TIR9	TIR10	TIR11	TIR12	TIR13	TIR14

THE NU AFRIKAN CALENDAR

SHAHIDI 529/2045-46/6281-82/2037-38						
Ashe	Ojobo	Iwa	Araorun	Ile	Itutu	Erinle
1	2	3	4	5	6	7
JA23	JA24	JA25	JA26	JA27	JA28	JA29
AMN7	AMN8	AMN9	AMN10	AMN11	AMN12	AMN13
TIR15	TIR16	TIR17	TIR18	TIR19	TIR20	TIR21
8	9	10	11	12	13	14
JA30	JA31	F1	F2	F3	F4	F5
AMN14	AMN15	AMN16	AMN17	AMN18	AMN19	AMN20
TIR22	TIR23	TIR24	TIR25	TIR26	TIR27	TIR28
15	16	17	18	19	20	21
F6	F7	F8	F9	F10	F 11	F12
AMN21	AMN22	AMN23	AMN24	AMN25	AMN26	AMN28
TIR29	TIR30	YEK1	YEK2	YEK3	YEK4	YEK5
22	23	24	25	26	27	28
F13	F14	F15	F16	F17	F18	F19
AMN29	AMN30	RNT1	RNT2	RNT3	RNT4	RNT5
YEK6	YEK7	YEK8	YEK9	YEK10	YEK11	YEK12
SHAHIDI 80						
F20						
RNT6						
YEK13						

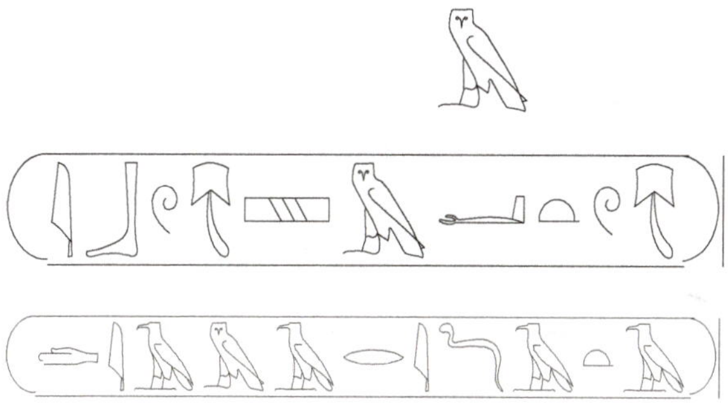

THE NU AFRIKAN CALENDAR

MAAFA 530 B.S.E./ASIENTO 2046-47 C.E.
KAM 6282-83 S.M./EC 2038-39 A.M. (ዓ.ም.)

AFU 530/2046-47/6282-83/2038-39

Ashe	Ojobo	Iwa	Araorun	Ile	Itutu	Erinle
1	2	3	4	5	6	7
F21	F22	F23	F24	F25	F26	F27
RNT7	RNT8	RNT9	RNT10	RNT11	RNT12	RNT13
YEK14	YEK15	YEK16	YEK17	YEK18	YEK19	YEK20
8	9	10	11	12	13	14
F28	M1	M2	M3	M4	M5	M6
RNT14	RNT15	RNT16	RNT17	RNT18	RNT19	RNT20
YEK21	YEK22	YEK23	YEK24	YEK25	YEK26	YEK27
15	16	17	18	19	20	21
M7	M8	M9	M10	M11	M12	M13
RNT21	RNT22	RNT23	RNT24	RNT25	RNT26	RNT27
YEK28	YEK29	YEK30	MAG1	MAG2	MAG3	MAG4
22	23	24	25	26	27	28
M14	M15	M16	M17	M18	M19	M20
RNT28	RNT29	RNT30	KHN1	KHN2	KHN3	KHN4
MAG5	MAG6	MAG7	MAG8	MAG9	MAG10	MAG11

KAWAIDA 530/2046-47/6282-83/2038-39

Ashe	Ojobo	Iwa	Araorun	Ile	Itutu	Erinle
1	2	3	4	5	6	7
A18	A19	A20	A21	A22	A23	A24
PIN3	PIN4	PIN5	PIN6	PIN7	PIN8	PIN9
MIY10	MIY11	MIY12	MIY13	MIY14	MIY15	MIY16
8	9	10	11	12	13	14
A25	A26	A27	A28	A29	A30	MA1
PIN10	PIN11	PIN12	PIN13	PIN14	PIN15	PIN16
MIY17	MIY18	MIY19	MIY20	MIY21	MIY22	MIY23
15	16	17	18	19	20	21
MA2	MA3	MA4	MA5	MA6	MA7	MA8
PIN17	PIN18	PIN19	PIN20	PIN21	PIN22	PIN23
MIY24	MIY25	MIY26	MIY27	MIY28	MIY29	MIY30
22	23	24	25	26	27	28
MA9	MA10	MA11	MA12	MA13	MA14	MA15
PIN24	PIN25	PIN26	PIN27	PIN28	PIN29	PIN30
GIN1	GIN2	GIN3	GIN4	GIN5	GIN6	GIN7

UHURU 530/2046-47/6282-83/2038-39

Ashe	Ojobo	Iwa	Araorun	Ile	Itutu	Erinle
1	2	3	4	5	6	7
M21	M22	M23	M24	M25	M26	M27
KHN5	KHN6	KHN7	KHN8	KHN9	KHN10	KHN11
MAG12	MAG13	MAG14	MAG15	MAG16	MAG17	MAG18
8	9	10	11	12	13	14
M28	M29	M30	M31	A1	A2	A3
KHN12	KHN13	KHN14	KHN15	KHN16	KHN17	KHN18
MAG19	MAG20	MAG21	MAG22	MAG23	MAG24	MAG25
15	16	17	18	19	20	21
A4	A5	A6	A7	A8	A9	A10
KHN19	KHN20	KHN21	KHN22	KHN23	KHN24	KHN25
MAG26	MAG27	MAG28	MAG29	MAG30	MIY1	MIY2
22	23	24	25	26	27	28
A11	A12	A13	A14	A15	A16	A17
KHN26	KHN27	KHN28	KHN29	KHN30	PIN1	PIN2
MIY3	MIY4	MIY5	MIY6	MIY7	MIY8	MIY9

MABABA WEUSI 530/2046-47/6282-83/2038-39

Ashe	Ojobo	Iwa	Araorun	Ile	Itutu	Erinle
1	2	3	4	5	6	7
MA16	MA17	MA18	MA19	MA20	MA21	MA22
APP1	APP2	APP3	APP4	APP5	APP6	APP7
GIN8	GIN9	GIN10	GIN11	GIN12	GIN13	GIN14
8	9	10	11	12	13	14
MA23	MA24	MA25	MA26	MA27	MA28	MA29
APP8	APP9	APP10	APP11	APP12	APP13	APP14
GIN15	GIN16	GIN17	GIN18	GIN19	GIN20	GIN21
15	16	17	18	19	20	21
MA30	MA31	J1	J2	J3	J4	J5
APP15	APP16	APP17	APP18	APP19	APP20	APP21
GIN22	GIN23	GIN24	GIN25	GIN26	GIN27	GIN28
22	23	24	25	26	27	28
J6	J7	J8	J9	J10	J11	J12
APP22	APP23	APP24	APP25	APP26	APP27	APP28
GIN29	GIN30	SAN1	SAN2	SAN3	SAN4	SAN5

THE NU AFRIKAN CALENDAR

MWANAMKE WEUSI 530/2046-47/6282-83/2038-39

Ashe	Ojobo	Iwa	Araorun	Ile	Itutu	Erinle
1	2	3	4	5	6	7
J13	J14	J15	J16	J17	J18	J19
APP29	APP30	MSR1	MSR2	MSR3	MSR4	MSR5
SAN6	SAN7	SAN8	SAN9	SAN10	SAN11	SAN12
8	9	10	11	12	13	14
J20	J21	J22	J23	J24	J25	J26
MSR6	MSR7	MSR8	MSR9	MSR10	MSR11	MSR12
SAN13	SAN14	SAN15	SAN16	SAN17	SAN18	SAN19
15	16	17	18	19	20	21
J27	J28	J29	J30	JU1	JU2	JU3
MSR13	MSR14	MSR15	MSR16	MSR17	MSR18	MSR19
SAN20	SAN21	SAN22	SAN23	SAN24	SAN25	SAN26
22	23	24	25	26	27	28
JU4	JU5	JU6	JU7	JU8	JU9	JU10
MSR20	MRS21	MSR22	MSR23	MSR24	MSR25	MSR26
SAN27	SAN28	SAN29	SAN30	HAM1	HAM2	HAM3

TAIFA WEUSI 530/2046-47/6282-83/2038-39

Ashe	Ojobo	Iwa	Araorun	Ile	Itutu	Erinle
1	2	3	4	5	6	7
AU8	AU9	AU10	AU11	AU12	AU13	AU14
THT20	THT21	THT22	THT23	THT24	THT25	THT26
NAH2	NAH3	NAH4	NAH5	NAH6	NAH7	NAH8
8	9	10	11	12	13	14
AU15	AU16	AU17	AU18	AU19	AU20	AU21
THT27	THT28	THT29	THT30	PAP1	PAP1	PAP2
NAH9	NAH10	NAH11	NAH12	NAH13	NAH14	NAH15
15	16	17	18	19	20	21
AU22	AU23	AU24	AU25	AU26	AU27	AU28
PAP3	PAP4	PAP5	PAP6	PAP7	PAP8	PAP9
NAH16	NAH17	NAH18	NAH19	NAH20	NAH21	NAH22
22	23	24	25	26	27	28
AU29	AU30	AU31	S1	S2	S3	S4
PAP10	PAP11	PAP12	PAP13	PAP14	PAP15	PAP16
NAH23	NAH24	NAH25	NAH26	NAH27	NAH28	NAH29

MTU WEUSI 530/2046-47/6282-83/2038-39

Ashe	Ojobo	Iwa	Araorun	Ile	Itutu	Erinle
1	2	3	4	5	6	7
JU11	JU12	JU13	JU14	JU15	JU16	JU17
MSR27	MSR28	MSR29	MSR30	*ASR1*	*HRU2*	*STH3*
HAM4	HAM5	HAM6	HAM7	HAM8	HAM9	HAM10
8	9	10	11	12	13	14
JU18	JU19	JU20	JU21	JU22	JU23	JU24
AST4	*NBH5*	THT1	THT2	THT3	THT4	THT5
HAM11	HAM12	HAM13	AM14	HAM15	HAM16	HAM17
15	16	17	18	19	20	21
JU25	JU26	JU27	JU28	JU29	JU30	JU31
THT6	THT7	THT8	THT9	THT10	THT11	THT12
HAM18	HAM19	HAM20	HAM21	HAM22	HAM23	HAM24
22	23	24	25	26	27	28
AU1	AU2	AU3	AU4	AU5	AU6	AU7
THT13	THT14	THT15	THT16	THT17	THT18	THT19
HAM25	HAM26	HAM27	HAM28	HAM29	HAM30	NAH1

UWEZO 530/2046-47/6282-83/2038-39

Ashe	Ojobo	Iwa	Araorun	Ile	Itutu	Erinle
1	2	3	4	5	6	7
S5	S6	S7	S8	S9	S10	S11
PAP17	PAP18	PAP19	PAP20	PAP21	PAP22	PAP23
NAH30	*PAG1*	*PAG2*	*PAG3*	*PAG4*	*PAG5*	MAS1
8	9	10	11	12	13	14
S12	S13	S14	S15	S16	S17	S18
PAP24	PAP25	PAP26	PAP27	PAP28	PAP29	PAP30
MAS2	MAS3	MAS4	MAS5	MAS6	MAS7	MAS8
15	16	17	18	19	20	21
S19	S20	S21	S22	S23	S24	S25
HTH1	HTH2	HTH3	HTH4	HTH5	HTH6	HTH7
MAS9	MAS10	MAS11	MAS12	MAS13	MAS14	MAS15
22	23	24	25	26	27	28
S26	S27	S28	S29	S30	O1	O2
HTH8	HTH9	HTH10	HTH11	HTH12	HTH13	HTH14
MAS16	MAS17	MAS18	MAS19	MAS20	MAS21	MAS22

THE NU AFRIKAN CALENDAR

EKUNDU 530/2046-47/6282-83/2038-39						
Ashe	Ojobo	Iwa	Araorun	Ile	Itutu	Erinle
1	2	3	4	5	6	7
O3	O4	O5	O6	O7	O8	O9
HTH15	HTH16	HTH17	HTH18	HTH19	HTH20	HTH21
MAS23	MAS24	MAS25	MAS26	MAS27	MAS28	MAS29
8	9	10	11	12	13	14
O10	O11	O12	O13	O14	O15	O16
HTH22	HTH23	HTH24	HTH25	HTH26	HTH27	HTH28
MAS30	THi1	THi2	THi3	THi4	THi5	THi6
15	16	17	18	19	20	21
O17	O18	O19	O20	O21	O22	O23
HTH29	HTH30	KHK1	KHK2	KHK3	KHK4	KHK5
THi7	THi8	THi9	THi10	THi11	THi12	THi13
22	23	24	25	26	27	28
O24	O25	O26	O27	O28	O29	O30
KHK6	KHK7	KHK8	KHK9	KHK10	KHK11	KHK12
THi14	THi15	THi16	THi17	THi18	THi19	THi20

KIJANI 530/2046-47/6282-83/2038-39						
Ashe	Ojobo	Iwa	Araorun	Ile	Itutu	Erinle
1	2	3	4	5	6	7
N28	N29	N30	D1	D2	D3	D4
TBT11	TBT12	TBT13	TBT14	TBT15	TBT16	TBT17
HID19	HID20	HID21	HID22	HID23	HID24	HID25
8	9	10	11	12	13	14
D5	D6	D7	D8	D9	D10	D11
TBT18	TBT19	TBT20	TBT21	TBT22	TBT23	TBT24
HID26	HID27	HID28	HID29	HID30	TAH1	TAH2
15	16	17	18	19	20	21
D12	D13	D14	D15	D16	D17	D18
TBT25	TBT26	TBT27	TBT28	TBT29	TBT30	MKH1
TAH3	TAH4	TAH5	TAH6	TAH7	TAH8	TAH9
22	23	24	25	26	27	28
D19	D20	D21	D22	D23	D24	D25
MKH2	MKH3	MKH4	MKH5	MKH6	MKH7	MKH8
TAH10	TAH11	TAH12	TAH13	TAH14	TAH15	TAH16

EUSI 530/2046-47/6282-83/2038-39						
Ashe	Ojobo	Iwa	Araorun	Ile	Itutu	Erinle
1	2	3	4	5	6	7
O31	N1	N2	N3	N4	N5	N6
KHK13	KHK14	KHK15	KHK16	KHK17	KHK18	KHK19
THi21	THi22	THi23	THi24	THi25	THi26	THi27
8	9	10	11	12	13	14
N7	N8	N9	N10	N11	N12	N13
KHK20	KHK21	KHK22	KHK23	KHK24	KHK25	KHK26
THi28	THi29	THi30	HID1	HID2	HID3	HID4
15	16	17	18	19	20	21
N14	N15	N16	N17	N18	N19	N20
KHK27	KHK28	KHK29	KHK30	TBT1	TBT2	TBT3
HID5	HID6	HID7	HID8	HID9	HID10	HID11
22	23	24	25	26	27	28
N21	N22	N23	N24	N25	N26	N27
TBT4	TBT5	TBT6	TBT7	TBT8	TBT9	TBT10
HID12	HID13	HID14	HID15	HID16	HID17	HID18

KWANZA 530/2046-47/6282-83/2038-39						
Umoja	Kujichagulia	Ujima	Ujamaa	Nia	Kuumba	Imani
1	2	3	4	5	6	7
D26	D27	D28	D29	D30	D31	JA1
MKH9	MKH10	MKH11	MKH12	MKH13	MKH14	MKH15
TAH17	TAH18	TAH19	TAH20	TAH21	TAH22	TAH23
8	9	10	11	12	13	14
JA2	JA3	JA4	JA5	JA6	JA7	JA8
MKH16	MKH17	MKH18	MKH19	MKH20	MKH21	MKH22
TAH24	TAH25	TAH26	TAH27	TAH28	TAH29	TAH30
15	16	17	18	19	20	21
JA9	JA10	JA11	JA12	JA13	JA14	JA15
MKH23	MKH24	MKH25	KMH26	MKH27	MKH28	MKH29
TIR1	TIR2	TIR3	TIR4	TIR5	TIR6	TIR7
22	23	24	25	26	27	28
JA16	JA17	JA18	JA19	JA20	JA21	JA22
MKH30	AMN1	AMN2	AMN3	AMN4	AMN5	AMN6
TIR8	TIR9	TIR10	TIR11	TIR12	TIR13	TIR14

THE NU AFRIKAN CALENDAR

SHAHIDI 530/2046-47/6282-83/2038-39						
Ashe	Ojobo	Iwa	Araorun	Ile	Itutu	Erinle
1	2	3	4	5	6	7
JA23	JA24	JA25	JA26	JA27	JA28	JA29
AMN7	AMN8	AMN9	AMN10	AMN11	AMN12	AMN13
TIR15	TIR16	TIR17	TIR18	TIR19	TIR20	TIR21
8	9	10	11	12	13	14
JA30	JA31	F1	F2	F3	F4	F5
AMN14	AMN15	AMN16	AMN17	AMN18	AMN19	AMN20
TIR22	TIR23	TIR24	TIR25	TIR26	TIR27	TIR28
15	16	17	18	19	20	21
F6	F7	F8	F9	F10	F 11	F12
AMN21	AMN22	AMN23	AMN24	AMN25	AMN26	AMN28
TIR29	TIR30	YEK1	YEK2	YEK3	YEK4	YEK5
22	23	24	25	26	27	28
F13	F14	F15	F16	F17	F18	F19
AMN29	AMN30	RNT1	RNT2	RNT3	RNT4	RNT5
YEK6	YEK7	YEK8	YEK9	YEK10	YEK11	YEK12
SHAHIDI 81						
F20						
RNT6						
YEK13						

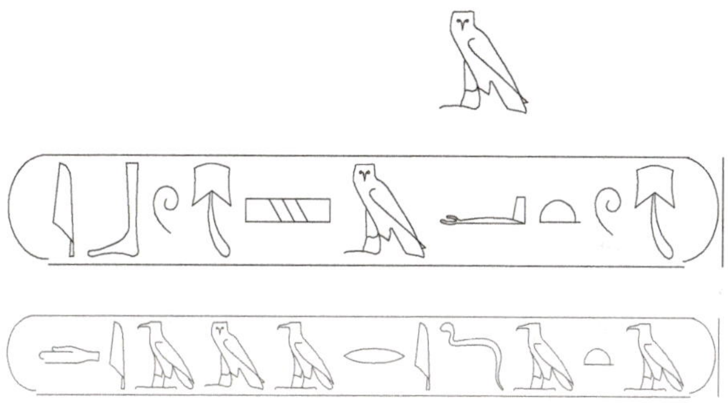

MAAFA 531 B.S.E./ASIENTO 2047-48 C.E./
KAM 6283-84 S.M./EC 2039-40 A.M. (ዓ.ም.)

AFU 531/2047-48/6283-84/2039-40

Ashe	Ojobo	Iwa	Araorun	Ile	Itutu	Erinle
1	2	3	4	5	6	7
F21	F22	F23	F24	F25	F26	F27
RNT7	RNT8	RNT9	RNT10	RNT11	RNT12	RNT13
YEK14	YEK15	YEK16	YEK17	YEK18	YEK19	YEK20
8	9	10	11	12	13	14
F28	M1	M2	M3	M4	M5	M6
RNT14	RNT15	RNT16	RNT17	RNT18	RNT19	RNT20
YEK21	YEK22	YEK23	YEK24	YEK25	YEK26	YEK27
15	16	17	18	19	20	21
M7	M8	M9	M10	M11	M12	M13
RNT21	RNT22	RNT23	RNT24	RNT25	RNT26	RNT27
YEK28	YEK 29	YEK30	MAG1	MAG2	MAG3	MAG4
22	23	24	25	26	27	28
M14	M15	M16	M17	M18	M19	M20
RNT28	RNT29	RNT30	KHN1	KHN2	KHN3	KHN4
MAG5	MAG6	MAG7	MAG8	MAG9	MAG10	MAG11

UHURU 531/2047-48/6283-84/2039-40

Ashe	Ojobo	Iwa	Araorun	Ile	Itutu	Erinle
1	2	3	4	5	6	7
M21	M22	M23	M24	M25	M26	M27
KHN5	KHN6	KHN7	KHN8	KHN9	KHN10	KHN11
MAG12	MAG13	MAG14	MAG15	MAG16	MAG17	MAG18
8	9	10	11	12	13	14
M28	M29	M30	M31	A1	A2	A3
KHN12	KHN13	KHN14	KHN15	KHN16	KHN17	KHN18
MAG19	MAG20	MAG21	MAG22	MAG23	MAG24	MAG25
15	16	17	18	19	20	21
A4	A5	A6	A7	A8	A9	A10
KHN19	KHN20	KHN21	KHN22	KHN23	KHN24	KHN25
MAG26	MAG27	MAG28	MAG29	MAG30	MIY1	MIY2
22	23	24	25	26	27	28
A11	A12	A13	A14	A15	A16	A17
KHN26	KHN27	KHN28	KHN29	KHN30	PIN1	PIN2
MIY3	MIY4	MIY5	MIY6	MIY7	MIY8	MIY9

KAWAIDA 531/2047-48/6283-84/2039-40

Ashe	Ojobo	Iwa	Araorun	Ile	Itutu	Erinle
1	2	3	4	5	6	7
A18	A19	A20	A21	A22	A23	A24
PIN3	PIN4	PIN5	PIN6	PIN7	PIN8	PIN9
MIY10	MIY11	MIY12	MIY13	MIY14	MIY15	MIY16
8	9	10	11	12	13	14
A25	A26	A27	A28	A29	A30	MA1
PIN10	PIN11	PIN12	PIN13	PIN14	PIN15	PIN16
MIY17	MIY18	MIY19	MIY20	MIY21	MIY22	MIY23
15	16	17	18	19	20	21
MA2	MA3	MA4	MA5	MA6	MA7	MA8
PIN17	PIN18	PIN19	PIN20	PIN21	PIN22	PIN23
MIY24	MIY25	MIY26	MIY27	MIY28	MIY29	MIY30
22	23	24	25	26	27	28
MA9	MA10	MA11	MA12	MA13	MA14	MA15
PIN24	PIN25	PIN26	PIN27	PIN28	PIN29	PIN30
GIN1	GIN2	GIN3	GIN4	GIN5	GIN6	GIN7

MABABA WEUSI 531/2047-48/6283-84/2039-40

Ashe	Ojobo	Iwa	Araorun	Ile	Itutu	Erinle
1	2	3	4	5	6	7
MA16	MA17	MA18	MA19	MA20	MA21	MA22
APP1	APP2	APP3	APP4	APP5	APP6	APP7
GIN8	GIN9	GIN10	GIN11	GIN12	GIN13	GIN14
8	9	10	11	12	13	14
MA23	MA24	MA25	MA26	MA27	MA28	MA29
APP8	APP9	APP10	APP11	APP12	APP13	APP14
GIN15	GIN16	GIN17	GIN18	GIN19	GIN20	GIN21
15	16	17	18	19	20	21
MA30	MA31	J1	J2	J3	J4	J5
APP15	APP16	APP17	APP18	APP19	APP20	APP21
GIN22	GIN23	GIN24	GIN25	GIN26	GIN27	GIN28
22	23	24	25	26	27	28
J6	J7	J8	J9	J10	J11	J12
APP22	APP23	APP24	APP25	APP26	APP27	APP28
GIN29	GIN30	SAN1	SAN2	SAN3	SAN4	SAN5

THE NU AFRIKAN CALENDAR

MWANAMKE WEUSI 531/2047-48/6283-84/2039-40

Ashe	Ojobo	Iwa	Araorun	Ile	Itutu	Erinle
1	2	3	4	5	6	7
J13	J14	J15	J16	J17	J18	J19
APP29	APP30	MSR1	MSR2	MSR3	MSR4	MSR5
SAN6	SAN7	SAN8	SAN9	SAN10	SAN11	SAN12
8	9	10	11	12	13	14
J20	J21	J22	J23	J24	J25	J26
MSR6	MSR7	MSR8	MSR9	MSR10	MSR11	MSR12
SAN13	SAN14	SAN15	SAN16	SAN17	SAN18	SAN19
15	16	17	18	19	20	21
J27	J28	J29	J30	JU1	JU2	JU3
MSR13	MSR14	MSR15	MSR16	MSR17	MSR18	MSR19
SAN20	SAN21	SAN22	SAN23	SAN24	SAN25	SAN26
22	23	24	25	26	27	28
JU4	JU5	JU6	JU7	JU8	JU9	JU10
MSR20	MRS21	MSR22	MSR23	MSR24	MSR25	MSR26
SAN27	SAN28	SAN29	SAN30	HAM1	HAM2	HAM3

TAIFA WEUSI 531/2047-48/6283-84/2039-40

Ashe	Ojobo	Iwa	Araorun	Ile	Itutu	Erinle
1	2	3	4	5	6	7
AU8	AU9	AU10	AU11	AU12	AU13	AU14
THT20	THT21	THT22	THT23	THT24	THT25	THT26
NAH2	NAH3	NAH4	NAH5	NAH6	NAH7	NAH8
8	9	10	11	12	13	14
AU15	AU16	AU17	AU18	AU19	AU20	AU21
THT27	THT28	THT29	THT30	PAP1	PAP1	PAP2
NAH9	NAH10	NAH11	NAH12	NAH13	NAH14	NAH15
15	16	17	18	19	20	21
AU22	AU23	AU24	AU25	AU26	AU27	AU28
PAP3	PAP4	PAP5	PAP6	PAP7	PAP8	PAP9
NAH16	NAH17	NAH18	NAH19	NAH20	NAH21	NAH22
22	23	24	25	26	27	28
AU29	AU30	AU31	S1	S2	S3	S4
PAP10	PAP11	PAP12	PAP13	PAP14	PAP15	PAP16
NAH23	NAH24	NAH25	NAH26	NAH27	NAH28	NAH29

MTU WEUSI 531/2047-48/6283-84/2039-40

Ashe	Ojobo	Iwa	Araorun	Ile	Itutu	Erinle
1	2	3	4	5	6	7
JU11	JU12	JU13	JU14	JU15	JU16	JU17
MSR27	MSR28	MSR29	MSR30	*ASR1*	*HRU2*	*STH3*
HAM4	HAM5	HAM6	HAM7	HAM8	HAM9	HAM10
8	9	10	11	12	13	14
JU18	JU19	JU20	JU21	JU22	JU23	JU24
AST4	*NBH5*	THT1	THT2	THT3	THT4	THT5
HAM11	HAM12	HAM13	HAM14	HAM15	HAM16	HAM17
15	16	17	18	19	20	21
JU25	JU26	JU27	JU28	JU29	JU30	JU31
THT6	THT7	THT8	THT9	THT10	THT11	THT12
HAM18	HAM19	HAM20	HAM21	HAM22	HAM23	HAM24
22	23	24	25	26	27	28
AU1	AU2	AU3	AU4	AU5	AU6	AU7
THT13	THT14	THT15	THT16	THT17	THT18	THT19
HAM25	HAM26	HAM27	HAM28	HAM29	HAM30	NAH1

UWEZO 531/2047-48/6283-84/2039-40

Ashe	Ojobo	Iwa	Araorun	Ile	Itutu	Erinle
1	2	3	4	5	6	7
S5	S6	S7	S8	S9	S10	S11
PAP17	PAP18	PAP19	PAP20	PAP21	PAP22	PAP23
NAH30	*QUA1*	*QUA2*	*QUA3*	*QUA4*	*QUA5*	*QUA6*
8	9	10	11	12	13	14
S12	S13	S14	S15	S16	S17	S18
PAP24	PAP25	PAP26	PAP27	PAP28	PAP29	PAP30
MAS1	MAS2	MAS3	MAS4	MAS5	MAS6	MAS7
15	16	17	18	19	20	21
S19	S20	S21	S22	S23	S24	S25
HTH1	HTH2	HTH3	HTH4	HTH5	HTH6	HTH7
MAS8	MAS9	MAS10	MAS11	MAS12	MAS13	MAS14
22	23	24	25	26	27	28
S26	S27	S28	S29	S30	O1	O2
HTH8	HTH9	HTH10	HTH11	HTH12	HTH13	HTH14
MAS15	MAS16	MAS17	MAS18	MAS19	MAS20	MAS21

THE NU AFRIKAN CALENDAR

EKUNDU 531/2047-48/6283-84/2039-40						
Ashe	Ojobo	Iwa	Araorun	Ile	Itutu	Erinle
1	2	3	4	5	6	7
O3	O4	O5	O6	O7	O8	O9
HTH15	HTH16	HTH17	HTH18	HTH19	HTH20	HTH21
MAS22	MAS23	MAS 24	MAS25	MAS26	MAS27	MAS28
8	9	10	11	12	13	14
O10	O11	O12	O13	O14	O15	O16
HTH22	HTH23	HTH24	HTH25	HTH26	HTH27	HTH28
MAS29	MAS30	THi1	THi2	THi3	THi4	THi5
15	16	17	18	19	20	21
O17	O18	O19	O20	O21	O22	O23
HTH29	HTH30	KHK1	KHK2	KHK3	KHK4	KHK5
THi6	THi7	THi8	THi9	THi10	THi11	THi12
22	23	24	25	26	27	28
O24	O25	O26	O27	O28	O29	O30
KHK6	KHK7	KHK8	KHK9	KHK10	KHK11	KHK12
THi13	THi14	THi15	THi16	THi17	THi18	THi19

KIJANI 531/2047-48/6283-84/2039-40						
Ashe	Ojobo	Iwa	Araorun	Ile	Itutu	Erinle
1	2	3	4	5	6	7
N28	N29	N30	D1	D2	D3	D4
TBT11	TBT12	TBT13	TBT14	TBT15	TBT16	TBT17
HID18	HID19	HID20	HID21	HID22	HID23	HID24
8	9	10	11	12	13	14
D5	D6	D7	D8	D9	D10	D11
TBT18	TBT19	TBT20	TBT21	TBT22	TBT23	TBT24
HID25	HID26	HID27	HID28	HID29	HID30	TAH1
15	16	17	18	19	20	21
D12	D13	D14	D15	D16	D17	D18
TBT25	TBT26	TBT27	TBT28	TBT29	TBT30	MKH1
TAH2	TAH3	TAH4	TAH5	TAH6	TAH7	TAH8
22	23	24	25	26	27	28
D19	D20	D21	D22	D23	D24	D25
MKH2	MKH3	MKH4	MKH5	MKH6	MKH7	MKH8
TAH9	TAH10	TAH11	TAH12	TAH13	TAH14	TAH15

EUSI 531/2047-48/6283-84/2039-40						
Ashe	Ojobo	Iwa	Araorun	Ile	Itutu	Erinle
1	2	3	4	5	6	7
O31	N1	N2	N3	N4	N5	N6
KHK13	KHK14	KHK15	KHK16	KHK17	KHK18	KHK19
THi20	THi21	THi22	THi23	THi24	THi25	THi26
8	9	10	11	12	13	14
N7	N8	N9	N10	N11	N12	N13
KHK20	KHK21	KHK22	KHK23	KHK24	KHK25	KHK26
THi27	THi28	THi29	THi30	HID1	HID2	HID3
15	16	17	18	19	20	21
N14	N15	N16	N17	N18	N19	N20
KHK27	KHK28	KHK29	KHK30	TBT1	TBT2	TBT3
HID4	HID5	HID6	HID7	HID8	HID9	HID10
22	23	24	25	26	27	28
N21	N22	N23	N24	N25	N26	N27
TBT4	TBT5	TBT6	TBT7	TBT8	TBT9	TBT10
HID11	HID12	HID13	HID14	HID15	HID16	HID17

KWANZA 531/2047-48/6283-84/2039-40						
Umoja	Kujichagulia	Ujima	Ujamaa	Nia	Kuumba	Imani
1	2	3	4	5	6	7
D26	D27	D28	D29	D30	D31	JA1
MKH9	MKH10	MKH11	MKH12	MKH13	MKH14	MKH1
TAH16	TAH17	TAH18	TAH19	TAH20	TAH21	TAH2
8	9	10	11	12	13	14
JA2	JA3	JA4	JA5	JA6	JA7	JA
MKH16	MKH17	MKH18	MKH19	MKH20	MKH21	MKH2
TAH23	TAH24	TAH25	TAH26	TAH27	TAH28	TAH29
15	16	17	18	19	20	21
JA9	JA10	JA11	JA12	JA13	JA14	JA1
MKH23	MKH24	MKH25	KMH26	MKH27	MKH28	MKH2
TAH30	TIR1	TIR2	TIR3	TIR4	TIR5	TIR6
22	23	24	25	26	27	28
JA16	JA17	JA18	JA19	JA20	JA21	JA2
MKH30	AMN1	AMN2	AMN3	AMN4	AMN5	AMN
TIR7	TIR8	TIR9	TIR10	TIR11	TIR12	TIR13

THE NU AFRIKAN CALENDAR

SHAHIDI 531/2047-48/6283-84/2039-40						
Ashe	Ojobo	Iwa	Araorun	Ile	Itutu	Erinle
1	2	3	4	5	6	7
JA23	JA24	JA25	JA26	JA27	JA28	JA29
AMN7	AMN8	AMN9	AMN10	AMN11	AMN12	AMN13
TIR14	TIR15	TIR16	TIR17	TIR18	TIR19	TIR20
8	9	10	11	12	13	14
JA30	JA31	F1	F2	F3	F4	F5
AMN14	AMN15	AMN16	AMN17	AMN18	AMN19	AMN20
TIR21	TIR22	TIR23	TIR24	TIR25	TIR26	TIR27
15	16	17	18	19	20	21
F6	F7	F8	F9	F10	F 11	F12
AMN21	AMN22	AMN23	AMN24	AMN25	AMN26	AMN28
TIR28	TIR29	TIR30	YEK1	YEK2	YEK3	YEK4
22	23	24	25	26	27	28
F13	F14	F15	F16	F17	F18	F19
AMN29	AMN30	RNT1	RNT2	RNT3	RNT4	RNT5
YEK5	YEK6	YEK7	YEK8	YEK9	YEK10	YEK11
SHAHIDI 82						
F20						
RNT6						
YEK12						

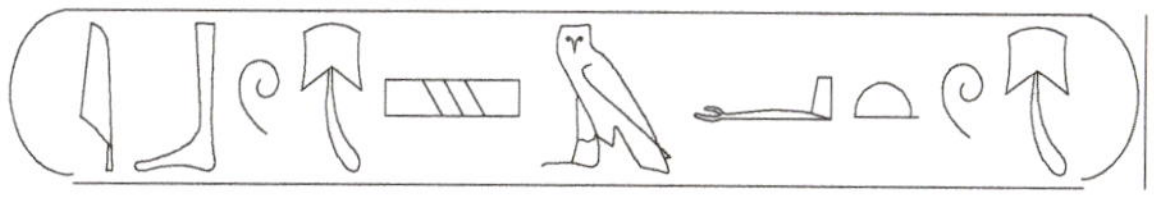

THE NU AFRIKAN CALENDAR

MAAFA 532 B.S.E./ASIENTO 2048-49 C.E./
KAM 6284-85 S.M./EC 2040-41 A.M. (ዓ.ም.)

AFU 532/2048-49/6284-85/2040-41

Ashe	Ojobo	Iwa	Araorun	Ile	Itutu	Erinle
1	2	3	4	5	6	7
F21	F22	F23	F24	F25	F26	F27
RNT7	RNT8	RNT9	RNT10	RNT11	RNT12	RNT13
YEK13	YEK14	YEK15	YEK16	YEK17	YEK18	YEK19
8	9	10	11	12	13	14
F28	F29	M1	M2	M3	M4	M5
RNT14	RNT15	RNT16	RNT17	RNT18	RNT19	RNT20
YEK 20	YEK21	YEK22	YEK23	YEK24	YEK25	YEK26
15	16	17	18	19	20	21
M6	M7	M8	M9	M10	M11	M12
RNT21	RNT22	RNT23	RNT24	RNT25	RNT26	RNT27
YEK 27	YEK28	YEK29	YEK30	MAG1	MAG2	MAG3
22	23	24	25	26	27	28
M13	M14	M15	M16	M17	M18	M19
RNT28	RNT29	RNT30	KHN1	KHN2	KHN3	KHN4
MAG4	MAG5	MAG6	MAG7	MAG8	MAG9	MAG10
SHARIFU 130						
M20						
KHN5						
MAG11						

KAWAIDA 532/2048-49/6284-85/2040-41

Ashe	Ojobo	Iwa	Araorun	Ile	Itutu	Erinle
1	2	3	4	5	6	7
A18	A19	A20	A21	A22	A23	A24
PIN4	PIN5	PIN6	PIN7	PIN8	PIN9	PIN10
MIY10	MIY11	MIY12	MIY13	MIY14	MIY15	MIY16
8	9	10	11	12	13	14
A25	A26	A27	A28	A29	A30	MA1
PIN11	PIN12	PIN13	PIN14	PIN15	PIN16	PIN17
MIY17	MIY18	MIY19	MIY20	MIY21	MIY22	MIY23
15	16	17	18	19	20	21
MA2	MA3	MA4	MA5	MA6	MA7	MA8
PIN18	PIN19	PIN20	PIN21	PIN22	PIN23	PIN24
MIY24	MIY25	MIY26	MIY27	MIY28	MIY29	MIY30
22	23	24	25	26	27	28
MA9	MA10	MA11	MA12	MA13	MA14	MA15
PIN25	PIN26	PIN27	PIN28	PIN29	PIN30	APP1
GIN1	GIN2	GIN3	GIN4	GIN5	GIN6	GIN7

UHURU 532/2048-49/6284-85/2040-41

Ashe	Ojobo	Iwa	Araorun	Ile	Itutu	Erinle
1	2	3	4	5	6	7
M21	M22	M23	M24	M25	M26	M27
KHN6	KHN7	KHN8	KHN9	KHN10	KHN11	KHN12
MAG12	MAG13	MAG14	MAG15	MAG16	MAG17	MAG18
8	9	10	11	12	13	14
M28	M29	M30	M31	A1	A2	A3
KHN13	KHN14	KHN15	KHN16	KHN17	KHN18	KHN19
MAG19	MAG20	MAG21	MAG22	MAG23	MAG24	MAG25
15	16	17	18	19	20	21
A4	A5	A6	A7	A8	A9	A10
KHN20	KHN21	KHN222	KHN23	KHN24	KHN25	KHN26
MAG26	MAG27	MAG28	MAG29	MAG30	MIY1	MIY2
22	23	24	25	26	27	28
A11	A12	A13	A14	A15	A16	A17
KHN27	KHN28	KHN29	KHN30	PIN1	PIN2	PIN3
MIY3	MIY4	MIY5	MIY6	MIY7	MIY8	MIY9

MABABA WEUSI 532/2048-49/6284-85/2040-41

Ashe	Ojobo	Iwa	Araorun	Ile	Itutu	Erinle
1	2	3	4	5	6	7
MA16	MA17	MA18	MA19	MA20	MA21	MA22
APP2	APP3	APP4	APP5	APP6	APP7	APP8
GIN8	GIN9	GIN10	GIN11	GIN12	GIN13	GIN14
8	9	10	11	12	13	14
MA23	MA24	MA25	MA26	MA27	MA28	MA29
APP9	APP10	APP11	APP12	APP13	APP14	APP15
GIN15	GIN16	GIN17	GIN18	GIN19	GIN20	GIN21
15	16	17	18	19	20	21
MA30	MA31	J1	J2	J3	J4	J5
APP16	APP17	APP18	APP19	APP20	APP21	APP22
GIN22	GIN23	GIN24	GIN25	GIN26	GIN27	GIN28
22	23	24	25	26	27	28
J6	J7	J8	J9	J10	J11	J12
APP23	APP24	APP25	APP26	APP27	APP28	APP29
GIN29	GIN30	SAN1	SAN2	SAN3	SAN4	SAN5

THE NU AFRIKAN CALENDAR

MWANAMKE WEUSI 532/2048-49/6284-85/2040-41						
Ashe	Ojobo	Iwa	Araorun	Ile	Itutu	Erinle
1	2	3	4	5	6	7
J13	J14	J15	J16	J17	J18	J19
APP30	MSR1	MSR2	MSR3	MSR4	MSR5	MSR6
SAN6	SAN7	SAN8	SAN9	SAN10	SAN11	SAN12
8	9	10	11	12	13	14
J20	J21	J22	J23	J24	J25	J26
MSR7	MSR8	MSR9	MSR10	MSR11	MSR12	MSR13
SAN13	SAN14	SAN15	SAN16	SAN17	SAN18	SAN19
15	16	17	18	19	20	21
J27	J28	J29	J30	JU1	JU2	JU3
MSR14	MSR15	MSR16	MSR17	MSR18	MSR19	MSR20
SAN20	SAN21	SAN22	SAN23	SAN24	SAN25	SAN26
22	23	24	25	26	27	28
JU4	JU5	JU6	JU7	JU8	JU9	JU10
MSR21	MRS22	MSR23	MSR24	MSR25	MSR26	MSR27
SAN27	SAN28	SAN29	SAN30	HAM1	HAM2	HAM3

TAIFA WEUSI 532/2048-49/6284-85/2040-41						
Ashe	Ojobo	Iwa	Araorun	Ile	Itutu	Erinle
1	2	3	4	5	6	7
AU8	AU9	AU10	AU11	AU12	AU13	AU14
THT20	THT21	THT22	THT23	THT24	THT25	THT26
NAH2	NAH3	NAH4	NAH5	NAH6	NAH7	NAH8
8	9	10	11	12	13	14
AU15	AU16	AU17	AU18	AU19	AU20	AU21
THT27	THT28	THT29	THT30	PAP1	PAP1	PAP2
NAH9	NAH10	NAH11	NAH12	NAH13	NAH14	NAH15
15	16	17	18	19	20	21
AU22	AU23	AU24	AU25	AU26	AU27	AU28
PAP3	PAP4	PAP5	PAP6	PAP7	PAP8	PAP9
NAH16	NAH17	NAH18	NAH19	NAH20	NAH21	NAH22
22	23	24	25	26	27	28
AU29	AU30	AU31	S1	S2	S3	S4
PAP10	PAP11	PAP12	PAP13	PAP14	PAP15	PAP16
NAH23	NAH24	NAH25	NAH26	NAH27	NAH28	NAH29

MTU WEUSI 532/2048-49/6284-85/2040-41						
Ashe	Ojobo	Iwa	Araorun	Ile	Itutu	Erinle
1	2	3	4	5	6	7
JU11	JU12	JU13	JU14	JU15	JU16	JU17
MSR28	MSR29	MSR30	*ASR1*	*HRU2*	*STH3*	*AST4*
HAM4	HAM5	HAM6	HAM7	HAM8	HAM9	HAM10
8	9	10	11	12	13	14
JU18	JU19	JU20	JU21	JU22	JU23	JU24
NBH5	*ANU6*	THT1	THT2	THT3	THT4	THT5
HAM11	HAM12	HAM13	HAM14	HAM15	HAM16	HAM17
15	16	17	18	19	20	21
JU25	JU26	JU27	JU28	JU29	JU30	JU31
THT6	THT7	THT8	THT9	THT10	THT11	THT12
HAM18	HAM19	HAM20	HAM21	HAM22	HAM23	HAM24
22	23	24	25	26	27	28
AU1	AU2	AU3	AU4	AU5	AU6	AU7
THT13	THT14	THT15	THT16	THT17	THT18	THT19
HAM25	HAM26	HAM27	HAM28	HAM29	HAM30	NAH1

UWEZO 532/2048-49/6284-85/2040-41						
Ashe	Ojobo	Iwa	Araorun	Ile	Itutu	Erinle
1	2	3	4	5	6	7
S5	S6	S7	S8	S9	S10	S11
PAP17	PAP18	PAP19	PAP20	PAP21	PAP22	PAP23
NAH30	*PAG1*	*PAG2*	*PAG3*	*PAG4*	*PAG5*	MAS1
8	9	10	11	12	13	14
S12	S13	S14	S15	S16	S17	S18
PAP24	PAP25	PAP26	PAP27	PAP28	PAP29	PAP30
MAS2	MAS3	MAS4	MAS5	MAS6	MAS7	MAS8
15	16	17	18	19	20	21
S19	S20	S21	S22	S23	S24	S25
HTH1	HTH2	HTH3	HTH4	HTH5	HTH6	HTH7
MAS9	MAS10	MAS11	MAS12	MAS13	MAS14	MAS15
22	23	24	25	26	27	28
S26	S27	S28	S29	S30	O1	O2
HTH8	HTH9	HTH10	HTH11	HTH12	HTH13	HTH14
MAS16	MAS17	MAS18	MAS19	MAS20	MAS21	MAS22

THE NU AFRIKAN CALENDAR

EKUNDU 532/2048-49/6284-85/2040-41						
Ashe	Ojobo	Iwa	Araorun	Ile	Itutu	Erinle
1	2	3	4	5	6	7
O3	O4	O5	O6	O7	O8	O9
HTH15	HTH16	HTH17	HTH18	HTH19	HTH20	HTH21
MAS23	MAS24	MAS25	MAS26	MAS27	MAS28	MAS29
8	9	10	11	12	13	14
O10	O11	O12	O13	O14	O15	O16
HTH22	HTH23	HTH24	HTH25	HTH26	HTH27	HTH28
MAS30	THi1	THi2	THi3	THi4	THi5	THi6
15	16	17	18	19	20	21
O17	O18	O19	O20	O21	O22	O23
HTH29	HTH30	KHK1	KHK2	KHK3	KHK4	KHK5
THi7	THi8	This9	THi10	THi11	THi12	THi13
22	23	24	25	26	27	28
O24	O25	O26	O27	O28	O29	O30
KHK6	KHK7	KHK8	KHK9	KHK10	KHK11	KHK12
Thi14	THi15	THi16	THi17	THi18	THi19	THi20

KIJANI 532/2048-49/6284-85/2040-41						
Ashe	Ojobo	Iwa	Araorun	Ile	Itutu	Erinle
1	2	3	4	5	6	7
N28	N29	N30	D1	D2	D3	D4
TBT11	TBT12	TBT13	TBT14	TBT15	TBT16	TBT17
HID19	HID20	HID21	HID22	HID23	HID24	HID25
8	9	10	11	12	13	14
D5	D6	D7	D8	D9	D10	D11
TBT18	TBT19	TBT20	TBT21	TBT22	TBT23	TBT24
HID26	HID27	HID28	HID29	HID30	TAH1	TAH2
15	16	17	18	19	20	21
D12	D13	D14	D15	D16	D17	D18
TBT25	TBT26	TBT27	TBT28	TBT29	TBT30	MKH1
TAH3	TAH4	TAH5	TAH6	TAH7	TAH8	TAH9
22	23	24	25	26	27	28
D19	D20	D21	D22	D23	D24	D25
MKH2	MKH3	MKH4	MKH5	MKH6	MKH7	MKH8
TAH10	TAH11	TAH12	TAH13	TAH14	TAH15	TAH16

EUSI 532/2048-49/6284-85/2040-41						
Ashe	Ojobo	Iwa	Araorun	Ile	Itutu	Erinle
1	2	3	4	5	6	7
O31	N1	N2	N3	N4	N5	N6
KHK13	KHK14	KHK15	KHK16	KHK17	KHK18	KHK19
THi21	THi22	THi23	THi24	THi25	THi26	THi27
8	9	10	11	12	13	14
N7	N8	N9	N10	N11	N12	N13
KHK20	KHK21	KHK22	KHK23	KHK24	KHK25	KHK26
THi28	THi29	THi30	HID1	HID2	HID3	HID4
15	16	17	18	19	20	21
N14	N15	N16	N17	N18	N19	N20
KHK27	KHK28	KHK29	KHK30	TBT1	TBT2	TBT3
HID5	HID6	HID7	HID8	HID9	HID10	HID11
22	23	24	25	26	27	28
N21	N22	N23	N24	N25	N26	N27
TBT4	TBT5	TBT6	TBT7	TBT8	TBT9	TBT10
HID12	HID13	HID14	HID15	HID16	HID17	HID18

KWANZA 532/2048-49/6284-85/2040-41						
Umoja	Kujichagulia	Ujima	Ujamaa	Nia	Kuumba	Imani
1	2	3	4	5	6	7
D26	D27	D28	D29	D30	D31	JA1
MKH9	MKH10	MKH11	MKH12	MKH13	MKH14	MKH1
TAH17	TAH18	TAH19	TAH20	TAH21	TAH22	TAH2
8	9	10	11	12	13	14
JA2	JA3	JA4	JA5	JA6	JA7	JA
MKH16	MKH17	MKH18	MKH19	MKH20	MKH21	MKH2
TAH24	TAH25	TAH26	TAH27	TAH28	TAH29	TAH3
15	16	17	18	19	20	21
JA9	JA10	JA11	JA12	JA13	JA14	JA1
MKH23	MKH24	MKH25	KMH26	MKH27	MKH28	MKH2
TIR1	TIR2	TIR3	TIR4	TIR5	TIR6	TIR7
22	23	24	25	26	27	28
JA16	JA17	JA18	JA19	JA20	JA21	JA2
MKH30	AMN1	AMN2	AMN3	AMN4	AMN5	AMN
TIR8	TIR9	TIR10	TIR11	TIR12	TIR13	TIR1

THE NU AFRIKAN CALENDAR

SHAHIDI 532/2048-49/6284-85/2040-41						
Ashe	Ojobo	Iwa	Araorun	Ile	Itutu	Erinle
1	2	3	4	5	6	7
JA23	JA24	JA25	JA26	JA27	JA28	JA29
AMN7	AMN8	AMN9	AMN10	AMN11	AMN12	AMN13
TIR15	TIR16	TIR17	TIR18	TIR19	TIR20	TIR21
8	9	10	11	12	13	14
JA30	JA31	F1	F2	F3	F4	F5
AMN14	AMN15	AMN16	AMN17	AMN18	AMN19	AMN20
TIR22	TIR23	TIR24	TIR25	TIR26	TIR27	TIR28
15	16	17	18	19	20	21
F6	F7	F8	F9	F10	F 11	F12
AMN21	AMN22	AMN23	AMN24	AMN25	AMN26	AMN28
TIR29	TIR30	YEK1	YEK2	YEK3	YEK4	YEK5
22	23	24	25	26	27	28
F13	F14	F15	F16	F17	F18	F19
AMN29	AMN30	RNT1	RNT2	RNT3	RNT4	RNT5
YEK6	YEK7	YEK8	YEK9	YEK10	YEK11	YEK12
SHAHIDI 83						
F20						
RNT6						
YEK13						

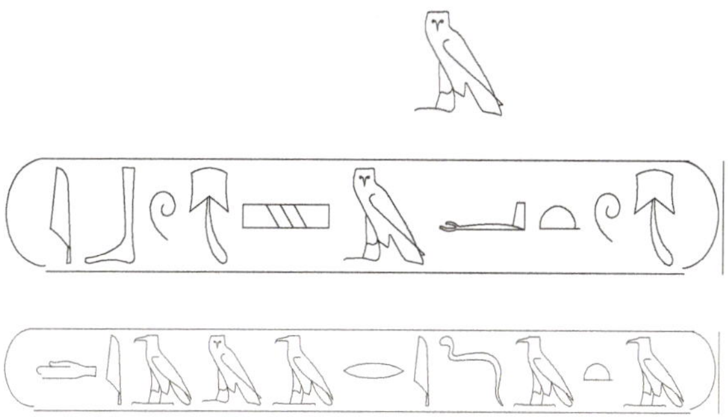

THE NU AFRIKAN CALENDAR

MAAFA 533 B.S.E./ASIENTO 2049-50 C.E.
KAM 6285-86 S.M./EC 2041-42 A.M. (ዓ.ም.)

AFU 533/2049-50/6285-86/2041-42						
Ashe	Ojobo	Iwa	Araorun	Ile	Itutu	Erinle
1	2	3	4	5	6	7
F21	F22	F23	F24	F25	F26	F27
RNT7	RNT8	RNT9	RNT10	RNT11	RNT12	RNT13
YEK14	YEK15	YEK16	YEK17	YEK18	YEK19	YEK20
8	9	10	11	12	13	14
F28	M1	M2	M3	M4	M5	M6
RNT14	RNT15	RNT16	RNT17	RNT18	RNT19	RNT20
YEK21	YEK22	YEK23	YEK24	YEK25	YEK26	YEK27
15	16	17	18	19	20	21
M7	M8	M9	M10	M11	M12	M13
RNT21	RNT22	RNT23	RNT24	RNT25	RNT26	RNT27
YEK28	YEK29	YEK30	MAG1	MAG2	MAG3	MAG4
22	23	24	25	26	27	28
M14	M15	M16	M17	M18	M19	M20
RNT28	RNT29	RNT30	KHN1	KHN2	KHN3	KHN4
MAG5	MAG6	MAG7	MAG8	MAG9	MAG10	MAG11

KAWAIDA 533/2049-50/6285-86/2041-42						
Ashe	Ojobo	Iwa	Araorun	Ile	Itutu	Erinle
1	2	3	4	5	6	7
A18	A19	A20	A21	A22	A23	A24
PIN3	PIN4	PIN5	PIN6	PIN7	PIN8	PIN9
MIY10	MIY11	MIY12	MIY13	MIY14	MIY15	MIY16
8	9	10	11	12	13	14
A25	A26	A27	A28	A29	A30	MA1
PIN10	PIN11	PIN12	PIN13	PIN14	PIN15	PIN16
MIY17	MIY18	MIY19	MIY20	MIY21	MIY22	MIY23
15	16	17	18	19	20	21
MA2	MA3	MA4	MA5	MA6	MA7	MA8
PIN17	PIN18	PIN19	PIN20	PIN21	PIN22	PIN23
MIY24	MIY25	MIY26	MIY27	MIY28	MIY29	MIY30
22	23	24	25	26	27	28
MA9	MA10	MA11	MA12	MA13	MA14	MA15
PIN24	PIN25	PIN26	PIN27	PIN28	PIN29	PIN30
GIN1	GIN2	GIN3	GIN4	GIN5	GIN6	GIN7

UHURU 533/2049-50/6285-86/2041-42						
Ashe	Ojobo	Iwa	Araorun	Ile	Itutu	Erinle
1	2	3	4	5	6	7
M21	M22	M23	M24	M25	M26	M27
KHN5	KHN6	KHN7	KHN8	KHN9	KHN10	KHN11
MAG12	MAG13	MAG14	MAG15	MAG16	MAG17	MAG18
8	9	10	11	12	13	14
M28	M29	M30	M31	A1	A2	A3
KHN12	KHN13	KHN14	KHN15	KHN16	KHN17	KHN18
MAG19	MAG20	MAG21	MAG22	MAG23	MAG24	MAG25
15	16	17	18	19	20	21
A4	A5	A6	A7	A8	A9	A10
KHN19	KHN20	KHN21	KHN22	KHN23	KHN24	KHN25
MAG26	MAG27	MAG28	MAG29	MAG30	MIY1	MIY2
22	23	24	25	26	27	28
A11	A12	A13	A14	A15	A16	A17
KHN26	KHN27	KHN28	KHN29	KHN30	PIN1	PIN2
MIY3	MIY4	MIY5	MIY6	MIY7	MIY8	MIY9

MABABA WEUSI 533/2049-50/6285-86/2041-42						
Ashe	Ojobo	Iwa	Araorun	Ile	Itutu	Erinle
1	2	3	4	5	6	7
MA16	MA17	MA18	MA19	MA20	MA21	MA22
APP1	APP2	APP3	APP4	APP5	APP6	APP7
GIN8	GIN9	GIN10	GIN11	GIN12	GIN13	GIN14
8	9	10	11	12	13	14
MA23	MA24	MA25	MA26	MA27	MA28	MA29
APP8	APP9	APP10	APP11	APP12	APP13	APP14
GIN15	GIN16	GIN17	GIN18	GIN19	GIN20	GIN21
15	16	17	18	19	20	21
MA30	MA31	J1	J2	J3	J4	J5
APP15	APP16	APP17	APP18	APP19	APP20	APP21
GIN22	GIN23	GIN24	GIN25	GIN26	GIN27	GIN28
22	23	24	25	26	27	28
J6	J7	J8	J9	J10	J11	J12
APP22	APP23	APP24	APP25	APP26	APP27	APP28
GIN29	GIN30	SAN1	SAN2	SAN3	SAN4	SAN5

THE NU AFRIKAN CALENDAR

MWANAMKE WEUSI 533/2049-50/6285-86/2041-42						
Ashe	Ojobo	Iwa	Araorun	Ile	Itutu	Erinle
1	2	3	4	5	6	7
J13	J14	J15	J16	J17	J18	J19
APP29	APP30	MSR1	MSR2	MSR3	MSR4	MSR5
SAN6	SAN7	SAN8	SAN9	SAN10	SAN11	SAN12
8	9	10	11	12	13	14
J20	J21	J22	J23	J24	J25	J26
MSR6	MSR7	MSR8	MSR9	MSR10	MSR11	MSR12
SAN13	SAN14	SAN15	SAN16	SAN17	SAN18	SAN19
15	16	17	18	19	20	21
J27	J28	J29	J30	JU1	JU2	JU3
MSR13	MSR14	MSR15	MSR16	MSR17	MSR18	MSR19
SAN20	SAN21	SAN22	SAN23	SAN24	SAN25	SAN26
22	23	24	25	26	27	28
JU4	JU5	JU6	JU7	JU8	JU9	JU10
MSR20	MRS21	MSR22	MSR23	MSR24	MSR25	MSR26
SAN27	SAN28	SAN29	SAN30	HAM1	HAM2	HAM3

TAIFA WEUSI 533/2049-50/6285-86/2041-42						
Ashe	Ojobo	Iwa	Araorun	Ile	Itutu	Erinle
1	2	3	4	5	6	7
AU8	AU9	AU10	AU11	AU12	AU13	AU14
THT20	THT21	THT22	THT23	THT24	THT25	THT26
NAH2	NAH3	NAH4	NAH5	NAH6	NAH7	NAH8
8	9	10	11	12	13	14
AU15	AU16	AU17	AU18	AU19	AU20	AU21
THT27	THT28	THT29	THT30	PAP1	PAP1	PAP2
NAH9	NAH10	NAH11	NAH12	NAH13	NAH14	NAH15
15	16	17	18	19	20	21
AU22	AU23	AU24	AU25	AU26	AU27	AU28
PAP3	PAP4	PAP5	PAP6	PAP7	PAP8	PAP9
NAH16	NAH17	NAH18	NAH19	NAH20	NAH21	NAH22
22	23	24	25	26	27	28
AU29	AU30	AU31	S1	S2	S3	S4
PAP10	PAP11	PAP12	PAP13	PAP14	PAP15	PAP16
NAH23	NAH24	NAH25	NAH26	NAH27	NAH28	NAH29

MTU WEUSI 533/2049-50/6285-86/2041-42						
Ashe	Ojobo	Iwa	Araorun	Ile	Itutu	Erinle
1	2	3	4	5	6	7
JU11	JU12	JU13	JU14	JU15	JU16	JU17
MSR27	MSR28	MSR29	MSR30	*ASR1*	*HRU2*	*STH3*
HAM4	HAM5	HAM6	HAM7	HAM8	HAM9	HAM10
8	9	10	11	12	13	14
JU18	JU19	JU20	JU21	JU22	JU23	JU24
AST4	*NBH5*	THT1	THT2	THT3	THT4	THT5
HAM11	HAM12	HAM13	AM14	HAM15	HAM16	HAM17
15	16	17	18	19	20	21
JU25	JU26	JU27	JU28	JU29	JU30	JU31
THT6	THT7	THT8	THT9	THT10	THT11	THT12
HAM18	HAM19	HAM20	HAM21	HAM22	HAM23	HAM24
22	23	24	25	26	27	28
AU1	AU2	AU3	AU4	AU5	AU6	AU7
THT13	THT14	THT15	THT16	THT17	THT18	THT19
HAM25	HAM26	HAM27	HAM28	HAM29	HAM30	NAH1

UWEZO 533/2049-50/6285-86/2041-42						
Ashe	Ojobo	Iwa	Araorun	Ile	Itutu	Erinle
1	2	3	4	5	6	7
S5	S6	S7	S8	S9	S10	S11
PAP17	PAP18	PAP19	PAP20	PAP21	PAP22	PAP23
NAH30	*PAG1*	*PAG2*	*PAG3*	*PAG4*	*PAG5*	MAS1
8	9	10	11	12	13	14
S12	S13	S14	S15	S16	S17	S18
PAP24	PAP25	PAP26	PAP27	PAP28	PAP29	PAP30
MAS2	MAS3	MAS4	MAS5	MAS6	MAS7	MAS8
15	16	17	18	19	20	21
S19	S20	S21	S22	S23	S24	S25
HTH1	HTH2	HTH3	HTH4	HTH5	HTH6	HTH7
MAS9	MAS10	MAS11	MAS12	MAS13	MAS14	MAS15
22	23	24	25	26	27	28
S26	S27	S28	S29	S30	O1	O2
HTH8	HTH9	HTH10	HTH11	HTH12	HTH13	HTH14
MAS16	MAS17	MAS18	MAS19	MAS20	MAS21	MAS22

THE NU AFRIKAN CALENDAR

EKUNDU 533/2049-50/6285-86/2041-42						
Ashe	Ojobo	Iwa	Araorun	Ile	Itutu	Erinle
1	2	3	4	5	6	7
O3	O4	O5	O6	O7	O8	O9
HTH15	HTH16	HTH17	HTH18	HTH19	HTH20	HTH21
MAS23	MAS24	MAS25	MAS26	MAS27	MAS28	MAS29
8	9	10	11	12	13	14
O10	O11	O12	O13	O14	O15	O16
HTH22	HTH23	HTH24	HTH25	HTH26	HTH27	HTH28
MAS30	THi1	THi2	THi3	THi4	THi5	THi6
15	16	17	18	19	20	21
O17	O18	O19	O20	O21	O22	O23
HTH29	HTH30	KHK1	KHK2	KHK3	KHK4	KHK5
THi7	THi8	THi9	THi10	THi11	THi12	THi13
22	23	24	25	26	27	28
O24	O25	O26	O27	O28	O29	O30
KHK6	KHK7	KHK8	KHK9	KHK10	KHK11	KHK12
THi14	THi15	THi16	THi17	THi18	THi19	THi20

KIJANI 533/2049-50/6285-86/2041-42						
Ashe	Ojobo	Iwa	Araorun	Ile	Itutu	Erinle
1	2	3	4	5	6	7
N28	N29	N30	D1	D2	D3	D4
TBT11	TBT12	TBT13	TBT14	TBT15	TBT16	TBT17
HID19	HID20	HID21	HID22	HID23	HID24	HID25
8	9	10	11	12	13	14
D5	D6	D7	D8	D9	D10	D11
TBT18	TBT19	TBT20	TBT21	TBT22	TBT23	TBT24
HID26	HID27	HID28	HID29	HID30	TAH1	TAH2
15	16	17	18	19	20	21
D12	D13	D14	D15	D16	D17	D18
TBT25	TBT26	TBT27	TBT28	TBT29	TBT30	MKH1
TAH3	TAH4	TAH5	TAH6	TAH7	TAH8	TAH9
22	23	24	25	26	27	28
D19	D20	D21	D22	D23	D24	D25
MKH2	MKH3	MKH4	MKH5	MKH6	MKH7	MKH8
TAH10	TAH11	TAH12	TAH13	TAH14	TAH15	TAH16

EUSI 533/2049-50/6285-86/2041-42						
Ashe	Ojobo	Iwa	Araorun	Ile	Itutu	Erinle
1	2	3	4	5	6	7
O31	N1	N2	N3	N4	N5	N6
KHK13	KHK14	KHK15	KHK16	KHK17	KHK18	KHK19
THi21	THi22	THi23	THi24	THi25	THi26	THi27
8	9	10	11	12	13	14
N7	N8	N9	N10	N11	N12	N13
KHK20	KHK21	KHK22	KHK23	KHK24	KHK25	KHK26
THi28	THi29	THi30	HID1	HID2	HID3	HID4
15	16	17	18	19	20	21
N14	N15	N16	N17	N18	N19	N20
KHK27	KHK28	KHK29	KHK30	TBT1	TBT2	TBT3
HID5	HID6	HID7	HID8	HID9	HID10	HID11
22	23	24	25	26	27	28
N21	N22	N23	N24	N25	N26	N27
TBT4	TBT5	TBT6	TBT7	TBT8	TBT9	TBT10
HID12	HID13	HID14	HID15	HID16	HID17	HID18

KWANZA 533/2049-50/6285-86/2041-42						
Umoja	Kujichagulia	Ujima	Ujamaa	Nia	Kuumba	Imani
1	2	3	4	5	6	7
D26	D27	D28	D29	D30	D31	JA1
MKH9	MKH10	MKH11	MKH12	MKH13	MKH14	MKH15
TAH17	TAH18	TAH19	TAH20	TAH21	TAH22	TAH23
8	9	10	11	12	13	14
JA2	JA3	JA4	JA5	JA6	JA7	JA8
MKH16	MKH17	MKH18	MKH19	MKH20	MKH21	MKH22
TAH24	TAH25	TAH26	TAH27	TAH28	TAH29	TAH30
15	16	17	18	19	20	21
JA9	JA10	JA11	JA12	JA13	JA14	JA15
MKH23	MKH24	MKH25	KMH26	MKH27	MKH28	MKH29
TIR1	TIR2	TIR3	TIR4	TIR5	TIR6	TIR7
22	23	24	25	26	27	28
JA16	JA17	JA18	JA19	JA20	JA21	JA22
MKH30	AMN1	AMN2	AMN3	AMN4	AMN5	AMN6
TIR8	TIR9	TIR10	TIR11	TIR12	TIR13	TIR14

THE NU AFRIKAN CALENDAR

SHAHIDI 533/2049-50/6285-86/2041-42						
Ashe	Ojobo	Iwa	Araorun	Ile	Itutu	Erinle
1	2	3	4	5	6	7
JA23	JA24	JA25	JA26	JA27	JA28	JA29
AMN7	AMN8	AMN9	AMN10	AMN11	AMN12	AMN13
TIR15	TIR16	TIR17	TIR18	TIR19	TIR20	TIR21
8	9	10	11	12	13	14
JA30	JA31	F1	F2	F3	F4	F5
AMN14	AMN15	AMN16	AMN17	AMN18	AMN19	AMN20
TIR22	TIR23	TIR24	TIR25	TIR26	TIR27	TIR28
15	16	17	18	19	20	21
F6	F7	F8	F9	F10	F 11	F12
AMN21	AMN22	AMN23	AMN24	AMN25	AMN26	AMN28
TIR29	TIR30	YEK1	YEK2	YEK3	YEK4	YEK5
22	23	24	25	26	27	28
F13	F14	F15	F16	F17	F18	F19
AMN29	AMN30	RNT1	RNT2	RNT3	RNT4	RNT5
YEK6	YEK7	YEK8	YEK9	YEK10	YEK11	YEK12
SHAHIDI 84						
F20						
RNT6						
YEK13						

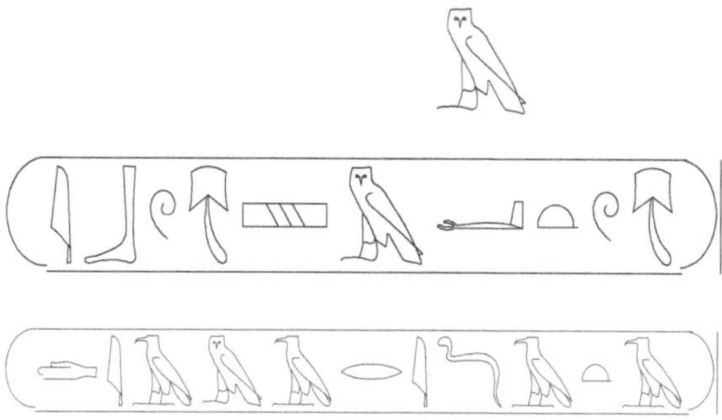

THE NU AFRIKAN CALENDAR

MAAFA 534 B.S.E./ASIENTO 2050-51 C.E.
KAM 6286-87 S.M./EC 2042-43 A.M. (ዓ.ም.)

AFU 534/2050-51/6286-87/2042-43						
Ashe	Ojobo	Iwa	Araorun	Ile	Itutu	Erinle
1	2	3	4	5	6	7
F21	F22	F23	F24	F25	F26	F27
RNT7	RNT8	RNT9	RNT10	RNT11	RNT12	RNT13
YEK14	YEK15	YEK16	YEK17	YEK18	YEK19	YEK20
8	9	10	11	12	13	14
F28	M1	M2	M3	M4	M5	M6
RNT14	RNT15	RNT16	RNT17	RNT18	RNT19	RNT20
YEK21	YEK22	YEK23	YEK24	YEK25	YEK26	YEK27
15	16	17	18	19	20	21
M7	M8	M9	M10	M11	M12	M13
RNT21	RNT22	RNT23	RNT24	RNT25	RNT26	RNT27
YEK28	YEK29	YEK30	MAG1	MAG2	MAG3	MAG4
22	23	24	25	26	27	28
M14	M15	M16	M17	M18	M19	M20
RNT28	RNT29	RNT30	KHN1	KHN2	KHN3	KHN4
MAG5	MAG6	MAG7	MAG8	MAG9	MAG10	MAG11

KAWAIDA 534/2050-51/6286-87/2042-43						
Ashe	Ojobo	Iwa	Araorun	Ile	Itutu	Erinle
1	2	3	4	5	6	7
A18	A19	A20	A21	A22	A23	A24
PIN3	PIN4	PIN5	PIN6	PIN7	PIN8	PIN9
MIY10	MIY11	MIY12	MIY13	MIY14	MIY15	MIY16
8	9	10	11	12	13	14
A25	A26	A27	A28	A29	A30	MA1
PIN10	PIN11	PIN12	PIN13	PIN14	PIN15	PIN16
MIY17	MIY18	MIY19	MIY20	MIY21	MIY22	MIY23
15	16	17	18	19	20	21
MA2	MA3	MA4	MA5	MA6	MA7	MA8
PIN17	PIN18	PIN19	PIN20	PIN21	PIN22	PIN23
MIY24	MIY25	MIY26	MIY27	MIY28	MIY29	MIY30
22	23	24	25	26	27	28
MA9	MA10	MA11	MA12	MA13	MA14	MA15
PIN24	PIN25	PIN26	PIN27	PIN28	PIN29	PIN30
GIN1	GIN2	GIN3	GIN4	GIN5	GIN6	GIN7

UHURU 534/2050-51/6286-87/2042-43						
Ashe	Ojobo	Iwa	Araorun	Ile	Itutu	Erinle
1	2	3	4	5	6	7
M21	M22	M23	M24	M25	M26	M27
KHN5	KHN6	KHN7	KHN8	KHN9	KHN10	KHN11
MAG12	MAG13	MAG14	MAG15	MAG16	MAG17	MAG18
8	9	10	11	12	13	14
M28	M29	M30	M31	A1	A2	A3
KHN12	KHN13	KHN14	KHN15	KHN16	KHN17	KHN18
MAG19	MAG20	MAG21	MAG22	MAG23	MAG24	MAG25
15	16	17	18	19	20	21
A4	A5	A6	A7	A8	A9	A10
KHN19	KHN20	KHN21	KHN22	KHN23	KHN24	KHN25
MAG26	MAG27	MAG28	MAG29	MAG30	MIY1	MIY2
22	23	24	25	26	27	28
A11	A12	A13	A14	A15	A16	A17
KHN26	KHN27	KHN28	KHN29	KHN30	PIN1	PIN2
MIY3	MIY4	MIY5	MIY6	MIY7	MIY8	MIY9

MABABA WEUSI 534/2050-51/6286-87/2042-43						
Ashe	Ojobo	Iwa	Araorun	Ile	Itutu	Erinle
1	2	3	4	5	6	7
MA16	MA17	MA18	MA19	MA20	MA21	MA22
APP1	APP2	APP3	APP4	APP5	APP6	APP7
GIN8	GIN9	GIN10	GIN11	GIN12	GIN13	GIN14
8	9	10	11	12	13	14
MA23	MA24	MA25	MA26	MA27	MA28	MA29
APP8	APP9	APP10	APP11	APP12	APP13	APP14
GIN15	GIN16	GIN17	GIN18	GIN19	GIN20	GIN21
15	16	17	18	19	20	21
MA30	MA31	J1	J2	J3	J4	J5
APP15	APP16	APP17	APP18	APP19	APP20	APP21
GIN22	GIN23	GIN24	GIN25	GIN26	GIN27	GIN28
22	23	24	25	26	27	28
J6	J7	J8	J9	J10	J11	J12
APP22	APP23	APP24	APP25	APP26	APP27	APP28
GIN29	GIN30	SAN1	SAN2	SAN3	SAN4	SAN5

THE NU AFRIKAN CALENDAR

MWANAMKE WEUSI 534/2050-51/6286-87/2042-43						
Ashe	Ojobo	Iwa	Araorun	Ile	Itutu	Erinle
1	2	3	4	5	6	7
J13	J14	J15	J16	J17	J18	J19
APP29	APP30	MSR1	MSR2	MSR3	MSR4	MSR5
SAN6	SAN7	SAN8	SAN9	SAN10	SAN11	SAN12
8	9	10	11	12	13	14
J20	J21	J22	J23	J24	J25	J26
MSR6	MSR7	MSR8	MSR9	MSR10	MSR11	MSR12
SAN13	SAN14	SAN15	SAN16	SAN17	SAN18	SAN19
15	16	17	18	19	20	21
J27	J28	J29	J30	JU1	JU2	JU3
MSR13	MSR14	MSR15	MSR16	MSR17	MSR18	MSR19
SAN20	SAN21	SAN22	SAN23	SAN24	SAN25	SAN26
22	23	24	25	26	27	28
JU4	JU5	JU6	JU7	JU8	JU9	JU10
MSR20	MRS21	MSR22	MSR23	MSR24	MSR25	MSR26
SAN27	SAN28	SAN29	SAN30	HAM1	HAM2	HAM3

TAIFA WEUSI 534/2050-51/6286-87/2042-43						
Ashe	Ojobo	Iwa	Araorun	Ile	Itutu	Erinle
1	2	3	4	5	6	7
AU8	AU9	AU10	AU11	AU12	AU13	AU14
THT20	THT21	THT22	THT23	THT24	THT25	THT26
NAH2	NAH3	NAH4	NAH5	NAH6	NAH7	NAH8
8	9	10	11	12	13	14
AU15	AU16	AU17	AU18	AU19	AU20	AU21
THT27	THT28	THT29	THT30	PAP1	PAP1	PAP2
NAH9	NAH10	NAH11	NAH12	NAH13	NAH14	NAH15
15	16	17	18	19	20	21
AU22	AU23	AU24	AU25	AU26	AU27	AU28
PAP3	PAP4	PAP5	PAP6	PAP7	PAP8	PAP9
NAH16	NAH17	NAH18	NAH19	NAH20	NAH21	NAH22
22	23	24	25	26	27	28
AU29	AU30	AU31	S1	S2	S3	S4
PAP10	PAP11	PAP12	PAP13	PAP14	PAP15	PAP16
NAH23	NAH24	NAH25	NAH26	NAH27	NAH28	NAH29

MTU WEUSI 534/2050-51/6286-87/2042-43						
Ashe	Ojobo	Iwa	Araorun	Ile	Itutu	Erinle
1	2	3	4	5	6	7
JU11	JU12	JU13	JU14	JU15	JU16	JU17
MSR27	MSR28	MSR29	MSR30	*ASR1*	*HRU2*	*STH3*
HAM4	HAM5	HAM6	HAM7	HAM8	HAM9	HAM10
8	9	10	11	12	13	14
JU18	JU19	JU20	JU21	JU22	JU23	JU24
AST4	*NBH5*	THT1	THT2	THT3	THT4	THT5
HAM11	HAM12	HAM13	AM14	HAM15	HAM16	HAM17
15	16	17	18	19	20	21
JU25	JU26	JU27	JU28	JU29	JU30	JU31
THT6	THT7	THT8	THT9	THT10	THT11	THT12
HAM18	HAM19	HAM20	HAM21	HAM22	HAM23	HAM24
22	23	24	25	26	27	28
AU1	AU2	AU3	AU4	AU5	AU6	AU7
THT13	THT14	THT15	THT16	THT17	THT18	THT19
HAM25	HAM26	HAM27	HAM28	HAM29	HAM30	NAH1

UWEZO 534/2050-51/6286-87/2042-43						
Ashe	Ojobo	Iwa	Araorun	Ile	Itutu	Erinle
1	2	3	4	5	6	7
S5	S6	S7	S8	S9	S10	S11
PAP17	PAP18	PAP19	PAP20	PAP21	PAP22	PAP23
NAH30	*PAG1*	*PAG2*	*PAG3*	*PAG4*	*PAG5*	MAS1
8	9	10	11	12	13	14
S12	S13	S14	S15	S16	S17	S18
PAP24	PAP25	PAP26	PAP27	PAP28	PAP29	PAP30
MAS2	MAS3	MAS4	MAS5	MAS6	MAS7	MAS8
15	16	17	18	19	20	21
S19	S20	S21	S22	S23	S24	S25
HTH1	HTH2	HTH3	HTH4	HTH5	HTH6	HTH7
MAS9	MAS10	MAS11	MAS12	MAS13	MAS14	MAS15
22	23	24	25	26	27	28
S26	S27	S28	S29	S30	O1	O2
HTH8	HTH9	HTH10	HTH11	HTH12	HTH13	HTH14
MAS16	MAS17	MAS18	MAS19	MAS20	MAS21	MAS22

THE NU AFRIKAN CALENDAR

EKUNDU 534/2050-51/6286-87/2042-43

Ashe	Ojobo	Iwa	Araorun	Ile	Itutu	Erinle
1	2	3	4	5	6	7
O3	O4	O5	O6	O7	O8	O9
HTH15	HTH16	HTH17	HTH18	HTH19	HTH20	HTH21
MAS23	MAS24	MAS25	MAS26	MAS27	MAS28	MAS29
8	9	10	11	12	13	14
O10	O11	O12	O13	O14	O15	O16
HTH22	HTH23	HTH24	HTH25	HTH26	HTH27	HTH28
MAS30	THi1	THi2	THi3	THi4	THi5	THi6
15	16	17	18	19	20	21
O17	O18	O19	O20	O21	O22	O23
HTH29	HTH30	KHK1	KHK2	KHK3	KHK4	KHK5
THi7	THi8	THi9	THi10	THi11	THi12	THi13
22	23	24	25	26	27	28
O24	O25	O26	O27	O28	O29	O30
KHK6	KHK7	KHK8	KHK9	KHK10	KHK11	KHK12
THi14	THi15	THi16	THi17	THi18	THi19	THi20

KIJANI 534/2050-51/6286-87/2042-43

Ashe	Ojobo	Iwa	Araorun	Ile	Itutu	Erinle
1	2	3	4	5	6	7
N28	N29	N30	D1	D2	D3	D4
TBT11	TBT12	TBT13	TBT14	TBT15	TBT16	TBT17
HID19	HID20	HID21	HID22	HID23	HID24	HID25
8	9	10	11	12	13	14
D5	D6	D7	D8	D9	D10	D11
TBT18	TBT19	TBT20	TBT21	TBT22	TBT23	TBT24
HID26	HID27	HID28	HID29	HID30	TAH1	TAH2
15	16	17	18	19	20	21
D12	D13	D14	D15	D16	D17	D18
TBT25	TBT26	TBT27	TBT28	TBT29	TBT30	MKH1
TAH3	TAH4	TAH5	TAH6	TAH7	TAH8	TAH9
22	23	24	25	26	27	28
D19	D20	D21	D22	D23	D24	D25
MKH2	MKH3	MKH4	MKH5	MKH6	MKH7	MKH8
TAH10	TAH11	TAH12	TAH13	TAH14	TAH15	TAH16

EUSI 534/2050-51/6286-87/2042-43

Ashe	Ojobo	Iwa	Araorun	Ile	Itutu	Erinle
1	2	3	4	5	6	7
O31	N1	N2	N3	N4	N5	N6
KHK13	KHK14	KHK15	KHK16	KHK17	KHK18	KHK19
THi21	THi22	THi23	THi24	THi25	THi26	THi27
8	9	10	11	12	13	14
N7	N8	N9	N10	N11	N12	N13
KHK20	KHK21	KHK22	KHK23	KHK24	KHK25	KHK26
THi28	THi29	THi30	HID1	HID2	HID3	HID4
15	16	17	18	19	20	21
N14	N15	N16	N17	N18	N19	N20
KHK27	KHK28	KHK29	KHK30	TBT1	TBT2	TBT3
HID5	HID6	HID7	HID8	HID9	HID10	HID11
22	23	24	25	26	27	28
N21	N22	N23	N24	N25	N26	N27
TBT4	TBT5	TBT6	TBT7	TBT8	TBT9	TBT10
HID12	HID13	HID14	HID15	HID16	HID17	HID18

KWANZA 534/2050-51/6286-87/2042-43

Umoja	Kujichagulia	Ujima	Ujamaa	Nia	Kuumba	Imani
1	2	3	4	5	6	7
D26	D27	D28	D29	D30	D31	JA1
MKH9	MKH10	MKH11	MKH12	MKH13	MKH14	MKH15
TAH17	TAH18	TAH19	TAH20	TAH21	TAH22	TAH23
8	9	10	11	12	13	14
JA2	JA3	JA4	JA5	JA6	JA7	JA8
MKH16	MKH17	MKH18	MKH19	MKH20	MKH21	MKH22
TAH24	TAH25	TAH26	TAH27	TAH28	TAH29	TAH30
15	16	17	18	19	20	21
JA9	JA10	JA11	JA12	JA13	JA14	JA15
MKH23	MKH24	MKH25	KMH26	MKH27	MKH28	MKH29
TIR1	TIR2	TIR3	TIR4	TIR5	TIR6	TIR7
22	23	24	25	26	27	28
JA16	JA17	JA18	JA19	JA20	JA21	JA22
MKH30	AMN1	AMN2	AMN3	AMN4	AMN5	AMN6
TIR8	TIR9	TIR10	TIR11	TIR12	TIR13	TIR14

THE NU AFRIKAN CALENDAR

SHAHIDI 534/2050-51/6286-87/2042-43						
Ashe	Ojobo	Iwa	Araorun	Ile	Itutu	Erinle
1	2	3	4	5	6	7
JA23	JA24	JA25	JA26	JA27	JA28	JA29
AMN7	AMN8	AMN9	AMN10	AMN11	AMN12	AMN13
TIR15	TIR16	TIR17	TIR18	TIR19	TIR20	TIR21
8	9	10	11	12	13	14
JA30	JA31	F1	F2	F3	F4	F5
AMN14	AMN15	AMN16	AMN17	AMN18	AMN19	AMN20
TIR22	TIR23	TIR24	TIR25	TIR26	TIR27	TIR28
15	16	17	18	19	20	21
F6	F7	F8	F9	F10	F 11	F12
AMN21	AMN22	AMN23	AMN24	AMN25	AMN26	AMN28
TIR29	TIR30	YEK1	YEK2	YEK3	YEK4	YEK5
22	23	24	25	26	27	28
F13	F14	F15	F16	F17	F18	F19
AMN29	AMN30	RNT1	RNT2	RNT3	RNT4	RNT5
YEK6	YEK7	YEK8	YEK9	YEK10	YEK11	YEK12
SHAHIDI 85						
F20						
RNT6						
YEK13						

THE NU AFRIKAN CALENDAR

MAAFA 535 B.S.E./ASIENTO 2051-52 C.E./
KAM 6287-88 S.M./EC 2043-44 A.M. (ዓ.ም.)

AFU 535/2051-52/6287-88/2043-44						
Ashe	Ojobo	Iwa	Araorun	Ile	Itutu	Erinle
1	2	3	4	5	6	7
F21	F22	F23	F24	F25	F26	F27
RNT7	RNT8	RNT9	RNT10	RNT11	RNT12	RNT13
YEK14	YEK15	YEK16	YEK17	YEK18	YEK19	YEK20
8	9	10	11	12	13	14
F28	M1	M2	M3	M4	M5	M6
RNT14	RNT15	RNT16	RNT17	RNT18	RNT19	RNT20
YEK21	YEK22	YEK23	YEK24	YEK25	YEK26	YEK27
15	16	17	18	19	20	21
M7	M8	M9	M10	M11	M12	M13
RNT21	RNT22	RNT23	RNT24	RNT25	RNT26	RNT27
YEK28	YEK 29	YEK30	MAG1	MAG2	MAG3	MAG4
22	23	24	25	26	27	28
M14	M15	M16	M17	M18	M19	M20
RNT28	RNT29	RNT30	KHN1	KHN2	KHN3	KHN4
MAG5	MAG6	MAG7	MAG8	MAG9	MAG10	MAG11

KAWAIDA 535/2051-52/6287-88/2043-44						
Ashe	Ojobo	Iwa	Araorun	Ile	Itutu	Erinle
1	2	3	4	5	6	7
A18	A19	A20	A21	A22	A23	A24
PIN3	PIN4	PIN5	PIN6	PIN7	PIN8	PIN9
MIY10	MIY11	MIY12	MIY13	MIY14	MIY15	MIY16
8	9	10	11	12	13	14
A25	A26	A27	A28	A29	A30	MA1
PIN10	PIN11	PIN12	PIN13	PIN14	PIN15	PIN16
MIY17	MIY18	MIY19	MIY20	MIY21	MIY22	MIY23
15	16	17	18	19	20	21
MA2	MA3	MA4	MA5	MA6	MA7	MA8
PIN17	PIN18	PIN19	PIN20	PIN21	PIN22	PIN23
MIY24	MIY25	MIY26	MIY27	MIY28	MIY29	MIY30
22	23	24	25	26	27	28
MA9	MA10	MA11	MA12	MA13	MA14	MA15
PIN24	PIN25	PIN26	PIN27	PIN28	PIN29	PIN30
GIN1	GIN2	GIN3	GIN4	GIN5	GIN6	GIN7

UHURU 535/2051-52/6287-88/2043-44						
Ashe	Ojobo	Iwa	Araorun	Ile	Itutu	Erinle
1	2	3	4	5	6	7
M21	M22	M23	M24	M25	M26	M27
KHN5	KHN6	KHN7	KHN8	KHN9	KHN10	KHN11
MAG12	MAG13	MAG14	MAG15	MAG16	MAG17	MAG18
8	9	10	11	12	13	14
M28	M29	M30	M31	A1	A2	A3
KHN12	KHN13	KHN14	KHN15	KHN16	KHN17	KHN18
MAG19	MAG20	MAG21	MAG22	MAG23	MAG24	MAG25
15	16	17	18	19	20	21
A4	A5	A6	A7	A8	A9	A10
KHN19	KHN20	KHN21	KHN22	KHN23	KHN24	KHN25
MAG26	MAG27	MAG28	MAG29	MAG30	MIY1	MIY2
22	23	24	25	26	27	28
A11	A12	A13	A14	A15	A16	A17
KHN26	KHN27	KHN28	KHN29	KHN30	PIN1	PIN2
MIY3	MIY4	MIY5	MIY6	MIY7	MIY8	MIY9

MABABA WEUSI 535/2051-52/6287-88/2043-44						
Ashe	Ojobo	Iwa	Araorun	Ile	Itutu	Erinle
1	2	3	4	5	6	7
MA16	MA17	MA18	MA19	MA20	MA21	MA22
APP1	APP2	APP3	APP4	APP5	APP6	APP7
GIN8	GIN9	GIN10	GIN11	GIN12	GIN13	GIN14
8	9	10	11	12	13	14
MA23	MA24	MA25	MA26	MA27	MA28	MA29
APP8	APP9	APP10	APP11	APP12	APP13	APP14
GIN15	GIN16	GIN17	GIN18	GIN19	GIN20	GIN21
15	16	17	18	19	20	21
MA30	MA31	J1	J2	J3	J4	J5
APP15	APP16	APP17	APP18	APP19	APP20	APP21
GIN22	GIN23	GIN24	GIN25	GIN26	GIN27	GIN28
22	23	24	25	26	27	28
J6	J7	J8	J9	J10	J11	J12
APP22	APP23	APP24	APP25	APP26	APP27	APP28
GIN29	GIN30	SAN1	SAN2	SAN3	SAN4	SAN5

THE NU AFRIKAN CALENDAR

MWANAMKE WEUSI 535/2051-52/6287-88/2043-44

Ashe	Ojobo	Iwa	Araorun	Ile	Itutu	Erinle
1	2	3	4	5	6	7
J13	J14	J15	J16	J17	J18	J19
APP29	APP30	MSR1	MSR2	MSR3	MSR4	MSR5
SAN6	SAN7	SAN8	SAN9	SAN10	SAN11	SAN12
8	9	10	11	12	13	14
J20	J21	J22	J23	J24	J25	J26
MSR6	MSR7	MSR8	MSR9	MSR10	MSR11	MSR12
SAN13	SAN14	SAN15	SAN16	SAN17	SAN18	SAN19
15	16	17	18	19	20	21
J27	J28	J29	J30	JU1	JU2	JU3
MSR13	MSR14	MSR15	MSR16	MSR17	MSR18	MSR19
SAN20	SAN21	SAN22	SAN23	SAN24	SAN25	SAN26
22	23	24	25	26	27	28
JU4	JU5	JU6	JU7	JU8	JU9	JU10
MSR20	MRS21	MSR22	MSR23	MSR24	MSR25	MSR26
SAN27	SAN28	SAN29	SAN30	HAM1	HAM2	HAM3

MTU WEUSI 535/2051-52/6287-88/2043-44

Ashe	Ojobo	Iwa	Araorun	Ile	Itutu	Erinle
1	2	3	4	5	6	7
JU11	JU12	JU13	JU14	JU15	JU16	JU17
MSR27	MSR28	MSR29	MSR30	*ASR1*	*HRU2*	*STH3*
HAM4	HAM5	HAM6	HAM7	HAM8	HAM9	HAM10
8	9	10	11	12	13	14
JU18	JU19	JU20	JU21	JU22	JU23	JU24
AST4	*NBH5*	THT1	THT2	THT3	THT4	THT5
HAM11	HAM12	HAM13	HAM14	HAM15	HAM16	HAM17
15	16	17	18	19	20	21
JU25	JU26	JU27	JU28	JU29	JU30	JU31
THT6	THT7	THT8	THT9	THT10	THT11	THT12
HAM18	HAM19	HAM20	HAM21	HAM22	HAM23	HAM24
22	23	24	25	26	27	28
AU1	AU2	AU3	AU4	AU5	AU6	AU7
THT13	THT14	THT15	THT16	THT17	THT18	THT19
HAM25	HAM26	HAM27	HAM28	HAM29	HAM30	NAH1

TAIFA WEUSI 535/2051-52/6287-88/2043-44

Ashe	Ojobo	Iwa	Araorun	Ile	Itutu	Erinle
1	2	3	4	5	6	7
AU8	AU9	AU10	AU11	AU12	AU13	AU14
THT20	THT21	THT22	THT23	THT24	THT25	THT26
NAH2	NAH3	NAH4	NAH5	NAH6	NAH7	NAH8
8	9	10	11	12	13	14
AU15	AU16	AU17	AU18	AU19	AU20	AU21
THT27	THT28	THT29	THT30	PAP1	PAP1	PAP2
NAH9	NAH10	NAH11	NAH12	NAH13	NAH14	NAH15
15	16	17	18	19	20	21
AU22	AU23	AU24	AU25	AU26	AU27	AU28
PAP3	PAP4	PAP5	PAP6	PAP7	PAP8	PAP9
NAH16	NAH17	NAH18	NAH19	NAH20	NAH21	NAH22
22	23	24	25	26	27	28
AU29	AU30	AU31	S1	S2	S3	S4
PAP10	PAP11	PAP12	PAP13	PAP14	PAP15	PAP16
NAH23	NAH24	NAH25	NAH26	NAH27	NAH28	NAH29

UWEZO 535/2051-52/6287-88/2043-44

Ashe	Ojobo	Iwa	Araorun	Ile	Itutu	Erinle
1	2	3	4	5	6	7
S5	S6	S7	S8	S9	S10	S11
PAP17	PAP18	PAP19	PAP20	PAP21	PAP22	PAP23
NAH30	*QUA1*	*QUA2*	*QUA3*	*QUA4*	*QUA5*	*QUA6*
8	9	10	11	12	13	14
S12	S13	S14	S15	S16	S17	S18
PAP24	PAP25	PAP26	PAP27	PAP28	PAP29	PAP30
MAS1	MAS2	MAS3	MAS4	MAS5	MAS6	MAS7
15	16	17	18	19	20	21
S19	S20	S21	S22	S23	S24	S25
HTH1	HTH2	HTH3	HTH4	HTH5	HTH6	HTH7
MAS8	MAS9	MAS10	MAS11	MAS12	MAS13	MAS14
22	23	24	25	26	27	28
S26	S27	S28	S29	S30	O1	O2
HTH8	HTH9	HTH10	HTH11	HTH12	HTH13	HTH14
MAS15	MAS16	MAS17	MAS18	MAS19	MAS20	MAS21

THE NU AFRIKAN CALENDAR

EKUNDU 535/2051-52/6287-88/2043-44						
Ashe	Ojobo	Iwa	Araorun	Ile	Itutu	Erinle
1	2	3	4	5	6	7
O3	O4	O5	O6	O7	O8	O9
HTH15	HTH16	HTH17	HTH18	HTH19	HTH20	HTH21
MAS22	MAS23	MAS 24	MAS25	MAS26	MAS27	MAS28
8	9	10	11	12	13	14
O10	O11	O12	O13	O14	O15	O16
HTH22	HTH23	HTH24	HTH25	HTH26	HTH27	HTH28
MAS29	MAS30	THi1	THi2	THi3	THi4	THi5
15	16	17	18	19	20	21
O17	O18	O19	O20	O21	O22	O23
HTH29	HTH30	KHK1	KHK2	KHK3	KHK4	KHK5
THi6	THi7	THi8	THi9	THi10	THi11	THi12
22	23	24	25	26	27	28
O24	O25	O26	O27	O28	O29	O30
KHK6	KHK7	KHK8	KHK9	KHK10	KHK11	KHK12
THi13	THi14	THi15	THi16	THi17	THi18	THi19

KIJANI 535/2051-52/6287-88/2043-44						
Ashe	Ojobo	Iwa	Araorun	Ile	Itutu	Erinle
1	2	3	4	5	6	7
N28	N29	N30	D1	D2	D3	D4
TBT11	TBT12	TBT13	TBT14	TBT15	TBT16	TBT17
HID18	HID19	HID20	HID21	HID22	HID23	HID24
8	9	10	11	12	13	14
D5	D6	D7	D8	D9	D10	D11
TBT18	TBT19	TBT20	TBT21	TBT22	TBT23	TBT24
HID25	HID26	HID27	HID28	HID29	HID30	TAH1
15	16	17	18	19	20	21
D12	D13	D14	D15	D16	D17	D18
TBT25	TBT26	TBT27	TBT28	TBT29	TBT30	MKH1
TAH2	TAH3	TAH4	TAH5	TAH6	TAH7	TAH8
22	23	24	25	26	27	28
D19	D20	D21	D22	D23	D24	D25
MKH2	MKH3	MKH4	MKH5	MKH6	MKH7	MKH8
TAH9	TAH10	TAH11	TAH12	TAH13	TAH14	TAH15

EUSI 535/2051-52/6287-88/2043-44						
Ashe	Ojobo	Iwa	Araorun	Ile	Itutu	Erinle
1	2	3	4	5	6	7
O31	N1	N2	N3	N4	N5	N6
KHK13	KHK14	KHK15	KHK16	KHK17	KHK18	KHK19
THi20	THi21	THi22	THi23	THi24	THi25	THi26
8	9	10	11	12	13	14
N7	N8	N9	N10	N11	N12	N13
KHK20	KHK21	KHK22	KHK23	KHK24	KHK25	KHK26
THi27	THi28	THi29	THi30	HID1	HID2	HID3
15	16	17	18	19	20	21
N14	N15	N16	N17	N18	N19	N20
KHK27	KHK28	KHK29	KHK30	TBT1	TBT2	TBT3
HID4	HID5	HID6	HID7	HID8	HID9	HID10
22	23	24	25	26	27	28
N21	N22	N23	N24	N25	N26	N27
TBT4	TBT5	TBT6	TBT7	TBT8	TBT9	TBT10
HID11	HID12	HID13	HID14	HID15	HID16	HID17

KWANZA 535/2051-52/6287-88/2043-44						
Umoja	Kujichagulia	Ujima	Ujamaa	Nia	Kuumba	Imani
1	2	3	4	5	6	7
D26	D27	D28	D29	D30	D31	JA1
MKH9	MKH10	MKH11	MKH12	MKH13	MKH14	MKH15
TAH16	TAH17	TAH18	TAH19	TAH20	TAH21	TAH22
8	9	10	11	12	13	14
JA2	JA3	JA4	JA5	JA6	JA7	JA8
MKH16	MKH17	MKH18	MKH19	MKH20	MKH21	MKH22
TAH23	TAH24	TAH25	TAH26	TAH27	TAH28	TAH29
15	16	17	18	19	20	21
JA9	JA10	JA11	JA12	JA13	JA14	JA15
MKH23	MKH24	MKH25	KMH26	MKH27	MKH28	MKH29
TAH30	TIR1	TIR2	TIR3	TIR4	TIR5	TIR6
22	23	24	25	26	27	28
JA16	JA17	JA18	JA19	JA20	JA21	JA22
MKH30	AMN1	AMN2	AMN3	AMN4	AMN5	AMN6
TIR7	TIR8	TIR9	TIR10	TIR11	TIR12	TIR13

THE NU AFRIKAN CALENDAR

SHAHIDI 535/2051-52/6287-88/2043-44						
Ashe	Ojobo	Iwa	Araorun	Ile	Itutu	Erinle
1	2	3	4	5	6	7
JA23	JA24	JA25	JA26	JA27	JA28	JA29
AMN7	AMN8	AMN9	AMN10	AMN11	AMN12	AMN13
TIR14	TIR15	TIR16	TIR17	TIR18	TIR19	TIR20
8	9	10	11	12	13	14
JA30	JA31	F1	F2	F3	F4	F5
AMN14	AMN15	AMN16	AMN17	AMN18	AMN19	AMN20
TIR21	TIR22	TIR23	TIR24	TIR25	TIR26	TIR27
15	16	17	18	19	20	21
F6	F7	F8	F9	F10	F 11	F12
AMN21	AMN22	AMN23	AMN24	AMN25	AMN26	AMN28
TIR28	TIR29	TIR30	YEK1	YEK2	YEK3	YEK4
22	23	24	25	26	27	28
F13	F14	F15	F16	F17	F18	F19
AMN29	AMN30	RNT1	RNT2	RNT3	RNT4	RNT5
YEK5	YEK6	YEK7	YEK8	YEK9	YEK10	YEK11
SHAHIDI 86						
F20						
RNT6						
YEK12						

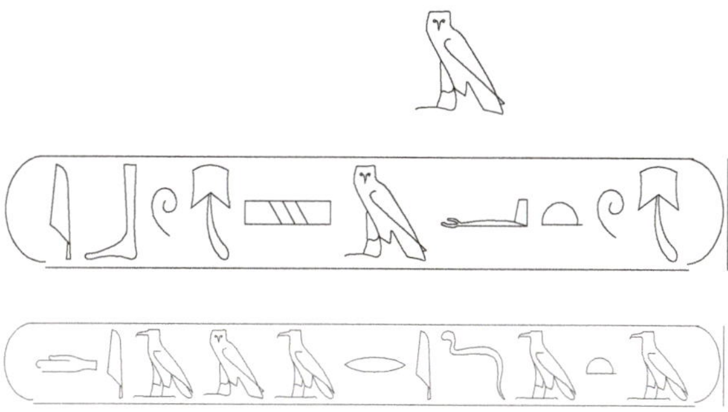

THE NU AFRIKAN CALENDAR

MAAFA 536 B.S.E./ASIENTO 2052-53 C.E./
KAM 6288-89 S.M./EC 2044-45 A.M. (ዓ.ም.)

AFU 536/2052-53/6288-89/2044-45

Ashe	Ojobo	Iwa	Araorun	Ile	Itutu	Erinle
1	2	3	4	5	6	7
F21	F22	F23	F24	F25	F26	F27
RNT7	RNT8	RNT9	RNT10	RNT11	RNT12	RNT13
YEK13	YEK14	YEK15	YEK16	YEK17	YEK18	YEK19
8	9	10	11	12	13	14
F28	F29	M1	M2	M3	M4	M5
RNT14	RNT15	RNT16	RNT17	RNT18	RNT19	RNT20
YEK 20	YEK21	YEK22	YEK23	YEK24	YEK25	YEK26
15	16	17	18	19	20	21
M6	M7	M8	M9	M10	M11	M12
RNT21	RNT22	RNT23	RNT24	RNT25	RNT26	RNT27
YEK 27	YEK28	YEK29	YEK30	MAG1	MAG2	MAG3
22	23	24	25	26	27	28
M13	M14	M15	M16	M17	M18	M19
RNT28	RNT29	RNT30	KHN1	KHN2	KHN3	KHN4
MAG4	MAG5	MAG6	MAG7	MAG8	MAG9	MAG10
SHARIFU 131						
M20						
KHN5						
MAG11						

KAWAIDA 536/2052-53/6288-89/2044-45

Ashe	Ojobo	Iwa	Araorun	Ile	Itutu	Erinle
1	2	3	4	5	6	7
A18	A19	A20	A21	A22	A23	A24
PIN4	PIN5	PIN6	PIN7	PIN8	PIN9	PIN10
MIY10	MIY11	MIY12	MIY13	MIY14	MIY15	MIY16
8	9	10	11	12	13	14
A25	A26	A27	A28	A29	A30	MA1
PIN11	PIN12	PIN13	PIN14	PIN15	PIN16	PIN17
MIY17	MIY18	MIY19	MIY20	MIY21	MIY22	MIY23
15	16	17	18	19	20	21
MA2	MA3	MA4	MA5	MA6	MA7	MA8
PIN18	PIN19	PIN20	PIN21	PIN22	PIN23	PIN24
MIY24	MIY25	MIY26	MIY27	MIY28	MIY29	MIY30
22	23	24	25	26	27	28
MA9	MA10	MA11	MA12	MA13	MA14	MA15
PIN25	PIN26	PIN27	PIN28	PIN29	PIN30	APP1
GIN1	GIN2	GIN3	GIN4	GIN5	GIN6	GIN7

UHURU 536/2052-53/6288-89/2044-45

Ashe	Ojobo	Iwa	Araorun	Ile	Itutu	Erinle
1	2	3	4	5	6	7
M21	M22	M23	M24	M25	M26	M27
KHN6	KHN7	KHN8	KHN9	KHN10	KHN11	KHN12
MAG12	MAG13	MAG14	MAG15	MAG16	MAG17	MAG18
8	9	10	11	12	13	14
M28	M29	M30	M31	A1	A2	A3
KHN13	KHN14	KHN15	KHN16	KHN17	KHN18	KHN19
MAG19	MAG20	MAG21	MAG22	MAG23	MAG24	MAG25
15	16	17	18	19	20	21
A4	A5	A6	A7	A8	A9	A10
KHN20	KHN21	KHN222	KHN23	KHN24	KHN25	KHN26
MAG26	MAG27	MAG28	MAG29	MAG30	MIY1	MIY2
22	23	24	25	26	27	28
A11	A12	A13	A14	A15	A16	A17
KHN27	KHN28	KHN29	KHN30	PIN1	PIN2	PIN3
MIY3	MIY4	MIY5	MIY6	MIY7	MIY8	MIY9

MABABA WEUSI 536/2052-53/6288-89/2044-45

Ashe	Ojobo	Iwa	Araorun	Ile	Itutu	Erinle
1	2	3	4	5	6	7
MA16	MA17	MA18	MA19	MA20	MA21	MA22
APP2	APP3	APP4	APP5	APP6	APP7	APP8
GIN8	GIN9	GIN10	GIN11	GIN12	GIN13	GIN14
8	9	10	11	12	13	14
MA23	MA24	MA25	MA26	MA27	MA28	MA29
APP9	APP10	APP11	APP12	APP13	APP14	APP15
GIN15	GIN16	GIN17	GIN18	GIN19	GIN20	GIN21
15	16	17	18	19	20	21
MA30	MA31	J1	J2	J3	J4	J5
APP16	APP17	APP18	APP19	APP20	APP21	APP22
GIN22	GIN23	GIN24	GIN25	GIN26	GIN27	GIN28
22	23	24	25	26	27	28
J6	J7	J8	J9	J10	J11	J12
APP23	APP24	APP25	APP26	APP27	APP28	APP29
GIN29	GIN30	SAN1	SAN2	SAN3	SAN4	SAN5

THE NU AFRIKAN CALENDAR

MWANAMKE WEUSI 536/2052-53/6288-89/2044-45						
Ashe	Ojobo	Iwa	Araorun	Ile	Itutu	Erinle
1	2	3	4	5	6	7
J13	J14	J15	J16	J17	J18	J19
APP30	MSR1	MSR2	MSR3	MSR4	MSR5	MSR6
SAN6	SAN7	SAN8	SAN9	SAN10	SAN11	SAN12
8	9	10	11	12	13	14
J20	J21	J22	J23	J24	J25	J26
MSR7	MSR8	MSR9	MSR10	MSR11	MSR12	MSR13
SAN13	SAN14	SAN15	SAN16	SAN17	SAN18	SAN19
15	16	17	18	19	20	21
J27	J28	J29	J30	JU1	JU2	JU3
MSR14	MSR15	MSR16	MSR17	MSR18	MSR19	MSR20
SAN20	SAN21	SAN22	SAN23	SAN24	SAN25	SAN26
22	23	24	25	26	27	28
JU4	JU5	JU6	JU7	JU8	JU9	JU10
MSR21	MRS22	MSR23	MSR24	MSR25	MSR26	MSR27
SAN27	SAN28	SAN29	SAN30	HAM1	HAM2	HAM3

TAIFA WEUSI 536/2052-53/6288-89/2044-45						
Ashe	Ojobo	Iwa	Araorun	Ile	Itutu	Erinle
1	2	3	4	5	6	7
AU8	AU9	AU10	AU11	AU12	AU13	AU14
THT20	THT21	THT22	THT23	THT24	THT25	THT26
NAH2	NAH3	NAH4	NAH5	NAH6	NAH7	NAH8
8	9	10	11	12	13	14
AU15	AU16	AU17	AU18	AU19	AU20	AU21
THT27	THT28	THT29	THT30	PAP1	PAP1	PAP2
NAH9	NAH10	NAH11	NAH12	NAH13	NAH14	NAH15
15	16	17	18	19	20	21
AU22	AU23	AU24	AU25	AU26	AU27	AU28
PAP3	PAP4	PAP5	PAP6	PAP7	PAP8	PAP9
NAH16	NAH17	NAH18	NAH19	NAH20	NAH21	NAH22
22	23	24	25	26	27	28
AU29	AU30	AU31	S1	S2	S3	S4
PAP10	PAP11	PAP12	PAP13	PAP14	PAP15	PAP16
NAH23	NAH24	NAH25	NAH26	NAH27	NAH28	NAH29

MTU WEUSI 536/2052-53/6288-89/2044-45						
Ashe	Ojobo	Iwa	Araorun	Ile	Itutu	Erinle
1	2	3	4	5	6	7
JU11	JU12	JU13	JU14	JU15	JU16	JU17
MSR28	MSR29	MSR30	*ASR1*	*HRU2*	*STH3*	*AST4*
HAM4	HAM5	HAM6	HAM7	HAM8	HAM9	HAM10
8	9	10	11	12	13	14
JU18	JU19	JU20	JU21	JU22	JU23	JU24
NBH5	*ANU6*	THT1	THT2	THT3	THT4	THT5
HAM11	HAM12	HAM13	HAM14	HAM15	HAM16	HAM17
15	16	17	18	19	20	21
JU25	JU26	JU27	JU28	JU29	JU30	JU31
THT6	THT7	THT8	THT9	THT10	THT11	THT12
HAM18	HAM19	HAM20	HAM21	HAM22	HAM23	HAM24
22	23	24	25	26	27	28
AU1	AU2	AU3	AU4	AU5	AU6	AU7
THT13	THT14	THT15	THT16	THT17	THT18	THT19
HAM25	HAM26	HAM27	HAM28	HAM29	HAM30	NAH1

UWEZO 536/2052-53/6288-89/2044-45						
Ashe	Ojobo	Iwa	Araorun	Ile	Itutu	Erinle
1	2	3	4	5	6	7
S5	S6	S7	S8	S9	S10	S11
PAP17	PAP18	PAP19	PAP20	PAP21	PAP22	PAP23
NAH30	*PAG1*	*PAG2*	*PAG3*	*PAG4*	*PAG5*	MAS1
8	9	10	11	12	13	14
S12	S13	S14	S15	S16	S17	S18
PAP24	PAP25	PAP26	PAP27	PAP28	PAP29	PAP30
MAS2	MAS3	MAS4	MAS5	MAS6	MAS7	MAS8
15	16	17	18	19	20	21
S19	S20	S21	S22	S23	S24	S25
HTH1	HTH2	HTH3	HTH4	HTH5	HTH6	HTH7
MAS9	MAS10	MAS11	MAS12	MAS13	MAS14	MAS15
22	23	24	25	26	27	28
S26	S27	S28	S29	S30	O1	O2
HTH8	HTH9	HTH10	HTH11	HTH12	HTH13	HTH14
MAS16	MAS17	MAS18	MAS19	MAS20	MAS21	MAS22

THE NU AFRIKAN CALENDAR

EKUNDU 536/2052-53/6288-89/2044-45						
Ashe	Ojobo	Iwa	Araorun	Ile	Itutu	Erinle
1	2	3	4	5	6	7
O3	O4	O5	O6	O7	O8	O9
HTH15	HTH16	HTH17	HTH18	HTH19	HTH20	HTH21
MAS23	MAS24	MAS25	MAS26	MAS27	MAS28	MAS29
8	9	10	11	12	13	14
O10	O11	O12	O13	O14	O15	O16
HTH22	HTH23	HTH24	HTH25	HTH26	HTH27	HTH28
MAS30	THi1	THi2	THi3	THi4	THi5	THi6
15	16	17	18	19	20	21
O17	O18	O19	O20	O21	O22	O23
HTH29	HTH30	KHK1	KHK2	KHK3	KHK4	KHK5
THi7	THi8	This9	THi10	THi11	THi12	THi13
22	23	24	25	26	27	28
O24	O25	O26	O27	O28	O29	O30
KHK6	KHK7	KHK8	KHK9	KHK10	KHK11	KHK12
Thi14	THi15	THi16	THi17	THi18	THi19	THi20

KIJANI 536/2052-53/6288-89/2044-45						
Ashe	Ojobo	Iwa	Araorun	Ile	Itutu	Erinle
1	2	3	4	5	6	7
N28	N29	N30	D1	D2	D3	D4
TBT11	TBT12	TBT13	TBT14	TBT15	TBT16	TBT17
HID19	HID20	HID21	HID22	HID23	HID24	HID25
8	9	10	11	12	13	14
D5	D6	D7	D8	D9	D10	D11
TBT18	TBT19	TBT20	TBT21	TBT22	TBT23	TBT24
HID26	HID27	HID28	HID29	HID30	TAH1	TAH2
15	16	17	18	19	20	21
D12	D13	D14	D15	D16	D17	D18
TBT25	TBT26	TBT27	TBT28	TBT29	TBT30	MKH1
TAH3	TAH4	TAH5	TAH6	TAH7	TAH8	TAH9
22	23	24	25	26	27	28
D19	D20	D21	D22	D23	D24	D25
MKH2	MKH3	MKH4	MKH5	MKH6	MKH7	MKH8
TAH10	TAH11	TAH12	TAH13	TAH14	TAH15	TAH16

EUSI 536/2052-53/6288-89/2044-45						
Ashe	Ojobo	Iwa	Araorun	Ile	Itutu	Erinle
1	2	3	4	5	6	7
O31	N1	N2	N3	N4	N5	N6
KHK13	KHK14	KHK15	KHK16	KHK17	KHK18	KHK19
THi21	THi22	THi23	THi24	THi25	THi26	THi27
8	9	10	11	12	13	14
N7	N8	N9	N10	N11	N12	N13
KHK20	KHK21	KHK22	KHK23	KHK24	KHK25	KHK26
THi28	THi29	THi30	HID1	HID2	HID3	HID4
15	16	17	18	19	20	21
N14	N15	N16	N17	N18	N19	N20
KHK27	KHK28	KHK29	KHK30	TBT1	TBT2	TBT3
HID5	HID6	HID7	HID8	HID9	HID10	HID11
22	23	24	25	26	27	28
N21	N22	N23	N24	N25	N26	N27
TBT4	TBT5	TBT6	TBT7	TBT8	TBT9	TBT10
HID12	HID13	HID14	HID15	HID16	HID17	HID18

KWANZA 536/2052-53/6288-89/2044-45						
Umoja	Kujichagulia	Ujima	Ujamaa	Nia	Kuumba	Imani
1	2	3	4	5	6	7
D26	D27	D28	D29	D30	D31	JA1
MKH9	MKH10	MKH11	MKH12	MKH13	MKH14	MKH15
TAH17	TAH18	TAH19	TAH20	TAH21	TAH22	TAH23
8	9	10	11	12	13	14
JA2	JA3	JA4	JA5	JA6	JA7	JA8
MKH16	MKH17	MKH18	MKH19	MKH20	MKH21	MKH22
TAH24	TAH25	TAH26	TAH27	TAH28	TAH29	TAH30
15	16	17	18	19	20	21
JA9	JA10	JA11	JA12	JA13	JA14	JA15
MKH23	MKH24	MKH25	KMH26	MKH27	MKH28	MKH29
TIR1	TIR2	TIR3	TIR4	TIR5	TIR6	TIR7
22	23	24	25	26	27	28
JA16	JA17	JA18	JA19	JA20	JA21	JA22
MKH30	AMN1	AMN2	AMN3	AMN4	AMN5	AMN6
TIR8	TIR9	TIR10	TIR11	TIR12	TIR13	TIR14

THE NU AFRIKAN CALENDAR

SHAHIDI 536/2052-53/6288-89/2044-45						
Ashe	Ojobo	Iwa	Araorun	Ile	Itutu	Erinle
1	2	3	4	5	6	7
JA23	JA24	JA25	JA26	JA27	JA28	JA29
AMN7	AMN8	AMN9	AMN10	AMN11	AMN12	AMN13
TIR15	TIR16	TIR17	TIR18	TIR19	TIR20	TIR21
8	9	10	11	12	13	14
JA30	JA31	F1	F2	F3	F4	F5
AMN14	AMN15	AMN16	AMN17	AMN18	AMN19	AMN20
TIR22	TIR23	TIR24	TIR25	TIR26	TIR27	TIR28
15	16	17	18	19	20	21
F6	F7	F8	F9	F10	F 11	F12
AMN21	AMN22	AMN23	AMN24	AMN25	AMN26	AMN28
TIR29	TIR30	YEK1	YEK2	YEK3	YEK4	YEK5
22	23	24	25	26	27	28
F13	F14	F15	F16	F17	F18	F19
AMN29	AMN30	RNT1	RNT2	RNT3	RNT4	RNT5
YEK6	YEK7	YEK8	YEK9	YEK10	YEK11	YEK12
SHAHIDI 87						
F20						
RNT6						
YEK13						

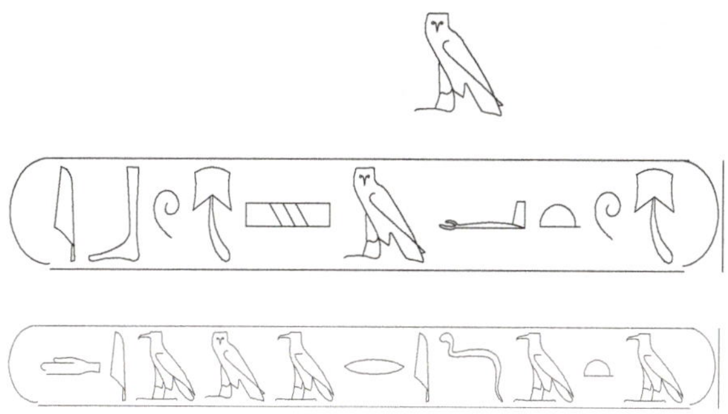

MAAFA 537 B.S.E./ASIENTO 2053-54 C.E.
KAM 6289-90 S.M./EC 2045-46 A.M. (ዓ.ም.)

AFU 537/2053-54/6289-90/2045-46						
she	Ojobo	Iwa	Araorun	Ile	Itutu	Erinle
1	2	3	4	5	6	7
21	F22	F23	F24	F25	F26	F27
NT7	RNT8	RNT9	RNT10	RNT11	RNT12	RNT13
EK14	YEK15	YEK16	YEK17	YEK18	YEK19	YEK20
	9	10	11	12	13	14
28	M1	M2	M3	M4	M5	M6
NT14	RNT15	RNT16	RNT17	RNT18	RNT19	RNT20
EK21	YEK22	YEK23	YEK24	YEK25	YEK26	YEK27
5	16	17	18	19	20	21
17	M8	M9	M10	M11	M12	M13
NT21	RNT22	RNT23	RNT24	RNT25	RNT26	RNT27
EK28	YEK29	YEK30	MAG1	MAG2	MAG3	MAG4
2	23	24	25	26	27	28
14	M15	M16	M17	M18	M19	M20
NT28	RNT29	RNT30	KHN1	KHN2	KHN3	KHN4
AG5	MAG6	MAG7	MAG8	MAG9	MAG10	MAG11

KAWAIDA 537/2053-54/6289-90/2045-46						
Ashe	Ojobo	Iwa	Araorun	Ile	Itutu	Erinle
1	2	3	4	5	6	7
A18	A19	A20	A21	A22	A23	A24
PIN3	PIN4	PIN5	PIN6	PIN7	PIN8	PIN9
MIY10	MIY11	MIY12	MIY13	MIY14	MIY15	MIY16
8	9	10	11	12	13	14
A25	A26	A27	A28	A29	A30	MA1
PIN10	PIN11	PIN12	PIN13	PIN14	PIN15	PIN16
MIY17	MIY18	MIY19	MIY20	MIY21	MIY22	MIY23
15	16	17	18	19	20	21
MA2	MA3	MA4	MA5	MA6	MA7	MA8
PIN17	PIN18	PIN19	PIN20	PIN21	PIN22	PIN23
MIY24	MIY25	MIY26	MIY27	MIY28	MIY29	MIY30
22	23	24	25	26	27	28
MA9	MA10	MA11	MA12	MA13	MA14	MA15
PIN24	PIN25	PIN26	PIN27	PIN28	PIN29	PIN30
GIN1	GIN2	GIN3	GIN4	GIN5	GIN6	GIN7

UHURU 537/2053-54/6289-90/2045-46						
she	Ojobo	Iwa	Araorun	Ile	Itutu	Erinle
	2	3	4	5	6	7
21	M22	M23	M24	M25	M26	M27
HN5	KHN6	KHN7	KHN8	KHN9	KHN10	KHN11
AG12	MAG13	MAG14	MAG15	MAG16	MAG17	MAG18
8	9	10	11	12	13	14
28	M29	M30	M31	A1	A2	A3
HN12	KHN13	KHN14	KHN15	KHN16	KHN17	KHN18
AG19	MAG20	MAG21	MAG22	MAG23	MAG24	MAG25
15	16	17	18	19	20	21
A4	A5	A6	A7	A8	A9	A10
HN19	KHN20	KHN21	KHN22	KHN23	KHN24	KHN25
AG26	MAG27	MAG28	MAG29	MAG30	MIY1	MIY2
22	23	24	25	26	27	28
11	A12	A13	A14	A15	A16	A17
HN26	KHN27	KHN28	KHN29	KHN30	PIN1	PIN2
IY3	MIY4	MIY5	MIY6	MIY7	MIY8	MIY9

MABABA WEUSI 537/2053-54/6289-90/2045-46						
Ashe	Ojobo	Iwa	Araorun	Ile	Itutu	Erinle
1	2	3	4	5	6	7
MA16	MA17	MA18	MA19	MA20	MA21	MA22
APP1	APP2	APP3	APP4	APP5	APP6	APP7
GIN8	GIN9	GIN10	GIN11	GIN12	GIN13	GIN14
8	9	10	11	12	13	14
MA23	MA24	MA25	MA26	MA27	MA28	MA29
APP8	APP9	APP10	APP11	APP12	APP13	APP14
GIN15	GIN16	GIN17	GIN18	GIN19	GIN20	GIN21
15	16	17	18	19	20	21
MA30	MA31	J1	J2	J3	J4	J5
APP15	APP16	APP17	APP18	APP19	APP20	APP21
GIN22	GIN23	GIN24	GIN25	GIN26	GIN27	GIN28
22	23	24	25	26	27	28
J6	J7	J8	J9	J10	J11	J12
APP22	APP23	APP24	APP25	APP26	APP27	APP28
GIN29	GIN30	SAN1	SAN2	SAN3	SAN4	SAN5

MWANAMKE WEUSI 537/2053-54/6289-90/2045-46						
Ashe	Ojobo	Iwa	Araorun	Ile	Itutu	Erinle
1	2	3	4	5	6	7
J13	J14	J15	J16	J17	J18	J19
APP29	APP30	MSR1	MSR2	MSR3	MSR4	MSR5
SAN6	SAN7	SAN8	SAN9	SAN10	SAN11	SAN12
8	9	10	11	12	13	14
J20	J21	J22	J23	J24	J25	J26
MSR6	MSR7	MSR8	MSR9	MSR10	MSR11	MSR12
SAN13	SAN14	SAN15	SAN16	SAN17	SAN18	SAN19
15	16	17	18	19	20	21
J27	J28	J29	J30	JU1	JU2	JU3
MSR13	MSR14	MSR15	MSR16	MSR17	MSR18	MSR19
SAN20	SAN21	SAN22	SAN23	SAN24	SAN25	SAN26
22	23	24	25	26	27	28
JU4	JU5	JU6	JU7	JU8	JU9	JU10
MSR20	MRS21	MSR22	MSR23	MSR24	MSR25	MSR26
SAN27	SAN28	SAN29	SAN30	HAM1	HAM2	HAM3

TAIFA WEUSI 537/2053-54/6289-90/2045-46						
Ashe	Ojobo	Iwa	Araorun	Ile	Itutu	Erinle
1	2	3	4	5	6	7
AU8	AU9	AU10	AU11	AU12	AU13	AU14
THT20	THT21	THT22	THT23	THT24	THT25	THT26
NAH2	NAH3	NAH4	NAH5	NAH6	NAH7	NAH8
8	9	10	11	12	13	14
AU15	AU16	AU17	AU18	AU19	AU20	AU21
THT27	THT28	THT29	THT30	PAP1	PAP1	PAP2
NAH9	NAH10	NAH11	NAH12	NAH13	NAH14	NAH15
15	16	17	18	19	20	21
AU22	AU23	AU24	AU25	AU26	AU27	AU28
PAP3	PAP4	PAP5	PAP6	PAP7	PAP8	PAP9
NAH16	NAH17	NAH18	NAH19	NAH20	NAH21	NAH22
22	23	24	25	26	27	28
AU29	AU30	AU31	S1	S2	S3	S4
PAP10	PAP11	PAP12	PAP13	PAP14	PAP15	PAP16
NAH23	NAH24	NAH25	NAH26	NAH27	NAH28	NAH29

MTU WEUSI 537/2053-54/6289-90/2045-46						
Ashe	Ojobo	Iwa	Araorun	Ile	Itutu	Erinle
1	2	3	4	5	6	7
JU11	JU12	JU13	JU14	JU15	JU16	JU17
MSR27	MSR28	MSR29	MSR30	*ASR1*	*HRU2*	*STH3*
HAM4	HAM5	HAM6	HAM7	HAM8	HAM9	HAM10
8	9	10	11	12	13	14
JU18	JU19	JU20	JU21	JU22	JU23	JU24
AST4	*NBH5*	THT1	THT2	THT3	THT4	THT5
HAM11	HAM12	HAM13	AM14	HAM15	HAM16	HAM17
15	16	17	18	19	20	21
JU25	JU26	JU27	JU28	JU29	JU30	JU31
THT6	THT7	THT8	THT9	THT10	THT11	THT12
HAM18	HAM19	HAM20	HAM21	HAM22	HAM23	HAM24
22	23	24	25	26	27	28
AU1	AU2	AU3	AU4	AU5	AU6	AU7
THT13	THT14	THT15	THT16	THT17	THT18	THT19
HAM25	HAM26	HAM27	HAM28	HAM29	HAM30	NAH1

UWEZO 537/2053-54/6289-90/2045-46						
Ashe	Ojobo	Iwa	Araorun	Ile	Itutu	Erinle
1	2	3	4	5	6	7
S5	S6	S7	S8	S9	S10	S11
PAP17	PAP18	PAP19	PAP20	PAP21	PAP22	PAP23
NAH30	*PAG1*	*PAG2*	*PAG3*	*PAG4*	*PAG5*	MAS1
8	9	10	11	12	13	14
S12	S13	S14	S15	S16	S17	S18
PAP24	PAP25	PAP26	PAP27	PAP28	PAP29	PAP30
MAS2	MAS3	MAS4	MAS5	MAS6	MAS7	MAS8
15	16	17	18	19	20	21
S19	S20	S21	S22	S23	S24	S25
HTH1	HTH2	HTH3	HTH4	HTH5	HTH6	HTH7
MAS9	MAS10	MAS11	MAS12	MAS13	MAS14	MAS15
22	23	24	25	26	27	28
S26	S27	S28	S29	S30	O1	O2
HTH8	HTH9	HTH10	HTH11	HTH12	HTH13	HTH14
MAS16	MAS17	MAS18	MAS19	MAS20	MAS21	MAS22

EKUNDU 537/2053-54/6289-90/2045-46						
Ashe	Ojobo	Iwa	Araorun	Ile	Itutu	Erinle
1	2	3	4	5	6	7
O3	O4	O5	O6	O7	O8	O9
HTH15	HTH16	HTH17	HTH18	HTH19	HTH20	HTH21
MAS23	MAS24	MAS25	MAS26	MAS27	MAS28	MAS29
8	9	10	11	12	13	14
O10	O11	O12	O13	O14	O15	O16
HTH22	HTH23	HTH24	HTH25	HTH26	HTH27	HTH28
MAS30	THi1	THi2	THi3	THi4	THi5	THi6
15	16	17	18	19	20	21
O17	O18	O19	O20	O21	O22	O23
HTH29	HTH30	KHK1	KHK2	KHK3	KHK4	KHK5
THi7	THi8	THi9	THi10	THi11	THi12	THi13
22	23	24	25	26	27	28
O24	O25	O26	O27	O28	O29	O30
KHK6	KHK7	KHK8	KHK9	KHK10	KHK11	KHK12
THi14	THi15	THi16	THi17	THi18	THi19	THi20

KIJANI 537/2053-54/6289-90/2045-46						
Ashe	Ojobo	Iwa	Araorun	Ile	Itutu	Erinle
1	2	3	4	5	6	7
N28	N29	N30	D1	D2	D3	D4
TBT11	TBT12	TBT13	TBT14	TBT15	TBT16	TBT17
HID19	HID20	HID21	HID22	HID23	HID24	HID25
8	9	10	11	12	13	14
D5	D6	D7	D8	D9	D10	D11
TBT18	TBT19	TBT20	TBT21	TBT22	TBT23	TBT24
HID26	HID27	HID28	HID29	HID30	TAH1	TAH2
15	16	17	18	19	20	21
D12	D13	D14	D15	D16	D17	D18
TBT25	TBT26	TBT27	TBT28	TBT29	TBT30	MKH1
TAH3	TAH4	TAH5	TAH6	TAH7	TAH8	TAH9
22	23	24	25	26	27	28
D19	D20	D21	D22	D23	D24	D25
MKH2	MKH3	MKH4	MKH5	MKH6	MKH7	MKH8
TAH10	TAH11	TAH12	TAH13	TAH14	TAH15	TAH16

EUSI 537/2053-54/6289-90/2045-46						
Ashe	Ojobo	Iwa	Araorun	Ile	Itutu	Erinle
1	2	3	4	5	6	7
O31	N1	N2	N3	N4	N5	N6
KHK13	KHK14	KHK15	KHK16	KHK17	KHK18	KHK19
THi21	THi22	THi23	THi24	THi25	THi26	THi27
8	9	10	11	12	13	14
N7	N8	N9	N10	N11	N12	N13
KHK20	KHK21	KHK22	KHK23	KHK24	KHK25	KHK26
THi28	THi29	THi30	HID1	HID2	HID3	HID4
15	16	17	18	19	20	21
N14	N15	N16	N17	N18	N19	N20
KHK27	KHK28	KHK29	KHK30	TBT1	TBT2	TBT3
HID5	HID6	HID7	HID8	HID9	HID10	HID11
22	23	24	25	26	27	28
N21	N22	N23	N24	N25	N26	N27
TBT4	TBT5	TBT6	TBT7	TBT8	TBT9	TBT10
HID12	HID13	HID14	HID15	HID16	HID17	HID18

KWANZA 537/2053-54/6289-90/2045-46						
Umoja	Kujichagulia	Ujima	Ujamaa	Nia	Kuumba	Imani
1	2	3	4	5	6	7
D26	D27	D28	D29	D30	D31	JA1
MKH9	MKH10	MKH11	MKH12	MKH13	MKH14	MKH15
TAH17	TAH18	TAH19	TAH20	TAH21	TAH22	TAH23
8	9	10	11	12	13	14
JA2	JA3	JA4	JA5	JA6	JA7	JA8
MKH16	MKH17	MKH18	MKH19	MKH20	MKH21	MKH22
TAH24	TAH25	TAH26	TAH27	TAH28	TAH29	TAH30
15	16	17	18	19	20	21
JA9	JA10	JA11	JA12	JA13	JA14	JA15
MKH23	MKH24	MKH25	KMH26	MKH27	MKH28	MKH29
TIR1	TIR2	TIR3	TIR4	TIR5	TIR6	TIR7
22	23	24	25	26	27	28
JA16	JA17	JA18	JA19	JA20	JA21	JA22
MKH30	AMN1	AMN2	AMN3	AMN4	AMN5	AMN6
TIR8	TIR9	TIR10	TIR11	TIR12	TIR13	TIR14

SHAHIDI 537/2053-54/6289-90/2045-46						
Ashe	Ojobo	Iwa	Araorun	Ile	Itutu	Erinle
1	2	3	4	5	6	7
JA23	JA24	JA25	JA26	JA27	JA28	JA29
AMN7	AMN8	AMN9	AMN10	AMN11	AMN12	AMN13
TIR15	TIR16	TIR17	TIR18	TIR19	TIR20	TIR21
8	9	10	11	12	13	14
JA30	JA31	F1	F2	F3	F4	F5
AMN14	AMN15	AMN16	AMN17	AMN18	AMN19	AMN20
TIR22	TIR23	TIR24	TIR25	TIR26	TIR27	TIR28
15	16	17	18	19	20	21
F6	F7	F8	F9	F10	F 11	F12
AMN21	AMN22	AMN23	AMN24	AMN25	AMN26	AMN28
TIR29	TIR30	YEK1	YEK2	YEK3	YEK4	YEK5
22	23	24	25	26	27	28
F13	F14	F15	F16	F17	F18	F19
AMN29	AMN30	RNT1	RNT2	RNT3	RNT4	RNT5
YEK6	YEK7	YEK8	YEK9	YEK10	YEK11	YEK12
SHAHIDI 88						
F20						
RNT6						
YEK13						

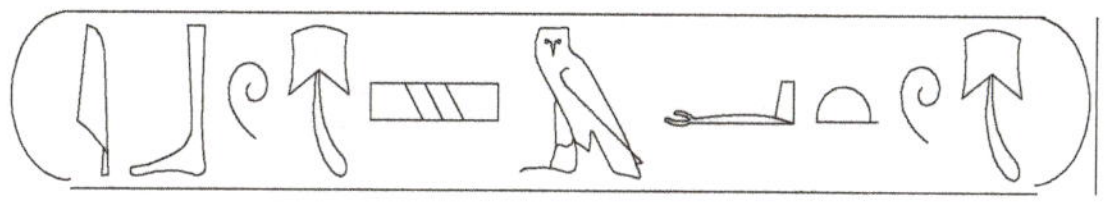

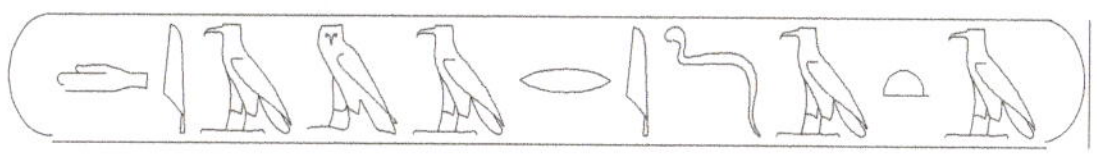

MAAFA 538 B.S.E./ASIENTO 2054-55 C.E.
KAM 6290-91 S.M./EC 2046-47 A.M. (ዓ.ም.)

AFU 538/2054-55/6290-91/2046-47						
shhe	Ojobo	Iwa	Araorun	Ile	Itutu	Erinle
1	2	3	4	5	6	7
21	F22	F23	F24	F25	F26	F27
NT7	RNT8	RNT9	RNT10	RNT11	RNT12	RNT13
EK14	YEK15	YEK16	YEK17	YEK18	YEK19	YEK20
	9	10	11	12	13	14
F28	M1	M2	M3	M4	M5	M6
NT14	RNT15	RNT16	RNT17	RNT18	RNT19	RNT20
EK21	YEK22	YEK23	YEK24	YEK25	YEK26	YEK27
5	16	17	18	19	20	21
M7	M8	M9	M10	M11	M12	M13
NT21	RNT22	RNT23	RNT24	RNT25	RNT26	RNT27
EK28	YEK29	YEK30	MAG1	MAG2	MAG3	MAG4
2	23	24	25	26	27	28
M14	M15	M16	M17	M18	M19	M20
NT28	RNT29	RNT30	KHN1	KHN2	KHN3	KHN4
AG5	MAG6	MAG7	MAG8	MAG9	MAG10	MAG11

KAWAIDA 538/2054-55/6290-91/2046-47						
Ashe	Ojobo	Iwa	Araorun	Ile	Itutu	Erinle
1	2	3	4	5	6	7
A18	A19	A20	A21	A22	A23	A24
PIN3	PIN4	PIN5	PIN6	PIN7	PIN8	PIN9
MIY10	MIY11	MIY12	MIY13	MIY14	MIY15	MIY16
8	9	10	11	12	13	14
A25	A26	A27	A28	A29	A30	MA1
PIN10	PIN11	PIN12	PIN13	PIN14	PIN15	PIN16
MIY17	MIY18	MIY19	MIY20	MIY21	MIY22	MIY23
15	16	17	18	19	20	21
MA2	MA3	MA4	MA5	MA6	MA7	MA8
PIN17	PIN18	PIN19	PIN20	PIN21	PIN22	PIN23
MIY24	MIY25	MIY26	MIY27	MIY28	MIY29	MIY30
22	23	24	25	26	27	28
MA9	MA10	MA11	MA12	MA13	MA14	MA15
PIN24	PIN25	PIN26	PIN27	PIN28	PIN29	PIN30
GIN1	GIN2	GIN3	GIN4	GIN5	GIN6	GIN7

UHURU 538/2054-55/6290-91/2046-47						
shhe	Ojobo	Iwa	Araorun	Ile	Itutu	Erinle
1	2	3	4	5	6	7
M21	M22	M23	M24	M25	M26	M27
HN5	KHN6	KHN7	KHN8	KHN9	KHN10	KHN11
AG12	MAG13	MAG14	MAG15	MAG16	MAG17	MAG18
8	9	10	11	12	13	14
M28	M29	M30	M31	A1	A2	A3
HN12	KHN13	KHN14	KHN15	KHN16	KHN17	KHN18
AG19	MAG20	MAG21	MAG22	MAG23	MAG24	MAG25
15	16	17	18	19	20	21
A4	A5	A6	A7	A8	A9	A10
HN19	KHN20	KHN21	KHN22	KHN23	KHN24	KHN25
AG26	MAG27	MAG28	MAG29	MAG30	MIY1	MIY2
22	23	24	25	26	27	28
A11	A12	A13	A14	A15	A16	A17
HN26	KHN27	KHN28	KHN29	KHN30	PIN1	PIN2
IY3	MIY4	MIY5	MIY6	MIY7	MIY8	MIY9

MABABA WEUSI 538/2054-55/6290-91/2046-47						
Ashe	Ojobo	Iwa	Araorun	Ile	Itutu	Erinle
1	2	3	4	5	6	7
MA16	MA17	MA18	MA19	MA20	MA21	MA22
APP1	APP2	APP3	APP4	APP5	APP6	APP7
GIN8	GIN9	GIN10	GIN11	GIN12	GIN13	GIN14
8	9	10	11	12	13	14
MA23	MA24	MA25	MA26	MA27	MA28	MA29
APP8	APP9	APP10	APP11	APP12	APP13	APP14
GIN15	GIN16	GIN17	GIN18	GIN19	GIN20	GIN21
15	16	17	18	19	20	21
MA30	MA31	J1	J2	J3	J4	J5
APP15	APP16	APP17	APP18	APP19	APP20	APP21
GIN22	GIN23	GIN24	GIN25	GIN26	GIN27	GIN28
22	23	24	25	26	27	28
J6	J7	J8	J9	J10	J11	J12
APP22	APP23	APP24	APP25	APP26	APP27	APP28
GIN29	GIN30	SAN1	SAN2	SAN3	SAN4	SAN5

MWANAMKE WEUSI 538/2054-55/6290-91/2046-47						
Ashe	Ojobo	Iwa	Araorun	Ile	Itutu	Erinle
1	2	3	4	5	6	7
J13	J14	J15	J16	J17	J18	J19
APP29	APP30	MSR1	MSR2	MSR3	MSR4	MSR5
SAN6	SAN7	SAN8	SAN9	SAN10	SAN11	SAN12
8	9	10	11	12	13	14
J20	J21	J22	J23	J24	J25	J26
MSR6	MSR7	MSR8	MSR9	MSR10	MSR11	MSR12
SAN13	SAN14	SAN15	SAN16	SAN17	SAN18	SAN19
15	16	17	18	19	20	21
J27	J28	J29	J30	JU1	JU2	JU3
MSR13	MSR14	MSR15	MSR16	MSR17	MSR18	MSR19
SAN20	SAN21	SAN22	SAN23	SAN24	SAN25	SAN26
22	23	24	25	26	27	28
JU4	JU5	JU6	JU7	JU8	JU9	JU10
MSR20	MRS21	MSR22	MSR23	MSR24	MSR25	MSR26
SAN27	SAN28	SAN29	SAN30	HAM1	HAM2	HAM3

TAIFA WEUSI 538/2054-55/6290-91/2046-47						
Ashe	Ojobo	Iwa	Araorun	Ile	Itutu	Erinle
1	2	3	4	5	6	7
AU8	AU9	AU10	AU11	AU12	AU13	AU14
THT20	THT21	THT22	THT23	THT24	THT25	THT26
NAH2	NAH3	NAH4	NAH5	NAH6	NAH7	NAH8
8	9	10	11	12	13	14
AU15	AU16	AU17	AU18	AU19	AU20	AU21
THT27	THT28	THT29	THT30	PAP1	PAP1	PAP2
NAH9	NAH10	NAH11	NAH12	NAH13	NAH14	NAH15
15	16	17	18	19	20	21
AU22	AU23	AU24	AU25	AU26	AU27	AU28
PAP3	PAP4	PAP5	PAP6	PAP7	PAP8	PAP9
NAH16	NAH17	NAH18	NAH19	NAH20	NAH21	NAH22
22	23	24	25	26	27	28
AU29	AU30	AU31	S1	S2	S3	S4
PAP10	PAP11	PAP12	PAP13	PAP14	PAP15	PAP16
NAH23	NAH24	NAH25	NAH26	NAH27	NAH28	NAH29

MTU WEUSI 538/2054-55/6290-91/2046-47						
Ashe	Ojobo	Iwa	Araorun	Ile	Itutu	Erinle
1	2	3	4	5	6	7
JU11	JU12	JU13	JU14	JU15	JU16	JU17
MSR27	MSR28	MSR29	MSR30	*ASR1*	*HRU2*	*STH3*
HAM4	HAM5	HAM6	HAM7	HAM8	HAM9	HAM10
8	9	10	11	12	13	14
JU18	JU19	JU20	JU21	JU22	JU23	JU24
AST4	*NBH5*	THT1	THT2	THT3	THT4	THT5
HAM11	HAM12	HAM13	AM14	HAM15	HAM16	HAM17
15	16	17	18	19	20	21
JU25	JU26	JU27	JU28	JU29	JU30	JU31
THT6	THT7	THT8	THT9	THT10	THT11	THT12
HAM18	HAM19	HAM20	HAM21	HAM22	HAM23	HAM24
22	23	24	25	26	27	28
AU1	AU2	AU3	AU4	AU5	AU6	AU7
THT13	THT14	THT15	THT16	THT17	THT18	THT19
HAM25	HAM26	HAM27	HAM28	HAM29	HAM30	NAH1

UWEZO 538/2054-55/6290-91/2046-47						
Ashe	Ojobo	Iwa	Araorun	Ile	Itutu	Erinle
1	2	3	4	5	6	7
S5	S6	S7	S8	S9	S10	S11
PAP17	PAP18	PAP19	PAP20	PAP21	PAP22	PAP23
NAH30	*PAG1*	*PAG2*	*PAG3*	*PAG4*	*PAG5*	MAS1
8	9	10	11	12	13	14
S12	S13	S14	S15	S16	S17	S18
PAP24	PAP25	PAP26	PAP27	PAP28	PAP29	PAP30
MAS2	MAS3	MAS4	MAS5	MAS6	MAS7	MAS8
15	16	17	18	19	20	21
S19	S20	S21	S22	S23	S24	S25
HTH1	HTH2	HTH3	HTH4	HTH5	HTH6	HTH7
MAS9	MAS10	MAS11	MAS12	MAS13	MAS14	MAS15
22	23	24	25	26	27	28
S26	S27	S28	S29	S30	O1	O2
HTH8	HTH9	HTH10	HTH11	HTH12	HTH13	HTH14
MAS16	MAS17	MAS18	MAS19	MAS20	MAS21	MAS22

EKUNDU 538/2054-55/6290-91/2046-47						
Ashe	Ojobo	Iwa	Araorun	Ile	Itutu	Erinle
1	2	3	4	5	6	7
O3	O4	O5	O6	O7	O8	O9
HTH15	HTH16	HTH17	HTH18	HTH19	HTH20	HTH21
MAS23	MAS24	MAS25	MAS26	MAS27	MAS28	MAS29
8	9	10	11	12	13	14
O10	O11	O12	O13	O14	O15	O16
HTH22	HTH23	HTH24	HTH25	HTH26	HTH27	HTH28
MAS30	THi1	THi2	THi3	THi4	THi5	THi6
15	16	17	18	19	20	21
O17	O18	O19	O20	O21	O22	O23
HTH29	HTH30	KHK1	KHK2	KHK3	KHK4	KHK5
THi7	THi8	THi9	THi10	THi11	THi12	THi13
22	23	24	25	26	27	28
O24	O25	O26	O27	O28	O29	O30
KHK6	KHK7	KHK8	KHK9	KHK10	KHK11	KHK12
THi14	THi15	THi16	THi17	THi18	THi19	THi20

KIJANI 538/2054-55/6290-91/2046-47						
Ashe	Ojobo	Iwa	Araorun	Ile	Itutu	Erinle
1	2	3	4	5	6	7
N28	N29	N30	D1	D2	D3	D4
TBT11	TBT12	TBT13	TBT14	TBT15	TBT16	TBT17
HID19	HID20	HID21	HID22	HID23	HID24	HID25
8	9	10	11	12	13	14
D5	D6	D7	D8	D9	D10	D11
TBT18	TBT19	TBT20	TBT21	TBT22	TBT23	TBT24
HID26	HID27	HID28	HID29	HID30	TAH1	TAH2
15	16	17	18	19	20	21
D12	D13	D14	D15	D16	D17	D18
TBT25	TBT26	TBT27	TBT28	TBT29	TBT30	MKH1
TAH3	TAH4	TAH5	TAH6	TAH7	TAH8	TAH9
22	23	24	25	26	27	28
D19	D20	D21	D22	D23	D24	D25
MKH2	MKH3	MKH4	MKH5	MKH6	MKH7	MKH8
TAH10	TAH11	TAH12	TAH13	TAH14	TAH15	TAH16

EUSI 538/2054-55/6290-91/2046-47						
Ashe	Ojobo	Iwa	Araorun	Ile	Itutu	Erinle
1	2	3	4	5	6	7
O31	N1	N2	N3	N4	N5	N6
KHK13	KHK14	KHK15	KHK16	KHK17	KHK18	KHK19
THi21	THi22	THi23	THi24	THi25	THi26	THi27
8	9	10	11	12	13	14
N7	N8	N9	N10	N11	N12	N13
KHK20	KHK21	KHK22	KHK23	KHK24	KHK25	KHK26
THi28	THi29	THi30	HID1	HID2	HID3	HID4
15	16	17	18	19	20	21
N14	N15	N16	N17	N18	N19	N20
KHK27	KHK28	KHK29	KHK30	TBT1	TBT2	TBT3
HID5	HID6	HID7	HID8	HID9	HID10	HID11
22	23	24	25	26	27	28
N21	N22	N23	N24	N25	N26	N27
TBT4	TBT5	TBT6	TBT7	TBT8	TBT9	TBT10
HID12	HID13	HID14	HID15	HID16	HID17	HID18

KWANZA 538/2054-55/6290-91/2046-47						
Umoja	Kujichagulia	Ujima	Ujamaa	Nia	Kuumba	Imani
1	2	3	4	5	6	7
D26	D27	D28	D29	D30	D31	JA1
MKH9	MKH10	MKH11	MKH12	MKH13	MKH14	MKH15
TAH17	TAH18	TAH19	TAH20	TAH21	TAH22	TAH23
8	9	10	11	12	13	14
JA2	JA3	JA4	JA5	JA6	JA7	JA8
MKH16	MKH17	MKH18	MKH19	MKH20	MKH21	MKH22
TAH24	TAH25	TAH26	TAH27	TAH28	TAH29	TAH30
15	16	17	18	19	20	21
JA9	JA10	JA11	JA12	JA13	JA14	JA15
MKH23	MKH24	MKH25	KMH26	MKH27	MKH28	MKH29
TIR1	TIR2	TIR3	TIR4	TIR5	TIR6	TIR7
22	23	24	25	26	27	28
JA16	JA17	JA18	JA19	JA20	JA21	JA22
MKH30	AMN1	AMN2	AMN3	AMN4	AMN5	AMN6
TIR8	TIR9	TIR10	TIR11	TIR12	TIR13	TIR14

SHAHIDI 538/2054-55/6290-91/2046-47						
Ashe	Ojobo	Iwa	Araorun	Ile	Itutu	Erinle
1	2	3	4	5	6	7
JA23	JA24	JA25	JA26	JA27	JA28	JA29
AMN7	AMN8	AMN9	AMN10	AMN11	AMN12	AMN13
TIR15	TIR16	TIR17	TIR18	TIR19	TIR20	TIR21
8	9	10	11	12	13	14
JA30	JA31	F1	F2	F3	F4	F5
AMN14	AMN15	AMN16	AMN17	AMN18	AMN19	AMN20
TIR22	TIR23	TIR24	TIR25	TIR26	TIR27	TIR28
15	16	17	18	19	20	21
F6	F7	F8	F9	F10	F 11	F12
AMN21	AMN22	AMN23	AMN24	AMN25	AMN26	AMN28
TIR29	TIR30	YEK1	YEK2	YEK3	YEK4	YEK5
22	23	24	25	26	27	28
F13	F14	F15	F16	F17	F18	F19
AMN29	AMN30	RNT1	RNT2	RNT3	RNT4	RNT5
YEK6	YEK7	YEK8	YEK9	YEK10	YEK11	YEK12
SHAHIDI 89						
F20						
RNT6						
YEK13						

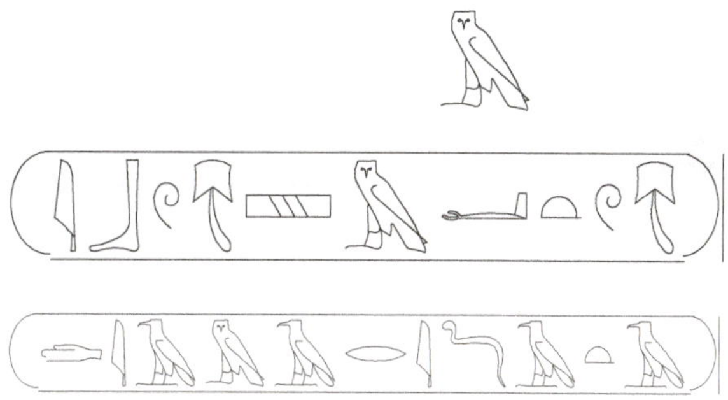

MAAFA 539 B.S.E./ASIENTO 2055-56 C.E./
KAM 6291-92 S.M./EC 2047-48 A.M. (ዓ.ም.)

AFU 539/2055-56/6291-92/2047-48

Ashe	Ojobo	Iwa	Araorun	Ile	Itutu	Erinle
1	2	3	4	5	6	7
F21	F22	F23	F24	F25	F26	F27
RNT7	RNT8	RNT9	RNT10	RNT11	RNT12	RNT13
YEK14	YEK15	YEK16	YEK17	YEK18	YEK19	YEK20
8	9	10	11	12	13	14
F28	M1	M2	M3	M4	M5	M6
RNT14	RNT15	RNT16	RNT17	RNT18	RNT19	RNT20
YEK21	YEK22	YEK23	YEK24	YEK25	YEK26	YEK27
15	16	17	18	19	20	21
M7	M8	M9	M10	M11	M12	M13
RNT21	RNT22	RNT23	RNT24	RNT25	RNT26	RNT27
YEK28	YEK 29	YEK30	MAG1	MAG2	MAG3	MAG4
22	23	24	25	26	27	28
M14	M15	M16	M17	M18	M19	M20
RNT28	RNT29	RNT30	KHN1	KHN2	KHN3	KHN4
MAG5	MAG6	MAG7	MAG8	MAG9	MAG10	MAG11

KAWAIDA 539/2055-56/6291-92/2047-48

Ashe	Ojobo	Iwa	Araorun	Ile	Itutu	Erinle
1	2	3	4	5	6	7
A18	A19	A20	A21	A22	A23	A24
PIN3	PIN4	PIN5	PIN6	PIN7	PIN8	PIN9
MIY10	MIY11	MIY12	MIY13	MIY14	MIY15	MIY16
8	9	10	11	12	13	14
A25	A26	A27	A28	A29	A30	MA1
PIN10	PIN11	PIN12	PIN13	PIN14	PIN15	PIN16
MIY17	MIY18	MIY19	MIY20	MIY21	MIY22	MIY23
15	16	17	18	19	20	21
MA2	MA3	MA4	MA5	MA6	MA7	MA8
PIN17	PIN18	PIN19	PIN20	PIN21	PIN22	PIN23
MIY24	MIY25	MIY26	MIY27	MIY28	MIY29	MIY30
22	23	24	25	26	27	28
MA9	MA10	MA11	MA12	MA13	MA14	MA15
PIN24	PIN25	PIN26	PIN27	PIN28	PIN29	PIN30
GIN1	GIN2	GIN3	GIN4	GIN5	GIN6	GIN7

UHURU 539/2055-56/6291-92/2047-48

Ashe	Ojobo	Iwa	Araorun	Ile	Itutu	Erinle
1	2	3	4	5	6	7
M21	M22	M23	M24	M25	M26	M27
KHN5	KHN6	KHN7	KHN8	KHN9	KHN10	KHN11
MAG12	MAG13	MAG14	MAG15	MAG16	MAG17	MAG18
8	9	10	11	12	13	14
M28	M29	M30	M31	A1	A2	A3
KHN12	KHN13	KHN14	KHN15	KHN16	KHN17	KHN18
MAG19	MAG20	MAG21	MAG22	MAG23	MAG24	MAG25
15	16	17	18	19	20	21
A4	A5	A6	A7	A8	A9	A10
KHN19	KHN20	KHN21	KHN22	KHN23	KHN24	KHN25
MAG26	MAG27	MAG28	MAG29	MAG30	MIY1	MIY2
22	23	24	25	26	27	28
A11	A12	A13	A14	A15	A16	A17
KHN26	KHN27	KHN28	KHN29	KHN30	PIN1	PIN2
MIY3	MIY4	MIY5	MIY6	MIY7	MIY8	MIY9

MABABA WEUSI 539/2055-56/6291-92/2047-48

Ashe	Ojobo	Iwa	Araorun	Ile	Itutu	Erinle
1	2	3	4	5	6	7
MA16	MA17	MA18	MA19	MA20	MA21	MA22
APP1	APP2	APP3	APP4	APP5	APP6	APP7
GIN8	GIN9	GIN10	GIN11	GIN12	GIN13	GIN14
8	9	10	11	12	13	14
MA23	MA24	MA25	MA26	MA27	MA28	MA29
APP8	APP9	APP10	APP11	APP12	APP13	APP14
GIN15	GIN16	GIN17	GIN18	GIN19	GIN20	GIN21
15	16	17	18	19	20	21
MA30	MA31	J1	J2	J3	J4	J5
APP15	APP16	APP17	APP18	APP19	APP20	APP21
GIN22	GIN23	GIN24	GIN25	GIN26	GIN27	GIN28
22	23	24	25	26	27	28
J6	J7	J8	J9	J10	J11	J12
APP22	APP23	APP24	APP25	APP26	APP27	APP28
GIN29	GIN30	SAN1	SAN2	SAN3	SAN4	SAN5

MWANAMKE WEUSI 539/2055-56/6291-92/2047-48						
Ashe	Ojobo	Iwa	Araorun	Ile	Itutu	Erinle
1	2	3	4	5	6	7
J13	J14	J15	J16	J17	J18	J19
APP29	APP30	MSR1	MSR2	MSR3	MSR4	MSR5
SAN6	SAN7	SAN8	SAN9	SAN10	SAN11	SAN12
8	9	10	11	12	13	14
J20	J21	J22	J23	J24	J25	J26
MSR6	MSR7	MSR8	MSR9	MSR10	MSR11	MSR12
SAN13	SAN14	SAN15	SAN16	SAN17	SAN18	SAN19
15	16	17	18	19	20	21
J27	J28	J29	J30	JU1	JU2	JU3
MSR13	MSR14	MSR15	MSR16	MSR17	MSR18	MSR19
SAN20	SAN21	SAN22	SAN23	SAN24	SAN25	SAN26
22	23	24	25	26	27	28
JU4	JU5	JU6	JU7	JU8	JU9	JU10
MSR20	MRS21	MSR22	MSR23	MSR24	MSR25	MSR26
SAN27	SAN28	SAN29	SAN30	HAM1	HAM2	HAM3

TAIFA WEUSI 539/2055-56/6291-92/2047-48						
Ashe	Ojobo	Iwa	Araorun	Ile	Itutu	Erinl
1	2	3	4	5	6	7
AU8	AU9	AU10	AU11	AU12	AU13	AU1
THT20	THT21	THT22	THT23	THT24	THT25	THT
NAH2	NAH3	NAH4	NAH5	NAH6	NAH7	NAH
8	9	10	11	12	13	14
AU15	AU16	AU17	AU18	AU19	AU20	AU21
THT27	THT28	THT29	THT30	PAP1	PAP1	PAP
NAH9	NAH10	NAH11	NAH12	NAH13	NAH14	NAH
15	16	17	18	19	20	21
AU22	AU23	AU24	AU25	AU26	AU27	AU2
PAP3	PAP4	PAP5	PAP6	PAP7	PAP8	PAP
NAH16	NAH17	NAH18	NAH19	NAH20	NAH21	NAH
22	23	24	25	26	27	28
AU29	AU30	AU31	S1	S2	S3	S4
PAP10	PAP11	PAP12	PAP13	PAP14	PAP15	PAP1
NAH23	NAH24	NAH25	NAH26	NAH27	NAH28	NAH

MTU WEUSI 539/2055-56/6291-92/2047-48						
Ashe	Ojobo	Iwa	Araorun	Ile	Itutu	Erinle
1	2	3	4	5	6	7
JU11	JU12	JU13	JU14	JU15	JU16	JU17
MSR27	MSR28	MSR29	MSR30	*ASR1*	*HRU2*	*STH3*
HAM4	HAM5	HAM6	HAM7	HAM8	HAM9	HAM10
8	9	10	11	12	13	14
JU18	JU19	JU20	JU21	JU22	JU23	JU24
AST4	*NBH5*	THT1	THT2	THT3	THT4	THT5
HAM11	HAM12	HAM13	HAM14	HAM15	HAM16	HAM17
15	16	17	18	19	20	21
JU25	JU26	JU27	JU28	JU29	JU30	JU31
THT6	THT7	THT8	THT9	THT10	THT11	THT12
HAM18	HAM19	HAM20	HAM21	HAM22	HAM23	HAM24
22	23	24	25	26	27	28
AU1	AU2	AU3	AU4	AU5	AU6	AU7
THT13	THT14	THT15	THT16	THT17	THT18	THT19
HAM25	HAM26	HAM27	HAM28	HAM29	HAM30	NAH1

UWEZO 539/2055-56/6291-92/2047-48						
Ashe	Ojobo	Iwa	Araorun	Ile	Itutu	Erinl
1	2	3	4	5	6	7
S5	S6	S7	S8	S9	S10	S11
PAP17	PAP18	PAP19	PAP20	PAP21	PAP22	PAP2
NAH30	*QUA1*	*QUA2*	*QUA3*	*QUA4*	*QUA5*	*QUA6*
8	9	10	11	12	13	14
S12	S13	S14	S15	S16	S17	S18
PAP24	PAP25	PAP26	PAP27	PAP28	PAP29	PAP3
MAS1	MAS2	MAS3	MAS4	MAS5	MAS6	MAS
15	16	17	18	19	20	21
S19	S20	S21	S22	S23	S24	S25
HTH1	HTH2	HTH3	HTH4	HTH5	HTH6	HTH
MAS8	MAS9	MAS10	MAS11	MAS12	MAS13	MAS1
22	23	24	25	26	27	28
S26	S27	S28	S29	S30	O1	O2
HTH8	HTH9	HTH10	HTH11	HTH12	HTH13	HTH
MAS15	MAS16	MAS17	MAS18	MAS19	MAS20	MAS

EKUNDU 539/2055-56/6291-92/2047-48

Ashe	Ojobo	Iwa	Araorun	Ile	Itutu	Erinle
1	2	3	4	5	6	7
O3	O4	O5	O6	O7	O8	O9
HTH15	HTH16	HTH17	HTH18	HTH19	HTH20	HTH21
MAS22	MAS23	MAS 24	MAS25	MAS26	MAS27	MAS28
8	9	10	11	12	13	14
O10	O11	O12	O13	O14	O15	O16
HTH22	HTH23	HTH24	HTH25	HTH26	HTH27	HTH28
MAS29	MAS30	THi1	THi2	THi3	THi4	THi5
15	16	17	18	19	20	21
O17	O18	O19	O20	O21	O22	O23
HTH29	HTH30	KHK1	KHK2	KHK3	KHK4	KHK5
THi6	THi7	THi8	THi9	THi10	THi11	THi12
22	23	24	25	26	27	28
O24	O25	O26	O27	O28	O29	O30
KHK6	KHK7	KHK8	KHK9	KHK10	KHK11	KHK12
THi13	THi14	THi15	THi16	THi17	THi18	THi19

KIJANI 539/2055-56/6291-92/2047-48

Ashe	Ojobo	Iwa	Araorun	Ile	Itutu	Erinle
1	2	3	4	5	6	7
N28	N29	N30	D1	D2	D3	D4
TBT11	TBT12	TBT13	TBT14	TBT15	TBT16	TBT17
HID18	HID19	HID20	HID21	HID22	HID23	HID24
8	9	10	11	12	13	14
D5	D6	D7	D8	D9	D10	D11
TBT18	TBT19	TBT20	TBT21	TBT22	TBT23	TBT24
HID25	HID26	HID27	HID28	HID29	HID30	TAH1
15	16	17	18	19	20	21
D12	D13	D14	D15	D16	D17	D18
TBT25	TBT26	TBT27	TBT28	TBT29	TBT30	MKH1
TAH2	TAH3	TAH4	TAH5	TAH6	TAH7	TAH8
22	23	24	25	26	27	28
D19	D20	D21	D22	D23	D24	D25
MKH2	MKH3	MKH4	MKH5	MKH6	MKH7	MKH8
TAH9	TAH10	TAH11	TAH12	TAH13	TAH14	TAH15

EUSI 539/2055-56/6291-92/2047-48

Ashe	Ojobo	Iwa	Araorun	Ile	Itutu	Erinle
1	2	3	4	5	6	7
O31	N1	N2	N3	N4	N5	N6
KHK13	KHK14	KHK15	KHK16	KHK17	KHK18	KHK19
THi20	THi21	THi22	THi23	THi24	THi25	THi26
8	9	10	11	12	13	14
N7	N8	N9	N10	N11	N12	N13
KHK20	KHK21	KHK22	KHK23	KHK24	KHK25	KHK26
THi27	THi28	THi29	THi30	HID1	HID2	HID3
15	16	17	18	19	20	21
N14	N15	N16	N17	N18	N19	N20
KHK27	KHK28	KHK29	KHK30	TBT1	TBT2	TBT3
HID4	HID5	HID6	HID7	HID8	HID9	HID10
22	23	24	25	26	27	28
N21	N22	N23	N24	N25	N26	N27
TBT4	TBT5	TBT6	TBT7	TBT8	TBT9	TBT10
HID11	HID12	HID13	HID14	HID15	HID16	HID17

KWANZA 539/2055-56/6291-92/2047-48

Umoja	Kujichagulia	Ujima	Ujamaa	Nia	Kuumba	Imani
1	2	3	4	5	6	7
D26	D27	D28	D29	D30	D31	JA1
MKH9	MKH10	MKH11	MKH12	MKH13	MKH14	MKH1
TAH16	TAH17	TAH18	TAH19	TAH20	TAH21	TAH2
8	9	10	11	12	13	14
JA2	JA3	JA4	JA5	JA6	JA7	JA
MKH16	MKH17	MKH18	MKH19	MKH20	MKH21	MKH2
TAH23	TAH24	TAH25	TAH26	TAH27	TAH28	TAH29
15	16	17	18	19	20	21
JA9	JA10	JA11	JA12	JA13	JA14	JA1
MKH23	MKH24	MKH25	KMH26	MKH27	MKH28	MKH2
TAH30	TIR1	TIR2	TIR3	TIR4	TIR5	TIR6
22	23	24	25	26	27	28
JA16	JA17	JA18	JA19	JA20	JA21	JA2
MKH30	AMN1	AMN2	AMN3	AMN4	AMN5	AMN
TIR7	TIR8	TIR9	TIR10	TIR11	TIR12	TIR13

SHAHIDI 539/2055-56/6291-92/2047-48						
Ashe	Ojobo	Iwa	Araorun	Ile	Itutu	Erinle
1	2	3	4	5	6	7
JA23	JA24	JA25	JA26	JA27	JA28	JA29
AMN7	AMN8	AMN9	AMN10	AMN11	AMN12	AMN13
TIR14	TIR15	TIR16	TIR17	TIR18	TIR19	TIR20
8	9	10	11	12	13	14
JA30	JA31	F1	F2	F3	F4	F5
AMN14	AMN15	AMN16	AMN17	AMN18	AMN19	AMN20
TIR21	TIR22	TIR23	TIR24	TIR25	TIR26	TIR27
15	16	17	18	19	20	21
F6	F7	F8	F9	F10	F 11	F12
AMN21	AMN22	AMN23	AMN24	AMN25	AMN26	AMN28
TIR28	TIR29	TIR30	YEK1	YEK2	YEK3	YEK4
22	23	24	25	26	27	28
F13	F14	F15	F16	F17	F18	F19
AMN29	AMN30	RNT1	RNT2	RNT3	RNT4	RNT5
YEK5	YEK6	YEK7	YEK8	YEK9	YEK10	YEK11
SHAHIDI 90						
F20						
RNT6						
YEK12						

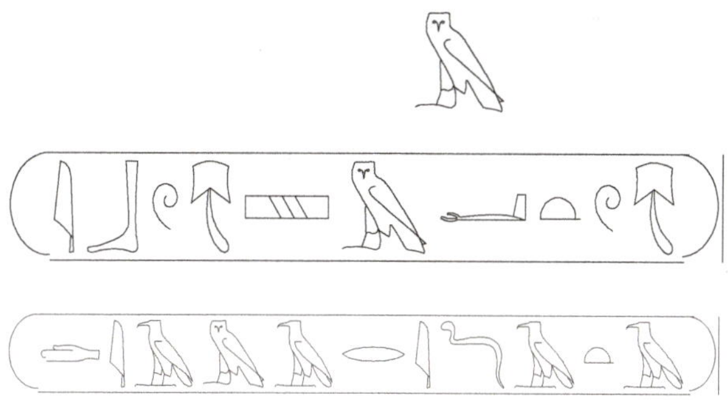

THE NU AFRIKAN CALENDAR

MAAFA 540 B.S.E./ASIENTO 2056-57 C.E./
KAM 6292-93 S.M./EC 2048-49 A.M. (ዓ.ም.)

AFU 540/2056-57/6292-93/2048-49						
Ashe	Ojobo	Iwa	Araorun	Ile	Itutu	Erinle
1	2	3	4	5	6	7
F21	F22	F23	F24	F25	F26	F27
RNT7	RNT8	RNT9	RNT10	RNT11	RNT12	RNT13
YEK13	YEK14	YEK15	YEK16	YEK17	YEK18	YEK19
8	9	10	11	12	13	14
F28	F29	M1	M2	M3	M4	M5
RNT14	RNT15	RNT16	RNT17	RNT18	RNT19	RNT20
YEK 20	YEK21	YEK22	YEK23	YEK24	YEK25	YEK26
15	16	17	18	19	20	21
M6	M7	M8	M9	M10	M11	M12
RNT21	RNT22	RNT23	RNT24	RNT25	RNT26	RNT27
YEK 27	YEK28	YEK29	YEK30	MAG1	MAG2	MAG3
22	23	24	25	26	27	28
M13	M14	M15	M16	M17	M18	M19
RNT28	RNT29	RNT30	KHN1	KHN2	KHN3	KHN4
MAG4	MAG5	MAG6	MAG7	MAG8	MAG9	MAG10
SHARIFU 132						
M20						
KHN5						
MAG11						

KAWAIDA 540/2056-57/6292-93/2048-49						
Ashe	Ojobo	Iwa	Araorun	Ile	Itutu	Erinle
1	2	3	4	5	6	7
A18	A19	A20	A21	A22	A23	A24
PIN4	PIN5	PIN6	PIN7	PIN8	PIN9	PIN10
MIY10	MIY11	MIY12	MIY13	MIY14	MIY15	MIY16
8	9	10	11	12	13	14
A25	A26	A27	A28	A29	A30	MA1
PIN11	PIN12	PIN13	PIN14	PIN15	PIN16	PIN17
MIY17	MIY18	MIY19	MIY20	MIY21	MIY22	MIY23
15	16	17	18	19	20	21
MA2	MA3	MA4	MA5	MA6	MA7	MA8
PIN18	PIN19	PIN20	PIN21	PIN22	PIN23	PIN24
MIY24	MIY25	MIY26	MIY27	MIY28	MIY29	MIY30
22	23	24	25	26	27	28
MA9	MA10	MA11	MA12	MA13	MA14	MA15
PIN25	PIN26	PIN27	PIN28	PIN29	PIN30	APP1
GIN1	GIN2	GIN3	GIN4	GIN5	GIN6	GIN7

UHURU 540/2056-57/6292-93/2048-49						
Ashe	Ojobo	Iwa	Araorun	Ile	Itutu	Erinle
1	2	3	4	5	6	7
M21	M22	M23	M24	M25	M26	M27
KHN6	KHN7	KHN8	KHN9	KHN10	KHN11	KHN12
MAG12	MAG13	MAG14	MAG15	MAG16	MAG17	MAG18
8	9	10	11	12	13	14
M28	M29	M30	M31	A1	A2	A3
KHN13	KHN14	KHN15	KHN16	KHN17	KHN18	KHN19
MAG19	MAG20	MAG21	MAG22	MAG23	MAG24	MAG25
15	16	17	18	19	20	21
A4	A5	A6	A7	A8	A9	A10
KHN20	KHN21	KHN222	KHN23	KHN24	KHN25	KHN26
MAG26	MAG27	MAG28	MAG29	MAG30	MIY1	MIY2
22	23	24	25	26	27	28
A11	A12	A13	A14	A15	A16	A17
KHN27	KHN28	KHN29	KHN30	PIN1	PIN2	PIN3
MIY3	MIY4	MIY5	MIY6	MIY7	MIY8	MIY9

MABABA WEUSI 540/2056-57/6292-93/2048-49						
Ashe	Ojobo	Iwa	Araorun	Ile	Itutu	Erinle
1	2	3	4	5	6	7
MA16	MA17	MA18	MA19	MA20	MA21	MA22
APP2	APP3	APP4	APP5	APP6	APP7	APP8
GIN8	GIN9	GIN10	GIN11	GIN12	GIN13	GIN14
8	9	10	11	12	13	14
MA23	MA24	MA25	MA26	MA27	MA28	MA29
APP9	APP10	APP11	APP12	APP13	APP14	APP15
GIN15	GIN16	GIN17	GIN18	GIN19	GIN20	GIN21
15	16	17	18	19	20	21
MA30	MA31	J1	J2	J3	J4	J5
APP16	APP17	APP18	APP19	APP20	APP21	APP22
GIN22	GIN23	GIN24	GIN25	GIN26	GIN27	GIN28
22	23	24	25	26	27	28
J6	J7	J8	J9	J10	J11	J12
APP23	APP24	APP25	APP26	APP27	APP28	APP29
GIN29	GIN30	SAN1	SAN2	SAN3	SAN4	SAN5

THE NU AFRIKAN CALENDAR

MWANAMKE WEUSI 540/2056-57/6292-93/2048-49

Ashe	Ojobo	Iwa	Araorun	Ile	Itutu	Erinle
1	2	3	4	5	6	7
J13	J14	J15	J16	J17	J18	J19
APP30	MSR1	MSR2	MSR3	MSR4	MSR5	MSR6
SAN6	SAN7	SAN8	SAN9	SAN10	SAN11	SAN12
8	9	10	11	12	13	14
J20	J21	J22	J23	J24	J25	J26
MSR7	MSR8	MSR9	MSR10	MSR11	MSR12	MSR13
SAN13	SAN14	SAN15	SAN16	SAN17	SAN18	SAN19
15	16	17	18	19	20	21
J27	J28	J29	J30	JU1	JU2	JU3
MSR14	MSR15	MSR16	MSR17	MSR18	MSR19	MSR20
SAN20	SAN21	SAN22	SAN23	SAN24	SAN25	SAN26
22	23	24	25	26	27	28
JU4	JU5	JU6	JU7	JU8	JU9	JU10
MSR21	MRS22	MSR23	MSR24	MSR25	MSR26	MSR27
SAN27	SAN28	SAN29	SAN30	HAM1	HAM2	HAM3

TAIFA WEUSI 540/2056-57/6292-93/2048-49

Ashe	Ojobo	Iwa	Araorun	Ile	Itutu	Erinle
1	2	3	4	5	6	7
AU8	AU9	AU10	AU11	AU12	AU13	AU14
THT20	THT21	THT22	THT23	THT24	THT25	THT26
NAH2	NAH3	NAH4	NAH5	NAH6	NAH7	NAH8
8	9	10	11	12	13	14
AU15	AU16	AU17	AU18	AU19	AU20	AU21
THT27	THT28	THT29	THT30	PAP1	PAP1	PAP2
NAH9	NAH10	NAH11	NAH12	NAH13	NAH14	NAH15
15	16	17	18	19	20	21
AU22	AU23	AU24	AU25	AU26	AU27	AU28
PAP3	PAP4	PAP5	PAP6	PAP7	PAP8	PAP9
NAH16	NAH17	NAH18	NAH19	NAH20	NAH21	NAH22
22	23	24	25	26	27	28
AU29	AU30	AU31	S1	S2	S3	S4
PAP10	PAP11	PAP12	PAP13	PAP14	PAP15	PAP16
NAH23	NAH24	NAH25	NAH26	NAH27	NAH28	NAH29

MTU WEUSI 540/2056-57/6292-93/2048-49

Ashe	Ojobo	Iwa	Araorun	Ile	Itutu	Erinle
1	2	3	4	5	6	7
JU11	JU12	JU13	JU14	JU15	JU16	JU17
MSR28	MSR29	MSR30	*ASR1*	*HRU2*	*STH3*	*AST4*
HAM4	HAM5	HAM6	HAM7	HAM8	HAM9	HAM10
8	9	10	11	12	13	14
JU18	JU19	JU20	JU21	JU22	JU23	JU24
NBH5	*ANU6*	THT1	THT2	THT3	THT4	THT5
HAM11	HAM12	HAM13	HAM14	HAM15	HAM16	HAM17
15	16	17	18	19	20	21
JU25	JU26	JU27	JU28	JU29	JU30	JU31
THT6	THT7	THT8	THT9	THT10	THT11	THT12
HAM18	HAM19	HAM20	HAM21	HAM22	HAM23	HAM24
22	23	24	25	26	27	28
AU1	AU2	AU3	AU4	AU5	AU6	AU7
THT13	THT14	THT15	THT16	THT17	THT18	THT19
HAM25	HAM26	HAM27	HAM28	HAM29	HAM30	NAH1

UWEZO 540/2056-57/6292-93/2048-49

Ashe	Ojobo	Iwa	Araorun	Ile	Itutu	Erinle
1	2	3	4	5	6	7
S5	S6	S7	S8	S9	S10	S11
PAP17	PAP18	PAP19	PAP20	PAP21	PAP22	PAP23
NAH30	*PAG1*	*PAG2*	*PAG3*	*PAG4*	*PAG5*	MAS1
8	9	10	11	12	13	14
S12	S13	S14	S15	S16	S17	S18
PAP24	PAP25	PAP26	PAP27	PAP28	PAP29	PAP30
MAS2	MAS3	MAS4	MAS5	MAS6	MAS7	MAS8
15	16	17	18	19	20	21
S19	S20	S21	S22	S23	S24	S25
HTH1	HTH2	HTH3	HTH4	HTH5	HTH6	HTH7
MAS9	MAS10	MAS11	MAS12	MAS13	MAS14	MAS15
22	23	24	25	26	27	28
S26	S27	S28	S29	S30	O1	O2
HTH8	HTH9	HTH10	HTH11	HTH12	HTH13	HTH14
MAS16	MAS17	MAS18	MAS19	MAS20	MAS21	MAS22

THE NU AFRIKAN CALENDAR

EKUNDU 540/2056-57/6292-93/2048-49						
Ashe	Ojobo	Iwa	Araorun	Ile	Itutu	Erinle
1	2	3	4	5	6	7
O3	O4	O5	O6	O7	O8	O9
HTH15	HTH16	HTH17	HTH18	HTH19	HTH20	HTH21
MAS23	MAS24	MAS25	MAS26	MAS27	MAS28	MAS29
8	9	10	11	12	13	14
O10	O11	O12	O13	O14	O15	O16
HTH22	HTH23	HTH24	HTH25	HTH26	HTH27	HTH28
MAS30	THi1	THi2	THi3	THi4	THi5	THi6
15	16	17	18	19	20	21
O17	O18	O19	O20	O21	O22	O23
HTH29	HTH30	KHK1	KHK2	KHK3	KHK4	KHK5
THi7	THi8	This9	THi10	THi11	THi12	THi13
22	23	24	25	26	27	28
O24	O25	O26	O27	O28	O29	O30
KHK6	KHK7	KHK8	KHK9	KHK10	KHK11	KHK12
Thi14	THi15	THi16	THi17	THi18	THi19	THi20

KIJANI 540/2056-57/6292-93/2048-49						
Ashe	Ojobo	Iwa	Araorun	Ile	Itutu	Erinle
1	2	3	4	5	6	7
N28	N29	N30	D1	D2	D3	D4
TBT11	TBT12	TBT13	TBT14	TBT15	TBT16	TBT17
HID19	HID20	HID21	HID22	HID23	HID24	HID25
8	9	10	11	12	13	14
D5	D6	D7	D8	D9	D10	D11
TBT18	TBT19	TBT20	TBT21	TBT22	TBT23	TBT24
HID26	HID27	HID28	HID29	HID30	TAH1	TAH2
15	16	17	18	19	20	21
D12	D13	D14	D15	D16	D17	D18
TBT25	TBT26	TBT27	TBT28	TBT29	TBT30	MKH1
TAH3	TAH4	TAH5	TAH6	TAH7	TAH8	TAH9
22	23	24	25	26	27	28
D19	D20	D21	D22	D23	D24	D25
MKH2	MKH3	MKH4	MKH5	MKH6	MKH7	MKH8
TAH10	TAH11	TAH12	TAH13	TAH14	TAH15	TAH16

EUSI 540/2056-57/6292-93/2048-49						
Ashe	Ojobo	Iwa	Araorun	Ile	Itutu	Erinle
1	2	3	4	5	6	7
O31	N1	N2	N3	N4	N5	N6
KHK13	KHK14	KHK15	KHK16	KHK17	KHK18	KHK19
THi21	THi22	THi23	THi24	THi25	THi26	THi27
8	9	10	11	12	13	14
N7	N8	N9	N10	N11	N12	N13
KHK20	KHK21	KHK22	KHK23	KHK24	KHK25	KHK26
THi28	THi29	THi30	HID1	HID2	HID3	HID4
15	16	17	18	19	20	21
N14	N15	N16	N17	N18	N19	N20
KHK27	KHK28	KHK29	KHK30	TBT1	TBT2	TBT3
HID5	HID6	HID7	HID8	HID9	HID10	HID11
22	23	24	25	26	27	28
N21	N22	N23	N24	N25	N26	N27
TBT4	TBT5	TBT6	TBT7	TBT8	TBT9	TBT10
HID12	HID13	HID14	HID15	HID16	HID17	HID18

KWANZA 540/2056-57/6292-93/2048-49						
Umoja	Kujichagulia	Ujima	Ujamaa	Nia	Kuumba	Imani
1	2	3	4	5	6	7
D26	D27	D28	D29	D30	D31	JA1
MKH9	MKH10	MKH11	MKH12	MKH13	MKH14	MKH1
TAH17	TAH18	TAH19	TAH20	TAH21	TAH22	TAH2
8	9	10	11	12	13	14
JA2	JA3	JA4	JA5	JA6	JA7	JA
MKH16	MKH17	MKH18	MKH19	MKH20	MKH21	MKH2
TAH24	TAH25	TAH26	TAH27	TAH28	TAH29	TAH3
15	16	17	18	19	20	21
JA9	JA10	JA11	JA12	JA13	JA14	JA1
MKH23	MKH24	MKH25	KMH26	MKH27	MKH28	MKH2
TIR1	TIR2	TIR3	TIR4	TIR5	TIR6	TIR7
22	23	24	25	26	27	28
JA16	JA17	JA18	JA19	JA20	JA21	JA2
MKH30	AMN1	AMN2	AMN3	AMN4	AMN5	AMN
TIR8	TIR9	TIR10	TIR11	TIR12	TIR13	TIR1

THE NU AFRIKAN CALENDAR

SHAHIDI 540/2056-57/6292-93/2048-49						
Ashe	Ojobo	Iwa	Araorun	Ile	Itutu	Erinle
1	2	3	4	5	6	7
JA23	JA24	JA25	JA26	JA27	JA28	JA29
AMN7	AMN8	AMN9	AMN10	AMN11	AMN12	AMN13
TIR15	TIR16	TIR17	TIR18	TIR19	TIR20	TIR21
8	9	10	11	12	13	14
JA30	JA31	F1	F2	F3	F4	F5
AMN14	AMN15	AMN16	AMN17	AMN18	AMN19	AMN20
TIR22	TIR23	TIR24	TIR25	TIR26	TIR27	TIR28
15	16	17	18	19	20	21
F6	F7	F8	F9	F10	F 11	F12
AMN21	AMN22	AMN23	AMN24	AMN25	AMN26	AMN28
TIR29	TIR30	YEK1	YEK2	YEK3	YEK4	YEK5
22	23	24	25	26	27	28
F13	F14	F15	F16	F17	F18	F19
AMN29	AMN30	RNT1	RNT2	RNT3	RNT4	RNT5
YEK6	YEK7	YEK8	YEK9	YEK10	YEK11	YEK12
SHAHIDI 91						
F20						
RNT6						
YEK13						

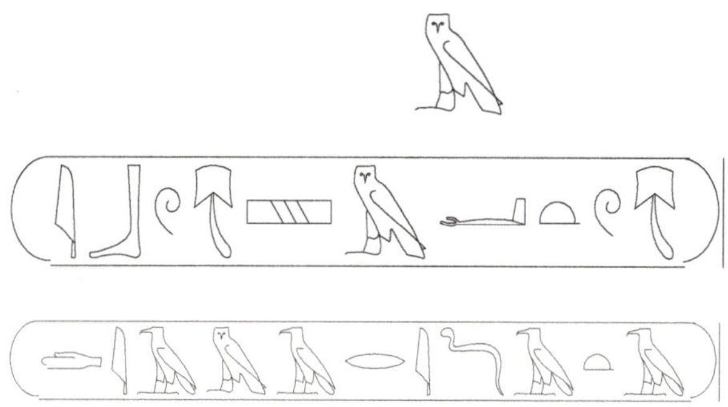

THE NU AFRIKAN CALENDAR

MAAFA 541 B.S.E./ASIENTO 2057-58 C.E.
KAM 6293-94 S.M./EC 2049-50 A.M. (ዓ.ም.)

AFU 541/2057-58/6293-94/2049-50

Ashe	Ojobo	Iwa	Araorun	Ile	Itutu	Erinle
1	2	3	4	5	6	7
F21	F22	F23	F24	F25	F26	F27
RNT7	RNT8	RNT9	RNT10	RNT11	RNT12	RNT13
YEK14	YEK15	YEK16	YEK17	YEK18	YEK19	YEK20
8	9	10	11	12	13	14
F28	M1	M2	M3	M4	M5	M6
RNT14	RNT15	RNT16	RNT17	RNT18	RNT19	RNT20
YEK21	YEK22	YEK23	YEK24	YEK25	YEK26	YEK27
15	16	17	18	19	20	21
M7	M8	M9	M10	M11	M12	M13
RNT21	RNT22	RNT23	RNT24	RNT25	RNT26	RNT27
YEK28	YEK29	YEK30	MAG1	MAG2	MAG3	MAG4
22	23	24	25	26	27	28
M14	M15	M16	M17	M18	M19	M20
RNT28	RNT29	RNT30	KHN1	KHN2	KHN3	KHN4
MAG5	MAG6	MAG7	MAG8	MAG9	MAG10	MAG11

KAWAIDA 541/2057-58/6293-94/2049-50

Ashe	Ojobo	Iwa	Araorun	Ile	Itutu	Erinle
1	2	3	4	5	6	7
A18	A19	A20	A21	A22	A23	A24
PIN3	PIN4	PIN5	PIN6	PIN7	PIN8	PIN9
MIY10	MIY11	MIY12	MIY13	MIY14	MIY15	MIY16
8	9	10	11	12	13	14
A25	A26	A27	A28	A29	A30	MA1
PIN10	PIN11	PIN12	PIN13	PIN14	PIN15	PIN16
MIY17	MIY18	MIY19	MIY20	MIY21	MIY22	MIY23
15	16	17	18	19	20	21
MA2	MA3	MA4	MA5	MA6	MA7	MA8
PIN17	PIN18	PIN19	PIN20	PIN21	PIN22	PIN23
MIY24	MIY25	MIY26	MIY27	MIY28	MIY29	MIY30
22	23	24	25	26	27	28
MA9	MA10	MA11	MA12	MA13	MA14	MA15
PIN24	PIN25	PIN26	PIN27	PIN28	PIN29	PIN30
GIN1	GIN2	GIN3	GIN4	GIN5	GIN6	GIN7

UHURU 541/2057-58/6293-94/2049-50

Ashe	Ojobo	Iwa	Araorun	Ile	Itutu	Erinle
1	2	3	4	5	6	7
M21	M22	M23	M24	M25	M26	M27
KHN5	KHN6	KHN7	KHN8	KHN9	KHN10	KHN11
MAG12	MAG13	MAG14	MAG15	MAG16	MAG17	MAG18
8	9	10	11	12	13	14
M28	M29	M30	M31	A1	A2	A3
KHN12	KHN13	KHN14	KHN15	KHN16	KHN17	KHN18
MAG19	MAG20	MAG21	MAG22	MAG23	MAG24	MAG25
15	16	17	18	19	20	21
A4	A5	A6	A7	A8	A9	A10
KHN19	KHN20	KHN21	KHN22	KHN23	KHN24	KHN25
MAG26	MAG27	MAG28	MAG29	MAG30	MIY1	MIY2
22	23	24	25	26	27	28
A11	A12	A13	A14	A15	A16	A17
KHN26	KHN27	KHN28	KHN29	KHN30	PIN1	PIN2
MIY3	MIY4	MIY5	MIY6	MIY7	MIY8	MIY9

MABABA WEUSI 541/2057-58/6293-94/2049-50

Ashe	Ojobo	Iwa	Araorun	Ile	Itutu	Erinle
1	2	3	4	5	6	7
MA16	MA17	MA18	MA19	MA20	MA21	MA22
APP1	APP2	APP3	APP4	APP5	APP6	APP7
GIN8	GIN9	GIN10	GIN11	GIN12	GIN13	GIN14
8	9	10	11	12	13	14
MA23	MA24	MA25	MA26	MA27	MA28	MA29
APP8	APP9	APP10	APP11	APP12	APP13	APP14
GIN15	GIN16	GIN17	GIN18	GIN19	GIN20	GIN21
15	16	17	18	19	20	21
MA30	MA31	J1	J2	J3	J4	J5
APP15	APP16	APP17	APP18	APP19	APP20	APP21
GIN22	GIN23	GIN24	GIN25	GIN26	GIN27	GIN28
22	23	24	25	26	27	28
J6	J7	J8	J9	J10	J11	J12
APP22	APP23	APP24	APP25	APP26	APP27	APP28
GIN29	GIN30	SAN1	SAN2	SAN3	SAN4	SAN5

THE NU AFRIKAN CALENDAR

MWANAMKE WEUSI 541/2057-58/6293-94/2049-50						
Ashe	Ojobo	Iwa	Araorun	Ile	Itutu	Erinle
1	2	3	4	5	6	7
J13	J14	J15	J16	J17	J18	J19
APP29	APP30	MSR1	MSR2	MSR3	MSR4	MSR5
SAN6	SAN7	SAN8	SAN9	SAN10	SAN11	SAN12
8	9	10	11	12	13	14
J20	J21	J22	J23	J24	J25	J26
MSR6	MSR7	MSR8	MSR9	MSR10	MSR11	MSR12
SAN13	SAN14	SAN15	SAN16	SAN17	SAN18	SAN19
15	16	17	18	19	20	21
J27	J28	J29	J30	JU1	JU2	JU3
MSR13	MSR14	MSR15	MSR16	MSR17	MSR18	MSR19
SAN20	SAN21	SAN22	SAN23	SAN24	SAN25	SAN26
22	23	24	25	26	27	28
JU4	JU5	JU6	JU7	JU8	JU9	JU10
MSR20	MRS21	MSR22	MSR23	MSR24	MSR25	MSR26
SAN27	SAN28	SAN29	SAN30	HAM1	HAM2	HAM3

TAIFA WEUSI 541/2057-58/6293-94/2049-50						
Ashe	Ojobo	Iwa	Araorun	Ile	Itutu	Erinle
1	2	3	4	5	6	7
AU8	AU9	AU10	AU11	AU12	AU13	AU14
THT20	THT21	THT22	THT23	THT24	THT25	THT26
NAH2	NAH3	NAH4	NAH5	NAH6	NAH7	NAH8
8	9	10	11	12	13	14
AU15	AU16	AU17	AU18	AU19	AU20	AU21
THT27	THT28	THT29	THT30	PAP1	PAP1	PAP2
NAH9	NAH10	NAH11	NAH12	NAH13	NAH14	NAH15
15	16	17	18	19	20	21
AU22	AU23	AU24	AU25	AU26	AU27	AU28
PAP3	PAP4	PAP5	PAP6	PAP7	PAP8	PAP9
NAH16	NAH17	NAH18	NAH19	NAH20	NAH21	NAH22
22	23	24	25	26	27	28
AU29	AU30	AU31	S1	S2	S3	S4
PAP10	PAP11	PAP12	PAP13	PAP14	PAP15	PAP16
NAH23	NAH24	NAH25	NAH26	NAH27	NAH28	NAH29

MTU WEUSI 541/2057-58/6293-94/2049-50						
Ashe	Ojobo	Iwa	Araorun	Ile	Itutu	Erinle
1	2	3	4	5	6	7
JU11	JU12	JU13	JU14	JU15	JU16	JU17
MSR27	MSR28	MSR29	MSR30	*ASR1*	*HRU2*	*STH3*
HAM4	HAM5	HAM6	HAM7	HAM8	HAM9	HAM10
8	9	10	11	12	13	14
JU18	JU19	JU20	JU21	JU22	JU23	JU24
AST4	*NBH5*	THT1	THT2	THT3	THT4	THT5
HAM11	HAM12	HAM13	AM14	HAM15	HAM16	HAM17
15	16	17	18	19	20	21
JU25	JU26	JU27	JU28	JU29	JU30	JU31
THT6	THT7	THT8	THT9	THT10	THT11	THT12
HAM18	HAM19	HAM20	HAM21	HAM22	HAM23	HAM24
22	23	24	25	26	27	28
AU1	AU2	AU3	AU4	AU5	AU6	AU7
THT13	THT14	THT15	THT16	THT17	THT18	THT19
HAM25	HAM26	HAM27	HAM28	HAM29	HAM30	NAH1

UWEZO 541/2057-58/6293-94/2049-50						
Ashe	Ojobo	Iwa	Araorun	Ile	Itutu	Erinle
1	2	3	4	5	6	7
S5	S6	S7	S8	S9	S10	S11
PAP17	PAP18	PAP19	PAP20	PAP21	PAP22	PAP23
NAH30	*PAG1*	*PAG2*	*PAG3*	*PAG4*	*PAG5*	MAS1
8	9	10	11	12	13	14
S12	S13	S14	S15	S16	S17	S18
PAP24	PAP25	PAP26	PAP27	PAP28	PAP29	PAP30
MAS2	MAS3	MAS4	MAS5	MAS6	MAS7	MAS8
15	16	17	18	19	20	21
S19	S20	S21	S22	S23	S24	S25
HTH1	HTH2	HTH3	HTH4	HTH5	HTH6	HTH7
MAS9	MAS10	MAS11	MAS12	MAS13	MAS14	MAS15
22	23	24	25	26	27	28
S26	S27	S28	S29	S30	O1	O2
HTH8	HTH9	HTH10	HTH11	HTH12	HTH13	HTH14
MAS16	MAS17	MAS18	MAS19	MAS20	MAS21	MAS22

THE NU AFRIKAN CALENDAR

EKUNDU 541/2057-58/6293-94/2049-50						
Ashe	Ojobo	Iwa	Araorun	Ile	Itutu	Erinle
1	2	3	4	5	6	7
O3	O4	O5	O6	O7	O8	O9
HTH15	HTH16	HTH17	HTH18	HTH19	HTH20	HTH21
MAS23	MAS24	MAS25	MAS26	MAS27	MAS28	MAS29
8	9	10	11	12	13	14
O10	O11	O12	O13	O14	O15	O16
HTH22	HTH23	HTH24	HTH25	HTH26	HTH27	HTH28
MAS30	THi1	THi2	THi3	THi4	THi5	THi6
15	16	17	18	19	20	21
O17	O18	O19	O20	O21	O22	O23
HTH29	HTH30	KHK1	KHK2	KHK3	KHK4	KHK5
THi7	THi8	THi9	THi10	THi11	THi12	THi13
22	23	24	25	26	27	28
O24	O25	O26	O27	O28	O29	O30
KHK6	KHK7	KHK8	KHK9	KHK10	KHK11	KHK12
THi14	THi15	THi16	THi17	THi18	THi19	THi20

KIJANI 541/2057-58/6293-94/2049-50						
Ashe	Ojobo	Iwa	Araorun	Ile	Itutu	Erinle
1	2	3	4	5	6	7
N28	N29	N30	D1	D2	D3	D4
TBT11	TBT12	TBT13	TBT14	TBT15	TBT16	TBT17
HID19	HID20	HID21	HID22	HID23	HID24	HID25
8	9	10	11	12	13	14
D5	D6	D7	D8	D9	D10	D11
TBT18	TBT19	TBT20	TBT21	TBT22	TBT23	TBT24
HID26	HID27	HID28	HID29	HID30	TAH1	TAH2
15	16	17	18	19	20	21
D12	D13	D14	D15	D16	D17	D18
TBT25	TBT26	TBT27	TBT28	TBT29	TBT30	MKH1
TAH3	TAH4	TAH5	TAH6	TAH7	TAH8	TAH9
22	23	24	25	26	27	28
D19	D20	D21	D22	D23	D24	D25
MKH2	MKH3	MKH4	MKH5	MKH6	MKH7	MKH8
TAH10	TAH11	TAH12	TAH13	TAH14	TAH15	TAH16

EUSI 541/2057-58/6293-94/2049-50						
Ashe	Ojobo	Iwa	Araorun	Ile	Itutu	Erinle
1	2	3	4	5	6	7
O31	N1	N2	N3	N4	N5	N6
KHK13	KHK14	KHK15	KHK16	KHK17	KHK18	KHK19
THi21	THi22	THi23	THi24	THi25	THi26	THi27
8	9	10	11	12	13	14
N7	N8	N9	N10	N11	N12	N13
KHK20	KHK21	KHK22	KHK23	KHK24	KHK25	KHK26
THi28	THi29	THi30	HID1	HID2	HID3	HID4
15	16	17	18	19	20	21
N14	N15	N16	N17	N18	N19	N20
KHK27	KHK28	KHK29	KHK30	TBT1	TBT2	TBT3
HID5	HID6	HID7	HID8	HID9	HID10	HID11
22	23	24	25	26	27	28
N21	N22	N23	N24	N25	N26	N27
TBT4	TBT5	TBT6	TBT7	TBT8	TBT9	TBT10
HID12	HID13	HID14	HID15	HID16	HID17	HID18

KWANZA 541/2057-58/6293-94/2049-50						
Umoja	Kujichagulia	Ujima	Ujamaa	Nia	Kuumba	Imani
1	2	3	4	5	6	7
D26	D27	D28	D29	D30	D31	JA1
MKH9	MKH10	MKH11	MKH12	MKH13	MKH14	MKH15
TAH17	TAH18	TAH19	TAH20	TAH21	TAH22	TAH23
8	9	10	11	12	13	14
JA2	JA3	JA4	JA5	JA6	JA7	JA8
MKH16	MKH17	MKH18	MKH19	MKH20	MKH21	MKH22
TAH24	TAH25	TAH26	TAH27	TAH28	TAH29	TAH30
15	16	17	18	19	20	21
JA9	JA10	JA11	JA12	JA13	JA14	JA15
MKH23	MKH24	MKH25	KMH26	MKH27	MKH28	MKH29
TIR1	TIR2	TIR3	TIR4	TIR5	TIR6	TIR7
22	23	24	25	26	27	28
JA16	JA17	JA18	JA19	JA20	JA21	JA22
MKH30	AMN1	AMN2	AMN3	AMN4	AMN5	AMN6
TIR8	TIR9	TIR10	TIR11	TIR12	TIR13	TIR14

SHAHIDI 541/2057-58/6293-94/2049-50						
Ashe	Ojobo	Iwa	Araorun	Ile	Itutu	Erinle
1	2	3	4	5	6	7
JA23	JA24	JA25	JA26	JA27	JA28	JA29
AMN7	AMN8	AMN9	AMN10	AMN11	AMN12	AMN13
TIR15	TIR16	TIR17	TIR18	TIR19	TIR20	TIR21
8	9	10	11	12	13	14
JA30	JA31	F1	F2	F3	F4	F5
AMN14	AMN15	AMN16	AMN17	AMN18	AMN19	AMN20
TIR22	TIR23	TIR24	TIR25	TIR26	TIR27	TIR28
15	16	17	18	19	20	21
F6	F7	F8	F9	F10	F 11	F12
AMN21	AMN22	AMN23	AMN24	AMN25	AMN26	AMN28
TIR29	TIR30	YEK1	YEK2	YEK3	YEK4	YEK5
22	23	24	25	26	27	28
F13	F14	F15	F16	F17	F18	F19
AMN29	AMN30	RNT1	RNT2	RNT3	RNT4	RNT5
YEK6	YEK7	YEK8	YEK9	YEK10	YEK11	YEK12
SHAHIDI 92						
F20						
RNT6						
YEK13						

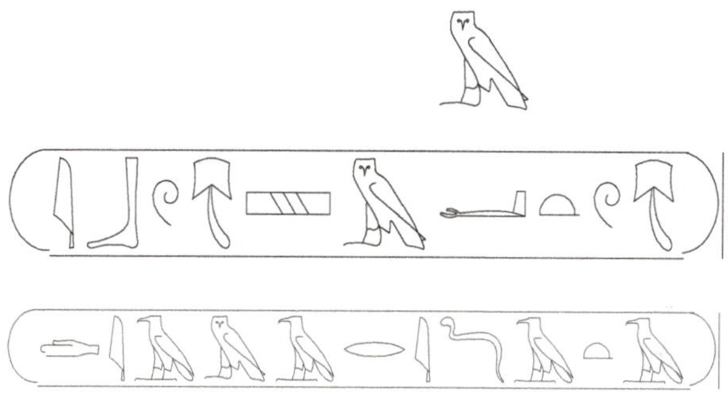

MAAFA 542 B.S.E./ASIENTO 2058-59 C.E.
KAM 6294-95 S.M./EC 2050-51 A.M. (ዓ.ም.)

AFU 542/2058-59/6294-95/2050-51

Ashe	Ojobo	Iwa	Araorun	Ile	Itutu	Erinle
1	2	3	4	5	6	7
21	F22	F23	F24	F25	F26	F27
NT7	RNT8	RNT9	RNT10	RNT11	RNT12	RNT13
EK14	YEK15	YEK16	YEK17	YEK18	YEK19	YEK20
	9	10	11	12	13	14
F28	M1	M2	M3	M4	M5	M6
NT14	RNT15	RNT16	RNT17	RNT18	RNT19	RNT20
EK21	YEK22	YEK23	YEK24	YEK25	YEK26	YEK27
5	16	17	18	19	20	21
M7	M8	M9	M10	M11	M12	M13
NT21	RNT22	RNT23	RNT24	RNT25	RNT26	RNT27
EK28	YEK29	YEK30	MAG1	MAG2	MAG3	MAG4
22	23	24	25	26	27	28
M14	M15	M16	M17	M18	M19	M20
NT28	RNT29	RNT30	KHN1	KHN2	KHN3	KHN4
AG5	MAG6	MAG7	MAG8	MAG9	MAG10	MAG11

KAWAIDA 542/2058-59/6294-95/2050-51

Ashe	Ojobo	Iwa	Araorun	Ile	Itutu	Erinle
1	2	3	4	5	6	7
A18	A19	A20	A21	A22	A23	A24
PIN3	PIN4	PIN5	PIN6	PIN7	PIN8	PIN9
MIY10	MIY11	MIY12	MIY13	MIY14	MIY15	MIY16
8	9	10	11	12	13	14
A25	A26	A27	A28	A29	A30	MA1
PIN10	PIN11	PIN12	PIN13	PIN14	PIN15	PIN16
MIY17	MIY18	MIY19	MIY20	MIY21	MIY22	MIY23
15	16	17	18	19	20	21
MA2	MA3	MA4	MA5	MA6	MA7	MA8
PIN17	PIN18	PIN19	PIN20	PIN21	PIN22	PIN23
MIY24	MIY25	MIY26	MIY27	MIY28	MIY29	MIY30
22	23	24	25	26	27	28
MA9	MA10	MA11	MA12	MA13	MA14	MA15
PIN24	PIN25	PIN26	PIN27	PIN28	PIN29	PIN30
GIN1	GIN2	GIN3	GIN4	GIN5	GIN6	GIN7

UHURU 542/2058-59/6294-95/2050-51

Ashe	Ojobo	Iwa	Araorun	Ile	Itutu	Erinle
1	2	3	4	5	6	7
M21	M22	M23	M24	M25	M26	M27
HN5	KHN6	KHN7	KHN8	KHN9	KHN10	KHN11
AG12	MAG13	MAG14	MAG15	MAG16	MAG17	MAG18
8	9	10	11	12	13	14
M28	M29	M30	M31	A1	A2	A3
HN12	KHN13	KHN14	KHN15	KHN16	KHN17	KHN18
AG19	MAG20	MAG21	MAG22	MAG23	MAG24	MAG25
15	16	17	18	19	20	21
A4	A5	A6	A7	A8	A9	A10
HN19	KHN20	KHN21	KHN22	KHN23	KHN24	KHN25
AG26	MAG27	MAG28	MAG29	MAG30	MIY1	MIY2
22	23	24	25	26	27	28
A11	A12	A13	A14	A15	A16	A17
HN26	KHN27	KHN28	KHN29	KHN30	PIN1	PIN2
IY3	MIY4	MIY5	MIY6	MIY7	MIY8	MIY9

MABABA WEUSI 542/2058-59/6294-95/2050-51

Ashe	Ojobo	Iwa	Araorun	Ile	Itutu	Erinle
1	2	3	4	5	6	7
MA16	MA17	MA18	MA19	MA20	MA21	MA22
APP1	APP2	APP3	APP4	APP5	APP6	APP7
GIN8	GIN9	GIN10	GIN11	GIN12	GIN13	GIN14
8	9	10	11	12	13	14
MA23	MA24	MA25	MA26	MA27	MA28	MA29
APP8	APP9	APP10	APP11	APP12	APP13	APP14
GIN15	GIN16	GIN17	GIN18	GIN19	GIN20	GIN21
15	16	17	18	19	20	21
MA30	MA31	J1	J2	J3	J4	J5
APP15	APP16	APP17	APP18	APP19	APP20	APP21
GIN22	GIN23	GIN24	GIN25	GIN26	GIN27	GIN28
22	23	24	25	26	27	28
J6	J7	J8	J9	J10	J11	J12
APP22	APP23	APP24	APP25	APP26	APP27	APP28
GIN29	GIN30	SAN1	SAN2	SAN3	SAN4	SAN5

THE NU AFRIKAN CALENDAR

MWANAMKE WEUSI 542/2058-59/6294-95/2050-51						
Ashe	Ojobo	Iwa	Araorun	Ile	Itutu	Erinle
1	2	3	4	5	6	7
J13	J14	J15	J16	J17	J18	J19
APP29	APP30	MSR1	MSR2	MSR3	MSR4	MSR5
SAN6	SAN7	SAN8	SAN9	SAN10	SAN11	SAN12
8	9	10	11	12	13	14
J20	J21	J22	J23	J24	J25	J26
MSR6	MSR7	MSR8	MSR9	MSR10	MSR11	MSR12
SAN13	SAN14	SAN15	SAN16	SAN17	SAN18	SAN19
15	16	17	18	19	20	21
J27	J28	J29	J30	JU1	JU2	JU3
MSR13	MSR14	MSR15	MSR16	MSR17	MSR18	MSR19
SAN20	SAN21	SAN22	SAN23	SAN24	SAN25	SAN26
22	23	24	25	26	27	28
JU4	JU5	JU6	JU7	JU8	JU9	JU10
MSR20	MRS21	MSR22	MSR23	MSR24	MSR25	MSR26
SAN27	SAN28	SAN29	SAN30	HAM1	HAM2	HAM3

TAIFA WEUSI 542/2058-59/6294-95/2050-51						
Ashe	Ojobo	Iwa	Araorun	Ile	Itutu	Erinle
1	2	3	4	5	6	7
AU8	AU9	AU10	AU11	AU12	AU13	AU14
THT20	THT21	THT22	THT23	THT24	THT25	THT26
NAH2	NAH3	NAH4	NAH5	NAH6	NAH7	NAH8
8	9	10	11	12	13	14
AU15	AU16	AU17	AU18	AU19	AU20	AU21
THT27	THT28	THT29	THT30	PAP1	PAP1	PAP2
NAH9	NAH10	NAH11	NAH12	NAH13	NAH14	NAH15
15	16	17	18	19	20	21
AU22	AU23	AU24	AU25	AU26	AU27	AU28
PAP3	PAP4	PAP5	PAP6	PAP7	PAP8	PAP9
NAH16	NAH17	NAH18	NAH19	NAH20	NAH21	NAH22
22	23	24	25	26	27	28
AU29	AU30	AU31	S1	S2	S3	S4
PAP10	PAP11	PAP12	PAP13	PAP14	PAP15	PAP16
NAH23	NAH24	NAH25	NAH26	NAH27	NAH28	NAH29

MTU WEUSI 542/2058-59/6294-95/2050-51						
Ashe	Ojobo	Iwa	Araorun	Ile	Itutu	Erinle
1	2	3	4	5	6	7
JU11	JU12	JU13	JU14	JU15	JU16	JU17
MSR27	MSR28	MSR29	MSR30	*ASR1*	*HRU2*	*STH3*
HAM4	HAM5	HAM6	HAM7	HAM8	HAM9	HAM10
8	9	10	11	12	13	14
JU18	JU19	JU20	JU21	JU22	JU23	JU24
AST4	*NBH5*	THT1	THT2	THT3	THT4	THT5
HAM11	HAM12	HAM13	AM14	HAM15	HAM16	HAM17
15	16	17	18	19	20	21
JU25	JU26	JU27	JU28	JU29	JU30	JU31
THT6	THT7	THT8	THT9	THT10	THT11	THT12
HAM18	HAM19	HAM20	HAM21	HAM22	HAM23	HAM24
22	23	24	25	26	27	28
AU1	AU2	AU3	AU4	AU5	AU6	AU7
THT13	THT14	THT15	THT16	THT17	THT18	THT19
HAM25	HAM26	HAM27	HAM28	HAM29	HAM30	NAH1

UWEZO 542/2058-59/6294-95/2050-51						
Ashe	Ojobo	Iwa	Araorun	Ile	Itutu	Erinle
1	2	3	4	5	6	7
S5	S6	S7	S8	S9	S10	S11
PAP17	PAP18	PAP19	PAP20	PAP21	PAP22	PAP23
NAH30	*PAG1*	*PAG2*	*PAG3*	*PAG4*	*PAG5*	MAS1
8	9	10	11	12	13	14
S12	S13	S14	S15	S16	S17	S18
PAP24	PAP25	PAP26	PAP27	PAP28	PAP29	PAP30
MAS2	MAS3	MAS4	MAS5	MAS6	MAS7	MAS8
15	16	17	18	19	20	21
S19	S20	S21	S22	S23	S24	S25
HTH1	HTH2	HTH3	HTH4	HTH5	HTH6	HTH7
MAS9	MAS10	MAS11	MAS12	MAS13	MAS14	MAS15
22	23	24	25	26	27	28
S26	S27	S28	S29	S30	O1	O2
HTH8	HTH9	HTH10	HTH11	HTH12	HTH13	HTH14
MAS16	MAS17	MAS18	MAS19	MAS20	MAS21	MAS22

EKUNDU 542/2058-59/6294-95/2050-51

Ashe	Ojobo	Iwa	Araorun	Ile	Itutu	Erinle
1	2	3	4	5	6	7
O3	O4	O5	O6	O7	O8	O9
HTH15	HTH16	HTH17	HTH18	HTH19	HTH20	HTH21
MAS23	MAS24	MAS25	MAS26	MAS27	MAS28	MAS29
8	9	10	11	12	13	14
O10	O11	O12	O13	O14	O15	O16
HTH22	HTH23	HTH24	HTH25	HTH26	HTH27	HTH28
MAS30	THi1	THi2	THi3	THi4	THi5	THi6
15	16	17	18	19	20	21
O17	O18	O19	O20	O21	O22	O23
HTH29	HTH30	KHK1	KHK2	KHK3	KHK4	KHK5
THi7	THi8	THi9	THi10	THi11	THi12	THi13
22	23	24	25	26	27	28
O24	O25	O26	O27	O28	O29	O30
KHK6	KHK7	KHK8	KHK9	KHK10	KHK11	KHK12
THi14	THi15	THi16	THi17	THi18	THi19	THi20

KIJANI 542/2058-59/6294-95/2050-51

Ashe	Ojobo	Iwa	Araorun	Ile	Itutu	Erinle
1	2	3	4	5	6	7
N28	N29	N30	D1	D2	D3	D4
TBT11	TBT12	TBT13	TBT14	TBT15	TBT16	TBT17
HID19	HID20	HID21	HID22	HID23	HID24	HID25
8	9	10	11	12	13	14
D5	D6	D7	D8	D9	D10	D11
TBT18	TBT19	TBT20	TBT21	TBT22	TBT23	TBT24
HID26	HID27	HID28	HID29	HID30	TAH1	TAH2
15	16	17	18	19	20	21
D12	D13	D14	D15	D16	D17	D18
TBT25	TBT26	TBT27	TBT28	TBT29	TBT30	MKH1
TAH3	TAH4	TAH5	TAH6	TAH7	TAH8	TAH9
22	23	24	25	26	27	28
D19	D20	D21	D22	D23	D24	D25
MKH2	MKH3	MKH4	MKH5	MKH6	MKH7	MKH8
TAH10	TAH11	TAH12	TAH13	TAH14	TAH15	TAH16

EUSI 542/2058-59/6294-95/2050-51

Ashe	Ojobo	Iwa	Araorun	Ile	Itutu	Erinle
1	2	3	4	5	6	7
O31	N1	N2	N3	N4	N5	N6
KHK13	KHK14	KHK15	KHK16	KHK17	KHK18	KHK19
THi21	THi22	THi23	THi24	THi25	THi26	THi27
8	9	10	11	12	13	14
N7	N8	N9	N10	N11	N12	N13
KHK20	KHK21	KHK22	KHK23	KHK24	KHK25	KHK26
THi28	THi29	THi30	HID1	HID2	HID3	HID4
15	16	17	18	19	20	21
N14	N15	N16	N17	N18	N19	N20
KHK27	KHK28	KHK29	KHK30	TBT1	TBT2	TBT3
HID5	HID6	HID7	HID8	HID9	HID10	HID11
22	23	24	25	26	27	28
N21	N22	N23	N24	N25	N26	N27
TBT4	TBT5	TBT6	TBT7	TBT8	TBT9	TBT10
HID12	HID13	HID14	HID15	HID16	HID17	HID18

KWANZA 542/2058-59/6294-95/2050-51

Umoja	Kujichagulia	Ujima	Ujamaa	Nia	Kuumba	Imani
1	2	3	4	5	6	7
D26	D27	D28	D29	D30	D31	JA1
MKH9	MKH10	MKH11	MKH12	MKH13	MKH14	MKH15
TAH17	TAH18	TAH19	TAH20	TAH21	TAH22	TAH23
8	9	10	11	12	13	14
JA2	JA3	JA4	JA5	JA6	JA7	JA8
MKH16	MKH17	MKH18	MKH19	MKH20	MKH21	MKH22
TAH24	TAH25	TAH26	TAH27	TAH28	TAH29	TAH30
15	16	17	18	19	20	21
JA9	JA10	JA11	JA12	JA13	JA14	JA15
MKH23	MKH24	MKH25	KMH26	MKH27	MKH28	MKH29
TIR1	TIR2	TIR3	TIR4	TIR5	TIR6	TIR7
22	23	24	25	26	27	28
JA16	JA17	JA18	JA19	JA20	JA21	JA22
MKH30	AMN1	AMN2	AMN3	AMN4	AMN5	AMN6
TIR8	TIR9	TIR10	TIR11	TIR12	TIR13	TIR14

SHAHIDI 542/2058-59/6294-95/2050-51						
Ashe	Ojobo	Iwa	Araorun	Ile	Itutu	Erinle
1	2	3	4	5	6	7
JA23	JA24	JA25	JA26	JA27	JA28	JA29
AMN7	AMN8	AMN9	AMN10	AMN11	AMN12	AMN13
TIR15	TIR16	TIR17	TIR18	TIR19	TIR20	TIR21
8	9	10	11	12	13	14
JA30	JA31	F1	F2	F3	F4	F5
AMN14	AMN15	AMN16	AMN17	AMN18	AMN19	AMN20
TIR22	TIR23	TIR24	TIR25	TIR26	TIR27	TIR28
15	16	17	18	19	20	21
F6	F7	F8	F9	F10	F 11	F12
AMN21	AMN22	AMN23	AMN24	AMN25	AMN26	AMN28
TIR29	TIR30	YEK1	YEK2	YEK3	YEK4	YEK5
22	23	24	25	26	27	28
F13	F14	F15	F16	F17	F18	F19
AMN29	AMN30	RNT1	RNT2	RNT3	RNT4	RNT5
YEK6	YEK7	YEK8	YEK9	YEK10	YEK11	YEK12
SHAHIDI 93						
F20						
RNT6						
YEK13						

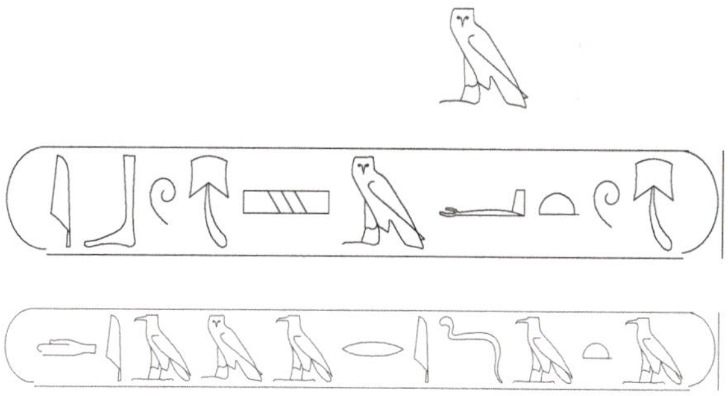

MAAFA 543 B.S.E./ASIENTO 2059-60 C.E./
KAM 6295-96 S.M./EC 2051-52 A.M. (ዓ.ም.)

AFU 543/2059-60/6295-96/2051-52						
Ashe	Ojobo	Iwa	Araorun	Ile	Itutu	Erinle
1	2	3	4	5	6	7
F21	F22	F23	F24	F25	F26	F27
RNT7	RNT8	RNT9	RNT10	RNT11	RNT12	RNT13
YEK14	YEK15	YEK16	YEK17	YEK18	YEK19	YEK20
8	9	10	11	12	13	14
F28	M1	M2	M3	M4	M5	M6
RNT14	RNT15	RNT16	RNT17	RNT18	RNT19	RNT20
YEK21	YEK22	YEK23	YEK24	YEK25	YEK26	YEK27
15	16	17	18	19	20	21
M7	M8	M9	M10	M11	M12	M13
RNT21	RNT22	RNT23	RNT24	RNT25	RNT26	RNT27
YEK28	YEK 29	YEK30	MAG1	MAG2	MAG3	MAG4
22	23	24	25	26	27	28
M14	M15	M16	M17	M18	M19	M20
RNT28	RNT29	RNT30	KHN1	KHN2	KHN3	KHN4
MAG5	MAG6	MAG7	MAG8	MAG9	MAG10	MAG11

UHURU 543/2059-60/6295-96/2051-52						
Ashe	Ojobo	Iwa	Araorun	Ile	Itutu	Erinle
1	2	3	4	5	6	7
M21	M22	M23	M24	M25	M26	M27
KHN5	KHN6	KHN7	KHN8	KHN9	KHN10	KHN11
MAG12	MAG13	MAG14	MAG15	MAG16	MAG17	MAG18
8	9	10	11	12	13	14
M28	M29	M30	M31	A1	A2	A3
KHN12	KHN13	KHN14	KHN15	KHN16	KHN17	KHN18
MAG19	MAG20	MAG21	MAG22	MAG23	MAG24	MAG25
15	16	17	18	19	20	21
A4	A5	A6	A7	A8	A9	A10
KHN19	KHN20	KHN21	KHN22	KHN23	KHN24	KHN25
MAG26	MAG27	MAG28	MAG29	MAG30	MIY1	MIY2
22	23	24	25	26	27	28
A11	A12	A13	A14	A15	A16	A17
KHN26	KHN27	KHN28	KHN29	KHN30	PIN1	PIN2
MIY3	MIY4	MIY5	MIY6	MIY7	MIY8	MIY9

KAWAIDA 543/2059-60/6295-96/2051-52						
Ashe	Ojobo	Iwa	Araorun	Ile	Itutu	Erinle
1	2	3	4	5	6	7
A18	A19	A20	A21	A22	A23	A24
PIN3	PIN4	PIN5	PIN6	PIN7	PIN8	PIN9
MIY10	MIY11	MIY12	MIY13	MIY14	MIY15	MIY16
8	9	10	11	12	13	14
A25	A26	A27	A28	A29	A30	MA1
PIN10	PIN11	PIN12	PIN13	PIN14	PIN15	PIN16
MIY17	MIY18	MIY19	MIY20	MIY21	MIY22	MIY23
15	16	17	18	19	20	21
MA2	MA3	MA4	MA5	MA6	MA7	MA8
PIN17	PIN18	PIN19	PIN20	PIN21	PIN22	PIN23
MIY24	MIY25	MIY26	MIY27	MIY28	MIY29	MIY30
22	23	24	25	26	27	28
MA9	MA10	MA11	MA12	MA13	MA14	MA15
PIN24	PIN25	PIN26	PIN27	PIN28	PIN29	PIN30
GIN1	GIN2	GIN3	GIN4	GIN5	GIN6	GIN7

MABABA WEUSI 543/2059-60/6295-96/2051-52						
Ashe	Ojobo	Iwa	Araorun	Ile	Itutu	Erinle
1	2	3	4	5	6	7
MA16	MA17	MA18	MA19	MA20	MA21	MA22
APP1	APP2	APP3	APP4	APP5	APP6	APP7
GIN8	GIN9	GIN10	GIN11	GIN12	GIN13	GIN14
8	9	10	11	12	13	14
MA23	MA24	MA25	MA26	MA27	MA28	MA29
APP8	APP9	APP10	APP11	APP12	APP13	APP14
GIN15	GIN16	GIN17	GIN18	GIN19	GIN20	GIN21
15	16	17	18	19	20	21
MA30	MA31	J1	J2	J3	J4	J5
APP15	APP16	APP17	APP18	APP19	APP20	APP21
GIN22	GIN23	GIN24	GIN25	GIN26	GIN27	GIN28
22	23	24	25	26	27	28
J6	J7	J8	J9	J10	J11	J12
APP22	APP23	APP24	APP25	APP26	APP27	APP28
GIN29	GIN30	SAN1	SAN2	SAN3	SAN4	SAN5

THE NU AFRIKAN CALENDAR

MWANAMKE WEUSI 543/2059-60/6295-96/2051-52

Ashe	Ojobo	Iwa	Araorun	Ile	Itutu	Erinle
1	2	3	4	5	6	7
J13	J14	J15	J16	J17	J18	J19
APP29	APP30	MSR1	MSR2	MSR3	MSR4	MSR5
SAN6	SAN7	SAN8	SAN9	SAN10	SAN11	SAN12
8	9	10	11	12	13	14
J20	J21	J22	J23	J24	J25	J26
MSR6	MSR7	MSR8	MSR9	MSR10	MSR11	MSR12
SAN13	SAN14	SAN15	SAN16	SAN17	SAN18	SAN19
15	16	17	18	19	20	21
J27	J28	J29	J30	JU1	JU2	JU3
MSR13	MSR14	MSR15	MSR16	MSR17	MSR18	MSR19
SAN20	SAN21	SAN22	SAN23	SAN24	SAN25	SAN26
22	23	24	25	26	27	28
JU4	JU5	JU6	JU7	JU8	JU9	JU10
MSR20	MRS21	MSR22	MSR23	MSR24	MSR25	MSR26
SAN27	SAN28	SAN29	SAN30	HAM1	HAM2	HAM3

TAIFA WEUSI 543/2059-60/6295-96/2051-52

Ashe	Ojobo	Iwa	Araorun	Ile	Itutu	Erinle
1	2	3	4	5	6	7
AU8	AU9	AU10	AU11	AU12	AU13	AU14
THT20	THT21	THT22	THT23	THT24	THT25	THT26
NAH2	NAH3	NAH4	NAH5	NAH6	NAH7	NAH8
8	9	10	11	12	13	14
AU15	AU16	AU17	AU18	AU19	AU20	AU21
THT27	THT28	THT29	THT30	PAP1	PAP1	PAP2
NAH9	NAH10	NAH11	NAH12	NAH13	NAH14	NAH15
15	16	17	18	19	20	21
AU22	AU23	AU24	AU25	AU26	AU27	AU28
PAP3	PAP4	PAP5	PAP6	PAP7	PAP8	PAP9
NAH16	NAH17	NAH18	NAH19	NAH20	NAH21	NAH22
22	23	24	25	26	27	28
AU29	AU30	AU31	S1	S2	S3	S4
PAP10	PAP11	PAP12	PAP13	PAP14	PAP15	PAP16
NAH23	NAH24	NAH25	NAH26	NAH27	NAH28	NAH29

MTU WEUSI 543/2059-60/6295-96/2051-52

Ashe	Ojobo	Iwa	Araorun	Ile	Itutu	Erinle
1	2	3	4	5	6	7
JU11	JU12	JU13	JU14	JU15	JU16	JU17
MSR27	MSR28	MSR29	MSR30	*ASR1*	*HRU2*	*STH3*
HAM4	HAM5	HAM6	HAM7	HAM8	HAM9	HAM10
8	9	10	11	12	13	14
JU18	JU19	JU20	JU21	JU22	JU23	JU24
AST4	*NBH5*	THT1	THT2	THT3	THT4	THT5
HAM11	HAM12	HAM13	HAM14	HAM15	HAM16	HAM17
15	16	17	18	19	20	21
JU25	JU26	JU27	JU28	JU29	JU30	JU31
THT6	THT7	THT8	THT9	THT10	THT11	THT12
HAM18	HAM19	HAM20	HAM21	HAM22	HAM23	HAM24
22	23	24	25	26	27	28
AU1	AU2	AU3	AU4	AU5	AU6	AU7
THT13	THT14	THT15	THT16	THT17	THT18	THT19
HAM25	HAM26	HAM27	HAM28	HAM29	HAM30	NAH1

UWEZO 543/2059-60/6295-96/2051-52

Ashe	Ojobo	Iwa	Araorun	Ile	Itutu	Erinle
1	2	3	4	5	6	7
S5	S6	S7	S8	S9	S10	S11
PAP17	PAP18	PAP19	PAP20	PAP21	PAP22	PAP23
NAH30	*QUA1*	*QUA2*	*QUA3*	*QUA4*	*QUA5*	*QUA6*
8	9	10	11	12	13	14
S12	S13	S14	S15	S16	S17	S18
PAP24	PAP25	PAP26	PAP27	PAP28	PAP29	PAP30
MAS1	MAS2	MAS3	MAS4	MAS5	MAS6	MAS7
15	16	17	18	19	20	21
S19	S20	S21	S22	S23	S24	S25
HTH1	HTH2	HTH3	HTH4	HTH5	HTH6	HTH7
MAS8	MAS9	MAS10	MAS11	MAS12	MAS13	MAS14
22	23	24	25	26	27	28
S26	S27	S28	S29	S30	O1	O2
HTH8	HTH9	HTH10	HTH11	HTH12	HTH13	HTH14
MAS15	MAS16	MAS17	MAS18	MAS19	MAS20	MAS21

THE NU AFRIKAN CALENDAR

EKUNDU 543/2059-60/6295-96/2051-52						
Ashe	Ojobo	Iwa	Araorun	Ile	Itutu	Erinle
1	2	3	4	5	6	7
O3	O4	O5	O6	O7	O8	O9
HTH15	HTH16	HTH17	HTH18	HTH19	HTH20	HTH21
MAS22	MAS23	MAS 24	MAS25	MAS26	MAS27	MAS28
8	9	10	11	12	13	14
O10	O11	O12	O13	O14	O15	O16
HTH22	HTH23	HTH24	HTH25	HTH26	HTH27	HTH28
MAS29	MAS30	THi1	THi2	THi3	THi4	THi5
15	16	17	18	19	20	21
O17	O18	O19	O20	O21	O22	O23
HTH29	HTH30	KHK1	KHK2	KHK3	KHK4	KHK5
THi6	THi7	THi8	THi9	THi10	THi11	THi12
22	23	24	25	26	27	28
O24	O25	O26	O27	O28	O29	O30
KHK6	KHK7	KHK8	KHK9	KHK10	KHK11	KHK12
THi13	THi14	THi15	THi16	THi17	THi18	THi19

KIJANI 543/2059-60/6295-96/2051-52						
Ashe	Ojobo	Iwa	Araorun	Ile	Itutu	Erinle
1	2	3	4	5	6	7
N28	N29	N30	D1	D2	D3	D4
TBT11	TBT12	TBT13	TBT14	TBT15	TBT16	TBT17
HID18	HID19	HID20	HID21	HID22	HID23	HID24
8	9	10	11	12	13	14
D5	D6	D7	D8	D9	D10	D11
TBT18	TBT19	TBT20	TBT21	TBT22	TBT23	TBT24
HID25	HID26	HID27	HID28	HID29	HID30	TAH1
15	16	17	18	19	20	21
D12	D13	D14	D15	D16	D17	D18
TBT25	TBT26	TBT27	TBT28	TBT29	TBT30	MKH1
TAH2	TAH3	TAH4	TAH5	TAH6	TAH7	TAH8
22	23	24	25	26	27	28
D19	D20	D21	D22	D23	D24	D25
MKH2	MKH3	MKH4	MKH5	MKH6	MKH7	MKH8
TAH9	TAH10	TAH11	TAH12	TAH13	TAH14	TAH15

EUSI 543/2059-60/6295-96/2051-52						
Ashe	Ojobo	Iwa	Araorun	Ile	Itutu	Erinle
1	2	3	4	5	6	7
O31	N1	N2	N3	N4	N5	N6
KHK13	KHK14	KHK15	KHK16	KHK17	KHK18	KHK19
THi20	THi21	THi22	THi23	THi24	THi25	THi26
8	9	10	11	12	13	14
N7	N8	N9	N10	N11	N12	N13
KHK20	KHK21	KHK22	KHK23	KHK24	KHK25	KHK26
THi27	THi28	THi29	THi30	HID1	HID2	HID3
15	16	17	18	19	20	21
N14	N15	N16	N17	N18	N19	N20
KHK27	KHK28	KHK29	KHK30	TBT1	TBT2	TBT3
HID4	HID5	HID6	HID7	HID8	HID9	HID10
22	23	24	25	26	27	28
N21	N22	N23	N24	N25	N26	N27
TBT4	TBT5	TBT6	TBT7	TBT8	TBT9	TBT10
HID11	HID12	HID13	HID14	HID15	HID16	HID17

KWANZA 543/2059-60/6295-96/2051-52						
Umoja	Kujichagulia	Ujima	Ujamaa	Nia	Kuumba	Imani
1	2	3	4	5	6	7
D26	D27	D28	D29	D30	D31	JA1
MKH9	MKH10	MKH11	MKH12	MKH13	MKH14	MKH1
TAH16	TAH17	TAH18	TAH19	TAH20	TAH21	TAH2
8	9	10	11	12	13	14
JA2	JA3	JA4	JA5	JA6	JA7	JA
MKH16	MKH17	MKH18	MKH19	MKH20	MKH21	MKH2
TAH23	TAH24	TAH25	TAH26	TAH27	TAH28	TAH2
15	16	17	18	19	20	21
JA9	JA10	JA11	JA12	JA13	JA14	JA1
MKH23	MKH24	MKH25	KMH26	MKH27	MKH28	MKH2
TAH30	TIR1	TIR2	TIR3	TIR4	TIR5	TIR6
22	23	24	25	26	27	28
JA16	JA17	JA18	JA19	JA20	JA21	JA2
MKH30	AMN1	AMN2	AMN3	AMN4	AMN5	AMN
TIR7	TIR8	TIR9	TIR10	TIR11	TIR12	TIR13

SHAHIDI 543/2059-60/6295-96/2051-52						
Ashe	Ojobo	Iwa	Araorun	Ile	Itutu	Erinle
1	2	3	4	5	6	7
JA23	JA24	JA25	JA26	JA27	JA28	JA29
AMN7	AMN8	AMN9	AMN10	AMN11	AMN12	AMN13
TIR14	TIR15	TIR16	TIR17	TIR18	TIR19	TIR20
8	9	10	11	12	13	14
JA30	JA31	F1	F2	F3	F4	F5
AMN14	AMN15	AMN16	AMN17	AMN18	AMN19	AMN20
TIR21	TIR22	TIR23	TIR24	TIR25	TIR26	TIR27
15	16	17	18	19	20	21
F6	F7	F8	F9	F10	F 11	F12
AMN21	AMN22	AMN23	AMN24	AMN25	AMN26	AMN28
TIR28	TIR29	TIR30	YEK1	YEK2	YEK3	YEK4
22	23	24	25	26	27	28
F13	F14	F15	F16	F17	F18	F19
AMN29	AMN30	RNT1	RNT2	RNT3	RNT4	RNT5
YEK5	YEK6	YEK7	YEK8	YEK9	YEK10	YEK11
SHAHIDI 94						
F20						
RNT6						
YEK12						

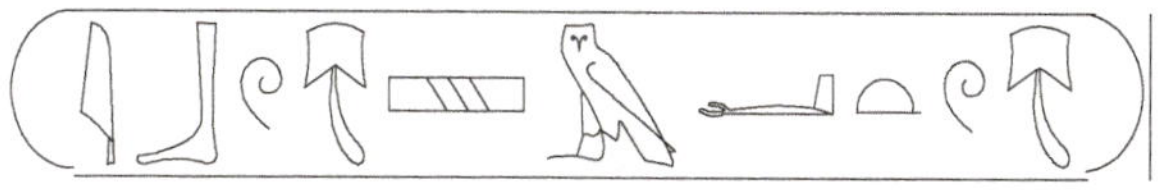

THE NU AFRIKAN CALENDAR

MAAFA 544 B.S.E./ASIENTO 2060-61 C.E./
KAM 6296-97 S.M./EC 2052-53 A.M. (ዓ.ም.)

AFU 544/2060-61/6296-97/2052-53						
Ashe	Ojobo	Iwa	Araorun	Ile	Itutu	Erinle
1	2	3	4	5	6	7
F21	F22	F23	F24	F25	F26	F27
RNT7	RNT8	RNT9	RNT10	RNT11	RNT12	RNT13
YEK13	YEK14	YEK15	YEK16	YEK17	YEK18	YEK19
8	9	10	11	12	13	14
F28	F29	M1	M2	M3	M4	M5
RNT14	RNT15	RNT16	RNT17	RNT18	RNT19	RNT20
YEK 20	YEK21	YEK22	YEK23	YEK24	YEK25	YEK26
15	16	17	18	19	20	21
M6	M7	M8	M9	M10	M11	M12
RNT21	RNT22	RNT23	RNT24	RNT25	RNT26	RNT27
YEK 27	YEK28	YEK29	YEK30	MAG1	MAG2	MAG3
22	23	24	25	26	27	28
M13	M14	M15	M16	M17	M18	M19
RNT28	RNT29	RNT30	KHN1	KHN2	KHN3	KHN4
MAG4	MAG5	MAG6	MAG7	MAG8	MAG9	MAG10
SHARIFU 133						
M20						
KHN5						
MAG11						

KAWAIDA 544/2060-61/6296-97/2052-53						
Ashe	Ojobo	Iwa	Araorun	Ile	Itutu	Erinle
1	2	3	4	5	6	7
A18	A19	A20	A21	A22	A23	A24
PIN4	PIN5	PIN6	PIN7	PIN8	PIN9	PIN10
MIY10	MIY11	MIY12	MIY13	MIY14	MIY15	MIY16
8	9	10	11	12	13	14
A25	A26	A27	A28	A29	A30	MA1
PIN11	PIN12	PIN13	PIN14	PIN15	PIN16	PIN17
MIY17	MIY18	MIY19	MIY20	MIY21	MIY22	MIY23
15	16	17	18	19	20	21
MA2	MA3	MA4	MA5	MA6	MA7	MA8
PIN18	PIN19	PIN20	PIN21	PIN22	PIN23	PIN24
MIY24	MIY25	MIY26	MIY27	MIY28	MIY29	MIY30
22	23	24	25	26	27	28
MA9	MA10	MA11	MA12	MA13	MA14	MA15
PIN25	PIN26	PIN27	PIN28	PIN29	PIN30	APP1
GIN1	GIN2	GIN3	GIN4	GIN5	GIN6	GIN7

UHURU 544/2060-61/6296-97/2052-53						
Ashe	Ojobo	Iwa	Araorun	Ile	Itutu	Erinle
1	2	3	4	5	6	7
M21	M22	M23	M24	M25	M26	M27
KHN6	KHN7	KHN8	KHN9	KHN10	KHN11	KHN12
MAG12	MAG13	MAG14	MAG15	MAG16	MAG17	MAG18
8	9	10	11	12	13	14
M28	M29	M30	M31	A1	A2	A3
KHN13	KHN14	KHN15	KHN16	KHN17	KHN18	KHN19
MAG19	MAG20	MAG21	MAG22	MAG23	MAG24	MAG25
15	16	17	18	19	20	21
A4	A5	A6	A7	A8	A9	A10
KHN20	KHN21	KHN222	KHN23	KHN24	KHN25	KHN26
MAG26	MAG27	MAG28	MAG29	MAG30	MIY1	MIY2
22	23	24	25	26	27	28
A11	A12	A13	A14	A15	A16	A17
KHN27	KHN28	KHN29	KHN30	PIN1	PIN2	PIN3
MIY3	MIY4	MIY5	MIY6	MIY7	MIY8	MIY9

MABABA WEUSI 544/2060-61/6296-97/2052-53						
Ashe	Ojobo	Iwa	Araorun	Ile	Itutu	Erinle
1	2	3	4	5	6	7
MA16	MA17	MA18	MA19	MA20	MA21	MA22
APP2	APP3	APP4	APP5	APP6	APP7	APP8
GIN8	GIN9	GIN10	GIN11	GIN12	GIN13	GIN14
8	9	10	11	12	13	14
MA23	MA24	MA25	MA26	MA27	MA28	MA29
APP9	APP10	APP11	APP12	APP13	APP14	APP15
GIN15	GIN16	GIN17	GIN18	GIN19	GIN20	GIN21
15	16	17	18	19	20	21
MA30	MA31	J1	J2	J3	J4	J5
APP16	APP17	APP18	APP19	APP20	APP21	APP22
GIN22	GIN23	GIN24	GIN25	GIN26	GIN27	GIN28
22	23	24	25	26	27	28
J6	J7	J8	J9	J10	J11	J12
APP23	APP24	APP25	APP26	APP27	APP28	APP29
GIN29	GIN30	SAN1	SAN2	SAN3	SAN4	SAN5

THE NU AFRIKAN CALENDAR

MWANAMKE WEUSI 544/2060-61/6296-97/2052-53

Ashe	Ojobo	Iwa	Araorun	Ile	Itutu	Erinle
1	2	3	4	5	6	7
J13	J14	J15	J16	J17	J18	J19
APP30	MSR1	MSR2	MSR3	MSR4	MSR5	MSR6
SAN6	SAN7	SAN8	SAN9	SAN10	SAN11	SAN12
8	9	10	11	12	13	14
J20	J21	J22	J23	J24	J25	J26
MSR7	MSR8	MSR9	MSR10	MSR11	MSR12	MSR13
SAN13	SAN14	SAN15	SAN16	SAN17	SAN18	SAN19
15	16	17	18	19	20	21
J27	J28	J29	J30	JU1	JU2	JU3
MSR14	MSR15	MSR16	MSR17	MSR18	MSR19	MSR20
SAN20	SAN21	SAN22	SAN23	SAN24	SAN25	SAN26
22	23	24	25	26	27	28
JU4	JU5	JU6	JU7	JU8	JU9	JU10
MSR21	MRS22	MSR23	MSR24	MSR25	MSR26	MSR27
SAN27	SAN28	SAN29	SAN30	HAM1	HAM2	HAM3

TAIFA WEUSI 544/2060-61/6296-97/2052-53

Ashe	Ojobo	Iwa	Araorun	Ile	Itutu	Erinle
1	2	3	4	5	6	7
AU8	AU9	AU10	AU11	AU12	AU13	AU14
THT20	THT21	THT22	THT23	THT24	THT25	THT26
NAH2	NAH3	NAH4	NAH5	NAH6	NAH7	NAH8
8	9	10	11	12	13	14
AU15	AU16	AU17	AU18	AU19	AU20	AU21
THT27	THT28	THT29	THT30	PAP1	PAP1	PAP2
NAH9	NAH10	NAH11	NAH12	NAH13	NAH14	NAH15
15	16	17	18	19	20	21
AU22	AU23	AU24	AU25	AU26	AU27	AU28
PAP3	PAP4	PAP5	PAP6	PAP7	PAP8	PAP9
NAH16	NAH17	NAH18	NAH19	NAH20	NAH21	NAH22
22	23	24	25	26	27	28
AU29	AU30	AU31	S1	S2	S3	S4
PAP10	PAP11	PAP12	PAP13	PAP14	PAP15	PAP16
NAH23	NAH24	NAH25	NAH26	NAH27	NAH28	NAH29

MTU WEUSI 544/2060-61/6296-97/2052-53

Ashe	Ojobo	Iwa	Araorun	Ile	Itutu	Erinle
1	2	3	4	5	6	7
JU11	JU12	JU13	JU14	JU15	JU16	JU17
MSR28	MSR29	MSR30	*ASR1*	*HRU2*	*STH3*	*AST4*
HAM4	HAM5	HAM6	HAM7	HAM8	HAM9	HAM10
8	9	10	11	12	13	14
JU18	JU19	JU20	JU21	JU22	JU23	JU24
NBH5	*ANU6*	THT1	THT2	THT3	THT4	THT5
HAM11	HAM12	HAM13	HAM14	HAM15	HAM16	HAM17
15	16	17	18	19	20	21
JU25	JU26	JU27	JU28	JU29	JU30	JU31
THT6	THT7	THT8	THT9	THT10	THT11	THT12
HAM18	HAM19	HAM20	HAM21	HAM22	HAM23	HAM24
22	23	24	25	26	27	28
AU1	AU2	AU3	AU4	AU5	AU6	AU7
THT13	THT14	THT15	THT16	THT17	THT18	THT19
HAM25	HAM26	HAM27	HAM28	HAM29	HAM30	NAH1

UWEZO 544/2060-61/6296-97/2052-53

Ashe	Ojobo	Iwa	Araorun	Ile	Itutu	Erinle
1	2	3	4	5	6	7
S5	S6	S7	S8	S9	S10	S11
PAP17	PAP18	PAP19	PAP20	PAP21	PAP22	PAP23
NAH30	*PAG1*	*PAG2*	*PAG3*	*PAG4*	*PAG5*	MAS1
8	9	10	11	12	13	14
S12	S13	S14	S15	S16	S17	S18
PAP24	PAP25	PAP26	PAP27	PAP28	PAP29	PAP30
MAS2	MAS3	MAS4	MAS5	MAS6	MAS7	MAS8
15	16	17	18	19	20	21
S19	S20	S21	S22	S23	S24	S25
HTH1	HTH2	HTH3	HTH4	HTH5	HTH6	HTH7
MAS9	MAS10	MAS11	MAS12	MAS13	MAS14	MAS15
22	23	24	25	26	27	28
S26	S27	S28	S29	S30	O1	O2
HTH8	HTH9	HTH10	HTH11	HTH12	HTH13	HTH14
MAS16	MAS17	MAS18	MAS19	MAS20	MAS21	MAS22

THE NU AFRIKAN CALENDAR

EKUNDU 544/2060-61/6296-97/2052-53						
Ashe	Ojobo	Iwa	Araorun	Ile	Itutu	Erinle
1	2	3	4	5	6	7
O3	O4	O5	O6	O7	O8	O9
HTH15	HTH16	HTH17	HTH18	HTH19	HTH20	HTH21
MAS23	MAS24	MAS25	MAS26	MAS27	MAS28	MAS29
8	9	10	11	12	13	14
O10	O11	O12	O13	O14	O15	O16
HTH22	HTH23	HTH24	HTH25	HTH26	HTH27	HTH28
MAS30	THi1	THi2	THi3	THi4	THi5	THi6
15	16	17	18	19	20	21
O17	O18	O19	O20	O21	O22	O23
HTH29	HTH30	KHK1	KHK2	KHK3	KHK4	KHK5
THi7	THi8	This9	THi10	THi11	THi12	THi13
22	23	24	25	26	27	28
O24	O25	O26	O27	O28	O29	O30
KHK6	KHK7	KHK8	KHK9	KHK10	KHK11	KHK12
Thi14	THi15	THi16	THi17	THi18	THi19	THi20

KIJANI 544/2060-61/6296-97/2052-53						
Ashe	Ojobo	Iwa	Araorun	Ile	Itutu	Erinle
1	2	3	4	5	6	7
N28	N29	N30	D1	D2	D3	D4
TBT11	TBT12	TBT13	TBT14	TBT15	TBT16	TBT17
HID19	HID20	HID21	HID22	HID23	HID24	HID25
8	9	10	11	12	13	14
D5	D6	D7	D8	D9	D10	D11
TBT18	TBT19	TBT20	TBT21	TBT22	TBT23	TBT24
HID26	HID27	HID28	HID29	HID30	TAH1	TAH2
15	16	17	18	19	20	21
D12	D13	D14	D15	D16	D17	D18
TBT25	TBT26	TBT27	TBT28	TBT29	TBT30	MKH1
TAH3	TAH4	TAH5	TAH6	TAH7	TAH8	TAH9
22	23	24	25	26	27	28
D19	D20	D21	D22	D23	D24	D25
MKH2	MKH3	MKH4	MKH5	MKH6	MKH7	MKH8
TAH10	TAH11	TAH12	TAH13	TAH14	TAH15	TAH16

EUSI 544/2060-61/6296-97/2052-53						
Ashe	Ojobo	Iwa	Araorun	Ile	Itutu	Erinle
1	2	3	4	5	6	7
O31	N1	N2	N3	N4	N5	N6
KHK13	KHK14	KHK15	KHK16	KHK17	KHK18	KHK19
THi21	THi22	THi23	THi24	THi25	THi26	THi27
8	9	10	11	12	13	14
N7	N8	N9	N10	N11	N12	N13
KHK20	KHK21	KHK22	KHK23	KHK24	KHK25	KHK26
THi28	THi29	THi30	HID1	HID2	HID3	HID4
15	16	17	18	19	20	21
N14	N15	N16	N17	N18	N19	N20
KHK27	KHK28	KHK29	KHK30	TBT1	TBT2	TBT3
HID5	HID6	HID7	HID8	HID9	HID10	HID11
22	23	24	25	26	27	28
N21	N22	N23	N24	N25	N26	N27
TBT4	TBT5	TBT6	TBT7	TBT8	TBT9	TBT10
HID12	HID13	HID14	HID15	HID16	HID17	HID18

KWANZA 544/2060-61/6296-97/2052-53						
Umoja	Kujichagulia	Ujima	Ujamaa	Nia	Kuumba	Imani
1	2	3	4	5	6	7
D26	D27	D28	D29	D30	D31	JA1
MKH9	MKH10	MKH11	MKH12	MKH13	MKH14	MKH15
TAH17	TAH18	TAH19	TAH20	TAH21	TAH22	TAH23
8	9	10	11	12	13	14
JA2	JA3	JA4	JA5	JA6	JA7	JA8
MKH16	MKH17	MKH18	MKH19	MKH20	MKH21	MKH22
TAH24	TAH25	TAH26	TAH27	TAH28	TAH29	TAH30
15	16	17	18	19	20	21
JA9	JA10	JA11	JA12	JA13	JA14	JA15
MKH23	MKH24	MKH25	KMH26	MKH27	MKH28	MKH29
TIR1	TIR2	TIR3	TIR4	TIR5	TIR6	TIR7
22	23	24	25	26	27	28
JA16	JA17	JA18	JA19	JA20	JA21	JA22
MKH30	AMN1	AMN2	AMN3	AMN4	AMN5	AMN6
TIR8	TIR9	TIR10	TIR11	TIR12	TIR13	TIR14

THE NU AFRIKAN CALENDAR

SHAHIDI 544/2060-61/6296-97/2052-53						
Ashe	Ojobo	Iwa	Araorun	Ile	Itutu	Erinle
1	2	3	4	5	6	7
JA23	JA24	JA25	JA26	JA27	JA28	JA29
AMN7	AMN8	AMN9	AMN10	AMN11	AMN12	AMN13
TIR15	TIR16	TIR17	TIR18	TIR19	TIR20	TIR21
8	9	10	11	12	13	14
JA30	JA31	F1	F2	F3	F4	F5
AMN14	AMN15	AMN16	AMN17	AMN18	AMN19	AMN20
TIR22	TIR23	TIR24	TIR25	TIR26	TIR27	TIR28
15	16	17	18	19	20	21
F6	F7	F8	F9	F10	F 11	F12
AMN21	AMN22	AMN23	AMN24	AMN25	AMN26	AMN28
TIR29	TIR30	YEK1	YEK2	YEK3	YEK4	YEK5
22	23	24	25	26	27	28
F13	F14	F15	F16	F17	F18	F19
AMN29	AMN30	RNT1	RNT2	RNT3	RNT4	RNT5
YEK6	YEK7	YEK8	YEK9	YEK10	YEK11	YEK12
SHAHIDI 95						
F20						
RNT6						
YEK13						

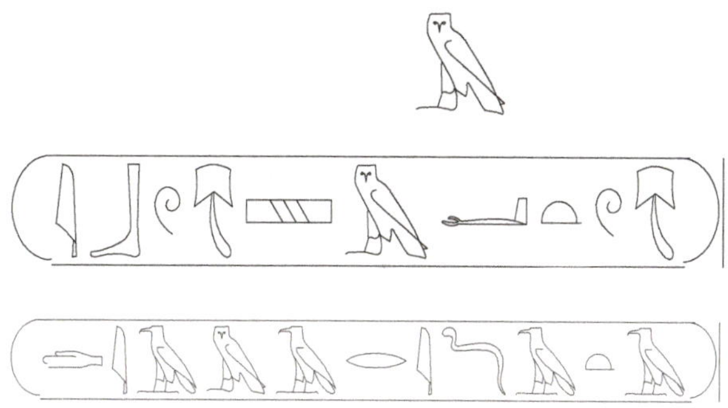

THE NU AFRIKAN CALENDAR

MAAFA 545 B.S.E./ASIENTO 2061-62 C.E.
KAM 6297-98 S.M./EC 2053-54 A.M. (ዓ.ም.)

AFU 545/2061-62/6297-98/2053-54						
Ashe	Ojobo	Iwa	Araorun	Ile	Itutu	Erinle
1	2	3	4	5	6	7
F21	F22	F23	F24	F25	F26	F27
RNT7	RNT8	RNT9	RNT10	RNT11	RNT12	RNT13
YEK14	YEK15	YEK16	YEK17	YEK18	YEK19	YEK20
8	9	10	11	12	13	14
F28	M1	M2	M3	M4	M5	M6
RNT14	RNT15	RNT16	RNT17	RNT18	RNT19	RNT20
YEK21	YEK22	YEK23	YEK24	YEK25	YEK26	YEK27
15	16	17	18	19	20	21
M7	M8	M9	M10	M11	M12	M13
RNT21	RNT22	RNT23	RNT24	RNT25	RNT26	RNT27
YEK28	YEK29	YEK30	MAG1	MAG2	MAG3	MAG4
22	23	24	25	26	27	28
M14	M15	M16	M17	M18	M19	M20
RNT28	RNT29	RNT30	KHN1	KHN2	KHN3	KHN4
MAG5	MAG6	MAG7	MAG8	MAG9	MAG10	MAG11

KAWAIDA 545/2061-62/6297-98/2053-54						
Ashe	Ojobo	Iwa	Araorun	Ile	Itutu	Erinle
1	2	3	4	5	6	7
A18	A19	A20	A21	A22	A23	A24
PIN3	PIN4	PIN5	PIN6	PIN7	PIN8	PIN9
MIY10	MIY11	MIY12	MIY13	MIY14	MIY15	MIY16
8	9	10	11	12	13	14
A25	A26	A27	A28	A29	A30	MA1
PIN10	PIN11	PIN12	PIN13	PIN14	PIN15	PIN16
MIY17	MIY18	MIY19	MIY20	MIY21	MIY22	MIY23
15	16	17	18	19	20	21
MA2	MA3	MA4	MA5	MA6	MA7	MA8
PIN17	PIN18	PIN19	PIN20	PIN21	PIN22	PIN23
MIY24	MIY25	MIY26	MIY27	MIY28	MIY29	MIY30
22	23	24	25	26	27	28
MA9	MA10	MA11	MA12	MA13	MA14	MA15
PIN24	PIN25	PIN26	PIN27	PIN28	PIN29	PIN30
GIN1	GIN2	GIN3	GIN4	GIN5	GIN6	GIN7

UHURU 545/2061-62/6297-98/2053-54						
Ashe	Ojobo	Iwa	Araorun	Ile	Itutu	Erinle
1	2	3	4	5	6	7
M21	M22	M23	M24	M25	M26	M27
KHN5	KHN6	KHN7	KHN8	KHN9	KHN10	KHN11
MAG12	MAG13	MAG14	MAG15	MAG16	MAG17	MAG18
8	9	10	11	12	13	14
M28	M29	M30	M31	A1	A2	A3
KHN12	KHN13	KHN14	KHN15	KHN16	KHN17	KHN18
MAG19	MAG20	MAG21	MAG22	MAG23	MAG24	MAG25
15	16	17	18	19	20	21
A4	A5	A6	A7	A8	A9	A10
KHN19	KHN20	KHN21	KHN22	KHN23	KHN24	KHN25
MAG26	MAG27	MAG28	MAG29	MAG30	MIY1	MIY2
22	23	24	25	26	27	28
A11	A12	A13	A14	A15	A16	A17
KHN26	KHN27	KHN28	KHN29	KHN30	PIN1	PIN2
MIY3	MIY4	MIY5	MIY6	MIY7	MIY8	MIY9

MABABA WEUSI 545/2061-62/6297-98/2053-54						
Ashe	Ojobo	Iwa	Araorun	Ile	Itutu	Erinle
1	2	3	4	5	6	7
MA16	MA17	MA18	MA19	MA20	MA21	MA22
APP1	APP2	APP3	APP4	APP5	APP6	APP7
GIN8	GIN9	GIN10	GIN11	GIN12	GIN13	GIN14
8	9	10	11	12	13	14
MA23	MA24	MA25	MA26	MA27	MA28	MA29
APP8	APP9	APP10	APP11	APP12	APP13	APP14
GIN15	GIN16	GIN17	GIN18	GIN19	GIN20	GIN21
15	16	17	18	19	20	21
MA30	MA31	J1	J2	J3	J4	J5
APP15	APP16	APP17	APP18	APP19	APP20	APP21
GIN22	GIN23	GIN24	GIN25	GIN26	GIN27	GIN28
22	23	24	25	26	27	28
J6	J7	J8	J9	J10	J11	J12
APP22	APP23	APP24	APP25	APP26	APP27	APP28
GIN29	GIN30	SAN1	SAN2	SAN3	SAN4	SAN5

THE NU AFRIKAN CALENDAR

MWANAMKE WEUSI 545/2061-62/6297-98/2053-54						
Ashe	Ojobo	Iwa	Araorun	Ile	Itutu	Erinle
1	2	3	4	5	6	7
J13	J14	J15	J16	J17	J18	J19
APP29	APP30	MSR1	MSR2	MSR3	MSR4	MSR5
SAN6	SAN7	SAN8	SAN9	SAN10	SAN11	SAN12
8	9	10	11	12	13	14
J20	J21	J22	J23	J24	J25	J26
MSR6	MSR7	MSR8	MSR9	MSR10	MSR11	MSR12
SAN13	SAN14	SAN15	SAN16	SAN17	SAN18	SAN19
15	16	17	18	19	20	21
J27	J28	J29	J30	JU1	JU2	JU3
MSR13	MSR14	MSR15	MSR16	MSR17	MSR18	MSR19
SAN20	SAN21	SAN22	SAN23	SAN24	SAN25	SAN26
22	23	24	25	26	27	28
JU4	JU5	JU6	JU7	JU8	JU9	JU10
MSR20	MRS21	MSR22	MSR23	MSR24	MSR25	MSR26
SAN27	SAN28	SAN29	SAN30	HAM1	HAM2	HAM3

TAIFA WEUSI 545/2061-62/6297-98/2053-54						
Ashe	Ojobo	Iwa	Araorun	Ile	Itutu	Erinle
1	2	3	4	5	6	7
AU8	AU9	AU10	AU11	AU12	AU13	AU14
THT20	THT21	THT22	THT23	THT24	THT25	THT26
NAH2	NAH3	NAH4	NAH5	NAH6	NAH7	NAH8
8	9	10	11	12	13	14
AU15	AU16	AU17	AU18	AU19	AU20	AU21
THT27	THT28	THT29	THT30	PAP1	PAP1	PAP2
NAH9	NAH10	NAH11	NAH12	NAH13	NAH14	NAH15
15	16	17	18	19	20	21
AU22	AU23	AU24	AU25	AU26	AU27	AU28
PAP3	PAP4	PAP5	PAP6	PAP7	PAP8	PAP9
NAH16	NAH17	NAH18	NAH19	NAH20	NAH21	NAH22
22	23	24	25	26	27	28
AU29	AU30	AU31	S1	S2	S3	S4
PAP10	PAP11	PAP12	PAP13	PAP14	PAP15	PAP16
NAH23	NAH24	NAH25	NAH26	NAH27	NAH28	NAH29

MTU WEUSI 545/2061-62/6297-98/2053-54						
Ashe	Ojobo	Iwa	Araorun	Ile	Itutu	Erinle
1	2	3	4	5	6	7
JU11	JU12	JU13	JU14	JU15	JU16	JU17
MSR27	MSR28	MSR29	MSR30	*ASR1*	*HRU2*	*STH3*
HAM4	HAM5	HAM6	HAM7	HAM8	HAM9	HAM10
8	9	10	11	12	13	14
JU18	JU19	JU20	JU21	JU22	JU23	JU24
AST4	*NBH5*	THT1	THT2	THT3	THT4	THT5
HAM11	HAM12	HAM13	AM14	HAM15	HAM16	HAM17
15	16	17	18	19	20	21
JU25	JU26	JU27	JU28	JU29	JU30	JU31
THT6	THT7	THT8	THT9	THT10	THT11	THT12
HAM18	HAM19	HAM20	HAM21	HAM22	HAM23	HAM24
22	23	24	25	26	27	28
AU1	AU2	AU3	AU4	AU5	AU6	AU7
THT13	THT14	THT15	THT16	THT17	THT18	THT19
HAM25	HAM26	HAM27	HAM28	HAM29	HAM30	NAH1

UWEZO 545/2061-62/6297-98/2053-54						
Ashe	Ojobo	Iwa	Araorun	Ile	Itutu	Erinle
1	2	3	4	5	6	7
S5	S6	S7	S8	S9	S10	S11
PAP17	PAP18	PAP19	PAP20	PAP21	PAP22	PAP23
NAH30	*PAG1*	*PAG2*	*PAG3*	*PAG4*	*PAG5*	MAS1
8	9	10	11	12	13	14
S12	S13	S14	S15	S16	S17	S18
PAP24	PAP25	PAP26	PAP27	PAP28	PAP29	PAP30
MAS2	MAS3	MAS4	MAS5	MAS6	MAS7	MAS8
15	16	17	18	19	20	21
S19	S20	S21	S22	S23	S24	S25
HTH1	HTH2	HTH3	HTH4	HTH5	HTH6	HTH7
MAS9	MAS10	MAS11	MAS12	MAS13	MAS14	MAS15
22	23	24	25	26	27	28
S26	S27	S28	S29	S30	O1	O2
HTH8	HTH9	HTH10	HTH11	HTH12	HTH13	HTH14
MAS16	MAS17	MAS18	MAS19	MAS20	MAS21	MAS22

THE NU AFRIKAN CALENDAR

EKUNDU 545/2061-62/6297-98/2053-54						
Ashe	Ojobo	Iwa	Araorun	Ile	Itutu	Erinle
1	2	3	4	5	6	7
O3	O4	O5	O6	O7	O8	O9
HTH15	HTH16	HTH17	HTH18	HTH19	HTH20	HTH21
MAS23	MAS24	MAS25	MAS26	MAS27	MAS28	MAS29
8	9	10	11	12	13	14
O10	O11	O12	O13	O14	O15	O16
HTH22	HTH23	HTH24	HTH25	HTH26	HTH27	HTH28
MAS30	THi1	THi2	THi3	THi4	THi5	THi6
15	16	17	18	19	20	21
O17	O18	O19	O20	O21	O22	O23
HTH29	HTH30	KHK1	KHK2	KHK3	KHK4	KHK5
THi7	THi8	THi9	THi10	THi11	THi12	THi13
22	23	24	25	26	27	28
O24	O25	O26	O27	O28	O29	O30
KHK6	KHK7	KHK8	KHK9	KHK10	KHK11	KHK12
THi14	THi15	THi16	THi17	THi18	THi19	THi20

KIJANI 545/2061-62/6297-98/2053-54						
Ashe	Ojobo	Iwa	Araorun	Ile	Itutu	Erinle
1	2	3	4	5	6	7
N28	N29	N30	D1	D2	D3	D4
TBT11	TBT12	TBT13	TBT14	TBT15	TBT16	TBT17
HID19	HID20	HID21	HID22	HID23	HID24	HID25
8	9	10	11	12	13	14
D5	D6	D7	D8	D9	D10	D11
TBT18	TBT19	TBT20	TBT21	TBT22	TBT23	TBT24
HID26	HID27	HID28	HID29	HID30	TAH1	TAH2
15	16	17	18	19	20	21
D12	D13	D14	D15	D16	D17	D18
TBT25	TBT26	TBT27	TBT28	TBT29	TBT30	MKH1
TAH3	TAH4	TAH5	TAH6	TAH7	TAH8	TAH9
22	23	24	25	26	27	28
D19	D20	D21	D22	D23	D24	D25
MKH2	MKH3	MKH4	MKH5	MKH6	MKH7	MKH8
TAH10	TAH11	TAH12	TAH13	TAH14	TAH15	TAH16

EUSI 545/2061-62/6297-98/2053-54						
Ashe	Ojobo	Iwa	Araorun	Ile	Itutu	Erinle
1	2	3	4	5	6	7
O31	N1	N2	N3	N4	N5	N6
KHK13	KHK14	KHK15	KHK16	KHK17	KHK18	KHK19
THi21	THi22	THi23	THi24	THi25	THi26	THi27
8	9	10	11	12	13	14
N7	N8	N9	N10	N11	N12	N13
KHK20	KHK21	KHK22	KHK23	KHK24	KHK25	KHK26
THi28	THi29	THi30	HID1	HID2	HID3	HID4
15	16	17	18	19	20	21
N14	N15	N16	N17	N18	N19	N20
KHK27	KHK28	KHK29	KHK30	TBT1	TBT2	TBT3
HID5	HID6	HID7	HID8	HID9	HID10	HID11
22	23	24	25	26	27	28
N21	N22	N23	N24	N25	N26	N27
TBT4	TBT5	TBT6	TBT7	TBT8	TBT9	TBT10
HID12	HID13	HID14	HID15	HID16	HID17	HID18

KWANZA 545/2061-62/6297-98/2053-54						
Umoja	Kujichagulia	Ujima	Ujamaa	Nia	Kuumba	Imani
1	2	3	4	5	6	7
D26	D27	D28	D29	D30	D31	JA1
MKH9	MKH10	MKH11	MKH12	MKH13	MKH14	MKH15
TAH17	TAH18	TAH19	TAH20	TAH21	TAH22	TAH23
8	9	10	11	12	13	14
JA2	JA3	JA4	JA5	JA6	JA7	JA8
MKH16	MKH17	MKH18	MKH19	MKH20	MKH21	MKH22
TAH24	TAH25	TAH26	TAH27	TAH28	TAH29	TAH30
15	16	17	18	19	20	21
JA9	JA10	JA11	JA12	JA13	JA14	JA15
MKH23	MKH24	MKH25	KMH26	MKH27	MKH28	MKH29
TIR1	TIR2	TIR3	TIR4	TIR5	TIR6	TIR7
22	23	24	25	26	27	28
JA16	JA17	JA18	JA19	JA20	JA21	JA22
MKH30	AMN1	AMN2	AMN3	AMN4	AMN5	AMN6
TIR8	TIR9	TIR10	TIR11	TIR12	TIR13	TIR14

THE NU AFRIKAN CALENDAR

SHAHIDI 545/2061-62/6297-98/2053-54						
Ashe	Ojobo	Iwa	Araorun	Ile	Itutu	Erinle
1	2	3	4	5	6	7
JA23	JA24	JA25	JA26	JA27	JA28	JA29
AMN7	AMN8	AMN9	AMN10	AMN11	AMN12	AMN13
TIR15	TIR16	TIR17	TIR18	TIR19	TIR20	TIR21
8	9	10	11	12	13	14
JA30	JA31	F1	F2	F3	F4	F5
AMN14	AMN15	AMN16	AMN17	AMN18	AMN19	AMN20
TIR22	TIR23	TIR24	TIR25	TIR26	TIR27	TIR28
15	16	17	18	19	20	21
F6	F7	F8	F9	F10	F 11	F12
AMN21	AMN22	AMN23	AMN24	AMN25	AMN26	AMN28
TIR29	TIR30	YEK1	YEK2	YEK3	YEK4	YEK5
22	23	24	25	26	27	28
F13	F14	F15	F16	F17	F18	F19
AMN29	AMN30	RNT1	RNT2	RNT3	RNT4	RNT5
YEK6	YEK7	YEK8	YEK9	YEK10	YEK11	YEK12
SHAHIDI 96						
F20						
RNT6						
YEK13						

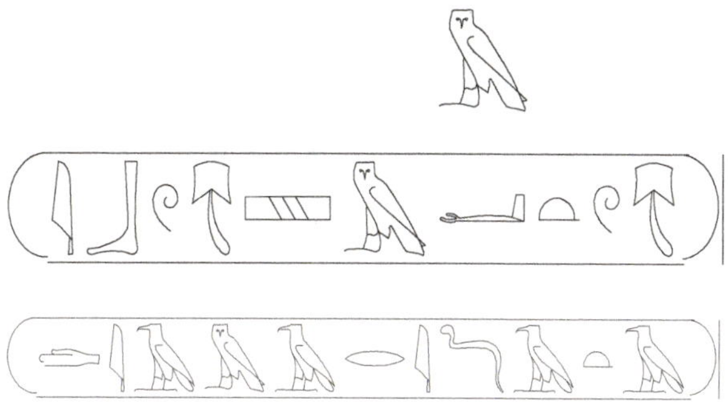

THE NU AFRIKAN CALENDAR

MAAFA 546 B.S.E./ASIENTO 2062-63 C.E.
KAM 6298-99 S.M./EC 2054-55 A.M. (ዓ.ም.)

AFU 546/2062-63/6298-99/2054-55

Ashe	Ojobo	Iwa	Araorun	Ile	Itutu	Erinle
1	2	3	4	5	6	7
F21	F22	F23	F24	F25	F26	F27
RNT7	RNT8	RNT9	RNT10	RNT11	RNT12	RNT13
YEK14	YEK15	YEK16	YEK17	YEK18	YEK19	YEK20
8	9	10	11	12	13	14
F28	M1	M2	M3	M4	M5	M6
RNT14	RNT15	RNT16	RNT17	RNT18	RNT19	RNT20
YEK21	YEK22	YEK23	YEK24	YEK25	YEK26	YEK27
15	16	17	18	19	20	21
M7	M8	M9	M10	M11	M12	M13
RNT21	RNT22	RNT23	RNT24	RNT25	RNT26	RNT27
YEK28	YEK29	YEK30	MAG1	MAG2	MAG3	MAG4
22	23	24	25	26	27	28
M14	M15	M16	M17	M18	M19	M20
RNT28	RNT29	RNT30	KHN1	KHN2	KHN3	KHN4
MAG5	MAG6	MAG7	MAG8	MAG9	MAG10	MAG11

UHURU 546/2062-63/6298-99/2054-55

Ashe	Ojobo	Iwa	Araorun	Ile	Itutu	Erinle
1	2	3	4	5	6	7
M21	M22	M23	M24	M25	M26	M27
KHN5	KHN6	KHN7	KHN8	KHN9	KHN10	KHN11
MAG12	MAG13	MAG14	MAG15	MAG16	MAG17	MAG18
8	9	10	11	12	13	14
M28	M29	M30	M31	A1	A2	A3
KHN12	KHN13	KHN14	KHN15	KHN16	KHN17	KHN18
MAG19	MAG20	MAG21	MAG22	MAG23	MAG24	MAG25
15	16	17	18	19	20	21
A4	A5	A6	A7	A8	A9	A10
KHN19	KHN20	KHN21	KHN22	KHN23	KHN24	KHN25
MAG26	MAG27	MAG28	MAG29	MAG30	MIY1	MIY2
22	23	24	25	26	27	28
A11	A12	A13	A14	A15	A16	A17
KHN26	KHN27	KHN28	KHN29	KHN30	PIN1	PIN2
MIY3	MIY4	MIY5	MIY6	MIY7	MIY8	MIY9

KAWAIDA 546/2062-63/6298-99/2054-55

Ashe	Ojobo	Iwa	Araorun	Ile	Itutu	Erinle
1	2	3	4	5	6	7
A18	A19	A20	A21	A22	A23	A24
PIN3	PIN4	PIN5	PIN6	PIN7	PIN8	PIN9
MIY10	MIY11	MIY12	MIY13	MIY14	MIY15	MIY16
8	9	10	11	12	13	14
A25	A26	A27	A28	A29	A30	MA1
PIN10	PIN11	PIN12	PIN13	PIN14	PIN15	PIN16
MIY17	MIY18	MIY19	MIY20	MIY21	MIY22	MIY23
15	16	17	18	19	20	21
MA2	MA3	MA4	MA5	MA6	MA7	MA8
PIN17	PIN18	PIN19	PIN20	PIN21	PIN22	PIN23
MIY24	MIY25	MIY26	MIY27	MIY28	MIY29	MIY30
22	23	24	25	26	27	28
MA9	MA10	MA11	MA12	MA13	MA14	MA15
PIN24	PIN25	PIN26	PIN27	PIN28	PIN29	PIN30
GIN1	GIN2	GIN3	GIN4	GIN5	GIN6	GIN7

MABABA WEUSI 546/2062-63/6298-99/2054-55

Ashe	Ojobo	Iwa	Araorun	Ile	Itutu	Erinle
1	2	3	4	5	6	7
MA16	MA17	MA18	MA19	MA20	MA21	MA22
APP1	APP2	APP3	APP4	APP5	APP6	APP7
GIN8	GIN9	GIN10	GIN11	GIN12	GIN13	GIN14
8	9	10	11	12	13	14
MA23	MA24	MA25	MA26	MA27	MA28	MA29
APP8	APP9	APP10	APP11	APP12	APP13	APP14
GIN15	GIN16	GIN17	GIN18	GIN19	GIN20	GIN21
15	16	17	18	19	20	21
MA30	MA31	J1	J2	J3	J4	J5
APP15	APP16	APP17	APP18	APP19	APP20	APP21
GIN22	GIN23	GIN24	GIN25	GIN26	GIN27	GIN28
22	23	24	25	26	27	28
J6	J7	J8	J9	J10	J11	J12
APP22	APP23	APP24	APP25	APP26	APP27	APP28
GIN29	GIN30	SAN1	SAN2	SAN3	SAN4	SAN5

THE NU AFRIKAN CALENDAR

MWANAMKE WEUSI 546/2062-63/6298-99/2054-55

Ashe	Ojobo	Iwa	Araorun	Ile	Itutu	Erinle
1	2	3	4	5	6	7
J13	J14	J15	J16	J17	J18	J19
APP29	APP30	MSR1	MSR2	MSR3	MSR4	MSR5
SAN6	SAN7	SAN8	SAN9	SAN10	SAN11	SAN12
8	9	10	11	12	13	14
J20	J21	J22	J23	J24	J25	J26
MSR6	MSR7	MSR8	MSR9	MSR10	MSR11	MSR12
SAN13	SAN14	SAN15	SAN16	SAN17	SAN18	SAN19
15	16	17	18	19	20	21
J27	J28	J29	J30	JU1	JU2	JU3
MSR13	MSR14	MSR15	MSR16	MSR17	MSR18	MSR19
SAN20	SAN21	SAN22	SAN23	SAN24	SAN25	SAN26
22	23	24	25	26	27	28
JU4	JU5	JU6	JU7	JU8	JU9	JU10
MSR20	MRS21	MSR22	MSR23	MSR24	MSR25	MSR26
SAN27	SAN28	SAN29	SAN30	HAM1	HAM2	HAM3

TAIFA WEUSI 546/2062-63/6298-99/2054-55

Ashe	Ojobo	Iwa	Araorun	Ile	Itutu	Erinle
1	2	3	4	5	6	7
AU8	AU9	AU10	AU11	AU12	AU13	AU14
THT20	THT21	THT22	THT23	THT24	THT25	THT26
NAH2	NAH3	NAH4	NAH5	NAH6	NAH7	NAH8
8	9	10	11	12	13	14
AU15	AU16	AU17	AU18	AU19	AU20	AU21
THT27	THT28	THT29	THT30	PAP1	PAP1	PAP2
NAH9	NAH10	NAH11	NAH12	NAH13	NAH14	NAH15
15	16	17	18	19	20	21
AU22	AU23	AU24	AU25	AU26	AU27	AU28
PAP3	PAP4	PAP5	PAP6	PAP7	PAP8	PAP9
NAH16	NAH17	NAH18	NAH19	NAH20	NAH21	NAH22
22	23	24	25	26	27	28
AU29	AU30	AU31	S1	S2	S3	S4
PAP10	PAP11	PAP12	PAP13	PAP14	PAP15	PAP16
NAH23	NAH24	NAH25	NAH26	NAH27	NAH28	NAH29

MTU WEUSI 546/2062-63/6298-99/2054-55

Ashe	Ojobo	Iwa	Araorun	Ile	Itutu	Erinle
1	2	3	4	5	6	7
JU11	JU12	JU13	JU14	JU15	JU16	JU17
MSR27	MSR28	MSR29	MSR30	*ASR1*	*HRU2*	*STH3*
HAM4	HAM5	HAM6	HAM7	HAM8	HAM9	HAM10
8	9	10	11	12	13	14
JU18	JU19	JU20	JU21	JU22	JU23	JU24
AST4	*NBH5*	THT1	THT2	THT3	THT4	THT5
HAM11	HAM12	HAM13	AM14	HAM15	HAM16	HAM17
15	16	17	18	19	20	21
JU25	JU26	JU27	JU28	JU29	JU30	JU31
THT6	THT7	THT8	THT9	THT10	THT11	THT12
HAM18	HAM19	HAM20	HAM21	HAM22	HAM23	HAM24
22	23	24	25	26	27	28
AU1	AU2	AU3	AU4	AU5	AU6	AU7
THT13	THT14	THT15	THT16	THT17	THT18	THT19
HAM25	HAM26	HAM27	HAM28	HAM29	HAM30	NAH1

UWEZO 546/2062-63/6298-99/2054-55

Ashe	Ojobo	Iwa	Araorun	Ile	Itutu	Erinle
1	2	3	4	5	6	7
S5	S6	S7	S8	S9	S10	S11
PAP17	PAP18	PAP19	PAP20	PAP21	PAP22	PAP23
NAH30	*PAG1*	*PAG2*	*PAG3*	*PAG4*	*PAG5*	MAS1
8	9	10	11	12	13	14
S12	S13	S14	S15	S16	S17	S18
PAP24	PAP25	PAP26	PAP27	PAP28	PAP29	PAP30
MAS2	MAS3	MAS4	MAS5	MAS6	MAS7	MAS8
15	16	17	18	19	20	21
S19	S20	S21	S22	S23	S24	S25
HTH1	HTH2	HTH3	HTH4	HTH5	HTH6	HTH7
MAS9	MAS10	MAS11	MAS12	MAS13	MAS14	MAS15
22	23	24	25	26	27	28
S26	S27	S28	S29	S30	O1	O2
HTH8	HTH9	HTH10	HTH11	HTH12	HTH13	HTH14
MAS16	MAS17	MAS18	MAS19	MAS20	MAS21	MAS22

EKUNDU 546/2062-63/6298-99/2054-55						
Ashe	Ojobo	Iwa	Araorun	Ile	Itutu	Erinle
1	2	3	4	5	6	7
O3	O4	O5	O6	O7	O8	O9
HTH15	HTH16	HTH17	HTH18	HTH19	HTH20	HTH21
MAS23	MAS24	MAS25	MAS26	MAS27	MAS28	MAS29
8	9	10	11	12	13	14
O10	O11	O12	O13	O14	O15	O16
HTH22	HTH23	HTH24	HTH25	HTH26	HTH27	HTH28
MAS30	THi1	THi2	THi3	THi4	THi5	THi6
15	16	17	18	19	20	21
O17	O18	O19	O20	O21	O22	O23
HTH29	HTH30	KHK1	KHK2	KHK3	KHK4	KHK5
THi7	THi8	THi9	THi10	THi11	THi12	THi13
22	23	24	25	26	27	28
O24	O25	O26	O27	O28	O29	O30
KHK6	KHK7	KHK8	KHK9	KHK10	KHK11	KHK12
THi14	THi15	THi16	THi17	THi18	THi19	THi20

KIJANI 546/2062-63/6298-99/2054-55						
Ashe	Ojobo	Iwa	Araorun	Ile	Itutu	Erinle
1	2	3	4	5	6	7
N28	N29	N30	D1	D2	D3	D4
TBT11	TBT12	TBT13	TBT14	TBT15	TBT16	TBT17
HID19	HID20	HID21	HID22	HID23	HID24	HID25
8	9	10	11	12	13	14
D5	D6	D7	D8	D9	D10	D11
TBT18	TBT19	TBT20	TBT21	TBT22	TBT23	TBT24
HID26	HID27	HID28	HID29	HID30	TAH1	TAH2
15	16	17	18	19	20	21
D12	D13	D14	D15	D16	D17	D18
TBT25	TBT26	TBT27	TBT28	TBT29	TBT30	MKH1
TAH3	TAH4	TAH5	TAH6	TAH7	TAH8	TAH9
22	23	24	25	26	27	28
D19	D20	D21	D22	D23	D24	D25
MKH2	MKH3	MKH4	MKH5	MKH6	MKH7	MKH8
TAH10	TAH11	TAH12	TAH13	TAH14	TAH15	TAH16

EUSI 546/2062-63/6298-99/2054-55						
Ashe	Ojobo	Iwa	Araorun	Ile	Itutu	Erinle
1	2	3	4	5	6	7
O31	N1	N2	N3	N4	N5	N6
KHK13	KHK14	KHK15	KHK16	KHK17	KHK18	KHK19
THi21	THi22	THi23	THi24	THi25	THi26	THi27
8	9	10	11	12	13	14
N7	N8	N9	N10	N11	N12	N13
KHK20	KHK21	KHK22	KHK23	KHK24	KHK25	KHK26
THi28	THi29	THi30	HID1	HID2	HID3	HID4
15	16	17	18	19	20	21
N14	N15	N16	N17	N18	N19	N20
KHK27	KHK28	KHK29	KHK30	TBT1	TBT2	TBT3
HID5	HID6	HID7	HID8	HID9	HID10	HID11
22	23	24	25	26	27	28
N21	N22	N23	N24	N25	N26	N27
TBT4	TBT5	TBT6	TBT7	TBT8	TBT9	TBT10
HID12	HID13	HID14	HID15	HID16	HID17	HID18

KWANZA 546/2062-63/6298-99/2054-55						
Umoja	Kujichagulia	Ujima	Ujamaa	Nia	Kuumba	Imani
1	2	3	4	5	6	7
D26	D27	D28	D29	D30	D31	JA1
MKH9	MKH10	MKH11	MKH12	MKH13	MKH14	MKH15
TAH17	TAH18	TAH19	TAH20	TAH21	TAH22	TAH23
8	9	10	11	12	13	14
JA2	JA3	JA4	JA5	JA6	JA7	JA8
MKH16	MKH17	MKH18	MKH19	MKH20	MKH21	MKH22
TAH24	TAH25	TAH26	TAH27	TAH28	TAH29	TAH30
15	16	17	18	19	20	21
JA9	JA10	JA11	JA12	JA13	JA14	JA15
MKH23	MKH24	MKH25	KMH26	MKH27	MKH28	MKH29
TIR1	TIR2	TIR3	TIR4	TIR5	TIR6	TIR7
22	23	24	25	26	27	28
JA16	JA17	JA18	JA19	JA20	JA21	JA22
MKH30	AMN1	AMN2	AMN3	AMN4	AMN5	AMN6
TIR8	TIR9	TIR10	TIR11	TIR12	TIR13	TIR14

SHAHIDI 546/2062-63/6298-99/2054-55						
Ashe	Ojobo	Iwa	Araorun	Ile	Itutu	Erinle
1	2	3	4	5	6	7
JA23	JA24	JA25	JA26	JA27	JA28	JA29
AMN7	AMN8	AMN9	AMN10	AMN11	AMN12	AMN13
TIR15	TIR16	TIR17	TIR18	TIR19	TIR20	TIR21
8	9	10	11	12	13	14
JA30	JA31	F1	F2	F3	F4	F5
AMN14	AMN15	AMN16	AMN17	AMN18	AMN19	AMN20
TIR22	TIR23	TIR24	TIR25	TIR26	TIR27	TIR28
15	16	17	18	19	20	21
F6	F7	F8	F9	F10	F 11	F12
AMN21	AMN22	AMN23	AMN24	AMN25	AMN26	AMN28
TIR29	TIR30	YEK1	YEK2	YEK3	YEK4	YEK5
22	23	24	25	26	27	28
F13	F14	F15	F16	F17	F18	F19
AMN29	AMN30	RNT1	RNT2	RNT3	RNT4	RNT5
YEK6	YEK7	YEK8	YEK9	YEK10	YEK11	YEK12
SHAHIDI 97						
F20						
RNT6						
YEK13						

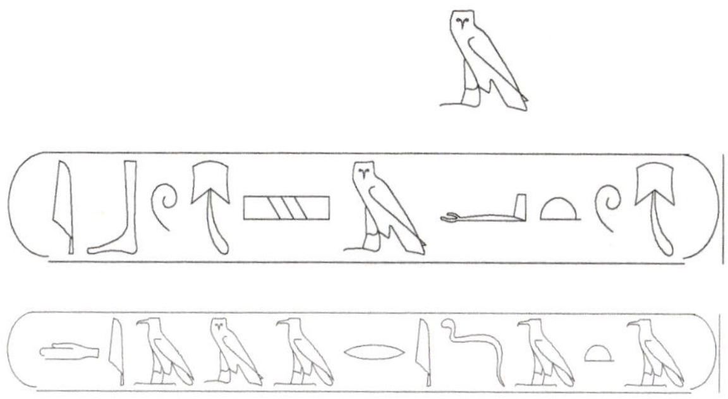

THE NU AFRIKAN CALENDAR

MAAFA 547 B.S.E./ASIENTO 2063-64 C.E./ KAM 6299-00 S.M./EC 2055-56 A.M. (ዓ.ም.)

AFU 547/2063-64/6299-00/2055-56						
Ashe	Ojobo	Iwa	Araorun	Ile	Itutu	Erinle
1	2	3	4	5	6	7
F21	F22	F23	F24	F25	F26	F27
RNT7	RNT8	RNT9	RNT10	RNT11	RNT12	RNT13
YEK14	YEK15	YEK16	YEK17	YEK18	YEK19	YEK20
8	9	10	11	12	13	14
F28	M1	M2	M3	M4	M5	M6
RNT14	RNT15	RNT16	RNT17	RNT18	RNT19	RNT20
YEK21	YEK22	YEK23	YEK24	YEK25	YEK26	YEK27
15	16	17	18	19	20	21
M7	M8	M9	M10	M11	M12	M13
RNT21	RNT22	RNT23	RNT24	RNT25	RNT26	RNT27
YEK28	YEK 29	YEK30	MAG1	MAG2	MAG3	MAG4
22	23	24	25	26	27	28
M14	M15	M16	M17	M18	M19	M20
RNT28	RNT29	RNT30	KHN1	KHN2	KHN3	KHN4
MAG5	MAG6	MAG7	MAG8	MAG9	MAG10	MAG11

UHURU 547/2063-64/6299-00/2055-56						
Ashe	Ojobo	Iwa	Araorun	Ile	Itutu	Erinle
1	2	3	4	5	6	7
M21	M22	M23	M24	M25	M26	M27
KHN5	KHN6	KHN7	KHN8	KHN9	KHN10	KHN11
MAG12	MAG13	MAG14	MAG15	MAG16	MAG17	MAG18
8	9	10	11	12	13	14
M28	M29	M30	M31	A1	A2	A3
KHN12	KHN13	KHN14	KHN15	KHN16	KHN17	KHN18
MAG19	MAG20	MAG21	MAG22	MAG23	MAG24	MAG25
15	16	17	18	19	20	21
A4	A5	A6	A7	A8	A9	A10
KHN19	KHN20	KHN21	KHN22	KHN23	KHN24	KHN25
MAG26	MAG27	MAG28	MAG29	MAG30	MIY1	MIY2
22	23	24	25	26	27	28
A11	A12	A13	A14	A15	A16	A17
KHN26	KHN27	KHN28	KHN29	KHN30	PIN1	PIN2
MIY3	MIY4	MIY5	MIY6	MIY7	MIY8	MIY9

KAWAIDA 547/2063-64/6299-00/2055-56						
Ashe	Ojobo	Iwa	Araorun	Ile	Itutu	Erinle
1	2	3	4	5	6	7
A18	A19	A20	A21	A22	A23	A24
PIN3	PIN4	PIN5	PIN6	PIN7	PIN8	PIN9
MIY10	MIY11	MIY12	MIY13	MIY14	MIY15	MIY1
8	9	10	11	12	13	14
A25	A26	A27	A28	A29	A30	MA1
PIN10	PIN11	PIN12	PIN13	PIN14	PIN15	PIN16
MIY17	MIY18	MIY19	MIY20	MIY21	MIY22	MIY23
15	16	17	18	19	20	21
MA2	MA3	MA4	MA5	MA6	MA7	MA8
PIN17	PIN18	PIN19	PIN20	PIN21	PIN22	PIN2
MIY24	MIY25	MIY26	MIY27	MIY28	MIY29	MIY3
22	23	24	25	26	27	28
MA9	MA10	MA11	MA12	MA13	MA14	MA1
PIN24	PIN25	PIN26	PIN27	PIN28	PIN29	PIN3
GIN1	GIN2	GIN3	GIN4	GIN5	GIN6	GIN7

MABABA WEUSI 547/2063-64/6299-00/2055-56						
Ashe	Ojobo	Iwa	Araorun	Ile	Itutu	Erinle
1	2	3	4	5	6	7
MA16	MA17	MA18	MA19	MA20	MA21	MA22
APP1	APP2	APP3	APP4	APP5	APP6	APP
GIN8	GIN9	GIN10	GIN11	GIN12	GIN13	GIN1
8	9	10	11	12	13	14
MA23	MA24	MA25	MA26	MA27	MA28	MA29
APP8	APP9	APP10	APP11	APP12	APP13	APP1
GIN15	GIN16	GIN17	GIN18	GIN19	GIN20	GIN21
15	16	17	18	19	20	21
MA30	MA31	J1	J2	J3	J4	J5
APP15	APP16	APP17	APP18	APP19	APP20	APP21
GIN22	GIN23	GIN24	GIN25	GIN26	GIN27	GIN28
22	23	24	25	26	27	28
J6	J7	J8	J9	J10	J11	J12
APP22	APP23	APP24	APP25	APP26	APP27	APP2
GIN29	GIN30	SAN1	SAN2	SAN3	SAN4	SAN5

THE NU AFRIKAN CALENDAR

MWANAMKE WEUSI 547/2063-64/6299-00/2055-56						
Ashe	Ojobo	Iwa	Araorun	Ile	Itutu	Erinle
1	2	3	4	5	6	7
J13	J14	J15	J16	J17	J18	J19
APP29	APP30	MSR1	MSR2	MSR3	MSR4	MSR5
SAN6	SAN7	SAN8	SAN9	SAN10	SAN11	SAN12
8	9	10	11	12	13	14
J20	J21	J22	J23	J24	J25	J26
MSR6	MSR7	MSR8	MSR9	MSR10	MSR11	MSR12
SAN13	SAN14	SAN15	SAN16	SAN17	SAN18	SAN19
15	16	17	18	19	20	21
J27	J28	J29	J30	JU1	JU2	JU3
MSR13	MSR14	MSR15	MSR16	MSR17	MSR18	MSR19
SAN20	SAN21	SAN22	SAN23	SAN24	SAN25	SAN26
22	23	24	25	26	27	28
JU4	JU5	JU6	JU7	JU8	JU9	JU10
MSR20	MRS21	MSR22	MSR23	MSR24	MSR25	MSR26
SAN27	SAN28	SAN29	SAN30	HAM1	HAM2	HAM3

MTU WEUSI 547/2063-64/6299-00/2055-56						
Ashe	Ojobo	Iwa	Araorun	Ile	Itutu	Erinle
1	2	3	4	5	6	7
JU11	JU12	JU13	JU14	JU15	JU16	JU17
MSR27	MSR28	MSR29	MSR30	*ASR1*	*HRU2*	*STH3*
HAM4	HAM5	HAM6	HAM7	HAM8	HAM9	HAM10
8	9	10	11	12	13	14
JU18	JU19	JU20	JU21	JU22	JU23	JU24
AST4	*NBH5*	THT1	THT2	THT3	THT4	THT5
HAM11	HAM12	HAM13	HAM14	HAM15	HAM16	HAM17
15	16	17	18	19	20	21
JU25	JU26	JU27	JU28	JU29	JU30	JU31
THT6	THT7	THT8	THT9	THT10	THT11	THT12
HAM18	HAM19	HAM20	HAM21	HAM22	HAM23	HAM24
22	23	24	25	26	27	28
AU1	AU2	AU3	AU4	AU5	AU6	AU7
THT13	THT14	THT15	THT16	THT17	THT18	THT19
HAM25	HAM26	HAM27	HAM28	HAM29	HAM30	NAH1

TAIFA WEUSI 547/2063-64/6299-00/2055-56						
Ashe	Ojobo	Iwa	Araorun	Ile	Itutu	Erinle
1	2	3	4	5	6	7
AU8	AU9	AU10	AU11	AU12	AU13	AU14
THT20	THT21	THT22	THT23	THT24	THT25	THT26
NAH2	NAH3	NAH4	NAH5	NAH6	NAH7	NAH8
8	9	10	11	12	13	14
AU15	AU16	AU17	AU18	AU19	AU20	AU21
THT27	THT28	THT29	THT30	PAP1	PAP1	PAP2
NAH9	NAH10	NAH11	NAH12	NAH13	NAH14	NAH15
15	16	17	18	19	20	21
AU22	AU23	AU24	AU25	AU26	AU27	AU28
PAP3	PAP4	PAP5	PAP6	PAP7	PAP8	PAP9
NAH16	NAH17	NAH18	NAH19	NAH20	NAH21	NAH22
22	23	24	25	26	27	28
AU29	AU30	AU31	S1	S2	S3	S4
PAP10	PAP11	PAP12	PAP13	PAP14	PAP15	PAP16
NAH23	NAH24	NAH25	NAH26	NAH27	NAH28	NAH29

UWEZO 547/2063-64/6299-00/2055-56						
Ashe	Ojobo	Iwa	Araorun	Ile	Itutu	Erinle
1	2	3	4	5	6	7
S5	S6	S7	S8	S9	S10	S11
PAP17	PAP18	PAP19	PAP20	PAP21	PAP22	PAP23
NAH30	*QUA1*	*QUA2*	*QUA3*	*QUA4*	*QUA5*	*QUA6*
8	9	10	11	12	13	14
S12	S13	S14	S15	S16	S17	S18
PAP24	PAP25	PAP26	PAP27	PAP28	PAP29	PAP30
MAS1	MAS2	MAS3	MAS4	MAS5	MAS6	MAS7
15	16	17	18	19	20	21
S19	S20	S21	S22	S23	S24	S25
HTH1	HTH2	HTH3	HTH4	HTH5	HTH6	HTH7
MAS8	MAS9	MAS10	MAS11	MAS12	MAS13	MAS14
22	23	24	25	26	27	28
S26	S27	S28	S29	S30	O1	O2
HTH8	HTH9	HTH10	HTH11	HTH12	HTH13	HTH14
MAS15	MAS16	MAS17	MAS18	MAS19	MAS20	MAS21

THE NU AFRIKAN CALENDAR

EKUNDU 547/2063-64/6299-00/2055-56						
Ashe	Ojobo	Iwa	Araorun	Ile	Itutu	Erinle
1	2	3	4	5	6	7
O3	O4	O5	O6	O7	O8	O9
HTH15	HTH16	HTH17	HTH18	HTH19	HTH20	HTH21
MAS22	MAS23	MAS 24	MAS25	MAS26	MAS27	MAS28
8	9	10	11	12	13	14
O10	O11	O12	O13	O14	O15	O16
HTH22	HTH23	HTH24	HTH25	HTH26	HTH27	HTH28
MAS29	MAS30	THi1	THi2	THi3	THi4	THi5
15	16	17	18	19	20	21
O17	O18	O19	O20	O21	O22	O23
HTH29	HTH30	KHK1	KHK2	KHK3	KHK4	KHK5
THi6	THi7	THi8	THi9	THi10	THi11	THi12
22	23	24	25	26	27	28
O24	O25	O26	O27	O28	O29	O30
KHK6	KHK7	KHK8	KHK9	KHK10	KHK11	KHK12
THi13	THi14	THi15	THi16	THi17	THi18	THi19

KIJANI 547/2063-64/6299-00/2055-56						
Ashe	Ojobo	Iwa	Araorun	Ile	Itutu	Erinle
1	2	3	4	5	6	7
N28	N29	N30	D1	D2	D3	D4
TBT11	TBT12	TBT13	TBT14	TBT15	TBT16	TBT17
HID18	HID19	HID20	HID21	HID22	HID23	HID24
8	9	10	11	12	13	14
D5	D6	D7	D8	D9	D10	D11
TBT18	TBT19	TBT20	TBT21	TBT22	TBT23	TBT24
HID25	HID26	HID27	HID28	HID29	HID30	TAH1
15	16	17	18	19	20	21
D12	D13	D14	D15	D16	D17	D18
TBT25	TBT26	TBT27	TBT28	TBT29	TBT30	MKH1
TAH2	TAH3	TAH4	TAH5	TAH6	TAH7	TAH8
22	23	24	25	26	27	28
D19	D20	D21	D22	D23	D24	D25
MKH2	MKH3	MKH4	MKH5	MKH6	MKH7	MKH8
TAH9	TAH10	TAH11	TAH12	TAH13	TAH14	TAH15

EUSI 547/2063-64/6299-00/2055-56						
Ashe	Ojobo	Iwa	Araorun	Ile	Itutu	Erinle
1	2	3	4	5	6	7
O31	N1	N2	N3	N4	N5	N6
KHK13	KHK14	KHK15	KHK16	KHK17	KHK18	KHK19
THi20	THi21	THi22	THi23	THi24	THi25	THi26
8	9	10	11	12	13	14
N7	N8	N9	N10	N11	N12	N13
KHK20	KHK21	KHK22	KHK23	KHK24	KHK25	KHK26
THi27	THi28	THi29	THi30	HID1	HID2	HID3
15	16	17	18	19	20	21
N14	N15	N16	N17	N18	N19	N20
KHK27	KHK28	KHK29	KHK30	TBT1	TBT2	TBT3
HID4	HID5	HID6	HID7	HID8	HID9	HID10
22	23	24	25	26	27	28
N21	N22	N23	N24	N25	N26	N27
TBT4	TBT5	TBT6	TBT7	TBT8	TBT9	TBT10
HID11	HID12	HID13	HID14	HID15	HID16	HID17

KWANZA 547/2063-64/6299-00/2055-56						
Umoja	Kujichagulia	Ujima	Ujamaa	Nia	Kuumba	Imani
1	2	3	4	5	6	7
D26	D27	D28	D29	D30	D31	JA1
MKH9	MKH10	MKH11	MKH12	MKH13	MKH14	MKH15
TAH16	TAH17	TAH18	TAH19	TAH20	TAH21	TAH22
8	9	10	11	12	13	14
JA2	JA3	JA4	JA5	JA6	JA7	JA8
MKH16	MKH17	MKH18	MKH19	MKH20	MKH21	MKH22
TAH23	TAH24	TAH25	TAH26	TAH27	TAH28	TAH29
15	16	17	18	19	20	21
JA9	JA10	JA11	JA12	JA13	JA14	JA15
MKH23	MKH24	MKH25	KMH26	MKH27	MKH28	MKH29
TAH30	TIR1	TIR2	TIR3	TIR4	TIR5	TIR6
22	23	24	25	26	27	28
JA16	JA17	JA18	JA19	JA20	JA21	JA22
MKH30	AMN1	AMN2	AMN3	AMN4	AMN5	AMN6
TIR7	TIR8	TIR9	TIR10	TIR11	TIR12	TIR13

SHAHIDI 547/2063-64/6299-00/2055-56						
Ashe	Ojobo	Iwa	Araorun	Ile	Itutu	Erinle
1	2	3	4	5	6	7
JA23	JA24	JA25	JA26	JA27	JA28	JA29
AMN7	AMN8	AMN9	AMN10	AMN11	AMN12	AMN13
TIR14	TIR15	TIR16	TIR17	TIR18	TIR19	TIR20
8	9	10	11	12	13	14
JA30	JA31	F1	F2	F3	F4	F5
AMN14	AMN15	AMN16	AMN17	AMN18	AMN19	AMN20
TIR21	TIR22	TIR23	TIR24	TIR25	TIR26	TIR27
15	16	17	18	19	20	21
F6	F7	F8	F9	F10	F 11	F12
AMN21	AMN22	AMN23	AMN24	AMN25	AMN26	AMN28
TIR28	TIR29	TIR30	YEK1	YEK2	YEK3	YEK4
22	23	24	25	26	27	28
F13	F14	F15	F16	F17	F18	F19
AMN29	AMN30	RNT1	RNT2	RNT3	RNT4	RNT5
YEK5	YEK6	YEK7	YEK8	YEK9	YEK10	YEK11
SHAHIDI 98						
F20						
RNT6						
YEK12						

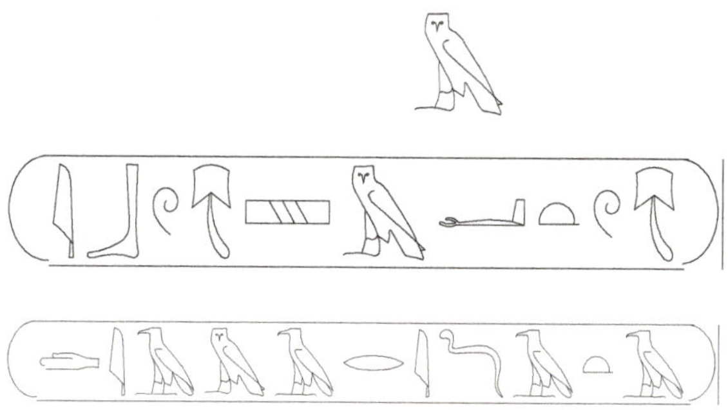

THE NU AFRIKAN CALENDAR

MAAFA 548 B.S.E./ASIENTO 2064-65 C.E./
KAM 6300-01 S.M./EC 2056-57 A.M. (ዓ.ም.)

AFU 548/2064-65/6300-01/2056-57						
Ashe	Ojobo	Iwa	Araorun	Ile	Itutu	Erinle
1	2	3	4	5	6	7
F21	F22	F23	F24	F25	F26	F27
RNT7	RNT8	RNT9	RNT10	RNT11	RNT12	RNT13
YEK13	YEK14	YEK15	YEK16	YEK17	YEK18	YEK19
8	9	10	11	12	13	14
F28	F29	M1	M2	M3	M4	M5
RNT14	RNT15	RNT16	RNT17	RNT18	RNT19	RNT20
YEK 20	YEK21	YEK22	YEK23	YEK24	YEK25	YEK26
15	16	17	18	19	20	21
M6	M7	M8	M9	M10	M11	M12
RNT21	RNT22	RNT23	RNT24	RNT25	RNT26	RNT27
YEK 27	YEK28	YEK29	YEK30	MAG1	MAG2	MAG3
22	23	24	25	26	27	28
M13	M14	M15	M16	M17	M18	M19
RNT28	RNT29	RNT30	KHN1	KHN2	KHN3	KHN4
MAG4	MAG5	MAG6	MAG7	MAG8	MAG9	MAG10
SHARIFU 134						
M20						
KHN5						
MAG11						

KAWAIDA 548/2064-65/6300-01/2056-57						
Ashe	Ojobo	Iwa	Araorun	Ile	Itutu	Erinle
1	2	3	4	5	6	7
A18	A19	A20	A21	A22	A23	A24
PIN4	PIN5	PIN6	PIN7	PIN8	PIN9	PIN10
MIY10	MIY11	MIY12	MIY13	MIY14	MIY15	MIY16
8	9	10	11	12	13	14
A25	A26	A27	A28	A29	A30	MA1
PIN11	PIN12	PIN13	PIN14	PIN15	PIN16	PIN17
MIY17	MIY18	MIY19	MIY20	MIY21	MIY22	MIY23
15	16	17	18	19	20	21
MA2	MA3	MA4	MA5	MA6	MA7	MA8
PIN18	PIN19	PIN20	PIN21	PIN22	PIN23	PIN24
MIY24	MIY25	MIY26	MIY27	MIY28	MIY29	MIY30
22	23	24	25	26	27	28
MA9	MA10	MA11	MA12	MA13	MA14	MA15
PIN25	PIN26	PIN27	PIN28	PIN29	PIN30	APP1
GIN1	GIN2	GIN3	GIN4	GIN5	GIN6	GIN7

UHURU 548/2064-65/6300-01/2056-57						
Ashe	Ojobo	Iwa	Araorun	Ile	Itutu	Erinle
1	2	3	4	5	6	7
M21	M22	M23	M24	M25	M26	M27
KHN6	KHN7	KHN8	KHN9	KHN10	KHN11	KHN12
MAG12	MAG13	MAG14	MAG15	MAG16	MAG17	MAG18
8	9	10	11	12	13	14
M28	M29	M30	M31	A1	A2	A3
KHN13	KHN14	KHN15	KHN16	KHN17	KHN18	KHN19
MAG19	MAG20	MAG21	MAG22	MAG23	MAG24	MAG25
15	16	17	18	19	20	21
A4	A5	A6	A7	A8	A9	A10
KHN20	KHN21	KHN222	KHN23	KHN24	KHN25	KHN26
MAG26	MAG27	MAG28	MAG29	MAG30	MIY1	MIY2
22	23	24	25	26	27	28
A11	A12	A13	A14	A15	A16	A17
KHN27	KHN28	KHN29	KHN30	PIN1	PIN2	PIN3
MIY3	MIY4	MIY5	MIY6	MIY7	MIY8	MIY9

MABABA WEUSI 548/2064-65/6300-01/2056-57						
Ashe	Ojobo	Iwa	Araorun	Ile	Itutu	Erinle
1	2	3	4	5	6	7
MA16	MA17	MA18	MA19	MA20	MA21	MA22
APP2	APP3	APP4	APP5	APP6	APP7	APP8
GIN8	GIN9	GIN10	GIN11	GIN12	GIN13	GIN14
8	9	10	11	12	13	14
MA23	MA24	MA25	MA26	MA27	MA28	MA29
APP9	APP10	APP11	APP12	APP13	APP14	APP15
GIN15	GIN16	GIN17	GIN18	GIN19	GIN20	GIN21
15	16	17	18	19	20	21
MA30	MA31	J1	J2	J3	J4	J5
APP16	APP17	APP18	APP19	APP20	APP21	APP22
GIN22	GIN23	GIN24	GIN25	GIN26	GIN27	GIN28
22	23	24	25	26	27	28
J6	J7	J8	J9	J10	J11	J12
APP23	APP24	APP25	APP26	APP27	APP28	APP29
GIN29	GIN30	SAN1	SAN2	SAN3	SAN4	SAN5

THE NU AFRIKAN CALENDAR

MWANAMKE WEUSI 548/2064-65/6300-01/2056-57						
Ashe	Ojobo	Iwa	Araorun	Ile	Itutu	Erinle
1	2	3	4	5	6	7
J13	J14	J15	J16	J17	J18	J19
APP30	MSR1	MSR2	MSR3	MSR4	MSR5	MSR6
SAN6	SAN7	SAN8	SAN9	SAN10	SAN11	SAN12
8	9	10	11	12	13	14
J20	J21	J22	J23	J24	J25	J26
MSR7	MSR8	MSR9	MSR10	MSR11	MSR12	MSR13
SAN13	SAN14	SAN15	SAN16	SAN17	SAN18	SAN19
15	16	17	18	19	20	21
J27	J28	J29	J30	JU1	JU2	JU3
MSR14	MSR15	MSR16	MSR17	MSR18	MSR19	MSR20
SAN20	SAN21	SAN22	SAN23	SAN24	SAN25	SAN26
22	23	24	25	26	27	28
JU4	JU5	JU6	JU7	JU8	JU9	JU10
MSR21	MRS22	MSR23	MSR24	MSR25	MSR26	MSR27
SAN27	SAN28	SAN29	SAN30	HAM1	HAM2	HAM3

TAIFA WEUSI 548/2064-65/6300-01/2056-57						
Ashe	Ojobo	Iwa	Araorun	Ile	Itutu	Erinle
1	2	3	4	5	6	7
AU8	AU9	AU10	AU11	AU12	AU13	AU14
THT20	THT21	THT22	THT23	THT24	THT25	THT26
NAH2	NAH3	NAH4	NAH5	NAH6	NAH7	NAH8
8	9	10	11	12	13	14
AU15	AU16	AU17	AU18	AU19	AU20	AU21
THT27	THT28	THT29	THT30	PAP1	PAP1	PAP2
NAH9	NAH10	NAH11	NAH12	NAH13	NAH14	NAH15
15	16	17	18	19	20	21
AU22	AU23	AU24	AU25	AU26	AU27	AU28
PAP3	PAP4	PAP5	PAP6	PAP7	PAP8	PAP9
NAH16	NAH17	NAH18	NAH19	NAH20	NAH21	NAH22
22	23	24	25	26	27	28
AU29	AU30	AU31	S1	S2	S3	S4
PAP10	PAP11	PAP12	PAP13	PAP14	PAP15	PAP16
NAH23	NAH24	NAH25	NAH26	NAH27	NAH28	NAH29

MTU WEUSI 548/2064-65/6300-01/2056-57						
Ashe	Ojobo	Iwa	Araorun	Ile	Itutu	Erinle
1	2	3	4	5	6	7
JU11	JU12	JU13	JU14	JU15	JU16	JU17
MSR28	MSR29	MSR30	*ASR1*	*HRU2*	*STH3*	*AST4*
HAM4	HAM5	HAM6	HAM7	HAM8	HAM9	HAM10
8	9	10	11	12	13	14
JU18	JU19	JU20	JU21	JU22	JU23	JU24
NBH5	*ANU6*	THT1	THT2	THT3	THT4	THT5
HAM11	HAM12	HAM13	HAM14	HAM15	HAM16	HAM17
15	16	17	18	19	20	21
JU25	JU26	JU27	JU28	JU29	JU30	JU31
THT6	THT7	THT8	THT9	THT10	THT11	THT12
HAM18	HAM19	HAM20	HAM21	HAM22	HAM23	HAM24
22	23	24	25	26	27	28
AU1	AU2	AU3	AU4	AU5	AU6	AU7
THT13	THT14	THT15	THT16	THT17	THT18	THT19
HAM25	HAM26	HAM27	HAM28	HAM29	HAM30	NAH1

UWEZO 548/2064-65/6300-01/2056-57						
Ashe	Ojobo	Iwa	Araorun	Ile	Itutu	Erinle
1	2	3	4	5	6	7
S5	S6	S7	S8	S9	S10	S11
PAP17	PAP18	PAP19	PAP20	PAP21	PAP22	PAP23
NAH30	*PAG1*	*PAG2*	*PAG3*	*PAG4*	*PAG5*	MAS1
8	9	10	11	12	13	14
S12	S13	S14	S15	S16	S17	S18
PAP24	PAP25	PAP26	PAP27	PAP28	PAP29	PAP30
MAS2	MAS3	MAS4	MAS5	MAS6	MAS7	MAS8
15	16	17	18	19	20	21
S19	S20	S21	S22	S23	S24	S25
HTH1	HTH2	HTH3	HTH4	HTH5	HTH6	HTH7
MAS9	MAS10	MAS11	MAS12	MAS13	MAS14	MAS15
22	23	24	25	26	27	28
S26	S27	S28	S29	S30	O1	O2
HTH8	HTH9	HTH10	HTH11	HTH12	HTH13	HTH14
MAS16	MAS17	MAS18	MAS19	MAS20	MAS21	MAS22

THE NU AFRIKAN CALENDAR

EKUNDU 548/2064-65/6300-01/2056-57						
Ashe	Ojobo	Iwa	Araorun	Ile	Itutu	Erinle
1	2	3	4	5	6	7
O3	O4	O5	O6	O7	O8	O9
HTH15	HTH16	HTH17	HTH18	HTH19	HTH20	HTH21
MAS23	MAS24	MAS25	MAS26	MAS27	MAS28	MAS29
8	9	10	11	12	13	14
O10	O11	O12	O13	O14	O15	O16
HTH22	HTH23	HTH24	HTH25	HTH26	HTH27	HTH28
MAS30	THi1	THi2	THi3	THi4	THi5	THi6
15	16	17	18	19	20	21
O17	O18	O19	O20	O21	O22	O23
HTH29	HTH30	KHK1	KHK2	KHK3	KHK4	KHK5
THi7	THi8	This9	THi10	THi11	THi12	THi13
22	23	24	25	26	27	28
O24	O25	O26	O27	O28	O29	O30
KHK6	KHK7	KHK8	KHK9	KHK10	KHK11	KHK12
Thi14	THi15	THi16	THi17	THi18	THi19	THi20

KIJANI 548/2064-65/6300-01/2056-57						
Ashe	Ojobo	Iwa	Araorun	Ile	Itutu	Erinle
1	2	3	4	5	6	7
N28	N29	N30	D1	D2	D3	D4
TBT11	TBT12	TBT13	TBT14	TBT15	TBT16	TBT17
HID19	HID20	HID21	HID22	HID23	HID24	HID25
8	9	10	11	12	13	14
D5	D6	D7	D8	D9	D10	D11
TBT18	TBT19	TBT20	TBT21	TBT22	TBT23	TBT24
HID26	HID27	HID28	HID29	HID30	TAH1	TAH2
15	16	17	18	19	20	21
D12	D13	D14	D15	D16	D17	D18
TBT25	TBT26	TBT27	TBT28	TBT29	TBT30	MKH1
TAH3	TAH4	TAH5	TAH6	TAH7	TAH8	TAH9
22	23	24	25	26	27	28
D19	D20	D21	D22	D23	D24	D25
MKH2	MKH3	MKH4	MKH5	MKH6	MKH7	MKH8
TAH10	TAH11	TAH12	TAH13	TAH14	TAH15	TAH16

EUSI 548/2064-65/6300-01/2056-57						
Ashe	Ojobo	Iwa	Araorun	Ile	Itutu	Erinle
1	2	3	4	5	6	7
O31	N1	N2	N3	N4	N5	N6
KHK13	KHK14	KHK15	KHK16	KHK17	KHK18	KHK19
THi21	THi22	THi23	THi24	THi25	THi26	THi27
8	9	10	11	12	13	14
N7	N8	N9	N10	N11	N12	N13
KHK20	KHK21	KHK22	KHK23	KHK24	KHK25	KHK26
THi28	THi29	THi30	HID1	HID2	HID3	HID4
15	16	17	18	19	20	21
N14	N15	N16	N17	N18	N19	N20
KHK27	KHK28	KHK29	KHK30	TBT1	TBT2	TBT3
HID5	HID6	HID7	HID8	HID9	HID10	HID11
22	23	24	25	26	27	28
N21	N22	N23	N24	N25	N26	N27
TBT4	TBT5	TBT6	TBT7	TBT8	TBT9	TBT10
HID12	HID13	HID14	HID15	HID16	HID17	HID18

KWANZA 548/2064-65/6300-01/2056-57						
Umoja	Kujichagulia	Ujima	Ujamaa	Nia	Kuumba	Imani
1	2	3	4	5	6	7
D26	D27	D28	D29	D30	D31	JA1
MKH9	MKH10	MKH11	MKH12	MKH13	MKH14	MKH15
TAH17	TAH18	TAH19	TAH20	TAH21	TAH22	TAH23
8	9	10	11	12	13	14
JA2	JA3	JA4	JA5	JA6	JA7	JA8
MKH16	MKH17	MKH18	MKH19	MKH20	MKH21	MKH22
TAH24	TAH25	TAH26	TAH27	TAH28	TAH29	TAH30
15	16	17	18	19	20	21
JA9	JA10	JA11	JA12	JA13	JA14	JA15
MKH23	MKH24	MKH25	KMH26	MKH27	MKH28	MKH29
TIR1	TIR2	TIR3	TIR4	TIR5	TIR6	TIR7
22	23	24	25	26	27	28
JA16	JA17	JA18	JA19	JA20	JA21	JA22
MKH30	AMN1	AMN2	AMN3	AMN4	AMN5	AMN6
TIR8	TIR9	TIR10	TIR11	TIR12	TIR13	TIR14

THE NU AFRIKAN CALENDAR

SHAHIDI 548/2064-65/6300-01/2056-57						
Ashe	Ojobo	Iwa	Araorun	Ile	Itutu	Erinle
1	2	3	4	5	6	7
JA23	JA24	JA25	JA26	JA27	JA28	JA29
AMN7	AMN8	AMN9	AMN10	AMN11	AMN12	AMN13
TIR15	TIR16	TIR17	TIR18	TIR19	TIR20	TIR21
8	9	10	11	12	13	14
JA30	JA31	F1	F2	F3	F4	F5
AMN14	AMN15	AMN16	AMN17	AMN18	AMN19	AMN20
TIR22	TIR23	TIR24	TIR25	TIR26	TIR27	TIR28
15	16	17	18	19	20	21
F6	F7	F8	F9	F10	F 11	F12
AMN21	AMN22	AMN23	AMN24	AMN25	AMN26	AMN28
TIR29	TIR30	YEK1	YEK2	YEK3	YEK4	YEK5
22	23	24	25	26	27	28
F13	F14	F15	F16	F17	F18	F19
AMN29	AMN30	RNT1	RNT2	RNT3	RNT4	RNT5
YEK6	YEK7	YEK8	YEK9	YEK10	YEK11	YEK12
SHAHIDI 99						
F20						
RNT6						
YEK13						

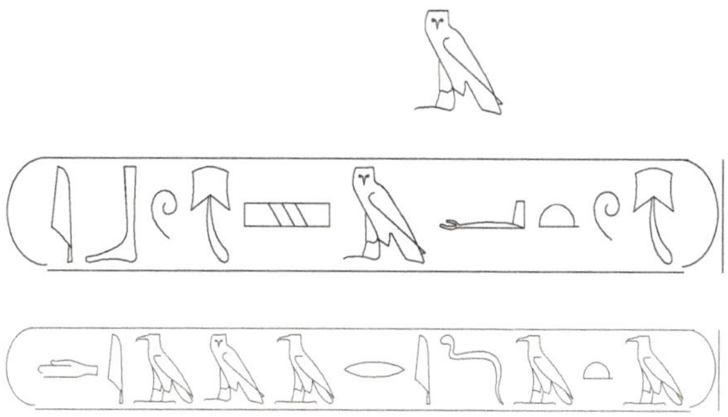

THE NU AFRIKAN CALENDAR

MAAFA 549 B.S.E./ASIENTO 2065-66 C.E.
KAM 6301-02 S.M./EC 2057-58 A.M. (ዓ.ም.)

AFU 549/2065-66/6301-02/2057-58						
Ashe	Ojobo	Iwa	Araorun	Ile	Itutu	Erinle
1	2	3	4	5	6	7
F21	F22	F23	F24	F25	F26	F27
RNT7	RNT8	RNT9	RNT10	RNT11	RNT12	RNT13
YEK14	YEK15	YEK16	YEK17	YEK18	YEK19	YEK20
8	9	10	11	12	13	14
F28	M1	M2	M3	M4	M5	M6
RNT14	RNT15	RNT16	RNT17	RNT18	RNT19	RNT20
YEK21	YEK22	YEK23	YEK24	YEK25	YEK26	YEK27
15	16	17	18	19	20	21
M7	M8	M9	M10	M11	M12	M13
RNT21	RNT22	RNT23	RNT24	RNT25	RNT26	RNT27
YEK28	YEK29	YEK30	MAG1	MAG2	MAG3	MAG4
22	23	24	25	26	27	28
M14	M15	M16	M17	M18	M19	M20
RNT28	RNT29	RNT30	KHN1	KHN2	KHN3	KHN4
MAG5	MAG6	MAG7	MAG8	MAG9	MAG10	MAG11

KAWAIDA 549/2065-66/6301-02/2057-58						
Ashe	Ojobo	Iwa	Araorun	Ile	Itutu	Erinle
1	2	3	4	5	6	7
A18	A19	A20	A21	A22	A23	A24
PIN3	PIN4	PIN5	PIN6	PIN7	PIN8	PIN9
MIY10	MIY11	MIY12	MIY13	MIY14	MIY15	MIY16
8	9	10	11	12	13	14
A25	A26	A27	A28	A29	A30	MA1
PIN10	PIN11	PIN12	PIN13	PIN14	PIN15	PIN16
MIY17	MIY18	MIY19	MIY20	MIY21	MIY22	MIY23
15	16	17	18	19	20	21
MA2	MA3	MA4	MA5	MA6	MA7	MA8
PIN17	PIN18	PIN19	PIN20	PIN21	PIN22	PIN23
MIY24	MIY25	MIY26	MIY27	MIY28	MIY29	MIY30
22	23	24	25	26	27	28
MA9	MA10	MA11	MA12	MA13	MA14	MA15
PIN24	PIN25	PIN26	PIN27	PIN28	PIN29	PIN30
GIN1	GIN2	GIN3	GIN4	GIN5	GIN6	GIN7

UHURU 549/2065-66/6301-02/2057-58						
Ashe	Ojobo	Iwa	Araorun	Ile	Itutu	Erinle
1	2	3	4	5	6	7
M21	M22	M23	M24	M25	M26	M27
KHN5	KHN6	KHN7	KHN8	KHN9	KHN10	KHN11
MAG12	MAG13	MAG14	MAG15	MAG16	MAG17	MAG18
8	9	10	11	12	13	14
M28	M29	M30	M31	A1	A2	A3
KHN12	KHN13	KHN14	KHN15	KHN16	KHN17	KHN18
MAG19	MAG20	MAG21	MAG22	MAG23	MAG24	MAG25
15	16	17	18	19	20	21
A4	A5	A6	A7	A8	A9	A10
KHN19	KHN20	KHN21	KHN22	KHN23	KHN24	KHN25
MAG26	MAG27	MAG28	MAG29	MAG30	MIY1	MIY2
22	23	24	25	26	27	28
A11	A12	A13	A14	A15	A16	A17
KHN26	KHN27	KHN28	KHN29	KHN30	PIN1	PIN2
MIY3	MIY4	MIY5	MIY6	MIY7	MIY8	MIY9

MABABA WEUSI 549/2065-66/6301-02/2057-58						
Ashe	Ojobo	Iwa	Araorun	Ile	Itutu	Erinle
1	2	3	4	5	6	7
MA16	MA17	MA18	MA19	MA20	MA21	MA22
APP1	APP2	APP3	APP4	APP5	APP6	APP7
GIN8	GIN9	GIN10	GIN11	GIN12	GIN13	GIN14
8	9	10	11	12	13	14
MA23	MA24	MA25	MA26	MA27	MA28	MA29
APP8	APP9	APP10	APP11	APP12	APP13	APP14
GIN15	GIN16	GIN17	GIN18	GIN19	GIN20	GIN21
15	16	17	18	19	20	21
MA30	MA31	J1	J2	J3	J4	J5
APP15	APP16	APP17	APP18	APP19	APP20	APP21
GIN22	GIN23	GIN24	GIN25	GIN26	GIN27	GIN28
22	23	24	25	26	27	28
J6	J7	J8	J9	J10	J11	J12
APP22	APP23	APP24	APP25	APP26	APP27	APP28
GIN29	GIN30	SAN1	SAN2	SAN3	SAN4	SAN5

THE NU AFRIKAN CALENDAR

MWANAMKE WEUSI 549/2065-66/6301-02/2057-58

Ashe	Ojobo	Iwa	Araorun	Ile	Itutu	Erinle
1	2	3	4	5	6	7
J13	J14	J15	J16	J17	J18	J19
APP29	APP30	MSR1	MSR2	MSR3	MSR4	MSR5
SAN6	SAN7	SAN8	SAN9	SAN10	SAN11	SAN12
8	9	10	11	12	13	14
J20	J21	J22	J23	J24	J25	J26
MSR6	MSR7	MSR8	MSR9	MSR10	MSR11	MSR12
SAN13	SAN14	SAN15	SAN16	SAN17	SAN18	SAN19
15	16	17	18	19	20	21
J27	J28	J29	J30	JU1	JU2	JU3
MSR13	MSR14	MSR15	MSR16	MSR17	MSR18	MSR19
SAN20	SAN21	SAN22	SAN23	SAN24	SAN25	SAN26
22	23	24	25	26	27	28
JU4	JU5	JU6	JU7	JU8	JU9	JU10
MSR20	MRS21	MSR22	MSR23	MSR24	MSR25	MSR26
SAN27	SAN28	SAN29	SAN30	HAM1	HAM2	HAM3

TAIFA WEUSI 549/2065-66/6301-02/2057-58

Ashe	Ojobo	Iwa	Araorun	Ile	Itutu	Erinle
1	2	3	4	5	6	7
AU8	AU9	AU10	AU11	AU12	AU13	AU14
THT20	THT21	THT22	THT23	THT24	THT25	THT26
NAH2	NAH3	NAH4	NAH5	NAH6	NAH7	NAH8
8	9	10	11	12	13	14
AU15	AU16	AU17	AU18	AU19	AU20	AU21
THT27	THT28	THT29	THT30	PAP1	PAP1	PAP2
NAH9	NAH10	NAH11	NAH12	NAH13	NAH14	NAH15
15	16	17	18	19	20	21
AU22	AU23	AU24	AU25	AU26	AU27	AU28
PAP3	PAP4	PAP5	PAP6	PAP7	PAP8	PAP9
NAH16	NAH17	NAH18	NAH19	NAH20	NAH21	NAH22
22	23	24	25	26	27	28
AU29	AU30	AU31	S1	S2	S3	S4
PAP10	PAP11	PAP12	PAP13	PAP14	PAP15	PAP16
NAH23	NAH24	NAH25	NAH26	NAH27	NAH28	NAH29

MTU WEUSI 549/2065-66/6301-02/2057-58

Ashe	Ojobo	Iwa	Araorun	Ile	Itutu	Erinle
1	2	3	4	5	6	7
JU11	JU12	JU13	JU14	JU15	JU16	JU17
MSR27	MSR28	MSR29	MSR30	*ASR1*	*HRU2*	*STH3*
HAM4	HAM5	HAM6	HAM7	HAM8	HAM9	HAM10
8	9	10	11	12	13	14
JU18	JU19	JU20	JU21	JU22	JU23	JU24
AST4	*NBH5*	THT1	THT2	THT3	THT4	THT5
HAM11	HAM12	HAM13	AM14	HAM15	HAM16	HAM17
15	16	17	18	19	20	21
JU25	JU26	JU27	JU28	JU29	JU30	JU31
THT6	THT7	THT8	THT9	THT10	THT11	THT12
HAM18	HAM19	HAM20	HAM21	HAM22	HAM23	HAM24
22	23	24	25	26	27	28
AU1	AU2	AU3	AU4	AU5	AU6	AU7
THT13	THT14	THT15	THT16	THT17	THT18	THT19
HAM25	HAM26	HAM27	HAM28	HAM29	HAM30	NAH1

UWEZO 549/2065-66/6301-02/2057-58

Ashe	Ojobo	Iwa	Araorun	Ile	Itutu	Erinle
1	2	3	4	5	6	7
S5	S6	S7	S8	S9	S10	S11
PAP17	PAP18	PAP19	PAP20	PAP21	PAP22	PAP23
NAH30	*PAG1*	*PAG2*	*PAG3*	*PAG4*	*PAG5*	MAS1
8	9	10	11	12	13	14
S12	S13	S14	S15	S16	S17	S18
PAP24	PAP25	PAP26	PAP27	PAP28	PAP29	PAP30
MAS2	MAS3	MAS4	MAS5	MAS6	MAS7	MAS8
15	16	17	18	19	20	21
S19	S20	S21	S22	S23	S24	S25
HTH1	HTH2	HTH3	HTH4	HTH5	HTH6	HTH7
MAS9	MAS10	MAS11	MAS12	MAS13	MAS14	MAS15
22	23	24	25	26	27	28
S26	S27	S28	S29	S30	O1	O2
HTH8	HTH9	HTH10	HTH11	HTH12	HTH13	HTH14
MAS16	MAS17	MAS18	MAS19	MAS20	MAS21	MAS22

THE NU AFRIKAN CALENDAR

EKUNDU 549/2065-66/6301-02/2057-58						
Ashe	Ojobo	Iwa	Araorun	Ile	Itutu	Erinle
1	2	3	4	5	6	7
O3	O4	O5	O6	O7	O8	O9
HTH15	HTH16	HTH17	HTH18	HTH19	HTH20	HTH21
MAS23	MAS24	MAS25	MAS26	MAS27	MAS28	MAS29
8	9	10	11	12	13	14
O10	O11	O12	O13	O14	O15	O16
HTH22	HTH23	HTH24	HTH25	HTH26	HTH27	HTH28
MAS30	THi1	THi2	THi3	THi4	THi5	THi6
15	16	17	18	19	20	21
O17	O18	O19	O20	O21	O22	O23
HTH29	HTH30	KHK1	KHK2	KHK3	KHK4	KHK5
THi7	THi8	THi9	THi10	THi11	THi12	THi13
22	23	24	25	26	27	28
O24	O25	O26	O27	O28	O29	O30
KHK6	KHK7	KHK8	KHK9	KHK10	KHK11	KHK12
THi14	THi15	THi16	THi17	THi18	THi19	THi20

KIJANI 549/2065-66/6301-02/2057-58						
Ashe	Ojobo	Iwa	Araorun	Ile	Itutu	Erinle
1	2	3	4	5	6	7
N28	N29	N30	D1	D2	D3	D4
TBT11	TBT12	TBT13	TBT14	TBT15	TBT16	TBT17
HID19	HID20	HID21	HID22	HID23	HID24	HID25
8	9	10	11	12	13	14
D5	D6	D7	D8	D9	D10	D11
TBT18	TBT19	TBT20	TBT21	TBT22	TBT23	TBT24
HID26	HID27	HID28	HID29	HID30	TAH1	TAH2
15	16	17	18	19	20	21
D12	D13	D14	D15	D16	D17	D18
TBT25	TBT26	TBT27	TBT28	TBT29	TBT30	MKH1
TAH3	TAH4	TAH5	TAH6	TAH7	TAH8	TAH9
22	23	24	25	26	27	28
D19	D20	D21	D22	D23	D24	D25
MKH2	MKH3	MKH4	MKH5	MKH6	MKH7	MKH8
TAH10	TAH11	TAH12	TAH13	TAH14	TAH15	TAH16

EUSI 549/2065-66/6301-02/2057-58						
Ashe	Ojobo	Iwa	Araorun	Ile	Itutu	Erinle
1	2	3	4	5	6	7
O31	N1	N2	N3	N4	N5	N6
KHK13	KHK14	KHK15	KHK16	KHK17	KHK18	KHK19
THi21	THi22	THi23	THi24	THi25	THi26	THi27
8	9	10	11	12	13	14
N7	N8	N9	N10	N11	N12	N13
KHK20	KHK21	KHK22	KHK23	KHK24	KHK25	KHK26
THi28	THi29	THi30	HID1	HID2	HID3	HID4
15	16	17	18	19	20	21
N14	N15	N16	N17	N18	N19	N20
KHK27	KHK28	KHK29	KHK30	TBT1	TBT2	TBT3
HID5	HID6	HID7	HID8	HID9	HID10	HID11
22	23	24	25	26	27	28
N21	N22	N23	N24	N25	N26	N27
TBT4	TBT5	TBT6	TBT7	TBT8	TBT9	TBT10
HID12	HID13	HID14	HID15	HID16	HID17	HID18

KWANZA 549/2065-66/6301-02/2057-58						
Umoja	Kujichagulia	Ujima	Ujamaa	Nia	Kuumba	Imani
1	2	3	4	5	6	7
D26	D27	D28	D29	D30	D31	JA1
MKH9	MKH10	MKH11	MKH12	MKH13	MKH14	MKH15
TAH17	TAH18	TAH19	TAH20	TAH21	TAH22	TAH23
8	9	10	11	12	13	14
JA2	JA3	JA4	JA5	JA6	JA7	JA8
MKH16	MKH17	MKH18	MKH19	MKH20	MKH21	MKH22
TAH24	TAH25	TAH26	TAH27	TAH28	TAH29	TAH30
15	16	17	18	19	20	21
JA9	JA10	JA11	JA12	JA13	JA14	JA15
MKH23	MKH24	MKH25	KMH26	MKH27	MKH28	MKH29
TIR1	TIR2	TIR3	TIR4	TIR5	TIR6	TIR7
22	23	24	25	26	27	28
JA16	JA17	JA18	JA19	JA20	JA21	JA22
MKH30	AMN1	AMN2	AMN3	AMN4	AMN5	AMN6
TIR8	TIR9	TIR10	TIR11	TIR12	TIR13	TIR14

THE NU AFRIKAN CALENDAR

SHAHIDI 549/2065-66/6301-02/2057-58						
Ashe	Ojobo	Iwa	Araorun	Ile	Itutu	Erinle
1	2	3	4	5	6	7
JA23	JA24	JA25	JA26	JA27	JA28	JA29
AMN7	AMN8	AMN9	AMN10	AMN11	AMN12	AMN13
TIR15	TIR16	TIR17	TIR18	TIR19	TIR20	TIR21
8	9	10	11	12	13	14
JA30	JA31	F1	F2	F3	F4	F5
AMN14	AMN15	AMN16	AMN17	AMN18	AMN19	AMN20
TIR22	TIR23	TIR24	TIR25	TIR26	TIR27	TIR28
15	16	17	18	19	20	21
F6	F7	F8	F9	F10	F 11	F12
AMN21	AMN22	AMN23	AMN24	AMN25	AMN26	AMN28
TIR29	TIR30	YEK1	YEK2	YEK3	YEK4	YEK5
22	23	24	25	26	27	28
F13	F14	F15	F16	F17	F18	F19
AMN29	AMN30	RNT1	RNT2	RNT3	RNT4	RNT5
YEK6	YEK7	YEK8	YEK9	YEK10	YEK11	YEK12
SHAHIDI 100						
F20						
RNT6						
YEK13						

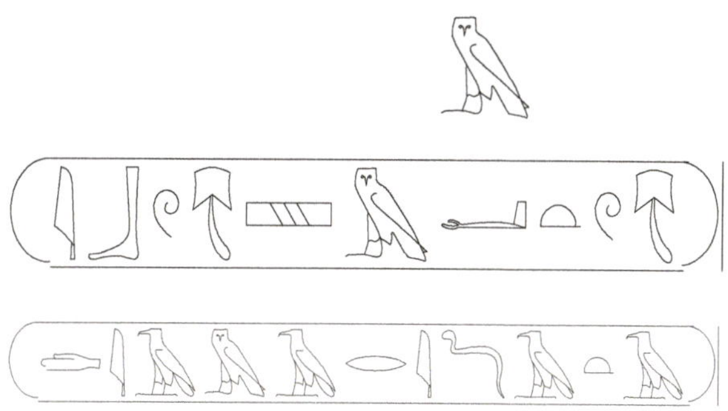

THE NU AFRIKAN CALENDAR

MAAFA 550 B.S.E./ASIENTO 2066-67 C.E.
KAM 6302-03 S.M./EC 2058-59 A.M. (ዓ.ም.)

AFU 550/2066-67/6302-03/2058-59						
Ashe	Ojobo	Iwa	Araorun	Ile	Itutu	Erinle
1	2	3	4	5	6	7
F21	F22	F23	F24	F25	F26	F27
RNT7	RNT8	RNT9	RNT10	RNT11	RNT12	RNT13
YEK14	YEK15	YEK16	YEK17	YEK18	YEK19	YEK20
8	9	10	11	12	13	14
F28	M1	M2	M3	M4	M5	M6
RNT14	RNT15	RNT16	RNT17	RNT18	RNT19	RNT20
YEK21	YEK22	YEK23	YEK24	YEK25	YEK26	YEK27
15	16	17	18	19	20	21
M7	M8	M9	M10	M11	M12	M13
RNT21	RNT22	RNT23	RNT24	RNT25	RNT26	RNT27
YEK28	YEK29	YEK30	MAG1	MAG2	MAG3	MAG4
22	23	24	25	26	27	28
M14	M15	M16	M17	M18	M19	M20
RNT28	RNT29	RNT30	KHN1	KHN2	KHN3	KHN4
MAG5	MAG6	MAG7	MAG8	MAG9	MAG10	MAG11

KAWAIDA 550/2066-67/6302-03/2058-59						
Ashe	Ojobo	Iwa	Araorun	Ile	Itutu	Erinle
1	2	3	4	5	6	7
A18	A19	A20	A21	A22	A23	A24
PIN3	PIN4	PIN5	PIN6	PIN7	PIN8	PIN9
MIY10	MIY11	MIY12	MIY13	MIY14	MIY15	MIY16
8	9	10	11	12	13	14
A25	A26	A27	A28	A29	A30	MA1
PIN10	PIN11	PIN12	PIN13	PIN14	PIN15	PIN16
MIY17	MIY18	MIY19	MIY20	MIY21	MIY22	MIY23
15	16	17	18	19	20	21
MA2	MA3	MA4	MA5	MA6	MA7	MA8
PIN17	PIN18	PIN19	PIN20	PIN21	PIN22	PIN23
MIY24	MIY25	MIY26	MIY27	MIY28	MIY29	MIY30
22	23	24	25	26	27	28
MA9	MA10	MA11	MA12	MA13	MA14	MA15
PIN24	PIN25	PIN26	PIN27	PIN28	PIN29	PIN30
GIN1	GIN2	GIN3	GIN4	GIN5	GIN6	GIN7

UHURU 550/2066-67/6302-03/2058-59						
Ashe	Ojobo	Iwa	Araorun	Ile	Itutu	Erinle
1	2	3	4	5	6	7
M21	M22	M23	M24	M25	M26	M27
KHN5	KHN6	KHN7	KHN8	KHN9	KHN10	KHN11
MAG12	MAG13	MAG14	MAG15	MAG16	MAG17	MAG18
8	9	10	11	12	13	14
M28	M29	M30	M31	A1	A2	A3
KHN12	KHN13	KHN14	KHN15	KHN16	KHN17	KHN18
MAG19	MAG20	MAG21	MAG22	MAG23	MAG24	MAG25
15	16	17	18	19	20	21
A4	A5	A6	A7	A8	A9	A10
KHN19	KHN20	KHN21	KHN22	KHN23	KHN24	KHN25
MAG26	MAG27	MAG28	MAG29	MAG30	MIY1	MIY2
22	23	24	25	26	27	28
A11	A12	A13	A14	A15	A16	A17
KHN26	KHN27	KHN28	KHN29	KHN30	PIN1	PIN2
MIY3	MIY4	MIY5	MIY6	MIY7	MIY8	MIY9

MABABA WEUSI 550/2066-67/6302-03/2058-59						
Ashe	Ojobo	Iwa	Araorun	Ile	Itutu	Erinle
1	2	3	4	5	6	7
MA16	MA17	MA18	MA19	MA20	MA21	MA22
APP1	APP2	APP3	APP4	APP5	APP6	APP7
GIN8	GIN9	GIN10	GIN11	GIN12	GIN13	GIN14
8	9	10	11	12	13	14
MA23	MA24	MA25	MA26	MA27	MA28	MA29
APP8	APP9	APP10	APP11	APP12	APP13	APP14
GIN15	GIN16	GIN17	GIN18	GIN19	GIN20	GIN21
15	16	17	18	19	20	21
MA30	MA31	J1	J2	J3	J4	J5
APP15	APP16	APP17	APP18	APP19	APP20	APP21
GIN22	GIN23	GIN24	GIN25	GIN26	GIN27	GIN28
22	23	24	25	26	27	28
J6	J7	J8	J9	J10	J11	J12
APP22	APP23	APP24	APP25	APP26	APP27	APP28
GIN29	GIN30	SAN1	SAN2	SAN3	SAN4	SAN5

THE NU AFRIKAN CALENDAR

MWANAMKE WEUSI 550/2066-67/6302-03/2058-59

Ashe	Ojobo	Iwa	Araorun	Ile	Itutu	Erinle
1	2	3	4	5	6	7
J13	J14	J15	J16	J17	J18	J19
APP29	APP30	MSR1	MSR2	MSR3	MSR4	MSR5
SAN6	SAN7	SAN8	SAN9	SAN10	SAN11	SAN12
8	9	10	11	12	13	14
J20	J21	J22	J23	J24	J25	J26
MSR6	MSR7	MSR8	MSR9	MSR10	MSR11	MSR12
SAN13	SAN14	SAN15	SAN16	SAN17	SAN18	SAN19
15	16	17	18	19	20	21
J27	J28	J29	J30	JU1	JU2	JU3
MSR13	MSR14	MSR15	MSR16	MSR17	MSR18	MSR19
SAN20	SAN21	SAN22	SAN23	SAN24	SAN25	SAN26
22	23	24	25	26	27	28
JU4	JU5	JU6	JU7	JU8	JU9	JU10
MSR20	MRS21	MSR22	MSR23	MSR24	MSR25	MSR26
SAN27	SAN28	SAN29	SAN30	HAM1	HAM2	HAM3

TAIFA WEUSI 550/2066-67/6302-03/2058-59

Ashe	Ojobo	Iwa	Araorun	Ile	Itutu	Erinle
1	2	3	4	5	6	7
AU8	AU9	AU10	AU11	AU12	AU13	AU14
THT20	THT21	THT22	THT23	THT24	THT25	THT26
NAH2	NAH3	NAH4	NAH5	NAH6	NAH7	NAH8
8	9	10	11	12	13	14
AU15	AU16	AU17	AU18	AU19	AU20	AU21
THT27	THT28	THT29	THT30	PAP1	PAP1	PAP2
NAH9	NAH10	NAH11	NAH12	NAH13	NAH14	NAH15
15	16	17	18	19	20	21
AU22	AU23	AU24	AU25	AU26	AU27	AU28
PAP3	PAP4	PAP5	PAP6	PAP7	PAP8	PAP9
NAH16	NAH17	NAH18	NAH19	NAH20	NAH21	NAH22
22	23	24	25	26	27	28
AU29	AU30	AU31	S1	S2	S3	S4
PAP10	PAP11	PAP12	PAP13	PAP14	PAP15	PAP16
NAH23	NAH24	NAH25	NAH26	NAH27	NAH28	NAH29

MTU WEUSI 550/2066-67/6302-03/2058-59

Ashe	Ojobo	Iwa	Araorun	Ile	Itutu	Erinle
1	2	3	4	5	6	7
JU11	JU12	JU13	JU14	JU15	JU16	JU17
MSR27	MSR28	MSR29	MSR30	*ASR1*	*HRU2*	*STH3*
HAM4	HAM5	HAM6	HAM7	HAM8	HAM9	HAM10
8	9	10	11	12	13	14
JU18	JU19	JU20	JU21	JU22	JU23	JU24
AST4	*NBH5*	THT1	THT2	THT3	THT4	THT5
HAM11	HAM12	HAM13	AM14	HAM15	HAM16	HAM17
15	16	17	18	19	20	21
JU25	JU26	JU27	JU28	JU29	JU30	JU31
THT6	THT7	THT8	THT9	THT10	THT11	THT12
HAM18	HAM19	HAM20	HAM21	HAM22	HAM23	HAM24
22	23	24	25	26	27	28
AU1	AU2	AU3	AU4	AU5	AU6	AU7
THT13	THT14	THT15	THT16	THT17	THT18	THT19
HAM25	HAM26	HAM27	HAM28	HAM29	HAM30	NAH1

UWEZO 550/2066-67/6302-03/2058-59

Ashe	Ojobo	Iwa	Araorun	Ile	Itutu	Erinle
1	2	3	4	5	6	7
S5	S6	S7	S8	S9	S10	S11
PAP17	PAP18	PAP19	PAP20	PAP21	PAP22	PAP23
NAH30	*PAG1*	*PAG2*	*PAG3*	*PAG4*	*PAG5*	MAS1
8	9	10	11	12	13	14
S12	S13	S14	S15	S16	S17	S18
PAP24	PAP25	PAP26	PAP27	PAP28	PAP29	PAP30
MAS2	MAS3	MAS4	MAS5	MAS6	MAS7	MAS8
15	16	17	18	19	20	21
S19	S20	S21	S22	S23	S24	S25
HTH1	HTH2	HTH3	HTH4	HTH5	HTH6	HTH7
MAS9	MAS10	MAS11	MAS12	MAS13	MAS14	MAS15
22	23	24	25	26	27	28
S26	S27	S28	S29	S30	O1	O2
HTH8	HTH9	HTH10	HTH11	HTH12	HTH13	HTH14
MAS16	MAS17	MAS18	MAS19	MAS20	MAS21	MAS22

EKUNDU 550/2066-67/6302-03/2058-59						
Ashe	Ojobo	Iwa	Araorun	Ile	Itutu	Erinle
1	2	3	4	5	6	7
O3	O4	O5	O6	O7	O8	O9
HTH15	HTH16	HTH17	HTH18	HTH19	HTH20	HTH21
MAS23	MAS24	MAS25	MAS26	MAS27	MAS28	MAS29
8	9	10	11	12	13	14
O10	O11	O12	O13	O14	O15	O16
HTH22	HTH23	HTH24	HTH25	HTH26	HTH27	HTH28
MAS30	THi1	THi2	THi3	THi4	THi5	THi6
15	16	17	18	19	20	21
O17	O18	O19	O20	O21	O22	O23
HTH29	HTH30	KHK1	KHK2	KHK3	KHK4	KHK5
THi7	THi8	THi9	THi10	THi11	THi12	THi13
22	23	24	25	26	27	28
O24	O25	O26	O27	O28	O29	O30
KHK6	KHK7	KHK8	KHK9	KHK10	KHK11	KHK12
THi14	THi15	THi16	THi17	THi18	THi19	THi20

KIJANI 550/2066-67/6302-03/2058-59						
Ashe	Ojobo	Iwa	Araorun	Ile	Itutu	Erinle
1	2	3	4	5	6	7
N28	N29	N30	D1	D2	D3	D4
TBT11	TBT12	TBT13	TBT14	TBT15	TBT16	TBT17
HID19	HID20	HID21	HID22	HID23	HID24	HID25
8	9	10	11	12	13	14
D5	D6	D7	D8	D9	D10	D11
TBT18	TBT19	TBT20	TBT21	TBT22	TBT23	TBT24
HID26	HID27	HID28	HID29	HID30	TAH1	TAH2
15	16	17	18	19	20	21
D12	D13	D14	D15	D16	D17	D18
TBT25	TBT26	TBT27	TBT28	TBT29	TBT30	MKH1
TAH3	TAH4	TAH5	TAH6	TAH7	TAH8	TAH9
22	23	24	25	26	27	28
D19	D20	D21	D22	D23	D24	D25
MKH2	MKH3	MKH4	MKH5	MKH6	MKH7	MKH8
TAH10	TAH11	TAH12	TAH13	TAH14	TAH15	TAH16

EUSI 550/2066-67/6302-03/2058-59						
Ashe	Ojobo	Iwa	Araorun	Ile	Itutu	Erinle
1	2	3	4	5	6	7
O31	N1	N2	N3	N4	N5	N6
KHK13	KHK14	KHK15	KHK16	KHK17	KHK18	KHK19
THi21	THi22	THi23	THi24	THi25	THi26	THi27
8	9	10	11	12	13	14
N7	N8	N9	N10	N11	N12	N13
KHK20	KHK21	KHK22	KHK23	KHK24	KHK25	KHK26
THi28	THi29	THi30	HID1	HID2	HID3	HID4
15	16	17	18	19	20	21
N14	N15	N16	N17	N18	N19	N20
KHK27	KHK28	KHK29	KHK30	TBT1	TBT2	TBT3
HID5	HID6	HID7	HID8	HID9	HID10	HID11
22	23	24	25	26	27	28
N21	N22	N23	N24	N25	N26	N27
TBT4	TBT5	TBT6	TBT7	TBT8	TBT9	TBT10
HID12	HID13	HID14	HID15	HID16	HID17	HID18

KWANZA 550/2066-67/6302-03/2058-59						
Umoja	Kujichagulia	Ujima	Ujamaa	Nia	Kuumba	Imani
1	2	3	4	5	6	7
D26	D27	D28	D29	D30	D31	JA1
MKH9	MKH10	MKH11	MKH12	MKH13	MKH14	MKH15
TAH17	TAH18	TAH19	TAH20	TAH21	TAH22	TAH23
8	9	10	11	12	13	14
JA2	JA3	JA4	JA5	JA6	JA7	JA8
MKH16	MKH17	MKH18	MKH19	MKH20	MKH21	MKH22
TAH24	TAH25	TAH26	TAH27	TAH28	TAH29	TAH30
15	16	17	18	19	20	21
JA9	JA10	JA11	JA12	JA13	JA14	JA15
MKH23	MKH24	MKH25	KMH26	MKH27	MKH28	MKH29
TIR1	TIR2	TIR3	TIR4	TIR5	TIR6	TIR7
22	23	24	25	26	27	28
JA16	JA17	JA18	JA19	JA20	JA21	JA22
MKH30	AMN1	AMN2	AMN3	AMN4	AMN5	AMN6
TIR8	TIR9	TIR10	TIR11	TIR12	TIR13	TIR14

SHAHIDI 550/2066-67/6302-03/2058-59						
Ashe	Ojobo	Iwa	Araorun	Ile	Itutu	Erinle
1	2	3	4	5	6	7
JA23	JA24	JA25	JA26	JA27	JA28	JA29
AMN7	AMN8	AMN9	AMN10	AMN11	AMN12	AMN13
TIR15	TIR16	TIR17	TIR18	TIR19	TIR20	TIR21
8	9	10	11	12	13	14
JA30	JA31	F1	F2	F3	F4	F5
AMN14	AMN15	AMN16	AMN17	AMN18	AMN19	AMN20
TIR22	TIR23	TIR24	TIR25	TIR26	TIR27	TIR28
15	16	17	18	19	20	21
F6	F7	F8	F9	F10	F 11	F12
AMN21	AMN22	AMN23	AMN24	AMN25	AMN26	AMN28
TIR29	TIR30	YEK1	YEK2	YEK3	YEK4	YEK5
22	23	24	25	26	27	28
F13	F14	F15	F16	F17	F18	F19
AMN29	AMN30	RNT1	RNT2	RNT3	RNT4	RNT5
YEK6	YEK7	YEK8	YEK9	YEK10	YEK11	YEK12
SHAHIDI 101						
F20						
RNT6						
YEK13						

MAAFA 551 B.S.E./ASIENTO 2067-68 C.E./
KAM 6303-04 S.M./EC 2059-60 A.M. (ዓ.ም.)

AFU 551/2067-68/6303-04/2059-60						
Ashe	Ojobo	Iwa	Araorun	Ile	Itutu	Erinle
1	2	3	4	5	6	7
F21	F22	F23	F24	F25	F26	F27
RNT7	RNT8	RNT9	RNT10	RNT11	RNT12	RNT13
YEK14	YEK15	YEK16	YEK17	YEK18	YEK19	YEK20
8	9	10	11	12	13	14
F28	M1	M2	M3	M4	M5	M6
RNT14	RNT15	RNT16	RNT17	RNT18	RNT19	RNT20
YEK21	YEK22	YEK23	YEK24	YEK25	YEK26	YEK27
15	16	17	18	19	20	21
M7	M8	M9	M10	M11	M12	M13
RNT21	RNT22	RNT23	RNT24	RNT25	RNT26	RNT27
YEK28	YEK 29	YEK30	MAG1	MAG2	MAG3	MAG4
22	23	24	25	26	27	28
M14	M15	M16	M17	M18	M19	M20
RNT28	RNT29	RNT30	KHN1	KHN2	KHN3	KHN4
MAG5	MAG6	MAG7	MAG8	MAG9	MAG10	MAG11

KAWAIDA 551/2067-68/6303-04/2059-60						
Ashe	Ojobo	Iwa	Araorun	Ile	Itutu	Erinle
1	2	3	4	5	6	7
A18	A19	A20	A21	A22	A23	A24
PIN3	PIN4	PIN5	PIN6	PIN7	PIN8	PIN9
MIY10	MIY11	MIY12	MIY13	MIY14	MIY15	MIY16
8	9	10	11	12	13	14
A25	A26	A27	A28	A29	A30	MA1
PIN10	PIN11	PIN12	PIN13	PIN14	PIN15	PIN16
MIY17	MIY18	MIY19	MIY20	MIY21	MIY22	MIY23
15	16	17	18	19	20	21
MA2	MA3	MA4	MA5	MA6	MA7	MA8
PIN17	PIN18	PIN19	PIN20	PIN21	PIN22	PIN23
MIY24	MIY25	MIY26	MIY27	MIY28	MIY29	MIY30
22	23	24	25	26	27	28
MA9	MA10	MA11	MA12	MA13	MA14	MA15
PIN24	PIN25	PIN26	PIN27	PIN28	PIN29	PIN30
GIN1	GIN2	GIN3	GIN4	GIN5	GIN6	GIN7

UHURU 551/2067-68/6303-04/2059-60						
Ashe	Ojobo	Iwa	Araorun	Ile	Itutu	Erinle
1	2	3	4	5	6	7
M21	M22	M23	M24	M25	M26	M27
KHN5	KHN6	KHN7	KHN8	KHN9	KHN10	KHN11
MAG12	MAG13	MAG14	MAG15	MAG16	MAG17	MAG18
8	9	10	11	12	13	14
M28	M29	M30	M31	A1	A2	A3
KHN12	KHN13	KHN14	KHN15	KHN16	KHN17	KHN18
MAG19	MAG20	MAG21	MAG22	MAG23	MAG24	MAG25
15	16	17	18	19	20	21
A4	A5	A6	A7	A8	A9	A10
KHN19	KHN20	KHN21	KHN22	KHN23	KHN24	KHN25
MAG26	MAG27	MAG28	MAG29	MAG30	MIY1	MIY2
22	23	24	25	26	27	28
A11	A12	A13	A14	A15	A16	A17
KHN26	KHN27	KHN28	KHN29	KHN30	PIN1	PIN2
MIY3	MIY4	MIY5	MIY6	MIY7	MIY8	MIY9

MABABA WEUSI 551/2067-68/6303-04/2059-60						
Ashe	Ojobo	Iwa	Araorun	Ile	Itutu	Erinle
1	2	3	4	5	6	7
MA16	MA17	MA18	MA19	MA20	MA21	MA22
APP1	APP2	APP3	APP4	APP5	APP6	APP7
GIN8	GIN9	GIN10	GIN11	GIN12	GIN13	GIN14
8	9	10	11	12	13	14
MA23	MA24	MA25	MA26	MA27	MA28	MA29
APP8	APP9	APP10	APP11	APP12	APP13	APP14
GIN15	GIN16	GIN17	GIN18	GIN19	GIN20	GIN21
15	16	17	18	19	20	21
MA30	MA31	J1	J2	J3	J4	J5
APP15	APP16	APP17	APP18	APP19	APP20	APP21
GIN22	GIN23	GIN24	GIN25	GIN26	GIN27	GIN28
22	23	24	25	26	27	28
J6	J7	J8	J9	J10	J11	J12
APP22	APP23	APP24	APP25	APP26	APP27	APP28
GIN29	GIN30	SAN1	SAN2	SAN3	SAN4	SAN5

THE NU AFRIKAN CALENDAR

MWANAMKE WEUSI 551/2067-68/6303-04/2059-60						
Ashe	Ojobo	Iwa	Araorun	Ile	Itutu	Erinle
1	2	3	4	5	6	7
J13	J14	J15	J16	J17	J18	J19
APP29	APP30	MSR1	MSR2	MSR3	MSR4	MSR5
SAN6	SAN7	SAN8	SAN9	SAN10	SAN11	SAN12
8	9	10	11	12	13	14
J20	J21	J22	J23	J24	J25	J26
MSR6	MSR7	MSR8	MSR9	MSR10	MSR11	MSR12
SAN13	SAN14	SAN15	SAN16	SAN17	SAN18	SAN19
15	16	17	18	19	20	21
J27	J28	J29	J30	JU1	JU2	JU3
MSR13	MSR14	MSR15	MSR16	MSR17	MSR18	MSR19
SAN20	SAN21	SAN22	SAN23	SAN24	SAN25	SAN26
22	23	24	25	26	27	28
JU4	JU5	JU6	JU7	JU8	JU9	JU10
MSR20	MRS21	MSR22	MSR23	MSR24	MSR25	MSR26
SAN27	SAN28	SAN29	SAN30	HAM1	HAM2	HAM3

TAIFA WEUSI 551/2067-68/6303-04/2059-60						
Ashe	Ojobo	Iwa	Araorun	Ile	Itutu	Erinle
1	2	3	4	5	6	7
AU8	AU9	AU10	AU11	AU12	AU13	AU14
THT20	THT21	THT22	THT23	THT24	THT25	THT26
NAH2	NAH3	NAH4	NAH5	NAH6	NAH7	NAH8
8	9	10	11	12	13	14
AU15	AU16	AU17	AU18	AU19	AU20	AU21
THT27	THT28	THT29	THT30	PAP1	PAP1	PAP2
NAH9	NAH10	NAH11	NAH12	NAH13	NAH14	NAH15
15	16	17	18	19	20	21
AU22	AU23	AU24	AU25	AU26	AU27	AU28
PAP3	PAP4	PAP5	PAP6	PAP7	PAP8	PAP9
NAH16	NAH17	NAH18	NAH19	NAH20	NAH21	NAH22
22	23	24	25	26	27	28
AU29	AU30	AU31	S1	S2	S3	S4
PAP10	PAP11	PAP12	PAP13	PAP14	PAP15	PAP16
NAH23	NAH24	NAH25	NAH26	NAH27	NAH28	NAH29

MTU WEUSI 551/2067-68/6303-04/2059-60						
Ashe	Ojobo	Iwa	Araorun	Ile	Itutu	Erinle
1	2	3	4	5	6	7
JU11	JU12	JU13	JU14	JU15	JU16	JU17
MSR27	MSR28	MSR29	MSR30	*ASR1*	*HRU2*	*STH3*
HAM4	HAM5	HAM6	HAM7	HAM8	HAM9	HAM10
8	9	10	11	12	13	14
JU18	JU19	JU20	JU21	JU22	JU23	JU24
AST4	*NBH5*	THT1	THT2	THT3	THT4	THT5
HAM11	HAM12	HAM13	HAM14	HAM15	HAM16	HAM17
15	16	17	18	19	20	21
JU25	JU26	JU27	JU28	JU29	JU30	JU31
THT6	THT7	THT8	THT9	THT10	THT11	THT12
HAM18	HAM19	HAM20	HAM21	HAM22	HAM23	HAM24
22	23	24	25	26	27	28
AU1	AU2	AU3	AU4	AU5	AU6	AU7
THT13	THT14	THT15	THT16	THT17	THT18	THT19
HAM25	HAM26	HAM27	HAM28	HAM29	HAM30	NAH1

UWEZO 551/2067-68/6303-04/2059-60						
Ashe	Ojobo	Iwa	Araorun	Ile	Itutu	Erinle
1	2	3	4	5	6	7
S5	S6	S7	S8	S9	S10	S11
PAP17	PAP18	PAP19	PAP20	PAP21	PAP22	PAP23
NAH30	*QUA1*	*QUA2*	*QUA3*	*QUA4*	*QUA5*	*QUA6*
8	9	10	11	12	13	14
S12	S13	S14	S15	S16	S17	S18
PAP24	PAP25	PAP26	PAP27	PAP28	PAP29	PAP30
MAS1	MAS2	MAS3	MAS4	MAS5	MAS6	MAS7
15	16	17	18	19	20	21
S19	S20	S21	S22	S23	S24	S25
HTH1	HTH2	HTH3	HTH4	HTH5	HTH6	HTH7
MAS8	MAS9	MAS10	MAS11	MAS12	MAS13	MAS14
S26	S27	S28	S29	S30	O1	O2
HTH8	HTH9	HTH10	HTH11	HTH12	HTH13	HTH14
MAS15	MAS16	MAS17	MAS18	MAS19	MAS20	MAS21

THE NU AFRIKAN CALENDAR

EKUNDU 551/2067-68/6303-04/2059-60						
Ashe	Ojobo	Iwa	Araorun	Ile	Itutu	Erinle
1	2	3	4	5	6	7
O3	O4	O5	O6	O7	O8	O9
HTH15	HTH16	HTH17	HTH18	HTH19	HTH20	HTH21
MAS22	MAS23	MAS 24	MAS25	MAS26	MAS27	MAS28
8	9	10	11	12	13	14
O10	O11	O12	O13	O14	O15	O16
HTH22	HTH23	HTH24	HTH25	HTH26	HTH27	HTH28
MAS29	MAS30	THi1	THi2	THi3	THi4	THi5
15	16	17	18	19	20	21
O17	O18	O19	O20	O21	O22	O23
HTH29	HTH30	KHK1	KHK2	KHK3	KHK4	KHK5
THi6	THi7	THi8	THi9	THi10	THi11	THi12
22	23	24	25	26	27	28
O24	O25	O26	O27	O28	O29	O30
KHK6	KHK7	KHK8	KHK9	KHK10	KHK11	KHK12
THi13	THi14	THi15	THi16	THi17	THi18	THi19

KIJANI 551/2067-68/6303-04/2059-60						
Ashe	Ojobo	Iwa	Araorun	Ile	Itutu	Erinle
1	2	3	4	5	6	7
N28	N29	N30	D1	D2	D3	D4
TBT11	TBT12	TBT13	TBT14	TBT15	TBT16	TBT17
HID18	HID19	HID20	HID21	HID22	HID23	HID24
8	9	10	11	12	13	14
D5	D6	D7	D8	D9	D10	D11
TBT18	TBT19	TBT20	TBT21	TBT22	TBT23	TBT24
HID25	HID26	HID27	HID28	HID29	HID30	TAH1
15	16	17	18	19	20	21
D12	D13	D14	D15	D16	D17	D18
TBT25	TBT26	TBT27	TBT28	TBT29	TBT30	MKH1
TAH2	TAH3	TAH4	TAH5	TAH6	TAH7	TAH8
22	23	24	25	26	27	28
D19	D20	D21	D22	D23	D24	D25
MKH2	MKH3	MKH4	MKH5	MKH6	MKH7	MKH8
TAH9	TAH10	TAH11	TAH12	TAH13	TAH14	TAH15

EUSI 551/2067-68/6303-04/2059-60						
Ashe	Ojobo	Iwa	Araorun	Ile	Itutu	Erinle
1	2	3	4	5	6	7
O31	N1	N2	N3	N4	N5	N6
KHK13	KHK14	KHK15	KHK16	KHK17	KHK18	KHK19
THi20	THi21	THi22	THi23	THi24	THi25	THi26
8	9	10	11	12	13	14
N7	N8	N9	N10	N11	N12	N13
KHK20	KHK21	KHK22	KHK23	KHK24	KHK25	KHK26
THi27	THi28	THi29	THi30	HID1	HID2	HID3
15	16	17	18	19	20	21
N14	N15	N16	N17	N18	N19	N20
KHK27	KHK28	KHK29	KHK30	TBT1	TBT2	TBT3
HID4	HID5	HID6	HID7	HID8	HID9	HID10
22	23	24	25	26	27	28
N21	N22	N23	N24	N25	N26	N27
TBT4	TBT5	TBT6	TBT7	TBT8	TBT9	TBT10
HID11	HID12	HID13	HID14	HID15	HID16	HID17

KWANZA 551/2067-68/6303-04/2059-60						
Umoja	Kujichagulia	Ujima	Ujamaa	Nia	Kuumba	Imani
1	2	3	4	5	6	7
D26	D27	D28	D29	D30	D31	JA1
MKH9	MKH10	MKH11	MKH12	MKH13	MKH14	MKH15
TAH16	TAH17	TAH18	TAH19	TAH20	TAH21	TAH22
8	9	10	11	12	13	14
JA2	JA3	JA4	JA5	JA6	JA7	JA8
MKH16	MKH17	MKH18	MKH19	MKH20	MKH21	MKH22
TAH23	TAH24	TAH25	TAH26	TAH27	TAH28	TAH29
15	16	17	18	19	20	21
JA9	JA10	JA11	JA12	JA13	JA14	JA15
MKH23	MKH24	MKH25	KMH26	MKH27	MKH28	MKH29
TAH30	TIR1	TIR2	TIR3	TIR4	TIR5	TIR6
22	23	24	25	26	27	28
JA16	JA17	JA18	JA19	JA20	JA21	JA22
MKH30	AMN1	AMN2	AMN3	AMN4	AMN5	AMN6
TIR7	TIR8	TIR9	TIR10	TIR11	TIR12	TIR13

SHAHIDI 551/2067-68/6303-04/2059-60						
Ashe	Ojobo	Iwa	Araorun	Ile	Itutu	Erinle
1	2	3	4	5	6	7
JA23	JA24	JA25	JA26	JA27	JA28	JA29
AMN7	AMN8	AMN9	AMN10	AMN11	AMN12	AMN13
TIR14	TIR15	TIR16	TIR17	TIR18	TIR19	TIR20
8	9	10	11	12	13	14
JA30	JA31	F1	F2	F3	F4	F5
AMN14	AMN15	AMN16	AMN17	AMN18	AMN19	AMN20
TIR21	TIR22	TIR23	TIR24	TIR25	TIR26	TIR27
15	16	17	18	19	20	21
F6	F7	F8	F9	F10	F 11	F12
AMN21	AMN22	AMN23	AMN24	AMN25	AMN26	AMN28
TIR28	TIR29	TIR30	YEK1	YEK2	YEK3	YEK4
22	23	24	25	26	27	28
F13	F14	F15	F16	F17	F18	F19
AMN29	AMN30	RNT1	RNT2	RNT3	RNT4	RNT5
YEK5	YEK6	YEK7	YEK8	YEK9	YEK10	YEK11
SHAHIDI 102						
F20						
RNT6						
YEK12						

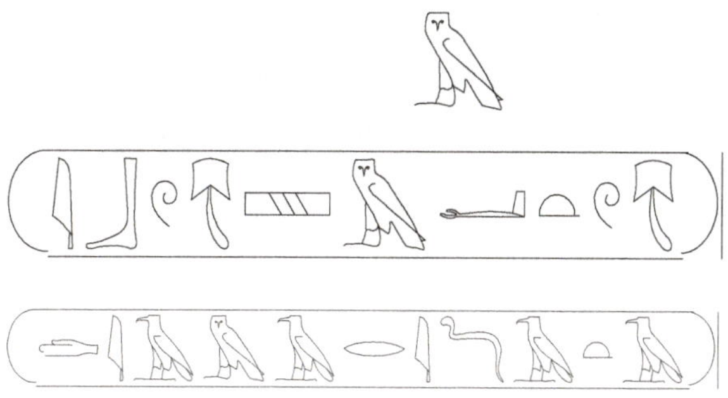

THE NU AFRIKAN CALENDAR

MAAFA 552 B.S.E./ASIENTO 2068-69 C.E./
KAM 6304-05 S.M./EC 2060-61 A.M. (ዓ.ም.)

AFU 552/2068-69/6304-05/2060-61						
Ashe	Ojobo	Iwa	Araorun	Ile	Itutu	Erinle
1	2	3	4	5	6	7
F21	F22	F23	F24	F25	F26	F27
RNT7	RNT8	RNT9	RNT10	RNT11	RNT12	RNT13
YEK13	YEK14	YEK15	YEK16	YEK17	YEK18	YEK19
8	9	10	11	12	13	14
F28	F29	M1	M2	M3	M4	M5
RNT14	RNT15	RNT16	RNT17	RNT18	RNT19	RNT20
YEK 20	YEK21	YEK22	YEK23	YEK24	YEK25	YEK26
15	16	17	18	19	20	21
M6	M7	M8	M9	M10	M11	M12
RNT21	RNT22	RNT23	RNT24	RNT25	RNT26	RNT27
YEK 27	YEK28	YEK29	YEK30	MAG1	MAG2	MAG3
22	23	24	25	26	27	28
M13	M14	M15	M16	M17	M18	M19
RNT28	RNT29	RNT30	KHN1	KHN2	KHN3	KHN4
MAG4	MAG5	MAG6	MAG7	MAG8	MAG9	MAG10
SHARIFU 135						
M20						
KHN5						
MAG11						

KAWAIDA 552/2068-69/6304-05/2060-61						
Ashe	Ojobo	Iwa	Araorun	Ile	Itutu	Erinle
1	2	3	4	5	6	7
A18	A19	A20	A21	A22	A23	A24
PIN4	PIN5	PIN6	PIN7	PIN8	PIN9	PIN10
MIY10	MIY11	MIY12	MIY13	MIY14	MIY15	MIY16
8	9	10	11	12	13	14
A25	A26	A27	A28	A29	A30	MA1
PIN11	PIN12	PIN13	PIN14	PIN15	PIN16	PIN17
MIY17	MIY18	MIY19	MIY20	MIY21	MIY22	MIY23
15	16	17	18	19	20	21
MA2	MA3	MA4	MA5	MA6	MA7	MA8
PIN18	PIN19	PIN20	PIN21	PIN22	PIN23	PIN24
MIY24	MIY25	MIY26	MIY27	MIY28	MIY29	MIY30
22	23	24	25	26	27	28
MA9	MA10	MA11	MA12	MA13	MA14	MA15
PIN25	PIN26	PIN27	PIN28	PIN29	PIN30	APP1
GIN1	GIN2	GIN3	GIN4	GIN5	GIN6	GIN7

UHURU 552/2068-69/6304-05/2060-61						
Ashe	Ojobo	Iwa	Araorun	Ile	Itutu	Erinle
1	2	3	4	5	6	7
M21	M22	M23	M24	M25	M26	M27
KHN6	KHN7	KHN8	KHN9	KHN10	KHN11	KHN12
MAG12	MAG13	MAG14	MAG15	MAG16	MAG17	MAG18
8	9	10	11	12	13	14
M28	M29	M30	M31	A1	A2	A3
KHN13	KHN14	KHN15	KHN16	KHN17	KHN18	KHN19
MAG19	MAG20	MAG21	MAG22	MAG23	MAG24	MAG25
15	16	17	18	19	20	21
A4	A5	A6	A7	A8	A9	A10
KHN20	KHN21	KHN222	KHN23	KHN24	KHN25	KHN26
MAG26	MAG27	MAG28	MAG29	MAG30	MIY1	MIY2
22	23	24	25	26	27	28
A11	A12	A13	A14	A15	A16	A17
KHN27	KHN28	KHN29	KHN30	PIN1	PIN2	PIN3
MIY3	MIY4	MIY5	MIY6	MIY7	MIY8	MIY9

MABABA WEUSI 552/2068-69/6304-05/2060-61						
Ashe	Ojobo	Iwa	Araorun	Ile	Itutu	Erinle
1	2	3	4	5	6	7
MA16	MA17	MA18	MA19	MA20	MA21	MA22
APP2	APP3	APP4	APP5	APP6	APP7	APP8
GIN8	GIN9	GIN10	GIN11	GIN12	GIN13	GIN14
8	9	10	11	12	13	14
MA23	MA24	MA25	MA26	MA27	MA28	MA29
APP9	APP10	APP11	APP12	APP13	APP14	APP15
GIN15	GIN16	GIN17	GIN18	GIN19	GIN20	GIN21
15	16	17	18	19	20	21
MA30	MA31	J1	J2	J3	J4	J5
APP16	APP17	APP18	APP19	APP20	APP21	APP22
GIN22	GIN23	GIN24	GIN25	GIN26	GIN27	GIN28
22	23	24	25	26	27	28
J6	J7	J8	J9	J10	J11	J12
APP23	APP24	APP25	APP26	APP27	APP28	APP29
GIN29	GIN30	SAN1	SAN2	SAN3	SAN4	SAN5

THE NU AFRIKAN CALENDAR

MWANAMKE WEUSI 552/2068-69/6304-05/2060-61

Ashe	Ojobo	Iwa	Araorun	Ile	Itutu	Erinle
1	2	3	4	5	6	7
J13	J14	J15	J16	J17	J18	J19
APP30	MSR1	MSR2	MSR3	MSR4	MSR5	MSR6
SAN6	SAN7	SAN8	SAN9	SAN10	SAN11	SAN12
8	9	10	11	12	13	14
J20	J21	J22	J23	J24	J25	J26
MSR7	MSR8	MSR9	MSR10	MSR11	MSR12	MSR13
SAN13	SAN14	SAN15	SAN16	SAN17	SAN18	SAN19
15	16	17	18	19	20	21
J27	J28	J29	J30	JU1	JU2	JU3
MSR14	MSR15	MSR16	MSR17	MSR18	MSR19	MSR20
SAN20	SAN21	SAN22	SAN23	SAN24	SAN25	SAN26
22	23	24	25	26	27	28
JU4	JU5	JU6	JU7	JU8	JU9	JU10
MSR21	MRS22	MSR23	MSR24	MSR25	MSR26	MSR27
SAN27	SAN28	SAN29	SAN30	HAM1	HAM2	HAM3

TAIFA WEUSI 552/2068-69/6304-05/2060-61

Ashe	Ojobo	Iwa	Araorun	Ile	Itutu	Erinle
1	2	3	4	5	6	7
AU8	AU9	AU10	AU11	AU12	AU13	AU14
THT20	THT21	THT22	THT23	THT24	THT25	THT26
NAH2	NAH3	NAH4	NAH5	NAH6	NAH7	NAH8
8	9	10	11	12	13	14
AU15	AU16	AU17	AU18	AU19	AU20	AU21
THT27	THT28	THT29	THT30	PAP1	PAP1	PAP2
NAH9	NAH10	NAH11	NAH12	NAH13	NAH14	NAH15
15	16	17	18	19	20	21
AU22	AU23	AU24	AU25	AU26	AU27	AU28
PAP3	PAP4	PAP5	PAP6	PAP7	PAP8	PAP9
NAH16	NAH17	NAH18	NAH19	NAH20	NAH21	NAH22
22	23	24	25	26	27	28
AU29	AU30	AU31	S1	S2	S3	S4
PAP10	PAP11	PAP12	PAP13	PAP14	PAP15	PAP16
NAH23	NAH24	NAH25	NAH26	NAH27	NAH28	NAH29

MTU WEUSI 552/2068-69/6304-05/2060-61

Ashe	Ojobo	Iwa	Araorun	Ile	Itutu	Erinle
1	2	3	4	5	6	7
JU11	JU12	JU13	JU14	JU15	JU16	JU17
MSR28	MSR29	MSR30	*ASR1*	*HRU2*	*STH3*	*AST4*
HAM4	HAM5	HAM6	HAM7	HAM8	HAM9	HAM10
8	9	10	11	12	13	14
JU18	JU19	JU20	JU21	JU22	JU23	JU24
NBH5	*ANU6*	THT1	THT2	THT3	THT4	THT5
HAM11	HAM12	HAM13	HAM14	HAM15	HAM16	HAM17
15	16	17	18	19	20	21
JU25	JU26	JU27	JU28	JU29	JU30	JU31
THT6	THT7	THT8	THT9	THT10	THT11	THT12
HAM18	HAM19	HAM20	HAM21	HAM22	HAM23	HAM24
22	23	24	25	26	27	28
AU1	AU2	AU3	AU4	AU5	AU6	AU7
THT13	THT14	THT15	THT16	THT17	THT18	THT19
HAM25	HAM26	HAM27	HAM28	HAM29	HAM30	NAH1

UWEZO 552/2068-69/6304-05/2060-61

Ashe	Ojobo	Iwa	Araorun	Ile	Itutu	Erinle
1	2	3	4	5	6	7
S5	S6	S7	S8	S9	S10	S11
PAP17	PAP18	PAP19	PAP20	PAP21	PAP22	PAP23
NAH30	*PAG1*	*PAG2*	*PAG3*	*PAG4*	*PAG5*	MAS1
8	9	10	11	12	13	14
S12	S13	S14	S15	S16	S17	S18
PAP24	PAP25	PAP26	PAP27	PAP28	PAP29	PAP30
MAS2	MAS3	MAS4	MAS5	MAS6	MAS7	MAS8
15	16	17	18	19	20	21
S19	S20	S21	S22	S23	S24	S25
HTH1	HTH2	HTH3	HTH4	HTH5	HTH6	HTH7
MAS9	MAS10	MAS11	MAS12	MAS13	MAS14	MAS15
22	23	24	25	26	27	28
S26	S27	S28	S29	S30	O1	O2
HTH8	HTH9	HTH10	HTH11	HTH12	HTH13	HTH14
MAS16	MAS17	MAS18	MAS19	MAS20	MAS21	MAS22

THE NU AFRIKAN CALENDAR

EKUNDU 552/2068-69/6304-05/2060-61						
Ashe	Ojobo	Iwa	Araorun	Ile	Itutu	Erinle
1	2	3	4	5	6	7
O3	O4	O5	O6	O7	O8	O9
HTH15	HTH16	HTH17	HTH18	HTH19	HTH20	HTH21
MAS23	MAS24	MAS25	MAS26	MAS27	MAS28	MAS29
8	9	10	11	12	13	14
O10	O11	O12	O13	O14	O15	O16
HTH22	HTH23	HTH24	HTH25	HTH26	HTH27	HTH28
MAS30	THi1	THi2	THi3	THi4	THi5	THi6
15	16	17	18	19	20	21
O17	O18	O19	O20	O21	O22	O23
HTH29	HTH30	KHK1	KHK2	KHK3	KHK4	KHK5
THi7	THi8	This9	THi10	THi11	THi12	THi13
22	23	24	25	26	27	28
O24	O25	O26	O27	O28	O29	O30
KHK6	KHK7	KHK8	KHK9	KHK10	KHK11	KHK12
Thi14	THi15	THi16	THi17	THi18	THi19	THi20

KIJANI 552/2068-69/6304-05/2060-61						
Ashe	Ojobo	Iwa	Araorun	Ile	Itutu	Erinle
1	2	3	4	5	6	7
N28	N29	N30	D1	D2	D3	D4
TBT11	TBT12	TBT13	TBT14	TBT15	TBT16	TBT17
HID19	HID20	HID21	HID22	HID23	HID24	HID25
8	9	10	11	12	13	14
D5	D6	D7	D8	D9	D10	D11
TBT18	TBT19	TBT20	TBT21	TBT22	TBT23	TBT24
HID26	HID27	HID28	HID29	HID30	TAH1	TAH2
15	16	17	18	19	20	21
D12	D13	D14	D15	D16	D17	D18
TBT25	TBT26	TBT27	TBT28	TBT29	TBT30	MKH1
TAH3	TAH4	TAH5	TAH6	TAH7	TAH8	TAH9
22	23	24	25	26	27	28
D19	D20	D21	D22	D23	D24	D25
MKH2	MKH3	MKH4	MKH5	MKH6	MKH7	MKH8
TAH10	TAH11	TAH12	TAH13	TAH14	TAH15	TAH16

EUSI 552/2068-69/6304-05/2060-61						
Ashe	Ojobo	Iwa	Araorun	Ile	Itutu	Erinle
1	2	3	4	5	6	7
O31	N1	N2	N3	N4	N5	N6
KHK13	KHK14	KHK15	KHK16	KHK17	KHK18	KHK19
THi21	THi22	THi23	THi24	THi25	THi26	THi27
8	9	10	11	12	13	14
N7	N8	N9	N10	N11	N12	N13
KHK20	KHK21	KHK22	KHK23	KHK24	KHK25	KHK26
THi28	THi29	THi30	HID1	HID2	HID3	HID4
15	16	17	18	19	20	21
N14	N15	N16	N17	N18	N19	N20
KHK27	KHK28	KHK29	KHK30	TBT1	TBT2	TBT3
HID5	HID6	HID7	HID8	HID9	HID10	HID11
22	23	24	25	26	27	28
N21	N22	N23	N24	N25	N26	N27
TBT4	TBT5	TBT6	TBT7	TBT8	TBT9	TBT10
HID12	HID13	HID14	HID15	HID16	HID17	HID18

KWANZA 552/2068-69/6304-05/2060-61						
Umoja	Kujichagulia	Ujima	Ujamaa	Nia	Kuumba	Imani
1	2	3	4	5	6	7
D26	D27	D28	D29	D30	D31	JA1
MKH9	MKH10	MKH11	MKH12	MKH13	MKH14	MKH15
TAH17	TAH18	TAH19	TAH20	TAH21	TAH22	TAH23
8	9	10	11	12	13	14
JA2	JA3	JA4	JA5	JA6	JA7	JA8
MKH16	MKH17	MKH18	MKH19	MKH20	MKH21	MKH22
TAH24	TAH25	TAH26	TAH27	TAH28	TAH29	TAH30
15	16	17	18	19	20	21
JA9	JA10	JA11	JA12	JA13	JA14	JA15
MKH23	MKH24	MKH25	KMH26	MKH27	MKH28	MKH29
TIR1	TIR2	TIR3	TIR4	TIR5	TIR6	TIR7
22	23	24	25	26	27	28
JA16	JA17	JA18	JA19	JA20	JA21	JA22
MKH30	AMN1	AMN2	AMN3	AMN4	AMN5	AMN6
TIR8	TIR9	TIR10	TIR11	TIR12	TIR13	TIR14

THE NU AFRIKAN CALENDAR

SHAHIDI 552/2068-69/6304-05/2060-61						
Ashe	Ojobo	Iwa	Araorun	Ile	Itutu	Erinle
1	2	3	4	5	6	7
JA23	JA24	JA25	JA26	JA27	JA28	JA29
AMN7	AMN8	AMN9	AMN10	AMN11	AMN12	AMN13
TIR15	TIR16	TIR17	TIR18	TIR19	TIR20	TIR21
8	9	10	11	12	13	14
JA30	JA31	F1	F2	F3	F4	F5
AMN14	AMN15	AMN16	AMN17	AMN18	AMN19	AMN20
TIR22	TIR23	TIR24	TIR25	TIR26	TIR27	TIR28
15	16	17	18	19	20	21
F6	F7	F8	F9	F10	F 11	F12
AMN21	AMN22	AMN23	AMN24	AMN25	AMN26	AMN28
TIR29	TIR30	YEK1	YEK2	YEK3	YEK4	YEK5
22	23	24	25	26	27	28
F13	F14	F15	F16	F17	F18	F19
AMN29	AMN30	RNT1	RNT2	RNT3	RNT4	RNT5
YEK6	YEK7	YEK8	YEK9	YEK10	YEK11	YEK12
SHAHIDI 103						
F20						
RNT6						
YEK13						

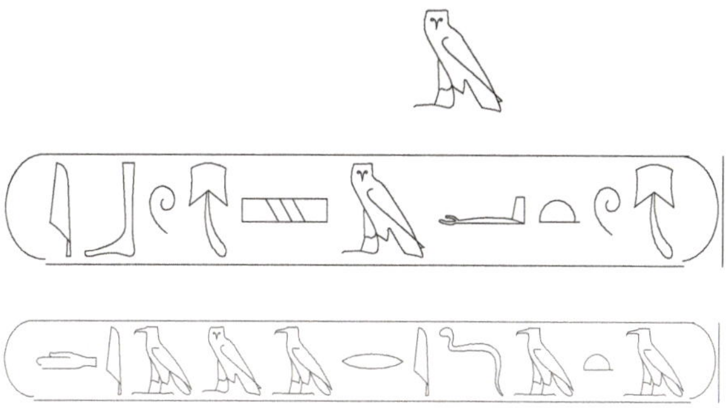

THE NU AFRIKAN CALENDAR

MAAFA 553 B.S.E./ASIENTO 2069-70 C.E.
KAM 6305-06 S.M./EC 2061-62 A.M. (ዓ.ም.)

AFU 553/2069-70/6305-06/2061-62

Ashe	Ojobo	Iwa	Araorun	Ile	Itutu	Erinle
1	2	3	4	5	6	7
F21	F22	F23	F24	F25	F26	F27
RNT7	RNT8	RNT9	RNT10	RNT11	RNT12	RNT13
YEK14	YEK15	YEK16	YEK17	YEK18	YEK19	YEK20
8	9	10	11	12	13	14
F28	M1	M2	M3	M4	M5	M6
RNT14	RNT15	RNT16	RNT17	RNT18	RNT19	RNT20
YEK21	YEK22	YEK23	YEK24	YEK25	YEK26	YEK27
15	16	17	18	19	20	21
M7	M8	M9	M10	M11	M12	M13
RNT21	RNT22	RNT23	RNT24	RNT25	RNT26	RNT27
YEK28	YEK29	YEK30	MAG1	MAG2	MAG3	MAG4
22	23	24	25	26	27	28
M14	M15	M16	M17	M18	M19	M20
RNT28	RNT29	RNT30	KHN1	KHN2	KHN3	KHN4
MAG5	MAG6	MAG7	MAG8	MAG9	MAG10	MAG11

KAWAIDA 553/2069-70/6305-06/2061-62

Ashe	Ojobo	Iwa	Araorun	Ile	Itutu	Erinle
1	2	3	4	5	6	7
A18	A19	A20	A21	A22	A23	A24
PIN3	PIN4	PIN5	PIN6	PIN7	PIN8	PIN9
MIY10	MIY11	MIY12	MIY13	MIY14	MIY15	MIY16
8	9	10	11	12	13	14
A25	A26	A27	A28	A29	A30	MA1
PIN10	PIN11	PIN12	PIN13	PIN14	PIN15	PIN16
MIY17	MIY18	MIY19	MIY20	MIY21	MIY22	MIY23
15	16	17	18	19	20	21
MA2	MA3	MA4	MA5	MA6	MA7	MA8
PIN17	PIN18	PIN19	PIN20	PIN21	PIN22	PIN23
MIY24	MIY25	MIY26	MIY27	MIY28	MIY29	MIY30
22	23	24	25	26	27	28
MA9	MA10	MA11	MA12	MA13	MA14	MA15
PIN24	PIN25	PIN26	PIN27	PIN28	PIN29	PIN30
GIN1	GIN2	GIN3	GIN4	GIN5	GIN6	GIN7

UHURU 553/2069-70/6305-06/2061-62

Ashe	Ojobo	Iwa	Araorun	Ile	Itutu	Erinle
1	2	3	4	5	6	7
M21	M22	M23	M24	M25	M26	M27
KHN5	KHN6	KHN7	KHN8	KHN9	KHN10	KHN11
MAG12	MAG13	MAG14	MAG15	MAG16	MAG17	MAG18
8	9	10	11	12	13	14
M28	M29	M30	M31	A1	A2	A3
KHN12	KHN13	KHN14	KHN15	KHN16	KHN17	KHN18
MAG19	MAG20	MAG21	MAG22	MAG23	MAG24	MAG25
15	16	17	18	19	20	21
A4	A5	A6	A7	A8	A9	A10
KHN19	KHN20	KHN21	KHN22	KHN23	KHN24	KHN25
MAG26	MAG27	MAG28	MAG29	MAG30	MIY1	MIY2
22	23	24	25	26	27	28
A11	A12	A13	A14	A15	A16	A17
KHN26	KHN27	KHN28	KHN29	KHN30	PIN1	PIN2
MIY3	MIY4	MIY5	MIY6	MIY7	MIY8	MIY9

MABABA WEUSI 553/2069-70/6305-06/2061-62

Ashe	Ojobo	Iwa	Araorun	Ile	Itutu	Erinle
1	2	3	4	5	6	7
MA16	MA17	MA18	MA19	MA20	MA21	MA22
APP1	APP2	APP3	APP4	APP5	APP6	APP7
GIN8	GIN9	GIN10	GIN11	GIN12	GIN13	GIN14
8	9	10	11	12	13	14
MA23	MA24	MA25	MA26	MA27	MA28	MA29
APP8	APP9	APP10	APP11	APP12	APP13	APP14
GIN15	GIN16	GIN17	GIN18	GIN19	GIN20	GIN21
15	16	17	18	19	20	21
MA30	MA31	J1	J2	J3	J4	J5
APP15	APP16	APP17	APP18	APP19	APP20	APP21
GIN22	GIN23	GIN24	GIN25	GIN26	GIN27	GIN28
22	23	24	25	26	27	28
J6	J7	J8	J9	J10	J11	J12
APP22	APP23	APP24	APP25	APP26	APP27	APP28
GIN29	GIN30	SAN1	SAN2	SAN3	SAN4	SAN5

THE NU AFRIKAN CALENDAR

MWANAMKE WEUSI 553/2069-70/6305-06/2061-62						
Ashe	Ojobo	Iwa	Araorun	Ile	Itutu	Erinle
1	2	3	4	5	6	7
J13	J14	J15	J16	J17	J18	J19
APP29	APP30	MSR1	MSR2	MSR3	MSR4	MSR5
SAN6	SAN7	SAN8	SAN9	SAN10	SAN11	SAN12
8	9	10	11	12	13	14
J20	J21	J22	J23	J24	J25	J26
MSR6	MSR7	MSR8	MSR9	MSR10	MSR11	MSR12
SAN13	SAN14	SAN15	SAN16	SAN17	SAN18	SAN19
15	16	17	18	19	20	21
J27	J28	J29	J30	JU1	JU2	JU3
MSR13	MSR14	MSR15	MSR16	MSR17	MSR18	MSR19
SAN20	SAN21	SAN22	SAN23	SAN24	SAN25	SAN26
22	23	24	25	26	27	28
JU4	JU5	JU6	JU7	JU8	JU9	JU10
MSR20	MRS21	MSR22	MSR23	MSR24	MSR25	MSR26
SAN27	SAN28	SAN29	SAN30	HAM1	HAM2	HAM3

TAIFA WEUSI 553/2069-70/6305-06/2061-62						
Ashe	Ojobo	Iwa	Araorun	Ile	Itutu	Erinle
1	2	3	4	5	6	7
AU8	AU9	AU10	AU11	AU12	AU13	AU14
THT20	THT21	THT22	THT23	THT24	THT25	THT26
NAH2	NAH3	NAH4	NAH5	NAH6	NAH7	NAH8
8	9	10	11	12	13	14
AU15	AU16	AU17	AU18	AU19	AU20	AU21
THT27	THT28	THT29	THT30	PAP1	PAP1	PAP2
NAH9	NAH10	NAH11	NAH12	NAH13	NAH14	NAH15
15	16	17	18	19	20	21
AU22	AU23	AU24	AU25	AU26	AU27	AU28
PAP3	PAP4	PAP5	PAP6	PAP7	PAP8	PAP9
NAH16	NAH17	NAH18	NAH19	NAH20	NAH21	NAH22
22	23	24	25	26	27	28
AU29	AU30	AU31	S1	S2	S3	S4
PAP10	PAP11	PAP12	PAP13	PAP14	PAP15	PAP16
NAH23	NAH24	NAH25	NAH26	NAH27	NAH28	NAH29

MTU WEUSI 553/2069-70/6305-06/2061-62						
Ashe	Ojobo	Iwa	Araorun	Ile	Itutu	Erinle
1	2	3	4	5	6	7
JU11	JU12	JU13	JU14	JU15	JU16	JU17
MSR27	MSR28	MSR29	MSR30	*ASR1*	*HRU2*	*STH3*
HAM4	HAM5	HAM6	HAM7	HAM8	HAM9	HAM10
8	9	10	11	12	13	14
JU18	JU19	JU20	JU21	JU22	JU23	JU24
AST4	*NBH5*	THT1	THT2	THT3	THT4	THT5
HAM11	HAM12	HAM13	AM14	HAM15	HAM16	HAM17
15	16	17	18	19	20	21
JU25	JU26	JU27	JU28	JU29	JU30	JU31
THT6	THT7	THT8	THT9	THT10	THT11	THT12
HAM18	HAM19	HAM20	HAM21	HAM22	HAM23	HAM24
22	23	24	25	26	27	28
AU1	AU2	AU3	AU4	AU5	AU6	AU7
THT13	THT14	THT15	THT16	THT17	THT18	THT19
HAM25	HAM26	HAM27	HAM28	HAM29	HAM30	NAH1

UWEZO 553/2069-70/6305-06/2061-62						
Ashe	Ojobo	Iwa	Araorun	Ile	Itutu	Erinle
1	2	3	4	5	6	7
S5	S6	S7	S8	S9	S10	S11
PAP17	PAP18	PAP19	PAP20	PAP21	PAP22	PAP23
NAH30	*PAG1*	*PAG2*	*PAG3*	*PAG4*	*PAG5*	MAS1
8	9	10	11	12	13	14
S12	S13	S14	S15	S16	S17	S18
PAP24	PAP25	PAP26	PAP27	PAP28	PAP29	PAP30
MAS2	MAS3	MAS4	MAS5	MAS6	MAS7	MAS8
15	16	17	18	19	20	21
S19	S20	S21	S22	S23	S24	S25
HTH1	HTH2	HTH3	HTH4	HTH5	HTH6	HTH7
MAS9	MAS10	MAS11	MAS12	MAS13	MAS14	MAS15
22	23	24	25	26	27	28
S26	S27	S28	S29	S30	O1	O2
HTH8	HTH9	HTH10	HTH11	HTH12	HTH13	HTH14
MAS16	MAS17	MAS18	MAS19	MAS20	MAS21	MAS22

THE NU AFRIKAN CALENDAR

EKUNDU 553/2069-70/6305-06/2061-62						
Ashe	Ojobo	Iwa	Araorun	Ile	Itutu	Erinle
1	2	3	4	5	6	7
O3	O4	O5	O6	O7	O8	O9
HTH15	HTH16	HTH17	HTH18	HTH19	HTH20	HTH21
MAS23	MAS24	MAS25	MAS26	MAS27	MAS28	MAS29
8	9	10	11	12	13	14
O10	O11	O12	O13	O14	O15	O16
HTH22	HTH23	HTH24	HTH25	HTH26	HTH27	HTH28
MAS30	THi1	THi2	THi3	THi4	THi5	THi6
15	16	17	18	19	20	21
O17	O18	O19	O20	O21	O22	O23
HTH29	HTH30	KHK1	KHK2	KHK3	KHK4	KHK5
THi7	THi8	THi9	THi10	THi11	THi12	THi13
22	23	24	25	26	27	28
O24	O25	O26	O27	O28	O29	O30
KHK6	KHK7	KHK8	KHK9	KHK10	KHK11	KHK12
THi14	THi15	THi16	THi17	THi18	THi19	THi20

KIJANI 553/2069-70/6305-06/2061-62						
Ashe	Ojobo	Iwa	Araorun	Ile	Itutu	Erinle
1	2	3	4	5	6	7
N28	N29	N30	D1	D2	D3	D4
TBT11	TBT12	TBT13	TBT14	TBT15	TBT16	TBT17
HID19	HID20	HID21	HID22	HID23	HID24	HID25
8	9	10	11	12	13	14
D5	D6	D7	D8	D9	D10	D11
TBT18	TBT19	TBT20	TBT21	TBT22	TBT23	TBT24
HID26	HID27	HID28	HID29	HID30	TAH1	TAH2
15	16	17	18	19	20	21
D12	D13	D14	D15	D16	D17	D18
TBT25	TBT26	TBT27	TBT28	TBT29	TBT30	MKH1
TAH3	TAH4	TAH5	TAH6	TAH7	TAH8	TAH9
22	23	24	25	26	27	28
D19	D20	D21	D22	D23	D24	D25
MKH2	MKH3	MKH4	MKH5	MKH6	MKH7	MKH8
TAH10	TAH11	TAH12	TAH13	TAH14	TAH15	TAH16

EUSI 553/2069-70/6305-06/2061-62						
Ashe	Ojobo	Iwa	Araorun	Ile	Itutu	Erinle
1	2	3	4	5	6	7
O31	N1	N2	N3	N4	N5	N6
KHK13	KHK14	KHK15	KHK16	KHK17	KHK18	KHK19
THi21	THi22	THi23	THi24	THi25	THi26	THi27
8	9	10	11	12	13	14
N7	N8	N9	N10	N11	N12	N13
KHK20	KHK21	KHK22	KHK23	KHK24	KHK25	KHK26
THi28	THi29	THi30	HID1	HID2	HID3	HID4
15	16	17	18	19	20	21
N14	N15	N16	N17	N18	N19	N20
KHK27	KHK28	KHK29	KHK30	TBT1	TBT2	TBT3
HID5	HID6	HID7	HID8	HID9	HID10	HID11
22	23	24	25	26	27	28
N21	N22	N23	N24	N25	N26	N27
TBT4	TBT5	TBT6	TBT7	TBT8	TBT9	TBT10
HID12	HID13	HID14	HID15	HID16	HID17	HID18

KWANZA 553/2069-70/6305-06/2061-62						
Umoja	Kujichagulia	Ujima	Ujamaa	Nia	Kuumba	Imani
1	2	3	4	5	6	7
D26	D27	D28	D29	D30	D31	JA1
MKH9	MKH10	MKH11	MKH12	MKH13	MKH14	MKH15
TAH17	TAH18	TAH19	TAH20	TAH21	TAH22	TAH23
8	9	10	11	12	13	14
JA2	JA3	JA4	JA5	JA6	JA7	JA8
MKH16	MKH17	MKH18	MKH19	MKH20	MKH21	MKH22
TAH24	TAH25	TAH26	TAH27	TAH28	TAH29	TAH30
15	16	17	18	19	20	21
JA9	JA10	JA11	JA12	JA13	JA14	JA15
MKH23	MKH24	MKH25	KMH26	MKH27	MKH28	MKH29
TIR1	TIR2	TIR3	TIR4	TIR5	TIR6	TIR7
22	23	24	25	26	27	28
JA16	JA17	JA18	JA19	JA20	JA21	JA22
MKH30	AMN1	AMN2	AMN3	AMN4	AMN5	AMN6
TIR8	TIR9	TIR10	TIR11	TIR12	TIR13	TIR14

THE NU AFRIKAN CALENDAR

SHAHIDI 553/2069-70/6305-06/2061-62						
Ashe	Ojobo	Iwa	Araorun	Ile	Itutu	Erinle
1	2	3	4	5	6	7
JA23	JA24	JA25	JA26	JA27	JA28	JA29
AMN7	AMN8	AMN9	AMN10	AMN11	AMN12	AMN13
TIR15	TIR16	TIR17	TIR18	TIR19	TIR20	TIR21
8	9	10	11	12	13	14
JA30	JA31	F1	F2	F3	F4	F5
AMN14	AMN15	AMN16	AMN17	AMN18	AMN19	AMN20
TIR22	TIR23	TIR24	TIR25	TIR26	TIR27	TIR28
15	16	17	18	19	20	21
F6	F7	F8	F9	F10	F 11	F12
AMN21	AMN22	AMN23	AMN24	AMN25	AMN26	AMN28
TIR29	TIR30	YEK1	YEK2	YEK3	YEK4	YEK5
22	23	24	25	26	27	28
F13	F14	F15	F16	F17	F18	F19
AMN29	AMN30	RNT1	RNT2	RNT3	RNT4	RNT5
YEK6	YEK7	YEK8	YEK9	YEK10	YEK11	YEK12
SHAHIDI 104						
F20						
RNT6						
YEK13						

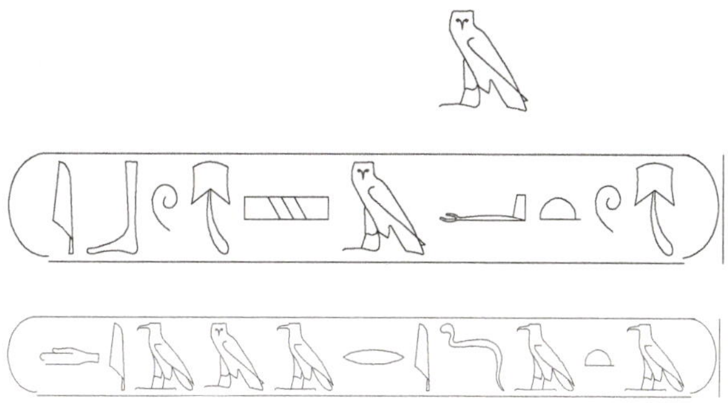

MAAFA 554 B.S.E./ASIENTO 2070-71 C.E.
KAM 6306-07 S.M./EC 2062-63 A.M. (ዓ.ም.)

AFU 554/2070-71/6306-07/2062-63						
Ashe	Ojobo	Iwa	Araorun	Ile	Itutu	Erinle
1	2	3	4	5	6	7
F21	F22	F23	F24	F25	F26	F27
RNT7	RNT8	RNT9	RNT10	RNT11	RNT12	RNT13
YEK14	YEK15	YEK16	YEK17	YEK18	YEK19	YEK20
8	9	10	11	12	13	14
F28	M1	M2	M3	M4	M5	M6
RNT14	RNT15	RNT16	RNT17	RNT18	RNT19	RNT20
YEK21	YEK22	YEK23	YEK24	YEK25	YEK26	YEK27
15	16	17	18	19	20	21
M7	M8	M9	M10	M11	M12	M13
RNT21	RNT22	RNT23	RNT24	RNT25	RNT26	RNT27
YEK28	YEK29	YEK30	MAG1	MAG2	MAG3	MAG4
22	23	24	25	26	27	28
M14	M15	M16	M17	M18	M19	M20
RNT28	RNT29	RNT30	KHN1	KHN2	KHN3	KHN4
MAG5	MAG6	MAG7	MAG8	MAG9	MAG10	MAG11

KAWAIDA 554/2070-71/6306-07/2062-63						
Ashe	Ojobo	Iwa	Araorun	Ile	Itutu	Erinle
1	2	3	4	5	6	7
A18	A19	A20	A21	A22	A23	A24
PIN3	PIN4	PIN5	PIN6	PIN7	PIN8	PIN9
MIY10	MIY11	MIY12	MIY13	MIY14	MIY15	MIY16
8	9	10	11	12	13	14
A25	A26	A27	A28	A29	A30	MA1
PIN10	PIN11	PIN12	PIN13	PIN14	PIN15	PIN16
MIY17	MIY18	MIY19	MIY20	MIY21	MIY22	MIY23
15	16	17	18	19	20	21
MA2	MA3	MA4	MA5	MA6	MA7	MA8
PIN17	PIN18	PIN19	PIN20	PIN21	PIN22	PIN23
MIY24	MIY25	MIY26	MIY27	MIY28	MIY29	MIY30
22	23	24	25	26	27	28
MA9	MA10	MA11	MA12	MA13	MA14	MA15
PIN24	PIN25	PIN26	PIN27	PIN28	PIN29	PIN30
GIN1	GIN2	GIN3	GIN4	GIN5	GIN6	GIN7

UHURU 554/2070-71/6306-07/2062-63						
Ashe	Ojobo	Iwa	Araorun	Ile	Itutu	Erinle
1	2	3	4	5	6	7
M21	M22	M23	M24	M25	M26	M27
KHN5	KHN6	KHN7	KHN8	KHN9	KHN10	KHN11
MAG12	MAG13	MAG14	MAG15	MAG16	MAG17	MAG18
8	9	10	11	12	13	14
M28	M29	M30	M31	A1	A2	A3
KHN12	KHN13	KHN14	KHN15	KHN16	KHN17	KHN18
MAG19	MAG20	MAG21	MAG22	MAG23	MAG24	MAG25
15	16	17	18	19	20	21
A4	A5	A6	A7	A8	A9	A10
KHN19	KHN20	KHN21	KHN22	KHN23	KHN24	KHN25
MAG26	MAG27	MAG28	MAG29	MAG30	MIY1	MIY2
22	23	24	25	26	27	28
A11	A12	A13	A14	A15	A16	A17
KHN26	KHN27	KHN28	KHN29	KHN30	PIN1	PIN2
MIY3	MIY4	MIY5	MIY6	MIY7	MIY8	MIY9

MABABA WEUSI 554/2070-71/6306-07/2062-63						
Ashe	Ojobo	Iwa	Araorun	Ile	Itutu	Erinle
1	2	3	4	5	6	7
MA16	MA17	MA18	MA19	MA20	MA21	MA22
APP1	APP2	APP3	APP4	APP5	APP6	APP7
GIN8	GIN9	GIN10	GIN11	GIN12	GIN13	GIN14
8	9	10	11	12	13	14
MA23	MA24	MA25	MA26	MA27	MA28	MA29
APP8	APP9	APP10	APP11	APP12	APP13	APP14
GIN15	GIN16	GIN17	GIN18	GIN19	GIN20	GIN21
15	16	17	18	19	20	21
MA30	MA31	J1	J2	J3	J4	J5
APP15	APP16	APP17	APP18	APP19	APP20	APP21
GIN22	GIN23	GIN24	GIN25	GIN26	GIN27	GIN28
22	23	24	25	26	27	28
J6	J7	J8	J9	J10	J11	J12
APP22	APP23	APP24	APP25	APP26	APP27	APP28
GIN29	GIN30	SAN1	SAN2	SAN3	SAN4	SAN5

THE NU AFRIKAN CALENDAR

MWANAMKE WEUSI 554/2070-71/6306-07/2062-63						
Ashe	Ojobo	Iwa	Araorun	Ile	Itutu	Erinle
1	2	3	4	5	6	7
J13	J14	J15	J16	J17	J18	J19
APP29	APP30	MSR1	MSR2	MSR3	MSR4	MSR5
SAN6	SAN7	SAN8	SAN9	SAN10	SAN11	SAN12
8	9	10	11	12	13	14
J20	J21	J22	J23	J24	J25	J26
MSR6	MSR7	MSR8	MSR9	MSR10	MSR11	MSR12
SAN13	SAN14	SAN15	SAN16	SAN17	SAN18	SAN19
15	16	17	18	19	20	21
J27	J28	J29	J30	JU1	JU2	JU3
MSR13	MSR14	MSR15	MSR16	MSR17	MSR18	MSR19
SAN20	SAN21	SAN22	SAN23	SAN24	SAN25	SAN26
22	23	24	25	26	27	28
JU4	JU5	JU6	JU7	JU8	JU9	JU10
MSR20	MRS21	MSR22	MSR23	MSR24	MSR25	MSR26
SAN27	SAN28	SAN29	SAN30	HAM1	HAM2	HAM3

TAIFA WEUSI 554/2070-71/6306-07/2062-63						
Ashe	Ojobo	Iwa	Araorun	Ile	Itutu	Erinle
1	2	3	4	5	6	7
AU8	AU9	AU10	AU11	AU12	AU13	AU14
THT20	THT21	THT22	THT23	THT24	THT25	THT26
NAH2	NAH3	NAH4	NAH5	NAH6	NAH7	NAH8
8	9	10	11	12	13	14
AU15	AU16	AU17	AU18	AU19	AU20	AU21
THT27	THT28	THT29	THT30	PAP1	PAP1	PAP2
NAH9	NAH10	NAH11	NAH12	NAH13	NAH14	NAH15
15	16	17	18	19	20	21
AU22	AU23	AU24	AU25	AU26	AU27	AU28
PAP3	PAP4	PAP5	PAP6	PAP7	PAP8	PAP9
NAH16	NAH17	NAH18	NAH19	NAH20	NAH21	NAH22
22	23	24	25	26	27	28
AU29	AU30	AU31	S1	S2	S3	S4
PAP10	PAP11	PAP12	PAP13	PAP14	PAP15	PAP16
NAH23	NAH24	NAH25	NAH26	NAH27	NAH28	NAH29

MTU WEUSI 554/2070-71/6306-07/2062-63						
Ashe	Ojobo	Iwa	Araorun	Ile	Itutu	Erinle
1	2	3	4	5	6	7
JU11	JU12	JU13	JU14	JU15	JU16	JU17
MSR27	MSR28	MSR29	MSR30	*ASR1*	*HRU2*	*STH3*
HAM4	HAM5	HAM6	HAM7	HAM8	HAM9	HAM10
8	9	10	11	12	13	14
JU18	JU19	JU20	JU21	JU22	JU23	JU24
AST4	*NBH5*	THT1	THT2	THT3	THT4	THT5
HAM11	HAM12	HAM13	AM14	HAM15	HAM16	HAM17
15	16	17	18	19	20	21
JU25	JU26	JU27	JU28	JU29	JU30	JU31
THT6	THT7	THT8	THT9	THT10	THT11	THT12
HAM18	HAM19	HAM20	HAM21	HAM22	HAM23	HAM24
22	23	24	25	26	27	28
AU1	AU2	AU3	AU4	AU5	AU6	AU7
THT13	THT14	THT15	THT16	THT17	THT18	THT19
HAM25	HAM26	HAM27	HAM28	HAM29	HAM30	NAH1

UWEZO 554/2070-71/6306-07/2062-63						
Ashe	Ojobo	Iwa	Araorun	Ile	Itutu	Erinle
1	2	3	4	5	6	7
S5	S6	S7	S8	S9	S10	S11
PAP17	PAP18	PAP19	PAP20	PAP21	PAP22	PAP23
NAH30	*PAG1*	*PAG2*	*PAG3*	*PAG4*	*PAG5*	MAS1
8	9	10	11	12	13	14
S12	S13	S14	S15	S16	S17	S18
PAP24	PAP25	PAP26	PAP27	PAP28	PAP29	PAP30
MAS2	MAS3	MAS4	MAS5	MAS6	MAS7	MAS8
15	16	17	18	19	20	21
S19	S20	S21	S22	S23	S24	S25
HTH1	HTH2	HTH3	HTH4	HTH5	HTH6	HTH7
MAS9	MAS10	MAS11	MAS12	MAS13	MAS14	MAS15
22	23	24	25	26	27	28
S26	S27	S28	S29	S30	O1	O2
HTH8	HTH9	HTH10	HTH11	HTH12	HTH13	HTH14
MAS16	MAS17	MAS18	MAS19	MAS20	MAS21	MAS22

THE NU AFRIKAN CALENDAR

EKUNDU 554/2070-71/6306-07/2062-63						
Ashe	Ojobo	Iwa	Araorun	Ile	Itutu	Erinle
1	2	3	4	5	6	7
O3	O4	O5	O6	O7	O8	O9
HTH15	HTH16	HTH17	HTH18	HTH19	HTH20	HTH21
MAS23	MAS24	MAS25	MAS26	MAS27	MAS28	MAS29
8	9	10	11	12	13	14
O10	O11	O12	O13	O14	O15	O16
HTH22	HTH23	HTH24	HTH25	HTH26	HTH27	HTH28
MAS30	THi1	THi2	THi3	THi4	THi5	THi6
15	16	17	18	19	20	21
O17	O18	O19	O20	O21	O22	O23
HTH29	HTH30	KHK1	KHK2	KHK3	KHK4	KHK5
THi7	THi8	THi9	THi10	THi11	THi12	THi13
22	23	24	25	26	27	28
O24	O25	O26	O27	O28	O29	O30
KHK6	KHK7	KHK8	KHK9	KHK10	KHK11	KHK12
THi14	THi15	THi16	THi17	THi18	THi19	THi20

KIJANI 554/2070-71/6306-07/2062-63						
Ashe	Ojobo	Iwa	Araorun	Ile	Itutu	Erinle
1	2	3	4	5	6	7
N28	N29	N30	D1	D2	D3	D4
TBT11	TBT12	TBT13	TBT14	TBT15	TBT16	TBT17
HID19	HID20	HID21	HID22	HID23	HID24	HID25
8	9	10	11	12	13	14
D5	D6	D7	D8	D9	D10	D11
TBT18	TBT19	TBT20	TBT21	TBT22	TBT23	TBT24
HID26	HID27	HID28	HID29	HID30	TAH1	TAH2
15	16	17	18	19	20	21
D12	D13	D14	D15	D16	D17	D18
TBT25	TBT26	TBT27	TBT28	TBT29	TBT30	MKH1
TAH3	TAH4	TAH5	TAH6	TAH7	TAH8	TAH9
22	23	24	25	26	27	28
D19	D20	D21	D22	D23	D24	D25
MKH2	MKH3	MKH4	MKH5	MKH6	MKH7	MKH8
TAH10	TAH11	TAH12	TAH13	TAH14	TAH15	TAH16

EUSI 554/2070-71/6306-07/2062-63						
Ashe	Ojobo	Iwa	Araorun	Ile	Itutu	Erinle
1	2	3	4	5	6	7
O31	N1	N2	N3	N4	N5	N6
KHK13	KHK14	KHK15	KHK16	KHK17	KHK18	KHK19
THi21	THi22	THi23	THi24	THi25	THi26	THi27
8	9	10	11	12	13	14
N7	N8	N9	N10	N11	N12	N13
KHK20	KHK21	KHK22	KHK23	KHK24	KHK25	KHK26
THi28	THi29	THi30	HID1	HID2	HID3	HID4
15	16	17	18	19	20	21
N14	N15	N16	N17	N18	N19	N20
KHK27	KHK28	KHK29	KHK30	TBT1	TBT2	TBT3
HID5	HID6	HID7	HID8	HID9	HID10	HID11
22	23	24	25	26	27	28
N21	N22	N23	N24	N25	N26	N27
TBT4	TBT5	TBT6	TBT7	TBT8	TBT9	TBT10
HID12	HID13	HID14	HID15	HID16	HID17	HID18

KWANZA 554/2070-71/6306-07/2062-63						
Umoja	Kujichagulia	Ujima	Ujamaa	Nia	Kuumba	Imani
1	2	3	4	5	6	7
D26	D27	D28	D29	D30	D31	JA1
MKH9	MKH10	MKH11	MKH12	MKH13	MKH14	MKH1
TAH17	TAH18	TAH19	TAH20	TAH21	TAH22	TAH23
8	9	10	11	12	13	14
JA2	JA3	JA4	JA5	JA6	JA7	JA8
MKH16	MKH17	MKH18	MKH19	MKH20	MKH21	MKH2
TAH24	TAH25	TAH26	TAH27	TAH28	TAH29	TAH30
15	16	17	18	19	20	21
JA9	JA10	JA11	JA12	JA13	JA14	JA15
MKH23	MKH24	MKH25	KMH26	MKH27	MKH28	MKH2
TIR1	TIR2	TIR3	TIR4	TIR5	TIR6	TIR7
22	23	24	25	26	27	28
JA16	JA17	JA18	JA19	JA20	JA21	JA22
MKH30	AMN1	AMN2	AMN3	AMN4	AMN5	AMN6
TIR8	TIR9	TIR10	TIR11	TIR12	TIR13	TIR14

SHAHIDI 554/2070-71/6306-07/2062-63						
Ashe	Ojobo	Iwa	Araorun	Ile	Itutu	Erinle
1	2	3	4	5	6	7
JA23	JA24	JA25	JA26	JA27	JA28	JA29
AMN7	AMN8	AMN9	AMN10	AMN11	AMN12	AMN13
TIR15	TIR16	TIR17	TIR18	TIR19	TIR20	TIR21
8	9	10	11	12	13	14
JA30	JA31	F1	F2	F3	F4	F5
AMN14	AMN15	AMN16	AMN17	AMN18	AMN19	AMN20
TIR22	TIR23	TIR24	TIR25	TIR26	TIR27	TIR28
15	16	17	18	19	20	21
F6	F7	F8	F9	F10	F 11	F12
AMN21	AMN22	AMN23	AMN24	AMN25	AMN26	AMN28
TIR29	TIR30	YEK1	YEK2	YEK3	YEK4	YEK5
22	23	24	25	26	27	28
F13	F14	F15	F16	F17	F18	F19
AMN29	AMN30	RNT1	RNT2	RNT3	RNT4	RNT5
YEK6	YEK7	YEK8	YEK9	YEK10	YEK11	YEK12
SHAHIDI 105						
F20						
RNT6						
YEK13						

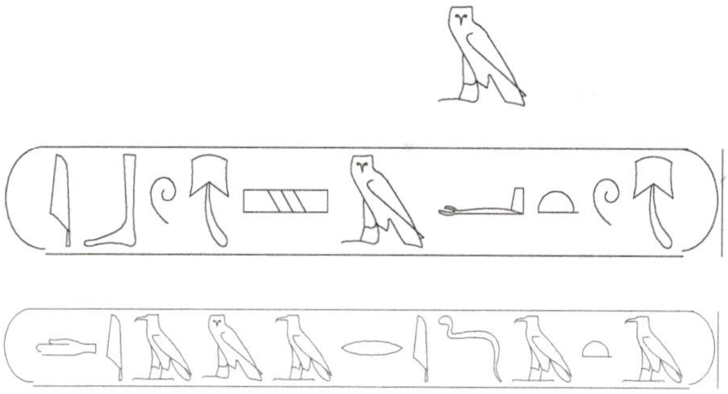

MAAFA 555 B.S.E./ASIENTO 2071-72 C.E./
KAM 6307-08 S.M./EC 2063-64 A.M. (ዓ.ም.)

AFU 555/2071-72/6307-08/2063-64						
Ashe	Ojobo	Iwa	Araorun	Ile	Itutu	Erinle
1	2	3	4	5	6	7
F21	F22	F23	F24	F25	F26	F27
RNT7	RNT8	RNT9	RNT10	RNT11	RNT12	RNT13
YEK14	YEK15	YEK16	YEK17	YEK18	YEK19	YEK20
8	9	10	11	12	13	14
F28	M1	M2	M3	M4	M5	M6
RNT14	RNT15	RNT16	RNT17	RNT18	RNT19	RNT20
YEK21	YEK22	YEK23	YEK24	YEK25	YEK26	YEK27
15	16	17	18	19	20	21
M7	M8	M9	M10	M11	M12	M13
RNT21	RNT22	RNT23	RNT24	RNT25	RNT26	RNT27
YEK28	YEK 29	YEK30	MAG1	MAG2	MAG3	MAG4
22	23	24	25	26	27	28
M14	M15	M16	M17	M18	M19	M20
RNT28	RNT29	RNT30	KHN1	KHN2	KHN3	KHN4
MAG5	MAG6	MAG7	MAG8	MAG9	MAG10	MAG11

KAWAIDA 555/2071-72/6307-08/2063-64						
Ashe	Ojobo	Iwa	Araorun	Ile	Itutu	Erinle
1	2	3	4	5	6	7
A18	A19	A20	A21	A22	A23	A24
PIN3	PIN4	PIN5	PIN6	PIN7	PIN8	PIN9
MIY10	MIY11	MIY12	MIY13	MIY14	MIY15	MIY1
8	9	10	11	12	13	14
A25	A26	A27	A28	A29	A30	MA1
PIN10	PIN11	PIN12	PIN13	PIN14	PIN15	PIN16
MIY17	MIY18	MIY19	MIY20	MIY21	MIY22	MIY23
15	16	17	18	19	20	21
MA2	MA3	MA4	MA5	MA6	MA7	MA8
PIN17	PIN18	PIN19	PIN20	PIN21	PIN22	PIN2
MIY24	MIY25	MIY26	MIY27	MIY28	MIY29	MIY3
22	23	24	25	26	27	28
MA9	MA10	MA11	MA12	MA13	MA14	MA1
PIN24	PIN25	PIN26	PIN27	PIN28	PIN29	PIN3
GIN1	GIN2	GIN3	GIN4	GIN5	GIN6	GIN7

UHURU 555/2071-72/6307-08/2063-64						
Ashe	Ojobo	Iwa	Araorun	Ile	Itutu	Erinle
1	2	3	4	5	6	7
M21	M22	M23	M24	M25	M26	M27
KHN5	KHN6	KHN7	KHN8	KHN9	KHN10	KHN11
MAG12	MAG13	MAG14	MAG15	MAG16	MAG17	MAG18
8	9	10	11	12	13	14
M28	M29	M30	M31	A1	A2	A3
KHN12	KHN13	KHN14	KHN15	KHN16	KHN17	KHN18
MAG19	MAG20	MAG21	MAG22	MAG23	MAG24	MAG25
15	16	17	18	19	20	21
A4	A5	A6	A7	A8	A9	A10
KHN19	KHN20	KHN21	KHN22	KHN23	KHN24	KHN25
MAG26	MAG27	MAG28	MAG29	MAG30	MIY1	MIY2
22	23	24	25	26	27	28
A11	A12	A13	A14	A15	A16	A17
KHN26	KHN27	KHN28	KHN29	KHN30	PIN1	PIN2
MIY3	MIY4	MIY5	MIY6	MIY7	MIY8	MIY9

MABABA WEUSI 555/2071-72/6307-08/2063-64						
Ashe	Ojobo	Iwa	Araorun	Ile	Itutu	Erinle
1	2	3	4	5	6	7
MA16	MA17	MA18	MA19	MA20	MA21	MA22
APP1	APP2	APP3	APP4	APP5	APP6	APP
GIN8	GIN9	GIN10	GIN11	GIN12	GIN13	GIN1
8	9	10	11	12	13	14
MA23	MA24	MA25	MA26	MA27	MA28	MA29
APP8	APP9	APP10	APP11	APP12	APP13	APP1
GIN15	GIN16	GIN17	GIN18	GIN19	GIN20	GIN21
15	16	17	18	19	20	21
MA30	MA31	J1	J2	J3	J4	J5
APP15	APP16	APP17	APP18	APP19	APP20	APP2
GIN22	GIN23	GIN24	GIN25	GIN26	GIN27	GIN28
22	23	24	25	26	27	28
J6	J7	J8	J9	J10	J11	J12
APP22	APP23	APP24	APP25	APP26	APP27	APP2
GIN29	GIN30	SAN1	SAN2	SAN3	SAN4	SAN5

THE NU AFRIKAN CALENDAR

MWANAMKE WEUSI 555/2071-72/6307-08/2063-64						
Ashe	Ojobo	Iwa	Araorun	Ile	Itutu	Erinle
1	2	3	4	5	6	7
J13	J14	J15	J16	J17	J18	J19
APP29	APP30	MSR1	MSR2	MSR3	MSR4	MSR5
SAN6	SAN7	SAN8	SAN9	SAN10	SAN11	SAN12
8	9	10	11	12	13	14
J20	J21	J22	J23	J24	J25	J26
MSR6	MSR7	MSR8	MSR9	MSR10	MSR11	MSR12
SAN13	SAN14	SAN15	SAN16	SAN17	SAN18	SAN19
15	16	17	18	19	20	21
J27	J28	J29	J30	JU1	JU2	JU3
MSR13	MSR14	MSR15	MSR16	MSR17	MSR18	MSR19
SAN20	SAN21	SAN22	SAN23	SAN24	SAN25	SAN26
22	23	24	25	26	27	28
JU4	JU5	JU6	JU7	JU8	JU9	JU10
MSR20	MRS21	MSR22	MSR23	MSR24	MSR25	MSR26
SAN27	SAN28	SAN29	SAN30	HAM1	HAM2	HAM3

TAIFA WEUSI 555/2071-72/6307-08/2063-64						
Ashe	Ojobo	Iwa	Araorun	Ile	Itutu	Erinle
1	2	3	4	5	6	7
AU8	AU9	AU10	AU11	AU12	AU13	AU14
THT20	THT21	THT22	THT23	THT24	THT25	THT26
NAH2	NAH3	NAH4	NAH5	NAH6	NAH7	NAH8
8	9	10	11	12	13	14
AU15	AU16	AU17	AU18	AU19	AU20	AU21
THT27	THT28	THT29	THT30	PAP1	PAP1	PAP2
NAH9	NAH10	NAH11	NAH12	NAH13	NAH14	NAH15
15	16	17	18	19	20	21
AU22	AU23	AU24	AU25	AU26	AU27	AU28
PAP3	PAP4	PAP5	PAP6	PAP7	PAP8	PAP9
NAH16	NAH17	NAH18	NAH19	NAH20	NAH21	NAH22
22	23	24	25	26	27	28
AU29	AU30	AU31	S1	S2	S3	S4
PAP10	PAP11	PAP12	PAP13	PAP14	PAP15	PAP16
NAH23	NAH24	NAH25	NAH26	NAH27	NAH28	NAH29

MTU WEUSI 555/2071-72/6307-08/2063-64						
Ashe	Ojobo	Iwa	Araorun	Ile	Itutu	Erinle
1	2	3	4	5	6	7
JU11	JU12	JU13	JU14	JU15	JU16	JU17
MSR27	MSR28	MSR29	MSR30	*ASR1*	*HRU2*	*STH3*
HAM4	HAM5	HAM6	HAM7	HAM8	HAM9	HAM10
8	9	10	11	12	13	14
JU18	JU19	JU20	JU21	JU22	JU23	JU24
AST4	*NBH5*	THT1	THT2	THT3	THT4	THT5
HAM11	HAM12	HAM13	HAM14	HAM15	HAM16	HAM17
15	16	17	18	19	20	21
JU25	JU26	JU27	JU28	JU29	JU30	JU31
THT6	THT7	THT8	THT9	THT10	THT11	THT12
HAM18	HAM19	HAM20	HAM21	HAM22	HAM23	HAM24
22	23	24	25	26	27	28
AU1	AU2	AU3	AU4	AU5	AU6	AU7
THT13	THT14	THT15	THT16	THT17	THT18	THT19
HAM25	HAM26	HAM27	HAM28	HAM29	HAM30	NAH1

UWEZO 555/2071-72/6307-08/2063-64						
Ashe	Ojobo	Iwa	Araorun	Ile	Itutu	Erinle
1	2	3	4	5	6	7
S5	S6	S7	S8	S9	S10	S11
PAP17	PAP18	PAP19	PAP20	PAP21	PAP22	PAP23
NAH30	*QUA1*	*QUA2*	*QUA3*	*QUA4*	*QUA5*	*QUA6*
8	9	10	11	12	13	14
S12	S13	S14	S15	S16	S17	S18
PAP24	PAP25	PAP26	PAP27	PAP28	PAP29	PAP30
MAS1	MAS2	MAS3	MAS4	MAS5	MAS6	MAS7
15	16	17	18	19	20	21
S19	S20	S21	S22	S23	S24	S25
HTH1	HTH2	HTH3	HTH4	HTH5	HTH6	HTH7
MAS8	MAS9	MAS10	MAS11	MAS12	MAS13	MAS14
22	23	24	25	26	27	28
S26	S27	S28	S29	S30	O1	O2
HTH8	HTH9	HTH10	HTH11	HTH12	HTH13	HTH14
MAS15	MAS16	MAS17	MAS18	MAS19	MAS20	MAS21

THE NU AFRIKAN CALENDAR

EKUNDU 555/2071-72/6307-08/2063-64						
Ashe	Ojobo	Iwa	Araorun	Ile	Itutu	Erinle
1	2	3	4	5	6	7
O3	O4	O5	O6	O7	O8	O9
HTH15	HTH16	HTH17	HTH18	HTH19	HTH20	HTH21
MAS22	MAS23	MAS 24	MAS25	MAS26	MAS27	MAS28
8	9	10	11	12	13	14
O10	O11	O12	O13	O14	O15	O16
HTH22	HTH23	HTH24	HTH25	HTH26	HTH27	HTH28
MAS29	MAS30	THi1	THi2	THi3	THi4	THi5
15	16	17	18	19	20	21
O17	O18	O19	O20	O21	O22	O23
HTH29	HTH30	KHK1	KHK2	KHK3	KHK4	KHK5
THi6	THi7	THi8	THi9	THi10	THi11	THi12
22	23	24	25	26	27	28
O24	O25	O26	O27	O28	O29	O30
KHK6	KHK7	KHK8	KHK9	KHK10	KHK11	KHK12
THi13	THi14	THi15	THi16	THi17	THi18	THi19

KIJANI 555/2071-72/6307-08/2063-64						
Ashe	Ojobo	Iwa	Araorun	Ile	Itutu	Erinle
1	2	3	4	5	6	7
N28	N29	N30	D1	D2	D3	D4
TBT11	TBT12	TBT13	TBT14	TBT15	TBT16	TBT17
HID18	HID19	HID20	HID21	HID22	HID23	HID24
8	9	10	11	12	13	14
D5	D6	D7	D8	D9	D10	D11
TBT18	TBT19	TBT20	TBT21	TBT22	TBT23	TBT24
HID25	HID26	HID27	HID28	HID29	HID30	TAH1
15	16	17	18	19	20	21
D12	D13	D14	D15	D16	D17	D18
TBT25	TBT26	TBT27	TBT28	TBT29	TBT30	MKH1
TAH2	TAH3	TAH4	TAH5	TAH6	TAH7	TAH8
22	23	24	25	26	27	28
D19	D20	D21	D22	D23	D24	D25
MKH2	MKH3	MKH4	MKH5	MKH6	MKH7	MKH8
TAH9	TAH10	TAH11	TAH12	TAH13	TAH14	TAH15

EUSI 555/2071-72/6307-08/2063-64						
Ashe	Ojobo	Iwa	Araorun	Ile	Itutu	Erinle
1	2	3	4	5	6	7
O31	N1	N2	N3	N4	N5	N6
KHK13	KHK14	KHK15	KHK16	KHK17	KHK18	KHK19
THi20	THi21	THi22	THi23	THi24	THi25	THi26
8	9	10	11	12	13	14
N7	N8	N9	N10	N11	N12	N13
KHK20	KHK21	KHK22	KHK23	KHK24	KHK25	KHK26
THi27	THi28	THi29	THi30	HID1	HID2	HID3
15	16	17	18	19	20	21
N14	N15	N16	N17	N18	N19	N20
KHK27	KHK28	KHK29	KHK30	TBT1	TBT2	TBT3
HID4	HID5	HID6	HID7	HID8	HID9	HID10
22	23	24	25	26	27	28
N21	N22	N23	N24	N25	N26	N27
TBT4	TBT5	TBT6	TBT7	TBT8	TBT9	TBT10
HID11	HID12	HID13	HID14	HID15	HID16	HID17

KWANZA 555/2071-72/6307-08/2063-64						
Umoja	Kujichagulia	Ujima	Ujamaa	Nia	Kuumba	Imani
1	2	3	4	5	6	7
D26	D27	D28	D29	D30	D31	JA1
MKH9	MKH10	MKH11	MKH12	MKH13	MKH14	MKH1
TAH16	TAH17	TAH18	TAH19	TAH20	TAH21	TAH2
8	9	10	11	12	13	14
JA2	JA3	JA4	JA5	JA6	JA7	JA
MKH16	MKH17	MKH18	MKH19	MKH20	MKH21	MKH2
TAH23	TAH24	TAH25	TAH26	TAH27	TAH28	TAH29
15	16	17	18	19	20	21
JA9	JA10	JA11	JA12	JA13	JA14	JA1
MKH23	MKH24	MKH25	KMH26	MKH27	MKH28	MKH2
TAH30	TIR1	TIR2	TIR3	TIR4	TIR5	TIR6
22	23	24	25	26	27	28
JA16	JA17	JA18	JA19	JA20	JA21	JA2
MKH30	AMN1	AMN2	AMN3	AMN4	AMN5	AMN
TIR7	TIR8	TIR9	TIR10	TIR11	TIR12	TIR13

SHAHIDI 555/2071-72/6307-08/2063-64						
Ashe	Ojobo	Iwa	Araorun	Ile	Itutu	Erinle
1	2	3	4	5	6	7
JA23	JA24	JA25	JA26	JA27	JA28	JA29
AMN7	AMN8	AMN9	AMN10	AMN11	AMN12	AMN13
TIR14	TIR15	TIR16	TIR17	TIR18	TIR19	TIR20
8	9	10	11	12	13	14
JA30	JA31	F1	F2	F3	F4	F5
AMN14	AMN15	AMN16	AMN17	AMN18	AMN19	AMN20
TIR21	TIR22	TIR23	TIR24	TIR25	TIR26	TIR27
15	16	17	18	19	20	21
F6	F7	F8	F9	F10	F 11	F12
AMN21	AMN22	AMN23	AMN24	AMN25	AMN26	AMN28
TIR28	TIR29	TIR30	YEK1	YEK2	YEK3	YEK4
22	23	24	25	26	27	28
F13	F14	F15	F16	F17	F18	F19
AMN29	AMN30	RNT1	RNT2	RNT3	RNT4	RNT5
YEK5	YEK6	YEK7	YEK8	YEK9	YEK10	YEK11
SHAHIDI 106						
F20						
RNT6						
YEK12						

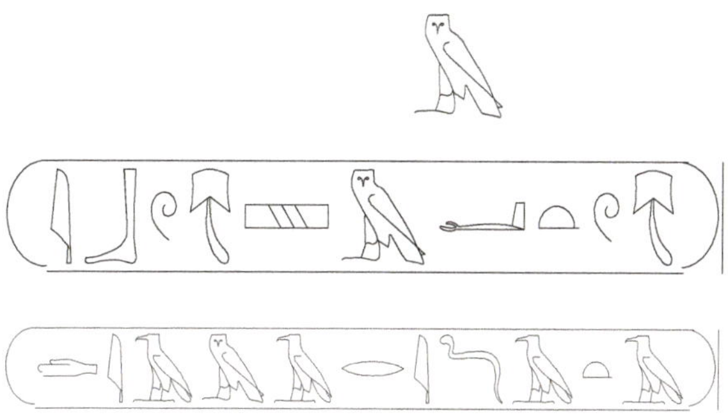

THE NU AFRIKAN CALENDAR

MAAFA 556 B.S.E./ASIENTO 2072-73 C.E./
KAM 6308-09 S.M./EC 2064-65 A.M. (ዓ.ም.)

AFU 556/2072-73/6308-09/2064-65						
Ashe	Ojobo	Iwa	Araorun	Ile	Itutu	Erinle
1	2	3	4	5	6	7
F21	F22	F23	F24	F25	F26	F27
RNT7	RNT8	RNT9	RNT10	RNT11	RNT12	RNT13
YEK13	YEK14	YEK15	YEK16	YEK17	YEK18	YEK19
8	9	10	11	12	13	14
F28	F29	M1	M2	M3	M4	M5
RNT14	RNT15	RNT16	RNT17	RNT18	RNT19	RNT20
YEK 20	YEK21	YEK22	YEK23	YEK24	YEK25	YEK26
15	16	17	18	19	20	21
M6	M7	M8	M9	M10	M11	M12
RNT21	RNT22	RNT23	RNT24	RNT25	RNT26	RNT27
YEK 27	YEK28	YEK29	YEK30	MAG1	MAG2	MAG3
22	23	24	25	26	27	28
M13	M14	M15	M16	M17	M18	M19
RNT28	RNT29	RNT30	KHN1	KHN2	KHN3	KHN4
MAG4	MAG5	MAG6	MAG7	MAG8	MAG9	MAG10
SHARIFU 136						
M20						
KHN5						
MAG11						

KAWAIDA 556/2072-73/6308-09/2064-65						
Ashe	Ojobo	Iwa	Araorun	Ile	Itutu	Erinle
1	2	3	4	5	6	7
A18	A19	A20	A21	A22	A23	A24
PIN4	PIN5	PIN6	PIN7	PIN8	PIN9	PIN10
MIY10	MIY11	MIY12	MIY13	MIY14	MIY15	MIY16
8	9	10	11	12	13	14
A25	A26	A27	A28	A29	A30	MA1
PIN11	PIN12	PIN13	PIN14	PIN15	PIN16	PIN17
MIY17	MIY18	MIY19	MIY20	MIY21	MIY22	MIY23
15	16	17	18	19	20	21
MA2	MA3	MA4	MA5	MA6	MA7	MA8
PIN18	PIN19	PIN20	PIN21	PIN22	PIN23	PIN24
MIY24	MIY25	MIY26	MIY27	MIY28	MIY29	MIY30
22	23	24	25	26	27	28
MA9	MA10	MA11	MA12	MA13	MA14	MA15
PIN25	PIN26	PIN27	PIN28	PIN29	PIN30	APP1
GIN1	GIN2	GIN3	GIN4	GIN5	GIN6	GIN7

UHURU 556/2072-73/6308-09/2064-65						
Ashe	Ojobo	Iwa	Araorun	Ile	Itutu	Erinle
1	2	3	4	5	6	7
M21	M22	M23	M24	M25	M26	M27
KHN6	KHN7	KHN8	KHN9	KHN10	KHN11	KHN12
MAG12	MAG13	MAG14	MAG15	MAG16	MAG17	MAG18
8	9	10	11	12	13	14
M28	M29	M30	M31	A1	A2	A3
KHN13	KHN14	KHN15	KHN16	KHN17	KHN18	KHN19
MAG19	MAG20	MAG21	MAG22	MAG23	MAG24	MAG25
15	16	17	18	19	20	21
A4	A5	A6	A7	A8	A9	A10
KHN20	KHN21	KHN222	KHN23	KHN24	KHN25	KHN26
MAG26	MAG27	MAG28	MAG29	MAG30	MIY1	MIY2
22	23	24	25	26	27	28
A11	A12	A13	A14	A15	A16	A17
KHN27	KHN28	KHN29	KHN30	PIN1	PIN2	PIN3
MIY3	MIY4	MIY5	MIY6	MIY7	MIY8	MIY9

MABABA WEUSI 556/2072-73/6308-09/2064-65						
Ashe	Ojobo	Iwa	Araorun	Ile	Itutu	Erinle
1	2	3	4	5	6	7
MA16	MA17	MA18	MA19	MA20	MA21	MA22
APP2	APP3	APP4	APP5	APP6	APP7	APP8
GIN8	GIN9	GIN10	GIN11	GIN12	GIN13	GIN14
8	9	10	11	12	13	14
MA23	MA24	MA25	MA26	MA27	MA28	MA29
APP9	APP10	APP11	APP12	APP13	APP14	APP15
GIN15	GIN16	GIN17	GIN18	GIN19	GIN20	GIN21
15	16	17	18	19	20	21
MA30	MA31	J1	J2	J3	J4	J5
APP16	APP17	APP18	APP19	APP20	APP21	APP22
GIN22	GIN23	GIN24	GIN25	GIN26	GIN27	GIN28
22	23	24	25	26	27	28
J6	J7	J8	J9	J10	J11	J12
APP23	APP24	APP25	APP26	APP27	APP28	APP29
GIN29	GIN30	SAN1	SAN2	SAN3	SAN4	SAN5

THE NU AFRIKAN CALENDAR

MWANAMKE WEUSI 556/2072-73/6308-09/2064-65						
Ashe	Ojobo	Iwa	Araorun	Ile	Itutu	Erinle
1	2	3	4	5	6	7
J13	J14	J15	J16	J17	J18	J19
APP30	MSR1	MSR2	MSR3	MSR4	MSR5	MSR6
SAN6	SAN7	SAN8	SAN9	SAN10	SAN11	SAN12
8	9	10	11	12	13	14
J20	J21	J22	J23	J24	J25	J26
MSR7	MSR8	MSR9	MSR10	MSR11	MSR12	MSR13
SAN13	SAN14	SAN15	SAN16	SAN17	SAN18	SAN19
15	16	17	18	19	20	21
J27	J28	J29	J30	JU1	JU2	JU3
MSR14	MSR15	MSR16	MSR17	MSR18	MSR19	MSR20
SAN20	SAN21	SAN22	SAN23	SAN24	SAN25	SAN26
22	23	24	25	26	27	28
JU4	JU5	JU6	JU7	JU8	JU9	JU10
MSR21	MRS22	MSR23	MSR24	MSR25	MSR26	MSR27
SAN27	SAN28	SAN29	SAN30	HAM1	HAM2	HAM3

TAIFA WEUSI 556/2072-73/6308-09/2064-65						
Ashe	Ojobo	Iwa	Araorun	Ile	Itutu	Erinle
1	2	3	4	5	6	7
AU8	AU9	AU10	AU11	AU12	AU13	AU14
THT20	THT21	THT22	THT23	THT24	THT25	THT26
NAH2	NAH3	NAH4	NAH5	NAH6	NAH7	NAH8
8	9	10	11	12	13	14
AU15	AU16	AU17	AU18	AU19	AU20	AU21
THT27	THT28	THT29	THT30	PAP1	PAP1	PAP2
NAH9	NAH10	NAH11	NAH12	NAH13	NAH14	NAH15
15	16	17	18	19	20	21
AU22	AU23	AU24	AU25	AU26	AU27	AU28
PAP3	PAP4	PAP5	PAP6	PAP7	PAP8	PAP9
NAH16	NAH17	NAH18	NAH19	NAH20	NAH21	NAH22
22	23	24	25	26	27	28
AU29	AU30	AU31	S1	S2	S3	S4
PAP10	PAP11	PAP12	PAP13	PAP14	PAP15	PAP16
NAH23	NAH24	NAH25	NAH26	NAH27	NAH28	NAH29

MTU WEUSI 556/2072-73/6308-09/2064-65						
Ashe	Ojobo	Iwa	Araorun	Ile	Itutu	Erinle
1	2	3	4	5	6	7
JU11	JU12	JU13	JU14	JU15	JU16	JU17
MSR28	MSR29	MSR30	*ASR1*	*HRU2*	*STH3*	*AST4*
HAM4	HAM5	HAM6	HAM7	HAM8	HAM9	HAM10
8	9	10	11	12	13	14
JU18	JU19	JU20	JU21	JU22	JU23	JU24
NBH5	*ANU6*	THT1	THT2	THT3	THT4	THT5
HAM11	HAM12	HAM13	HAM14	HAM15	HAM16	HAM17
15	16	17	18	19	20	21
JU25	JU26	JU27	JU28	JU29	JU30	JU31
THT6	THT7	THT8	THT9	THT10	THT11	THT12
HAM18	HAM19	HAM20	HAM21	HAM22	HAM23	HAM24
22	23	24	25	26	27	28
AU1	AU2	AU3	AU4	AU5	AU6	AU7
THT13	THT14	THT15	THT16	THT17	THT18	THT19
HAM25	HAM26	HAM27	HAM28	HAM29	HAM30	NAH1

UWEZO 556/2072-73/6308-09/2064-65						
Ashe	Ojobo	Iwa	Araorun	Ile	Itutu	Erinle
1	2	3	4	5	6	7
S5	S6	S7	S8	S9	S10	S11
PAP17	PAP18	PAP19	PAP20	PAP21	PAP22	PAP23
NAH30	*PAG1*	*PAG2*	*PAG3*	*PAG4*	*PAG5*	MAS1
8	9	10	11	12	13	14
S12	S13	S14	S15	S16	S17	S18
PAP24	PAP25	PAP26	PAP27	PAP28	PAP29	PAP30
MAS2	MAS3	MAS4	MAS5	MAS6	MAS7	MAS8
15	16	17	18	19	20	21
S19	S20	S21	S22	S23	S24	S25
HTH1	HTH2	HTH3	HTH4	HTH5	HTH6	HTH7
MAS9	MAS10	MAS11	MAS12	MAS13	MAS14	MAS15
22	23	24	25	26	27	28
S26	S27	S28	S29	S30	O1	O2
HTH8	HTH9	HTH10	HTH11	HTH12	HTH13	HTH14
MAS16	MAS17	MAS18	MAS19	MAS20	MAS21	MAS22

EKUNDU 556/2072-73/6308-09/2064-65						
Ashe	Ojobo	Iwa	Araorun	Ile	Itutu	Erinle
1	2	3	4	5	6	7
O3	O4	O5	O6	O7	O8	O9
HTH15	HTH16	HTH17	HTH18	HTH19	HTH20	HTH21
MAS23	MAS24	MAS25	MAS26	MAS27	MAS28	MAS29
8	9	10	11	12	13	14
O10	O11	O12	O13	O14	O15	O16
HTH22	HTH23	HTH24	HTH25	HTH26	HTH27	HTH28
MAS30	THi1	THi2	THi3	THi4	THi5	THi6
15	16	17	18	19	20	21
O17	O18	O19	O20	O21	O22	O23
HTH29	HTH30	KHK1	KHK2	KHK3	KHK4	KHK5
THi7	THi8	This9	THi10	THi11	THi12	THi13
22	23	24	25	26	27	28
O24	O25	O26	O27	O28	O29	O30
KHK6	KHK7	KHK8	KHK9	KHK10	KHK11	KHK12
Thi14	THi15	THi16	THi17	THi18	THi19	THi20

KIJANI 556/2072-73/6308-09/2064-65						
Ashe	Ojobo	Iwa	Araorun	Ile	Itutu	Erinle
1	2	3	4	5	6	7
N28	N29	N30	D1	D2	D3	D4
TBT11	TBT12	TBT13	TBT14	TBT15	TBT16	TBT17
HID19	HID20	HID21	HID22	HID23	HID24	HID25
8	9	10	11	12	13	14
D5	D6	D7	D8	D9	D10	D11
TBT18	TBT19	TBT20	TBT21	TBT22	TBT23	TBT24
HID26	HID27	HID28	HID29	HID30	TAH1	TAH2
15	16	17	18	19	20	21
D12	D13	D14	D15	D16	D17	D18
TBT25	TBT26	TBT27	TBT28	TBT29	TBT30	MKH1
TAH3	TAH4	TAH5	TAH6	TAH7	TAH8	TAH9
22	23	24	25	26	27	28
D19	D20	D21	D22	D23	D24	D25
MKH2	MKH3	MKH4	MKH5	MKH6	MKH7	MKH8
TAH10	TAH11	TAH12	TAH13	TAH14	TAH15	TAH16

EUSI 556/2072-73/6308-09/2064-65						
Ashe	Ojobo	Iwa	Araorun	Ile	Itutu	Erinle
1	2	3	4	5	6	7
O31	N1	N2	N3	N4	N5	N6
KHK13	KHK14	KHK15	KHK16	KHK17	KHK18	KHK19
THi21	THi22	THi23	THi24	THi25	THi26	THi27
8	9	10	11	12	13	14
N7	N8	N9	N10	N11	N12	N13
KHK20	KHK21	KHK22	KHK23	KHK24	KHK25	KHK26
THi28	THi29	THi30	HID1	HID2	HID3	HID4
15	16	17	18	19	20	21
N14	N15	N16	N17	N18	N19	N20
KHK27	KHK28	KHK29	KHK30	TBT1	TBT2	TBT3
HID5	HID6	HID7	HID8	HID9	HID10	HID11
22	23	24	25	26	27	28
N21	N22	N23	N24	N25	N26	N27
TBT4	TBT5	TBT6	TBT7	TBT8	TBT9	TBT10
HID12	HID13	HID14	HID15	HID16	HID17	HID18

KWANZA 556/2072-73/6308-09/2064-65						
Umoja	Kujichagulia	Ujima	Ujamaa	Nia	Kuumba	Imani
1	2	3	4	5	6	7
D26	D27	D28	D29	D30	D31	JA1
MKH9	MKH10	MKH11	MKH12	MKH13	MKH14	MKH15
TAH17	TAH18	TAH19	TAH20	TAH21	TAH22	TAH23
8	9	10	11	12	13	14
JA2	JA3	JA4	JA5	JA6	JA7	JA8
MKH16	MKH17	MKH18	MKH19	MKH20	MKH21	MKH22
TAH24	TAH25	TAH26	TAH27	TAH28	TAH29	TAH30
15	16	17	18	19	20	21
JA9	JA10	JA11	JA12	JA13	JA14	JA15
MKH23	MKH24	MKH25	KMH26	MKH27	MKH28	MKH29
TIR1	TIR2	TIR3	TIR4	TIR5	TIR6	TIR7
22	23	24	25	26	27	28
JA16	JA17	JA18	JA19	JA20	JA21	JA22
MKH30	AMN1	AMN2	AMN3	AMN4	AMN5	AMN6
TIR8	TIR9	TIR10	TIR11	TIR12	TIR13	TIR14

THE NU AFRIKAN CALENDAR

SHAHIDI 556/2072-73/6308-09/2064-65						
Ashe	Ojobo	Iwa	Araorun	Ile	Itutu	Erinle
1	2	3	4	5	6	7
JA23	JA24	JA25	JA26	JA27	JA28	JA29
AMN7	AMN8	AMN9	AMN10	AMN11	AMN12	AMN13
TIR15	TIR16	TIR17	TIR18	TIR19	TIR20	TIR21
8	9	10	11	12	13	14
JA30	JA31	F1	F2	F3	F4	F5
AMN14	AMN15	AMN16	AMN17	AMN18	AMN19	AMN20
TIR22	TIR23	TIR24	TIR25	TIR26	TIR27	TIR28
15	16	17	18	19	20	21
F6	F7	F8	F9	F10	F 11	F12
AMN21	AMN22	AMN23	AMN24	AMN25	AMN26	AMN28
TIR29	TIR30	YEK1	YEK2	YEK3	YEK4	YEK5
22	23	24	25	26	27	28
F13	F14	F15	F16	F17	F18	F19
AMN29	AMN30	RNT1	RNT2	RNT3	RNT4	RNT5
YEK6	YEK7	YEK8	YEK9	YEK10	YEK11	YEK12
SHAHIDI 107						
F20						
RNT6						
YEK13						

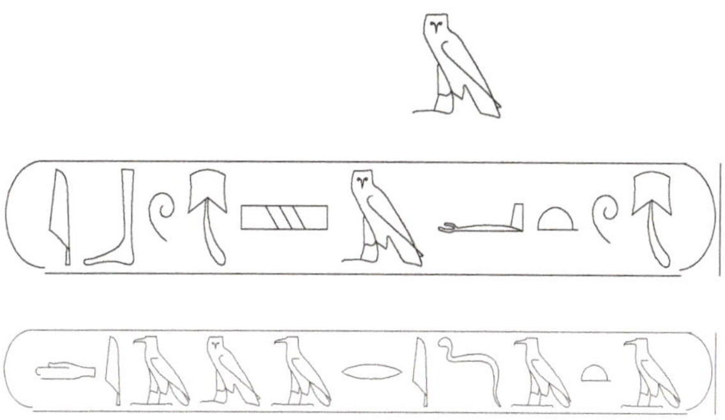

THE NU AFRIKAN CALENDAR

MAAFA 557 B.S.E./ASIENTO 2073-74 C.E.
KAM 6209-10 S.M./EC 2065-66 A.M. (ዓ.ም.)

AFU 557/2073-74/6309-10/2065-66

Ashe	Ojobo	Iwa	Araorun	Ile	Itutu	Erinle
1	2	3	4	5	6	7
F21	F22	F23	F24	F25	F26	F27
RNT7	RNT8	RNT9	RNT10	RNT11	RNT12	RNT13
YEK14	YEK15	YEK16	YEK17	YEK18	YEK19	YEK20
8	9	10	11	12	13	14
F28	M1	M2	M3	M4	M5	M6
RNT14	RNT15	RNT16	RNT17	RNT18	RNT19	RNT20
YEK21	YEK22	YEK23	YEK24	YEK25	YEK26	YEK27
15	16	17	18	19	20	21
M7	M8	M9	M10	M11	M12	M13
RNT21	RNT22	RNT23	RNT24	RNT25	RNT26	RNT27
YEK28	YEK29	YEK30	MAG1	MAG2	MAG3	MAG4
22	23	24	25	26	27	28
M14	M15	M16	M17	M18	M19	M20
RNT28	RNT29	RNT30	KHN1	KHN2	KHN3	KHN4
MAG5	MAG6	MAG7	MAG8	MAG9	MAG10	MAG11

KAWAIDA 557/2073-74/6309-10/2065-66

Ashe	Ojobo	Iwa	Araorun	Ile	Itutu	Erinle
1	2	3	4	5	6	7
A18	A19	A20	A21	A22	A23	A24
PIN3	PIN4	PIN5	PIN6	PIN7	PIN8	PIN9
MIY10	MIY11	MIY12	MIY13	MIY14	MIY15	MIY16
8	9	10	11	12	13	14
A25	A26	A27	A28	A29	A30	MA1
PIN10	PIN11	PIN12	PIN13	PIN14	PIN15	PIN16
MIY17	MIY18	MIY19	MIY20	MIY21	MIY22	MIY23
15	16	17	18	19	20	21
MA2	MA3	MA4	MA5	MA6	MA7	MA8
PIN17	PIN18	PIN19	PIN20	PIN21	PIN22	PIN23
MIY24	MIY25	MIY26	MIY27	MIY28	MIY29	MIY30
22	23	24	25	26	27	28
MA9	MA10	MA11	MA12	MA13	MA14	MA15
PIN24	PIN25	PIN26	PIN27	PIN28	PIN29	PIN30
GIN1	GIN2	GIN3	GIN4	GIN5	GIN6	GIN7

UHURU 557/2073-74/6309-10/2065-66

Ashe	Ojobo	Iwa	Araorun	Ile	Itutu	Erinle
1	2	3	4	5	6	7
M21	M22	M23	M24	M25	M26	M27
KHN5	KHN6	KHN7	KHN8	KHN9	KHN10	KHN11
MAG12	MAG13	MAG14	MAG15	MAG16	MAG17	MAG18
8	9	10	11	12	13	14
M28	M29	M30	M31	A1	A2	A3
KHN12	KHN13	KHN14	KHN15	KHN16	KHN17	KHN18
MAG19	MAG20	MAG21	MAG22	MAG23	MAG24	MAG25
15	16	17	18	19	20	21
A4	A5	A6	A7	A8	A9	A10
KHN19	KHN20	KHN21	KHN22	KHN23	KHN24	KHN25
MAG26	MAG27	MAG28	MAG29	MAG30	MIY1	MIY2
22	23	24	25	26	27	28
A11	A12	A13	A14	A15	A16	A17
KHN26	KHN27	KHN28	KHN29	KHN30	PIN1	PIN2
MIY3	MIY4	MIY5	MIY6	MIY7	MIY8	MIY9

MABABA WEUSI 557/2073-74/6309-10/2065-66

Ashe	Ojobo	Iwa	Araorun	Ile	Itutu	Erinle
1	2	3	4	5	6	7
MA16	MA17	MA18	MA19	MA20	MA21	MA22
APP1	APP2	APP3	APP4	APP5	APP6	APP7
GIN8	GIN9	GIN10	GIN11	GIN12	GIN13	GIN14
8	9	10	11	12	13	14
MA23	MA24	MA25	MA26	MA27	MA28	MA29
APP8	APP9	APP10	APP11	APP12	APP13	APP14
GIN15	GIN16	GIN17	GIN18	GIN19	GIN20	GIN21
15	16	17	18	19	20	21
MA30	MA31	J1	J2	J3	J4	J5
APP15	APP16	APP17	APP18	APP19	APP20	APP21
GIN22	GIN23	GIN24	GIN25	GIN26	GIN27	GIN28
22	23	24	25	26	27	28
J6	J7	J8	J9	J10	J11	J12
APP22	APP23	APP24	APP25	APP26	APP27	APP28
GIN29	GIN30	SAN1	SAN2	SAN3	SAN4	SAN5

THE NU AFRIKAN CALENDAR

MWANAMKE WEUSI 557/2073-74/6309-10/2065-66

Ashe	Ojobo	Iwa	Araorun	Ile	Itutu	Erinle
1	2	3	4	5	6	7
J13	J14	J15	J16	J17	J18	J19
APP29	APP30	MSR1	MSR2	MSR3	MSR4	MSR5
SAN6	SAN7	SAN8	SAN9	SAN10	SAN11	SAN12
8	9	10	11	12	13	14
J20	J21	J22	J23	J24	J25	J26
MSR6	MSR7	MSR8	MSR9	MSR10	MSR11	MSR12
SAN13	SAN14	SAN15	SAN16	SAN17	SAN18	SAN19
15	16	17	18	19	20	21
J27	J28	J29	J30	JU1	JU2	JU3
MSR13	MSR14	MSR15	MSR16	MSR17	MSR18	MSR19
SAN20	SAN21	SAN22	SAN23	SAN24	SAN25	SAN26
22	23	24	25	26	27	28
JU4	JU5	JU6	JU7	JU8	JU9	JU10
MSR20	MRS21	MSR22	MSR23	MSR24	MSR25	MSR26
SAN27	SAN28	SAN29	SAN30	HAM1	HAM2	HAM3

MTU WEUSI 557/2073-74/6309-10/2065-66

Ashe	Ojobo	Iwa	Araorun	Ile	Itutu	Erinle
1	2	3	4	5	6	7
JU11	JU12	JU13	JU14	JU15	JU16	JU17
MSR27	MSR28	MSR29	MSR30	*ASR1*	*HRU2*	*STH3*
HAM4	HAM5	HAM6	HAM7	HAM8	HAM9	HAM10
8	9	10	11	12	13	14
JU18	JU19	JU20	JU21	JU22	JU23	JU24
AST4	*NBH5*	THT1	THT2	THT3	THT4	THT5
HAM11	HAM12	HAM13	AM14	HAM15	HAM16	HAM17
15	16	17	18	19	20	21
JU25	JU26	JU27	JU28	JU29	JU30	JU31
THT6	THT7	THT8	THT9	THT10	THT11	THT12
HAM18	HAM19	HAM20	HAM21	HAM22	HAM23	HAM24
22	23	24	25	26	27	28
AU1	AU2	AU3	AU4	AU5	AU6	AU7
THT13	THT14	THT15	THT16	THT17	THT18	THT19
HAM25	HAM26	HAM27	HAM28	HAM29	HAM30	NAH1

TAIFA WEUSI 557/2073-74/6309-10/2065-66

Ashe	Ojobo	Iwa	Araorun	Ile	Itutu	Erinle
1	2	3	4	5	6	7
AU8	AU9	AU10	AU11	AU12	AU13	AU14
THT20	THT21	THT22	THT23	THT24	THT25	THT26
NAH2	NAH3	NAH4	NAH5	NAH6	NAH7	NAH8
8	9	10	11	12	13	14
AU15	AU16	AU17	AU18	AU19	AU20	AU21
THT27	THT28	THT29	THT30	PAP1	PAP1	PAP2
NAH9	NAH10	NAH11	NAH12	NAH13	NAH14	NAH15
15	16	17	18	19	20	21
AU22	AU23	AU24	AU25	AU26	AU27	AU28
PAP3	PAP4	PAP5	PAP6	PAP7	PAP8	PAP9
NAH16	NAH17	NAH18	NAH19	NAH20	NAH21	NAH22
22	23	24	25	26	27	28
AU29	AU30	AU31	S1	S2	S3	S4
PAP10	PAP11	PAP12	PAP13	PAP14	PAP15	PAP16
NAH23	NAH24	NAH25	NAH26	NAH27	NAH28	NAH29

UWEZO 557/2073-74/6309-10/2065-66

Ashe	Ojobo	Iwa	Araorun	Ile	Itutu	Erinle
1	2	3	4	5	6	7
S5	S6	S7	S8	S9	S10	S11
PAP17	PAP18	PAP19	PAP20	PAP21	PAP22	PAP23
NAH30	*PAG1*	*PAG2*	*PAG3*	*PAG4*	*PAG5*	MAS1
8	9	10	11	12	13	14
S12	S13	S14	S15	S16	S17	S18
PAP24	PAP25	PAP26	PAP27	PAP28	PAP29	PAP30
MAS2	MAS3	MAS4	MAS5	MAS6	MAS7	MAS8
15	16	17	18	19	20	21
S19	S20	S21	S22	S23	S24	S25
HTH1	HTH2	HTH3	HTH4	HTH5	HTH6	HTH7
MAS9	MAS10	MAS11	MAS12	MAS13	MAS14	MAS15
22	23	24	25	26	27	28
S26	S27	S28	S29	S30	O1	O2
HTH8	HTH9	HTH10	HTH11	HTH12	HTH13	HTH14
MAS16	MAS17	MAS18	MAS19	MAS20	MAS21	MAS22

THE NU AFRIKAN CALENDAR

EKUNDU 557/2073-74/6309-10/2065-66						
Ashe	Ojobo	Iwa	Araorun	Ile	Itutu	Erinle
1	2	3	4	5	6	7
O3	O4	O5	O6	O7	O8	O9
HTH15	HTH16	HTH17	HTH18	HTH19	HTH20	HTH21
MAS23	MAS24	MAS25	MAS26	MAS27	MAS28	MAS29
8	9	10	11	12	13	14
O10	O11	O12	O13	O14	O15	O16
HTH22	HTH23	HTH24	HTH25	HTH26	HTH27	HTH28
MAS30	THi1	THi2	THi3	THi4	THi5	THi6
15	16	17	18	19	20	21
O17	O18	O19	O20	O21	O22	O23
HTH29	HTH30	KHK1	KHK2	KHK3	KHK4	KHK5
THi7	THi8	THi9	THi10	THi11	THi12	THi13
22	23	24	25	26	27	28
O24	O25	O26	O27	O28	O29	O30
KHK6	KHK7	KHK8	KHK9	KHK10	KHK11	KHK12
THi14	THi15	THi16	THi17	THi18	THi19	THi20

KIJANI 557/2073-74/6309-10/2065-66						
Ashe	Ojobo	Iwa	Araorun	Ile	Itutu	Erinle
1	2	3	4	5	6	7
N28	N29	N30	D1	D2	D3	D4
TBT11	TBT12	TBT13	TBT14	TBT15	TBT16	TBT17
HID19	HID20	HID21	HID22	HID23	HID24	HID25
8	9	10	11	12	13	14
D5	D6	D7	D8	D9	D10	D11
TBT18	TBT19	TBT20	TBT21	TBT22	TBT23	TBT24
HID26	HID27	HID28	HID29	HID30	TAH1	TAH2
15	16	17	18	19	20	21
D12	D13	D14	D15	D16	D17	D18
TBT25	TBT26	TBT27	TBT28	TBT29	TBT30	MKH1
TAH3	TAH4	TAH5	TAH6	TAH7	TAH8	TAH9
22	23	24	25	26	27	28
D19	D20	D21	D22	D23	D24	D25
MKH2	MKH3	MKH4	MKH5	MKH6	MKH7	MKH8
TAH10	TAH11	TAH12	TAH13	TAH14	TAH15	TAH16

EUSI 557/2073-74/6309-10/2065-66						
Ashe	Ojobo	Iwa	Araorun	Ile	Itutu	Erinle
1	2	3	4	5	6	7
O31	N1	N2	N3	N4	N5	N6
KHK13	KHK14	KHK15	KHK16	KHK17	KHK18	KHK19
THi21	THi22	THi23	THi24	THi25	THi26	THi27
8	9	10	11	12	13	14
N7	N8	N9	N10	N11	N12	N13
KHK20	KHK21	KHK22	KHK23	KHK24	KHK25	KHK26
THi28	THi29	THi30	HID1	HID2	HID3	HID4
15	16	17	18	19	20	21
N14	N15	N16	N17	N18	N19	N20
KHK27	KHK28	KHK29	KHK30	TBT1	TBT2	TBT3
HID5	HID6	HID7	HID8	HID9	HID10	HID11
22	23	24	25	26	27	28
N21	N22	N23	N24	N25	N26	N27
TBT4	TBT5	TBT6	TBT7	TBT8	TBT9	TBT10
HID12	HID13	HID14	HID15	HID16	HID17	HID18

KWANZA 557/2073-74/6309-10/2065-66						
Umoja	Kujichagulia	Ujima	Ujamaa	Nia	Kuumba	Imani
1	2	3	4	5	6	7
D26	D27	D28	D29	D30	D31	JA1
MKH9	MKH10	MKH11	MKH12	MKH13	MKH14	MKH15
TAH17	TAH18	TAH19	TAH20	TAH21	TAH22	TAH23
8	9	10	11	12	13	14
JA2	JA3	JA4	JA5	JA6	JA7	JA8
MKH16	MKH17	MKH18	MKH19	MKH20	MKH21	MKH22
TAH24	TAH25	TAH26	TAH27	TAH28	TAH29	TAH30
15	16	17	18	19	20	21
JA9	JA10	JA11	JA12	JA13	JA14	JA15
MKH23	MKH24	MKH25	KMH26	MKH27	MKH28	MKH29
TIR1	TIR2	TIR3	TIR4	TIR5	TIR6	TIR7
22	23	24	25	26	27	28
JA16	JA17	JA18	JA19	JA20	JA21	JA22
MKH30	AMN1	AMN2	AMN3	AMN4	AMN5	AMN6
TIR8	TIR9	TIR10	TIR11	TIR12	TIR13	TIR14

THE NU AFRIKAN CALENDAR

SHAHIDI 557/2073-74/6309-10/2065-66						
Ashe	Ojobo	Iwa	Araorun	Ile	Itutu	Erinle
1	2	3	4	5	6	7
JA23	JA24	JA25	JA26	JA27	JA28	JA29
AMN7	AMN8	AMN9	AMN10	AMN11	AMN12	AMN13
TIR15	TIR16	TIR17	TIR18	TIR19	TIR20	TIR21
8	9	10	11	12	13	14
JA30	JA31	F1	F2	F3	F4	F5
AMN14	AMN15	AMN16	AMN17	AMN18	AMN19	AMN20
TIR22	TIR23	TIR24	TIR25	TIR26	TIR27	TIR28
15	16	17	18	19	20	21
F6	F7	F8	F9	F10	F 11	F12
AMN21	AMN22	AMN23	AMN24	AMN25	AMN26	AMN28
TIR29	TIR30	YEK1	YEK2	YEK3	YEK4	YEK5
22	23	24	25	26	27	28
F13	F14	F15	F16	F17	F18	F19
AMN29	AMN30	RNT1	RNT2	RNT3	RNT4	RNT5
YEK6	YEK7	YEK8	YEK9	YEK10	YEK11	YEK12
SHAHIDI 108						
F20						
RNT6						
YEK13						

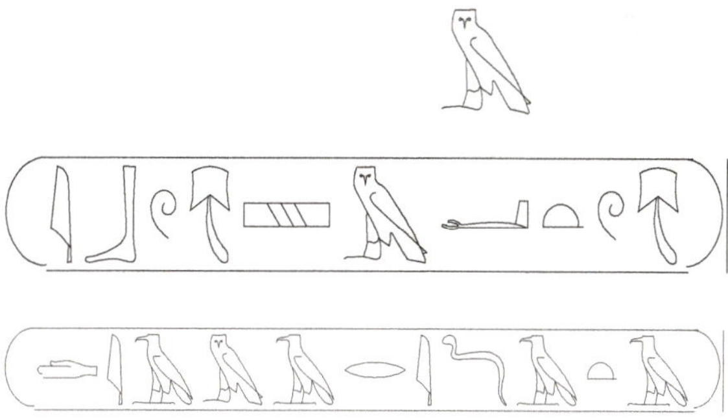

THE NU AFRIKAN CALENDAR

MAAFA 558 B.S.E./ASIENTO 2074-75 C.E.
KAM 6210-11 S.M./EC 2066-67 A.M. (ዓ.ም.)

AFU 558/2074-75/6310-11/2066-67

Ashe	Ojobo	Iwa	Araorun	Ile	Itutu	Erinle
1	2	3	4	5	6	7
F21	F22	F23	F24	F25	F26	F27
RNT7	RNT8	RNT9	RNT10	RNT11	RNT12	RNT13
YEK14	YEK15	YEK16	YEK17	YEK18	YEK19	YEK20
8	9	10	11	12	13	14
F28	M1	M2	M3	M4	M5	M6
RNT14	RNT15	RNT16	RNT17	RNT18	RNT19	RNT20
YEK21	YEK22	YEK23	YEK24	YEK25	YEK26	YEK27
15	16	17	18	19	20	21
M7	M8	M9	M10	M11	M12	M13
RNT21	RNT22	RNT23	RNT24	RNT25	RNT26	RNT27
YEK28	YEK29	YEK30	MAG1	MAG2	MAG3	MAG4
22	23	24	25	26	27	28
M14	M15	M16	M17	M18	M19	M20
RNT28	RNT29	RNT30	KHN1	KHN2	KHN3	KHN4
MAG5	MAG6	MAG7	MAG8	MAG9	MAG10	MAG11

KAWAIDA 558/2074-75/6310-11/2066-67

Ashe	Ojobo	Iwa	Araorun	Ile	Itutu	Erinle
1	2	3	4	5	6	7
A18	A19	A20	A21	A22	A23	A24
PIN3	PIN4	PIN5	PIN6	PIN7	PIN8	PIN9
MIY10	MIY11	MIY12	MIY13	MIY14	MIY15	MIY16
8	9	10	11	12	13	14
A25	A26	A27	A28	A29	A30	MA1
PIN10	PIN11	PIN12	PIN13	PIN14	PIN15	PIN16
MIY17	MIY18	MIY19	MIY20	MIY21	MIY22	MIY23
15	16	17	18	19	20	21
MA2	MA3	MA4	MA5	MA6	MA7	MA8
PIN17	PIN18	PIN19	PIN20	PIN21	PIN22	PIN23
MIY24	MIY25	MIY26	MIY27	MIY28	MIY29	MIY30
22	23	24	25	26	27	28
MA9	MA10	MA11	MA12	MA13	MA14	MA15
PIN24	PIN25	PIN26	PIN27	PIN28	PIN29	PIN30
GIN1	GIN2	GIN3	GIN4	GIN5	GIN6	GIN7

UHURU 558/2074-75/6310-11/2066-67

Ashe	Ojobo	Iwa	Araorun	Ile	Itutu	Erinle
1	2	3	4	5	6	7
M21	M22	M23	M24	M25	M26	M27
KHN5	KHN6	KHN7	KHN8	KHN9	KHN10	KHN11
MAG12	MAG13	MAG14	MAG15	MAG16	MAG17	MAG18
8	9	10	11	12	13	14
M28	M29	M30	M31	A1	A2	A3
KHN12	KHN13	KHN14	KHN15	KHN16	KHN17	KHN18
MAG19	MAG20	MAG21	MAG22	MAG23	MAG24	MAG25
15	16	17	18	19	20	21
A4	A5	A6	A7	A8	A9	A10
KHN19	KHN20	KHN21	KHN22	KHN23	KHN24	KHN25
MAG26	MAG27	MAG28	MAG29	MAG30	MIY1	MIY2
22	23	24	25	26	27	28
A11	A12	A13	A14	A15	A16	A17
KHN26	KHN27	KHN28	KHN29	KHN30	PIN1	PIN2
MIY3	MIY4	MIY5	MIY6	MIY7	MIY8	MIY9

MABABA WEUSI 558/2074-75/6310-11/2066-67

Ashe	Ojobo	Iwa	Araorun	Ile	Itutu	Erinle
1	2	3	4	5	6	7
MA16	MA17	MA18	MA19	MA20	MA21	MA22
APP1	APP2	APP3	APP4	APP5	APP6	APP7
GIN8	GIN9	GIN10	GIN11	GIN12	GIN13	GIN14
8	9	10	11	12	13	14
MA23	MA24	MA25	MA26	MA27	MA28	MA29
APP8	APP9	APP10	APP11	APP12	APP13	APP14
GIN15	GIN16	GIN17	GIN18	GIN19	GIN20	GIN21
15	16	17	18	19	20	21
MA30	MA31	J1	J2	J3	J4	J5
APP15	APP16	APP17	APP18	APP19	APP20	APP21
GIN22	GIN23	GIN24	GIN25	GIN26	GIN27	GIN28
22	23	24	25	26	27	28
J6	J7	J8	J9	J10	J11	J12
APP22	APP23	APP24	APP25	APP26	APP27	APP28
GIN29	GIN30	SAN1	SAN2	SAN3	SAN4	SAN5

MWANAMKE WEUSI 558/2074-75/6310-11/2066-67						
Ashe	Ojobo	Iwa	Araorun	Ile	Itutu	Erinle
1	2	3	4	5	6	7
J13	J14	J15	J16	J17	J18	J19
APP29	APP30	MSR1	MSR2	MSR3	MSR4	MSR5
SAN6	SAN7	SAN8	SAN9	SAN10	SAN11	SAN12
8	9	10	11	12	13	14
J20	J21	J22	J23	J24	J25	J26
MSR6	MSR7	MSR8	MSR9	MSR10	MSR11	MSR12
SAN13	SAN14	SAN15	SAN16	SAN17	SAN18	SAN19
15	16	17	18	19	20	21
J27	J28	J29	J30	JU1	JU2	JU3
MSR13	MSR14	MSR15	MSR16	MSR17	MSR18	MSR19
SAN20	SAN21	SAN22	SAN23	SAN24	SAN25	SAN26
22	23	24	25	26	27	28
JU4	JU5	JU6	JU7	JU8	JU9	JU10
MSR20	MRS21	MSR22	MSR23	MSR24	MSR25	MSR26
SAN27	SAN28	SAN29	SAN30	HAM1	HAM2	HAM3

TAIFA WEUSI 558/2074-75/6310-11/2066-67						
Ashe	Ojobo	Iwa	Araorun	Ile	Itutu	Erinle
1	2	3	4	5	6	7
AU8	AU9	AU10	AU11	AU12	AU13	AU14
THT20	THT21	THT22	THT23	THT24	THT25	THT26
NAH2	NAH3	NAH4	NAH5	NAH6	NAH7	NAH8
8	9	10	11	12	13	14
AU15	AU16	AU17	AU18	AU19	AU20	AU21
THT27	THT28	THT29	THT30	PAP1	PAP1	PAP2
NAH9	NAH10	NAH11	NAH12	NAH13	NAH14	NAH15
15	16	17	18	19	20	21
AU22	AU23	AU24	AU25	AU26	AU27	AU28
PAP3	PAP4	PAP5	PAP6	PAP7	PAP8	PAP9
NAH16	NAH17	NAH18	NAH19	NAH20	NAH21	NAH22
22	23	24	25	26	27	28
AU29	AU30	AU31	S1	S2	S3	S4
PAP10	PAP11	PAP12	PAP13	PAP14	PAP15	PAP16
NAH23	NAH24	NAH25	NAH26	NAH27	NAH28	NAH29

MTU WEUSI 558/2074-75/6310-11/2066-67						
Ashe	Ojobo	Iwa	Araorun	Ile	Itutu	Erinle
1	2	3	4	5	6	7
JU11	JU12	JU13	JU14	JU15	JU16	JU17
MSR27	MSR28	MSR29	MSR30	*ASR1*	*HRU2*	*STH3*
HAM4	HAM5	HAM6	HAM7	HAM8	HAM9	HAM10
8	9	10	11	12	13	14
JU18	JU19	JU20	JU21	JU22	JU23	JU24
AST4	*NBH5*	THT1	THT2	THT3	THT4	THT5
HAM11	HAM12	HAM13	AM14	HAM15	HAM16	HAM17
15	16	17	18	19	20	21
JU25	JU26	JU27	JU28	JU29	JU30	JU31
THT6	THT7	THT8	THT9	THT10	THT11	THT12
HAM18	HAM19	HAM20	HAM21	HAM22	HAM23	HAM24
22	23	24	25	26	27	28
AU1	AU2	AU3	AU4	AU5	AU6	AU7
THT13	THT14	THT15	THT16	THT17	THT18	THT19
HAM25	HAM26	HAM27	HAM28	HAM29	HAM30	NAH1

UWEZO 558/2074-75/6310-11/2066-67						
Ashe	Ojobo	Iwa	Araorun	Ile	Itutu	Erinle
1	2	3	4	5	6	7
S5	S6	S7	S8	S9	S10	S11
PAP17	PAP18	PAP19	PAP20	PAP21	PAP22	PAP23
NAH30	*PAG1*	*PAG2*	*PAG3*	*PAG4*	*PAG5*	MAS1
8	9	10	11	12	13	14
S12	S13	S14	S15	S16	S17	S18
PAP24	PAP25	PAP26	PAP27	PAP28	PAP29	PAP30
MAS2	MAS3	MAS4	MAS5	MAS6	MAS7	MAS8
15	16	17	18	19	20	21
S19	S20	S21	S22	S23	S24	S25
HTH1	HTH2	HTH3	HTH4	HTH5	HTH6	HTH7
MAS9	MAS10	MAS11	MAS12	MAS13	MAS14	MAS15
22	23	24	25	26	27	28
S26	S27	S28	S29	S30	O1	O2
HTH8	HTH9	HTH10	HTH11	HTH12	HTH13	HTH14
MAS16	MAS17	MAS18	MAS19	MAS20	MAS21	MAS22

THE NU AFRIKAN CALENDAR

EKUNDU 558/2074-75/6310-11/2066-67						
Ashe	Ojobo	Iwa	Araorun	Ile	Itutu	Erinle
1	2	3	4	5	6	7
O3	O4	O5	O6	O7	O8	O9
HTH15	HTH16	HTH17	HTH18	HTH19	HTH20	HTH21
MAS23	MAS24	MAS25	MAS26	MAS27	MAS28	MAS29
8	9	10	11	12	13	14
O10	O11	O12	O13	O14	O15	O16
HTH22	HTH23	HTH24	HTH25	HTH26	HTH27	HTH28
MAS30	THi1	THi2	THi3	THi4	THi5	THi6
15	16	17	18	19	20	21
O17	O18	O19	O20	O21	O22	O23
HTH29	HTH30	KHK1	KHK2	KHK3	KHK4	KHK5
THi7	THi8	THi9	THi10	THi11	THi12	THi13
22	23	24	25	26	27	28
O24	O25	O26	O27	O28	O29	O30
KHK6	KHK7	KHK8	KHK9	KHK10	KHK11	KHK12
THi14	THi15	THi16	THi17	THi18	THi19	THi20

KIJANI 558/2074-75/6310-11/2066-67						
Ashe	Ojobo	Iwa	Araorun	Ile	Itutu	Erinle
1	2	3	4	5	6	7
N28	N29	N30	D1	D2	D3	D4
TBT11	TBT12	TBT13	TBT14	TBT15	TBT16	TBT17
HID19	HID20	HID21	HID22	HID23	HID24	HID25
8	9	10	11	12	13	14
D5	D6	D7	D8	D9	D10	D11
TBT18	TBT19	TBT20	TBT21	TBT22	TBT23	TBT24
HID26	HID27	HID28	HID29	HID30	TAH1	TAH2
15	16	17	18	19	20	21
D12	D13	D14	D15	D16	D17	D18
TBT25	TBT26	TBT27	TBT28	TBT29	TBT30	MKH1
TAH3	TAH4	TAH5	TAH6	TAH7	TAH8	TAH9
22	23	24	25	26	27	28
D19	D20	D21	D22	D23	D24	D25
MKH2	MKH3	MKH4	MKH5	MKH6	MKH7	MKH8
TAH10	TAH11	TAH12	TAH13	TAH14	TAH15	TAH16

EUSI 558/2074-75/6310-11/2066-67						
Ashe	Ojobo	Iwa	Araorun	Ile	Itutu	Erinle
1	2	3	4	5	6	7
O31	N1	N2	N3	N4	N5	N6
KHK13	KHK14	KHK15	KHK16	KHK17	KHK18	KHK19
THi21	THi22	THi23	THi24	THi25	THi26	THi27
8	9	10	11	12	13	14
N7	N8	N9	N10	N11	N12	N13
KHK20	KHK21	KHK22	KHK23	KHK24	KHK25	KHK26
THi28	THi29	THi30	HID1	HID2	HID3	HID4
15	16	17	18	19	20	21
N14	N15	N16	N17	N18	N19	N20
KHK27	KHK28	KHK29	KHK30	TBT1	TBT2	TBT3
HID5	HID6	HID7	HID8	HID9	HID10	HID11
22	23	24	25	26	27	28
N21	N22	N23	N24	N25	N26	N27
TBT4	TBT5	TBT6	TBT7	TBT8	TBT9	TBT10
HID12	HID13	HID14	HID15	HID16	HID17	HID18

KWANZA 558/2074-75/6310-11/2066-67						
Umoja	Kujichagulia	Ujima	Ujamaa	Nia	Kuumba	Imani
1	2	3	4	5	6	7
D26	D27	D28	D29	D30	D31	JA1
MKH9	MKH10	MKH11	MKH12	MKH13	MKH14	MKH15
TAH17	TAH18	TAH19	TAH20	TAH21	TAH22	TAH23
8	9	10	11	12	13	14
JA2	JA3	JA4	JA5	JA6	JA7	JA8
MKH16	MKH17	MKH18	MKH19	MKH20	MKH21	MKH22
TAH24	TAH25	TAH26	TAH27	TAH28	TAH29	TAH30
15	16	17	18	19	20	21
JA9	JA10	JA11	JA12	JA13	JA14	JA15
MKH23	MKH24	MKH25	KMH26	MKH27	MKH28	MKH29
TIR1	TIR2	TIR3	TIR4	TIR5	TIR6	TIR7
22	23	24	25	26	27	28
JA16	JA17	JA18	JA19	JA20	JA21	JA22
MKH30	AMN1	AMN2	AMN3	AMN4	AMN5	AMN6
TIR8	TIR9	TIR10	TIR11	TIR12	TIR13	TIR14

SHAHIDI 558/2074-75/6310-11/2066-67						
Ashe	Ojobo	Iwa	Araorun	Ile	Itutu	Erinle
1	2	3	4	5	6	7
JA23	JA24	JA25	JA26	JA27	JA28	JA29
AMN7	AMN8	AMN9	AMN10	AMN11	AMN12	AMN13
TIR15	TIR16	TIR17	TIR18	TIR19	TIR20	TIR21
8	9	10	11	12	13	14
JA30	JA31	F1	F2	F3	F4	F5
AMN14	AMN15	AMN16	AMN17	AMN18	AMN19	AMN20
TIR22	TIR23	TIR24	TIR25	TIR26	TIR27	TIR28
15	16	17	18	19	20	21
F6	F7	F8	F9	F10	F 11	F12
AMN21	AMN22	AMN23	AMN24	AMN25	AMN26	AMN28
TIR29	TIR30	YEK1	YEK2	YEK3	YEK4	YEK5
22	23	24	25	26	27	28
F13	F14	F15	F16	F17	F18	F19
AMN29	AMN30	RNT1	RNT2	RNT3	RNT4	RNT5
YEK6	YEK7	YEK8	YEK9	YEK10	YEK11	YEK12
SHAHIDI 109						
F20						
RNT6						
YEK13						

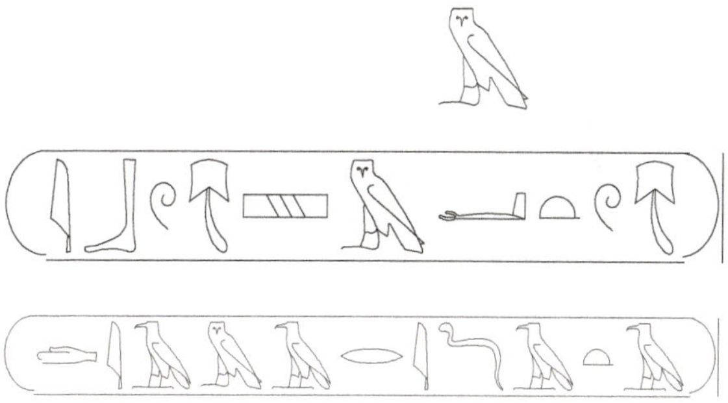

MAAFA 559 B.S.E./ASIENTO 2075-76 C.E./
KAM 6311-12 S.M./EC 2067-68 A.M. (ዓ.ም.)

AFU 559/2075-76/6311-12/2067-68

Ashe	Ojobo	Iwa	Araorun	Ile	Itutu	Erinle
1	2	3	4	5	6	7
F21	F22	F23	F24	F25	F26	F27
RNT7	RNT8	RNT9	RNT10	RNT11	RNT12	RNT13
YEK14	YEK15	YEK16	YEK17	YEK18	YEK19	YEK20
8	9	10	11	12	13	14
F28	M1	M2	M3	M4	M5	M6
RNT14	RNT15	RNT16	RNT17	RNT18	RNT19	RNT20
YEK21	YEK22	YEK23	YEK24	YEK25	YEK26	YEK27
15	16	17	18	19	20	21
M7	M8	M9	M10	M11	M12	M13
RNT21	RNT22	RNT23	RNT24	RNT25	RNT26	RNT27
YEK28	YEK 29	YEK30	MAG1	MAG2	MAG3	MAG4
22	23	24	25	26	27	28
M14	M15	M16	M17	M18	M19	M20
RNT28	RNT29	RNT30	KHN1	KHN2	KHN3	KHN4
MAG5	MAG6	MAG7	MAG8	MAG9	MAG10	MAG11

UHURU 559/2075-76/6311-12/2067-68

Ashe	Ojobo	Iwa	Araorun	Ile	Itutu	Erinle
1	2	3	4	5	6	7
M21	M22	M23	M24	M25	M26	M27
KHN5	KHN6	KHN7	KHN8	KHN9	KHN10	KHN11
MAG12	MAG13	MAG14	MAG15	MAG16	MAG17	MAG18
8	9	10	11	12	13	14
M28	M29	M30	M31	A1	A2	A3
KHN12	KHN13	KHN14	KHN15	KHN16	KHN17	KHN18
MAG19	MAG20	MAG21	MAG22	MAG23	MAG24	MAG25
15	16	17	18	19	20	21
A4	A5	A6	A7	A8	A9	A10
KHN19	KHN20	KHN21	KHN22	KHN23	KHN24	KHN25
MAG26	MAG27	MAG28	MAG29	MAG30	MIY1	MIY2
22	23	24	25	26	27	28
A11	A12	A13	A14	A15	A16	A17
KHN26	KHN27	KHN28	KHN29	KHN30	PIN1	PIN2
MIY3	MIY4	MIY5	MIY6	MIY7	MIY8	MIY9

KAWAIDA 559/2075-76/6311-12/2067-68

Ashe	Ojobo	Iwa	Araorun	Ile	Itutu	Erinle
1	2	3	4	5	6	7
A18	A19	A20	A21	A22	A23	A24
PIN3	PIN4	PIN5	PIN6	PIN7	PIN8	PIN9
MIY10	MIY11	MIY12	MIY13	MIY14	MIY15	MIY16
8	9	10	11	12	13	14
A25	A26	A27	A28	A29	A30	MA1
PIN10	PIN11	PIN12	PIN13	PIN14	PIN15	PIN16
MIY17	MIY18	MIY19	MIY20	MIY21	MIY22	MIY23
15	16	17	18	19	20	21
MA2	MA3	MA4	MA5	MA6	MA7	MA8
PIN17	PIN18	PIN19	PIN20	PIN21	PIN22	PIN23
MIY24	MIY25	MIY26	MIY27	MIY28	MIY29	MIY30
22	23	24	25	26	27	28
MA9	MA10	MA11	MA12	MA13	MA14	MA15
PIN24	PIN25	PIN26	PIN27	PIN28	PIN29	PIN30
GIN1	GIN2	GIN3	GIN4	GIN5	GIN6	GIN7

MABABA WEUSI 559/2075-76/6311-12/2067-68

Ashe	Ojobo	Iwa	Araorun	Ile	Itutu	Erinle
1	2	3	4	5	6	7
MA16	MA17	MA18	MA19	MA20	MA21	MA22
APP1	APP2	APP3	APP4	APP5	APP6	APP7
GIN8	GIN9	GIN10	GIN11	GIN12	GIN13	GIN14
8	9	10	11	12	13	14
MA23	MA24	MA25	MA26	MA27	MA28	MA29
APP8	APP9	APP10	APP11	APP12	APP13	APP14
GIN15	GIN16	GIN17	GIN18	GIN19	GIN20	GIN21
15	16	17	18	19	20	21
MA30	MA31	J1	J2	J3	J4	J5
APP15	APP16	APP17	APP18	APP19	APP20	APP21
GIN22	GIN23	GIN24	GIN25	GIN26	GIN27	GIN28
22	23	24	25	26	27	28
J6	J7	J8	J9	J10	J11	J12
APP22	APP23	APP24	APP25	APP26	APP27	APP28
GIN29	GIN30	SAN1	SAN2	SAN3	SAN4	SAN5

THE NU AFRIKAN CALENDAR

MWANAMKE WEUSI 559/2075-76/6311-12/2067-68						
Ashe	Ojobo	Iwa	Araorun	Ile	Itutu	Erinle
1	2	3	4	5	6	7
J13	J14	J15	J16	J17	J18	J19
APP29	APP30	MSR1	MSR2	MSR3	MSR4	MSR5
SAN6	SAN7	SAN8	SAN9	SAN10	SAN11	SAN12
8	9	10	11	12	13	14
J20	J21	J22	J23	J24	J25	J26
MSR6	MSR7	MSR8	MSR9	MSR10	MSR11	MSR12
SAN13	SAN14	SAN15	SAN16	SAN17	SAN18	SAN19
15	16	17	18	19	20	21
J27	J28	J29	J30	JU1	JU2	JU3
MSR13	MSR14	MSR15	MSR16	MSR17	MSR18	MSR19
SAN20	SAN21	SAN22	SAN23	SAN24	SAN25	SAN26
22	23	24	25	26	27	28
JU4	JU5	JU6	JU7	JU8	JU9	JU10
MSR20	MRS21	MSR22	MSR23	MSR24	MSR25	MSR26
SAN27	SAN28	SAN29	SAN30	HAM1	HAM2	HAM3

MTU WEUSI 559/2075-76/6311-12/2067-68						
Ashe	Ojobo	Iwa	Araorun	Ile	Itutu	Erinle
1	2	3	4	5	6	7
JU11	JU12	JU13	JU14	JU15	JU16	JU17
MSR27	MSR28	MSR29	MSR30	*ASR1*	*HRU2*	*STH3*
HAM4	HAM5	HAM6	HAM7	HAM8	HAM9	HAM10
8	9	10	11	12	13	14
JU18	JU19	JU20	JU21	JU22	JU23	JU24
AST4	*NBH5*	THT1	THT2	THT3	THT4	THT5
HAM11	HAM12	HAM13	HAM14	HAM15	HAM16	HAM17
15	16	17	18	19	20	21
JU25	JU26	JU27	JU28	JU29	JU30	JU31
THT6	THT7	THT8	THT9	THT10	THT11	THT12
HAM18	HAM19	HAM20	HAM21	HAM22	HAM23	HAM24
22	23	24	25	26	27	28
AU1	AU2	AU3	AU4	AU5	AU6	AU7
THT13	THT14	THT15	THT16	THT17	THT18	THT19
HAM25	HAM26	HAM27	HAM28	HAM29	HAM30	NAH1

TAIFA WEUSI 559/2075-76/6311-12/2067-68						
Ashe	Ojobo	Iwa	Araorun	Ile	Itutu	Erinle
1	2	3	4	5	6	7
AU8	AU9	AU10	AU11	AU12	AU13	AU14
THT20	THT21	THT22	THT23	THT24	THT25	THT26
NAH2	NAH3	NAH4	NAH5	NAH6	NAH7	NAH8
8	9	10	11	12	13	14
AU15	AU16	AU17	AU18	AU19	AU20	AU21
THT27	THT28	THT29	THT30	PAP1	PAP1	PAP2
NAH9	NAH10	NAH11	NAH12	NAH13	NAH14	NAH15
15	16	17	18	19	20	21
AU22	AU23	AU24	AU25	AU26	AU27	AU28
PAP3	PAP4	PAP5	PAP6	PAP7	PAP8	PAP9
NAH16	NAH17	NAH18	NAH19	NAH20	NAH21	NAH22
22	23	24	25	26	27	28
AU29	AU30	AU31	S1	S2	S3	S4
PAP10	PAP11	PAP12	PAP13	PAP14	PAP15	PAP16
NAH23	NAH24	NAH25	NAH26	NAH27	NAH28	NAH29

UWEZO 559/2075-76/6311-12/2067-68						
Ashe	Ojobo	Iwa	Araorun	Ile	Itutu	Erinle
1	2	3	4	5	6	7
S5	S6	S7	S8	S9	S10	S11
PAP17	PAP18	PAP19	PAP20	PAP21	PAP22	PAP23
NAH30	*QUA1*	*QUA2*	*QUA3*	*QUA4*	*QUA5*	*QUA6*
8	9	10	11	12	13	14
S12	S13	S14	S15	S16	S17	S18
PAP24	PAP25	PAP26	PAP27	PAP28	PAP29	PAP30
MAS1	MAS2	MAS3	MAS4	MAS5	MAS6	MAS7
15	16	17	18	19	20	21
S19	S20	S21	S22	S23	S24	S25
HTH1	HTH2	HTH3	HTH4	HTH5	HTH6	HTH7
MAS8	MAS9	MAS10	MAS11	MAS12	MAS13	MAS14
22	23	24	25	26	27	28
S26	S27	S28	S29	S30	O1	O2
HTH8	HTH9	HTH10	HTH11	HTH12	HTH13	HTH14
MAS15	MAS16	MAS17	MAS18	MAS19	MAS20	MAS21

THE NU AFRIKAN CALENDAR

EKUNDU 559/2075-76/6311-12/2067-68						
Ashe	Ojobo	Iwa	Araorun	Ile	Itutu	Erinle
1	2	3	4	5	6	7
O3	O4	O5	O6	O7	O8	O9
HTH15	HTH16	HTH17	HTH18	HTH19	HTH20	HTH21
MAS22	MAS23	MAS 24	MAS25	MAS26	MAS27	MAS28
8	9	10	11	12	13	14
O10	O11	O12	O13	O14	O15	O16
HTH22	HTH23	HTH24	HTH25	HTH26	HTH27	HTH28
MAS29	MAS30	THi1	THi2	THi3	THi4	THi5
15	16	17	18	19	20	21
O17	O18	O19	O20	O21	O22	O23
HTH29	HTH30	KHK1	KHK2	KHK3	KHK4	KHK5
THi6	THi7	THi8	THi9	THi10	THi11	THi12
22	23	24	25	26	27	28
O24	O25	O26	O27	O28	O29	O30
KHK6	KHK7	KHK8	KHK9	KHK10	KHK11	KHK12
THi13	THi14	THi15	THi16	THi17	THi18	THi19

KIJANI 559/2075-76/6311-12/2067-68						
Ashe	Ojobo	Iwa	Araorun	Ile	Itutu	Erinle
1	2	3	4	5	6	7
N28	N29	N30	D1	D2	D3	D4
TBT11	TBT12	TBT13	TBT14	TBT15	TBT16	TBT17
HID18	HID19	HID20	HID21	HID22	HID23	HID24
8	9	10	11	12	13	14
D5	D6	D7	D8	D9	D10	D11
TBT18	TBT19	TBT20	TBT21	TBT22	TBT23	TBT24
HID25	HID26	HID27	HID28	HID29	HID30	TAH1
15	16	17	18	19	20	21
D12	D13	D14	D15	D16	D17	D18
TBT25	TBT26	TBT27	TBT28	TBT29	TBT30	MKH1
TAH2	TAH3	TAH4	TAH5	TAH6	TAH7	TAH8
22	23	24	25	26	27	28
D19	D20	D21	D22	D23	D24	D25
MKH2	MKH3	MKH4	MKH5	MKH6	MKH7	MKH8
TAH9	TAH10	TAH11	TAH12	TAH13	TAH14	TAH15

EUSI 559/2075-76/6311-12/2067-68						
Ashe	Ojobo	Iwa	Araorun	Ile	Itutu	Erinle
1	2	3	4	5	6	7
O31	N1	N2	N3	N4	N5	N6
KHK13	KHK14	KHK15	KHK16	KHK17	KHK18	KHK19
THi20	THi21	THi22	THi23	THi24	THi25	THi26
8	9	10	11	12	13	14
N7	N8	N9	N10	N11	N12	N13
KHK20	KHK21	KHK22	KHK23	KHK24	KHK25	KHK26
THi27	THi28	THi29	THi30	HID1	HID2	HID3
15	16	17	18	19	20	21
N14	N15	N16	N17	N18	N19	N20
KHK27	KHK28	KHK29	KHK30	TBT1	TBT2	TBT3
HID4	HID5	HID6	HID7	HID8	HID9	HID10
22	23	24	25	26	27	28
N21	N22	N23	N24	N25	N26	N27
TBT4	TBT5	TBT6	TBT7	TBT8	TBT9	TBT10
HID11	HID12	HID13	HID14	HID15	HID16	HID17

KWANZA 559/2075-76/6311-12/2067-68						
Umoja	Kujichagulia	Ujima	Ujamaa	Nia	Kuumba	Imani
1	2	3	4	5	6	7
D26	D27	D28	D29	D30	D31	JA1
MKH9	MKH10	MKH11	MKH12	MKH13	MKH14	MKH1
TAH16	TAH17	TAH18	TAH19	TAH20	TAH21	TAH2
8	9	10	11	12	13	14
JA2	JA3	JA4	JA5	JA6	JA7	JA
MKH16	MKH17	MKH18	MKH19	MKH20	MKH21	MKH2
TAH23	TAH24	TAH25	TAH26	TAH27	TAH28	TAH29
15	16	17	18	19	20	21
JA9	JA10	JA11	JA12	JA13	JA14	JA1
MKH23	MKH24	MKH25	KMH26	MKH27	MKH28	MKH2
TAH30	TIR1	TIR2	TIR3	TIR4	TIR5	TIR6
22	23	24	25	26	27	28
JA16	JA17	JA18	JA19	JA20	JA21	JA2
MKH30	AMN1	AMN2	AMN3	AMN4	AMN5	AMN
TIR7	TIR8	TIR9	TIR10	TIR11	TIR12	TIR13

SHAHIDI 559/2075-76/6311-12/2067-68						
Ashe	Ojobo	Iwa	Araorun	Ile	Itutu	Erinle
1	2	3	4	5	6	7
JA23	JA24	JA25	JA26	JA27	JA28	JA29
AMN7	AMN8	AMN9	AMN10	AMN11	AMN12	AMN13
TIR14	TIR15	TIR16	TIR17	TIR18	TIR19	TIR20
8	9	10	11	12	13	14
JA30	JA31	F1	F2	F3	F4	F5
AMN14	AMN15	AMN16	AMN17	AMN18	AMN19	AMN20
TIR21	TIR22	TIR23	TIR24	TIR25	TIR26	TIR27
15	16	17	18	19	20	21
F6	F7	F8	F9	F10	F 11	F12
AMN21	AMN22	AMN23	AMN24	AMN25	AMN26	AMN28
TIR28	TIR29	TIR30	YEK1	YEK2	YEK3	YEK4
22	23	24	25	26	27	28
F13	F14	F15	F16	F17	F18	F19
AMN29	AMN30	RNT1	RNT2	RNT3	RNT4	RNT5
YEK5	YEK6	YEK7	YEK8	YEK9	YEK10	YEK11
SHAHIDI 110						
F20						
RNT6						
YEK12						

THE NU AFRIKAN CALENDAR

MAAFA 560 B.S.E./ASIENTO 2076-77 C.E./
KAM 6312-13 S.M./EC 2068-69 A.M. (ዓ.ም.)

AFU 560/2076-77/6312-13/2068-69

Ashe	Ojobo	Iwa	Araorun	Ile	Itutu	Erinle
1	2	3	4	5	6	7
F21	F22	F23	F24	F25	F26	F27
RNT7	RNT8	RNT9	RNT10	RNT11	RNT12	RNT13
YEK13	YEK14	YEK15	YEK16	YEK17	YEK18	YEK19
8	9	10	11	12	13	14
F28	F29	M1	M2	M3	M4	M5
RNT14	RNT15	RNT16	RNT17	RNT18	RNT19	RNT20
YEK 20	YEK21	YEK22	YEK23	YEK24	YEK25	YEK26
15	16	17	18	19	20	21
M6	M7	M8	M9	M10	M11	M12
RNT21	RNT22	RNT23	RNT24	RNT25	RNT26	RNT27
YEK 27	YEK28	YEK29	YEK30	MAG1	MAG2	MAG3
22	23	24	25	26	27	28
M13	M14	M15	M16	M17	M18	M19
RNT28	RNT29	RNT30	KHN1	KHN2	KHN3	KHN4
MAG4	MAG5	MAG6	MAG7	MAG8	MAG9	MAG10
SHARIFU 137						
M20						
KHN5						
MAG11						

KAWAIDA 560/2076-77/6312-13/2068-69

Ashe	Ojobo	Iwa	Araorun	Ile	Itutu	Erinle
1	2	3	4	5	6	7
A18	A19	A20	A21	A22	A23	A24
PIN4	PIN5	PIN6	PIN7	PIN8	PIN9	PIN10
MIY10	MIY11	MIY12	MIY13	MIY14	MIY15	MIY16
8	9	10	11	12	13	14
A25	A26	A27	A28	A29	A30	MA1
PIN11	PIN12	PIN13	PIN14	PIN15	PIN16	PIN17
MIY17	MIY18	MIY19	MIY20	MIY21	MIY22	MIY23
15	16	17	18	19	20	21
MA2	MA3	MA4	MA5	MA6	MA7	MA8
PIN18	PIN19	PIN20	PIN21	PIN22	PIN23	PIN24
MIY24	MIY25	MIY26	MIY27	MIY28	MIY29	MIY30
22	23	24	25	26	27	28
MA9	MA10	MA11	MA12	MA13	MA14	MA15
PIN25	PIN26	PIN27	PIN28	PIN29	PIN30	APP1
GIN1	GIN2	GIN3	GIN4	GIN5	GIN6	GIN7

UHURU 560/2076-77/6312-13/2068-69

Ashe	Ojobo	Iwa	Araorun	Ile	Itutu	Erinle
1	2	3	4	5	6	7
M21	M22	M23	M24	M25	M26	M27
KHN6	KHN7	KHN8	KHN9	KHN10	KHN11	KHN12
MAG12	MAG13	MAG14	MAG15	MAG16	MAG17	MAG18
8	9	10	11	12	13	14
M28	M29	M30	M31	A1	A2	A3
KHN13	KHN14	KHN15	KHN16	KHN17	KHN18	KHN19
MAG19	MAG20	MAG21	MAG22	MAG23	MAG24	MAG25
15	16	17	18	19	20	21
A4	A5	A6	A7	A8	A9	A10
KHN20	KHN21	KHN222	KHN23	KHN24	KHN25	KHN26
MAG26	MAG27	MAG28	MAG29	MAG30	MIY1	MIY2
22	23	24	25	26	27	28
A11	A12	A13	A14	A15	A16	A17
KHN27	KHN28	KHN29	KHN30	PIN1	PIN2	PIN3
MIY3	MIY4	MIY5	MIY6	MIY7	MIY8	MIY9

MABABA WEUSI 560/2076-77/6312-13/2068-69

Ashe	Ojobo	Iwa	Araorun	Ile	Itutu	Erinle
1	2	3	4	5	6	7
MA16	MA17	MA18	MA19	MA20	MA21	MA22
APP2	APP3	APP4	APP5	APP6	APP7	APP8
GIN8	GIN9	GIN10	GIN11	GIN12	GIN13	GIN14
8	9	10	11	12	13	14
MA23	MA24	MA25	MA26	MA27	MA28	MA29
APP9	APP10	APP11	APP12	APP13	APP14	APP15
GIN15	GIN16	GIN17	GIN18	GIN19	GIN20	GIN21
15	16	17	18	19	20	21
MA30	MA31	J1	J2	J3	J4	J5
APP16	APP17	APP18	APP19	APP20	APP21	APP22
GIN22	GIN23	GIN24	GIN25	GIN26	GIN27	GIN28
22	23	24	25	26	27	28
J6	J7	J8	J9	J10	J11	J12
APP23	APP24	APP25	APP26	APP27	APP28	APP29
GIN29	GIN30	SAN1	SAN2	SAN3	SAN4	SAN5

THE NU AFRIKAN CALENDAR

MWANAMKE WEUSI 560/2076-77/6312-13/2068-69						
Ashe	Ojobo	Iwa	Araorun	Ile	Itutu	Erinle
1	2	3	4	5	6	7
J13	J14	J15	J16	J17	J18	J19
APP30	MSR1	MSR2	MSR3	MSR4	MSR5	MSR6
SAN6	SAN7	SAN8	SAN9	SAN10	SAN11	SAN12
8	9	10	11	12	13	14
J20	J21	J22	J23	J24	J25	J26
MSR7	MSR8	MSR9	MSR10	MSR11	MSR12	MSR13
SAN13	SAN14	SAN15	SAN16	SAN17	SAN18	SAN19
15	16	17	18	19	20	21
J27	J28	J29	J30	JU1	JU2	JU3
MSR14	MSR15	MSR16	MSR17	MSR18	MSR19	MSR20
SAN20	SAN21	SAN22	SAN23	SAN24	SAN25	SAN26
22	23	24	25	26	27	28
JU4	JU5	JU6	JU7	JU8	JU9	JU10
MSR21	MRS22	MSR23	MSR24	MSR25	MSR26	MSR27
SAN27	SAN28	SAN29	SAN30	HAM1	HAM2	HAM3

TAIFA WEUSI 560/2076-77/6312-13/2068-69						
Ashe	Ojobo	Iwa	Araorun	Ile	Itutu	Erinle
1	2	3	4	5	6	7
AU8	AU9	AU10	AU11	AU12	AU13	AU14
THT20	THT21	THT22	THT23	THT24	THT25	THT26
NAH2	NAH3	NAH4	NAH5	NAH6	NAH7	NAH8
8	9	10	11	12	13	14
AU15	AU16	AU17	AU18	AU19	AU20	AU21
THT27	THT28	THT29	THT30	PAP1	PAP1	PAP2
NAH9	NAH10	NAH11	NAH12	NAH13	NAH14	NAH15
15	16	17	18	19	20	21
AU22	AU23	AU24	AU25	AU26	AU27	AU28
PAP3	PAP4	PAP5	PAP6	PAP7	PAP8	PAP9
NAH16	NAH17	NAH18	NAH19	NAH20	NAH21	NAH22
22	23	24	25	26	27	28
AU29	AU30	AU31	S1	S2	S3	S4
PAP10	PAP11	PAP12	PAP13	PAP14	PAP15	PAP16
NAH23	NAH24	NAH25	NAH26	NAH27	NAH28	NAH29

MTU WEUSI 560/2076-77/6312-13/2068-69						
Ashe	Ojobo	Iwa	Araorun	Ile	Itutu	Erinle
1	2	3	4	5	6	7
JU11	JU12	JU13	JU14	JU15	JU16	JU17
MSR28	MSR29	MSR30	*ASR1*	*HRU2*	*STH3*	*AST4*
HAM4	HAM5	HAM6	HAM7	HAM8	HAM9	HAM10
8	9	10	11	12	13	14
JU18	JU19	JU20	JU21	JU22	JU23	JU24
NBH5	*ANU6*	THT1	THT2	THT3	THT4	THT5
HAM11	HAM12	HAM13	HAM14	HAM15	HAM16	HAM17
15	16	17	18	19	20	21
JU25	JU26	JU27	JU28	JU29	JU30	JU31
THT6	THT7	THT8	THT9	THT10	THT11	THT12
HAM18	HAM19	HAM20	HAM21	HAM22	HAM23	HAM24
22	23	24	25	26	27	28
AU1	AU2	AU3	AU4	AU5	AU6	AU7
THT13	THT14	THT15	THT16	THT17	THT18	THT19
HAM25	HAM26	HAM27	HAM28	HAM29	HAM30	NAH1

UWEZO 560/2076-77/6312-13/2068-69						
Ashe	Ojobo	Iwa	Araorun	Ile	Itutu	Erinle
1	2	3	4	5	6	7
S5	S6	S7	S8	S9	S10	S11
PAP17	PAP18	PAP19	PAP20	PAP21	PAP22	PAP23
NAH30	*PAG1*	*PAG2*	*PAG3*	*PAG4*	*PAG5*	MAS1
8	9	10	11	12	13	14
S12	S13	S14	S15	S16	S17	S18
PAP24	PAP25	PAP26	PAP27	PAP28	PAP29	PAP30
MAS2	MAS3	MAS4	MAS5	MAS6	MAS7	MAS8
15	16	17	18	19	20	21
S19	S20	S21	S22	S23	S24	S25
HTH1	HTH2	HTH3	HTH4	HTH5	HTH6	HTH7
MAS9	MAS10	MAS11	MAS12	MAS13	MAS14	MAS15
22	23	24	25	26	27	28
S26	S27	S28	S29	S30	O1	O2
HTH8	HTH9	HTH10	HTH11	HTH12	HTH13	HTH14
MAS16	MAS17	MAS18	MAS19	MAS20	MAS21	MAS22

THE NU AFRIKAN CALENDAR

EKUNDU 560/2076-77/6312-13/2068-69						
Ashe	Ojobo	Iwa	Araorun	Ile	Itutu	Erinle
1	2	3	4	5	6	7
O3	O4	O5	O6	O7	O8	O9
HTH15	HTH16	HTH17	HTH18	HTH19	HTH20	HTH21
MAS23	MAS24	MAS25	MAS26	MAS27	MAS28	MAS29
8	9	10	11	12	13	14
O10	O11	O12	O13	O14	O15	O16
HTH22	HTH23	HTH24	HTH25	HTH26	HTH27	HTH28
MAS30	THi1	THi2	THi3	THi4	THi5	THi6
15	16	17	18	19	20	21
O17	O18	O19	O20	O21	O22	O23
HTH29	HTH30	KHK1	KHK2	KHK3	KHK4	KHK5
THi7	THi8	This9	THi10	THi11	THi12	THi13
22	23	24	25	26	27	28
O24	O25	O26	O27	O28	O29	O30
KHK6	KHK7	KHK8	KHK9	KHK10	KHK11	KHK12
Thi14	THi15	THi16	THi17	THi18	THi19	THi20

KIJANI 560/2076-77/6312-13/2068-69						
Ashe	Ojobo	Iwa	Araorun	Ile	Itutu	Erinle
1	2	3	4	5	6	7
N28	N29	N30	D1	D2	D3	D4
TBT11	TBT12	TBT13	TBT14	TBT15	TBT16	TBT17
HID19	HID20	HID21	HID22	HID23	HID24	HID25
8	9	10	11	12	13	14
D5	D6	D7	D8	D9	D10	D11
TBT18	TBT19	TBT20	TBT21	TBT22	TBT23	TBT24
HID26	HID27	HID28	HID29	HID30	TAH1	TAH2
15	16	17	18	19	20	21
D12	D13	D14	D15	D16	D17	D18
TBT25	TBT26	TBT27	TBT28	TBT29	TBT30	MKH1
TAH3	TAH4	TAH5	TAH6	TAH7	TAH8	TAH9
22	23	24	25	26	27	28
D19	D20	D21	D22	D23	D24	D25
MKH2	MKH3	MKH4	MKH5	MKH6	MKH7	MKH8
TAH10	TAH11	TAH12	TAH13	TAH14	TAH15	TAH16

EUSI 560/2076-77/6312-13/2068-69						
Ashe	Ojobo	Iwa	Araorun	Ile	Itutu	Erinle
1	2	3	4	5	6	7
O31	N1	N2	N3	N4	N5	N6
KHK13	KHK14	KHK15	KHK16	KHK17	KHK18	KHK19
THi21	THi22	THi23	THi24	THi25	THi26	THi27
8	9	10	11	12	13	14
N7	N8	N9	N10	N11	N12	N13
KHK20	KHK21	KHK22	KHK23	KHK24	KHK25	KHK26
THi28	THi29	THi30	HID1	HID2	HID3	HID4
15	16	17	18	19	20	21
N14	N15	N16	N17	N18	N19	N20
KHK27	KHK28	KHK29	KHK30	TBT1	TBT2	TBT3
HID5	HID6	HID7	HID8	HID9	HID10	HID11
22	23	24	25	26	27	28
N21	N22	N23	N24	N25	N26	N27
TBT4	TBT5	TBT6	TBT7	TBT8	TBT9	TBT10
HID12	HID13	HID14	HID15	HID16	HID17	HID18

KWANZA 560/2076-77/6312-13/2068-69						
Umoja	Kujichagulia	Ujima	Ujamaa	Nia	Kuumba	Imani
1	2	3	4	5	6	7
D26	D27	D28	D29	D30	D31	JA1
MKH9	MKH10	MKH11	MKH12	MKH13	MKH14	MKH15
TAH17	TAH18	TAH19	TAH20	TAH21	TAH22	TAH23
8	9	10	11	12	13	14
JA2	JA3	JA4	JA5	JA6	JA7	JA8
MKH16	MKH17	MKH18	MKH19	MKH20	MKH21	MKH22
TAH24	TAH25	TAH26	TAH27	TAH28	TAH29	TAH30
15	16	17	18	19	20	21
JA9	JA10	JA11	JA12	JA13	JA14	JA15
MKH23	MKH24	MKH25	KMH26	MKH27	MKH28	MKH29
TIR1	TIR2	TIR3	TIR4	TIR5	TIR6	TIR7
22	23	24	25	26	27	28
JA16	JA17	JA18	JA19	JA20	JA21	JA22
MKH30	AMN1	AMN2	AMN3	AMN4	AMN5	AMN6
TIR8	TIR9	TIR10	TIR11	TIR12	TIR13	TIR14

THE NU AFRIKAN CALENDAR

SHAHIDI 560/2076-77/6312-13/2068-69						
Ashe	Ojobo	Iwa	Araorun	Ile	Itutu	Erinle
1	2	3	4	5	6	7
JA23	JA24	JA25	JA26	JA27	JA28	JA29
AMN7	AMN8	AMN9	AMN10	AMN11	AMN12	AMN13
TIR15	TIR16	TIR17	TIR18	TIR19	TIR20	TIR21
8	9	10	11	12	13	14
JA30	JA31	F1	F2	F3	F4	F5
AMN14	AMN15	AMN16	AMN17	AMN18	AMN19	AMN20
TIR22	TIR23	TIR24	TIR25	TIR26	TIR27	TIR28
15	16	17	18	19	20	21
F6	F7	F8	F9	F10	F 11	F12
AMN21	AMN22	AMN23	AMN24	AMN25	AMN26	AMN28
TIR29	TIR30	YEK1	YEK2	YEK3	YEK4	YEK5
22	23	24	25	26	27	28
F13	F14	F15	F16	F17	F18	F19
AMN29	AMN30	RNT1	RNT2	RNT3	RNT4	RNT5
YEK6	YEK7	YEK8	YEK9	YEK10	YEK11	YEK12
SHAHIDI 111						
F20						
RNT6						
YEK13						

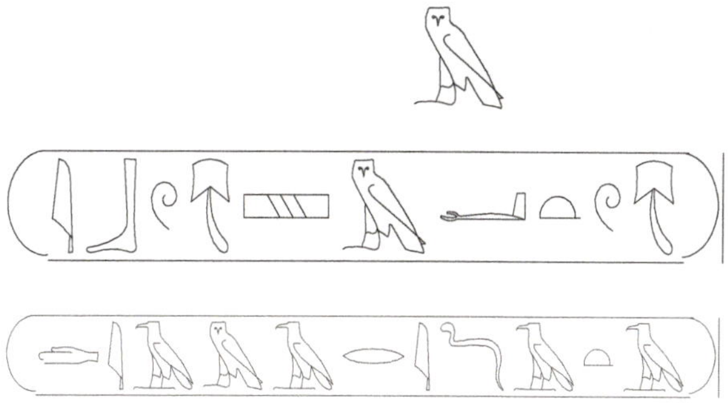

THE NU AFRIKAN CALENDAR

MAAFA 561 B.S.E./ASIENTO 2077-78 C.E.
KAM 6313-14 S.M./EC 2069-70 A.M. (ዓ.ም.)

AFU 561/2077-78/6313-14/2069-70

Ashe	Ojobo	Iwa	Araorun	Ile	Itutu	Erinle
1	2	3	4	5	6	7
F21	F22	F23	F24	F25	F26	F27
RNT7	RNT8	RNT9	RNT10	RNT11	RNT12	RNT13
YEK14	YEK15	YEK16	YEK17	YEK18	YEK19	YEK20
8	9	10	11	12	13	14
F28	M1	M2	M3	M4	M5	M6
RNT14	RNT15	RNT16	RNT17	RNT18	RNT19	RNT20
YEK21	YEK22	YEK23	YEK24	YEK25	YEK26	YEK27
15	16	17	18	19	20	21
M7	M8	M9	M10	M11	M12	M13
RNT21	RNT22	RNT23	RNT24	RNT25	RNT26	RNT27
YEK28	YEK29	YEK30	MAG1	MAG2	MAG3	MAG4
22	23	24	25	26	27	28
M14	M15	M16	M17	M18	M19	M20
RNT28	RNT29	RNT30	KHN1	KHN2	KHN3	KHN4
MAG5	MAG6	MAG7	MAG8	MAG9	MAG10	MAG11

KAWAIDA 561/2077-78/6313-14/2069-70

Ashe	Ojobo	Iwa	Araorun	Ile	Itutu	Erinle
1	2	3	4	5	6	7
A18	A19	A20	A21	A22	A23	A24
PIN3	PIN4	PIN5	PIN6	PIN7	PIN8	PIN9
MIY10	MIY11	MIY12	MIY13	MIY14	MIY15	MIY16
8	9	10	11	12	13	14
A25	A26	A27	A28	A29	A30	MA1
PIN10	PIN11	PIN12	PIN13	PIN14	PIN15	PIN16
MIY17	MIY18	MIY19	MIY20	MIY21	MIY22	MIY23
15	16	17	18	19	20	21
MA2	MA3	MA4	MA5	MA6	MA7	MA8
PIN17	PIN18	PIN19	PIN20	PIN21	PIN22	PIN23
MIY24	MIY25	MIY26	MIY27	MIY28	MIY29	MIY30
22	23	24	25	26	27	28
MA9	MA10	MA11	MA12	MA13	MA14	MA15
PIN24	PIN25	PIN26	PIN27	PIN28	PIN29	PIN30
GIN1	GIN2	GIN3	GIN4	GIN5	GIN6	GIN7

UHURU 561/2077-78/6313-14/2069-70

Ashe	Ojobo	Iwa	Araorun	Ile	Itutu	Erinle
1	2	3	4	5	6	7
M21	M22	M23	M24	M25	M26	M27
KHN5	KHN6	KHN7	KHN8	KHN9	KHN10	KHN11
MAG12	MAG13	MAG14	MAG15	MAG16	MAG17	MAG18
8	9	10	11	12	13	14
M28	M29	M30	M31	A1	A2	A3
KHN12	KHN13	KHN14	KHN15	KHN16	KHN17	KHN18
MAG19	MAG20	MAG21	MAG22	MAG23	MAG24	MAG25
15	16	17	18	19	20	21
A4	A5	A6	A7	A8	A9	A10
KHN19	KHN20	KHN21	KHN22	KHN23	KHN24	KHN25
MAG26	MAG27	MAG28	MAG29	MAG30	MIY1	MIY2
22	23	24	25	26	27	28
A11	A12	A13	A14	A15	A16	A17
KHN26	KHN27	KHN28	KHN29	KHN30	PIN1	PIN2
MIY3	MIY4	MIY5	MIY6	MIY7	MIY8	MIY9

MABABA WEUSI 561/2077-78/6313-14/2069-70

Ashe	Ojobo	Iwa	Araorun	Ile	Itutu	Erinle
1	2	3	4	5	6	7
MA16	MA17	MA18	MA19	MA20	MA21	MA22
APP1	APP2	APP3	APP4	APP5	APP6	APP7
GIN8	GIN9	GIN10	GIN11	GIN12	GIN13	GIN14
8	9	10	11	12	13	14
MA23	MA24	MA25	MA26	MA27	MA28	MA29
APP8	APP9	APP10	APP11	APP12	APP13	APP14
GIN15	GIN16	GIN17	GIN18	GIN19	GIN20	GIN21
15	16	17	18	19	20	21
MA30	MA31	J1	J2	J3	J4	J5
APP15	APP16	APP17	APP18	APP19	APP20	APP21
GIN22	GIN23	GIN24	GIN25	GIN26	GIN27	GIN28
22	23	24	25	26	27	28
J6	J7	J8	J9	J10	J11	J12
APP22	APP23	APP24	APP25	APP26	APP27	APP28
GIN29	GIN30	SAN1	SAN2	SAN3	SAN4	SAN5

THE NU AFRIKAN CALENDAR

MWANAMKE WEUSI 561/2077-78/6313-14/2069-70						
Ashe	Ojobo	Iwa	Araorun	Ile	Itutu	Erinle
1	2	3	4	5	6	7
J13	J14	J15	J16	J17	J18	J19
APP29	APP30	MSR1	MSR2	MSR3	MSR4	MSR5
SAN6	SAN7	SAN8	SAN9	SAN10	SAN11	SAN12
8	9	10	11	12	13	14
J20	J21	J22	J23	J24	J25	J26
MSR6	MSR7	MSR8	MSR9	MSR10	MSR11	MSR12
SAN13	SAN14	SAN15	SAN16	SAN17	SAN18	SAN19
15	16	17	18	19	20	21
J27	J28	J29	J30	JU1	JU2	JU3
MSR13	MSR14	MSR15	MSR16	MSR17	MSR18	MSR19
SAN20	SAN21	SAN22	SAN23	SAN24	SAN25	SAN26
22	23	24	25	26	27	28
JU4	JU5	JU6	JU7	JU8	JU9	JU10
MSR20	MRS21	MSR22	MSR23	MSR24	MSR25	MSR26
SAN27	SAN28	SAN29	SAN30	HAM1	HAM2	HAM3

TAIFA WEUSI 561/2077-78/6313-14/2069-70						
Ashe	Ojobo	Iwa	Araorun	Ile	Itutu	Erinle
1	2	3	4	5	6	7
AU8	AU9	AU10	AU11	AU12	AU13	AU14
THT20	THT21	THT22	THT23	THT24	THT25	THT26
NAH2	NAH3	NAH4	NAH5	NAH6	NAH7	NAH8
8	9	10	11	12	13	14
AU15	AU16	AU17	AU18	AU19	AU20	AU21
THT27	THT28	THT29	THT30	PAP1	PAP1	PAP2
NAH9	NAH10	NAH11	NAH12	NAH13	NAH14	NAH15
15	16	17	18	19	20	21
AU22	AU23	AU24	AU25	AU26	AU27	AU28
PAP3	PAP4	PAP5	PAP6	PAP7	PAP8	PAP9
NAH16	NAH17	NAH18	NAH19	NAH20	NAH21	NAH22
22	23	24	25	26	27	28
AU29	AU30	AU31	S1	S2	S3	S4
PAP10	PAP11	PAP12	PAP13	PAP14	PAP15	PAP16
NAH23	NAH24	NAH25	NAH26	NAH27	NAH28	NAH29

MTU WEUSI 561/2077-78/6313-14/2069-70						
Ashe	Ojobo	Iwa	Araorun	Ile	Itutu	Erinle
1	2	3	4	5	6	7
JU11	JU12	JU13	JU14	JU15	JU16	JU17
MSR27	MSR28	MSR29	MSR30	*ASR1*	*HRU2*	*STH3*
HAM4	HAM5	HAM6	HAM7	HAM8	HAM9	HAM10
8	9	10	11	12	13	14
JU18	JU19	JU20	JU21	JU22	JU23	JU24
AST4	*NBH5*	THT1	THT2	THT3	THT4	THT5
HAM11	HAM12	HAM13	AM14	HAM15	HAM16	HAM17
15	16	17	18	19	20	21
JU25	JU26	JU27	JU28	JU29	JU30	JU31
THT6	THT7	THT8	THT9	THT10	THT11	THT12
HAM18	HAM19	HAM20	HAM21	HAM22	HAM23	HAM24
22	23	24	25	26	27	28
AU1	AU2	AU3	AU4	AU5	AU6	AU7
THT13	THT14	THT15	THT16	THT17	THT18	THT19
HAM25	HAM26	HAM27	HAM28	HAM29	HAM30	NAH1

UWEZO 561/2077-78/6313-14/2069-70						
Ashe	Ojobo	Iwa	Araorun	Ile	Itutu	Erinle
1	2	3	4	5	6	7
S5	S6	S7	S8	S9	S10	S11
PAP17	PAP18	PAP19	PAP20	PAP21	PAP22	PAP23
NAH30	*PAG1*	*PAG2*	*PAG3*	*PAG4*	*PAG5*	MAS1
8	9	10	11	12	13	14
S12	S13	S14	S15	S16	S17	S18
PAP24	PAP25	PAP26	PAP27	PAP28	PAP29	PAP30
MAS2	MAS3	MAS4	MAS5	MAS6	MAS7	MAS8
15	16	17	18	19	20	21
S19	S20	S21	S22	S23	S24	S25
HTH1	HTH2	HTH3	HTH4	HTH5	HTH6	HTH7
MAS9	MAS10	MAS11	MAS12	MAS13	MAS14	MAS15
22	23	24	25	26	27	28
S26	S27	S28	S29	S30	O1	O2
HTH8	HTH9	HTH10	HTH11	HTH12	HTH13	HTH14
MAS16	MAS17	MAS18	MAS19	MAS20	MAS21	MAS22

THE NU AFRIKAN CALENDAR

EKUNDU 561/2077-78/6313-14/2069-70						
Ashe	Ojobo	Iwa	Araorun	Ile	Itutu	Erinle
1	2	3	4	5	6	7
O3	O4	O5	O6	O7	O8	O9
HTH15	HTH16	HTH17	HTH18	HTH19	HTH20	HTH21
MAS23	MAS24	MAS25	MAS26	MAS27	MAS28	MAS29
8	9	10	11	12	13	14
O10	O11	O12	O13	O14	O15	O16
HTH22	HTH23	HTH24	HTH25	HTH26	HTH27	HTH28
MAS30	THi1	THi2	THi3	THi4	THi5	THi6
15	16	17	18	19	20	21
O17	O18	O19	O20	O21	O22	O23
HTH29	HTH30	KHK1	KHK2	KHK3	KHK4	KHK5
THi7	THi8	THi9	THi10	THi11	THi12	THi13
22	23	24	25	26	27	28
O24	O25	O26	O27	O28	O29	O30
KHK6	KHK7	KHK8	KHK9	KHK10	KHK11	KHK12
THi14	THi15	THi16	THi17	THi18	THi19	THi20

KIJANI 561/2077-78/6313-14/2069-70						
Ashe	Ojobo	Iwa	Araorun	Ile	Itutu	Erinle
1	2	3	4	5	6	7
N28	N29	N30	D1	D2	D3	D4
TBT11	TBT12	TBT13	TBT14	TBT15	TBT16	TBT17
HID19	HID20	HID21	HID22	HID23	HID24	HID25
8	9	10	11	12	13	14
D5	D6	D7	D8	D9	D10	D11
TBT18	TBT19	TBT20	TBT21	TBT22	TBT23	TBT24
HID26	HID27	HID28	HID29	HID30	TAH1	TAH2
15	16	17	18	19	20	21
D12	D13	D14	D15	D16	D17	D18
TBT25	TBT26	TBT27	TBT28	TBT29	TBT30	MKH1
TAH3	TAH4	TAH5	TAH6	TAH7	TAH8	TAH9
22	23	24	25	26	27	28
D19	D20	D21	D22	D23	D24	D25
MKH2	MKH3	MKH4	MKH5	MKH6	MKH7	MKH8
TAH10	TAH11	TAH12	TAH13	TAH14	TAH15	TAH16

EUSI 561/2077-78/6313-14/2069-70						
Ashe	Ojobo	Iwa	Araorun	Ile	Itutu	Erinle
1	2	3	4	5	6	7
O31	N1	N2	N3	N4	N5	N6
KHK13	KHK14	KHK15	KHK16	KHK17	KHK18	KHK19
THi21	THi22	THi23	THi24	THi25	THi26	THi27
8	9	10	11	12	13	14
N7	N8	N9	N10	N11	N12	N13
KHK20	KHK21	KHK22	KHK23	KHK24	KHK25	KHK26
THi28	THi29	THi30	HID1	HID2	HID3	HID4
15	16	17	18	19	20	21
N14	N15	N16	N17	N18	N19	N20
KHK27	KHK28	KHK29	KHK30	TBT1	TBT2	TBT3
HID5	HID6	HID7	HID8	HID9	HID10	HID11
22	23	24	25	26	27	28
N21	N22	N23	N24	N25	N26	N27
TBT4	TBT5	TBT6	TBT7	TBT8	TBT9	TBT10
HID12	HID13	HID14	HID15	HID16	HID17	HID18

KWANZA 561/2077-78/6313-14/2069-70						
Umoja	Kujichagulia	Ujima	Ujamaa	Nia	Kuumba	Imani
1	2	3	4	5	6	7
D26	D27	D28	D29	D30	D31	JA1
MKH9	MKH10	MKH11	MKH12	MKH13	MKH14	MKH15
TAH17	TAH18	TAH19	TAH20	TAH21	TAH22	TAH23
8	9	10	11	12	13	14
JA2	JA3	JA4	JA5	JA6	JA7	JA8
MKH16	MKH17	MKH18	MKH19	MKH20	MKH21	MKH22
TAH24	TAH25	TAH26	TAH27	TAH28	TAH29	TAH30
15	16	17	18	19	20	21
JA9	JA10	JA11	JA12	JA13	JA14	JA15
MKH23	MKH24	MKH25	KMH26	MKH27	MKH28	MKH29
TIR1	TIR2	TIR3	TIR4	TIR5	TIR6	TIR7
22	23	24	25	26	27	28
JA16	JA17	JA18	JA19	JA20	JA21	JA22
MKH30	AMN1	AMN2	AMN3	AMN4	AMN5	AMN6
TIR8	TIR9	TIR10	TIR11	TIR12	TIR13	TIR14

THE NU AFRIKAN CALENDAR

SHAHIDI 561/2077-78/6313-14/2069-70						
Ashe	Ojobo	Iwa	Araorun	Ile	Itutu	Erinle
1	2	3	4	5	6	7
JA23	JA24	JA25	JA26	JA27	JA28	JA29
AMN7	AMN8	AMN9	AMN10	AMN11	AMN12	AMN13
TIR15	TIR16	TIR17	TIR18	TIR19	TIR20	TIR21
8	9	10	11	12	13	14
JA30	JA31	F1	F2	F3	F4	F5
AMN14	AMN15	AMN16	AMN17	AMN18	AMN19	AMN20
TIR22	TIR23	TIR24	TIR25	TIR26	TIR27	TIR28
15	16	17	18	19	20	21
F6	F7	F8	F9	F10	F 11	F12
AMN21	AMN22	AMN23	AMN24	AMN25	AMN26	AMN28
TIR29	TIR30	YEK1	YEK2	YEK3	YEK4	YEK5
22	23	24	25	26	27	28
F13	F14	F15	F16	F17	F18	F19
AMN29	AMN30	RNT1	RNT2	RNT3	RNT4	RNT5
YEK6	YEK7	YEK8	YEK9	YEK10	YEK11	YEK12
SHAHIDI 112						
F20						
RNT6						
YEK13						

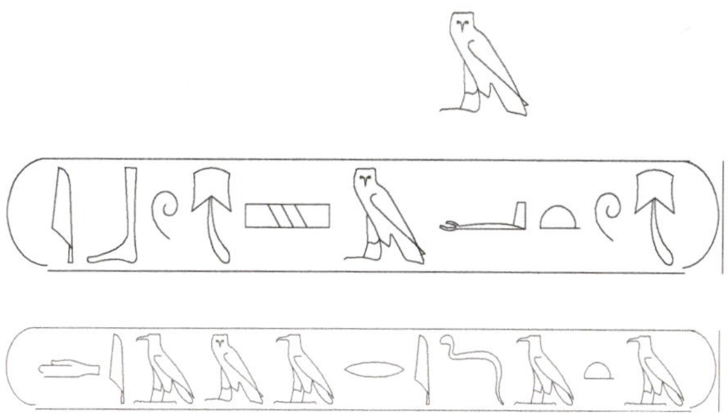

MAAFA 562 B.S.E./ASIENTO 2078-79 C.E.
KAM 6314-15 S.M./EC 2070-71 A.M. (ዓ.ም.)

AFU 562/2078-79/6314-15/2070-71						
Ashe	Ojobo	Iwa	Araorun	Ile	Itutu	Erinle
1	2	3	4	5	6	7
F21	F22	F23	F24	F25	F26	F27
RNT7	RNT8	RNT9	RNT10	RNT11	RNT12	RNT13
YEK14	YEK15	YEK16	YEK17	YEK18	YEK19	YEK20
8	9	10	11	12	13	14
F28	M1	M2	M3	M4	M5	M6
RNT14	RNT15	RNT16	RNT17	RNT18	RNT19	RNT20
YEK21	YEK22	YEK23	YEK24	YEK25	YEK26	YEK27
15	16	17	18	19	20	21
M7	M8	M9	M10	M11	M12	M13
RNT21	RNT22	RNT23	RNT24	RNT25	RNT26	RNT27
YEK28	YEK29	YEK30	MAG1	MAG2	MAG3	MAG4
22	23	24	25	26	27	28
M14	M15	M16	M17	M18	M19	M20
RNT28	RNT29	RNT30	KHN1	KHN2	KHN3	KHN4
MAG5	MAG6	MAG7	MAG8	MAG9	MAG10	MAG11

KAWAIDA 562/2078-79/6314-15/2070-71						
Ashe	Ojobo	Iwa	Araorun	Ile	Itutu	Erinle
1	2	3	4	5	6	7
A18	A19	A20	A21	A22	A23	A24
PIN3	PIN4	PIN5	PIN6	PIN7	PIN8	PIN9
MIY10	MIY11	MIY12	MIY13	MIY14	MIY15	MIY16
8	9	10	11	12	13	14
A25	A26	A27	A28	A29	A30	MA1
PIN10	PIN11	PIN12	PIN13	PIN14	PIN15	PIN16
MIY17	MIY18	MIY19	MIY20	MIY21	MIY22	MIY23
15	16	17	18	19	20	21
MA2	MA3	MA4	MA5	MA6	MA7	MA8
PIN17	PIN18	PIN19	PIN20	PIN21	PIN22	PIN23
MIY24	MIY25	MIY26	MIY27	MIY28	MIY29	MIY30
22	23	24	25	26	27	28
MA9	MA10	MA11	MA12	MA13	MA14	MA15
PIN24	PIN25	PIN26	PIN27	PIN28	PIN29	PIN30
GIN1	GIN2	GIN3	GIN4	GIN5	GIN6	GIN7

UHURU 562/2078-79/6314-15/2070-71						
Ashe	Ojobo	Iwa	Araorun	Ile	Itutu	Erinle
1	2	3	4	5	6	7
M21	M22	M23	M24	M25	M26	M27
KHN5	KHN6	KHN7	KHN8	KHN9	KHN10	KHN11
MAG12	MAG13	MAG14	MAG15	MAG16	MAG17	MAG18
8	9	10	11	12	13	14
M28	M29	M30	M31	A1	A2	A3
KHN12	KHN13	KHN14	KHN15	KHN16	KHN17	KHN18
MAG19	MAG20	MAG21	MAG22	MAG23	MAG24	MAG25
15	16	17	18	19	20	21
A4	A5	A6	A7	A8	A9	A10
KHN19	KHN20	KHN21	KHN22	KHN23	KHN24	KHN25
MAG26	MAG27	MAG28	MAG29	MAG30	MIY1	MIY2
22	23	24	25	26	27	28
A11	A12	A13	A14	A15	A16	A17
KHN26	KHN27	KHN28	KHN29	KHN30	PIN1	PIN2
MIY3	MIY4	MIY5	MIY6	MIY7	MIY8	MIY9

MABABA WEUSI 562/2078-79/6314-15/2070-71						
Ashe	Ojobo	Iwa	Araorun	Ile	Itutu	Erinle
1	2	3	4	5	6	7
MA16	MA17	MA18	MA19	MA20	MA21	MA22
APP1	APP2	APP3	APP4	APP5	APP6	APP7
GIN8	GIN9	GIN10	GIN11	GIN12	GIN13	GIN14
8	9	10	11	12	13	14
MA23	MA24	MA25	MA26	MA27	MA28	MA29
APP8	APP9	APP10	APP11	APP12	APP13	APP14
GIN15	GIN16	GIN17	GIN18	GIN19	GIN20	GIN21
15	16	17	18	19	20	21
MA30	MA31	J1	J2	J3	J4	J5
APP15	APP16	APP17	APP18	APP19	APP20	APP21
GIN22	GIN23	GIN24	GIN25	GIN26	GIN27	GIN28
22	23	24	25	26	27	28
J6	J7	J8	J9	J10	J11	J12
APP22	APP23	APP24	APP25	APP26	APP27	APP28
GIN29	GIN30	SAN1	SAN2	SAN3	SAN4	SAN5

THE NU AFRIKAN CALENDAR

MWANAMKE WEUSI 562/2078-79/6314-15/2070-71						
Ashe	Ojobo	Iwa	Araorun	Ile	Itutu	Erinle
1	2	3	4	5	6	7
J13	J14	J15	J16	J17	J18	J19
APP29	APP30	MSR1	MSR2	MSR3	MSR4	MSR5
SAN6	SAN7	SAN8	SAN9	SAN10	SAN11	SAN12
8	9	10	11	12	13	14
J20	J21	J22	J23	J24	J25	J26
MSR6	MSR7	MSR8	MSR9	MSR10	MSR11	MSR12
SAN13	SAN14	SAN15	SAN16	SAN17	SAN18	SAN19
15	16	17	18	19	20	21
J27	J28	J29	J30	JU1	JU2	JU3
MSR13	MSR14	MSR15	MSR16	MSR17	MSR18	MSR19
SAN20	SAN21	SAN22	SAN23	SAN24	SAN25	SAN26
22	23	24	25	26	27	28
JU4	JU5	JU6	JU7	JU8	JU9	JU10
MSR20	MRS21	MSR22	MSR23	MSR24	MSR25	MSR26
SAN27	SAN28	SAN29	SAN30	HAM1	HAM2	HAM3

TAIFA WEUSI 562/2078-79/6314-15/2070-71						
Ashe	Ojobo	Iwa	Araorun	Ile	Itutu	Erinle
1	2	3	4	5	6	7
AU8	AU9	AU10	AU11	AU12	AU13	AU14
THT20	THT21	THT22	THT23	THT24	THT25	THT26
NAH2	NAH3	NAH4	NAH5	NAH6	NAH7	NAH8
8	9	10	11	12	13	14
AU15	AU16	AU17	AU18	AU19	AU20	AU21
THT27	THT28	THT29	THT30	PAP1	PAP1	PAP2
NAH9	NAH10	NAH11	NAH12	NAH13	NAH14	NAH15
15	16	17	18	19	20	21
AU22	AU23	AU24	AU25	AU26	AU27	AU28
PAP3	PAP4	PAP5	PAP6	PAP7	PAP8	PAP9
NAH16	NAH17	NAH18	NAH19	NAH20	NAH21	NAH22
22	23	24	25	26	27	28
AU29	AU30	AU31	S1	S2	S3	S4
PAP10	PAP11	PAP12	PAP13	PAP14	PAP15	PAP16
NAH23	NAH24	NAH25	NAH26	NAH27	NAH28	NAH29

MTU WEUSI 562/2078-79/6314-15/2070-71						
Ashe	Ojobo	Iwa	Araorun	Ile	Itutu	Erinle
1	2	3	4	5	6	7
JU11	JU12	JU13	JU14	JU15	JU16	JU17
MSR27	MSR28	MSR29	MSR30	*ASR1*	*HRU2*	*STH3*
HAM4	HAM5	HAM6	HAM7	HAM8	HAM9	HAM10
8	9	10	11	12	13	14
JU18	JU19	JU20	JU21	JU22	JU23	JU24
AST4	*NBH5*	THT1	THT2	THT3	THT4	THT5
HAM11	HAM12	HAM13	AM14	HAM15	HAM16	HAM17
15	16	17	18	19	20	21
JU25	JU26	JU27	JU28	JU29	JU30	JU31
THT6	THT7	THT8	THT9	THT10	THT11	THT12
HAM18	HAM19	HAM20	HAM21	HAM22	HAM23	HAM24
22	23	24	25	26	27	28
AU1	AU2	AU3	AU4	AU5	AU6	AU7
THT13	THT14	THT15	THT16	THT17	THT18	THT19
HAM25	HAM26	HAM27	HAM28	HAM29	HAM30	NAH1

UWEZO 562/2078-79/6314-15/2070-71						
Ashe	Ojobo	Iwa	Araorun	Ile	Itutu	Erinle
1	2	3	4	5	6	7
S5	S6	S7	S8	S9	S10	S11
PAP17	PAP18	PAP19	PAP20	PAP21	PAP22	PAP23
NAH30	*PAG1*	*PAG2*	*PAG3*	*PAG4*	*PAG5*	MAS1
8	9	10	11	12	13	14
S12	S13	S14	S15	S16	S17	S18
PAP24	PAP25	PAP26	PAP27	PAP28	PAP29	PAP30
MAS2	MAS3	MAS4	MAS5	MAS6	MAS7	MAS8
15	16	17	18	19	20	21
S19	S20	S21	S22	S23	S24	S25
HTH1	HTH2	HTH3	HTH4	HTH5	HTH6	HTH7
MAS9	MAS10	MAS11	MAS12	MAS13	MAS14	MAS15
22	23	24	25	26	27	28
S26	S27	S28	S29	S30	O1	O2
HTH8	HTH9	HTH10	HTH11	HTH12	HTH13	HTH14
MAS16	MAS17	MAS18	MAS19	MAS20	MAS21	MAS22

THE NU AFRIKAN CALENDAR

EKUNDU 562/2078-79/6314-15/2070-71

Ashe	Ojobo	Iwa	Araorun	Ile	Itutu	Erinle
1	2	3	4	5	6	7
O3	O4	O5	O6	O7	O8	O9
HTH15	HTH16	HTH17	HTH18	HTH19	HTH20	HTH21
MAS23	MAS24	MAS25	MAS26	MAS27	MAS28	MAS29
8	9	10	11	12	13	14
O10	O11	O12	O13	O14	O15	O16
HTH22	HTH23	HTH24	HTH25	HTH26	HTH27	HTH28
MAS30	THi1	THi2	THi3	THi4	THi5	THi6
15	16	17	18	19	20	21
O17	O18	O19	O20	O21	O22	O23
HTH29	HTH30	KHK1	KHK2	KHK3	KHK4	KHK5
THi7	THi8	THi9	THi10	THi11	THi12	THi13
22	23	24	25	26	27	28
O24	O25	O26	O27	O28	O29	O30
KHK6	KHK7	KHK8	KHK9	KHK10	KHK11	KHK12
THi14	THi15	THi16	THi17	THi18	THi19	THi20

KIJANI 562/2078-79/6314-15/2070-71

Ashe	Ojobo	Iwa	Araorun	Ile	Itutu	Erinle
1	2	3	4	5	6	7
N28	N29	N30	D1	D2	D3	D4
TBT11	TBT12	TBT13	TBT14	TBT15	TBT16	TBT17
HID19	HID20	HID21	HID22	HID23	HID24	HID25
8	9	10	11	12	13	14
D5	D6	D7	D8	D9	D10	D11
TBT18	TBT19	TBT20	TBT21	TBT22	TBT23	TBT24
HID26	HID27	HID28	HID29	HID30	TAH1	TAH2
15	16	17	18	19	20	21
D12	D13	D14	D15	D16	D17	D18
TBT25	TBT26	TBT27	TBT28	TBT29	TBT30	MKH1
TAH3	TAH4	TAH5	TAH6	TAH7	TAH8	TAH9
22	23	24	25	26	27	28
D19	D20	D21	D22	D23	D24	D25
MKH2	MKH3	MKH4	MKH5	MKH6	MKH7	MKH8
TAH10	TAH11	TAH12	TAH13	TAH14	TAH15	TAH16

EUSI 562/2078-79/6314-15/2070-71

Ashe	Ojobo	Iwa	Araorun	Ile	Itutu	Erinle
1	2	3	4	5	6	7
O31	N1	N2	N3	N4	N5	N6
KHK13	KHK14	KHK15	KHK16	KHK17	KHK18	KHK19
THi21	THi22	THi23	THi24	THi25	THi26	THi27
8	9	10	11	12	13	14
N7	N8	N9	N10	N11	N12	N13
KHK20	KHK21	KHK22	KHK23	KHK24	KHK25	KHK26
THi28	THi29	THi30	HID1	HID2	HID3	HID4
15	16	17	18	19	20	21
N14	N15	N16	N17	N18	N19	N20
KHK27	KHK28	KHK29	KHK30	TBT1	TBT2	TBT3
HID5	HID6	HID7	HID8	HID9	HID10	HID11
22	23	24	25	26	27	28
N21	N22	N23	N24	N25	N26	N27
TBT4	TBT5	TBT6	TBT7	TBT8	TBT9	TBT10
HID12	HID13	HID14	HID15	HID16	HID17	HID18

KWANZA 562/2078-79/6314-15/2070-71

Umoja	Kujichagulia	Ujima	Ujamaa	Nia	Kuumba	Imani
1	2	3	4	5	6	7
D26	D27	D28	D29	D30	D31	JA1
MKH9	MKH10	MKH11	MKH12	MKH13	MKH14	MKH15
TAH17	TAH18	TAH19	TAH20	TAH21	TAH22	TAH23
8	9	10	11	12	13	14
JA2	JA3	JA4	JA5	JA6	JA7	JA8
MKH16	MKH17	MKH18	MKH19	MKH20	MKH21	MKH22
TAH24	TAH25	TAH26	TAH27	TAH28	TAH29	TAH30
15	16	17	18	19	20	21
JA9	JA10	JA11	JA12	JA13	JA14	JA15
MKH23	MKH24	MKH25	KMH26	MKH27	MKH28	MKH29
TIR1	TIR2	TIR3	TIR4	TIR5	TIR6	TIR7
22	23	24	25	26	27	28
JA16	JA17	JA18	JA19	JA20	JA21	JA22
MKH30	AMN1	AMN2	AMN3	AMN4	AMN5	AMN6
TIR8	TIR9	TIR10	TIR11	TIR12	TIR13	TIR14

SHAHIDI 562/2078-79/6314-15/2070-71						
Ashe	Ojobo	Iwa	Araorun	Ile	Itutu	Erinle
1	2	3	4	5	6	7
JA23	JA24	JA25	JA26	JA27	JA28	JA29
AMN7	AMN8	AMN9	AMN10	AMN11	AMN12	AMN13
TIR15	TIR16	TIR17	TIR18	TIR19	TIR20	TIR21
8	9	10	11	12	13	14
JA30	JA31	F1	F2	F3	F4	F5
AMN14	AMN15	AMN16	AMN17	AMN18	AMN19	AMN20
TIR22	TIR23	TIR24	TIR25	TIR26	TIR27	TIR28
15	16	17	18	19	20	21
F6	F7	F8	F9	F10	F 11	F12
AMN21	AMN22	AMN23	AMN24	AMN25	AMN26	AMN28
TIR29	TIR30	YEK1	YEK2	YEK3	YEK4	YEK5
22	23	24	25	26	27	28
F13	F14	F15	F16	F17	F18	F19
AMN29	AMN30	RNT1	RNT2	RNT3	RNT4	RNT5
YEK6	YEK7	YEK8	YEK9	YEK10	YEK11	YEK12
SHAHIDI 113						
F20						
RNT6						
YEK13						

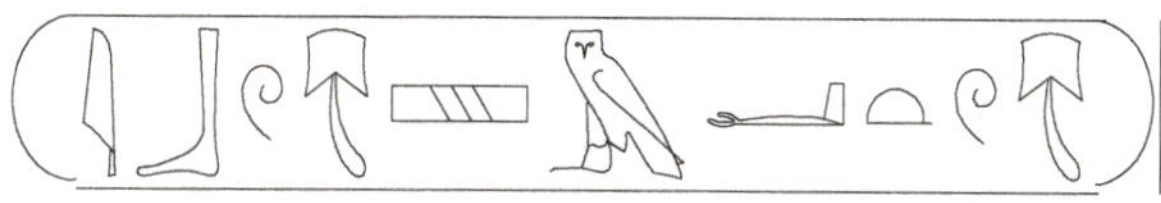

THE NU AFRIKAN CALENDAR

MAAFA 563 B.S.E./ASIENTO 2079-80 C.E./
KAM 6315-16 S.M./EC 2071-72 A.M. (ዓ.ም.)

AFU 563/2079-80/6315-16/2071-72						
Ashe	Ojobo	Iwa	Araorun	Ile	Itutu	Erinle
1	2	3	4	5	6	7
F21	F22	F23	F24	F25	F26	F27
RNT7	RNT8	RNT9	RNT10	RNT11	RNT12	RNT13
YEK14	YEK15	YEK16	YEK17	YEK18	YEK19	YEK20
8	9	10	11	12	13	14
F28	M1	M2	M3	M4	M5	M6
RNT14	RNT15	RNT16	RNT17	RNT18	RNT19	RNT20
YEK21	YEK22	YEK23	YEK24	YEK25	YEK26	YEK27
15	16	17	18	19	20	21
M7	M8	M9	M10	M11	M12	M13
RNT21	RNT22	RNT23	RNT24	RNT25	RNT26	RNT27
YEK28	YEK 29	YEK30	MAG1	MAG2	MAG3	MAG4
22	23	24	25	26	27	28
M14	M15	M16	M17	M18	M19	M20
RNT28	RNT29	RNT30	KHN1	KHN2	KHN3	KHN4
MAG5	MAG6	MAG7	MAG8	MAG9	MAG10	MAG11

KAWAIDA 563/2079-80/6315-16/2071-72						
Ashe	Ojobo	Iwa	Araorun	Ile	Itutu	Erinle
1	2	3	4	5	6	7
A18	A19	A20	A21	A22	A23	A24
PIN3	PIN4	PIN5	PIN6	PIN7	PIN8	PIN9
MIY10	MIY11	MIY12	MIY13	MIY14	MIY15	MIY16
8	9	10	11	12	13	14
A25	A26	A27	A28	A29	A30	MA1
PIN10	PIN11	PIN12	PIN13	PIN14	PIN15	PIN16
MIY17	MIY18	MIY19	MIY20	MIY21	MIY22	MIY23
15	16	17	18	19	20	21
MA2	MA3	MA4	MA5	MA6	MA7	MA8
PIN17	PIN18	PIN19	PIN20	PIN21	PIN22	PIN23
MIY24	MIY25	MIY26	MIY27	MIY28	MIY29	MIY30
22	23	24	25	26	27	28
MA9	MA10	MA11	MA12	MA13	MA14	MA15
PIN24	PIN25	PIN26	PIN27	PIN28	PIN29	PIN30
GIN1	GIN2	GIN3	GIN4	GIN5	GIN6	GIN7

UHURU 563/2079-80/6315-16/2071-72						
Ashe	Ojobo	Iwa	Araorun	Ile	Itutu	Erinle
1	2	3	4	5	6	7
M21	M22	M23	M24	M25	M26	M27
KHN5	KHN6	KHN7	KHN8	KHN9	KHN10	KHN11
MAG12	MAG13	MAG14	MAG15	MAG16	MAG17	MAG18
8	9	10	11	12	13	14
M28	M29	M30	M31	A1	A2	A3
KHN12	KHN13	KHN14	KHN15	KHN16	KHN17	KHN18
MAG19	MAG20	MAG21	MAG22	MAG23	MAG24	MAG25
15	16	17	18	19	20	21
A4	A5	A6	A7	A8	A9	A10
KHN19	KHN20	KHN21	KHN22	KHN23	KHN24	KHN25
MAG26	MAG27	MAG28	MAG29	MAG30	MIY1	MIY2
22	23	24	25	26	27	28
A11	A12	A13	A14	A15	A16	A17
KHN26	KHN27	KHN28	KHN29	KHN30	PIN1	PIN2
MIY3	MIY4	MIY5	MIY6	MIY7	MIY8	MIY9

MABABA WEUSI 563/2079-80/6315-16/2071-72						
Ashe	Ojobo	Iwa	Araorun	Ile	Itutu	Erinle
1	2	3	4	5	6	7
MA16	MA17	MA18	MA19	MA20	MA21	MA22
APP1	APP2	APP3	APP4	APP5	APP6	APP7
GIN8	GIN9	GIN10	GIN11	GIN12	GIN13	GIN14
8	9	10	11	12	13	14
MA23	MA24	MA25	MA26	MA27	MA28	MA29
APP8	APP9	APP10	APP11	APP12	APP13	APP14
GIN15	GIN16	GIN17	GIN18	GIN19	GIN20	GIN21
15	16	17	18	19	20	21
MA30	MA31	J1	J2	J3	J4	J5
APP15	APP16	APP17	APP18	APP19	APP20	APP21
GIN22	GIN23	GIN24	GIN25	GIN26	GIN27	GIN28
22	23	24	25	26	27	28
J6	J7	J8	J9	J10	J11	J12
APP22	APP23	APP24	APP25	APP26	APP27	APP28
GIN29	GIN30	SAN1	SAN2	SAN3	SAN4	SAN5

THE NU AFRIKAN CALENDAR

MWANAMKE WEUSI 563/2079-80/6315-16/2071-72

Ashe	Ojobo	Iwa	Araorun	Ile	Itutu	Erinle
1	2	3	4	5	6	7
J13	J14	J15	J16	J17	J18	J19
APP29	APP30	MSR1	MSR2	MSR3	MSR4	MSR5
SAN6	SAN7	SAN8	SAN9	SAN10	SAN11	SAN12
8	9	10	11	12	13	14
J20	J21	J22	J23	J24	J25	J26
MSR6	MSR7	MSR8	MSR9	MSR10	MSR11	MSR12
SAN13	SAN14	SAN15	SAN16	SAN17	SAN18	SAN19
15	16	17	18	19	20	21
J27	J28	J29	J30	JU1	JU2	JU3
MSR13	MSR14	MSR15	MSR16	MSR17	MSR18	MSR19
SAN20	SAN21	SAN22	SAN23	SAN24	SAN25	SAN26
22	23	24	25	26	27	28
JU4	JU5	JU6	JU7	JU8	JU9	JU10
MSR20	MRS21	MSR22	MSR23	MSR24	MSR25	MSR26
SAN27	SAN28	SAN29	SAN30	HAM1	HAM2	HAM3

TAIFA WEUSI 563/2079-80/6315-16/2071-72

Ashe	Ojobo	Iwa	Araorun	Ile	Itutu	Erinle
1	2	3	4	5	6	7
AU8	AU9	AU10	AU11	AU12	AU13	AU14
THT20	THT21	THT22	THT23	THT24	THT25	THT26
NAH2	NAH3	NAH4	NAH5	NAH6	NAH7	NAH8
8	9	10	11	12	13	14
AU15	AU16	AU17	AU18	AU19	AU20	AU21
THT27	THT28	THT29	THT30	PAP1	PAP1	PAP2
NAH9	NAH10	NAH11	NAH12	NAH13	NAH14	NAH15
15	16	17	18	19	20	21
AU22	AU23	AU24	AU25	AU26	AU27	AU28
PAP3	PAP4	PAP5	PAP6	PAP7	PAP8	PAP9
NAH16	NAH17	NAH18	NAH19	NAH20	NAH21	NAH22
22	23	24	25	26	27	28
AU29	AU30	AU31	S1	S2	S3	S4
PAP10	PAP11	PAP12	PAP13	PAP14	PAP15	PAP16
NAH23	NAH24	NAH25	NAH26	NAH27	NAH28	NAH29

MTU WEUSI 563/2079-80/6315-16/2071-72

Ashe	Ojobo	Iwa	Araorun	Ile	Itutu	Erinle
1	2	3	4	5	6	7
JU11	JU12	JU13	JU14	JU15	JU16	JU17
MSR27	MSR28	MSR29	MSR30	*ASR1*	*HRU2*	*STH3*
HAM4	HAM5	HAM6	HAM7	HAM8	HAM9	HAM10
8	9	10	11	12	13	14
JU18	JU19	JU20	JU21	JU22	JU23	JU24
AST4	*NBH5*	THT1	THT2	THT3	THT4	THT5
HAM11	HAM12	HAM13	HAM14	HAM15	HAM16	HAM17
15	16	17	18	19	20	21
JU25	JU26	JU27	JU28	JU29	JU30	JU31
THT6	THT7	THT8	THT9	THT10	THT11	THT12
HAM18	HAM19	HAM20	HAM21	HAM22	HAM23	HAM24
22	23	24	25	26	27	28
AU1	AU2	AU3	AU4	AU5	AU6	AU7
THT13	THT14	THT15	THT16	THT17	THT18	THT19
HAM25	HAM26	HAM27	HAM28	HAM29	HAM30	NAH1

UWEZO 563/2079-80/6315-16/2071-72

Ashe	Ojobo	Iwa	Araorun	Ile	Itutu	Erinle
1	2	3	4	5	6	7
S5	S6	S7	S8	S9	S10	S11
PAP17	PAP18	PAP19	PAP20	PAP21	PAP22	PAP23
NAH30	*QUA1*	*QUA2*	*QUA3*	*QUA4*	*QUA5*	*QUA6*
8	9	10	11	12	13	14
S12	S13	S14	S15	S16	S17	S18
PAP24	PAP25	PAP26	PAP27	PAP28	PAP29	PAP30
MAS1	MAS2	MAS3	MAS4	MAS5	MAS6	MAS7
15	16	17	18	19	20	21
S19	S20	S21	S22	S23	S24	S25
HTH1	HTH2	HTH3	HTH4	HTH5	HTH6	HTH7
MAS8	MAS9	MAS10	MAS11	MAS12	MAS13	MAS14
22	23	24	25	26	27	28
S26	S27	S28	S29	S30	O1	O2
HTH8	HTH9	HTH10	HTH11	HTH12	HTH13	HTH14
MAS15	MAS16	MAS17	MAS18	MAS19	MAS20	MAS21

THE NU AFRIKAN CALENDAR

EKUNDU 563/2079-80/6315-16/2071-72						
Ashe	Ojobo	Iwa	Araorun	Ile	Itutu	Erinle
1	2	3	4	5	6	7
O3	O4	O5	O6	O7	O8	O9
HTH15	HTH16	HTH17	HTH18	HTH19	HTH20	HTH21
MAS22	MAS23	MAS 24	MAS25	MAS26	MAS27	MAS28
8	9	10	11	12	13	14
O10	O11	O12	O13	O14	O15	O16
HTH22	HTH23	HTH24	HTH25	HTH26	HTH27	HTH28
MAS29	MAS30	THi1	THi2	THi3	THi4	THi5
15	16	17	18	19	20	21
O17	O18	O19	O20	O21	O22	O23
HTH29	HTH30	KHK1	KHK2	KHK3	KHK4	KHK5
THi6	THi7	THi8	THi9	THi10	THi11	THi12
22	23	24	25	26	27	28
O24	O25	O26	O27	O28	O29	O30
KHK6	KHK7	KHK8	KHK9	KHK10	KHK11	KHK12
THi13	THi14	THi15	THi16	THi17	THi18	THi19

KIJANI 563/2079-80/6315-16/2071-72						
Ashe	Ojobo	Iwa	Araorun	Ile	Itutu	Erinle
1	2	3	4	5	6	7
N28	N29	N30	D1	D2	D3	D4
TBT11	TBT12	TBT13	TBT14	TBT15	TBT16	TBT17
HID18	HID19	HID20	HID21	HID22	HID23	HID24
8	9	10	11	12	13	14
D5	D6	D7	D8	D9	D10	D11
TBT18	TBT19	TBT20	TBT21	TBT22	TBT23	TBT24
HID25	HID26	HID27	HID28	HID29	HID30	TAH1
15	16	17	18	19	20	21
D12	D13	D14	D15	D16	D17	D18
TBT25	TBT26	TBT27	TBT28	TBT29	TBT30	MKH1
TAH2	TAH3	TAH4	TAH5	TAH6	TAH7	TAH8
22	23	24	25	26	27	28
D19	D20	D21	D22	D23	D24	D25
MKH2	MKH3	MKH4	MKH5	MKH6	MKH7	MKH8
TAH9	TAH10	TAH11	TAH12	TAH13	TAH14	TAH15

EUSI 563/2079-80/6315-16/2071-72						
Ashe	Ojobo	Iwa	Araorun	Ile	Itutu	Erinle
1	2	3	4	5	6	7
O31	N1	N2	N3	N4	N5	N6
KHK13	KHK14	KHK15	KHK16	KHK17	KHK18	KHK19
THi20	THi21	THi22	THi23	THi24	THi25	THi26
8	9	10	11	12	13	14
N7	N8	N9	N10	N11	N12	N13
KHK20	KHK21	KHK22	KHK23	KHK24	KHK25	KHK26
THi27	THi28	THi29	THi30	HID1	HID2	HID3
15	16	17	18	19	20	21
N14	N15	N16	N17	N18	N19	N20
KHK27	KHK28	KHK29	KHK30	TBT1	TBT2	TBT3
HID4	HID5	HID6	HID7	HID8	HID9	HID10
22	23	24	25	26	27	28
N21	N22	N23	N24	N25	N26	N27
TBT4	TBT5	TBT6	TBT7	TBT8	TBT9	TBT10
HID11	HID12	HID13	HID14	HID15	HID16	HID17

KWANZA 563/2079-80/6315-16/2071-72						
Umoja	Kujichagulia	Ujima	Ujamaa	Nia	Kuumba	Imani
1	2	3	4	5	6	7
D26	D27	D28	D29	D30	D31	JA1
MKH9	MKH10	MKH11	MKH12	MKH13	MKH14	MKH15
TAH16	TAH17	TAH18	TAH19	TAH20	TAH21	TAH22
8	9	10	11	12	13	14
JA2	JA3	JA4	JA5	JA6	JA7	JA8
MKH16	MKH17	MKH18	MKH19	MKH20	MKH21	MKH22
TAH23	TAH24	TAH25	TAH26	TAH27	TAH28	TAH29
15	16	17	18	19	20	21
JA9	JA10	JA11	JA12	JA13	JA14	JA15
MKH23	MKH24	MKH25	KMH26	MKH27	MKH28	MKH29
TAH30	TIR1	TIR2	TIR3	TIR4	TIR5	TIR6
22	23	24	25	26	27	28
JA16	JA17	JA18	JA19	JA20	JA21	JA22
MKH30	AMN1	AMN2	AMN3	AMN4	AMN5	AMN6
TIR7	TIR8	TIR9	TIR10	TIR11	TIR12	TIR13

SHAHIDI 563/2079-80/6315-16/2071-72						
Ashe	Ojobo	Iwa	Araorun	Ile	Itutu	Erinle
1	2	3	4	5	6	7
JA23	JA24	JA25	JA26	JA27	JA28	JA29
AMN7	AMN8	AMN9	AMN10	AMN11	AMN12	AMN13
TIR14	TIR15	TIR16	TIR17	TIR18	TIR19	TIR20
8	9	10	11	12	13	14
JA30	JA31	F1	F2	F3	F4	F5
AMN14	AMN15	AMN16	AMN17	AMN18	AMN19	AMN20
TIR21	TIR22	TIR23	TIR24	TIR25	TIR26	TIR27
15	16	17	18	19	20	21
F6	F7	F8	F9	F10	F 11	F12
AMN21	AMN22	AMN23	AMN24	AMN25	AMN26	AMN28
TIR28	TIR29	TIR30	YEK1	YEK2	YEK3	YEK4
22	23	24	25	26	27	28
F13	F14	F15	F16	F17	F18	F19
AMN29	AMN30	RNT1	RNT2	RNT3	RNT4	RNT5
YEK5	YEK6	YEK7	YEK8	YEK9	YEK10	YEK11
SHAHIDI 114						
F20						
RNT6						
YEK12						

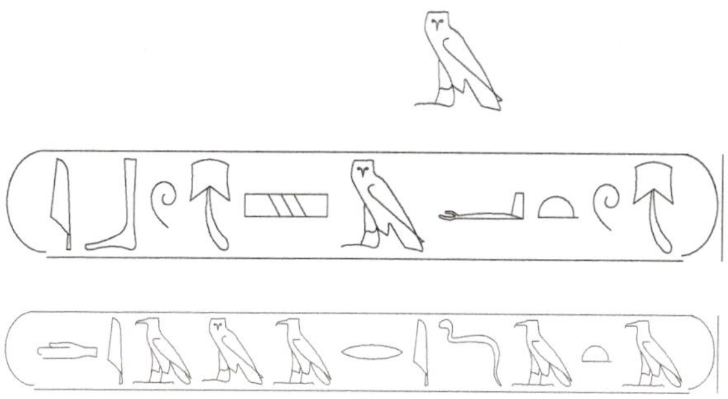

THE NU AFRIKAN CALENDAR

MAAFA 564 B.S.E./ASIENTO 2080-81 C.E./
KAM 6316-17 S.M./EC 2072-73 A.M. (ዓ.ም.)

AFU 564/2080-81/6316-17/2072-73

Ashe	Ojobo	Iwa	Araorun	Ile	Itutu	Erinle
1	2	3	4	5	6	7
F21	F22	F23	F24	F25	F26	F27
RNT7	RNT8	RNT9	RNT10	RNT11	RNT12	RNT13
YEK13	YEK14	YEK15	YEK16	YEK17	YEK18	YEK19
8	9	10	11	12	13	14
F28	F29	M1	M2	M3	M4	M5
RNT14	RNT15	RNT16	RNT17	RNT18	RNT19	RNT20
YEK 20	YEK21	YEK22	YEK23	YEK24	YEK25	YEK26
15	16	17	18	19	20	21
M6	M7	M8	M9	M10	M11	M12
RNT21	RNT22	RNT23	RNT24	RNT25	RNT26	RNT27
YEK 27	YEK28	YEK29	YEK30	MAG1	MAG2	MAG3
22	23	24	25	26	27	28
M13	M14	M15	M16	M17	M18	M19
RNT28	RNT29	RNT30	KHN1	KHN2	KHN3	KHN4
MAG4	MAG5	MAG6	MAG7	MAG8	MAG9	MAG10
SHARIFU 138						
M20						
KHN5						
MAG11						

KAWAIDA 564/2080-81/6316-17/2072-73

Ashe	Ojobo	Iwa	Araorun	Ile	Itutu	Erinle
1	2	3	4	5	6	7
A18	A19	A20	A21	A22	A23	A24
PIN4	PIN5	PIN6	PIN7	PIN8	PIN9	PIN10
MIY10	MIY11	MIY12	MIY13	MIY14	MIY15	MIY16
8	9	10	11	12	13	14
A25	A26	A27	A28	A29	A30	MA1
PIN11	PIN12	PIN13	PIN14	PIN15	PIN16	PIN17
MIY17	MIY18	MIY19	MIY20	MIY21	MIY22	MIY23
15	16	17	18	19	20	21
MA2	MA3	MA4	MA5	MA6	MA7	MA8
PIN18	PIN19	PIN20	PIN21	PIN22	PIN23	PIN24
MIY24	MIY25	MIY26	MIY27	MIY28	MIY29	MIY30
22	23	24	25	26	27	28
MA9	MA10	MA11	MA12	MA13	MA14	MA15
PIN25	PIN26	PIN27	PIN28	PIN29	PIN30	APP1
GIN1	GIN2	GIN3	GIN4	GIN5	GIN6	GIN7

UHURU 564/2080-81/6316-17/2072-73

Ashe	Ojobo	Iwa	Araorun	Ile	Itutu	Erinle
1	2	3	4	5	6	7
M21	M22	M23	M24	M25	M26	M27
KHN6	KHN7	KHN8	KHN9	KHN10	KHN11	KHN12
MAG12	MAG13	MAG14	MAG15	MAG16	MAG17	MAG18
8	9	10	11	12	13	14
M28	M29	M30	M31	A1	A2	A3
KHN13	KHN14	KHN15	KHN16	KHN17	KHN18	KHN19
MAG19	MAG20	MAG21	MAG22	MAG23	MAG24	MAG25
15	16	17	18	19	20	21
A4	A5	A6	A7	A8	A9	A10
KHN20	KHN21	KHN222	KHN23	KHN24	KHN25	KHN26
MAG26	MAG27	MAG28	MAG29	MAG30	MIY1	MIY2
22	23	24	25	26	27	28
A11	A12	A13	A14	A15	A16	A17
KHN27	KHN28	KHN29	KHN30	PIN1	PIN2	PIN3
MIY3	MIY4	MIY5	MIY6	MIY7	MIY8	MIY9

MABABA WEUSI 564/2080-81/6316-17/2072-73

Ashe	Ojobo	Iwa	Araorun	Ile	Itutu	Erinle
1	2	3	4	5	6	7
MA16	MA17	MA18	MA19	MA20	MA21	MA22
APP2	APP3	APP4	APP5	APP6	APP7	APP8
GIN8	GIN9	GIN10	GIN11	GIN12	GIN13	GIN14
8	9	10	11	12	13	14
MA23	MA24	MA25	MA26	MA27	MA28	MA29
APP9	APP10	APP11	APP12	APP13	APP14	APP15
GIN15	GIN16	GIN17	GIN18	GIN19	GIN20	GIN21
15	16	17	18	19	20	21
MA30	MA31	J1	J2	J3	J4	J5
APP16	APP17	APP18	APP19	APP20	APP21	APP22
GIN22	GIN23	GIN24	GIN25	GIN26	GIN27	GIN28
22	23	24	25	26	27	28
J6	J7	J8	J9	J10	J11	J12
APP23	APP24	APP25	APP26	APP27	APP28	APP29
GIN29	GIN30	SAN1	SAN2	SAN3	SAN4	SAN5

THE NU AFRIKAN CALENDAR

MWANAMKE WEUSI 564/2080-81/6316-17/2072-73						
Ashe	Ojobo	Iwa	Araorun	Ile	Itutu	Erinle
1	2	3	4	5	6	7
J13	J14	J15	J16	J17	J18	J19
APP30	MSR1	MSR2	MSR3	MSR4	MSR5	MSR6
SAN6	SAN7	SAN8	SAN9	SAN10	SAN11	SAN12
8	9	10	11	12	13	14
J20	J21	J22	J23	J24	J25	J26
MSR7	MSR8	MSR9	MSR10	MSR11	MSR12	MSR13
SAN13	SAN14	SAN15	SAN16	SAN17	SAN18	SAN19
15	16	17	18	19	20	21
J27	J28	J29	J30	JU1	JU2	JU3
MSR14	MSR15	MSR16	MSR17	MSR18	MSR19	MSR20
SAN20	SAN21	SAN22	SAN23	SAN24	SAN25	SAN26
22	23	24	25	26	27	28
JU4	JU5	JU6	JU7	JU8	JU9	JU10
MSR21	MRS22	MSR23	MSR24	MSR25	MSR26	MSR27
SAN27	SAN28	SAN29	SAN30	HAM1	HAM2	HAM3

TAIFA WEUSI 564/2080-81/6316-17/2072-73						
Ashe	Ojobo	Iwa	Araorun	Ile	Itutu	Erinle
1	2	3	4	5	6	7
AU8	AU9	AU10	AU11	AU12	AU13	AU14
THT20	THT21	THT22	THT23	THT24	THT25	THT26
NAH2	NAH3	NAH4	NAH5	NAH6	NAH7	NAH8
8	9	10	11	12	13	14
AU15	AU16	AU17	AU18	AU19	AU20	AU21
THT27	THT28	THT29	THT30	PAP1	PAP1	PAP2
NAH9	NAH10	NAH11	NAH12	NAH13	NAH14	NAH15
15	16	17	18	19	20	21
AU22	AU23	AU24	AU25	AU26	AU27	AU28
PAP3	PAP4	PAP5	PAP6	PAP7	PAP8	PAP9
NAH16	NAH17	NAH18	NAH19	NAH20	NAH21	NAH22
22	23	24	25	26	27	28
AU29	AU30	AU31	S1	S2	S3	S4
PAP10	PAP11	PAP12	PAP13	PAP14	PAP15	PAP16
NAH23	NAH24	NAH25	NAH26	NAH27	NAH28	NAH29

MTU WEUSI 564/2080-81/6316-17/2072-73						
Ashe	Ojobo	Iwa	Araorun	Ile	Itutu	Erinle
1	2	3	4	5	6	7
JU11	JU12	JU13	JU14	JU15	JU16	JU17
MSR28	MSR29	MSR30	*ASR1*	*HRU2*	*STH3*	*AST4*
HAM4	HAM5	HAM6	HAM7	HAM8	HAM9	HAM10
8	9	10	11	12	13	14
JU18	JU19	JU20	JU21	JU22	JU23	JU24
NBH5	*ANU6*	THT1	THT2	THT3	THT4	THT5
HAM11	HAM12	HAM13	HAM14	HAM15	HAM16	HAM17
15	16	17	18	19	20	21
JU25	JU26	JU27	JU28	JU29	JU30	JU31
THT6	THT7	THT8	THT9	THT10	THT11	THT12
HAM18	HAM19	HAM20	HAM21	HAM22	HAM23	HAM24
22	23	24	25	26	27	28
AU1	AU2	AU3	AU4	AU5	AU6	AU7
THT13	THT14	THT15	THT16	THT17	THT18	THT19
HAM25	HAM26	HAM27	HAM28	HAM29	HAM30	NAH1

UWEZO 564/2080-81/6316-17/2072-73						
Ashe	Ojobo	Iwa	Araorun	Ile	Itutu	Erinle
1	2	3	4	5	6	7
S5	S6	S7	S8	S9	S10	S11
PAP17	PAP18	PAP19	PAP20	PAP21	PAP22	PAP23
NAH30	*PAG1*	*PAG2*	*PAG3*	*PAG4*	*PAG5*	MAS1
8	9	10	11	12	13	14
S12	S13	S14	S15	S16	S17	S18
PAP24	PAP25	PAP26	PAP27	PAP28	PAP29	PAP30
MAS2	MAS3	MAS4	MAS5	MAS6	MAS7	MAS8
15	16	17	18	19	20	21
S19	S20	S21	S22	S23	S24	S25
HTH1	HTH2	HTH3	HTH4	HTH5	HTH6	HTH7
MAS9	MAS10	MAS11	MAS12	MAS13	MAS14	MAS15
22	23	24	25	26	27	28
S26	S27	S28	S29	S30	O1	O2
HTH8	HTH9	HTH10	HTH11	HTH12	HTH13	HTH14
MAS16	MAS17	MAS18	MAS19	MAS20	MAS21	MAS22

THE NU AFRIKAN CALENDAR

EKUNDU 564/2080-81/6316-17/2072-73						
Ashe	Ojobo	Iwa	Araorun	Ile	Itutu	Erinle
1	2	3	4	5	6	7
O3	O4	O5	O6	O7	O8	O9
HTH15	HTH16	HTH17	HTH18	HTH19	HTH20	HTH21
MAS23	MAS24	MAS25	MAS26	MAS27	MAS28	MAS29
8	9	10	11	12	13	14
O10	O11	O12	O13	O14	O15	O16
HTH22	HTH23	HTH24	HTH25	HTH26	HTH27	HTH28
MAS30	THi1	THi2	THi3	THi4	THi5	THi6
15	16	17	18	19	20	21
O17	O18	O19	O20	O21	O22	O23
HTH29	HTH30	KHK1	KHK2	KHK3	KHK4	KHK5
THi7	THi8	This9	THi10	THi11	THi12	THi13
22	23	24	25	26	27	28
O24	O25	O26	O27	O28	O29	O30
KHK6	KHK7	KHK8	KHK9	KHK10	KHK11	KHK12
Thi14	THi15	THi16	THi17	THi18	THi19	THi20

KIJANI 564/2080-81/6316-17/2072-73						
Ashe	Ojobo	Iwa	Araorun	Ile	Itutu	Erinle
1	2	3	4	5	6	7
N28	N29	N30	D1	D2	D3	D4
TBT11	TBT12	TBT13	TBT14	TBT15	TBT16	TBT17
HID19	HID20	HID21	HID22	HID23	HID24	HID25
8	9	10	11	12	13	14
D5	D6	D7	D8	D9	D10	D11
TBT18	TBT19	TBT20	TBT21	TBT22	TBT23	TBT24
HID26	HID27	HID28	HID29	HID30	TAH1	TAH2
15	16	17	18	19	20	21
D12	D13	D14	D15	D16	D17	D18
TBT25	TBT26	TBT27	TBT28	TBT29	TBT30	MKH1
TAH3	TAH4	TAH5	TAH6	TAH7	TAH8	TAH9
22	23	24	25	26	27	28
D19	D20	D21	D22	D23	D24	D25
MKH2	MKH3	MKH4	MKH5	MKH6	MKH7	MKH8
TAH10	TAH11	TAH12	TAH13	TAH14	TAH15	TAH16

EUSI 564/2080-81/6316-17/2072-73						
Ashe	Ojobo	Iwa	Araorun	Ile	Itutu	Erinle
1	2	3	4	5	6	7
O31	N1	N2	N3	N4	N5	N6
KHK13	KHK14	KHK15	KHK16	KHK17	KHK18	KHK19
THi21	THi22	THi23	THi24	THi25	THi26	THi27
8	9	10	11	12	13	14
N7	N8	N9	N10	N11	N12	N13
KHK20	KHK21	KHK22	KHK23	KHK24	KHK25	KHK26
THi28	THi29	THi30	HID1	HID2	HID3	HID4
15	16	17	18	19	20	21
N14	N15	N16	N17	N18	N19	N20
KHK27	KHK28	KHK29	KHK30	TBT1	TBT2	TBT3
HID5	HID6	HID7	HID8	HID9	HID10	HID11
22	23	24	25	26	27	28
N21	N22	N23	N24	N25	N26	N27
TBT4	TBT5	TBT6	TBT7	TBT8	TBT9	TBT10
HID12	HID13	HID14	HID15	HID16	HID17	HID18

KWANZA 564/2080-81/6316-17/2072-73						
Umoja	Kujichagulia	Ujima	Ujamaa	Nia	Kuumba	Imani
1	2	3	4	5	6	7
D26	D27	D28	D29	D30	D31	JA1
MKH9	MKH10	MKH11	MKH12	MKH13	MKH14	MKH15
TAH17	TAH18	TAH19	TAH20	TAH21	TAH22	TAH23
8	9	10	11	12	13	14
JA2	JA3	JA4	JA5	JA6	JA7	JA8
MKH16	MKH17	MKH18	MKH19	MKH20	MKH21	MKH22
TAH24	TAH25	TAH26	TAH27	TAH28	TAH29	TAH30
15	16	17	18	19	20	21
JA9	JA10	JA11	JA12	JA13	JA14	JA15
MKH23	MKH24	MKH25	KMH26	MKH27	MKH28	MKH29
TIR1	TIR2	TIR3	TIR4	TIR5	TIR6	TIR7
22	23	24	25	26	27	28
JA16	JA17	JA18	JA19	JA20	JA21	JA22
MKH30	AMN1	AMN2	AMN3	AMN4	AMN5	AMN6
TIR8	TIR9	TIR10	TIR11	TIR12	TIR13	TIR14

THE NU AFRIKAN CALENDAR

SHAHIDI 564/2080-81/6316-17/2072-73						
Ashe	Ojobo	Iwa	Araorun	Ile	Itutu	Erinle
1	2	3	4	5	6	7
JA23	JA24	JA25	JA26	JA27	JA28	JA29
AMN7	AMN8	AMN9	AMN10	AMN11	AMN12	AMN13
TIR15	TIR16	TIR17	TIR18	TIR19	TIR20	TIR21
8	9	10	11	12	13	14
JA30	JA31	F1	F2	F3	F4	F5
AMN14	AMN15	AMN16	AMN17	AMN18	AMN19	AMN20
TIR22	TIR23	TIR24	TIR25	TIR26	TIR27	TIR28
15	16	17	18	19	20	21
F6	F7	F8	F9	F10	F 11	F12
AMN21	AMN22	AMN23	AMN24	AMN25	AMN26	AMN28
TIR29	TIR30	YEK1	YEK2	YEK3	YEK4	YEK5
22	23	24	25	26	27	28
F13	F14	F15	F16	F17	F18	F19
AMN29	AMN30	RNT1	RNT2	RNT3	RNT4	RNT5
YEK6	YEK7	YEK8	YEK9	YEK10	YEK11	YEK12
SHAHIDI 115						
F20						
RNT6						
YEK13						

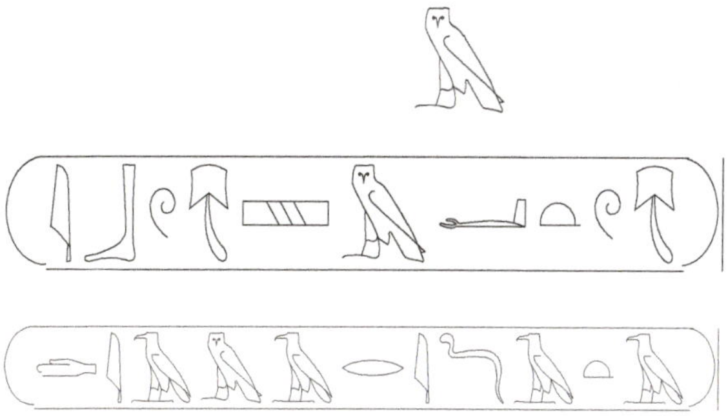

THE NU AFRIKAN CALENDAR

MAAFA 565 B.S.E./ASIENTO 2081-82 C.E.
KAM 6317-18 S.M./EC 2073-74 A.M. (ዓ.ም.)

AFU 565/2081-82/6317-18/2073-74						
Ashe	Ojobo	Iwa	Araorun	Ile	Itutu	Erinle
1	2	3	4	5	6	7
F21	F22	F23	F24	F25	F26	F27
RNT7	RNT8	RNT9	RNT10	RNT11	RNT12	RNT13
YEK14	YEK15	YEK16	YEK17	YEK18	YEK19	YEK20
8	9	10	11	12	13	14
F28	M1	M2	M3	M4	M5	M6
RNT14	RNT15	RNT16	RNT17	RNT18	RNT19	RNT20
YEK21	YEK22	YEK23	YEK24	YEK25	YEK26	YEK27
15	16	17	18	19	20	21
M7	M8	M9	M10	M11	M12	M13
RNT21	RNT22	RNT23	RNT24	RNT25	RNT26	RNT27
YEK28	YEK29	YEK30	MAG1	MAG2	MAG3	MAG4
22	23	24	25	26	27	28
M14	M15	M16	M17	M18	M19	M20
RNT28	RNT29	RNT30	KHN1	KHN2	KHN3	KHN4
MAG5	MAG6	MAG7	MAG8	MAG9	MAG10	MAG11

KAWAIDA 565/2081-82/6317-18/2073-74						
Ashe	Ojobo	Iwa	Araorun	Ile	Itutu	Erinle
1	2	3	4	5	6	7
A18	A19	A20	A21	A22	A23	A24
PIN3	PIN4	PIN5	PIN6	PIN7	PIN8	PIN9
MIY10	MIY11	MIY12	MIY13	MIY14	MIY15	MIY16
8	9	10	11	12	13	14
A25	A26	A27	A28	A29	A30	MA1
PIN10	PIN11	PIN12	PIN13	PIN14	PIN15	PIN16
MIY17	MIY18	MIY19	MIY20	MIY21	MIY22	MIY23
15	16	17	18	19	20	21
MA2	MA3	MA4	MA5	MA6	MA7	MA8
PIN17	PIN18	PIN19	PIN20	PIN21	PIN22	PIN23
MIY24	MIY25	MIY26	MIY27	MIY28	MIY29	MIY30
22	23	24	25	26	27	28
MA9	MA10	MA11	MA12	MA13	MA14	MA15
PIN24	PIN25	PIN26	PIN27	PIN28	PIN29	PIN30
GIN1	GIN2	GIN3	GIN4	GIN5	GIN6	GIN7

UHURU 565/2081-82/6317-18/2073-74						
Ashe	Ojobo	Iwa	Araorun	Ile	Itutu	Erinle
1	2	3	4	5	6	7
M21	M22	M23	M24	M25	M26	M27
KHN5	KHN6	KHN7	KHN8	KHN9	KHN10	KHN11
MAG12	MAG13	MAG14	MAG15	MAG16	MAG17	MAG18
8	9	10	11	12	13	14
M28	M29	M30	M31	A1	A2	A3
KHN12	KHN13	KHN14	KHN15	KHN16	KHN17	KHN18
MAG19	MAG20	MAG21	MAG22	MAG23	MAG24	MAG25
15	16	17	18	19	20	21
A4	A5	A6	A7	A8	A9	A10
KHN19	KHN20	KHN21	KHN22	KHN23	KHN24	KHN25
MAG26	MAG27	MAG28	MAG29	MAG30	MIY1	MIY2
22	23	24	25	26	27	28
A11	A12	A13	A14	A15	A16	A17
KHN26	KHN27	KHN28	KHN29	KHN30	PIN1	PIN2
MIY3	MIY4	MIY5	MIY6	MIY7	MIY8	MIY9

MABABA WEUSI 565/2081-82/6317-18/2073-74						
Ashe	Ojobo	Iwa	Araorun	Ile	Itutu	Erinle
1	2	3	4	5	6	7
MA16	MA17	MA18	MA19	MA20	MA21	MA22
APP1	APP2	APP3	APP4	APP5	APP6	APP7
GIN8	GIN9	GIN10	GIN11	GIN12	GIN13	GIN14
8	9	10	11	12	13	14
MA23	MA24	MA25	MA26	MA27	MA28	MA29
APP8	APP9	APP10	APP11	APP12	APP13	APP14
GIN15	GIN16	GIN17	GIN18	GIN19	GIN20	GIN21
15	16	17	18	19	20	21
MA30	MA31	J1	J2	J3	J4	J5
APP15	APP16	APP17	APP18	APP19	APP20	APP21
GIN22	GIN23	GIN24	GIN25	GIN26	GIN27	GIN28
22	23	24	25	26	27	28
J6	J7	J8	J9	J10	J11	J12
APP22	APP23	APP24	APP25	APP26	APP27	APP28
GIN29	GIN30	SAN1	SAN2	SAN3	SAN4	SAN5

THE NU AFRIKAN CALENDAR

MWANAMKE WEUSI 565/2081-82/6317-18/2073-74						
Ashe	Ojobo	Iwa	Araorun	Ile	Itutu	Erinle
1	2	3	4	5	6	7
J13	J14	J15	J16	J17	J18	J19
APP29	APP30	MSR1	MSR2	MSR3	MSR4	MSR5
SAN6	SAN7	SAN8	SAN9	SAN10	SAN11	SAN12
8	9	10	11	12	13	14
J20	J21	J22	J23	J24	J25	J26
MSR6	MSR7	MSR8	MSR9	MSR10	MSR11	MSR12
SAN13	SAN14	SAN15	SAN16	SAN17	SAN18	SAN19
15	16	17	18	19	20	21
J27	J28	J29	J30	JU1	JU2	JU3
MSR13	MSR14	MSR15	MSR16	MSR17	MSR18	MSR19
SAN20	SAN21	SAN22	SAN23	SAN24	SAN25	SAN26
22	23	24	25	26	27	28
JU4	JU5	JU6	JU7	JU8	JU9	JU10
MSR20	MRS21	MSR22	MSR23	MSR24	MSR25	MSR26
SAN27	SAN28	SAN29	SAN30	HAM1	HAM2	HAM3

TAIFA WEUSI 565/2081-82/6317-18/2073-74						
Ashe	Ojobo	Iwa	Araorun	Ile	Itutu	Erinle
1	2	3	4	5	6	7
AU8	AU9	AU10	AU11	AU12	AU13	AU14
THT20	THT21	THT22	THT23	THT24	THT25	THT26
NAH2	NAH3	NAH4	NAH5	NAH6	NAH7	NAH8
8	9	10	11	12	13	14
AU15	AU16	AU17	AU18	AU19	AU20	AU21
THT27	THT28	THT29	THT30	PAP1	PAP1	PAP2
NAH9	NAH10	NAH11	NAH12	NAH13	NAH14	NAH15
15	16	17	18	19	20	21
AU22	AU23	AU24	AU25	AU26	AU27	AU28
PAP3	PAP4	PAP5	PAP6	PAP7	PAP8	PAP9
NAH16	NAH17	NAH18	NAH19	NAH20	NAH21	NAH22
22	23	24	25	26	27	28
AU29	AU30	AU31	S1	S2	S3	S4
PAP10	PAP11	PAP12	PAP13	PAP14	PAP15	PAP16
NAH23	NAH24	NAH25	NAH26	NAH27	NAH28	NAH29

MTU WEUSI 565/2081-82/6317-18/2073-74						
Ashe	Ojobo	Iwa	Araorun	Ile	Itutu	Erinle
1	2	3	4	5	6	7
JU11	JU12	JU13	JU14	JU15	JU16	JU17
MSR27	MSR28	MSR29	MSR30	*ASR1*	*HRU2*	*STH3*
HAM4	HAM5	HAM6	HAM7	HAM8	HAM9	HAM10
8	9	10	11	12	13	14
JU18	JU19	JU20	JU21	JU22	JU23	JU24
AST4	*NBH5*	THT1	THT2	THT3	THT4	THT5
HAM11	HAM12	HAM13	AM14	HAM15	HAM16	HAM17
15	16	17	18	19	20	21
JU25	JU26	JU27	JU28	JU29	JU30	JU31
THT6	THT7	THT8	THT9	THT10	THT11	THT12
HAM18	HAM19	HAM20	HAM21	HAM22	HAM23	HAM24
22	23	24	25	26	27	28
AU1	AU2	AU3	AU4	AU5	AU6	AU7
THT13	THT14	THT15	THT16	THT17	THT18	THT19
HAM25	HAM26	HAM27	HAM28	HAM29	HAM30	NAH1

UWEZO 565/2081-82/6317-18/2073-74						
Ashe	Ojobo	Iwa	Araorun	Ile	Itutu	Erinle
1	2	3	4	5	6	7
S5	S6	S7	S8	S9	S10	S11
PAP17	PAP18	PAP19	PAP20	PAP21	PAP22	PAP23
NAH30	*PAG1*	*PAG2*	*PAG3*	*PAG4*	*PAG5*	MAS1
8	9	10	11	12	13	14
S12	S13	S14	S15	S16	S17	S18
PAP24	PAP25	PAP26	PAP27	PAP28	PAP29	PAP30
MAS2	MAS3	MAS4	MAS5	MAS6	MAS7	MAS8
15	16	17	18	19	20	21
S19	S20	S21	S22	S23	S24	S25
HTH1	HTH2	HTH3	HTH4	HTH5	HTH6	HTH7
MAS9	MAS10	MAS11	MAS12	MAS13	MAS14	MAS15
22	23	24	25	26	27	28
S26	S27	S28	S29	S30	O1	O2
HTH8	HTH9	HTH10	HTH11	HTH12	HTH13	HTH14
MAS16	MAS17	MAS18	MAS19	MAS20	MAS21	MAS22

THE NU AFRIKAN CALENDAR

EKUNDU 565/2081-82/6317-18/2073-74						
Ashe	Ojobo	Iwa	Araorun	Ile	Itutu	Erinle
1	2	3	4	5	6	7
O3	O4	O5	O6	O7	O8	O9
HTH15	HTH16	HTH17	HTH18	HTH19	HTH20	HTH21
MAS23	MAS24	MAS25	MAS26	MAS27	MAS28	MAS29
8	9	10	11	12	13	14
O10	O11	O12	O13	O14	O15	O16
HTH22	HTH23	HTH24	HTH25	HTH26	HTH27	HTH28
MAS30	THi1	THi2	THi3	THi4	THi5	THi6
15	16	17	18	19	20	21
O17	O18	O19	O20	O21	O22	O23
HTH29	HTH30	KHK1	KHK2	KHK3	KHK4	KHK5
THi7	THi8	THi9	THi10	THi11	THi12	THi13
22	23	24	25	26	27	28
O24	O25	O26	O27	O28	O29	O30
KHK6	KHK7	KHK8	KHK9	KHK10	KHK11	KHK12
THi14	THi15	THi16	THi17	THi18	THi19	THi20

KIJANI 565/2081-82/6317-18/2073-74						
Ashe	Ojobo	Iwa	Araorun	Ile	Itutu	Erinle
1	2	3	4	5	6	7
N28	N29	N30	D1	D2	D3	D4
TBT11	TBT12	TBT13	TBT14	TBT15	TBT16	TBT17
HID19	HID20	HID21	HID22	HID23	HID24	HID25
8	9	10	11	12	13	14
D5	D6	D7	D8	D9	D10	D11
TBT18	TBT19	TBT20	TBT21	TBT22	TBT23	TBT24
HID26	HID27	HID28	HID29	HID30	TAH1	TAH2
15	16	17	18	19	20	21
D12	D13	D14	D15	D16	D17	D18
TBT25	TBT26	TBT27	TBT28	TBT29	TBT30	MKH1
TAH3	TAH4	TAH5	TAH6	TAH7	TAH8	TAH9
22	23	24	25	26	27	28
D19	D20	D21	D22	D23	D24	D25
MKH2	MKH3	MKH4	MKH5	MKH6	MKH7	MKH8
TAH10	TAH11	TAH12	TAH13	TAH14	TAH15	TAH16

EUSI 565/2081-82/6317-18/2073-74						
Ashe	Ojobo	Iwa	Araorun	Ile	Itutu	Erinle
1	2	3	4	5	6	7
O31	N1	N2	N3	N4	N5	N6
KHK13	KHK14	KHK15	KHK16	KHK17	KHK18	KHK19
THi21	THi22	THi23	THi24	THi25	THi26	THi27
8	9	10	11	12	13	14
N7	N8	N9	N10	N11	N12	N13
KHK20	KHK21	KHK22	KHK23	KHK24	KHK25	KHK26
THi28	THi29	THi30	HID1	HID2	HID3	HID4
15	16	17	18	19	20	21
N14	N15	N16	N17	N18	N19	N20
KHK27	KHK28	KHK29	KHK30	TBT1	TBT2	TBT3
HID5	HID6	HID7	HID8	HID9	HID10	HID11
22	23	24	25	26	27	28
N21	N22	N23	N24	N25	N26	N27
TBT4	TBT5	TBT6	TBT7	TBT8	TBT9	TBT10
HID12	HID13	HID14	HID15	HID16	HID17	HID18

KWANZA 565/2081-82/6317-18/2073-74						
Umoja	Kujichagulia	Ujima	Ujamaa	Nia	Kuumba	Imani
1	2	3	4	5	6	7
D26	D27	D28	D29	D30	D31	JA1
MKH9	MKH10	MKH11	MKH12	MKH13	MKH14	MKH15
TAH17	TAH18	TAH19	TAH20	TAH21	TAH22	TAH23
8	9	10	11	12	13	14
JA2	JA3	JA4	JA5	JA6	JA7	JA8
MKH16	MKH17	MKH18	MKH19	MKH20	MKH21	MKH22
TAH24	TAH25	TAH26	TAH27	TAH28	TAH29	TAH30
15	16	17	18	19	20	21
JA9	JA10	JA11	JA12	JA13	JA14	JA15
MKH23	MKH24	MKH25	KMH26	MKH27	MKH28	MKH29
TIR1	TIR2	TIR3	TIR4	TIR5	TIR6	TIR7
22	23	24	25	26	27	28
JA16	JA17	JA18	JA19	JA20	JA21	JA22
MKH30	AMN1	AMN2	AMN3	AMN4	AMN5	AMN6
TIR8	TIR9	TIR10	TIR11	TIR12	TIR13	TIR14

THE NU AFRIKAN CALENDAR

SHAHIDI 565/2081-82/6317-18/2073-74						
Ashe	Ojobo	Iwa	Araorun	Ile	Itutu	Erinle
1	2	3	4	5	6	7
JA23	JA24	JA25	JA26	JA27	JA28	JA29
AMN7	AMN8	AMN9	AMN10	AMN11	AMN12	AMN13
TIR15	TIR16	TIR17	TIR18	TIR19	TIR20	TIR21
8	9	10	11	12	13	14
JA30	JA31	F1	F2	F3	F4	F5
AMN14	AMN15	AMN16	AMN17	AMN18	AMN19	AMN20
TIR22	TIR23	TIR24	TIR25	TIR26	TIR27	TIR28
15	16	17	18	19	20	21
F6	F7	F8	F9	F10	F 11	F12
AMN21	AMN22	AMN23	AMN24	AMN25	AMN26	AMN28
TIR29	TIR30	YEK1	YEK2	YEK3	YEK4	YEK5
22	23	24	25	26	27	28
F13	F14	F15	F16	F17	F18	F19
AMN29	AMN30	RNT1	RNT2	RNT3	RNT4	RNT5
YEK6	YEK7	YEK8	YEK9	YEK10	YEK11	YEK12
SHAHIDI 116						
F20						
RNT6						
YEK13						

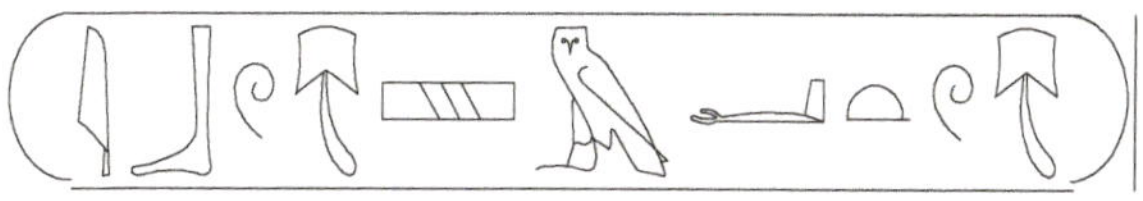

THE NU AFRIKAN CALENDAR

MAAFA 566 B.S.E./ASIENTO 2082-83 C.E.
KAM 6318-19 S.M./EC 2074-75 A.M. (ዓ.ም.)

AFU 566/2082-83/6318-19/2074-75

Ashe	Ojobo	Iwa	Araorun	Ile	Itutu	Erinle
1	2	3	4	5	6	7
F21	F22	F23	F24	F25	F26	F27
RNT7	RNT8	RNT9	RNT10	RNT11	RNT12	RNT13
YEK14	YEK15	YEK16	YEK17	YEK18	YEK19	YEK20
8	9	10	11	12	13	14
F28	M1	M2	M3	M4	M5	M6
RNT14	RNT15	RNT16	RNT17	RNT18	RNT19	RNT20
YEK21	YEK22	YEK23	YEK24	YEK25	YEK26	YEK27
15	16	17	18	19	20	21
M7	M8	M9	M10	M11	M12	M13
RNT21	RNT22	RNT23	RNT24	RNT25	RNT26	RNT27
YEK28	YEK29	YEK30	MAG1	MAG2	MAG3	MAG4
22	23	24	25	26	27	28
M14	M15	M16	M17	M18	M19	M20
RNT28	RNT29	RNT30	KHN1	KHN2	KHN3	KHN4
MAG5	MAG6	MAG7	MAG8	MAG9	MAG10	MAG11

KAWAIDA 566/2082-83/6318-19/2074-75

Ashe	Ojobo	Iwa	Araorun	Ile	Itutu	Erinle
1	2	3	4	5	6	7
A18	A19	A20	A21	A22	A23	A24
PIN3	PIN4	PIN5	PIN6	PIN7	PIN8	PIN9
MIY10	MIY11	MIY12	MIY13	MIY14	MIY15	MIY16
8	9	10	11	12	13	14
A25	A26	A27	A28	A29	A30	MA1
PIN10	PIN11	PIN12	PIN13	PIN14	PIN15	PIN16
MIY17	MIY18	MIY19	MIY20	MIY21	MIY22	MIY23
15	16	17	18	19	20	21
MA2	MA3	MA4	MA5	MA6	MA7	MA8
PIN17	PIN18	PIN19	PIN20	PIN21	PIN22	PIN23
MIY24	MIY25	MIY26	MIY27	MIY28	MIY29	MIY30
22	23	24	25	26	27	28
MA9	MA10	MA11	MA12	MA13	MA14	MA15
PIN24	PIN25	PIN26	PIN27	PIN28	PIN29	PIN30
GIN1	GIN2	GIN3	GIN4	GIN5	GIN6	GIN7

UHURU 566/2082-83/6318-19/2074-75

Ashe	Ojobo	Iwa	Araorun	Ile	Itutu	Erinle
1	2	3	4	5	6	7
M21	M22	M23	M24	M25	M26	M27
KHN5	KHN6	KHN7	KHN8	KHN9	KHN10	KHN11
MAG12	MAG13	MAG14	MAG15	MAG16	MAG17	MAG18
8	9	10	11	12	13	14
M28	M29	M30	M31	A1	A2	A3
KHN12	KHN13	KHN14	KHN15	KHN16	KHN17	KHN18
MAG19	MAG20	MAG21	MAG22	MAG23	MAG24	MAG25
15	16	17	18	19	20	21
A4	A5	A6	A7	A8	A9	A10
KHN19	KHN20	KHN21	KHN22	KHN23	KHN24	KHN25
MAG26	MAG27	MAG28	MAG29	MAG30	MIY1	MIY2
22	23	24	25	26	27	28
A11	A12	A13	A14	A15	A16	A17
KHN26	KHN27	KHN28	KHN29	KHN30	PIN1	PIN2
MIY3	MIY4	MIY5	MIY6	MIY7	MIY8	MIY9

MABABA WEUSI 566/2082-83/6318-19/2074-75

Ashe	Ojobo	Iwa	Araorun	Ile	Itutu	Erinle
1	2	3	4	5	6	7
MA16	MA17	MA18	MA19	MA20	MA21	MA22
APP1	APP2	APP3	APP4	APP5	APP6	APP7
GIN8	GIN9	GIN10	GIN11	GIN12	GIN13	GIN14
8	9	10	11	12	13	14
MA23	MA24	MA25	MA26	MA27	MA28	MA29
APP8	APP9	APP10	APP11	APP12	APP13	APP14
GIN15	GIN16	GIN17	GIN18	GIN19	GIN20	GIN21
15	16	17	18	19	20	21
MA30	MA31	J1	J2	J3	J4	J5
APP15	APP16	APP17	APP18	APP19	APP20	APP21
GIN22	GIN23	GIN24	GIN25	GIN26	GIN27	GIN28
22	23	24	25	26	27	28
J6	J7	J8	J9	J10	J11	J12
APP22	APP23	APP24	APP25	APP26	APP27	APP28
GIN29	GIN30	SAN1	SAN2	SAN3	SAN4	SAN5

THE NU AFRIKAN CALENDAR

MWANAMKE WEUSI 566/2082-83/6318-19/2074-75

Ashe	Ojobo	Iwa	Araorun	Ile	Itutu	Erinle
1	2	3	4	5	6	7
J13	J14	J15	J16	J17	J18	J19
APP29	APP30	MSR1	MSR2	MSR3	MSR4	MSR5
SAN6	SAN7	SAN8	SAN9	SAN10	SAN11	SAN12
8	9	10	11	12	13	14
J20	J21	J22	J23	J24	J25	J26
MSR6	MSR7	MSR8	MSR9	MSR10	MSR11	MSR12
SAN13	SAN14	SAN15	SAN16	SAN17	SAN18	SAN19
15	16	17	18	19	20	21
J27	J28	J29	J30	JU1	JU2	JU3
MSR13	MSR14	MSR15	MSR16	MSR17	MSR18	MSR19
SAN20	SAN21	SAN22	SAN23	SAN24	SAN25	SAN26
22	23	24	25	26	27	28
JU4	JU5	JU6	JU7	JU8	JU9	JU10
MSR20	MRS21	MSR22	MSR23	MSR24	MSR25	MSR26
SAN27	SAN28	SAN29	SAN30	HAM1	HAM2	HAM3

TAIFA WEUSI 566/2082-83/6318-19/2074-75

Ashe	Ojobo	Iwa	Araorun	Ile	Itutu	Erinle
1	2	3	4	5	6	7
AU8	AU9	AU10	AU11	AU12	AU13	AU14
THT20	THT21	THT22	THT23	THT24	THT25	THT26
NAH2	NAH3	NAH4	NAH5	NAH6	NAH7	NAH8
8	9	10	11	12	13	14
AU15	AU16	AU17	AU18	AU19	AU20	AU21
THT27	THT28	THT29	THT30	PAP1	PAP1	PAP2
NAH9	NAH10	NAH11	NAH12	NAH13	NAH14	NAH15
15	16	17	18	19	20	21
AU22	AU23	AU24	AU25	AU26	AU27	AU28
PAP3	PAP4	PAP5	PAP6	PAP7	PAP8	PAP9
NAH16	NAH17	NAH18	NAH19	NAH20	NAH21	NAH22
22	23	24	25	26	27	28
AU29	AU30	AU31	S1	S2	S3	S4
PAP10	PAP11	PAP12	PAP13	PAP14	PAP15	PAP16
NAH23	NAH24	NAH25	NAH26	NAH27	NAH28	NAH29

MTU WEUSI 566/2082-83/6318-19/2074-75

Ashe	Ojobo	Iwa	Araorun	Ile	Itutu	Erinle
1	2	3	4	5	6	7
JU11	JU12	JU13	JU14	JU15	JU16	JU17
MSR27	MSR28	MSR29	MSR30	*ASR1*	*HRU2*	*STH3*
HAM4	HAM5	HAM6	HAM7	HAM8	HAM9	HAM10
8	9	10	11	12	13	14
JU18	JU19	JU20	JU21	JU22	JU23	JU24
AST4	*NBH5*	THT1	THT2	THT3	THT4	THT5
HAM11	HAM12	HAM13	AM14	HAM15	HAM16	HAM17
15	16	17	18	19	20	21
JU25	JU26	JU27	JU28	JU29	JU30	JU31
THT6	THT7	THT8	THT9	THT10	THT11	THT12
HAM18	HAM19	HAM20	HAM21	HAM22	HAM23	HAM24
22	23	24	25	26	27	28
AU1	AU2	AU3	AU4	AU5	AU6	AU7
THT13	THT14	THT15	THT16	THT17	THT18	THT19
HAM25	HAM26	HAM27	HAM28	HAM29	HAM30	NAH1

UWEZO 566/2082-83/6318-19/2074-75

Ashe	Ojobo	Iwa	Araorun	Ile	Itutu	Erinle
1	2	3	4	5	6	7
S5	S6	S7	S8	S9	S10	S11
PAP17	PAP18	PAP19	PAP20	PAP21	PAP22	PAP23
NAH30	*PAG1*	*PAG2*	*PAG3*	*PAG4*	*PAG5*	MAS1
8	9	10	11	12	13	14
S12	S13	S14	S15	S16	S17	S18
PAP24	PAP25	PAP26	PAP27	PAP28	PAP29	PAP30
MAS2	MAS3	MAS4	MAS5	MAS6	MAS7	MAS8
15	16	17	18	19	20	21
S19	S20	S21	S22	S23	S24	S25
HTH1	HTH2	HTH3	HTH4	HTH5	HTH6	HTH7
MAS9	MAS10	MAS11	MAS12	MAS13	MAS14	MAS15
22	23	24	25	26	27	28
S26	S27	S28	S29	S30	O1	O2
HTH8	HTH9	HTH10	HTH11	HTH12	HTH13	HTH14
MAS16	MAS17	MAS18	MAS19	MAS20	MAS21	MAS22

THE NU AFRIKAN CALENDAR

EKUNDU 566/2082-83/6318-19/2074-75						
Ashe	Ojobo	Iwa	Araorun	Ile	Itutu	Erinle
1	2	3	4	5	6	7
O3	O4	O5	O6	O7	O8	O9
HTH15	HTH16	HTH17	HTH18	HTH19	HTH20	HTH21
MAS23	MAS24	MAS25	MAS26	MAS27	MAS28	MAS29
8	9	10	11	12	13	14
O10	O11	O12	O13	O14	O15	O16
HTH22	HTH23	HTH24	HTH25	HTH26	HTH27	HTH28
MAS30	THi1	THi2	THi3	THi4	THi5	THi6
15	16	17	18	19	20	21
O17	O18	O19	O20	O21	O22	O23
HTH29	HTH30	KHK1	KHK2	KHK3	KHK4	KHK5
THi7	THi8	THi9	THi10	THi11	THi12	THi13
22	23	24	25	26	27	28
O24	O25	O26	O27	O28	O29	O30
KHK6	KHK7	KHK8	KHK9	KHK10	KHK11	KHK12
THi14	THi15	THi16	THi17	THi18	THi19	THi20

KIJANI 566/2082-83/6318-19/2074-75						
Ashe	Ojobo	Iwa	Araorun	Ile	Itutu	Erinle
1	2	3	4	5	6	7
N28	N29	N30	D1	D2	D3	D4
TBT11	TBT12	TBT13	TBT14	TBT15	TBT16	TBT17
HID19	HID20	HID21	HID22	HID23	HID24	HID25
8	9	10	11	12	13	14
D5	D6	D7	D8	D9	D10	D11
TBT18	TBT19	TBT20	TBT21	TBT22	TBT23	TBT24
HID26	HID27	HID28	HID29	HID30	TAH1	TAH2
15	16	17	18	19	20	21
D12	D13	D14	D15	D16	D17	D18
TBT25	TBT26	TBT27	TBT28	TBT29	TBT30	MKH1
TAH3	TAH4	TAH5	TAH6	TAH7	TAH8	TAH9
22	23	24	25	26	27	28
D19	D20	D21	D22	D23	D24	D25
MKH2	MKH3	MKH4	MKH5	MKH6	MKH7	MKH8
TAH10	TAH11	TAH12	TAH13	TAH14	TAH15	TAH16

EUSI 566/2082-83/6318-19/2074-75						
Ashe	Ojobo	Iwa	Araorun	Ile	Itutu	Erinle
1	2	3	4	5	6	7
O31	N1	N2	N3	N4	N5	N6
KHK13	KHK14	KHK15	KHK16	KHK17	KHK18	KHK19
THi21	THi22	THi23	THi24	THi25	THi26	THi27
8	9	10	11	12	13	14
N7	N8	N9	N10	N11	N12	N13
KHK20	KHK21	KHK22	KHK23	KHK24	KHK25	KHK26
THi28	THi29	THi30	HID1	HID2	HID3	HID4
15	16	17	18	19	20	21
N14	N15	N16	N17	N18	N19	N20
KHK27	KHK28	KHK29	KHK30	TBT1	TBT2	TBT3
HID5	HID6	HID7	HID8	HID9	HID10	HID11
22	23	24	25	26	27	28
N21	N22	N23	N24	N25	N26	N27
TBT4	TBT5	TBT6	TBT7	TBT8	TBT9	TBT10
HID12	HID13	HID14	HID15	HID16	HID17	HID18

KWANZA 566/2082-83/6318-19/2074-75						
Umoja	Kujichagulia	Ujima	Ujamaa	Nia	Kuumba	Imani
1	2	3	4	5	6	7
D26	D27	D28	D29	D30	D31	JA1
MKH9	MKH10	MKH11	MKH12	MKH13	MKH14	MKH15
TAH17	TAH18	TAH19	TAH20	TAH21	TAH22	TAH23
8	9	10	11	12	13	14
JA2	JA3	JA4	JA5	JA6	JA7	JA8
MKH16	MKH17	MKH18	MKH19	MKH20	MKH21	MKH22
TAH24	TAH25	TAH26	TAH27	TAH28	TAH29	TAH30
15	16	17	18	19	20	21
JA9	JA10	JA11	JA12	JA13	JA14	JA15
MKH23	MKH24	MKH25	KMH26	MKH27	MKH28	MKH29
TIR1	TIR2	TIR3	TIR4	TIR5	TIR6	TIR7
22	23	24	25	26	27	28
JA16	JA17	JA18	JA19	JA20	JA21	JA22
MKH30	AMN1	AMN2	AMN3	AMN4	AMN5	AMN6
TIR8	TIR9	TIR10	TIR11	TIR12	TIR13	TIR14

THE NU AFRIKAN CALENDAR

SHAHIDI 566/2082-83/6318-19/2074-75						
Ashe	Ojobo	Iwa	Araorun	Ile	Itutu	Erinle
1	2	3	4	5	6	7
JA23	JA24	JA25	JA26	JA27	JA28	JA29
AMN7	AMN8	AMN9	AMN10	AMN11	AMN12	AMN13
TIR15	TIR16	TIR17	TIR18	TIR19	TIR20	TIR21
8	9	10	11	12	13	14
JA30	JA31	F1	F2	F3	F4	F5
AMN14	AMN15	AMN16	AMN17	AMN18	AMN19	AMN20
TIR22	TIR23	TIR24	TIR25	TIR26	TIR27	TIR28
15	16	17	18	19	20	21
F6	F7	F8	F9	F10	F 11	F12
AMN21	AMN22	AMN23	AMN24	AMN25	AMN26	AMN28
TIR29	TIR30	YEK1	YEK2	YEK3	YEK4	YEK5
22	23	24	25	26	27	28
F13	F14	F15	F16	F17	F18	F19
AMN29	AMN30	RNT1	RNT2	RNT3	RNT4	RNT5
YEK6	YEK7	YEK8	YEK9	YEK10	YEK11	YEK12
SHAHIDI 117						
F20						
RNT6						
YEK13						

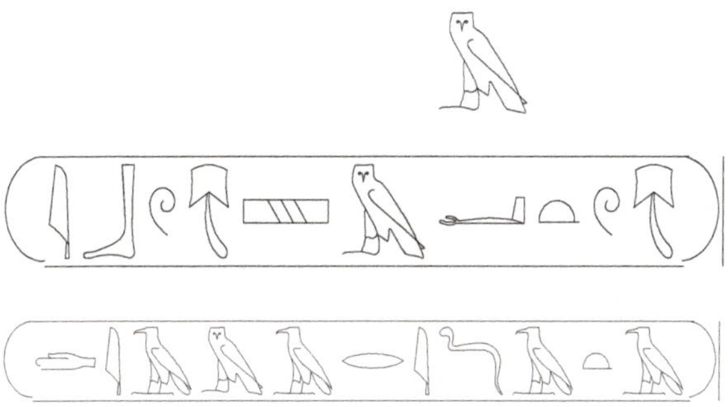

THE NU AFRIKAN CALENDAR

MAAFA 567 B.S.E./ASIENTO 2083-84 C.E./
KAM 6319-20 S.M./EC 2075-76 A.M. (ዓ.ም.)

AFU 567/2083-84/6319-20/2075-76						
Ashe	Ojobo	Iwa	Araorun	Ile	Itutu	Erinle
1	2	3	4	5	6	7
F21	F22	F23	F24	F25	F26	F27
RNT7	RNT8	RNT9	RNT10	RNT11	RNT12	RNT13
YEK14	YEK15	YEK16	YEK17	YEK18	YEK19	YEK20
8	9	10	11	12	13	14
F28	M1	M2	M3	M4	M5	M6
RNT14	RNT15	RNT16	RNT17	RNT18	RNT19	RNT20
YEK21	YEK22	YEK23	YEK24	YEK25	YEK26	YEK27
15	16	17	18	19	20	21
M7	M8	M9	M10	M11	M12	M13
RNT21	RNT22	RNT23	RNT24	RNT25	RNT26	RNT27
YEK28	YEK 29	YEK30	MAG1	MAG2	MAG3	MAG4
22	23	24	25	26	27	28
M14	M15	M16	M17	M18	M19	M20
RNT28	RNT29	RNT30	KHN1	KHN2	KHN3	KHN4
MAG5	MAG6	MAG7	MAG8	MAG9	MAG10	MAG11

UHURU 567/2083-84/6319-20/2075-76						
Ashe	Ojobo	Iwa	Araorun	Ile	Itutu	Erinle
1	2	3	4	5	6	7
M21	M22	M23	M24	M25	M26	M27
KHN5	KHN6	KHN7	KHN8	KHN9	KHN10	KHN11
MAG12	MAG13	MAG14	MAG15	MAG16	MAG17	MAG18
8	9	10	11	12	13	14
M28	M29	M30	M31	A1	A2	A3
KHN12	KHN13	KHN14	KHN15	KHN16	KHN17	KHN18
MAG19	MAG20	MAG21	MAG22	MAG23	MAG24	MAG25
15	16	17	18	19	20	21
A4	A5	A6	A7	A8	A9	A10
KHN19	KHN20	KHN21	KHN22	KHN23	KHN24	KHN25
MAG26	MAG27	MAG28	MAG29	MAG30	MIY1	MIY2
22	23	24	25	26	27	28
A11	A12	A13	A14	A15	A16	A17
KHN26	KHN27	KHN28	KHN29	KHN30	PIN1	PIN2
MIY3	MIY4	MIY5	MIY6	MIY7	MIY8	MIY9

KAWAIDA 567/2083-84/6319-20/2075-76						
Ashe	Ojobo	Iwa	Araorun	Ile	Itutu	Erinle
1	2	3	4	5	6	7
A18	A19	A20	A21	A22	A23	A24
PIN3	PIN4	PIN5	PIN6	PIN7	PIN8	PIN9
MIY10	MIY11	MIY12	MIY13	MIY14	MIY15	MIY16
8	9	10	11	12	13	14
A25	A26	A27	A28	A29	A30	MA1
PIN10	PIN11	PIN12	PIN13	PIN14	PIN15	PIN16
MIY17	MIY18	MIY19	MIY20	MIY21	MIY22	MIY23
15	16	17	18	19	20	21
MA2	MA3	MA4	MA5	MA6	MA7	MA8
PIN17	PIN18	PIN19	PIN20	PIN21	PIN22	PIN23
MIY24	MIY25	MIY26	MIY27	MIY28	MIY29	MIY30
22	23	24	25	26	27	28
MA9	MA10	MA11	MA12	MA13	MA14	MA15
PIN24	PIN25	PIN26	PIN27	PIN28	PIN29	PIN30
GIN1	GIN2	GIN3	GIN4	GIN5	GIN6	GIN7

MABABA WEUSI 567/2083-84/6319-20/2075-76						
Ashe	Ojobo	Iwa	Araorun	Ile	Itutu	Erinle
1	2	3	4	5	6	7
MA16	MA17	MA18	MA19	MA20	MA21	MA22
APP1	APP2	APP3	APP4	APP5	APP6	APP7
GIN8	GIN9	GIN10	GIN11	GIN12	GIN13	GIN14
8	9	10	11	12	13	14
MA23	MA24	MA25	MA26	MA27	MA28	MA29
APP8	APP9	APP10	APP11	APP12	APP13	APP14
GIN15	GIN16	GIN17	GIN18	GIN19	GIN20	GIN21
15	16	17	18	19	20	21
MA30	MA31	J1	J2	J3	J4	J5
APP15	APP16	APP17	APP18	APP19	APP20	APP21
GIN22	GIN23	GIN24	GIN25	GIN26	GIN27	GIN28
22	23	24	25	26	27	28
J6	J7	J8	J9	J10	J11	J12
APP22	APP23	APP24	APP25	APP26	APP27	APP28
GIN29	GIN30	SAN1	SAN2	SAN3	SAN4	SAN5

THE NU AFRIKAN CALENDAR

MWANAMKE WEUSI 567/2083-84/6319-20/2075-76

Ashe	Ojobo	Iwa	Araorun	Ile	Itutu	Erinle
1	2	3	4	5	6	7
J13	J14	J15	J16	J17	J18	J19
APP29	APP30	MSR1	MSR2	MSR3	MSR4	MSR5
SAN6	SAN7	SAN8	SAN9	SAN10	SAN11	SAN12
8	9	10	11	12	13	14
J20	J21	J22	J23	J24	J25	J26
MSR6	MSR7	MSR8	MSR9	MSR10	MSR11	MSR12
SAN13	SAN14	SAN15	SAN16	SAN17	SAN18	SAN19
15	16	17	18	19	20	21
J27	J28	J29	J30	JU1	JU2	JU3
MSR13	MSR14	MSR15	MSR16	MSR17	MSR18	MSR19
SAN20	SAN21	SAN22	SAN23	SAN24	SAN25	SAN26
22	23	24	25	26	27	28
JU4	JU5	JU6	JU7	JU8	JU9	JU10
MSR20	MRS21	MSR22	MSR23	MSR24	MSR25	MSR26
SAN27	SAN28	SAN29	SAN30	HAM1	HAM2	HAM3

TAIFA WEUSI 567/2083-84/6319-20/2075-76

Ashe	Ojobo	Iwa	Araorun	Ile	Itutu	Erinle
1	2	3	4	5	6	7
AU8	AU9	AU10	AU11	AU12	AU13	AU14
THT20	THT21	THT22	THT23	THT24	THT25	THT26
NAH2	NAH3	NAH4	NAH5	NAH6	NAH7	NAH8
8	9	10	11	12	13	14
AU15	AU16	AU17	AU18	AU19	AU20	AU21
THT27	THT28	THT29	THT30	PAP1	PAP1	PAP2
NAH9	NAH10	NAH11	NAH12	NAH13	NAH14	NAH15
15	16	17	18	19	20	21
AU22	AU23	AU24	AU25	AU26	AU27	AU28
PAP3	PAP4	PAP5	PAP6	PAP7	PAP8	PAP9
NAH16	NAH17	NAH18	NAH19	NAH20	NAH21	NAH22
22	23	24	25	26	27	28
AU29	AU30	AU31	S1	S2	S3	S4
PAP10	PAP11	PAP12	PAP13	PAP14	PAP15	PAP16
NAH23	NAH24	NAH25	NAH26	NAH27	NAH28	NAH29

MTU WEUSI 567/2083-84/6319-20/2075-76

Ashe	Ojobo	Iwa	Araorun	Ile	Itutu	Erinle
1	2	3	4	5	6	7
JU11	JU12	JU13	JU14	JU15	JU16	JU17
MSR27	MSR28	MSR29	MSR30	*ASR1*	*HRU2*	*STH3*
HAM4	HAM5	HAM6	HAM7	HAM8	HAM9	HAM10
8	9	10	11	12	13	14
JU18	JU19	JU20	JU21	JU22	JU23	JU24
AST4	*NBH5*	THT1	THT2	THT3	THT4	THT5
HAM11	HAM12	HAM13	HAM14	HAM15	HAM16	HAM17
15	16	17	18	19	20	21
JU25	JU26	JU27	JU28	JU29	JU30	JU31
THT6	THT7	THT8	THT9	THT10	THT11	THT12
HAM18	HAM19	HAM20	HAM21	HAM22	HAM23	HAM24
22	23	24	25	26	27	28
AU1	AU2	AU3	AU4	AU5	AU6	AU7
THT13	THT14	THT15	THT16	THT17	THT18	THT19
HAM25	HAM26	HAM27	HAM28	HAM29	HAM30	NAH1

UWEZO 567/2083-84/6319-20/2075-76

Ashe	Ojobo	Iwa	Araorun	Ile	Itutu	Erinle
1	2	3	4	5	6	7
S5	S6	S7	S8	S9	S10	S11
PAP17	PAP18	PAP19	PAP20	PAP21	PAP22	PAP23
NAH30	*QUA1*	*QUA2*	*QUA3*	*QUA4*	*QUA5*	*QUA6*
8	9	10	11	12	13	14
S12	S13	S14	S15	S16	S17	S18
PAP24	PAP25	PAP26	PAP27	PAP28	PAP29	PAP30
MAS1	MAS2	MAS3	MAS4	MAS5	MAS6	MAS7
15	16	17	18	19	20	21
S19	S20	S21	S22	S23	S24	S25
HTH1	HTH2	HTH3	HTH4	HTH5	HTH6	HTH7
MAS8	MAS9	MAS10	MAS11	MAS12	MAS13	MAS14
22	23	24	25	26	27	28
S26	S27	S28	S29	S30	O1	O2
HTH8	HTH9	HTH10	HTH11	HTH12	HTH13	HTH14
MAS15	MAS16	MAS17	MAS18	MAS19	MAS20	MAS21

THE NU AFRIKAN CALENDAR

EKUNDU 567/2083-84/6319-20/2075-76						
Ashe	Ojobo	Iwa	Araorun	Ile	Itutu	Erinle
1	2	3	4	5	6	7
O3	O4	O5	O6	O7	O8	O9
HTH15	HTH16	HTH17	HTH18	HTH19	HTH20	HTH21
MAS22	MAS23	MAS 24	MAS25	MAS26	MAS27	MAS28
8	9	10	11	12	13	14
O10	O11	O12	O13	O14	O15	O16
HTH22	HTH23	HTH24	HTH25	HTH26	HTH27	HTH28
MAS29	MAS30	THi1	THi2	THi3	THi4	THi5
15	16	17	18	19	20	21
O17	O18	O19	O20	O21	O22	O23
HTH29	HTH30	KHK1	KHK2	KHK3	KHK4	KHK5
THi6	THi7	THi8	THi9	THi10	THi11	THi12
22	23	24	25	26	27	28
O24	O25	O26	O27	O28	O29	O30
KHK6	KHK7	KHK8	KHK9	KHK10	KHK11	KHK12
THi13	THi14	THi15	THi16	THi17	THi18	THi19

KIJANI 567/2083-84/6319-20/2075-76						
Ashe	Ojobo	Iwa	Araorun	Ile	Itutu	Erinle
1	2	3	4	5	6	7
N28	N29	N30	D1	D2	D3	D4
TBT11	TBT12	TBT13	TBT14	TBT15	TBT16	TBT17
HID18	HID19	HID20	HID21	HID22	HID23	HID24
8	9	10	11	12	13	14
D5	D6	D7	D8	D9	D10	D11
TBT18	TBT19	TBT20	TBT21	TBT22	TBT23	TBT24
HID25	HID26	HID27	HID28	HID29	HID30	TAH1
15	16	17	18	19	20	21
D12	D13	D14	D15	D16	D17	D18
TBT25	TBT26	TBT27	TBT28	TBT29	TBT30	MKH1
TAH2	TAH3	TAH4	TAH5	TAH6	TAH7	TAH8
22	23	24	25	26	27	28
D19	D20	D21	D22	D23	D24	D25
MKH2	MKH3	MKH4	MKH5	MKH6	MKH7	MKH8
TAH9	TAH10	TAH11	TAH12	TAH13	TAH14	TAH15

EUSI 567/2083-84/6319-20/2075-76						
Ashe	Ojobo	Iwa	Araorun	Ile	Itutu	Erinle
1	2	3	4	5	6	7
O31	N1	N2	N3	N4	N5	N6
KHK13	KHK14	KHK15	KHK16	KHK17	KHK18	KHK19
THi20	THi21	THi22	THi23	THi24	THi25	THi26
8	9	10	11	12	13	14
N7	N8	N9	N10	N11	N12	N13
KHK20	KHK21	KHK22	KHK23	KHK24	KHK25	KHK26
THi27	THi28	THi29	THi30	HID1	HID2	HID3
15	16	17	18	19	20	21
N14	N15	N16	N17	N18	N19	N20
KHK27	KHK28	KHK29	KHK30	TBT1	TBT2	TBT3
HID4	HID5	HID6	HID7	HID8	HID9	HID10
22	23	24	25	26	27	28
N21	N22	N23	N24	N25	N26	N27
TBT4	TBT5	TBT6	TBT7	TBT8	TBT9	TBT10
HID11	HID12	HID13	HID14	HID15	HID16	HID17

KWANZA 567/2083-84/6319-20/2075-76						
Umoja	Kujichagulia	Ujima	Ujamaa	Nia	Kuumba	Imani
1	2	3	4	5	6	7
D26	D27	D28	D29	D30	D31	JA1
MKH9	MKH10	MKH11	MKH12	MKH13	MKH14	MKH1
TAH16	TAH17	TAH18	TAH19	TAH20	TAH21	TAH2
8	9	10	11	12	13	14
JA2	JA3	JA4	JA5	JA6	JA7	JA
MKH16	MKH17	MKH18	MKH19	MKH20	MKH21	MKH2
TAH23	TAH24	TAH25	TAH26	TAH27	TAH28	TAH29
15	16	17	18	19	20	21
JA9	JA10	JA11	JA12	JA13	JA14	JA1
MKH23	MKH24	MKH25	KMH26	MKH27	MKH28	MKH2
TAH30	TIR1	TIR2	TIR3	TIR4	TIR5	TIR6
22	23	24	25	26	27	28
JA16	JA17	JA18	JA19	JA20	JA21	JA2
MKH30	AMN1	AMN2	AMN3	AMN4	AMN5	AMN
TIR7	TIR8	TIR9	TIR10	TIR11	TIR12	TIR13

SHAHIDI 567/2083-84/6319-20/2075-76						
Ashe	Ojobo	Iwa	Araorun	Ile	Itutu	Erinle
1	2	3	4	5	6	7
JA23	JA24	JA25	JA26	JA27	JA28	JA29
AMN7	AMN8	AMN9	AMN10	AMN11	AMN12	AMN13
TIR14	TIR15	TIR16	TIR17	TIR18	TIR19	TIR20
8	9	10	11	12	13	14
JA30	JA31	F1	F2	F3	F4	F5
AMN14	AMN15	AMN16	AMN17	AMN18	AMN19	AMN20
TIR21	TIR22	TIR23	TIR24	TIR25	TIR26	TIR27
15	16	17	18	19	20	21
F6	F7	F8	F9	F10	F 11	F12
AMN21	AMN22	AMN23	AMN24	AMN25	AMN26	AMN28
TIR28	TIR29	TIR30	YEK1	YEK2	YEK3	YEK4
22	23	24	25	26	27	28
F13	F14	F15	F16	F17	F18	F19
AMN29	AMN30	RNT1	RNT2	RNT3	RNT4	RNT5
YEK5	YEK6	YEK7	YEK8	YEK9	YEK10	YEK11
SHAHIDI 118						
F20						
RNT6						
YEK12						

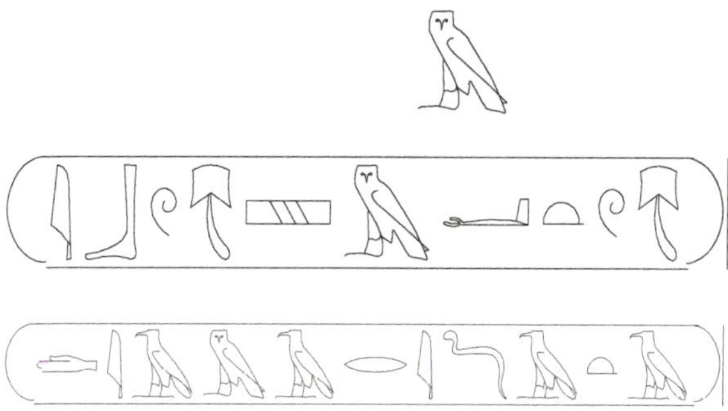

THE NU AFRIKAN CALENDAR

MAAFA 568 B.S.E./ASIENTO 2084-85 C.E./
KAM 6320-21 S.M./EC 2076-77 A.M. (ዓ.ም.)

AFU 568/2084-85/6320-21/2076-77

Ashe	Ojobo	Iwa	Araorun	Ile	Itutu	Erinle
1	2	3	4	5	6	7
F21	F22	F23	F24	F25	F26	F27
RNT7	RNT8	RNT9	RNT10	RNT11	RNT12	RNT13
YEK13	YEK14	YEK15	YEK16	YEK17	YEK18	YEK19
8	9	10	11	12	13	14
F28	F29	M1	M2	M3	M4	M5
RNT14	RNT15	RNT16	RNT17	RNT18	RNT19	RNT20
YEK 20	YEK21	YEK22	YEK23	YEK24	YEK25	YEK26
15	16	17	18	19	20	21
M6	M7	M8	M9	M10	M11	M12
RNT21	RNT22	RNT23	RNT24	RNT25	RNT26	RNT27
YEK 27	YEK28	YEK29	YEK30	MAG1	MAG2	MAG3
22	23	24	25	26	27	28
M13	M14	M15	M16	M17	M18	M19
RNT28	RNT29	RNT30	KHN1	KHN2	KHN3	KHN4
MAG4	MAG5	MAG6	MAG7	MAG8	MAG9	MAG10
SHARIFU 139						
M20						
KHN5						
MAG11						

KAWAIDA 568/2084-85/6320-21/2076-77

Ashe	Ojobo	Iwa	Araorun	Ile	Itutu	Erinle
1	2	3	4	5	6	7
A18	A19	A20	A21	A22	A23	A24
PIN4	PIN5	PIN6	PIN7	PIN8	PIN9	PIN10
MIY10	MIY11	MIY12	MIY13	MIY14	MIY15	MIY16
8	9	10	11	12	13	14
A25	A26	A27	A28	A29	A30	MA1
PIN11	PIN12	PIN13	PIN14	PIN15	PIN16	PIN17
MIY17	MIY18	MIY19	MIY20	MIY21	MIY22	MIY23
15	16	17	18	19	20	21
MA2	MA3	MA4	MA5	MA6	MA7	MA8
PIN18	PIN19	PIN20	PIN21	PIN22	PIN23	PIN24
MIY24	MIY25	MIY26	MIY27	MIY28	MIY29	MIY30
22	23	24	25	26	27	28
MA9	MA10	MA11	MA12	MA13	MA14	MA15
PIN25	PIN26	PIN27	PIN28	PIN29	PIN30	APP1
GIN1	GIN2	GIN3	GIN4	GIN5	GIN6	GIN7

UHURU 568/2084-85/6320-21/2076-77

Ashe	Ojobo	Iwa	Araorun	Ile	Itutu	Erinle
1	2	3	4	5	6	7
M21	M22	M23	M24	M25	M26	M27
KHN6	KHN7	KHN8	KHN9	KHN10	KHN11	KHN12
MAG12	MAG13	MAG14	MAG15	MAG16	MAG17	MAG18
8	9	10	11	12	13	14
M28	M29	M30	M31	A1	A2	A3
KHN13	KHN14	KHN15	KHN16	KHN17	KHN18	KHN19
MAG19	MAG20	MAG21	MAG22	MAG23	MAG24	MAG25
15	16	17	18	19	20	21
A4	A5	A6	A7	A8	A9	A10
KHN20	KHN21	KHN222	KHN23	KHN24	KHN25	KHN26
MAG26	MAG27	MAG28	MAG29	MAG30	MIY1	MIY2
22	23	24	25	26	27	28
A11	A12	A13	A14	A15	A16	A17
KHN27	KHN28	KHN29	KHN30	PIN1	PIN2	PIN3
MIY3	MIY4	MIY5	MIY6	MIY7	MIY8	MIY9

MABABA WEUSI 568/2084-85/6320-21/2076-77

Ashe	Ojobo	Iwa	Araorun	Ile	Itutu	Erinle
1	2	3	4	5	6	7
MA16	MA17	MA18	MA19	MA20	MA21	MA22
APP2	APP3	APP4	APP5	APP6	APP7	APP8
GIN8	GIN9	GIN10	GIN11	GIN12	GIN13	GIN14
8	9	10	11	12	13	14
MA23	MA24	MA25	MA26	MA27	MA28	MA29
APP9	APP10	APP11	APP12	APP13	APP14	APP15
GIN15	GIN16	GIN17	GIN18	GIN19	GIN20	GIN21
15	16	17	18	19	20	21
MA30	MA31	J1	J2	J3	J4	J5
APP16	APP17	APP18	APP19	APP20	APP21	APP22
GIN22	GIN23	GIN24	GIN25	GIN26	GIN27	GIN28
22	23	24	25	26	27	28
J6	J7	J8	J9	J10	J11	J12
APP23	APP24	APP25	APP26	APP27	APP28	APP29
GIN29	GIN30	SAN1	SAN2	SAN3	SAN4	SAN5

THE NU AFRIKAN CALENDAR

MWANAMKE WEUSI 568/2084-85/6320-21/2076-77						
Ashe	Ojobo	Iwa	Araorun	Ile	Itutu	Erinle
1	2	3	4	5	6	7
J13	J14	J15	J16	J17	J18	J19
APP30	MSR1	MSR2	MSR3	MSR4	MSR5	MSR6
SAN6	SAN7	SAN8	SAN9	SAN10	SAN11	SAN12
8	9	10	11	12	13	14
J20	J21	J22	J23	J24	J25	J26
MSR7	MSR8	MSR9	MSR10	MSR11	MSR12	MSR13
SAN13	SAN14	SAN15	SAN16	SAN17	SAN18	SAN19
15	16	17	18	19	20	21
J27	J28	J29	J30	JU1	JU2	JU3
MSR14	MSR15	MSR16	MSR17	MSR18	MSR19	MSR20
SAN20	SAN21	SAN22	SAN23	SAN24	SAN25	SAN26
22	23	24	25	26	27	28
JU4	JU5	JU6	JU7	JU8	JU9	JU10
MSR21	MRS22	MSR23	MSR24	MSR25	MSR26	MSR27
SAN27	SAN28	SAN29	SAN30	HAM1	HAM2	HAM3

TAIFA WEUSI 568/2084-85/6320-21/2076-77						
Ashe	Ojobo	Iwa	Araorun	Ile	Itutu	Erinle
1	2	3	4	5	6	7
AU8	AU9	AU10	AU11	AU12	AU13	AU14
THT20	THT21	THT22	THT23	THT24	THT25	THT26
NAH2	NAH3	NAH4	NAH5	NAH6	NAH7	NAH8
8	9	10	11	12	13	14
AU15	AU16	AU17	AU18	AU19	AU20	AU21
THT27	THT28	THT29	THT30	PAP1	PAP1	PAP2
NAH9	NAH10	NAH11	NAH12	NAH13	NAH14	NAH15
15	16	17	18	19	20	21
AU22	AU23	AU24	AU25	AU26	AU27	AU28
PAP3	PAP4	PAP5	PAP6	PAP7	PAP8	PAP9
NAH16	NAH17	NAH18	NAH19	NAH20	NAH21	NAH22
22	23	24	25	26	27	28
AU29	AU30	AU31	S1	S2	S3	S4
PAP10	PAP11	PAP12	PAP13	PAP14	PAP15	PAP16
NAH23	NAH24	NAH25	NAH26	NAH27	NAH28	NAH29

MTU WEUSI 568/2084-85/6320-21/2076-77						
Ashe	Ojobo	Iwa	Araorun	Ile	Itutu	Erinle
1	2	3	4	5	6	7
JU11	JU12	JU13	JU14	JU15	JU16	JU17
MSR28	MSR29	MSR30	*ASR1*	*HRU2*	*STH3*	*AST4*
HAM4	HAM5	HAM6	HAM7	HAM8	HAM9	HAM10
8	9	10	11	12	13	14
JU18	JU19	JU20	JU21	JU22	JU23	JU24
NBH5	*ANU6*	THT1	THT2	THT3	THT4	THT5
HAM11	HAM12	HAM13	HAM14	HAM15	HAM16	HAM17
15	16	17	18	19	20	21
JU25	JU26	JU27	JU28	JU29	JU30	JU31
THT6	THT7	THT8	THT9	THT10	THT11	THT12
HAM18	HAM19	HAM20	HAM21	HAM22	HAM23	HAM24
22	23	24	25	26	27	28
AU1	AU2	AU3	AU4	AU5	AU6	AU7
THT13	THT14	THT15	THT16	THT17	THT18	THT19
HAM25	HAM26	HAM27	HAM28	HAM29	HAM30	NAH1

UWEZO 568/2084-85/6320-21/2076-77						
Ashe	Ojobo	Iwa	Araorun	Ile	Itutu	Erinle
1	2	3	4	5	6	7
S5	S6	S7	S8	S9	S10	S11
PAP17	PAP18	PAP19	PAP20	PAP21	PAP22	PAP23
NAH30	*PAG1*	*PAG2*	*PAG3*	*PAG4*	*PAG5*	MAS1
8	9	10	11	12	13	14
S12	S13	S14	S15	S16	S17	S18
PAP24	PAP25	PAP26	PAP27	PAP28	PAP29	PAP30
MAS2	MAS3	MAS4	MAS5	MAS6	MAS7	MAS8
15	16	17	18	19	20	21
S19	S20	S21	S22	S23	S24	S25
HTH1	HTH2	HTH3	HTH4	HTH5	HTH6	HTH7
MAS9	MAS10	MAS11	MAS12	MAS13	MAS14	MAS15
22	23	24	25	26	27	28
S26	S27	S28	S29	S30	O1	O2
HTH8	HTH9	HTH10	HTH11	HTH12	HTH13	HTH14
MAS16	MAS17	MAS18	MAS19	MAS20	MAS21	MAS22

THE NU AFRIKAN CALENDAR

EKUNDU 568/2084-85/6320-21/2076-77						
Ashe	Ojobo	Iwa	Araorun	Ile	Itutu	Erinle
1	2	3	4	5	6	7
O3	O4	O5	O6	O7	O8	O9
HTH15	HTH16	HTH17	HTH18	HTH19	HTH20	HTH21
MAS23	MAS24	MAS25	MAS26	MAS27	MAS28	MAS29
8	9	10	11	12	13	14
O10	O11	O12	O13	O14	O15	O16
HTH22	HTH23	HTH24	HTH25	HTH26	HTH27	HTH28
MAS30	THi1	THi2	THi3	THi4	THi5	THi6
15	16	17	18	19	20	21
O17	O18	O19	O20	O21	O22	O23
HTH29	HTH30	KHK1	KHK2	KHK3	KHK4	KHK5
THi7	THi8	This9	THi10	THi11	THi12	THi13
22	23	24	25	26	27	28
O24	O25	O26	O27	O28	O29	O30
KHK6	KHK7	KHK8	KHK9	KHK10	KHK11	KHK12
Thi14	THi15	THi16	THi17	THi18	THi19	THi20

KIJANI 568/2084-85/6320-21/2076-77						
Ashe	Ojobo	Iwa	Araorun	Ile	Itutu	Erinle
1	2	3	4	5	6	7
N28	N29	N30	D1	D2	D3	D4
TBT11	TBT12	TBT13	TBT14	TBT15	TBT16	TBT17
HID19	HID20	HID21	HID22	HID23	HID24	HID25
8	9	10	11	12	13	14
D5	D6	D7	D8	D9	D10	D11
TBT18	TBT19	TBT20	TBT21	TBT22	TBT23	TBT24
HID26	HID27	HID28	HID29	HID30	TAH1	TAH2
15	16	17	18	19	20	21
D12	D13	D14	D15	D16	D17	D18
TBT25	TBT26	TBT27	TBT28	TBT29	TBT30	MKH1
TAH3	TAH4	TAH5	TAH6	TAH7	TAH8	TAH9
22	23	24	25	26	27	28
D19	D20	D21	D22	D23	D24	D25
MKH2	MKH3	MKH4	MKH5	MKH6	MKH7	MKH8
TAH10	TAH11	TAH12	TAH13	TAH14	TAH15	TAH16

EUSI 568/2084-85/6320-21/2076-77						
Ashe	Ojobo	Iwa	Araorun	Ile	Itutu	Erinle
1	2	3	4	5	6	7
O31	N1	N2	N3	N4	N5	N6
KHK13	KHK14	KHK15	KHK16	KHK17	KHK18	KHK19
THi21	THi22	THi23	THi24	THi25	THi26	THi27
8	9	10	11	12	13	14
N7	N8	N9	N10	N11	N12	N13
KHK20	KHK21	KHK22	KHK23	KHK24	KHK25	KHK26
THi28	THi29	THi30	HID1	HID2	HID3	HID4
15	16	17	18	19	20	21
N14	N15	N16	N17	N18	N19	N20
KHK27	KHK28	KHK29	KHK30	TBT1	TBT2	TBT3
HID5	HID6	HID7	HID8	HID9	HID10	HID11
22	23	24	25	26	27	28
N21	N22	N23	N24	N25	N26	N27
TBT4	TBT5	TBT6	TBT7	TBT8	TBT9	TBT10
HID12	HID13	HID14	HID15	HID16	HID17	HID18

KWANZA 568/2084-85/6320-21/2076-77						
Umoja	Kujichagulia	Ujima	Ujamaa	Nia	Kuumba	Iman
1	2	3	4	5	6	7
D26	D27	D28	D29	D30	D31	JA1
MKH9	MKH10	MKH11	MKH12	MKH13	MKH14	MKH
TAH17	TAH18	TAH19	TAH20	TAH21	TAH22	TAH
8	9	10	11	12	13	14
JA2	JA3	JA4	JA5	JA6	JA7	J
MKH16	MKH17	MKH18	MKH19	MKH20	MKH21	MKH
TAH24	TAH25	TAH26	TAH27	TAH28	TAH29	TAH
15	16	17	18	19	20	21
JA9	JA10	JA11	JA12	JA13	JA14	JA
MKH23	MKH24	MKH25	KMH26	MKH27	MKH28	MKH
TIR1	TIR2	TIR3	TIR4	TIR5	TIR6	TIR
22	23	24	25	26	27	28
JA16	JA17	JA18	JA19	JA20	JA21	JA
MKH30	AMN1	AMN2	AMN3	AMN4	AMN5	AM
TIR8	TIR9	TIR10	TIR11	TIR12	TIR13	TIR

THE NU AFRIKAN CALENDAR

SHAHIDI 568/2084-85/6320-21/2076-77						
Ashe	**Ojobo**	**Iwa**	**Araorun**	**Ile**	**Itutu**	**Erinle**
1	2	3	4	5	6	7
JA23	JA24	JA25	JA26	JA27	JA28	JA29
AMN7	AMN8	AMN9	AMN10	AMN11	AMN12	AMN13
TIR15	TIR16	TIR17	TIR18	TIR19	TIR20	TIR21
8	9	10	11	12	13	14
JA30	JA31	F1	F2	F3	F4	F5
AMN14	AMN15	AMN16	AMN17	AMN18	AMN19	AMN20
TIR22	TIR23	TIR24	TIR25	TIR26	TIR27	TIR28
15	16	17	18	19	20	21
F6	F7	F8	F9	F10	F 11	F12
AMN21	AMN22	AMN23	AMN24	AMN25	AMN26	AMN28
TIR29	TIR30	YEK1	YEK2	YEK3	YEK4	YEK5
22	23	24	25	26	27	28
F13	F14	F15	F16	F17	F18	F19
AMN29	AMN30	RNT1	RNT2	RNT3	RNT4	RNT5
YEK6	YEK7	YEK8	YEK9	YEK10	YEK11	YEK12

SHAHIDI 119
F20
RNT6
YEK13

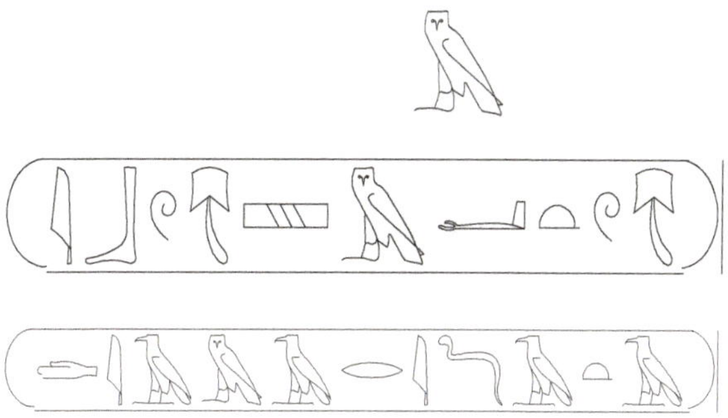

THE NU AFRIKAN CALENDAR

MAAFA 569 B.S.E./ASIENTO 2085-86 C.E.
KAM 6321-22 S.M./EC 2077-78 A.M. (ዓ.ም.)

AFU 569/2085-86/6321-22/2077-78						
Ashe	Ojobo	Iwa	Araorun	Ile	Itutu	Erinle
1	2	3	4	5	6	7
F21	F22	F23	F24	F25	F26	F27
RNT7	RNT8	RNT9	RNT10	RNT11	RNT12	RNT13
YEK14	YEK15	YEK16	YEK17	YEK18	YEK19	YEK20
8	9	10	11	12	13	14
F28	M1	M2	M3	M4	M5	M6
RNT14	RNT15	RNT16	RNT17	RNT18	RNT19	RNT20
YEK21	YEK22	YEK23	YEK24	YEK25	YEK26	YEK27
15	16	17	18	19	20	21
M7	M8	M9	M10	M11	M12	M13
RNT21	RNT22	RNT23	RNT24	RNT25	RNT26	RNT27
YEK28	YEK29	YEK30	MAG1	MAG2	MAG3	MAG4
22	23	24	25	26	27	28
M14	M15	M16	M17	M18	M19	M20
RNT28	RNT29	RNT30	KHN1	KHN2	KHN3	KHN4
MAG5	MAG6	MAG7	MAG8	MAG9	MAG10	MAG11

KAWAIDA 569/2085-86/6321-22/2077-78						
Ashe	Ojobo	Iwa	Araorun	Ile	Itutu	Erinle
1	2	3	4	5	6	7
A18	A19	A20	A21	A22	A23	A24
PIN3	PIN4	PIN5	PIN6	PIN7	PIN8	PIN9
MIY10	MIY11	MIY12	MIY13	MIY14	MIY15	MIY16
8	9	10	11	12	13	14
A25	A26	A27	A28	A29	A30	MA1
PIN10	PIN11	PIN12	PIN13	PIN14	PIN15	PIN16
MIY17	MIY18	MIY19	MIY20	MIY21	MIY22	MIY23
15	16	17	18	19	20	21
MA2	MA3	MA4	MA5	MA6	MA7	MA8
PIN17	PIN18	PIN19	PIN20	PIN21	PIN22	PIN23
MIY24	MIY25	MIY26	MIY27	MIY28	MIY29	MIY30
22	23	24	25	26	27	28
MA9	MA10	MA11	MA12	MA13	MA14	MA15
PIN24	PIN25	PIN26	PIN27	PIN28	PIN29	PIN30
GIN1	GIN2	GIN3	GIN4	GIN5	GIN6	GIN7

UHURU 569/2085-86/6321-22/2077-78						
Ashe	Ojobo	Iwa	Araorun	Ile	Itutu	Erinle
1	2	3	4	5	6	7
M21	M22	M23	M24	M25	M26	M27
KHN5	KHN6	KHN7	KHN8	KHN9	KHN10	KHN11
MAG12	MAG13	MAG14	MAG15	MAG16	MAG17	MAG18
8	9	10	11	12	13	14
M28	M29	M30	M31	A1	A2	A3
KHN12	KHN13	KHN14	KHN15	KHN16	KHN17	KHN18
MAG19	MAG20	MAG21	MAG22	MAG23	MAG24	MAG25
15	16	17	18	19	20	21
A4	A5	A6	A7	A8	A9	A10
KHN19	KHN20	KHN21	KHN22	KHN23	KHN24	KHN25
MAG26	MAG27	MAG28	MAG29	MAG30	MIY1	MIY2
22	23	24	25	26	27	28
A11	A12	A13	A14	A15	A16	A17
KHN26	KHN27	KHN28	KHN29	KHN30	PIN1	PIN2
MIY3	MIY4	MIY5	MIY6	MIY7	MIY8	MIY9

MABABA WEUSI 569/2085-86/6321-22/2077-78						
Ashe	Ojobo	Iwa	Araorun	Ile	Itutu	Erinle
1	2	3	4	5	6	7
MA16	MA17	MA18	MA19	MA20	MA21	MA22
APP1	APP2	APP3	APP4	APP5	APP6	APP7
GIN8	GIN9	GIN10	GIN11	GIN12	GIN13	GIN14
8	9	10	11	12	13	14
MA23	MA24	MA25	MA26	MA27	MA28	MA29
APP8	APP9	APP10	APP11	APP12	APP13	APP14
GIN15	GIN16	GIN17	GIN18	GIN19	GIN20	GIN21
15	16	17	18	19	20	21
MA30	MA31	J1	J2	J3	J4	J5
APP15	APP16	APP17	APP18	APP19	APP20	APP21
GIN22	GIN23	GIN24	GIN25	GIN26	GIN27	GIN28
22	23	24	25	26	27	28
J6	J7	J8	J9	J10	J11	J12
APP22	APP23	APP24	APP25	APP26	APP27	APP28
GIN29	GIN30	SAN1	SAN2	SAN3	SAN4	SAN5

THE NU AFRIKAN CALENDAR

MWANAMKE WEUSI 569/2085-86/6321-22/2077-78						
Ashe	Ojobo	Iwa	Araorun	Ile	Itutu	Erinle
1	2	3	4	5	6	7
J13	J14	J15	J16	J17	J18	J19
APP29	APP30	MSR1	MSR2	MSR3	MSR4	MSR5
SAN6	SAN7	SAN8	SAN9	SAN10	SAN11	SAN12
8	9	10	11	12	13	14
J20	J21	J22	J23	J24	J25	J26
MSR6	MSR7	MSR8	MSR9	MSR10	MSR11	MSR12
SAN13	SAN14	SAN15	SAN16	SAN17	SAN18	SAN19
15	16	17	18	19	20	21
J27	J28	J29	J30	JU1	JU2	JU3
MSR13	MSR14	MSR15	MSR16	MSR17	MSR18	MSR19
SAN20	SAN21	SAN22	SAN23	SAN24	SAN25	SAN26
22	23	24	25	26	27	28
JU4	JU5	JU6	JU7	JU8	JU9	JU10
MSR20	MRS21	MSR22	MSR23	MSR24	MSR25	MSR26
SAN27	SAN28	SAN29	SAN30	HAM1	HAM2	HAM3

TAIFA WEUSI 569/2085-86/6321-22/2077-78						
Ashe	Ojobo	Iwa	Araorun	Ile	Itutu	Erinle
1	2	3	4	5	6	7
AU8	AU9	AU10	AU11	AU12	AU13	AU14
THT20	THT21	THT22	THT23	THT24	THT25	THT26
NAH2	NAH3	NAH4	NAH5	NAH6	NAH7	NAH8
8	9	10	11	12	13	14
AU15	AU16	AU17	AU18	AU19	AU20	AU21
THT27	THT28	THT29	THT30	PAP1	PAP1	PAP2
NAH9	NAH10	NAH11	NAH12	NAH13	NAH14	NAH15
15	16	17	18	19	20	21
AU22	AU23	AU24	AU25	AU26	AU27	AU28
PAP3	PAP4	PAP5	PAP6	PAP7	PAP8	PAP9
NAH16	NAH17	NAH18	NAH19	NAH20	NAH21	NAH22
22	23	24	25	26	27	28
AU29	AU30	AU31	S1	S2	S3	S4
PAP10	PAP11	PAP12	PAP13	PAP14	PAP15	PAP16
NAH23	NAH24	NAH25	NAH26	NAH27	NAH28	NAH29

MTU WEUSI 569/2085-86/6321-22/2077-78						
Ashe	Ojobo	Iwa	Araorun	Ile	Itutu	Erinle
1	2	3	4	5	6	7
JU11	JU12	JU13	JU14	JU15	JU16	JU17
MSR27	MSR28	MSR29	MSR30	*ASR1*	*HRU2*	*STH3*
HAM4	HAM5	HAM6	HAM7	HAM8	HAM9	HAM10
8	9	10	11	12	13	14
JU18	JU19	JU20	JU21	JU22	JU23	JU24
AST4	*NBH5*	THT1	THT2	THT3	THT4	THT5
HAM11	HAM12	HAM13	AM14	HAM15	HAM16	HAM17
15	16	17	18	19	20	21
JU25	JU26	JU27	JU28	JU29	JU30	JU31
THT6	THT7	THT8	THT9	THT10	THT11	THT12
HAM18	HAM19	HAM20	HAM21	HAM22	HAM23	HAM24
22	23	24	25	26	27	28
AU1	AU2	AU3	AU4	AU5	AU6	AU7
THT13	THT14	THT15	THT16	THT17	THT18	THT19
HAM25	HAM26	HAM27	HAM28	HAM29	HAM30	NAH1

UWEZO 569/2085-86/6321-22/2077-78						
Ashe	Ojobo	Iwa	Araorun	Ile	Itutu	Erinle
1	2	3	4	5	6	7
S5	S6	S7	S8	S9	S10	S11
PAP17	PAP18	PAP19	PAP20	PAP21	PAP22	PAP23
NAH30	*PAG1*	*PAG2*	*PAG3*	*PAG4*	*PAG5*	MAS1
8	9	10	11	12	13	14
S12	S13	S14	S15	S16	S17	S18
PAP24	PAP25	PAP26	PAP27	PAP28	PAP29	PAP30
MAS2	MAS3	MAS4	MAS5	MAS6	MAS7	MAS8
15	16	17	18	19	20	21
S19	S20	S21	S22	S23	S24	S25
HTH1	HTH2	HTH3	HTH4	HTH5	HTH6	HTH7
MAS9	MAS10	MAS11	MAS12	MAS13	MAS14	MAS15
22	23	24	25	26	27	28
S26	S27	S28	S29	S30	O1	O2
HTH8	HTH9	HTH10	HTH11	HTH12	HTH13	HTH14
MAS16	MAS17	MAS18	MAS19	MAS20	MAS21	MAS22

THE NU AFRIKAN CALENDAR

EKUNDU 569/2085-86/6321-22/2077-78						
Ashe	Ojobo	Iwa	Araorun	Ile	Itutu	Erinle
1	2	3	4	5	6	7
O3	O4	O5	O6	O7	O8	O9
HTH15	HTH16	HTH17	HTH18	HTH19	HTH20	HTH21
MAS23	MAS24	MAS25	MAS26	MAS27	MAS28	MAS29
8	9	10	11	12	13	14
O10	O11	O12	O13	O14	O15	O16
HTH22	HTH23	HTH24	HTH25	HTH26	HTH27	HTH28
MAS30	THi1	THi2	THi3	THi4	THi5	THi6
15	16	17	18	19	20	21
O17	O18	O19	O20	O21	O22	O23
HTH29	HTH30	KHK1	KHK2	KHK3	KHK4	KHK5
THi7	THi8	THi9	THi10	THi11	THi12	THi13
22	23	24	25	26	27	28
O24	O25	O26	O27	O28	O29	O30
KHK6	KHK7	KHK8	KHK9	KHK10	KHK11	KHK12
THi14	THi15	THi16	THi17	THi18	THi19	THi20

KIJANI 569/2085-86/6321-22/2077-78						
Ashe	Ojobo	Iwa	Araorun	Ile	Itutu	Erinle
1	2	3	4	5	6	7
N28	N29	N30	D1	D2	D3	D4
TBT11	TBT12	TBT13	TBT14	TBT15	TBT16	TBT17
HID19	HID20	HID21	HID22	HID23	HID24	HID25
8	9	10	11	12	13	14
D5	D6	D7	D8	D9	D10	D11
TBT18	TBT19	TBT20	TBT21	TBT22	TBT23	TBT24
HID26	HID27	HID28	HID29	HID30	TAH1	TAH2
15	16	17	18	19	20	21
D12	D13	D14	D15	D16	D17	D18
TBT25	TBT26	TBT27	TBT28	TBT29	TBT30	MKH1
TAH3	TAH4	TAH5	TAH6	TAH7	TAH8	TAH9
22	23	24	25	26	27	28
D19	D20	D21	D22	D23	D24	D25
MKH2	MKH3	MKH4	MKH5	MKH6	MKH7	MKH8
TAH10	TAH11	TAH12	TAH13	TAH14	TAH15	TAH16

EUSI 569/2085-86/6321-22/2077-78						
Ashe	Ojobo	Iwa	Araorun	Ile	Itutu	Erinle
1	2	3	4	5	6	7
O31	N1	N2	N3	N4	N5	N6
KHK13	KHK14	KHK15	KHK16	KHK17	KHK18	KHK19
THi21	THi22	THi23	THi24	THi25	THi26	THi27
8	9	10	11	12	13	14
N7	N8	N9	N10	N11	N12	N13
KHK20	KHK21	KHK22	KHK23	KHK24	KHK25	KHK26
THi28	THi29	THi30	HID1	HID2	HID3	HID4
15	16	17	18	19	20	21
N14	N15	N16	N17	N18	N19	N20
KHK27	KHK28	KHK29	KHK30	TBT1	TBT2	TBT3
HID5	HID6	HID7	HID8	HID9	HID10	HID11
22	23	24	25	26	27	28
N21	N22	N23	N24	N25	N26	N27
TBT4	TBT5	TBT6	TBT7	TBT8	TBT9	TBT10
HID12	HID13	HID14	HID15	HID16	HID17	HID18

KWANZA 569/2085-86/6321-22/2077-78						
Umoja	Kujichagulia	Ujima	Ujamaa	Nia	Kuumba	Imani
1	2	3	4	5	6	7
D26	D27	D28	D29	D30	D31	JA1
MKH9	MKH10	MKH11	MKH12	MKH13	MKH14	MKH15
TAH17	TAH18	TAH19	TAH20	TAH21	TAH22	TAH23
8	9	10	11	12	13	14
JA2	JA3	JA4	JA5	JA6	JA7	JA8
MKH16	MKH17	MKH18	MKH19	MKH20	MKH21	MKH22
TAH24	TAH25	TAH26	TAH27	TAH28	TAH29	TAH30
15	16	17	18	19	20	21
JA9	JA10	JA11	JA12	JA13	JA14	JA15
MKH23	MKH24	MKH25	KMH26	MKH27	MKH28	MKH29
TIR1	TIR2	TIR3	TIR4	TIR5	TIR6	TIR7
22	23	24	25	26	27	28
JA16	JA17	JA18	JA19	JA20	JA21	JA22
MKH30	AMN1	AMN2	AMN3	AMN4	AMN5	AMN6
TIR8	TIR9	TIR10	TIR11	TIR12	TIR13	TIR14

THE NU AFRIKAN CALENDAR

SHAHIDI 569/2085-86/6321-22/2077-78						
Ashe	Ojobo	Iwa	Araorun	Ile	Itutu	Erinle
1	2	3	4	5	6	7
JA23	JA24	JA25	JA26	JA27	JA28	JA29
AMN7	AMN8	AMN9	AMN10	AMN11	AMN12	AMN13
TIR15	TIR16	TIR17	TIR18	TIR19	TIR20	TIR21
8	9	10	11	12	13	14
JA30	JA31	F1	F2	F3	F4	F5
AMN14	AMN15	AMN16	AMN17	AMN18	AMN19	AMN20
TIR22	TIR23	TIR24	TIR25	TIR26	TIR27	TIR28
15	16	17	18	19	20	21
F6	F7	F8	F9	F10	F 11	F12
AMN21	AMN22	AMN23	AMN24	AMN25	AMN26	AMN28
TIR29	TIR30	YEK1	YEK2	YEK3	YEK4	YEK5
22	23	24	25	26	27	28
F13	F14	F15	F16	F17	F18	F19
AMN29	AMN30	RNT1	RNT2	RNT3	RNT4	RNT5
YEK6	YEK7	YEK8	YEK9	YEK10	YEK11	YEK12
SHAHIDI 120						
F20						
RNT6						
YEK13						

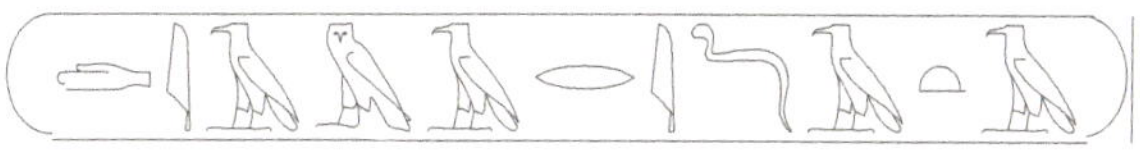

THE NU AFRIKAN CALENDAR

MAAFA 570 B.S.E./ASIENTO 2086-87 C.E.
KAM 6322-23 S.M./EC 2078-79 A.M. (ዓ.ም.)

AFU 570/2086-87/6322-23/2078-79

Ashe	Ojobo	Iwa	Araorun	Ile	Itutu	Erinle
1	2	3	4	5	6	7
F21	F22	F23	F24	F25	F26	F27
RNT7	RNT8	RNT9	RNT10	RNT11	RNT12	RNT13
YEK14	YEK15	YEK16	YEK17	YEK18	YEK19	YEK20
8	9	10	11	12	13	14
F28	M1	M2	M3	M4	M5	M6
RNT14	RNT15	RNT16	RNT17	RNT18	RNT19	RNT20
YEK21	YEK22	YEK23	YEK24	YEK25	YEK26	YEK27
15	16	17	18	19	20	21
M7	M8	M9	M10	M11	M12	M13
RNT21	RNT22	RNT23	RNT24	RNT25	RNT26	RNT27
YEK28	YEK29	YEK30	MAG1	MAG2	MAG3	MAG4
22	23	24	25	26	27	28
M14	M15	M16	M17	M18	M19	M20
RNT28	RNT29	RNT30	KHN1	KHN2	KHN3	KHN4
MAG5	MAG6	MAG7	MAG8	MAG9	MAG10	MAG11

KAWAIDA 570/2086-87/6322-23/2078-79

Ashe	Ojobo	Iwa	Araorun	Ile	Itutu	Erinle
1	2	3	4	5	6	7
A18	A19	A20	A21	A22	A23	A24
PIN3	PIN4	PIN5	PIN6	PIN7	PIN8	PIN9
MIY10	MIY11	MIY12	MIY13	MIY14	MIY15	MIY16
8	9	10	11	12	13	14
A25	A26	A27	A28	A29	A30	MA1
PIN10	PIN11	PIN12	PIN13	PIN14	PIN15	PIN16
MIY17	MIY18	MIY19	MIY20	MIY21	MIY22	MIY23
15	16	17	18	19	20	21
MA2	MA3	MA4	MA5	MA6	MA7	MA8
PIN17	PIN18	PIN19	PIN20	PIN21	PIN22	PIN23
MIY24	MIY25	MIY26	MIY27	MIY28	MIY29	MIY30
22	23	24	25	26	27	28
MA9	MA10	MA11	MA12	MA13	MA14	MA15
PIN24	PIN25	PIN26	PIN27	PIN28	PIN29	PIN30
GIN1	GIN2	GIN3	GIN4	GIN5	GIN6	GIN7

UHURU 570/2086-87/6322-23/2078-79

Ashe	Ojobo	Iwa	Araorun	Ile	Itutu	Erinle
1	2	3	4	5	6	7
M21	M22	M23	M24	M25	M26	M27
KHN5	KHN6	KHN7	KHN8	KHN9	KHN10	KHN11
MAG12	MAG13	MAG14	MAG15	MAG16	MAG17	MAG18
8	9	10	11	12	13	14
M28	M29	M30	M31	A1	A2	A3
KHN12	KHN13	KHN14	KHN15	KHN16	KHN17	KHN18
MAG19	MAG20	MAG21	MAG22	MAG23	MAG24	MAG25
15	16	17	18	19	20	21
A4	A5	A6	A7	A8	A9	A10
KHN19	KHN20	KHN21	KHN22	KHN23	KHN24	KHN25
MAG26	MAG27	MAG28	MAG29	MAG30	MIY1	MIY2
22	23	24	25	26	27	28
A11	A12	A13	A14	A15	A16	A17
KHN26	KHN27	KHN28	KHN29	KHN30	PIN1	PIN2
MIY3	MIY4	MIY5	MIY6	MIY7	MIY8	MIY9

MABABA WEUSI 570/2086-87/6322-23/2078-79

Ashe	Ojobo	Iwa	Araorun	Ile	Itutu	Erinle
1	2	3	4	5	6	7
MA16	MA17	MA18	MA19	MA20	MA21	MA22
APP1	APP2	APP3	APP4	APP5	APP6	APP7
GIN8	GIN9	GIN10	GIN11	GIN12	GIN13	GIN14
8	9	10	11	12	13	14
MA23	MA24	MA25	MA26	MA27	MA28	MA29
APP8	APP9	APP10	APP11	APP12	APP13	APP14
GIN15	GIN16	GIN17	GIN18	GIN19	GIN20	GIN21
15	16	17	18	19	20	21
MA30	MA31	J1	J2	J3	J4	J5
APP15	APP16	APP17	APP18	APP19	APP20	APP21
GIN22	GIN23	GIN24	GIN25	GIN26	GIN27	GIN28
22	23	24	25	26	27	28
J6	J7	J8	J9	J10	J11	J12
APP22	APP23	APP24	APP25	APP26	APP27	APP28
GIN29	GIN30	SAN1	SAN2	SAN3	SAN4	SAN5

THE NU AFRIKAN CALENDAR

MWANAMKE WEUSI 570/2086-87/6322-23/2078-79						
Ashe	Ojobo	Iwa	Araorun	Ile	Itutu	Erinle
1	2	3	4	5	6	7
J13	J14	J15	J16	J17	J18	J19
APP29	APP30	MSR1	MSR2	MSR3	MSR4	MSR5
SAN6	SAN7	SAN8	SAN9	SAN10	SAN11	SAN12
8	9	10	11	12	13	14
J20	J21	J22	J23	J24	J25	J26
MSR6	MSR7	MSR8	MSR9	MSR10	MSR11	MSR12
SAN13	SAN14	SAN15	SAN16	SAN17	SAN18	SAN19
15	16	17	18	19	20	21
J27	J28	J29	J30	JU1	JU2	JU3
MSR13	MSR14	MSR15	MSR16	MSR17	MSR18	MSR19
SAN20	SAN21	SAN22	SAN23	SAN24	SAN25	SAN26
22	23	24	25	26	27	28
JU4	JU5	JU6	JU7	JU8	JU9	JU10
MSR20	MRS21	MSR22	MSR23	MSR24	MSR25	MSR26
SAN27	SAN28	SAN29	SAN30	HAM1	HAM2	HAM3

TAIFA WEUSI 570/2086-87/6322-23/2078-79						
Ashe	Ojobo	Iwa	Araorun	Ile	Itutu	Erinle
1	2	3	4	5	6	7
AU8	AU9	AU10	AU11	AU12	AU13	AU14
THT20	THT21	THT22	THT23	THT24	THT25	THT26
NAH2	NAH3	NAH4	NAH5	NAH6	NAH7	NAH8
8	9	10	11	12	13	14
AU15	AU16	AU17	AU18	AU19	AU20	AU21
THT27	THT28	THT29	THT30	PAP1	PAP1	PAP2
NAH9	NAH10	NAH11	NAH12	NAH13	NAH14	NAH15
15	16	17	18	19	20	21
AU22	AU23	AU24	AU25	AU26	AU27	AU28
PAP3	PAP4	PAP5	PAP6	PAP7	PAP8	PAP9
NAH16	NAH17	NAH18	NAH19	NAH20	NAH21	NAH22
22	23	24	25	26	27	28
AU29	AU30	AU31	S1	S2	S3	S4
PAP10	PAP11	PAP12	PAP13	PAP14	PAP15	PAP16
NAH23	NAH24	NAH25	NAH26	NAH27	NAH28	NAH29

MTU WEUSI 570/2086-87/6322-23/2078-79						
Ashe	Ojobo	Iwa	Araorun	Ile	Itutu	Erinle
1	2	3	4	5	6	7
JU11	JU12	JU13	JU14	JU15	JU16	JU17
MSR27	MSR28	MSR29	MSR30	*ASR1*	*HRU2*	*STH3*
HAM4	HAM5	HAM6	HAM7	HAM8	HAM9	HAM10
8	9	10	11	12	13	14
JU18	JU19	JU20	JU21	JU22	JU23	JU24
AST4	*NBH5*	THT1	THT2	THT3	THT4	THT5
HAM11	HAM12	HAM13	AM14	HAM15	HAM16	HAM17
15	16	17	18	19	20	21
JU25	JU26	JU27	JU28	JU29	JU30	JU31
THT6	THT7	THT8	THT9	THT10	THT11	THT12
HAM18	HAM19	HAM20	HAM21	HAM22	HAM23	HAM24
22	23	24	25	26	27	28
AU1	AU2	AU3	AU4	AU5	AU6	AU7
THT13	THT14	THT15	THT16	THT17	THT18	THT19
HAM25	HAM26	HAM27	HAM28	HAM29	HAM30	NAH1

UWEZO 570/2086-87/6322-23/2078-79						
Ashe	Ojobo	Iwa	Araorun	Ile	Itutu	Erinle
1	2	3	4	5	6	7
S5	S6	S7	S8	S9	S10	S11
PAP17	PAP18	PAP19	PAP20	PAP21	PAP22	PAP23
NAH30	*PAG1*	*PAG2*	*PAG3*	*PAG4*	*PAG5*	MAS1
8	9	10	11	12	13	14
S12	S13	S14	S15	S16	S17	S18
PAP24	PAP25	PAP26	PAP27	PAP28	PAP29	PAP30
MAS2	MAS3	MAS4	MAS5	MAS6	MAS7	MAS8
15	16	17	18	19	20	21
S19	S20	S21	S22	S23	S24	S25
HTH1	HTH2	HTH3	HTH4	HTH5	HTH6	HTH7
MAS9	MAS10	MAS11	MAS12	MAS13	MAS14	MAS15
22	23	24	25	26	27	28
S26	S27	S28	S29	S30	O1	O2
HTH8	HTH9	HTH10	HTH11	HTH12	HTH13	HTH14
MAS16	MAS17	MAS18	MAS19	MAS20	MAS21	MAS22

THE NU AFRIKAN CALENDAR

EKUNDU 570/2086-87/6322-23/2078-79						
Ashe	Ojobo	Iwa	Araorun	Ile	Itutu	Erinle
1	2	3	4	5	6	7
O3	O4	O5	O6	O7	O8	O9
HTH15	HTH16	HTH17	HTH18	HTH19	HTH20	HTH21
MAS23	MAS24	MAS25	MAS26	MAS27	MAS28	MAS29
8	9	10	11	12	13	14
O10	O11	O12	O13	O14	O15	O16
HTH22	HTH23	HTH24	HTH25	HTH26	HTH27	HTH28
MAS30	THi1	THi2	THi3	THi4	THi5	THi6
15	16	17	18	19	20	21
O17	O18	O19	O20	O21	O22	O23
HTH29	HTH30	KHK1	KHK2	KHK3	KHK4	KHK5
THi7	THi8	THi9	THi10	THi11	THi12	THi13
22	23	24	25	26	27	28
O24	O25	O26	O27	O28	O29	O30
KHK6	KHK7	KHK8	KHK9	KHK10	KHK11	KHK12
THi14	THi15	THi16	THi17	THi18	THi19	THi20

KIJANI 570/2086-87/6322-23/2078-79						
Ashe	Ojobo	Iwa	Araorun	Ile	Itutu	Erinle
1	2	3	4	5	6	7
N28	N29	N30	D1	D2	D3	D4
TBT11	TBT12	TBT13	TBT14	TBT15	TBT16	TBT17
HID19	HID20	HID21	HID22	HID23	HID24	HID25
8	9	10	11	12	13	14
D5	D6	D7	D8	D9	D10	D11
TBT18	TBT19	TBT20	TBT21	TBT22	TBT23	TBT24
HID26	HID27	HID28	HID29	HID30	TAH1	TAH2
15	16	17	18	19	20	21
D12	D13	D14	D15	D16	D17	D18
TBT25	TBT26	TBT27	TBT28	TBT29	TBT30	MKH1
TAH3	TAH4	TAH5	TAH6	TAH7	TAH8	TAH9
22	23	24	25	26	27	28
D19	D20	D21	D22	D23	D24	D25
MKH2	MKH3	MKH4	MKH5	MKH6	MKH7	MKH8
TAH10	TAH11	TAH12	TAH13	TAH14	TAH15	TAH16

EUSI 570/2086-87/6322-23/2078-79						
Ashe	Ojobo	Iwa	Araorun	Ile	Itutu	Erinle
1	2	3	4	5	6	7
O31	N1	N2	N3	N4	N5	N6
KHK13	KHK14	KHK15	KHK16	KHK17	KHK18	KHK19
THi21	THi22	THi23	THi24	THi25	THi26	THi27
8	9	10	11	12	13	14
N7	N8	N9	N10	N11	N12	N13
KHK20	KHK21	KHK22	KHK23	KHK24	KHK25	KHK26
THi28	THi29	THi30	HID1	HID2	HID3	HID4
15	16	17	18	19	20	21
N14	N15	N16	N17	N18	N19	N20
KHK27	KHK28	KHK29	KHK30	TBT1	TBT2	TBT3
HID5	HID6	HID7	HID8	HID9	HID10	HID11
22	23	24	25	26	27	28
N21	N22	N23	N24	N25	N26	N27
TBT4	TBT5	TBT6	TBT7	TBT8	TBT9	TBT10
HID12	HID13	HID14	HID15	HID16	HID17	HID18

KWANZA 570/2086-87/6322-23/2078-79						
Umoja	Kujichagulia	Ujima	Ujamaa	Nia	Kuumba	Imani
1	2	3	4	5	6	7
D26	D27	D28	D29	D30	D31	JA1
MKH9	MKH10	MKH11	MKH12	MKH13	MKH14	MKH15
TAH17	TAH18	TAH19	TAH20	TAH21	TAH22	TAH23
8	9	10	11	12	13	14
JA2	JA3	JA4	JA5	JA6	JA7	JA8
MKH16	MKH17	MKH18	MKH19	MKH20	MKH21	MKH22
TAH24	TAH25	TAH26	TAH27	TAH28	TAH29	TAH30
15	16	17	18	19	20	21
JA9	JA10	JA11	JA12	JA13	JA14	JA15
MKH23	MKH24	MKH25	KMH26	MKH27	MKH28	MKH29
TIR1	TIR2	TIR3	TIR4	TIR5	TIR6	TIR7
22	23	24	25	26	27	28
JA16	JA17	JA18	JA19	JA20	JA21	JA22
MKH30	AMN1	AMN2	AMN3	AMN4	AMN5	AMN6
TIR8	TIR9	TIR10	TIR11	TIR12	TIR13	TIR14

SHAHIDI 570/2086-87/6322-23/2078-79						
Ashe	Ojobo	Iwa	Araorun	Ile	Itutu	Erinle
1	2	3	4	5	6	7
JA23	JA24	JA25	JA26	JA27	JA28	JA29
AMN7	AMN8	AMN9	AMN10	AMN11	AMN12	AMN13
TIR15	TIR16	TIR17	TIR18	TIR19	TIR20	TIR21
8	9	10	11	12	13	14
JA30	JA31	F1	F2	F3	F4	F5
AMN14	AMN15	AMN16	AMN17	AMN18	AMN19	AMN20
TIR22	TIR23	TIR24	TIR25	TIR26	TIR27	TIR28
15	16	17	18	19	20	21
F6	F7	F8	F9	F10	F 11	F12
AMN21	AMN22	AMN23	AMN24	AMN25	AMN26	AMN28
TIR29	TIR30	YEK1	YEK2	YEK3	YEK4	YEK5
22	23	24	25	26	27	28
F13	F14	F15	F16	F17	F18	F19
AMN29	AMN30	RNT1	RNT2	RNT3	RNT4	RNT5
YEK6	YEK7	YEK8	YEK9	YEK10	YEK11	YEK12
SHAHIDI 121						
F20						
RNT6						
YEK13						

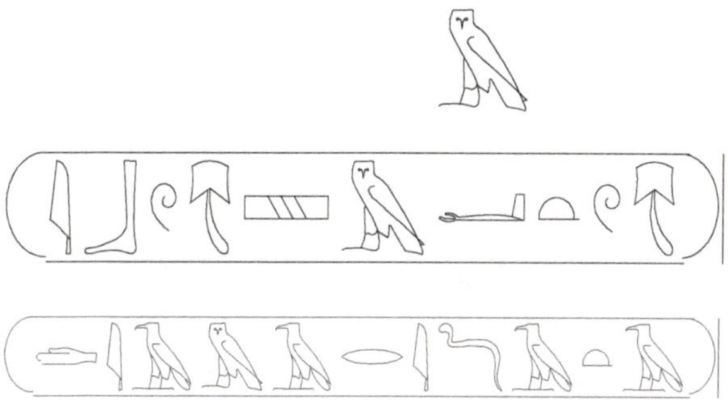

THE NU AFRIKAN CALENDAR

MAAFA 571 B.S.E./ASIENTO 2087-88 C.E./
KAM 6323-24 S.M./EC 2079-80 A.M. (ዓ.ም.)

AFU 571/2087-88/6323-24/2079-80						
Ashe	Ojobo	Iwa	Araorun	Ile	Itutu	Erinle
1	2	3	4	5	6	7
F21	F22	F23	F24	F25	F26	F27
RNT7	RNT8	RNT9	RNT10	RNT11	RNT12	RNT13
YEK14	YEK15	YEK16	YEK17	YEK18	YEK19	YEK20
8	9	10	11	12	13	14
F28	M1	M2	M3	M4	M5	M6
RNT14	RNT15	RNT16	RNT17	RNT18	RNT19	RNT20
YEK21	YEK22	YEK23	YEK24	YEK25	YEK26	YEK27
15	16	17	18	19	20	21
M7	M8	M9	M10	M11	M12	M13
RNT21	RNT22	RNT23	RNT24	RNT25	RNT26	RNT27
YEK28	YEK 29	YEK30	MAG1	MAG2	MAG3	MAG4
22	23	24	25	26	27	28
M14	M15	M16	M17	M18	M19	M20
RNT28	RNT29	RNT30	KHN1	KHN2	KHN3	KHN4
MAG5	MAG6	MAG7	MAG8	MAG9	MAG10	MAG11

UHURU 571/2087-88/6323-24/2079-80						
Ashe	Ojobo	Iwa	Araorun	Ile	Itutu	Erinle
1	2	3	4	5	6	7
M21	M22	M23	M24	M25	M26	M27
KHN5	KHN6	KHN7	KHN8	KHN9	KHN10	KHN11
MAG12	MAG13	MAG14	MAG15	MAG16	MAG17	MAG18
8	9	10	11	12	13	14
M28	M29	M30	M31	A1	A2	A3
KHN12	KHN13	KHN14	KHN15	KHN16	KHN17	KHN18
MAG19	MAG20	MAG21	MAG22	MAG23	MAG24	MAG25
15	16	17	18	19	20	21
A4	A5	A6	A7	A8	A9	A10
KHN19	KHN20	KHN21	KHN22	KHN23	KHN24	KHN25
MAG26	MAG27	MAG28	MAG29	MAG30	MIY1	MIY2
22	23	24	25	26	27	28
A11	A12	A13	A14	A15	A16	A17
KHN26	KHN27	KHN28	KHN29	KHN30	PIN1	PIN2
MIY3	MIY4	MIY5	MIY6	MIY7	MIY8	MIY9

KAWAIDA 571/2087-88/6323-24/2079-80						
Ashe	Ojobo	Iwa	Araorun	Ile	Itutu	Erinle
1	2	3	4	5	6	7
A18	A19	A20	A21	A22	A23	A24
PIN3	PIN4	PIN5	PIN6	PIN7	PIN8	PIN9
MIY10	MIY11	MIY12	MIY13	MIY14	MIY15	MIY16
8	9	10	11	12	13	14
A25	A26	A27	A28	A29	A30	MA1
PIN10	PIN11	PIN12	PIN13	PIN14	PIN15	PIN16
MIY17	MIY18	MIY19	MIY20	MIY21	MIY22	MIY23
15	16	17	18	19	20	21
MA2	MA3	MA4	MA5	MA6	MA7	MA8
PIN17	PIN18	PIN19	PIN20	PIN21	PIN22	PIN23
MIY24	MIY25	MIY26	MIY27	MIY28	MIY29	MIY30
22	23	24	25	26	27	28
MA9	MA10	MA11	MA12	MA13	MA14	MA15
PIN24	PIN25	PIN26	PIN27	PIN28	PIN29	PIN30
GIN1	GIN2	GIN3	GIN4	GIN5	GIN6	GIN7

MABABA WEUSI 571/2087-88/6323-24/2079-80						
Ashe	Ojobo	Iwa	Araorun	Ile	Itutu	Erinle
1	2	3	4	5	6	7
MA16	MA17	MA18	MA19	MA20	MA21	MA22
APP1	APP2	APP3	APP4	APP5	APP6	APP7
GIN8	GIN9	GIN10	GIN11	GIN12	GIN13	GIN14
8	9	10	11	12	13	14
MA23	MA24	MA25	MA26	MA27	MA28	MA29
APP8	APP9	APP10	APP11	APP12	APP13	APP14
GIN15	GIN16	GIN17	GIN18	GIN19	GIN20	GIN21
15	16	17	18	19	20	21
MA30	MA31	J1	J2	J3	J4	J5
APP15	APP16	APP17	APP18	APP19	APP20	APP21
GIN22	GIN23	GIN24	GIN25	GIN26	GIN27	GIN28
22	23	24	25	26	27	28
J6	J7	J8	J9	J10	J11	J12
APP22	APP23	APP24	APP25	APP26	APP27	APP28
GIN29	GIN30	SAN1	SAN2	SAN3	SAN4	SAN5

THE NU AFRIKAN CALENDAR

MWANAMKE WEUSI 571/2087-88/6323-24/2079-80

Ashe	Ojobo	Iwa	Araorun	Ile	Itutu	Erinle
1	2	3	4	5	6	7
J13	J14	J15	J16	J17	J18	J19
APP29	APP30	MSR1	MSR2	MSR3	MSR4	MSR5
SAN6	SAN7	SAN8	SAN9	SAN10	SAN11	SAN12
8	9	10	11	12	13	14
J20	J21	J22	J23	J24	J25	J26
MSR6	MSR7	MSR8	MSR9	MSR10	MSR11	MSR12
SAN13	SAN14	SAN15	SAN16	SAN17	SAN18	SAN19
15	16	17	18	19	20	21
J27	J28	J29	J30	JU1	JU2	JU3
MSR13	MSR14	MSR15	MSR16	MSR17	MSR18	MSR19
SAN20	SAN21	SAN22	SAN23	SAN24	SAN25	SAN26
22	23	24	25	26	27	28
JU4	JU5	JU6	JU7	JU8	JU9	JU10
MSR20	MRS21	MSR22	MSR23	MSR24	MSR25	MSR26
SAN27	SAN28	SAN29	SAN30	HAM1	HAM2	HAM3

TAIFA WEUSI 571/2087-88/6323-24/2079-80

Ashe	Ojobo	Iwa	Araorun	Ile	Itutu	Erinle
1	2	3	4	5	6	7
AU8	AU9	AU10	AU11	AU12	AU13	AU14
THT20	THT21	THT22	THT23	THT24	THT25	THT26
NAH2	NAH3	NAH4	NAH5	NAH6	NAH7	NAH8
8	9	10	11	12	13	14
AU15	AU16	AU17	AU18	AU19	AU20	AU21
THT27	THT28	THT29	THT30	PAP1	PAP1	PAP2
NAH9	NAH10	NAH11	NAH12	NAH13	NAH14	NAH15
15	16	17	18	19	20	21
AU22	AU23	AU24	AU25	AU26	AU27	AU28
PAP3	PAP4	PAP5	PAP6	PAP7	PAP8	PAP9
NAH16	NAH17	NAH18	NAH19	NAH20	NAH21	NAH22
22	23	24	25	26	27	28
AU29	AU30	AU31	S1	S2	S3	S4
PAP10	PAP11	PAP12	PAP13	PAP14	PAP15	PAP16
NAH23	NAH24	NAH25	NAH26	NAH27	NAH28	NAH29

MTU WEUSI 571/2087-88/6323-24/2079-80

Ashe	Ojobo	Iwa	Araorun	Ile	Itutu	Erinle
1	2	3	4	5	6	7
JU11	JU12	JU13	JU14	JU15	JU16	JU17
MSR27	MSR28	MSR29	MSR30	*ASR1*	*HRU2*	*STH3*
HAM4	HAM5	HAM6	HAM7	HAM8	HAM9	HAM10
8	9	10	11	12	13	14
JU18	JU19	JU20	JU21	JU22	JU23	JU24
AST4	*NBH5*	THT1	THT2	THT3	THT4	THT5
HAM11	HAM12	HAM13	HAM14	HAM15	HAM16	HAM17
15	16	17	18	19	20	21
JU25	JU26	JU27	JU28	JU29	JU30	JU31
THT6	THT7	THT8	THT9	THT10	THT11	THT12
HAM18	HAM19	HAM20	HAM21	HAM22	HAM23	HAM24
22	23	24	25	26	27	28
AU1	AU2	AU3	AU4	AU5	AU6	AU7
THT13	THT14	THT15	THT16	THT17	THT18	THT19
HAM25	HAM26	HAM27	HAM28	HAM29	HAM30	NAH1

UWEZO 571/2087-88/6323-24/2079-80

Ashe	Ojobo	Iwa	Araorun	Ile	Itutu	Erinle
1	2	3	4	5	6	7
S5	S6	S7	S8	S9	S10	S11
PAP17	PAP18	PAP19	PAP20	PAP21	PAP22	PAP23
NAH30	*QUA1*	*QUA2*	*QUA3*	*QUA4*	*QUA5*	*QUA6*
8	9	10	11	12	13	14
S12	S13	S14	S15	S16	S17	S18
PAP24	PAP25	PAP26	PAP27	PAP28	PAP29	PAP30
MAS1	MAS2	MAS3	MAS4	MAS5	MAS6	MAS7
15	16	17	18	19	20	21
S19	S20	S21	S22	S23	S24	S25
HTH1	HTH2	HTH3	HTH4	HTH5	HTH6	HTH7
MAS8	MAS9	MAS10	MAS11	MAS12	MAS13	MAS14
22	23	24	25	26	27	28
S26	S27	S28	S29	S30	O1	O2
HTH8	HTH9	HTH10	HTH11	HTH12	HTH13	HTH14
MAS15	MAS16	MAS17	MAS18	MAS19	MAS20	MAS21

THE NU AFRIKAN CALENDAR

EKUNDU 571/2087-88/6323-24/2079-80						
Ashe	Ojobo	Iwa	Araorun	Ile	Itutu	Erinle
1	2	3	4	5	6	7
O3	O4	O5	O6	O7	O8	O9
HTH15	HTH16	HTH17	HTH18	HTH19	HTH20	HTH21
MAS22	MAS23	MAS 24	MAS25	MAS26	MAS27	MAS28
8	9	10	11	12	13	14
O10	O11	O12	O13	O14	O15	O16
HTH22	HTH23	HTH24	HTH25	HTH26	HTH27	HTH28
MAS29	MAS30	THi1	THi2	THi3	THi4	THi5
15	16	17	18	19	20	21
O17	O18	O19	O20	O21	O22	O23
HTH29	HTH30	KHK1	KHK2	KHK3	KHK4	KHK5
THi6	THi7	THi8	THi9	THi10	THi11	THi12
22	23	24	25	26	27	28
O24	O25	O26	O27	O28	O29	O30
KHK6	KHK7	KHK8	KHK9	KHK10	KHK11	KHK12
THi13	THi14	THi15	THi16	THi17	THi18	THi19

KIJANI 571/2087-88/6323-24/2079-80						
Ashe	Ojobo	Iwa	Araorun	Ile	Itutu	Erinle
1	2	3	4	5	6	7
N28	N29	N30	D1	D2	D3	D4
TBT11	TBT12	TBT13	TBT14	TBT15	TBT16	TBT17
HID18	HID19	HID20	HID21	HID22	HID23	HID24
8	9	10	11	12	13	14
D5	D6	D7	D8	D9	D10	D11
TBT18	TBT19	TBT20	TBT21	TBT22	TBT23	TBT24
HID25	HID26	HID27	HID28	HID29	HID30	TAH1
15	16	17	18	19	20	21
D12	D13	D14	D15	D16	D17	D18
TBT25	TBT26	TBT27	TBT28	TBT29	TBT30	MKH1
TAH2	TAH3	TAH4	TAH5	TAH6	TAH7	TAH8
22	23	24	25	26	27	28
D19	D20	D21	D22	D23	D24	D25
MKH2	MKH3	MKH4	MKH5	MKH6	MKH7	MKH8
TAH9	TAH10	TAH11	TAH12	TAH13	TAH14	TAH15

EUSI 571/2087-88/6323-24/2079-80						
Ashe	Ojobo	Iwa	Araorun	Ile	Itutu	Erinle
1	2	3	4	5	6	7
O31	N1	N2	N3	N4	N5	N6
KHK13	KHK14	KHK15	KHK16	KHK17	KHK18	KHK19
THi20	THi21	THi22	THi23	THi24	THi25	THi26
8	9	10	11	12	13	14
N7	N8	N9	N10	N11	N12	N13
KHK20	KHK21	KHK22	KHK23	KHK24	KHK25	KHK26
THi27	THi28	THi29	THi30	HID1	HID2	HID3
15	16	17	18	19	20	21
N14	N15	N16	N17	N18	N19	N20
KHK27	KHK28	KHK29	KHK30	TBT1	TBT2	TBT3
HID4	HID5	HID6	HID7	HID8	HID9	HID10
22	23	24	25	26	27	28
N21	N22	N23	N24	N25	N26	N27
TBT4	TBT5	TBT6	TBT7	TBT8	TBT9	TBT10
HID11	HID12	HID13	HID14	HID15	HID16	HID17

KWANZA 571/2087-88/6323-24/2079-80						
Umoja	Kujichagulia	Ujima	Ujamaa	Nia	Kuumba	Imani
1	2	3	4	5	6	7
D26	D27	D28	D29	D30	D31	JA1
MKH9	MKH10	MKH11	MKH12	MKH13	MKH14	MKH15
TAH16	TAH17	TAH18	TAH19	TAH20	TAH21	TAH22
8	9	10	11	12	13	14
JA2	JA3	JA4	JA5	JA6	JA7	JA8
MKH16	MKH17	MKH18	MKH19	MKH20	MKH21	MKH22
TAH23	TAH24	TAH25	TAH26	TAH27	TAH28	TAH29
15	16	17	18	19	20	21
JA9	JA10	JA11	JA12	JA13	JA14	JA15
MKH23	MKH24	MKH25	KMH26	MKH27	MKH28	MKH29
TAH30	TIR1	TIR2	TIR3	TIR4	TIR5	TIR6
22	23	24	25	26	27	28
JA16	JA17	JA18	JA19	JA20	JA21	JA22
MKH30	AMN1	AMN2	AMN3	AMN4	AMN5	AMN6
TIR7	TIR8	TIR9	TIR10	TIR11	TIR12	TIR13

SHAHIDI 571/2087-88/6323-24/2079-80						
Ashe	Ojobo	Iwa	Araorun	Ile	Itutu	Erinle
1	2	3	4	5	6	7
JA23	JA24	JA25	JA26	JA27	JA28	JA29
AMN7	AMN8	AMN9	AMN10	AMN11	AMN12	AMN13
TIR14	TIR15	TIR16	TIR17	TIR18	TIR19	TIR20
8	9	10	11	12	13	14
JA30	JA31	F1	F2	F3	F4	F5
AMN14	AMN15	AMN16	AMN17	AMN18	AMN19	AMN20
TIR21	TIR22	TIR23	TIR24	TIR25	TIR26	TIR27
15	16	17	18	19	20	21
F6	F7	F8	F9	F10	F 11	F12
AMN21	AMN22	AMN23	AMN24	AMN25	AMN26	AMN28
TIR28	TIR29	TIR30	YEK1	YEK2	YEK3	YEK4
22	23	24	25	26	27	28
F13	F14	F15	F16	F17	F18	F19
AMN29	AMN30	RNT1	RNT2	RNT3	RNT4	RNT5
YEK5	YEK6	YEK7	YEK8	YEK9	YEK10	YEK11
SHAHIDI 122						
F20						
RNT6						
YEK12						

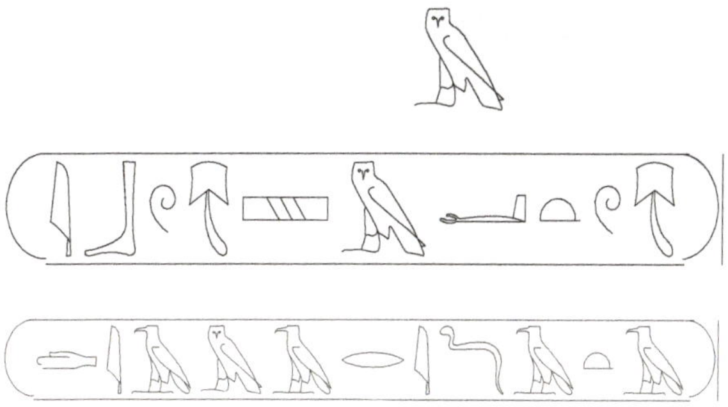

THE NU AFRIKAN CALENDAR

MAAFA 572 B.S.E./ASIENTO 2088-89 C.E./
KAM 6324-25 S.M./EC 2080-81 A.M. (ዓ.ም.)

AFU 572/2088-89/6324-25/2080-81						
Ashe	Ojobo	Iwa	Araorun	Ile	Itutu	Erinle
1	2	3	4	5	6	7
F21	F22	F23	F24	F25	F26	F27
RNT7	RNT8	RNT9	RNT10	RNT11	RNT12	RNT13
YEK13	YEK14	YEK15	YEK16	YEK17	YEK18	YEK19
8	9	10	11	12	13	14
F28	F29	M1	M2	M3	M4	M5
RNT14	RNT15	RNT16	RNT17	RNT18	RNT19	RNT20
YEK 20	YEK21	YEK22	YEK23	YEK24	YEK25	YEK26
15	16	17	18	19	20	21
M6	M7	M8	M9	M10	M11	M12
RNT21	RNT22	RNT23	RNT24	RNT25	RNT26	RNT27
YEK 27	YEK28	YEK29	YEK30	MAG1	MAG2	MAG3
22	23	24	25	26	27	28
M13	M14	M15	M16	M17	M18	M19
RNT28	RNT29	RNT30	KHN1	KHN2	KHN3	KHN4
MAG4	MAG5	MAG6	MAG7	MAG8	MAG9	MAG10
SHARIFU 140						
M20						
KHN5						
MAG11						

KAWAIDA 572/2088-89/6324-25/2080-81						
Ashe	Ojobo	Iwa	Araorun	Ile	Itutu	Erinle
1	2	3	4	5	6	7
A18	A19	A20	A21	A22	A23	A24
PIN4	PIN5	PIN6	PIN7	PIN8	PIN9	PIN10
MIY10	MIY11	MIY12	MIY13	MIY14	MIY15	MIY16
8	9	10	11	12	13	14
A25	A26	A27	A28	A29	A30	MA1
PIN11	PIN12	PIN13	PIN14	PIN15	PIN16	PIN17
MIY17	MIY18	MIY19	MIY20	MIY21	MIY22	MIY23
15	16	17	18	19	20	21
MA2	MA3	MA4	MA5	MA6	MA7	MA8
PIN18	PIN19	PIN20	PIN21	PIN22	PIN23	PIN24
MIY24	MIY25	MIY26	MIY27	MIY28	MIY29	MIY30
22	23	24	25	26	27	28
MA9	MA10	MA11	MA12	MA13	MA14	MA15
PIN25	PIN26	PIN27	PIN28	PIN29	PIN30	APP1
GIN1	GIN2	GIN3	GIN4	GIN5	GIN6	GIN7

UHURU 572/2088-89/6324-25/2080-81						
Ashe	Ojobo	Iwa	Araorun	Ile	Itutu	Erinle
1	2	3	4	5	6	7
M21	M22	M23	M24	M25	M26	M27
KHN6	KHN7	KHN8	KHN9	KHN10	KHN11	KHN12
MAG12	MAG13	MAG14	MAG15	MAG16	MAG17	MAG18
8	9	10	11	12	13	14
M28	M29	M30	M31	A1	A2	A3
KHN13	KHN14	KHN15	KHN16	KHN17	KHN18	KHN19
MAG19	MAG20	MAG21	MAG22	MAG23	MAG24	MAG25
15	16	17	18	19	20	21
A4	A5	A6	A7	A8	A9	A10
KHN20	KHN21	KHN222	KHN23	KHN24	KHN25	KHN26
MAG26	MAG27	MAG28	MAG29	MAG30	MIY1	MIY2
22	23	24	25	26	27	28
A11	A12	A13	A14	A15	A16	A17
KHN27	KHN28	KHN29	KHN30	PIN1	PIN2	PIN3
MIY3	MIY4	MIY5	MIY6	MIY7	MIY8	MIY9

MABABA WEUSI 572/2088-89/6324-25/2080-81						
Ashe	Ojobo	Iwa	Araorun	Ile	Itutu	Erinle
1	2	3	4	5	6	7
MA16	MA17	MA18	MA19	MA20	MA21	MA22
APP2	APP3	APP4	APP5	APP6	APP7	APP8
GIN8	GIN9	GIN10	GIN11	GIN12	GIN13	GIN14
8	9	10	11	12	13	14
MA23	MA24	MA25	MA26	MA27	MA28	MA29
APP9	APP10	APP11	APP12	APP13	APP14	APP15
GIN15	GIN16	GIN17	GIN18	GIN19	GIN20	GIN21
15	16	17	18	19	20	21
MA30	MA31	J1	J2	J3	J4	J5
APP16	APP17	APP18	APP19	APP20	APP21	APP22
GIN22	GIN23	GIN24	GIN25	GIN26	GIN27	GIN28
22	23	24	25	26	27	28
J6	J7	J8	J9	J10	J11	J12
APP23	APP24	APP25	APP26	APP27	APP28	APP29
GIN29	GIN30	SAN1	SAN2	SAN3	SAN4	SAN5

THE NU AFRIKAN CALENDAR

MWANAMKE WEUSI 572/2088-89/6324-25/2080-81

Ashe	Ojobo	Iwa	Araorun	Ile	Itutu	Erinle
1	2	3	4	5	6	7
J13	J14	J15	J16	J17	J18	J19
APP30	MSR1	MSR2	MSR3	MSR4	MSR5	MSR6
SAN6	SAN7	SAN8	SAN9	SAN10	SAN11	SAN12
8	9	10	11	12	13	14
J20	J21	J22	J23	J24	J25	J26
MSR7	MSR8	MSR9	MSR10	MSR11	MSR12	MSR13
SAN13	SAN14	SAN15	SAN16	SAN17	SAN18	SAN19
15	16	17	18	19	20	21
J27	J28	J29	J30	JU1	JU2	JU3
MSR14	MSR15	MSR16	MSR17	MSR18	MSR19	MSR20
SAN20	SAN21	SAN22	SAN23	SAN24	SAN25	SAN26
22	23	24	25	26	27	28
JU4	JU5	JU6	JU7	JU8	JU9	JU10
MSR21	MRS22	MSR23	MSR24	MSR25	MSR26	MSR27
SAN27	SAN28	SAN29	SAN30	HAM1	HAM2	HAM3

TAIFA WEUSI 572/2088-89/6324-25/2080-81

Ashe	Ojobo	Iwa	Araorun	Ile	Itutu	Erinle
1	2	3	4	5	6	7
AU8	AU9	AU10	AU11	AU12	AU13	AU14
THT20	THT21	THT22	THT23	THT24	THT25	THT26
NAH2	NAH3	NAH4	NAH5	NAH6	NAH7	NAH8
8	9	10	11	12	13	14
AU15	AU16	AU17	AU18	AU19	AU20	AU21
THT27	THT28	THT29	THT30	PAP1	PAP1	PAP2
NAH9	NAH10	NAH11	NAH12	NAH13	NAH14	NAH15
15	16	17	18	19	20	21
AU22	AU23	AU24	AU25	AU26	AU27	AU28
PAP3	PAP4	PAP5	PAP6	PAP7	PAP8	PAP9
NAH16	NAH17	NAH18	NAH19	NAH20	NAH21	NAH22
22	23	24	25	26	27	28
AU29	AU30	AU31	S1	S2	S3	S4
PAP10	PAP11	PAP12	PAP13	PAP14	PAP15	PAP16
NAH23	NAH24	NAH25	NAH26	NAH27	NAH28	NAH29

MTU WEUSI 572/2088-89/6324-25/2080-81

Ashe	Ojobo	Iwa	Araorun	Ile	Itutu	Erinle
1	2	3	4	5	6	7
JU11	JU12	JU13	JU14	JU15	JU16	JU17
MSR28	MSR29	MSR30	*ASR1*	*HRU2*	*STH3*	*AST4*
HAM4	HAM5	HAM6	HAM7	HAM8	HAM9	HAM10
8	9	10	11	12	13	14
JU18	JU19	JU20	JU21	JU22	JU23	JU24
NBH5	*ANU6*	THT1	THT2	THT3	THT4	THT5
HAM11	HAM12	HAM13	HAM14	HAM15	HAM16	HAM17
15	16	17	18	19	20	21
JU25	JU26	JU27	JU28	JU29	JU30	JU31
THT6	THT7	THT8	THT9	THT10	THT11	THT12
HAM18	HAM19	HAM20	HAM21	HAM22	HAM23	HAM24
22	23	24	25	26	27	28
AU1	AU2	AU3	AU4	AU5	AU6	AU7
THT13	THT14	THT15	THT16	THT17	THT18	THT19
HAM25	HAM26	HAM27	HAM28	HAM29	HAM30	NAH1

UWEZO 572/2088-89/6324-25/2080-81

Ashe	Ojobo	Iwa	Araorun	Ile	Itutu	Erinle
1	2	3	4	5	6	7
S5	S6	S7	S8	S9	S10	S11
PAP17	PAP18	PAP19	PAP20	PAP21	PAP22	PAP23
NAH30	*PAG1*	*PAG2*	*PAG3*	*PAG4*	*PAG5*	MAS1
8	9	10	11	12	13	14
S12	S13	S14	S15	S16	S17	S18
PAP24	PAP25	PAP26	PAP27	PAP28	PAP29	PAP30
MAS2	MAS3	MAS4	MAS5	MAS6	MAS7	MAS8
15	16	17	18	19	20	21
S19	S20	S21	S22	S23	S24	S25
HTH1	HTH2	HTH3	HTH4	HTH5	HTH6	HTH7
MAS9	MAS10	MAS11	MAS12	MAS13	MAS14	MAS15
22	23	24	25	26	27	28
S26	S27	S28	S29	S30	O1	O2
HTH8	HTH9	HTH10	HTH11	HTH12	HTH13	HTH14
MAS16	MAS17	MAS18	MAS19	MAS20	MAS21	MAS22

THE NU AFRIKAN CALENDAR

EKUNDU 572/2088-89/6324-25/2080-81						
Ashe	Ojobo	Iwa	Araorun	Ile	Itutu	Erinle
1	2	3	4	5	6	7
O3	O4	O5	O6	O7	O8	O9
HTH15	HTH16	HTH17	HTH18	HTH19	HTH20	HTH21
MAS23	MAS24	MAS25	MAS26	MAS27	MAS28	MAS29
8	9	10	11	12	13	14
O10	O11	O12	O13	O14	O15	O16
HTH22	HTH23	HTH24	HTH25	HTH26	HTH27	HTH28
MAS30	THi1	THi2	THi3	THi4	THi5	THi6
15	16	17	18	19	20	21
O17	O18	O19	O20	O21	O22	O23
HTH29	HTH30	KHK1	KHK2	KHK3	KHK4	KHK5
THi7	THi8	This9	THi10	THi11	THi12	THi13
22	23	24	25	26	27	28
O24	O25	O26	O27	O28	O29	O30
KHK6	KHK7	KHK8	KHK9	KHK10	KHK11	KHK12
Thi14	THi15	THi16	THi17	THi18	THi19	THi20

KIJANI 572/2088-89/6324-25/2080-81						
Ashe	Ojobo	Iwa	Araorun	Ile	Itutu	Erinle
1	2	3	4	5	6	7
N28	N29	N30	D1	D2	D3	D4
TBT11	TBT12	TBT13	TBT14	TBT15	TBT16	TBT17
HID19	HID20	HID21	HID22	HID23	HID24	HID25
8	9	10	11	12	13	14
D5	D6	D7	D8	D9	D10	D11
TBT18	TBT19	TBT20	TBT21	TBT22	TBT23	TBT24
HID26	HID27	HID28	HID29	HID30	TAH1	TAH2
15	16	17	18	19	20	21
D12	D13	D14	D15	D16	D17	D18
TBT25	TBT26	TBT27	TBT28	TBT29	TBT30	MKH1
TAH3	TAH4	TAH5	TAH6	TAH7	TAH8	TAH9
22	23	24	25	26	27	28
D19	D20	D21	D22	D23	D24	D25
MKH2	MKH3	MKH4	MKH5	MKH6	MKH7	MKH8
TAH10	TAH11	TAH12	TAH13	TAH14	TAH15	TAH16

EUSI 572/2088-89/6324-25/2080-81						
Ashe	Ojobo	Iwa	Araorun	Ile	Itutu	Erinle
1	2	3	4	5	6	7
O31	N1	N2	N3	N4	N5	N6
KHK13	KHK14	KHK15	KHK16	KHK17	KHK18	KHK19
THi21	THi22	THi23	THi24	THi25	THi26	THi27
8	9	10	11	12	13	14
N7	N8	N9	N10	N11	N12	N13
KHK20	KHK21	KHK22	KHK23	KHK24	KHK25	KHK26
THi28	THi29	THi30	HID1	HID2	HID3	HID4
15	16	17	18	19	20	21
N14	N15	N16	N17	N18	N19	N20
KHK27	KHK28	KHK29	KHK30	TBT1	TBT2	TBT3
HID5	HID6	HID7	HID8	HID9	HID10	HID11
22	23	24	25	26	27	28
N21	N22	N23	N24	N25	N26	N27
TBT4	TBT5	TBT6	TBT7	TBT8	TBT9	TBT10
HID12	HID13	HID14	HID15	HID16	HID17	HID18

KWANZA 572/2088-89/6324-25/2080-81						
Umoja	Kujichagulia	Ujima	Ujamaa	Nia	Kuumba	Imani
1	2	3	4	5	6	7
D26	D27	D28	D29	D30	D31	JA1
MKH9	MKH10	MKH11	MKH12	MKH13	MKH14	MKH1
TAH17	TAH18	TAH19	TAH20	TAH21	TAH22	TAH2
8	9	10	11	12	13	14
JA2	JA3	JA4	JA5	JA6	JA7	JA
MKH16	MKH17	MKH18	MKH19	MKH20	MKH21	MKH2
TAH24	TAH25	TAH26	TAH27	TAH28	TAH29	TAH3
15	16	17	18	19	20	21
JA9	JA10	JA11	JA12	JA13	JA14	JA1
MKH23	MKH24	MKH25	KMH26	MKH27	MKH28	MKH2
TIR1	TIR2	TIR3	TIR4	TIR5	TIR6	TIR7
22	23	24	25	26	27	28
JA16	JA17	JA18	JA19	JA20	JA21	JA2
MKH30	AMN1	AMN2	AMN3	AMN4	AMN5	AMN
TIR8	TIR9	TIR10	TIR11	TIR12	TIR13	TIR1

THE NU AFRIKAN CALENDAR

SHAHIDI 572/2088-89/6324-25/2080-81						
Ashe	Ojobo	Iwa	Araorun	Ile	Itutu	Erinle
1	2	3	4	5	6	7
JA23	JA24	JA25	JA26	JA27	JA28	JA29
AMN7	AMN8	AMN9	AMN10	AMN11	AMN12	AMN13
TIR15	TIR16	TIR17	TIR18	TIR19	TIR20	TIR21
8	9	10	11	12	13	14
JA30	JA31	F1	F2	F3	F4	F5
AMN14	AMN15	AMN16	AMN17	AMN18	AMN19	AMN20
TIR22	TIR23	TIR24	TIR25	TIR26	TIR27	TIR28
15	16	17	18	19	20	21
F6	F7	F8	F9	F10	F 11	F12
AMN21	AMN22	AMN23	AMN24	AMN25	AMN26	AMN28
TIR29	TIR30	YEK1	YEK2	YEK3	YEK4	YEK5
22	23	24	25	26	27	28
F13	F14	F15	F16	F17	F18	F19
AMN29	AMN30	RNT1	RNT2	RNT3	RNT4	RNT5
YEK6	YEK7	YEK8	YEK9	YEK10	YEK11	YEK12
SHAHIDI 123						
F20						
RNT6						
YEK13						

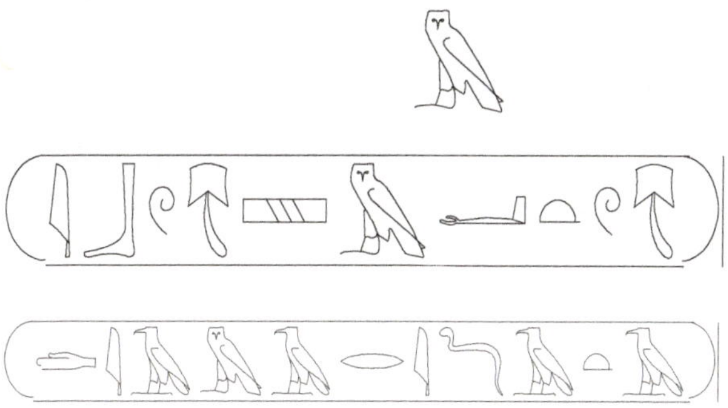

THE NU AFRIKAN CALENDAR

MAAFA 573 B.S.E./ASIENTO 2089-90 C.E.
KAM 6325-26 S.M./EC 2081-82 A.M. (ዓ.ም.)

AFU 573/2089-90/6325-26/2081-82

Ashe	Ojobo	Iwa	Araorun	Ile	Itutu	Erinle
1	2	3	4	5	6	7
F21	F22	F23	F24	F25	F26	F27
RNT7	RNT8	RNT9	RNT10	RNT11	RNT12	RNT13
YEK14	YEK15	YEK16	YEK17	YEK18	YEK19	YEK20
8	9	10	11	12	13	14
F28	M1	M2	M3	M4	M5	M6
RNT14	RNT15	RNT16	RNT17	RNT18	RNT19	RNT20
YEK21	YEK22	YEK23	YEK24	YEK25	YEK26	YEK27
15	16	17	18	19	20	21
M7	M8	M9	M10	M11	M12	M13
RNT21	RNT22	RNT23	RNT24	RNT25	RNT26	RNT27
YEK28	YEK29	YEK30	MAG1	MAG2	MAG3	MAG4
22	23	24	25	26	27	28
M14	M15	M16	M17	M18	M19	M20
RNT28	RNT29	RNT30	KHN1	KHN2	KHN3	KHN4
MAG5	MAG6	MAG7	MAG8	MAG9	MAG10	MAG11

KAWAIDA 573/2089-90/6325-26/2081-82

Ashe	Ojobo	Iwa	Araorun	Ile	Itutu	Erinle
1	2	3	4	5	6	7
A18	A19	A20	A21	A22	A23	A24
PIN3	PIN4	PIN5	PIN6	PIN7	PIN8	PIN9
MIY10	MIY11	MIY12	MIY13	MIY14	MIY15	MIY16
8	9	10	11	12	13	14
A25	A26	A27	A28	A29	A30	MA1
PIN10	PIN11	PIN12	PIN13	PIN14	PIN15	PIN16
MIY17	MIY18	MIY19	MIY20	MIY21	MIY22	MIY23
15	16	17	18	19	20	21
MA2	MA3	MA4	MA5	MA6	MA7	MA8
PIN17	PIN18	PIN19	PIN20	PIN21	PIN22	PIN23
MIY24	MIY25	MIY26	MIY27	MIY28	MIY29	MIY30
22	23	24	25	26	27	28
MA9	MA10	MA11	MA12	MA13	MA14	MA15
PIN24	PIN25	PIN26	PIN27	PIN28	PIN29	PIN30
GIN1	GIN2	GIN3	GIN4	GIN5	GIN6	GIN7

UHURU 573/2089-90/6325-26/2081-82

Ashe	Ojobo	Iwa	Araorun	Ile	Itutu	Erinle
1	2	3	4	5	6	7
M21	M22	M23	M24	M25	M26	M27
KHN5	KHN6	KHN7	KHN8	KHN9	KHN10	KHN11
MAG12	MAG13	MAG14	MAG15	MAG16	MAG17	MAG18
8	9	10	11	12	13	14
M28	M29	M30	M31	A1	A2	A3
KHN12	KHN13	KHN14	KHN15	KHN16	KHN17	KHN18
MAG19	MAG20	MAG21	MAG22	MAG23	MAG24	MAG25
15	16	17	18	19	20	21
A4	A5	A6	A7	A8	A9	A10
KHN19	KHN20	KHN21	KHN22	KHN23	KHN24	KHN25
MAG26	MAG27	MAG28	MAG29	MAG30	MIY1	MIY2
22	23	24	25	26	27	28
A11	A12	A13	A14	A15	A16	A17
KHN26	KHN27	KHN28	KHN29	KHN30	PIN1	PIN2
MIY3	MIY4	MIY5	MIY6	MIY7	MIY8	MIY9

MABABA WEUSI 573/2089-90/6325-26/2081-82

Ashe	Ojobo	Iwa	Araorun	Ile	Itutu	Erinle
1	2	3	4	5	6	7
MA16	MA17	MA18	MA19	MA20	MA21	MA22
APP1	APP2	APP3	APP4	APP5	APP6	APP7
GIN8	GIN9	GIN10	GIN11	GIN12	GIN13	GIN14
8	9	10	11	12	13	14
MA23	MA24	MA25	MA26	MA27	MA28	MA29
APP8	APP9	APP10	APP11	APP12	APP13	APP14
GIN15	GIN16	GIN17	GIN18	GIN19	GIN20	GIN21
15	16	17	18	19	20	21
MA30	MA31	J1	J2	J3	J4	J5
APP15	APP16	APP17	APP18	APP19	APP20	APP21
GIN22	GIN23	GIN24	GIN25	GIN26	GIN27	GIN28
22	23	24	25	26	27	28
J6	J7	J8	J9	J10	J11	J12
APP22	APP23	APP24	APP25	APP26	APP27	APP28
GIN29	GIN30	SAN1	SAN2	SAN3	SAN4	SAN5

THE NU AFRIKAN CALENDAR

MWANAMKE WEUSI 573/2089-90/6325-26/2081-82						
Ashe	Ojobo	Iwa	Araorun	Ile	Itutu	Erinle
1	2	3	4	5	6	7
J13	J14	J15	J16	J17	J18	J19
APP29	APP30	MSR1	MSR2	MSR3	MSR4	MSR5
SAN6	SAN7	SAN8	SAN9	SAN10	SAN11	SAN12
8	9	10	11	12	13	14
J20	J21	J22	J23	J24	J25	J26
MSR6	MSR7	MSR8	MSR9	MSR10	MSR11	MSR12
SAN13	SAN14	SAN15	SAN16	SAN17	SAN18	SAN19
15	16	17	18	19	20	21
J27	J28	J29	J30	JU1	JU2	JU3
MSR13	MSR14	MSR15	MSR16	MSR17	MSR18	MSR19
SAN20	SAN21	SAN22	SAN23	SAN24	SAN25	SAN26
22	23	24	25	26	27	28
JU4	JU5	JU6	JU7	JU8	JU9	JU10
MSR20	MRS21	MSR22	MSR23	MSR24	MSR25	MSR26
SAN27	SAN28	SAN29	SAN30	HAM1	HAM2	HAM3

MTU WEUSI 573/2089-90/6325-26/2081-82						
Ashe	Ojobo	Iwa	Araorun	Ile	Itutu	Erinle
1	2	3	4	5	6	7
JU11	JU12	JU13	JU14	JU15	JU16	JU17
MSR27	MSR28	MSR29	MSR30	*ASR1*	*HRU2*	*STH3*
HAM4	HAM5	HAM6	HAM7	HAM8	HAM9	HAM10
8	9	10	11	12	13	14
JU18	JU19	JU20	JU21	JU22	JU23	JU24
AST4	*NBH5*	THT1	THT2	THT3	THT4	THT5
HAM11	HAM12	HAM13	AM14	HAM15	HAM16	HAM17
15	16	17	18	19	20	21
JU25	JU26	JU27	JU28	JU29	JU30	JU31
THT6	THT7	THT8	THT9	THT10	THT11	THT12
HAM18	HAM19	HAM20	HAM21	HAM22	HAM23	HAM24
22	23	24	25	26	27	28
AU1	AU2	AU3	AU4	AU5	AU6	AU7
THT13	THT14	THT15	THT16	THT17	THT18	THT19
HAM25	HAM26	HAM27	HAM28	HAM29	HAM30	NAH1

TAIFA WEUSI 573/2089-90/6325-26/2081-82						
Ashe	Ojobo	Iwa	Araorun	Ile	Itutu	Erinle
1	2	3	4	5	6	7
AU8	AU9	AU10	AU11	AU12	AU13	AU14
THT20	THT21	THT22	THT23	THT24	THT25	THT26
NAH2	NAH3	NAH4	NAH5	NAH6	NAH7	NAH8
8	9	10	11	12	13	14
AU15	AU16	AU17	AU18	AU19	AU20	AU21
THT27	THT28	THT29	THT30	PAP1	PAP1	PAP2
NAH9	NAH10	NAH11	NAH12	NAH13	NAH14	NAH15
15	16	17	18	19	20	21
AU22	AU23	AU24	AU25	AU26	AU27	AU28
PAP3	PAP4	PAP5	PAP6	PAP7	PAP8	PAP9
NAH16	NAH17	NAH18	NAH19	NAH20	NAH21	NAH22
22	23	24	25	26	27	28
AU29	AU30	AU31	S1	S2	S3	S4
PAP10	PAP11	PAP12	PAP13	PAP14	PAP15	PAP16
NAH23	NAH24	NAH25	NAH26	NAH27	NAH28	NAH29

UWEZO 573/2089-90/6325-26/2081-82						
Ashe	Ojobo	Iwa	Araorun	Ile	Itutu	Erinle
1	2	3	4	5	6	7
S5	S6	S7	S8	S9	S10	S11
PAP17	PAP18	PAP19	PAP20	PAP21	PAP22	PAP23
NAH30	*PAG1*	*PAG2*	*PAG3*	*PAG4*	*PAG5*	MAS1
8	9	10	11	12	13	14
S12	S13	S14	S15	S16	S17	S18
PAP24	PAP25	PAP26	PAP27	PAP28	PAP29	PAP30
MAS2	MAS3	MAS4	MAS5	MAS6	MAS7	MAS8
15	16	17	18	19	20	21
S19	S20	S21	S22	S23	S24	S25
HTH1	HTH2	HTH3	HTH4	HTH5	HTH6	HTH7
MAS9	MAS10	MAS11	MAS12	MAS13	MAS14	MAS15
22	23	24	25	26	27	28
S26	S27	S28	S29	S30	O1	O2
HTH8	HTH9	HTH10	HTH11	HTH12	HTH13	HTH14
MAS16	MAS17	MAS18	MAS19	MAS20	MAS21	MAS22

THE NU AFRIKAN CALENDAR

EKUNDU 573/2089-90/6325-26/2081-82						
Ashe	Ojobo	Iwa	Araorun	Ile	Itutu	Erinle
1	2	3	4	5	6	7
O3	O4	O5	O6	O7	O8	O9
HTH15	HTH16	HTH17	HTH18	HTH19	HTH20	HTH21
MAS23	MAS24	MAS25	MAS26	MAS27	MAS28	MAS29
8	9	10	11	12	13	14
O10	O11	O12	O13	O14	O15	O16
HTH22	HTH23	HTH24	HTH25	HTH26	HTH27	HTH28
MAS30	THi1	THi2	THi3	THi4	THi5	THi6
15	16	17	18	19	20	21
O17	O18	O19	O20	O21	O22	O23
HTH29	HTH30	KHK1	KHK2	KHK3	KHK4	KHK5
THi7	THi8	THi9	THi10	THi11	THi12	THi13
22	23	24	25	26	27	28
O24	O25	O26	O27	O28	O29	O30
KHK6	KHK7	KHK8	KHK9	KHK10	KHK11	KHK12
THi14	THi15	THi16	THi17	THi18	THi19	THi20

EUSI 573/2089-90/6325-26/2081-82						
Ashe	Ojobo	Iwa	Araorun	Ile	Itutu	Erinle
1	2	3	4	5	6	7
O31	N1	N2	N3	N4	N5	N6
KHK13	KHK14	KHK15	KHK16	KHK17	KHK18	KHK19
THi21	THi22	THi23	THi24	THi25	THi26	THi27
8	9	10	11	12	13	14
N7	N8	N9	N10	N11	N12	N13
KHK20	KHK21	KHK22	KHK23	KHK24	KHK25	KHK26
THi28	THi29	THi30	HID1	HID2	HID3	HID4
15	16	17	18	19	20	21
N14	N15	N16	N17	N18	N19	N20
KHK27	KHK28	KHK29	KHK30	TBT1	TBT2	TBT3
HID5	HID6	HID7	HID8	HID9	HID10	HID11
22	23	24	25	26	27	28
N21	N22	N23	N24	N25	N26	N27
TBT4	TBT5	TBT6	TBT7	TBT8	TBT9	TBT10
HID12	HID13	HID14	HID15	HID16	HID17	HID18

KIJANI 573/2089-90/6325-26/2081-82						
Ashe	Ojobo	Iwa	Araorun	Ile	Itutu	Erinle
1	2	3	4	5	6	7
N28	N29	N30	D1	D2	D3	D4
TBT11	TBT12	TBT13	TBT14	TBT15	TBT16	TBT17
HID19	HID20	HID21	HID22	HID23	HID24	HID25
8	9	10	11	12	13	14
D5	D6	D7	D8	D9	D10	D11
TBT18	TBT19	TBT20	TBT21	TBT22	TBT23	TBT24
HID26	HID27	HID28	HID29	HID30	TAH1	TAH2
15	16	17	18	19	20	21
D12	D13	D14	D15	D16	D17	D18
TBT25	TBT26	TBT27	TBT28	TBT29	TBT30	MKH1
TAH3	TAH4	TAH5	TAH6	TAH7	TAH8	TAH9
22	23	24	25	26	27	28
D19	D20	D21	D22	D23	D24	D25
MKH2	MKH3	MKH4	MKH5	MKH6	MKH7	MKH8
TAH10	TAH11	TAH12	TAH13	TAH14	TAH15	TAH16

KWANZA 573/2089-90/6325-26/2081-82						
Umoja	Kujichagulia	Ujima	Ujamaa	Nia	Kuumba	Imani
1	2	3	4	5	6	7
D26	D27	D28	D29	D30	D31	JA1
MKH9	MKH10	MKH11	MKH12	MKH13	MKH14	MKH15
TAH17	TAH18	TAH19	TAH20	TAH21	TAH22	TAH23
8	9	10	11	12	13	14
JA2	JA3	JA4	JA5	JA6	JA7	JA8
MKH16	MKH17	MKH18	MKH19	MKH20	MKH21	MKH22
TAH24	TAH25	TAH26	TAH27	TAH28	TAH29	TAH30
15	16	17	18	19	20	21
JA9	JA10	JA11	JA12	JA13	JA14	JA15
MKH23	MKH24	MKH25	KMH26	MKH27	MKH28	MKH29
TIR1	TIR2	TIR3	TIR4	TIR5	TIR6	TIR7
22	23	24	25	26	27	28
JA16	JA17	JA18	JA19	JA20	JA21	JA22
MKH30	AMN1	AMN2	AMN3	AMN4	AMN5	AMN6
TIR8	TIR9	TIR10	TIR11	TIR12	TIR13	TIR14

THE NU AFRIKAN CALENDAR

SHAHIDI 573/2089-90/6325-26/2081-82						
Ashe	Ojobo	Iwa	Araorun	Ile	Itutu	Erinle
1	2	3	4	5	6	7
JA23	JA24	JA25	JA26	JA27	JA28	JA29
AMN7	AMN8	AMN9	AMN10	AMN11	AMN12	AMN13
TIR15	TIR16	TIR17	TIR18	TIR19	TIR20	TIR21
8	9	10	11	12	13	14
JA30	JA31	F1	F2	F3	F4	F5
AMN14	AMN15	AMN16	AMN17	AMN18	AMN19	AMN20
TIR22	TIR23	TIR24	TIR25	TIR26	TIR27	TIR28
15	16	17	18	19	20	21
F6	F7	F8	F9	F10	F 11	F12
AMN21	AMN22	AMN23	AMN24	AMN25	AMN26	AMN28
TIR29	TIR30	YEK1	YEK2	YEK3	YEK4	YEK5
22	23	24	25	26	27	28
F13	F14	F15	F16	F17	F18	F19
AMN29	AMN30	RNT1	RNT2	RNT3	RNT4	RNT5
YEK6	YEK7	YEK8	YEK9	YEK10	YEK11	YEK12
SHAHIDI 124						
F20						
RNT6						
YEK13						

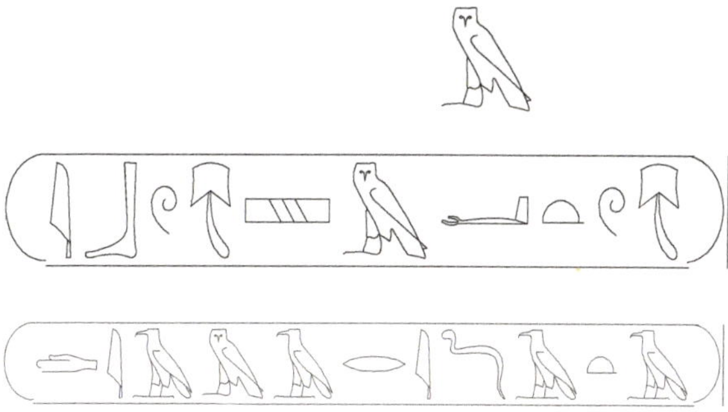

THE NU AFRIKAN CALENDAR

MAAFA 574 B.S.E./ASIENTO 2090-91 C.E.
KAM 6326-27 S.M./EC 2082-83 A.M. (ዓ.ም.)

AFU 574/2090-91/6326-27/2082-83						
Ashe	Ojobo	Iwa	Araorun	Ile	Itutu	Erinle
1	2	3	4	5	6	7
F21	F22	F23	F24	F25	F26	F27
RNT7	RNT8	RNT9	RNT10	RNT11	RNT12	RNT13
YEK14	YEK15	YEK16	YEK17	YEK18	YEK19	YEK20
8	9	10	11	12	13	14
F28	M1	M2	M3	M4	M5	M6
RNT14	RNT15	RNT16	RNT17	RNT18	RNT19	RNT20
YEK21	YEK22	YEK23	YEK24	YEK25	YEK26	YEK27
15	16	17	18	19	20	21
M7	M8	M9	M10	M11	M12	M13
RNT21	RNT22	RNT23	RNT24	RNT25	RNT26	RNT27
YEK28	YEK29	YEK30	MAG1	MAG2	MAG3	MAG4
22	23	24	25	26	27	28
M14	M15	M16	M17	M18	M19	M20
RNT28	RNT29	RNT30	KHN1	KHN2	KHN3	KHN4
MAG5	MAG6	MAG7	MAG8	MAG9	MAG10	MAG11

KAWAIDA 574/2090-91/6326-27/2082-83						
Ashe	Ojobo	Iwa	Araorun	Ile	Itutu	Erinle
1	2	3	4	5	6	7
A18	A19	A20	A21	A22	A23	A24
PIN3	PIN4	PIN5	PIN6	PIN7	PIN8	PIN9
MIY10	MIY11	MIY12	MIY13	MIY14	MIY15	MIY16
8	9	10	11	12	13	14
A25	A26	A27	A28	A29	A30	MA1
PIN10	PIN11	PIN12	PIN13	PIN14	PIN15	PIN16
MIY17	MIY18	MIY19	MIY20	MIY21	MIY22	MIY23
15	16	17	18	19	20	21
MA2	MA3	MA4	MA5	MA6	MA7	MA8
PIN17	PIN18	PIN19	PIN20	PIN21	PIN22	PIN23
MIY24	MIY25	MIY26	MIY27	MIY28	MIY29	MIY30
22	23	24	25	26	27	28
MA9	MA10	MA11	MA12	MA13	MA14	MA15
PIN24	PIN25	PIN26	PIN27	PIN28	PIN29	PIN30
GIN1	GIN2	GIN3	GIN4	GIN5	GIN6	GIN7

UHURU 574/2090-91/6326-27/2082-83						
Ashe	Ojobo	Iwa	Araorun	Ile	Itutu	Erinle
1	2	3	4	5	6	7
M21	M22	M23	M24	M25	M26	M27
KHN5	KHN6	KHN7	KHN8	KHN9	KHN10	KHN11
MAG12	MAG13	MAG14	MAG15	MAG16	MAG17	MAG18
8	9	10	11	12	13	14
M28	M29	M30	M31	A1	A2	A3
KHN12	KHN13	KHN14	KHN15	KHN16	KHN17	KHN18
MAG19	MAG20	MAG21	MAG22	MAG23	MAG24	MAG25
15	16	17	18	19	20	21
A4	A5	A6	A7	A8	A9	A10
KHN19	KHN20	KHN21	KHN22	KHN23	KHN24	KHN25
MAG26	MAG27	MAG28	MAG29	MAG30	MIY1	MIY2
22	23	24	25	26	27	28
A11	A12	A13	A14	A15	A16	A17
KHN26	KHN27	KHN28	KHN29	KHN30	PIN1	PIN2
MIY3	MIY4	MIY5	MIY6	MIY7	MIY8	MIY9

MABABA WEUSI 574/2090-91/6326-27/2082-83						
Ashe	Ojobo	Iwa	Araorun	Ile	Itutu	Erinle
1	2	3	4	5	6	7
MA16	MA17	MA18	MA19	MA20	MA21	MA22
APP1	APP2	APP3	APP4	APP5	APP6	APP7
GIN8	GIN9	GIN10	GIN11	GIN12	GIN13	GIN14
8	9	10	11	12	13	14
MA23	MA24	MA25	MA26	MA27	MA28	MA29
APP8	APP9	APP10	APP11	APP12	APP13	APP14
GIN15	GIN16	GIN17	GIN18	GIN19	GIN20	GIN21
15	16	17	18	19	20	21
MA30	MA31	J1	J2	J3	J4	J5
APP15	APP16	APP17	APP18	APP19	APP20	APP21
GIN22	GIN23	GIN24	GIN25	GIN26	GIN27	GIN28
22	23	24	25	26	27	28
J6	J7	J8	J9	J10	J11	J12
APP22	APP23	APP24	APP25	APP26	APP27	APP28
GIN29	GIN30	SAN1	SAN2	SAN3	SAN4	SAN5

THE NU AFRIKAN CALENDAR

MWANAMKE WEUSI 574/2090-91/6326-27/2082-83						
Ashe	Ojobo	Iwa	Araorun	Ile	Itutu	Erinle
1	2	3	4	5	6	7
J13	J14	J15	J16	J17	J18	J19
APP29	APP30	MSR1	MSR2	MSR3	MSR4	MSR5
SAN6	SAN7	SAN8	SAN9	SAN10	SAN11	SAN12
8	9	10	11	12	13	14
J20	J21	J22	J23	J24	J25	J26
MSR6	MSR7	MSR8	MSR9	MSR10	MSR11	MSR12
SAN13	SAN14	SAN15	SAN16	SAN17	SAN18	SAN19
15	16	17	18	19	20	21
J27	J28	J29	J30	JU1	JU2	JU3
MSR13	MSR14	MSR15	MSR16	MSR17	MSR18	MSR19
SAN20	SAN21	SAN22	SAN23	SAN24	SAN25	SAN26
22	23	24	25	26	27	28
JU4	JU5	JU6	JU7	JU8	JU9	JU10
MSR20	MRS21	MSR22	MSR23	MSR24	MSR25	MSR26
SAN27	SAN28	SAN29	SAN30	HAM1	HAM2	HAM3

TAIFA WEUSI 574/2090-91/6326-27/2082-83						
Ashe	Ojobo	Iwa	Araorun	Ile	Itutu	Erinle
1	2	3	4	5	6	7
AU8	AU9	AU10	AU11	AU12	AU13	AU14
THT20	THT21	THT22	THT23	THT24	THT25	THT26
NAH2	NAH3	NAH4	NAH5	NAH6	NAH7	NAH8
8	9	10	11	12	13	14
AU15	AU16	AU17	AU18	AU19	AU20	AU21
THT27	THT28	THT29	THT30	PAP1	PAP1	PAP2
NAH9	NAH10	NAH11	NAH12	NAH13	NAH14	NAH15
15	16	17	18	19	20	21
AU22	AU23	AU24	AU25	AU26	AU27	AU28
PAP3	PAP4	PAP5	PAP6	PAP7	PAP8	PAP9
NAH16	NAH17	NAH18	NAH19	NAH20	NAH21	NAH22
22	23	24	25	26	27	28
AU29	AU30	AU31	S1	S2	S3	S4
PAP10	PAP11	PAP12	PAP13	PAP14	PAP15	PAP16
NAH23	NAH24	NAH25	NAH26	NAH27	NAH28	NAH29

MTU WEUSI 574/2090-91/6326-27/2082-83						
Ashe	Ojobo	Iwa	Araorun	Ile	Itutu	Erinle
1	2	3	4	5	6	7
JU11	JU12	JU13	JU14	JU15	JU16	JU17
MSR27	MSR28	MSR29	MSR30	*ASR1*	*HRU2*	*STH3*
HAM4	HAM5	HAM6	HAM7	HAM8	HAM9	HAM10
8	9	10	11	12	13	14
JU18	JU19	JU20	JU21	JU22	JU23	JU24
AST4	*NBH5*	THT1	THT2	THT3	THT4	THT5
HAM11	HAM12	HAM13	AM14	HAM15	HAM16	HAM17
15	16	17	18	19	20	21
JU25	JU26	JU27	JU28	JU29	JU30	JU31
THT6	THT7	THT8	THT9	THT10	THT11	THT12
HAM18	HAM19	HAM20	HAM21	HAM22	HAM23	HAM24
22	23	24	25	26	27	28
AU1	AU2	AU3	AU4	AU5	AU6	AU7
THT13	THT14	THT15	THT16	THT17	THT18	THT19
HAM25	HAM26	HAM27	HAM28	HAM29	HAM30	NAH1

UWEZO 574/2090-91/6326-27/2082-83						
Ashe	Ojobo	Iwa	Araorun	Ile	Itutu	Erinle
1	2	3	4	5	6	7
S5	S6	S7	S8	S9	S10	S11
PAP17	PAP18	PAP19	PAP20	PAP21	PAP22	PAP23
NAH30	*PAG1*	*PAG2*	*PAG3*	*PAG4*	*PAG5*	MAS1
8	9	10	11	12	13	14
S12	S13	S14	S15	S16	S17	S18
PAP24	PAP25	PAP26	PAP27	PAP28	PAP29	PAP30
MAS2	MAS3	MAS4	MAS5	MAS6	MAS7	MAS8
15	16	17	18	19	20	21
S19	S20	S21	S22	S23	S24	S25
HTH1	HTH2	HTH3	HTH4	HTH5	HTH6	HTH7
MAS9	MAS10	MAS11	MAS12	MAS13	MAS14	MAS15
22	23	24	25	26	27	28
S26	S27	S28	S29	S30	O1	O2
HTH8	HTH9	HTH10	HTH11	HTH12	HTH13	HTH14
MAS16	MAS17	MAS18	MAS19	MAS20	MAS21	MAS22

EKUNDU 574/2090-91/6326-27/2082-83

Ashe	Ojobo	Iwa	Araorun	Ile	Itutu	Erinle
1	2	3	4	5	6	7
O3	O4	O5	O6	O7	O8	O9
HTH15	HTH16	HTH17	HTH18	HTH19	HTH20	HTH21
MAS23	MAS24	MAS25	MAS26	MAS27	MAS28	MAS29
8	9	10	11	12	13	14
O10	O11	O12	O13	O14	O15	O16
HTH22	HTH23	HTH24	HTH25	HTH26	HTH27	HTH28
MAS30	THi1	THi2	THi3	THi4	THi5	THi6
15	16	17	18	19	20	21
O17	O18	O19	O20	O21	O22	O23
HTH29	HTH30	KHK1	KHK2	KHK3	KHK4	KHK5
THi7	THi8	THi9	THi10	THi11	THi12	THi13
22	23	24	25	26	27	28
O24	O25	O26	O27	O28	O29	O30
KHK6	KHK7	KHK8	KHK9	KHK10	KHK11	KHK12
THi14	THi15	THi16	THi17	THi18	THi19	THi20

KIJANI 574/2090-91/6326-27/2082-83

Ashe	Ojobo	Iwa	Araorun	Ile	Itutu	Erinle
1	2	3	4	5	6	7
N28	N29	N30	D1	D2	D3	D4
TBT11	TBT12	TBT13	TBT14	TBT15	TBT16	TBT17
HID19	HID20	HID21	HID22	HID23	HID24	HID25
8	9	10	11	12	13	14
D5	D6	D7	D8	D9	D10	D11
TBT18	TBT19	TBT20	TBT21	TBT22	TBT23	TBT24
HID26	HID27	HID28	HID29	HID30	TAH1	TAH2
15	16	17	18	19	20	21
D12	D13	D14	D15	D16	D17	D18
TBT25	TBT26	TBT27	TBT28	TBT29	TBT30	MKH1
TAH3	TAH4	TAH5	TAH6	TAH7	TAH8	TAH9
22	23	24	25	26	27	28
D19	D20	D21	D22	D23	D24	D25
MKH2	MKH3	MKH4	MKH5	MKH6	MKH7	MKH8
TAH10	TAH11	TAH12	TAH13	TAH14	TAH15	TAH16

EUSI 574/2090-91/6326-27/2082-83

Ashe	Ojobo	Iwa	Araorun	Ile	Itutu	Erinle
1	2	3	4	5	6	7
O31	N1	N2	N3	N4	N5	N6
KHK13	KHK14	KHK15	KHK16	KHK17	KHK18	KHK19
THi21	THi22	THi23	THi24	THi25	THi26	THi27
8	9	10	11	12	13	14
N7	N8	N9	N10	N11	N12	N13
KHK20	KHK21	KHK22	KHK23	KHK24	KHK25	KHK26
THi28	THi29	THi30	HID1	HID2	HID3	HID4
15	16	17	18	19	20	21
N14	N15	N16	N17	N18	N19	N20
KHK27	KHK28	KHK29	KHK30	TBT1	TBT2	TBT3
HID5	HID6	HID7	HID8	HID9	HID10	HID11
22	23	24	25	26	27	28
N21	N22	N23	N24	N25	N26	N27
TBT4	TBT5	TBT6	TBT7	TBT8	TBT9	TBT10
HID12	HID13	HID14	HID15	HID16	HID17	HID18

KWANZA 574/2090-91/6326-27/2082-83

Umoja	Kujichagulia	Ujima	Ujamaa	Nia	Kuumba	Imani
1	2	3	4	5	6	7
D26	D27	D28	D29	D30	D31	JA1
MKH9	MKH10	MKH11	MKH12	MKH13	MKH14	MKH15
TAH17	TAH18	TAH19	TAH20	TAH21	TAH22	TAH23
8	9	10	11	12	13	14
JA2	JA3	JA4	JA5	JA6	JA7	JA8
MKH16	MKH17	MKH18	MKH19	MKH20	MKH21	MKH22
TAH24	TAH25	TAH26	TAH27	TAH28	TAH29	TAH30
15	16	17	18	19	20	21
JA9	JA10	JA11	JA12	JA13	JA14	JA15
MKH23	MKH24	MKH25	KMH26	MKH27	MKH28	MKH29
TIR1	TIR2	TIR3	TIR4	TIR5	TIR6	TIR7
22	23	24	25	26	27	28
JA16	JA17	JA18	JA19	JA20	JA21	JA22
MKH30	AMN1	AMN2	AMN3	AMN4	AMN5	AMN6
TIR8	TIR9	TIR10	TIR11	TIR12	TIR13	TIR14

SHAHIDI 574/2090-91/6326-27/2082-83						
Ashe	Ojobo	Iwa	Araorun	Ile	Itutu	Erinle
1	2	3	4	5	6	7
JA23	JA24	JA25	JA26	JA27	JA28	JA29
AMN7	AMN8	AMN9	AMN10	AMN11	AMN12	AMN13
TIR15	TIR16	TIR17	TIR18	TIR19	TIR20	TIR21
8	9	10	11	12	13	14
JA30	JA31	F1	F2	F3	F4	F5
AMN14	AMN15	AMN16	AMN17	AMN18	AMN19	AMN20
TIR22	TIR23	TIR24	TIR25	TIR26	TIR27	TIR28
15	16	17	18	19	20	21
F6	F7	F8	F9	F10	F 11	F12
AMN21	AMN22	AMN23	AMN24	AMN25	AMN26	AMN28
TIR29	TIR30	YEK1	YEK2	YEK3	YEK4	YEK5
22	23	24	25	26	27	28
F13	F14	F15	F16	F17	F18	F19
AMN29	AMN30	RNT1	RNT2	RNT3	RNT4	RNT5
YEK6	YEK7	YEK8	YEK9	YEK10	YEK11	YEK12
SHAHIDI 125						
F20						
RNT6						
YEK13						

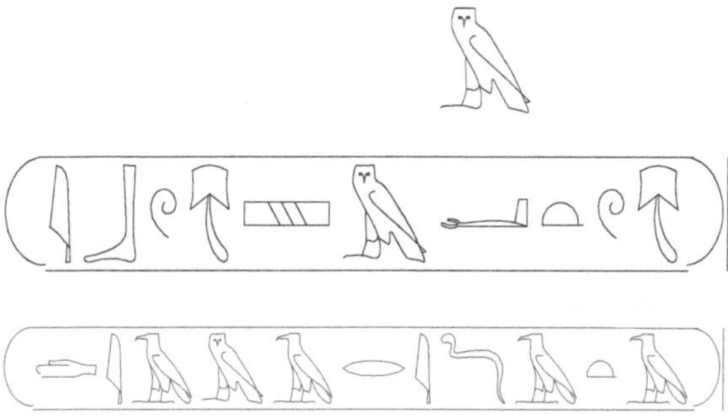

THE NU AFRIKAN CALENDAR

MAAFA 575 B.S.E./ASIENTO 2091-92 C.E./
KAM 6327-28 S.M./EC 2083-84 A.M. (ዓ.ም.)

AFU 575/2091-92/6327-28/2083-84						
Ashe	Ojobo	Iwa	Araorun	Ile	Itutu	Erinle
1	2	3	4	5	6	7
F21	F22	F23	F24	F25	F26	F27
RNT7	RNT8	RNT9	RNT10	RNT11	RNT12	RNT13
YEK14	YEK15	YEK16	YEK17	YEK18	YEK19	YEK20
8	9	10	11	12	13	14
F28	M1	M2	M3	M4	M5	M6
RNT14	RNT15	RNT16	RNT17	RNT18	RNT19	RNT20
YEK21	YEK22	YEK23	YEK24	YEK25	YEK26	YEK27
15	16	17	18	19	20	21
M7	M8	M9	M10	M11	M12	M13
RNT21	RNT22	RNT23	RNT24	RNT25	RNT26	RNT27
YEK28	YEK 29	YEK30	MAG1	MAG2	MAG3	MAG4
22	23	24	25	26	27	28
M14	M15	M16	M17	M18	M19	M20
RNT28	RNT29	RNT30	KHN1	KHN2	KHN3	KHN4
MAG5	MAG6	MAG7	MAG8	MAG9	MAG10	MAG11

UHURU 575/2091-92/6327-28/2083-84						
Ashe	Ojobo	Iwa	Araorun	Ile	Itutu	Erinle
1	2	3	4	5	6	7
M21	M22	M23	M24	M25	M26	M27
KHN5	KHN6	KHN7	KHN8	KHN9	KHN10	KHN11
MAG12	MAG13	MAG14	MAG15	MAG16	MAG17	MAG18
8	9	10	11	12	13	14
M28	M29	M30	M31	A1	A2	A3
KHN12	KHN13	KHN14	KHN15	KHN16	KHN17	KHN18
MAG19	MAG20	MAG21	MAG22	MAG23	MAG24	MAG25
15	16	17	18	19	20	21
A4	A5	A6	A7	A8	A9	A10
KHN19	KHN20	KHN21	KHN22	KHN23	KHN24	KHN25
MAG26	MAG27	MAG28	MAG29	MAG30	MIY1	MIY2
22	23	24	25	26	27	28
A11	A12	A13	A14	A15	A16	A17
KHN26	KHN27	KHN28	KHN29	KHN30	PIN1	PIN2
MIY3	MIY4	MIY5	MIY6	MIY7	MIY8	MIY9

KAWAIDA 575/2091-92/6327-28/2083-84						
Ashe	Ojobo	Iwa	Araorun	Ile	Itutu	Erinle
1	2	3	4	5	6	7
A18	A19	A20	A21	A22	A23	A24
PIN3	PIN4	PIN5	PIN6	PIN7	PIN8	PIN9
MIY10	MIY11	MIY12	MIY13	MIY14	MIY15	MIY1
8	9	10	11	12	13	14
A25	A26	A27	A28	A29	A30	MA1
PIN10	PIN11	PIN12	PIN13	PIN14	PIN15	PIN16
MIY17	MIY18	MIY19	MIY20	MIY21	MIY22	MIY23
15	16	17	18	19	20	21
MA2	MA3	MA4	MA5	MA6	MA7	MA8
PIN17	PIN18	PIN19	PIN20	PIN21	PIN22	PIN2
MIY24	MIY25	MIY26	MIY27	MIY28	MIY29	MIY3
22	23	24	25	26	27	28
MA9	MA10	MA11	MA12	MA13	MA14	MA1
PIN24	PIN25	PIN26	PIN27	PIN28	PIN29	PIN3
GIN1	GIN2	GIN3	GIN4	GIN5	GIN6	GIN7

MABABA WEUSI 575/2091-92/6327-28/2083-84						
Ashe	Ojobo	Iwa	Araorun	Ile	Itutu	Erinle
1	2	3	4	5	6	7
MA16	MA17	MA18	MA19	MA20	MA21	MA22
APP1	APP2	APP3	APP4	APP5	APP6	APP
GIN8	GIN9	GIN10	GIN11	GIN12	GIN13	GIN1
8	9	10	11	12	13	14
MA23	MA24	MA25	MA26	MA27	MA28	MA29
APP8	APP9	APP10	APP11	APP12	APP13	APP1
GIN15	GIN16	GIN17	GIN18	GIN19	GIN20	GIN21
15	16	17	18	19	20	21
MA30	MA31	J1	J2	J3	J4	J5
APP15	APP16	APP17	APP18	APP19	APP20	APP21
GIN22	GIN23	GIN24	GIN25	GIN26	GIN27	GIN28
22	23	24	25	26	27	28
J6	J7	J8	J9	J10	J11	J12
APP22	APP23	APP24	APP25	APP26	APP27	APP2
GIN29	GIN30	SAN1	SAN2	SAN3	SAN4	SAN5

THE NU AFRIKAN CALENDAR

MWANAMKE WEUSI 575/2091-92/6327-28/2083-84						
Ashe	Ojobo	Iwa	Araorun	Ile	Itutu	Erinle
1	2	3	4	5	6	7
J13	J14	J15	J16	J17	J18	J19
APP29	APP30	MSR1	MSR2	MSR3	MSR4	MSR5
SAN6	SAN7	SAN8	SAN9	SAN10	SAN11	SAN12
8	9	10	11	12	13	14
J20	J21	J22	J23	J24	J25	J26
MSR6	MSR7	MSR8	MSR9	MSR10	MSR11	MSR12
SAN13	SAN14	SAN15	SAN16	SAN17	SAN18	SAN19
15	16	17	18	19	20	21
J27	J28	J29	J30	JU1	JU2	JU3
MSR13	MSR14	MSR15	MSR16	MSR17	MSR18	MSR19
SAN20	SAN21	SAN22	SAN23	SAN24	SAN25	SAN26
22	23	24	25	26	27	28
JU4	JU5	JU6	JU7	JU8	JU9	JU10
MSR20	MRS21	MSR22	MSR23	MSR24	MSR25	MSR26
SAN27	SAN28	SAN29	SAN30	HAM1	HAM2	HAM3

TAIFA WEUSI 575/2091-92/6327-28/2083-84						
Ashe	Ojobo	Iwa	Araorun	Ile	Itutu	Erinle
1	2	3	4	5	6	7
AU8	AU9	AU10	AU11	AU12	AU13	AU14
THT20	THT21	THT22	THT23	THT24	THT25	THT26
NAH2	NAH3	NAH4	NAH5	NAH6	NAH7	NAH8
8	9	10	11	12	13	14
AU15	AU16	AU17	AU18	AU19	AU20	AU21
THT27	THT28	THT29	THT30	PAP1	PAP1	PAP2
NAH9	NAH10	NAH11	NAH12	NAH13	NAH14	NAH15
15	16	17	18	19	20	21
AU22	AU23	AU24	AU25	AU26	AU27	AU28
PAP3	PAP4	PAP5	PAP6	PAP7	PAP8	PAP9
NAH16	NAH17	NAH18	NAH19	NAH20	NAH21	NAH22
22	23	24	25	26	27	28
AU29	AU30	AU31	S1	S2	S3	S4
PAP10	PAP11	PAP12	PAP13	PAP14	PAP15	PAP16
NAH23	NAH24	NAH25	NAH26	NAH27	NAH28	NAH29

MTU WEUSI 575/2091-92/6327-28/2083-84						
Ashe	Ojobo	Iwa	Araorun	Ile	Itutu	Erinle
1	2	3	4	5	6	7
JU11	JU12	JU13	JU14	JU15	JU16	JU17
MSR27	MSR28	MSR29	MSR30	*ASR1*	*HRU2*	*STH3*
HAM4	HAM5	HAM6	HAM7	HAM8	HAM9	HAM10
8	9	10	11	12	13	14
JU18	JU19	JU20	JU21	JU22	JU23	JU24
AST4	*NBH5*	THT1	THT2	THT3	THT4	THT5
HAM11	HAM12	HAM13	HAM14	HAM15	HAM16	HAM17
15	16	17	18	19	20	21
JU25	JU26	JU27	JU28	JU29	JU30	JU31
THT6	THT7	THT8	THT9	THT10	THT11	THT12
HAM18	HAM19	HAM20	HAM21	HAM22	HAM23	HAM24
22	23	24	25	26	27	28
AU1	AU2	AU3	AU4	AU5	AU6	AU7
THT13	THT14	THT15	THT16	THT17	THT18	THT19
HAM25	HAM26	HAM27	HAM28	HAM29	HAM30	NAH1

UWEZO 575/2091-92/6327-28/2083-84						
Ashe	Ojobo	Iwa	Araorun	Ile	Itutu	Erinle
1	2	3	4	5	6	7
S5	S6	S7	S8	S9	S10	S11
PAP17	PAP18	PAP19	PAP20	PAP21	PAP22	PAP23
NAH30	*QUA1*	*QUA2*	*QUA3*	*QUA4*	*QUA5*	*QUA6*
8	9	10	11	12	13	14
S12	S13	S14	S15	S16	S17	S18
PAP24	PAP25	PAP26	PAP27	PAP28	PAP29	PAP30
MAS1	MAS2	MAS3	MAS4	MAS5	MAS6	MAS7
15	16	17	18	19	20	21
S19	S20	S21	S22	S23	S24	S25
HTH1	HTH2	HTH3	HTH4	HTH5	HTH6	HTH7
MAS8	MAS9	MAS10	MAS11	MAS12	MAS13	MAS14
22	23	24	25	26	27	28
S26	S27	S28	S29	S30	O1	O2
HTH8	HTH9	HTH10	HTH11	HTH12	HTH13	HTH14
MAS15	MAS16	MAS17	MAS18	MAS19	MAS20	MAS21

THE NU AFRIKAN CALENDAR

EKUNDU 575/2091-92/6327-28/2083-84						
Ashe	Ojobo	Iwa	Araorun	Ile	Itutu	Erinle
1	2	3	4	5	6	7
O3	O4	O5	O6	O7	O8	O9
HTH15	HTH16	HTH17	HTH18	HTH19	HTH20	HTH21
MAS22	MAS23	MAS 24	MAS25	MAS26	MAS27	MAS28
8	9	10	11	12	13	14
O10	O11	O12	O13	O14	O15	O16
HTH22	HTH23	HTH24	HTH25	HTH26	HTH27	HTH28
MAS29	MAS30	THi1	THi2	THi3	THi4	THi5
15	16	17	18	19	20	21
O17	O18	O19	O20	O21	O22	O23
HTH29	HTH30	KHK1	KHK2	KHK3	KHK4	KHK5
THi6	THi7	THi8	THi9	THi10	THi11	THi12
22	23	24	25	26	27	28
O24	O25	O26	O27	O28	O29	O30
KHK6	KHK7	KHK8	KHK9	KHK10	KHK11	KHK12
THi13	THi14	THi15	THi16	THi17	THi18	THi19

KIJANI 575/2091-92/6327-28/2083-84						
Ashe	Ojobo	Iwa	Araorun	Ile	Itutu	Erinle
1	2	3	4	5	6	7
N28	N29	N30	D1	D2	D3	D4
TBT11	TBT12	TBT13	TBT14	TBT15	TBT16	TBT17
HID18	HID19	HID20	HID21	HID22	HID23	HID24
8	9	10	11	12	13	14
D5	D6	D7	D8	D9	D10	D11
TBT18	TBT19	TBT20	TBT21	TBT22	TBT23	TBT24
HID25	HID26	HID27	HID28	HID29	HID30	TAH1
15	16	17	18	19	20	21
D12	D13	D14	D15	D16	D17	D18
TBT25	TBT26	TBT27	TBT28	TBT29	TBT30	MKH1
TAH2	TAH3	TAH4	TAH5	TAH6	TAH7	TAH8
22	23	24	25	26	27	28
D19	D20	D21	D22	D23	D24	D25
MKH2	MKH3	MKH4	MKH5	MKH6	MKH7	MKH8
TAH9	TAH10	TAH11	TAH12	TAH13	TAH14	TAH15

EUSI 575/2091-92/6327-28/2083-84						
Ashe	Ojobo	Iwa	Araorun	Ile	Itutu	Erinle
1	2	3	4	5	6	7
O31	N1	N2	N3	N4	N5	N6
KHK13	KHK14	KHK15	KHK16	KHK17	KHK18	KHK19
THi20	THi21	THi22	THi23	THi24	THi25	THi26
8	9	10	11	12	13	14
N7	N8	N9	N10	N11	N12	N13
KHK20	KHK21	KHK22	KHK23	KHK24	KHK25	KHK26
THi27	THi28	THi29	THi30	HID1	HID2	HID3
15	16	17	18	19	20	21
N14	N15	N16	N17	N18	N19	N20
KHK27	KHK28	KHK29	KHK30	TBT1	TBT2	TBT3
HID4	HID5	HID6	HID7	HID8	HID9	HID10
22	23	24	25	26	27	28
N21	N22	N23	N24	N25	N26	N27
TBT4	TBT5	TBT6	TBT7	TBT8	TBT9	TBT10
HID11	HID12	HID13	HID14	HID15	HID16	HID17

KWANZA 575/2091-92/6327-28/2083-84						
Umoja	Kujichagulia	Ujima	Ujamaa	Nia	Kuumba	Imani
1	2	3	4	5	6	7
D26	D27	D28	D29	D30	D31	JA1
MKH9	MKH10	MKH11	MKH12	MKH13	MKH14	MKH1
TAH16	TAH17	TAH18	TAH19	TAH20	TAH21	TAH2
8	9	10	11	12	13	14
JA2	JA3	JA4	JA5	JA6	JA7	JA
MKH16	MKH17	MKH18	MKH19	MKH20	MKH21	MKH2
TAH23	TAH24	TAH25	TAH26	TAH27	TAH28	TAH29
15	16	17	18	19	20	21
JA9	JA10	JA11	JA12	JA13	JA14	JA1
MKH23	MKH24	MKH25	KMH26	MKH27	MKH28	MKH2
TAH30	TIR1	TIR2	TIR3	TIR4	TIR5	TIR6
22	23	24	25	26	27	28
JA16	JA17	JA18	JA19	JA20	JA21	JA2
MKH30	AMN1	AMN2	AMN3	AMN4	AMN5	AMN
TIR7	TIR8	TIR9	TIR10	TIR11	TIR12	TIR13

THE NU AFRIKAN CALENDAR

SHAHIDI 575/2091-92/6327-28/2083-84						
Ashe	Ojobo	Iwa	Araorun	Ile	Itutu	Erinle
1	2	3	4	5	6	7
JA23	JA24	JA25	JA26	JA27	JA28	JA29
AMN7	AMN8	AMN9	AMN10	AMN11	AMN12	AMN13
TIR14	TIR15	TIR16	TIR17	TIR18	TIR19	TIR20
8	9	10	11	12	13	14
JA30	JA31	F1	F2	F3	F4	F5
AMN14	AMN15	AMN16	AMN17	AMN18	AMN19	AMN20
TIR21	TIR22	TIR23	TIR24	TIR25	TIR26	TIR27
15	16	17	18	19	20	21
F6	F7	F8	F9	F10	F 11	F12
AMN21	AMN22	AMN23	AMN24	AMN25	AMN26	AMN28
TIR28	TIR29	TIR30	YEK1	YEK2	YEK3	YEK4
22	23	24	25	26	27	28
F13	F14	F15	F16	F17	F18	F19
AMN29	AMN30	RNT1	RNT2	RNT3	RNT4	RNT5
YEK5	YEK6	YEK7	YEK8	YEK9	YEK10	YEK11
SHAHIDI 126						
F20						
RNT6						
YEK12						

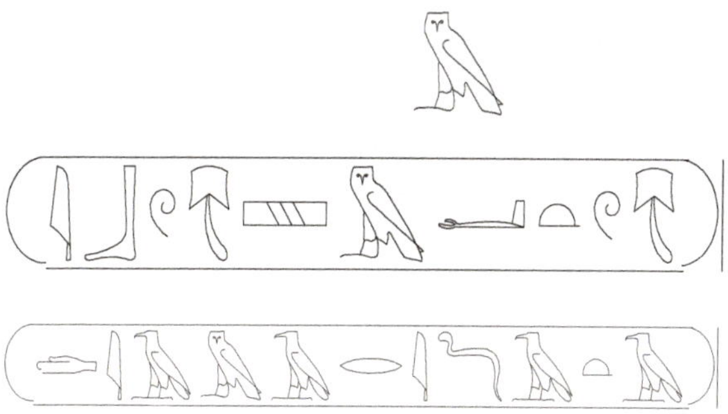

THE NU AFRIKAN CALENDAR

MAAFA 576 B.S.E./ASIENTO 2092-93 C.E./ KAM 6328-29 S.M./EC 2084-85 A.M. (ዓ.ም.)

AFU 576/2092-93/6328-29/2084-85						
Ashe	**Ojobo**	**Iwa**	**Araorun**	**Ile**	**Itutu**	**Erinle**
1	2	3	4	5	6	7
F21	F22	F23	F24	F25	F26	F27
RNT7	RNT8	RNT9	RNT10	RNT11	RNT12	RNT13
YEK13	YEK14	YEK15	YEK16	YEK17	YEK18	YEK19
8	9	10	11	12	13	14
F28	F29	M1	M2	M3	M4	M5
RNT14	RNT15	RNT16	RNT17	RNT18	RNT19	RNT20
YEK 20	YEK21	YEK22	YEK23	YEK24	YEK25	YEK26
15	16	17	18	19	20	21
M6	M7	M8	M9	M10	M11	M12
RNT21	RNT22	RNT23	RNT24	RNT25	RNT26	RNT27
YEK 27	YEK28	YEK29	YEK30	MAG1	MAG2	MAG3
22	23	24	25	26	27	28
M13	M14	M15	M16	M17	M18	M19
RNT28	RNT29	RNT30	KHN1	KHN2	KHN3	KHN4
MAG4	MAG5	MAG6	MAG7	MAG8	MAG9	MAG10
SHARIFU 141						
M20						
KHN5						
MAG11						

KAWAIDA 576/2092-93/6328-29/2084-85						
Ashe	**Ojobo**	**Iwa**	**Araorun**	**Ile**	**Itutu**	**Erinle**
1	2	3	4	5	6	7
A18	A19	A20	A21	A22	A23	A24
PIN4	PIN5	PIN6	PIN7	PIN8	PIN9	PIN10
MIY10	MIY11	MIY12	MIY13	MIY14	MIY15	MIY16
8	9	10	11	12	13	14
A25	A26	A27	A28	A29	A30	MA1
PIN11	PIN12	PIN13	PIN14	PIN15	PIN16	PIN17
MIY17	MIY18	MIY19	MIY20	MIY21	MIY22	MIY23
15	16	17	18	19	20	21
MA2	MA3	MA4	MA5	MA6	MA7	MA8
PIN18	PIN19	PIN20	PIN21	PIN22	PIN23	PIN24
MIY24	MIY25	MIY26	MIY27	MIY28	MIY29	MIY30
22	23	24	25	26	27	28
MA9	MA10	MA11	MA12	MA13	MA14	MA15
PIN25	PIN26	PIN27	PIN28	PIN29	PIN30	APP1
GIN1	GIN2	GIN3	GIN4	GIN5	GIN6	GIN7

UHURU 576/2092-93/6328-29/2084-85						
Ashe	**Ojobo**	**Iwa**	**Araorun**	**Ile**	**Itutu**	**Erinle**
1	2	3	4	5	6	7
M21	M22	M23	M24	M25	M26	M27
KHN6	KHN7	KHN8	KHN9	KHN10	KHN11	KHN12
MAG12	MAG13	MAG14	MAG15	MAG16	MAG17	MAG18
8	9	10	11	12	13	14
M28	M29	M30	M31	A1	A2	A3
KHN13	KHN14	KHN15	KHN16	KHN17	KHN18	KHN19
MAG19	MAG20	MAG21	MAG22	MAG23	MAG24	MAG25
15	16	17	18	19	20	21
A4	A5	A6	A7	A8	A9	A10
KHN20	KHN21	KHN222	KHN23	KHN24	KHN25	KHN26
MAG26	MAG27	MAG28	MAG29	MAG30	MIY1	MIY2
22	23	24	25	26	27	28
A11	A12	A13	A14	A15	A16	A17
KHN27	KHN28	KHN29	KHN30	PIN1	PIN2	PIN3
MIY3	MIY4	MIY5	MIY6	MIY7	MIY8	MIY9

MABABA WEUSI 576/2092-93/6328-29/2084-85						
Ashe	**Ojobo**	**Iwa**	**Araorun**	**Ile**	**Itutu**	**Erinle**
1	2	3	4	5	6	7
MA16	MA17	MA18	MA19	MA20	MA21	MA22
APP2	APP3	APP4	APP5	APP6	APP7	APP8
GIN8	GIN9	GIN10	GIN11	GIN12	GIN13	GIN14
8	9	10	11	12	13	14
MA23	MA24	MA25	MA26	MA27	MA28	MA29
APP9	APP10	APP11	APP12	APP13	APP14	APP15
GIN15	GIN16	GIN17	GIN18	GIN19	GIN20	GIN21
15	16	17	18	19	20	21
MA30	MA31	J1	J2	J3	J4	J5
APP16	APP17	APP18	APP19	APP20	APP21	APP22
GIN22	GIN23	GIN24	GIN25	GIN26	GIN27	GIN28
22	23	24	25	26	27	28
J6	J7	J8	J9	J10	J11	J12
APP23	APP24	APP25	APP26	APP27	APP28	APP29
GIN29	GIN30	SAN1	SAN2	SAN3	SAN4	SAN5

THE NU AFRIKAN CALENDAR

MWANAMKE WEUSI 576/2092-93/6328-29/2084-85

Ashe	Ojobo	Iwa	Araorun	Ile	Itutu	Erinle
1	2	3	4	5	6	7
J13	J14	J15	J16	J17	J18	J19
APP30	MSR1	MSR2	MSR3	MSR4	MSR5	MSR6
SAN6	SAN7	SAN8	SAN9	SAN10	SAN11	SAN12
8	9	10	11	12	13	14
J20	J21	J22	J23	J24	J25	J26
MSR7	MSR8	MSR9	MSR10	MSR11	MSR12	MSR13
SAN13	SAN14	SAN15	SAN16	SAN17	SAN18	SAN19
15	16	17	18	19	20	21
J27	J28	J29	J30	JU1	JU2	JU3
MSR14	MSR15	MSR16	MSR17	MSR18	MSR19	MSR20
SAN20	SAN21	SAN22	SAN23	SAN24	SAN25	SAN26
22	23	24	25	26	27	28
JU4	JU5	JU6	JU7	JU8	JU9	JU10
MSR21	MRS22	MSR23	MSR24	MSR25	MSR26	MSR27
SAN27	SAN28	SAN29	SAN30	HAM1	HAM2	HAM3

TAIFA WEUSI 576/2092-93/6328-29/2084-85

Ashe	Ojobo	Iwa	Araorun	Ile	Itutu	Erinle
1	2	3	4	5	6	7
AU8	AU9	AU10	AU11	AU12	AU13	AU14
THT20	THT21	THT22	THT23	THT24	THT25	THT26
NAH2	NAH3	NAH4	NAH5	NAH6	NAH7	NAH8
8	9	10	11	12	13	14
AU15	AU16	AU17	AU18	AU19	AU20	AU21
THT27	THT28	THT29	THT30	PAP1	PAP1	PAP2
NAH9	NAH10	NAH11	NAH12	NAH13	NAH14	NAH15
15	16	17	18	19	20	21
AU22	AU23	AU24	AU25	AU26	AU27	AU28
PAP3	PAP4	PAP5	PAP6	PAP7	PAP8	PAP9
NAH16	NAH17	NAH18	NAH19	NAH20	NAH21	NAH22
22	23	24	25	26	27	28
AU29	AU30	AU31	S1	S2	S3	S4
PAP10	PAP11	PAP12	PAP13	PAP14	PAP15	PAP16
NAH23	NAH24	NAH25	NAH26	NAH27	NAH28	NAH29

MTU WEUSI 576/2092-93/6328-29/2084-85

Ashe	Ojobo	Iwa	Araorun	Ile	Itutu	Erinle
1	2	3	4	5	6	7
JU11	JU12	JU13	JU14	JU15	JU16	JU17
MSR28	MSR29	MSR30	*ASR1*	*HRU2*	*STH3*	*AST4*
HAM4	HAM5	HAM6	HAM7	HAM8	HAM9	HAM10
8	9	10	11	12	13	14
JU18	JU19	JU20	JU21	JU22	JU23	JU24
NBH5	*ANU6*	THT1	THT2	THT3	THT4	THT5
HAM11	HAM12	HAM13	HAM14	HAM15	HAM16	HAM17
15	16	17	18	19	20	21
JU25	JU26	JU27	JU28	JU29	JU30	JU31
THT6	THT7	THT8	THT9	THT10	THT11	THT12
HAM18	HAM19	HAM20	HAM21	HAM22	HAM23	HAM24
22	23	24	25	26	27	28
AU1	AU2	AU3	AU4	AU5	AU6	AU7
THT13	THT14	THT15	THT16	THT17	THT18	THT19
HAM25	HAM26	HAM27	HAM28	HAM29	HAM30	NAH1

UWEZO 576/2092-93/6328-29/2084-85

Ashe	Ojobo	Iwa	Araorun	Ile	Itutu	Erinle
1	2	3	4	5	6	7
S5	S6	S7	S8	S9	S10	S11
PAP17	PAP18	PAP19	PAP20	PAP21	PAP22	PAP23
NAH30	*PAG1*	*PAG2*	*PAG3*	*PAG4*	*PAG5*	MAS1
8	9	10	11	12	13	14
S12	S13	S14	S15	S16	S17	S18
PAP24	PAP25	PAP26	PAP27	PAP28	PAP29	PAP30
MAS2	MAS3	MAS4	MAS5	MAS6	MAS7	MAS8
15	16	17	18	19	20	21
S19	S20	S21	S22	S23	S24	S25
HTH1	HTH2	HTH3	HTH4	HTH5	HTH6	HTH7
MAS9	MAS10	MAS11	MAS12	MAS13	MAS14	MAS15
22	23	24	25	26	27	28
S26	S27	S28	S29	S30	O1	O2
HTH8	HTH9	HTH10	HTH11	HTH12	HTH13	HTH14
MAS16	MAS17	MAS18	MAS19	MAS20	MAS21	MAS22

THE NU AFRIKAN CALENDAR

EKUNDU 576/2092-93/6328-29/2084-85

Ashe	Ojobo	Iwa	Araorun	Ile	Itutu	Erinle
1	2	3	4	5	6	7
O3	O4	O5	O6	O7	O8	O9
HTH15	HTH16	HTH17	HTH18	HTH19	HTH20	HTH21
MAS23	MAS24	MAS25	MAS26	MAS27	MAS28	MAS29
8	9	10	11	12	13	14
O10	O11	O12	O13	O14	O15	O16
HTH22	HTH23	HTH24	HTH25	HTH26	HTH27	HTH28
MAS30	THi1	THi2	THi3	THi4	THi5	THi6
15	16	17	18	19	20	21
O17	O18	O19	O20	O21	O22	O23
HTH29	HTH30	KHK1	KHK2	KHK3	KHK4	KHK5
THi7	THi8	This9	THi10	THi11	THi12	THi13
22	23	24	25	26	27	28
O24	O25	O26	O27	O28	O29	O30
KHK6	KHK7	KHK8	KHK9	KHK10	KHK11	KHK12
Thi14	THi15	THi16	THi17	THi18	THi19	THi20

KIJANI 576/2092-93/6328-29/2084-85

Ashe	Ojobo	Iwa	Araorun	Ile	Itutu	Erinle
1	2	3	4	5	6	7
N28	N29	N30	D1	D2	D3	D4
TBT11	TBT12	TBT13	TBT14	TBT15	TBT16	TBT17
HID19	HID20	HID21	HID22	HID23	HID24	HID25
8	9	10	11	12	13	14
D5	D6	D7	D8	D9	D10	D11
TBT18	TBT19	TBT20	TBT21	TBT22	TBT23	TBT24
HID26	HID27	HID28	HID29	HID30	TAH1	TAH2
15	16	17	18	19	20	21
D12	D13	D14	D15	D16	D17	D18
TBT25	TBT26	TBT27	TBT28	TBT29	TBT30	MKH1
TAH3	TAH4	TAH5	TAH6	TAH7	TAH8	TAH9
22	23	24	25	26	27	28
D19	D20	D21	D22	D23	D24	D25
MKH2	MKH3	MKH4	MKH5	MKH6	MKH7	MKH8
TAH10	TAH11	TAH12	TAH13	TAH14	TAH15	TAH16

EUSI 576/2092-93/6328-29/2084-85

Ashe	Ojobo	Iwa	Araorun	Ile	Itutu	Erinle
1	2	3	4	5	6	7
O31	N1	N2	N3	N4	N5	N6
KHK13	KHK14	KHK15	KHK16	KHK17	KHK18	KHK19
THi21	THi22	THi23	THi24	THi25	THi26	THi27
8	9	10	11	12	13	14
N7	N8	N9	N10	N11	N12	N13
KHK20	KHK21	KHK22	KHK23	KHK24	KHK25	KHK26
THi28	THi29	THi30	HID1	HID2	HID3	HID4
15	16	17	18	19	20	21
N14	N15	N16	N17	N18	N19	N20
KHK27	KHK28	KHK29	KHK30	TBT1	TBT2	TBT3
HID5	HID6	HID7	HID8	HID9	HID10	HID11
22	23	24	25	26	27	28
N21	N22	N23	N24	N25	N26	N27
TBT4	TBT5	TBT6	TBT7	TBT8	TBT9	TBT10
HID12	HID13	HID14	HID15	HID16	HID17	HID18

KWANZA 576/2092-93/6328-29/2084-85

Umoja	Kujichagulia	Ujima	Ujamaa	Nia	Kuumba	Imani
1	2	3	4	5	6	7
D26	D27	D28	D29	D30	D31	JA1
MKH9	MKH10	MKH11	MKH12	MKH13	MKH14	MKH1
TAH17	TAH18	TAH19	TAH20	TAH21	TAH22	TAH2
8	9	10	11	12	13	14
JA2	JA3	JA4	JA5	JA6	JA7	JA
MKH16	MKH17	MKH18	MKH19	MKH20	MKH21	MKH2
TAH24	TAH25	TAH26	TAH27	TAH28	TAH29	TAH3
15	16	17	18	19	20	21
JA9	JA10	JA11	JA12	JA13	JA14	JA1
MKH23	MKH24	MKH25	KMH26	MKH27	MKH28	MKH2
TIR1	TIR2	TIR3	TIR4	TIR5	TIR6	TIR7
22	23	24	25	26	27	28
JA16	JA17	JA18	JA19	JA20	JA21	JA2
MKH30	AMN1	AMN2	AMN3	AMN4	AMN5	AMN
TIR8	TIR9	TIR10	TIR11	TIR12	TIR13	TIR1

THE NU AFRIKAN CALENDAR

SHAHIDI 576/2092-93/6328-29/2084-85						
Ashe	Ojobo	Iwa	Araorun	Ile	Itutu	Erinle
1	2	3	4	5	6	7
JA23	JA24	JA25	JA26	JA27	JA28	JA29
AMN7	AMN8	AMN9	AMN10	AMN11	AMN12	AMN13
TIR15	TIR16	TIR17	TIR18	TIR19	TIR20	TIR21
8	9	10	11	12	13	14
JA30	JA31	F1	F2	F3	F4	F5
AMN14	AMN15	AMN16	AMN17	AMN18	AMN19	AMN20
TIR22	TIR23	TIR24	TIR25	TIR26	TIR27	TIR28
15	16	17	18	19	20	21
F6	F7	F8	F9	F10	F 11	F12
AMN21	AMN22	AMN23	AMN24	AMN25	AMN26	AMN28
TIR29	TIR30	YEK1	YEK2	YEK3	YEK4	YEK5
22	23	24	25	26	27	28
F13	F14	F15	F16	F17	F18	F19
AMN29	AMN30	RNT1	RNT2	RNT3	RNT4	RNT5
YEK6	YEK7	YEK8	YEK9	YEK10	YEK11	YEK12
SHAHIDI 127						
F20						
RNT6						
YEK13						

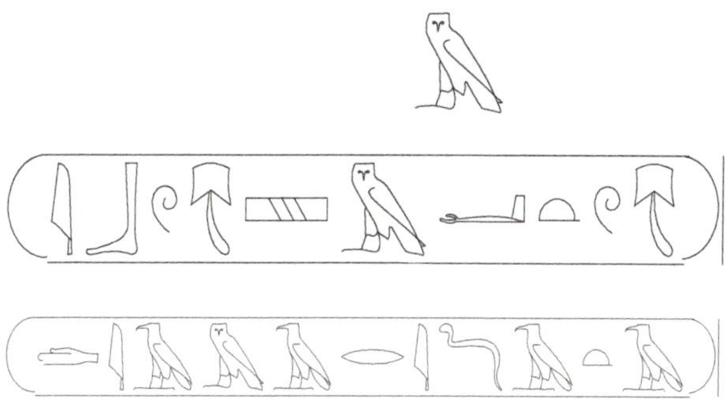

THE NU AFRIKAN CALENDAR

MAAFA 577 B.S.E./ASIENTO 2093-94 C.E.
KAM 6329-30 S.M./EC 2085-86 A.M. (ዓ.ም.)

AFU 577/2093-94/6329-30/2085-86

Ashe	Ojobo	Iwa	Araorun	Ile	Itutu	Erinle
1	2	3	4	5	6	7
F21	F22	F23	F24	F25	F26	F27
RNT7	RNT8	RNT9	RNT10	RNT11	RNT12	RNT13
YEK14	YEK15	YEK16	YEK17	YEK18	YEK19	YEK20
8	9	10	11	12	13	14
F28	M1	M2	M3	M4	M5	M6
RNT14	RNT15	RNT16	RNT17	RNT18	RNT19	RNT20
YEK21	YEK22	YEK23	YEK24	YEK25	YEK26	YEK27
15	16	17	18	19	20	21
M7	M8	M9	M10	M11	M12	M13
RNT21	RNT22	RNT23	RNT24	RNT25	RNT26	RNT27
YEK28	YEK29	YEK30	MAG1	MAG2	MAG3	MAG4
22	23	24	25	26	27	28
M14	M15	M16	M17	M18	M19	M20
RNT28	RNT29	RNT30	KHN1	KHN2	KHN3	KHN4
MAG5	MAG6	MAG7	MAG8	MAG9	MAG10	MAG11

KAWAIDA 577/2093-94/6329-30/2085-86

Ashe	Ojobo	Iwa	Araorun	Ile	Itutu	Erinle
1	2	3	4	5	6	7
A18	A19	A20	A21	A22	A23	A24
PIN3	PIN4	PIN5	PIN6	PIN7	PIN8	PIN9
MIY10	MIY11	MIY12	MIY13	MIY14	MIY15	MIY16
8	9	10	11	12	13	14
A25	A26	A27	A28	A29	A30	MA1
PIN10	PIN11	PIN12	PIN13	PIN14	PIN15	PIN16
MIY17	MIY18	MIY19	MIY20	MIY21	MIY22	MIY23
15	16	17	18	19	20	21
MA2	MA3	MA4	MA5	MA6	MA7	MA8
PIN17	PIN18	PIN19	PIN20	PIN21	PIN22	PIN23
MIY24	MIY25	MIY26	MIY27	MIY28	MIY29	MIY30
22	23	24	25	26	27	28
MA9	MA10	MA11	MA12	MA13	MA14	MA15
PIN24	PIN25	PIN26	PIN27	PIN28	PIN29	PIN30
GIN1	GIN2	GIN3	GIN4	GIN5	GIN6	GIN7

UHURU 577/2093-94/6329-30/2085-86

Ashe	Ojobo	Iwa	Araorun	Ile	Itutu	Erinle
1	2	3	4	5	6	7
M21	M22	M23	M24	M25	M26	M27
KHN5	KHN6	KHN7	KHN8	KHN9	KHN10	KHN11
MAG12	MAG13	MAG14	MAG15	MAG16	MAG17	MAG18
8	9	10	11	12	13	14
M28	M29	M30	M31	A1	A2	A3
KHN12	KHN13	KHN14	KHN15	KHN16	KHN17	KHN18
MAG19	MAG20	MAG21	MAG22	MAG23	MAG24	MAG25
15	16	17	18	19	20	21
A4	A5	A6	A7	A8	A9	A10
KHN19	KHN20	KHN21	KHN22	KHN23	KHN24	KHN25
MAG26	MAG27	MAG28	MAG29	MAG30	MIY1	MIY2
22	23	24	25	26	27	28
A11	A12	A13	A14	A15	A16	A17
KHN26	KHN27	KHN28	KHN29	KHN30	PIN1	PIN2
MIY3	MIY4	MIY5	MIY6	MIY7	MIY8	MIY9

MABABA WEUSI 577/2093-94/6329-30/2085-86

Ashe	Ojobo	Iwa	Araorun	Ile	Itutu	Erinle
1	2	3	4	5	6	7
MA16	MA17	MA18	MA19	MA20	MA21	MA22
APP1	APP2	APP3	APP4	APP5	APP6	APP7
GIN8	GIN9	GIN10	GIN11	GIN12	GIN13	GIN14
8	9	10	11	12	13	14
MA23	MA24	MA25	MA26	MA27	MA28	MA29
APP8	APP9	APP10	APP11	APP12	APP13	APP14
GIN15	GIN16	GIN17	GIN18	GIN19	GIN20	GIN21
15	16	17	18	19	20	21
MA30	MA31	J1	J2	J3	J4	J5
APP15	APP16	APP17	APP18	APP19	APP20	APP21
GIN22	GIN23	GIN24	GIN25	GIN26	GIN27	GIN28
22	23	24	25	26	27	28
J6	J7	J8	J9	J10	J11	J12
APP22	APP23	APP24	APP25	APP26	APP27	APP28
GIN29	GIN30	SAN1	SAN2	SAN3	SAN4	SAN5

THE NU AFRIKAN CALENDAR

MWANAMKE WEUSI 577/2093-94/6329-30/2085-86						
Ashe	Ojobo	Iwa	Araorun	Ile	Itutu	Erinle
1	2	3	4	5	6	7
J13	J14	J15	J16	J17	J18	J19
APP29	APP30	MSR1	MSR2	MSR3	MSR4	MSR5
SAN6	SAN7	SAN8	SAN9	SAN10	SAN11	SAN12
8	9	10	11	12	13	14
J20	J21	J22	J23	J24	J25	J26
MSR6	MSR7	MSR8	MSR9	MSR10	MSR11	MSR12
SAN13	SAN14	SAN15	SAN16	SAN17	SAN18	SAN19
15	16	17	18	19	20	21
J27	J28	J29	J30	JU1	JU2	JU3
MSR13	MSR14	MSR15	MSR16	MSR17	MSR18	MSR19
SAN20	SAN21	SAN22	SAN23	SAN24	SAN25	SAN26
22	23	24	25	26	27	28
JU4	JU5	JU6	JU7	JU8	JU9	JU10
MSR20	MRS21	MSR22	MSR23	MSR24	MSR25	MSR26
SAN27	SAN28	SAN29	SAN30	HAM1	HAM2	HAM3

TAIFA WEUSI 577/2093-94/6329-30/2085-86						
Ashe	Ojobo	Iwa	Araorun	Ile	Itutu	Erinle
1	2	3	4	5	6	7
AU8	AU9	AU10	AU11	AU12	AU13	AU14
THT20	THT21	THT22	THT23	THT24	THT25	THT26
NAH2	NAH3	NAH4	NAH5	NAH6	NAH7	NAH8
8	9	10	11	12	13	14
AU15	AU16	AU17	AU18	AU19	AU20	AU21
THT27	THT28	THT29	THT30	PAP1	PAP1	PAP2
NAH9	NAH10	NAH11	NAH12	NAH13	NAH14	NAH15
15	16	17	18	19	20	21
AU22	AU23	AU24	AU25	AU26	AU27	AU28
PAP3	PAP4	PAP5	PAP6	PAP7	PAP8	PAP9
NAH16	NAH17	NAH18	NAH19	NAH20	NAH21	NAH22
22	23	24	25	26	27	28
AU29	AU30	AU31	S1	S2	S3	S4
PAP10	PAP11	PAP12	PAP13	PAP14	PAP15	PAP16
NAH23	NAH24	NAH25	NAH26	NAH27	NAH28	NAH29

MTU WEUSI 577/2093-94/6329-30/2085-86						
Ashe	Ojobo	Iwa	Araorun	Ile	Itutu	Erinle
1	2	3	4	5	6	7
JU11	JU12	JU13	JU14	JU15	JU16	JU17
MSR27	MSR28	MSR29	MSR30	*ASR1*	*HRU2*	*STH3*
HAM4	HAM5	HAM6	HAM7	HAM8	HAM9	HAM10
8	9	10	11	12	13	14
JU18	JU19	JU20	JU21	JU22	JU23	JU24
AST4	*NBH5*	THT1	THT2	THT3	THT4	THT5
HAM11	HAM12	HAM13	AM14	HAM15	HAM16	HAM17
15	16	17	18	19	20	21
JU25	JU26	JU27	JU28	JU29	JU30	JU31
THT6	THT7	THT8	THT9	THT10	THT11	THT12
HAM18	HAM19	HAM20	HAM21	HAM22	HAM23	HAM24
22	23	24	25	26	27	28
AU1	AU2	AU3	AU4	AU5	AU6	AU7
THT13	THT14	THT15	THT16	THT17	THT18	THT19
HAM25	HAM26	HAM27	HAM28	HAM29	HAM30	NAH1

UWEZO 577/2093-94/6329-30/2085-86						
Ashe	Ojobo	Iwa	Araorun	Ile	Itutu	Erinle
1	2	3	4	5	6	7
S5	S6	S7	S8	S9	S10	S11
PAP17	PAP18	PAP19	PAP20	PAP21	PAP22	PAP23
NAH30	*PAG1*	*PAG2*	*PAG3*	*PAG4*	*PAG5*	MAS1
8	9	10	11	12	13	14
S12	S13	S14	S15	S16	S17	S18
PAP24	PAP25	PAP26	PAP27	PAP28	PAP29	PAP30
MAS2	MAS3	MAS4	MAS5	MAS6	MAS7	MAS8
15	16	17	18	19	20	21
S19	S20	S21	S22	S23	S24	S25
HTH1	HTH2	HTH3	HTH4	HTH5	HTH6	HTH7
MAS9	MAS10	MAS11	MAS12	MAS13	MAS14	MAS15
22	23	24	25	26	27	28
S26	S27	S28	S29	S30	O1	O2
HTH8	HTH9	HTH10	HTH11	HTH12	HTH13	HTH14
MAS16	MAS17	MAS18	MAS19	MAS20	MAS21	MAS22

THE NU AFRIKAN CALENDAR

EKUNDU 577/2093-94/6329-30/2085-86						
Ashe	Ojobo	Iwa	Araorun	Ile	Itutu	Erinle
1	2	3	4	5	6	7
O3	O4	O5	O6	O7	O8	O9
HTH15	HTH16	HTH17	HTH18	HTH19	HTH20	HTH21
MAS23	MAS24	MAS25	MAS26	MAS27	MAS28	MAS29
8	9	10	11	12	13	14
O10	O11	O12	O13	O14	O15	O16
HTH22	HTH23	HTH24	HTH25	HTH26	HTH27	HTH28
MAS30	THi1	THi2	THi3	THi4	THi5	THi6
15	16	17	18	19	20	21
O17	O18	O19	O20	O21	O22	O23
HTH29	HTH30	KHK1	KHK2	KHK3	KHK4	KHK5
THi7	THi8	THi9	THi10	THi11	THi12	THi13
22	23	24	25	26	27	28
O24	O25	O26	O27	O28	O29	O30
KHK6	KHK7	KHK8	KHK9	KHK10	KHK11	KHK12
THi14	THi15	THi16	THi17	THi18	THi19	THi20

KIJANI 577/2093-94/6329-30/2085-86						
Ashe	Ojobo	Iwa	Araorun	Ile	Itutu	Erinle
1	2	3	4	5	6	7
N28	N29	N30	D1	D2	D3	D4
TBT11	TBT12	TBT13	TBT14	TBT15	TBT16	TBT17
HID19	HID20	HID21	HID22	HID23	HID24	HID25
8	9	10	11	12	13	14
D5	D6	D7	D8	D9	D10	D11
TBT18	TBT19	TBT20	TBT21	TBT22	TBT23	TBT24
HID26	HID27	HID28	HID29	HID30	TAH1	TAH2
15	16	17	18	19	20	21
D12	D13	D14	D15	D16	D17	D18
TBT25	TBT26	TBT27	TBT28	TBT29	TBT30	MKH1
TAH3	TAH4	TAH5	TAH6	TAH7	TAH8	TAH9
22	23	24	25	26	27	28
D19	D20	D21	D22	D23	D24	D25
MKH2	MKH3	MKH4	MKH5	MKH6	MKH7	MKH8
TAH10	TAH11	TAH12	TAH13	TAH14	TAH15	TAH16

EUSI 577/2093-94/6329-30/2085-86						
Ashe	Ojobo	Iwa	Araorun	Ile	Itutu	Erinle
1	2	3	4	5	6	7
O31	N1	N2	N3	N4	N5	N6
KHK13	KHK14	KHK15	KHK16	KHK17	KHK18	KHK19
THi21	THi22	THi23	THi24	THi25	THi26	THi27
8	9	10	11	12	13	14
N7	N8	N9	N10	N11	N12	N13
KHK20	KHK21	KHK22	KHK23	KHK24	KHK25	KHK26
THi28	THi29	THi30	HID1	HID2	HID3	HID4
15	16	17	18	19	20	21
N14	N15	N16	N17	N18	N19	N20
KHK27	KHK28	KHK29	KHK30	TBT1	TBT2	TBT3
HID5	HID6	HID7	HID8	HID9	HID10	HID11
22	23	24	25	26	27	28
N21	N22	N23	N24	N25	N26	N27
TBT4	TBT5	TBT6	TBT7	TBT8	TBT9	TBT10
HID12	HID13	HID14	HID15	HID16	HID17	HID18

KWANZA 577/2093-94/6329-30/2085-86						
Umoja	Kujichagulia	Ujima	Ujamaa	Nia	Kuumba	Imani
1	2	3	4	5	6	7
D26	D27	D28	D29	D30	D31	JA1
MKH9	MKH10	MKH11	MKH12	MKH13	MKH14	MKH15
TAH17	TAH18	TAH19	TAH20	TAH21	TAH22	TAH23
8	9	10	11	12	13	14
JA2	JA3	JA4	JA5	JA6	JA7	JA8
MKH16	MKH17	MKH18	MKH19	MKH20	MKH21	MKH22
TAH24	TAH25	TAH26	TAH27	TAH28	TAH29	TAH30
15	16	17	18	19	20	21
JA9	JA10	JA11	JA12	JA13	JA14	JA15
MKH23	MKH24	MKH25	KMH26	MKH27	MKH28	MKH29
TIR1	TIR2	TIR3	TIR4	TIR5	TIR6	TIR7
22	23	24	25	26	27	28
JA16	JA17	JA18	JA19	JA20	JA21	JA22
MKH30	AMN1	AMN2	AMN3	AMN4	AMN5	AMN6
TIR8	TIR9	TIR10	TIR11	TIR12	TIR13	TIR14

THE NU AFRIKAN CALENDAR

SHAHIDI 577/2093-94/6329-30/2085-86						
Ashe	Ojobo	Iwa	Araorun	Ile	Itutu	Erinle
1	2	3	4	5	6	7
JA23	JA24	JA25	JA26	JA27	JA28	JA29
AMN7	AMN8	AMN9	AMN10	AMN11	AMN12	AMN13
TIR15	TIR16	TIR17	TIR18	TIR19	TIR20	TIR21
8	9	10	11	12	13	14
JA30	JA31	F1	F2	F3	F4	F5
AMN14	AMN15	AMN16	AMN17	AMN18	AMN19	AMN20
TIR22	TIR23	TIR24	TIR25	TIR26	TIR27	TIR28
15	16	17	18	19	20	21
F6	F7	F8	F9	F10	F 11	F12
AMN21	AMN22	AMN23	AMN24	AMN25	AMN26	AMN28
TIR29	TIR30	YEK1	YEK2	YEK3	YEK4	YEK5
22	23	24	25	26	27	28
F13	F14	F15	F16	F17	F18	F19
AMN29	AMN30	RNT1	RNT2	RNT3	RNT4	RNT5
YEK6	YEK7	YEK8	YEK9	YEK10	YEK11	YEK12
SHAHIDI 128						
F20						
RNT6						
YEK13						

THE NU AFRIKAN CALENDAR

MAAFA 578 B.S.E./ASIENTO 2094-95 C.E.
KAM 6330-31 S.M./EC 2086-87 A.M. (ዓ.ም.)

AFU 578/2094-95/6330-31/2086-87

Ashe	Ojobo	Iwa	Araorun	Ile	Itutu	Erinle
1	2	3	4	5	6	7
F21	F22	F23	F24	F25	F26	F27
RNT7	RNT8	RNT9	RNT10	RNT11	RNT12	RNT13
YEK14	YEK15	YEK16	YEK17	YEK18	YEK19	YEK20
8	9	10	11	12	13	14
F28	M1	M2	M3	M4	M5	M6
RNT14	RNT15	RNT16	RNT17	RNT18	RNT19	RNT20
YEK21	YEK22	YEK23	YEK24	YEK25	YEK26	YEK27
15	16	17	18	19	20	21
M7	M8	M9	M10	M11	M12	M13
RNT21	RNT22	RNT23	RNT24	RNT25	RNT26	RNT27
YEK28	YEK29	YEK30	MAG1	MAG2	MAG3	MAG4
22	23	24	25	26	27	28
M14	M15	M16	M17	M18	M19	M20
RNT28	RNT29	RNT30	KHN1	KHN2	KHN3	KHN4
MAG5	MAG6	MAG7	MAG8	MAG9	MAG10	MAG11

KAWAIDA 578/2094-95/6330-31/2086-87

Ashe	Ojobo	Iwa	Araorun	Ile	Itutu	Erinle
1	2	3	4	5	6	7
A18	A19	A20	A21	A22	A23	A24
PIN3	PIN4	PIN5	PIN6	PIN7	PIN8	PIN9
MIY10	MIY11	MIY12	MIY13	MIY14	MIY15	MIY16
8	9	10	11	12	13	14
A25	A26	A27	A28	A29	A30	MA1
PIN10	PIN11	PIN12	PIN13	PIN14	PIN15	PIN16
MIY17	MIY18	MIY19	MIY20	MIY21	MIY22	MIY23
15	16	17	18	19	20	21
MA2	MA3	MA4	MA5	MA6	MA7	MA8
PIN17	PIN18	PIN19	PIN20	PIN21	PIN22	PIN23
MIY24	MIY25	MIY26	MIY27	MIY28	MIY29	MIY30
22	23	24	25	26	27	28
MA9	MA10	MA11	MA12	MA13	MA14	MA15
PIN24	PIN25	PIN26	PIN27	PIN28	PIN29	PIN30
GIN1	GIN2	GIN3	GIN4	GIN5	GIN6	GIN7

UHURU 578/2094-95/6330-31/2086-87

Ashe	Ojobo	Iwa	Araorun	Ile	Itutu	Erinle
1	2	3	4	5	6	7
M21	M22	M23	M24	M25	M26	M27
KHN5	KHN6	KHN7	KHN8	KHN9	KHN10	KHN11
MAG12	MAG13	MAG14	MAG15	MAG16	MAG17	MAG18
8	9	10	11	12	13	14
M28	M29	M30	M31	A1	A2	A3
KHN12	KHN13	KHN14	KHN15	KHN16	KHN17	KHN18
MAG19	MAG20	MAG21	MAG22	MAG23	MAG24	MAG25
15	16	17	18	19	20	21
A4	A5	A6	A7	A8	A9	A10
KHN19	KHN20	KHN21	KHN22	KHN23	KHN24	KHN25
MAG26	MAG27	MAG28	MAG29	MAG30	MIY1	MIY2
22	23	24	25	26	27	28
A11	A12	A13	A14	A15	A16	A17
KHN26	KHN27	KHN28	KHN29	KHN30	PIN1	PIN2
MIY3	MIY4	MIY5	MIY6	MIY7	MIY8	MIY9

MABABA WEUSI 578/2094-95/6330-31/2086-87

Ashe	Ojobo	Iwa	Araorun	Ile	Itutu	Erinle
1	2	3	4	5	6	7
MA16	MA17	MA18	MA19	MA20	MA21	MA22
APP1	APP2	APP3	APP4	APP5	APP6	APP7
GIN8	GIN9	GIN10	GIN11	GIN12	GIN13	GIN14
8	9	10	11	12	13	14
MA23	MA24	MA25	MA26	MA27	MA28	MA29
APP8	APP9	APP10	APP11	APP12	APP13	APP14
GIN15	GIN16	GIN17	GIN18	GIN19	GIN20	GIN21
15	16	17	18	19	20	21
MA30	MA31	J1	J2	J3	J4	J5
APP15	APP16	APP17	APP18	APP19	APP20	APP21
GIN22	GIN23	GIN24	GIN25	GIN26	GIN27	GIN28
22	23	24	25	26	27	28
J6	J7	J8	J9	J10	J11	J12
APP22	APP23	APP24	APP25	APP26	APP27	APP28
GIN29	GIN30	SAN1	SAN2	SAN3	SAN4	SAN5

THE NU AFRIKAN CALENDAR

MWANAMKE WEUSI 578/2094-95/6330-31/2086-87						
Ashe	Ojobo	Iwa	Araorun	Ile	Itutu	Erinle
1	2	3	4	5	6	7
J13	J14	J15	J16	J17	J18	J19
APP29	APP30	MSR1	MSR2	MSR3	MSR4	MSR5
SAN6	SAN7	SAN8	SAN9	SAN10	SAN11	SAN12
8	9	10	11	12	13	14
J20	J21	J22	J23	J24	J25	J26
MSR6	MSR7	MSR8	MSR9	MSR10	MSR11	MSR12
SAN13	SAN14	SAN15	SAN16	SAN17	SAN18	SAN19
15	16	17	18	19	20	21
J27	J28	J29	J30	JU1	JU2	JU3
MSR13	MSR14	MSR15	MSR16	MSR17	MSR18	MSR19
SAN20	SAN21	SAN22	SAN23	SAN24	SAN25	SAN26
22	23	24	25	26	27	28
JU4	JU5	JU6	JU7	JU8	JU9	JU10
MSR20	MRS21	MSR22	MSR23	MSR24	MSR25	MSR26
SAN27	SAN28	SAN29	SAN30	HAM1	HAM2	HAM3

TAIFA WEUSI 578/2094-95/6330-31/2086-87						
Ashe	Ojobo	Iwa	Araorun	Ile	Itutu	Erinle
1	2	3	4	5	6	7
AU8	AU9	AU10	AU11	AU12	AU13	AU14
THT20	THT21	THT22	THT23	THT24	THT25	THT26
NAH2	NAH3	NAH4	NAH5	NAH6	NAH7	NAH8
8	9	10	11	12	13	14
AU15	AU16	AU17	AU18	AU19	AU20	AU21
THT27	THT28	THT29	THT30	PAP1	PAP1	PAP2
NAH9	NAH10	NAH11	NAH12	NAH13	NAH14	NAH15
15	16	17	18	19	20	21
AU22	AU23	AU24	AU25	AU26	AU27	AU28
PAP3	PAP4	PAP5	PAP6	PAP7	PAP8	PAP9
NAH16	NAH17	NAH18	NAH19	NAH20	NAH21	NAH22
22	23	24	25	26	27	28
AU29	AU30	AU31	S1	S2	S3	S4
PAP10	PAP11	PAP12	PAP13	PAP14	PAP15	PAP16
NAH23	NAH24	NAH25	NAH26	NAH27	NAH28	NAH29

MTU WEUSI 578/2094-95/6330-31/2086-87						
Ashe	Ojobo	Iwa	Araorun	Ile	Itutu	Erinle
1	2	3	4	5	6	7
JU11	JU12	JU13	JU14	JU15	JU16	JU17
MSR27	MSR28	MSR29	MSR30	*ASR1*	*HRU2*	*STH3*
HAM4	HAM5	HAM6	HAM7	HAM8	HAM9	HAM10
8	9	10	11	12	13	14
JU18	JU19	JU20	JU21	JU22	JU23	JU24
AST4	*NBH5*	THT1	THT2	THT3	THT4	THT5
HAM11	HAM12	HAM13	AM14	HAM15	HAM16	HAM17
15	16	17	18	19	20	21
JU25	JU26	JU27	JU28	JU29	JU30	JU31
THT6	THT7	THT8	THT9	THT10	THT11	THT12
HAM18	HAM19	HAM20	HAM21	HAM22	HAM23	HAM24
22	23	24	25	26	27	28
AU1	AU2	AU3	AU4	AU5	AU6	AU7
THT13	THT14	THT15	THT16	THT17	THT18	THT19
HAM25	HAM26	HAM27	HAM28	HAM29	HAM30	NAH1

UWEZO 578/2094-95/6330-31/2086-87						
Ashe	Ojobo	Iwa	Araorun	Ile	Itutu	Erinle
1	2	3	4	5	6	7
S5	S6	S7	S8	S9	S10	S11
PAP17	PAP18	PAP19	PAP20	PAP21	PAP22	PAP23
NAH30	*PAG1*	*PAG2*	*PAG3*	*PAG4*	*PAG5*	MAS1
8	9	10	11	12	13	14
S12	S13	S14	S15	S16	S17	S18
PAP24	PAP25	PAP26	PAP27	PAP28	PAP29	PAP30
MAS2	MAS3	MAS4	MAS5	MAS6	MAS7	MAS8
15	16	17	18	19	20	21
S19	S20	S21	S22	S23	S24	S25
HTH1	HTH2	HTH3	HTH4	HTH5	HTH6	HTH7
MAS9	MAS10	MAS11	MAS12	MAS13	MAS14	MAS15
22	23	24	25	26	27	28
S26	S27	S28	S29	S30	O1	O2
HTH8	HTH9	HTH10	HTH11	HTH12	HTH13	HTH14
MAS16	MAS17	MAS18	MAS19	MAS20	MAS21	MAS22

THE NU AFRIKAN CALENDAR

EKUNDU 578/2094-95/6330-31/2086-87						
Ashe	Ojobo	Iwa	Araorun	Ile	Itutu	Erinle
1	2	3	4	5	6	7
O3	O4	O5	O6	O7	O8	O9
HTH15	HTH16	HTH17	HTH18	HTH19	HTH20	HTH21
MAS23	MAS24	MAS25	MAS26	MAS27	MAS28	MAS29
8	9	10	11	12	13	14
O10	O11	O12	O13	O14	O15	O16
HTH22	HTH23	HTH24	HTH25	HTH26	HTH27	HTH28
MAS30	THi1	THi2	THi3	THi4	THi5	THi6
15	16	17	18	19	20	21
O17	O18	O19	O20	O21	O22	O23
HTH29	HTH30	KHK1	KHK2	KHK3	KHK4	KHK5
THi7	THi8	THi9	THi10	THi11	THi12	THi13
22	23	24	25	26	27	28
O24	O25	O26	O27	O28	O29	O30
KHK6	KHK7	KHK8	KHK9	KHK10	KHK11	KHK12
THi14	THi15	THi16	THi17	THi18	THi19	THi20

KIJANI 578/2094-95/6330-31/2086-87						
Ashe	Ojobo	Iwa	Araorun	Ile	Itutu	Erinle
1	2	3	4	5	6	7
N28	N29	N30	D1	D2	D3	D4
TBT11	TBT12	TBT13	TBT14	TBT15	TBT16	TBT17
HID19	HID20	HID21	HID22	HID23	HID24	HID25
8	9	10	11	12	13	14
D5	D6	D7	D8	D9	D10	D11
TBT18	TBT19	TBT20	TBT21	TBT22	TBT23	TBT24
HID26	HID27	HID28	HID29	HID30	TAH1	TAH2
15	16	17	18	19	20	21
D12	D13	D14	D15	D16	D17	D18
TBT25	TBT26	TBT27	TBT28	TBT29	TBT30	MKH1
TAH3	TAH4	TAH5	TAH6	TAH7	TAH8	TAH9
22	23	24	25	26	27	28
D19	D20	D21	D22	D23	D24	D25
MKH2	MKH3	MKH4	MKH5	MKH6	MKH7	MKH8
TAH10	TAH11	TAH12	TAH13	TAH14	TAH15	TAH16

EUSI 578/2094-95/6330-31/2086-87						
Ashe	Ojobo	Iwa	Araorun	Ile	Itutu	Erinle
1	2	3	4	5	6	7
O31	N1	N2	N3	N4	N5	N6
KHK13	KHK14	KHK15	KHK16	KHK17	KHK18	KHK19
THi21	THi22	THi23	THi24	THi25	THi26	THi27
8	9	10	11	12	13	14
N7	N8	N9	N10	N11	N12	N13
KHK20	KHK21	KHK22	KHK23	KHK24	KHK25	KHK26
THi28	THi29	THi30	HID1	HID2	HID3	HID4
15	16	17	18	19	20	21
N14	N15	N16	N17	N18	N19	N20
KHK27	KHK28	KHK29	KHK30	TBT1	TBT2	TBT3
HID5	HID6	HID7	HID8	HID9	HID10	HID11
22	23	24	25	26	27	28
N21	N22	N23	N24	N25	N26	N27
TBT4	TBT5	TBT6	TBT7	TBT8	TBT9	TBT10
HID12	HID13	HID14	HID15	HID16	HID17	HID18

KWANZA 578/2094-95/6330-31/2086-87						
Umoja	Kujichagulia	Ujima	Ujamaa	Nia	Kuumba	Imani
1	2	3	4	5	6	7
D26	D27	D28	D29	D30	D31	JA1
MKH9	MKH10	MKH11	MKH12	MKH13	MKH14	MKH15
TAH17	TAH18	TAH19	TAH20	TAH21	TAH22	TAH23
8	9	10	11	12	13	14
JA2	JA3	JA4	JA5	JA6	JA7	JA8
MKH16	MKH17	MKH18	MKH19	MKH20	MKH21	MKH22
TAH24	TAH25	TAH26	TAH27	TAH28	TAH29	TAH30
15	16	17	18	19	20	21
JA9	JA10	JA11	JA12	JA13	JA14	JA15
MKH23	MKH24	MKH25	KMH26	MKH27	MKH28	MKH29
TIR1	TIR2	TIR3	TIR4	TIR5	TIR6	TIR7
22	23	24	25	26	27	28
JA16	JA17	JA18	JA19	JA20	JA21	JA22
MKH30	AMN1	AMN2	AMN3	AMN4	AMN5	AMN6
TIR8	TIR9	TIR10	TIR11	TIR12	TIR13	TIR14

SHAHIDI 578/2094-95/6330-31/2086-87						
Ashe	Ojobo	Iwa	Araorun	Ile	Itutu	Erinle
1	2	3	4	5	6	7
JA23	JA24	JA25	JA26	JA27	JA28	JA29
AMN7	AMN8	AMN9	AMN10	AMN11	AMN12	AMN13
TIR15	TIR16	TIR17	TIR18	TIR19	TIR20	TIR21
8	9	10	11	12	13	14
JA30	JA31	F1	F2	F3	F4	F5
AMN14	AMN15	AMN16	AMN17	AMN18	AMN19	AMN20
TIR22	TIR23	TIR24	TIR25	TIR26	TIR27	TIR28
15	16	17	18	19	20	21
F6	F7	F8	F9	F10	F 11	F12
AMN21	AMN22	AMN23	AMN24	AMN25	AMN26	AMN28
TIR29	TIR30	YEK1	YEK2	YEK3	YEK4	YEK5
22	23	24	25	26	27	28
F13	F14	F15	F16	F17	F18	F19
AMN29	AMN30	RNT1	RNT2	RNT3	RNT4	RNT5
YEK6	YEK7	YEK8	YEK9	YEK10	YEK11	YEK12
SHAHIDI 129						
F20						
RNT6						
YEK13						

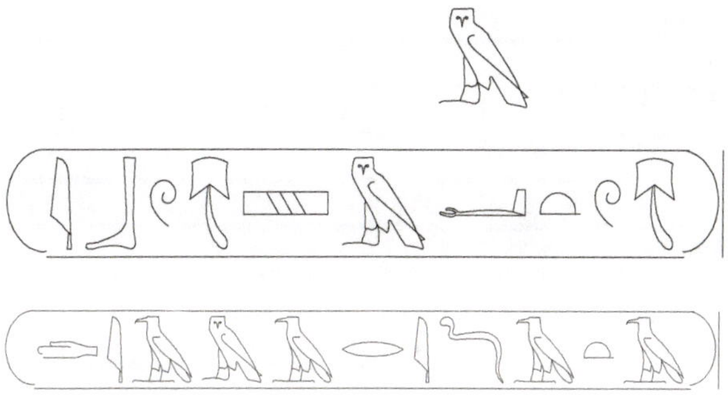

MAAFA 579 B.S.E./ASIENTO 2095-96 C.E./

KAM 6331-32 S.M./EC 2087-88 A.M. (ዓ.ም.)

AFU 579/2095-96/6331-32/2087-88						
Ashe	Ojobo	Iwa	Araorun	Ile	Itutu	Erinle
1	2	3	4	5	6	7
F21	F22	F23	F24	F25	F26	F27
RNT7	RNT8	RNT9	RNT10	RNT11	RNT12	RNT13
YEK14	YEK15	YEK16	YEK17	YEK18	YEK19	YEK20
8	9	10	11	12	13	14
F28	M1	M2	M3	M4	M5	M6
RNT14	RNT15	RNT16	RNT17	RNT18	RNT19	RNT20
YEK21	YEK22	YEK23	YEK24	YEK25	YEK26	YEK27
15	16	17	18	19	20	21
M7	M8	M9	M10	M11	M12	M13
RNT21	RNT22	RNT23	RNT24	RNT25	RNT26	RNT27
YEK28	YEK 29	YEK30	MAG1	MAG2	MAG3	MAG4
22	23	24	25	26	27	28
M14	M15	M16	M17	M18	M19	M20
RNT28	RNT29	RNT30	KHN1	KHN2	KHN3	KHN4
MAG5	MAG6	MAG7	MAG8	MAG9	MAG10	MAG11

UHURU 579/2095-96/6331-32/2087-88						
Ashe	Ojobo	Iwa	Araorun	Ile	Itutu	Erinle
1	2	3	4	5	6	7
M21	M22	M23	M24	M25	M26	M27
KHN5	KHN6	KHN7	KHN8	KHN9	KHN10	KHN11
MAG12	MAG13	MAG14	MAG15	MAG16	MAG17	MAG18
8	9	10	11	12	13	14
M28	M29	M30	M31	A1	A2	A3
KHN12	KHN13	KHN14	KHN15	KHN16	KHN17	KHN18
MAG19	MAG20	MAG21	MAG22	MAG23	MAG24	MAG25
15	16	17	18	19	20	21
A4	A5	A6	A7	A8	A9	A10
KHN19	KHN20	KHN21	KHN22	KHN23	KHN24	KHN25
MAG26	MAG27	MAG28	MAG29	MAG30	MIY1	MIY2
22	23	24	25	26	27	28
A11	A12	A13	A14	A15	A16	A17
KHN26	KHN27	KHN28	KHN29	KHN30	PIN1	PIN2
MIY3	MIY4	MIY5	MIY6	MIY7	MIY8	MIY9

KAWAIDA 579/2095-96/6331-32/2087-88						
Ashe	Ojobo	Iwa	Araorun	Ile	Itutu	Erinle
1	2	3	4	5	6	7
A18	A19	A20	A21	A22	A23	A24
PIN3	PIN4	PIN5	PIN6	PIN7	PIN8	PIN9
MIY10	MIY11	MIY12	MIY13	MIY14	MIY15	MIY16
8	9	10	11	12	13	14
A25	A26	A27	A28	A29	A30	MA1
PIN10	PIN11	PIN12	PIN13	PIN14	PIN15	PIN16
MIY17	MIY18	MIY19	MIY20	MIY21	MIY22	MIY23
15	16	17	18	19	20	21
MA2	MA3	MA4	MA5	MA6	MA7	MA8
PIN17	PIN18	PIN19	PIN20	PIN21	PIN22	PIN23
MIY24	MIY25	MIY26	MIY27	MIY28	MIY29	MIY30
22	23	24	25	26	27	28
MA9	MA10	MA11	MA12	MA13	MA14	MA15
PIN24	PIN25	PIN26	PIN27	PIN28	PIN29	PIN30
GIN1	GIN2	GIN3	GIN4	GIN5	GIN6	GIN7

MABABA WEUSI 579/2095-96/6331-32/2087-88						
Ashe	Ojobo	Iwa	Araorun	Ile	Itutu	Erinle
1	2	3	4	5	6	7
MA16	MA17	MA18	MA19	MA20	MA21	MA22
APP1	APP2	APP3	APP4	APP5	APP6	APP7
GIN8	GIN9	GIN10	GIN11	GIN12	GIN13	GIN14
8	9	10	11	12	13	14
MA23	MA24	MA25	MA26	MA27	MA28	MA29
APP8	APP9	APP10	APP11	APP12	APP13	APP14
GIN15	GIN16	GIN17	GIN18	GIN19	GIN20	GIN21
15	16	17	18	19	20	21
MA30	MA31	J1	J2	J3	J4	J5
APP15	APP16	APP17	APP18	APP19	APP20	APP21
GIN22	GIN23	GIN24	GIN25	GIN26	GIN27	GIN28
22	23	24	25	26	27	28
J6	J7	J8	J9	J10	J11	J12
APP22	APP23	APP24	APP25	APP26	APP27	APP28
GIN29	GIN30	SAN1	SAN2	SAN3	SAN4	SAN5

THE NU AFRIKAN CALENDAR

MWANAMKE WEUSI 579/2095-96/6331-32/2087-88

Ashe	Ojobo	Iwa	Araorun	Ile	Itutu	Erinle
1	2	3	4	5	6	7
J13	J14	J15	J16	J17	J18	J19
APP29	APP30	MSR1	MSR2	MSR3	MSR4	MSR5
SAN6	SAN7	SAN8	SAN9	SAN10	SAN11	SAN12
8	9	10	11	12	13	14
J20	J21	J22	J23	J24	J25	J26
MSR6	MSR7	MSR8	MSR9	MSR10	MSR11	MSR12
SAN13	SAN14	SAN15	SAN16	SAN17	SAN18	SAN19
15	16	17	18	19	20	21
J27	J28	J29	J30	JU1	JU2	JU3
MSR13	MSR14	MSR15	MSR16	MSR17	MSR18	MSR19
SAN20	SAN21	SAN22	SAN23	SAN24	SAN25	SAN26
22	23	24	25	26	27	28
JU4	JU5	JU6	JU7	JU8	JU9	JU10
MSR20	MRS21	MSR22	MSR23	MSR24	MSR25	MSR26
SAN27	SAN28	SAN29	SAN30	HAM1	HAM2	HAM3

TAIFA WEUSI 579/2095-96/6331-32/2087-88

Ashe	Ojobo	Iwa	Araorun	Ile	Itutu	Erinle
1	2	3	4	5	6	7
AU8	AU9	AU10	AU11	AU12	AU13	AU14
THT20	THT21	THT22	THT23	THT24	THT25	THT26
NAH2	NAH3	NAH4	NAH5	NAH6	NAH7	NAH8
8	9	10	11	12	13	14
AU15	AU16	AU17	AU18	AU19	AU20	AU21
THT27	THT28	THT29	THT30	PAP1	PAP1	PAP2
NAH9	NAH10	NAH11	NAH12	NAH13	NAH14	NAH15
15	16	17	18	19	20	21
AU22	AU23	AU24	AU25	AU26	AU27	AU28
PAP3	PAP4	PAP5	PAP6	PAP7	PAP8	PAP9
NAH16	NAH17	NAH18	NAH19	NAH20	NAH21	NAH22
22	23	24	25	26	27	28
AU29	AU30	AU31	S1	S2	S3	S4
PAP10	PAP11	PAP12	PAP13	PAP14	PAP15	PAP16
NAH23	NAH24	NAH25	NAH26	NAH27	NAH28	NAH29

MTU WEUSI 579/2095-96/6331-32/2087-88

Ashe	Ojobo	Iwa	Araorun	Ile	Itutu	Erinle
1	2	3	4	5	6	7
JU11	JU12	JU13	JU14	JU15	JU16	JU17
MSR27	MSR28	MSR29	MSR30	*ASR1*	*HRU2*	*STH3*
HAM4	HAM5	HAM6	HAM7	HAM8	HAM9	HAM10
8	9	10	11	12	13	14
JU18	JU19	JU20	JU21	JU22	JU23	JU24
AST4	*NBH5*	THT1	THT2	THT3	THT4	THT5
HAM11	HAM12	HAM13	HAM14	HAM15	HAM16	HAM17
15	16	17	18	19	20	21
JU25	JU26	JU27	JU28	JU29	JU30	JU31
THT6	THT7	THT8	THT9	THT10	THT11	THT12
HAM18	HAM19	HAM20	HAM21	HAM22	HAM23	HAM24
22	23	24	25	26	27	28
AU1	AU2	AU3	AU4	AU5	AU6	AU7
THT13	THT14	THT15	THT16	THT17	THT18	THT19
HAM25	HAM26	HAM27	HAM28	HAM29	HAM30	NAH1

UWEZO 579/2095-96/6331-32/2087-88

Ashe	Ojobo	Iwa	Araorun	Ile	Itutu	Erinle
1	2	3	4	5	6	7
S5	S6	S7	S8	S9	S10	S11
PAP17	PAP18	PAP19	PAP20	PAP21	PAP22	PAP23
NAH30	*QUA1*	*QUA2*	*QUA3*	*QUA4*	*QUA5*	*QUA6*
8	9	10	11	12	13	14
S12	S13	S14	S15	S16	S17	S18
PAP24	PAP25	PAP26	PAP27	PAP28	PAP29	PAP30
MAS1	MAS2	MAS3	MAS4	MAS5	MAS6	MAS7
15	16	17	18	19	20	21
S19	S20	S21	S22	S23	S24	S25
HTH1	HTH2	HTH3	HTH4	HTH5	HTH6	HTH7
MAS8	MAS9	MAS10	MAS11	MAS12	MAS13	MAS14
22	23	24	25	26	27	28
S26	S27	S28	S29	S30	O1	O2
HTH8	HTH9	HTH10	HTH11	HTH12	HTH13	HTH14
MAS15	MAS16	MAS17	MAS18	MAS19	MAS20	MAS21

THE NU AFRIKAN CALENDAR

EKUNDU 579/2095-96/6331-32/2087-88						
Ashe	Ojobo	Iwa	Araorun	Ile	Itutu	Erinle
1	2	3	4	5	6	7
O3	O4	O5	O6	O7	O8	O9
HTH15	HTH16	HTH17	HTH18	HTH19	HTH20	HTH21
MAS22	MAS23	MAS 24	MAS25	MAS26	MAS27	MAS28
8	9	10	11	12	13	14
O10	O11	O12	O13	O14	O15	O16
HTH22	HTH23	HTH24	HTH25	HTH26	HTH27	HTH28
MAS29	MAS30	THi1	THi2	THi3	THi4	THi5
15	16	17	18	19	20	21
O17	O18	O19	O20	O21	O22	O23
HTH29	HTH30	KHK1	KHK2	KHK3	KHK4	KHK5
THi6	THi7	THi8	THi9	THi10	THi11	THi12
22	23	24	25	26	27	28
O24	O25	O26	O27	O28	O29	O30
KHK6	KHK7	KHK8	KHK9	KHK10	KHK11	KHK12
THi13	THi14	THi15	THi16	THi17	THi18	THi19

KIJANI 579/2095-96/6331-32/2087-88						
Ashe	Ojobo	Iwa	Araorun	Ile	Itutu	Erinle
1	2	3	4	5	6	7
N28	N29	N30	D1	D2	D3	D4
TBT11	TBT12	TBT13	TBT14	TBT15	TBT16	TBT17
HID18	HID19	HID20	HID21	HID22	HID23	HID24
8	9	10	11	12	13	14
D5	D6	D7	D8	D9	D10	D11
TBT18	TBT19	TBT20	TBT21	TBT22	TBT23	TBT24
HID25	HID26	HID27	HID28	HID29	HID30	TAH1
15	16	17	18	19	20	21
D12	D13	D14	D15	D16	D17	D18
TBT25	TBT26	TBT27	TBT28	TBT29	TBT30	MKH1
TAH2	TAH3	TAH4	TAH5	TAH6	TAH7	TAH8
22	23	24	25	26	27	28
D19	D20	D21	D22	D23	D24	D25
MKH2	MKH3	MKH4	MKH5	MKH6	MKH7	MKH8
TAH9	TAH10	TAH11	TAH12	TAH13	TAH14	TAH15

EUSI 579/2095-96/6331-32/2087-88						
Ashe	Ojobo	Iwa	Araorun	Ile	Itutu	Erinle
1	2	3	4	5	6	7
O31	N1	N2	N3	N4	N5	N6
KHK13	KHK14	KHK15	KHK16	KHK17	KHK18	KHK19
THi20	THi21	THi22	THi23	THi24	THi25	THi26
8	9	10	11	12	13	14
N7	N8	N9	N10	N11	N12	N13
KHK20	KHK21	KHK22	KHK23	KHK24	KHK25	KHK26
THi27	THi28	THi29	THi30	HID1	HID2	HID3
15	16	17	18	19	20	21
N14	N15	N16	N17	N18	N19	N20
KHK27	KHK28	KHK29	KHK30	TBT1	TBT2	TBT3
HID4	HID5	HID6	HID7	HID8	HID9	HID10
22	23	24	25	26	27	28
N21	N22	N23	N24	N25	N26	N27
TBT4	TBT5	TBT6	TBT7	TBT8	TBT9	TBT10
HID11	HID12	HID13	HID14	HID15	HID16	HID17

KWANZA 579/2095-96/6331-32/2087-88						
Umoja	Kujichagulia	Ujima	Ujamaa	Nia	Kuumba	Imani
1	2	3	4	5	6	7
D26	D27	D28	D29	D30	D31	JA1
MKH9	MKH10	MKH11	MKH12	MKH13	MKH14	MKH15
TAH16	TAH17	TAH18	TAH19	TAH20	TAH21	TAH22
8	9	10	11	12	13	14
JA2	JA3	JA4	JA5	JA6	JA7	JA8
MKH16	MKH17	MKH18	MKH19	MKH20	MKH21	MKH22
TAH23	TAH24	TAH25	TAH26	TAH27	TAH28	TAH29
15	16	17	18	19	20	21
JA9	JA10	JA11	JA12	JA13	JA14	JA15
MKH23	MKH24	MKH25	KMH26	MKH27	MKH28	MKH29
TAH30	TIR1	TIR2	TIR3	TIR4	TIR5	TIR6
22	23	24	25	26	27	28
JA16	JA17	JA18	JA19	JA20	JA21	JA22
MKH30	AMN1	AMN2	AMN3	AMN4	AMN5	AMN6
TIR7	TIR8	TIR9	TIR10	TIR11	TIR12	TIR13

SHAHIDI 579/2095-96/6331-32/2087-88						
Ashe	Ojobo	Iwa	Araorun	Ile	Itutu	Erinle
1	2	3	4	5	6	7
JA23	JA24	JA25	JA26	JA27	JA28	JA29
AMN7	AMN8	AMN9	AMN10	AMN11	AMN12	AMN13
TIR14	TIR15	TIR16	TIR17	TIR18	TIR19	TIR20
8	9	10	11	12	13	14
JA30	JA31	F1	F2	F3	F4	F5
AMN14	AMN15	AMN16	AMN17	AMN18	AMN19	AMN20
TIR21	TIR22	TIR23	TIR24	TIR25	TIR26	TIR27
15	16	17	18	19	20	21
F6	F7	F8	F9	F10	F 11	F12
AMN21	AMN22	AMN23	AMN24	AMN25	AMN26	AMN28
TIR28	TIR29	TIR30	YEK1	YEK2	YEK3	YEK4
22	23	24	25	26	27	28
F13	F14	F15	F16	F17	F18	F19
AMN29	AMN30	RNT1	RNT2	RNT3	RNT4	RNT5
YEK5	YEK6	YEK7	YEK8	YEK9	YEK10	YEK11
SHAHIDI 130						
F20						
RNT6						
YEK12						

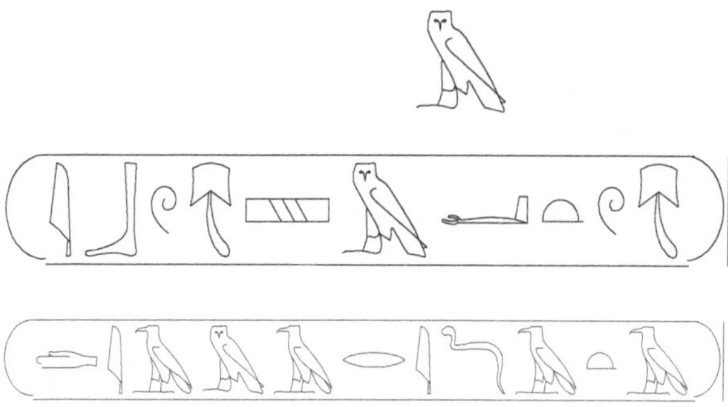

MAAFA 580 B.S.E./ASIENTO 2096-97 C.E.
KAM 6332-33 S.M./E.C. 2088-89 a.m.

AFU 580/2096-97/6332-33/2088-89

e	Ojobo	Iwa	Araorun	Ile	Itutu	Erinle
	2	3	4	5	6	7
	F22	F23	F24	F25	F26	F27
7	RNT8	RNT9	RNT10	RNT11	RNT12	RNT13
14	YEK15	YEK16	YEK17	YEK18	YEK19	YEK20
	9	10	11	12	13	14
	M1	M2	M3	M4	M5	M6
14	RNT15	RNT16	RNT17	RNT18	RNT19	RNT20
21	YEK22	YEK23	YEK24	YEK25	YEK26	YEK27
	16	17	18	19	20	21
	M8	M9	M10	M11	M12	M13
T21	RNT22	RNT23	RNT24	RNT25	RNT26	RNT27
28	YEK 29	YEK30	MAG1	MAG2	MAG3	MAG4
	23	24	25	26	27	28
4	M15	M16	M17	M18	M19	M20
T28	RNT29	RNT30	KHN1	KHN2	KHN3	KHN4
G5	MAG6	MAG7	MAG8	MAG9	MAG10	MAG11

KWAIDA 580/2096-97/6332-33/2088-89

Ashe	Ojobo	Iwa	Araorun	Ile	Itutu	Erinle
1	2	3	4	5	6	7
A18	A19	A20	A21	A22	A23	A24
PIN3	PIN4	PIN5	PIN6	PIN7	PIN8	PIN9
MIY10	MIY11	MIY12	MIY13	MIY14	MIY15	MIY16
8	9	10	11	12	13	14
A25	A26	A27	A28	A29	A30	MA1
PIN10	PIN11	PIN12	PIN13	PIN14	PIN15	PIN16
MIY17	MIY18	MIY19	MIY20	MIY21	MIY22	MIY23
15	16	17	18	19	20	21
MA2	MA3	MA4	MA5	MA6	MA7	MA8
PIN17	PIN18	PIN19	PIN20	PIN21	PIN22	PIN23
MIY24	MIY25	MIY26	MIY27	MIY28	MIY29	MIY30
22	23	24	25	26	27	28
MA9	MA10	MA11	MA12	MA13	MA14	MA15
PIN24	PIN25	PIN26	PIN27	PIN28	PIN29	PIN30
GIN1	GIN2	GIN3	GIN4	GIN5	GIN6	GIN7

UHURU 580/2096-97/6332-33/2088-89

ie	Ojobo	Iwa	Araorun	Ile	Itutu	Erinle
	2	3	4	5	6	7
1	M22	M23	M24	M25	M26	M27
N5	KHN6	KHN7	KHN8	KHN9	KHN10	KHN11
G12	MAG13	MAG14	MAG15	MAG16	MAG17	MAG18
	9	10	11	12	13	14
8	M29	M30	M31	A1	A2	A3
N12	KHN13	KHN14	KHN15	KHN16	KHN17	KHN18
G19	MAG20	MAG21	MAG22	MAG23	MAG24	MAG25
	16	17	18	19	20	21
4	A5	A6	A7	A8	A9	A10
N19	KHN20	KHN21	KHN22	KHN23	KHN24	KHN25
G26	MAG27	MAG28	MAG29	MAG30	MIY1	MIY2
	23	24	25	26	27	28
1	A12	A13	A14	A15	A16	A17
N26	KHN27	KHN28	KHN29	KHN30	PIN1	PIN2
Y3	MIY4	MIY5	MIY6	MIY7	MIY8	MIY9

MABABA WEUSI 580/2096-97/6332-33/2088-89

Ashe	Ojobo	Iwa	Araorun	Ile	Itutu	Erinle
1	2	3	4	5	6	7
MA16	MA17	MA18	MA19	MA20	MA21	MA22
APP1	APP2	APP3	APP4	APP5	APP6	APP7
GIN8	GIN9	GIN10	GIN11	GIN12	GIN13	GIN14
8	9	10	11	12	13	14
MA23	MA24	MA25	MA26	MA27	MA28	MA29
APP8	APP9	APP10	APP11	APP12	APP13	APP14
GIN15	GIN16	GIN17	GIN18	GIN19	GIN20	GIN21
15	16	17	18	19	20	21
MA30	MA31	J1	J2	J3	J4	J5
APP15	APP16	APP17	APP18	APP19	APP20	APP21
GIN22	GIN23	GIN24	GIN25	GIN26	GIN27	GIN28
22	23	24	25	26	27	28
J6	J7	J8	J9	J10	J11	J12
APP22	APP23	APP24	APP25	APP26	APP27	APP28
GIN29	GIN30	SAN1	SAN2	SAN3	SAN4	SAN5

THE NU AFRIKAN CALENDAR

MWANAMKE WEUSI 580/2096-97/6332-33/2088-89						
Ashe	Ojobo	Iwa	Araorun	Ile	Itutu	Erinle
1	2	3	4	5	6	7
J13	J14	J15	J16	J17	J18	J19
APP30	MSR1	MSR2	MSR3	MSR4	MSR5	MSR6
SAN6	SAN7	SAN8	SAN9	SAN10	SAN11	SAN12
8	9	10	11	12	13	14
J20	J21	J22	J23	J24	J25	J26
MSR7	MSR8	MSR9	MSR10	MSR11	MSR12	MSR13
SAN13	SAN14	SAN15	SAN16	SAN17	SAN18	SAN19
15	16	17	18	19	20	21
J27	J28	J29	J30	JU1	JU2	JU3
MSR14	MSR15	MSR16	MSR17	MSR18	MSR19	MSR20
SAN20	SAN21	SAN22	SAN23	SAN24	SAN25	SAN26
22	23	24	25	26	27	28
JU4	JU5	JU6	JU7	JU8	JU9	JU10
MSR21	MRS22	MSR23	MSR24	MSR25	MSR26	MSR27
SAN27	SAN28	SAN29	SAN30	HAM1	HAM2	HAM3

TAIFA WEUSI 580/2096-97/6332-33/2088-89						
Ashe	Ojobo	Iwa	Araorun	Ile	Itutu	Erinle
1	2	3	4	5	6	7
AU8	AU9	AU10	AU11	AU12	AU13	AU14
THT20	THT21	THT22	THT23	THT24	THT25	THT26
NAH2	NAH3	NAH4	NAH5	NAH6	NAH7	NAH8
8	9	10	11	12	13	14
AU15	AU16	AU17	AU18	AU19	AU20	AU21
THT27	THT28	THT29	THT30	PAP1	PAP1	PAP2
NAH9	NAH10	NAH11	NAH12	NAH13	NAH14	NAH15
15	16	17	18	19	20	21
AU22	AU23	AU24	AU25	AU26	AU27	AU28
PAP3	PAP4	PAP5	PAP6	PAP7	PAP8	PAP9
NAH16	NAH17	NAH18	NAH19	NAH20	NAH21	NAH22
22	23	24	25	26	27	28
AU29	AU30	AU31	S1	S2	S3	S4
PAP10	PAP11	PAP12	PAP13	PAP14	PAP15	PAP16
NAH23	NAH24	NAH25	NAH26	NAH27	NAH28	NAH29

MTU WEUSI 580/2096-97/6332-33/2088-89						
Ashe	Ojobo	Iwa	Araorun	Ile	Itutu	Erinle
1	2	3	4	5	6	7
JU11	JU12	JU13	JU14	JU15	JU16	JU17
MSR28	MSR29	MSR30	*ASR1*	*HRU2*	*STH3*	*AST4*
HAM4	HAM5	HAM6	HAM7	HAM8	HAM9	HAM10
8	9	10	11	12	13	14
JU18	JU19	JU20	JU21	JU22	JU23	JU24
NBH5	*ANU6*	THT1	THT2	THT3	THT4	THT5
HAM11	HAM12	HAM13	HAM14	HAM15	HAM16	HAM17
15	16	17	18	19	20	21
JU25	JU26	JU27	JU28	JU29	JU30	JU31
THT6	THT7	THT8	THT9	THT10	THT11	THT12
HAM18	HAM19	HAM20	HAM21	HAM22	HAM23	HAM24
22	23	24	25	26	27	28
AU1	AU2	AU3	AU4	AU5	AU6	AU7
THT13	THT14	THT15	THT16	THT17	THT18	THT19
HAM25	HAM26	HAM27	HAM28	HAM29	HAM30	NAH1

UWEZO 580/2096-97/6332-33/2088-89						
Ashe	Ojobo	Iwa	Araorun	Ile	Itutu	Erinle
1	2	3	4	5	6	7
S5	S6	S7	S8	S9	S10	S11
PAP17	PAP18	PAP19	PAP20	PAP21	PAP22	PAP23
NAH30	*PAG1*	*PAG2*	*PAG3*	*PAG4*	*PAG5*	MAS1
8	9	10	11	12	13	14
S12	S13	S14	S15	S16	S17	S18
PAP24	PAP25	PAP26	PAP27	PAP28	PAP29	PAP30
MAS2	MAS3	MAS4	MAS5	MAS6	MAS7	MAS8
15	16	17	18	19	20	21
S19	S20	S21	S22	S23	S24	S25
HTH1	HTH2	HTH3	HTH4	HTH5	HTH6	HTH7
MAS9	MAS10	MAS11	MAS12	MAS13	MAS14	MAS15
22	23	24	25	26	27	28
S26	S27	S28	S29	S30	O1	O2
HTH8	HTH9	HTH10	HTH11	HTH12	HTH13	HTH14
MAS16	MAS17	MAS18	MAS19	MAS20	MAS21	MAS22

THE NU AFRIKAN CALENDAR

EKUNDU 580/2096-97/6332-33/2088-89						
Ashe	Ojobo	Iwa	Araorun	Ile	Itutu	Erinle
1	2	3	4	5	6	7
O3	O4	O5	O6	O7	O8	O9
HTH15	HTH16	HTH17	HTH18	HTH19	HTH20	HTH21
MAS23	MAS24	MAS25	MAS26	MAS27	MAS28	MAS29
8	9	10	11	12	13	14
O10	O11	O12	O13	O14	O15	O16
HTH22	HTH23	HTH24	HTH25	HTH26	HTH27	HTH28
MAS30	THi1	THi2	THi3	THi4	THi5	THi6
15	16	17	18	19	20	21
O17	O18	O19	O20	O21	O22	O23
HTH29	HTH30	KHK1	KHK2	KHK3	KHK4	KHK5
THi7	THi8	This9	THi10	THi11	THi12	THi13
22	23	24	25	26	27	28
O24	O25	O26	O27	O28	O29	O30
KHK6	KHK7	KHK8	KHK9	KHK10	KHK11	KHK12
Thi14	THi15	THi16	THi17	THi18	THi19	THi20

KIJANI 580/2096-97/6332-33/2088-89						
Ashe	Ojobo	Iwa	Araorun	Ile	Itutu	Erinle
1	2	3	4	5	6	7
N28	N29	N30	D1	D2	D3	D4
TBT11	TBT12	TBT13	TBT14	TBT15	TBT16	TBT17
HID19	HID20	HID21	HID22	HID23	HID24	HID25
8	9	10	11	12	13	14
D5	D6	D7	D8	D9	D10	D11
TBT18	TBT19	TBT20	TBT21	TBT22	TBT23	TBT24
HID26	HID27	HID28	HID29	HID30	TAH1	TAH2
15	16	17	18	19	20	21
D12	D13	D14	D15	D16	D17	D18
TBT25	TBT26	TBT27	TBT28	TBT29	TBT30	MKH1
TAH3	TAH4	TAH5	TAH6	TAH7	TAH8	TAH9
22	23	24	25	26	27	28
D19	D20	D21	D22	D23	D24	D25
MKH2	MKH3	MKH4	MKH5	MKH6	MKH7	MKH8
TAH10	TAH11	TAH12	TAH13	TAH14	TAH15	TAH16

EUSI 580/2096-97/6332-33/2088-89						
Ashe	Ojobo	Iwa	Araorun	Ile	Itutu	Erinle
1	2	3	4	5	6	7
O31	N1	N2	N3	N4	N5	N6
KHK13	KHK14	KHK15	KHK16	KHK17	KHK18	KHK19
THi21	THi22	THi23	THi24	THi25	THi26	THi27
8	9	10	11	12	13	14
N7	N8	N9	N10	N11	N12	N13
KHK20	KHK21	KHK22	KHK23	KHK24	KHK25	KHK26
THi28	THi29	THi30	HID1	HID2	HID3	HID4
15	16	17	18	19	20	21
N14	N15	N16	N17	N18	N19	N20
KHK27	KHK28	KHK29	KHK30	TBT1	TBT2	TBT3
HID5	HID6	HID7	HID8	HID9	HID10	HID11
22	23	24	25	26	27	28
N21	N22	N23	N24	N25	N26	N27
TBT4	TBT5	TBT6	TBT7	TBT8	TBT9	TBT10
HID12	HID13	HID14	HID15	HID16	HID17	HID18

KWANZA 580/2096-97/6332-33/2088-89						
Umoja	Kujichagulia	Ujima	Ujamaa	Nia	Kuumba	Imani
1	2	3	4	5	6	7
D26	D27	D28	D29	D30	D31	JA1
MKH9	MKH10	MKH11	MKH12	MKH13	MKH14	MKH15
TAH17	TAH18	TAH19	TAH20	TAH21	TAH22	TAH23
8	9	10	11	12	13	14
JA2	JA3	JA4	JA5	JA6	JA7	JA8
MKH16	MKH17	MKH18	MKH19	MKH20	MKH21	MKH22
TAH24	TAH25	TAH26	TAH27	TAH28	TAH29	TAH30
15	16	17	18	19	20	21
JA9	JA10	JA11	JA12	JA13	JA14	JA15
MKH23	MKH24	MKH25	KMH26	MKH27	MKH28	MKH29
TIR1	TIR2	TIR3	TIR4	TIR5	TIR6	TIR7
22	23	24	25	26	27	28
JA16	JA17	JA18	JA19	JA20	JA21	JA22
MKH30	AMN1	AMN2	AMN3	AMN4	AMN5	AMN6
TIR8	TIR9	TIR10	TIR11	TIR12	TIR13	TIR14

THE NU AFRIKAN CALENDAR

SHAHIDI 580/2096-97/6332-33/2088-89						
Ashe	Ojobo	Iwa	Araorun	Ile	Itutu	Erinle
1	2	3	4	5	6	7
JA23	JA24	JA25	JA26	JA27	JA28	JA29
AMN7	AMN8	AMN9	AMN10	AMN11	AMN12	AMN13
TIR15	TIR16	TIR17	TIR18	TIR19	TIR20	TIR21
8	9	10	11	12	13	14
JA30	JA31	F1	F2	F3	F4	F5
AMN14	AMN15	AMN16	AMN17	AMN18	AMN19	AMN20
TIR22	TIR23	TIR24	TIR25	TIR26	TIR27	TIR28
15	16	17	18	19	20	21
F6	F7	F8	F9	F10	F 11	F12
AMN21	AMN22	AMN23	AMN24	AMN25	AMN26	AMN28
TIR29	TIR30	YEK1	YEK2	YEK3	YEK4	YEK5
22	23	24	25	26	27	28
F13	F14	F15	F16	F17	F18	F19
AMN29	AMN30	RNT1	RNT2	RNT3	RNT4	RNT5
YEK6	YEK7	YEK8	YEK9	YEK10	YEK11	YEK12
SHAHIDI 131						
F20						
RNT6						
YEK13						

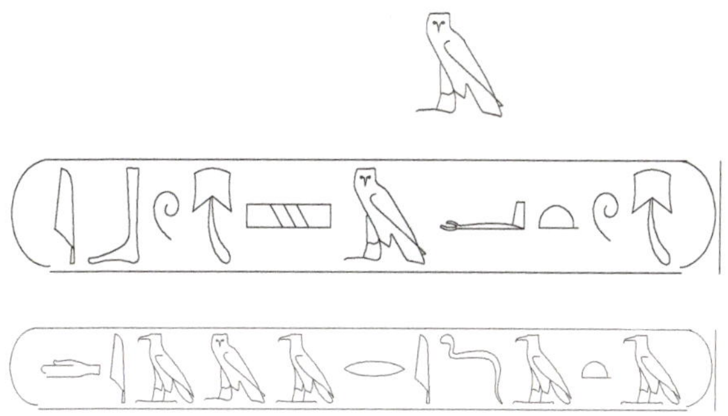

THE NU AFRIKAN CALENDAR

MAAFA 581 B.S.E./ASIENTO 2097-98 C.E.
KAM 6333-34 S.M./EC 2089-90 A.M. (ዓ.ም.)

AFU 581/2097-98/6333-34/2089-90						
Ashe	Ojobo	Iwa	Araorun	Ile	Itutu	Erinle
1	2	3	4	5	6	7
F21	F22	F23	F24	F25	F26	F27
RNT7	RNT8	RNT9	RNT10	RNT11	RNT12	RNT13
YEK14	YEK15	YEK16	YEK17	YEK18	YEK19	YEK20
8	9	10	11	12	13	14
F28	M1	M2	M3	M4	M5	M6
RNT14	RNT15	RNT16	RNT17	RNT18	RNT19	RNT20
YEK21	YEK22	YEK23	YEK24	YEK25	YEK26	YEK27
15	16	17	18	19	20	21
M7	M8	M9	M10	M11	M12	M13
RNT21	RNT22	RNT23	RNT24	RNT25	RNT26	RNT27
YEK28	YEK29	YEK30	MAG1	MAG2	MAG3	MAG4
22	23	24	25	26	27	28
M14	M15	M16	M17	M18	M19	M20
RNT28	RNT29	RNT30	KHN1	KHN2	KHN3	KHN4
MAG5	MAG6	MAG7	MAG8	MAG9	MAG10	MAG11

KAWAIDA 581/2097-98/6333-34/2089-90						
Ashe	Ojobo	Iwa	Araorun	Ile	Itutu	Erinle
1	2	3	4	5	6	7
A18	A19	A20	A21	A22	A23	A24
PIN3	PIN4	PIN5	PIN6	PIN7	PIN8	PIN9
MIY10	MIY11	MIY12	MIY13	MIY14	MIY15	MIY16
8	9	10	11	12	13	14
A25	A26	A27	A28	A29	A30	MA1
PIN10	PIN11	PIN12	PIN13	PIN14	PIN15	PIN16
MIY17	MIY18	MIY19	MIY20	MIY21	MIY22	MIY23
15	16	17	18	19	20	21
MA2	MA3	MA4	MA5	MA6	MA7	MA8
PIN17	PIN18	PIN19	PIN20	PIN21	PIN22	PIN23
MIY24	MIY25	MIY26	MIY27	MIY28	MIY29	MIY30
22	23	24	25	26	27	28
MA9	MA10	MA11	MA12	MA13	MA14	MA15
PIN24	PIN25	PIN26	PIN27	PIN28	PIN29	PIN30
GIN1	GIN2	GIN3	GIN4	GIN5	GIN6	GIN7

UHURU 581/2097-98/6333-34/2089-90						
Ashe	Ojobo	Iwa	Araorun	Ile	Itutu	Erinle
1	2	3	4	5	6	7
M21	M22	M23	M24	M25	M26	M27
KHN5	KHN6	KHN7	KHN8	KHN9	KHN10	KHN11
MAG12	MAG13	MAG14	MAG15	MAG16	MAG17	MAG18
8	9	10	11	12	13	14
M28	M29	M30	M31	A1	A2	A3
KHN12	KHN13	KHN14	KHN15	KHN16	KHN17	KHN18
MAG19	MAG20	MAG21	MAG22	MAG23	MAG24	MAG25
15	16	17	18	19	20	21
A4	A5	A6	A7	A8	A9	A10
KHN19	KHN20	KHN21	KHN22	KHN23	KHN24	KHN25
MAG26	MAG27	MAG28	MAG29	MAG30	MIY1	MIY2
22	23	24	25	26	27	28
A11	A12	A13	A14	A15	A16	A17
KHN26	KHN27	KHN28	KHN29	KHN30	PIN1	PIN2
MIY3	MIY4	MIY5	MIY6	MIY7	MIY8	MIY9

MABABA WEUSI 581/2097-98/6333-34/2089-90						
Ashe	Ojobo	Iwa	Araorun	Ile	Itutu	Erinle
1	2	3	4	5	6	7
MA16	MA17	MA18	MA19	MA20	MA21	MA22
APP1	APP2	APP3	APP4	APP5	APP6	APP7
GIN8	GIN9	GIN10	GIN11	GIN12	GIN13	GIN14
8	9	10	11	12	13	14
MA23	MA24	MA25	MA26	MA27	MA28	MA29
APP8	APP9	APP10	APP11	APP12	APP13	APP14
GIN15	GIN16	GIN17	GIN18	GIN19	GIN20	GIN21
15	16	17	18	19	20	21
MA30	MA31	J1	J2	J3	J4	J5
APP15	APP16	APP17	APP18	APP19	APP20	APP21
GIN22	GIN23	GIN24	GIN25	GIN26	GIN27	GIN28
22	23	24	25	26	27	28
J6	J7	J8	J9	J10	J11	J12
APP22	APP23	APP24	APP25	APP26	APP27	APP28
GIN29	GIN30	SAN1	SAN2	SAN3	SAN4	SAN5

THE NU AFRIKAN CALENDAR

MWANAMKE WEUSI 581/2097-98/6333-34/2089-90						
Ashe	Ojobo	Iwa	Araorun	Ile	Itutu	Erinle
1	2	3	4	5	6	7
J13	J14	J15	J16	J17	J18	J19
APP29	APP30	MSR1	MSR2	MSR3	MSR4	MSR5
SAN6	SAN7	SAN8	SAN9	SAN10	SAN11	SAN12
8	9	10	11	12	13	14
J20	J21	J22	J23	J24	J25	J26
MSR6	MSR7	MSR8	MSR9	MSR10	MSR11	MSR12
SAN13	SAN14	SAN15	SAN16	SAN17	SAN18	SAN19
15	16	17	18	19	20	21
J27	J28	J29	J30	JU1	JU2	JU3
MSR13	MSR14	MSR15	MSR16	MSR17	MSR18	MSR19
SAN20	SAN21	SAN22	SAN23	SAN24	SAN25	SAN26
22	23	24	25	26	27	28
JU4	JU5	JU6	JU7	JU8	JU9	JU10
MSR20	MRS21	MSR22	MSR23	MSR24	MSR25	MSR26
SAN27	SAN28	SAN29	SAN30	HAM1	HAM2	HAM3

TAIFA WEUSI 581/2097-98/6333-34/2089-90						
Ashe	Ojobo	Iwa	Araorun	Ile	Itutu	Erinle
1	2	3	4	5	6	7
AU8	AU9	AU10	AU11	AU12	AU13	AU14
THT20	THT21	THT22	THT23	THT24	THT25	THT26
NAH2	NAH3	NAH4	NAH5	NAH6	NAH7	NAH8
8	9	10	11	12	13	14
AU15	AU16	AU17	AU18	AU19	AU20	AU21
THT27	THT28	THT29	THT30	PAP1	PAP1	PAP2
NAH9	NAH10	NAH11	NAH12	NAH13	NAH14	NAH15
15	16	17	18	19	20	21
AU22	AU23	AU24	AU25	AU26	AU27	AU28
PAP3	PAP4	PAP5	PAP6	PAP7	PAP8	PAP9
NAH16	NAH17	NAH18	NAH19	NAH20	NAH21	NAH22
22	23	24	25	26	27	28
AU29	AU30	AU31	S1	S2	S3	S4
PAP10	PAP11	PAP12	PAP13	PAP14	PAP15	PAP16
NAH23	NAH24	NAH25	NAH26	NAH27	NAH28	NAH29

MTU WEUSI 581/2097-98/6333-34/2089-90						
Ashe	Ojobo	Iwa	Araorun	Ile	Itutu	Erinle
1	2	3	4	5	6	7
JU11	JU12	JU13	JU14	JU15	JU16	JU17
MSR27	MSR28	MSR29	MSR30	*ASR1*	*HRU2*	*STH3*
HAM4	HAM5	HAM6	HAM7	HAM8	HAM9	HAM10
8	9	10	11	12	13	14
JU18	JU19	JU20	JU21	JU22	JU23	JU24
AST4	*NBH5*	THT1	THT2	THT3	THT4	THT5
HAM11	HAM12	HAM13	AM14	HAM15	HAM16	HAM17
15	16	17	18	19	20	21
JU25	JU26	JU27	JU28	JU29	JU30	JU31
THT6	THT7	THT8	THT9	THT10	THT11	THT12
HAM18	HAM19	HAM20	HAM21	HAM22	HAM23	HAM24
22	23	24	25	26	27	28
AU1	AU2	AU3	AU4	AU5	AU6	AU7
THT13	THT14	THT15	THT16	THT17	THT18	THT19
HAM25	HAM26	HAM27	HAM28	HAM29	HAM30	NAH1

UWEZO 581/2097-98/6333-34/2089-90						
Ashe	Ojobo	Iwa	Araorun	Ile	Itutu	Erinle
1	2	3	4	5	6	7
S5	S6	S7	S8	S9	S10	S11
PAP17	PAP18	PAP19	PAP20	PAP21	PAP22	PAP23
NAH30	*PAG1*	*PAG2*	*PAG3*	*PAG4*	*PAG5*	MAS1
8	9	10	11	12	13	14
S12	S13	S14	S15	S16	S17	S18
PAP24	PAP25	PAP26	PAP27	PAP28	PAP29	PAP30
MAS2	MAS3	MAS4	MAS5	MAS6	MAS7	MAS8
15	16	17	18	19	20	21
S19	S20	S21	S22	S23	S24	S25
HTH1	HTH2	HTH3	HTH4	HTH5	HTH6	HTH7
MAS9	MAS10	MAS11	MAS12	MAS13	MAS14	MAS15
22	23	24	25	26	27	28
S26	S27	S28	S29	S30	O1	O2
HTH8	HTH9	HTH10	HTH11	HTH12	HTH13	HTH14
MAS16	MAS17	MAS18	MAS19	MAS20	MAS21	MAS22

THE NU AFRIKAN CALENDAR

EKUNDU 581/2097-98/6333-34/2089-90						
Ashe	Ojobo	Iwa	Araorun	Ile	Itutu	Erinle
1	2	3	4	5	6	7
O3	O4	O5	O6	O7	O8	O9
HTH15	HTH16	HTH17	HTH18	HTH19	HTH20	HTH21
MAS23	MAS24	MAS25	MAS26	MAS27	MAS28	MAS29
8	9	10	11	12	13	14
O10	O11	O12	O13	O14	O15	O16
HTH22	HTH23	HTH24	HTH25	HTH26	HTH27	HTH28
MAS30	THi1	THi2	THi3	THi4	THi5	THi6
15	16	17	18	19	20	21
O17	O18	O19	O20	O21	O22	O23
HTH29	HTH30	KHK1	KHK2	KHK3	KHK4	KHK5
THi7	THi8	THi9	THi10	THi11	THi12	THi13
22	23	24	25	26	27	28
O24	O25	O26	O27	O28	O29	O30
KHK6	KHK7	KHK8	KHK9	KHK10	KHK11	KHK12
THi14	THi15	THi16	THi17	THi18	THi19	THi20

KIJANI 581/2097-98/6333-34/2089-90						
Ashe	Ojobo	Iwa	Araorun	Ile	Itutu	Erinle
1	2	3	4	5	6	7
N28	N29	N30	D1	D2	D3	D4
TBT11	TBT12	TBT13	TBT14	TBT15	TBT16	TBT17
HID19	HID20	HID21	HID22	HID23	HID24	HID25
8	9	10	11	12	13	14
D5	D6	D7	D8	D9	D10	D11
TBT18	TBT19	TBT20	TBT21	TBT22	TBT23	TBT24
HID26	HID27	HID28	HID29	HID30	TAH1	TAH2
15	16	17	18	19	20	21
D12	D13	D14	D15	D16	D17	D18
TBT25	TBT26	TBT27	TBT28	TBT29	TBT30	MKH1
TAH3	TAH4	TAH5	TAH6	TAH7	TAH8	TAH9
22	23	24	25	26	27	28
D19	D20	D21	D22	D23	D24	D25
MKH2	MKH3	MKH4	MKH5	MKH6	MKH7	MKH8
TAH10	TAH11	TAH12	TAH13	TAH14	TAH15	TAH16

EUSI 581/2097-98/6333-34/2089-90						
Ashe	Ojobo	Iwa	Araorun	Ile	Itutu	Erinle
1	2	3	4	5	6	7
O31	N1	N2	N3	N4	N5	N6
KHK13	KHK14	KHK15	KHK16	KHK17	KHK18	KHK19
THi21	THi22	THi23	THi24	THi25	THi26	THi27
8	9	10	11	12	13	14
N7	N8	N9	N10	N11	N12	N13
KHK20	KHK21	KHK22	KHK23	KHK24	KHK25	KHK26
THi28	THi29	THi30	HID1	HID2	HID3	HID4
15	16	17	18	19	20	21
N14	N15	N16	N17	N18	N19	N20
KHK27	KHK28	KHK29	KHK30	TBT1	TBT2	TBT3
HID5	HID6	HID7	HID8	HID9	HID10	HID11
22	23	24	25	26	27	28
N21	N22	N23	N24	N25	N26	N27
TBT4	TBT5	TBT6	TBT7	TBT8	TBT9	TBT10
HID12	HID13	HID14	HID15	HID16	HID17	HID18

KWANZA 581/2097-98/6333-34/2089-90						
Umoja	Kujichagulia	Ujima	Ujamaa	Nia	Kuumba	Imani
1	2	3	4	5	6	7
D26	D27	D28	D29	D30	D31	JA1
MKH9	MKH10	MKH11	MKH12	MKH13	MKH14	MKH15
TAH17	TAH18	TAH19	TAH20	TAH21	TAH22	TAH23
8	9	10	11	12	13	14
JA2	JA3	JA4	JA5	JA6	JA7	JA8
MKH16	MKH17	MKH18	MKH19	MKH20	MKH21	MKH22
TAH24	TAH25	TAH26	TAH27	TAH28	TAH29	TAH30
15	16	17	18	19	20	21
JA9	JA10	JA11	JA12	JA13	JA14	JA15
MKH23	MKH24	MKH25	KMH26	MKH27	MKH28	MKH29
TIR1	TIR2	TIR3	TIR4	TIR5	TIR6	TIR7
22	23	24	25	26	27	28
JA16	JA17	JA18	JA19	JA20	JA21	JA22
MKH30	AMN1	AMN2	AMN3	AMN4	AMN5	AMN6
TIR8	TIR9	TIR10	TIR11	TIR12	TIR13	TIR14

THE NU AFRIKAN CALENDAR

SHAHIDI 581/2097-98/6333-34/2089-90						
Ashe	Ojobo	Iwa	Araorun	Ile	Itutu	Erinle
1	2	3	4	5	6	7
JA23	JA24	JA25	JA26	JA27	JA28	JA29
AMN7	AMN8	AMN9	AMN10	AMN11	AMN12	AMN13
TIR15	TIR16	TIR17	TIR18	TIR19	TIR20	TIR21
8	9	10	11	12	13	14
JA30	JA31	F1	F2	F3	F4	F5
AMN14	AMN15	AMN16	AMN17	AMN18	AMN19	AMN20
TIR22	TIR23	TIR24	TIR25	TIR26	TIR27	TIR28
15	16	17	18	19	20	21
F6	F7	F8	F9	F10	F 11	F12
AMN21	AMN22	AMN23	AMN24	AMN25	AMN26	AMN28
TIR29	TIR30	YEK1	YEK2	YEK3	YEK4	YEK5
22	23	24	25	26	27	28
F13	F14	F15	F16	F17	F18	F19
AMN29	AMN30	RNT1	RNT2	RNT3	RNT4	RNT5
YEK6	YEK7	YEK8	YEK9	YEK10	YEK11	YEK12
SHAHIDI 132						
F20						
RNT6						
YEK13						

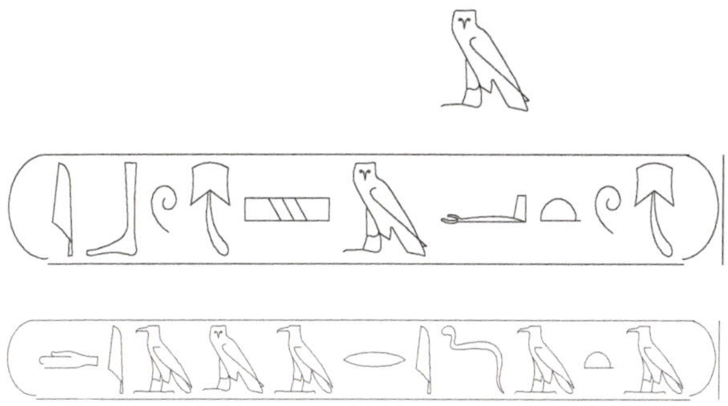

THE NU AFRIKAN CALENDAR

MAAFA 582 B.S.E./ASIENTO 2098-99 C.E.
KAM 6334-35 S.M./EC 2090-91 A.M. (ዓ.ም.)

AFU 582/2098-99/6334-35/2090-91

Ashe	Ojobo	Iwa	Araorun	Ile	Itutu	Erinle
1	2	3	4	5	6	7
F21	F22	F23	F24	F25	F26	F27
RNT7	RNT8	RNT9	RNT10	RNT11	RNT12	RNT13
YEK14	YEK15	YEK16	YEK17	YEK18	YEK19	YEK20
8	9	10	11	12	13	14
F28	M1	M2	M3	M4	M5	M6
RNT14	RNT15	RNT16	RNT17	RNT18	RNT19	RNT20
YEK21	YEK22	YEK23	YEK24	YEK25	YEK26	YEK27
15	16	17	18	19	20	21
M7	M8	M9	M10	M11	M12	M13
RNT21	RNT22	RNT23	RNT24	RNT25	RNT26	RNT27
YEK28	YEK29	YEK30	MAG1	MAG2	MAG3	MAG4
22	23	24	25	26	27	28
M14	M15	M16	M17	M18	M19	M20
RNT28	RNT29	RNT30	KHN1	KHN2	KHN3	KHN4
MAG5	MAG6	MAG7	MAG8	MAG9	MAG10	MAG11

KAWAIDA 582/2098-99/6334-35/2090-91

Ashe	Ojobo	Iwa	Araorun	Ile	Itutu	Erinle
1	2	3	4	5	6	7
A18	A19	A20	A21	A22	A23	A24
PIN3	PIN4	PIN5	PIN6	PIN7	PIN8	PIN9
MIY10	MIY11	MIY12	MIY13	MIY14	MIY15	MIY16
8	9	10	11	12	13	14
A25	A26	A27	A28	A29	A30	MA1
PIN10	PIN11	PIN12	PIN13	PIN14	PIN15	PIN16
MIY17	MIY18	MIY19	MIY20	MIY21	MIY22	MIY23
15	16	17	18	19	20	21
MA2	MA3	MA4	MA5	MA6	MA7	MA8
PIN17	PIN18	PIN19	PIN20	PIN21	PIN22	PIN23
MIY24	MIY25	MIY26	MIY27	MIY28	MIY29	MIY30
22	23	24	25	26	27	28
MA9	MA10	MA11	MA12	MA13	MA14	MA15
PIN24	PIN25	PIN26	PIN27	PIN28	PIN29	PIN30
GIN1	GIN2	GIN3	GIN4	GIN5	GIN6	GIN7

UHURU 582/2098-99/6334-35/2090-91

Ashe	Ojobo	Iwa	Araorun	Ile	Itutu	Erinle
1	2	3	4	5	6	7
M21	M22	M23	M24	M25	M26	M27
KHN5	KHN6	KHN7	KHN8	KHN9	KHN10	KHN11
MAG12	MAG13	MAG14	MAG15	MAG16	MAG17	MAG18
8	9	10	11	12	13	14
M28	M29	M30	M31	A1	A2	A3
KHN12	KHN13	KHN14	KHN15	KHN16	KHN17	KHN18
MAG19	MAG20	MAG21	MAG22	MAG23	MAG24	MAG25
15	16	17	18	19	20	21
A4	A5	A6	A7	A8	A9	A10
KHN19	KHN20	KHN21	KHN22	KHN23	KHN24	KHN25
MAG26	MAG27	MAG28	MAG29	MAG30	MIY1	MIY2
22	23	24	25	26	27	28
A11	A12	A13	A14	A15	A16	A17
KHN26	KHN27	KHN28	KHN29	KHN30	PIN1	PIN2
MIY3	MIY4	MIY5	MIY6	MIY7	MIY8	MIY9

MABABA WEUSI 582/2098-99/6334-35/2090-91

Ashe	Ojobo	Iwa	Araorun	Ile	Itutu	Erinle
1	2	3	4	5	6	7
MA16	MA17	MA18	MA19	MA20	MA21	MA22
APP1	APP2	APP3	APP4	APP5	APP6	APP7
GIN8	GIN9	GIN10	GIN11	GIN12	GIN13	GIN14
8	9	10	11	12	13	14
MA23	MA24	MA25	MA26	MA27	MA28	MA29
APP8	APP9	APP10	APP11	APP12	APP13	APP14
GIN15	GIN16	GIN17	GIN18	GIN19	GIN20	GIN21
15	16	17	18	19	20	21
MA30	MA31	J1	J2	J3	J4	J5
APP15	APP16	APP17	APP18	APP19	APP20	APP21
GIN22	GIN23	GIN24	GIN25	GIN26	GIN27	GIN28
22	23	24	25	26	27	28
J6	J7	J8	J9	J10	J11	J12
APP22	APP23	APP24	APP25	APP26	APP27	APP28
GIN29	GIN30	SAN1	SAN2	SAN3	SAN4	SAN5

THE NU AFRIKAN CALENDAR

EKUNDU 582/2098-99/6334-35/2090-91

Ashe	Ojobo	Iwa	Araorun	Ile	Itutu	Erinle
1	2	3	4	5	6	7
O3	O4	O5	O6	O7	O8	O9
HTH15	HTH16	HTH17	HTH18	HTH19	HTH20	HTH21
MAS23	MAS24	MAS25	MAS26	MAS27	MAS28	MAS29
8	9	10	11	12	13	14
O10	O11	O12	O13	O14	O15	O16
HTH22	HTH23	HTH24	HTH25	HTH26	HTH27	HTH28
MAS30	THi1	THi2	THi3	THi4	THi5	THi6
15	16	17	18	19	20	21
O17	O18	O19	O20	O21	O22	O23
HTH29	HTH30	KHK1	KHK2	KHK3	KHK4	KHK5
THi7	THi8	THi9	THi10	THi11	THi12	THi13
22	23	24	25	26	27	28
O24	O25	O26	O27	O28	O29	O30
KHK6	KHK7	KHK8	KHK9	KHK10	KHK11	KHK12
THi14	THi15	THi16	THi17	THi18	THi19	THi20

KIJANI 582/2098-99/6334-35/2090-91

Ashe	Ojobo	Iwa	Araorun	Ile	Itutu	Erinle
1	2	3	4	5	6	7
N28	N29	N30	D1	D2	D3	D4
TBT11	TBT12	TBT13	TBT14	TBT15	TBT16	TBT17
HID19	HID20	HID21	HID22	HID23	HID24	HID25
8	9	10	11	12	13	14
D5	D6	D7	D8	D9	D10	D11
TBT18	TBT19	TBT20	TBT21	TBT22	TBT23	TBT24
HID26	HID27	HID28	HID29	HID30	TAH1	TAH2
15	16	17	18	19	20	21
D12	D13	D14	D15	D16	D17	D18
TBT25	TBT26	TBT27	TBT28	TBT29	TBT30	MKH1
TAH3	TAH4	TAH5	TAH6	TAH7	TAH8	TAH9
22	23	24	25	26	27	28
D19	D20	D21	D22	D23	D24	D25
MKH2	MKH3	MKH4	MKH5	MKH6	MKH7	MKH8
TAH10	TAH11	TAH12	TAH13	TAH14	TAH15	TAH16

EUSI 582/2098-99/6334-35/2090-91

Ashe	Ojobo	Iwa	Araorun	Ile	Itutu	Erinle
1	2	3	4	5	6	7
O31	N1	N2	N3	N4	N5	N6
KHK13	KHK14	KHK15	KHK16	KHK17	KHK18	KHK19
THi21	THi22	THi23	THi24	THi25	THi26	THi27
8	9	10	11	12	13	14
N7	N8	N9	N10	N11	N12	N13
KHK20	KHK21	KHK22	KHK23	KHK24	KHK25	KHK26
THi28	THi29	THi30	HID1	HID2	HID3	HID4
15	16	17	18	19	20	21
N14	N15	N16	N17	N18	N19	N20
KHK27	KHK28	KHK29	KHK30	TBT1	TBT2	TBT3
HID5	HID6	HID7	HID8	HID9	HID10	HID11
22	23	24	25	26	27	28
N21	N22	N23	N24	N25	N26	N27
TBT4	TBT5	TBT6	TBT7	TBT8	TBT9	TBT10
HID12	HID13	HID14	HID15	HID16	HID17	HID18

KWANZA 582/2098-99/6334-35/2090-91

Umoja	Kujichagulia	Ujima	Ujamaa	Nia	Kuumba	Imani
1	2	3	4	5	6	7
D26	D27	D28	D29	D30	D31	JA1
MKH9	MKH10	MKH11	MKH12	MKH13	MKH14	MKH1
TAH17	TAH18	TAH19	TAH20	TAH21	TAH22	TAH23
8	9	10	11	12	13	14
JA2	JA3	JA4	JA5	JA6	JA7	JA8
MKH16	MKH17	MKH18	MKH19	MKH20	MKH21	MKH2
TAH24	TAH25	TAH26	TAH27	TAH28	TAH29	TAH30
15	16	17	18	19	20	21
JA9	JA10	JA11	JA12	JA13	JA14	JA15
MKH23	MKH24	MKH25	KMH26	MKH27	MKH28	MKH2
TIR1	TIR2	TIR3	TIR4	TIR5	TIR6	TIR7
22	23	24	25	26	27	28
JA16	JA17	JA18	JA19	JA20	JA21	JA22
MKH30	AMN1	AMN2	AMN3	AMN4	AMN5	AMN6
TIR8	TIR9	TIR10	TIR11	TIR12	TIR13	TIR14

THE NU AFRIKAN CALENDAR

SHAHIDI 582/2098-99/6334-35/2090-91						
Ashe	Ojobo	Iwa	Araorun	Ile	Itutu	Erinle
1	2	3	4	5	6	7
JA23	JA24	JA25	JA26	JA27	JA28	JA29
AMN7	AMN8	AMN9	AMN10	AMN11	AMN12	AMN13
TIR15	TIR16	TIR17	TIR18	TIR19	TIR20	TIR21
8	9	10	11	12	13	14
JA30	JA31	F1	F2	F3	F4	F5
AMN14	AMN15	AMN16	AMN17	AMN18	AMN19	AMN20
TIR22	TIR23	TIR24	TIR25	TIR26	TIR27	TIR28
15	16	17	18	19	20	21
F6	F7	F8	F9	F10	F 11	F12
AMN21	AMN22	AMN23	AMN24	AMN25	AMN26	AMN28
TIR29	TIR30	YEK1	YEK2	YEK3	YEK4	YEK5
22	23	24	25	26	27	28
F13	F14	F15	F16	F17	F18	F19
AMN29	AMN30	RNT1	RNT2	RNT3	RNT4	RNT5
YEK6	YEK7	YEK8	YEK9	YEK10	YEK11	YEK12
SHAHIDI 133						
F20						
RNT6						
YEK13						

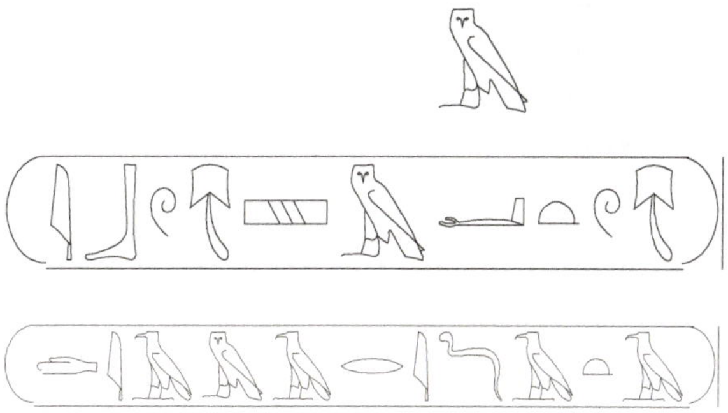

THE NU AFRIKAN CALENDAR

MAAFA 583 B.S.E./ASIENTO 2099-00 C.E.
KAM 6335-36 S.M./EC 2091-92 A.M. (ዓ.ም.)

AFU 583/2099-00/6335-36/2091-92						
Ashe	Ojobo	Iwa	Araorun	Ile	Itutu	Erinle
1	2	3	4	5	6	7
F21	F22	F23	F24	F25	F26	F27
RNT7	RNT8	RNT9	RNT10	RNT11	RNT12	RNT13
YEK14	YEK15	YEK16	YEK17	YEK18	YEK19	YEK20
8	9	10	11	12	13	14
F28	M1	M2	M3	M4	M5	M6
RNT14	RNT15	RNT16	RNT17	RNT18	RNT19	RNT20
YEK21	YEK22	YEK23	YEK24	YEK25	YEK26	YEK27
15	16	17	18	19	20	21
M7	M8	M9	M10	M11	M12	M13
RNT21	RNT22	RNT23	RNT24	RNT25	RNT26	RNT27
YEK28	YEK29	YEK30	MAG1	MAG2	MAG3	MAG4
22	23	24	25	26	27	28
M14	M15	M16	M17	M18	M19	M20
RNT28	RNT29	RNT30	KHN1	KHN2	KHN3	KHN4
MAG5	MAG6	MAG7	MAG8	MAG9	MAG10	MAG11

KAWAIDA 583/2099-00/6335-36/2091-92						
Ashe	Ojobo	Iwa	Araorun	Ile	Itutu	Erinle
1	2	3	4	5	6	7
A18	A19	A20	A21	A22	A23	A24
PIN3	PIN4	PIN5	PIN6	PIN7	PIN8	PIN9
MIY10	MIY11	MIY12	MIY13	MIY14	MIY15	MIY16
8	9	10	11	12	13	14
A25	A26	A27	A28	A29	A30	MA1
PIN10	PIN11	PIN12	PIN13	PIN14	PIN15	PIN16
MIY17	MIY18	MIY19	MIY20	MIY21	MIY22	MIY23
15	16	17	18	19	20	21
MA2	MA3	MA4	MA5	MA6	MA7	MA8
PIN17	PIN18	PIN19	PIN20	PIN21	PIN22	PIN23
MIY24	MIY25	MIY26	MIY27	MIY28	MIY29	MIY30
22	23	24	25	26	27	28
MA9	MA10	MA11	MA12	MA13	MA14	MA15
PIN24	PIN25	PIN26	PIN27	PIN28	PIN29	PIN30
GIN1	GIN2	GIN3	GIN4	GIN5	GIN6	GIN7

UHURU 583/2099-00/6335-36/2091-92						
Ashe	Ojobo	Iwa	Araorun	Ile	Itutu	Erinle
1	2	3	4	5	6	7
M21	M22	M23	M24	M25	M26	M27
KHN5	KHN6	KHN7	KHN8	KHN9	KHN10	KHN11
MAG12	MAG13	MAG14	MAG15	MAG16	MAG17	MAG18
8	9	10	11	12	13	14
M28	M29	M30	M31	A1	A2	A3
KHN12	KHN13	KHN14	KHN15	KHN16	KHN17	KHN18
MAG19	MAG20	MAG21	MAG22	MAG23	MAG24	MAG25
15	16	17	18	19	20	21
A4	A5	A6	A7	A8	A9	A10
KHN19	KHN20	KHN21	KHN22	KHN23	KHN24	KHN25
MAG26	MAG27	MAG28	MAG29	MAG30	MIY1	MIY2
22	23	24	25	26	27	28
A11	A12	A13	A14	A15	A16	A17
KHN26	KHN27	KHN28	KHN29	KHN30	PIN1	PIN2
MIY3	MIY4	MIY5	MIY6	MIY7	MIY8	MIY9

MABABA WEUSI 583/2099-00/6335-36/2091-92						
Ashe	Ojobo	Iwa	Araorun	Ile	Itutu	Erinle
1	2	3	4	5	6	7
MA16	MA17	MA18	MA19	MA20	MA21	MA22
APP1	APP2	APP3	APP4	APP5	APP6	APP7
GIN8	GIN9	GIN10	GIN11	GIN12	GIN13	GIN14
8	9	10	11	12	13	14
MA23	MA24	MA25	MA26	MA27	MA28	MA29
APP8	APP9	APP10	APP11	APP12	APP13	APP14
GIN15	GIN16	GIN17	GIN18	GIN19	GIN20	GIN21
15	16	17	18	19	20	21
MA30	MA31	J1	J2	J3	J4	J5
APP15	APP16	APP17	APP18	APP19	APP20	APP21
GIN22	GIN23	GIN24	GIN25	GIN26	GIN27	GIN28
22	23	24	25	26	27	28
J6	J7	J8	J9	J10	J11	J12
APP22	APP23	APP24	APP25	APP26	APP27	APP28
GIN29	GIN30	SAN1	SAN2	SAN3	SAN4	SAN5

THE NU AFRIKAN CALENDAR

MWANAMKE WEUSI 583/2099-00/6335-36/2091-92						
Ashe	Ojobo	Iwa	Araorun	Ile	Itutu	Erinle
1	2	3	4	5	6	7
J13	J14	J15	J16	J17	J18	J19
APP29	APP30	MSR1	MSR2	MSR3	MSR4	MSR5
SAN6	SAN7	SAN8	SAN9	SAN10	SAN11	SAN12
8	9	10	11	12	13	14
J20	J21	J22	J23	J24	J25	J26
MSR6	MSR7	MSR8	MSR9	MSR10	MSR11	MSR12
SAN13	SAN14	SAN15	SAN16	SAN17	SAN18	SAN19
15	16	17	18	19	20	21
J27	J28	J29	J30	JU1	JU2	JU3
MSR13	MSR14	MSR15	MSR16	MSR17	MSR18	MSR19
SAN20	SAN21	SAN22	SAN23	SAN24	SAN25	SAN26
22	23	24	25	26	27	28
JU4	JU5	JU6	JU7	JU8	JU9	JU10
MSR20	MRS21	MSR22	MSR23	MSR24	MSR25	MSR26
SAN27	SAN28	SAN29	SAN30	HAM1	HAM2	HAM3

TAIFA WEUSI 583/2099-00/6335-36/2091-92						
Ashe	Ojobo	Iwa	Araorun	Ile	Itutu	Erinle
1	2	3	4	5	6	7
AU8	AU9	AU10	AU11	AU12	AU13	AU14
THT20	THT21	THT22	THT23	THT24	THT25	THT26
NAH2	NAH3	NAH4	NAH5	NAH6	NAH7	NAH8
8	9	10	11	12	13	14
AU15	AU16	AU17	AU18	AU19	AU20	AU21
THT27	THT28	THT29	THT30	PAP1	PAP1	PAP2
NAH9	NAH10	NAH11	NAH12	NAH13	NAH14	NAH15
15	16	17	18	19	20	21
AU22	AU23	AU24	AU25	AU26	AU27	AU28
PAP3	PAP4	PAP5	PAP6	PAP7	PAP8	PAP9
NAH16	NAH17	NAH18	NAH19	NAH20	NAH21	NAH22
22	23	24	25	26	27	28
AU29	AU30	AU31	S1	S2	S3	S4
PAP10	PAP11	PAP12	PAP13	PAP14	PAP15	PAP16
NAH23	NAH24	NAH25	NAH26	NAH27	NAH28	NAH29

MTU WEUSI 583/2099-00/6335-36/2091-92						
Ashe	Ojobo	Iwa	Araorun	Ile	Itutu	Erinle
1	2	3	4	5	6	7
JU11	JU12	JU13	JU14	JU15	JU16	JU17
MSR27	MSR28	MSR29	MSR30	*ASR1*	*HRU2*	*STH3*
HAM4	HAM5	HAM6	HAM7	HAM8	HAM9	HAM10
8	9	10	11	12	13	14
JU18	JU19	JU20	JU21	JU22	JU23	JU24
AST4	*NBH5*	THT1	THT2	THT3	THT4	THT5
HAM11	HAM12	HAM13	AM14	HAM15	HAM16	HAM17
15	16	17	18	19	20	21
JU25	JU26	JU27	JU28	JU29	JU30	JU31
THT6	THT7	THT8	THT9	THT10	THT11	THT12
HAM18	HAM19	HAM20	HAM21	HAM22	HAM23	HAM24
22	23	24	25	26	27	28
AU1	AU2	AU3	AU4	AU5	AU6	AU7
THT13	THT14	THT15	THT16	THT17	THT18	THT19
HAM25	HAM26	HAM27	HAM28	HAM29	HAM30	NAH1

UWEZO 583/2099-00/6335-36/2091-92						
Ashe	Ojobo	Iwa	Araorun	Ile	Itutu	Erinle
1	2	3	4	5	6	7
S5	S6	S7	S8	S9	S10	S11
PAP17	PAP18	PAP19	PAP20	PAP21	PAP22	PAP23
NAH30	*PAG1*	*PAG2*	*PAG3*	*PAG4*	*PAG5*	MAS1
8	9	10	11	12	13	14
S12	S13	S14	S15	S16	S17	S18
PAP24	PAP25	PAP26	PAP27	PAP28	PAP29	PAP30
MAS2	MAS3	MAS4	MAS5	MAS6	MAS7	MAS8
15	16	17	18	19	20	21
S19	S20	S21	S22	S23	S24	S25
HTH1	HTH2	HTH3	HTH4	HTH5	HTH6	HTH7
MAS9	MAS10	MAS11	MAS12	MAS13	MAS14	MAS15
22	23	24	25	26	27	28
S26	S27	S28	S29	S30	O1	O2
HTH8	HTH9	HTH10	HTH11	HTH12	HTH13	HTH14
MAS16	MAS17	MAS18	MAS19	MAS20	MAS21	MAS22

THE NU AFRIKAN CALENDAR

EKUNDU 583/2099-00/6335-36/2091-92						
Ashe	Ojobo	Iwa	Araorun	Ile	Itutu	Erinle
1	2	3	4	5	6	7
O3	O4	O5	O6	O7	O8	O9
HTH15	HTH16	HTH17	HTH18	HTH19	HTH20	HTH21
MAS23	MAS24	MAS25	MAS26	MAS27	MAS28	MAS29
8	9	10	11	12	13	14
O10	O11	O12	O13	O14	O15	O16
HTH22	HTH23	HTH24	HTH25	HTH26	HTH27	HTH28
MAS30	THi1	THi2	THi3	THi4	THi5	THi6
15	16	17	18	19	20	21
O17	O18	O19	O20	O21	O22	O23
HTH29	HTH30	KHK1	KHK2	KHK3	KHK4	KHK5
THi7	THi8	THi9	THi10	THi11	THi12	THi13
22	23	24	25	26	27	28
O24	O25	O26	O27	O28	O29	O30
KHK6	KHK7	KHK8	KHK9	KHK10	KHK11	KHK12
THi14	THi15	THi16	THi17	THi18	THi19	THi20

KIJANI 583/2099-00/6335-36/2091-92						
Ashe	Ojobo	Iwa	Araorun	Ile	Itutu	Erinle
1	2	3	4	5	6	7
N28	N29	N30	D1	D2	D3	D4
TBT11	TBT12	TBT13	TBT14	TBT15	TBT16	TBT17
HID19	HID20	HID21	HID22	HID23	HID24	HID25
8	9	10	11	12	13	14
D5	D6	D7	D8	D9	D10	D11
TBT18	TBT19	TBT20	TBT21	TBT22	TBT23	TBT24
HID26	HID27	HID28	HID29	HID30	TAH1	TAH2
15	16	17	18	19	20	21
D12	D13	D14	D15	D16	D17	D18
TBT25	TBT26	TBT27	TBT28	TBT29	TBT30	MKH1
TAH3	TAH4	TAH5	TAH6	TAH7	TAH8	TAH9
22	23	24	25	26	27	28
D19	D20	D21	D22	D23	D24	D25
MKH2	MKH3	MKH4	MKH5	MKH6	MKH7	MKH8
TAH10	TAH11	TAH12	TAH13	TAH14	TAH15	TAH16

EUSI 583/2099-00/6335-36/2091-92						
Ashe	Ojobo	Iwa	Araorun	Ile	Itutu	Erinle
1	2	3	4	5	6	7
O31	N1	N2	N3	N4	N5	N6
KHK13	KHK14	KHK15	KHK16	KHK17	KHK18	KHK19
THi21	THi22	THi23	THi24	THi25	THi26	THi27
8	9	10	11	12	13	14
N7	N8	N9	N10	N11	N12	N13
KHK20	KHK21	KHK22	KHK23	KHK24	KHK25	KHK26
THi28	THi29	THi30	HID1	HID2	HID3	HID4
15	16	17	18	19	20	21
N14	N15	N16	N17	N18	N19	N20
KHK27	KHK28	KHK29	KHK30	TBT1	TBT2	TBT3
HID5	HID6	HID7	HID8	HID9	HID10	HID11
22	23	24	25	26	27	28
N21	N22	N23	N24	N25	N26	N27
TBT4	TBT5	TBT6	TBT7	TBT8	TBT9	TBT10
HID12	HID13	HID14	HID15	HID16	HID17	HID18

KWANZA 583/2099-00/6335-36/2091-92						
Umoja	Kujichagulia	Ujima	Ujamaa	Nia	Kuumba	Imani
1	2	3	4	5	6	7
D26	D27	D28	D29	D30	D31	JA1
MKH9	MKH10	MKH11	MKH12	MKH13	MKH14	MKH15
TAH17	TAH18	TAH19	TAH20	TAH21	TAH22	TAH23
8	9	10	11	12	13	14
JA2	JA3	JA4	JA5	JA6	JA7	JA8
MKH16	MKH17	MKH18	MKH19	MKH20	MKH21	MKH22
TAH24	TAH25	TAH26	TAH27	TAH28	TAH29	TAH30
15	16	17	18	19	20	21
JA9	JA10	JA11	JA12	JA13	JA14	JA15
MKH23	MKH24	MKH25	KMH26	MKH27	MKH28	MKH29
TIR1	TIR2	TIR3	TIR4	TIR5	TIR6	TIR7
22	23	24	25	26	27	28
JA16	JA17	JA18	JA19	JA20	JA21	JA22
MKH30	AMN1	AMN2	AMN3	AMN4	AMN5	AMN6
TIR8	TIR9	TIR10	TIR11	TIR12	TIR13	TIR14

THE NU AFRIKAN CALENDAR

SHAHIDI 583/2099-00/6335-36/2091-92						
Ashe	Ojobo	Iwa	Araorun	Ile	Itutu	Erinle
1	2	3	4	5	6	7
JA23	JA24	JA25	JA26	JA27	JA28	JA29
AMN7	AMN8	AMN9	AMN10	AMN11	AMN12	AMN13
TIR15	TIR16	TIR17	TIR18	TIR19	TIR20	TIR21
8	9	10	11	12	13	14
JA30	JA31	F1	F2	F3	F4	F5
AMN14	AMN15	AMN16	AMN17	AMN18	AMN19	AMN20
TIR22	TIR23	TIR24	TIR25	TIR26	TIR27	TIR28
15	16	17	18	19	20	21
F6	F7	F8	F9	F10	F 11	F12
AMN21	AMN22	AMN23	AMN24	AMN25	AMN26	AMN28
TIR29	TIR30	YEK1	YEK2	YEK3	YEK4	YEK5
22	23	24	25	26	27	28
F13	F14	F15	F16	F17	F18	F19
AMN29	AMN30	RNT1	RNT2	RNT3	RNT4	RNT5
YEK6	YEK7	YEK8	YEK9	YEK10	YEK11	YEK12
SHAHIDI 134						
F20						
RNT6						
YEK13						

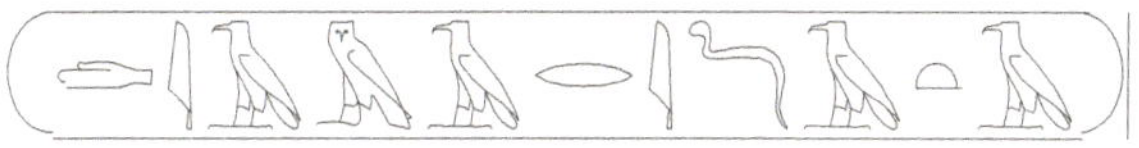

THE NU AFRIKAN CALENDAR

MAAFA 584 B.S.E./ASIENTO 2100-01 C.E.
KAM 6336-37 S.M./EC 2092-93 A.M. (ዓ.ም.)

AFU 584/2100-01/6336-37/2092-93

Ashe	Ojobo	Iwa	Araorun	Ile	Itutu	Erinle
1	2	3	4	5	6	7
F21	F22	F23	F24	F25	F26	F27
RNT7	RNT8	RNT9	RNT10	RNT11	RNT12	RNT13
YEK14	YEK15	YEK16	YEK17	YEK18	YEK19	YEK20
8	9	10	11	12	13	14
F28	M1	M2	M3	M4	M5	M6
RNT14	RNT15	RNT16	RNT17	RNT18	RNT19	RNT20
YEK21	YEK22	YEK23	YEK24	YEK25	YEK26	YEK27
15	16	17	18	19	20	21
M7	M8	M9	M10	M11	M12	M13
RNT21	RNT22	RNT23	RNT24	RNT25	RNT26	RNT27
YEK28	YEK29	YEK30	MAG1	MAG2	MAG3	MAG4
22	23	24	25	26	27	28
M14	M15	M16	M17	M18	M19	M20
RNT28	RNT29	RNT30	KHN1	KHN2	KHN3	KHN4
MAG5	MAG6	MAG7	MAG8	MAG9	MAG10	MAG11

KAWAIDA 584/2100-01/6336-37/2092-93

Ashe	Ojobo	Iwa	Araorun	Ile	Itutu	Erinle
1	2	3	4	5	6	7
A18	A19	A20	A21	A22	A23	A24
PIN3	PIN4	PIN5	PIN6	PIN7	PIN8	PIN9
MIY10	MIY11	MIY12	MIY13	MIY14	MIY15	MIY16
8	9	10	11	12	13	14
A25	A26	A27	A28	A29	A30	MA1
PIN10	PIN11	PIN12	PIN13	PIN14	PIN15	PIN16
MIY17	MIY18	MIY19	MIY20	MIY21	MIY22	MIY23
15	16	17	18	19	20	21
MA2	MA3	MA4	MA5	MA6	MA7	MA8
PIN17	PIN18	PIN19	PIN20	PIN21	PIN22	PIN23
MIY24	MIY25	MIY26	MIY27	MIY28	MIY29	MIY30
22	23	24	25	26	27	28
MA9	MA10	MA11	MA12	MA13	MA14	MA15
PIN24	PIN25	PIN26	PIN27	PIN28	PIN29	PIN30
GIN1	GIN2	GIN3	GIN4	GIN5	GIN6	GIN7

UHURU 584/2100-01/6336-37/2092-93

Ashe	Ojobo	Iwa	Araorun	Ile	Itutu	Erinle
1	2	3	4	5	6	7
M21	M22	M23	M24	M25	M26	M27
KHN5	KHN6	KHN7	KHN8	KHN9	KHN10	KHN11
MAG12	MAG13	MAG14	MAG15	MAG16	MAG17	MAG18
8	9	10	11	12	13	14
M28	M29	M30	M31	A1	A2	A3
KHN12	KHN13	KHN14	KHN15	KHN16	KHN17	KHN18
MAG19	MAG20	MAG21	MAG22	MAG23	MAG24	MAG25
15	16	17	18	19	20	21
A4	A5	A6	A7	A8	A9	A10
KHN19	KHN20	KHN21	KHN22	KHN23	KHN24	KHN25
MAG26	MAG27	MAG28	MAG29	MAG30	MIY1	MIY2
22	23	24	25	26	27	28
A11	A12	A13	A14	A15	A16	A17
KHN26	KHN27	KHN28	KHN29	KHN30	PIN1	PIN2
MIY3	MIY4	MIY5	MIY6	MIY7	MIY8	MIY9

MABABA WEUSI 584/2100-01/6336-37/2092-93

Ashe	Ojobo	Iwa	Araorun	Ile	Itutu	Erinle
1	2	3	4	5	6	7
MA16	MA17	MA18	MA19	MA20	MA21	MA22
APP1	APP2	APP3	APP4	APP5	APP6	APP7
GIN8	GIN9	GIN10	GIN11	GIN12	GIN13	GIN14
8	9	10	11	12	13	14
MA23	MA24	MA25	MA26	MA27	MA28	MA29
APP8	APP9	APP10	APP11	APP12	APP13	APP14
GIN15	GIN16	GIN17	GIN18	GIN19	GIN20	GIN21
15	16	17	18	19	20	21
MA30	MA31	J1	J2	J3	J4	J5
APP15	APP16	APP17	APP18	APP19	APP20	APP21
GIN22	GIN23	GIN24	GIN25	GIN26	GIN27	GIN28
22	23	24	25	26	27	28
J6	J7	J8	J9	J10	J11	J12
APP22	APP23	APP24	APP25	APP26	APP27	APP28
GIN29	GIN30	SAN1	SAN2	SAN3	SAN4	SAN5

THE NU AFRIKAN CALENDAR

MWANAMKE WEUSI 584/2100-01/6336-37/2092-93

Ashe	Ojobo	Iwa	Araorun	Ile	Itutu	Erinle
1	2	3	4	5	6	7
J13	J14	J15	J16	J17	J18	J19
APP29	APP30	MSR1	MSR2	MSR3	MSR4	MSR5
SAN6	SAN7	SAN8	SAN9	SAN10	SAN11	SAN12
8	9	10	11	12	13	14
J20	J21	J22	J23	J24	J25	J26
MSR6	MSR7	MSR8	MSR9	MSR10	MSR11	MSR12
SAN13	SAN14	SAN15	SAN16	SAN17	SAN18	SAN19
15	16	17	18	19	20	21
J27	J28	J29	J30	JU1	JU2	JU3
MSR13	MSR14	MSR15	MSR16	MSR17	MSR18	MSR19
SAN20	SAN21	SAN22	SAN23	SAN24	SAN25	SAN26
22	23	24	25	26	27	28
JU4	JU5	JU6	JU7	JU8	JU9	JU10
MSR20	MRS21	MSR22	MSR23	MSR24	MSR25	MSR26
SAN27	SAN28	SAN29	SAN30	HAM1	HAM2	HAM3

TAIFA WEUSI 584/2100-01/6336-37/2092-93

Ashe	Ojobo	Iwa	Araorun	Ile	Itutu	Erinle
1	2	3	4	5	6	7
AU8	AU9	AU10	AU11	AU12	AU13	AU14
THT20	THT21	THT22	THT23	THT24	THT25	THT26
NAH2	NAH3	NAH4	NAH5	NAH6	NAH7	NAH8
8	9	10	11	12	13	14
AU15	AU16	AU17	AU18	AU19	AU20	AU21
THT27	THT28	THT29	THT30	PAP1	PAP1	PAP2
NAH9	NAH10	NAH11	NAH12	NAH13	NAH14	NAH15
15	16	17	18	19	20	21
AU22	AU23	AU24	AU25	AU26	AU27	AU28
PAP3	PAP4	PAP5	PAP6	PAP7	PAP8	PAP9
NAH16	NAH17	NAH18	NAH19	NAH20	NAH21	NAH22
22	23	24	25	26	27	28
AU29	AU30	AU31	S1	S2	S3	S4
PAP10	PAP11	PAP12	PAP13	PAP14	PAP15	PAP16
NAH23	NAH24	NAH25	NAH26	NAH27	NAH28	NAH29

MTU WEUSI 584/2100-01/6336-37/2092-93

Ashe	Ojobo	Iwa	Araorun	Ile	Itutu	Erinle
1	2	3	4	5	6	7
JU11	JU12	JU13	JU14	JU15	JU16	JU17
MSR27	MSR28	MSR29	MSR30	*ASR1*	*HRU2*	*STH3*
HAM4	HAM5	HAM6	HAM7	HAM8	HAM9	HAM10
8	9	10	11	12	13	14
JU18	JU19	JU20	JU21	JU22	JU23	JU24
AST4	*NBH5*	THT1	THT2	THT3	THT4	THT5
HAM11	HAM12	HAM13	AM14	HAM15	HAM16	HAM17
15	16	17	18	19	20	21
JU25	JU26	JU27	JU28	JU29	JU30	JU31
THT6	THT7	THT8	THT9	THT10	THT11	THT12
HAM18	HAM19	HAM20	HAM21	HAM22	HAM23	HAM24
22	23	24	25	26	27	28
AU1	AU2	AU3	AU4	AU5	AU6	AU7
THT13	THT14	THT15	THT16	THT17	THT18	THT19
HAM25	HAM26	HAM27	HAM28	HAM29	HAM30	NAH1

UWEZO 584/2100-01/6336-37/2092-93

Ashe	Ojobo	Iwa	Araorun	Ile	Itutu	Erinle
1	2	3	4	5	6	7
S5	S6	S7	S8	S9	S10	S11
PAP17	PAP18	PAP19	PAP20	PAP21	PAP22	PAP23
NAH30	*PAG1*	*PAG2*	*PAG3*	*PAG4*	*PAG5*	MAS1
8	9	10	11	12	13	14
S12	S13	S14	S15	S16	S17	S18
PAP24	PAP25	PAP26	PAP27	PAP28	PAP29	PAP30
MAS2	MAS3	MAS4	MAS5	MAS6	MAS7	MAS8
15	16	17	18	19	20	21
S19	S20	S21	S22	S23	S24	S25
HTH1	HTH2	HTH3	HTH4	HTH5	HTH6	HTH7
MAS9	MAS10	MAS11	MAS12	MAS13	MAS14	MAS15
22	23	24	25	26	27	28
S26	S27	S28	S29	S30	O1	O2
HTH8	HTH9	HTH10	HTH11	HTH12	HTH13	HTH14
MAS16	MAS17	MAS18	MAS19	MAS20	MAS21	MAS22

THE NU AFRIKAN CALENDAR

EKUNDU 584/2100-01/6336-37/2092-93						
Ashe	Ojobo	Iwa	Araorun	Ile	Itutu	Erinle
1	2	3	4	5	6	7
O3	O4	O5	O6	O7	O8	O9
HTH15	HTH16	HTH17	HTH18	HTH19	HTH20	HTH21
MAS23	MAS24	MAS25	MAS26	MAS27	MAS28	MAS29
8	9	10	11	12	13	14
O10	O11	O12	O13	O14	O15	O16
HTH22	HTH23	HTH24	HTH25	HTH26	HTH27	HTH28
MAS30	THi1	THi2	THi3	THi4	THi5	THi6
15	16	17	18	19	20	21
O17	O18	O19	O20	O21	O22	O23
HTH29	HTH30	KHK1	KHK2	KHK3	KHK4	KHK5
THi7	THi8	THi9	THi10	THi11	THi12	THi13
22	23	24	25	26	27	28
O24	O25	O26	O27	O28	O29	O30
KHK6	KHK7	KHK8	KHK9	KHK10	KHK11	KHK12
THi14	THi15	THi16	THi17	THi18	THi19	THi20

KIJANI 584/2100-01/6336-37/2092-93						
Ashe	Ojobo	Iwa	Araorun	Ile	Itutu	Erinle
1	2	3	4	5	6	7
N28	N29	N30	D1	D2	D3	D4
TBT11	TBT12	TBT13	TBT14	TBT15	TBT16	TBT17
HID19	HID20	HID21	HID22	HID23	HID24	HID25
8	9	10	11	12	13	14
D5	D6	D7	D8	D9	D10	D11
TBT18	TBT19	TBT20	TBT21	TBT22	TBT23	TBT24
HID26	HID27	HID28	HID29	HID30	TAH1	TAH2
15	16	17	18	19	20	21
D12	D13	D14	D15	D16	D17	D18
TBT25	TBT26	TBT27	TBT28	TBT29	TBT30	MKH1
TAH3	TAH4	TAH5	TAH6	TAH7	TAH8	TAH9
22	23	24	25	26	27	28
D19	D20	D21	D22	D23	D24	D25
MKH2	MKH3	MKH4	MKH5	MKH6	MKH7	MKH8
TAH10	TAH11	TAH12	TAH13	TAH14	TAH15	TAH16

EUSI 584/2100-01/6336-37/2092-93						
Ashe	Ojobo	Iwa	Araorun	Ile	Itutu	Erinle
1	2	3	4	5	6	7
O31	N1	N2	N3	N4	N5	N6
KHK13	KHK14	KHK15	KHK16	KHK17	KHK18	KHK19
THi21	THi22	THi23	THi24	THi25	THi26	THi27
8	9	10	11	12	13	14
N7	N8	N9	N10	N11	N12	N13
KHK20	KHK21	KHK22	KHK23	KHK24	KHK25	KHK26
THi28	THi29	THi30	HID1	HID2	HID3	HID4
15	16	17	18	19	20	21
N14	N15	N16	N17	N18	N19	N20
KHK27	KHK28	KHK29	KHK30	TBT1	TBT2	TBT3
HID5	HID6	HID7	HID8	HID9	HID10	HID11
22	23	24	25	26	27	28
N21	N22	N23	N24	N25	N26	N27
TBT4	TBT5	TBT6	TBT7	TBT8	TBT9	TBT10
HID12	HID13	HID14	HID15	HID16	HID17	HID18

KWANZA 584/2100-01/6336-37/2092-93						
Umoja	Kujichagulia	Ujima	Ujamaa	Nia	Kuumba	Imani
1	2	3	4	5	6	7
D26	D27	D28	D29	D30	D31	JA1
MKH9	MKH10	MKH11	MKH12	MKH13	MKH14	MKH15
TAH17	TAH18	TAH19	TAH20	TAH21	TAH22	TAH23
8	9	10	11	12	13	14
JA2	JA3	JA4	JA5	JA6	JA7	JA8
MKH16	MKH17	MKH18	MKH19	MKH20	MKH21	MKH22
TAH24	TAH25	TAH26	TAH27	TAH28	TAH29	TAH30
15	16	17	18	19	20	21
JA9	JA10	JA11	JA12	JA13	JA14	JA15
MKH23	MKH24	MKH25	KMH26	MKH27	MKH28	MKH29
TIR1	TIR2	TIR3	TIR4	TIR5	TIR6	TIR7
22	23	24	25	26	27	28
JA16	JA17	JA18	JA19	JA20	JA21	JA22
MKH30	AMN1	AMN2	AMN3	AMN4	AMN5	AMN6
TIR8	TIR9	TIR10	TIR11	TIR12	TIR13	TIR14

THE NU AFRIKAN CALENDAR

SHAHIDI 584/2100-01/6336-37/2092-93						
Ashe	Ojobo	Iwa	Araorun	Ile	Itutu	Erinle
1	2	3	4	5	6	7
JA23	JA24	JA25	JA26	JA27	JA28	JA29
AMN7	AMN8	AMN9	AMN10	AMN11	AMN12	AMN13
TIR15	TIR16	TIR17	TIR18	TIR19	TIR20	TIR21
8	9	10	11	12	13	14
JA30	JA31	F1	F2	F3	F4	F5
AMN14	AMN15	AMN16	AMN17	AMN18	AMN19	AMN20
TIR22	TIR23	TIR24	TIR25	TIR26	TIR27	TIR28
15	16	17	18	19	20	21
F6	F7	F8	F9	F10	F 11	F12
AMN21	AMN22	AMN23	AMN24	AMN25	AMN26	AMN28
TIR29	TIR30	YEK1	YEK2	YEK3	YEK4	YEK5
22	23	24	25	26	27	28
F13	F14	F15	F16	F17	F18	F19
AMN29	AMN30	RNT1	RNT2	RNT3	RNT4	RNT5
YEK6	YEK7	YEK8	YEK9	YEK10	YEK11	YEK12
SHAHIDI 135						
F20						
RNT6						
YEK13						

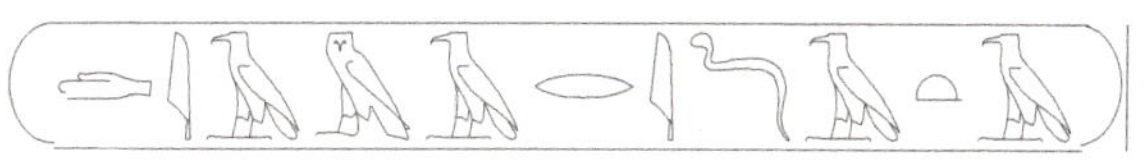

THE NU AFRIKAN CALENDAR

MAAFA 585 B.S.E./ASIENTO 2101-02 C.E.
KAM 6337-38S.M./EC 2093-94 A.M. (ዓ.ም.)

AFU 585/2101-02/6337-38/2093-94						
Ashe	Ojobo	Iwa	Araorun	Ile	Itutu	Erinle
1	2	3	4	5	6	7
F21	F22	F23	F24	F25	F26	F27
RNT7	RNT8	RNT9	RNT10	RNT11	RNT12	RNT13
YEK14	YEK15	YEK16	YEK17	YEK18	YEK19	YEK20
8	9	10	11	12	13	14
F28	M1	M2	M3	M4	M5	M6
RNT14	RNT15	RNT16	RNT17	RNT18	RNT19	RNT20
YEK21	YEK22	YEK23	YEK24	YEK25	YEK26	YEK27
15	16	17	18	19	20	21
M7	M8	M9	M10	M11	M12	M13
RNT21	RNT22	RNT23	RNT24	RNT25	RNT26	RNT27
YEK28	YEK29	YEK30	MAG1	MAG2	MAG3	MAG4
22	23	24	25	26	27	28
M14	M15	M16	M17	M18	M19	M20
RNT28	RNT29	RNT30	KHN1	KHN2	KHN3	KHN4
MAG5	MAG6	MAG7	MAG8	MAG9	MAG10	MAG11

KAWAIDA 585/2101-02/6337-38/2093-94						
Ashe	Ojobo	Iwa	Araorun	Ile	Itutu	Erinle
1	2	3	4	5	6	7
A18	A19	A20	A21	A22	A23	A24
PIN3	PIN4	PIN5	PIN6	PIN7	PIN8	PIN9
MIY10	MIY11	MIY12	MIY13	MIY14	MIY15	MIY16
8	9	10	11	12	13	14
A25	A26	A27	A28	A29	A30	MA1
PIN10	PIN11	PIN12	PIN13	PIN14	PIN15	PIN16
MIY17	MIY18	MIY19	MIY20	MIY21	MIY22	MIY23
15	16	17	18	19	20	21
MA2	MA3	MA4	MA5	MA6	MA7	MA8
PIN17	PIN18	PIN19	PIN20	PIN21	PIN22	PIN23
MIY24	MIY25	MIY26	MIY27	MIY28	MIY29	MIY30
22	23	24	25	26	27	28
MA9	MA10	MA11	MA12	MA13	MA14	MA15
PIN24	PIN25	PIN26	PIN27	PIN28	PIN29	PIN30
GIN1	GIN2	GIN3	GIN4	GIN5	GIN6	GIN7

UHURU 585/2101-02/6337-38/2093-94						
Ashe	Ojobo	Iwa	Araorun	Ile	Itutu	Erinle
1	2	3	4	5	6	7
M21	M22	M23	M24	M25	M26	M27
KHN5	KHN6	KHN7	KHN8	KHN9	KHN10	KHN11
MAG12	MAG13	MAG14	MAG15	MAG16	MAG17	MAG18
8	9	10	11	12	13	14
M28	M29	M30	M31	A1	A2	A3
KHN12	KHN13	KHN14	KHN15	KHN16	KHN17	KHN18
MAG19	MAG20	MAG21	MAG22	MAG23	MAG24	MAG25
15	16	17	18	19	20	21
A4	A5	A6	A7	A8	A9	A10
KHN19	KHN20	KHN21	KHN22	KHN23	KHN24	KHN25
MAG26	MAG27	MAG28	MAG29	MAG30	MIY1	MIY2
22	23	24	25	26	27	28
A11	A12	A13	A14	A15	A16	A17
KHN26	KHN27	KHN28	KHN29	KHN30	PIN1	PIN2
MIY3	MIY4	MIY5	MIY6	MIY7	MIY8	MIY9

MABABA WEUSI 585/2101-02/6337-38/2093-94						
Ashe	Ojobo	Iwa	Araorun	Ile	Itutu	Erinle
1	2	3	4	5	6	7
MA16	MA17	MA18	MA19	MA20	MA21	MA22
APP1	APP2	APP3	APP4	APP5	APP6	APP7
GIN8	GIN9	GIN10	GIN11	GIN12	GIN13	GIN14
8	9	10	11	12	13	14
MA23	MA24	MA25	MA26	MA27	MA28	MA29
APP8	APP9	APP10	APP11	APP12	APP13	APP14
GIN15	GIN16	GIN17	GIN18	GIN19	GIN20	GIN21
15	16	17	18	19	20	21
MA30	MA31	J1	J2	J3	J4	J5
APP15	APP16	APP17	APP18	APP19	APP20	APP21
GIN22	GIN23	GIN24	GIN25	GIN26	GIN27	GIN28
22	23	24	25	26	27	28
J6	J7	J8	J9	J10	J11	J12
APP22	APP23	APP24	APP25	APP26	APP27	APP28
GIN29	GIN30	SAN1	SAN2	SAN3	SAN4	SAN5

THE NU AFRIKAN CALENDAR

MWANAMKE WEUSI 585/2101-02/6337-38/2093-94

Ashe	Ojobo	Iwa	Araorun	Ile	Itutu	Erinle
1	2	3	4	5	6	7
J13	J14	J15	J16	J17	J18	J19
APP29	APP30	MSR1	MSR2	MSR3	MSR4	MSR5
SAN6	SAN7	SAN8	SAN9	SAN10	SAN11	SAN12
8	9	10	11	12	13	14
J20	J21	J22	J23	J24	J25	J26
MSR6	MSR7	MSR8	MSR9	MSR10	MSR11	MSR12
SAN13	SAN14	SAN15	SAN16	SAN17	SAN18	SAN19
15	16	17	18	19	20	21
J27	J28	J29	J30	JU1	JU2	JU3
MSR13	MSR14	MSR15	MSR16	MSR17	MSR18	MSR19
SAN20	SAN21	SAN22	SAN23	SAN24	SAN25	SAN26
22	23	24	25	26	27	28
JU4	JU5	JU6	JU7	JU8	JU9	JU10
MSR20	MRS21	MSR22	MSR23	MSR24	MSR25	MSR26
SAN27	SAN28	SAN29	SAN30	HAM1	HAM2	HAM3

TAIFA WEUSI 585/2101-02/6337-38/2093-94

Ashe	Ojobo	Iwa	Araorun	Ile	Itutu	Erinle
1	2	3	4	5	6	7
AU8	AU9	AU10	AU11	AU12	AU13	AU14
THT20	THT21	THT22	THT23	THT24	THT25	THT26
NAH2	NAH3	NAH4	NAH5	NAH6	NAH7	NAH8
8	9	10	11	12	13	14
AU15	AU16	AU17	AU18	AU19	AU20	AU21
THT27	THT28	THT29	THT30	PAP1	PAP1	PAP2
NAH9	NAH10	NAH11	NAH12	NAH13	NAH14	NAH15
15	16	17	18	19	20	21
AU22	AU23	AU24	AU25	AU26	AU27	AU28
PAP3	PAP4	PAP5	PAP6	PAP7	PAP8	PAP9
NAH16	NAH17	NAH18	NAH19	NAH20	NAH21	NAH22
22	23	24	25	26	27	28
AU29	AU30	AU31	S1	S2	S3	S4
PAP10	PAP11	PAP12	PAP13	PAP14	PAP15	PAP16
NAH23	NAH24	NAH25	NAH26	NAH27	NAH28	NAH29

MTU WEUSI 585/2101-02/6337-38/2093-94

Ashe	Ojobo	Iwa	Araorun	Ile	Itutu	Erinle
1	2	3	4	5	6	7
JU11	JU12	JU13	JU14	JU15	JU16	JU17
MSR27	MSR28	MSR29	MSR30	*ASR1*	*HRU2*	*STH3*
HAM4	HAM5	HAM6	HAM7	HAM8	HAM9	HAM10
8	9	10	11	12	13	14
JU18	JU19	JU20	JU21	JU22	JU23	JU24
AST4	*NBH5*	THT1	THT2	THT3	THT4	THT5
HAM11	HAM12	HAM13	AM14	HAM15	HAM16	HAM17
15	16	17	18	19	20	21
JU25	JU26	JU27	JU28	JU29	JU30	JU31
THT6	THT7	THT8	THT9	THT10	THT11	THT12
HAM18	HAM19	HAM20	HAM21	HAM22	HAM23	HAM24
22	23	24	25	26	27	28
AU1	AU2	AU3	AU4	AU5	AU6	AU7
THT13	THT14	THT15	THT16	THT17	THT18	THT19
HAM25	HAM26	HAM27	HAM28	HAM29	HAM30	NAH1

UWEZO 585/2101-02/6337-38/2093-94

Ashe	Ojobo	Iwa	Araorun	Ile	Itutu	Erinle
1	2	3	4	5	6	7
S5	S6	S7	S8	S9	S10	S11
PAP17	PAP18	PAP19	PAP20	PAP21	PAP22	PAP23
NAH30	*PAG1*	*PAG2*	*PAG3*	*PAG4*	*PAG5*	MAS1
8	9	10	11	12	13	14
S12	S13	S14	S15	S16	S17	S18
PAP24	PAP25	PAP26	PAP27	PAP28	PAP29	PAP30
MAS2	MAS3	MAS4	MAS5	MAS6	MAS7	MAS8
15	16	17	18	19	20	21
S19	S20	S21	S22	S23	S24	S25
HTH1	HTH2	HTH3	HTH4	HTH5	HTH6	HTH7
MAS9	MAS10	MAS11	MAS12	MAS13	MAS14	MAS15
22	23	24	25	26	27	28
S26	S27	S28	S29	S30	O1	O2
HTH8	HTH9	HTH10	HTH11	HTH12	HTH13	HTH14
MAS16	MAS17	MAS18	MAS19	MAS20	MAS21	MAS22

THE NU AFRIKAN CALENDAR

EKUNDU 585/2101-02/6337-38/2093-94

Ashe	Ojobo	Iwa	Araorun	Ile	Itutu	Erinle
1	2	3	4	5	6	7
O3	O4	O5	O6	O7	O8	O9
HTH15	HTH16	HTH17	HTH18	HTH19	HTH20	HTH21
MAS23	MAS24	MAS25	MAS26	MAS27	MAS28	MAS29
8	9	10	11	12	13	14
O10	O11	O12	O13	O14	O15	O16
HTH22	HTH23	HTH24	HTH25	HTH26	HTH27	HTH28
MAS30	THi1	THi2	THi3	THi4	THi5	THi6
15	16	17	18	19	20	21
O17	O18	O19	O20	O21	O22	O23
HTH29	HTH30	KHK1	KHK2	KHK3	KHK4	KHK5
THi7	THi8	THi9	THi10	THi11	THi12	THi13
22	23	24	25	26	27	28
O24	O25	O26	O27	O28	O29	O30
KHK6	KHK7	KHK8	KHK9	KHK10	KHK11	KHK12
THi14	THi15	THi16	THi17	THi18	THi19	THi20

KIJANI 585/2101-02/6337-38/2093-94

Ashe	Ojobo	Iwa	Araorun	Ile	Itutu	Erinle
1	2	3	4	5	6	7
N28	N29	N30	D1	D2	D3	D4
TBT11	TBT12	TBT13	TBT14	TBT15	TBT16	TBT17
HID19	HID20	HID21	HID22	HID23	HID24	HID25
8	9	10	11	12	13	14
D5	D6	D7	D8	D9	D10	D11
TBT18	TBT19	TBT20	TBT21	TBT22	TBT23	TBT24
HID26	HID27	HID28	HID29	HID30	TAH1	TAH2
15	16	17	18	19	20	21
D12	D13	D14	D15	D16	D17	D18
TBT25	TBT26	TBT27	TBT28	TBT29	TBT30	MKH1
TAH3	TAH4	TAH5	TAH6	TAH7	TAH8	TAH9
22	23	24	25	26	27	28
D19	D20	D21	D22	D23	D24	D25
MKH2	MKH3	MKH4	MKH5	MKH6	MKH7	MKH8
TAH10	TAH11	TAH12	TAH13	TAH14	TAH15	TAH16

EUSI 585/2101-02/6337-38/2093-94

Ashe	Ojobo	Iwa	Araorun	Ile	Itutu	Erinle
1	2	3	4	5	6	7
O31	N1	N2	N3	N4	N5	N6
KHK13	KHK14	KHK15	KHK16	KHK17	KHK18	KHK19
THi21	THi22	THi23	THi24	THi25	THi26	THi27
8	9	10	11	12	13	14
N7	N8	N9	N10	N11	N12	N13
KHK20	KHK21	KHK22	KHK23	KHK24	KHK25	KHK26
THi28	THi29	THi30	HID1	HID2	HID3	HID4
15	16	17	18	19	20	21
N14	N15	N16	N17	N18	N19	N20
KHK27	KHK28	KHK29	KHK30	TBT1	TBT2	TBT3
HID5	HID6	HID7	HID8	HID9	HID10	HID11
22	23	24	25	26	27	28
N21	N22	N23	N24	N25	N26	N27
TBT4	TBT5	TBT6	TBT7	TBT8	TBT9	TBT10
HID12	HID13	HID14	HID15	HID16	HID17	HID18

KWANZA 585/2101-02/6337-38/2093-94

Umoja	Kujichagulia	Ujima	Ujamaa	Nia	Kuumba	Imani
1	2	3	4	5	6	7
D26	D27	D28	D29	D30	D31	JA1
MKH9	MKH10	MKH11	MKH12	MKH13	MKH14	MKH15
TAH17	TAH18	TAH19	TAH20	TAH21	TAH22	TAH23
8	9	10	11	12	13	14
JA2	JA3	JA4	JA5	JA6	JA7	JA8
MKH16	MKH17	MKH18	MKH19	MKH20	MKH21	MKH22
TAH24	TAH25	TAH26	TAH27	TAH28	TAH29	TAH30
15	16	17	18	19	20	21
JA9	JA10	JA11	JA12	JA13	JA14	JA15
MKH23	MKH24	MKH25	KMH26	MKH27	MKH28	MKH29
TIR1	TIR2	TIR3	TIR4	TIR5	TIR6	TIR7
22	23	24	25	26	27	28
JA16	JA17	JA18	JA19	JA20	JA21	JA22
MKH30	AMN1	AMN2	AMN3	AMN4	AMN5	AMN6
TIR8	TIR9	TIR10	TIR11	TIR12	TIR13	TIR14

THE NU AFRIKAN CALENDAR

SHAHIDI 585/2101-02/6337-38/2093-94						
Ashe	Ojobo	Iwa	Araorun	Ile	Itutu	Erinle
1	2	3	4	5	6	7
JA23	JA24	JA25	JA26	JA27	JA28	JA29
AMN7	AMN8	AMN9	AMN10	AMN11	AMN12	AMN13
TIR15	TIR16	TIR17	TIR18	TIR19	TIR20	TIR21
8	9	10	11	12	13	14
JA30	JA31	F1	F2	F3	F4	F5
AMN14	AMN15	AMN16	AMN17	AMN18	AMN19	AMN20
TIR22	TIR23	TIR24	TIR25	TIR26	TIR27	TIR28
15	16	17	18	19	20	21
F6	F7	F8	F9	F10	F 11	F12
AMN21	AMN22	AMN23	AMN24	AMN25	AMN26	AMN28
TIR29	TIR30	YEK1	YEK2	YEK3	YEK4	YEK5
22	23	24	25	26	27	28
F13	F14	F15	F16	F17	F18	F19
AMN29	AMN30	RNT1	RNT2	RNT3	RNT4	RNT5
YEK6	YEK7	YEK8	YEK9	YEK10	YEK11	YEK12
SHAHIDI 136						
F20						
RNT6						
YEK13						

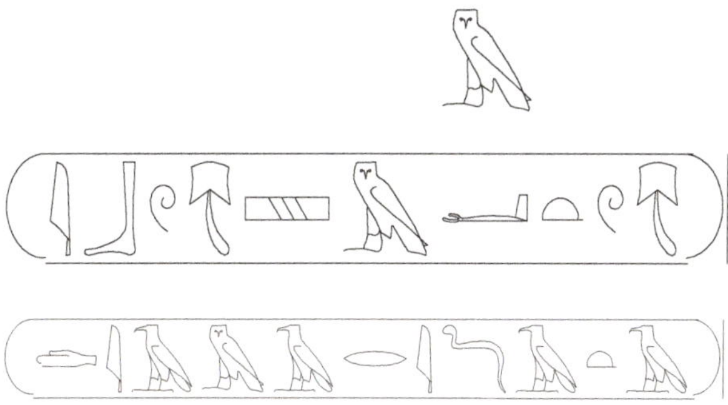

THE NU AFRIKAN CALENDAR

MAAFA 586 B.S.E./ASIENTO 2102-03 C.E.
KAM 6338-39 S.M./EC 2094-95 A.M. (ዓ.ም.)

AFU 586/2102-03/6338-39/2094-95						
Ashe	Ojobo	Iwa	Araorun	Ile	Itutu	Erinle
1	2	3	4	5	6	7
F21	F22	F23	F24	F25	F26	F27
RNT7	RNT8	RNT9	RNT10	RNT11	RNT12	RNT13
YEK14	YEK15	YEK16	YEK17	YEK18	YEK19	YEK20
8	9	10	11	12	13	14
F28	M1	M2	M3	M4	M5	M6
RNT14	RNT15	RNT16	RNT17	RNT18	RNT19	RNT20
YEK21	YEK22	YEK23	YEK24	YEK25	YEK26	YEK27
15	16	17	18	19	20	21
M7	M8	M9	M10	M11	M12	M13
RNT21	RNT22	RNT23	RNT24	RNT25	RNT26	RNT27
YEK28	YEK29	YEK30	MAG1	MAG2	MAG3	MAG4
22	23	24	25	26	27	28
M14	M15	M16	M17	M18	M19	M20
RNT28	RNT29	RNT30	KHN1	KHN2	KHN3	KHN4
MAG5	MAG6	MAG7	MAG8	MAG9	MAG10	MAG11

KAWAIDA 586/2102-03/6338-39/2094-95						
Ashe	Ojobo	Iwa	Araorun	Ile	Itutu	Erinle
1	2	3	4	5	6	7
A18	A19	A20	A21	A22	A23	A24
PIN3	PIN4	PIN5	PIN6	PIN7	PIN8	PIN9
MIY10	MIY11	MIY12	MIY13	MIY14	MIY15	MIY16
8	9	10	11	12	13	14
A25	A26	A27	A28	A29	A30	MA1
PIN10	PIN11	PIN12	PIN13	PIN14	PIN15	PIN16
MIY17	MIY18	MIY19	MIY20	MIY21	MIY22	MIY23
15	16	17	18	19	20	21
MA2	MA3	MA4	MA5	MA6	MA7	MA8
PIN17	PIN18	PIN19	PIN20	PIN21	PIN22	PIN23
MIY24	MIY25	MIY26	MIY27	MIY28	MIY29	MIY30
22	23	24	25	26	27	28
MA9	MA10	MA11	MA12	MA13	MA14	MA15
PIN24	PIN25	PIN26	PIN27	PIN28	PIN29	PIN30
GIN1	GIN2	GIN3	GIN4	GIN5	GIN6	GIN7

UHURU 586/2102-03/6338-39/2094-95						
Ashe	Ojobo	Iwa	Araorun	Ile	Itutu	Erinle
1	2	3	4	5	6	7
M21	M22	M23	M24	M25	M26	M27
KHN5	KHN6	KHN7	KHN8	KHN9	KHN10	KHN11
MAG12	MAG13	MAG14	MAG15	MAG16	MAG17	MAG18
8	9	10	11	12	13	14
M28	M29	M30	M31	A1	A2	A3
KHN12	KHN13	KHN14	KHN15	KHN16	KHN17	KHN18
MAG19	MAG20	MAG21	MAG22	MAG23	MAG24	MAG25
15	16	17	18	19	20	21
A4	A5	A6	A7	A8	A9	A10
KHN19	KHN20	KHN21	KHN22	KHN23	KHN24	KHN25
MAG26	MAG27	MAG28	MAG29	MAG30	MIY1	MIY2
22	23	24	25	26	27	28
A11	A12	A13	A14	A15	A16	A17
KHN26	KHN27	KHN28	KHN29	KHN30	PIN1	PIN2
MIY3	MIY4	MIY5	MIY6	MIY7	MIY8	MIY9

MABABA WEUSI 586/2102-03/6338-39/2094-95						
Ashe	Ojobo	Iwa	Araorun	Ile	Itutu	Erinle
1	2	3	4	5	6	7
MA16	MA17	MA18	MA19	MA20	MA21	MA22
APP1	APP2	APP3	APP4	APP5	APP6	APP7
GIN8	GIN9	GIN10	GIN11	GIN12	GIN13	GIN14
8	9	10	11	12	13	14
MA23	MA24	MA25	MA26	MA27	MA28	MA29
APP8	APP9	APP10	APP11	APP12	APP13	APP14
GIN15	GIN16	GIN17	GIN18	GIN19	GIN20	GIN21
15	16	17	18	19	20	21
MA30	MA31	J1	J2	J3	J4	J5
APP15	APP16	APP17	APP18	APP19	APP20	APP21
GIN22	GIN23	GIN24	GIN25	GIN26	GIN27	GIN28
22	23	24	25	26	27	28
J6	J7	J8	J9	J10	J11	J12
APP22	APP23	APP24	APP25	APP26	APP27	APP28
GIN29	GIN30	SAN1	SAN2	SAN3	SAN4	SAN5

THE NU AFRIKAN CALENDAR

MWANAMKE WEUSI 586/2102-03/6338-39/2094-95

Ashe	Ojobo	Iwa	Araorun	Ile	Itutu	Erinle
1	2	3	4	5	6	7
J13	J14	J15	J16	J17	J18	J19
APP29	APP30	MSR1	MSR2	MSR3	MSR4	MSR5
SAN6	SAN7	SAN8	SAN9	SAN10	SAN11	SAN12
8	9	10	11	12	13	14
J20	J21	J22	J23	J24	J25	J26
MSR6	MSR7	MSR8	MSR9	MSR10	MSR11	MSR12
SAN13	SAN14	SAN15	SAN16	SAN17	SAN18	SAN19
15	16	17	18	19	20	21
J27	J28	J29	J30	JU1	JU2	JU3
MSR13	MSR14	MSR15	MSR16	MSR17	MSR18	MSR19
SAN20	SAN21	SAN22	SAN23	SAN24	SAN25	SAN26
22	23	24	25	26	27	28
JU4	JU5	JU6	JU7	JU8	JU9	JU10
MSR20	MRS21	MSR22	MSR23	MSR24	MSR25	MSR26
SAN27	SAN28	SAN29	SAN30	HAM1	HAM2	HAM3

TAIFA WEUSI 586/2102-03/6338-39/2094-95

Ashe	Ojobo	Iwa	Araorun	Ile	Itutu	Erinle
1	2	3	4	5	6	7
AU8	AU9	AU10	AU11	AU12	AU13	AU14
THT20	THT21	THT22	THT23	THT24	THT25	THT26
NAH2	NAH3	NAH4	NAH5	NAH6	NAH7	NAH8
8	9	10	11	12	13	14
AU15	AU16	AU17	AU18	AU19	AU20	AU21
THT27	THT28	THT29	THT30	PAP1	PAP1	PAP2
NAH9	NAH10	NAH11	NAH12	NAH13	NAH14	NAH15
15	16	17	18	19	20	21
AU22	AU23	AU24	AU25	AU26	AU27	AU28
PAP3	PAP4	PAP5	PAP6	PAP7	PAP8	PAP9
NAH16	NAH17	NAH18	NAH19	NAH20	NAH21	NAH22
22	23	24	25	26	27	28
AU29	AU30	AU31	S1	S2	S3	S4
PAP10	PAP11	PAP12	PAP13	PAP14	PAP15	PAP16
NAH23	NAH24	NAH25	NAH26	NAH27	NAH28	NAH29

MTU WEUSI 586/2102-03/6338-39/2094-95

Ashe	Ojobo	Iwa	Araorun	Ile	Itutu	Erinle
1	2	3	4	5	6	7
JU11	JU12	JU13	JU14	JU15	JU16	JU17
MSR27	MSR28	MSR29	MSR30	*ASR1*	*HRU2*	*STH3*
HAM4	HAM5	HAM6	HAM7	HAM8	HAM9	HAM10
8	9	10	11	12	13	14
JU18	JU19	JU20	JU21	JU22	JU23	JU24
AST4	*NBH5*	THT1	THT2	THT3	THT4	THT5
HAM11	HAM12	HAM13	AM14	HAM15	HAM16	HAM17
15	16	17	18	19	20	21
JU25	JU26	JU27	JU28	JU29	JU30	JU31
THT6	THT7	THT8	THT9	THT10	THT11	THT12
HAM18	HAM19	HAM20	HAM21	HAM22	HAM23	HAM24
22	23	24	25	26	27	28
AU1	AU2	AU3	AU4	AU5	AU6	AU7
THT13	THT14	THT15	THT16	THT17	THT18	THT19
HAM25	HAM26	HAM27	HAM28	HAM29	HAM30	NAH1

UWEZO 586/2102-03/6338-39/2094-95

Ashe	Ojobo	Iwa	Araorun	Ile	Itutu	Erinle
1	2	3	4	5	6	7
S5	S6	S7	S8	S9	S10	S11
PAP17	PAP18	PAP19	PAP20	PAP21	PAP22	PAP23
NAH30	*PAG1*	*PAG2*	*PAG3*	*PAG4*	*PAG5*	MAS1
8	9	10	11	12	13	14
S12	S13	S14	S15	S16	S17	S18
PAP24	PAP25	PAP26	PAP27	PAP28	PAP29	PAP30
MAS2	MAS3	MAS4	MAS5	MAS6	MAS7	MAS8
15	16	17	18	19	20	21
S19	S20	S21	S22	S23	S24	S25
HTH1	HTH2	HTH3	HTH4	HTH5	HTH6	HTH7
MAS9	MAS10	MAS11	MAS12	MAS13	MAS14	MAS15
22	23	24	25	26	27	28
S26	S27	S28	S29	S30	O1	O2
HTH8	HTH9	HTH10	HTH11	HTH12	HTH13	HTH14
MAS16	MAS17	MAS18	MAS19	MAS20	MAS21	MAS22

THE NU AFRIKAN CALENDAR

EKUNDU 586/2102-03/6338-39/2094-95						
Ashe	Ojobo	Iwa	Araorun	Ile	Itutu	Erinle
1	2	3	4	5	6	7
O3	O4	O5	O6	O7	O8	O9
HTH15	HTH16	HTH17	HTH18	HTH19	HTH20	HTH21
MAS23	MAS24	MAS25	MAS26	MAS27	MAS28	MAS29
8	9	10	11	12	13	14
O10	O11	O12	O13	O14	O15	O16
HTH22	HTH23	HTH24	HTH25	HTH26	HTH27	HTH28
MAS30	THi1	THi2	THi3	THi4	THi5	THi6
15	16	17	18	19	20	21
O17	O18	O19	O20	O21	O22	O23
HTH29	HTH30	KHK1	KHK2	KHK3	KHK4	KHK5
THi7	THi8	THi9	THi10	THi11	THi12	THi13
22	23	24	25	26	27	28
O24	O25	O26	O27	O28	O29	O30
KHK6	KHK7	KHK8	KHK9	KHK10	KHK11	KHK12
THi14	THi15	THi16	THi17	THi18	THi19	THi20

KIJANI 586/2102-03/6338-39/2094-95						
Ashe	Ojobo	Iwa	Araorun	Ile	Itutu	Erinle
1	2	3	4	5	6	7
N28	N29	N30	D1	D2	D3	D4
TBT11	TBT12	TBT13	TBT14	TBT15	TBT16	TBT17
HID19	HID20	HID21	HID22	HID23	HID24	HID25
8	9	10	11	12	13	14
D5	D6	D7	D8	D9	D10	D11
TBT18	TBT19	TBT20	TBT21	TBT22	TBT23	TBT24
HID26	HID27	HID28	HID29	HID30	TAH1	TAH2
15	16	17	18	19	20	21
D12	D13	D14	D15	D16	D17	D18
TBT25	TBT26	TBT27	TBT28	TBT29	TBT30	MKH1
TAH3	TAH4	TAH5	TAH6	TAH7	TAH8	TAH9
22	23	24	25	26	27	28
D19	D20	D21	D22	D23	D24	D25
MKH2	MKH3	MKH4	MKH5	MKH6	MKH7	MKH8
TAH10	TAH11	TAH12	TAH13	TAH14	TAH15	TAH16

EUSI 586/2102-03/6338-39/2094-95						
Ashe	Ojobo	Iwa	Araorun	Ile	Itutu	Erinle
1	2	3	4	5	6	7
O31	N1	N2	N3	N4	N5	N6
KHK13	KHK14	KHK15	KHK16	KHK17	KHK18	KHK19
THi21	THi22	THi23	THi24	THi25	THi26	THi27
8	9	10	11	12	13	14
N7	N8	N9	N10	N11	N12	N13
KHK20	KHK21	KHK22	KHK23	KHK24	KHK25	KHK26
THi28	THi29	THi30	HID1	HID2	HID3	HID4
15	16	17	18	19	20	21
N14	N15	N16	N17	N18	N19	N20
KHK27	KHK28	KHK29	KHK30	TBT1	TBT2	TBT3
HID5	HID6	HID7	HID8	HID9	HID10	HID11
22	23	24	25	26	27	28
N21	N22	N23	N24	N25	N26	N27
TBT4	TBT5	TBT6	TBT7	TBT8	TBT9	TBT10
HID12	HID13	HID14	HID15	HID16	HID17	HID18

KWANZA 586/2102-03/6338-39/2094-95						
Umoja	Kujichagulia	Ujima	Ujamaa	Nia	Kuumba	Imani
1	2	3	4	5	6	7
D26	D27	D28	D29	D30	D31	JA1
MKH9	MKH10	MKH11	MKH12	MKH13	MKH14	MKH15
TAH17	TAH18	TAH19	TAH20	TAH21	TAH22	TAH23
8	9	10	11	12	13	14
JA2	JA3	JA4	JA5	JA6	JA7	JA8
MKH16	MKH17	MKH18	MKH19	MKH20	MKH21	MKH22
TAH24	TAH25	TAH26	TAH27	TAH28	TAH29	TAH30
15	16	17	18	19	20	21
JA9	JA10	JA11	JA12	JA13	JA14	JA15
MKH23	MKH24	MKH25	KMH26	MKH27	MKH28	MKH29
TIR1	TIR2	TIR3	TIR4	TIR5	TIR6	TIR7
22	23	24	25	26	27	28
JA16	JA17	JA18	JA19	JA20	JA21	JA22
MKH30	AMN1	AMN2	AMN3	AMN4	AMN5	AMN6
TIR8	TIR9	TIR10	TIR11	TIR12	TIR13	TIR14

THE NU AFRIKAN CALENDAR

SHAHIDI 586/2102-03/6338-39/2094-95						
Ashe	Ojobo	Iwa	Araorun	Ile	Itutu	Erinle
1	2	3	4	5	6	7
JA23	JA24	JA25	JA26	JA27	JA28	JA29
AMN7	AMN8	AMN9	AMN10	AMN11	AMN12	AMN13
TIR15	TIR16	TIR17	TIR18	TIR19	TIR20	TIR21
8	9	10	11	12	13	14
JA30	JA31	F1	F2	F3	F4	F5
AMN14	AMN15	AMN16	AMN17	AMN18	AMN19	AMN20
TIR22	TIR23	TIR24	TIR25	TIR26	TIR27	TIR28
15	16	17	18	19	20	21
F6	F7	F8	F9	F10	F 11	F12
AMN21	AMN22	AMN23	AMN24	AMN25	AMN26	AMN28
TIR29	TIR30	YEK1	YEK2	YEK3	YEK4	YEK5
22	23	24	25	26	27	28
F13	F14	F15	F16	F17	F18	F19
AMN29	AMN30	RNT1	RNT2	RNT3	RNT4	RNT5
YEK6	YEK7	YEK8	YEK9	YEK10	YEK11	YEK12
SHAHIDI 137						
F20						
RNT6						
YEK13						

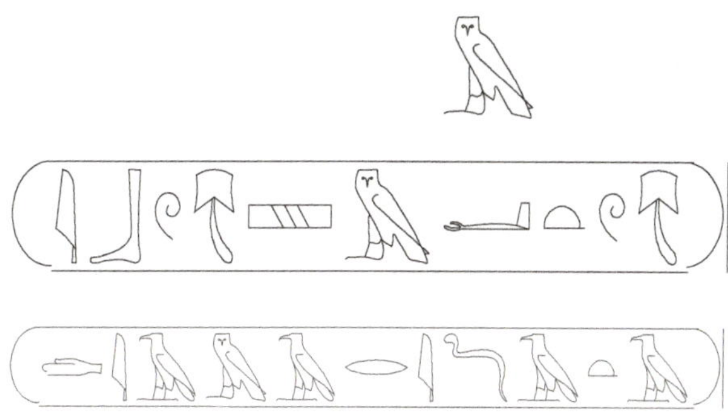

MAAFA 587 B.S.E./ASIENTO 2103-04 C.E./
KAM 6339-40 S.M./EC 2095-96 A.M. (ዓ.ም.)

AFU 587/2103-04/6339-40/2095-96

Ashe	Ojobo	Iwa	Araorun	Ile	Itutu	Erinle
1	2	3	4	5	6	7
F21	F22	F23	F24	F25	F26	F27
RNT7	RNT8	RNT9	RNT10	RNT11	RNT12	RNT13
YEK14	YEK15	YEK16	YEK17	YEK18	YEK19	YEK20
8	9	10	11	12	13	14
F28	M1	M2	M3	M4	M5	M6
RNT14	RNT15	RNT16	RNT17	RNT18	RNT19	RNT20
YEK21	YEK22	YEK23	YEK24	YEK25	YEK26	YEK27
15	16	17	18	19	20	21
M7	M8	M9	M10	M11	M12	M13
RNT21	RNT22	RNT23	RNT24	RNT25	RNT26	RNT27
YEK28	YEK 29	YEK30	MAG1	MAG2	MAG3	MAG4
22	23	24	25	26	27	28
M14	M15	M16	M17	M18	M19	M20
RNT28	RNT29	RNT30	KHN1	KHN2	KHN3	KHN4
MAG5	MAG6	MAG7	MAG8	MAG9	MAG10	MAG11

UHURU 587/2103-04/6339-40/2095-96

Ashe	Ojobo	Iwa	Araorun	Ile	Itutu	Erinle
1	2	3	4	5	6	7
M21	M22	M23	M24	M25	M26	M27
KHN5	KHN6	KHN7	KHN8	KHN9	KHN10	KHN11
MAG12	MAG13	MAG14	MAG15	MAG16	MAG17	MAG18
8	9	10	11	12	13	14
M28	M29	M30	M31	A1	A2	A3
KHN12	KHN13	KHN14	KHN15	KHN16	KHN17	KHN18
MAG19	MAG20	MAG21	MAG22	MAG23	MAG24	MAG25
15	16	17	18	19	20	21
A4	A5	A6	A7	A8	A9	A10
KHN19	KHN20	KHN21	KHN22	KHN23	KHN24	KHN25
MAG26	MAG27	MAG28	MAG29	MAG30	MIY1	MIY2
22	23	24	25	26	27	28
A11	A12	A13	A14	A15	A16	A17
KHN26	KHN27	KHN28	KHN29	KHN30	PIN1	PIN2
MIY3	MIY4	MIY5	MIY6	MIY7	MIY8	MIY9

KAWAIDA 587/2103-04/6339-40/2095-96

Ashe	Ojobo	Iwa	Araorun	Ile	Itutu	Erinle
1	2	3	4	5	6	7
A18	A19	A20	A21	A22	A23	A24
PIN3	PIN4	PIN5	PIN6	PIN7	PIN8	PIN9
MIY10	MIY11	MIY12	MIY13	MIY14	MIY15	MIY16
8	9	10	11	12	13	14
A25	A26	A27	A28	A29	A30	MA1
PIN10	PIN11	PIN12	PIN13	PIN14	PIN15	PIN16
MIY17	MIY18	MIY19	MIY20	MIY21	MIY22	MIY23
15	16	17	18	19	20	21
MA2	MA3	MA4	MA5	MA6	MA7	MA8
PIN17	PIN18	PIN19	PIN20	PIN21	PIN22	PIN23
MIY24	MIY25	MIY26	MIY27	MIY28	MIY29	MIY30
22	23	24	25	26	27	28
MA9	MA10	MA11	MA12	MA13	MA14	MA15
PIN24	PIN25	PIN26	PIN27	PIN28	PIN29	PIN30
GIN1	GIN2	GIN3	GIN4	GIN5	GIN6	GIN7

MABABA WEUSI 587/2103-04/6339-40/2095-96

Ashe	Ojobo	Iwa	Araorun	Ile	Itutu	Erinle
1	2	3	4	5	6	7
MA16	MA17	MA18	MA19	MA20	MA21	MA22
APP1	APP2	APP3	APP4	APP5	APP6	APP7
GIN8	GIN9	GIN10	GIN11	GIN12	GIN13	GIN14
8	9	10	11	12	13	14
MA23	MA24	MA25	MA26	MA27	MA28	MA29
APP8	APP9	APP10	APP11	APP12	APP13	APP14
GIN15	GIN16	GIN17	GIN18	GIN19	GIN20	GIN21
15	16	17	18	19	20	21
MA30	MA31	J1	J2	J3	J4	J5
APP15	APP16	APP17	APP18	APP19	APP20	APP21
GIN22	GIN23	GIN24	GIN25	GIN26	GIN27	GIN28
22	23	24	25	26	27	28
J6	J7	J8	J9	J10	J11	J12
APP22	APP23	APP24	APP25	APP26	APP27	APP28
GIN29	GIN30	SAN1	SAN2	SAN3	SAN4	SAN5

THE NU AFRIKAN CALENDAR

MWANAMKE WEUSI 587/2103-04/6339-40/2095-96						
Ashe	Ojobo	Iwa	Araorun	Ile	Itutu	Erinle
1	2	3	4	5	6	7
J13	J14	J15	J16	J17	J18	J19
APP29	APP30	MSR1	MSR2	MSR3	MSR4	MSR5
SAN6	SAN7	SAN8	SAN9	SAN10	SAN11	SAN12
8	9	10	11	12	13	14
J20	J21	J22	J23	J24	J25	J26
MSR6	MSR7	MSR8	MSR9	MSR10	MSR11	MSR12
SAN13	SAN14	SAN15	SAN16	SAN17	SAN18	SAN19
15	16	17	18	19	20	21
J27	J28	J29	J30	JU1	JU2	JU3
MSR13	MSR14	MSR15	MSR16	MSR17	MSR18	MSR19
SAN20	SAN21	SAN22	SAN23	SAN24	SAN25	SAN26
22	23	24	25	26	27	28
JU4	JU5	JU6	JU7	JU8	JU9	JU10
MSR20	MRS21	MSR22	MSR23	MSR24	MSR25	MSR26
SAN27	SAN28	SAN29	SAN30	HAM1	HAM2	HAM3

TAIFA WEUSI 587/2103-04/6339-40/2095-96						
Ashe	Ojobo	Iwa	Araorun	Ile	Itutu	Erinle
1	2	3	4	5	6	7
AU8	AU9	AU10	AU11	AU12	AU13	AU14
THT20	THT21	THT22	THT23	THT24	THT25	THT26
NAH2	NAH3	NAH4	NAH5	NAH6	NAH7	NAH8
8	9	10	11	12	13	14
AU15	AU16	AU17	AU18	AU19	AU20	AU21
THT27	THT28	THT29	THT30	PAP1	PAP1	PAP2
NAH9	NAH10	NAH11	NAH12	NAH13	NAH14	NAH15
15	16	17	18	19	20	21
AU22	AU23	AU24	AU25	AU26	AU27	AU28
PAP3	PAP4	PAP5	PAP6	PAP7	PAP8	PAP9
NAH16	NAH17	NAH18	NAH19	NAH20	NAH21	NAH22
22	23	24	25	26	27	28
AU29	AU30	AU31	S1	S2	S3	S4
PAP10	PAP11	PAP12	PAP13	PAP14	PAP15	PAP16
NAH23	NAH24	NAH25	NAH26	NAH27	NAH28	NAH29

MTU WEUSI 587/2103-04/6339-40/2095-96						
Ashe	Ojobo	Iwa	Araorun	Ile	Itutu	Erinle
1	2	3	4	5	6	7
JU11	JU12	JU13	JU14	JU15	JU16	JU17
MSR27	MSR28	MSR29	MSR30	*ASR1*	*HRU2*	*STH3*
HAM4	HAM5	HAM6	HAM7	HAM8	HAM9	HAM10
8	9	10	11	12	13	14
JU18	JU19	JU20	JU21	JU22	JU23	JU24
AST4	*NBH5*	THT1	THT2	THT3	THT4	THT5
HAM11	HAM12	HAM13	HAM14	HAM15	HAM16	HAM17
15	16	17	18	19	20	21
JU25	JU26	JU27	JU28	JU29	JU30	JU31
THT6	THT7	THT8	THT9	THT10	THT11	THT12
HAM18	HAM19	HAM20	HAM21	HAM22	HAM23	HAM24
22	23	24	25	26	27	28
AU1	AU2	AU3	AU4	AU5	AU6	AU7
THT13	THT14	THT15	THT16	THT17	THT18	THT19
HAM25	HAM26	HAM27	HAM28	HAM29	HAM30	NAH1

UWEZO 587/2103-04/6339-40/2095-96						
Ashe	Ojobo	Iwa	Araorun	Ile	Itutu	Erinle
1	2	3	4	5	6	7
S5	S6	S7	S8	S9	S10	S11
PAP17	PAP18	PAP19	PAP20	PAP21	PAP22	PAP23
NAH30	*QUA1*	*QUA2*	*QUA3*	*QUA4*	*QUA5*	*QUA6*
8	9	10	11	12	13	14
S12	S13	S14	S15	S16	S17	S18
PAP24	PAP25	PAP26	PAP27	PAP28	PAP29	PAP30
MAS1	MAS2	MAS3	MAS4	MAS5	MAS6	MAS7
15	16	17	18	19	20	21
S19	S20	S21	S22	S23	S24	S25
HTH1	HTH2	HTH3	HTH4	HTH5	HTH6	HTH7
MAS8	MAS9	MAS10	MAS11	MAS12	MAS13	MAS14
22	23	24	25	26	27	28
S26	S27	S28	S29	S30	O1	O2
HTH8	HTH9	HTH10	HTH11	HTH12	HTH13	HTH14
MAS15	MAS16	MAS17	MAS18	MAS19	MAS20	MAS21

THE NU AFRIKAN CALENDAR

EKUNDU 587/2103-04/6339-40/2095-96						
Ashe	Ojobo	Iwa	Araorun	Ile	Itutu	Erinle
1	2	3	4	5	6	7
O3	O4	O5	O6	O7	O8	O9
HTH15	HTH16	HTH17	HTH18	HTH19	HTH20	HTH21
MAS22	MAS23	MAS 24	MAS25	MAS26	MAS27	MAS28
8	9	10	11	12	13	14
O10	O11	O12	O13	O14	O15	O16
HTH22	HTH23	HTH24	HTH25	HTH26	HTH27	HTH28
MAS29	MAS30	THi1	THi2	THi3	THi4	THi5
15	16	17	18	19	20	21
O17	O18	O19	O20	O21	O22	O23
HTH29	HTH30	KHK1	KHK2	KHK3	KHK4	KHK5
THi6	THi7	THi8	THi9	THi10	THi11	THi12
22	23	24	25	26	27	28
O24	O25	O26	O27	O28	O29	O30
KHK6	KHK7	KHK8	KHK9	KHK10	KHK11	KHK12
THi13	THi14	THi15	THi16	THi17	THi18	THi19

KIJANI 587/2103-04/6339-40/2095-96						
Ashe	Ojobo	Iwa	Araorun	Ile	Itutu	Erinle
1	2	3	4	5	6	7
N28	N29	N30	D1	D2	D3	D4
TBT11	TBT12	TBT13	TBT14	TBT15	TBT16	TBT17
HID18	HID19	HID20	HID21	HID22	HID23	HID24
8	9	10	11	12	13	14
D5	D6	D7	D8	D9	D10	D11
TBT18	TBT19	TBT20	TBT21	TBT22	TBT23	TBT24
HID25	HID26	HID27	HID28	HID29	HID30	TAH1
15	16	17	18	19	20	21
D12	D13	D14	D15	D16	D17	D18
TBT25	TBT26	TBT27	TBT28	TBT29	TBT30	MKH1
TAH2	TAH3	TAH4	TAH5	TAH6	TAH7	TAH8
22	23	24	25	26	27	28
D19	D20	D21	D22	D23	D24	D25
MKH2	MKH3	MKH4	MKH5	MKH6	MKH7	MKH8
TAH9	TAH10	TAH11	TAH12	TAH13	TAH14	TAH15

EUSI 587/2103-04/6339-40/2095-96						
Ashe	Ojobo	Iwa	Araorun	Ile	Itutu	Erinle
1	2	3	4	5	6	7
O31	N1	N2	N3	N4	N5	N6
KHK13	KHK14	KHK15	KHK16	KHK17	KHK18	KHK19
THi20	THi21	THi22	THi23	THi24	THi25	THi26
8	9	10	11	12	13	14
N7	N8	N9	N10	N11	N12	N13
KHK20	KHK21	KHK22	KHK23	KHK24	KHK25	KHK26
THi27	THi28	THi29	THi30	HID1	HID2	HID3
15	16	17	18	19	20	21
N14	N15	N16	N17	N18	N19	N20
KHK27	KHK28	KHK29	KHK30	TBT1	TBT2	TBT3
HID4	HID5	HID6	HID7	HID8	HID9	HID10
22	23	24	25	26	27	28
N21	N22	N23	N24	N25	N26	N27
TBT4	TBT5	TBT6	TBT7	TBT8	TBT9	TBT10
HID11	HID12	HID13	HID14	HID15	HID16	HID17

KWANZA 587/2103-04/6339-40/2095-96						
Umoja	Kujichagulia	Ujima	Ujamaa	Nia	Kuumba	Imani
1	2	3	4	5	6	7
D26	D27	D28	D29	D30	D31	JA1
MKH9	MKH10	MKH11	MKH12	MKH13	MKH14	MKH1
TAH16	TAH17	TAH18	TAH19	TAH20	TAH21	TAH2
8	9	10	11	12	13	14
JA2	JA3	JA4	JA5	JA6	JA7	JA
MKH16	MKH17	MKH18	MKH19	MKH20	MKH21	MKH2
TAH23	TAH24	TAH25	TAH26	TAH27	TAH28	TAH29
15	16	17	18	19	20	21
JA9	JA10	JA11	JA12	JA13	JA14	JA1
MKH23	MKH24	MKH25	KMH26	MKH27	MKH28	MKH2
TAH30	TIR1	TIR2	TIR3	TIR4	TIR5	TIR6
22	23	24	25	26	27	28
JA16	JA17	JA18	JA19	JA20	JA21	JA2
MKH30	AMN1	AMN2	AMN3	AMN4	AMN5	AMN
TIR7	TIR8	TIR9	TIR10	TIR11	TIR12	TIR13

THE NU AFRIKAN CALENDAR

SHAHIDI 587/2103-04/6339-40/2095-96						
Ashe	Ojobo	Iwa	Araorun	Ile	Itutu	Erinle
1	2	3	4	5	6	7
JA23	JA24	JA25	JA26	JA27	JA28	JA29
AMN7	AMN8	AMN9	AMN10	AMN11	AMN12	AMN13
TIR14	TIR15	TIR16	TIR17	TIR18	TIR19	TIR20
8	9	10	11	12	13	14
JA30	JA31	F1	F2	F3	F4	F5
AMN14	AMN15	AMN16	AMN17	AMN18	AMN19	AMN20
TIR21	TIR22	TIR23	TIR24	TIR25	TIR26	TIR27
15	16	17	18	19	20	21
F6	F7	F8	F9	F10	F 11	F12
AMN21	AMN22	AMN23	AMN24	AMN25	AMN26	AMN28
TIR28	TIR29	TIR30	YEK1	YEK2	YEK3	YEK4
22	23	24	25	26	27	28
F13	F14	F15	F16	F17	F18	F19
AMN29	AMN30	RNT1	RNT2	RNT3	RNT4	RNT5
YEK5	YEK6	YEK7	YEK8	YEK9	YEK10	YEK11
SHAHIDI 138						
F20						
RNT6						
YEK12						

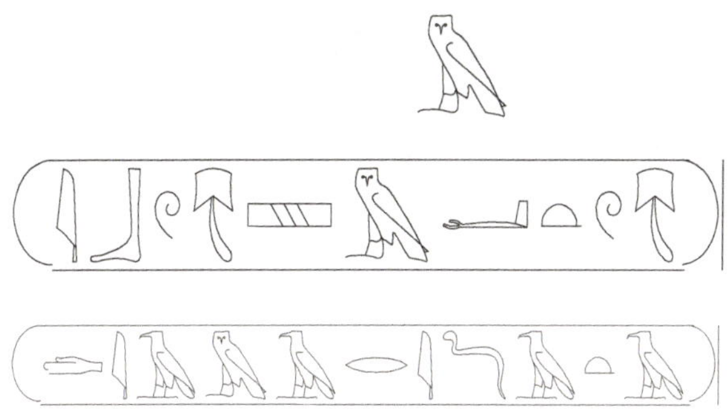

THE NU AFRIKAN CALENDAR

MAAFA 588 B.S.E./ASIENTO 2104-05 C.E./
KAM 6340-41 S.M./EC 2096-97 A.M. (ዓ.ም.)

AFU 588/2104-05/6340-41/2096-97						
Ashe	Ojobo	Iwa	Araorun	Ile	Itutu	Erinle
1	2	3	4	5	6	7
F21	F22	F23	F24	F25	F26	F27
RNT7	RNT8	RNT9	RNT10	RNT11	RNT12	RNT13
YEK13	YEK14	YEK15	YEK16	YEK17	YEK18	YEK19
8	9	10	11	12	13	14
F28	F29	M1	M2	M3	M4	M5
RNT14	RNT15	RNT16	RNT17	RNT18	RNT19	RNT20
YEK 20	YEK21	YEK22	YEK23	YEK24	YEK25	YEK26
15	16	17	18	19	20	21
M6	M7	M8	M9	M10	M11	M12
RNT21	RNT22	RNT23	RNT24	RNT25	RNT26	RNT27
YEK 27	YEK28	YEK29	YEK30	MAG1	MAG2	MAG3
22	23	24	25	26	27	28
M13	M14	M15	M16	M17	M18	M19
RNT28	RNT29	RNT30	KHN1	KHN2	KHN3	KHN4
MAG4	MAG5	MAG6	MAG7	MAG8	MAG9	MAG10
SHARIFU 143						
M20						
KHN5						
MAG11						

KAWAIDA 588/2104-05/6340-41/2096-97						
Ashe	Ojobo	Iwa	Araorun	Ile	Itutu	Erinle
1	2	3	4	5	6	7
A18	A19	A20	A21	A22	A23	A24
PIN4	PIN5	PIN6	PIN7	PIN8	PIN9	PIN10
MIY10	MIY11	MIY12	MIY13	MIY14	MIY15	MIY16
8	9	10	11	12	13	14
A25	A26	A27	A28	A29	A30	MA1
PIN11	PIN12	PIN13	PIN14	PIN15	PIN16	PIN17
MIY17	MIY18	MIY19	MIY20	MIY21	MIY22	MIY23
15	16	17	18	19	20	21
MA2	MA3	MA4	MA5	MA6	MA7	MA8
PIN18	PIN19	PIN20	PIN21	PIN22	PIN23	PIN24
MIY24	MIY25	MIY26	MIY27	MIY28	MIY29	MIY30
22	23	24	25	26	27	28
MA9	MA10	MA11	MA12	MA13	MA14	MA15
PIN25	PIN26	PIN27	PIN28	PIN29	PIN30	APP1
GIN1	GIN2	GIN3	GIN4	GIN5	GIN6	GIN7

UHURU 588/2104-05/6340-41/2096-97						
Ashe	Ojobo	Iwa	Araorun	Ile	Itutu	Erinle
1	2	3	4	5	6	7
M21	M22	M23	M24	M25	M26	M27
KHN6	KHN7	KHN8	KHN9	KHN10	KHN11	KHN12
MAG12	MAG13	MAG14	MAG15	MAG16	MAG17	MAG18
8	9	10	11	12	13	14
M28	M29	M30	M31	A1	A2	A3
KHN13	KHN14	KHN15	KHN16	KHN17	KHN18	KHN19
MAG19	MAG20	MAG21	MAG22	MAG23	MAG24	MAG25
15	16	17	18	19	20	21
A4	A5	A6	A7	A8	A9	A10
KHN20	KHN21	KHN222	KHN23	KHN24	KHN25	KHN26
MAG26	MAG27	MAG28	MAG29	MAG30	MIY1	MIY2
22	23	24	25	26	27	28
A11	A12	A13	A14	A15	A16	A17
KHN27	KHN28	KHN29	KHN30	PIN1	PIN2	PIN3
MIY3	MIY4	MIY5	MIY6	MIY7	MIY8	MIY9

MABABA WEUSI 588/2104-05/6340-41/2096-97						
Ashe	Ojobo	Iwa	Araorun	Ile	Itutu	Erinle
1	2	3	4	5	6	7
MA16	MA17	MA18	MA19	MA20	MA21	MA22
APP2	APP3	APP4	APP5	APP6	APP7	APP8
GIN8	GIN9	GIN10	GIN11	GIN12	GIN13	GIN14
8	9	10	11	12	13	14
MA23	MA24	MA25	MA26	MA27	MA28	MA29
APP9	APP10	APP11	APP12	APP13	APP14	APP15
GIN15	GIN16	GIN17	GIN18	GIN19	GIN20	GIN21
15	16	17	18	19	20	21
MA30	MA31	J1	J2	J3	J4	J5
APP16	APP17	APP18	APP19	APP20	APP21	APP22
GIN22	GIN23	GIN24	GIN25	GIN26	GIN27	GIN28
22	23	24	25	26	27	28
J6	J7	J8	J9	J10	J11	J12
APP23	APP24	APP25	APP26	APP27	APP28	APP29
GIN29	GIN30	SAN1	SAN2	SAN3	SAN4	SAN5

THE NU AFRIKAN CALENDAR

MWANAMKE WEUSI 588/2104-05/6340-41/2096-97						
Ashe	Ojobo	Iwa	Araorun	Ile	Itutu	Erinle
1	2	3	4	5	6	7
J13	J14	J15	J16	J17	J18	J19
APP30	MSR1	MSR2	MSR3	MSR4	MSR5	MSR6
SAN6	SAN7	SAN8	SAN9	SAN10	SAN11	SAN12
8	9	10	11	12	13	14
J20	J21	J22	J23	J24	J25	J26
MSR7	MSR8	MSR9	MSR10	MSR11	MSR12	MSR13
SAN13	SAN14	SAN15	SAN16	SAN17	SAN18	SAN19
15	16	17	18	19	20	21
J27	J28	J29	J30	JU1	JU2	JU3
MSR14	MSR15	MSR16	MSR17	MSR18	MSR19	MSR20
SAN20	SAN21	SAN22	SAN23	SAN24	SAN25	SAN26
22	23	24	25	26	27	28
JU4	JU5	JU6	JU7	JU8	JU9	JU10
MSR21	MRS22	MSR23	MSR24	MSR25	MSR26	MSR27
SAN27	SAN28	SAN29	SAN30	HAM1	HAM2	HAM3

TAIFA WEUSI 588/2104-05/6340-41/2096-97						
Ashe	Ojobo	Iwa	Araorun	Ile	Itutu	Erinle
1	2	3	4	5	6	7
AU8	AU9	AU10	AU11	AU12	AU13	AU14
THT20	THT21	THT22	THT23	THT24	THT25	THT26
NAH2	NAH3	NAH4	NAH5	NAH6	NAH7	NAH8
8	9	10	11	12	13	14
AU15	AU16	AU17	AU18	AU19	AU20	AU21
THT27	THT28	THT29	THT30	PAP1	PAP1	PAP2
NAH9	NAH10	NAH11	NAH12	NAH13	NAH14	NAH15
15	16	17	18	19	20	21
AU22	AU23	AU24	AU25	AU26	AU27	AU28
PAP3	PAP4	PAP5	PAP6	PAP7	PAP8	PAP9
NAH16	NAH17	NAH18	NAH19	NAH20	NAH21	NAH22
22	23	24	25	26	27	28
AU29	AU30	AU31	S1	S2	S3	S4
PAP10	PAP11	PAP12	PAP13	PAP14	PAP15	PAP16
NAH23	NAH24	NAH25	NAH26	NAH27	NAH28	NAH29

MTU WEUSI 588/2104-05/6340-41/2096-97						
Ashe	Ojobo	Iwa	Araorun	Ile	Itutu	Erinle
1	2	3	4	5	6	7
JU11	JU12	JU13	JU14	JU15	JU16	JU17
MSR28	MSR29	MSR30	*ASR1*	*HRU2*	*STH3*	*AST4*
HAM4	HAM5	HAM6	HAM7	HAM8	HAM9	HAM10
8	9	10	11	12	13	14
JU18	JU19	JU20	JU21	JU22	JU23	JU24
NBH5	*ANU6*	THT1	THT2	THT3	THT4	THT5
HAM11	HAM12	HAM13	HAM14	HAM15	HAM16	HAM17
15	16	17	18	19	20	21
JU25	JU26	JU27	JU28	JU29	JU30	JU31
THT6	THT7	THT8	THT9	THT10	THT11	THT12
HAM18	HAM19	HAM20	HAM21	HAM22	HAM23	HAM24
22	23	24	25	26	27	28
AU1	AU2	AU3	AU4	AU5	AU6	AU7
THT13	THT14	THT15	THT16	THT17	THT18	THT19
HAM25	HAM26	HAM27	HAM28	HAM29	HAM30	NAH1

UWEZO 588/2104-05/6340-41/2096-97						
Ashe	Ojobo	Iwa	Araorun	Ile	Itutu	Erinle
1	2	3	4	5	6	7
S5	S6	S7	S8	S9	S10	S11
PAP17	PAP18	PAP19	PAP20	PAP21	PAP22	PAP23
NAH30	*PAG1*	*PAG2*	*PAG3*	*PAG4*	*PAG5*	MAS1
8	9	10	11	12	13	14
S12	S13	S14	S15	S16	S17	S18
PAP24	PAP25	PAP26	PAP27	PAP28	PAP29	PAP30
MAS2	MAS3	MAS4	MAS5	MAS6	MAS7	MAS8
15	16	17	18	19	20	21
S19	S20	S21	S22	S23	S24	S25
HTH1	HTH2	HTH3	HTH4	HTH5	HTH6	HTH7
MAS9	MAS10	MAS11	MAS12	MAS13	MAS14	MAS15
22	23	24	25	26	27	28
S26	S27	S28	S29	S30	O1	O2
HTH8	HTH9	HTH10	HTH11	HTH12	HTH13	HTH14
MAS16	MAS17	MAS18	MAS19	MAS20	MAS21	MAS22

THE NU AFRIKAN CALENDAR

EKUNDU 588/2104-05/6340-41/2096-97						
Ashe	Ojobo	Iwa	Araorun	Ile	Itutu	Erinle
1	2	3	4	5	6	7
O3	O4	O5	O6	O7	O8	O9
HTH15	HTH16	HTH17	HTH18	HTH19	HTH20	HTH21
MAS23	MAS24	MAS25	MAS26	MAS27	MAS28	MAS29
8	9	10	11	12	13	14
O10	O11	O12	O13	O14	O15	O16
HTH22	HTH23	HTH24	HTH25	HTH26	HTH27	HTH28
MAS30	THi1	THi2	THi3	THi4	THi5	THi6
15	16	17	18	19	20	21
O17	O18	O19	O20	O21	O22	O23
HTH29	HTH30	KHK1	KHK2	KHK3	KHK4	KHK5
THi7	THi8	This9	THi10	THi11	THi12	THi13
22	23	24	25	26	27	28
O24	O25	O26	O27	O28	O29	O30
KHK6	KHK7	KHK8	KHK9	KHK10	KHK11	KHK12
Thi14	THi15	THi16	THi17	THi18	THi19	THi20

KIJANI 588/2104-05/6340-41/2096-97						
Ashe	Ojobo	Iwa	Araorun	Ile	Itutu	Erinle
1	2	3	4	5	6	7
N28	N29	N30	D1	D2	D3	D4
TBT11	TBT12	TBT13	TBT14	TBT15	TBT16	TBT17
HID19	HID20	HID21	HID22	HID23	HID24	HID25
8	9	10	11	12	13	14
D5	D6	D7	D8	D9	D10	D11
TBT18	TBT19	TBT20	TBT21	TBT22	TBT23	TBT24
HID26	HID27	HID28	HID29	HID30	TAH1	TAH2
15	16	17	18	19	20	21
D12	D13	D14	D15	D16	D17	D18
TBT25	TBT26	TBT27	TBT28	TBT29	TBT30	MKH1
TAH3	TAH4	TAH5	TAH6	TAH7	TAH8	TAH9
22	23	24	25	26	27	28
D19	D20	D21	D22	D23	D24	D25
MKH2	MKH3	MKH4	MKH5	MKH6	MKH7	MKH8
TAH10	TAH11	TAH12	TAH13	TAH14	TAH15	TAH16

EUSI 588/2104-05/6340-41/2096-97						
Ashe	Ojobo	Iwa	Araorun	Ile	Itutu	Erinle
1	2	3	4	5	6	7
O31	N1	N2	N3	N4	N5	N6
KHK13	KHK14	KHK15	KHK16	KHK17	KHK18	KHK19
THi21	THi22	THi23	THi24	THi25	THi26	THi27
8	9	10	11	12	13	14
N7	N8	N9	N10	N11	N12	N13
KHK20	KHK21	KHK22	KHK23	KHK24	KHK25	KHK26
THi28	THi29	THi30	HID1	HID2	HID3	HID4
15	16	17	18	19	20	21
N14	N15	N16	N17	N18	N19	N20
KHK27	KHK28	KHK29	KHK30	TBT1	TBT2	TBT3
HID5	HID6	HID7	HID8	HID9	HID10	HID11
22	23	24	25	26	27	28
N21	N22	N23	N24	N25	N26	N27
TBT4	TBT5	TBT6	TBT7	TBT8	TBT9	TBT10
HID12	HID13	HID14	HID15	HID16	HID17	HID18

KWANZA 588/2104-05/6340-41/2096-97						
Umoja	Kujichagulia	Ujima	Ujamaa	Nia	Kuumba	Imani
1	2	3	4	5	6	7
D26	D27	D28	D29	D30	D31	JA1
MKH9	MKH10	MKH11	MKH12	MKH13	MKH14	MKH1
TAH17	TAH18	TAH19	TAH20	TAH21	TAH22	TAH2
8	9	10	11	12	13	14
JA2	JA3	JA4	JA5	JA6	JA7	JA
MKH16	MKH17	MKH18	MKH19	MKH20	MKH21	MKH2
TAH24	TAH25	TAH26	TAH27	TAH28	TAH29	TAH3
15	16	17	18	19	20	21
JA9	JA10	JA11	JA12	JA13	JA14	JA1
MKH23	MKH24	MKH25	KMH26	MKH27	MKH28	MKH2
TIR1	TIR2	TIR3	TIR4	TIR5	TIR6	TIR7
22	23	24	25	26	27	28
JA16	JA17	JA18	JA19	JA20	JA21	JA2
MKH30	AMN1	AMN2	AMN3	AMN4	AMN5	AMN
TIR8	TIR9	TIR10	TIR11	TIR12	TIR13	TIR1

THE NU AFRIKAN CALENDAR

SHAHIDI 588/2104-05/6340-41/2096-97						
Ashe	Ojobo	Iwa	Araorun	Ile	Itutu	Erinle
1	2	3	4	5	6	7
JA23	JA24	JA25	JA26	JA27	JA28	JA29
AMN7	AMN8	AMN9	AMN10	AMN11	AMN12	AMN13
TIR15	TIR16	TIR17	TIR18	TIR19	TIR20	TIR21
8	9	10	11	12	13	14
JA30	JA31	F1	F2	F3	F4	F5
AMN14	AMN15	AMN16	AMN17	AMN18	AMN19	AMN20
TIR22	TIR23	TIR24	TIR25	TIR26	TIR27	TIR28
15	16	17	18	19	20	21
F6	F7	F8	F9	F10	F 11	F12
AMN21	AMN22	AMN23	AMN24	AMN25	AMN26	AMN28
TIR29	TIR30	YEK1	YEK2	YEK3	YEK4	YEK5
22	23	24	25	26	27	28
F13	F14	F15	F16	F17	F18	F19
AMN29	AMN30	RNT1	RNT2	RNT3	RNT4	RNT5
YEK6	YEK7	YEK8	YEK9	YEK10	YEK11	YEK12
SHAHIDI 139						
F20						
RNT6						
YEK13						

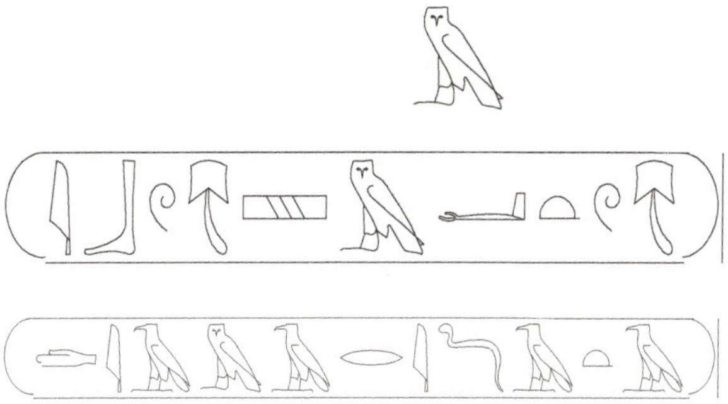

THE NU AFRIKAN CALENDAR

MAAFA 583 B.S.E./ASIENTO 2099-00 C.E.
KAM 6335-36 S.M./EC 2091-92 A.M. (ዓ.ም.)

AFU 583/2099-00/6335-36/2091-92						
Ashe	Ojobo	Iwa	Araorun	Ile	Itutu	Erinle
1	2	3	4	5	6	7
F21	F22	F23	F24	F25	F26	F27
RNT7	RNT8	RNT9	RNT10	RNT11	RNT12	RNT13
YEK14	YEK15	YEK16	YEK17	YEK18	YEK19	YEK20
8	9	10	11	12	13	14
F28	M1	M2	M3	M4	M5	M6
RNT14	RNT15	RNT16	RNT17	RNT18	RNT19	RNT20
YEK21	YEK22	YEK23	YEK24	YEK25	YEK26	YEK27
15	16	17	18	19	20	21
M7	M8	M9	M10	M11	M12	M13
RNT21	RNT22	RNT23	RNT24	RNT25	RNT26	RNT27
YEK28	YEK29	YEK30	MAG1	MAG2	MAG3	MAG4
22	23	24	25	26	27	28
M14	M15	M16	M17	M18	M19	M20
RNT28	RNT29	RNT30	KHN1	KHN2	KHN3	KHN4
MAG5	MAG6	MAG7	MAG8	MAG9	MAG10	MAG11

KAWAIDA 583/2099-00/6335-36/2091-92						
Ashe	Ojobo	Iwa	Araorun	Ile	Itutu	Erinle
1	2	3	4	5	6	7
A18	A19	A20	A21	A22	A23	A24
PIN3	PIN4	PIN5	PIN6	PIN7	PIN8	PIN9
MIY10	MIY11	MIY12	MIY13	MIY14	MIY15	MIY16
8	9	10	11	12	13	14
A25	A26	A27	A28	A29	A30	MA1
PIN10	PIN11	PIN12	PIN13	PIN14	PIN15	PIN16
MIY17	MIY18	MIY19	MIY20	MIY21	MIY22	MIY23
15	16	17	18	19	20	21
MA2	MA3	MA4	MA5	MA6	MA7	MA8
PIN17	PIN18	PIN19	PIN20	PIN21	PIN22	PIN23
MIY24	MIY25	MIY26	MIY27	MIY28	MIY29	MIY30
22	23	24	25	26	27	28
MA9	MA10	MA11	MA12	MA13	MA14	MA15
PIN24	PIN25	PIN26	PIN27	PIN28	PIN29	PIN30
GIN1	GIN2	GIN3	GIN4	GIN5	GIN6	GIN7

UHURU 583/2099-00/6335-36/2091-92						
Ashe	Ojobo	Iwa	Araorun	Ile	Itutu	Erinle
1	2	3	4	5	6	7
M21	M22	M23	M24	M25	M26	M27
KHN5	KHN6	KHN7	KHN8	KHN9	KHN10	KHN11
MAG12	MAG13	MAG14	MAG15	MAG16	MAG17	MAG18
8	9	10	11	12	13	14
M28	M29	M30	M31	A1	A2	A3
KHN12	KHN13	KHN14	KHN15	KHN16	KHN17	KHN18
MAG19	MAG20	MAG21	MAG22	MAG23	MAG24	MAG25
15	16	17	18	19	20	21
A4	A5	A6	A7	A8	A9	A10
KHN19	KHN20	KHN21	KHN22	KHN23	KHN24	KHN25
MAG26	MAG27	MAG28	MAG29	MAG30	MIY1	MIY2
22	23	24	25	26	27	28
A11	A12	A13	A14	A15	A16	A17
KHN26	KHN27	KHN28	KHN29	KHN30	PIN1	PIN2
MIY3	MIY4	MIY5	MIY6	MIY7	MIY8	MIY9

MABABA WEUSI 583/2099-00/6335-36/2091-92						
Ashe	Ojobo	Iwa	Araorun	Ile	Itutu	Erinle
1	2	3	4	5	6	7
MA16	MA17	MA18	MA19	MA20	MA21	MA22
APP1	APP2	APP3	APP4	APP5	APP6	APP7
GIN8	GIN9	GIN10	GIN11	GIN12	GIN13	GIN14
8	9	10	11	12	13	14
MA23	MA24	MA25	MA26	MA27	MA28	MA29
APP8	APP9	APP10	APP11	APP12	APP13	APP14
GIN15	GIN16	GIN17	GIN18	GIN19	GIN20	GIN21
15	16	17	18	19	20	21
MA30	MA31	J1	J2	J3	J4	J5
APP15	APP16	APP17	APP18	APP19	APP20	APP21
GIN22	GIN23	GIN24	GIN25	GIN26	GIN27	GIN28
22	23	24	25	26	27	28
J6	J7	J8	J9	J10	J11	J12
APP22	APP23	APP24	APP25	APP26	APP27	APP28
GIN29	GIN30	SAN1	SAN2	SAN3	SAN4	SAN5

THE NU AFRIKAN CALENDAR

MWANAMKE WEUSI 583/2099-00/6335-36/2091-92						
Ashe	Ojobo	Iwa	Araorun	Ile	Itutu	Erinle
1	2	3	4	5	6	7
J13	J14	J15	J16	J17	J18	J19
APP29	APP30	MSR1	MSR2	MSR3	MSR4	MSR5
SAN6	SAN7	SAN8	SAN9	SAN10	SAN11	SAN12
8	9	10	11	12	13	14
J20	J21	J22	J23	J24	J25	J26
MSR6	MSR7	MSR8	MSR9	MSR10	MSR11	MSR12
SAN13	SAN14	SAN15	SAN16	SAN17	SAN18	SAN19
15	16	17	18	19	20	21
J27	J28	J29	J30	JU1	JU2	JU3
MSR13	MSR14	MSR15	MSR16	MSR17	MSR18	MSR19
SAN20	SAN21	SAN22	SAN23	SAN24	SAN25	SAN26
22	23	24	25	26	27	28
JU4	JU5	JU6	JU7	JU8	JU9	JU10
MSR20	MRS21	MSR22	MSR23	MSR24	MSR25	MSR26
SAN27	SAN28	SAN29	SAN30	HAM1	HAM2	HAM3

TAIFA WEUSI 583/2099-00/6335-36/2091-92						
Ashe	Ojobo	Iwa	Araorun	Ile	Itutu	Erinle
1	2	3	4	5	6	7
AU8	AU9	AU10	AU11	AU12	AU13	AU14
THT20	THT21	THT22	THT23	THT24	THT25	THT26
NAH2	NAH3	NAH4	NAH5	NAH6	NAH7	NAH8
8	9	10	11	12	13	14
AU15	AU16	AU17	AU18	AU19	AU20	AU21
THT27	THT28	THT29	THT30	PAP1	PAP1	PAP2
NAH9	NAH10	NAH11	NAH12	NAH13	NAH14	NAH15
15	16	17	18	19	20	21
AU22	AU23	AU24	AU25	AU26	AU27	AU28
PAP3	PAP4	PAP5	PAP6	PAP7	PAP8	PAP9
NAH16	NAH17	NAH18	NAH19	NAH20	NAH21	NAH22
22	23	24	25	26	27	28
AU29	AU30	AU31	S1	S2	S3	S4
PAP10	PAP11	PAP12	PAP13	PAP14	PAP15	PAP16
NAH23	NAH24	NAH25	NAH26	NAH27	NAH28	NAH29

MTU WEUSI 583/2099-00/6335-36/2091-92						
Ashe	Ojobo	Iwa	Araorun	Ile	Itutu	Erinle
1	2	3	4	5	6	7
JU11	JU12	JU13	JU14	JU15	JU16	JU17
MSR27	MSR28	MSR29	MSR30	*ASR1*	*HRU2*	*STH3*
HAM4	HAM5	HAM6	HAM7	HAM8	HAM9	HAM10
8	9	10	11	12	13	14
JU18	JU19	JU20	JU21	JU22	JU23	JU24
AST4	*NBH5*	THT1	THT2	THT3	THT4	THT5
HAM11	HAM12	HAM13	AM14	HAM15	HAM16	HAM17
15	16	17	18	19	20	21
JU25	JU26	JU27	JU28	JU29	JU30	JU31
THT6	THT7	THT8	THT9	THT10	THT11	THT12
HAM18	HAM19	HAM20	HAM21	HAM22	HAM23	HAM24
22	23	24	25	26	27	28
AU1	AU2	AU3	AU4	AU5	AU6	AU7
THT13	THT14	THT15	THT16	THT17	THT18	THT19
HAM25	HAM26	HAM27	HAM28	HAM29	HAM30	NAH1

UWEZO 583/2099-00/6335-36/2091-92						
Ashe	Ojobo	Iwa	Araorun	Ile	Itutu	Erinle
1	2	3	4	5	6	7
S5	S6	S7	S8	S9	S10	S11
PAP17	PAP18	PAP19	PAP20	PAP21	PAP22	PAP23
NAH30	*PAG1*	*PAG2*	*PAG3*	*PAG4*	*PAG5*	MAS1
8	9	10	11	12	13	14
S12	S13	S14	S15	S16	S17	S18
PAP24	PAP25	PAP26	PAP27	PAP28	PAP29	PAP30
MAS2	MAS3	MAS4	MAS5	MAS6	MAS7	MAS8
15	16	17	18	19	20	21
S19	S20	S21	S22	S23	S24	S25
HTH1	HTH2	HTH3	HTH4	HTH5	HTH6	HTH7
MAS9	MAS10	MAS11	MAS12	MAS13	MAS14	MAS15
22	23	24	25	26	27	28
S26	S27	S28	S29	S30	O1	O2
HTH8	HTH9	HTH10	HTH11	HTH12	HTH13	HTH14
MAS16	MAS17	MAS18	MAS19	MAS20	MAS21	MAS22

THE NU AFRIKAN CALENDAR

EKUNDU 583/2099-00/6335-36/2091-92

Ashe	Ojobo	Iwa	Araorun	Ile	Itutu	Erinle
1	2	3	4	5	6	7
O3	O4	O5	O6	O7	O8	O9
HTH15	HTH16	HTH17	HTH18	HTH19	HTH20	HTH21
MAS23	MAS24	MAS25	MAS26	MAS27	MAS28	MAS29
8	9	10	11	12	13	14
O10	O11	O12	O13	O14	O15	O16
HTH22	HTH23	HTH24	HTH25	HTH26	HTH27	HTH28
MAS30	THi1	THi2	THi3	THi4	THi5	THi6
15	16	17	18	19	20	21
O17	O18	O19	O20	O21	O22	O23
HTH29	HTH30	KHK1	KHK2	KHK3	KHK4	KHK5
THi7	THi8	THi9	THi10	THi11	THi12	THi13
22	23	24	25	26	27	28
O24	O25	O26	O27	O28	O29	O30
KHK6	KHK7	KHK8	KHK9	KHK10	KHK11	KHK12
THi14	THi15	THi16	THi17	THi18	THi19	THi20

KIJANI 583/2099-00/6335-36/2091-92

Ashe	Ojobo	Iwa	Araorun	Ile	Itutu	Erinle
1	2	3	4	5	6	7
N28	N29	N30	D1	D2	D3	D4
TBT11	TBT12	TBT13	TBT14	TBT15	TBT16	TBT17
HID19	HID20	HID21	HID22	HID23	HID24	HID25
8	9	10	11	12	13	14
D5	D6	D7	D8	D9	D10	D11
TBT18	TBT19	TBT20	TBT21	TBT22	TBT23	TBT24
HID26	HID27	HID28	HID29	HID30	TAH1	TAH2
15	16	17	18	19	20	21
D12	D13	D14	D15	D16	D17	D18
TBT25	TBT26	TBT27	TBT28	TBT29	TBT30	MKH1
TAH3	TAH4	TAH5	TAH6	TAH7	TAH8	TAH9
22	23	24	25	26	27	28
D19	D20	D21	D22	D23	D24	D25
MKH2	MKH3	MKH4	MKH5	MKH6	MKH7	MKH8
TAH10	TAH11	TAH12	TAH13	TAH14	TAH15	TAH16

EUSI 583/2099-00/6335-36/2091-92

Ashe	Ojobo	Iwa	Araorun	Ile	Itutu	Erinle
1	2	3	4	5	6	7
O31	N1	N2	N3	N4	N5	N6
KHK13	KHK14	KHK15	KHK16	KHK17	KHK18	KHK19
THi21	THi22	THi23	THi24	THi25	THi26	THi27
8	9	10	11	12	13	14
N7	N8	N9	N10	N11	N12	N13
KHK20	KHK21	KHK22	KHK23	KHK24	KHK25	KHK26
THi28	THi29	THi30	HID1	HID2	HID3	HID4
15	16	17	18	19	20	21
N14	N15	N16	N17	N18	N19	N20
KHK27	KHK28	KHK29	KHK30	TBT1	TBT2	TBT3
HID5	HID6	HID7	HID8	HID9	HID10	HID11
22	23	24	25	26	27	28
N21	N22	N23	N24	N25	N26	N27
TBT4	TBT5	TBT6	TBT7	TBT8	TBT9	TBT10
HID12	HID13	HID14	HID15	HID16	HID17	HID18

KWANZA 583/2099-00/6335-36/2091-92

Umoja	Kujichagulia	Ujima	Ujamaa	Nia	Kuumba	Imani
1	2	3	4	5	6	7
D26	D27	D28	D29	D30	D31	JA1
MKH9	MKH10	MKH11	MKH12	MKH13	MKH14	MKH1
TAH17	TAH18	TAH19	TAH20	TAH21	TAH22	TAH23
8	9	10	11	12	13	14
JA2	JA3	JA4	JA5	JA6	JA7	JA8
MKH16	MKH17	MKH18	MKH19	MKH20	MKH21	MKH2
TAH24	TAH25	TAH26	TAH27	TAH28	TAH29	TAH30
15	16	17	18	19	20	21
JA9	JA10	JA11	JA12	JA13	JA14	JA15
MKH23	MKH24	MKH25	KMH26	MKH27	MKH28	MKH2
TIR1	TIR2	TIR3	TIR4	TIR5	TIR6	TIR7
22	23	24	25	26	27	28
JA16	JA17	JA18	JA19	JA20	JA21	JA22
MKH30	AMN1	AMN2	AMN3	AMN4	AMN5	AMN6
TIR8	TIR9	TIR10	TIR11	TIR12	TIR13	TIR14

THE NU AFRIKAN CALENDAR

SHAHIDI 583/2099-00/6335-36/2091-92						
Ashe	Ojobo	Iwa	Araorun	Ile	Itutu	Erinle
1	2	3	4	5	6	7
JA23	JA24	JA25	JA26	JA27	JA28	JA29
AMN7	AMN8	AMN9	AMN10	AMN11	AMN12	AMN13
TIR15	TIR16	TIR17	TIR18	TIR19	TIR20	TIR21
8	9	10	11	12	13	14
JA30	JA31	F1	F2	F3	F4	F5
AMN14	AMN15	AMN16	AMN17	AMN18	AMN19	AMN20
TIR22	TIR23	TIR24	TIR25	TIR26	TIR27	TIR28
15	16	17	18	19	20	21
F6	F7	F8	F9	F10	F 11	F12
AMN21	AMN22	AMN23	AMN24	AMN25	AMN26	AMN28
TIR29	TIR30	YEK1	YEK2	YEK3	YEK4	YEK5
22	23	24	25	26	27	28
F13	F14	F15	F16	F17	F18	F19
AMN29	AMN30	RNT1	RNT2	RNT3	RNT4	RNT5
YEK6	YEK7	YEK8	YEK9	YEK10	YEK11	YEK12
SHAHIDI 134						
F20						
RNT6						
YEK13						

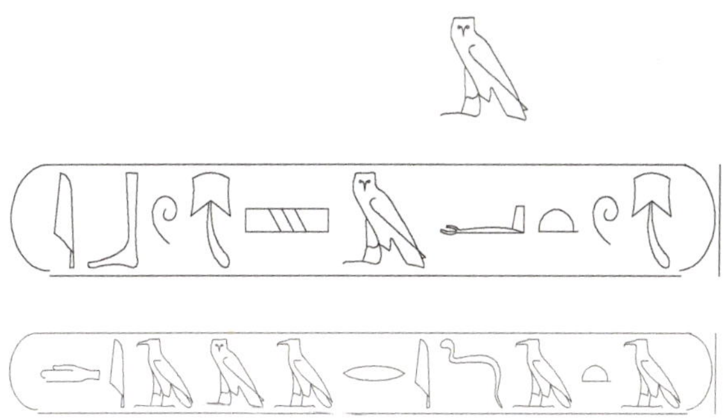

THE NU AFRIKAN CALENDAR

MAAFA 584 B.S.E./ASIENTO 2100-01 C.E.
KAM 6336-37 S.M./EC 2092-93 A.M. (ዓ.ም.)

AFU 584/2100-01/6336-37/2092-93

Ashe	Ojobo	Iwa	Araorun	Ile	Itutu	Erinle
1	2	3	4	5	6	7
F21	F22	F23	F24	F25	F26	F27
RNT7	RNT8	RNT9	RNT10	RNT11	RNT12	RNT13
YEK14	YEK15	YEK16	YEK17	YEK18	YEK19	YEK20
8	9	10	11	12	13	14
F28	M1	M2	M3	M4	M5	M6
RNT14	RNT15	RNT16	RNT17	RNT18	RNT19	RNT20
YEK21	YEK22	YEK23	YEK24	YEK25	YEK26	YEK27
15	16	17	18	19	20	21
M7	M8	M9	M10	M11	M12	M13
RNT21	RNT22	RNT23	RNT24	RNT25	RNT26	RNT27
YEK28	YEK29	YEK30	MAG1	MAG2	MAG3	MAG4
22	23	24	25	26	27	28
M14	M15	M16	M17	M18	M19	M20
RNT28	RNT29	RNT30	KHN1	KHN2	KHN3	KHN4
MAG5	MAG6	MAG7	MAG8	MAG9	MAG10	MAG11

KAWAIDA 584/2100-01/6336-37/2092-93

Ashe	Ojobo	Iwa	Araorun	Ile	Itutu	Erinle
1	2	3	4	5	6	7
A18	A19	A20	A21	A22	A23	A24
PIN3	PIN4	PIN5	PIN6	PIN7	PIN8	PIN9
MIY10	MIY11	MIY12	MIY13	MIY14	MIY15	MIY16
8	9	10	11	12	13	14
A25	A26	A27	A28	A29	A30	MA1
PIN10	PIN11	PIN12	PIN13	PIN14	PIN15	PIN16
MIY17	MIY18	MIY19	MIY20	MIY21	MIY22	MIY23
15	16	17	18	19	20	21
MA2	MA3	MA4	MA5	MA6	MA7	MA8
PIN17	PIN18	PIN19	PIN20	PIN21	PIN22	PIN23
MIY24	MIY25	MIY26	MIY27	MIY28	MIY29	MIY30
22	23	24	25	26	27	28
MA9	MA10	MA11	MA12	MA13	MA14	MA15
PIN24	PIN25	PIN26	PIN27	PIN28	PIN29	PIN30
GIN1	GIN2	GIN3	GIN4	GIN5	GIN6	GIN7

UHURU 584/2100-01/6336-37/2092-93

Ashe	Ojobo	Iwa	Araorun	Ile	Itutu	Erinle
1	2	3	4	5	6	7
M21	M22	M23	M24	M25	M26	M27
KHN5	KHN6	KHN7	KHN8	KHN9	KHN10	KHN11
MAG12	MAG13	MAG14	MAG15	MAG16	MAG17	MAG18
8	9	10	11	12	13	14
M28	M29	M30	M31	A1	A2	A3
KHN12	KHN13	KHN14	KHN15	KHN16	KHN17	KHN18
MAG19	MAG20	MAG21	MAG22	MAG23	MAG24	MAG25
15	16	17	18	19	20	21
A4	A5	A6	A7	A8	A9	A10
KHN19	KHN20	KHN21	KHN22	KHN23	KHN24	KHN25
MAG26	MAG27	MAG28	MAG29	MAG30	MIY1	MIY2
22	23	24	25	26	27	28
A11	A12	A13	A14	A15	A16	A17
KHN26	KHN27	KHN28	KHN29	KHN30	PIN1	PIN2
MIY3	MIY4	MIY5	MIY6	MIY7	MIY8	MIY9

MABABA WEUSI 584/2100-01/6336-37/2092-93

Ashe	Ojobo	Iwa	Araorun	Ile	Itutu	Erinle
1	2	3	4	5	6	7
MA16	MA17	MA18	MA19	MA20	MA21	MA22
APP1	APP2	APP3	APP4	APP5	APP6	APP7
GIN8	GIN9	GIN10	GIN11	GIN12	GIN13	GIN14
8	9	10	11	12	13	14
MA23	MA24	MA25	MA26	MA27	MA28	MA29
APP8	APP9	APP10	APP11	APP12	APP13	APP14
GIN15	GIN16	GIN17	GIN18	GIN19	GIN20	GIN21
15	16	17	18	19	20	21
MA30	MA31	J1	J2	J3	J4	J5
APP15	APP16	APP17	APP18	APP19	APP20	APP21
GIN22	GIN23	GIN24	GIN25	GIN26	GIN27	GIN28
22	23	24	25	26	27	28
J6	J7	J8	J9	J10	J11	J12
APP22	APP23	APP24	APP25	APP26	APP27	APP28
GIN29	GIN30	SAN1	SAN2	SAN3	SAN4	SAN5

THE NU AFRIKAN CALENDAR

MWANAMKE WEUSI 584/2100-01/6336-37/2092-93						
Ashe	Ojobo	Iwa	Araorun	Ile	Itutu	Erinle
1	2	3	4	5	6	7
J13	J14	J15	J16	J17	J18	J19
APP29	APP30	MSR1	MSR2	MSR3	MSR4	MSR5
SAN6	SAN7	SAN8	SAN9	SAN10	SAN11	SAN12
8	9	10	11	12	13	14
J20	J21	J22	J23	J24	J25	J26
MSR6	MSR7	MSR8	MSR9	MSR10	MSR11	MSR12
SAN13	SAN14	SAN15	SAN16	SAN17	SAN18	SAN19
15	16	17	18	19	20	21
J27	J28	J29	J30	JU1	JU2	JU3
MSR13	MSR14	MSR15	MSR16	MSR17	MSR18	MSR19
SAN20	SAN21	SAN22	SAN23	SAN24	SAN25	SAN26
22	23	24	25	26	27	28
JU4	JU5	JU6	JU7	JU8	JU9	JU10
MSR20	MRS21	MSR22	MSR23	MSR24	MSR25	MSR26
SAN27	SAN28	SAN29	SAN30	HAM1	HAM2	HAM3

TAIFA WEUSI 584/2100-01/6336-37/2092-93						
Ashe	Ojobo	Iwa	Araorun	Ile	Itutu	Erinle
1	2	3	4	5	6	7
AU8	AU9	AU10	AU11	AU12	AU13	AU14
THT20	THT21	THT22	THT23	THT24	THT25	THT26
NAH2	NAH3	NAH4	NAH5	NAH6	NAH7	NAH8
8	9	10	11	12	13	14
AU15	AU16	AU17	AU18	AU19	AU20	AU21
THT27	THT28	THT29	THT30	PAP1	PAP1	PAP2
NAH9	NAH10	NAH11	NAH12	NAH13	NAH14	NAH15
15	16	17	18	19	20	21
AU22	AU23	AU24	AU25	AU26	AU27	AU28
PAP3	PAP4	PAP5	PAP6	PAP7	PAP8	PAP9
NAH16	NAH17	NAH18	NAH19	NAH20	NAH21	NAH22
22	23	24	25	26	27	28
AU29	AU30	AU31	S1	S2	S3	S4
PAP10	PAP11	PAP12	PAP13	PAP14	PAP15	PAP16
NAH23	NAH24	NAH25	NAH26	NAH27	NAH28	NAH29

MTU WEUSI 584/2100-01/6336-37/2092-93						
Ashe	Ojobo	Iwa	Araorun	Ile	Itutu	Erinle
1	2	3	4	5	6	7
JU11	JU12	JU13	JU14	JU15	JU16	JU17
MSR27	MSR28	MSR29	MSR30	*ASR1*	*HRU2*	*STH3*
HAM4	HAM5	HAM6	HAM7	HAM8	HAM9	HAM10
8	9	10	11	12	13	14
JU18	JU19	JU20	JU21	JU22	JU23	JU24
AST4	*NBH5*	THT1	THT2	THT3	THT4	THT5
HAM11	HAM12	HAM13	AM14	HAM15	HAM16	HAM17
15	16	17	18	19	20	21
JU25	JU26	JU27	JU28	JU29	JU30	JU31
THT6	THT7	THT8	THT9	THT10	THT11	THT12
HAM18	HAM19	HAM20	HAM21	HAM22	HAM23	HAM24
22	23	24	25	26	27	28
AU1	AU2	AU3	AU4	AU5	AU6	AU7
THT13	THT14	THT15	THT16	THT17	THT18	THT19
HAM25	HAM26	HAM27	HAM28	HAM29	HAM30	NAH1

UWEZO 584/2100-01/6336-37/2092-93						
Ashe	Ojobo	Iwa	Araorun	Ile	Itutu	Erinle
1	2	3	4	5	6	7
S5	S6	S7	S8	S9	S10	S11
PAP17	PAP18	PAP19	PAP20	PAP21	PAP22	PAP23
NAH30	*PAG1*	*PAG2*	*PAG3*	*PAG4*	*PAG5*	MAS1
8	9	10	11	12	13	14
S12	S13	S14	S15	S16	S17	S18
PAP24	PAP25	PAP26	PAP27	PAP28	PAP29	PAP30
MAS2	MAS3	MAS4	MAS5	MAS6	MAS7	MAS8
15	16	17	18	19	20	21
S19	S20	S21	S22	S23	S24	S25
HTH1	HTH2	HTH3	HTH4	HTH5	HTH6	HTH7
MAS9	MAS10	MAS11	MAS12	MAS13	MAS14	MAS15
22	23	24	25	26	27	28
S26	S27	S28	S29	S30	O1	O2
HTH8	HTH9	HTH10	HTH11	HTH12	HTH13	HTH14
MAS16	MAS17	MAS18	MAS19	MAS20	MAS21	MAS22

THE NU AFRIKAN CALENDAR

EKUNDU 584/2100-01/6336-37/2092-93

Ashe	Ojobo	Iwa	Araorun	Ile	Itutu	Erinle
1	2	3	4	5	6	7
O3	O4	O5	O6	O7	O8	O9
HTH15	HTH16	HTH17	HTH18	HTH19	HTH20	HTH21
MAS23	MAS24	MAS25	MAS26	MAS27	MAS28	MAS29
8	9	10	11	12	13	14
O10	O11	O12	O13	O14	O15	O16
HTH22	HTH23	HTH24	HTH25	HTH26	HTH27	HTH28
MAS30	THi1	THi2	THi3	THi4	THi5	THi6
15	16	17	18	19	20	21
O17	O18	O19	O20	O21	O22	O23
HTH29	HTH30	KHK1	KHK2	KHK3	KHK4	KHK5
THi7	THi8	THi9	THi10	THi11	THi12	THi13
22	23	24	25	26	27	28
O24	O25	O26	O27	O28	O29	O30
KHK6	KHK7	KHK8	KHK9	KHK10	KHK11	KHK12
THi14	THi15	THi16	THi17	THi18	THi19	THi20

EUSI 584/2100-01/6336-37/2092-93

Ashe	Ojobo	Iwa	Araorun	Ile	Itutu	Erinle
1	2	3	4	5	6	7
O31	N1	N2	N3	N4	N5	N6
KHK13	KHK14	KHK15	KHK16	KHK17	KHK18	KHK19
THi21	THi22	THi23	THi24	THi25	THi26	THi27
8	9	10	11	12	13	14
N7	N8	N9	N10	N11	N12	N13
KHK20	KHK21	KHK22	KHK23	KHK24	KHK25	KHK26
THi28	THi29	THi30	HID1	HID2	HID3	HID4
15	16	17	18	19	20	21
N14	N15	N16	N17	N18	N19	N20
KHK27	KHK28	KHK29	KHK30	TBT1	TBT2	TBT3
HID5	HID6	HID7	HID8	HID9	HID10	HID11
22	23	24	25	26	27	28
N21	N22	N23	N24	N25	N26	N27
TBT4	TBT5	TBT6	TBT7	TBT8	TBT9	TBT10
HID12	HID13	HID14	HID15	HID16	HID17	HID18

KIJANI 584/2100-01/6336-37/2092-93

Ashe	Ojobo	Iwa	Araorun	Ile	Itutu	Erinle
1	2	3	4	5	6	7
N28	N29	N30	D1	D2	D3	D4
TBT11	TBT12	TBT13	TBT14	TBT15	TBT16	TBT17
HID19	HID20	HID21	HID22	HID23	HID24	HID25
8	9	10	11	12	13	14
D5	D6	D7	D8	D9	D10	D11
TBT18	TBT19	TBT20	TBT21	TBT22	TBT23	TBT24
HID26	HID27	HID28	HID29	HID30	TAH1	TAH2
15	16	17	18	19	20	21
D12	D13	D14	D15	D16	D17	D18
TBT25	TBT26	TBT27	TBT28	TBT29	TBT30	MKH1
TAH3	TAH4	TAH5	TAH6	TAH7	TAH8	TAH9
22	23	24	25	26	27	28
D19	D20	D21	D22	D23	D24	D25
MKH2	MKH3	MKH4	MKH5	MKH6	MKH7	MKH8
TAH10	TAH11	TAH12	TAH13	TAH14	TAH15	TAH16

KWANZA 584/2100-01/6336-37/2092-93

Umoja	Kujichagulia	Ujima	Ujamaa	Nia	Kuumba	Imani
1	2	3	4	5	6	7
D26	D27	D28	D29	D30	D31	JA1
MKH9	MKH10	MKH11	MKH12	MKH13	MKH14	MKH1
TAH17	TAH18	TAH19	TAH20	TAH21	TAH22	TAH23
8	9	10	11	12	13	14
JA2	JA3	JA4	JA5	JA6	JA7	JA8
MKH16	MKH17	MKH18	MKH19	MKH20	MKH21	MKH2
TAH24	TAH25	TAH26	TAH27	TAH28	TAH29	TAH30
15	16	17	18	19	20	21
JA9	JA10	JA11	JA12	JA13	JA14	JA15
MKH23	MKH24	MKH25	KMH26	MKH27	MKH28	MKH2
TIR1	TIR2	TIR3	TIR4	TIR5	TIR6	TIR7
22	23	24	25	26	27	28
JA16	JA17	JA18	JA19	JA20	JA21	JA22
MKH30	AMN1	AMN2	AMN3	AMN4	AMN5	AMN6
TIR8	TIR9	TIR10	TIR11	TIR12	TIR13	TIR14

THE NU AFRIKAN CALENDAR

SHAHIDI 584/2100-01/6336-37/2092-93						
Ashe	Ojobo	Iwa	Araorun	Ile	Itutu	Erinle
1	2	3	4	5	6	7
JA23	JA24	JA25	JA26	JA27	JA28	JA29
AMN7	AMN8	AMN9	AMN10	AMN11	AMN12	AMN13
TIR15	TIR16	TIR17	TIR18	TIR19	TIR20	TIR21
8	9	10	11	12	13	14
JA30	JA31	F1	F2	F3	F4	F5
AMN14	AMN15	AMN16	AMN17	AMN18	AMN19	AMN20
TIR22	TIR23	TIR24	TIR25	TIR26	TIR27	TIR28
15	16	17	18	19	20	21
F6	F7	F8	F9	F10	F 11	F12
AMN21	AMN22	AMN23	AMN24	AMN25	AMN26	AMN28
TIR29	TIR30	YEK1	YEK2	YEK3	YEK4	YEK5
22	23	24	25	26	27	28
F13	F14	F15	F16	F17	F18	F19
AMN29	AMN30	RNT1	RNT2	RNT3	RNT4	RNT5
YEK6	YEK7	YEK8	YEK9	YEK10	YEK11	YEK12
SHAHIDI 135						
F20						
RNT6						
YEK13						

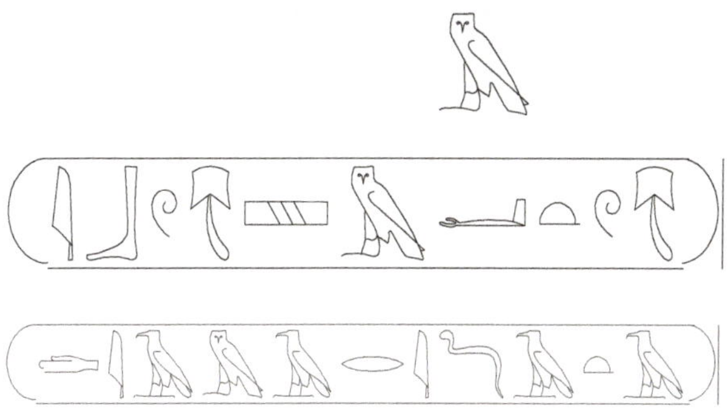

THE NU AFRIKAN CALENDAR

MAAFA 585 B.S.E./ASIENTO 2101-02 C.E.
KAM 6337-38S.M./EC 2093-94 A.M. (ዓ.ም.)

AFU 585/2101-02/6337-38/2093-94

Ashe	Ojobo	Iwa	Araorun	Ile	Itutu	Erinle
1	2	3	4	5	6	7
F21	F22	F23	F24	F25	F26	F27
RNT7	RNT8	RNT9	RNT10	RNT11	RNT12	RNT13
YEK14	YEK15	YEK16	YEK17	YEK18	YEK19	YEK20
8	9	10	11	12	13	14
F28	M1	M2	M3	M4	M5	M6
RNT14	RNT15	RNT16	RNT17	RNT18	RNT19	RNT20
YEK21	YEK22	YEK23	YEK24	YEK25	YEK26	YEK27
15	16	17	18	19	20	21
M7	M8	M9	M10	M11	M12	M13
RNT21	RNT22	RNT23	RNT24	RNT25	RNT26	RNT27
YEK28	YEK29	YEK30	MAG1	MAG2	MAG3	MAG4
22	23	24	25	26	27	28
M14	M15	M16	M17	M18	M19	M20
RNT28	RNT29	RNT30	KHN1	KHN2	KHN3	KHN4
MAG5	MAG6	MAG7	MAG8	MAG9	MAG10	MAG11

KAWAIDA 585/2101-02/6337-38/2093-94

Ashe	Ojobo	Iwa	Araorun	Ile	Itutu	Erinle
1	2	3	4	5	6	7
A18	A19	A20	A21	A22	A23	A24
PIN3	PIN4	PIN5	PIN6	PIN7	PIN8	PIN9
MIY10	MIY11	MIY12	MIY13	MIY14	MIY15	MIY16
8	9	10	11	12	13	14
A25	A26	A27	A28	A29	A30	MA1
PIN10	PIN11	PIN12	PIN13	PIN14	PIN15	PIN16
MIY17	MIY18	MIY19	MIY20	MIY21	MIY22	MIY23
15	16	17	18	19	20	21
MA2	MA3	MA4	MA5	MA6	MA7	MA8
PIN17	PIN18	PIN19	PIN20	PIN21	PIN22	PIN23
MIY24	MIY25	MIY26	MIY27	MIY28	MIY29	MIY30
22	23	24	25	26	27	28
MA9	MA10	MA11	MA12	MA13	MA14	MA15
PIN24	PIN25	PIN26	PIN27	PIN28	PIN29	PIN30
GIN1	GIN2	GIN3	GIN4	GIN5	GIN6	GIN7

UHURU 585/2101-02/6337-38/2093-94

Ashe	Ojobo	Iwa	Araorun	Ile	Itutu	Erinle
1	2	3	4	5	6	7
M21	M22	M23	M24	M25	M26	M27
KHN5	KHN6	KHN7	KHN8	KHN9	KHN10	KHN11
MAG12	MAG13	MAG14	MAG15	MAG16	MAG17	MAG18
8	9	10	11	12	13	14
M28	M29	M30	M31	A1	A2	A3
KHN12	KHN13	KHN14	KHN15	KHN16	KHN17	KHN18
MAG19	MAG20	MAG21	MAG22	MAG23	MAG24	MAG25
15	16	17	18	19	20	21
A4	A5	A6	A7	A8	A9	A10
KHN19	KHN20	KHN21	KHN22	KHN23	KHN24	KHN25
MAG26	MAG27	MAG28	MAG29	MAG30	MIY1	MIY2
22	23	24	25	26	27	28
A11	A12	A13	A14	A15	A16	A17
KHN26	KHN27	KHN28	KHN29	KHN30	PIN1	PIN2
MIY3	MIY4	MIY5	MIY6	MIY7	MIY8	MIY9

MABABA WEUSI 585/2101-02/6337-38/2093-94

Ashe	Ojobo	Iwa	Araorun	Ile	Itutu	Erinle
1	2	3	4	5	6	7
MA16	MA17	MA18	MA19	MA20	MA21	MA22
APP1	APP2	APP3	APP4	APP5	APP6	APP7
GIN8	GIN9	GIN10	GIN11	GIN12	GIN13	GIN14
8	9	10	11	12	13	14
MA23	MA24	MA25	MA26	MA27	MA28	MA29
APP8	APP9	APP10	APP11	APP12	APP13	APP14
GIN15	GIN16	GIN17	GIN18	GIN19	GIN20	GIN21
15	16	17	18	19	20	21
MA30	MA31	J1	J2	J3	J4	J5
APP15	APP16	APP17	APP18	APP19	APP20	APP21
GIN22	GIN23	GIN24	GIN25	GIN26	GIN27	GIN28
22	23	24	25	26	27	28
J6	J7	J8	J9	J10	J11	J12
APP22	APP23	APP24	APP25	APP26	APP27	APP28
GIN29	GIN30	SAN1	SAN2	SAN3	SAN4	SAN5

THE NU AFRIKAN CALENDAR

MWANAMKE WEUSI 585/2101-02/6337-38/2093-94						
Ashe	Ojobo	Iwa	Araorun	Ile	Itutu	Erinle
1	2	3	4	5	6	7
J13	J14	J15	J16	J17	J18	J19
APP29	APP30	MSR1	MSR2	MSR3	MSR4	MSR5
SAN6	SAN7	SAN8	SAN9	SAN10	SAN11	SAN12
8	9	10	11	12	13	14
J20	J21	J22	J23	J24	J25	J26
MSR6	MSR7	MSR8	MSR9	MSR10	MSR11	MSR12
SAN13	SAN14	SAN15	SAN16	SAN17	SAN18	SAN19
15	16	17	18	19	20	21
J27	J28	J29	J30	JU1	JU2	JU3
MSR13	MSR14	MSR15	MSR16	MSR17	MSR18	MSR19
SAN20	SAN21	SAN22	SAN23	SAN24	SAN25	SAN26
22	23	24	25	26	27	28
JU4	JU5	JU6	JU7	JU8	JU9	JU10
MSR20	MRS21	MSR22	MSR23	MSR24	MSR25	MSR26
SAN27	SAN28	SAN29	SAN30	HAM1	HAM2	HAM3

TAIFA WEUSI 585/2101-02/6337-38/2093-94						
Ashe	Ojobo	Iwa	Araorun	Ile	Itutu	Erinle
1	2	3	4	5	6	7
AU8	AU9	AU10	AU11	AU12	AU13	AU14
THT20	THT21	THT22	THT23	THT24	THT25	THT26
NAH2	NAH3	NAH4	NAH5	NAH6	NAH7	NAH8
8	9	10	11	12	13	14
AU15	AU16	AU17	AU18	AU19	AU20	AU21
THT27	THT28	THT29	THT30	PAP1	PAP1	PAP2
NAH9	NAH10	NAH11	NAH12	NAH13	NAH14	NAH15
15	16	17	18	19	20	21
AU22	AU23	AU24	AU25	AU26	AU27	AU28
PAP3	PAP4	PAP5	PAP6	PAP7	PAP8	PAP9
NAH16	NAH17	NAH18	NAH19	NAH20	NAH21	NAH22
22	23	24	25	26	27	28
AU29	AU30	AU31	S1	S2	S3	S4
PAP10	PAP11	PAP12	PAP13	PAP14	PAP15	PAP16
NAH23	NAH24	NAH25	NAH26	NAH27	NAH28	NAH29

MTU WEUSI 585/2101-02/6337-38/2093-94						
Ashe	Ojobo	Iwa	Araorun	Ile	Itutu	Erinle
1	2	3	4	5	6	7
JU11	JU12	JU13	JU14	JU15	JU16	JU17
MSR27	MSR28	MSR29	MSR30	*ASR1*	*HRU2*	*STH3*
HAM4	HAM5	HAM6	HAM7	HAM8	HAM9	HAM10
8	9	10	11	12	13	14
JU18	JU19	JU20	JU21	JU22	JU23	JU24
AST4	*NBH5*	THT1	THT2	THT3	THT4	THT5
HAM11	HAM12	HAM13	AM14	HAM15	HAM16	HAM17
15	16	17	18	19	20	21
JU25	JU26	JU27	JU28	JU29	JU30	JU31
THT6	THT7	THT8	THT9	THT10	THT11	THT12
HAM18	HAM19	HAM20	HAM21	HAM22	HAM23	HAM24
22	23	24	25	26	27	28
AU1	AU2	AU3	AU4	AU5	AU6	AU7
THT13	THT14	THT15	THT16	THT17	THT18	THT19
HAM25	HAM26	HAM27	HAM28	HAM29	HAM30	NAH1

UWEZO 585/2101-02/6337-38/2093-94						
Ashe	Ojobo	Iwa	Araorun	Ile	Itutu	Erinle
1	2	3	4	5	6	7
S5	S6	S7	S8	S9	S10	S11
PAP17	PAP18	PAP19	PAP20	PAP21	PAP22	PAP23
NAH30	*PAG1*	*PAG2*	*PAG3*	*PAG4*	*PAG5*	MAS1
8	9	10	11	12	13	14
S12	S13	S14	S15	S16	S17	S18
PAP24	PAP25	PAP26	PAP27	PAP28	PAP29	PAP30
MAS2	MAS3	MAS4	MAS5	MAS6	MAS7	MAS8
15	16	17	18	19	20	21
S19	S20	S21	S22	S23	S24	S25
HTH1	HTH2	HTH3	HTH4	HTH5	HTH6	HTH7
MAS9	MAS10	MAS11	MAS12	MAS13	MAS14	MAS15
22	23	24	25	26	27	28
S26	S27	S28	S29	S30	O1	O2
HTH8	HTH9	HTH10	HTH11	HTH12	HTH13	HTH14
MAS16	MAS17	MAS18	MAS19	MAS20	MAS21	MAS22

THE NU AFRIKAN CALENDAR

EKUNDU 585/2101-02/6337-38/2093-94						
Ashe	Ojobo	Iwa	Araorun	Ile	Itutu	Erinle
1	2	3	4	5	6	7
O3	O4	O5	O6	O7	O8	O9
HTH15	HTH16	HTH17	HTH18	HTH19	HTH20	HTH21
MAS23	MAS24	MAS25	MAS26	MAS27	MAS28	MAS29
8	9	10	11	12	13	14
O10	O11	O12	O13	O14	O15	O16
HTH22	HTH23	HTH24	HTH25	HTH26	HTH27	HTH28
MAS30	THi1	THi2	THi3	THi4	THi5	THi6
15	16	17	18	19	20	21
O17	O18	O19	O20	O21	O22	O23
HTH29	HTH30	KHK1	KHK2	KHK3	KHK4	KHK5
THi7	THi8	THi9	THi10	THi11	THi12	THi13
22	23	24	25	26	27	28
O24	O25	O26	O27	O28	O29	O30
KHK6	KHK7	KHK8	KHK9	KHK10	KHK11	KHK12
THi14	THi15	THi16	THi17	THi18	THi19	THi20

KIJANI 585/2101-02/6337-38/2093-94						
Ashe	Ojobo	Iwa	Araorun	Ile	Itutu	Erinle
1	2	3	4	5	6	7
N28	N29	N30	D1	D2	D3	D4
TBT11	TBT12	TBT13	TBT14	TBT15	TBT16	TBT17
HID19	HID20	HID21	HID22	HID23	HID24	HID25
8	9	10	11	12	13	14
D5	D6	D7	D8	D9	D10	D11
TBT18	TBT19	TBT20	TBT21	TBT22	TBT23	TBT24
HID26	HID27	HID28	HID29	HID30	TAH1	TAH2
15	16	17	18	19	20	21
D12	D13	D14	D15	D16	D17	D18
TBT25	TBT26	TBT27	TBT28	TBT29	TBT30	MKH1
TAH3	TAH4	TAH5	TAH6	TAH7	TAH8	TAH9
22	23	24	25	26	27	28
D19	D20	D21	D22	D23	D24	D25
MKH2	MKH3	MKH4	MKH5	MKH6	MKH7	MKH8
TAH10	TAH11	TAH12	TAH13	TAH14	TAH15	TAH16

EUSI 585/2101-02/6337-38/2093-94						
Ashe	Ojobo	Iwa	Araorun	Ile	Itutu	Erinle
1	2	3	4	5	6	7
O31	N1	N2	N3	N4	N5	N6
KHK13	KHK14	KHK15	KHK16	KHK17	KHK18	KHK19
THi21	THi22	THi23	THi24	THi25	THi26	THi27
8	9	10	11	12	13	14
N7	N8	N9	N10	N11	N12	N13
KHK20	KHK21	KHK22	KHK23	KHK24	KHK25	KHK26
THi28	THi29	THi30	HID1	HID2	HID3	HID4
15	16	17	18	19	20	21
N14	N15	N16	N17	N18	N19	N20
KHK27	KHK28	KHK29	KHK30	TBT1	TBT2	TBT3
HID5	HID6	HID7	HID8	HID9	HID10	HID11
22	23	24	25	26	27	28
N21	N22	N23	N24	N25	N26	N27
TBT4	TBT5	TBT6	TBT7	TBT8	TBT9	TBT10
HID12	HID13	HID14	HID15	HID16	HID17	HID18

KWANZA 585/2101-02/6337-38/2093-94						
Umoja	Kujichagulia	Ujima	Ujamaa	Nia	Kuumba	Imani
1	2	3	4	5	6	7
D26	D27	D28	D29	D30	D31	JA1
MKH9	MKH10	MKH11	MKH12	MKH13	MKH14	MKH15
TAH17	TAH18	TAH19	TAH20	TAH21	TAH22	TAH23
8	9	10	11	12	13	14
JA2	JA3	JA4	JA5	JA6	JA7	JA8
MKH16	MKH17	MKH18	MKH19	MKH20	MKH21	MKH22
TAH24	TAH25	TAH26	TAH27	TAH28	TAH29	TAH30
15	16	17	18	19	20	21
JA9	JA10	JA11	JA12	JA13	JA14	JA15
MKH23	MKH24	MKH25	KMH26	MKH27	MKH28	MKH29
TIR1	TIR2	TIR3	TIR4	TIR5	TIR6	TIR7
22	23	24	25	26	27	28
JA16	JA17	JA18	JA19	JA20	JA21	JA22
MKH30	AMN1	AMN2	AMN3	AMN4	AMN5	AMN6
TIR8	TIR9	TIR10	TIR11	TIR12	TIR13	TIR14

THE NU AFRIKAN CALENDAR

SHAHIDI 585/2101-02/6337-38/2093-94						
Ashe	Ojobo	Iwa	Araorun	Ile	Itutu	Erinle
1	2	3	4	5	6	7
JA23	JA24	JA25	JA26	JA27	JA28	JA29
AMN7	AMN8	AMN9	AMN10	AMN11	AMN12	AMN13
TIR15	TIR16	TIR17	TIR18	TIR19	TIR20	TIR21
8	9	10	11	12	13	14
JA30	JA31	F1	F2	F3	F4	F5
AMN14	AMN15	AMN16	AMN17	AMN18	AMN19	AMN20
TIR22	TIR23	TIR24	TIR25	TIR26	TIR27	TIR28
15	16	17	18	19	20	21
F6	F7	F8	F9	F10	F 11	F12
AMN21	AMN22	AMN23	AMN24	AMN25	AMN26	AMN28
TIR29	TIR30	YEK1	YEK2	YEK3	YEK4	YEK5
22	23	24	25	26	27	28
F13	F14	F15	F16	F17	F18	F19
AMN29	AMN30	RNT1	RNT2	RNT3	RNT4	RNT5
YEK6	YEK7	YEK8	YEK9	YEK10	YEK11	YEK12
SHAHIDI 136						
F20						
RNT6						
YEK13						

THE NU AFRIKAN CALENDAR

MAAFA 586 B.S.E./ASIENTO 2102-03 C.E.
KAM 6338-39 S.M./EC 2094-95 A.M. (ዓ.ም.)

AFU 586/2102-03/6338-39/2094-95						
Ashe	Ojobo	Iwa	Araorun	Ile	Itutu	Erinle
1	2	3	4	5	6	7
F21	F22	F23	F24	F25	F26	F27
RNT7	RNT8	RNT9	RNT10	RNT11	RNT12	RNT13
YEK14	YEK15	YEK16	YEK17	YEK18	YEK19	YEK20
8	9	10	11	12	13	14
F28	M1	M2	M3	M4	M5	M6
RNT14	RNT15	RNT16	RNT17	RNT18	RNT19	RNT20
YEK21	YEK22	YEK23	YEK24	YEK25	YEK26	YEK27
15	16	17	18	19	20	21
M7	M8	M9	M10	M11	M12	M13
RNT21	RNT22	RNT23	RNT24	RNT25	RNT26	RNT27
YEK28	YEK29	YEK30	MAG1	MAG2	MAG3	MAG4
22	23	24	25	26	27	28
M14	M15	M16	M17	M18	M19	M20
RNT28	RNT29	RNT30	KHN1	KHN2	KHN3	KHN4
MAG5	MAG6	MAG7	MAG8	MAG9	MAG10	MAG11

KAWAIDA 586/2102-03/6338-39/2094-95						
Ashe	Ojobo	Iwa	Araorun	Ile	Itutu	Erinle
1	2	3	4	5	6	7
A18	A19	A20	A21	A22	A23	A24
PIN3	PIN4	PIN5	PIN6	PIN7	PIN8	PIN9
MIY10	MIY11	MIY12	MIY13	MIY14	MIY15	MIY16
8	9	10	11	12	13	14
A25	A26	A27	A28	A29	A30	MA1
PIN10	PIN11	PIN12	PIN13	PIN14	PIN15	PIN16
MIY17	MIY18	MIY19	MIY20	MIY21	MIY22	MIY23
15	16	17	18	19	20	21
MA2	MA3	MA4	MA5	MA6	MA7	MA8
PIN17	PIN18	PIN19	PIN20	PIN21	PIN22	PIN23
MIY24	MIY25	MIY26	MIY27	MIY28	MIY29	MIY30
22	23	24	25	26	27	28
MA9	MA10	MA11	MA12	MA13	MA14	MA15
PIN24	PIN25	PIN26	PIN27	PIN28	PIN29	PIN30
GIN1	GIN2	GIN3	GIN4	GIN5	GIN6	GIN7

UHURU 586/2102-03/6338-39/2094-95						
Ashe	Ojobo	Iwa	Araorun	Ile	Itutu	Erinle
1	2	3	4	5	6	7
M21	M22	M23	M24	M25	M26	M27
KHN5	KHN6	KHN7	KHN8	KHN9	KHN10	KHN11
MAG12	MAG13	MAG14	MAG15	MAG16	MAG17	MAG18
8	9	10	11	12	13	14
M28	M29	M30	M31	A1	A2	A3
KHN12	KHN13	KHN14	KHN15	KHN16	KHN17	KHN18
MAG19	MAG20	MAG21	MAG22	MAG23	MAG24	MAG25
15	16	17	18	19	20	21
A4	A5	A6	A7	A8	A9	A10
KHN19	KHN20	KHN21	KHN22	KHN23	KHN24	KHN25
MAG26	MAG27	MAG28	MAG29	MAG30	MIY1	MIY2
22	23	24	25	26	27	28
A11	A12	A13	A14	A15	A16	A17
KHN26	KHN27	KHN28	KHN29	KHN30	PIN1	PIN2
MIY3	MIY4	MIY5	MIY6	MIY7	MIY8	MIY9

MABABA WEUSI 586/2102-03/6338-39/2094-95						
Ashe	Ojobo	Iwa	Araorun	Ile	Itutu	Erinle
1	2	3	4	5	6	7
MA16	MA17	MA18	MA19	MA20	MA21	MA22
APP1	APP2	APP3	APP4	APP5	APP6	APP7
GIN8	GIN9	GIN10	GIN11	GIN12	GIN13	GIN14
8	9	10	11	12	13	14
MA23	MA24	MA25	MA26	MA27	MA28	MA29
APP8	APP9	APP10	APP11	APP12	APP13	APP14
GIN15	GIN16	GIN17	GIN18	GIN19	GIN20	GIN21
15	16	17	18	19	20	21
MA30	MA31	J1	J2	J3	J4	J5
APP15	APP16	APP17	APP18	APP19	APP20	APP21
GIN22	GIN23	GIN24	GIN25	GIN26	GIN27	GIN28
22	23	24	25	26	27	28
J6	J7	J8	J9	J10	J11	J12
APP22	APP23	APP24	APP25	APP26	APP27	APP28
GIN29	GIN30	SAN1	SAN2	SAN3	SAN4	SAN5

THE NU AFRIKAN CALENDAR

MWANAMKE WEUSI 586/2102-03/6338-39/2094-95

Ashe	Ojobo	Iwa	Araorun	Ile	Itutu	Erinle
1	2	3	4	5	6	7
J13	J14	J15	J16	J17	J18	J19
APP29	APP30	MSR1	MSR2	MSR3	MSR4	MSR5
SAN6	SAN7	SAN8	SAN9	SAN10	SAN11	SAN12
8	9	10	11	12	13	14
J20	J21	J22	J23	J24	J25	J26
MSR6	MSR7	MSR8	MSR9	MSR10	MSR11	MSR12
SAN13	SAN14	SAN15	SAN16	SAN17	SAN18	SAN19
15	16	17	18	19	20	21
J27	J28	J29	J30	JU1	JU2	JU3
MSR13	MSR14	MSR15	MSR16	MSR17	MSR18	MSR19
SAN20	SAN21	SAN22	SAN23	SAN24	SAN25	SAN26
22	23	24	25	26	27	28
JU4	JU5	JU6	JU7	JU8	JU9	JU10
MSR20	MRS21	MSR22	MSR23	MSR24	MSR25	MSR26
SAN27	SAN28	SAN29	SAN30	HAM1	HAM2	HAM3

TAIFA WEUSI 586/2102-03/6338-39/2094-95

Ashe	Ojobo	Iwa	Araorun	Ile	Itutu	Erinle
1	2	3	4	5	6	7
AU8	AU9	AU10	AU11	AU12	AU13	AU14
THT20	THT21	THT22	THT23	THT24	THT25	THT26
NAH2	NAH3	NAH4	NAH5	NAH6	NAH7	NAH8
8	9	10	11	12	13	14
AU15	AU16	AU17	AU18	AU19	AU20	AU21
THT27	THT28	THT29	THT30	PAP1	PAP1	PAP2
NAH9	NAH10	NAH11	NAH12	NAH13	NAH14	NAH15
15	16	17	18	19	20	21
AU22	AU23	AU24	AU25	AU26	AU27	AU28
PAP3	PAP4	PAP5	PAP6	PAP7	PAP8	PAP9
NAH16	NAH17	NAH18	NAH19	NAH20	NAH21	NAH22
22	23	24	25	26	27	28
AU29	AU30	AU31	S1	S2	S3	S4
PAP10	PAP11	PAP12	PAP13	PAP14	PAP15	PAP16
NAH23	NAH24	NAH25	NAH26	NAH27	NAH28	NAH29

MTU WEUSI 586/2102-03/6338-39/2094-95

Ashe	Ojobo	Iwa	Araorun	Ile	Itutu	Erinle
1	2	3	4	5	6	7
JU11	JU12	JU13	JU14	JU15	JU16	JU17
MSR27	MSR28	MSR29	MSR30	*ASR1*	*HRU2*	*STH3*
HAM4	HAM5	HAM6	HAM7	HAM8	HAM9	HAM10
8	9	10	11	12	13	14
JU18	JU19	JU20	JU21	JU22	JU23	JU24
AST4	*NBH5*	THT1	THT2	THT3	THT4	THT5
HAM11	HAM12	HAM13	AM14	HAM15	HAM16	HAM17
15	16	17	18	19	20	21
JU25	JU26	JU27	JU28	JU29	JU30	JU31
THT6	THT7	THT8	THT9	THT10	THT11	THT12
HAM18	HAM19	HAM20	HAM21	HAM22	HAM23	HAM24
22	23	24	25	26	27	28
AU1	AU2	AU3	AU4	AU5	AU6	AU7
THT13	THT14	THT15	THT16	THT17	THT18	THT19
HAM25	HAM26	HAM27	HAM28	HAM29	HAM30	NAH1

UWEZO 586/2102-03/6338-39/2094-95

Ashe	Ojobo	Iwa	Araorun	Ile	Itutu	Erinle
1	2	3	4	5	6	7
S5	S6	S7	S8	S9	S10	S11
PAP17	PAP18	PAP19	PAP20	PAP21	PAP22	PAP23
NAH30	*PAG1*	*PAG2*	*PAG3*	*PAG4*	*PAG5*	MAS1
8	9	10	11	12	13	14
S12	S13	S14	S15	S16	S17	S18
PAP24	PAP25	PAP26	PAP27	PAP28	PAP29	PAP30
MAS2	MAS3	MAS4	MAS5	MAS6	MAS7	MAS8
15	16	17	18	19	20	21
S19	S20	S21	S22	S23	S24	S25
HTH1	HTH2	HTH3	HTH4	HTH5	HTH6	HTH7
MAS9	MAS10	MAS11	MAS12	MAS13	MAS14	MAS15
22	23	24	25	26	27	28
S26	S27	S28	S29	S30	O1	O2
HTH8	HTH9	HTH10	HTH11	HTH12	HTH13	HTH14
MAS16	MAS17	MAS18	MAS19	MAS20	MAS21	MAS22

THE NU AFRIKAN CALENDAR

EKUNDU 586/2102-03/6338-39/2094-95						
Ashe	Ojobo	Iwa	Araorun	Ile	Itutu	Erinle
1	2	3	4	5	6	7
O3	O4	O5	O6	O7	O8	O9
HTH15	HTH16	HTH17	HTH18	HTH19	HTH20	HTH21
MAS23	MAS24	MAS25	MAS26	MAS27	MAS28	MAS29
8	9	10	11	12	13	14
O10	O11	O12	O13	O14	O15	O16
HTH22	HTH23	HTH24	HTH25	HTH26	HTH27	HTH28
MAS30	THi1	THi2	THi3	THi4	THi5	THi6
15	16	17	18	19	20	21
O17	O18	O19	O20	O21	O22	O23
HTH29	HTH30	KHK1	KHK2	KHK3	KHK4	KHK5
THi7	THi8	THi9	THi10	THi11	THi12	THi13
22	23	24	25	26	27	28
O24	O25	O26	O27	O28	O29	O30
KHK6	KHK7	KHK8	KHK9	KHK10	KHK11	KHK12
THi14	THi15	THi16	THi17	THi18	THi19	THi20

KIJANI 586/2102-03/6338-39/2094-95						
Ashe	Ojobo	Iwa	Araorun	Ile	Itutu	Erinle
1	2	3	4	5	6	7
N28	N29	N30	D1	D2	D3	D4
TBT11	TBT12	TBT13	TBT14	TBT15	TBT16	TBT17
HID19	HID20	HID21	HID22	HID23	HID24	HID25
8	9	10	11	12	13	14
D5	D6	D7	D8	D9	D10	D11
TBT18	TBT19	TBT20	TBT21	TBT22	TBT23	TBT24
HID26	HID27	HID28	HID29	HID30	TAH1	TAH2
15	16	17	18	19	20	21
D12	D13	D14	D15	D16	D17	D18
TBT25	TBT26	TBT27	TBT28	TBT29	TBT30	MKH1
TAH3	TAH4	TAH5	TAH6	TAH7	TAH8	TAH9
22	23	24	25	26	27	28
D19	D20	D21	D22	D23	D24	D25
MKH2	MKH3	MKH4	MKH5	MKH6	MKH7	MKH8
TAH10	TAH11	TAH12	TAH13	TAH14	TAH15	TAH16

EUSI 586/2102-03/6338-39/2094-95						
Ashe	Ojobo	Iwa	Araorun	Ile	Itutu	Erinle
1	2	3	4	5	6	7
O31	N1	N2	N3	N4	N5	N6
KHK13	KHK14	KHK15	KHK16	KHK17	KHK18	KHK19
THi21	THi22	THi23	THi24	THi25	THi26	THi27
8	9	10	11	12	13	14
N7	N8	N9	N10	N11	N12	N13
KHK20	KHK21	KHK22	KHK23	KHK24	KHK25	KHK26
THi28	THi29	THi30	HID1	HID2	HID3	HID4
15	16	17	18	19	20	21
N14	N15	N16	N17	N18	N19	N20
KHK27	KHK28	KHK29	KHK30	TBT1	TBT2	TBT3
HID5	HID6	HID7	HID8	HID9	HID10	HID11
22	23	24	25	26	27	28
N21	N22	N23	N24	N25	N26	N27
TBT4	TBT5	TBT6	TBT7	TBT8	TBT9	TBT10
HID12	HID13	HID14	HID15	HID16	HID17	HID18

KWANZA 586/2102-03/6338-39/2094-95						
Umoja	Kujichagulia	Ujima	Ujamaa	Nia	Kuumba	Imani
1	2	3	4	5	6	7
D26	D27	D28	D29	D30	D31	JA1
MKH9	MKH10	MKH11	MKH12	MKH13	MKH14	MKH15
TAH17	TAH18	TAH19	TAH20	TAH21	TAH22	TAH23
8	9	10	11	12	13	14
JA2	JA3	JA4	JA5	JA6	JA7	JA8
MKH16	MKH17	MKH18	MKH19	MKH20	MKH21	MKH22
TAH24	TAH25	TAH26	TAH27	TAH28	TAH29	TAH30
15	16	17	18	19	20	21
JA9	JA10	JA11	JA12	JA13	JA14	JA15
MKH23	MKH24	MKH25	KMH26	MKH27	MKH28	MKH29
TIR1	TIR2	TIR3	TIR4	TIR5	TIR6	TIR7
22	23	24	25	26	27	28
JA16	JA17	JA18	JA19	JA20	JA21	JA22
MKH30	AMN1	AMN2	AMN3	AMN4	AMN5	AMN6
TIR8	TIR9	TIR10	TIR11	TIR12	TIR13	TIR14

THE NU AFRIKAN CALENDAR

SHAHIDI 586/2102-03/6338-39/2094-95						
Ashe	**Ojobo**	**Iwa**	**Araorun**	**Ile**	**Itutu**	**Erinle**
1	2	3	4	5	6	7
JA23	JA24	JA25	JA26	JA27	JA28	JA29
AMN7	AMN8	AMN9	AMN10	AMN11	AMN12	AMN13
TIR15	TIR16	TIR17	TIR18	TIR19	TIR20	TIR21
8	9	10	11	12	13	14
JA30	JA31	F1	F2	F3	F4	F5
AMN14	AMN15	AMN16	AMN17	AMN18	AMN19	AMN20
TIR22	TIR23	TIR24	TIR25	TIR26	TIR27	TIR28
15	16	17	18	19	20	21
F6	F7	F8	F9	F10	F 11	F12
AMN21	AMN22	AMN23	AMN24	AMN25	AMN26	AMN28
TIR29	TIR30	YEK1	YEK2	YEK3	YEK4	YEK5
22	23	24	25	26	27	28
F13	F14	F15	F16	F17	F18	F19
AMN29	AMN30	RNT1	RNT2	RNT3	RNT4	RNT5
YEK6	YEK7	YEK8	YEK9	YEK10	YEK11	YEK12
SHAHIDI 137						
F20						
RNT6						
YEK13						

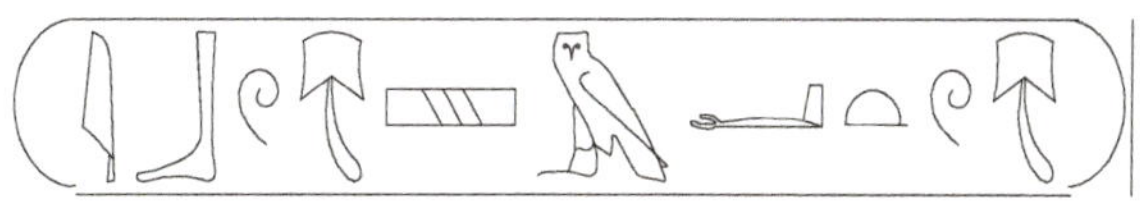

MAAFA 587 B.S.E./ASIENTO 2103-04 C.E./
KAM 6339-40 S.M./EC 2095-96 A.M. (ዓ.ም.)

AFU 587/2103-04/6339-40/2095-96						
Ashe	Ojobo	Iwa	Araorun	Ile	Itutu	Erinle
1	2	3	4	5	6	7
F21	F22	F23	F24	F25	F26	F27
RNT7	RNT8	RNT9	RNT10	RNT11	RNT12	RNT13
YEK14	YEK15	YEK16	YEK17	YEK18	YEK19	YEK20
8	9	10	11	12	13	14
F28	M1	M2	M3	M4	M5	M6
RNT14	RNT15	RNT16	RNT17	RNT18	RNT19	RNT20
YEK21	YEK22	YEK23	YEK24	YEK25	YEK26	YEK27
15	16	17	18	19	20	21
M7	M8	M9	M10	M11	M12	M13
RNT21	RNT22	RNT23	RNT24	RNT25	RNT26	RNT27
YEK28	YEK 29	YEK30	MAG1	MAG2	MAG3	MAG4
22	23	24	25	26	27	28
M14	M15	M16	M17	M18	M19	M20
RNT28	RNT29	RNT30	KHN1	KHN2	KHN3	KHN4
MAG5	MAG6	MAG7	MAG8	MAG9	MAG10	MAG11

UHURU 587/2103-04/6339-40/2095-96						
Ashe	Ojobo	Iwa	Araorun	Ile	Itutu	Erinle
1	2	3	4	5	6	7
M21	M22	M23	M24	M25	M26	M27
KHN5	KHN6	KHN7	KHN8	KHN9	KHN10	KHN11
MAG12	MAG13	MAG14	MAG15	MAG16	MAG17	MAG18
8	9	10	11	12	13	14
M28	M29	M30	M31	A1	A2	A3
KHN12	KHN13	KHN14	KHN15	KHN16	KHN17	KHN18
MAG19	MAG20	MAG21	MAG22	MAG23	MAG24	MAG25
15	16	17	18	19	20	21
A4	A5	A6	A7	A8	A9	A10
KHN19	KHN20	KHN21	KHN22	KHN23	KHN24	KHN25
MAG26	MAG27	MAG28	MAG29	MAG30	MIY1	MIY2
22	23	24	25	26	27	28
A11	A12	A13	A14	A15	A16	A17
KHN26	KHN27	KHN28	KHN29	KHN30	PIN1	PIN2
MIY3	MIY4	MIY5	MIY6	MIY7	MIY8	MIY9

KAWAIDA 587/2103-04/6339-40/2095-96						
Ashe	Ojobo	Iwa	Araorun	Ile	Itutu	Erinle
1	2	3	4	5	6	7
A18	A19	A20	A21	A22	A23	A24
PIN3	PIN4	PIN5	PIN6	PIN7	PIN8	PIN9
MIY10	MIY11	MIY12	MIY13	MIY14	MIY15	MIY16
8	9	10	11	12	13	14
A25	A26	A27	A28	A29	A30	MA1
PIN10	PIN11	PIN12	PIN13	PIN14	PIN15	PIN16
MIY17	MIY18	MIY19	MIY20	MIY21	MIY22	MIY23
15	16	17	18	19	20	21
MA2	MA3	MA4	MA5	MA6	MA7	MA8
PIN17	PIN18	PIN19	PIN20	PIN21	PIN22	PIN23
MIY24	MIY25	MIY26	MIY27	MIY28	MIY29	MIY30
22	23	24	25	26	27	28
MA9	MA10	MA11	MA12	MA13	MA14	MA15
PIN24	PIN25	PIN26	PIN27	PIN28	PIN29	PIN30
GIN1	GIN2	GIN3	GIN4	GIN5	GIN6	GIN7

MABABA WEUSI 587/2103-04/6339-40/2095-96						
Ashe	Ojobo	Iwa	Araorun	Ile	Itutu	Erinle
1	2	3	4	5	6	7
MA16	MA17	MA18	MA19	MA20	MA21	MA22
APP1	APP2	APP3	APP4	APP5	APP6	APP7
GIN8	GIN9	GIN10	GIN11	GIN12	GIN13	GIN14
8	9	10	11	12	13	14
MA23	MA24	MA25	MA26	MA27	MA28	MA29
APP8	APP9	APP10	APP11	APP12	APP13	APP14
GIN15	GIN16	GIN17	GIN18	GIN19	GIN20	GIN21
15	16	17	18	19	20	21
MA30	MA31	J1	J2	J3	J4	J5
APP15	APP16	APP17	APP18	APP19	APP20	APP21
GIN22	GIN23	GIN24	GIN25	GIN26	GIN27	GIN28
22	23	24	25	26	27	28
J6	J7	J8	J9	J10	J11	J12
APP22	APP23	APP24	APP25	APP26	APP27	APP28
GIN29	GIN30	SAN1	SAN2	SAN3	SAN4	SAN5

THE NU AFRIKAN CALENDAR

MWANAMKE WEUSI 587/2103-04/6339-40/2095-96						
Ashe	Ojobo	Iwa	Araorun	Ile	Itutu	Erinle
1	2	3	4	5	6	7
J13	J14	J15	J16	J17	J18	J19
APP29	APP30	MSR1	MSR2	MSR3	MSR4	MSR5
SAN6	SAN7	SAN8	SAN9	SAN10	SAN11	SAN12
8	9	10	11	12	13	14
J20	J21	J22	J23	J24	J25	J26
MSR6	MSR7	MSR8	MSR9	MSR10	MSR11	MSR12
SAN13	SAN14	SAN15	SAN16	SAN17	SAN18	SAN19
15	16	17	18	19	20	21
J27	J28	J29	J30	JU1	JU2	JU3
MSR13	MSR14	MSR15	MSR16	MSR17	MSR18	MSR19
SAN20	SAN21	SAN22	SAN23	SAN24	SAN25	SAN26
22	23	24	25	26	27	28
JU4	JU5	JU6	JU7	JU8	JU9	JU10
MSR20	MRS21	MSR22	MSR23	MSR24	MSR25	MSR26
SAN27	SAN28	SAN29	SAN30	HAM1	HAM2	HAM3

TAIFA WEUSI 587/2103-04/6339-40/2095-96						
Ashe	Ojobo	Iwa	Araorun	Ile	Itutu	Erinle
1	2	3	4	5	6	7
AU8	AU9	AU10	AU11	AU12	AU13	AU14
THT20	THT21	THT22	THT23	THT24	THT25	THT26
NAH2	NAH3	NAH4	NAH5	NAH6	NAH7	NAH8
8	9	10	11	12	13	14
AU15	AU16	AU17	AU18	AU19	AU20	AU21
THT27	THT28	THT29	THT30	PAP1	PAP1	PAP2
NAH9	NAH10	NAH11	NAH12	NAH13	NAH14	NAH15
15	16	17	18	19	20	21
AU22	AU23	AU24	AU25	AU26	AU27	AU28
PAP3	PAP4	PAP5	PAP6	PAP7	PAP8	PAP9
NAH16	NAH17	NAH18	NAH19	NAH20	NAH21	NAH22
22	23	24	25	26	27	28
AU29	AU30	AU31	S1	S2	S3	S4
PAP10	PAP11	PAP12	PAP13	PAP14	PAP15	PAP16
NAH23	NAH24	NAH25	NAH26	NAH27	NAH28	NAH29

MTU WEUSI 587/2103-04/6339-40/2095-96						
Ashe	Ojobo	Iwa	Araorun	Ile	Itutu	Erinle
1	2	3	4	5	6	7
JU11	JU12	JU13	JU14	JU15	JU16	JU17
MSR27	MSR28	MSR29	MSR30	*ASR1*	*HRU2*	*STH3*
HAM4	HAM5	HAM6	HAM7	HAM8	HAM9	HAM10
8	9	10	11	12	13	14
JU18	JU19	JU20	JU21	JU22	JU23	JU24
AST4	*NBH5*	THT1	THT2	THT3	THT4	THT5
HAM11	HAM12	HAM13	HAM14	HAM15	HAM16	HAM17
15	16	17	18	19	20	21
JU25	JU26	JU27	JU28	JU29	JU30	JU31
THT6	THT7	THT8	THT9	THT10	THT11	THT12
HAM18	HAM19	HAM20	HAM21	HAM22	HAM23	HAM24
22	23	24	25	26	27	28
AU1	AU2	AU3	AU4	AU5	AU6	AU7
THT13	THT14	THT15	THT16	THT17	THT18	THT19
HAM25	HAM26	HAM27	HAM28	HAM29	HAM30	NAH1

UWEZO 587/2103-04/6339-40/2095-96						
Ashe	Ojobo	Iwa	Araorun	Ile	Itutu	Erinle
1	2	3	4	5	6	7
S5	S6	S7	S8	S9	S10	S11
PAP17	PAP18	PAP19	PAP20	PAP21	PAP22	PAP23
NAH30	*QUA1*	*QUA2*	*QUA3*	*QUA4*	*QUA5*	*QUA6*
8	9	10	11	12	13	14
S12	S13	S14	S15	S16	S17	S18
PAP24	PAP25	PAP26	PAP27	PAP28	PAP29	PAP30
MAS1	MAS2	MAS3	MAS4	MAS5	MAS6	MAS7
15	16	17	18	19	20	21
S19	S20	S21	S22	S23	S24	S25
HTH1	HTH2	HTH3	HTH4	HTH5	HTH6	HTH7
MAS8	MAS9	MAS10	MAS11	MAS12	MAS13	MAS14
22	23	24	25	26	27	28
S26	S27	S28	S29	S30	O1	O2
HTH8	HTH9	HTH10	HTH11	HTH12	HTH13	HTH14
MAS15	MAS16	MAS17	MAS18	MAS19	MAS20	MAS21

THE NU AFRIKAN CALENDAR

EKUNDU 587/2103-04/6339-40/2095-96						
Ashe	Ojobo	Iwa	Araorun	Ile	Itutu	Erinle
1	2	3	4	5	6	7
O3	O4	O5	O6	O7	O8	O9
HTH15	HTH16	HTH17	HTH18	HTH19	HTH20	HTH21
MAS22	MAS23	MAS 24	MAS25	MAS26	MAS27	MAS28
8	9	10	11	12	13	14
O10	O11	O12	O13	O14	O15	O16
HTH22	HTH23	HTH24	HTH25	HTH26	HTH27	HTH28
MAS29	MAS30	THi1	THi2	THi3	THi4	THi5
15	16	17	18	19	20	21
O17	O18	O19	O20	O21	O22	O23
HTH29	HTH30	KHK1	KHK2	KHK3	KHK4	KHK5
THi6	THi7	THi8	THi9	THi10	THi11	THi12
22	23	24	25	26	27	28
O24	O25	O26	O27	O28	O29	O30
KHK6	KHK7	KHK8	KHK9	KHK10	KHK11	KHK12
THi13	THi14	THi15	THi16	THi17	THi18	THi19

KIJANI 587/2103-04/6339-40/2095-96						
Ashe	Ojobo	Iwa	Araorun	Ile	Itutu	Erinle
1	2	3	4	5	6	7
N28	N29	N30	D1	D2	D3	D4
TBT11	TBT12	TBT13	TBT14	TBT15	TBT16	TBT17
HID18	HID19	HID20	HID21	HID22	HID23	HID24
8	9	10	11	12	13	14
D5	D6	D7	D8	D9	D10	D11
TBT18	TBT19	TBT20	TBT21	TBT22	TBT23	TBT24
HID25	HID26	HID27	HID28	HID29	HID30	TAH1
15	16	17	18	19	20	21
D12	D13	D14	D15	D16	D17	D18
TBT25	TBT26	TBT27	TBT28	TBT29	TBT30	MKH1
TAH2	TAH3	TAH4	TAH5	TAH6	TAH7	TAH8
22	23	24	25	26	27	28
D19	D20	D21	D22	D23	D24	D25
MKH2	MKH3	MKH4	MKH5	MKH6	MKH7	MKH8
TAH9	TAH10	TAH11	TAH12	TAH13	TAH14	TAH15

EUSI 587/2103-04/6339-40/2095-96						
Ashe	Ojobo	Iwa	Araorun	Ile	Itutu	Erinle
1	2	3	4	5	6	7
O31	N1	N2	N3	N4	N5	N6
KHK13	KHK14	KHK15	KHK16	KHK17	KHK18	KHK19
THi20	THi21	THi22	THi23	THi24	THi25	THi26
8	9	10	11	12	13	14
N7	N8	N9	N10	N11	N12	N13
KHK20	KHK21	KHK22	KHK23	KHK24	KHK25	KHK26
THi27	THi28	THi29	THi30	HID1	HID2	HID3
15	16	17	18	19	20	21
N14	N15	N16	N17	N18	N19	N20
KHK27	KHK28	KHK29	KHK30	TBT1	TBT2	TBT3
HID4	HID5	HID6	HID7	HID8	HID9	HID10
22	23	24	25	26	27	28
N21	N22	N23	N24	N25	N26	N27
TBT4	TBT5	TBT6	TBT7	TBT8	TBT9	TBT10
HID11	HID12	HID13	HID14	HID15	HID16	HID17

KWANZA 587/2103-04/6339-40/2095-96						
Umoja	Kujichagulia	Ujima	Ujamaa	Nia	Kuumba	Imani
1	2	3	4	5	6	7
D26	D27	D28	D29	D30	D31	JA1
MKH9	MKH10	MKH11	MKH12	MKH13	MKH14	MKH15
TAH16	TAH17	TAH18	TAH19	TAH20	TAH21	TAH22
8	9	10	11	12	13	14
JA2	JA3	JA4	JA5	JA6	JA7	JA8
MKH16	MKH17	MKH18	MKH19	MKH20	MKH21	MKH22
TAH23	TAH24	TAH25	TAH26	TAH27	TAH28	TAH29
15	16	17	18	19	20	21
JA9	JA10	JA11	JA12	JA13	JA14	JA15
MKH23	MKH24	MKH25	KMH26	MKH27	MKH28	MKH29
TAH30	TIR1	TIR2	TIR3	TIR4	TIR5	TIR6
22	23	24	25	26	27	28
JA16	JA17	JA18	JA19	JA20	JA21	JA22
MKH30	AMN1	AMN2	AMN3	AMN4	AMN5	AMN6
TIR7	TIR8	TIR9	TIR10	TIR11	TIR12	TIR13

THE NU AFRIKAN CALENDAR

MWANAMKE WEUSI 587/2103-04/6339-40/2095-96

Ashe	Ojobo	Iwa	Araorun	Ile	Itutu	Erinle
1	2	3	4	5	6	7
J13	J14	J15	J16	J17	J18	J19
APP29	APP30	MSR1	MSR2	MSR3	MSR4	MSR5
SAN6	SAN7	SAN8	SAN9	SAN10	SAN11	SAN12
8	9	10	11	12	13	14
J20	J21	J22	J23	J24	J25	J26
MSR6	MSR7	MSR8	MSR9	MSR10	MSR11	MSR12
SAN13	SAN14	SAN15	SAN16	SAN17	SAN18	SAN19
15	16	17	18	19	20	21
J27	J28	J29	J30	JU1	JU2	JU3
MSR13	MSR14	MSR15	MSR16	MSR17	MSR18	MSR19
SAN20	SAN21	SAN22	SAN23	SAN24	SAN25	SAN26
22	23	24	25	26	27	28
JU4	JU5	JU6	JU7	JU8	JU9	JU10
MSR20	MRS21	MSR22	MSR23	MSR24	MSR25	MSR26
SAN27	SAN28	SAN29	SAN30	HAM1	HAM2	HAM3

TAIFA WEUSI 587/2103-04/6339-40/2095-96

Ashe	Ojobo	Iwa	Araorun	Ile	Itutu	Erinle
1	2	3	4	5	6	7
AU8	AU9	AU10	AU11	AU12	AU13	AU14
THT20	THT21	THT22	THT23	THT24	THT25	THT26
NAH2	NAH3	NAH4	NAH5	NAH6	NAH7	NAH8
8	9	10	11	12	13	14
AU15	AU16	AU17	AU18	AU19	AU20	AU21
THT27	THT28	THT29	THT30	PAP1	PAP1	PAP2
NAH9	NAH10	NAH11	NAH12	NAH13	NAH14	NAH15
15	16	17	18	19	20	21
AU22	AU23	AU24	AU25	AU26	AU27	AU28
PAP3	PAP4	PAP5	PAP6	PAP7	PAP8	PAP9
NAH16	NAH17	NAH18	NAH19	NAH20	NAH21	NAH22
22	23	24	25	26	27	28
AU29	AU30	AU31	S1	S2	S3	S4
PAP10	PAP11	PAP12	PAP13	PAP14	PAP15	PAP16
NAH23	NAH24	NAH25	NAH26	NAH27	NAH28	NAH29

MTU WEUSI 587/2103-04/6339-40/2095-96

Ashe	Ojobo	Iwa	Araorun	Ile	Itutu	Erinle
1	2	3	4	5	6	7
JU11	JU12	JU13	JU14	JU15	JU16	JU17
MSR27	MSR28	MSR29	MSR30	*ASR1*	*HRU2*	*STH3*
HAM4	HAM5	HAM6	HAM7	HAM8	HAM9	HAM10
8	9	10	11	12	13	14
JU18	JU19	JU20	JU21	JU22	JU23	JU24
AST4	*NBH5*	THT1	THT2	THT3	THT4	THT5
HAM11	HAM12	HAM13	HAM14	HAM15	HAM16	HAM17
15	16	17	18	19	20	21
JU25	JU26	JU27	JU28	JU29	JU30	JU31
THT6	THT7	THT8	THT9	THT10	THT11	THT12
HAM18	HAM19	HAM20	HAM21	HAM22	HAM23	HAM24
22	23	24	25	26	27	28
AU1	AU2	AU3	AU4	AU5	AU6	AU7
THT13	THT14	THT15	THT16	THT17	THT18	THT19
HAM25	HAM26	HAM27	HAM28	HAM29	HAM30	NAH1

UWEZO 587/2103-04/6339-40/2095-96

Ashe	Ojobo	Iwa	Araorun	Ile	Itutu	Erinle
1	2	3	4	5	6	7
S5	S6	S7	S8	S9	S10	S11
PAP17	PAP18	PAP19	PAP20	PAP21	PAP22	PAP23
NAH30	*QUA1*	*QUA2*	*QUA3*	*QUA4*	*QUA5*	*QUA6*
8	9	10	11	12	13	14
S12	S13	S14	S15	S16	S17	S18
PAP24	PAP25	PAP26	PAP27	PAP28	PAP29	PAP30
MAS1	MAS2	MAS3	MAS4	MAS5	MAS6	MAS7
15	16	17	18	19	20	21
S19	S20	S21	S22	S23	S24	S25
HTH1	HTH2	HTH3	HTH4	HTH5	HTH6	HTH7
MAS8	MAS9	MAS10	MAS11	MAS12	MAS13	MAS14
22	23	24	25	26	27	28
S26	S27	S28	S29	S30	O1	O2
HTH8	HTH9	HTH10	HTH11	HTH12	HTH13	HTH14
MAS15	MAS16	MAS17	MAS18	MAS19	MAS20	MAS21

THE NU AFRIKAN CALENDAR

EKUNDU 587/2103-04/6339-40/2095-96						
Ashe	Ojobo	Iwa	Araorun	Ile	Itutu	Erinle
1	2	3	4	5	6	7
O3	O4	O5	O6	O7	O8	O9
HTH15	HTH16	HTH17	HTH18	HTH19	HTH20	HTH21
MAS22	MAS23	MAS 24	MAS25	MAS26	MAS27	MAS28
8	9	10	11	12	13	14
O10	O11	O12	O13	O14	O15	O16
HTH22	HTH23	HTH24	HTH25	HTH26	HTH27	HTH28
MAS29	MAS30	THi1	THi2	THi3	THi4	THi5
15	16	17	18	19	20	21
O17	O18	O19	O20	O21	O22	O23
HTH29	HTH30	KHK1	KHK2	KHK3	KHK4	KHK5
THi6	THi7	THi8	THi9	THi10	THi11	THi12
22	23	24	25	26	27	28
O24	O25	O26	O27	O28	O29	O30
KHK6	KHK7	KHK8	KHK9	KHK10	KHK11	KHK12
THi13	THi14	THi15	THi16	THi17	THi18	THi19

KIJANI 587/2103-04/6339-40/2095-96						
Ashe	Ojobo	Iwa	Araorun	Ile	Itutu	Erinle
1	2	3	4	5	6	7
N28	N29	N30	D1	D2	D3	D4
TBT11	TBT12	TBT13	TBT14	TBT15	TBT16	TBT17
HID18	HID19	HID20	HID21	HID22	HID23	HID24
8	9	10	11	12	13	14
D5	D6	D7	D8	D9	D10	D11
TBT18	TBT19	TBT20	TBT21	TBT22	TBT23	TBT24
HID25	HID26	HID27	HID28	HID29	HID30	TAH1
15	16	17	18	19	20	21
D12	D13	D14	D15	D16	D17	D18
TBT25	TBT26	TBT27	TBT28	TBT29	TBT30	MKH1
TAH2	TAH3	TAH4	TAH5	TAH6	TAH7	TAH8
22	23	24	25	26	27	28
D19	D20	D21	D22	D23	D24	D25
MKH2	MKH3	MKH4	MKH5	MKH6	MKH7	MKH8
TAH9	TAH10	TAH11	TAH12	TAH13	TAH14	TAH15

EUSI 587/2103-04/6339-40/2095-96						
Ashe	Ojobo	Iwa	Araorun	Ile	Itutu	Erinle
1	2	3	4	5	6	7
O31	N1	N2	N3	N4	N5	N6
KHK13	KHK14	KHK15	KHK16	KHK17	KHK18	KHK19
THi20	THi21	THi22	THi23	THi24	THi25	THi26
8	9	10	11	12	13	14
N7	N8	N9	N10	N11	N12	N13
KHK20	KHK21	KHK22	KHK23	KHK24	KHK25	KHK26
THi27	THi28	THi29	THi30	HID1	HID2	HID3
15	16	17	18	19	20	21
N14	N15	N16	N17	N18	N19	N20
KHK27	KHK28	KHK29	KHK30	TBT1	TBT2	TBT3
HID4	HID5	HID6	HID7	HID8	HID9	HID10
22	23	24	25	26	27	28
N21	N22	N23	N24	N25	N26	N27
TBT4	TBT5	TBT6	TBT7	TBT8	TBT9	TBT10
HID11	HID12	HID13	HID14	HID15	HID16	HID17

KWANZA 587/2103-04/6339-40/2095-96						
Umoja	Kujichagulia	Ujima	Ujamaa	Nia	Kuumba	Imani
1	2	3	4	5	6	7
D26	D27	D28	D29	D30	D31	JA1
MKH9	MKH10	MKH11	MKH12	MKH13	MKH14	MKH15
TAH16	TAH17	TAH18	TAH19	TAH20	TAH21	TAH22
8	9	10	11	12	13	14
JA2	JA3	JA4	JA5	JA6	JA7	JA8
MKH16	MKH17	MKH18	MKH19	MKH20	MKH21	MKH22
TAH23	TAH24	TAH25	TAH26	TAH27	TAH28	TAH29
15	16	17	18	19	20	21
JA9	JA10	JA11	JA12	JA13	JA14	JA15
MKH23	MKH24	MKH25	KMH26	MKH27	MKH28	MKH29
TAH30	TIR1	TIR2	TIR3	TIR4	TIR5	TIR6
22	23	24	25	26	27	28
JA16	JA17	JA18	JA19	JA20	JA21	JA22
MKH30	AMN1	AMN2	AMN3	AMN4	AMN5	AMN6
TIR7	TIR8	TIR9	TIR10	TIR11	TIR12	TIR13

THE NU AFRIKAN CALENDAR

SHAHIDI 587/2103-04/6339-40/2095-96						
Ashe	Ojobo	Iwa	Araorun	Ile	Itutu	Erinle
1	2	3	4	5	6	7
JA23	JA24	JA25	JA26	JA27	JA28	JA29
AMN7	AMN8	AMN9	AMN10	AMN11	AMN12	AMN13
TIR14	TIR15	TIR16	TIR17	TIR18	TIR19	TIR20
8	9	10	11	12	13	14
JA30	JA31	F1	F2	F3	F4	F5
AMN14	AMN15	AMN16	AMN17	AMN18	AMN19	AMN20
TIR21	TIR22	TIR23	TIR24	TIR25	TIR26	TIR27
15	16	17	18	19	20	21
F6	F7	F8	F9	F10	F 11	F12
AMN21	AMN22	AMN23	AMN24	AMN25	AMN26	AMN28
TIR28	TIR29	TIR30	YEK1	YEK2	YEK3	YEK4
22	23	24	25	26	27	28
F13	F14	F15	F16	F17	F18	F19
AMN29	AMN30	RNT1	RNT2	RNT3	RNT4	RNT5
YEK5	YEK6	YEK7	YEK8	YEK9	YEK10	YEK11
SHAHIDI 138						
F20						
RNT6						
YEK12						

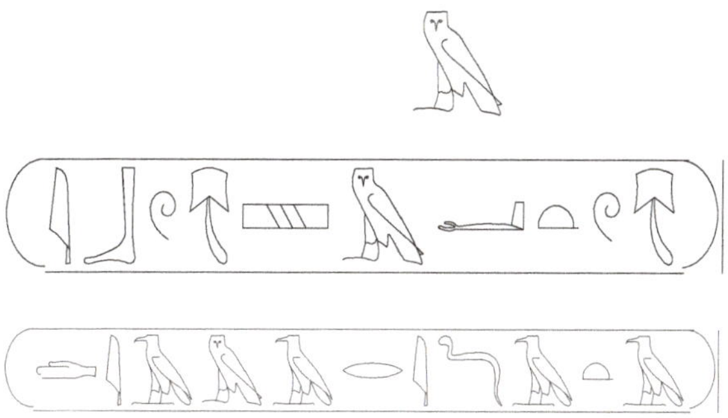

THE NU AFRIKAN CALENDAR

MAAFA 588 B.S.E./ASIENTO 2104-05 C.E./
KAM 6340-41 S.M./EC 2096-97 A.M. (ዓ.ም.)

AFU 588/2104-05/6340-41/2096-97						
Ashe	Ojobo	Iwa	Araorun	Ile	Itutu	Erinle
1	2	3	4	5	6	7
F21	F22	F23	F24	F25	F26	F27
RNT7	RNT8	RNT9	RNT10	RNT11	RNT12	RNT13
YEK13	YEK14	YEK15	YEK16	YEK17	YEK18	YEK19
8	9	10	11	12	13	14
F28	F29	M1	M2	M3	M4	M5
RNT14	RNT15	RNT16	RNT17	RNT18	RNT19	RNT20
YEK 20	YEK21	YEK22	YEK23	YEK24	YEK25	YEK26
15	16	17	18	19	20	21
M6	M7	M8	M9	M10	M11	M12
RNT21	RNT22	RNT23	RNT24	RNT25	RNT26	RNT27
YEK 27	YEK28	YEK29	YEK30	MAG1	MAG2	MAG3
22	23	24	25	26	27	28
M13	M14	M15	M16	M17	M18	M19
RNT28	RNT29	RNT30	KHN1	KHN2	KHN3	KHN4
MAG4	MAG5	MAG6	MAG7	MAG8	MAG9	MAG10
SHARIFU 143						
M20						
KHN5						
MAG11						

KAWAIDA 588/2104-05/6340-41/2096-97						
Ashe	Ojobo	Iwa	Araorun	Ile	Itutu	Erinle
1	2	3	4	5	6	7
A18	A19	A20	A21	A22	A23	A24
PIN4	PIN5	PIN6	PIN7	PIN8	PIN9	PIN10
MIY10	MIY11	MIY12	MIY13	MIY14	MIY15	MIY16
8	9	10	11	12	13	14
A25	A26	A27	A28	A29	A30	MA1
PIN11	PIN12	PIN13	PIN14	PIN15	PIN16	PIN17
MIY17	MIY18	MIY19	MIY20	MIY21	MIY22	MIY23
15	16	17	18	19	20	21
MA2	MA3	MA4	MA5	MA6	MA7	MA8
PIN18	PIN19	PIN20	PIN21	PIN22	PIN23	PIN24
MIY24	MIY25	MIY26	MIY27	MIY28	MIY29	MIY3
22	23	24	25	26	27	28
MA9	MA10	MA11	MA12	MA13	MA14	MA15
PIN25	PIN26	PIN27	PIN28	PIN29	PIN30	APP1
GIN1	GIN2	GIN3	GIN4	GIN5	GIN6	GIN7

UHURU 588/2104-05/6340-41/2096-97						
Ashe	Ojobo	Iwa	Araorun	Ile	Itutu	Erinle
1	2	3	4	5	6	7
M21	M22	M23	M24	M25	M26	M27
KHN6	KHN7	KHN8	KHN9	KHN10	KHN11	KHN12
MAG12	MAG13	MAG14	MAG15	MAG16	MAG17	MAG18
8	9	10	11	12	13	14
M28	M29	M30	M31	A1	A2	A3
KHN13	KHN14	KHN15	KHN16	KHN17	KHN18	KHN19
MAG19	MAG20	MAG21	MAG22	MAG23	MAG24	MAG25
15	16	17	18	19	20	21
A4	A5	A6	A7	A8	A9	A10
KHN20	KHN21	KHN222	KHN23	KHN24	KHN25	KHN26
MAG26	MAG27	MAG28	MAG29	MAG30	MIY1	MIY2
22	23	24	25	26	27	28
A11	A12	A13	A14	A15	A16	A17
KHN27	KHN28	KHN29	KHN30	PIN1	PIN2	PIN3
MIY3	MIY4	MIY5	MIY6	MIY7	MIY8	MIY9

MABABA WEUSI 588/2104-05/6340-41/2096-97						
Ashe	Ojobo	Iwa	Araorun	Ile	Itutu	Erinle
1	2	3	4	5	6	7
MA16	MA17	MA18	MA19	MA20	MA21	MA22
APP2	APP3	APP4	APP5	APP6	APP7	APP8
GIN8	GIN9	GIN10	GIN11	GIN12	GIN13	GIN14
8	9	10	11	12	13	14
MA23	MA24	MA25	MA26	MA27	MA28	MA29
APP9	APP10	APP11	APP12	APP13	APP14	APP15
GIN15	GIN16	GIN17	GIN18	GIN19	GIN20	GIN21
15	16	17	18	19	20	21
MA30	MA31	J1	J2	J3	J4	J5
APP16	APP17	APP18	APP19	APP20	APP21	APP2
GIN22	GIN23	GIN24	GIN25	GIN26	GIN27	GIN2
22	23	24	25	26	27	28
J6	J7	J8	J9	J10	J11	J12
APP23	APP24	APP25	APP26	APP27	APP28	APP29
GIN29	GIN30	SAN1	SAN2	SAN3	SAN4	SAN5

THE NU AFRIKAN CALENDAR

MWANAMKE WEUSI 588/2104-05/6340-41/2096-97

Ashe	Ojobo	Iwa	Araorun	Ile	Itutu	Erinle
1	2	3	4	5	6	7
J13	J14	J15	J16	J17	J18	J19
APP29	APP30	MSR1	MSR2	MSR3	MSR4	MSR5
SAN6	SAN7	SAN8	SAN9	SAN10	SAN11	SAN12
8	9	10	11	12	13	14
J20	J21	J22	J23	J24	J25	J26
MSR6	MSR7	MSR8	MSR9	MSR10	MSR11	MSR12
SAN13	SAN14	SAN15	SAN16	SAN17	SAN18	SAN19
15	16	17	18	19	20	21
J27	J28	J29	J30	JU1	JU2	JU3
MSR13	MSR14	MSR15	MSR16	MSR17	MSR18	MSR19
SAN20	SAN21	SAN22	SAN23	SAN24	SAN25	SAN26
22	23	24	25	26	27	28
JU4	JU5	JU6	JU7	JU8	JU9	JU10
MSR20	MRS21	MSR22	MSR23	MSR24	MSR25	MSR26
SAN27	SAN28	SAN29	SAN30	HAM1	HAM2	HAM3

TAIFA WEUSI 588/2104-05/6340-41/2096-97

Ashe	Ojobo	Iwa	Araorun	Ile	Itutu	Erinle
1	2	3	4	5	6	7
AU8	AU9	AU10	AU11	AU12	AU13	AU14
THT20	THT21	THT22	THT23	THT24	THT25	THT2
NAH2	NAH3	NAH4	NAH5	NAH6	NAH7	NAH8
8	9	10	11	12	13	14
AU15	AU16	AU17	AU18	AU19	AU20	AU21
THT27	THT28	THT29	THT30	PAP1	PAP1	PAP2
NAH9	NAH10	NAH11	NAH12	NAH13	NAH14	NAH15
15	16	17	18	19	20	21
AU22	AU23	AU24	AU25	AU26	AU27	AU28
PAP3	PAP4	PAP5	PAP6	PAP7	PAP8	PAP9
NAH16	NAH17	NAH18	NAH19	NAH20	NAH21	NAH2
22	23	24	25	26	27	28
AU29	AU30	AU31	S1	S2	S3	S4
PAP10	PAP11	PAP12	PAP13	PAP14	PAP15	PAP16
NAH23	NAH24	NAH25	NAH26	NAH27	NAH28	NAH29

MTU WEUSI 588/2104-05/6340-41/2096-97

Ashe	Ojobo	Iwa	Araorun	Ile	Itutu	Erinle
1	2	3	4	5	6	7
JU11	JU12	JU13	JU14	JU15	JU16	JU17
MSR27	MSR28	MSR29	MSR30	*ASR1*	*HRU2*	*STH3*
HAM4	HAM5	HAM6	HAM7	HAM8	HAM9	HAM10
8	9	10	11	12	13	14
JU18	JU19	JU20	JU21	JU22	JU23	JU24
AST4	*NBH5*	THT1	THT2	THT3	THT4	THT5
HAM11	HAM12	HAM13	AM14	HAM15	HAM16	HAM17
15	16	17	18	19	20	21
JU25	JU26	JU27	JU28	JU29	JU30	JU31
THT6	THT7	THT8	THT9	THT10	THT11	THT12
HAM18	HAM19	HAM20	HAM21	HAM22	HAM23	HAM24
22	23	24	25	26	27	28
AU1	AU2	AU3	AU4	AU5	AU6	AU7
THT13	THT14	THT15	THT16	THT17	THT18	THT19
HAM25	HAM26	HAM27	HAM28	HAM29	HAM30	NAH1

UWEZO 588/2104-05/6340-41/2096-97

Ashe	Ojobo	Iwa	Araorun	Ile	Itutu	Erinle
1	2	3	4	5	6	7
S5	S6	S7	S8	S9	S10	S11
PAP17	PAP18	PAP19	PAP20	PAP21	PAP22	PAP23
NAH30	*PAG1*	*PAG2*	*PAG3*	*PAG4*	*PAG5*	MAS1
8	9	10	11	12	13	14
S12	S13	S14	S15	S16	S17	S18
PAP24	PAP25	PAP26	PAP27	PAP28	PAP29	PAP30
MAS2	MAS3	MAS4	MAS5	MAS6	MAS7	MAS8
15	16	17	18	19	20	21
S19	S20	S21	S22	S23	S24	S25
HTH1	HTH2	HTH3	HTH4	HTH5	HTH6	HTH7
MAS9	MAS10	MAS11	MAS12	MAS13	MAS14	MAS15
22	23	24	25	26	27	28
S26	S27	S28	S29	S30	O1	O2
HTH8	HTH9	HTH10	HTH11	HTH12	HTH13	HTH14
MAS16	MAS17	MAS18	MAS19	MAS20	MAS21	MAS22

THE NU AFRIKAN CALENDAR

EKUNDU 588/2104-05/6340-41/2096-97						
Ashe	Ojobo	Iwa	Araorun	Ile	Itutu	Erinle
1	2	3	4	5	6	7
O3	O4	O5	O6	O7	O8	O9
HTH15	HTH16	HTH17	HTH18	HTH19	HTH20	HTH21
MAS23	MAS24	MAS25	MAS26	MAS27	MAS28	MAS29
8	9	10	11	12	13	14
O10	O11	O12	O13	O14	O15	O16
HTH22	HTH23	HTH24	HTH25	HTH26	HTH27	HTH28
MAS30	THi1	THi2	THi3	THi4	THi5	THi6
15	16	17	18	19	20	21
O17	O18	O19	O20	O21	O22	O23
HTH29	HTH30	KHK1	KHK2	KHK3	KHK4	KHK5
THi7	THi8	This9	THi10	THi11	THi12	THi13
22	23	24	25	26	27	28
O24	O25	O26	O27	O28	O29	O30
KHK6	KHK7	KHK8	KHK9	KHK10	KHK11	KHK12
Thi14	THi15	THi16	THi17	THi18	THi19	THi20

KIJANI 588/2104-05/6340-41/2096-97						
Ashe	Ojobo	Iwa	Araorun	Ile	Itutu	Erinle
1	2	3	4	5	6	7
N28	N29	N30	D1	D2	D3	D4
TBT11	TBT12	TBT13	TBT14	TBT15	TBT16	TBT17
HID19	HID20	HID21	HID22	HID23	HID24	HID25
8	9	10	11	12	13	14
D5	D6	D7	D8	D9	D10	D11
TBT18	TBT19	TBT20	TBT21	TBT22	TBT23	TBT24
HID26	HID27	HID28	HID29	HID30	TAH1	TAH2
15	16	17	18	19	20	21
D12	D13	D14	D15	D16	D17	D18
TBT25	TBT26	TBT27	TBT28	TBT29	TBT30	MKH1
TAH3	TAH4	TAH5	TAH6	TAH7	TAH8	TAH9
22	23	24	25	26	27	28
D19	D20	D21	D22	D23	D24	D25
MKH2	MKH3	MKH4	MKH5	MKH6	MKH7	MKH8
TAH10	TAH11	TAH12	TAH13	TAH14	TAH15	TAH16

EUSI 588/2104-05/6340-41/2096-97						
Ashe	Ojobo	Iwa	Araorun	Ile	Itutu	Erinle
1	2	3	4	5	6	7
O31	N1	N2	N3	N4	N5	N6
KHK13	KHK14	KHK15	KHK16	KHK17	KHK18	KHK19
THi21	THi22	THi23	THi24	THi25	THi26	THi27
8	9	10	11	12	13	14
N7	N8	N9	N10	N11	N12	N13
KHK20	KHK21	KHK22	KHK23	KHK24	KHK25	KHK26
THi28	THi29	THi30	HID1	HID2	HID3	HID4
15	16	17	18	19	20	21
N14	N15	N16	N17	N18	N19	N20
KHK27	KHK28	KHK29	KHK30	TBT1	TBT2	TBT3
HID5	HID6	HID7	HID8	HID9	HID10	HID11
22	23	24	25	26	27	28
N21	N22	N23	N24	N25	N26	N27
TBT4	TBT5	TBT6	TBT7	TBT8	TBT9	TBT10
HID12	HID13	HID14	HID15	HID16	HID17	HID18

KWANZA 588/2104-05/6340-41/2096-97						
Umoja	Kujichagulia	Ujima	Ujamaa	Nia	Kuumba	Imani
1	2	3	4	5	6	7
D26	D27	D28	D29	D30	D31	JA1
MKH9	MKH10	MKH11	MKH12	MKH13	MKH14	MKH1
TAH17	TAH18	TAH19	TAH20	TAH21	TAH22	TAH2
8	9	10	11	12	13	14
JA2	JA3	JA4	JA5	JA6	JA7	JA
MKH16	MKH17	MKH18	MKH19	MKH20	MKH21	MKH2
TAH24	TAH25	TAH26	TAH27	TAH28	TAH29	TAH3
15	16	17	18	19	20	21
JA9	JA10	JA11	JA12	JA13	JA14	JA1
MKH23	MKH24	MKH25	KMH26	MKH27	MKH28	MKH2
TIR1	TIR2	TIR3	TIR4	TIR5	TIR6	TIR7
22	23	24	25	26	27	28
JA16	JA17	JA18	JA19	JA20	JA21	JA2
MKH30	AMN1	AMN2	AMN3	AMN4	AMN5	AMN
TIR8	TIR9	TIR10	TIR11	TIR12	TIR13	TIR1

www.ingramcontent.com/pod-product-compliance
Lightning Source LLC
LaVergne TN
LVHW070943210926
848482LV00002B/76
* 9 7 9 8 9 8 7 9 7 6 6 6 1 *